Photonics

THE OXFORD SERIES IN ELECTRICAL AND COMPUTER ENGINEERING

Adel S. Sedra, Series Editor

Allen and Holberg, *CMOS Analog Circuit Design, 2nd edition*
Bobrow, *Elementary Linear Circuit Analysis, 2nd edition*
Bobrow, *Fundamentals of Electrical Engineering, 2nd edition*
Burns and Roberts, *An Introduction to Mixed-Signal IC Test and Measurement*
Campbell, *The Science and Engineering of Microelectronic Fabrication, 2nd edition*
Chen, *Digital Signal Processing*
Chen, *Linear System Theory and Design, 3rd edition*
Chen, *Signals and Systems, 3rd edition*
Comer, *Digital Logic and State Machine Design, 3rd edition*
Comer, *Microprocessor-Based System Design*
Cooper and McGillem, *Probabilistic Methods of Signal and System Analysis, 3rd edition*
DeCarlo and Lin, *Linear Circuit Analysis, 2nd edition*
Dimitrijev, *Principles of Semiconductor Devices*
Dimitrijev, *Understanding Semiconductor Devices*
Fortney, *Principles of Electronics: Analog & Digital*
Franco, *Electric Circuits Fundamentals*
Ghausi, *Electronic Devices and Circuits: Discrete and Integrated*
Guru and Hiziroğlu, *Electric Machinery and Transformers, 3rd edition*
Guru and Warrier, *Electrical Circuits: Analysis and Design*
Houts, *Signal Analysis in Linear Systems*
Jones, *Introduction to Optical Fiber Communication Systems*
Krein, *Elements of Power Electronics*
Kuo, *Digital Control Systems, 3rd edition*
Lathi, *Linear Systems and Signals, 2nd edition*
Lathi, *Modern Digital and Analog Communication Systems, 3rd edition*
Lathi, *Signal Processing and Linear Systems*
Martin, *Digital Integrated Circuit Design*
Miner, *Lines and Electromagnetic Fields for Engineers*
Parhami, *Computer Architecture*
Parhami, *Computer Arithmetic*
Roberts and Sedra, *SPICE, 2nd edition*
Roulston, *An Introduction to the Physics of Semiconductor Devices*
Sadiku, *Elements of Electromagnetics, 3rd edition*
Santina, Stubberud, and Hostetter, *Digital Control System Design, 2nd edition*
Sarma, *Introduction to Electrical Engineering*
Schaumann and Van Valkenburg, *Design of Analog Filters*
Schwarz and Oldham, *Electrical Engineering: An Introduction, 2nd edition*
Sedra and Smith, *Microelectronic Circuits, 5th edition*
Stefani, Savant, Shahian, and Hostetter, *Design of Feedback Control Systems, 4th edition*
Tsividis, *Operation and Modeling of the MOS Transistor, 2nd edition*
Van Valkenburg, *Analog Filter Design*
Warner and Grung, *Semiconductor Device Electronics*
Wolovich, *Automatic Control Systems*
Yariv and Yeh, *Photonics: Optical Electronics in Modern Communications, 6th edition*
Żak, *Systems and Control*

Photonics

Optical Electronics in Modern Communications

SIXTH EDITION

Amnon Yariv
California Institute of Technology

Pochi Yeh
University of California, Santa Barbara

New York Oxford
OXFORD UNIVERSITY PRESS
2007

Oxford University Press, Inc., publishes works that further Oxford University's
objective of excellence in research, scholarship, and education.

Oxford New York
Auckland Cape Town Dar es Salaam Hong Kong Karachi
Kuala Lumpur Madrid Melbourne Mexico City Nairobi
New Delhi Shanghai Taipei Toronto

With offices in
Argentina Austria Brazil Chile Czech Republic France Greece
Guatemala Hungary Italy Japan Poland Portugal Singapore
South Korea Switzerland Thailand Turkey Ukraine Vietnam

Published by Oxford University Press, Inc.
198 Madison Avenue, New York, New York 10016
http://www.oup.com

Library of Congress Cataloging-in-Publication Data

Yariv, Amnon.
Photonics : optical electronics in modern communications / Amnon Yariv,
Pochi Yeh.—6th ed.
p. cm.—(The Oxford series in electrical and computer engineering)
Rev. ed of: Optical electronics in modern communications. 5th ed., c1997.
Includes bibliographical references and index.
ISBN-13: 978-0-19-517946-0
ISBN 0-19-517946-3
1. Photonics. I. Yeh, Pochi. II. Yariv, Amnon. Optical electronics in modern
communications. III. Title. IV. Series.

TA1520.Y37 2006
621.382′7—dc22

2005047270

Printing number: 9 8 7 6 5 4 3 2 1

Printed in the United States of America
on acid-free paper

Contents

Preface

The field of photonics, sometimes referred to as optical electronics, has continued to evolve vigorously during the last decade, thus justifying a major updating of the last (fifth) edition. The present edition has a broader theoretical underpinning and includes new and important subjects.

The book continues the tradition of introducing basic principles in a systematic self-contained treatment with minimal reliance on outside sources. It describes the physics and methodology of operation of the basic optoelectronic components of importance to optical communications and optical electronics. The book is intended to serve both as a text for students in electrical engineering and applied physics as well as a reference book for engineers and scientists working in those fields.

The present edition reflects two major efforts on our part: (1) the addition of new topics related to technology development in optical electronics and communications (and the omission of some less important topics) and (2) the refinement and improvement of materials already in the fifth edition. In the revision process, we decided to tailor the new edition for students, researchers, and engineers in the area of optical communications who are interested in learning how to generate and manipulate optical radiation and how to put this knowledge to work in analyzing and designing photonic components for the transmission of information. The presentation and inclusion of topics also reflect comments and suggestions from many anonymous reviewers and instructors.

Specifically, the main new features of this edition are:

1. The introduction of Stokes parameters and the Poincaré sphere for the representation of polarization states in birefringent optical networks.
2. The use of Fermat's principle for the derivation of rays, beam propagation, and the Fresnel diffraction integral.
3. The use of matrix methods for treating wave propagation in coupled resonator optical waveguides (CROWs).
4. Matrix treatment of multicavity etalons and multilayer structures.
5. Matrix treatment of mode coupling and supermodes in mode-locked lasers.
6. Chromatic dispersion, polarization mode dispersion (PMD) in fibers, and their compensation.
7. Nonlinear optical effects in fibers: self-phase modulation, cross-phase modulation, stimulated Brillouin scattering (SBS) and stimulated Raman scattering (SRS) in fibers, optical four-wave mixing, and spectral reversal (phase conjugation) in fibers.
8. Electroabsorption and waveguide electro-optic Mach–Zehnder modulators.
9. Periodic layered media, fiber Bragg gratings and photonic crystals, and Bragg reflection waveguides.
10. Optical amplifiers: semiconductor optical amplifiers, erbium-doped fiber amplifiers, and Raman amplifiers.

As in the earlier editions, we assume a basic background in electromagnetic theory and familiarity with Maxwell's equations and electromagnetic wave propagation in the bulk and in waveguides. An elementary acquaintance with quantum mechanics is recommended but may be acquired en route.

A generous use of numerical examples is intended to help bridge the gap between theory and applications.

The authors thank their students and colleagues as well as the many reviewers and lecturers whose comments constituted an important factor in the revision.

Pasadena, California
Santa Barbara, California

Amnon Yariv
Pochi Yeh

CHAPTER 1

ELECTROMAGNETIC FIELDS AND WAVES

1.0 INTRODUCTION

This book deals with many important subjects in optical electronics and their applications in modern optical communications, where optical waves are employed as carriers of information for local and long distance transmission. In this chapter, we review some of the most important and basic properties of electromagnetic radiation. These background materials are included for completeness and as a ready source of reference.

We begin by reviewing Maxwell's equations, which govern the propagation of optical waves in various media, including free space, optical crystals, periodic media, optical fibers, and waveguides. We then describe the boundary conditions on the electric and magnetic field vectors. One of the most important contributions of Maxwell's equations is the prediction of the existence of electromagnetic waves in free space. We discuss the storage as well as the transport of energy that accompany the propagation of electromagnetic waves. These are followed by a derivation of the wave equation and an analysis of the monochromatic plane waves and some of their important properties. We also discuss the polarization states of optical waves and their representation in terms of Jones vectors, Stokes parameters, and the Poincaré sphere. Comprehensive coverage of the propagation of optical waves in anisotropic media is included. This is followed by a detailed discussion of the Jones calculus, which is a powerful technique for analyzing the propagation of optical waves in birefringent systems. In the end, we present a brief discussion of an elementary theory of coherence.

1.1 MAXWELL'S EQUATIONS AND BOUNDARY CONDITIONS

In the great work *Treatise of Electricity and Magnetism*, the Scottish physicist James Clerk Maxwell published in 1873 his original finding of the electromagnetic theory of light. His theory has led to many important discoveries, including the existence of electromagnetic waves. Based on his theory, all electric, magnetic, electromagnetic, and optical phenomena are governed by the same fundamental laws of electromagnetism. These laws are written mathematically in terms of the Maxwell equations (in MKS units):

$$\nabla \times \mathbf{E} + \frac{\partial \mathbf{B}}{\partial t} = 0 \tag{1.1-1}$$

$$\nabla \times \mathbf{H} - \frac{\partial \mathbf{D}}{\partial t} = \mathbf{J} \tag{1.1-2}$$

$$\nabla \cdot \mathbf{D} = \rho \tag{1.1-3}$$

$$\nabla \cdot \mathbf{B} = 0 \tag{1.1-4}$$

In these equations, **E** and **H** are the electric field vector (in volts per meter) and magnetic field vector (in amperes per meter), respectively. These two field vectors are often employed to describe an electromagnetic field or an optical wave. The quantities **D** and **B** are called the electric displacement vector (in coulombs per square meter) and magnetic induction vector (in webers per square meter), respectively. These two vectors are introduced to include the effect of the electromagnetic field on matter. The quantities ρ and **J** are the electric charge density (in coulombs per cubic meter) and electric current density vector (in amperes per square meter), respectively. The electric charge and current may be considered the source of the electromagnetic radiation, represented by the **E** and **H** vector fields. These four equations completely determine the electromagnetic field and are the fundamental equations of the theory of such fields, that is, of electrodynamics.

In the field of optical electronics and optical communications, one often deals with the transmission and propagation of electromagnetic radiation in regions of space where both charge density and current density are zero. In fact, if we set $\rho = 0$ and $\mathbf{J} = 0$ in Maxwell's equations, we find that nonzero solutions exist. This means that an electromagnetic field can exist even in the absence of any charges or currents. Electromagnetic fields occurring in media in the absence of charges are called electromagnetic waves. Maxwell's equations (Equations (1.1-1) to (1.1-4)) consist of 8 scalar equations that relate a total of 12 variables, 3 for each of the 4 field vectors **E**, **H**, **D**, and **B**. They cannot be solved uniquely unless the relationships between **B** and **H** as well as that between **E** and **D** are known. To obtain a unique determination of the field vectors, Maxwell's equations must be supplemented by the so-called constitutive equations (or material equations),

$$\mathbf{D} = \varepsilon \mathbf{E} = \varepsilon_0 \mathbf{E} + \mathbf{P} \tag{1.1-5}$$

$$\mathbf{B} = \mu \mathbf{H} = \mu_0 \mathbf{H} + \mu_0 \mathbf{M} \tag{1.1-6}$$

where the constitutive parameters ε and μ are tensors of rank 2 and are known as the dielectric tensor (or permittivity tensor) and the permeability tensor, respectively; **P** and **M** are electric and magnetic polarizations, respectively. In vector notation, the dielectric tensor and the permeability tensor are often written in terms of 3×3 matrices. When an electromagnetic field is present in matter, the electric field can perturb the motion of electrons and produce a distribution of charge separation. This leads to a dipole polarization per unit volume. Analogously, the magnetic field can also produce a magnetization in materials having a permeability that is different from μ_0. The constant ε_0 is called the permittivity of a vacuum and has a value of 8.854×10^{-12} F/m. The constant μ_0 is known as the permeability of a vacuum. It has, by definition, the exact value of $4\pi \times 10^{-7}$ H/m.

If the material medium is isotropic, both ε and μ reduce to scalars. For most applications in optical electronics and optical communications, the quantities ε and μ can be assumed to be independent of the field strengths. However, if the fields are sufficiently strong, such as obtained, for example, by focusing a laser beam or by applying a strong dc electric field to an electro-optic crystal, the dependence of these quantities on **E** and **H** must be considered. These nonlinear optical effects will be discussed later in this book.

Boundary Conditions

Maxwell's equations can be solved in regions where both ε and μ are continuous. In optical electronics and optical communications, one often deals with situations in which the physical properties (characterized by ε and μ) change abruptly across one or more surfaces. This occurs, for example, in dielectric waveguides, which consist of transparent layers of different dielectric constants (or refractive indices). Although the physical properties may change abruptly across the dielectric interfaces, there exist continuity relationships of some of the components of the field vectors at the dielectric boundary. These continuity conditions can be derived directly from Maxwell's equations.

We consider a boundary surface separating two media with different dielectric permittivity and permeability (medium 1 and medium 2). To obtain the boundary conditions for **B** and **D**, we construct a thin cylinder over a unit area of the surface, as shown in Figure 1.1a. The end faces of the cylinder are parallel to the surface. We now apply the Gauss divergence theorem

$$\iiint \nabla \cdot \mathbf{F}\, dV = \iint \mathbf{F} \cdot d\mathbf{S} \tag{1.1-7}$$

to both sides of Equations (1.1-3) and (1.1-4). The surface integral reduces, in the limit as the height of the cylinder approaches zero, to an integral over the end faces only. This leads to

$$\begin{aligned} \mathbf{n} \cdot (\mathbf{B}_2 - \mathbf{B}_1) &= 0 \\ \mathbf{n} \cdot (\mathbf{D}_2 - \mathbf{D}_1) &= \sigma \end{aligned} \tag{1.1-8}$$

where **n** is the unit vector normal to the surface directed from medium 1 into medium 2, $\mathbf{B}_1$ and $\mathbf{D}_1$ are the field vectors in medium 1 in the immediate vicinity of the boundary, $\mathbf{B}_2$ and $\mathbf{D}_2$ are the field vectors in medium 2 in the immediate vicinity of the boundary, and σ is the surface charge density (in coulombs per square meter). These boundary conditions are often written

$$\begin{aligned} B_{2n} &= B_{1n} \\ D_{2n} - D_{1n} &= \sigma \end{aligned} \tag{1.1-9}$$

where $B_{1n} = \mathbf{B}_1 \cdot \mathbf{n}$, $B_{2n} = \mathbf{B}_2 \cdot \mathbf{n}$, $D_{1n} = \mathbf{D}_1 \cdot \mathbf{n}$, and $D_{2n} = \mathbf{D}_2 \cdot \mathbf{n}$. In other words, the normal component of the magnetic induction is always continuous, and the difference between the normal components of the electric displacement vector **D** is equal in magnitude to the surface charge density σ.

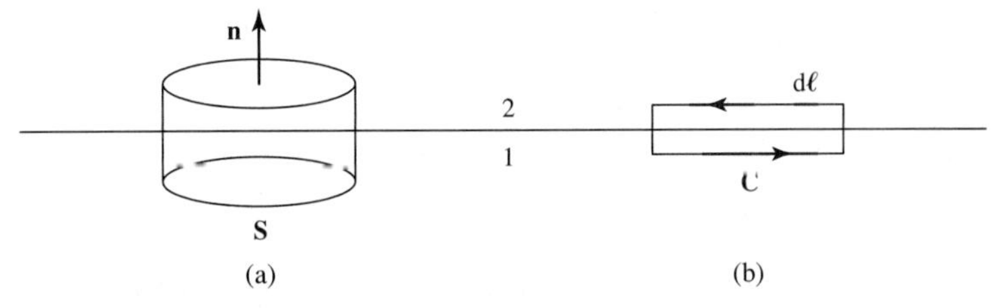

Figure 1.1 Boundary between medium 1 and medium 2. **n** is the unit vector normal to the surface directed from medium 1 into medium 2. $d\ell$ is a differential element along the rectangular contour C.

For the field vectors **E** and **H**, we draw a rectangular contour with two long sides parallel to the surface of discontinuity, as shown in Figure 1.1b. We now apply the Stokes theorem

$$\iint \nabla \times \mathbf{F} \cdot d\mathbf{S} = \int \mathbf{F} \cdot d\ell \tag{1.1-10}$$

to both sides of Equations (1.1-1) and (1.1-2). The contour integral reduces, in the limit as the width of the rectangle approaches zero, to an integral over the two sides only. This leads to

$$\begin{aligned} \mathbf{n} \times (\mathbf{E}_2 - \mathbf{E}_1) &= 0 \\ \mathbf{n} \times (\mathbf{H}_2 - \mathbf{H}_1) &= \mathbf{K} \end{aligned} \tag{1.1-11}$$

where **K** is the surface current density (in amperes per meter), and $\mathbf{E}_1$, $\mathbf{H}_1$, $\mathbf{E}_2$, and $\mathbf{H}_2$ are the field vectors in the immediate vicinity of the boundary in media 1 and 2, respectively. Again, the boundary conditions for the electric field and magnetic field vectors (1.1-11) are often written as

$$\begin{aligned} \mathbf{E}_{2t} &= \mathbf{E}_{1t} \\ \mathbf{H}_{2t} - \mathbf{H}_{1t} &= \mathbf{K} \end{aligned} \tag{1.1-12}$$

where the subscript t means the tangential component of the field vector. (Note: The tangential components of these field vectors to the boundary surface are still vectors in the tangential plane of the surface.) In other words, the tangential component of the electric field vector **E** is always continuous at the boundary surface, and the difference between the tangential components of the magnetic field vector **H** is equal to the surface current density **K**.

In many areas of optical electronics and optical communications, one often deals with situations in which the surface charge density σ and the surface current density **K** both vanish. This occurs, for example, at the boundary between core and cladding of dielectric waveguides. It follows that, in such cases, the tangential components of **E** and **H**, and the normal components of **D** and **B** are continuous across the interface separating media 1 and 2. These boundary conditions are important in solving many wave propagation problems in optical electronics and optical communications, including guided waves in optical fibers and dielectric slab waveguides.

1.2 ENERGY DENSITY AND POYNTING VECTOR

It has been known for some time that light carries energy with it and is a form of electromagnetic radiation. The sun delivers an enormous amount of energy (with an intensity of about 1 kW/m^2 at sea level) to the planet Earth everyday. As we indicated, the first and most conspicuous success of Maxwell's theory was the prediction of the existence of electromagnetic waves and the transport of energy. We now consider two of the most important aspects of electrodynamics: the energy stored with an electromagnetic wave and the power flow associated with an electromagnetic wave. To derive the energy density and the power flow, we consider the conservation of energy in a small volume element in space. The work done per unit volume by an electromagnetic field is $\mathbf{J} \cdot \mathbf{E}$, which may also be considered as the energy dissipation per unit volume. This energy dissipation must be connected with the net decrease of energy density and the power flow out of the volume. According to Equation (1.1-2), the work done by the electromagnetic wave can be written

$$\mathbf{J} \cdot \mathbf{E} = \mathbf{E} \cdot (\nabla \times \mathbf{H}) - \mathbf{E} \cdot \frac{\partial \mathbf{D}}{\partial t} \tag{1.2-1}$$

If we now employ the vector identity

$$\nabla \cdot (\mathbf{E} \times \mathbf{H}) = \mathbf{H} \cdot (\nabla \times \mathbf{E}) - \mathbf{E} \cdot (\nabla \times \mathbf{H}) \tag{1.2-2}$$

and use Equation (1.1-1), the right side of Equation (1.2-1) becomes

$$\mathbf{J} \cdot \mathbf{E} = -\nabla \cdot (\mathbf{E} \times \mathbf{H}) - \mathbf{H} \cdot \frac{\partial \mathbf{B}}{\partial t} - \mathbf{E} \cdot \frac{\partial \mathbf{D}}{\partial t} \tag{1.2-3}$$

If we now further assume that the material medium involved is linear in its electromagnetic properties (i.e., ε and μ are independent of the field strength), Equation (1.2-3) can be written

$$\frac{\partial U}{\partial t} + \nabla \cdot \mathbf{S} = -\mathbf{J} \cdot \mathbf{E} \tag{1.2-4}$$

where U and $\mathbf{S}$ are defined as

$$U = \tfrac{1}{2}(\mathbf{E} \cdot \mathbf{D} + \mathbf{B} \cdot \mathbf{H}) \tag{1.2-5}$$

$$\mathbf{S} = \mathbf{E} \times \mathbf{H} \tag{1.2-6}$$

The scalar U represents the energy density of the electromagnetic fields and has the dimension of joules per cubic meter. The vector $\mathbf{S}$, representing the energy flux, is called the Poynting vector and has the dimensions of joules per square meter per second. It is consistent to view $|\mathbf{S}|$ as the power per unit area (watts per square meter) carried by the field in the direction of $\mathbf{S}$. The quantity $\nabla \cdot \mathbf{S}$ thus represents the net electromagnetic power outflow per unit volume. Equation (1.2-4) is known as the continuity equation or Poynting's theorem. It represents the conservation of energy for the electromagnetic field. The conservation laws for the linear momentum of electromagnetic fields can be obtained in a similar way (see Problem 1.5).

In a region where there's no electric current ($\mathbf{J} = 0$), the continuity equation becomes

$$\frac{\partial U}{\partial t} + \nabla \cdot \mathbf{S} = 0 \tag{1.2-7}$$

Thus, in a current-free region, the decrease of electromagnetic energy density in a volume is a result of the outflow of electromagnetic energy through the surface of the volume.

Dipolar Dissipation

Using the material equations (1.1-5) and (1.1-6), the work done per unit volume by an electromagnetic field, $\mathbf{J} \cdot \mathbf{E}$ (or Equation (1.2-3)), can be written

$$\mathbf{J} \cdot \mathbf{E} = -\nabla \cdot (\mathbf{E} \times \mathbf{H}) - \mu_0 \mathbf{H} \cdot \frac{\partial \mathbf{H}}{\partial t} - \varepsilon_0 \mathbf{E} \cdot \frac{\partial \mathbf{E}}{\partial t} - \mu_0 \mathbf{H} \cdot \frac{\partial \mathbf{M}}{\partial t} - \mathbf{E} \cdot \frac{\partial \mathbf{P}}{\partial t} \tag{1.2-8}$$

Of particular interest in this book is the last term,

$$\mathbf{E} \cdot \frac{\partial \mathbf{P}}{\partial t} \tag{1.2-9}$$

which represents the power per unit volume expended by the field on the electric dipoles. This power goes into an increase in the electrostatic energy stored by the dipoles as well as into supplying the dissipation that may accompany the change in **P**.

1.3 MONOCHROMATIC FIELDS AND COMPLEX-FUNCTION FORMALISM

In optical electronics and optical communications, one often deals with monochromatic light, which serves as a carrier of information. It is known that monochromatic light has a unique angular frequency of oscillation. The field vectors of monochromatic light are sinusoidal functions of time. For the purpose of simplifying the algebraic manipulations, we introduce the complex-function (analytic function) formalism. To illustrate this, we consider, as an example, the function

$$a(t) = |A| \cos(\omega t + \alpha) \tag{1.3-1}$$

where ω is the angular frequency (in units of radian per second), $|A|$ is the amplitude, and α is the phase. If we define the complex amplitude of $a(t)$ by

$$A = |A| e^{i\alpha} \tag{1.3-2}$$

then we can rewrite Equation (1.3-1) as

$$a(t) = \mathrm{Re}[Ae^{i\omega t}] \tag{1.3-3}$$

In complex-function formalism, we will often represent $a(t)$ as

$$a(t) = Ae^{i\omega t} \tag{1.3-4}$$

instead of by Equation (1.3-1) or (1.3-3). This, of course, is not strictly correct so that when this happens it is always understood that what is meant by Equation (1.3-4) is the real part of $A \exp(i\omega t)$. It is important to note that the angular frequency ω is related to the real frequency ν (in units of hertz) by the following equation:

$$\omega = 2\pi\nu \tag{1.3-5}$$

In most situations, when linear operations such as differentiation and integration are involved, the replacement of Equation (1.3-3) by the complex form (1.3-4) poses no problems. The exceptions are cases that involve the product (or powers) of monochromatic field vectors such as energy density and Poynting's vector. In these cases, we must use the real form of the field vectors.

To illustrate the case where the distinction between the real and complex forms is not necessary, consider the problem of taking the derivative of $a(t)$. Using the real form (1.3-1), we obtain

$$\frac{d}{dt} a(t) = \frac{d}{dt} |A| \cos(\omega t + \alpha) = -\omega |A| \sin(\omega t + \alpha) \tag{1.3-6}$$

If we use instead the complex form (1.3-4), we get

$$\frac{d}{dt} a(t) = \frac{d}{dt} Ae^{i\omega t} = i\omega Ae^{i\omega t} \tag{1.3-7}$$

Taking, as agreed, the real part of the last expression and using Equation (1.3-2), we obtain Equation (1.3-6). In fact, the complex form is most convenient when linear operations are involved.

As an example of a case in which we have to use the real form of the function, consider the product of two sinusoidal functions $a(t)$ and $b(t)$, where

$$a(t) = |A| \cos(\omega t + \alpha) = \mathrm{Re}[Ae^{i\omega t}] \tag{1.3-8}$$

$$b(t) = |B| \cos(\omega t + \beta) = \mathrm{Re}[Be^{i\omega t}] \tag{1.3-9}$$

with $A = |A|e^{i\alpha}$ and $B = |B|e^{i\beta}$. Using the real functions, we get

$$a(t)b(t) = \frac{|A||B|}{2}[\cos(2\omega t + \alpha + \beta) + \cos(\alpha - \beta)] \tag{1.3-10}$$

If we were to evaluate the product $a(t)b(t)$ using the complex form of the functions, we would get

$$a(t)b(t) = ABe^{i2\omega t} = |A||B|e^{i(2\omega t+\alpha+\beta)} \tag{1.3-11}$$

Comparing with the last result, Equation (1.3-10) using the real form, we notice that the time-independent term $\cos(\alpha - \beta)$ in parentheses is missing. In addition, the real part of Equation (1.3-11) is a factor of 2 larger than the corresponding term in Equation (1.3-10). Thus, the use of the complex form led to an erroneous result. Generally, the product of the real parts of two complex numbers may not be equal to the real part of the product of these two complex numbers. In other words, if x and y are two arbitrary complex numbers, the following is generally true:

$$\mathrm{Re}[x] \cdot \mathrm{Re}[y] \neq \mathrm{Re}[xy] \tag{1.3-12}$$

Time Averaging of Sinusoidal Products

In optical electronics, one often deals with optical waves with field vectors that are fast varying functions of time. For example, the period of a monochromatic light with a wavelength of $\lambda = 1.5$ μm is $T = \lambda/c = 0.5 \times 10^{-14}$ s. Most optical detectors are unable to respond at such a short period of time. One often considers the time-averaged values rather than the instantaneous values of many physical quantities such as Poynting's vector and energy density. It is frequently necessary to find the time average of the product of two sinusoidal functions of the same frequency:

$$\langle a(t)b(t)\rangle = \frac{1}{T}\int_0^T |A| \cos(\omega t + \alpha)\, |B| \cos(\omega t + \beta)\, dt \tag{1.3-13}$$

where $a(t)$ and $b(t)$ are given by Equations (1.3-8) and (1.3-9), and the angular brackets denote time averaging; $T = 2\pi/\omega$ is the period of oscillation. Since the integrand in the above equation is periodic with a period T, the averaging can be performed over any time period that is an integral number of T. By using Equation (1.3-10), we obtain directly

$$\langle a(t)b(t)\rangle = \tfrac{1}{2}|A||B| \cos(\alpha - \beta) \tag{1.3-14}$$

This last result can be written in terms of the complex amplitudes A and B, defined immediately following Equation (1.3-9), as

$$\langle a(t)b(t)\rangle = \tfrac{1}{2}\mathrm{Re}[A^*B] \tag{1.3-15}$$

or in terms of the analytic form of $a(t)$ and $b(t)$ directly as

$$\langle \mathrm{Re}[a(t)]\mathrm{Re}[b(t)]\rangle = \tfrac{1}{2}\mathrm{Re}[a(t)^*b(t)] \tag{1.3-16}$$

where the asterisk superscript indicates the complex conjugate. The time dependence on the right side of Equation (1.3-16) disappears because both $a(t)$ and $b(t)$ have the same sinusoidal time dependence $\exp(i\omega t)$. These two results, Equations (1.3-15) and (1.3-16), are important and will find frequent use throughout the book. Equation (1.3-16) is also useful when the frequencies of the two functions are slightly different.

By using the complex formalism (or analytic representation) for the field vectors **E**, **H**, **D**, and **B**, the time-averaged Poynting vector (1.2-6) and the energy density (1.2-5) for sinusoidally varying fields are given by

$$\mathbf{S} = \tfrac{1}{2}\mathrm{Re}[\mathbf{E} \times \mathbf{H}^*] \tag{1.3-17}$$

and

$$U = \tfrac{1}{4}\mathrm{Re}[\mathbf{E} \cdot \mathbf{D}^* + \mathbf{H} \cdot \mathbf{B}^*] \tag{1.3-18}$$

respectively. The dipolar dissipation (1.2-9) for sinusoidally varying fields can be written

$$P_D = \tfrac{1}{2}\mathrm{Re}\left(\mathbf{E}^* \cdot \frac{\partial \mathbf{P}}{\partial t}\right) \tag{1.3-19}$$

If an electromagnetic field is written

$$\begin{aligned} \mathbf{E} &= \mathbf{E}_0 e^{i\omega t} \\ \mathbf{H} &= \mathbf{H}_0 e^{i\omega t} \end{aligned} \tag{1.3-20}$$

and the polarization **P** is written

$$\mathbf{P} = \varepsilon_0 \chi \mathbf{E}_0 e^{i\omega t} \tag{1.3-21}$$

where χ is the electric susceptibility coefficient, then the dipolar dissipation can be written

$$P_D = \tfrac{1}{2}\mathrm{Re}(i\omega\varepsilon_0\chi\mathbf{E}_0 \cdot \mathbf{E}_0^*) \tag{1.3-22}$$

In general, χ is a complex number that leads to a phase delay between the polarization **P** and the electric field **E**. The phase delay depends on the imaginary part of χ. If the material response to the electric field is instantaneous (i.e., no phase delay), then χ is a real number. This medium is lossless according to Equation (1.3-22). If the complex susceptibility is written as

$$\chi = \chi' - i\chi'' \tag{1.3-23}$$

where χ' is the real part and χ'' is the imaginary part, then the dipolar dissipation can be written

$$P_D = \tfrac{1}{2}\omega\varepsilon_0\mathbf{E}_0 \cdot \mathbf{E}_0^*\chi'' \tag{1.3-24}$$

We note that the imaginary part of the electric susceptibility is responsible for the dipolar dissipation.

1.4 WAVE EQUATIONS AND MONOCHROMATIC PLANE WAVES

Two of the most important results of Maxwell's equations are the wave equations and the existence of electromagnetic waves that are solutions to them. Here we derive the wave equations in isotropic material media and then consider the propagation of electromagnetic plane waves in homogeneous and isotropic media, where both ε and μ are scalar constants. Vacuum is, of course, the best example of such a "medium." Glasses with uniform compositions are material media that can be treated as homogeneous and isotropic.

To derive the wave equations, we start from Maxwell's equations (1.1-1) to (1.1-4). We limit our attention to regions where both charge density ρ and current density $\mathbf{J}$ vanish. Using the constitutive equations $\mathbf{D} = \varepsilon\mathbf{E}$ and $\mathbf{B} = \mu\mathbf{H}$, the first two of Maxwell's equations can be written

$$\nabla \times \mathbf{E} + \mu \frac{\partial \mathbf{H}}{\partial t} = 0 \tag{1.4-1}$$

$$\nabla \times \mathbf{H} - \varepsilon \frac{\partial \mathbf{E}}{\partial t} = 0 \tag{1.4-2}$$

We now take a curl operation of Equation (1.4-1) and eliminate $\nabla \times \mathbf{H}$ by using Equation (1.4-2). This leads to

$$\nabla \times (\nabla \times \mathbf{E}) + \mu\varepsilon \frac{\partial^2}{\partial t^2} \mathbf{E} = 0 \tag{1.4-3}$$

The first term in the above equation can be expanded as

$$\nabla \times (\nabla \times \mathbf{E}) = \nabla(\nabla \cdot \mathbf{E}) - \nabla^2 \mathbf{E} \tag{1.4-4}$$

Using Equation (1.1-3) and $\mathbf{D} = \varepsilon\mathbf{E}$, we find that for homogeneous and isotropic media, the first term on the right side of Equation (1.4-4) is zero. Thus, Equation (1.4-3) becomes

$$\nabla^2 \mathbf{E} - \mu\varepsilon \frac{\partial^2}{\partial t^2} \mathbf{E} = 0 \tag{1.4-5}$$

This is the wave equation for the electric field vector $\mathbf{E}$ in homogeneous and isotropic media. A similar equation for the magnetic field vector $\mathbf{H}$ can be written

$$\nabla^2 \mathbf{H} - \mu\varepsilon \frac{\partial^2}{\partial t^2} \mathbf{H} = 0 \tag{1.4-6}$$

These are the standard electromagnetic wave equations. They are satisfied by the well-known monochromatic plane wave solution

$$\psi = A e^{i(\omega t - \mathbf{k} \cdot \mathbf{r})} \tag{1.4-7}$$

where A is a constant and is often called the amplitude. In Equation (1.4-7), the angular frequency ω is related to the magnitude of the wavevector $\mathbf{k}$ by

$$|\mathbf{k}| = \omega \sqrt{\mu\varepsilon} \tag{1.4-8}$$

and ψ can be any Cartesian component of $\mathbf{E}$ and $\mathbf{H}$.

We now examine the meaning of this solution. The expression (1.4-7) represents the field component as a function of time and space. At any point in space, the field is a sinusoidal function of time. In addition, at each given moment, the field is a sinusoidal function of space. It is clear that the field has the same value for coordinates $\mathbf{r}$ and times t, which satisfy

$$\omega t - \mathbf{k} \cdot \mathbf{r} = \text{constant} \tag{1.4-9}$$

where the constant is arbitrary and determines the field value. The above equation determines a plane normal to the wavevector $\mathbf{k}$ at any instant of time. This plane is called a surface of constant phase. The surfaces of constant phases are often referred to as wavefronts. The electromagnetic wave represented by Equation (1.4-7) is called a plane wave because all the wavefronts are planar. It is easily seen (from Equation (1.4-9)) that the surfaces of constant phase travel in the direction of **k** with a velocity whose magnitude is

$$v = \frac{\omega}{|\mathbf{k}|} \tag{1.4-10}$$

This is the phase velocity of the wave. If we set $t = 0$ in Equation (1.4-7) and examine the spatial variation of the wave amplitude, the separation between two neighboring field peaks (i.e., the wavelength) is

$$\lambda' = \frac{2\pi}{k} = 2\pi\frac{v}{\omega} \tag{1.4-11}$$

where $k = |\mathbf{k}|$ and the prime indicates the wavelength of light inside the medium. In optical electronics and optical communications, λ is reserved for the wavelength of light in a vacuum.

The value of the phase velocity is a property of the medium and can be expressed in terms of the dielectric constant ε and the magnetic permeability μ. From Equations (1.4-8) and (1.4-10), we obtain

$$v = \frac{1}{\sqrt{\mu\varepsilon}} \tag{1.4-12}$$

The phase velocity of the electromagnetic radiation in a vacuum is

$$c = \frac{1}{\sqrt{\mu_0\varepsilon_0}} = 299{,}792{,}458 \text{ m/s} \tag{1.4-13}$$

whereas in material media it has the value

$$v = \frac{c}{n} \tag{1.4-14}$$

where

$$n = \sqrt{\frac{\mu\varepsilon}{\mu_0\varepsilon_0}} \tag{1.4-15}$$

Most transparent media in optical electronics are nonmagnetic and have a magnetic permeability μ_0. In that case, $n = \sqrt{\varepsilon/\varepsilon_0}$ and is called the index of refraction of the medium. Table 1.1 lists the index of refraction of some common optical electronic materials. We must keep in mind, however, that ε, and therefore n, of a nonmagnetic material ($\mu = \mu_0$) are functions of the frequency ω. The variation of n with frequency ω gives rise to the well-known phenomenon of chromatic dispersion in optics. In a dispersive medium, the phase velocity of a light wave depends on the frequency ω. For example, red light travels faster than blue light in most optical glasses. The physical origin of the refractive index and its dispersion is discussed in some detail in Chapter 5.

We now turn our attention to the vector nature of the electromagnetic field and the requirements of satisfying Maxwell's equations. Using the complex formalism, we write the electromagnetic fields of the monochromatic plane wave in the forms

TABLE 1.1 Index of Refraction of Selected Materials

	Wavelength (μm)									
Material	0.488	0.5	0.5145	0.6328	1.0	1.3	1.55	3.0	5.0	10.6
As-S glass	2.786	2.77	2.75	2.606	2.478	2.449	2.437	2.416	2.407	2.378
BaF_2	1.478	1.4779	1.477	1.473	1.4686	1.467	1.466	1.4611	1.451	1.3928
$Bi_{12}GeO_{20}$			2.55	2.54						
$Bi_{12}SiO_{20}$				2.54						
$Bi_{12}TiO_{20}$				2.55						
CaF_2	1.437	1.4366	1.4362	1.433	1.429	1.427	1.426	1.418	1.399	1.2803
CdS, n_o			2.747	2.46	2.334			2.279	2.266	2.226
CdSe, n_o					2.55	2.50	2.48	2.454	2.446	2.43
CdTe					2.84			2.81	2.77	2.69
CsBr	1.712	1.710	1.707	1.694	1.679	1.675	1.673	1.670	1.668	1.663
CsI	1.810	1.806	1.802	1.781	1.757	1.751	1.749	1.744	1.742	1.739
CuBr	2.201	2.184	2.167	2.102						
CuCl	2.019	2.01	2.002	1.9613	1.924	1.92	1.91	1.903	1.901	1.90
CuI	2.437	2.422	2.405	2.321						
GaAs					3.5	3.41	3.38	3.35	3.29	3.135
GaP			3.66	3.38	3.17	3.07	3.05	2.97	2.94	2.90
GaSb								3.898	3.824	3.843
Ge								4.045	4.0163	4.0029
InAs									3.46	3.42
InP					3.327	3.21	3.17	3.11	3.08	3.05
InSb										3.95
Intran 1 (MgF_2)					1.378	1.377	1.376	1.364	1.337	
Intran 2 (ZnS)					2.291			2.256	2.245	2.1902
Intran 3 (CdF_2)					1.429			1.418	1.399	1.2817
Intran 4 (ZnSe)					2.485			2.44	2.432	2.4034
Intran 5 (MgO)					1.723			1.692	1.637	
KBr	1.572	1.57	1.568	1.558	1.5443	1.541	1.540	1.536	1.534	1.525
KCl	1.498	1.497	1.496	1.488	1.4799	1.478	1.477	1.474	1.471	1.454
KI	1.686	1.684	1.68	1.661	1.64	1.636	1.634	1.6284	1.626	1.6191
LiF	1.395	1.394	1.394	1.392	1.387	1.385	1.383	1.367	1.327	1.05
MgF_2		1.3703			1.363					
MgF_2 crystal, n_o	1.380	1.3798	1.379	1.377	1.374	1.372	1.371			
MgF_2 crystal, n_e	1.392	1.3917	1.391	1.389	1.385	1.384	1.382			
MgO		1.76			1.7237			1.6922	1.6262	
PbF_2	1.786	1.782	1.779	1.761	1.742			1.724	1.708	1.625
Quartz, crystal, n_o	1.55	1.549	1.548	1.543	1.535	1.531	1.528	1.50	1.417	
Ruby, Al_2O_3, n_o	1.776	1.775	1.774	1.766						
Sapphire, Al_2O_3, n_o	1.776	1.775	1.774	1.766	1.756	1.745	1.736	1.7115	1.6239	
Si						3.502	3.476	3.432	3.422	3.418
SiC, crystal	2.698	2.691	2.682	2.64	2.583					
SiO_2, fused silica		1.462	1.462	1.457	1.45	1.447	1.444	1.4185		
$SrTiO_3$, crystal	2.489	2.477	2.461	2.389	2.316			2.231	2.122	
Ta_2O_5	2.27	2.26	2.25	2.16	2.03	1.99				
TiO_2, rutile, n_o	2.731	2.712	2.694	2.59	2.484					
ZnO, crystal, n_o	2.064	2.051	2.044	1.99	1.944	1.932	1.928	1.9072		
ZnS		2.42	2.409	2.351	2.32	2.28	2.27	2.26	2.25	
ZnS, sphalerite	2.433	2.421	2.406	2.354	2.293	2.28	2.27			
ZnS wurtzite, n_o	2.43	2.421	2.41	2.352	2.301					
ZnSe		2.7	2.7	2.59	2.48	2.47	2.46	2.44	2.43	2.392
ZnTe				2.984	2.79	2.75	2.73	2.71	2.70	2.70

$$\mathbf{E} = \mathbf{u}_1 E_0 e^{i(\omega t - \mathbf{k}\cdot\mathbf{r})} \tag{1.4-16}$$

$$\mathbf{H} = \mathbf{u}_2 H_0 e^{i(\omega t - \mathbf{k}\cdot\mathbf{r})} \tag{1.4-17}$$

where $\mathbf{u}_1$ and $\mathbf{u}_2$ are two constant unit vectors, and E_0 and H_0 are complex amplitudes that are constant in space and time. In a homogeneous charge-free medium, the divergence Maxwell equations are $\nabla \cdot \mathbf{E} = 0$ and $\nabla \cdot \mathbf{H} = 0$, which, when applied to Equations (1.4-16) and (1.4-17), result in

$$\mathbf{u}_1 \cdot \mathbf{k} = \mathbf{u}_2 \cdot \mathbf{k} = 0 \tag{1.4-18}$$

This means that **E** and **H** are both perpendicular to the direction of propagation. For this reason, electromagnetic waves are said to be transverse waves. The transverse condition (1.4-18) holds for all four field vectors of the plane wave in homogeneous and isotropic media. In a general anisotropic charge-free medium, only the field vectors **D** and **B** of a plane wave are perpendicular to the direction of propagation.

The curl Maxwell equations provide further restrictions on the field vectors. These are obtained by substituting Equations (1.4-16) and (1.4-17) into Equation (1.1-1) and are given by

$$\mathbf{u}_2 = \frac{\mathbf{k} \times \mathbf{u}_1}{|\mathbf{k}|} \tag{1.4-19}$$

and

$$H_0 = \frac{E_0}{\eta}, \quad \eta = \sqrt{\frac{\mu}{\varepsilon}} \tag{1.4-20}$$

This shows that the triad ($\mathbf{u}_1$, $\mathbf{u}_2$, $\mathbf{k}$) forms a set of mutually orthogonal vectors and that the field vectors **E** and **H** are in phase and in constant ratio provided ε and μ are real. The parameter η in Equation (1.4-20) has the dimension of resistance and is called the impedance of space. In a vacuum,

$$\eta_0 = \sqrt{\mu_0/\varepsilon_0} = 377\ \Omega$$

The plane wave we have just derived is a transverse wave propagating in the direction **k**. It represents a time-averaged energy flux given, according to Poynting's theorem (1.3-17), by

$$\mathbf{S} = \frac{1}{2\eta}|E_0|^2 \mathbf{u}_3 = \frac{\mathbf{k}}{2\omega\mu}|E_0|^2 = \frac{\mathbf{k}}{2\omega\mu}|\mathbf{E}|^2 \tag{1.4-21}$$

where $\mathbf{u}_3 = \mathbf{u}_1 \times \mathbf{u}_2$ is a unit vector in the direction of **k**. The time-averaged energy density is

$$U = \tfrac{1}{2}\varepsilon|E_0|^2 = \tfrac{1}{2}\varepsilon|\mathbf{E}|^2 \tag{1.4-22}$$

We note from Equations (1.4-21) and (1.4-22) that the energy flux is directed along the direction of propagation. We also note from Equations (1.4-13) and (1.4-15) that we have the relation

$$S = vU \tag{1.4-23}$$

This equation simply states that the energy is flowing at a speed $v = c/n$ along the direction of propagation. It is important to note that this is valid only in media without dispersion. When chromatic dispersion is present, the energy is flowing at a speed known as the group velocity.

1.5 CHROMATIC DISPERSION AND GROUP VELOCITY

In the previous section we discussed the plane wave solutions to Maxwell's equations and studied some of the basic properties. Only monochromatic waves with a definite frequency and wavenumber were treated. In the field of optical communications, short pulses of laser radiation are employed to transmit information. The finite duration of an optical pulse results in a finite spread of frequencies or, equivalently, wavelengths. The propagation of an optical pulse in a linear medium can be described in terms of the propagation of an appropriate superposition of plane waves with different frequencies, since Maxwell's equations are linear. In a linear medium, the induced polarization is proportional to the electric field. If the medium is dispersive, the phase velocity depends on the frequency. This is known as the chromatic dispersion. As a result of the chromatic dispersion, different frequency components of the pulse propagate with different speeds. This can lead to a change of the shape or even a spreading of the pulse. In addition, the velocity of energy flow of an optical pulse in a dispersive medium may be different from the phase velocity. This is a subtle subject and requires careful investigation.

These effects on the propagation of an optical pulse due to the chromatic dispersion can be described by representing the pulse as a sum of many monochromatic plane wave components. Each of these plane wave components is a solution of Maxwell's equations. Since the equations are linear, the sum is also a solution of Maxwell's equations. In general, we must replace the summation by an integration. For the sake of clarity in introducing the basic concepts, we will only consider the case of a beam of polarized light which is uniform in the transverse directions. If $A(k)$ denotes the amplitude of the monochromatic plane wave component with wavenumber k, then the electric field of the pulse can be written

$$E(z, t) = \int_{-\infty}^{\infty} A(k) e^{i[\omega(k)t - kz]}\, dk \tag{1.5-1}$$

where the angular frequency is written $\omega(k)$ to show its dependence on k. This integral satisfies Maxwell's equations, since the integrand is a basic monochromatic plane wave solution to the same equations. Note that if we view $E(z, t)$ at some instant of time, say, $t = 0$, then $A(k)$ is formally the Fourier transform of $E(z, 0)$. In other words,

$$E(z, 0) = \int_{-\infty}^{\infty} A(k) e^{-ikz}\, dk \tag{1.5-2}$$

We will then refer to $|A(k)|^2$ as the Fourier spectrum (or power spectrum) of $E(z, t)$.

The relation between ω and k (known as the dispersion relation) is given by Equation (1.4-8) for the electromagnetic field. In the following discussion, we assume that both k and $\omega(k)$ are real. A typical optical pulse and its Fourier spectrum are shown in Figure 1.2.

A beam of polarized optical pulse is usually characterized by its center frequency ω_0 (or its corresponding wavenumber k_0), and the frequency spread $\Delta\omega$ around ω_0 (or the corresponding spread in wavenumber Δk). Typically, $A(k)$ is sharply peaked around k_0 (i.e., $\Delta k \ll k_0$). To study the propagation of such an optical pulse in space and time, we expand $\omega(k)$ around the value k_0 in terms of a Taylor series,

$$\omega(k) = \omega_0 + \left(\frac{d\omega}{dk}\right)_{\omega_0} (k - k_0) + \cdots \tag{1.5-3}$$

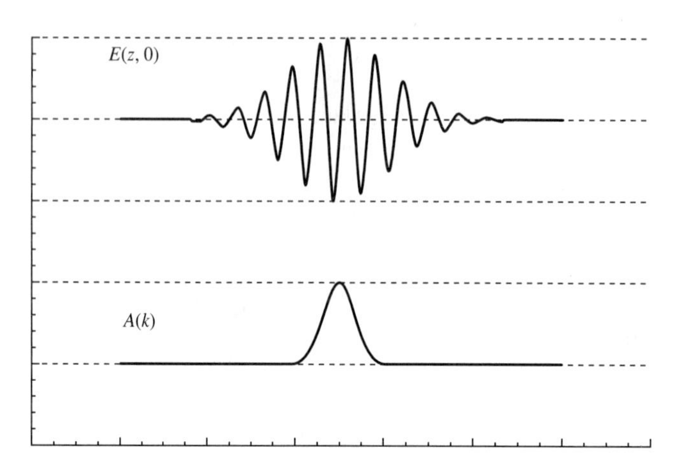

Figure 1.2 An optical pulse of finite extent and its Fourier spectrum in wavenumber (k) space.

and then substitute for ω from Equation (1.5-3) into Equation (1.5-1). Thus Equation (1.5-2) can be written

$$E(z,t) = e^{i(\omega_0 t - k_0 z)} \int_{-\infty}^{\infty} A(k) \exp\left\{ i\left[\left(\frac{d\omega}{dk}\right)_{\omega_0} t - z \right](k - k_0) \right\} dk \tag{1.5-4}$$

where we neglect the higher-order terms in $(k - k_0)$. The integral in Equation (1.5-4) is a function of the composite variable

$$\xi = \left[z - \left(\frac{d\omega}{dk}\right)_{\omega_0} t \right] \tag{1.5-5}$$

only and may be called the envelope function $V(\xi)$. Thus the amplitude of the optical pulse can be written

$$E(z,t) = e^{i(\omega_0 t - k_0 z)} V\left[z - \left(\frac{d\omega}{dk}\right)_{\omega_0} t \right] \tag{1.5-6}$$

This shows that, apart from an overall phase factor, the optical pulse travels along undistorted in shape, with a velocity

$$v_g = \left(\frac{d\omega}{dk}\right)_{\omega_0} \tag{1.5-7}$$

which is called the group velocity of the pulse. Figure 1.3 shows the propagation of an optical pulse in a dispersive medium in this approximation. The approximation is legitimate only when the distribution is sharply peaked at k_0 and the frequency ω is a smoothly varying function of k around k_0. If the electromagnetic energy density of an optical pulse is associated with the absolute square of the amplitude, it is obvious that in this approximation the group velocity represents the speed of transport of energy. We note that, in the case of a pulse, the phase velocity and group velocity are, in general, different. The phase velocity, which is usually greater than the group velocity, has the same significance as in the case of a plane wave. It is the velocity needed to stay on top of a given wave crest or valley.

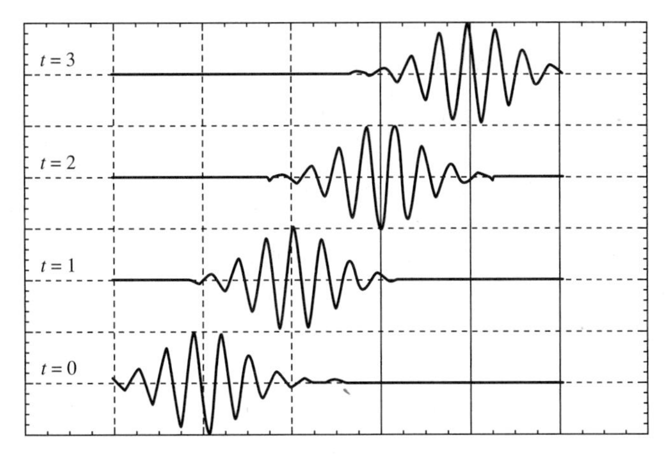

Figure 1.3 Propagation of an optical pulse.

In optical communications, optical pulses are often propagating in silica fibers, which exhibit chromatic dispersion. The dispersion is usually described by the index of refraction, $n(\omega)$, as a function of frequency (or wavelength), and the relation between ω and k is given by

$$k = n(\omega)\frac{\omega}{c} \tag{1.5-8}$$

where c is the speed of light in vacuum. The phase velocity is

$$v_p = \frac{c}{n(\omega)} \tag{1.5-9}$$

and is greater than or smaller than c depending on whether $n(\omega)$ is smaller or larger than unity. From Equations (1.5-7) and (1.5-8), the group velocity is given by

$$v_g = \frac{c}{n + \omega(dn/d\omega)} \tag{1.5-10}$$

For normal dispersion, $dn/d\omega > 0$, the group velocity is less than the phase velocity. In regions of anomalous dispersion, however, $dn/d\omega$ can become large and negative. Then the group velocity differs greatly from the phase velocity and can be greater than c. The latter case occurs only when $dn/d\omega$ is very negative, which is equivalent to a rapid variation of ω as a function of k, thus making the approximation illegitimate. In most transparent media, the materials exhibit normal dispersion. Anomalous dispersion occurs usually in the spectral regime within the absorption bands.

Equation (1.5-10) is often written, equivalently, in terms of wavelength as

$$v_g = \frac{c}{n - \lambda(dn/d\lambda)} \tag{1.5-11}$$

Group Velocity Dispersion and Pulse Spreading

We have seen in an earlier discussion that an optical pulse propagating in a dispersive medium can remain undistorted in shape provided the approximation (1.5-3) is valid. In high-speed optical communications, the pulse width can be on the order of picoseconds. In this case, the amplitude $A(k)$ is no longer sharply peaked around the central wavenumber k_0. Consequently, the next higher-order term

$$\frac{1}{2}\left(\frac{d^2\omega}{dk^2}\right)_{\omega_0}(k-k_0)^2 \tag{1.5-12}$$

cannot be neglected. The pulse will no longer remain unchanged and will, in general, spread as the pulse travels. The spreading of the pulses in optical communication networks can lead to a degradation or loss of the signal transmission.

The spreading of the optical pulse can be accounted for by noting that the group velocity v_g may not be the same for each spectral component of the pulse. This is known as the group velocity dispersion (GVD). If Δk is the width of the spread in wavenumbers, then the group velocity spread is on the order of

$$\Delta v_g = \left(\frac{d^2\omega}{dk^2}\right)_{\omega_0}\Delta k \tag{1.5-13}$$

As the optical pulse propagates, one expects a spread in position on the order of $\Delta v_g\,T$, where T is the time of flight of a reference frequency component. The group velocity spread is often written

$$\Delta v_g = \left(\frac{dv_g}{d\omega}\right)_{\omega_0}\Delta\omega \tag{1.5-14}$$

where $\Delta\omega$ is the width of spread of the angular frequency, or equivalently,

$$\Delta v_g = \left(\frac{dv_g}{d\lambda}\right)_{\omega_0}\Delta\lambda \tag{1.5-15}$$

where $\Delta\lambda$ is the width of spread of the wavelength.

In optical communications, signals are represented in terms of a sequence of pulses. Each of the pulses has a well-defined temporal duration. For example, in a 40 Gb/s transmission system, the pulse duration is around 25 ps. The pulse spreading in terms of picosecond is an important measurement of the signal integrity. To minimize the bit errors in transmission, the pulse spreading is often required to be less than a fraction of the pulse duration. Thus, in fiber-optic communications, the group velocity dispersion is often characterized by a parameter

$$D = \frac{1}{L}\frac{dT}{d\lambda} \tag{1.5-16}$$

where T is the pulse transmission time through a length L of the fiber. Physically, D is a measure of the pulse spreading (in units of seconds) per unit bandwidth per unit length of the medium of transmission. Using $T = L/v_g$, it can be shown that

$$T = L\left(\frac{n}{c} + \frac{\omega}{c}\frac{dn}{d\omega}\right) = L\left(\frac{n}{c} - \frac{\lambda}{c}\frac{dn}{d\lambda}\right) \tag{1.5-17}$$

and

$$D = -\frac{1}{c\lambda}\left(\lambda^2\frac{d^2n}{d\lambda^2}\right) \tag{1.5-18}$$

or equivalently,

$$D = -\frac{2\pi c}{\lambda^2}\frac{d^2k}{d\omega^2} \tag{1.5-19}$$

We note that the group velocity dispersion is related to the second-order derivative of the index of refraction with respect to frequency (or wavelength).

EXAMPLE: DISPERSION IN SILICA (SiO_2)

Most optical fibers for communications are made of silica glass. Pure silica glass exhibits extremely low absorption loss. The material, however, exhibits a chromatic dispersion. An accurate empirical expression of the refractive index of fused silica can be written

$$n(\lambda) = c_0 + c_1\lambda^2 + c_2\lambda^4 + \frac{c_3}{(\lambda^2 - a)} + \frac{c_4}{(\lambda^2 - a)^2} + \frac{c_5}{(\lambda^2 - a)^3} \tag{1.5-20}$$

where

$$\begin{aligned} c_0 &= 1.4508554 \\ c_1 &= -0.0031268 \\ c_2 &= -0.0000381 \\ c_3 &= 0.0030270 \\ c_4 &= -0.0000779 \\ c_5 &= 0.0000018 \\ a &= 0.035 \end{aligned}$$

and λ is in units of micrometers.

Figure 1.4 shows the index of refraction of silica glass as a function of wavelength. We note that the index of refraction is a decreasing function of wavelength. This is typical in the normal dispersion regime.

Using Equation (1.5-11), we obtain the group velocity of optical propagation in silica glass. Figure 1.5 shows v_g/c as a function of wavelength. It is clear that the group velocity depends

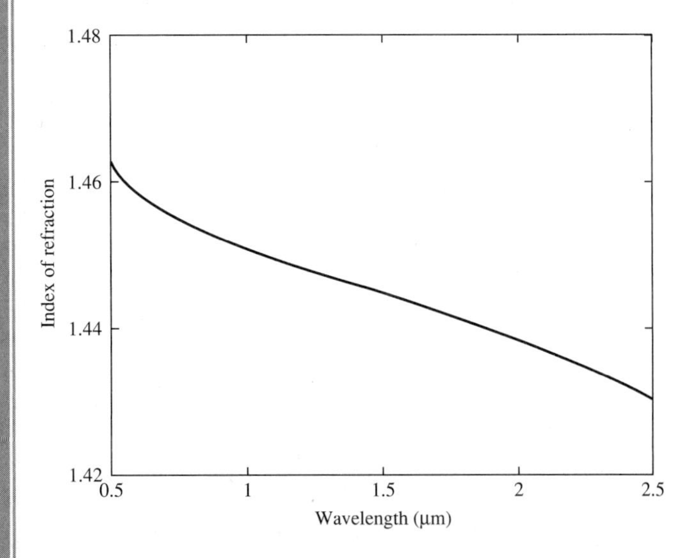

Figure 1.4 The index of refraction of pure silica glass as a function of wavelength.

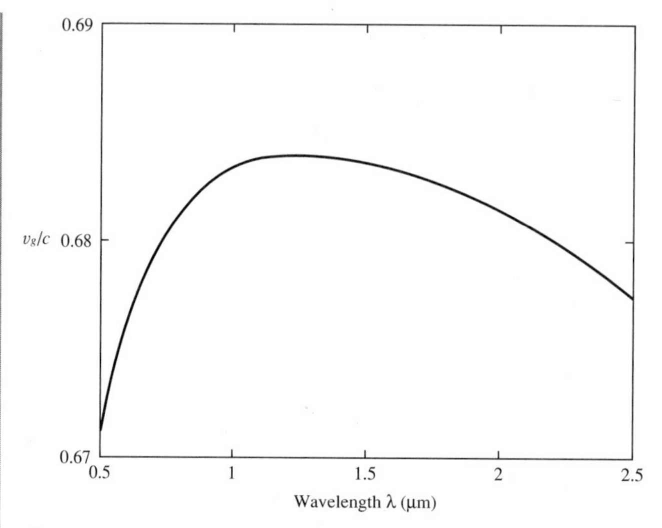

Figure 1.5 Group velocity in units of c in silica glass as a function of wavelength.

on the wavelength. The group velocity of light in pure silica exhibits a maximum at a wavelength near $\lambda = 1.275$ μm. For optical transmission of a pulse in the spectral regime around $\lambda = 1.55$ μm, the short-wavelength components travel faster than the long-wavelength components of the optical pulse. The disparity in the group velocity leads to pulse broadening and a descending frequency chirp in the pulse.

Using Equation (1.5-18), we obtain the group velocity dispersion in silica glass. Figure 1.6 shows the group velocity dispersion D as a function of wavelength. We note that D becomes zero at a wavelength near $\lambda = 1.275$ μm. The group velocity dispersion due to bulk silica glass is around 20 ps/nm-km in the spectral regime around $\lambda = 1500$ nm.

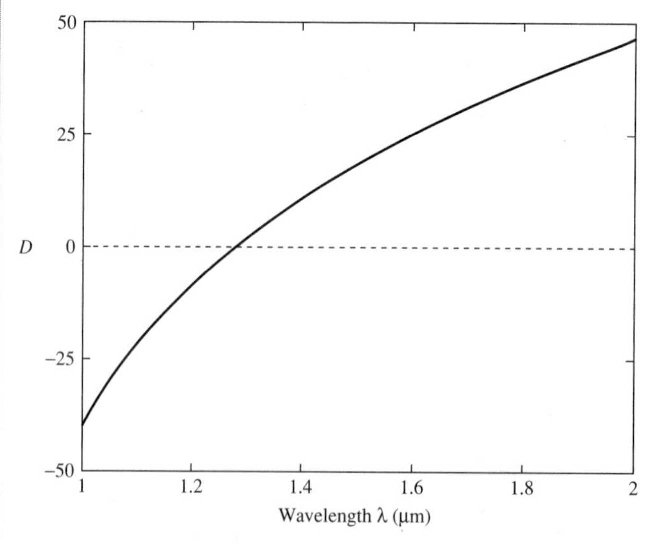

Figure 1.6 Group velocity dispersion of silica glass in units of ps/nm-km as a function of wavelength.

1.6 POLARIZATION STATES AND REPRESENTATIONS (STOKES PARAMETERS AND POINCARÉ SPHERE)

In the electromagnetic theory of light, beams of light are represented by electromagnetic waves propagating in space. A beam of light is often represented by its electric field vector, which oscillates in time and space as the beam propagates. In an isotropic medium, the direction of oscillation (or vibration) is always perpendicular to the direction of propagation. Being a transverse wave, there are two independent directions of oscillation. In an isotropic medium (e.g., glass, vacuum), the two mutually independent directions of oscillation can be chosen arbitrarily. If these two components of oscillation are totally uncorrelated, the resultant direction of oscillation is random and the beam of light is said to be unpolarized. For all thermal sources in nature, the direction of vibration is random. A beam of light is said to be linearly polarized if its electric field vector vibrates in one particular direction. In this section, we describe the polarization states of light and various ways of representing the states of polarization.

The main objective of this book is to provide comprehensive coverage of optical electronics and its application in optical communications. Many of the components in optical communications involve the use of polarized light and anisotropic optical media. Here we introduce the concept of polarization by considering a beam of monochromatic plane waves propagating in an isotropic and homogeneous medium. The beam can be represented by its electric field $\mathbf{E}(\mathbf{r}, t)$, which can be written

$$\mathbf{E} = \mathbf{A} \cos(\omega t - \mathbf{k} \cdot \mathbf{r}) \tag{1.6-1}$$

where ω is the angular frequency, $\mathbf{k}$ is the wavevector, and $\mathbf{A}$ is a constant vector representing the amplitude. The magnitude of the wavevector k is related to the frequency by the following equation:

$$k = n\frac{\omega}{c} = n\frac{2\pi}{\lambda} \tag{1.6-2}$$

where n is the index of refraction of the medium, c is the speed of light in vacuum, and λ is the wavelength of light in vacuum. For materials with some absorption, the index of refraction is a complex number. Reflecting the transverse nature, the electric field vector is always perpendicular to the direction of propagation,

$$\mathbf{k} \cdot \mathbf{E} = 0 \tag{1.6-3}$$

For mathematical simplicity, the monochromatic plane wave in Equation (1.6-1) is often written

$$\mathbf{E} = \mathbf{A} \exp[i(\omega t - \mathbf{k} \cdot \mathbf{r})] \tag{1.6-4}$$

with the understanding that only the real part of the right side represents the actual electric field. This is the analytic representation described in Section 1.3. The polarization state of a beam of monochromatic light is specified by its electric field vector $\mathbf{E}(\mathbf{r}, t)$. The time evolution of the electric field vector is exactly sinusoidal; that is, the electric field must oscillate at a definite frequency. For the purpose of describing various representations of the polarization states, we consider propagation along the z axis. Being a transverse wave, the electric field vector must lie in the xy plane. The two mutually independent components of the electric field vector can be written

$$E_x = A_x \cos(\omega t - kz + \delta_x)$$
$$E_y = A_y \cos(\omega t - kz + \delta_y) \tag{1.6-5}$$

where we have used two independent and positive amplitudes A_x and A_y and have added two independent phases δ_x and δ_y to reflect the mutual independence of the two components. With the amplitudes being positive, the phase angles are defined in the range $-\pi < \delta_{x,y} \leq \pi$. Since the x component and the y component of the electric field vector can oscillate independently at a definite frequency, one must consider the effect produced by the vector addition of these two oscillating orthogonal components. The problem of superposing two independent oscillations at right angles to each other and with the same frequency is well known and is completely analogous to the classical motion of a two-dimensional harmonic oscillator. The orbit of a general motion is an ellipse, which corresponds to oscillations in which the x and y components are not in phase. For optical waves, this corresponds to elliptic polarization states. There are, of course, many special cases that are important in optics. These include linear polarization states and circular polarization states.

We start with the discussion of two special cases of interest. Without loss of generality, we consider the time evolution of the electric field vector at the origin $z = 0$. According to Equation (1.6-5), the electric field components are written

$$E_x = A_x \cos(\omega t + \delta_x)$$
$$E_y = A_y \cos(\omega t + \delta_y) \tag{1.6-6}$$

We define a relative phase as

$$\delta = \delta_y - \delta_x \tag{1.6-7}$$

where, again, δ is limited to the region $-\pi < \delta \leq \pi$.

Linear Polarization States

A beam of light is said to be linearly polarized if the electric field vector vibrates in a constant direction (in the xy plane). This occurs when the two components of oscillation are in phase ($\delta = \delta_y - \delta_x = 0$) or π out of phase ($\delta = \delta_y - \delta_x = \pi$); that is,

$$\delta = \delta_y - \delta_x = 0 \quad \text{or} \quad \pi \tag{1.6-8}$$

In this case, the electric field vector vibrates sinusoidally along a constant direction in the xy plane defined by the ratio of the two components,

$$\frac{E_y}{E_x} = \frac{A_y}{A_x} \quad \text{or} \quad -\frac{A_y}{A_x} \tag{1.6-9}$$

Since the amplitudes A_x and A_y are independent, the electric field vector of linearly polarized light can vibrate along any direction in the xy plane. Linearly polarized light is often called plane polarized light. If we examine the space evolution of the electric field vector at a fixed point in time (say, $t = 0$), the components of the electric field vector can be written

$$E_x = A_x \cos(-kz + \delta_x)$$
$$E_y = A_y \cos(-kz + \delta_y) \tag{1.6-10}$$

with $\delta = \delta_y - \delta_x = 0$ or π. We note that the sinusoidal curve traced by the components in space is confined in a plane defined by Equation (1.6-9). The vibration of the electric field vector is confined in this plane. Thus the beam of light is said to be plane polarized. The terms plane polarized light and linearly polarized light are interchangeable. Linear polarization states are most widely used in optics because of their simplicity and ease of preparation.

Circular Polarization States

The other special case of importance is that of the circular polarization state. A beam of light is said to be circularly polarized if the electric field vector undergoes uniform rotation in the xy plane. This occurs when $A_x = A_y$ and

$$\delta = \delta_y - \delta_x = \pm\pi/2 \tag{1.6-11}$$

According to our convention, the beam of light is right-handed circularly polarized when $\delta = -\pi/2$, which corresponds to a counterclockwise rotation of the electric field vector in the xy plane; and left-handed circularly polarized when $\delta = \pi/2$, which corresponds to a clockwise rotation of the electric field vector in the xy plane. Our convention for labeling right-handed and left-handed polarization is consistent with the terminology of modern physics in which a photon with a right-handed circular polarization has a positive angular momentum along the direction of propagation (see Problem 1.36). However, some optics books adopt the opposite convention. The opposite convention arises from the description of the evolution of the electric field vector in space (see Problem 1.16).

It is interesting to note that the conditions of equal amplitude and $\pm\pi/2$ phase shift for circular polarization states are valid in any set of perpendicular coordinates in the xy plane. In other words, when the electric field vector of a circularly polarized light is decomposed into any two mutually perpendicular components, the amplitudes are always equal and the phase shift is always $\pm\pi/2$.

Elliptic Polarization States

A beam of light is said to be elliptically polarized if the curve traced by the end point of the electric field vector is an ellipse (in the xy plane). This is the most general case of a polarized light. Both linear polarization states and circular polarization states are special cases of elliptic polarization states. At a given point in space (say, $z = 0$), Equation (1.6-5) is a parametric representation of an ellipse traced by the end point of the electric field vector. The equation of the ellipse can be obtained by eliminating ωt in Equation (1.6-6). After several steps of elementary algebra, we obtain

$$\left(\frac{E_x}{A_x}\right)^2 + \left(\frac{E_y}{A_y}\right)^2 - 2\frac{\cos\delta}{A_x A_y} E_x E_y = \sin^2\delta \tag{1.6-12}$$

Equation (1.6-12) is an equation of a conic. From Equation (1.6-6) it is obvious that this conic section is confined in a rectangular region with sides parallel to the coordinate axes and whose lengths are $2A_x$ and $2A_y$. Therefore, the curve must be an ellipse. Thus, we find that the polarization states of light are, in general, elliptical. A complete description of an elliptical polarization state includes the orientation of the ellipse relative to the coordinate axes and the shape and sense of revolution of the electric field vector. In general, the principal axes of the ellipse are not in the x and y directions. By using a transformation (rotation) of the coordinate

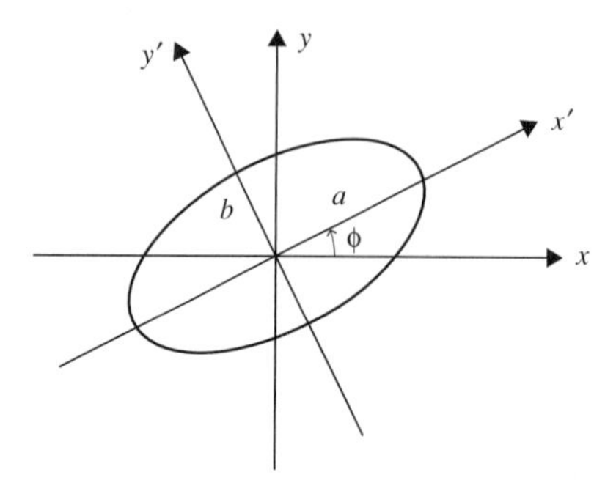

Figure 1.7 Polarization ellipse.

system, we are able to diagonalize Equation (1.6-12). Let x' and y' be the new set of axes along the principal axes of the ellipse. Then the equation of the ellipse in this new coordinate system becomes

$$\left(\frac{E_{x'}}{a}\right)^2 + \left(\frac{E_{y'}}{b}\right)^2 = 1 \tag{1.6-13}$$

where a and b are the lengths of the principal semiaxes of the ellipse, and $E_{x'}$ and $E_{y'}$ are the components of the electric field vector in this principal coordinate system.

Let ϕ be the angle between the x' axis and the x axis (see Figure 1.7). Then the lengths of the principal axes are given by

$$\begin{aligned} a^2 &= A_x^2 \cos^2\phi + A_y^2 \sin^2\phi + 2A_xA_y \cos\delta \cos\phi \sin\phi \\ b^2 &= A_x^2 \sin^2\phi + A_y^2 \cos^2\phi - 2A_xA_y \cos\delta \cos\phi \sin\phi \end{aligned} \tag{1.6-14}$$

The angle ϕ can be expressed in terms of A_x, A_y, and $\cos\delta$ as

$$\tan 2\phi = \frac{2A_xA_y}{A_x^2 - A_y^2}\cos\delta \tag{1.6-15}$$

It is important to note that $\phi + \pi/2$ is also a solution, if ϕ is a solution of the equation. The sense of revolution of an elliptical polarization is determined by the sign of $\sin\delta$. The end point of the electric vector will revolve in a clockwise direction if $\sin\delta > 0$ and in a counterclockwise direction if $\sin\delta < 0$. Figure 1.8 illustrates how the polarization ellipse changes with varying phase difference δ.

The ellipticity of a polarization ellipse is defined as

$$e = \pm\frac{b}{a} \tag{1.6-16}$$

where a and b are the lengths of the principal semiaxes. The ellipticity is taken as positive when the rotation of the electric field vector is right-handed and negative otherwise. With this definition, $e = \pm 1$ for circularly polarized light.

As discussed earlier, light is linearly polarized when the tip of the electric field vector **E** moves along a straight line. When it describes an ellipse, the light is elliptically polarized. When it describes a circle, the light is circularly polarized. If the end point of the electric field vector is seen to move in a counterclockwise direction by an observer facing the approaching wave, the field is said to possess right-handed polarization. Figure 1.8 also illustrates the sense of revolution of the ellipse.

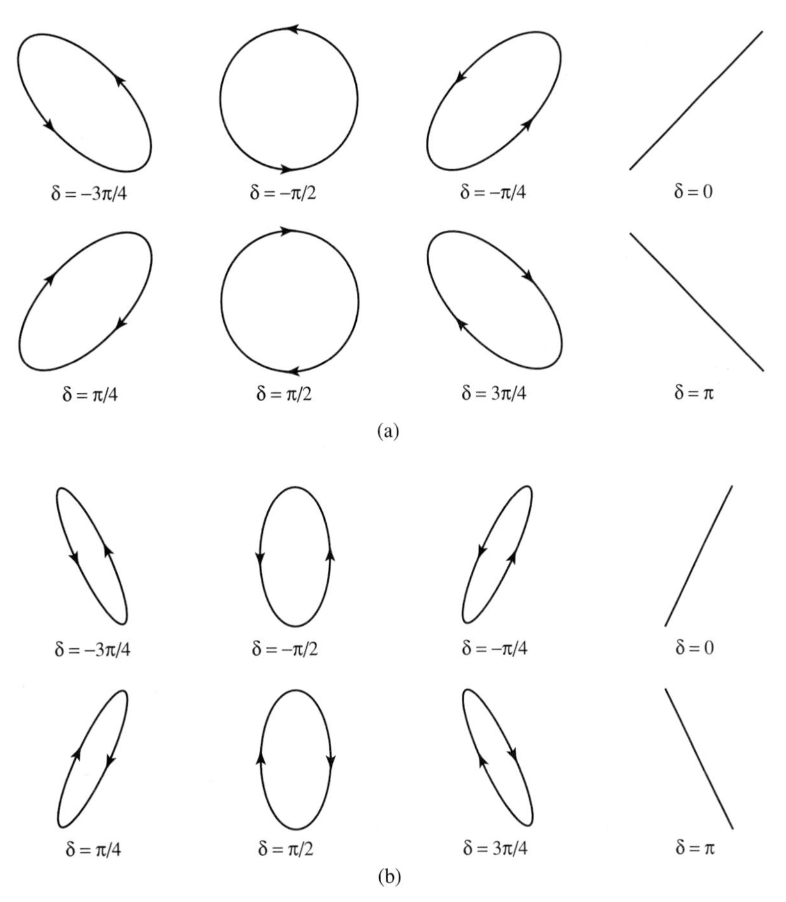

Figure 1.8 Polarization ellipses at various phase angles δ: (a) $E_x = \cos(\omega t - kz)$, $E_y = \cos(\omega t - kz + \delta)$; (b) $E_x = \frac{1}{2}\cos(\omega t - kz)$, $E_y = \cos(\omega t - kz + \delta)$.

An elliptic polarization state can always be decomposed into two mutually orthogonal components. The relative phase shift between these two components can be anywhere between $-\pi$ and π. However, in the principal coordinate system, the relative phase shift between the two orthogonal components is always $-\pi/2$ or $\pi/2$, depending on the sense of revolution.

Complex-Number Representation

From the previous discussion we found how the polarization state of a beam of light can be described in terms of the amplitudes and the phase angles of the x and y components of the electric field vector. In fact, all the information about the polarization state of a wave is contained in the complex amplitude $\mathbf{A}$ of the plane wave (Equation (1.6-4)). Therefore, a complex number χ defined as

$$\chi = e^{i\delta} \tan \psi = \frac{A_y}{A_x} e^{i(\delta_y - \delta_x)} \tag{1.6-17}$$

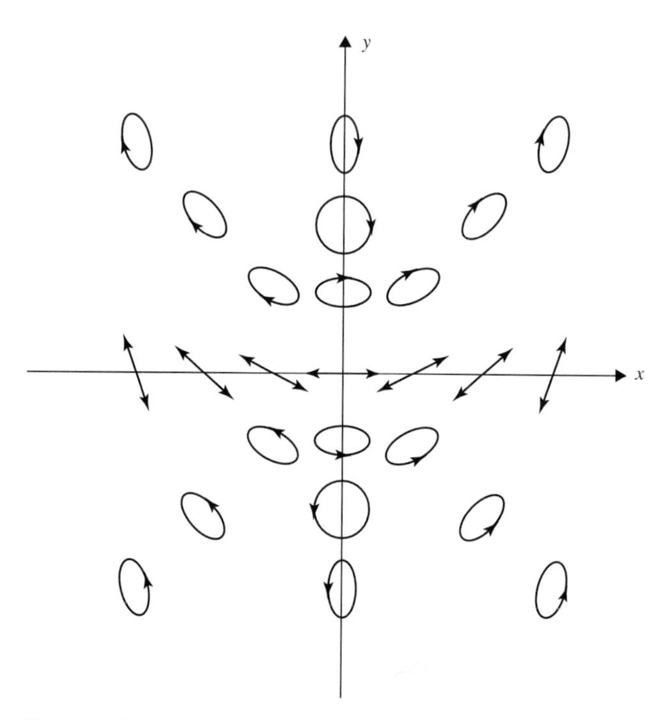

Figure 1.9 Complex-number representation of polarization states. Each point on the complex plane represents a unique state of polarization. The origin corresponds to a linearly polarized state along the x direction. Point (1, 0) represents a linearly polarized state with an azimuth angle of 45°, (0, 1) represents a left-handed circularly (LHC) polarized state, and (0, −1) represents a right-handed circularly (RHC) polarized state.

is sufficient to describe the polarization states. The angle ψ is defined to be between 0 and $\pi/2$. A complete description of the ellipse of polarization, which includes the orientation, sense of revolution, and ellipticity (see Equation (1.6-16)), can be expressed in terms of δ and ψ. Figure 1.9 illustrates various different polarization states in the complex plane. It can be seen from the figure that all the right-handed elliptical polarization states are in the lower half of the plane, whereas the left-handed elliptical polarization states are in the upper half of the plane. The origin corresponds to a linear polarization state with direction of oscillation parallel to the x axis. Thus each point on the complex plane represents a unique polarization state. Each point on the x axis represents a linearly polarized state with different azimuth angles of oscillation. Only two points $(0, \pm 1)$ correspond to circular polarization. Each point of the rest of the complex plane corresponds to a unique elliptical polarization state.

The inclination angle ϕ and the ellipticity angle θ ($\theta \equiv \tan^{-1} e$) of the polarization ellipse corresponding to a given complex number χ are given by

$$\tan 2\phi = \frac{2\mathrm{Re}[\chi]}{1 - |\chi|^2} = \tan 2\psi \cos \delta \tag{1.6-18}$$

and

$$\sin 2\theta = -\frac{2\mathrm{Im}[\chi]}{1 + |\chi|^2} = -\sin 2\psi \sin \delta \tag{1.6-19}$$

Jones Vector Representation

The Jones vector, introduced in 1941 by R. C. Jones [1], is very convenient for description of the polarization state of a plane wave. In this representation, the plane wave (1.6-4) is expressed in terms of its complex amplitudes as a column vector

$$\mathbf{J} = \begin{bmatrix} A_x e^{i\delta_x} \\ A_y e^{i\delta_y} \end{bmatrix} \tag{1.6-20}$$

Note that the Jones vector is a complex vector; that is, its elements are complex numbers. **J** is not a vector in the real physical space; rather, it is a vector in an abstract mathematical space. To obtain, as an example, the real x component of the electric field, we must perform the operation $E_x(t) = \mathrm{Re}[J_x e^{i\omega t}] = \mathrm{Re}[A_x e^{i(\omega t+\delta_x)}]$.

The Jones vector contains complete information about the amplitudes and phases of the electric-field-vector components. It thus specifies the polarization state of the wave uniquely. If we are only interested in the polarization state of the wave, it is convenient to use the normalized Jones vector, which satisfies the condition

$$\mathbf{J}^* \cdot \mathbf{J} = 1 \tag{1.6-21}$$

where the asterisk (∗) denotes complex conjugation. Thus a beam of linearly polarized light with the electric field vector oscillating along a given direction can be represented by the Jones vector

$$\begin{bmatrix} \cos\psi \\ \sin\psi \end{bmatrix} \tag{1.6-22}$$

where ψ is the azimuth angle of the oscillation direction with respect to the x axis. The state of polarization that is orthogonal to the state represented by Equation (1.6-22) can be obtained by substituting $\psi + \pi/2$ for ψ, leading to a Jones vector

$$\begin{bmatrix} -\sin\psi \\ \cos\psi \end{bmatrix} \tag{1.6-23}$$

The special case, when $\psi = 0$, represents linearly polarized waves whose electric field vector oscillates along the coordinate axes. These Jones vectors are given by

$$\hat{\mathbf{x}} = \begin{bmatrix} 1 \\ 0 \end{bmatrix} \quad \text{and} \quad \hat{\mathbf{y}} = \begin{bmatrix} 0 \\ 1 \end{bmatrix} \tag{1.6-24}$$

Jones vectors for the right- and left-handed circularly polarized light waves are given by

$$\mathbf{R} = \frac{1}{\sqrt{2}} \begin{bmatrix} 1 \\ -i \end{bmatrix} \tag{1.6-25}$$

$$\mathbf{L} = \frac{1}{\sqrt{2}} \begin{bmatrix} 1 \\ i \end{bmatrix} \tag{1.6-26}$$

These two states of circular polarizations are mutually orthogonal in the sense that

$$\mathbf{R}^* \cdot \mathbf{L} = 0 \tag{1.6-27}$$

Since the Jones vector is a column matrix of rank 2, any pair of orthogonal Jones vectors can be used as a basis of the mathematical space spanned by all the Jones vectors. Any polarization state can be represented as a superposition of two mutually orthogonal polarization states $\hat{\mathbf{x}}$ and $\hat{\mathbf{y}}$, or **R** and **L**. In particular, we can resolve the basic linear polarization states $\hat{\mathbf{x}}$ and $\hat{\mathbf{y}}$ into two circular polarization states **R** and **L** and vice versa. These relations are given by

$$\mathbf{R} = \frac{1}{\sqrt{2}}(\hat{\mathbf{x}} - i\hat{\mathbf{y}}) \tag{1.6-28}$$

$$\mathbf{L} = \frac{1}{\sqrt{2}}(\hat{\mathbf{x}} + i\hat{\mathbf{y}}) \tag{1.6-29}$$

$$\hat{\mathbf{x}} = \frac{1}{\sqrt{2}}(\mathbf{R} + \mathbf{L}) \tag{1.6-30}$$

$$\hat{\mathbf{y}} = \frac{i}{\sqrt{2}}(\mathbf{R} - \mathbf{L}) \tag{1.6-31}$$

Circular polarization states are seen to consist of linear oscillations along the x and y directions with equal amplitude $1/\sqrt{2}$, but with a phase difference of $\pi/2$. Similarly, a linear polarization state can be viewed as a superposition of two oppositely sensed circular polarization states.

We have so far discussed only the Jones vectors of some simple special cases of polarization. It is easy to show that a general elliptic polarization state can be represented by the following Jones vector:

$$\mathbf{J}(\psi, \delta) = \begin{bmatrix} \cos\psi \\ e^{i\delta}\sin\psi \end{bmatrix} \tag{1.6-32}$$

This Jones vector represents the same polarization state as the one represented by the complex number $\chi = e^{i\delta}\tan\psi$. Table 1.2 shows the Jones vectors of some typical polarization states.

The most important application of Jones vectors is in conjunction with the Jones calculus. This is a powerful technique used for studying the propagation of plane waves with arbitrary states of polarization through an arbitrary sequence of birefringent elements and polarizers. This topic will be considered in some detail later in this chapter.

Stokes Parameters and Partially Polarized Light

By virtue of its nature, a monochromatic plane wave must be polarized; that is, the end point of its electric field vector at each point in space must trace out periodically an ellipse or one of its special forms, such as a circle or a straight line. However, if the light is not absolutely monochromatic, the amplitudes and relative phase δ between the x and y components can both vary with time, and the electric field vector will first vibrate in one ellipse and then in another. As a result, the polarization state of a polychromatic plane wave may be constantly changing. If the polarization state changes more rapidly than the speed of observation, we say the light is partially polarized or unpolarized depending on the time-averaged behavior of the polarization state. In optical electronics, one often deals with light having oscillation frequencies of about $10^{14}\ \mathrm{s}^{-1}$, whereas the polarization state may change in a time period of 10^{-8} s depending on the nature of the light source.

We will limit ourselves to the case of quasi-monochromatic waves, whose frequency spectrum is confined to a narrow bandwidth $\Delta\omega$ (i.e., $\Delta\omega \ll \omega$). Such a wave can still be described by Equation (1.6-4), provided we relax the constancy condition of the amplitude **A**. Now ω denotes the center frequency, and the complex amplitude **A** is a function of time. Because the

TABLE 1.2 Various Representations of Polarization States

Polarization Ellipse	Jones Vector	(δ, ψ)	(ϕ, θ)	Stokes Vector
	$\begin{bmatrix} 1 \\ 0 \end{bmatrix}$	(0, 0)	(0, 0)	$\begin{bmatrix} 1 \\ 1 \\ 0 \\ 0 \end{bmatrix}$
	$\begin{bmatrix} 0 \\ 1 \end{bmatrix}$	$(0, \pi/2)$	$(\pi/2, 0)$	$\begin{bmatrix} 1 \\ -1 \\ 0 \\ 0 \end{bmatrix}$
	$\frac{1}{\sqrt{2}}\begin{bmatrix} 1 \\ 1 \end{bmatrix}$	$(0, \pi/4)$	$(\pi/4, 0)$	$\begin{bmatrix} 1 \\ 0 \\ 1 \\ 0 \end{bmatrix}$
	$\frac{1}{\sqrt{2}}\begin{bmatrix} 1 \\ -1 \end{bmatrix}$	$(\pi, \pi/4)$	$(-\pi/4, 0)$	$\begin{bmatrix} 1 \\ 0 \\ -1 \\ 0 \end{bmatrix}$
	$\frac{1}{\sqrt{2}}\begin{bmatrix} 1 \\ -i \end{bmatrix}$	$(-\pi/2, \pi/4)$	$(0, \pi/4)$	$\begin{bmatrix} 1 \\ 0 \\ 0 \\ -1 \end{bmatrix}$
	$\frac{1}{\sqrt{2}}\begin{bmatrix} 1 \\ i \end{bmatrix}$	$(\pi/2, \pi/4)$	$(0, -\pi/4)$	$\begin{bmatrix} 1 \\ 0 \\ 0 \\ 1 \end{bmatrix}$

bandwidth is narrow, $\mathbf{A}(t)$ may change only by a relatively small amount in a time interval $1/\Delta\omega$, and in this sense it is a slowly varying function of time. However, if the time constant of the detector, τ_D, is greater than $1/\Delta\omega$, $\mathbf{A}(t)$ may change significantly in a time interval τ_D. Although the amplitudes and phases are irregularly varying functions of time, certain correlations may exist among them.

To describe the polarization state of this type of radiation, we introduce the following time-averaged quantities:

$$\begin{aligned} S_0 &= \langle\langle A_x^2 + A_y^2 \rangle\rangle \\ S_1 &= \langle\langle A_x^2 - A_y^2 \rangle\rangle \\ S_2 &= 2\langle\langle A_x A_y \cos\delta \rangle\rangle \\ S_3 &= 2\langle\langle A_x A_y \sin\delta \rangle\rangle \end{aligned} \tag{1.6-33}$$

where the amplitudes A_x and A_y and the relative phase δ are assumed to be time dependent, and the double brackets denote averages performed over a time interval τ_D that is the characteristic time constant of the detection process. These four quantities are known as the *Stokes parameters* of a quasi-monochromatic plane wave. Note that all four quantities have the same dimension of intensity. It can be shown that the Stokes parameters satisfy the relation

$$S_1^2 + S_2^2 + S_3^2 \le S_0^2 \tag{1.6-34}$$

where the equality sign holds only for polarized waves.

It is a simple exercise to compute, from the definitions, various Stokes parameters of principal interest. Consider, for example, unpolarized light. There is no preference between A_x and A_y; consequently, $\langle\langle A_x^2 + A_y^2 \rangle\rangle$ reduces to $2\langle\langle A_x^2 \rangle\rangle$, and $\langle\langle A_x^2 - A_y^2 \rangle\rangle$ reduces to zero. The other quantities also reduce to zero because δ is a random function of time. If the field is normalized such that $S_0 = 1$, the Stokes vector representation of an unpolarized light wave is (1, 0, 0, 0). Similar reasoning shows that a horizontally polarized beam can be represented by the Stokes vector (1, 1, 0, 0), and a vertically polarized beam can be represented by (1, −1, 0, 0). Right-handed circularly polarized light ($\delta = -\pi/2$) is represented by (1, 0, 0, −1), and the left-handed circularly polarized light ($\delta = \pi/2$) is represented by (1, 0, 0, 1). From the definition, none of the parameters can be greater than the first S_0, which is normalized to 1. Therefore, each of the others lies in the range from −1 to 1. If the beam is entirely unpolarized, $S_1 = S_2 = S_3 = 0$. If it is completely polarized, $S_1^2 + S_2^2 + S_3^2 = 1$. The degree of polarization is therefore defined as

$$\gamma = \frac{(S_1^2 + S_2^2 + S_3^2)^{1/2}}{S_0} \tag{1.6-35}$$

According to Equation (1.6-35), this parameter γ is a real number between 0 and 1. It is thus very useful in describing the partially polarized light. The polarization preference of a partially polarized light can be seen directly from the sign of the parameters S_1, S_2, and S_3.

The parameter S_1 describes the linear polarization along the x or y axis; the probability that the light is linearly polarized along the x axis is $\frac{1}{2}(1 + S_1)$ and along the y axis, $\frac{1}{2}(1 - S_1)$. Thus the values $S_1 = 1, -1$ correspond to complete polarization in these directions. The parameter S_2 describes the linear polarization along directions at angles $\phi = \pm 45°$ to the x axis. The probability that the light is linearly polarized along these directions is, respectively, $\frac{1}{2}(1 + S_2)$ and $\frac{1}{2}(1 - S_2)$. Thus the values $S_2 = 1, -1$ correspond to complete polarization in these directions. Finally, the parameter S_3 represents the degree of circular polarization; the probability that the light wave has right-handed circular polarization is $\frac{1}{2}(1 - S_3)$, and left-handed circular polarization, $\frac{1}{2}(1 + S_3)$.

The Stokes parameters for a polarized light with a complex representation $\chi = e^{i\delta} \tan \psi$ are given by (according to Equation (1.6-17))

$$\begin{aligned} S_0 &= 1 \\ S_1 &= \cos 2\psi \\ S_2 &= \sin 2\psi \cos \delta \\ S_3 &= \sin 2\psi \sin \delta \end{aligned} \tag{1.6-36}$$

According to our convention, a positive S_3 corresponds to left-hand elliptical polarization ($\sin \delta > 0$, clockwise revolution).

Poincaré Sphere

Although the Stokes parameters are introduced for describing partially polarized light, they are also convenient parameters to describe the polarization states of polarized light. For polarized light, the Stokes parameters S_1, S_2, and S_3 can also be employed to represent the polarization states. Since $S_0 = 1$, all points with coordinate (S_1, S_2, S_3) are confined on the surface of a unit sphere in three-dimensional (3-D) space. This sphere is known as the Poincaré

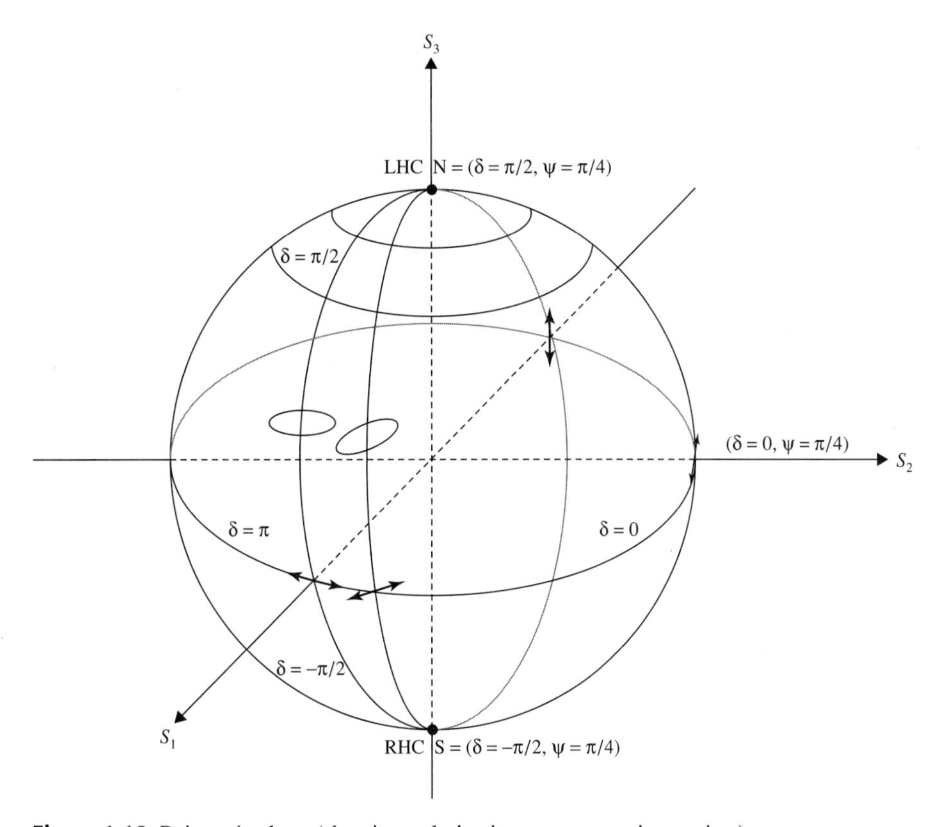

Figure 1.10 Poincaré sphere (showing polarization states at various points).

sphere (see Figure 1.10). Each point on the surface of the sphere represents a unique polarization state. For example, the north pole (0, 0, 1) corresponds to a left-handed circularly (LHC) polarized state, whereas the south pole (0, 0, −1) corresponds to a right-handed circularly (RHC) polarized state. Point (1, 0, 0) corresponds to a linear polarization state parallel to the horizontal direction, whereas point (−1, 0, 0) corresponds to a linear polarization state parallel to the vertical direction. In fact, all points on the equator correspond to a unique linear polarization state. The rest of the points correspond to elliptical polarization states. It is interesting to note that any pair of antipodal points on the Poincaré sphere (two points on the opposite side of the center of the sphere) correspond to states with orthogonal polarization.

According to Equations (1.6-18) and (1.6-19), we have

$$\tan 2\phi = S_2/S_1 \quad \text{and} \quad \sin 2\theta = -S_3 \tag{1.6-37}$$

where ϕ is the inclination angle of the polarization ellipse and θ is the ellipticity angle defined as $\tan^{-1}e$. Generally, S_2/S_1 = constant represents a vertical plane containing the poles. Since both S_1 and S_2 are confined on the surface of the sphere for polarized light, S_2/S_1 = constant actually represents a meridian—a half-circle connecting the north and south poles. According to Equation (1.6-37), ϕ is a constant on the meridian. Thus each meridian represents a class of elliptical polarization states with the same inclination angle ϕ, but with different ellipticities. In addition, S_3 = constant represents a circle on the sphere parallel to the equatorial plane. According to Equation (1.6-37), θ is a constant on this circle (parallel or latitude). Thus each parallel (latitude) represents a class of elliptical polarization states with the same ellipticity $e = \tan\theta$, but with different inclination angles. The Poincaré sphere is particularly useful in optical birefringent networks where wave plates are employed to change the polarization state of light. It is also particularly useful in the description of the evolution of polarization states in optical fibers when polarization mode dispersion (PMD) is present.

Consider two different points on the Poincaré sphere. Each point represents a polarization state. Let the Stokes vectors be written

$$\mathbf{S}_a = (1, S_{a1}, S_{a2}, S_{a3})$$
$$\mathbf{S}_b = (1, S_{b1}, S_{b2}, S_{b3}) \tag{1.6-38}$$

Using Equation (1.6-33), it can easily be shown that

$$\mathbf{S}_a \cdot \mathbf{S}_b = 2|\mathbf{J}_a^* \cdot \mathbf{J}_b|^2 \tag{1.6-39}$$

where $\mathbf{J}_a$ and $\mathbf{J}_b$ are the corresponding Jones vectors. For polarized light, the first component of the Stokes vectors is 1, so it is convenient to define three-component unit vectors, consisting of the three components S_1, S_2, and S_3 of the Stokes vectors as

$$\mathbf{s}_a = (S_{a1}, S_{a2}, S_{a3})$$
$$\mathbf{s}_b = (S_{b1}, S_{b2}, S_{b3}) \tag{1.6-40}$$

These three-component unit vectors are real vectors in Poincaré space. The tips of these unit vectors correspond to points on the Poincaré sphere. With this definition, Equation (1.6-39) can be written

$$|\mathbf{J}_a^* \cdot \mathbf{J}_b|^2 = \tfrac{1}{2}\mathbf{S}_a \cdot \mathbf{S}_b = \tfrac{1}{2}(1 + \mathbf{s}_a \cdot \mathbf{s}_b) \tag{1.6-41}$$

For a pair of antipodal points, $\mathbf{s}_a \cdot \mathbf{s}_b = -1$, the two polarization states form an orthogonal pair.

Stokes vectors and the Poincaré sphere will be employed later in this book to study polarization mode dispersion in optical fibers. It will be shown in Section 7.5 that the polarization transformation by a wave plate can easily be represented by a simple rotation on the Poincaré sphere. Let P be the input polarization state, P' be the output polarization state, and R be the polarization state of the slow mode of the wave plate. The output polarization state P' is obtained by rotating P through an angle Γ around the axis OR, where O is the origin of the sphere and Γ is the phase retardation of the wave plate (see Figure 7.9).

1.7 ELECTROMAGNETIC PROPAGATION IN ANISOTROPIC MEDIA (CRYSTALS)

In anisotropic media such as lithium niobate, quartz, nematic liquid crystals, and calcite, the propagation of electromagnetic radiation is determined by the dielectric tensor ε_{ij} that links the displacement vector and the electric field vector,

$$D_i = \varepsilon_{ij} E_j \tag{1.7-1}$$

where the convention of summation over repeated indices is observed. In nonmagnetic and transparent materials, this tensor is real and symmetric (see Problem 1.37):

$$\varepsilon_{ij} = \varepsilon_{ji} \tag{1.7-2}$$

The magnitude of these nine tensor elements depends, of course, on the choice of the x, y, and z axes relative to the crystal structure. Because of its real and symmetric nature, it is always possible to find three mutually orthogonal axes in such a way that the off-diagonal elements vanish, leaving

$$\varepsilon = \varepsilon_0 \begin{bmatrix} n_x^2 & 0 & 0 \\ 0 & n_y^2 & 0 \\ 0 & 0 & n_z^2 \end{bmatrix} = \begin{bmatrix} \varepsilon_x & 0 & 0 \\ 0 & \varepsilon_y & 0 \\ 0 & 0 & \varepsilon_z \end{bmatrix} \tag{1.7-3}$$

where ε_x, ε_y, and ε_z are the principal dielectric constants and n_x, n_y, and n_z are the principal indices of refraction. These directions (x, y, and z) are called the principal dielectric axes of the crystal. According to Equations (1.7-1) and (1.7-3), a plane wave propagating along the z axis can have two phase velocities, depending on its state of polarization. Specifically, the phase velocity is c/n_x for x-polarized light and c/n_y for y-polarized light. Generally, there are two normal modes of polarization for each direction of propagation.

It is important to note that the dielectric tensor (or dielectric constants) is a function of the frequency (or wavelength) of the electromagnetic field. This is known as dispersion. In the regime of optical waves, the frequencies are in the range of 10^{14}/s. We often use the refractive indices to describe the propagation of waves in optical media.

Plane Waves in Homogeneous Media and Normal Surface

To study such propagation along a general direction, we assume a monochromatic plane wave with an electric field vector

$$\mathbf{E} \exp[i(\omega t - \mathbf{k} \cdot \mathbf{r})] \tag{1.7-4}$$

and a magnetic field vector

$$\mathbf{H} \exp[i(\omega t - \mathbf{k} \cdot \mathbf{r})] \tag{1.7-5}$$

where $\mathbf{k}$ is the wavevector $\mathbf{k} = (\omega/c)n\mathbf{s}$, with $\mathbf{s}$ as a unit vector in the direction of propagation. The phase velocity c/n, or equivalently the refractive index n, is to be determined. Substituting $\mathbf{E}$ and $\mathbf{H}$ from Equations (1.7-4) and (1.7-5), respectively, into Maxwell's equations (1.1-1), (1.1-2) and (1.7-1) gives

$$\mathbf{k} \times \mathbf{E} = \omega\mu\mathbf{H} \tag{1.7-6}$$

$$\mathbf{k} \times \mathbf{H} = -\omega\varepsilon\mathbf{E} = -\omega\mathbf{D} \tag{1.7-7}$$

By eliminating $\mathbf{H}$ from Equations (1.7-6) and (1.7-7), we obtain

$$\mathbf{k} \times (\mathbf{k} \times \mathbf{E}) + \omega^2\mu\varepsilon\mathbf{E} = 0 \tag{1.7-8}$$

This equation will now be used to solve for the eigenvectors $\mathbf{E}$ and the corresponding eigenvalues n.

In the principal coordinate system, the dielectric tensor ε is given by Equation (1.7-3). Equation (1.7-8) can be written

$$\begin{bmatrix} \omega^2\mu\varepsilon_x - k_y^2 - k_z^2 & k_x k_y & k_x k_z \\ k_y k_x & \omega^2\mu\varepsilon_y - k_x^2 - k_z^2 & k_y k_z \\ k_z k_x & k_z k_y & \omega^2\mu\varepsilon_z - k_x^2 - k_y^2 \end{bmatrix} \begin{bmatrix} E_x \\ E_y \\ E_z \end{bmatrix} = 0 \tag{1.7-9}$$

where we recall that $\varepsilon_x = \varepsilon_0 n_x^2$, $\varepsilon_y = \varepsilon_0 n_y^2$, and $\varepsilon_z = \varepsilon_0 n_z^2$.

For nontrivial solutions to exist, the determinant of the matrix in Equation (1.7-9) must vanish. This leads to a relation between ω and $\mathbf{k}$:

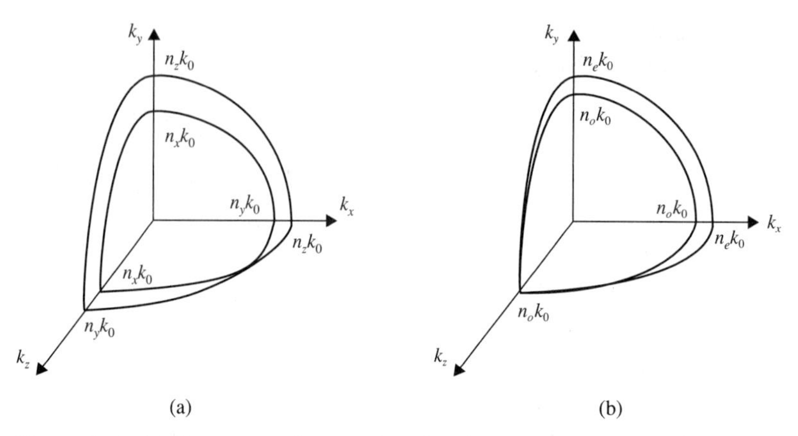

Figure 1.11 (a) One octant of a normal surface in momentum space with $n_x < n_y < n_z$. k_x, k_y, and k_z are in units of $k_0 = \omega/c$. (b) One octant of a normal surface in momentum space with $n_o = n_x = n_y < n_z = n_e$, $k_0 = \omega/c$.

$$\det \begin{vmatrix} \omega^2\mu\varepsilon_x - k_y^2 - k_z^2 & k_x k_y & k_x k_z \\ k_y k_x & \omega^2\mu\varepsilon_y - k_x^2 - k_z^2 & k_y k_z \\ k_z k_x & k_z k_y & \omega^2\mu\varepsilon_z - k_x^2 - k_y^2 \end{vmatrix} = 0 \qquad (1.7\text{-}10)$$

At a given frequency ω, this equation represents a three-dimensional surface in **k** space (momentum space). This surface, known as the *normal surface*, consists of two shells. These two shells, in general, have four points in common. The two lines that go through the origin and these points are known as the optic axes. Figure 1.11 shows one octant of a general normal surface. Given a direction of propagation, there are in general two k values that are the intersections of the direction of propagation **s** and the normal surface. These two k values correspond to two different phase velocities (ω/k) of the waves propagating along the chosen direction. The directions of the electric field vector associated with these propagations can also be obtained from Equation (1.7-9) and are given by

$$\begin{bmatrix} \dfrac{k_x}{k^2 - \omega^2\mu\varepsilon_x} \\ \dfrac{k_y}{k^2 - \omega^2\mu\varepsilon_y} \\ \dfrac{k_z}{k^2 - \omega^2\mu\varepsilon_z} \end{bmatrix} \qquad (1.7\text{-}11)$$

provided that the denominators do not vanish. In the case when $n_o = n_x = n_y < n_z = n_e$, the normal surface consists of a sphere and an ellipsoid of revolution with the z axis being the axis of symmetry.

For propagation in the direction of the optic axes, there is only one value of k and consequently only one phase velocity. There are, however, two independent directions of polarization. Equations (1.7-10) and (1.7-11) are often written in terms of the direction cosines of the wavevector. By using the relation $\mathbf{k} = (\omega/c)n\mathbf{s}$ for the plane wave given by Equation (1.7-4), Equations (1.7-10) and (1.7-11) can be written as

$$\frac{s_x^2}{n^2 - n_x^2} + \frac{s_y^2}{n^2 - n_y^2} + \frac{s_z^2}{n^2 - n_z^2} = \frac{1}{n^2} \qquad (1.7\text{-}12)$$

and

$$\begin{bmatrix} \dfrac{s_x}{n^2 - n_x^2} \\ \dfrac{s_y}{n^2 - n_y^2} \\ \dfrac{s_z}{n^2 - n_z^2} \end{bmatrix} \tag{1.7-13}$$

respectively, where we have used $\varepsilon_x = \varepsilon_0 n_x^2$, $\varepsilon_y = \varepsilon_0 n_y^2$, and $\varepsilon_z = \varepsilon_0 n_z^2$.

Equation (1.7-12) is known as *Fresnel's equation of wave normals* and can be solved for the eigenvalues of index of refraction, and Equation (1.7-13) gives the directions of polarization. Equation (1.7-12) is a quadratic equation in n^2. Therefore, for each direction of propagation (a set of s_x, s_y, s_z), two solutions for n^2 can be obtained from Equation (1.7-12). To complete the solution of the problem, we use the values of n^2, one at a time, in Equation (1.7-13). This gives us the polarizations (electric field vectors) of these waves. It can be seen that in a nonabsorbing medium these normal modes are linearly polarized since all the components are real in Equation (1.7-13). Let $\mathbf{E}_1$ and $\mathbf{E}_2$ be the electric field vectors and $\mathbf{D}_1$ and $\mathbf{D}_2$ be the displacement vectors of the linearly polarized normal modes associated with n_1^2 and n_2^2, respectively. Maxwell's equation $\nabla \cdot \mathbf{D} = 0$ requires that $\mathbf{D}_1$ and $\mathbf{D}_2$ are orthogonal to $\mathbf{s}$. Since $\mathbf{D}_1 \cdot \mathbf{D}_2 = 0$ (the proof of this orthogonal relation is left as a problem for students), the three vectors $\mathbf{D}_1$, $\mathbf{D}_2$, and $\mathbf{s}$ form an orthogonal triad. According to Equations (1.7-6) and (1.7-7), $\mathbf{D}$ and $\mathbf{H}$ are both perpendicular to the direction of propagation $\mathbf{s}$. Consequently, the direction of energy flow as given by the Poynting vector $\mathbf{E} \times \mathbf{H}$ is, in general, not collinear with the direction of propagation $\mathbf{s}$. Since $\mathbf{D}$, $\mathbf{E}$, and $\mathbf{k}$ are all orthogonal to $\mathbf{H}$, they must lie in the same plane.

Orthogonality of Normal Modes (Eigenmodes)

It can be shown that $\mathbf{D}$, $\mathbf{E}$, and $\mathbf{s}$ all lie in the same plane. In addition, these field vectors satisfy the following relations (see Problem 1.27):

$$\begin{gathered} \mathbf{D}_1 \cdot \mathbf{D}_2 = 0 \\ \mathbf{D}_1 \cdot \mathbf{E}_2 = 0 \\ \mathbf{D}_2 \cdot \mathbf{E}_1 = 0 \\ \mathbf{s} \cdot \mathbf{D}_1 = \mathbf{s} \cdot \mathbf{D}_2 = 0 \end{gathered} \tag{1.7-14}$$

The electric field vectors $\mathbf{E}_1$ and $\mathbf{E}_2$ are, in general, not orthogonal. The general orthogonality relation of the eigenmodes of propagation is often written

$$\mathbf{s} \cdot (\mathbf{E}_1 \times \mathbf{H}_2) = 0 \tag{1.7-15}$$

This latter relation shows that the power flow in a lossless anisotropic medium along the direction of propagation is the sum of the power carried by each mode individually.

To summarize: Along an arbitrary direction of propagation $\mathbf{s}$, there can exist two independent plane wave, linearly polarized propagation modes. These modes have phase velocities $\pm(c/n_1)$ and $\pm(c/n_2)$, where n_1^2 and n_2^2 are the two solutions of Fresnel's equation (1.7-12). The electric field vectors of these two normal modes are given by Equation (1.7-11) or (1.7-13).

Although Equation (1.7-11) provides a general explicit expression for the polarization states (**E** vector) of the normal modes in a general anisotropic medium, one must be careful in applying the equation for propagation along the principal axes or principal planes when the denominators in Equation (1.7-11) can become zero. For propagation along these special directions, it is often easier to obtain the normal modes directly from the wave equation (1.7-9).

Classification of Media

We have shown above that the normal surface contains a good deal of information about the wave propagation in anisotropic media. The normal surface is uniquely determined by the principal indices of refraction n_x, n_y, n_z. In the general case when the three principal indices n_x, n_y, n_z are all different, there are two optical axes. In this case, the medium is said to be biaxial. In many optical electronic materials (e.g., $LiNbO_3$ crystals, nematic liquid crystals) it happens that two of the principal indices are equal, in which case the equation for the normal surface (Equation (1.7-10) or (1.7-12)) can be factored according to

$$\left(\frac{k_x^2 + k_y^2}{n_e^2} + \frac{k_z^2}{n_o^2} - \frac{\omega^2}{c^2}\right)\left(\frac{k^2}{n_o^2} - \frac{\omega^2}{c^2}\right) = 0 \tag{1.7-16}$$

where $n_o^2 = \varepsilon_x/\varepsilon_0 = \varepsilon_y/\varepsilon_0$ and $n_e^2 = \varepsilon_z/\varepsilon_0$.

The normal surface in this case consists of a sphere and an ellipsoid of revolution (see Figure 1.11b). These two sheets of the normal surface touch at two points on the z axis. The z axis is therefore the only optic axis, and the medium is said to be uniaxial. If all three principal indices are equal, the two sheets of normal surface degenerate to a single sphere, and the medium is optically isotropic.

In a biaxial medium, the principal coordinate axes can be labeled in such a way that the three principal indices are in the following order:

$$n_x < n_y < n_z \tag{1.7-17}$$

In this convention, the optical axes lie in the xz plane. Cross sections of the normal surfaces with the xz plane are shown in Figure 1.12a. In a uniaxial medium, the index of refraction that corresponds to the two equal elements, $n_o^2 = \varepsilon_x/\varepsilon_0 = \varepsilon_y/\varepsilon_0$, is called the ordinary index n_o; the other index, corresponding to ε_z, is called the extraordinary index n_e. If $n_o < n_e$, the medium is said to be positive, whereas if $n_o > n_e$, it is said to be negative. Most liquid crystals with rodlike molecules are positive uniaxial media with $n_o < n_e$. Intersections of the normal surfaces with the xz plane are again shown in Figures 1.12b and 1.12c. The optic axis corresponds to the principal axis, which has a unique index of refraction. Table 1.3 lists some examples of solid crystals with their indices of refraction.

The Index Ellipsoid

The surface of constant energy density U_e in **D** space can be written

$$\frac{D_x^2}{\varepsilon_x} + \frac{D_y^2}{\varepsilon_y} + \frac{D_z^2}{\varepsilon_z} = 2U_e \tag{1.7-18}$$

where ε_x, ε_y, and ε_z are the principal dielectric constants. If we replace $\mathbf{D}/\sqrt{2U_e}$ by $\mathbf{r}$ and use the principal refractive indices $n_i^2 = \varepsilon_i/\varepsilon_0$ ($i = x, y, z$), the last equation can be written

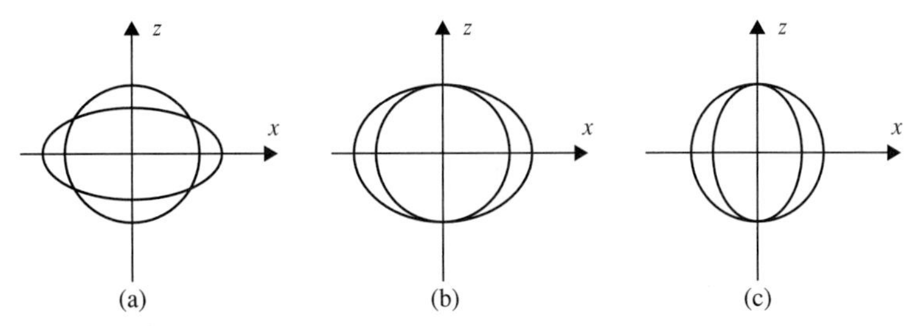

Figure 1.12 Intersection of the normal surface with the xz plane for (a) biaxial media (with $n_x < n_y < n_z$), (b) uniaxial media with positive birefringence ($n_o < n_e$), and (c) uniaxial media with negative birefringence ($n_e < n_o$).

TABLE 1.3 Refractive Indices[a] of Some Typical Solid Crystals

Isotropic	Fluorite		1.392	
	Sodium chloride, NaCl		1.544	
	Diamond, C		2.417	
	CdTe		2.69	
	GaAs		3.40	
	Ge		3.40	
	InP		3.61	
	GaP		3.73	
Uniaxial		n_o		n_e
Positive	MgF_2	1.378		1.390
	Quartz, SiO_2	1.544		1.553
	Beryllium oxide, BeO	1.717		1.732
	$La_3Ga_5SiO_{14}$	1.90		1.91
	ZnO	1.94		1.96
	SnO_2	2.01		2.10
	YVO_4	1.96		2.16
	$LiTaO_3$	2.183		2.188
	ZnS	2.354		2.358
	Rutile, TiO_2	2.616		2.903
Negative	KDP, KH_2PO_4	1.507		1.467
	ADP, $(NH_4)\ H_2PO_4$	1.522		1.478
	Beryl, $Be_3Al_2(SiO_3)_6$	1.598		1.590
	Sodium nitrate, $NaNO_3$	1.587		1.366
	Calcite, $CaCO_3$	1.658		1.486
	β-BaB_2O_4 (BBO)	1.67		1.55
	Sapphire, Al_2O_3	1.768		1.760
	Lithium niobate, $LiNbO_3$	2.300		2.208
	$PbMoO_3$	2.40		2.27
	Proustite, Ag_3AsS_3	3.019		2.739
Biaxial		n_x	n_y	n_z
	Gypsum	1.520	1.523	1.530
	Feldspar	1.522	1.526	1.530
	Mica	1.552	1.582	1.588
	Topaz	1.619	1.620	1.627
	Sodium nitrite, $NaNO_2$	1.344	1.411	1.651
	$YAlO_3$	1.923	1.938	1.947
	SbSI	2.7	1.7	3.8

[a] The refractive indices of most materials depend on the wavelength (dispersion). The listed numbers are typical values.

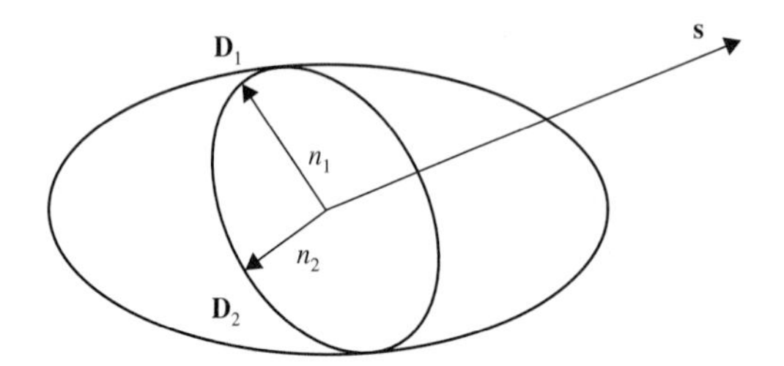

Figure 1.13 Method of index of ellipsoid. The inner ellipse is the intersection of the index ellipsoid with the plane that is normal to **s** and passes through the center of the ellipsoid.

$$\frac{x^2}{n_x^2} + \frac{y^2}{n_y^2} + \frac{z^2}{n_z^2} = 1 \tag{1.7-19}$$

This is the equation of a general ellipsoid with major axes parallel to the principal axes of the crystal, whose respective lengths are $2n_x$, $2n_y$, and $2n_z$. The ellipsoid is known as the *index ellipsoid* or, sometimes, as the *optical indicatrix*. The index ellipsoid is used mainly to find the **D** vectors and the two corresponding indices of refraction of the normal modes of propagation along a given direction of propagation **s** in a crystal. This is done by means of the following prescription: Find the intersection ellipse between the index ellipsoid and a plane through the origin that is normal to the direction of propagation **s**. The two axes of the intersection ellipse are equal in length to $2n_1$ and $2n_2$, where n_1 and n_2 are the two indices of refraction associated with the normal modes of propagation, that is, the solutions of Equation (1.7-12). These two axes are parallel, respectively, to the **D** vectors of the normal modes of propagation (see Figure 1.13). A formal proof of such a procedure is given in References [2, 3].

1.8 PLANE WAVES IN UNIAXIALLY ANISOTROPIC MEDIA—PHASE RETARDATION

Many of the optical electronic materials (e.g., $LiNbO_3$ crystal) are optically uniaxial. In addition, most liquid crystal displays involve the use of nematic liquid crystals. A planar nematic slab is a good example of a homogeneous uniaxial liquid crystal. Thus the propagation of optical waves in uniaxially anisotropic media deserves special attention. For the purpose of discussion, we rewrite the normal surface of a uniaxial medium as

$$\left(\frac{k_x^2 + k_y^2}{n_e^2} + \frac{k_z^2}{n_o^2} - \frac{\omega^2}{c^2}\right)\left(\frac{k^2}{n_o^2} - \frac{\omega^2}{c^2}\right) = 0 \tag{1.8-1}$$

We note that the normal surface consists of two parts. The sphere gives the relation between ω and **k** of the ordinary (O) wave. The ellipsoid of revolution gives the similar relation for the extraordinary (E) wave. These two surfaces touch at two points on the z axis. The eigen refractive indices associated with these two modes of propagation are given by

$$O \text{ wave:} \quad n = n_o \tag{1.8-2}$$

$$E \text{ wave:} \quad \frac{1}{n^2} = \frac{\cos^2\theta}{n_o^2} + \frac{\sin^2\theta}{n_e^2} \tag{1.8-3}$$

where θ is the angle between the direction of propagation and the c axis (the crystal z axis). For propagation along the optic axis (c axis), the eigen refractive indices of both modes are n_o, according to Equations (1.8-2) and (1.8-3).

The electric field vector of the O wave cannot be obtained directly from Equation (1.7-13) because of the vanishing denominators. It can easily be obtained from Equation (1.7-9). By using $\varepsilon_x = \varepsilon_y = \varepsilon_0 n_o^2$, $\varepsilon_z = \varepsilon_0 n_e^2$, and $\mathbf{k}_o = (\omega/c) n_o \mathbf{s}$, Equation (1.7-9) can be written

$$\begin{bmatrix} s_x^2 & s_x s_y & s_x s_z \\ s_y s_x & s_y^2 & s_y s_z \\ s_z s_x & s_z s_y & (n_e/n_o)^2 - (s_x^2 + s_y^2) \end{bmatrix} \begin{bmatrix} E_x \\ E_y \\ E_z \end{bmatrix} = 0 \tag{1.8-4}$$

A simple inspection of this equation yields the following direction of polarization:

$$O \text{ wave:} \quad \mathbf{E} = \begin{bmatrix} s_y \\ -s_x \\ 0 \end{bmatrix} \tag{1.8-5}$$

where we recall that s_x, s_y, s_z are directional cosines of the direction of propagation. The electric field vector of the E wave can be obtained from Equation (1.7-13) and is given by

$$E \text{ wave:} \quad \mathbf{E} = \begin{bmatrix} \dfrac{s_x}{n^2 - n_o^2} \\ \dfrac{s_y}{n^2 - n_o^2} \\ \dfrac{s_z}{n^2 - n_e^2} \end{bmatrix} \tag{1.8-6}$$

where n is given by Equation (1.8-3). The corresponding wavevector is $\mathbf{k}_e = (\omega/c) n\mathbf{s}$.

Notice that the electric field vector of the O wave is perpendicular to the plane formed by the wavevector $\mathbf{k}_o$ and the c axis, whereas the electric field vector of the E wave is not exactly perpendicular to the wavevector $\mathbf{k}_e$. However, the deviation from 90° is very small. This small angle between the field vectors $\mathbf{E}$ and $\mathbf{D}$ is also the angle between the phase velocity and the group velocity (see Problem 1.30). Therefore, for practical purposes, we may assume that the electric field is transverse to the direction of propagation. The displacement vectors $\mathbf{D}$ of the normal modes are exactly perpendicular to the wavevectors $\mathbf{k}_o$ and $\mathbf{k}_e$, respectively, and can be written

$$O \text{ wave:} \quad \mathbf{D}_o = \frac{\mathbf{k}_o \times \mathbf{c}}{|\mathbf{k}_o \times \mathbf{c}|} \tag{1.8-7}$$

$$E \text{ wave:} \quad \mathbf{D}_e = \frac{\mathbf{D}_o \times \mathbf{k}_e}{|\mathbf{D}_o \times \mathbf{k}_e|} \tag{1.8-8}$$

where $\mathbf{c}$ is a unit vector parallel to the c axis of the crystal.

Let (θ, ϕ) be the angle of propagation in spherical coordinates. The unit vector $\mathbf{s}$ can be written

$$\mathbf{s} = \begin{bmatrix} \sin\theta \cos\phi \\ \sin\theta \sin\phi \\ \cos\theta \end{bmatrix} \tag{1.8-9}$$

Using Equation (1.8-9), the normal modes for **E** can be written

$$O \text{ wave:} \quad \mathbf{E}_o = \begin{bmatrix} \sin\phi \\ -\cos\phi \\ 0 \end{bmatrix} \tag{1.8-10}$$

$$E \text{ wave:} \quad \mathbf{E}_e = \begin{bmatrix} n_e^2 \cos\theta\cos\phi \\ n_e^2 \cos\theta\sin\phi \\ -n_o^2 \sin\theta \end{bmatrix} \tag{1.8-11}$$

We note that these two modes are mutually orthogonal. The normal modes for **D** can be written

$$O \text{ wave:} \quad \mathbf{D}_o = \begin{bmatrix} \sin\phi \\ -\cos\phi \\ 0 \end{bmatrix} \tag{1.8-12}$$

$$E \text{ wave:} \quad \mathbf{D}_e = \begin{bmatrix} \cos\theta\cos\phi \\ \cos\theta\sin\phi \\ -\sin\theta \end{bmatrix} \tag{1.8-13}$$

These two **D** vectors and **s** are also mutually orthogonal. The above results can also be obtained by using the index ellipsoid.

If inside the uniaxial medium a polarized light is generated that is to propagate along a direction **s**, the displacement vector of this light can always be written as a linear combination of these two normal modes; that is,

$$\mathbf{D} = C_o\mathbf{D}_o \exp(-i\mathbf{k}_o \cdot \mathbf{r}) + C_e\mathbf{D}_e \exp(-i\mathbf{k}_e \cdot \mathbf{r}) \tag{1.8-14}$$

where C_o and C_e are constants and $\mathbf{k}_o$ and $\mathbf{k}_e$ are the wavevectors that are, in general, different (Equation (1.8-2)). As the light propagates inside the medium, a phase retardation between these two components is built up due to the difference in their phase velocities. Such a phase retardation between the two components leads to a new polarization state. Thus birefringent plates can be used to alter the polarization state of light. For a parallel plate with thickness d, the phase retardation can be written

$$\Gamma = (k_{ez} - k_{oz})d \tag{1.8-15}$$

where k_{ez} and k_{oz} are the z components of the wavevectors and the z axis is perpendicular to the surface of the plates.

Optical Rotatory Power and Faraday Rotation

In addition to linear birefringence in crystals, optical electronic materials (e.g., quartz crystal) exist in which the normal modes of propagation are circularly polarized. These media are said to be circularly birefringent or optically active (OA). When a beam of linearly polarized light traverses through these materials, the plane of polarization is rotated by an angle that is proportional to the distance of propagation. It was first observed in 1811 that quartz crystal rotates the plane of polarization of a beam of linearly polarized light traversing through it in the direction of the optical axis. The plane of polarization rotates 21.7 degrees per mm for sodium light ($\lambda = 589$ nm) at 20 °C. Both right-handed and left-handed rotations exist in different specimens. The optical activity of liquids was first observed in sugar solutions in 1815.

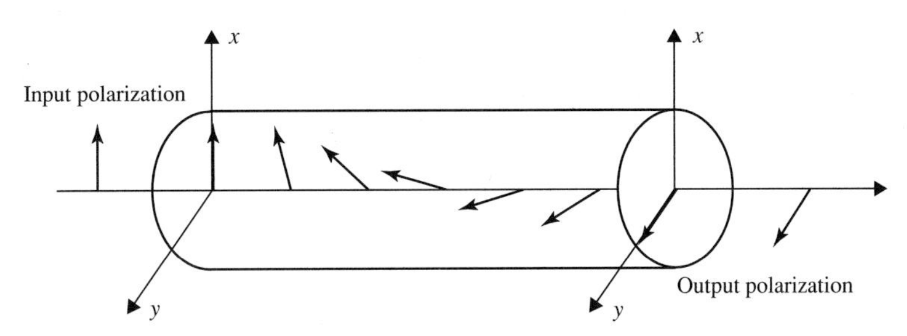

Figure 1.14 Rotation of the plane of polarization by an optically active medium.

It is known that fused quartz (silica) is optically inactive. The optical rotatory power in solids is due to the crystalline structure and the arrangement of molecules in the crystal. In liquids, the molecular orientations are random. Thus the cause of optical rotatory power in liquids must lie in the structure of the molecules themselves. Virtually all molecules in optically active liquids possess an asymmetric atom of carbon, nitrogen, or sulfur. In addition, all these liquids have a twin substance (isomer) with an opposite rotatory power. The existence of mirror-image isomers is a result of the presence of one or more asymmetric carbon atoms in the compound. Thus, these molecular structures can have left- and right-hand (chiral) forms. These forms are conventionally designated as dextro (D) and levo (L) because they compare to each other structurally as do the right and left hands when the molecule is compared with its mirror-image isomer. Optical rotatory power in liquids is due to the asymmetric structural arrangement (e.g., helical structure) of atoms in the randomly oriented molecules.

Figure 1.14 illustrates the rotation of the plane of polarization by an optically active medium. The amount of rotation is proportional to the path length of light in the medium. Conventionally, the rotatory power of a medium is given in degrees per mm; that is, the specific rotatory power is defined as the amount of rotation per unit length.

The sense of rotation in optically active materials bears a fixed relation to the propagation wavevector (**k**) of the beam of light, so if the beam of light is made to traverse the medium once in each of two opposite directions, as in the case of reflection from the right end face in Figure 1.14, the net rotation is zero. A substance is called dextrorotatory (or right-handed) if the sense of rotation of the plane of polarization is counterclockwise as viewed by an observer facing the approaching light beam. If the sense of rotation is clockwise, the substance is called levorotatory (or left-handed). Quartz occurs in both right-handed and left-handed crystalline forms. Many other substances are now known to exhibit optical activity; these include cinnabar, sodium chlorate, turpentine, sugar, strychnine sulfate, tellurium, selenium, and silver thiogallate ($AgGaS_2$). Many liquid crystal materials and organic compounds also exhibit optical rotatory power. The specific rotatory powers of some optically active media are given in Table 1.4.

Fresnel first recognized in 1825 that the optical activity arises from circular double refraction, in which the eigenwaves of propagation (i.e., the independent plane wave solutions of Maxwell's equations) are right and left circularly polarized waves. Let n_r and n_l be the refractive indices associated with these two waves, and assume that the waves are propagating in the $+z$ direction. It can be shown that the specific rotatory power is given by [2, 3]

$$\rho = \frac{\pi}{\lambda}(n_l - n_r) \tag{1.8-16}$$

TABLE 1.4 Optical Rotatory Powers of Some Solids

	λ (Å)	ρ (degree/mm)
Quartz	4000	49
	4550	37
	5000	31
	5500	26
	6000	22
	6500	17
$AgGaS_2$	4850	950
	4900	700
	4950	600
	5000	500
	5050	430
Se	7500	180
	10,000	30
Te	(6 μm)	40
	(10 μm)	15
TeO_2	3698	587
	4382	271
	5300	143
	6328	87
	10,000	30

The optical rotation is right-handed (counterclockwise) if $n_r < n_l$. Thus the plane of polarization turns in the same sense as the circularly polarized wave, which travels with the greater phase velocity. It can be shown that the polarization ellipse of a beam of elliptically polarized light will rotate the same angle $\omega z(n_l - n_r)/2c$ with its shape remaining unchanged (see Problem 1.18).

Faraday Rotation

In 1845, Faraday discovered that isotropic substances rotate the plane of polarization of linearly polarized light when placed in a strong magnetic field. When the field is reversed, the rotation is reversed. This is known as the Faraday effect. The rotation is proportional to the length of the optical path, and to the component of the magnetic field along the direction of propagation of the light. In an optically active medium, the direction of rotation bears a fixed relation to the direction of propagation, so that if a beam of light is reflected back on itself, the net rotation is zero. Such a medium is said to be reciprocal. In the Faraday effect, however, the rotation bears a fixed relation to the magnetic field **B**, so that reflection back on itself doubles the rotation. Such a phenomenon is known as nonreciprocal.

The specific rotation (i.e., rotation per unit length) of a Faraday cell is often written

$$\rho = VB \tag{1.8-17}$$

where B is the component of magnetic field along the direction of propagation and V is a constant known as the Verdet constant.

The Faraday effect originates from the effect of the static magnetic field on the motion of electrons via the Lorentz force. Faraday rotation has been observed in many solids, liquids, and even gases. A few values of the Verdet constant are given in Table 1.5.

TABLE 1.5 Value of the Verdet Constant at $\lambda = 5893$ Å

Substance	T (°C)	V (deg/G-mm)[a]
Water	20	2.18×10^{-5}
Fluorite		1.5×10^{-6}
Diamond		2.0×10^{-5}
Glass (crown)	18	2.68×10^{-5}
Glass (flint)		5.28×10^{-5}
Carbon disulfide (CS_2)	20	7.05×10^{-5}
Phosphorus	33	2.21×10^{-4}
Sodium chloride		6.0×10^{-5}
MBBA	20	6.67×10^{-5}

[a] 1 gauss = 10^{-4} tesla.

From the atomic point of view, the Faraday effect is related to the Zeeman effect. As a result of interaction between the orbiting electrons and the magnetic field, each electron energy level is split into several sublevels. By virtue of conservation of angular momentum, RHC polarized light and LHC polarized light that propagate along the direction of the magnetic field will interact with different sets of the sublevels. Thus the medium exhibits circular birefringence in the presence of the magnetic field, leading to a rotation of the polarization vector.

Faraday Isolator and Optical Circulator

Faraday effect plays an important role in nonreciprocal optical devices such as isolators and circulators. These nonreciprocal optical devices now form an integral part of most optical communication networks employing semiconductor lasers and amplifiers. Most of these lasers are extremely sensitive to even small amounts of reflected light, which cause instabilities in their output power and frequency characteristics. Optical isolators are made of Faraday cells in conjunction with polarizers that virtually remove all reflected light. Figure 1.15 shows the principle of operation of an optical isolator.

In an optical isolator as shown in Figure 1.15, a beam of polarized light will undergo a 45° rotation of the direction of propagation. Upon reflection, the same beam of light will undergo an additional 45° rotation, resulting in a net rotation of 90°. Using polarizers, it is possible to eliminate reflected light. As we know, reflected light can cause stability problems when the light enters the laser cavity. Optical isolators play an important role in optical networks to ensure stable operation of lasers. Figures 1.15c and 1.15d show an optical arrangement of an isolator for a beam of unpolarized light. A beam of unpolarized light is split into two polarized components by a polarizing beam splitter (PBS). These two polarized components are rotated by 90° by a combination of a 45° Faraday cell (FC) and a 45° optical rotator (OA). After rotation, these two components are recombined at an output PBS. For reflected light as shown in Figure 1.15d, the net rotation through the combination of FC and OA is zero. As a result, the reflected light is directed to a new output port.

1.9 JONES MATRIX METHOD

In Section 1.8, we discussed the propagation of plane waves in homogeneous and anisotropic media. Any plane wave propagation can easily be described in terms of a superposition (or linear combination) of the propagation of two orthogonally polarized components. Many

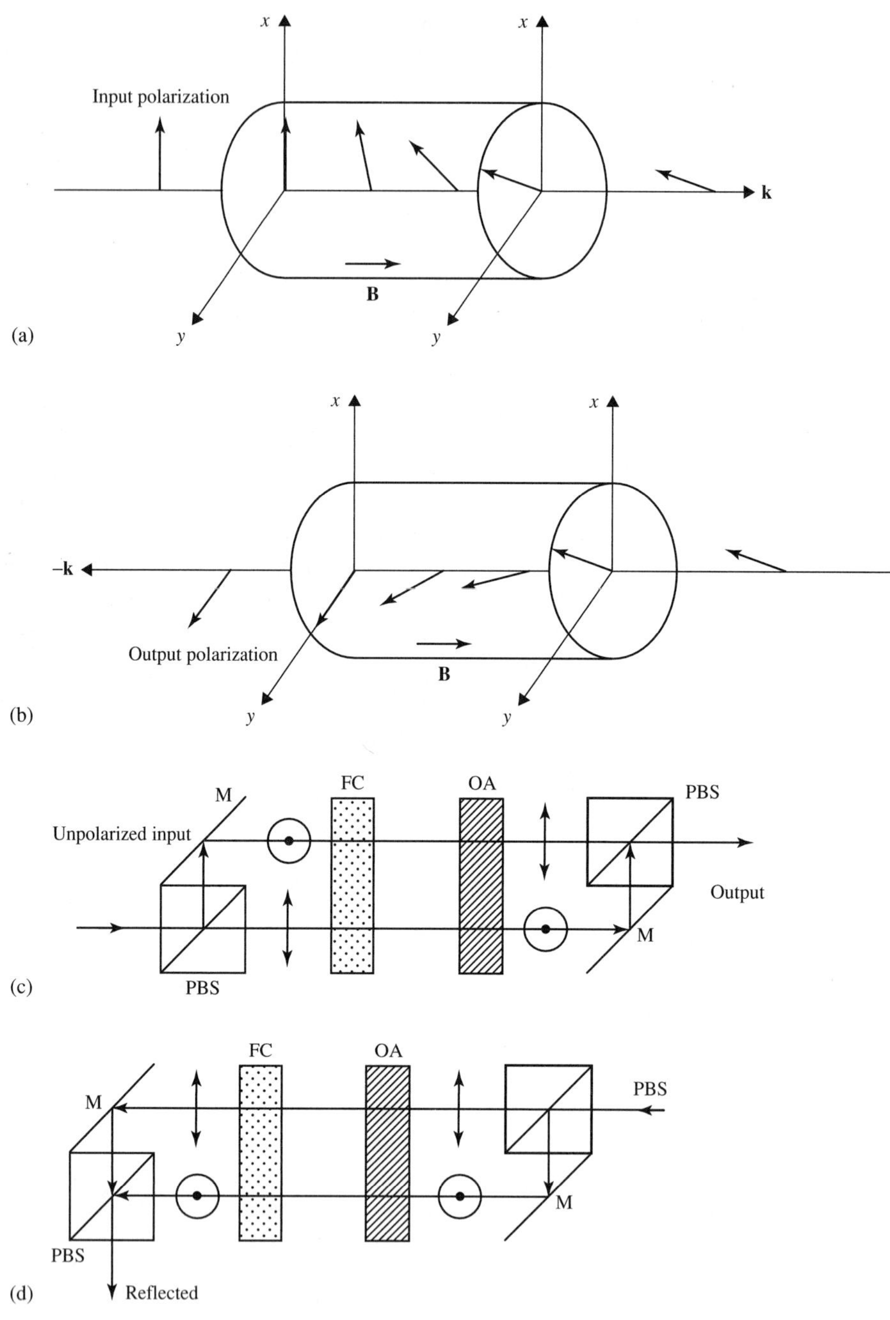

Figure 1.15 (a) A beam of linearly polarized light is rotated 45° by a Faraday rotator. (b) The retroreflected beam is rotated by an additional 45° by the Faraday rotator. (c) Schematic drawing of an optical circulator that also functions as an isolator. (d) Schematic drawing of the same circulator showing the reflected light is directed at a new output port. M = mirror, PBS = polarizing beam splitter, FC = Faraday cell, OA = optical rotator.

optical systems consist of a sequence of birefringent media. In the case of an imperfect optical fiber (e.g., silica fiber with an elliptical core), the birefringence may depend on the position along the fiber. In a situation like this, a systematic method is needed to treat the propagation of light through a series of birefringent elements. In this section, we describe such a powerful technique, known as the Jones matrix method [1].

We have shown in Section 1.8 that light propagation in a birefringent medium consists of a linear superposition of two normal modes. These normal modes have well-defined phase velocities and directions of polarization. The birefringent media may be either uniaxial or biaxial. However, most commonly used materials in optical electronics are uniaxial. In a uniaxial medium, these normal modes are the ordinary and the extraordinary waves. The directions of polarization for these normal modes are mutually orthogonal and are called the "slow" and "fast" axes of the medium for that direction of propagation. In traditional birefringent optics, retardation plates are usually cut in such a way that the c axis lies in the plane of the plate surfaces. In this case the propagation direction of normally incident light is perpendicular to the c axis.

Retardation plates (also called wave plates) and birefringent fibers are polarization-state converters, or transformers. The polarization state of a light beam can be converted to any other polarization state by means of a suitable retardation plate. In formulating the Jones matrix method, we assume that there is no reflection of light from either surface of the plate and the light is totally transmitted through the plate surfaces. In practice, there is reflection, though most retardation plates are coated so as to reduce the surface reflection loss. The Fresnel reflections at the plate surfaces not only decrease the transmitted intensity but also produce a fine structure (ripple) of the spectral transmittance because of multiple-reflection interference (Fabry–Perot effect). Referring to Figure 1.16, we consider an incident beam of light with a polarization state described by the Jones vector

$$\mathbf{V} = \begin{bmatrix} V_x \\ V_y \end{bmatrix} \tag{1.9-1}$$

where V_x and V_y are two complex numbers. The x and y axes are fixed laboratory axes. To determine how the light propagates in the retardation plate, we need to decompose the light into a linear combination of the "fast" and "slow" normal modes of the medium. This is done by the coordinate transformation

$$\begin{bmatrix} V_s \\ V_f \end{bmatrix} = \begin{bmatrix} \cos\psi & \sin\psi \\ -\sin\psi & \cos\psi \end{bmatrix} \begin{bmatrix} V_x \\ V_y \end{bmatrix} \equiv R(\psi) \begin{bmatrix} V_x \\ V_y \end{bmatrix} \tag{1.9-2}$$

where $R(\psi)$ is called the coordinate rotation matrix and V_s and V_f are the slow and fast components, respectively, of the polarization vector $\mathbf{V}$. The slow and fast axes are fixed in the medium. These two components are normal modes of the retardation plate and will propagate with their own phase velocities and polarizations. The azimuth ψ is defined as the angle between the sf coordinate and the xy coordinate (see Figure 1.16) with the z axis as the axis of the coordinate rotation. Because of the difference in phase velocity, one component is retarded relative to the other. This retardation changes the polarization state of the emerging beam. Table 1.6 lists the Jones vectors of various polarization states.

Let n_s and n_f be the refractive indices associated with the propagation of the slow and fast components, respectively. The polarization state of the emerging beam in the medium sf coordinate system is given by

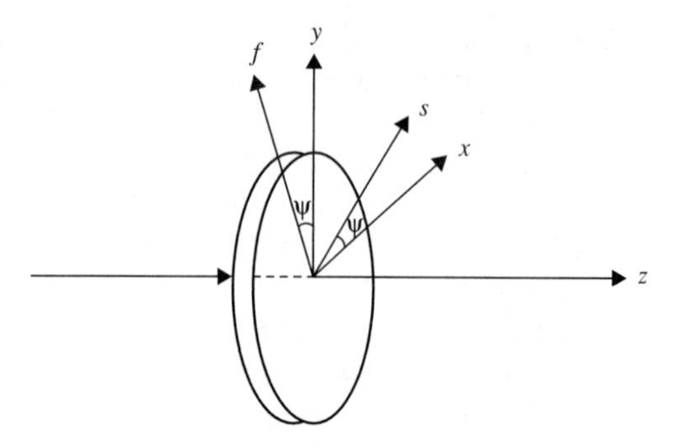

Figure 1.16 A wave plate (or retardation plate) with azimuth angle ψ. The s axis is in the direction of polarization of the slow mode, whereas the f axis is in the direction of polarization of the fast mode. The azimuth ψ is defined as the angle between the s axis and the x axis.

TABLE 1.6 Jones Vectors

Polarization State	Jones Vector
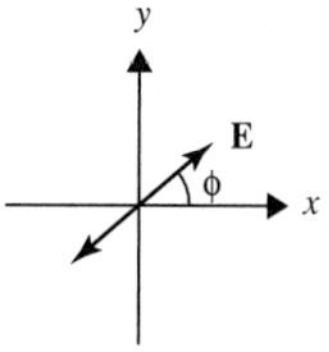	$\begin{bmatrix} \cos\phi \\ \sin\phi \end{bmatrix}$
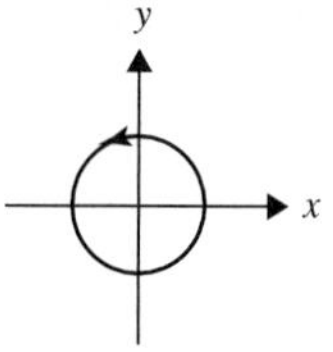	$\frac{1}{\sqrt{2}}\begin{bmatrix} 1 \\ -i \end{bmatrix}$
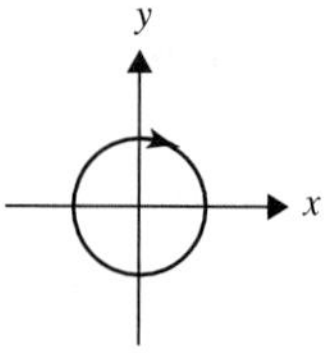	$\frac{1}{\sqrt{2}}\begin{bmatrix} 1 \\ i \end{bmatrix}$
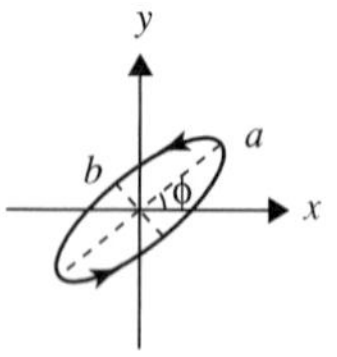	$\begin{bmatrix} a\cos\phi + ib\sin\phi \\ a\sin\phi - ib\cos\phi \end{bmatrix}$
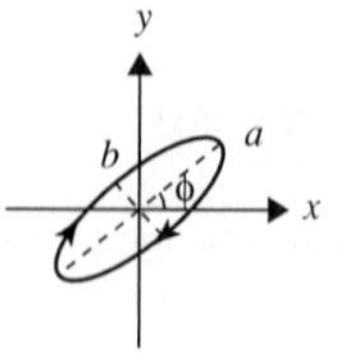	$\begin{bmatrix} a\cos\phi - ib\sin\phi \\ a\sin\phi + ib\cos\phi \end{bmatrix}$

$$\begin{bmatrix} V'_s \\ V'_f \end{bmatrix} = \begin{bmatrix} \exp\left(-in_s \frac{2\pi}{\lambda} d\right) & 0 \\ 0 & \exp\left(-in_f \frac{2\pi}{\lambda} d\right) \end{bmatrix} \begin{bmatrix} V_s \\ V_f \end{bmatrix} \tag{1.9-3}$$

where d is the thickness of the plate and λ is the wavelength of the light beam. The phase retardation is the difference of the exponents in Equation (1.9-3) and is defined as

$$\Gamma = \frac{2\pi}{\lambda}(n_s - n_f)d \tag{1.9-4}$$

Notice that the phase retardation Γ is a measure of the relative change in phase as a result of the propagation, not the absolute change. The birefringence of a typical retardation plate is small; that is, $|n_s - n_f| \ll n_s, n_f$. Consequently, the absolute change in phase caused by the plate may be hundreds of times greater than the phase retardati on. Let ϕ be the mean absolute phase change,

$$\phi = \tfrac{1}{2}(n_s + n_f)\frac{2\pi}{\lambda} d \tag{1.9-5}$$

Then Equation (1.9-3) can be written in terms of ϕ and Γ as

$$\begin{bmatrix} V'_s \\ V'_f \end{bmatrix} = e^{-i\phi} \begin{bmatrix} e^{-i\Gamma/2} & 0 \\ 0 & e^{i\Gamma/2} \end{bmatrix} \begin{bmatrix} V_s \\ V_f \end{bmatrix} \tag{1.9-6}$$

The Jones vector of the polarization state of the emerging beam in the xy coordinate is found by transforming back from the sf coordinate system:

$$\begin{bmatrix} V'_x \\ V'_y \end{bmatrix} = \begin{bmatrix} \cos\psi & -\sin\psi \\ \sin\psi & \cos\psi \end{bmatrix} \begin{bmatrix} V'_s \\ V'_f \end{bmatrix} \tag{1.9-7}$$

By combining Equations (1.9-1), (1.9-6), and (1.9-7), we can write the transformation due to the retardation plate as

$$\begin{bmatrix} V'_x \\ V'_y \end{bmatrix} = R(-\psi) W_0 R(\psi) \begin{bmatrix} V_x \\ V_y \end{bmatrix} \tag{1.9-8}$$

where $R(\psi)$ is the coordinate rotation matrix and W_0 is the Jones matrix for the retardation plate. These are given, respectively, by

$$R(\psi) = \begin{bmatrix} \cos\psi & \sin\psi \\ -\sin\psi & \cos\psi \end{bmatrix} \tag{1.9-9}$$

and

$$W_0 = e^{\;i\phi} \begin{bmatrix} e^{-i\Gamma/2} & 0 \\ 0 & e^{i\Gamma/2} \end{bmatrix} \tag{1.9-10}$$

The phase factor $e^{-i\phi}$ can be neglected if interference effects due to multiple reflections are not important, or not observable. A retardation plate is characterized by its phase retardation Γ and its azimuth angle ψ and is represented by the product of three matrices (Equation (1.9-8)):

TABLE 1.7 Jones Matrices

Optical Element	Jones Matrices
Wave plates	$\Gamma \equiv \frac{2\pi}{\lambda}(n_e - n_o)d$
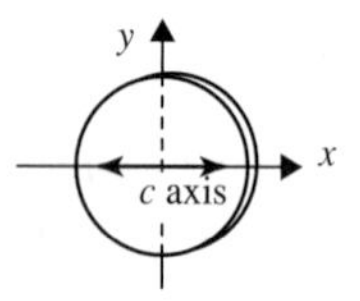	$\begin{bmatrix} e^{-i\Gamma/2} & 0 \\ 0 & e^{i\Gamma/2} \end{bmatrix}$
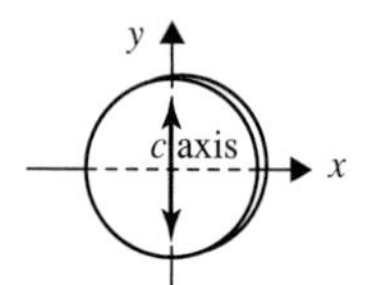	$\begin{bmatrix} e^{i\Gamma/2} & 0 \\ 0 & e^{-i\Gamma/2} \end{bmatrix}$
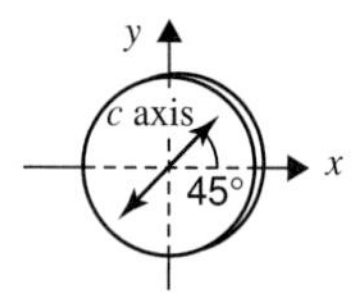	$\begin{bmatrix} \cos(\Gamma/2) & -i\sin(\Gamma/2) \\ -i\sin(\Gamma/2) & \cos(\Gamma/2) \end{bmatrix}$
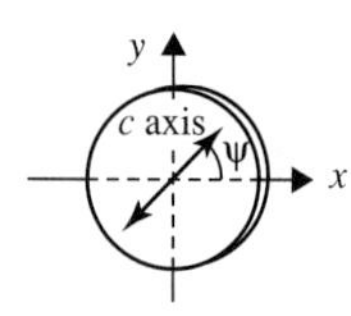	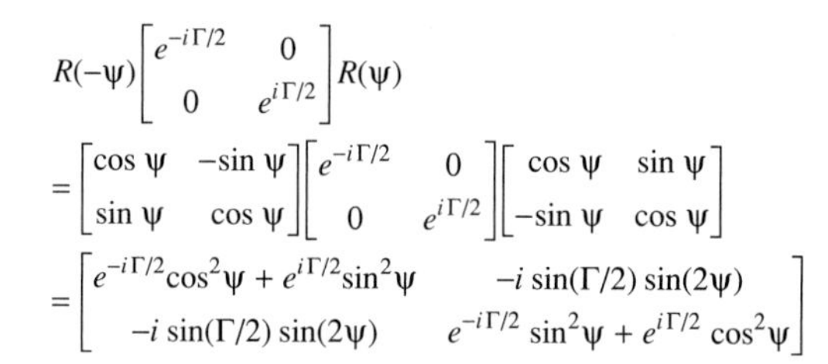 $R(-\psi)\begin{bmatrix} e^{-i\Gamma/2} & 0 \\ 0 & e^{i\Gamma/2} \end{bmatrix}R(\psi)$ $= \begin{bmatrix} \cos\psi & -\sin\psi \\ \sin\psi & \cos\psi \end{bmatrix}\begin{bmatrix} e^{-i\Gamma/2} & 0 \\ 0 & e^{i\Gamma/2} \end{bmatrix}\begin{bmatrix} \cos\psi & \sin\psi \\ -\sin\psi & \cos\psi \end{bmatrix}$ $= \begin{bmatrix} e^{-i\Gamma/2}\cos^2\psi + e^{i\Gamma/2}\sin^2\psi & -i\sin(\Gamma/2)\sin(2\psi) \\ -i\sin(\Gamma/2)\sin(2\psi) & e^{-i\Gamma/2}\sin^2\psi + e^{i\Gamma/2}\cos^2\psi \end{bmatrix}$
Polarizers	
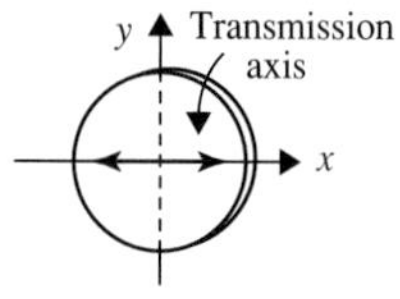 	$\begin{bmatrix} 1 & 0 \\ 0 & 0 \end{bmatrix}$
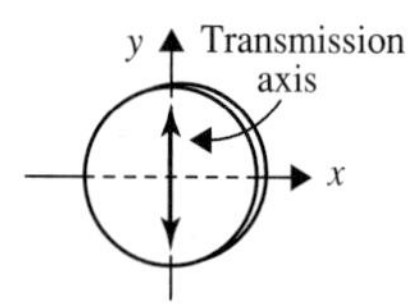	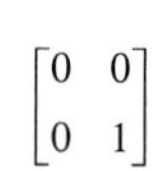 $\begin{bmatrix} 0 & 0 \\ 0 & 1 \end{bmatrix}$
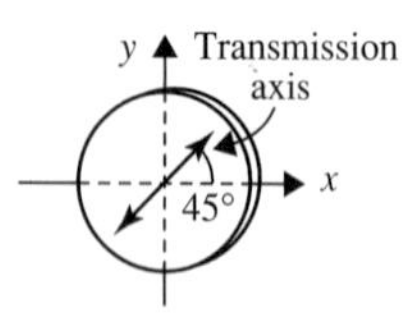	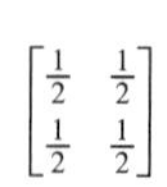 $\begin{bmatrix} \frac{1}{2} & \frac{1}{2} \\ \frac{1}{2} & \frac{1}{2} \end{bmatrix}$
y Transmission axis ψ x	$R(-\psi)\begin{bmatrix} 1 & 0 \\ 0 & 0 \end{bmatrix}R(\psi)$ $= \begin{bmatrix} \cos\psi & -\sin\psi \\ \sin\psi & \cos\psi \end{bmatrix}\begin{bmatrix} 1 & 0 \\ 0 & 0 \end{bmatrix}\begin{bmatrix} \cos\psi & \sin\psi \\ -\sin\psi & \cos\psi \end{bmatrix}$ $= \begin{bmatrix} \cos\psi & -\sin\psi \\ \sin\psi & \cos\psi \end{bmatrix}\begin{bmatrix} \cos\psi & \sin\psi \\ 0 & 0 \end{bmatrix} = \begin{vmatrix} \cos^2\psi & \cos\psi\sin\psi \\ \sin\psi\cos\psi & \sin^2\psi \end{vmatrix}$

$$W = R(-\psi)W_0R(\psi) \tag{1.9-11}$$

Note that the Jones matrix of a wave plate is a unitary matrix; that is,

$$W^{\dagger}W = 1$$

where the dagger (†) means Hermitian conjugate. The passage of a beam of polarized light through a wave plate is mathematically described as a unitary transformation. Many physical properties are invariant under unitary transformations; these include the orthogonal relationship between the Jones vectors and the magnitude of the Jones vectors. Thus if the polarization states of two beams are mutually orthogonal, they will remain orthogonal after passing through an arbitrary wave plate.

The Jones matrix of an ideal homogeneous linear sheet polarizer oriented with its transmission axis parallel to the laboratory x axis is

$$P_0 = e^{-i\phi}\begin{bmatrix} 1 & 0 \\ 0 & 0 \end{bmatrix} \tag{1.9-12}$$

where ϕ is the absolute phase accumulated due to the finite optical thickness of the polarizer. A sheet polarizer is made of an anisotropic medium in which one of the normal modes suffers attenuation due to material absorption. In this case, the y component is absorbed by the sheet polarizer. The Jones matrix of a polarizer rotated by an angle ψ about the z axis is given by

$$P = R(-\psi)P_0R(\psi) \tag{1.9-13}$$

Thus, if we neglect the absolute phase ϕ, the Jones matrix representations of the polarizers transmitting light with electric field vectors parallel to the x and y axes, respectively, are given by

$$P_x = \begin{bmatrix} 1 & 0 \\ 0 & 0 \end{bmatrix} \quad \text{and} \quad P_y = \begin{bmatrix} 0 & 0 \\ 0 & 1 \end{bmatrix}$$

To find the effect of a train of retardation plates and polarizers on the polarization state of a polarized light beam, we write down the Jones vector of the incident beam, and then write down the Jones matrices of the various elements. The Jones vector of the emerging beam is obtained by carrying out the matrix multiplication in sequence. Table 1.7 lists the Jones matrices of wave plates and polarizers at selected orientations.

EXAMPLE: A HALF-WAVE RETARDATION PLATE

A half-wave plate has a phase retardation of $\Gamma = \pi$. According, to Equation (1.9-4), an a-cut (surface is perpendicular to a axis) uniaxial plate will act as a half-wave plate provided the thickness is $d = \lambda/(2|n_e - n_o|)$. We will determine the effect of a half-wave plate on the polarization state of a transmitted light beam. The azimuth angle of the wave plate is taken as 45°, and the incident beam is vertically polarized. The Jones vector for the incident beam can be written

$$\mathbf{V} = \begin{bmatrix} 0 \\ 1 \end{bmatrix} \tag{1.9-14}$$

and the Jones matrix for the half-wave plate is obtained by using Equations (1.9-9) to (1.9-11):

$$W = \frac{1}{\sqrt{2}}\begin{bmatrix} 1 & -1 \\ 1 & 1 \end{bmatrix}\begin{bmatrix} -i & 0 \\ 0 & i \end{bmatrix}\frac{1}{\sqrt{2}}\begin{bmatrix} 1 & 1 \\ -1 & 1 \end{bmatrix} = \begin{bmatrix} 0 & -i \\ -i & 0 \end{bmatrix} \tag{1.9-15}$$

The Jones vector for the emerging beam is obtained by multiplying Equations (1.9-15) and (1.9-14); the result is

$$\mathbf{V}' = \begin{bmatrix} -i \\ 0 \end{bmatrix} = -i\begin{bmatrix} 1 \\ 0 \end{bmatrix} \tag{1.9-16}$$

This represents a beam of horizontally polarized light. The effect of the half-wave plate is to rotate the polarization by 90°. It can be shown that for a general azimuth angle ψ, the half-wave plate will rotate the polarization by an angle 2ψ (see Problem 1.19). In other words, linearly polarized light remains linearly polarized, except that the plane of polarization is rotated by an angle of 2ψ.

When the beam of incident light is circularly polarized, a half-wave plate will convert a beam of right-handed circularly polarized light into a beam of left-handed circularly polarized light and vice versa, regardless of the azimuth angle. The proof is left as an exercise (see Problem 1.19). Figure 1.17 illustrates the effect of a half-wave plate.

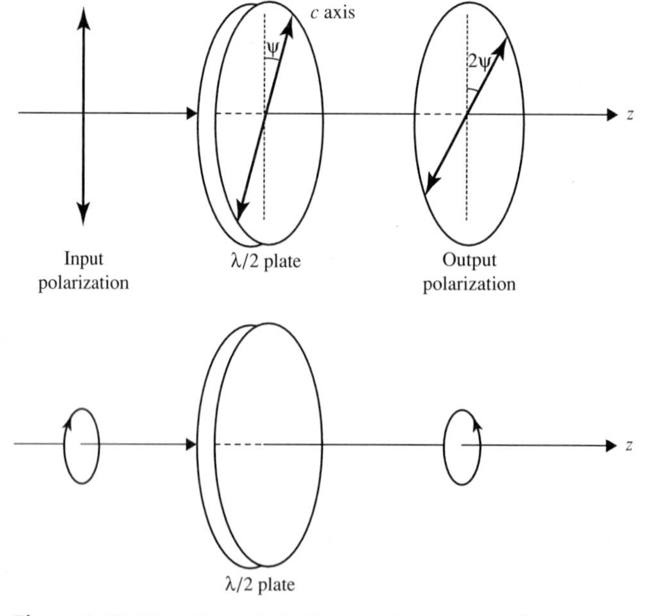

Figure 1.17 The effect of a half-wave plate on the polarization state of a beam.

EXAMPLE: A QUARTER-WAVE PLATE

A quarter-wave plate has a phase retardation of $\Gamma = \pi/2$. If the plate is made of an *a*-cut (or *b*-cut) uniaxially anisotropic medium, the thickness is $d = \lambda/(4|n_e - n_o|)$ (or odd multiples thereof). Suppose again that the azimuth angle of the plate is $\psi = 45°$ and the incident beam is vertically polarized. The Jones vector for the incident beam is given again by Equation (1.9-14). The Jones matrix for this quarter-wave plate, according to Equation (1.9-11), is

$$W = \frac{1}{\sqrt{2}}\begin{bmatrix} 1 & -1 \\ 1 & 1 \end{bmatrix}\begin{bmatrix} e^{-i\pi/4} & 0 \\ 0 & e^{i\pi/4} \end{bmatrix}\frac{1}{\sqrt{2}}\begin{bmatrix} 1 & 1 \\ -1 & 1 \end{bmatrix} = \frac{1}{\sqrt{2}}\begin{bmatrix} 1 & -i \\ -i & 1 \end{bmatrix} \tag{1.9-17}$$

The Jones vector for the emerging beam is obtained by multiplying Equations (1.9-17) and (1.9-14) and is given by

$$\mathbf{V}' = \frac{1}{\sqrt{2}}\begin{bmatrix} -i \\ 1 \end{bmatrix} = \frac{-i}{\sqrt{2}}\begin{bmatrix} 1 \\ i \end{bmatrix} \tag{1.9-18}$$

This represents a beam of left-handed circularly polarized light. The effect of a 45° oriented quarter-wave plate is to convert a beam of vertically polarized light into a beam of left-handed circularly polarized light. If the incident beam is horizontally polarized, the emerging beam will be right-handed circularly polarized. In general, a quarter-wave plate can convert a linearly polarized light into an elliptically polarized light and vice versa. The effect of the quarter-wave plate is illustrated in Figure 1.18.

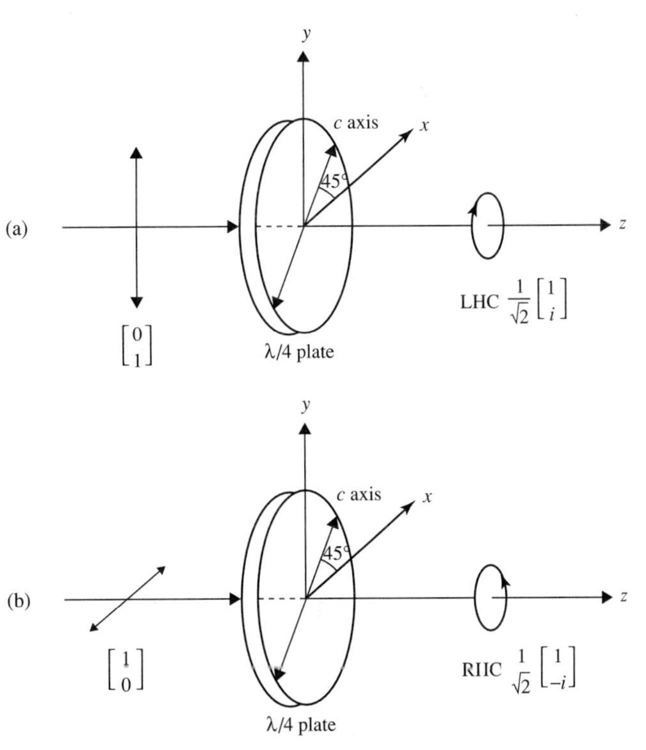

Figure 1.18 The effect of a quarter-wave plate (assuming $\Gamma = k(n_e - n_o)d = \pi/2$) on the polarization state of a linearly polarized beam. LHE = left-handed elliptically, RHE = right-handed elliptically.

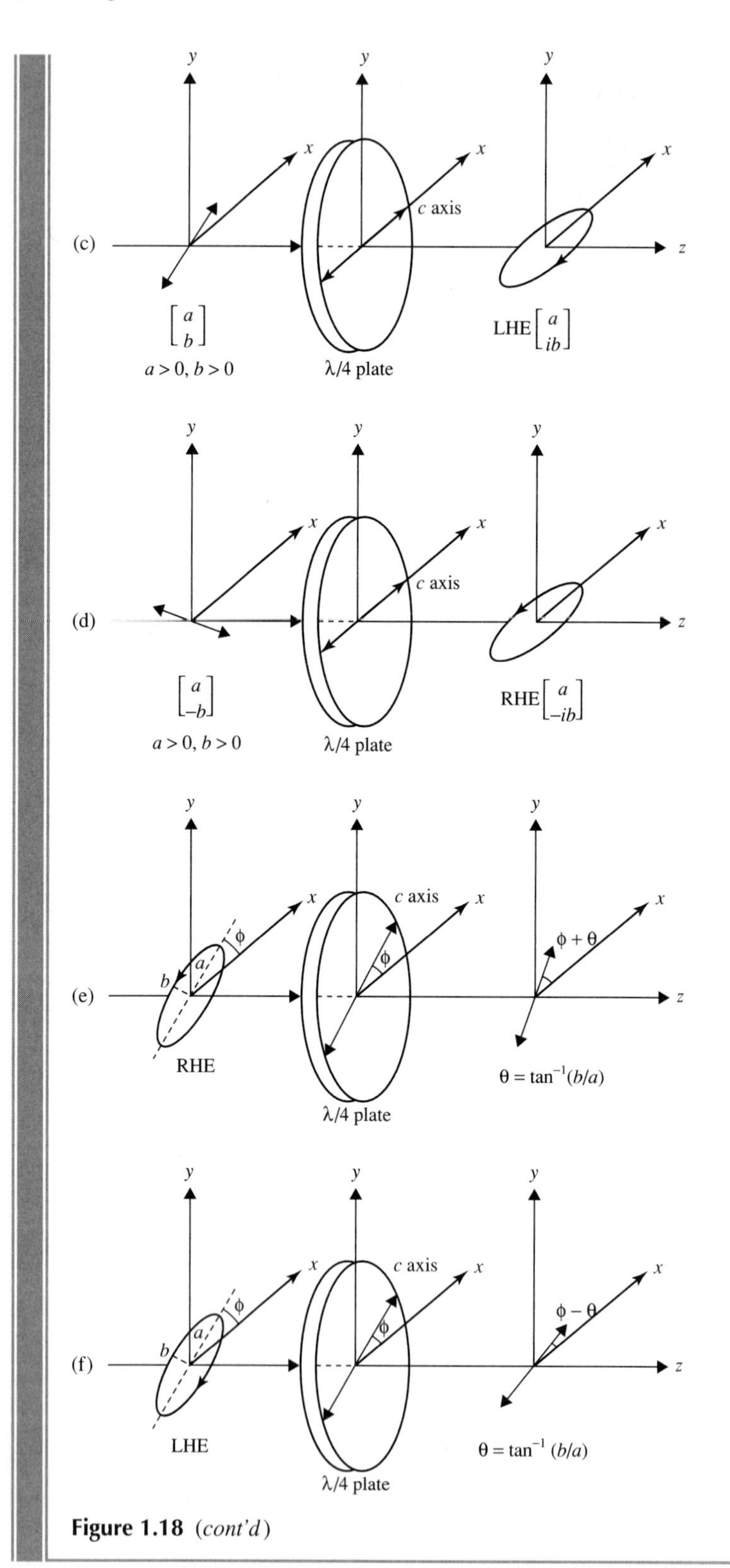

Figure 1.18 (*cont'd*)

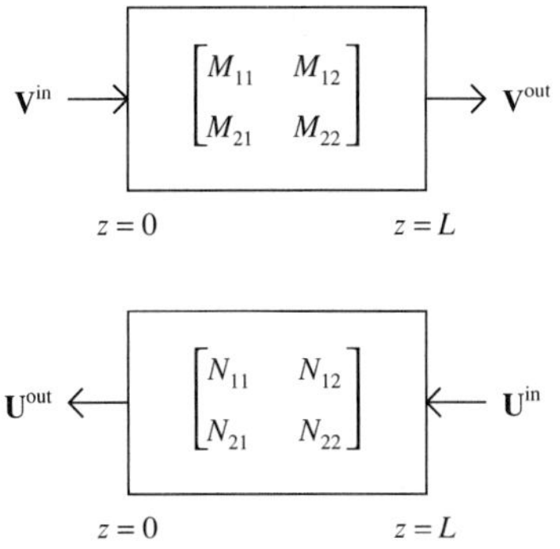

$$\mathbf{V}^{\text{in}} \rightarrow \begin{bmatrix} M_{11} & M_{12} \\ M_{21} & M_{22} \end{bmatrix} \rightarrow \mathbf{V}^{\text{out}}$$

$z = 0$ $z = L$

$$\mathbf{U}^{\text{out}} \leftarrow \begin{bmatrix} N_{11} & N_{12} \\ N_{21} & N_{22} \end{bmatrix} \leftarrow \mathbf{U}^{\text{in}}$$

$z = 0$ $z = L$

Figure 1.19 Schematic drawing of input–output relationships and definition of Jones matrices M and N.

General Properties of the Jones Matrix

Here, we consider some general properties of the Jones matrix. Specifically, we consider the Jones matrix under the transformation of retroreflection, mirror reflection, and time reversal and then we discuss the principle of reciprocity. We also consider the unimodular and unitary nature of the Jones matrix. The results are extremely useful in understanding the transmission properties of birefringent networks. We consider a general birefringent system (between $z = 0$ and $z = L$), which consists of a series of anisotropic wave plates (e.g., wave plates, birefringent fibers, LC cells) with arbitrary orientations of the optical axes. Referring to Figure 1.19, we consider the situation of incidence from both ends of a general birefringent system. For the purpose of discussion, we define two Jones matrices as shown in the figure.

1. Incidence from the left side ($z = 0$): The relationship between the input Jones vector $\mathbf{V}^{\text{in}}$ at ($z = 0$) and the output Jones vector $\mathbf{V}^{\text{out}}$ at ($z = L$) is written

$$\begin{bmatrix} V_x^{\text{out}} \\ V_y^{\text{out}} \end{bmatrix} = \begin{bmatrix} M_{11} & M_{12} \\ M_{21} & M_{22} \end{bmatrix} \begin{bmatrix} V_x^{\text{in}} \\ V_y^{\text{in}} \end{bmatrix} \tag{1.9-19}$$

   where M is the Jones matrix for incidence from the left side.

2. Incidence from the right side ($z = L$): The relationship between the input Jones vector $\mathbf{U}^{\text{in}}$ at ($z = L$) and the output Jones vector $\mathbf{U}^{\text{out}}$ at ($z = 0$) is written

$$\begin{bmatrix} U_x^{\text{out}} \\ U_y^{\text{out}} \end{bmatrix} = \begin{bmatrix} N_{11} & N_{12} \\ N_{21} & N_{22} \end{bmatrix} \begin{bmatrix} U_x^{\text{in}} \\ U_y^{\text{in}} \end{bmatrix} \tag{1.9-20}$$

   where N is the Jones matrix for incidence from the right side. Here we assume that the beam of light retraces the path defined by the incidence from $z = 0$ in Case 1. In other words, Case 2 can be obtained by a retroreflection of the beam of light in Case 1.

Based on the above definitions, the following are fundamental properties of these Jones matrices.

Time Reversal Symmetry

Consider the situation when the time is reversed. The output beam will retrace the beam path, propagate through the birefringent system, and become the phase conjugate of the input beam. In this case, $\mathbf{U}^{in} = (\mathbf{V}^{out})^*$. As a result of the time reversal symmetry, $\mathbf{U}^{out} = (\mathbf{V}^{in})^*$. This leads to

$$NM^* = 1 \tag{1.9-21}$$

Principle of Reciprocity

Based on the fundamental principle of reciprocity in physics and the definition of the Jones matrices in Equations (1.9-19) and (1.9-20), we have

$$\begin{aligned} N_{11} &= M_{11} \\ N_{22} &= M_{22} \\ N_{21} &= M_{12} \\ N_{12} &= M_{21} \end{aligned} \tag{1.9-22}$$

In other words, N is the transpose of M,

$$N = \tilde{M} \tag{1.9-23}$$

where the notation ~ indicates a transpose operation. This property is particularly useful in optical systems that involve the use of a reflector (mirror).

Using Equations (1.9-21), (1.9-22), and (1.9-23), it can be shown that both N and M are unitary matrices; that is,

$$\begin{aligned} M^\dagger M &= 1 \\ N^\dagger N &= 1 \end{aligned} \tag{1.9-24}$$

where $N^\dagger$ and $M^\dagger$ are the Hermitian conjugates of N and M, respectively. This proves the unitary nature of the Jones matrix. Thus, if the Jones matrix M is written

$$M = \begin{bmatrix} A & B \\ C & D \end{bmatrix} \tag{1.9-25}$$

where A, B, C, and D are the matrix elements, then the inverse of M can be written

$$M^{-1} = M^\dagger = \begin{bmatrix} A^* & C^* \\ B^* & D^* \end{bmatrix} \tag{1.9-26}$$

Furthermore, the Jones matrix is unimodular based on the definition in Equation (1.9-11), provided all the birefringent plates are lossless. In other words,

$$\det(M) = AD - BC = 1 \tag{1.9-27}$$

Thus the inverse of the Jones matrix can also be written

$$M^{-1} = \begin{bmatrix} D & -B \\ -C & A \end{bmatrix} \tag{1.9-28}$$

Using Equations (1.9-26) and (1.9-28), we obtain the following relationships for the matrix elements:

$$\begin{aligned} C &= -B^* \\ D &= A^* \end{aligned} \tag{1.9-29}$$

These relationships are very useful in simplifying calculations involving use of the Jones matrix method. Based on the relationships in Equation (1.9-29), the Jones matrix can be written

$$M = \begin{bmatrix} A & B \\ -B^* & A^* \end{bmatrix} \tag{1.9-30}$$

We note that the collection of all Jones matrices forms a mathematical group. In other words, the relationships remain true after the multiplication of any two Jones matrices.

Intensity Transmission Spectrum

So far our development of the Jones matrix method was concerned with the polarization state of the light beam. In many cases, we need to determine the transmitted intensity. A narrowband filter, for example, transmits radiation only in a small spectral regime and rejects (or absorbs) radiation at other wavelengths. To change the intensity of the transmitted beam, an analyzer is usually required. An analyzer is basically a polarizer. It is called an analyzer simply because of its location in the optical system. In most birefringent optical systems, a polarizer is placed in front of the system in order to "prepare" a beam of polarized light. A second polarizer (analyzer) is placed at the output to analyze the polarization state of the emerging beam. Because the phase retardation of each wave plate is wavelength dependent, the polarization state of the emerging beam depends on the wavelength of the light. A polarizer at the rear will cause the overall transmitted intensity to be wavelength dependent.

The Jones vector representation of a beam of light contains information about not only the polarization state but also the intensity of light. Let us now consider the light beam after it passes through the polarizer. Its electric vector can be written as a Jones vector

$$\mathbf{E} = \begin{bmatrix} E_x \\ E_y \end{bmatrix} \tag{1.9-31}$$

where E_x, E_y are the components in the xy coordinate. The intensity is calculated as follows:

$$I = \mathbf{E}^\dagger \cdot \mathbf{E} = |E_x|^2 + |E_y|^2 \tag{1.9-32}$$

where the dagger indicates the Hermitian conjugate. If the Jones vector of the emerging beam after it passes through the analyzer is written

$$\mathbf{E}' = \begin{bmatrix} E'_x \\ E'_y \end{bmatrix} \tag{1.9-33}$$

the transmittance of the birefringent optical system is calculated as

$$T = \frac{|E'_x|^2 + |E'_y|^2}{|E_x|^2 + |E_y|^2} \tag{1.9-34}$$

EXAMPLE: A BIREFRINGENT PLATE SANDWICHED BETWEEN PARALLEL POLARIZERS

Referring to Figure 1.20, we consider a birefringent plate sandwiched between a pair of parallel polarizers. The plate is oriented so that the "slow" and "fast" axes are at 45° with respect to the transmission axes of polarizers. Let the birefringence be $n_e - n_o$ and the plate thickness be d. The phase retardation is then given by

$$\Gamma = \frac{2\pi}{\lambda}(n_e - n_o)d \tag{1.9-35}$$

and the corresponding Jones matrix is, according to Equation (1.9-11) (or Table 1.7),

$$W = \begin{bmatrix} \cos(\Gamma/2) & -i\sin(\Gamma/2) \\ -i\sin(\Gamma/2) & \cos(\Gamma/2) \end{bmatrix} \tag{1.9-36}$$

Let the incident beam be unpolarized, so that after it passes through the front polarizer, the electric field vector can be represented by the following Jones vector:

$$\frac{1}{\sqrt{2}}\begin{bmatrix} 0 \\ 1 \end{bmatrix} \tag{1.9-37}$$

where we assume that the intensity of the incident beam is unity and only half of the intensity passes through the polarizer where the transmission axis is parallel to the y axis. The Jones vector representation of the electric field vector of the transmitted beam is obtained as follows:

$$\begin{aligned} \mathbf{E}' &= \begin{bmatrix} 0 & 0 \\ 0 & 1 \end{bmatrix}\begin{bmatrix} \cos(\Gamma/2) & -i\sin(\Gamma/2) \\ -i\sin(\Gamma/2) & \cos(\Gamma/2) \end{bmatrix}\frac{1}{\sqrt{2}}\begin{bmatrix} 0 \\ 1 \end{bmatrix} \\ &= \frac{1}{\sqrt{2}}\begin{bmatrix} 0 \\ \cos(\Gamma/2) \end{bmatrix} \end{aligned} \tag{1.9-38}$$

The transmitted beam is vertically (y) polarized with an intensity given by

$$I = \tfrac{1}{2}\cos^2(\Gamma/2) = \tfrac{1}{2}\cos^2\left[\pi(n_e - n_o)d/\lambda\right] \tag{1.9-39}$$

It can be seen from Equation (1.9-39) that the transmitted intensity is a sinusoidal function of the wavenumber and peaks at $\lambda = (n_e - n_o)d, (n_e - n_o)d/2, (n_e - n_o)d/3, \ldots$. These wavelengths correspond to $\Gamma = 2\pi, 4\pi, 6\pi, \ldots$. In other words, maximum transmission occurs when the plate is an integral number of full waves. The wavenumber separation between transmission maxima increases with decreasing plate thickness.

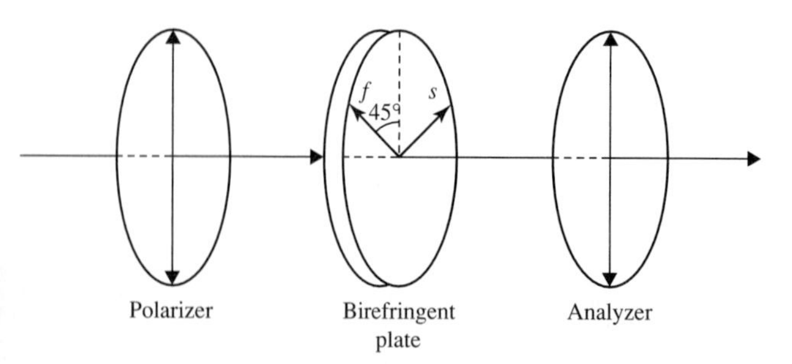

Figure 1.20 A birefringent plate sandwiched between a pair of parallel polarizers.

EXAMPLE: A BIREFRINGENT PLATE SANDWICHED BETWEEN A PAIR OF CROSSED POLARIZERS

If we rotate the analyzer shown in Figure 1.20 by 90°, then the input and output polarizers are crossed. The transmitted beam for this case is obtained as follows:

$$\mathbf{E}' = \begin{bmatrix} 1 & 0 \\ 0 & 0 \end{bmatrix} \begin{bmatrix} \cos(\Gamma/2) & -i\sin(\Gamma/2) \\ -i\sin(\Gamma/2) & \cos(\Gamma/2) \end{bmatrix} \frac{1}{\sqrt{2}} \begin{bmatrix} 0 \\ 1 \end{bmatrix}$$

$$= \frac{-i}{\sqrt{2}} \begin{bmatrix} \sin(\Gamma/2) \\ 0 \end{bmatrix} \qquad (1.9\text{-}40)$$

The transmitted beam is horizontally (x) polarized with an intensity given by

$$I = \tfrac{1}{2}\sin^2(\Gamma/2) = \tfrac{1}{2}\sin^2[\pi(n_e - n_o)d/\lambda] \qquad (1.9\text{-}41)$$

This is again a sinusoidal function of the wavenumber. The transmission spectrum consists of a series of maxima at $\lambda = 2(n_e - n_o)d$, $2(n_e - n_o)d/3, \ldots$. These wavelengths correspond to phase retardations of π, 3π, $5\pi, \ldots$, that is, when the wave plate becomes a half-wave plate or an odd integral multiple of a half-wave plate.

EXAMPLE: A BIREFRINGENT PLATE SANDWICHED BETWEEN A PAIR OF POLARIZERS

If we rotate the analyzer shown in Figure 1.20 so that the transmission axis of the analyzer forms an angle ψ with respect to the x axis, then the input and output polarizers are neither parallel nor crossed. The transmitted beam for this case is obtained as follows:

$$\mathbf{E}' = \begin{bmatrix} \cos^2\psi & \cos\psi\sin\psi \\ \sin\psi\cos\psi & \sin^2\psi \end{bmatrix} \begin{bmatrix} \cos(\Gamma/2) & -i\sin(\Gamma/2) \\ -i\sin(\Gamma/2) & \cos(\Gamma/2) \end{bmatrix} \frac{1}{\sqrt{2}} \begin{bmatrix} 0 \\ 1 \end{bmatrix}$$

$$= \frac{-i\cos\psi\sin(\Gamma/2) + \sin\psi\cos(\Gamma/2)}{\sqrt{2}} \begin{bmatrix} \cos\psi \\ \sin\psi \end{bmatrix} \qquad (1.9\text{-}42)$$

The transmitted beam is polarized in the same direction as the transmission axis of the analyzer with an intensity

$$I = \tfrac{1}{2}\cos^2\psi\sin^2(\Gamma/2) + \tfrac{1}{2}\sin^2\psi\cos^2(\Gamma/2) \qquad (1.9\text{-}43)$$

where

$$\Gamma = \frac{2\pi(n_e - n_o)d}{\lambda}$$

We note that Equation (1.9-43) reduces to Equation (1.9-39) when $\psi = \pi/2$ (i.e., when the analyzer is parallel to the polarizer) and Equation (1.9-43) reduces to Equation (1.9-41) when $\psi = 0$ (i.e., when the analyzer is perpendicular to the polarizer).

1.10 ELEMENTARY THEORY OF COHERENCE

It is known that the principle of superposition applies to the electromagnetic field. This principle states that the total field due to all sources is the sum of the field due to each source. The reason why this is true is that Maxwell's equations, which govern the electromagnetic field, are linear differential equations, provided both ε and μ are independent of the field strength. Consider now two linearly polarized plane waves of the same frequency ω. Let the electric fields be

$$\begin{aligned} \mathbf{E}_1 &= \mathbf{A}_1 \exp[i(\omega t - \mathbf{k}_1 \cdot \mathbf{r} + \phi_1)] \\ \mathbf{E}_2 &= \mathbf{A}_2 \exp[i(\omega t - \mathbf{k}_2 \cdot \mathbf{r} + \phi_2)] \end{aligned} \tag{1.10-1}$$

where $\mathbf{A}_1$ and $\mathbf{A}_2$ are the amplitudes and are real vectors because of the assumption of linear polarization states. The real quantities ϕ_1 and ϕ_2 are the phases, and $\mathbf{k}_1$ and $\mathbf{k}_2$ are the wavevectors. We assume that the amplitudes $\mathbf{A}_1$ and $\mathbf{A}_2$ are constants. The sources of these two planes waves are said to be mutually coherent if the phase difference $\phi_1 - \phi_2$ is constant. On the other hand, if the quantity $\phi_1 - \phi_2$ varies with time in a random fashion, the sources of these two plane waves are said to be mutually incoherent. To define the degree of mutual coherence of these two waves (1.10-1), we can examine the interference pattern formed by these two waves. We have shown earlier that the time-averaged intensity of electromagnetic radiation at any point in space is proportional to the square of the amplitude of the electric field. Thus, aside from a constant factor, the time-averaged intensity distribution of the interference pattern formed by the superposition of these two waves can be written

$$I = |\mathbf{E}_1 + \mathbf{E}_2|^2 = I_1 + I_2 + 2\mathbf{A}_1 \cdot \mathbf{A}_2 \cos(\mathbf{K} \cdot \mathbf{r} - \phi) \tag{1.10-2}$$

where $I_1 = |\mathbf{E}_1|^2$, $I_2 = |\mathbf{E}_2|^2$, $\mathbf{K} = \mathbf{k}_1 - \mathbf{k}_2$, and $\phi = \phi_1 - \phi_2$. The first two terms are the time-averaged intensities of the two waves, respectively. The third term is the interference term and contains information about the degree of mutual coherence. Note that Equation (1.10-2) is a time-averaged result, averaging over a period of $2\pi/\omega$. If we limit ourselves to the case of quasi-monochromatic plane waves, all the phases ϕ, ϕ_1, and ϕ_2 are considered constant during such a small time interval $2\pi/\omega$, which is on the order of 10^{-15} s for visible light. If these two waves are of the same polarization state, then the intensity of the interference pattern can be written

$$I = |\mathbf{E}_1 + \mathbf{E}_2|^2 = I_1 + I_2 + 2\sqrt{I_1 I_2}\,\cos(\mathbf{K} \cdot \mathbf{r} - \phi) \tag{1.10-3}$$

If the two waves are mutually coherent (i.e., ϕ = constant), the interference pattern is stationary with a spatial period of

$$\Lambda = \frac{2\pi}{|\mathbf{K}|} = \frac{\lambda}{2\sin(\theta/2)} \tag{1.10-4}$$

where θ is the angle between the two wavevectors, and λ is the wavelength of the light (see Figure 1.21).

If the two waves are not mutually coherent, the phase ϕ varies with time. As a result, the time-averaged intensity in Equation (1.10-3) fluctuates rapidly with time. If ϕ varies significantly over a time period τ_D, the characteristic time constant associated with the detection process, the fluctuation is too fast for the detector to respond, and the detected intensity is averaged over a time interval τ_D. As a result of this time averaging, the visibility of the fringes decreases. For mutually incoherent sources, the phase ϕ varies with time in a

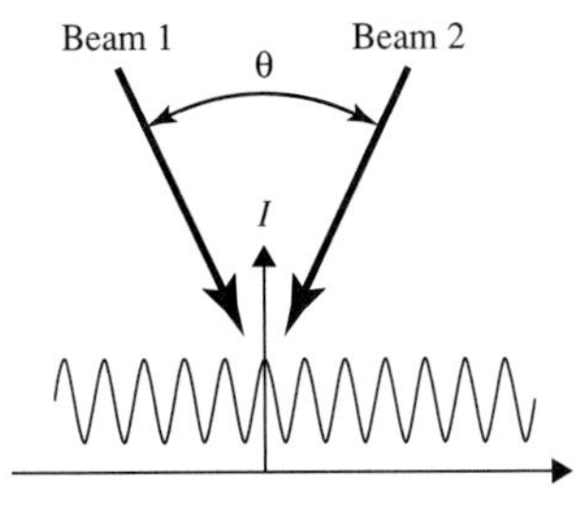

Figure 1.21 Intensity distribution of a fringe pattern due to the interference of two beams.

random fashion between 0 and 2π, and the averaging of the third term in Equation (1.10-3) is obviously zero. This leads to the disappearance of the fringe pattern. Thus, for many practical purposes, the degree of mutual coherence between the two waves (1.10-1) can be defined as

$$\gamma_{12} = \left(\frac{\langle\langle \mathbf{E}_1^* \cdot \mathbf{E}_2 \rangle\rangle}{\langle\langle \mathbf{E}_1^* \cdot \mathbf{E}_1 \rangle\rangle^{1/2} \langle\langle \mathbf{E}_2^* \cdot \mathbf{E}_2 \rangle\rangle^{1/2}} \right)_{\mathbf{r}=0} = \langle\langle e^{i(\phi_2-\phi_1)} \rangle\rangle = \frac{1}{\tau_D} \int_0^{\tau_D} e^{i(\phi_2-\phi_1)}\, dt \qquad (1.10\text{-}5)$$

where the double brackets indicate time averaging over a time interval τ_D, the characteristic time constant of the detection system. The parameter is, in general, complex and can be written

$$\gamma_{12} = |\gamma_{12}| \exp(i\alpha) \qquad (1.10\text{-}6)$$

where α is a real constant.

We now evaluate the detection of the time-averaged intensity. From Equation (1.10-2), we obtain

$$\langle\langle I \rangle\rangle = |\mathbf{E}_1 + \mathbf{E}_2|^2 = I_1 + I_2 + 2\mathbf{A}_1 \cdot \mathbf{A}_2 \langle\langle \cos(\mathbf{K} \cdot \mathbf{r} - \phi) \rangle\rangle \qquad (1.10\text{-}7)$$

where we assumed $\mathbf{A}_1$ and $\mathbf{A}_2$ are constants.

Using $\cos(\mathbf{K} \cdot \mathbf{r} - \phi) = \cos(\mathbf{K} \cdot \mathbf{r}) \cos \phi + \sin(\mathbf{K} \cdot \mathbf{r}) \sin \phi$, and Equation (1.10-5), we obtain

$$\langle\langle I \rangle\rangle = |\mathbf{E}_1 + \mathbf{E}_2|^2 = I_1 + I_2 + 2\mathbf{A}_1 \cdot \mathbf{A}_2 |\gamma_{12}| \cos(\mathbf{K} \cdot \mathbf{r} - \alpha) \qquad (1.10\text{-}8)$$

Thus an interference fringe pattern will result if $|\gamma_{12}|$, called the degree of mutual coherence, has a value other than zero. In other words, a periodic variation of intensity exists provided the mutual coherence of the two beams is nonzero. By definition (1.10-5), the magnitude of $|\gamma_{12}|$ is always between zero and one. In terms of $|\gamma_{12}|$, we have the following types of mutual coherence:

Complete mutual coherence:	$|\gamma_{12}| = 1$
Partial coherence:	$0 < |\gamma_{12}| < 1$
Complete mutual incoherence:	$|\gamma_{12}| = 0$

Fringe Visibility

In a pattern of interference fringes, the intensity varies between two limits I_{max} and I_{min}. From the above discussion, we note that these two limits are governed by

$$I_{\max} = I_1 + I_2 + 2|\mathbf{A}_1 \cdot \mathbf{A}_2||\gamma_{12}| \tag{1.10-9}$$

$$I_{\min} = I_1 + I_2 - 2|\mathbf{A}_1 \cdot \mathbf{A}_2||\gamma_{12}| \tag{1.10-10}$$

The fringe visibility V is defined as the ratio

$$V = \frac{I_{\max} - I_{\min}}{I_{\max} + I_{\min}} \tag{1.10-11}$$

For beams with the same polarization state (i.e., $\mathbf{A}_1$ is parallel to $\mathbf{A}_2$), it follows that

$$V = \frac{2\sqrt{I_1 I_2}}{I_1 + I_2}|\gamma_{12}| \tag{1.10-12}$$

The fringe visibility V is always between 0 and 1. In particular, if $I_1 = I_2$, then

$$V = |\gamma_{12}| \tag{1.10-13}$$

In other words, the interference fringe visibility is equal to the degree of mutual coherence. In the case of complete mutual coherence ($|\gamma_{12}| = 1$), the interference fringes have the maximum visibility of unity; whereas for beams of complete incoherence, $|\gamma_{12}| = 0$, the fringe visibility is zero—in other words, there are no interference fringes at all.

Coherence Time and Coherence Length

In addition to the mutual coherence between two beams, we can define the coherence of electromagnetic radiation itself. Consider a quasi-monochromatic wave of the form

$$E = A(t) \exp[i(\omega t - \mathbf{k}_1 \cdot \mathbf{r})] \tag{1.10-14}$$

where $A(t)$ is a time-varying complex amplitude. In the two-beam interference discussed earlier, the two electric fields E_1 and E_2 may originate from the same source given by Equation (1.10-14). This happens in many interferometers, including the Michelson interferometer, the Mach–Zehnder interferometer, and Young's double-slit interferometer. In these interference experiments, the two beams of light are only different in their optical paths. Referring to Figure 1.21, we assume that the two beams are

$$E_1 = A(t) \exp[i(\omega t - \mathbf{k}_1 \cdot \mathbf{r})] \tag{1.10-15}$$

$$E_2 = A(t + \tau) \exp[i(\omega t + \omega\tau - \mathbf{k}_2 \cdot \mathbf{r})] \tag{1.10-16}$$

where τ is the time delay. We can see that these two waves originate from the same source (1.10-14). The degree of mutual coherence between these two waves is a measure of the self-coherence of the wave (1.10-14). Thus self-coherence $\gamma(\tau)$ is defined as

$$\gamma(\tau) = \frac{\langle\langle E^*(t)\, E(t+\tau)\rangle\rangle}{\langle\langle E^*(t)\, E(t)\rangle\rangle^{1/2}\langle\langle E^*(t+\tau)\, E(t+\tau)\rangle\rangle^{1/2}} \tag{1.10-17}$$

Assume that all quantities are stationary; that is, the time average is independent of the choice of the origin. Then the self-coherence function $\gamma(\tau)$ becomes

$$\gamma(\tau) = \frac{\langle\langle E^*(t)\, E(t+\tau)\rangle\rangle}{\langle\langle E^*(t)\, E(t)\rangle\rangle} \tag{1.10-18}$$

The self-coherence function defined earlier also satisfies the condition

$$0 \le |\gamma(\tau)| \le 1 \tag{1.10-19}$$

For monochromatic plane waves, $A(t)$ is a constant amplitude, and the self-coherence function is unity for any τ. Such a wave is called coherent radiation. For most polychromatic waves, $|\gamma(\tau)| = 1$ only occurs when $\tau = 0$, and $|\gamma(\tau)|$ approaches zero when τ becomes large. Such waves are called partially coherent radiation. If $|\gamma(\tau)|$ is a monotonically decreasing function of τ and, in addition, if $|\gamma(\tau)|$ drops to virtually zero for $\tau > \tau_c$, then we say that τ_c is the coherence time of the wave. Thus, in the two-beam interference experiment, if the path difference between the two beams does not exceed the value

$$l_c = c\tau_c \tag{1.10-20}$$

interference fringes appear. The quantity l_c is known as the coherence length of the wave in Equation (1.10-14).

For many practical purposes, the time interval for averaging in Equation (1.10-17) or (1.10-18) can be taken as infinity. Thus the self-coherence function can be considered as the normalized autocorrelation function of the electromagnetic field. A basic theorem of stochastic theory, the so-called Wiener–Khintchine theorem, states that the power spectrum (or spectral density) of the electromagnetic field and the autocorrelation function form a Fourier transform pair. Therefore $\gamma(\tau)$, as given by Equation (1.10-18), is the normalized Fourier transform of the power spectrum of the electromagnetic field. Let $\Delta\omega$ be the width of the power spectral distribution of the field. Then, according to the theory of Fourier transform, we have

$$\tau_c = \frac{2\pi}{\Delta\omega} = \frac{1}{\Delta f} \tag{1.10-21}$$

which states that the coherence time of quasi-monochromatic radiation is given by the inverse of its spectral linewidth Δf.

PROBLEMS

1.1 Conservation of charges requires that the charge density at any point in space be related to the current density in that neighborhood by a continuity equation

$$\frac{\partial \rho}{\partial t} + \nabla \cdot \mathbf{J} = 0$$

Derive this equation from Maxwell's equation. If we integrate this equation over a volume element, the first term represents the change of total electric charge per unit time within the volume element; the second term represents the amount of charge flowing out of the volume element per unit time.

1.2 Derive the boundary conditions for the normal components of **B** and **D**, by using an application of the Gauss divergence theorem to Maxwell's equations.

1.3 Derive the boundary conditions that apply to the tangential components of **E** and **H** by using an application of the Stokes theorem to Maxwell's equations.

1.4 Let $\mathbf{E} = \mathbf{E}_0 \exp(i\omega t)$ and $\mathbf{H} = \mathbf{H}_0 \exp(i\omega t)$ be solutions to Maxwell's equations.

(a) Show that $\mathbf{E}^*$ and $\mathbf{H}^*$ also satisfy Maxwell's equations. Note that $\mathbf{E}^*$, $\mathbf{H}^*$ and **E**, **H** are, in fact, the same field since only the real parts have physical meaning.

(b) Show that the conjugate waves

$$\mathbf{E}_c = \mathbf{E}_0^* \exp(i\omega t) \quad \text{and} \quad \mathbf{H}_c = \mathbf{H}_0^* \exp(i\omega t)$$

also satisfy Maxwell's equations, provided the medium is lossless (i.e., ε and μ are real tensors.)

1.5 The electromagnetic-field momentum density and Maxwell's stress tensor are given by

$$\mathbf{P} = \mu\varepsilon(\mathbf{E} \times \mathbf{H})$$

$$T_{ij} = \varepsilon E_i E_j + \mu H_i H_j - \tfrac{1}{2}\delta_{ij}(\varepsilon E^2 + \mu H^2)$$

Derive the hydrodynamical equation of motion

$$\frac{\partial \mathbf{P}}{\partial t} = \nabla \cdot \mathbf{T} - \mathbf{F}$$

where **F** is the Lorentz force exerted on a distribution of charges and currents by the electromagnetic field, $\mathbf{F} = \rho\mathbf{E} + \mathbf{J} \times \mathbf{B}$.

1.6 Consider a charge particle with a charge q under the influence of an electric field E.

(a) Show that the work done per unit time by the field on the particle can be written work/time = $\mathbf{E} \cdot q\mathbf{v} = \mathbf{E} \cdot q\, d\mathbf{x}/dt$, where $\mathbf{v} = d\mathbf{x}/dt$ is the velocity of the charged particle.

(b) Consider a material medium that consists of a large number of charged particles. Show that the total work done per unit time is

$$\frac{dW}{dt} = \mathbf{E} \cdot \sum_i q_i \frac{d\mathbf{x}_i}{dt}$$

where i stands for the ith particle, and the summation is over all charged particles.

(c) The polarization **P** is defined as the dipole moment per unit volume. Show that

$$\frac{dW}{dt}\frac{1}{\text{Volume}} = \mathbf{E} \cdot \frac{d\mathbf{P}}{dt}$$

1.7 Consider the propagation of an optical pulse in a homogeneous medium of length L.

(a) Show that the time of propagation is

$$\tau = L\left(\frac{n}{c} + \frac{\omega}{c}\frac{dn}{d\omega}\right) = L\left(\frac{n}{c} - \frac{\lambda}{c}\frac{dn}{d\lambda}\right)$$

(b) Show that the group velocity dispersion defined as

$$D = \frac{d}{d\lambda}\left(\frac{\tau}{L}\right)$$

can be written

$$D = -\frac{1}{c\lambda}\left(\lambda^2 \frac{d^2 n}{d\lambda^2}\right)$$

Note that the term in parentheses is dimensionless. The parameter D has a dimension of s/m-m. For practical application, D is often expressed in units of ps/nm-km.

Consider the superposition of two waves. Let

$$v_1 = \frac{\omega_1}{k_1}, \quad v_2 = \frac{\omega_2}{k_2}, \quad v_g = \frac{\omega_2 - \omega_1}{k_2 - k_1}$$

Show that

$$(v_2 - v_g)(v_1 - v_g) > 0$$

In other words, the group velocity is either greater than both v_1 and v_2, or less than both v_1 and v_2. In normally dispersive media, the group velocity is always less than both v_1 and v_2.

1.8 Using Equation (1.5-20), plot the following in the spectral regime from $\lambda = 0.5$ to $2.0\ \mu$m.

(a) Plot n versus λ.
(b) Plot group velocity (in units of c) versus λ.
(c) Plot D (in units of ps/nm-km) versus λ.

1.9 Derive the polarization ellipse, Equation (1.6-12).

1.10 Derive the major and minor axes, and the inclination angle of the polarization ellipse, Equations (1.6-14) and (1.6-15), respectively.

1.11 Show that the end point of the electric vector of an elliptically polarized light beam will revolve in a clockwise direction if $\sin\delta > 0$ and in a counterclockwise direction if $\sin\delta < 0$.

1.12

(a) Find a polarization state that is orthogonal to the polarization state

$$\mathbf{J}(\psi, \delta) = \begin{bmatrix} \cos\psi \\ e^{i\delta}\sin\psi \end{bmatrix}$$

Answer: $\begin{bmatrix} \sin\psi \\ e^{i(\pi+\delta)}\cos\psi \end{bmatrix}$

(b) Show that the major axes of the ellipses of two mutually orthogonal polarization states are perpendicular to each other and the senses of revolution are opposite.

1.13 Derive the inclination angle and the ellipticity angle of the polarization ellipse, Equations (1.6-18) and (1.6-19), respectively.

1.14 Consider two monochromatic plane waves of the form

$$\mathbf{E}_a(z, t) = \text{Re}[\mathbf{A}e^{i(\omega t - kz)}] \quad \text{and} \quad \mathbf{E}_b(z, t) = \text{Re}[\mathbf{B}e^{i(\omega t - kz)}]$$

The polarization states of these two waves are orthogonal; that is, $\mathbf{A}^* \cdot \mathbf{B} = 0$.

(a) Let δ_a and δ_b be the phase angles defined in Equation (1.6-5). Show that

$$\delta_a - \delta_b = \pm\pi$$

(b) Since δ_a and δ_b are both in the range $-\pi < \delta \le \pi$, show that

$$\delta_a\delta_b \le 0$$

(c) Let χ_a and χ_b be the complex numbers representing the polarization states of these two waves. Show that

$$\chi_a^* \chi_b = -1$$

(d) Show that the major axes of the polarization ellipses are mutually orthogonal and the ellipticities are of the same magnitude with opposite signs.

1.15 Show that any polarization state can be converted into a linearly polarized state by using a quarter-wave plate. Describe your approach.

1.16 Draw the end point of the electric field vector in space for right-handed circularly polarized light at a given point in time (say, $t = 0$). Show that the locus is a left-handed helix. (This has been the traditional way of defining the handedness of circular polarization states. The advantage of this convention is that the handedness of the space helix is independent of the direction of viewing.)

(a) Show that a linear combination of two beams of circularly polarized light with equal amplitude and opposite handedness is a beam of plane polarized light. What determines the plane of polarization of the resultant beam?
(b) Show that the combination of two beams of elliptically polarized light is in general another beam of elliptically polarized light. Write down the conditions for which the resultant beam shall be plane polarized or circularly polarized.

1.17

(a) Show that it is impossible to distinguish between a beam of unpolarized light and a beam of circularly polarized light by using linear polarizers. Generally, wave plates and polarizers are needed to determine the degree of polarization.
(b) Given a beam of completely polarized light (say, elliptically polarized), show that the ellipticity and the orientation of the principal axes can be measured by using a single linear polarizer.

1.18 Use the following basis

$$\mathbf{E}_1 = (a, ib) \quad \text{and} \quad \mathbf{E}_2 = (b, -ia)$$

where a and b are real.

(a) Let $\mathbf{E} = (\cos\psi, \sin\psi) = c_1\mathbf{E}_1 + c_2\mathbf{E}_2$. Find c_1 and c_2.
(b) Show that a beam of elliptically polarized light will maintain its polarization ellipse after transmitting through an optically active medium. The angle of rotation of the major axis of the polarization ellipse is the same as that of a beam of linearly polarized light.

1.19 A half-wave plate has a phase retardation of $\Gamma = \pi$. Assume that the plate is oriented so that the azimuth angle (i.e., the angle between the x axis and the slow axis of the plate) is ψ.

(a) Find the polarization state of the transmitted beam, assuming that the incident beam is linearly polarized in the y direction.
(b) Show that a half-wave plate will convert right-handed circularly polarized light into left-handed circularly polarized light, and vice versa, regardless of the azimuth angle of the plate.
(c) E7 is a nematic liquid crystal with $n_o = 1.52$ and $n_e = 1.75$ at $\lambda = 577$ nm. Find the half-wave plate thickness at this wavelength, assuming the plate is made in such a way that the surfaces are parallel to the directors (i.e., a-plate).

1.20 A quarter-wave plate has a phase retardation of $\Gamma = \pi/2$. Assume that the plate is oriented in a direction with azimuth angle ψ.

(a) Find the polarization state of the transmitted beam, assuming that the incident beam is linearly polarized in the y direction.
(b) If the polarization state resulting from (a) is represented by a complex number on the complex plane, show that the locus of these points as ψ varies from 0 to $\pi/2$ is a branch of a hyperbola. Obtain the equation of the hyperbola.
(c) ZLI-1646 is a nematic liquid crystal with $n_o = 1.478$ and $n_e = 1.558$ at $\lambda = 589$ nm. Find the thickness of an a-plate at this wavelength.

1.21 A wave plate is characterized by its phase retardation Γ and azimuth angle ψ.

(a) Find the polarization state of the emerging beam, assuming that the incident beam is polarized in the y direction.
(b) Use a complex number to represent the resulting polarization state obtained in (a).
(c) The polarization state of the emerging beam is represented by a point in the complex plane. Show that the transformed polarization state can be anywhere in the complex plane, provided Γ can be varied from 0 to 2π and ψ can be varied from 0 to $\pi/2$. Physically, this means that any polarization state can be produced from linearly polarized light, provided a proper wave plate is available.
(d) Show that the locus of these points in the complex plane obtained by rotating a wave plate from $\psi = 0$ to $\psi = \pi/2$ is a hyperbola. Derive the equation of this hyperbola.

(e) Show that the Jones matrix W of a wave plate is unitary; that is,

$$W^{\dagger}W = 1$$

where the dagger indicates Hermitian conjugation.

(f) Let $\mathbf{V}_1'$ and $\mathbf{V}_2'$ be the transformed Jones vectors from $\mathbf{V}_1$ and $\mathbf{V}_2$, respectively. Show that if $\mathbf{V}_1$ and $\mathbf{V}_2$ are orthogonal, so are $\mathbf{V}_1'$ and $\mathbf{V}_2'$.

1.22 An ideal polarizer can be considered as a projection operator that acts on the incident polarization state and projects the polarization vector along the transmission axis of the polarizer.

(a) If we neglect the absolute phase factor in Equation (1.9-12), show that

$$P_o^2 = P_0 \quad \text{and} \quad P^2 = P$$

Operators satisfying these conditions are called projection operators in linear algebra.

(b) Show that if $\mathbf{E}_1$ is the amplitude of the electric field, the amplitude of the beam after it passes through the polarizer is given by

$$\mathbf{p}(\mathbf{p} \cdot \mathbf{E}_1)$$

where $\mathbf{p}$ is the unit vector along the transmission axis of the polarizer.

(c) If the incident beam is vertically polarized (i.e., $\mathbf{E}_1 = \hat{\mathbf{y}}E_0$), the polarizer transmission axis is in the x direction (i.e., $\mathbf{p} = \hat{\mathbf{x}}$). The transmitted beam has zero amplitude, since $\hat{\mathbf{x}} \cdot \hat{\mathbf{y}} = 0$. However, if a second polarizer is placed in front of the first polarizer and is oriented at 45° with respect to it, the transmitted amplitude is not zero. Find this amplitude.

(d) Consider a series of N polarizers with the first one oriented at $\psi_1 = \pi/2N$, the second one at $\psi_2 = 2(\pi/2N)$, the third one at $\psi_3 = 3(\pi/2N), \ldots,$ and the Nth one at $N(\pi/2N)$. Let the incident beam be horizontally polarized. Show that the transmitted beam is vertically polarized with an amplitude of

$$[\cos(\pi/2N)]^N$$

Evaluate the amplitude for $N = 1, 2, 3, \ldots, 10$. Show that in the limit of $N \to \infty$, the amplitude becomes one. In other words, a series of polarizers oriented like a fan can rotate the polarization of the light without attenuation.

1.23 In solar physics, the distribution of hydrogen in the solar corona is measured by photographing at the wavelength of the H_α line ($\lambda = 6563$ Å). To enhance the signal-to-noise ratio, a filter of extremely narrow bandwidth (~ 1 Å) is required. The polarization filter devised by Lyot and Öhman consists of a set of birefringent plates separated by parallel polarizers. The plate thicknesses are in geometric progression, that is, d, $2d$, $4d$, $8d, \ldots$. All the plates are oriented at an azimuth angle of 45° [4–7].

(a) Show that if n_o and n_e are the refractive indices of the plates, the transmittance of the whole stack of N plates is given by

$$T = \tfrac{1}{2}\cos^2 x \cos^2 2x \cos^2 4x \cdots \cos^2 2^{N-1}x$$

with

$$x = \frac{\pi d(n_e - n_o)}{\lambda} = \frac{\pi d(n_e - n_o)\nu}{c}$$

(b) Show that the transmission can be written

$$T = \frac{1}{2}\left(\frac{\sin 2^N x}{2^N \sin x}\right)^2$$

(c) Show that the transmission bandwidth (FWHM, full width at half-maximum) of the whole system is governed by that of the bands of the thickest plate, that is,

$$\Delta\nu_{1/2} \sim \frac{c}{2^N d(n_e - n_o)}$$

and the free spectral range $\Delta\nu$ is governed by that of the bands of the thinnest plate, that is,

$$\Delta\nu \sim \frac{c}{d(n_e - n_o)}$$

The finesse F of the system, defined as $\Delta\nu/\Delta\nu_{1/2}$, is then

$$F \sim 2^N$$

(d) Design a filter with a bandwidth of 1 Å at the H_α line, using quartz as the birefringent material. Assume that $n_o = 1.5416$ and $n_e = 1.5506$ at $\lambda = 6563$ Å. Find the required thickness of the thickest plate.

(e) Show that according to (b) the bandwidth (FWHM) is given by

$$\Delta\nu_{1/2} = 0.886\frac{c}{2^N d(n_e - n_o)}$$

1.24 In the examples in Section 1.9, we assumed that the c axis of the a-plate is oriented at 45° to the transmission axis of the polarizer. Show that if the orientation is at a general angle θ, the transmission can be written

$$T = \tfrac{1}{2}\sin^2 2\theta \sin^2(\Gamma/2)$$

which reduces to Equation (1.9-41) when $\theta = 45°$.

1.25 It can be shown that a general birefringent network (e.g., a sequence of wave plates) is equivalent to a wave plate and a polarization rotator. A polarization rotator will rotate the polarization ellipse of an input beam by an angle without changing the shape of the ellipse. The Jones matrix of a polarization rotator can be written

$$A(\rho) = \begin{bmatrix} \cos\rho & -\sin\rho \\ \sin\rho & \cos\rho \end{bmatrix}$$

where ρ is the angle of rotation. We note that a polarization rotation of ρ is equivalent to a coordinate rotation of $-\rho$.

(a) Find the Jones matrix of a birefringent wave plate with phase retardation Γ and azimuth angle ψ followed by a polarization rotator. Show that the Jones matrix is unitary.

(b) The Jones matrix of a general birefringent network is unitary. Based on the result in (a), an arbitrary birefringent network is equivalent to a phase retardation wave plate followed by a polarization rotator. Let the Jones matrix of the general birefringent network be written

$$M = \begin{bmatrix} a+ib & c+id \\ -c+id & a-ib \end{bmatrix}$$

where a, b, c, and d are real. Show that the phase retardation Γ, azimuth angle ψ, and rotation angle ρ of the equivalent network (a wave plate and a rotator) can be written

$$\cos^2(\Gamma/2) = a^2 + c^2, \quad \sin^2(\Gamma/2) = b^2 + d^2$$

$$\tan(\rho + 2\psi) = d/b, \quad \tan\rho = -c/a$$

Based on the above results in (a) and (b), an input beam of linearly polarized light with electric field vector oriented along the principal axes of the equivalent wave plate (ψ or $\psi + \pi/2$) will be transformed into an output beam of linearly polarized light.

1.26 Given a general birefringent network, there exist two linearly polarized input states such that the corresponding output polarization states are linear. These two input polarization states are mutually orthogonal. So are the two output polarization states.

(a) A beam of linearly polarized light is transformed into a beam of elliptically polarized light by using a wave plate (uniaxial) of phase retardation Γ. Let θ be the angle between the polarization vector and the x axis, and let the c axis (slow axis) of the plate be parallel to the x axis. Show that as θ varies from 0 to π, the output polarization state traces out a great circle on the Poincaré sphere. This circle can be obtained by rotating the equator around the S_1 axis by an angle Γ. Show that as Γ varies from 0 to 2π, the output polarization state traces out a circle on the Poincaré sphere. This circle can be obtained by rotating the initial polarization state on the equator around the S_1 axis. For an arbitrary phase retardation Γ, the output polarization state can be obtained by rotating the initial polarization state on the equator around the S_1 axis by an angle Γ.

(b) If a polarization rotator (ρ) is placed after the wave plate in Problem 1.25(a), show that as θ varies from 0 to π, the output polarization state also traces out a great circle on the Poincaré sphere. This great circle is obtained by rotating the great circle in (a) around the polar axis by an angle 2ρ. Based on the result, any input polarization state of a beam of monochromatic light can be transformed into any output polarization state by using the combination of a single wave plate and a polarization rotator.

(c) Using the equivalent circuit in Problem 1.25, show that given a general birefringent network, there exist two linearly polarized input states such that the output polarization states are linear. These two polarization states are mutually orthogonal. Find the polarization states of the input and output beams.

The existence of these linearly polarized states can be proved geometrically as follows. Given an input beam of linearly polarized light, an output beam of elliptically polarized light is obtained. By varying the angle of polarization of the input linearly polarized state from 0 to π, the output polarization state will trace out a great circle on the Poincaré sphere. This great circle either is the equator itself or will intersect with the equator at two points. These two points correspond to output states of linear polarization.

1.27

(a) Derive the expression for the eigenpolarization of the electric field vector in Equations (1.7-11) and (1.7-13).

(b) Use the relation $\mathbf{D} = \varepsilon\mathbf{E}$ and obtain an expression for the corresponding eigenpolarization of the displacement vector $\mathbf{D}$.

(c) Let n_1 and n_2 be the solution of the Fresnel equation (1.7-12), and $\mathbf{E}_1$, $\mathbf{E}_2$, $\mathbf{D}_1$, and $\mathbf{D}_2$ the corresponding eigen-field-vectors. Evaluate $\mathbf{E}_1 \cdot \mathbf{E}_2$ and $\mathbf{D}_1 \cdot \mathbf{D}_2$ and show that $\mathbf{D}_1$ and $\mathbf{D}_2$ are always mutually orthogonal, whereas $\mathbf{E}_1$ and $\mathbf{E}_2$ are only mutually orthogonal in a uniaxial or an isotropic medium.

(d) Show that $\mathbf{E}_1 \cdot \mathbf{D}_2 = 0$ and $\mathbf{E}_2 \cdot \mathbf{D}_1 = 0$.

1.28

(a) Derive the Fresnel equation (1.7-12) directly from Equation (1.7-10).
(b) Show that the Fresnel equation (1.7-12) is a quadratic equation in n^2; that is,

$$An^4 + Bn^2 + C = 0$$

and obtain expressions for A, B, and C.
(c) Show that $B^2 - 4AC > 0$ for the case of pure dielectrics with real ε_x, ε_y, ε_z.
(d) Derive Equation (1.7-16) from Equation (1.7-10) for the case of uniaxial crystals.
(e) Show that in an isotropic medium, Equation (1.7-10) reduces to

$$k^2 - \frac{\varepsilon}{\varepsilon_0}\left(\frac{\omega}{c}\right)^2 = 0$$

1.29 Show that the group velocity for a wave packet propagation in an anisotropic medium also represents the transport of energy, that is, $\mathbf{v}_g = \mathbf{v}_e$. Show that the equality $\mathbf{v}_g = \mathbf{v}_e$ also holds for complex field amplitudes **E** and **H**.

1.30

(a) Derive an expression for the group velocity of the extraordinary wave in a uniaxial crystal as a function of the polar angle θ of the propagation vector.
(b) Derive an expression for the angle α between the phase velocity and the group velocity. This angle is also the angle between the field vectors **E** and **D**.
(c) Show that $\alpha = 0$ when $\theta = 0, \pi/2$. Find the angle θ at which α is maximized, and obtain an expression for α_{max}. Calculate this angle α_{max} for ZLI-1646 with $n_o = 1.478$ and $n_e = 1.558$.
(d) Show that for $n_o \cong n_e$ the maximum angular separation α_{max} occurs at $\theta \cong 45°$, and show that α_{max} is proportional to $|n_o - n_e|$.

1.31

(a) Show that the normal surface Equation (1.7-10) can also be written

$$\frac{k_x^2}{k^2 - \omega^2\mu\varepsilon_x} + \frac{k_y^2}{k^2 - \omega^2\mu\varepsilon_y} + \frac{k_z^2}{k^2 - \omega^2\mu\varepsilon_z} = 1$$

where $k^2 = k_x^2 + k_y^2 + k_z^2$.
(b) Find the normal vector to the normal surface by taking the gradient of the above equation and show that the normal vector is perpendicular to the eigenvectors in Equation (1.7-11). This proves that the eigenvectors for **E** are tangent to the normal surface.

1.32 The phenomenon of double refraction in an anisotropic crystal may be utilized to produce polarized light. Consider a light beam incident on a plane boundary from the inside of a calcite crystal ($n_o = 1.658$, $n_e = 1.486$). Suppose that the c axis of the crystal is normal to the plane of incidence.

(a) Find the range of the internal angle of incidence such that the ordinary wave is totally reflected. The transmitted wave is thus completely polarized.
(b) Use the basic principle described in (a) to design a calcite Glan prism shown in Figure P1.32.

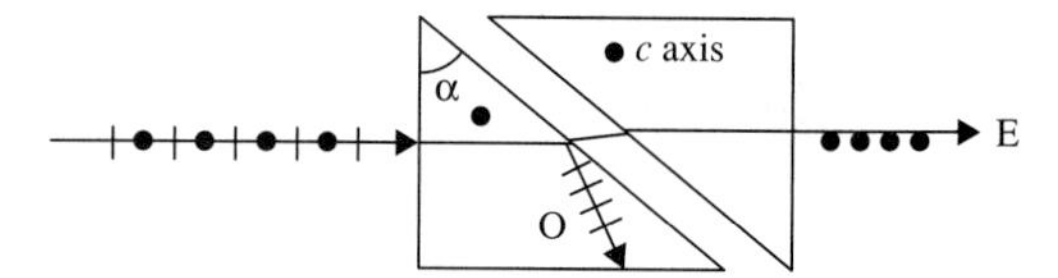

Figure P1.32

Find the range of the apex angle α.

1.33 Dichroic polarizers are materials whose absorption properties and reflection properties are strongly dependent on the direction of vibration of the electric field. If the two absorption coefficients are very different, a thin sheet of the material will be sufficient to transform unpolarized light into a nearly linearly polarized light.

(a) Let α_1 and α_2 be the two absorption coefficients corresponding to two independent polarizations. Derive an expression for the ratio of the two transmitted components as a function of the thickness of the medium.
(b) Show that, strictly speaking, the normal modes of propagation are no longer linearly polarized in the presence of absorption.

1.34 The wave equation (1.7-8) can be written

$$\mathbf{s} \times \left(\mathbf{s} \times \frac{\varepsilon_0}{\varepsilon}\mathbf{D}\right) = -\frac{1}{n^2}\mathbf{D}$$

where **s** is a unit vector in the direction of propagation. Let $\mathbf{D}_1$, $\mathbf{D}_2$ be the normalized eigenvectors with eigenvalues $1/n_1^2$, $1/n_2^2$, respectively. We assume that $\varepsilon_0/\varepsilon$ is a Hermitian tensor.

(a) Show that

$$\left(\frac{\varepsilon_0}{\varepsilon}\right)_{11} \equiv \mathbf{D}_1^* \cdot \frac{\varepsilon_0}{\varepsilon}\mathbf{D}_1 \equiv \frac{1}{n_1^2},$$

$$\left(\frac{\varepsilon_0}{\varepsilon}\right)_{22} \equiv \mathbf{D}_2^* \cdot \frac{\varepsilon_0}{\varepsilon}\mathbf{D}_2 \equiv \frac{1}{n_2^2},$$

$$\left(\frac{\varepsilon_0}{\varepsilon}\right)_{12} \equiv \mathbf{D}_1^* \cdot \frac{\varepsilon_0}{\varepsilon}\mathbf{D}_2 = 0$$

(b) Show that $\mathbf{D}_1^* \cdot \mathbf{D}_2 = 0$.

1.35

(a) Show that the power flow in the direction of propagation is given by

$$\mathbf{S} \cdot \mathbf{s} = \frac{1}{2}\frac{c}{\varepsilon_0}\left(\frac{1}{n_1^3}|\mathbf{D}_1|^2 + \frac{1}{n_2^3}|\mathbf{D}_2|^2\right)$$

where $\mathbf{D}_1$ and $\mathbf{D}_2$ are the amplitudes of the displacement fields of the eigenmodes and n_1 and n_2 are the corresponding refractive indices.

(b) Show that

$$\mathbf{S} \cdot \mathbf{s} = \frac{1}{2\mu c}(n_1|\mathbf{A}_1|^2 + n_2|\mathbf{A}_2|^2)$$

where $\mathbf{A}_1$ and $\mathbf{A}_2$ are the transverse parts of the electric field vectors.

(c) Show that the total power flow along the direction of propagation is a constant of integration; that is,

$$\frac{d}{d\xi}(n_1|\mathbf{A}_1|^2 + n_2|\mathbf{A}_2|^2) = 0$$

where ξ is the distance along the direction of propagation $\mathbf{s}$ (i.e., $\xi = \mathbf{s} \cdot \mathbf{r}$).

1.36 Consider a beam of circularly polarized light with a finite transverse dimension.

(a) Let the transverse dimension be much greater than the wavelength λ. Show that the electric field and magnetic field vector can be written approximately

$$E(x, y, z, t) \approx E_0(x, y)(\hat{\mathbf{x}} - i\hat{\mathbf{y}}) - \frac{i}{k}\left(\frac{\partial E_0}{\partial x} - i\frac{\partial E_0}{\partial y}\right)\hat{\mathbf{z}}e^{i(\omega t - kz)}$$

$$H(x, y, z, t) \approx i\frac{k}{\omega\mu}E(x, y, z, t)$$

(b) Using the angular momentum defined as $\mathbf{L} = \mathbf{r} \times \mathbf{P}$, where $\mathbf{P}$ is the linear momentum defined in Problem 1.5, calculate the time-averaged component of the angular momentum along the direction of propagation ($+z$). Show that this component of the angular momentum is $\hbar$ provided the energy of the electromagnetic wave is normalized to $\hbar\omega$.

(c) Show that the transverse components of the angular momentum vanish.

1.37 According to Problem 1.6, the work done per unit volume by applying an electric field in a dielectric medium is $W = \int \mathbf{E} \cdot d\mathbf{P}$, where $\mathbf{P}$ is the polarization. In a linear medium $P_i = \varepsilon_0\chi_{ij}E_j$. Carry out the integration from (0, 0) to (E_{10}, E_{20}) in a two-dimensional case by using two different paths.

(a) Path A: First from (0, 0) to $(E_{10}, 0)$, and then from $(E_{10}, 0)$ to (E_{10}, E_{20}). Show that the integration yields the following result:

$$W = \tfrac{1}{2}\varepsilon_0\chi_{11}E_{10}^2 + \tfrac{1}{2}\varepsilon_0\chi_{22}E_{20}^2 + \varepsilon_0\chi_{12}E_{10}E_{20}$$

(b) Path B: First from (0, 0) to $(0, E_{20})$, and then from $(0, E_{20})$ to (E_{10}, E_{20}). Show that the integration yields the following result:

$$W = \tfrac{1}{2}\varepsilon_0\chi_{11}E_{10}^2 + \tfrac{1}{2}\varepsilon_0\chi_{22}E_{20}^2 + \varepsilon_0\chi_{21}E_{10}E_{20}$$

In a lossless medium, the integration should be independent of the path. This leads to $\chi_{12} = \chi_{21}$. A similar analysis in the three-dimensional case will yield $\chi_{ij} = \chi_{ji}$.

REFERENCES

1. Jones, R. C., New calculus for the treatment of optical systems. *J. Opt. Soc. Am.* **31**:488 (1941).
2. Yariv, A., and P. Yeh, *Optical Waves in Crystals*. Wiley, New York, 1984.
3. Yeh, P., *Optical Waves in Layered Media*. Wiley, New York, 1988.
4. Lyot, B., Optical apparatus with wide field using interference of polarized light. *C. R. Acad. Sci. (Paris)* **197**:1593 (1933).
5. Lyot, B., Filter monochromatique polarisant et ses applications en physique solaire. *Ann. Astrophys.* **7**:31 (1944).
6. Öhman, Y., A new monochromator. *Nature* **41**:157 (1938).
7. Öhman, Y., On some new birefringent filter for solar research. *Ark. Astron.* **2**:165 (1958).

CHAPTER 2

RAYS AND OPTICAL BEAMS

2.0 INTRODUCTION

In the last chapter we described plane waves in homogeneous media. In this chapter we consider the propagation of rays and beams through a variety of media that are often employed in optical electronic systems. The media include homogeneous and inhomogeneous materials, thin lenses, gradient index (GRIN) lenses, dielectric interfaces, and mirrors. In geometrical optics, the propagation of optical waves can be described approximately by using the concept of rays. This is valid provided the beam diameter is much larger than the wavelength, and the diffraction can be neglected. The rays travel in straight lines in homogeneous media and obey Fermat's principle of least time in inhomogeneous media. A good understanding of the propagation of rays makes it possible to trace their trajectories when they are passing through various optical media. We find that the passage of rays through these optical elements can be described by simple 2×2 matrices. In this chapter, we also describe the propagation of Gaussian beams in various optical media, as well as the integral representation of the propagation of a general paraxial beam of light. It will be shown that simple ray matrices can be employed to describe the propagation of spherical waves and Gaussian beams, such as those characteristic of the output of lasers. Furthermore, the ray matrices are also useful for the propagation of a general paraxial beam through various optical media.

2.1 RAY MATRICES

Referring to Figure 2.1, we consider the passing of a paraxial ray through a thin lens of focal length f. In most optical systems that involve lenses made of homogeneous optical materials with spherical surfaces, the axis of symmetry (z axis) is often called the optical axis. Paraxial rays are defined as those rays whose angular deviation from the optical axis is small enough that the sine and tangent of the angle can be approximated by the angle itself. For such a system with cylindrical symmetry, we consider a class of rays whose trajectories can be conveniently described by $r(z)$, where r is the distance measured from the optical axis. These are known as meridional rays whose trajectories are confined in a plane through the optical axis.

For a thin lens, the output ray and the input ray are related by the following simple equations:

$$r_{\text{out}} = r_{\text{in}} \tag{2.1-1}$$

$$r'_{\text{out}} = r'_{\text{in}} - \frac{r_{\text{in}}}{f} \tag{2.1-2}$$

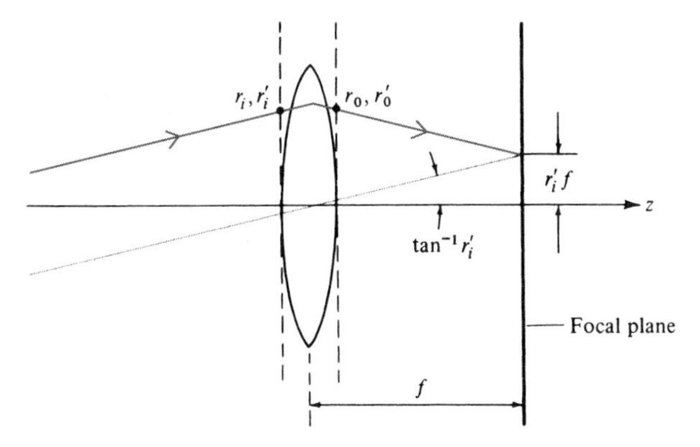

Figure 2.1 Deflection of a ray by a thin lens.

where r_{in} and r_{out} are the positions of the ray measured from the optical axis, and r'_{in} and r'_{out} are the slopes of the ray relative to the optical axis. In the paraxial approximation, the slopes are actually the ray angles measured relative to the optical axis. We note that the ray position is unchanged, reflecting the definition of a thin lens. The slope of the output ray is changed due to the presence of the thin lens. Equation (2.1-2) can easily be derived by considering the focusing nature of the thin lens. In other words, parallel rays intersect at the focal plane after passing through a thin lens.

Equations (2.1-1) and (2.1-2) can be represented conveniently by a simple 2×2 matrix equation

$$\begin{bmatrix} r_{\text{out}} \\ r'_{\text{out}} \end{bmatrix} = \begin{bmatrix} 1 & 0 \\ -\dfrac{1}{f} & 1 \end{bmatrix} \begin{bmatrix} r_{\text{in}} \\ r'_{\text{in}} \end{bmatrix} \tag{2.1-3}$$

where $f > 0$ for a convergent lens and is negative for a divergent one. The ray matrices for a number of optical elements are shown in Table 2.1.

According to Table 2.1, propagation through a homogeneous medium with length d can be described as

$$\begin{bmatrix} r_{\text{out}} \\ r'_{\text{out}} \end{bmatrix} = \begin{bmatrix} 1 & d \\ 0 & 1 \end{bmatrix} \begin{bmatrix} r_{\text{in}} \\ r'_{\text{in}} \end{bmatrix} \tag{2.1-4}$$

In the matrix approach, each ray is represented by a ray vector, while the optical elements are represented by ray matrices. The ray vector is a two-element column vector that consists of the ray distance from the axis $r(z)$ and the ray angle $r'(z)$, at any position z on the axis. In an optical system consisting of a series of optical elements as shown in Table 2.1, the output ray vector can easily be obtained by multiplying the ray matrices in sequence with the input ray vector.

To illustrate this, we consider the propagation of a ray through a periodic lens system, which consists of lenses of focal lengths f_1 and f_2 separated by a distance d as shown in Figure 2.2. Such an optical system will be shown later in this book to be formally equivalent to the problem of Gaussian beam propagation inside an optical resonator with mirrors of radii of curvature $R_1 = 2f_1$ and $R_2 = 2f_2$ that are separated by a distance d.

TABLE 2.1 Ray Matrices for Some Optical Elements and Media.

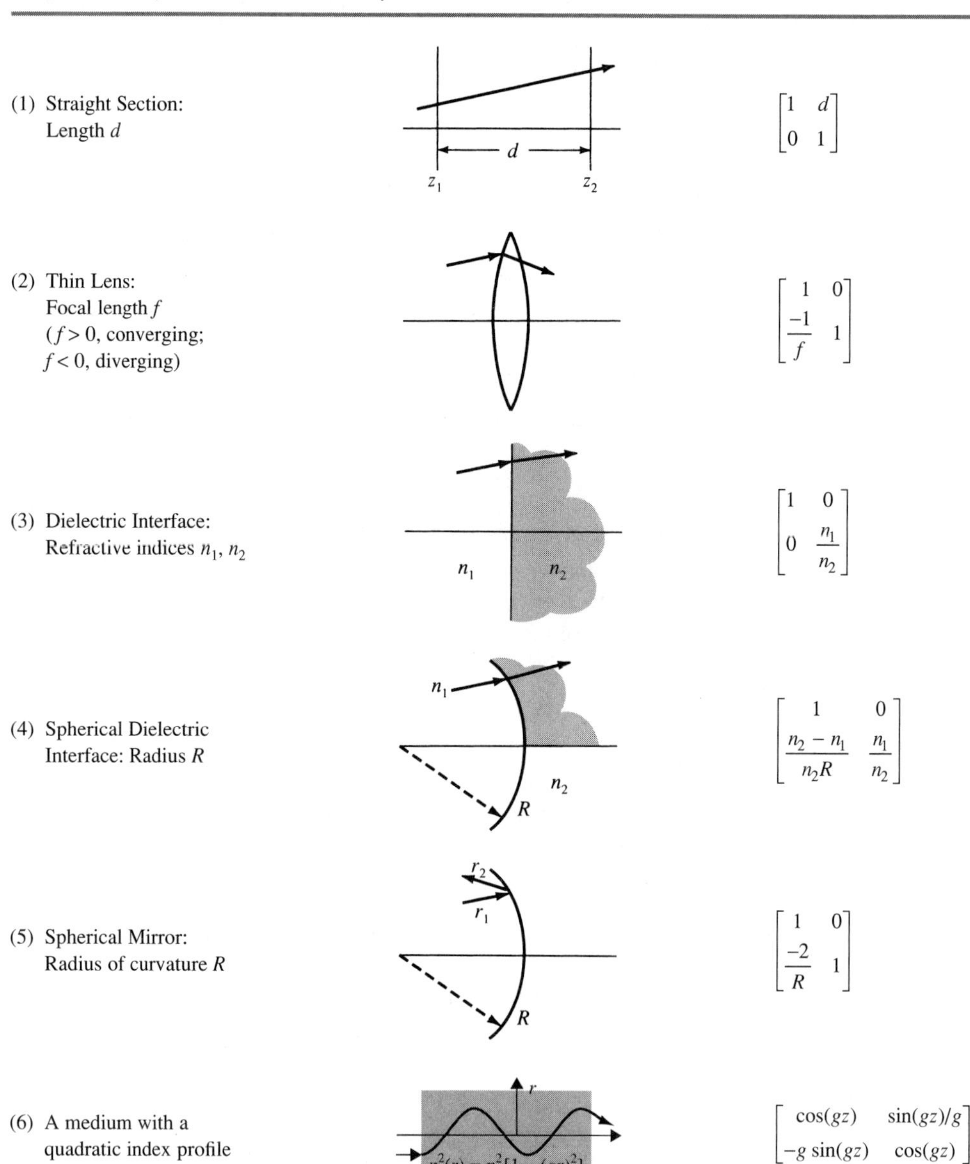

Element	Ray matrix
(1) Straight Section: Length d	$\begin{bmatrix} 1 & d \\ 0 & 1 \end{bmatrix}$
(2) Thin Lens: Focal length f ($f > 0$, converging; $f < 0$, diverging)	$\begin{bmatrix} 1 & 0 \\ \dfrac{-1}{f} & 1 \end{bmatrix}$
(3) Dielectric Interface: Refractive indices n_1, n_2	$\begin{bmatrix} 1 & 0 \\ 0 & \dfrac{n_1}{n_2} \end{bmatrix}$
(4) Spherical Dielectric Interface: Radius R	$\begin{bmatrix} 1 & 0 \\ \dfrac{n_2 - n_1}{n_2 R} & \dfrac{n_1}{n_2} \end{bmatrix}$
(5) Spherical Mirror: Radius of curvature R	$\begin{bmatrix} 1 & 0 \\ \dfrac{-2}{R} & 1 \end{bmatrix}$
(6) A medium with a quadratic index profile	$\begin{bmatrix} \cos(gz) & \sin(gz)/g \\ -g\sin(gz) & \cos(gz) \end{bmatrix}$

The section between planes s and $s + 1$ can be considered as a unit cell of the optical system. The matrix relating the ray vector at the output of a unit cell to that at the input is the product of the ray matrices of the elements in sequence. In other words,

$$\begin{bmatrix} r_{s+1} \\ r'_{s+1} \end{bmatrix} = \begin{bmatrix} 1 & 0 \\ -1/f_1 & 1 \end{bmatrix} \begin{bmatrix} 1 & d \\ 0 & 1 \end{bmatrix} \begin{bmatrix} 1 & 0 \\ -1/f_2 & 1 \end{bmatrix} \begin{bmatrix} 1 & d \\ 0 & 1 \end{bmatrix} \begin{bmatrix} r_s \\ r'_s \end{bmatrix} \tag{2.1-5}$$

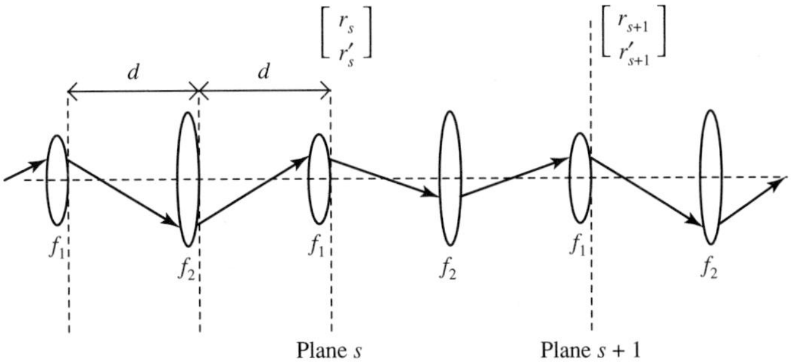

Figure 2.2 A periodic lens waveguide consisting of lenses with focal lengths f_1 and f_2.

where we note the rightmost ray matrix corresponds to the free space between plane s and the lens with focal length f_2 in the unit cell. By carrying out the matrix multiplication in Equation (2.1-5), we obtain

$$\begin{bmatrix} r_{s+1} \\ r'_{s+1} \end{bmatrix} = \begin{bmatrix} A & B \\ C & D \end{bmatrix} \begin{bmatrix} r_s \\ r'_s \end{bmatrix} \tag{2.1-6}$$

or equivalently,

$$\begin{aligned} r_{s+1} &= Ar_s + Br'_s \\ r'_{s+1} &= Cr_s + Dr'_s \end{aligned} \tag{2.1-7}$$

where A, B, C, and D are the matrix elements given by

$$\begin{aligned} A &= 1 - \frac{d}{f_2} \\ B &= d\left(2 - \frac{d}{f_2}\right) \\ C &= -\left[\frac{1}{f_1} + \frac{1}{f_2}\left(1 - \frac{d}{f_1}\right)\right] \\ D &= -\left[\frac{d}{f_1} - \left(1 - \frac{d}{f_1}\right)\left(1 - \frac{d}{f_2}\right)\right] \end{aligned} \tag{2.1-8}$$

It is important to note that the ray matrix for the unit cell is a unimodular matrix. In other words,

$$AD - BC = 1 \tag{2.1-9}$$

The result (Equation (2.1-6)) can now be applied to the propagation through N unit cells, say, between planes 0 and N. We obtain

$$\begin{bmatrix} r_N \\ r'_N \end{bmatrix} = \begin{bmatrix} A & B \\ C & D \end{bmatrix}^N \begin{bmatrix} r_0 \\ r'_0 \end{bmatrix} \tag{2.1-10}$$

The Nth power of a unimodular matrix can be simplified by the following matrix identity (Chebyshev's identity, see Problem 2.18):

$$\begin{bmatrix} A & B \\ C & D \end{bmatrix}^N = \begin{bmatrix} AU_{N-1} - U_{N-2} & BU_{N-1} \\ CU_{N-1} & DU_{N-1} - U_{N-2} \end{bmatrix} \tag{2.1-11}$$

where

$$U_N = \frac{\sin(N+1)\theta}{\sin\theta} \tag{2.1-12}$$

with

$$\theta = \cos^{-1}\left(\frac{A+D}{2}\right) \tag{2.1-13}$$

If a ray is launched at plane 0 with finite r_0 and r'_0, the ray will remain confined by the lens system along the optical axis provided the ray parameters r_N and r'_N remain finite as N approaches infinity. By examining Equations (2.1-10)–(2.1-13), we find that confined propagation of a ray requires a real θ, or equivalently,

$$\left|\frac{A+D}{2}\right| \leq 1 \tag{2.1-14}$$

In terms of the system parameters (Equation (2.1-8)), the condition for confined propagation becomes

$$-1 \leq 1 - \frac{d}{f_1} - \frac{d}{f_2} + \frac{d^2}{2f_1f_2} \leq 1 \tag{2.1-15}$$

or equivalently,

$$0 \leq \left(1 - \frac{d}{2f_1}\right)\left(1 - \frac{d}{2f_2}\right) \leq 1 \tag{2.1-16}$$

If, on the other hand, the confinement condition Equation (2.1-14) is violated, then the parameter θ becomes a purely imaginary number, $\theta = iq$, where q is a real number. In this case, U_N becomes

$$U_N = \frac{\sinh(N+1)q}{\sinh q} \tag{2.1-17}$$

and thus the ray parameters r_N and r'_N will grow exponentially as a function of N. This leads to unconfined propagation.

Identical Lens Waveguide—A Special Case

Consider the special case when $f_1 = f_2 = f$ in Figure 2.2. This is the simplest lens waveguide. The confinement condition Equation (2.1-16) becomes

$$\left(1 - \frac{d}{2f}\right)^2 \leq 1 \tag{2.1-18}$$

Since d is always positive, the condition is satisfied only when $f > 0$. In other words, in the simplest lens waveguide, confinement is possible only when the lenses are convergent lenses. The confinement condition can further be simplified as

$$0 \le d \le 4f \tag{2.1-19}$$

The same problem can also be solved directly. In this case, the ray parameters from one plane to the next one are given by

$$\begin{bmatrix} r_{s+1} \\ r'_{s+1} \end{bmatrix} = \begin{bmatrix} 1 & 0 \\ -1/f & 1 \end{bmatrix} \begin{bmatrix} 1 & d \\ 0 & 1 \end{bmatrix} \begin{bmatrix} r_s \\ r'_s \end{bmatrix} = \begin{bmatrix} A & B \\ C & D \end{bmatrix} \begin{bmatrix} r_s \\ r'_s \end{bmatrix} \tag{2.1-20}$$

By carrying out the matrix multiplication in Equation (2.1-20), we obtain

$$\begin{aligned} A &= 1 \\ B &= d \\ C &= -1/f \\ D &= 1 - d/f \end{aligned} \tag{2.1-21}$$

The confinement condition is given by, according to Equation (2.1-14)

$$-1 \le 1 - \frac{d}{2f} \le 1 \tag{2.1-22}$$

which is equivalent to Equations (2.1-18) and (2.1-19).

The beam radius (or ray location) at the mth lens can be written, according to Equations (2.1-10)–(2.1-12)

$$r_m = r_{\max} \sin(m\theta + \alpha) \tag{2.1-23}$$

with

$$\cos\theta = 1 - d/2f \tag{2.1-24}$$

Using Equations (2.1-10)–(2.1-12) and the explicit expressions of A, B, C, and D, we obtain

$$(r_{\max})^2 = \frac{4f}{4f - d}(r_0^2 + dr_0 r'_0 + dfr'^2_0) \tag{2.1-25}$$

$$\tan\alpha = \frac{\sqrt{4f/d - 1}}{(1 + 2fr'_0/r_0)} \tag{2.1-26}$$

where m corresponds to the plane immediately to the right of the mth lens. The derivation of the last two equations is left as an exercise.

The stability criteria can be demonstrated experimentally by tracing the behavior of a laser beam as it propagates down a sequence of lenses spaced uniformly. One can easily note the rapid escape of the beam once condition (2.1-19) or (2.1-22) is violated.

Rays in Optical Resonators

Consider the bouncing of rays between two curved mirrors that form an optical resonator. Since the reflection at a mirror with a radius of curvature R is equivalent, except for the folding of the path, to passage through a lens with a focal length of $f = R/2$, we can use the ray matrix formulation to describe the propagation of a ray inside the resonator, which consists of two mirrors of radii R_1 and R_2. Figure 2.3 shows such an optical resonator and its equivalent optical lens system.

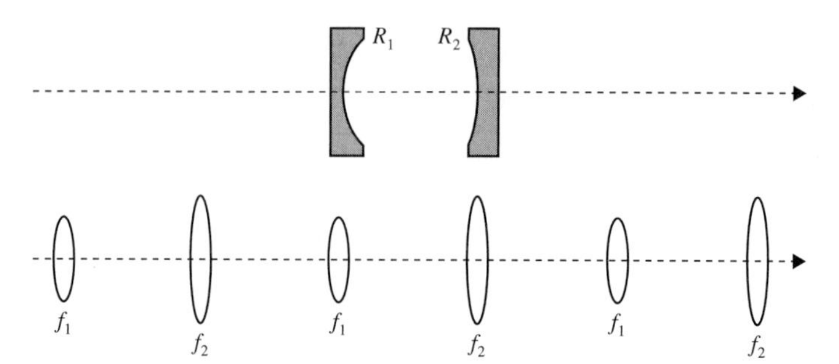

Figure 2.3 An optical resonator and its equivalent optical lens system. R_1 and R_2 are radii of curvature of the mirrors: $f_1 = R_1/2$ and $f_2 = R_2/2$.

For an optical resonator with two concave mirrors, the confinement condition becomes

$$0 \leq \left(1 - \frac{d}{R_1}\right)\left(1 - \frac{d}{R_2}\right) \leq 1 \tag{2.1-27}$$

where d is the separation between the mirrors, and R_1 and R_2 are the radii of curvature of the mirrors.

2.2 SKEW RAYS AND REENTRANT RAYS

Skew Rays

So far, we have only considered a particular class of rays (the meridional rays) in which each ray trajectory is confined in a plane that passes through the optical axis. In general, a ray can propagate with a three-dimensional trajectory along the optical axis (e.g., a helical trajectory). These are the so-called skew rays. For these rays, the x and y coordinates are independent variables. A complete description of the ray requires $x(z)$, $x'(z)$, $y(z)$, and $y'(z)$. In the lens waveguide described in the previous section, the propagation of the rays (from plane 0 to plane N) can be written

$$\begin{bmatrix} x_N \\ x'_N \end{bmatrix} = \begin{bmatrix} A & B \\ C & D \end{bmatrix}^N \begin{bmatrix} x_0 \\ x'_0 \end{bmatrix} \tag{2.2-1}$$

$$\begin{bmatrix} y_N \\ y'_N \end{bmatrix} = \begin{bmatrix} A & B \\ C & D \end{bmatrix}^N \begin{bmatrix} y_0 \\ y'_0 \end{bmatrix} \tag{2.2-2}$$

We note that both components (x and y) of the ray parameters obey the same ray matrix for propagation in the lens system. Using Equations (2.1-11)–(2.1-13), the ray position at the Nth plane can be written

$$x_N = x_{\max} \sin(N\theta + \alpha_x) \tag{2.2-3}$$

$$y_N = y_{\max} \sin(N\theta + \alpha_y) \tag{2.2-4}$$

where N is an integer, and $x_{\max}$, $y_{\max}$, α_x, and α_y are constants. These constants depend on the initial ray parameters at the 0th plane. In general, the two phases, α_x and α_y are different.

According to Equations (2.2-3) and (2.2-4), the locus of the points (x_N, y_N) lies on an ellipse for confined propagation when θ is a real number. These are the skew rays. The ray coordinates are confined in a rectangle of $2x_{\max}$ by $2y_{\max}$. In the special case when $\alpha_x = \alpha_y$, the ray trajectory becomes confined in a meridional plane through the optical axis. This corresponds to a meridional ray. We note that θ is a system parameter that is independent of the ray coordinates.

Reentrant Rays

Based on Equations (2.2-3) and (2.2-4), the ray coordinates in a periodic lens waveguide are sinusoidal functions of $N\theta$. Under the appropriate conditions (system parameters), the ray may reproduce itself after certain periods of propagation. If the system parameters are such that $\theta/2\pi$ is a rational number, then an integer N can be found such that

$$N\theta = 2M\pi \tag{2.2-5}$$

where M is also an integer. When this happens, the ray matrix after N periods becomes, according to Equations (2.1-11) and (2.1-12),

$$\begin{bmatrix} A & B \\ C & D \end{bmatrix}^N = \begin{bmatrix} 1 & 0 \\ 0 & 1 \end{bmatrix}$$

and thus the ray retraces its trajectory after N periods.

EXAMPLE: REENTRANT RAYS IN SYMMETRIC CONFOCAL RESONATORS

Consider an optical resonator made of two identical concave mirrors of radius of curvature R. If the separation d equals the radius of curvature R ($d = R$), then these two mirrors share a common focal point. Such resonators are called confocal resonators. For such resonators, $f_1 = f_2 = R/2$ in the equivalent lens waveguide, the system parameter θ is, according to Equation (2.1-8),

$$\theta = \cos^{-1}\left(\frac{A+D}{2}\right) = \cos^{-1}(-1) = \pi$$

The rays inside the resonator will thus repeat after two periods in the equivalent lens waveguide. Since each period in the lens waveguide is equivalent to a round-trip in the optical resonator, the ray trajectory will repeat after two round-trips in the resonator. It is important to note that the periodic retracing of the ray trajectory is independent of the initial launching condition (r_0, r_0').

2.3 RAYS IN LENSLIKE MEDIA

We consider in this section the propagation of a ray in an optically inhomogeneous medium, particularly a lenslike medium. In the realm of geometrical optics, a light ray travels between two points along a trajectory such that the time taken is the least. This is known as *Fermat's principle of least time*. In terms of calculus of variation, this is written

$$\delta \int_{P_1}^{P_2} \frac{n}{c}\, ds = 0 \tag{2.3-1}$$

where P_1 and P_2 are the two points, c is speed of light in vacuum, n is the index of refraction as a function of position, and ds is a differential element along the ray path. In other words, the optical length $\int n\, ds$ of the path traversed by a ray between two points is stationary, that is, either a maximum or a minimum.

We represent the ray trajectory in parametric form as

$$x = x(t), \quad y = y(t), \quad z = z(t) \tag{2.3-2}$$

where t is an arbitrary parameter. Equation (2.3-1) can then be written

$$\delta \int_{t_1}^{t_2} n(x, y, z)\sqrt{x'^2 + y'^2 + z'^2}\, dt = 0 \tag{2.3-3}$$

where x', y', and z' are derivatives with respect to the parameter t, and we have ignored the constant c. If we define a functional

$$F(x, x', y, y', z, z', t) = n(x, y, z)\sqrt{x'^2 + y'^2 + z'^2} \tag{2.3-4}$$

then the Euler–Lagrange equations following the condition of Equation (2.3-1) can be written

$$\frac{\partial F}{\partial x} = \frac{d}{dt}\left(\frac{\partial F}{\partial x'}\right), \quad \frac{\partial F}{\partial y} = \frac{d}{dt}\left(\frac{\partial F}{\partial y'}\right), \quad \frac{\partial F}{\partial z} = \frac{d}{dt}\left(\frac{\partial F}{\partial z'}\right) \tag{2.3-5}$$

In other words, the ray trajectory with the stationary optical length must satisfy the Euler–Lagrange equations (2.3-5). Using

$$ds = dt\sqrt{x'^2 + y'^2 + z'^2} \tag{2.3-6}$$

the above Euler–Lagrange equations can be written

$$\frac{\partial n}{\partial x} = \frac{d}{ds}\left(n\frac{dx}{ds}\right), \quad \frac{\partial n}{\partial y} = \frac{d}{ds}\left(n\frac{dy}{ds}\right), \quad \frac{\partial n}{\partial z} = \frac{d}{ds}\left(n\frac{dz}{ds}\right) \tag{2.3-7}$$

These three equations can be written as the following vector equation:

$$\frac{d}{ds}\left(n\frac{d\mathbf{r}}{ds}\right) = \boldsymbol{\nabla} n \tag{2.3-8}$$

where we recall that s is the arc length along the ray trajectory measured from P_1 and $\mathbf{r}$ is the vector representation of (x, y, z). This is the ray equation. The left side is a measure of the bending of the ray, while the right side is the gradient of the index of refraction. Thus the ray always bends toward the high-index region as it propagates. The ray equation has successfully been employed to explain many atmospheric phenomena, including mirage and green flash. In this chapter, we will investigate the propagation of a ray in a quadratic index medium (or lenslike medium).

We now consider the propagation of a ray in a lenslike medium whose index of refraction n varies according to

$$n^2(x, y) = n_0^2[1 - g^2(x^2 + y^2)] \quad \text{with} \quad g^2(x^2 + y^2) \ll 1 \tag{2.3-9}$$

where n_0 is the refractive index at the axis of symmetry and g is a real constant characteristic of the medium. Since the phase delay of a wave propagating through a section dz of a medium with an index of refraction n is $(2\pi\, dz/\lambda)n$, it follows directly that a thin slab of the medium described by Equation (2.3-9) will act as a thin lens, introducing a phase shift proportional to r^2. The medium has a maximum index of refraction at the axis of symmetry ($r = 0$).

The index of refraction decreases in the radial direction away from the axis. This type of medium is also known as gradient index (GRIN) medium, or quadratic index medium. This equation can also be viewed as consisting of the lowest terms in the Taylor series expansion of $n(x, y)$ for a continuous optical medium with cylindrical symmetry. Using cylindrical coordinates with

$$r = \sqrt{x^2 + y^2}$$

Equation (2.3-9) can be written

$$n^2(r) = n_0^2[1 - (gr)^2] \tag{2.3-10}$$

For paraxial rays, we may replace d/ds by d/dz in Equation (2.3-8). Using Equation (2.3-9) for the refractive index, we obtain

$$\frac{d^2\mathbf{r}}{dz^2} + g^2\mathbf{r} = 0 \quad \text{with} \quad \mathbf{r} = (x, y) \tag{2.3-11}$$

This is a differential equation for the ray trajectory in a lenslike medium. For meridional rays, the above equation can be written

$$\frac{d^2r}{d^2z} + g^2r = 0 \quad \text{with} \quad r = \sqrt{x^2 + y^2} \tag{2.3-12}$$

If at the input plane $z = 0$, the ray has a position r_0 and slope r_0', the ray trajectory can be written, according to Equation (2.3-12),

$$\begin{aligned} r(z) &= r_0 \cos(gz) + r_0' \sin(gz)/g \\ r'(z) &= -r_0 g \sin(gz) + r_0' \cos(gz) \end{aligned} \tag{2.3-13}$$

The ray trajectory can also be written as a matrix equation,

$$\begin{bmatrix} r \\ r' \end{bmatrix} = \begin{bmatrix} \cos(gz) & \sin(gz)/g \\ -g\sin(gz) & \cos(gz) \end{bmatrix} \begin{bmatrix} r_0 \\ r_0' \end{bmatrix} \tag{2.3-14}$$

According to Equation (2.3-13), the ray oscillates back and forth across the optical axis, as shown in Figure 2.4. This is in agreement with the fact that rays tend to bend toward the high-index region as they propagate. In fact, we note that the ray trajectory is a periodic function of z with a period

$$\Lambda = \frac{2\pi}{g} \tag{2.3-15}$$

This period is also known as the pitch.

As a result of the periodic behavior, sections of such a lenslike medium can be used for imaging purposes. These sections are called GRIN-rod lenses, or simply rod lenses. To illustrate this, let us consider the situation when a bundle of parallel rays are incident on such a medium at $z = 0$. All the incident rays are parallel to the z axis. The paths of these rays are represented by

$$r(z) = r_0 \cos(gz) \tag{2.3-16}$$

where r_0 is the point of incidence for each ray. After traversing a distance d such that

$$gd = \pi/2 \tag{2.3-17}$$

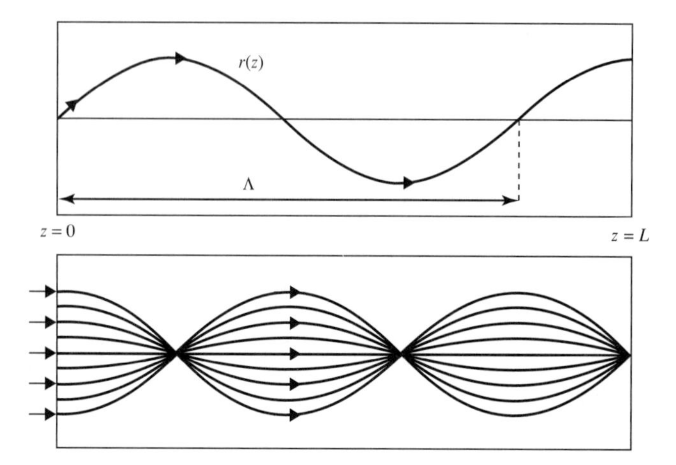

Figure 2.4 Path of rays in a medium with a quadratic index variation.

all the rays collapse into a point at the axis. If this is called a focal point, then the focal length is

$$f = \pi/2g = \Lambda/4 \tag{2.3-18}$$

where $\Lambda = 2\pi/g$ is the period. If the GRIN-rod is long enough, the bundle of rays will diverge and then converge again. In fact, the focusing will repeat itself indefinitely at an interval of π/g. Figure 2.4 illustrates the ray paths inside the medium.

If the length of the medium L is not an exact integral number of f, then the bundle of rays will converge upon emerging at $z = L$ to a common focus at a distance

$$h = \frac{1}{n_0 g \tan(gL)} \tag{2.3-19}$$

from the exit plane. The factor n_0 accounts for the refraction at the boundary, assuming the medium at $z > L$ posseses an index of $n = 1$, and a small angle of incidence. The derivation of Equation (2.3-19) is left as an exercise (Problem 2.3). If h in Equation (2.3-19) is negative, then the bundle of rays are actually divergent upon emerging at $z = L$. The diverging beam appears to originate from a point source located at $z = L + h$ (note that $h < 0$), inside the medium. In this case, the section of the medium acts like a negative lens.

There are physical situations that give rise to the quadratic index variation. These include the following:

1. Propagation of laser beams with Gaussian-like intensity profile in a slightly absorbing medium. As a result of the absorption, the temperature tends to increase more at the high-intensity region. This leads to a quadratic temperature profile. If the medium exhibits a thermo-optical coefficient of $dn/dT > 0$, then a lenslike medium is produced by the presence of the laser beam. A negative lens effect is produced if $dn/dT < 0$.
2. Propagation of laser beams with Gaussian intensity profile in Kerr media. As a result of the nonlinear response, the local index of refraction increases with the intensity, according to

$$n(x, y, z) = n_0 + n_2 I(x, y, z)$$

where n_0 is a constant and n_2 is known as the Kerr coefficient. For a Gaussian beam, the intensity has a quadratic profile near the axis of the beam. If $n_2 > 0$, then a lenslike medium is produced by the presence of the laser beam. A negative lens effect is produced if $n_2 < 0$. The Kerr effect will be discussed further in Chapter 8, as well as in Chapter 14.

3. Gradient index glass fibers produced by ion-exchange diffusion. The ion-exchange diffusions inside a hollow glass tube create an index distribution profile along the radial direction after the tube collapses into a solid fiber. These optical fibers can be employed for transmitting images in medical examinations.

The rays described in the preceding are all in meridional planes that contain the z axis. There are other types of rays that propagate out of the meridional planes. These are the so-called skew rays. Typical examples of skew rays take a helical path around the z axis (see examples in Problem 2.19).

2.4 WAVE EQUATION IN QUADRATIC INDEX MEDIA AND BEAMS

In geometrical optics, a bundle of parallel rays can be focused into a point by using a thin lens or a GRIN-rod lens. In reality, the focused spot has a finite size. Furthermore, the effect of diffraction is completely ignored in geometrical optics. In this section, we consider the propagation of optical beams by taking into account the wave nature of electromagnetic radiation. Particularly, we are interested in optical beams in which the flow of energy is predominantly along one direction. We are also interested in optical beams whose wavefronts are spherical. Later in this chapter, we will show that optical beams with spherical wavefronts can propagate in optical resonators made of spherical mirrors with minimum propagation loss.

In Chapter 1, we obtained the following equation for the propagation of an electromagnetic wave in isotropic media:

$$\nabla \times (\nabla \times \mathbf{E}) + \mu\varepsilon \frac{\partial^2}{\partial t^2}\mathbf{E} = 0 \tag{2.4-1}$$

In this section, we will find a solution of this equation in the form of an optical beam. The first term in the above equation can be expanded as

$$\nabla \times (\nabla \times \mathbf{E}) = \nabla(\nabla \cdot \mathbf{E}) - \nabla^2\mathbf{E} \tag{2.4-2}$$

Using Equation (1.1-3) and $\mathbf{D} = \varepsilon\mathbf{E}$, we obtain

$$\nabla^2\mathbf{E} - \mu\varepsilon \frac{\partial^2}{\partial t^2}\mathbf{E} = -\nabla\left(\frac{1}{\varepsilon}\mathbf{E} \cdot \nabla\varepsilon\right) \tag{2.4-3}$$

The term on the right side of Equation (2.4-3) can be neglected provided the fractional change of ε in one wavelength is $\ll 1$. In this case, the wave equation for a harmonic field with a time dependence of $\exp(i\omega t)$ becomes

$$\nabla^2\mathbf{E} + k^2(\mathbf{r})\mathbf{E} = 0 \tag{2.4-4}$$

where

$$k^2(\mathbf{r}) = \omega^2\mu\varepsilon(\mathbf{r})\left(1 - i\frac{\sigma(\mathbf{r})}{\omega\varepsilon}\right) \tag{2.4-5}$$

thus allowing for a possible dependence of ε on position $\mathbf{r}$. We have also taken k as a complex number to allow for the possibility of losses ($\sigma > 0$) or gain ($\sigma < 0$) in the medium. According to the discussion in Chapter 1, the field amplitude of a plane wave $E_0 \exp[(i\omega t - kz)]$ with a complex wavenumber k will either grow or decay exponentially.

Equation (2.4-4) is known as the *Helmholtz equation*. The most widely encountered type of laser beam in optical electronics is one where the intensity distribution at planes normal to the propagation direction is Gaussian, and the wavefronts are spherical. The medium in which the laser beam will propagate often possesses a cylindrical symmetry with respect to the beam axis. In particular, we will consider a lenslike medium whose index of refraction varies according to

$$n^2(x, y) = \frac{\varepsilon(r)}{\varepsilon_0} = n_0^2[1 - g^2(x^2 + y^2)] \quad \text{with} \quad g^2(x^2 + y^2) \ll 1 \tag{2.4-6}$$

where n_0 is the refractive index at the axis of symmetry, and g is a real constant characteristic of the medium. The special case of $g = 0$ corresponds to a homogeneous medium.

We limit our derivation to the solution of the field with cylindrical symmetry, so that the Laplacian ∇^2 can be written

$$\nabla^2 = \nabla_t^2 + \frac{\partial^2}{\partial z^2} = \frac{\partial^2}{\partial r^2} + \frac{1}{r}\frac{\partial}{\partial r} + \frac{\partial^2}{\partial z^2} \tag{2.4-7}$$

The kind of propagation we are considering is that of a nearly plane wave in which the flow of energy is predominantly along a single (e.g., z axis) direction, so that we may limit our derivation to a single transverse field component of $\mathbf{E}$. Taking $\mathbf{E}$ as

$$\mathbf{E} = \hat{\mathbf{a}}\psi(r, z)e^{-ikz} \tag{2.4-8}$$

where $\hat{\mathbf{a}}$ is a unit vector in the xy plane representing the polarization state, we obtain from Equations (2.4-4) and (2.4-7) in a few steps

$$\nabla_t^2\psi - 2ik\psi' - k^2g^2r^2\psi = 0 \tag{2.4-9}$$

where $\psi' = \partial\psi/\partial z$, and where we assume that the variation of the field amplitude is slow enough that

$$\psi'' \ll k\psi', k^2\psi \tag{2.4-10}$$

This approximation is known as the *slowly varying amplitude* (SVA) approximation. The approximation is legitimate provided the transverse dimension of the beam is much larger than the wavelength (see Problem 2.22). Since we are looking for a beam amplitude with cylindrical symmetry, it is convenient to introduce two complex functions $P(z)$ and $q(z)$ such that ψ is in the form of

$$\psi = \exp\left[-i\left(P(z) + \frac{k}{2q(z)}r^2\right)\right] \tag{2.4-11}$$

This solution can be viewed as an optical wave with a phase of $P(z)$ and a spherical wavefront with a radius of curvature of $q(z)$. The problem at hand is to find $P(z)$ and $q(z)$ so that Equation (2.4-11) is a solution of Equation (2.4-9).

We now substitute ψ into Equation (2.4-9). Using Equation (2.4-7), we obtain

$$-\left(\frac{k}{q}\right)^2 r^2 - 2i\left(\frac{k}{q}\right) - k^2r^2\left(\frac{1}{q}\right)' - 2kP' - k^2g^2r^2 = 0 \tag{2.4-12}$$

where the prime indicates differentiation with respect to z. If Equation (2.4-12) is to hold for all r, the coefficient of the different powers of r must be equal to zero. This leads to

$$\left(\frac{1}{q}\right)^2 + \left(\frac{1}{q}\right)' + g^2 = 0 \quad \text{and} \quad P' = -\frac{i}{q} \tag{2.4-13}$$

The Helmholtz equation for ψ is thus reduced to Equation (2.4-13) for the functions $P(z)$ and $q(z)$. The first differential equation can be solved for $q(z)$. The other beam parameter $P(z)$ can then be integrated by using the second equation. In practice, a general solution for the nonlinear differential equation (2.4-13) is not available. In the next section, we consider the solution for the case of homogeneous media where $g = 0$.

2.5 GAUSSIAN BEAMS IN HOMOGENEOUS MEDIA

If the medium is homogeneous, we can, according to Equation (2.4-6), put $g = 0$, and Equation (2.4-13) becomes

$$\left(\frac{1}{q}\right)^2 + \left(\frac{1}{q}\right)' = 0 \quad \text{and} \quad P' = -\frac{i}{q} \tag{2.5-1}$$

where we recall that the prime indicates a differentiation with respect to z. We now introduce a function $u(z)$ by the relation

$$\frac{1}{q} = \frac{1}{u}\frac{du}{dz} \tag{2.5-2}$$

We obtain directly from Equation (2.5-1)

$$\frac{d^2u}{dz^2} = 0 \tag{2.5-3}$$

The general solution for Equation (2.5-3) can be written

$$u(z) = az + b \tag{2.5-4}$$

where a and b are arbitrary constants. The beam parameter $q(z)$ can then be written, according to Equation (2.5-2),

$$\frac{1}{q} = \frac{a}{az + b} \tag{2.5-5}$$

or equivalently,

$$q(z) = z + q_0 \tag{2.5-6}$$

where q_0 is a constant ($q_0 = q(0) = b/a$). The other beam parameter $P(z)$ can be obtained from Equations (2.5-1) and (2.5-6):

$$P' = -\frac{i}{q} = -\frac{i}{z + q_0} \tag{2.5-7}$$

so that, upon integration, one obtains

$$P(z) = -i \ln\left(1 + \frac{z}{q_0}\right) \tag{2.5-8}$$

where the arbitrary constant is taken to be zero. This is legitimate, as the constant will only modify the phase of the field solution (2.4-11). This is equivalent to a mere shift of the time origin.

Combining Equations (2.5-6) and (2.5-8) in Equation (2.4-11), we obtain the following cylindrically symmetric solution of the Helmholtz equation:

$$\psi = \frac{q_0}{q_0 + z} \exp\left(-i \frac{k}{2(q_0 + z)} r^2\right) \tag{2.5-9}$$

We recall that q_0 is an arbitrary complex constant. For a physically realizable beam with its energy concentrated around the axis, the field amplitude must approach zero as $r \to \infty$. This requires a choice of a complex q_0 with a positive imaginary part. We can lump the real part of q_0 with z by selecting a new origin of the z axis. This leads to a pure imaginary q_0. We can now reexpress q_0 in terms of a new constant ω_0:

$$q_0 = i \frac{\pi \omega_0^2 n}{\lambda} = i \frac{k \omega_0^2}{2} \equiv i z_0 \tag{2.5-10}$$

where n is the index of refraction of the medium and z_0 is the magnitude of q_0. To see the physical meaning of ω_0, we examine the field amplitude at $z = 0$,

$$\psi(z = 0) = \exp\left(-\frac{r^2}{\omega_0^2}\right) \tag{2.5-11}$$

We note the beam amplitude has a Gaussian distribution with a spot size of ω_0. So ω_0 is a measure of the beam spot at $z = 0$.

For $z \neq 0$, we can rewrite the beam amplitude as, according to Equations (2.5-9) and (2.5-10),

$$\psi = \frac{1}{1 - iz/z_0} \exp\left(-\frac{r^2}{\omega_0^2 (1 - iz/z_0)}\right) \tag{2.5-12}$$

or equivalently,

$$\psi = \frac{1}{\sqrt{1 + (z/z_0)^2}} \exp\left(i \tan^{-1} \frac{z}{z_0}\right) \exp\left(-\frac{r^2}{\omega_0^2 [1 + (z/z_0)^2]} (1 + iz/z_0)\right) \tag{2.5-13}$$

If we define the parameters

$$\omega^2(z) = \omega_0^2 \left(1 + \frac{z^2}{z_0^2}\right) = \omega_0^2 \left[1 + \left(\frac{\lambda z}{\pi \omega_0^2 n}\right)^2\right] \tag{2.5-14}$$

$$R(z) = z\left(1 + \frac{z_0^2}{z^2}\right) = z\left[1 + \left(\frac{\pi \omega_0^2 n}{\lambda z}\right)^2\right] \tag{2.5-15}$$

so that

$$\frac{1}{q(z)} = \frac{1}{z + i z_0} = \frac{1}{R(z)} - i \frac{\lambda}{\pi \omega^2(z) n} \tag{2.5-16}$$

and

$$\eta(z) = \tan^{-1}\left(\frac{z}{z_0}\right) = \tan^{-1}\left(\frac{\lambda z}{\pi\omega_0^2 n}\right) \tag{2.5-17}$$

then we can combine Equations (2.5-9) and (2.5-11)–(2.5-13) and, recalling that $E = \psi \exp(-ikz)$, obtain

$$E(x, y, z) = E_0 \frac{\omega_0}{\omega(z)} \exp\left(-i[kz - \eta(z)] - i\frac{kr^2}{2q(z)}\right)$$

or equivalently,

$$E(x, y, z) = E_0 \frac{\omega_0}{\omega(z)} \exp\left[-i[kz - \eta(z)] - r^2\left(\frac{1}{\omega^2(z)} + \frac{ik}{2R(z)}\right)\right] \tag{2.5-18}$$

where we recall that $k = 2\pi n/\lambda$. This is our basic result. We refer to it as the fundamental Gaussian-beam solution, since we have excluded the more complicated solutions of Equation (2.4-4) (i.e., those with azimuthal variation) by limiting ourselves to transverse dependence involving $r = \sqrt{x^2 + y^2}$ only. These higher-order modes will be discussed separately.

From Equation (2.5-18) the parameter $\omega(z)$, which evolves according to Equation (2.5-14), is the distance r at which the field amplitude is down by a factor $1/e$ compared to its value on the axis. We will consequently refer to it as the beam *spot size*. The parameter ω_0 is the minimum spot size. It is the beam spot size at the plane $z = 0$. The parameter R in Equation (2.5-18) is the radius of curvature of the very nearly spherical wavefronts at z. We can verify this statement by deriving the radius of curvature of the constant phase surfaces (wavefronts) or, more simply, by considering the form of a spherical wave emitted by a point radiator placed at $z = 0$. It is given by

$$\begin{aligned} E &\propto \frac{1}{R}e^{-ikR} = \frac{1}{R}\exp(-ik\sqrt{x^2 + y^2 + z^2}) \\ &\approx \frac{1}{R}\exp\left(-ikz - ik\frac{x^2 + y^2}{2R}\right), \quad x^2 + y^2 \ll z^2 \end{aligned} \tag{2.5-19}$$

since z is equal to R, the radius of curvature of the spherical wave. Comparing Equations (2.5-19) and (2.5-18), we identify R as the radius of curvature of the Gaussian beam, with the exception of the immediate vicinity of the plane $z = 0$. In arriving at the above conclusion, we ignored the phase factor $\eta(z)$, which is much smaller than kz for most practical situations. The complex parameter q is often referred to as the Gaussian beam parameter, or the complex radius of curvature of the Gaussian beam.

The convention regarding the sign of $R(z)$ is that it is negative if the center of curvature occurs at a position on the right side of z (i.e., $z' > z$) and vice versa. The form of the fundamental Gaussian beam is, according to Equation (2.5-18), uniquely determined once its minimum spot size ω_0 and its location—that is, the plane $z = 0$—are specified. The spot size $\omega(z)$ and radius of curvature R at any plane z are then found from Equations (2.5-14) and (2.5-15). Some of these characteristics are displayed in Figure 2.5. The hyperbolas shown in this figure correspond to the ray direction and are intersections of planes that include the z axis and the hyperboloids

$$x^2 + y^2 = \text{const. } \omega^2(z) \tag{2.5-20}$$

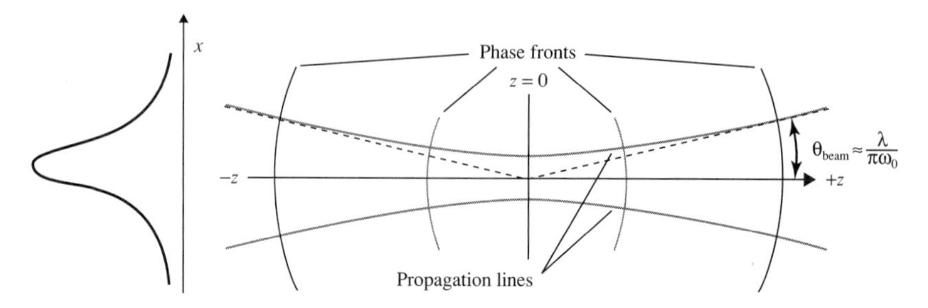

Figure 2.5 Propagating Gaussian beam.

These hyperbolas correspond to the local direction of energy propagation. The spherical surfaces shown have radii of curvature given by Equation (2.5-15). For large z the hyperboloids $x^2 + y^2 = \omega^2(z)$ are asymptotic to the cone

$$r = \sqrt{x^2 + y^2} = \frac{\lambda}{\pi\omega_0 n} z \tag{2.5-21}$$

whose half-apex angle, which we take as a measure of the angular beam spread, is

$$\theta_{\text{beam}} = \tan^{-1}\left(\frac{\lambda}{\pi\omega_0 n}\right) \cong \frac{\lambda}{\pi\omega_0 n} \quad \text{for} \quad \theta_{\text{beam}} \ll \pi \tag{2.5-22}$$

This last result is a rigorous manifestation of wave diffraction according to which a wave that is confined in the transverse direction to an aperture of radius ω_0 will spread (diffract) in the far field ($z \gg \pi\omega_0^2 n/\lambda$) according to Equation (2.5-22).

EXAMPLE: GAUSSIAN BEAM

The outputs of laser resonators are Gaussian beams. Consider a Gaussian beam with a beam spot size of $\omega_0 = 1$ mm at the laser output, and a wavelength of $\lambda = 1.06\ \mu\text{m}$. According to Equation (2.5-10), the confocal parameter z_0 is

$$z_0 = \frac{\pi\omega_0^2 n}{\lambda} = \frac{\pi\omega_0^2}{\lambda} = 3\ \text{m}$$

Assume that the minimum spot size ω_0 is located at the output of the laser. The beam spot size at a distance of 10 m from the output becomes

$$\omega(z) = \omega_0\left(1 + \frac{z^2}{z_0^2}\right)^{1/2} = \omega_0\left[1 + \left(\frac{\lambda z}{\pi\omega_0^2}\right)^2\right]^{1/2} = 3.5\ \text{mm}$$

The radius of curvature of the wavefront at $z = 10$ m becomes

$$R(z) = z\left(1 + \frac{z_0^2}{z^2}\right) = z\left[1 + \left(\frac{\pi\omega_0^2}{\lambda z}\right)^2\right] = 10.9\ \text{m}$$

The beam divergence angle is given by

$$\theta_{\text{beam}} = \tan^{-1}\left(\frac{\lambda}{\pi\omega_0}\right) \cong \frac{\lambda}{\pi\omega_0} = 0.00034\ \text{radian} = 0.02^\circ$$

2.6 FUNDAMENTAL GAUSSIAN BEAM IN A LENSLIKE MEDIUM—THE *ABCD* LAW

The Gaussian beam solution obtained in the previous section is a solution of the wave equation in homogeneous media. In this section, we return to the general case of a lenslike medium so that $g \neq 0$. The P and q functions of Equation (2.4-11) obey, according to Equation (2.4-13),

$$\left(\frac{1}{q}\right)^2 + \left(\frac{1}{q}\right)' + g^2 = 0 \quad \text{and} \quad P' = -\frac{i}{q} \tag{2.6-1}$$

If we introduce a function u defined by

$$\frac{1}{q} = \frac{1}{u}\frac{du}{dz} \tag{2.6-2}$$

we obtain from Equation (2.6-1)

$$\frac{d^2u}{dz^2} + g^2u = 0 \tag{2.6-3}$$

This is a differential equation for simple harmonic oscillators. The general solutions can be written

$$\begin{aligned} u(z) &= a \sin gz + b \cos gz \\ u'(z) &= ag \cos gz - bg \sin gz \end{aligned} \tag{2.6-4}$$

where a and b are arbitrary constants.

Using Equation (2.6-4) in Equation (2.6-2) and expressing the result in terms of an input value q_0, we obtain the following result for the complex beam parameter (complex radius of curvature) $q(z)$:

$$q(z) = \frac{\cos(gz)q_0 + \sin(gz)/g}{-\sin(gz)gq_0 + \cos(gz)} \tag{2.6-5}$$

This is the solution of Equation (2.6-1) that describes the evolution of the complex beam parameter $q(z)$, with an initial complex beam parameter of q_0 at $z = 0$. The physical significance of $q(z)$ in this case can be extracted from Equation (2.4-11). We focus on the part of $\psi(r, z)$ that involves r. Equation (2.4-11) can thus be written

$$\psi \propto e^{-ikr^2/2q(z)} \tag{2.6-6}$$

If we express the real and imaginary parts of $q(z)$ by means of

$$\frac{1}{q(z)} = \frac{1}{R(z)} - i\frac{\lambda}{\pi n \omega^2(z)} \tag{2.6-7}$$

we obtain

$$\psi \propto \exp\left(\frac{-r^2}{\omega^2(z)} - i\frac{kr^2}{2R(z)}\right) \tag{2.6-8}$$

so that $\omega(z)$ is the beam spot size and R its radius of curvature, as in the case of a homogeneous medium, which is described by Equation (2.5-18). For the special case of a homogeneous medium ($g = 0$), Equation (2.6-5) reduces to Equation (2.5-4).

Transformation of the Gaussian Beam—The *ABCD* Law

We have derived the transformation law of a Gaussian beam (Equation (2.6-5)) propagating through a lenslike medium that is characterized by g. We note first by comparing Equation (2.6-5) to Table 2.1(6) and to Equation (2.3-14) that the transformation of the Gaussian beam parameter (or the complex radius of curvature) can be described by

$$q_2 = \frac{Aq_1 + B}{Cq_1 + D} \tag{2.6-9}$$

where A, B, C, and D are the elements of the ray matrix that relate the ray (r, r') at plane 2 to the ray at plane 1. It follows immediately that the propagation through, or reflection from, any of the elements shown in Table 2.1 also obeys Equation (2.6-9), since these elements can all be viewed as special cases of a lenslike medium. This can also be seen by comparing Equation (2.6-3) to Equation (2.3-12). We note that the equation for the ray parameter $r(z)$ is identical to that of the function $u(z)$, defined in Equation (2.6-2). Equation (2.6-9) is known as a *bilinear transformation* (or *Möbius transformation*). For real matrix elements A, B, C, and D, it can be shown that q_2 is in the upper complex plane if q_1 is in the upper complex plane. In other words, a confined Gaussian beam remains a confined Gaussian beam provided the matrix elements are all real. For lenslike media, all matrix elements are real.

For future reference we note that by applying Equation (2.6-9) to a thin lens of focal length f we obtain from (2.6-9) and Table 2.1(2)

$$\frac{1}{q_2} = \frac{1}{q_1} - \frac{1}{f} \tag{2.6-10}$$

so that using Equation (2.6-7)

$$\omega_2 = \omega_1$$
$$\frac{1}{R_2} = \frac{1}{R_1} - \frac{1}{f} \tag{2.6-11}$$

These results apply, as well, to reflection from a mirror with a radius of curvature R if we replace f by $R/2$. Consider next the propagation of a Gaussian beam through two lenslike media that are adjacent to each other. The ray matrix describing the first one is (A_1, B_1, C_1, D_1) while that of the second one is (A_2, B_2, C_2, D_2). Taking the input beam parameter as q_1 and the output beam parameter as q_3 we have from Equation (2.6-9)

$$q_2 = \frac{A_1 q_1 + B_1}{C_1 q_1 + D_1}$$

for the beam parameter at the output of medium 1 and

$$q_3 = \frac{A_2 q_2 + B_2}{C_2 q_2 + D_2}$$

and after combining the last two equations,

$$q_3 = \frac{A_T q_1 + B_T}{C_T q_1 + D_T} \tag{2.6-12}$$

where A_T, B_T, C_T, D_T are the elements of the matrix relating the output plane (3) to the input plant (1); that is,

$$\begin{bmatrix} A_T & B_T \\ C_T & D_T \end{bmatrix} = \begin{bmatrix} A_2 & B_2 \\ C_2 & D_2 \end{bmatrix} \begin{bmatrix} A_1 & B_1 \\ C_1 & D_1 \end{bmatrix} \tag{2.6-13}$$

It follows by induction that Equation (2.6-9) applies to the propagation of a Gaussian beam through any arbitrary number of lenslike media and elements. The matrix with elements (A_T, B_T, C_T, D_T) is the ordered product of the matrices characterizing the individual members of the chain. The great power of the *ABCD* law is that it enables us to trace the Gaussian beam parameter $q(z)$ through a complicated sequence of lenslike elements. The beam radius $R(z)$ and spot size $\omega(z)$ at any plane z can be recovered through the use of Equation (2.6-7). The application of this method will be made clear by the following example.

EXAMPLE: GAUSSIAN BEAM FOCUSING

As an illustration of the application of the *ABCD* law, we consider the case of a Gaussian beam with a minimum beam spot size ω_{01} that is incident at a thin lens of focal length f, as shown in Figure 2.6. Let the waist of the incident beam be located at distance d_1 in front of the lens. We will find the location d_2 of the waist of the output beam and the beam spot size ω_{02} at that point.

At the input plane 1 (location of beam waist of incident beam), $\omega = \omega_{01}$ and $R_1 = \infty$, so that

$$\frac{1}{q_1} = \frac{1}{R_1} - i\frac{\lambda}{\pi\omega_{01}^2 n} = -i\frac{\lambda}{\pi\omega_{01}^2 n} \equiv \frac{1}{iz_1}$$

Similarly, at plane 2 (location of output beam waist), $\omega = \omega_{02}$ and $R_2 = \infty$, so the complex Gaussian beam parameter can be written

$$\frac{1}{q_2} = \frac{1}{R_2} - i\frac{\lambda}{\pi\omega_{02}^2 n} = -i\frac{\lambda}{\pi\omega_{02}^2 n} \equiv \frac{1}{iz_2}$$

These two beam parameters are related by the *ABCD* law:

$$q_2 = \frac{Aq_1 + B}{Cq_1 + D}$$

where A, B, C, and D are elements of the following matrix:

$$\begin{bmatrix} A & B \\ C & D \end{bmatrix} = \begin{bmatrix} 1 & d_2 \\ 0 & 1 \end{bmatrix} \begin{bmatrix} 1 & 0 \\ -1/f & 1 \end{bmatrix} \begin{bmatrix} 1 & d_1 \\ 0 & 1 \end{bmatrix}$$

By carrying out the matrix multiplications, we obtain

$$A = 1 - \frac{d_2}{f}, \quad B = d_1 + d_2 - \frac{d_1 d_2}{f}$$

$$C = -\frac{1}{f}, \quad D = 1 - \frac{d_1}{f}$$

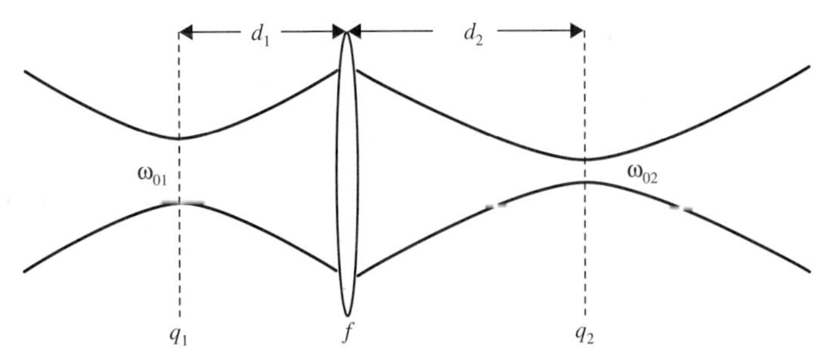

Figure 2.6 Focusing of a Gaussian beam.

Using the transformation formula, we have

$$iz_2 = \frac{Aiz_1 + B}{Ciz_1 + D} = \frac{(Aiz_1 + B)(-izC + D)}{C^2 z_1^2 + D^2} = \frac{ACz_1^2 + BD + iz_1}{C^2 z_1^2 + D^2}$$

where we have used the unimodular relationship $AD - BC = 1$. Since A, B, C, D, z_1, and z_2 are all real, we obtain

$$ACz_1^2 + BD = 0 \tag{2.6-14}$$

and

$$z_2 = \frac{z_1}{C^2 z_1^2 + D^2} \tag{2.6-15}$$

Using $AD - BC = 1$ and eliminating z_1^2 from the above two equations, we obtain

$$z_2 = \frac{A}{D} z_1 = \frac{d_2 - f}{d_1 - f} z_1$$

or equivalently,

$$\omega_{02}^2 = \frac{d_2 - f}{d_1 - f} \omega_{01}^2 \tag{2.6-16}$$

The location of the output beam waist can be found by using Equation (2.6-14). We obtain

$$z_1^2 = -\frac{BD}{AC} = \frac{(d_1 - f)(d_1 f + d_2 f - d_1 d_2)}{(d_2 - f)}$$

or equivalently,

$$(d_2 - f) = \frac{f^2}{z_1^2 + (d_1 - f)^2}(d_1 - f) \tag{2.6-17}$$

The minimum spot size of the output beam can now be written, according to Equations (2.6-16) and (2.6-17),

$$\omega_{02}^2 = \frac{f^2}{z_1^2 + (d_1 - f)^2} \omega_{01}^2 \tag{2.6-18}$$

The last two expressions are for the spot size and location of the beam waist.

We now examine some limiting cases of interest:

(a) For a point source with z_1 approaching zero, the image location is given by, according to Equation (2.6-17),

$$\frac{1}{d_1} + \frac{1}{d_2} = \frac{1}{f} \tag{2.6-19}$$

which is the familiar lens formula.

(b) For an input Gaussian beam of infinite spot size (plane wave with z_1 approaching infinity), the output beam waist is located at the focal point ($d_2 = f$).

(c) For an input Gaussian beam with its waist located at the front focal point ($d_1 = f$), the output beam waist is located at the rear focal point ($d_2 = f$). The output beam spot size at the waist is related to the input beam spot size by the following relationship:

$$z_1 z_2 = f^2 \tag{2.6-20}$$

or equivalently,

$$\omega_{01}^2 \omega_{02}^2 = \frac{\lambda^2}{\pi^2 n^2} f^2 \tag{2.6-21}$$

(d) For an input Gaussian beam with its waist located at the lens ($d_1 = 0$), the output beam waist and location are given by

$$d_2 = \frac{f}{1 + (f/\pi\omega_{01}^2 n/\lambda)^2} = \frac{f}{1 + (f/z_1)^2} \tag{2.6-22}$$

and

$$\frac{\omega_{02}}{\omega_{01}} = \frac{f\lambda/\pi\omega_{01}^2 n}{\sqrt{1 + (f\lambda/\pi\omega_{01}^2 n)^2}} = \frac{f/z_1}{\sqrt{1 + (f/z_1)^2}} \tag{2.6-23}$$

The confocal beam parameter

$$z_1 \equiv \frac{\pi\omega_{01}^2 n}{\lambda}$$

is, according to Equation (2.5-14), the distance from the waist at which the input beam spot size increases by $\sqrt{2}$ and is a convenient measure of the convergence of the input beam. The smaller z_1, the "stronger" the convergence.

According to Equation (2.6-11), the radius of curvature of a beam is decreased by $1/f$, after passing through a thin lens, while the spot size remains unchanged. Using Equation (2.6-8), the input–output relationship can thus be written

$$E_{\text{out}}(x, y) = E_{\text{in}}(x, y) \exp\left(+i\frac{kr^2}{2f}\right) = E_{\text{in}}(x, y) \exp\left(+i\frac{k(x^2 + y^2)}{2f}\right) \tag{2.6-24}$$

where f is the focal length of the thin lens, and $k = 2\pi n/\lambda$.

2.7 GAUSSIAN BEAMS IN LENS WAVEGUIDE

As another example of the application of the *ABCD* law, we consider the propagation of a Gaussian beam through a sequence of thin lenses, as shown in Figure 2.2. The matrix relating a ray in plane $s = N$ to that in plane $s = 0$ is

$$\begin{bmatrix} A_T & B_T \\ C_T & D_T \end{bmatrix} = \begin{bmatrix} A & B \\ C & D \end{bmatrix}^N \tag{2.7-1}$$

where (A, B, C, D) are the elements of the matrix for propagation through a single two-lens, unit cell ($\Delta s = 1$) and is given by Equation (2.1-6). We can use the matrix identity of Equation (2.1-11) for the Nth power of a unimodular matrix to obtain

$$\begin{aligned} A_T &= \frac{A\sin(N\theta) - \sin[(N-1)\theta]}{\sin\theta} \\ B_T &= \frac{B\sin(N\theta)}{\sin\theta} \\ C_T &= \frac{C\sin(N\theta)}{\sin\theta} \\ D_T &= \frac{D\sin(N\theta) - \sin[(N-1)\theta]}{\sin\theta} \end{aligned} \tag{2.7-2}$$

where

$$\cos\theta = \tfrac{1}{2}(A + D) = \left(1 - \frac{d}{f_2} - \frac{d}{f_1} + \frac{d^2}{2f_1f_2}\right) \tag{2.7-3}$$

and then use Equation (2.7-2) in (2.6-9) with the result

$$q_N = \frac{\{A\sin(N\theta) - \sin[(N-1)\theta]\}q_0 + B\sin(N\theta)}{C\sin(N\theta)q_0 + D\sin(N\theta) - \sin[(N-1)\theta]} \tag{2.7-4}$$

We now examine the behavior of q_N as N becomes infinity. For the case $|\cos\theta| > 1$, the sine functions on the right side will yield growing exponentials. This leads to a limiting value on the right side. It can be shown that this limiting value is a real number (see Problem 2.7). A Gaussian beam with a real beam parameter q is an unconfined beam. For the case $|\cos\theta| \leq 1$, the right side remains oscillatory as a function of N. The bilinear transformation with real constants will maintain the confinement of the beam. The condition for the confinement of the Gaussian beam by the lens sequence is thus $|\cos\theta| \leq 1$, or

$$0 \leq \left(1 - \frac{d}{2f_1}\right)\left(1 - \frac{d}{2f_2}\right) \leq 1 \tag{2.7-5}$$

that is, the same as condition (2.1-16) for stable ray propagation.

2.8 HIGH-ORDER GAUSSIAN BEAM MODES IN A HOMOGENEOUS MEDIUM

The Gaussian mode treated up to this point has a field variation that depends only on the axial distance z and the distance r from the axis. If we do not impose the condition $\partial/\partial\phi = 0$ (where ϕ is the azimuthal angle in a cylindrical coordinate system (r, ϕ, z)) and take $g = 0$, the wave equation (2.4-4) has solutions in the form of

$$\begin{aligned} E_{l,m}(x, y, z) &= E_0\frac{\omega_0}{\omega(z)}H_l\left(\sqrt{2}\frac{x}{\omega(z)}\right)H_m\left(\sqrt{2}\frac{y}{\omega(z)}\right)\times\exp\left(-ik\frac{x^2+y^2}{2q(z)} - ikz + i(l+m+1)\eta\right) \\ &= E_0\frac{\omega_0}{\omega(z)}H_l\left(\sqrt{2}\frac{x}{\omega(z)}\right)H_m\left(\sqrt{2}\frac{y}{\omega(z)}\right)\times\exp\left(-\frac{x^2+y^2}{\omega^2(z)} - ik\frac{x^2+y^2}{2R(z)} - ikz + i(l+m+1)\eta\right) \end{aligned} \tag{2.8-1}$$

where H_l is the Hermite polynomial of order l, and $\omega(z)$, $R(z)$, $q(z)$, and η are defined as in Equations (2.5-14)–(2.5-17) [1]. We note for future reference that the phase shift on the axis is

$$\begin{aligned} \theta &= kz - (l+m+1)\tan^{-1}\left(\frac{z}{z_0}\right) \\ z_0 &= \frac{\pi\omega_0^2 n}{\lambda} \end{aligned} \tag{2.8-2}$$

The transverse variation of the electric field along x (or y) is seen to be of the form $H_l(\xi)\exp(-\xi^2/2)$, where $\xi = \sqrt{2}x/\omega$. This function has been studied extensively, since it also corresponds to the quantum mechanical wavefunction $u_l(\xi)$ of the harmonic oscillator [2]. The Hermite–Gaussian beams form a complete set of orthogonal functions. They satisfy the following orthogonal relationship:

$$\iint E^*_{l_1m_1}(x, y, z)\, E_{l_2m_2}(x, y, z)\, dx\, dy = \delta_{(l_1m_1)(l_2m_2)} \tag{2.8-3}$$

These Hermite–Gaussian functions can be used as a basis to expand any arbitrary paraxial optical beam $E(x, y, z)$ in the form

$$E(x, y, z) = \sum_l \sum_m c_{lm} E_{l,m}(x, y, z) \tag{2.8-4}$$

where c_{lm} are constants that can be obtained by using the orthogonal relationship.

An alternative but equally valid family of solutions to the paraxial wave equation can be written in cylindrical rather than rectangular coordinates. They are the so-called Laguerre–Gaussian beams of the form

$$E_{p,m}(r, \phi, z) = E_0 \frac{\omega_0}{\omega(z)} \left(\frac{\sqrt{2}r}{\omega(z)}\right)^{|m|} L_p^{|m|}\left(\frac{2r^2}{\omega^2(z)}\right) \exp\left[-ik\frac{r^2}{2q(z)} - ikz + im\phi + i(2p + |m| + 1)\eta\right]$$

$$= E_0 \frac{\omega_0}{\omega(z)} \left(\frac{\sqrt{2}r}{\omega(z)}\right)^{|m|} L_p^{|m|}\left(\frac{2r^2}{\omega^2(z)}\right) \exp\left[-\frac{r^2}{\omega^2(z)} - ik\frac{r^2}{2R(z)} - ikz + im\phi + i(2p + |m| + 1)\eta\right] \tag{2.8-5}$$

where L_p^m functions are the associated Laguerre polynomials [3]. In the above solutions, the integer $p \geq 0$ is the radial index and the integer m is the azimuthal index. The Laguerre–Gaussian beams are also mutually orthogonal and are particularly useful for problems with cylindrical symmetry. As a result of the cylindrical symmetry, the radial dependence of the Laguerre–Gaussian beams with $\exp(\pm im\phi)$ are identical. In addition, the intensity patterns of Laguerre–Gaussian beams are cylindrically symmetric (i.e., no ϕ dependence), according to Equation (2.8-5). We also note the fundamental Laguerre–Gaussian beam ($p = 0$, $m = 0$) is exactly the same as the fundamental Hermite–Gaussian beam ($l = 0$, $m = 0$).

Some low-order Hermite–Gaussian functions normalized to represent the same amount of total beam power are shown in Figure 2.7. Photographs of actual field patterns are shown in Figure 2.8. Note that the first four correspond to the intensity $|u_l(\xi)|^2$ plots ($l = 0, 1, 2, 3$) of Figure 2.7.

The following is a list of the first few Hermite polynomials and associated Laguerre polynomials.

$$H_0(x) = 1$$

$$H_1(x) = 2x$$

$$H_2(x) = 4x^2 - 2$$

$$H_3(x) = 8x^3 - 12x$$

$$H_4(x) = 16x^4 - 48x^2 + 12$$

$$L_0^m(x) = 1$$

$$L_1^m(x) = -x + (m + 1)$$

$$L_2^m(x) = \tfrac{1}{2}[x^2 - 2(m + 2)x + (m + 1)(m + 2)]$$

$$L_3^m(x) = \tfrac{1}{6}[-x^3 + 3(m + 3)x^2 - 3(m + 2)(m + 3)x + (m + 1)(m + 2)(m + 3)]$$

These polynomials are provided as a ready source of reference.

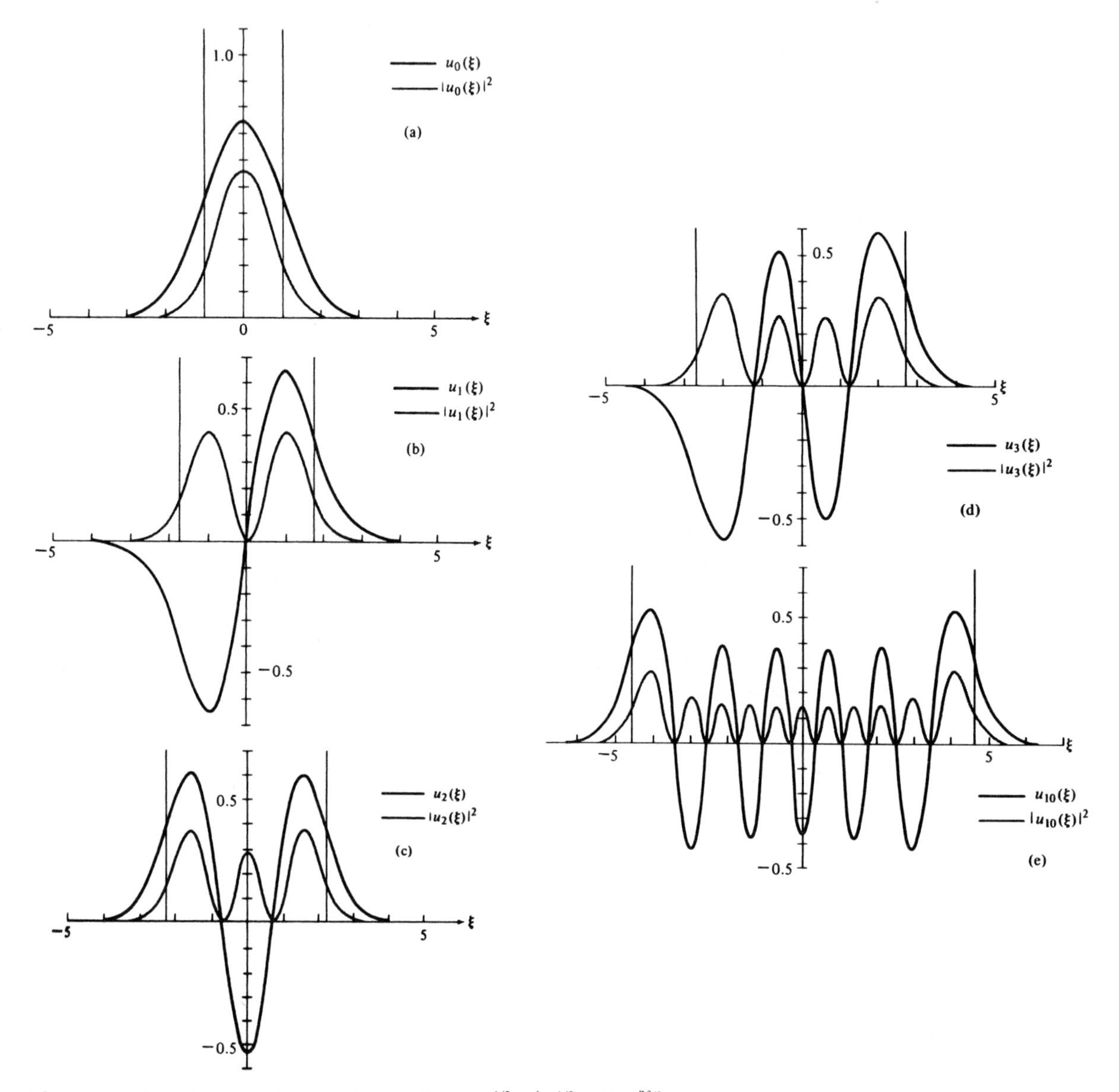

Figure 2.7 Hermite–Gaussian functions $u_l(\xi) = (\pi^{1/2}l!2^l)^{-1/2}H_l(\xi)e^{-\xi^2/2}$ corresponding to higher-order beam solutions (Equation (2.8-1)). The curves are normalized so as to represent a fixed amount of total beam power in all the modes $\left(\int_{-\infty}^{\infty} u_l^2(\xi)\, d\xi = 1\right)$. The solid curves are the functions $u_l(\xi)$ for $l = 0, 1, 2, 3$, and 10. The curves limited to the upper half-space are $u_l^2(\xi)$.

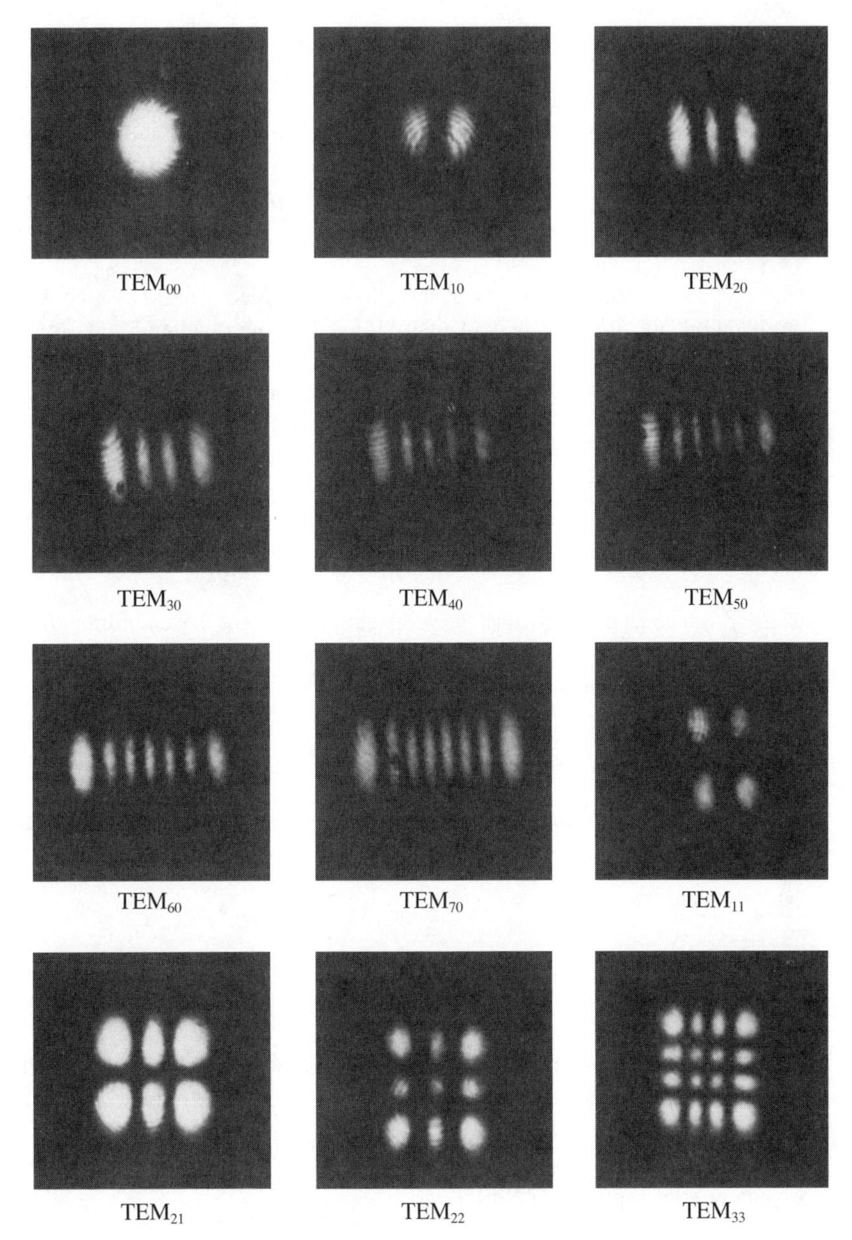

Figure 2.8 Intensity photographs of some low-order Gaussian beam modes. (After Reference [4].)

2.9 GAUSSIAN BEAM MODES IN QUADRATIC INDEX MEDIA

In Section 2.6 we treated the propagation of a circularly symmetric Gaussian beam in a lenslike medium by using the Gaussian beam solution of the homogeneous medium. Since the Gaussian beam solution of the homogeneous medium is not a solution in lenslike media, this treatment is possible provided we allow the complex Gaussian beam parameter q to vary as a function of z according to Equation (2.6-1). Thus, in this approach, the complex beam parameter q of the Gaussian beam of the homogeneous medium will vary as a function of z according to Equation (2.6-4).

Here we describe a different approach. In this approach, we find directly the Gaussian beam solution of the lenslike medium. In other words, we are looking for the modes of propagation in a lenslike medium whose index of refraction can be described by

$$n^2(r) = n_0^2(1 - g^2r^2) \qquad r^2 = x^2 + y^2 \tag{2.9-1}$$

where n_0 is the refractive index at the axis of symmetry and g is a constant. For modes of propagation, the transverse field distribution remains unchanged as the modes propagate in the medium. For these modes of propagation the complex beam parameter q is a constant.

The vector wave equation (2.4-4) for lenslike media takes the form

$$\nabla^2\mathbf{E} + k^2(1 - g^2r^2)\mathbf{E} = 0 \tag{2.9-2}$$

where $k = n_0\omega/c$ is the wavenumber. We consider some (scalar) component E of the last equation and assume a solution in the form

$$E(x, y) = \psi(x, y)\exp(-i\beta z) \tag{2.9-3}$$

where β is a constant, known as the propagation constant. By examining the index profile and the wave equation, we note that the wave equation in this case is separable. Taking $\psi(x, y) = U(x)V(y)$, the wave equation (2.9-2) becomes

$$\frac{1}{U}\frac{\partial^2 U}{\partial x^2} + \frac{1}{V}\frac{\partial^2 V}{\partial y^2} + k^2 - k^2g^2(x^2 + y^2) - \beta^2 = 0 \tag{2.9-4}$$

where U and V are one-dimensional functions. By grouping x-dependent terms and y-dependent terms in Equation (2.9-4), we obtain

$$\frac{1}{U}\frac{\partial^2 U}{\partial x^2} + (k^2 - \beta^2 - k^2g^2x^2) = C \tag{2.9-5}$$

$$\frac{1}{V}\frac{\partial^2 V}{\partial y^2} - k^2g^2y^2 = -C \tag{2.9-6}$$

where C is some constant. We note that the differential equation (2.9-2) is now separated into two. Each of these equations can be solved separately. Defining a variable ξ by

$$\xi = \alpha y, \quad \alpha \equiv \sqrt{gk} \tag{2.9-7}$$

we can rewrite Equation (2.9-6) as

$$\frac{d^2V}{d\xi^2} + \left(\frac{C}{\alpha^2} - \xi^2\right)V = 0 \tag{2.9-8}$$

This is a well-known differential equation and is identical to the Schrödinger equation of the harmonic oscillator [2]. The solution and the corresponding eigenvalues can be written

$$\frac{C}{\alpha^2} = (2m + 1), \quad m = 1, 2, 3, \ldots \tag{2.9-9}$$

$$V_m(\xi) = H_m(\xi)e^{-\xi^2/2} \tag{2.9-10}$$

where H_m is the Hermite polynomial of order m.

We now repeat the procedure with Equation (2.9-5). Substituting

$$\zeta = \alpha x$$

we obtain

$$\frac{\partial^2 U}{\partial \zeta^2} + \left(\frac{k^2 - \beta^2 - C}{\alpha^2} - \zeta^2 \right) U = 0$$

The eigenvalues and eigenfunctions are given by

$$\frac{k^2 - \beta^2 - C}{\alpha^2} = (2l + 1), \quad l = 1, 2, 3, \ldots \tag{2.9-11}$$

and

$$U_l(\zeta) = H_l(\zeta) e^{-\xi^2/2} \tag{2.9-12}$$

The total solution for the wavefunction ψ is thus

$$\psi(x, y) = H_l\left(\frac{\sqrt{2}x}{\omega_0}\right) H_m\left(\frac{\sqrt{2}y}{\omega_0}\right) e^{-(x^2+y^2)/\omega_0^2}$$

where the "spot size" ω_0 is, according to Equation (2.9-7),

$$\omega_0 = \frac{\sqrt{2}}{\alpha} = \sqrt{\frac{2}{gk}} = \sqrt{\frac{\lambda}{g n_0 \pi}} \tag{2.9-13}$$

This beam spot size can also be obtained directly from Equation (2.4-13) by assuming $q' = 0$. In other words, the complex beam parameter of the modes of propagation is a constant given by $q = i/g$.

The complex field of the mode of propagation is

$$\begin{aligned} E_{l,m}(x, y, z) &= \psi_{l,m}(x, y) e^{-i\beta_{l,m} z} \\ &= E_0 H_l\left(\sqrt{2}\frac{x}{\omega_0}\right) H_m\left(\sqrt{2}\frac{y}{\omega_0}\right) \exp\left(-\frac{x^2 + y^2}{\omega_0^2}\right) \exp(-i\beta_{l,m} z) \end{aligned} \tag{2.9-14}$$

where E_0 is a constant. The propagation constant $\beta_{l,m}$ of the l, m mode is obtained from Equations (2.9-9) and (2.9-11):

$$\beta_{l,m} = k\left(1 - \frac{2}{k} g(l + m + 1)\right)^{1/2} \tag{2.9-15}$$

Two features of the mode solutions are noteworthy. (1) Unlike the homogeneous medium solution ($g = 0$), the mode spot size ω is independent of z. This can be explained by the focusing action of the index variation ($g > 0$), which counteracts the natural tendency of a confined beam to diffract (spread). In the case of an index of refraction that increases with r ($g < 0$), it follows from Equations (2.9-13) and (2.9-14) that $\omega^2 < 0$ and no confined solutions exist. The index profile in this case leads to defocusing, thus reinforcing the diffraction of the beam. (2) The dependence of β on the mode indices l, m causes the different modes to have phase velocities $v_{l,m} = \omega/\beta_{l,m}$ as well as group velocities $(v_g)_{l,m} = d\omega/d\beta_{l,m}$ that depend on l and m.

Let us consider modal dispersion (i.e., the dependence on l and m) of the group velocity of mode

$$(v_g)_{l,m} = \frac{d\omega}{d\beta_{l,m}} \tag{2.9-16}$$

If the index variation is small so that

$$\frac{1}{k} g(l + m + 1) \ll 1 \tag{2.9-17}$$

we can approximate Equation (2.9-15) as

$$\beta_{l,m} \cong k - g(l + m + 1) - \frac{g^2}{2k}(l + m + 1)^2 \tag{2.9-18}$$

so that, according to Equation (2.9-16),

$$(v_g)_{l,m} = \frac{c/n_0}{\left(1 + \frac{g^2}{2k^2}(l + m + 1)^2\right)} \tag{2.9-19}$$

We have shown in Chapter 1 that optical pulses propagate at the group velocity v_g. The effect of the group velocity dispersion on pulse propagation in lenslike media is considered next. We will also treat the effect of group velocity dispersion ($dv_g/d\omega \neq 0$) on pulse broadening.

Pulse Spreading in Quadratic Index Glass Fibers

Glass fibers with quadratic index profiles (2.9-1) can be employed for optical communications systems [5, 6]. The signals are coded into trains of optical pulses and the information capacity is thus fundamentally limited by the number of pulses that can be transmitted per unit time [7, 8]. There are two ways in which the dispersion limits the pulse repetition rate: modal dispersion and group velocity dispersion.

Modal Dispersion

If the optical pulses fed into the input end of the fiber excite a large number of modes (this will be the case if the input light is strongly focused so that the "rays" subtend a large angle), then each mode will travel with a group velocity $(v_g)_{l,m}$, as given by Equation (2.9-19). If all the modes from (0, 0) to ($l_{\max}$, $m_{\max}$) are excited, the output pulse at $z = L$ will broaden to

$$\Delta\tau \cong L\left(\frac{1}{(v_g)_{l_{\max},m_{\max}}} - \frac{1}{(v_g)_{0,0}}\right) \tag{2.9-20}$$

We can use Equation (2.9-19) to obtain

$$\Delta\tau \cong \frac{n_0 g^2 L}{2ck^2}[(l_{\max} + m_{\max} + 1)^2 - 1] \tag{2.9-21}$$

The maximum number of pulses per second that can be transmitted without serious overlap of adjacent output pulses is thus $f_{\max} \sim 1/\Delta\tau$. High data rate transmission will thus require the use of single-mode excitation, which can be achieved by the use of coherent single-mode laser excitation [6–8, 9].

EXAMPLE: MAXIMUM PULSE RATE

Consider a 1 km long quadratic index fiber with $n_0 = 1.5$ and $g^2 = 3.4 \times 10^3\ \text{cm}^{-2}$. Let the input optical pulses at $\lambda = 1.0\ \mu\text{m}$ excite the modes up to $l_{\text{max}} = m_{\text{max}} = 30$. Substitution in Equation (2.9-21) gives

$$\Delta\tau = 3.6 \times 10^{-9}\ \text{s}$$

and $f_{\text{max}} \sim (\Delta\tau)^{-1} = 2.8 \times 10^8$ pulses per second for the maximum pulse rate.

Group Velocity Dispersion

The pulse spreading (2.9-21) due to multimode excitation can be eliminated if one were to excite a single mode, say, l, m only. In this case, pulse spreading would still result from the dependence of $(v_g)_{l,m}$ on frequency. This spreading can be explained by the fact that a pulse with a spectral width $\Delta\omega$ will spread in a distance L by

$$\Delta\tau = \Delta\left(\frac{L}{v_g}\right) \approx L\left|\frac{d}{d\omega}\left(\frac{1}{v_g}\right)\right|\Delta\omega = \frac{L}{v_g^2}\left|\frac{dv_g}{d\omega}\right|\Delta\omega \tag{2.9-22}$$

If the pulse is derived, say, by gating, from a coherent continuous source with a negligible spectral width, the pulse spectral width is related to the pulse duration τ by $\Delta\omega \sim 2\pi/\tau$ and Equation (2.9-22) becomes

$$\Delta\tau \approx \frac{2\pi L}{v_g^2 \tau}\left|\frac{dv_g}{d\omega}\right| \tag{2.9-23}$$

If the source bandwidth $\Delta\omega$ exceeds $2\pi/\tau$, then we need to replace $\Delta\omega$ in Equation (2.9-22) by the actual spectral width $\Delta\omega_s$. A rigorous treatment of this subject is reserved for Chapter 7.

In the situation when $g(l + m + 1) \ll k$, so that we can neglect the second-order term in Equation (2.9-18), we have

$$\beta_{l,m} \cong k - g(l + m + 1) \tag{2.9-24}$$

We note that the propagation constants of the modes are equally spaced with a spacing of g. It can easily be shown that the field distribution in the multimode waveguide with equally spaced propagation constants will reproduce itself periodically. In other words,

$$E(x, y, z + 2\pi/g) = E(x, y, z) \tag{2.9-25}$$

This result is consistent with the periodic solutions of rays in lenslike media.

2.10 PROPAGATION IN MEDIA WITH A QUADRATIC GAIN PROFILE

In many laser media the gain is a strong function of position. This variation can be due to a variety of causes, among them: (1) the radial distribution of energetic electrons in the plasma region of gas lasers [10], (2) the variation of pumping intensity in solid state lasers, and (3) the dependence of the degree of gain saturation on the radial position in the beam. We can account for an optical medium with quadratic gain (or loss) variation by taking the complex propagation constant $k(r)$ in Equation (2.4-5) as

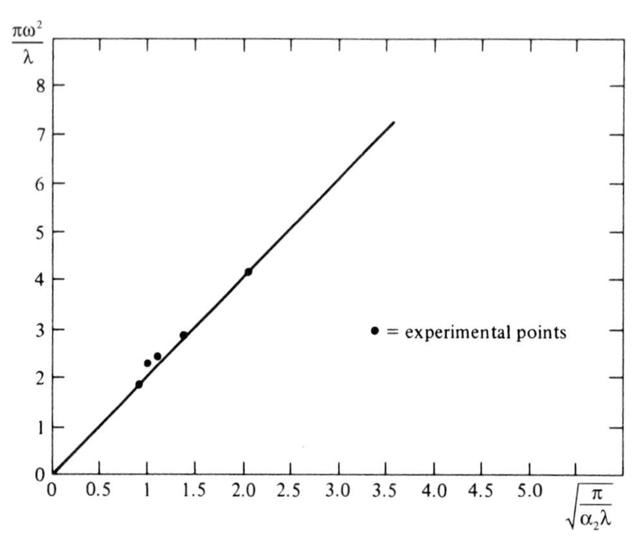

Figure 2.9 Theoretical curve showing the dependence of beam radius on quadratic gain constant α_2. Experimental points were obtained in a xenon 3.39 μm laser in which α_2 was varied by controlling the unsaturated laser gain. (After Reference [11].)

$$k(r) = k \pm i(\alpha_0 - \tfrac{1}{2}\alpha_2 r^2) \tag{2.10-1}$$

where α_0 and α_2 are constants. The plus (minus) sign applies to the case of gain (loss). The expression used previously for the quadratic index variation can be employed here, provided we treat the constant g in Equation (2.4-6) as a complex number and we assume $g^2r^2 \ll 1$. In other words, Equation (2.10-1) can be obtained from Equation (2.4-6) if we take $g^2 = i\alpha_2/k$. Using this value in Equation (2.4-13) to obtain the steady-state ($(1/q)' = 0$) solution of the complex beam radius q yields

$$\frac{1}{q} = -ig = -i\sqrt{\frac{i\alpha_2}{k}} \tag{2.10-2}$$

The steady-state beam radius and spot size are obtained from Equations (2.6-7) and (2.10-2):

$$\omega_0^2 = 2\sqrt{\frac{\lambda}{\pi n\alpha_2}}$$
$$R = 2\sqrt{\frac{\pi n}{\lambda\alpha_2}} \tag{2.10-3}$$

where n is the real part of the index of refraction. We thus find that the steady-state solution corresponds to a beam with a constant spot size but with a finite radius of curvature. The general (non-steady-state) behavior of the Gaussian beam in a quadratic gain medium is described in terms of its complex beam parameter $q(z)$ given by Equation (2.6-5), where $g^2 = i\alpha_2/k$.

Experimental data showing a decrease of the beam spot size with increasing gain parameter α_2 in agreement with Equation (2.10-3) are shown in Figure 2.9. "Steady state" here refers not to the intensity, which according to Equation (2.10-1) is growing or decaying with z, but to the beam radius of curvature and spot size.

2.11 ELLIPTIC GAUSSIAN BEAMS

All the beam solutions considered up to this point have one feature in common. The field drops off as in Equation (2.5-18), according to

$$E_{l,m} \propto \exp\left(-\frac{x^2+y^2}{\omega^2(z)}\right) \tag{2.11-1}$$

so that the locus in the xy plane of the points where the field is down by a factor of $1/e$ from its value on the axis is a circle of radius $\omega(z)$. We will refer to such beams as *circular Gaussian beams.*

The wave equation (2.4-9) also admits solutions in which the variation in the x and y directions is characterized by

$$E_{l,m} \propto \exp\left(-\frac{x^2}{\omega_x^2(z)}-\frac{y^2}{\omega_y^2(z)}\right) \tag{2.11-2}$$

with $\omega_x \neq \omega_y$. Such beams, which we name *elliptic Gaussian*, result, for example, when a circular Gaussian beam passes through a cylindrical lens or when a laser beam emerges from an astigmatic resonator—that is, one whose mirrors possess different radii of curvature in the zy and zx planes.

We will not repeat the whole derivation for this case, but will indicate the main steps. Instead of Equation (2.4-11), we assume a solution

$$\psi = \exp\left[-i\left(P(z)+\frac{k}{2q_x(z)}x^2+\frac{k}{2q_y(z)}y^2\right)\right] \tag{2.11-3}$$

where q_x and q_y are the complex Gaussian beam parameters. We also generalize Equation (2.4-6) and assume an astigmatic quadratic index medium with the following index profile:

$$n^2(x,y) = n_0^2(1-g_x^2x^2-g_y^2y^2)$$

where g_x and g_y are constants.

Using Equation (2.11-3) and following the approach in Section 2.4, we obtain

$$\left(\frac{1}{q_x}\right)^2+\left(\frac{1}{q_x}\right)'+g_x^2=0, \qquad \left(\frac{1}{q_y}\right)^2+\left(\frac{1}{q_y}\right)'+g_y^2=0 \tag{2.11-4}$$

and

$$\frac{dP}{dz} = -\frac{i}{2}\left(\frac{1}{q_x}+\frac{1}{q_y}\right) \tag{2.11-5}$$

where the prime indicates a differentiation with respect to z. In the case of a homogeneous ($g_x = g_y = 0$) beam we obtain, as in Equation (2.5-6),

$$q_x(z) = z + C_x \tag{2.11-6}$$

where C_x is an arbitrary constant of integration. We find it useful to write C_x as

$$C_x = -z_x + q_{0x} \tag{2.11-7}$$

where z_x is real and q_{0x} is imaginary. The physical significance of these two constants will become clear in what follows. A similar result with $x \to y$ is obtained for $q_y(z)$. Using the solutions of $q_x(z)$ and $q_y(z)$ in Equation (2.11-5) gives

$$P = -\frac{i}{2}\left[\ln\left(1 + \frac{z - z_x}{q_{0x}}\right) + \ln\left(1 + \frac{z - z_y}{q_{0y}}\right)\right]$$

Proceeding straightforwardly, as in the derivation connecting Equations (2.5-6)–(2.5-18), we find

$$E(x, y, z) = E_0 \frac{\sqrt{\omega_{0x}\omega_{0y}}}{\sqrt{\omega_x(z)\omega_y(z)}} \exp\left(-i[kz - \eta(z)] - \frac{ikx^2}{2q_x(z)} - \frac{iky^2}{2q_y(z)}\right)$$

$$= E_0 \frac{\sqrt{\omega_{0x}\omega_{0y}}}{\sqrt{\omega_x(z)\omega_y(z)}} \exp\left[-i[kz - \eta(z)] - x^2\left(\frac{1}{\omega_x^2(z)} + \frac{ik}{2R_x(z)}\right) - y^2\left(\frac{1}{\omega_y^2(z)} + \frac{ik}{2R_y(z)}\right)\right] \tag{2.11-8}$$

where

$$q_{0x} = i\frac{\pi\omega_{0x}^2 n}{\lambda}$$

$$\omega_x^2(z) = \omega_{0x}^2\left[1 + \left(\frac{\lambda(z - z_x)}{\pi\omega_{0x}^2 n}\right)^2\right] \tag{2.11-9}$$

$$R_x(z) = (z - z_x)\left[1 + \left(\frac{\pi\omega_{0x}^2 n}{\lambda(z - z_x)}\right)^2\right]$$

where $n = n_0$ is the index of refraction of the medium. Similar expressions for q_{0y}, ω_y, and R_y can be obtained by $x \to y$.

The phase delay $\eta(z)$ in Equation (2.11-8) is now given by

$$\eta(z) = \tfrac{1}{2}\tan^{-1}\left(\frac{\lambda(z - z_x)}{\pi\omega_{0x}^2 n}\right) + \tfrac{1}{2}\tan^{-1}\left(\frac{\lambda(z - z_y)}{\pi\omega_{0y}^2 n}\right) \tag{2.11-10}$$

It follows that *all* the results derived for the case of circular Gaussian beams apply, separately, to the xz and to the yz behavior of the elliptic Gaussian beam. For the purpose of analysis, the elliptic beam can be considered as two independent "beams." The position of the waist is not necessarily the same for these two beams. It occurs at $z = z_x$ for the xz beam and at $z = z_y$ for the yz beam in the example of Figure 2.10, where z_x and z_y are arbitrary. It also follows from the similarity between Equations (2.11-4) and (2.4-11) that the *ABCD* transformation law (2.6-9) can be applied separately to $q_x(z)$ and $q_y(z)$, which, according to Equation (2.11-8), are given by

$$\frac{1}{q_x(z)} = \frac{1}{R_x(z)} - i\frac{\lambda}{\pi n\omega_x^2(z)}$$

$$\frac{1}{q_y(z)} = \frac{1}{R_y(z)} - i\frac{\lambda}{\pi n\omega_y^2(z)} \tag{2.11-11}$$

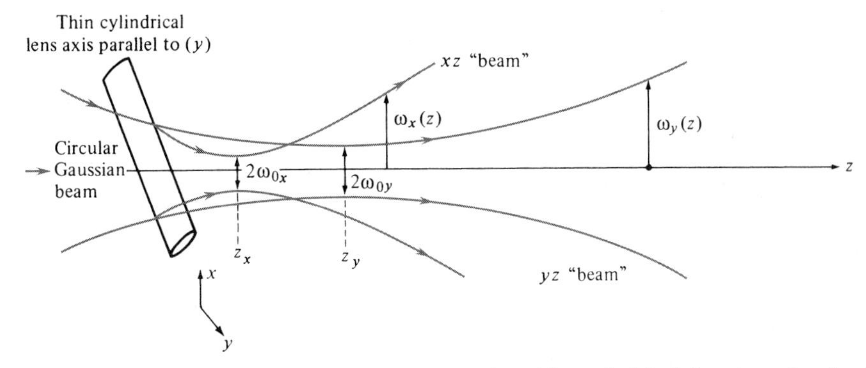

Figure 2.10 Illustration of an elliptic beam produced by cylindrical focusing of a circular Gaussian beam.

Elliptic Gaussian Beams in a Quadratic Lenslike Medium

Here we consider the mode of propagation in an astigmatic medium whose index of refraction is given by

$$n^2(x, y) = n_0^2(1 - g_x^2x^2 - g_y^2y^2) \tag{2.11-12}$$

where n_0, g_x, and g_y are constants. These modes of propagation have well-defined transverse wavefunctions that remain the same as they propagate. The derivation is identical to that presented in Section 2.9, resulting in the following wavefunctions for elliptic Gaussian beams:

$$E_{l,m}(r) = E_0 e^{-i\beta_{l,m}z} H_l\left(\sqrt{2}\frac{x}{\omega_x}\right) H_m\left(\sqrt{2}\frac{y}{\omega_y}\right) \exp\left(-\frac{x^2}{\omega_x^2} - \frac{y^2}{\omega_y^2}\right) \tag{2.11-13}$$

where E_0 is a constant and

$$\omega_x = \sqrt{\frac{2}{g_x k}} = \sqrt{\frac{\lambda}{g_x n_0 \pi}}, \quad \omega_y = \sqrt{\frac{2}{g_y k}} = \sqrt{\frac{\lambda}{g_y n_0 \pi}} \tag{2.11-14}$$

$$\beta_{l,m} = k\left\{1 - \frac{2}{k}\left[g_x\left(l + \frac{1}{2}\right) + g_y\left(m + \frac{1}{2}\right)\right]\right\}^{1/2}, \quad m, l = 0, 1, 2, 3, \ldots \tag{2.11-15}$$

Stripe-geometry GaAs-AlGaAs semiconductor lasers have an astigmatic quadratic-index profile approximately described by Equation (2.11-12) with $g_x \neq g_y$. Figure 2.11 shows the intensity patterns of the elliptic Gaussian beam modes of these lasers with various stripe widths (ω = 10 μm–50 μm). The principles of operation of semiconductor lasers will be discussed in Chapters 15 and 16.

2.12 BEAM PROPAGATION AND DIFFRACTION INTEGRAL

In previous sections, we considered the propagation of rays and Gaussian beams in homogeneous media. Their propagation can easily be described by using *ABCD* matrices. Here, we consider the propagation of a general beam of light with a finite transverse dimension. We will first describe the beam propagation by decomposing the beam into a superposition of plane waves. By virtue of the linearity of the wave equation, all optical wave of finite

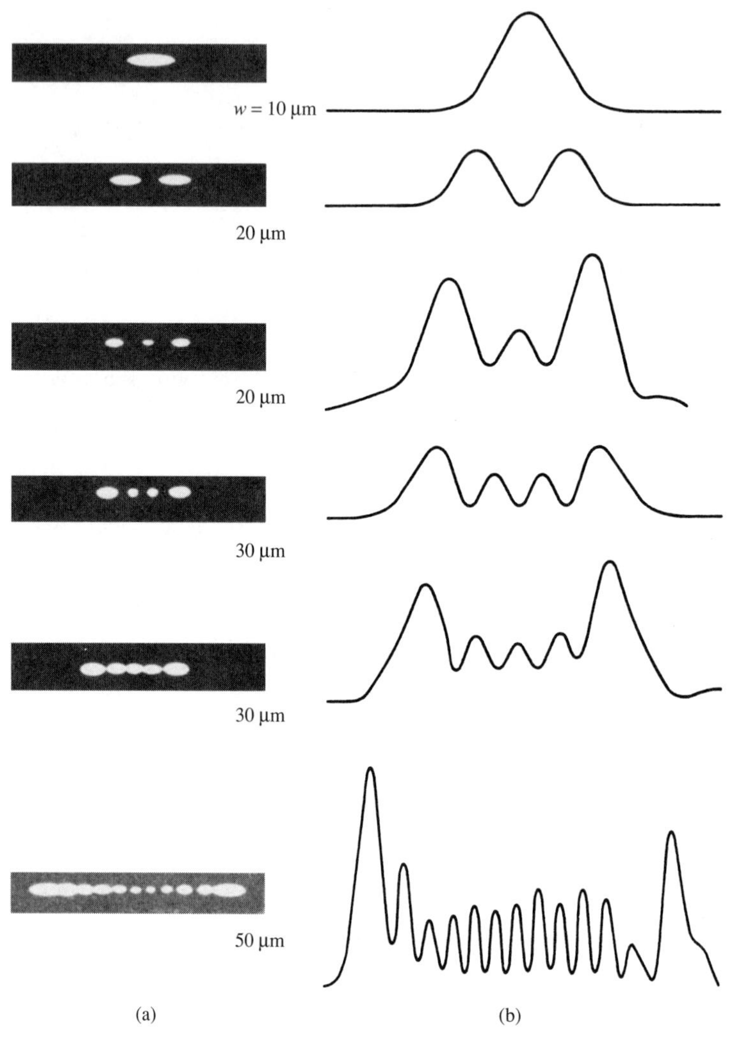

Figure 2.11 (a) Near-field and (b) far-field intensity distributions of the output of stripe contact GaAs–GaAlAs lasers. (After Reference [12].)

extent can be expressed in terms of a linear combination of plane waves. The propagation can then be described by the propagation of all these plane wave components. This approach leads to an integral representation of the propagation, known as the *Fresnel–Kirchhoff diffraction integral*. Near the end of this section, we will show that *ABCD* matrices can also be applied to the Fresnel–Kirchhoff diffraction integral for the propagation of a general optical beam in lenslike media.

Referring to Figure 2.12, let us consider the propagation of a general beam of light along the z axis. Let $E(x, y)$ be the amplitude of an optical beam at $z = 0$. The amplitude can be written as a Fourier integral of plane waves,

$$E(x, y) = \iint A(k_x, k_y) \exp(-ik_x x - ik_y y)\, dk_x\, dk_y \quad (z = 0) \tag{2.12-1}$$

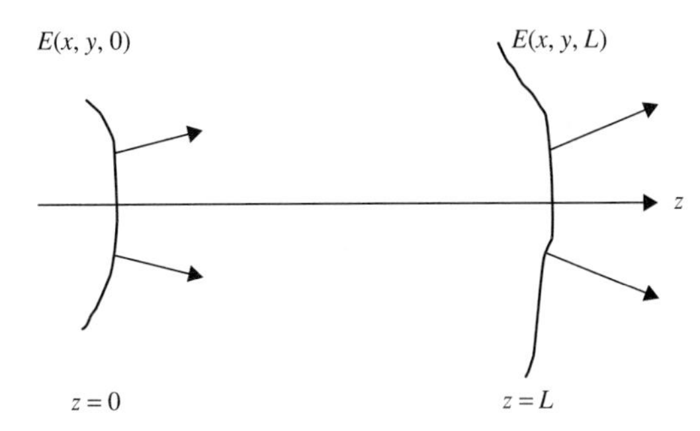

Figure 2.12 Propagation of a general beam from $z = 0$ to $z = L$.

where $A(k_x, k_y)$ is the amplitude of the (k_x, k_y) plane wave component. We assume that the beam is propagating along the z axis so that the amplitude $A(k_x, k_y)$ is centered around (0, 0) in the (k_x, k_y) plane. The amplitude distribution $A(k_x, k_y)$ is often called the Fourier transform of $E(x, y)$. Given the Fourier transform $A(k_x, k_y)$, the field amplitude at $z > 0$ can be written

$$E(x, y, z) = \iint A(k_x, k_y) \exp(-ik_x x - ik_y y) \exp(-ik_z z)\, dk_x\, dk_y \tag{2.12-2}$$

where k_z is the z component of the wavevector. This equation can be viewed as a linear summation of all the plane wave components. The components of the wavevector must satisfy

$$k_x^2 + k_y^2 + k_z^2 = \left(\frac{\omega}{c} n\right)^2 \equiv k^2 \tag{2.12-3}$$

where n is the index of refraction, ω is the angular frequency, and k is the wavenumber of light in the medium. For a well-confined beam or a paraxial wave, the Fourier transform $A(k_x, k_y)$ is significant only in the regime where

$$k_x, k_y \ll k \tag{2.12-4}$$

In this case, the z component k_z can be written

$$k_z = k - \frac{k_x^2 + k_y^2}{2k} \tag{2.12-5}$$

Substituting Equation (2.12-5) into Equation (2.12-2), we obtain

$$E(x, y, z) = \iint A(k_x, k_y)\, e^{-ikz + i[(k_x^2+k_y^2)/2k]z} \exp(-ik_x x - ik_y y)\, dk_x\, dk_y \tag{2.12-6}$$

By comparing with Equation (2.12-1), we note that the Fourier transform of $E(x, y, z)$ can be written

$$A'(k_x, k_y) = H(k_x, k_y) A(k_x, k_y) = e^{-ikz + i[(k_x^2+k_y^2)/2k]z} A(k_x, k_y) \tag{2.12-7}$$

where H is known as the *transfer function*. The transfer function for propagation over a distance z in space is thus

$$H(k_x, k_y) = e^{-ikz + i[(k_x^2+k_y^2)/2k]z} \tag{2.12-8}$$

By inverting Equation (2.12-1), the Fourier transform of $E(x, y)$ can be written

$$A(k_x, k_y) = \frac{1}{(2\pi)^2} \iint E(x, y) \exp(ik_x x + ik_y y)\, dx\, dy \tag{2.12-9}$$

To find the field amplitude at $z > 0$, we can substitute the Fourier transform of Equation (2.12-9) into Equation (2.12-6). We obtain

$$E(x, y, z) = \frac{1}{(2\pi)^2} \iint dx'\, dy' \iint E(x', y') e^{-ik_x(x-x')-ik_y(y-y')} e^{-ikz+i[(k_x^2+k_y^2)/2k]z}\, dk_x\, dk_y \quad (2.12\text{-}10)$$

We now carry out the integration over k_x and k_y. This leads to

$$E(x, y, z) = \frac{i}{\lambda z} e^{-ikz} \iint E(x', y') e^{-ik[(x-x')^2+(y-y')^2]/2z}\, dx'\, dy' \quad (2.12\text{-}11)$$

This is known as the Fresnel–Kirchhoff diffraction integral and is often written

$$E(x, y, L) = \frac{i}{\lambda L} e^{-ikL} \iint E(x', y') e^{-ik[(x-x')^2+(y-y')^2]/2L}\, dx'\, dy' \quad (2.12\text{-}12)$$

where L is the distance of propagation from $z = 0$. Notice that this equation is consistent with the *Huygens principle* in which the field at point (x, y) is a sum of all spherical waves emitted from source point (x', y'). In arriving at the above equations, we have used the following integral:

$$\int_{-\infty}^{\infty} \exp(-\alpha x^2 - \beta x)\, dx = \sqrt{\frac{\pi}{\alpha}} \exp\left(\frac{\beta^2}{4\alpha}\right) \quad (2.12\text{-}13)$$

Thus, given a field of $E(x', y')$ at the input plane $z = 0$, the field $E(x, y, L)$ at the output plane $z = L$ can be obtained by carrying out the integral (2.12-12) over the source plane (x', y').

EXAMPLE: FRESNEL–KIRCHHOFF DIFFRACTION

Consider an optical beam with a Gaussian beam profile at $z = 0$. In other words, the field amplitude can be written

$$E(x, y, 0) = E_0 \exp[-(x^2 + y^2)/a^2]$$

where a is a constant. Physically, a is a measure of the beam size.

Substituting the above expression into Equation (2.12-11), and carrying out the integration with the help of Equation (2.12-13), we obtain

$$E(x, y, z) = \frac{E_0 e^{-ikz}}{1 - i\dfrac{\lambda z}{\pi a^2 n}} \exp\left(-i\frac{k}{2R}(x^2 + y^2)\right) \exp\left(-\frac{x^2 + y^2}{\omega^2}\right)$$

with

$$R = z + \frac{1}{z}\left(\frac{\pi a^2 n}{\lambda}\right)^2, \quad \omega^2 = a^2\left[1 + \left(\frac{z\lambda}{\pi a^2 n}\right)^2\right]$$

We note that these results are in agreement with the propagation of a Gaussian beam.

The Fresnel–Kirchhoff diffraction integral (2.12-12) is now reproduced here with a slight change of notation:

$$f_1(x_1, y_1) = \frac{ik}{2\pi L} \exp(-ikL) \iint f_0(x_0, y_0) \exp\left(-\frac{ik}{2L}[(x_1 - x_0)^2 + (y_1 - y_0)^2]\right) dx_0\, dy_0 \quad (2.12\text{-}14)$$

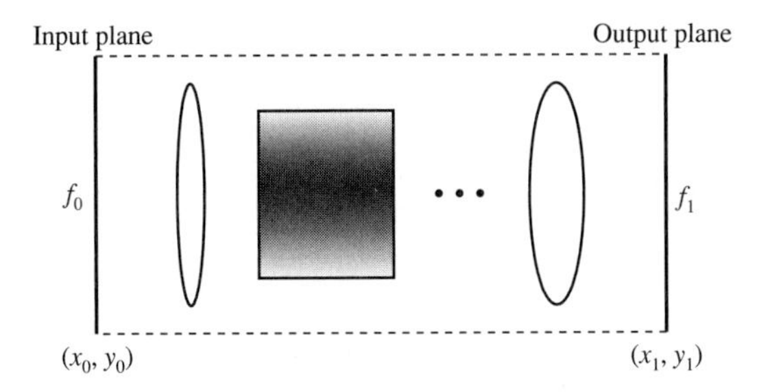

Figure 2.13 A cascade of lenslike (paraxial) elements with an overall *ABCD* matrix.

which relates the field $f_1(x_1, y_1)$ at $z = L$ to the source field $f_0(x_0, y_0)$ at $z = 0$ in the case of a homogeneous and isotropic medium. Equation (2.12-14) is often written in the following form:

$$f_1(x_1, y_1) = \frac{ik}{2\pi L}\iint f_0(x_0, y_0) \exp[-ik\rho(x_0, y_0; x_1, y_1)]\, dx_0\, dy_0 \tag{2.12-15}$$

where

$$\rho(x_0, y_0; x_1, y_1) = L + \frac{1}{2L}(x_1 - x_0)^2 + \frac{1}{2L}(y_1 - y_0)^2 \tag{2.12-16}$$

is the distance between the source point (x_0, y_0) and the observation point (x_1, y_1) in free space. We recall that the above is valid provided $\sqrt{(x_1 - x_0)^2 + (y_1 - y_0)^2} \ll L$. This result can be extended to the case where the intervening medium consists of a cascade of paraxial and axially symmetric components, which can be described by an overall *ABCD* matrix. An example of such a system is sketched in Figure 2.13.

The result is

$$f_1(x_1, y_1) = \frac{ik_0}{2\pi B}\exp(-ik_0 L)\iint f_0(x_0, y_0)$$
$$\exp\left(-\frac{ik_0}{2B}[A(x_0^2 + y_0^2) - 2x_0x_1 - 2y_0y_1 + D(x_1^2 + y_1^2)]\right) dx_0\, dy_0 \tag{2.12-17}$$

$$k_0 = \frac{2\pi}{\lambda} = \frac{\omega}{c} = \text{wavenumber in vacuum}$$

$$L = \sum_i n_i L_i = \text{opticla path length along the axis}$$

This important result can be proved as follows by using Fermat's principle of least time [3, 13–14]. Referring to Figure 2.14, we consider propagation through an optical system that can be represented by an overall *ABCD* matrix. Let X_0 and X_1 be general points at the input plane and output plane, respectively. Without loss of generality, we further assume that the medium before the input and the medium after the output are vacuum with a unity refractive index. Let the optical path between X_0 and X_1 be written $\rho(x_0, x_1)$. We extrapolate the rays at points X_0 and X_1. This leads to intersections with the optical axis at P_0 and P_1.

According to Fermat's principle, the optical path $P_0Q_0Q_1P_1$ along the axis should be the same as that of $P_0X_0X_1P_1$. Let the radii of curvature be

$$R_0 = P_0Q_0$$
$$R_1 = Q_1P_1 \tag{2.12-18}$$

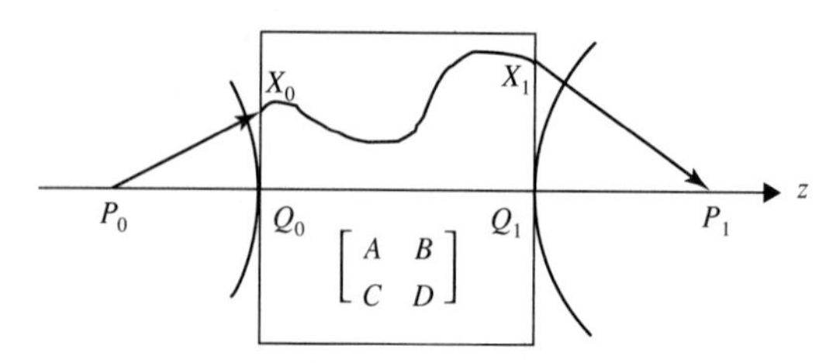

Figure 2.14 Schematic drawing showing the propagation of a beam through a general optical system described by a ray matrix with elements A, B, C, and D. This figure is used in the proof of the Fresnel–Kirchhoff integral (2.12-17) using Fermat's principle.

then the optical paths can be written

$$P_0Q_0Q_1P_1 = R_0 + \sum_i n_i L_i + R_1 \tag{2.12-19}$$

$$P_0X_0X_1P_1 = R_0 + \frac{x_0^2}{2R_0} + \rho(x_0, x_1) + R_1 + \frac{x_1^2}{2R_1} \tag{2.12-20}$$

By using the above equations and $P_0Q_0Q_1P_1 = P_0X_0X_1P_1$, we obtain

$$\rho(x_0, x_1) = \sum_i n_i L_i - \frac{x_0^2}{2R_0} - \frac{x_1^2}{2R_1} \tag{2.12-21}$$

where R_0 and R_1 are related to the ray slopes as

$$x_0' = \frac{x_0}{R_0} \quad \text{and} \quad x_1' = \frac{x_1}{R_1} \tag{2.12-22}$$

Furthermore, the ray parameters (including the slopes) can be written

$$\begin{aligned} x_1 &= Ax_0 + Bx_0' \\ x_1' &= Cx_0 + Dx_0' \end{aligned} \tag{2.12-23}$$

By eliminating R_0, R_1, and the ray slopes, we obtain

$$\rho(x_0, x_1) = \sum_i n_i L_i + \frac{1}{2B}(Ax_0^2 + Dx_1^2 - 2x_0x_1) \tag{2.12-24}$$

The extension to the general case of from point (x_0, y_0) to point (x_1, y_1) leads to

$$\rho(x_0, y_0; x_1, y_1) = \sum_i n_i L_i + \frac{1}{2B}[A(x_0^2 + y_0^2) + D(x_1^2 + y_1^2) - 2x_0x_1 - 2y_0y_1] \tag{2.12-25}$$

Substituting Equation (2.12-25) into Equation (2.12-15), we obtain the general expression (2.12-17). Note that the factor $(ik/2\pi L)$ is replaced by $(ik/2\pi B)$. This is essential to conserve power and to make the general result agree with the free-space result when the *ABCD* system consists only of free space. The matrix element B plays the same role as the distance of propagation $L = z_1 - z_0$. Thus an arbitrary optical wave can propagate through a sequence of paraxial optical systems, including all diffraction effects, by using knowledge only of the overall *ABCD* coefficients of the optical system between the input and output plane.

To demonstrate the power of this general result (2.12-17), we will use it to derive the imaging condition of a generalized cascade of optical elements described by an overall *ABCD* matrix. By rearranging and manipulating the exponent of Equation (2.12-17), we can rewrite it as

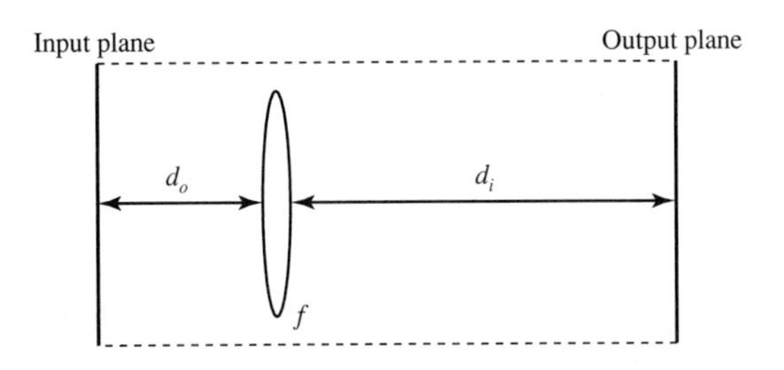

Figure 2.15 A single lens imaging system.

$$f_1(x_1, y_1) = \frac{ik_0}{2\pi B} \exp\left[-ik_0 L - \frac{ik_0}{2B}\left(D - \frac{1}{A}\right)(x_1^2 + y_1^2)\right]$$

$$\times \iint f_0(x_0, y_0) \exp\left\{-\frac{ik_0}{2B}\left[A\left(x_0 - \frac{x_1}{A}\right)^2 + A\left(y_0 - \frac{y_1}{A}\right)^2\right]\right\} dx_0\, dy_0 \quad (2.12\text{-}26)$$

It is a straightforward matter to show [15] that

$$\lim_{B\to 0} \sqrt{\frac{i}{2\pi B}} \exp\left(-i\frac{x^2}{2B}\right) = \delta(x) \quad (2.12\text{-}27)$$

where $\delta(x)$ is the Dirac delta function. Using Equation (2.12-27) in (2.12-26), we obtain

$$\lim_{B\to 0} f_1(x_1, y_1) = \frac{\exp(-ik_0 L)}{A} f_0\left(\frac{x_1}{A}, \frac{y_1}{A}\right) \exp\left(-i\frac{k_0(DA-1)}{2AB}(x_1^2 + y_1^2)\right) \quad (2.12\text{-}28)$$

Using the property $AD - BC = 1$, which holds when the individual A, B, C, D matrices of the cascade possess unity determinants, we rewrite Equation (2.12-28) as

$$\lim_{B\to 0} f_1(x_1, y_1) = \frac{\exp(-ik_0 L)}{A} \exp\left(-i\frac{k_0 C}{2A}(x_1^2 + y_1^2)\right) \times f_0\left(\frac{x_1}{A}, \frac{y_1}{A}\right) \quad (2.12\text{-}29)$$

This is the main result. It shows that in the limit of $B \to 0$, $f_1(x_1, y_1)$ is magnified by a factor A to $f_0(x_1/A, y_1/A)$ as well as multiplied by a quadratic phase factor (in x_1 and y_1). We thus associate $B = 0$ with the generalized imaging condition and A with the magnification. We leave it as a (simple) exercise to prove that in the case of imaging by a single thin lens, as shown in Figure 2.15, the imaging condition $B \to 0$ leads to the familiar lens formula

$$\frac{1}{f} = \frac{1}{d_o} + \frac{1}{d_i} \quad (2.12\text{-}30)$$

while the magnification is $A = -d_i/d_o$, where d_o and d_i are positions of the object and image, respectively.

The Fraunhofer Approximation

The diffraction integral (2.12-14) can be further simplified in the so-called Fraunhofer regime, where

$$k(x_0^2 + y_0^2) \ll 2L \quad (2.12\text{-}31)$$

for all points (x_0, y_0) of the source field. In this case, the quadratic phase function $\exp[-ik(x_0^2 + y_0^2)/2L]$ is approximately unity over the entire source field (aperture). The Fresnel–Kirchoff integral becomes, according to Equation (2.12-14),

$$f_1(x_1, y_1) = \frac{ik}{2\pi L}\exp[-ikL - ik(x_1^2 + y_1^2)/2L]\iint f_0(x_0, y_0)\exp\left(\frac{ik}{L}(x_1x_0 + y_1y_0)\right) dx_0\, dy_0 \quad (2.12\text{-}32)$$

We note that the field distribution $f_1(x_1, y_1)$ at $z = L$ is proportional to the Fourier transform of the source field distribution $f_0(x_0, y_0)$, provided condition (2.12-31) is satisfied. This is the regime of the Fraunhofer diffraction. The spatial region defined by Equation (2.12-31) is often called the far field. Thus, aside from a multiplicative phase factor, the field distribution in the far field is simply the Fourier transform of the source field.

EXAMPLE: FAR FIELD OF A CIRCLE APERTURE

Consider the passage of light with $\lambda = 1$ μm through a circular aperture of radius $a = 1$ cm. The far field according to Equation (2.12-31) is

$$L \gg k(x_0^2 + y_0^2)_{max}/2 = \pi a^2/\lambda = 314 \text{ m}$$

In the situation when the far field is indeed very far from the aperture, the far-field distribution can often be observed with the help of an imaging lens.

PROBLEMS

2.1 The unit cell matrix of a periodic lens waveguide is a unimodular matrix ($AD - BC = 1$). Solve the following eigenvalue problem:

$$\begin{bmatrix} A & B \\ C & D \end{bmatrix}\begin{bmatrix} r \\ r' \end{bmatrix} = \lambda \begin{bmatrix} r \\ r' \end{bmatrix}$$

(a) Show that the eigenvalues λ of the equation are $\lambda = e^{\pm i\theta}$ with θ given by Equation (2.1-13).

(b) Show that the eigenvectors are

$$\begin{bmatrix} r \\ r' \end{bmatrix}_1 = \begin{bmatrix} B \\ e^{i\theta} - A \end{bmatrix} \quad \text{and} \quad \begin{bmatrix} r \\ r' \end{bmatrix}_2 = \begin{bmatrix} B \\ e^{-i\theta} - A \end{bmatrix}$$

These two eigenvectors are independent provided B is finite and θ is not an integral multiple of π (i.e., $\theta \neq m\pi$). Although they are independent, show that these two eigenvectors are, in general, not orthogonal.

(c) A matrix M can be found such that the ray matrix is diagonalized,

$$M^{-1}\begin{bmatrix} A & B \\ C & D \end{bmatrix} M = \begin{bmatrix} e^{i\theta} & 0 \\ 0 & e^{-i\theta} \end{bmatrix}$$

Show that M can be written

$$M = \begin{bmatrix} B & B \\ e^{i\theta} - A & e^{-i\theta} - A \end{bmatrix}$$

The determinant of matrix M is $-2iB \sin\theta$. The inverse matrix exists provided B is finite and θ is not an integral multiple of π (i.e., $\theta \neq m\pi$). Find the inverse matrix M^{-1}.

2.2 Given two independent eigenvectors, it is possible to express an arbitrary ray vector in terms of a linear combination of the eigenvectors. In other words, we can write

$$\begin{bmatrix} r_0 \\ r_0' \end{bmatrix} = \alpha_0 \begin{bmatrix} B \\ e^{i\theta} - A \end{bmatrix} + \beta_0 \begin{bmatrix} B \\ e^{-i\theta} - A \end{bmatrix}$$

where α_0 and β_0 are constants.

(a) Find α_0 and β_0 in terms of r_0 and r_0'. Since both r_0 and r_0' are real, α_0 and β_0 form a complex conjugate pair.

(b) Show that the ray vector at the mth period can be written

$$\begin{bmatrix} r_m \\ r'_m \end{bmatrix} = \alpha_0 e^{im\theta} \begin{bmatrix} B \\ e^{i\theta} - A \end{bmatrix} + \beta_0 e^{-im\theta} \begin{bmatrix} B \\ e^{-i\theta} - A \end{bmatrix}$$

and show that a general expression for the ray location can be written in the form (2.1-23):

$$r_m = r_{\max} \sin(m\theta + \alpha)$$

(c) Derive Equation (2.1-23) directly from Equations (2.1-10) and (2.1-11).
(d) Derive Equations (2.1-24) through (2.1-26).

2.3 Derive Equation (2.3-19).

2.4 Consider a thin convex lens made of a homogeneous medium with spherical surfaces. Let the radii of curvatures of the surfaces be R_1 and R_2. Show that for a plane wave incident on a thin lens, the input–output relationship can be written

$$E_{\text{out}}(x, y) = E_{\text{in}}(x, y) \exp\left(ik \frac{x^2 + y^2}{2f} \right) \exp(-i\phi)$$

where f is the focal length of the thin lens. Show that f is given by

$$\frac{1}{f} = (n-1)\left(\frac{1}{R_1} + \frac{1}{R_2} \right)$$

What is the physical meaning of ϕ?

2.5 Show that a lenslike medium occupying the region $0 \le z \le l$ will image a point on the axis at $z < 0$ onto a single point. (If the image point occurs at $z < l$, the image is virtual.)

2.6 Derive the ray matrices of Table 2.1.

2.7 For confined Gaussian beams, the complex beam parameters q must be in the upper complex plane.

(a) Using the *ABCD* law of Gaussian beam transformation (2.6-9),

$$q_2 = \frac{Aq_1 + B}{Cq_1 + D}$$

Show that if q_1 is in the upper complex plane, so is q_2, provided A, B, C, and D are all real and $AD - BC = 1$. [Hint: Rationalize the denominator.]

(b) Using (a), show that q_N is always in the lower plane, if q_0 is in the lower complex plane.

$$q_N = \frac{\{A \sin(N\theta) - \sin[(N-1)\theta]\}q_0 + B \sin(N\theta)}{C \sin(N\theta) q_0 + D \sin(N\theta) - \sin[(N-1)\theta]}$$

provided θ is real.

(c) Show that for $|\cos\theta| > 1$, q_N approaches a real number when N becomes infinity. [Hint: First show the existence of a limit. Then show that the limit must satisfy $q = (Aq + B)/(Cq + D)$.]

2.8 Consider the incidence of a Gaussian beam normally on a solid prism with an index of refraction n (as shown in Figure P2.8).

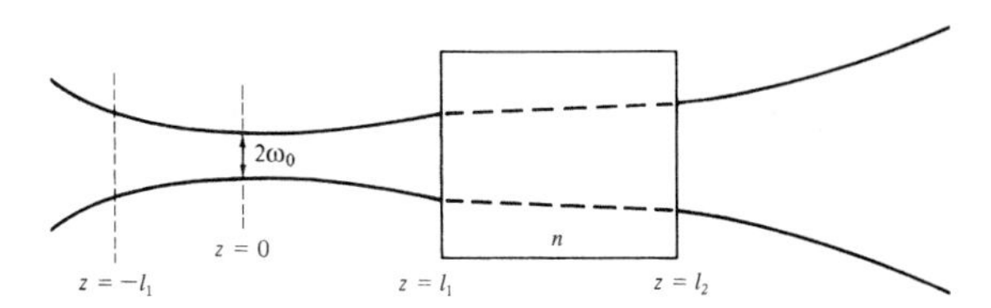

Figure P2.8

(a) What is the far-field diffraction angle of the output beam?
(b) Assume that the prism is moved to the left until its input face is at $z = -l_1$. What is the new beam waist and what is its location? (Assume that the crystal is long enough that the beam waist is inside the crystal.)

2.9 A Gaussian beam with a waist located at $z = 0$, and with a wavelength λ, is incident on a lens placed at $z = l$ as shown in Figure P2.9. Find the lens focal length, f, so that the output beam has a waist located at a given location $z = L$. Show that (given l and L) two solutions for the focal length exist. Sketch the beam behavior for each of these solutions.

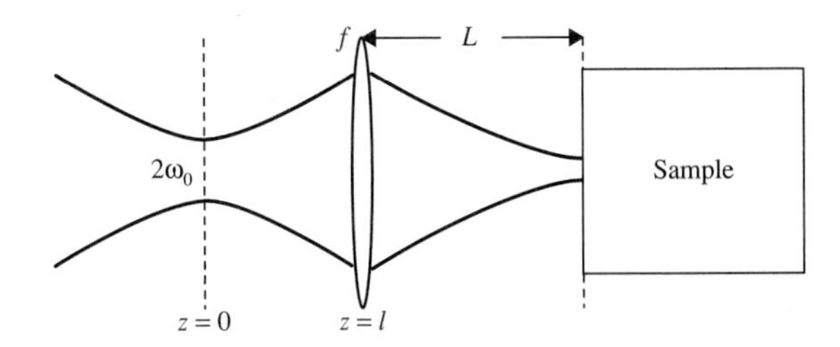

Figure P2.9

2.10 Complete all the missing steps in the derivation of Section 2.11.

2.11 Find the beam spot size and the maximum number of pulses per second that can be carried by an optical beam ($\lambda = 1\ \mu\text{m}$) propagating in a quadratic index glass fiber with $n = 1.5$, $g^2 = 3.3 \times 10^2\ \text{cm}^{-2}$: (a) in the case of a single-mode excitation $l = m = 0$ and (b) in the case where all the modes with $l, m < 5$ are excited. Using dispersion data of any typical commercial glass and taking $g^2 = 3.3 \times 10^2$ cm^{-2}, $l_{max} = m_{max} = 30$, compare the relative contributions of modal and glass dispersion to pulse broadening.

2.12 Consider a thick lens with radii of curvature R_1 and R_2 on its entrance and exit surfaces, an index of refraction n, and a thickness d.

(a) Obtain the *ABCD* matrix of the lens.
(b) What is its focal distance for light incident from the left?

2.13 Show that

$$\lim_{B \to 0} \sqrt{\frac{i}{2\pi B}} \exp\left(-i\frac{x^2}{2B}\right) = \delta(x)$$

where $\delta(x)$ is the Dirac δ function.

2.14 Show that, in the case of imaging by a thin lens, the generalized imaging condition $B = 0$ leads to Equation (2.12-30).

2.15 Show that Snell's law is consistent with Fermat's principle of least time.

2.16 Derive Snell's law directly from the ray equation.

2.17 The index of refraction of the atmosphere is given approximately as

$$n = 1 + \Delta n \exp[-\alpha(r - R)], \quad R < r$$

where Δn is approximately 0.0003, r is the distance measured from the center of the Earth, R is the radius of Earth, and α is approximately 1/7.8 km. For sunset and sunrise, the rays are nearly horizontal.

(a) Define β as the angle between the tangent vector of the ray and the horizon (x axis). Show that the ray equation can be written approximately

$$\frac{d}{dx}(n\beta) = \hat{\mathbf{y}} \cdot \nabla n = \frac{dn}{dr}\cos\theta$$

where r is the radial coordinate of the ray and y is the vertical axis with $r^2 = R^2 + x^2$.

(b) Define $r = R + z$, with R = Earth's radius. The ray bending can be obtained by integrating the above equation from $z = 0$ to infinity. Show that

$$\beta_\infty = \int_0^\infty \left(\frac{dn}{dr}\right) \frac{R}{\sqrt{(R+z)^2 - R^2}} dz$$

(c) Show that

$$\beta_\infty = -\Delta n \sqrt{\frac{\alpha R \pi}{2}}$$

[Hint: Neglect the z^2 term in the denominator ($z^2 \ll 2Rz$). This is legitimate as $R = 6400$ km and (dn/dr) is almost zero for altitudes above 100 km.] Using $\alpha = 1/7.8$ km, $R = 6400$ km, and $\Delta n = 0.0003$, $\beta_\infty = 0.011 = 0.6°$.

2.18

(a) Show that the inverse matrix M^{-1} of Problem 2.1 can be written

$$M^{-1} = \frac{i}{2B\sin\theta}\begin{bmatrix} e^{-i\theta} - A & -B \\ A - e^{i\theta} & B \end{bmatrix}$$

(b) Show that

$$\begin{bmatrix} A & B \\ C & D \end{bmatrix}^N = M \begin{bmatrix} e^{iN\theta} & 0 \\ 0 & e^{-iN\theta} \end{bmatrix} M^{-1}$$

(c) Carry out the matrix multiplications and derive the matrix identity (2.1-11).

2.19 Rays that are not confined to meridional planes exist in lenslike media, described by Equation (2.3-10). These rays are the so-called skew rays.

(a) Starting from the ray equation (2.3-8) or (2.3-11), show that rays with helical paths exist. In other words, show that the solution

$$r = a, \quad \theta = bz$$

satisfies the ray equation, where a and b are constants. Note that the helical pitch is $2\pi/b$.

(b) Show that b is related to the helical radius a by

$$b^2 = \frac{2n_0^2 g^2}{n^2 - n_0^2 g^2 a^2} = \frac{2n_0^2 g^2}{n_0^2 - 2n_0^2 g^2 a^2}$$

where n is the index of refraction at $r = a$.

(c) Show that circular paths also exist. Show that the radius of such a circular path satisfies the equation

$$n^2 = n_0^2 g^2 r^2$$

Note that the circular path may be viewed as a degenerate helix with zero pitch or an infinite b.

2.20 The lenslike medium described by Equation (2.3-9) can be shown to be a limiting case of the periodic lens waveguide consisting of identical lenses separated by an equal distance of d. Let $z = Nd$. Consider the limit when $d \to 0$ and $f \to \infty$ such that $df = 1/g^2$.

(a) Show that $\theta \to \sqrt{d/f}$ and $N\theta \to gz$.
(b) Using $A = 1$, $B = d$, $C = -1/f$, and $D = 1 - d/f$, show that in the limit when $N \to \infty$

$$\begin{bmatrix} A & B \\ C & D \end{bmatrix}^N = \begin{bmatrix} \cos(gz) & \sin(gz)/g \\ -g\sin(gz) & \cos(gz) \end{bmatrix}$$

So the lenslike medium can be viewed as a distributed case of a lens waveguide.

2.21 For the following problems, do the cases with $p = 0$, 1 and $m = 0$, 1, 2.

(a) Plot the intensity pattern of the first few low-order modes of Laguerre–Gaussian beams.
(b) Plot the sum of (p, m) and $(p, -m)$ modes.
(c) Express the first few low-order modes of Laguerre–Gaussian beams in terms of the linear combination of Hermite–Gaussian beams.

2.22

(a) Using the fundamental Gaussian beam in homogeneous media as an example, show that

$$k(\partial\psi/\partial z) = \left[-\frac{k}{q} + \frac{ik(kr^2/2)}{q^2}\right]\psi,$$

$$\partial^2\psi/\partial z^2 = \left[\frac{2}{q^2} - \frac{4i(kr^2/2)}{q^3} - \frac{(kr^2/2)^2}{q^4}\right]\psi$$

(b) Show that within the Gaussian beam, $(kr^2/2q) \leq 1$, the SVA is justified provided

$$\omega_0 \gg \lambda$$

that is, the beam spot size is much bigger than the wavelength.

(c) Show that the SVA approximation is equivalent to paraxial ray approximation.

REFERENCES

1. Marcuse, D., *Light Transmission Optics*. Van Nostrand, Princeton, NJ, 1972.
2. Yariv, A., *Quantum Electronics*, 2nd ed. Wiley, New York, 1975, Section 2.2.
3. Siegman, A. E., *Lasers*, University Science Books, Mill Valley, CA, 1986, pp. 779–782.
4. Kogelnik, H., and W. Rigrod, *Proc. IRE* **50**:230 (1962).
5. Kawakami, S., and J. Nishizawa, An optical waveguide with the optimum distribution of the refractive index with reference to waveform distortion. *IEEE Trans. Microwave Theory Technique* **10**:814 (1968).
6. Miller, S. E., E. A. J. Marcatili, and T. Li, Research toward optical fiber transmission systems. *Proc. IEEE* **61**:1703 (1973).
7. Cohen, L. G., and H. M. Presby, Shuttle pulse measurement of pulse spreading in a low loss graded index fiber. *Appl. Opt.* **14**:1361 (1975).
8. Bloom, D. M., L. F. Mollenauer, Chinlon Lin, D. W. Taylor, and A. M. DelGaudio, Direct demonstration of distortionless picosecond-pulse propagation in kilometer-length optical fibers. *Opt. Lett.* **4**:297 (1979).
9. Suematsu, Y., Long wavelength optical fiber communication. *Proc. IEEE* **71**:692 (1983).
10. Bennett, W. R., Inversion mechanisms in gas lasers. *Appl. Opt. Suppl. Chem. Lasers* **2**:3 (1965).
11. Casperson, L., and A. Yariv, The Gaussian mode in optical resonators with a radial gain profile. *Appl. Phys. Lett.* **12**:355 (1968).
12. Zachos, T. H., Gaussian beams from GaAs junction lasers. *Appl. Phys. Lett.* **12**:318 (1969).
13. Baues, P., Huygens' principle in inhomogeneous isotropic media. *Optoelectronics* **1**:37 (1969).
14. Collins, S. A., Lens-system diffraction integral written in terms of matrix optics. *J. Opt. Soc. Am.* **60**:1168 (1970).
15. Yariv, A., Imaging of coherent fields through lenslike systems. *Opt. Lett.* **19**:1607 (1994).

CHAPTER 3

GUIDED WAVES IN DIELECTRIC SLABS AND FIBERS

3.0 INTRODUCTION

So far, we have discussed wave propagation in free space, including the propagation of plane waves and beams. As a result of diffraction, a beam of light with a finite cross section will spread as it propagates in free space. Lenses or gradient index media may be employed at appropriate locations to focus the beams. Generally, dielectric media of high refractive index can be employed to confine the propagation of the beam. In this chapter, we will show that both dielectric slabs and circular fibers can support confined electromagnetic propagation. These modes of propagation are the so-called guided waves (or guided modes), and the structures that support guided waves are called waveguides. In this chapter, we first discuss the propagation of guided waves in dielectric slabs. As we know, any light beam with a finite transverse dimension will diverge as it propagates in a homogeneous medium. This divergence disappears in guiding dielectric structures under the appropriate conditions. In dielectric waveguides, the transverse dimension of these modes of propagation is determined by the dielectric waveguide.

We shall derive first the properties of guided modes in a dielectric slab structure. Optical modes are presented as the solution of the eigenvalue equation, which is derived from Maxwell's equations subject to the boundary conditions imposed by waveguide geometry. Both transverse electric (TE) and transverse magnetic (TM) modes of propagation are derived. The physics of confined propagation is explained in terms of the total internal reflection of plane waves from the dielectric interfaces. After discussion of the slab waveguides, we will cover the important subject of guided waves in circular fibers. We will then introduce the simple theory of effective index, which is useful in understanding waveguiding in two-dimensional structures. In the last part of the chapter, we will take up the subject of signal corruption due to chromatic dispersion in optical fibers, and the signal attenuation due to scattering and absorption. The subject of coupling between waveguides and the method of dispersion compensation in fiber transmission will be discussed later in this book.

3.1 TE AND TM CONFINED MODES IN SYMMETRIC SLAB WAVEGUIDES

Dielectric slabs are the simplest optical waveguides. Figure 3.1 shows a typical example of a slab waveguide. It consists of a thin dielectric layer (called the guiding layer, or simply the core)

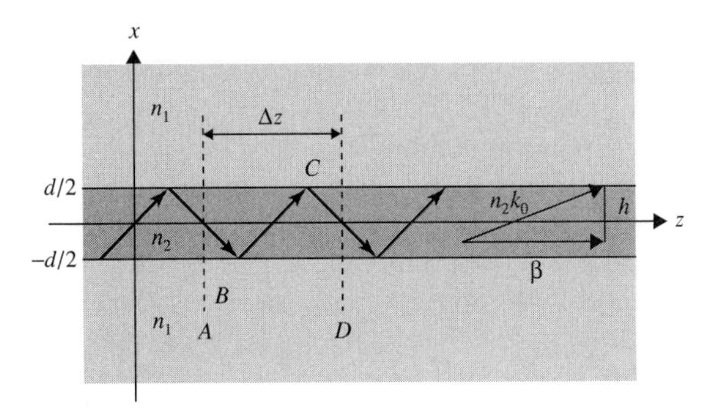

Figure 3.1 Schematic drawing of a symmetric slab waveguide. The waveguide consists of a guiding layer of thickness d with a refractive index n_2, surrounded by media of refractive index $n_1 \cdot k_0 = \omega/c$.

sandwiched between two semi-infinite bounding media (clad). Generally, the index of refraction of the guiding layer must be greater than those of the surrounding media. In addition, the thickness of the guiding layer is typically on the order of a wavelength. In symmetric slab waveguides, the two bounding media are identical. The simplest example will be a thin glass film (or layer) immersed in air or another fluid (or solid) of a lower index of refraction.

The following equation describes the index profile of a symmetric dielectric waveguide:

$$n(x) = \begin{cases} n_2, & |x| < d/2 \\ n_1, & \text{otherwise} \end{cases} \tag{3.1-1}$$

where d is the thickness of the guiding layer (core), n_2 is the index of refraction of the core, and n_1 is the index of refraction of the bounding media. To support guided modes, n_2 must be greater than n_1. The problem at hand is to find these guided modes.

The electromagnetic treatment of such a problem is relatively easy because the medium is homogeneous in each segment of the dielectric structure. In addition, the solutions of Maxwell's equations in homogeneous media are simply plane waves. Thus all we need to do is write the plane wave solutions for each segment and then match the boundary conditions at the interfaces to ensure the continuity of the wavefunction. We now consider the propagation of monochromatic radiation along the z axis. Maxwell's equation can be written in the form

$$\nabla \times \mathbf{H} = i\omega\varepsilon_0 n^2 \mathbf{E}, \qquad \nabla \times \mathbf{E} = -i\omega\mu\mathbf{H} \tag{3.1-2}$$

where n is the refractive index profile given in Equation (3.1-1). Since the whole structure is homogeneous along the z axis, solutions to the wave equations (3.1-2) can be taken as

$$\begin{aligned} \mathbf{E}(x, t) &= \mathbf{E}_m(x) \exp[i(\omega t - \beta z)] \\ \mathbf{H}(x, t) &= \mathbf{H}_m(x) \exp[i(\omega t - \beta z)] \end{aligned} \tag{3.1-3}$$

where β is the z component of the wavevectors and is known as the propagation constant to be determined from Maxwell's equations, and $\mathbf{E}_m(x)$ and $\mathbf{H}_m(x)$ are wavefunctions of the guided modes, with subscript m being an integer (called the mode number). For a layered dielectric structure that consists of homogeneous and isotropic materials, the wave equation can be obtained by eliminating $\mathbf{H}$ from Equartion (3.1-2):

$$\left(\frac{\partial^2}{\partial x^2} + \frac{\partial^2}{\partial y^2}\right)\mathbf{E}(x, y) + [k_0^2 n^2(r) - \beta^2]\mathbf{E}(x, y) = 0 \tag{3.1-4}$$

where $k_0 = \omega/c$ is the wavenumber in vacuum. Note that the preceding equation is not valid at the dielectric interfaces for TM waves (the $\mathbf{H}$ vector is transverse to the xz plane). We will solve for this equation separately in each segment of the structure and then match the tangential components of the field vectors at each interface. In addition to the continuity conditions at the interfaces, another important boundary condition for guided modes is that the field amplitudes be zero at infinity. A discussion of the general properties of the solutions of Equation (3.1-4) will be given later in this book.

We now proceed with the solution of Equation (3.1-4) in each segment of the dielectric structure. The propagation constant β is a very important parameter because it determines whether the field varies sinusoidally or exponentially. For confined modes, the field amplitude must fall off exponentially outside the guide structure. Consequently, the quantity $(n\omega/c)^2 - \beta^2$ in Equation (3.1-4) must be negative for $|x| > \frac{1}{2}d$. In other words, the propagation constant β of a confined mode must be such that

$$\beta > \frac{n_1\omega}{c} \tag{3.1-5}$$

where we recall that n_1 is the index of refraction of the bounding media. On the other hand, the continuity of the field requires that the magnitude of the field $\mathbf{E}_m(x)$ attain a maximum value. The existence of a maximum requires that the Laplacian of the field be negative. In other words, the propagation constant of a confined mode must be such that

$$\beta < \frac{n_2\omega}{c} \tag{3.1-6}$$

Thus we will find confined modes whose propagation constant satisfies these conditions, Equations (3.1-5) and (3.1-6). The modes can also be classified as either TE or TM modes. The TE modes have their electric field perpendicular to the xz plane (plane of incidence, or plane of propagation) and thus have only the field components E_y, H_x, and H_z. The TM modes have the field components H_y, E_x, and E_z.

Guided TE Modes

The electric field amplitude of the guided TE modes can be written in the form

$$E_y(x, z, t) = E_m(x) \exp[i(\omega t - \beta z)] \tag{3.1-7}$$

In a manner very similar to the wavefunction of a particle in a square-well potential, the mode function $E_m(x)$ is taken as

$$E_m(x) = \begin{cases} A \sin hx + B \cos hx, & |x| < \frac{1}{2}d \\ C \exp(-qx), & x > \frac{1}{2}d \\ D \exp(qx), & x < -\frac{1}{2}d \end{cases} \tag{3.1-8}$$

where A, B, C, and D are constants, and the parameters h and q are related to the propagation constant by

$$\begin{aligned} h &= \left[\left(\frac{n_2\omega}{c}\right)^2 - \beta^2\right]^{1/2} \\ q &= \left[\beta^2 - \left(\frac{n_1\omega}{c}\right)^2\right]^{1/2} \end{aligned} \tag{3.1-9}$$

The parameter h may be considered as the transverse component of the wavevector in the guiding layer. To be acceptable solutions, the tangential component of the electric and magnetic fields, E_y and H_z, must be continuous at the interfaces. Since $H_z = (i/\omega\mu)(\partial E_y/\partial x)$, we must match the magnitude as well as the slope of the TE mode functions $E_m(x)$ at the interfaces. This leads to

$$A \sin(\tfrac{1}{2}hd) + B \cos(\tfrac{1}{2}hd) = C \exp(-\tfrac{1}{2}qd)$$

$$hA \cos(\tfrac{1}{2}hd) - hB \sin(\tfrac{1}{2}hd) = -qC \exp(-\tfrac{1}{2}qd)$$

$$-A \sin(\tfrac{1}{2}hd) + B \cos(\tfrac{1}{2}hd) = D \exp(-\tfrac{1}{2}qd)$$

$$hA \cos(\tfrac{1}{2}hd) + hB \sin(\tfrac{1}{2}hd) = qD \exp(-\tfrac{1}{2}qd)$$

from which we obtain

$$2A \sin(\tfrac{1}{2}hd) = (C - D) \exp(-\tfrac{1}{2}qd) \tag{3.1-10}$$

$$2hA \cos(\tfrac{1}{2}hd) = -q(C - D) \exp(-\tfrac{1}{2}qd) \tag{3.1-11}$$

$$2B \cos(\tfrac{1}{2}hd) = (C + D) \exp(-\tfrac{1}{2}qd) \tag{3.1-12}$$

$$2hB \sin(\tfrac{1}{2}hd) = q(C + D) \exp(-\tfrac{1}{2}qd) \tag{3.1-13}$$

By examining the above equations, we find there are two sets of solutions.

(a) Symmetric modes ($A = 0$ and $C = D$): Equations (3.1-12) and (3.1-13) yield

$$h \tan(\tfrac{1}{2}hd) = q \quad \text{(for symmetric TE modes)} \tag{3.1-14}$$

(b) Antisymmetric modes ($B = 0$ and $C = -D$): Equations (3.1-10) and (3.1-11) give

$$h \cot(\tfrac{1}{2}hd) = -q \quad \text{(for antisymmetric TE modes)} \tag{3.1-15}$$

Note that both Equations (3.1-14) and (3.1-15) cannot be satisfied simultaneously since the elimination of q would lead to a pure imaginary h and a negative q. However, these two equations can be combined into a single equation (see Problem 3.14):

$$\tan(hd) = \frac{2hq}{h^2 - q^2} \tag{3.1-16}$$

The solutions of TE modes may thus be divided into two classes. For the first class,

$$A = 0, \quad C = D, \quad h \tan(\tfrac{1}{2}hd) = q \tag{3.1-17}$$

and for the second class,

$$B = 0, \quad C = -D, \quad h \cot(\tfrac{1}{2}hd) = -q \tag{3.1-18}$$

Note that the solutions in the first class have symmetric wavefunctions, whereas those of the second class have antisymmetric wavefunctions.

The propagation constants of the TE modes are found from a numerical or graphical solution of Equations (3.1-17) and (3.1-18), with the definition of h and q given by Equation (3.1-9). A very simple and well-known graphic solution is described here, since it clearly shows the way in which the number of TE modes depends on both the thickness d and the difference of indices of refraction. By putting $u = \tfrac{1}{2}hd$ and $v = \tfrac{1}{2}qd$, Equation (3.1-17) becomes $u \tan u = v$, with

$$u^2 + v^2 = (n_2^2 - n_1^2)\left(\frac{\omega d}{2c}\right)^2 = (n_2^2 - n_1^2)\left(\frac{\pi d}{\lambda}\right)^2 \equiv V^2 \tag{3.1-19}$$

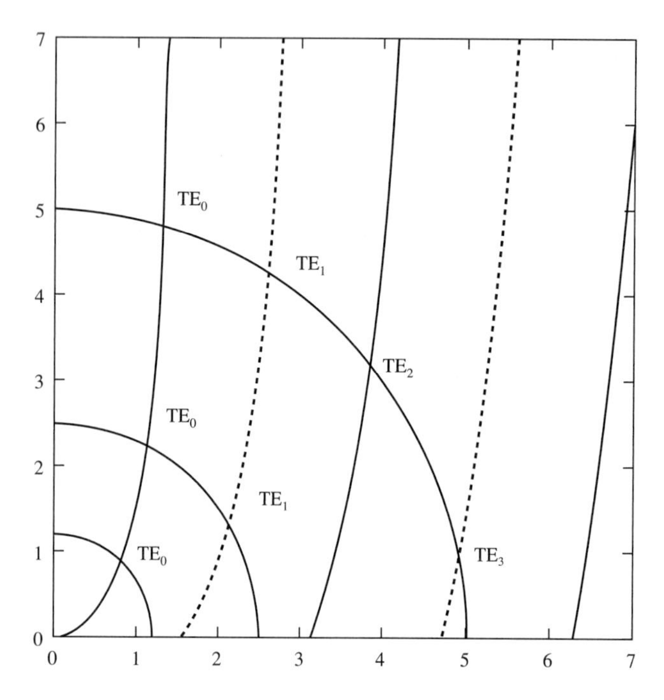

Figure 3.2 Graphic solution of Equations (3.1-17) and (3.1-18) for three values of V. Solid curves are $v = u \tan u$, and the dotted curves are $v = -u \cot u$.

Since u and v are restricted to positive values, the propagation constants may be found in this case from the intersection of both the curve $v = u \tan u$ and a circle of known radius $V = (n_2^2 - n_1^2)^{1/2}(\pi d/\lambda)$ in the first quadrant of the uv plane. A similar graphic construction for the solution of Equation (3.1-18) can be obtained by plotting $v = -u \cot u$ and the circle on the uv plane. Figure 3.2 shows such a graphic method for three values of V. For $V = 1.2$, there is only one solution—the TE_0 mode. There are two solutions (TE_0 and TE_1) when $V = 2.5$ and four solutions when $V = 5$. Note that the number of solutions depends on the value of V.

From Figure 3.2, it is clear that the number of confined TE modes depends on the magnitude of the parameter V. For V between zero and $\frac{1}{2}\pi$, there is just one TE mode of the first class. The first mode of the second class appears when the parameter V is greater than $\frac{1}{2}\pi$. As this parameter V increases, confined modes appear successively, first of one class and then of the other. Figure 3.3 plots the wavefunctions of a symmetrical slab waveguide with $n_2 = 1.6$, $n_1 = 1.5$, $d = 5$ μm, and $\lambda = 1.55$ μm. According to Equation (3.1-19), the parameter $V = 5.64$. This waveguide supports four TE modes. It is not difficult to see from Figure 3.3 that, when ordered according to the propagation constant β, the mth wavefunction has $m - 1$ nodes. We also notice that the wavefunctions are either symmetric or antisymmetric with respect to the origin $x = 0$. It follows from the discussion earlier that the wavefunctions are divided into two classes (see Equations (3.1-17) and (3.1-18)). This division is a direct consequence of the fact that the index profile $n(x)$ is symmetric about $x = 0$.

Knowledge that the solution possesses a definite symmetry sometimes simplifies the determination of the propagation constant, since we need only find the solution for positive x. Even solutions have zero slope and odd solutions have zero value at the origin $x = 0$. Thus the wavefunction of the even solutions can be written as $\cos(hx)$, whereas those of the odd solutions can be written as $\sin(hx)$. Both types of solutions decay exponentially in the region $|x| > \frac{1}{2}d$. The solutions are then obtained by matching the value and the slope at $|x| = \frac{1}{2}d$.

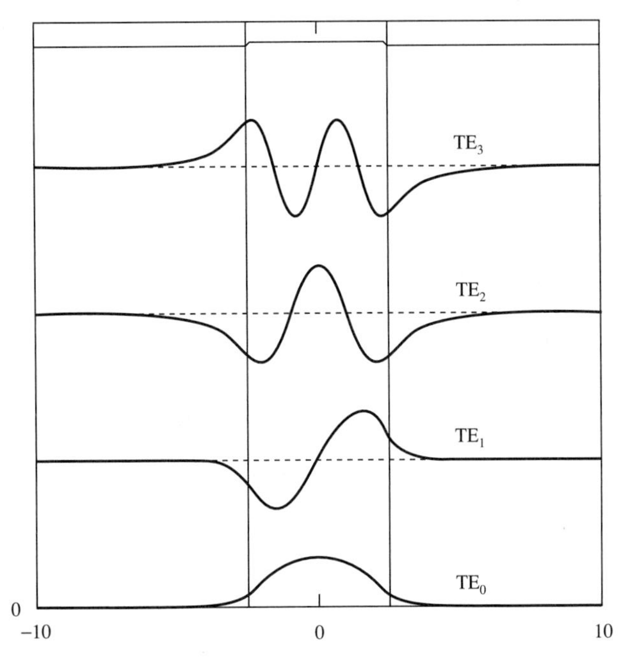

Figure 3.3 Wavefunctions of a symmetrical slab waveguide with $n_2 = 1.6$ and $n_1 = 1.5$. The thickness of the core is equal to $d = 5$ μm, and $\lambda = 1.55$ μm. The normalized propagation constants are 1.5946, 1.5785, 1.5521, and 1.5175. The fundamental mode has the largest propagation constant. The confinement factors (fraction of energy inside the core) for the modes are $\Gamma = 0.9914,\ 0.9631, 0.9033$ and 0.7511. The fundamental mode has the highest confinement.

For the purpose of describing and comparing the confined modes, it is convenient to define the normalized propagation constant as

$$\bar{\beta} = \frac{\beta}{\omega/c} \tag{3.1-20}$$

Such a normalized propagation is often called the effective index of refraction of the mode, n_{eff}, and is related to the phase velocity of the mode:

$$\bar{\beta} = n_{\text{eff}} = \frac{c}{v_p} \tag{3.1-21}$$

where v_p is the phase velocity of the mode, $v_p = \omega/\beta$. Thus, for confined modes, the normalized propagation constant $\bar{\beta}$ or the effective index n_{eff} is between n_2 and n_1.

Guided TM Modes

We now consider the TM modes whose magnetic field vector is perpendicular to the plane of propagation (xz plane). The derivation of the confined TM modes is similar in principle to that of the TE modes. The field amplitudes are written

$$\begin{aligned}
H_y(x, z, t) &= H_m(x) \exp[i(\omega t - \beta z)] \\
E_x(x, z, t) &= \frac{i}{\omega\mu}\frac{\partial}{\partial z} H_y \\
E_z(x, z, t) &= -\frac{i}{\omega\mu}\frac{\partial}{\partial x} H_y
\end{aligned} \tag{3.1-22}$$

The wavefunction $H_m(x)$ is

$$H_m(x) = \begin{cases} A \sin hx + B \cos hx, & |x| < \frac{1}{2}d \\ C \exp(-qx), & x > \frac{1}{2}d \\ D \exp(qx), & x < -\frac{1}{2}d \end{cases} \tag{3.1-23}$$

where A, B, C, and D are constants, and the parameters h and q are given by Equation (3.1-9).

The continuity of H_y and E_z at the two interfaces $x = \pm\frac{1}{2}d$ leads, in a manner similar to Equations (3.1-14) and (3.1-15), to the following eigenvalue equation:

$$\begin{aligned} h \tan(\tfrac{1}{2}hd) &= \frac{n_2^2}{n_1^2}q \quad \text{for even solutions} \\ h \cot(\tfrac{1}{2}hd) &= -\frac{n_2^2}{n_1^2}q \quad \text{for odd solutions} \end{aligned} \tag{3.1-24}$$

These two equations can also be combined into a single equation,

$$\tan(hd) = \frac{2h\bar{q}}{h^2 - \bar{q}^2} \tag{3.1-25}$$

where

$$\bar{q} = \frac{n_2^2}{n_1^2}q \tag{3.1-26}$$

Equation (3.1-24) can also be solved by using the graphic method described earlier. Figure 3.4 shows the dispersion relation (effective index n_{eff} versus normalized frequency V) of a typical symmetric waveguide.

The frequency at which $q = 0$ is a cutoff frequency. For a mode with $q = 0$, the field is no longer exponentially decaying in the cladding region and the propagation is no longer confined. Referring to Figure 3.4, we note that TE_0 and TM_0 modes have no cutoff frequency. In other words, these two modes are always confined in a symmetric waveguide. $V = \pi/2$ is

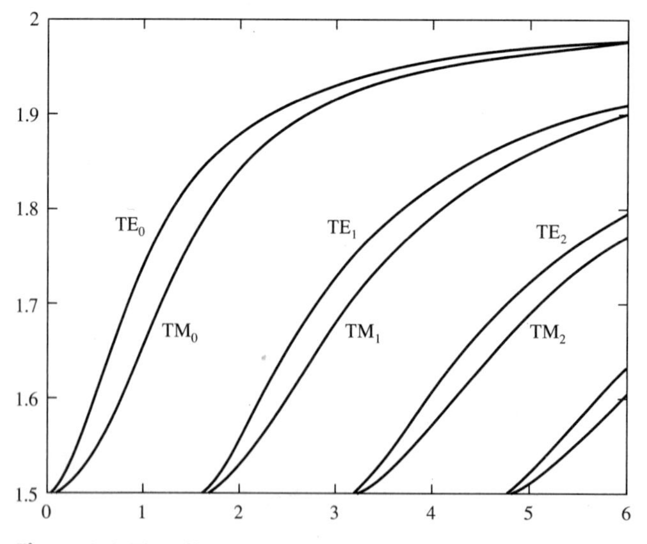

Figure 3.4 The effective index n_{eff} versus normalized frequency V of a typical symmetric waveguide with $n_1 = 1.5$ and $n_2 = 2.0$.

the cutoff frequency for TE_1 and TM_1 modes. For frequency in the range $V < \pi/2$, the waveguide can only support TE_0 and TM_0 modes. At a higher frequency in the range $\pi/2 < V < \pi$, additional modes TE_1 and TM_1 can be supported by the waveguide.

Geometric Optics Treatment

In the preceding, we solved the wave equation for the TE and TM modes of a symmetric slab waveguide. These modes can also be derived by using geometric optics. This is possible because the waveguide consists of layers of homogeneous dielectric materials. Wave propagation in each of the homogeneous regions can be represented by a superposition of two plane waves. One of the plane waves may be considered as the incident wave, whereas the other may be viewed as the reflected one. According to Equations (3.1-5) and (3.1-6), the plane waves that represent the confined modes experience total internal reflection at both interfaces $|x| = \frac{1}{2}d$. It seems that the condition of total internal reflection at both interfaces was enough to assure the confinement of energy in the guiding layer. However, it turns out that total internal reflection is only a necessary condition. In other words, not all rays trapped by internal reflection constitute a mode. A mode, by definition, must have a unique propagation constant and a well-defined field amplitude at each point in space and time. Consider the representation of a mode $E_m(x) \exp[i(\omega t - \beta z)]$ by a zigzagging plane wave of the form

$$E(x, z, t) = E_0(x) \exp[i(\omega t - \beta z - hx)] \tag{3.1-27}$$

where E_0 is a constant, and h is the transverse wavenumber. Let the plane wave travel a distance Δz in a time Δt in one zigzag $ABCD$ (see Figure 3.1). If we follow the ray path $ABCD$ and add the phase shifts due to propagation and total reflections at points B and C, we obtain a total phase shift of

$$\omega\, \Delta t - \beta\, \Delta z - 2hd + 2\phi \tag{3.1-28}$$

where ϕ is the phase shift upon total internal reflection at one of the interfaces. A mode of the form $E_m(x) \exp[i(\omega t - \beta z)]$ propagating from point A to point D will gain a phase shift of $\omega\, \Delta t - \beta\, \Delta z$. Therefore a total reflecting zigzag ray will become a mode only when the extra transverse phase shift is an integral multiple of 2π; that is,

$$-2hd + 2\phi = -2m\pi \tag{3.1-29}$$

where m is an integer. The minus sign preceding $2m\pi$ is chosen such that m corresponds to the TE_m and TM_m modes of the waveguide. This phase shift, 2ϕ, upon total internal reflection for TE or TM waves, can be expressed conveniently in terms of h and q as [1, 2]

$$\phi = \begin{cases} 2\tan^{-1}\left(\dfrac{q}{h}\right), & \text{TE} \\[2ex] 2\tan^{-1}\left(\dfrac{n_2^2 q}{n_1^2 h}\right), & \text{TM} \end{cases} \tag{3.1-30}$$

These phase shifts are limited between 0 and π. Therefore the fundamental mode corresponds to the situation when $2hd = 2\phi$ (i.e., $m = 0$). Higher-order modes involve larger hd. The integer m in Equation (3.1-27) therefore assumes only nonnegative values, that is, $m = 0, 1, 2, 3, \ldots$. Condition (3.1-29) is equivalent to the eigenvalue Equations (3.1-16) and (3.1-25).

Consider the example shown in Figure 3.3. This waveguide has $n_1 = 1.5$, $n_2 = 1.6$, $d = 5.0$ μm, $\lambda = 1.550$ μm, with a normalized frequency of $V = 5.6425$. There are four confined TE

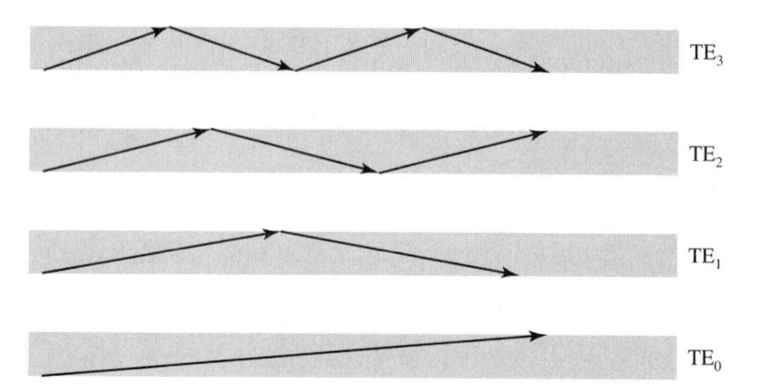

Figure 3.5 Zigzag ray representation of the four TE modes of a symmetrical slab waveguide.

modes in this slab waveguide. The normalized propagation constants $\beta/(\omega/c)$ for these modes are 1.5946, 1.5785, 1.5521, and 1.5175. These normalized propagation constants correspond to the ray incident angles of 85.29°, 80.60°, 75.94°, and 71.52°, in that order. Note that all these angles are greater than the critical angle 69.64°. The fundamental mode ($m = 0$) always has the largest propagation constant, and the angle of incidence with the normalized propagation constant approaches n_2. The highest order mode has the least normalized propagation constant approaching n_1 and the smallest angle of incidence approaching the critical angle. Figure 3.5 shows the zigzag rays corresponding to the modes previously discussed.

Ray optics offers a convenient way of obtaining the mode condition of a slab waveguide. In a more complicated waveguide structure that involves multilayers or inhomogeneous layers, it is very difficult to use the ray optics approach. In addition, the ray optics approach only yields the mode condition. Nothing about the field distribution and the mode orthogonality is obtained.

3.2 TE AND TM CONFINED MODES IN ASYMMETRIC SLAB WAVEGUIDES

The symmetric slab waveguides described in the previous section are useful for introducing the concept of confined propagation because of their simplicity in mathematics. In practice, however, most slab waveguides are not symmetric. In fact, most of the guiding layers are so thin that a substrate is necessary for support of the structure. In this section, we treat the confined modes in asymmetric slab waveguides. The symmetric slab waveguides described in the last section may be viewed as a special case of the asymmetric ones.

Referring to Figure 3.6, we now consider the propagation of confined modes in an asymmetric slab waveguide whose index profile is given by

$$n(x) = \begin{cases} n_1, & 0 < x \\ n_2, & -t < x < 0 \\ n_3, & x < -t \end{cases} \tag{3.2-1}$$

where t is the thickness of the guiding layer, and the index of refraction of the guiding media, n_2, is greater than those of the bounding media, n_1 and n_3. Without loss of generality, we assume that $n_1 < n_3 < n_2$. To be a satisfactory wavefunction, a solution to Maxwell's wave equation must be continuous, single valued, and finite throughout the space. We will first examine the physical nature of the solution as a function of the propagation constant β at a fixed frequency ω. For $\beta > n_2\omega/c$, it follows directly from Equation (3.1-4) that $(1/E)(\partial^2 E/\partial x^2) > 0$ everywhere,

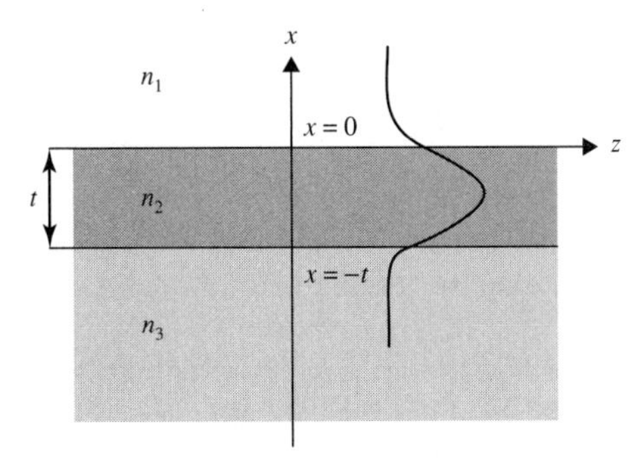

Figure 3.6 Schematic drawing of an asymmetric slab waveguide with a core index of n_2.

and $E(x)$ is exponential at all three regions of the waveguides. If we take $E(x) = \exp(-qx)$, which decays to zero at $x = +\infty$, because of the need to match both $E(x)$ and its derivatives at the two interfaces, the resulting field distribution is infinite at $x = -\infty$, as shown in Figure 3.7a. Such a solution would correspond to a field of infinite energy. It is not physically realizable and thus does not correspond to a real wave.

For $n_3(\omega/c) < \beta < n_2(\omega/c)$, as in Figures 3.7b and 3.7c, it follows from Equation (3.1-4) that the solution is sinusoidal in the core ($-t < x < 0$), since $(1/E)(\partial^2 E/\partial x^2) < 0$, but is exponential

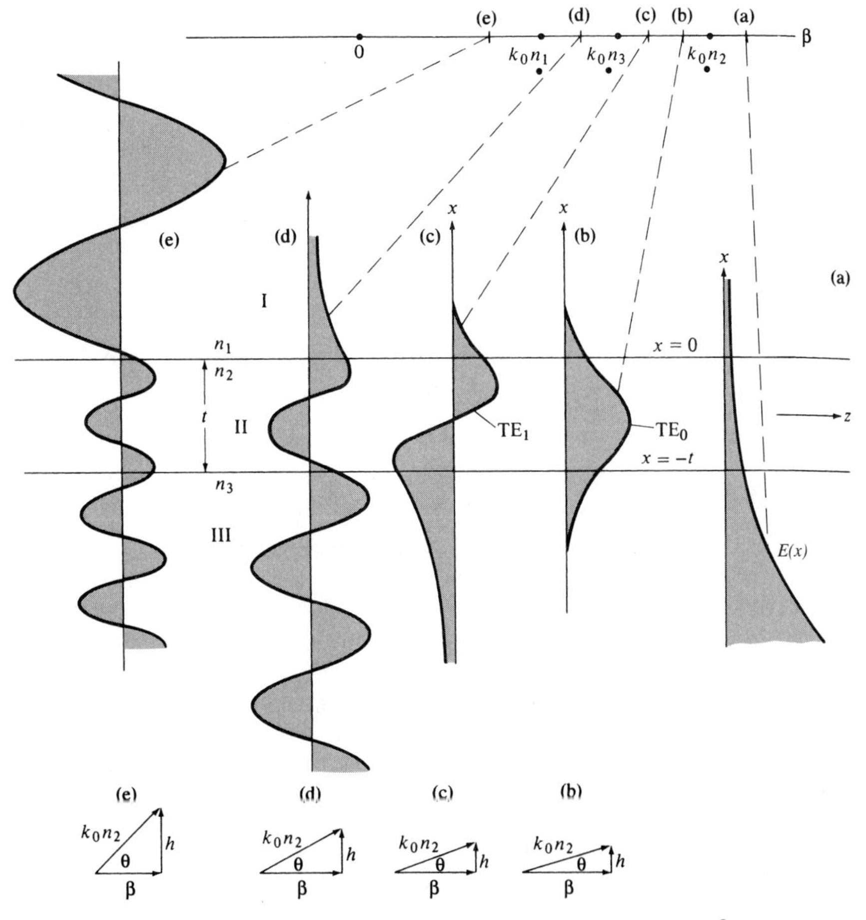

Figure 3.7 Typical field distributions corresponding to different values of β.

in the bounding media. This makes it possible to have a solution $E(x)$ that satisfies the boundary conditions while decaying exponentially in the regions $x < -t$ and $x > 0$. Two such solutions are shown in Figures 3.7b and 3.7c. The energy carried by these modes, as represented by the Poynting vector, is confined to the vicinity of the guiding layer, and, consequently, we will refer to them as confined or guided modes. In fact, only a small fraction of the energy is flowing outside the guiding layer. From the preceding discussion, it follows that a necessary condition for their existence is that $n_1(\omega/c), n_3(\omega/c) < \beta < n_2(\omega/c)$, so that confined modes are possible only when $n_2 > n_1, n_3$; that is, the inner layer (core) possesses the highest index of refraction. In some basic sense, the confined modes in this regime are reminiscent of quantized states of an electron in a potential well, in which the electron is trapped by the potential well.

Mode solutions for $n_1(\omega/c) < \beta < n_3(\omega/c)$ (regime (d) in Figure 3.7) correspond, according to Equation (3.1-4), to exponential behavior in the region $x > 0$ and to sinusoidal behavior in the regions $x < 0$, as illustrated in Figure 3.7d. In this regime, almost all the energy is flowing in the substrate. We will refer to these modes as substrate radiation modes. For $0 < \beta < n_1(\omega/c)$, as in Figure 3.7e, the solution for $E(x)$ becomes sinusoidal in all three regions. These are the so-called radiation modes of the waveguides.

A solution of Equation (3.1-4), subject to the boundary conditions at the interfaces given in what follows, shows that while in regimes (d) and (e) of Figure 3.7, β is a continuous variable, the values of allowed β in the propagation regime $n_3(\omega/c) < \beta < n_2(\omega/c)$ are discrete. The number of confined modes depends on the thickness t, the frequency, and the indices of refraction n_1, n_2, n_3. At a given wavelength, the number of confined modes increases from zero with increasing t. At some cutoff thickness t, the mode TE_0 becomes confined. Further increases in the layer thickness will allow TE_1 to exist as well, and so on.

We now turn our attention to solution of the wave Equation (3.1-4) for the dielectric waveguide sketched in Figure 3.6. The derivation will be limited to the guided modes, which, according to Figure 3.7, have a propagation constant β such that

$$n_3\frac{\omega}{c} < \beta < n_2\frac{\omega}{c}$$

where $n_1 < n_3$. This guide can, in the general case, support a finite number of confined TE modes with field components E_y, H_x, and H_z and TM modes with components H_y, E_x, and E_z. We will derive the mode wavefunction and the corresponding propagation constants.

Guided TE Modes

For TE modes, the electric field vector is perpendicular to the plane of incidence (xz plane). Thus E_y is the only component of the electric field vector. The field component E_y of the TE mode can be taken in the form

$$E_y(x, y, z, t) = E_m(x)e^{i(\omega t-\beta z)} \tag{3.2-2}$$

where β is the propagation constant and $E_m(x)$ is the wavefunction of the mth mode.

The function $E_m(x)$ assumes the following forms in each of the three regions:

$$E_m = \begin{cases} C\exp(-qx), & 0 \le x \\ C\left(\cos(hx) - \dfrac{q}{h}\sin(hx)\right), & -t \le x \le 0 \\ C\left(\cos(ht) + \dfrac{q}{h}\sin(ht)\right)\exp[p(x+t)], & x \le -t \end{cases} \tag{3.2-3}$$

where C is a normalization constant and h, q, and p are given by

$$h = \left[\left(\frac{n_2\omega}{c}\right)^2 - \beta^2\right]^{1/2}, \quad q = \left[\beta^2 - \left(\frac{n_1\omega}{c}\right)^2\right]^{1/2}, \quad p = \left[\beta^2 - \left(\frac{n_3\omega}{c}\right)^2\right]^{1/2} \tag{3.2-4}$$

These relationships are obtained by substituting Equation (3.2-3) into Equation (3.1-4).

The boundary conditions require that E_y and $H_z = (i/\omega\mu)(\partial E_y/\partial x)$ be continuous at both $x = 0$ and $x = -t$. The wavefunction in Equation (3.2-3) has been chosen such that E_y is continuous at both interfaces as well as $\partial E_y/\partial x$ at $x = 0$. By imposing the continuity requirements on $\partial E_y/\partial x$ at $x = -t$, we get, from Equation (3.2-3),

$$h \sin(ht) - q \cos(ht) = p\left(\cos(ht) + \frac{q}{h}\sin(ht)\right)$$

or

$$\tan(ht) = \frac{p + q}{h(1 - pq/h^2)} \tag{3.2-5}$$

This is the so-called mode condition. The propagation constant β of a guided TE mode must satisfy this condition. Given a set of refractive indices n_1, n_2, and n_3 of a slab waveguide, Equation (3.2-5) in general yields a finite number of solutions for β provided the thickness t is large enough. These modes are mutually orthogonal.

The normalization constant C is chosen so that the field $E_m(x)$ in Equation (3.2-3) corresponds to a power flow of 1 W (per unit width in the y direction) along the z axis in the mode. A mode for which $E_y = AE_m(x)$ will thus correspond to a power flow of $|A|^2$ W/m. The normalization condition is given by

$$S_z = \tfrac{1}{2}\int \text{Re}[\mathbf{E} \times \mathbf{H}^*]_z \, dx = 1$$

or equivalently,

$$-\tfrac{1}{2}\int_{-\infty}^{\infty} E_y H_x^* \, dx = \frac{\beta_m}{2\omega\mu}\int_{-\infty}^{\infty} [E_m(x)]^2 \, dx = 1 \tag{3.2-6}$$

where m denotes the mth confined TE mode (corresponding to the mth eigenvalue of Equation (3.2-5)] and $H_x = -i(\omega\mu)^{-1}\partial E_y/\partial z$. Substitution of Equation (3.2-3) for the wavefunction in Equation (3.2-6) and carrying out the integration lead to, after a few steps of algebraic manipulation,

$$C_m = 2h_m\left(\frac{\omega\mu}{|\beta_m|[t + (1/q_m) + (1/p_m)](h_m^2 + q_m^2)}\right)^{1/2} \tag{3.2-7}$$

The orthonormalization of the modes can be written

$$\int_{-\infty}^{\infty} E_m E_l \, dx = \frac{2\omega\mu}{|\beta_m|}\delta_{l,m} \tag{3.2-8}$$

where $\delta_{l,m}$ is a Kronecker delta.

Guided TM Modes

For TM modes, the magnetic field vector is perpendicular to the plane of incidence (xz plane). The derivation of the confined TM modes is similar in principle to that of the TE modes. The field components are

$$
\begin{aligned}
H_y(x, z, t) &= H_m(x)e^{i(\omega t-\beta z)} \\
E_x(x, z, t) &= \frac{i}{\omega\varepsilon}\frac{\partial H_y}{\partial z} = \frac{\beta}{\omega\varepsilon}H_m(x)e^{i(\omega t-\beta z)} \\
E_z(x, z, t) &= -\frac{i}{\omega\varepsilon}\frac{\partial H_y}{\partial x}
\end{aligned}
\tag{3.2-9}
$$

The function $H_m(x)$ assumes the following forms in each of the three regions:

$$
H_m(x) = \begin{cases} -C\left(\dfrac{h}{\bar{q}}\cos(ht) + \sin(ht)\right)e^{p(x+t)}, & x \le -t \\ C\left(-\dfrac{h}{\bar{q}}\cos(hx) + \sin(hx)\right), & -t \le x \le 0 \\ -\dfrac{h}{\bar{q}}Ce^{-qx}, & 0 \le x \end{cases}
\tag{3.2-10}
$$

where C is a normalization constant, h, q, and p are given by Equation (3.2-4), and $\bar{q}$ is defined as follows.

The continuity of H_y and E_z at the two interfaces leads, in a manner similar to Equation (3.2-5), to the eigenvalue equation

$$
\tan(ht) = \frac{h(\bar{p} + \bar{q})}{(h^2 - \bar{p}\bar{q})}
\tag{3.2-11}
$$

where

$$
\bar{p} \equiv \frac{n_2^2}{n_3^2}p \quad \text{and} \quad \bar{q} \equiv \frac{n_2^2}{n_1^2}q
$$

The normalization constant C is again chosen so that the field represented by Equations (3.2-9) and (3.2-10) carries 1 W of power flow along the z axis per unit width in the y direction. Thus we have

$$
\frac{1}{2}\int_{-\infty}^{\infty} H_y E_x^* \, dx = \frac{\beta_m}{2\omega}\int_{-\infty}^{\infty}\frac{H_m^2(x)}{\varepsilon(x)}dx = 1
$$

or, using $n^2(x) = \varepsilon(x)/\varepsilon_0$,

$$
\int_{-\infty}^{\infty}\frac{|H_m(x)|^2}{n^2(x)}dx = \frac{2\omega\varepsilon_0}{\beta_m}
\tag{3.2-12}
$$

Carrying out the integration using Equation (3.2-10) gives

$$
\begin{aligned}
C_m &= 2\sqrt{\frac{\omega\varepsilon_0}{|\beta_m| t_{\text{eff}}}} \\
t_{\text{eff}} &= \frac{\bar{q}^2 + h^2}{\bar{q}^2}\left(\frac{t}{n_2^2} + \frac{q^2 + h^2}{\bar{q}^2 + h^2}\frac{1}{n_1^2 q} + \frac{p^2 + h^2}{\bar{p}^2 + h^2}\frac{1}{n_3^2 p}\right)
\end{aligned}
\tag{3.2-13}
$$

The TM modes are also mutually orthogonal. In fact, all TE and TM modes are mutually orthogonal. The general orthonormality property will be discussed later in this book.

Figure 3.8 illustrates the dependence of propagation constants β on the waveguide thickness (t/λ) for a waveguide with $n_2 = 2.00$, $n_3 = 1.70$, and $n_1 = 1.0$. According to this figure,

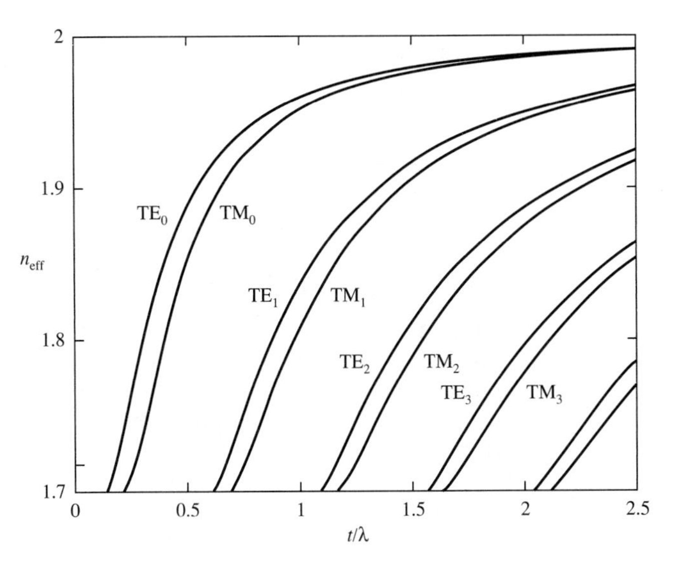

Figure 3.8 Effective index versus thickness/wavelength for the confined modes of an asymmetric waveguide with $n_1 = 1.0$, $n_2 = 2.0$, and $n_3 = 1.7$.

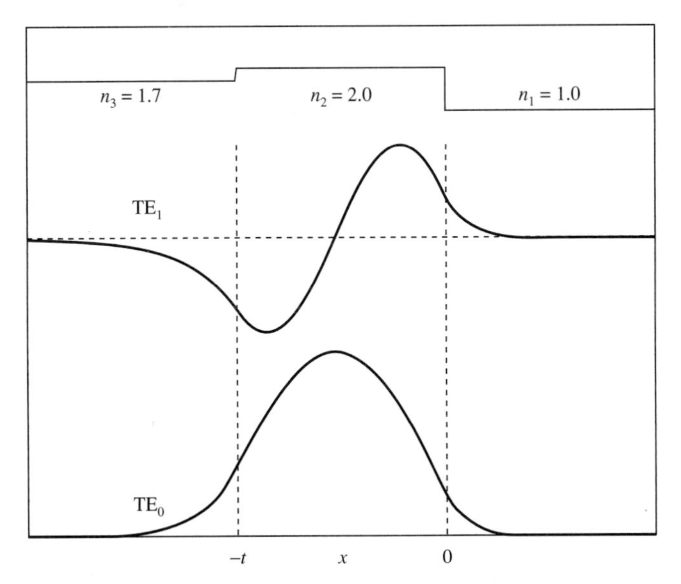

Figure 3.9 Field distribution of the two TE modes of a waveguide with $n_2 = 2.0$, $n_3 = 1.70$, $n_1 = 1.0$, and $t/\lambda = 1.0$. The top curve shows the index profile of the waveguide.

such a waveguide can support two TE and two TM modes when the core thickness is equal to the wavelength (i.e., $t = \lambda$). In fact, at $t/\lambda = 1$, the normalized propagation constants (or effective index) for the TE modes are $n_{\text{eff}} = 1.9594$, 1.8375. Wavefunctions of these two TE modes are plotted in Figure 3.9. We note that the wavefunctions are no longer symmetric. The wavefunctions are sinusoidal in the core and evanescent in the cladding regions.

Generally, a mode becomes confined above a certain (cutoff) value of t/λ. At the cutoff value, $p = 0$ and $\beta = n_3\omega/c$, and the mode extends to $x = -\infty$. According to the mode conditions (3.2-5), (3.2-11), and (3.2-4), cutoff values of t/λ for TE and TM modes are given, respectively, by

$$\left(\frac{t}{\lambda}\right)_{\rm TE} = \frac{1}{2\pi\sqrt{n_2^2 - n_3^2}}\left[m\pi + \tan^{-1}\left(\frac{n_3^2 - n_1^2}{n_2^2 - n_3^2}\right)^{1/2}\right]$$
$$\left(\frac{t}{\lambda}\right)_{\rm TM} = \frac{1}{2\pi\sqrt{n_2^2 - n_3^2}}\left[m\pi + \tan^{-1}\frac{n_2^2}{n_1^2}\left(\frac{n_3^2 - n_1^2}{n_2^2 - n_3^2}\right)^{1/2}\right] \tag{3.2-14}$$

where m is an integer ($m = 0, 1, 2, 3, \ldots$) that denotes the mth confined TE (or TM) mode. Note that the cutoff thickness of the TM_m mode is always larger than that of the TE_m mode because $n_1 < n_2$. For values of t/λ slightly above the cutoff value, $p \geq 0$, and the mode is poorly confined. As the values of t/λ increase, so does the value of p, and the mode becomes increasingly confined to layer 2. This is reflected in the normalized propagation constant or the effective index, $\beta\lambda/2\pi$, which at cutoff is equal to n_3 and for large t/λ approaches n_2. In a symmetric waveguide ($n_1 = n_3$), the lowest order modes TE_0 and TM_0 have no cutoff and are confined for all values of t/λ. The confinement is, however, poor when t/λ becomes small.

The total number of confined modes that can be supported by a waveguide depends on the value of t/λ. To study the number of confined modes, we defined the parameter

$$V \equiv \frac{\pi t}{\lambda}\sqrt{n_2^2 - n_3^2} \tag{3.2-15}$$

Let us now consider what happens about the TE modes in a given waveguide (i.e., fixed n_1, n_2, n_3, and t) as the wavelength of the light decreases gradually, assuming that the medium remains transparent and the indices of refraction n_1, n_2, and n_3 do not vary significantly. Since $\omega/c = 2\pi/\lambda$, the effect of decreasing the wavelength is to increase the value of ω/c. At long wavelengths (low frequencies), such that

$$0 < V < \frac{1}{2}\tan^{-1}\left(\frac{n_3^2 - n_1^2}{n_2^2 - n_3^2}\right)^{1/2} \tag{3.2-16}$$

the value of t/λ is below the cutoff value, and no confined mode exists in the waveguide. As the wavelength decreases such that

$$\frac{1}{2}\tan^{-1}\left(\frac{n_3^2 - n_1^2}{n_2^2 - n_3^2}\right)^{1/2} < V < \frac{\pi}{2} + \frac{1}{2}\tan^{-1}\left(\frac{n_3^2 - n_1^2}{n_2^2 - n_3^2}\right)^{1/2} \tag{3.2-17}$$

one solution exists to the mode condition (3.2-5). The mode is designated as TE_0 and has a transverse h parameter falling within the range

$$0 < ht < \pi$$

so that it has no zero crossings in the interior of the guiding layer ($-t < x < 0$). When the wavelength decreases further such that the parameter falls within the range

$$\frac{\pi}{2} + \frac{1}{2}\tan^{-1}\left(\frac{n_3^2 - n_1^2}{n_2^2 - n_3^2}\right)^{1/2} < V < \pi + \frac{1}{2}\tan^{-1}\left(\frac{n_3^2 - n_1^2}{n_2^2 - n_3^2}\right)^{1/2} \tag{3.2-18}$$

the mode condition (3.2-5) yields two solutions. One corresponds to a value of $ht < \pi$ and is thus that of the lowest order TE_0 mode. In the second mode,

$$\pi < ht < 2\pi \tag{3.2-19}$$

and, consequently, it has one zero crossing (i.e., the point where $E_y = 0$) in the guiding region ($-t < x < 0$). This is the so-called TE_1 mode. Both of these modes correspond to the same frequency and can thus be excited simultaneously by the same input field. We notice, however, that the TE_0 mode has a larger value of p (i.e., $p_0 > p_1$) and is therefore more tightly confined to the guiding slab. It follows from Equation (3.2-4) that $\beta_0 > \beta_1$, so that the phase velocity $v_0 = \omega/\beta_0$ of the TE_0 mode is smaller than that of the TE_1 mode. We can now generalize and state that the mth mode (TE or TM) satisfies

$$(m-1)\pi < ht < m\pi \tag{3.2-20}$$

and has $m - 1$ zero crossings in the guiding layer ($-t < x < 0$). The general features of TM modes are similar to those of TE modes, except that the corresponding values of p are somewhat smaller, indicating a lesser degree of confinement. A larger fraction of the total TM mode power thus propagates in the outer medium compared to a TE mode of the same order. This point is taken up in Problem 3.2.

Geometric Optics Treatment

In a manner similar to that leading to Equation (3.1-29), geometric optics can also be used to obtain the mode condition. Following the procedure described in Section 3.1, we obtain

$$-2ht + \phi_{21} + \phi_{23} = -2m\pi \tag{3.2-21}$$

where m is an integer, ϕ_{21} is the phase shift upon total reflection from the interface $x = 0$, and ϕ_{23} is the phase shift upon total reflection from the interface $x = -t$. These phase shifts can be conveniently written in terms of p, q, and h as [1, 2]

$$\phi_{21} = \begin{cases} 2\tan^{-1}\left(\dfrac{q}{h}\right), & \text{TE} \\[2ex] 2\tan^{-1}\left(\dfrac{n_2^2 q}{n_1^2 h}\right), & \text{TM} \end{cases} \tag{3.2-22}$$

$$\phi_{23} = \begin{cases} 2\tan^{-1}\left(\dfrac{p}{h}\right), & \text{TE} \\[2ex] 2\tan^{-1}\left(\dfrac{n_2^2 p}{n_3^2 h}\right), & \text{TM} \end{cases} \tag{3.2-23}$$

The phase shifts in the preceding equations are again limited between 0 and π. Therefore the fundamental modes correspond to the situation when $m = 0$. Substitution of Equations (3.2-22) and (3.2-23) into Equation (3.2-21) will lead to the mode condition in Equations (3.2-5) and (3.2-11).

Mode Confinement Factor

As shown in Figures 3.3 and 3.9, the electric field and magnetic field of confined modes are finite in the cladding regions. Although the power associated with the mode is propagating along the z direction, part of the electromagnetic field energy is actually propagating outside the guiding layer. Let Γ_i be the fraction of power flowing in medium i ($i = 1, 2, 3$) of the slab waveguide. By definition, we have

$$\Gamma_i = \frac{\mathrm{Re}\int_i (\mathbf{E} \times \mathbf{H}^*) \cdot \mathbf{z}\, dx}{\mathrm{Re}\int_{-\infty}^{\infty} (\mathbf{E} \times \mathbf{H}^*) \cdot \mathbf{z}\, dx}, \quad i = 1, 2, 3 \tag{3.2-24}$$

where the integral in the numerator is carried out over the ith layer.

Γ_2 is the fraction of mode power flowing in the guiding layer n_2 and is often called the mode confinement factor. The mode functions derived earlier in this section can be employed for calculation of the confinement factor. This is left for the student as an exercise (see Problem 3.2).

3.3 STEP-INDEX CIRCULAR DIELECTRIC WAVEGUIDES (LINEARLY POLARIZED MODES IN OPTICAL FIBERS)

Confined propagation can also occur in circular dielectric waveguides (e.g., silica glass fiber). By virtue of its transmission capacity, silica glass fiber has become the most important medium for long-distance, high-data-rate optical communications. It has caused what can be called, with very little exaggeration, a revolution in the art and practice of communication. This success is due mostly to the prediction [3] and realization [4] of low-loss propagation of confined optical modes in such fibers once the concentration of the absorbing impurities has been reduced to insignificance. The technology of optical communications in fibers can be stated, in simple terms, as that of feeding optical pulses at a maximal rate into one end of a fiber and retrieving them at the other end. The length of the fiber may vary from a few meters in the case of computer interconnect applications, to thousands of kilometers, in the case of transoceanic submarine cables. The main goal of a communication system is to receive the pulses at the output end with minimal loss of energy, minimal spread of pulses, and minimal contamination by noise. Nature, naturally, throws up many obstacles in the path of anyone trying to achieve these modest goals. Some of nature's tricks include diffraction; group velocity dispersion, which causes pulse spreading; and a variety of nonlinear scattering mechanisms. Much of this book is devoted to the understanding of these phenomena and, using this knowledge, to devising strategies for optimal transmission. A good understanding of the propagation of optical waves in fibers is very important. This section is devoted to the confined modes of propagation in fibers with a circular cross section. We will also investigate the problem of group velocity dispersion and propagation attenuation. The topics of detection and noise will be taken up in Chapters 10 and 11.

Referring to Figure 3.10, we consider a circular dielectric waveguide with a step-index profile,

$$n(r) = \begin{cases} n_1, & 0 < r < a \\ n_2, & a < r < b \end{cases} \tag{3.3-1a}$$

where a is the core radius and b is the clad radius.

It consists of a core of refractive index n_1 and radius a, and a cladding of refractive index n_2 and radius b. The radius b of the cladding is usually chosen to be large enough so that the

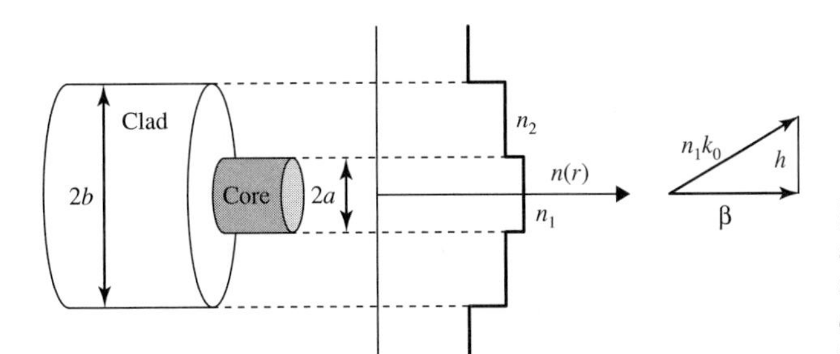

Figure 3.10 Schematic drawing of a circular dielectric waveguide with a clad radius b and a core radius $a \ll b$. $n(r)$ is the index profile of the fiber. $k_0 = \omega/c$.

field of confined modes is virtually zero at $r = b$. A typical optical fiber that supports a single mode of propagation at $\lambda = 1.55$ μm has a core radius of around 5 μm and a clad radius of greater than 100 μm. In the calculation below we will put $b = \infty$ for mathematical simplicity; this is a legitimate assumption in most waveguides, as far as confined modes are concerned.

Unlike the modes of propagation in dielectric slab waveguides, the modes of propagation in circular waveguides are in general a mixture of TE and TM waves. The exact solutions of these hybrid modes (*EH* and *HE* modes) require solving the wave equations in cylindrical coordinates and are very complicated. They are given in Appendixes A and B. In what follows, we will describe approximate solutions of the circular waveguides. A good approximation of the field components and mode condition can be obtained in most fibers whose core refractive index is only slightly higher than that of the cladding medium. As a matter of fact, the index difference $(n_1 - n_2)$ is of the order of 10^{-3} in most single-mode fibers. Assuming that

$$n_1 - n_2 \ll 1 \tag{3.3-1b}$$

it can be shown that the continuity condition on the tangential components of **H** at the interface $(r = a)$ between the core (n_1) and the clad (n_2) becomes identical to that of the tangential components of the field vector **E**. This leads to a tremendous simplification in matching the field components at the core–clad interface. Thus we may use Cartesian components of the field vectors without introducing much complexity in solving the wave equation. This simplified solution of the linearly polarized modes for the circular fibers using the assumption (3.3-1b) is due to Gloge [5]. In the limit (3.3-1b), all the transverse wavenumbers (h, q) of the confined modes $(n_2k_0 < \beta < n_1k_0)$ are much smaller compared to the propagation constant β; that is,

$$q, h \ll \beta \tag{3.3-2a}$$

where q and h are given by

$$h = \sqrt{n_1^2k_0^2 - \beta^2} \quad \text{and} \quad q = \sqrt{\beta^2 - n_2^2k_0^2} \tag{3.3-2b}$$

with k_0 being the wavenumber of light in vacuum $(k_0 = \omega/c)$.

We now start by solving the wave equation for the transverse Cartesian field components E_x, E_y, H_x, and H_y. These field components also satisfy the wave equations (A-1). For a step-index circular dielectric waveguide, the general solutions are Bessel functions as given in Appendix A. The tangential components (E_z, E_ϕ, H_z, and H_ϕ) must be continuous at the core–clad boundary. We now look for solutions where either the x or y component of the electric field vanishes. In other words, we look for either x-polarized or y-polarized modes of propagation. Since E_ϕ can be expressed in terms of E_x and E_y as

$$E_\phi = -E_x \sin\phi + E_y \cos\phi \tag{3.3-3}$$

it is apparent that the E_ϕ component is simply proportional to either E_x or E_y. Thus the continuity of E_ϕ becomes equivalent to the continuity of E_x or E_y in these new solutions. Take the E field of a y-polarized solution of the form

$$E_x = 0 \tag{3.3-4}$$

$$E_y = \begin{cases} AJ_l(hr)e^{il\phi}\exp[i(\omega t - \beta z)], & r < a \\ BK_l(qr)e^{il\phi}\exp[i(\omega t - \beta z)], & r > a \end{cases} \tag{3.3-5}$$

where A and B are constants, and J_l and K_l are Bessel functions. For confined modes $(n_2k_0 < \beta < n_1k_0)$, the electric wavefunction is oscillatory inside the core and evanescent in the cladding region. The Bessel functions $K_l(qr)$ decay exponentially in the cladding region along the radial direction. We assume that $E_z \ll E_y$. The magnetic field components are then given, according to Appendix A, by

$$\begin{aligned} H_x &= \frac{-i}{\omega\mu}\frac{\partial}{\partial z}E_y = \frac{-\beta}{\omega\mu}E_y \\ H_y &\approx 0 \\ H_z &= \frac{i}{\omega\mu}\frac{\partial}{\partial x}E_y \end{aligned} \tag{3.3-6}$$

The longitudinal component of the electric field vector $\mathbf{E}$ is related to H_x according to the Maxwell equation $\nabla \times \mathbf{H} = \varepsilon\, \partial\mathbf{E}/\partial t$:

$$E_z = \frac{i}{\omega\varepsilon}\frac{\partial}{\partial y}H_x = \frac{-i\beta}{\omega^2\mu\varepsilon}\frac{\partial}{\partial y}E_y \tag{3.3-7}$$

where we used Equation (3.3-6) in arriving at the last equation. We note that the field components E_x and H_y are zero in this solution. The other four field components can be expressed in terms of E_y. In order to calculate H_z and E_z, we need to carry out the differentiation with respect to x and y, respectively, according to Equations (3.3-6) and (3.3-7). Since E_y is of the form (3.3-5), we need the relations

$$\frac{\partial}{\partial x} = \frac{\partial r}{\partial x}\frac{\partial}{\partial r} + \frac{\partial\phi}{\partial x}\frac{\partial}{\partial\phi} \tag{3.3-8}$$

and

$$\frac{\partial}{\partial y} = \frac{\partial r}{\partial y}\frac{\partial}{\partial r} + \frac{\partial\phi}{\partial y}\frac{\partial}{\partial\phi} \tag{3.3-9}$$

By using the definition of r and ϕ

$$r = (x^2 + y^2)^{1/2} \tag{3.3-10}$$

$$\phi = \tan^{-1}\left(\frac{y}{x}\right) \tag{3.3-11}$$

we obtain

$$\frac{\partial r}{\partial x} = \frac{x}{r} = \cos\phi \tag{3.3-12}$$

$$\frac{\partial r}{\partial y} = \frac{y}{r} = \sin\phi \tag{3.3-13}$$

$$\frac{\partial \phi}{\partial x} = -\frac{y}{r^2} = -\frac{1}{r}\sin\phi \tag{3.3-14}$$

and

$$\frac{\partial \phi}{\partial y} = \frac{x}{r^2} = \frac{1}{r}\cos\phi \tag{3.3-15}$$

We now substitute Equation (3.3-5) for E_y in Equations (3.3-6) and (3.3-7) and carry out the differentiation, using Equations (3.3-8)–(3.3-15). After some laborious algebra and using the following functional relations of the Bessel function,

$$\begin{aligned} J_l'(x) &= \tfrac{1}{2}[J_{l-1}(x) - J_{l+1}(x)] \\ K_l'(x) &= -\tfrac{1}{2}[K_{l-1}(x) + K_{l+1}(x)] \end{aligned} \tag{3.3-16}$$

$$\begin{aligned} \frac{l}{x}J_l(x) &= \tfrac{1}{2}[J_{l-1}(x) + J_{l+1}(x)] \\ \frac{l}{x}K_l(x) &= -\tfrac{1}{2}[K_{l-1}(x) - K_{l+1}(x)] \end{aligned} \tag{3.3-17}$$

we obtain the following expressions for the field components:

Core ($r < a$):

$$\begin{aligned} E_x &= 0 \\ E_y &= AJ_l(hr)e^{il\phi}\exp[i(\omega t - \beta z)] \\ E_z &= \frac{h}{\beta}\frac{A}{2}[J_{l+1}(hr)e^{i(l+1)\phi} + J_{l-1}(hr)e^{i(l-1)\phi}]\exp[i(\omega t - \beta z)] \\ H_x &= -\frac{\beta}{\omega\mu}AJ_l(hr)e^{il\phi}\exp[i(\omega t - \beta z)] \\ H_y &\cong 0 \\ H_z &= -\frac{ih}{\omega\mu}\frac{A}{2}[J_{l+1}(hr)e^{i(l+1)\phi} - J_{l-1}(hr)e^{i(l-1)\phi}]\exp[i(\omega t - \beta z)] \end{aligned} \tag{3.3-18}$$

Cladding ($r > a$):

$$\begin{aligned} E_x &= 0 \\ E_y &= BK_l(qr)e^{il\phi}\exp[i(\omega t - \beta z)] \\ E_z &= \frac{q}{\beta}\frac{B}{2}[K_{l+1}(qr)e^{i(l+1)\phi} - K_{l-1}(qr)e^{i(l-1)\phi}]\exp[i(\omega t - \beta z)] \\ H_x &= -\frac{\beta}{\omega\mu}BK_l(qr)e^{il\phi}\exp[i(\omega t - \beta z)] \\ H_y &\cong 0 \\ H_z &= -\frac{iq}{\omega\mu}\frac{B}{2}[K_{l+1}(qr)e^{i(l+1)\phi} + K_{l-1}(qr)e^{i(l-1)\phi}]\exp[i(\omega t - \beta z)] \end{aligned} \tag{3.3-19}$$

In arriving at Equations (3.3-18) and (3.3-19), we have also used $\beta = n_1 k_0 \sim n_2 k_0$, since $n_2 k_0 < \beta < n_1 k_0$ and $n_1 - n_2 \ll 1$. Note that E_y and H_x are the dominant field components because, in the limit (3.3-1b), $h, q \ll \beta$. In other words, the field is essentially transverse. The constant B is given by

$$B = \frac{AJ_l(ha)}{K_l(qa)} \tag{3.3-20}$$

to ensure the continuity of E_y ($E_\phi \propto E_y$) at the core boundary $r = a$. The constant A is then determined by the normalization condition.

The field solution (3.3-18) and (3.3-19) is a y-polarized wave ($E_x = 0$). For a complete field description, we also need the mode with the orthogonal polarization (i.e., an x-polarized wave). The field components E_x and E_y of this orthogonal mode are taken as

$$E_x = \begin{cases} AJ_l(hr)e^{il\phi} \exp[i(\omega t - \beta z)], & r < a \\ BK_l(qr)e^{il\phi} \exp[i(\omega t - \beta z)], & r > a \end{cases} \tag{3.3-21}$$

$$E_y = 0 \tag{3.3-22}$$

and the other field components are, according to Maxwell's equations,

$$\begin{aligned} E_z &= \frac{-i}{\omega\varepsilon}\frac{\partial}{\partial x}H_y = \frac{-i\beta}{\omega^2\mu\varepsilon}\frac{\partial}{\partial x}E_x \\ H_x &\approx 0 \\ H_y &= \frac{i}{\omega\mu}\frac{\partial}{\partial z}E_x = \frac{\beta}{\omega\mu}E_x \\ H_z &= \frac{-i}{\omega\mu}\frac{\partial}{\partial y}E_x \end{aligned} \tag{3.3-23}$$

where we have assumed that $E_z \ll E_x$. We note that $E_y = 0$ and $H_x \sim 0$ in this solution. By substituting Equation (3.3-21) for E_x in Equation (3.3-23) and carrying out the differentiation, using Equations (3.3-8)–(3.3-15), we obtain, again after some laborious algebra and using the relations (3.3-16) and (3.3-17), the following expressions for the field amplitudes:

Core ($r < a$):

$$\begin{aligned} E_x &= AJ_l(hr)e^{il\phi} \exp[i(\omega t - \beta z)] \\ E_y &= 0 \\ E_z &= i\frac{h}{\beta}\frac{A}{2}[J_{l+1}(hr)e^{i(l+1)\phi} - J_{l-1}(hr)e^{i(l-1)\phi}] \exp[i(\omega t - \beta z)] \\ H_x &\cong 0 \\ H_y &= \frac{\beta}{\omega\mu}AJ_l(hr)e^{il\phi} \exp[i(\omega t - \beta z)] \\ H_z &= \frac{h}{\omega\mu}\frac{A}{2}[J_{l+1}(hr)e^{i(l+1)\phi} + J_{l-1}(hr)e^{i(l-1)\phi}] \exp[i(\omega t - \beta z)] \end{aligned} \tag{3.3-24}$$

Cladding ($r > a$):

$$
\begin{aligned}
E_x &= BK_l(qr)e^{il\phi} \exp[i(\omega t - \beta z)] \\
E_y &= 0 \\
E_z &= i\frac{q}{\beta}\frac{B}{2}[K_{l+1}(qr)e^{i(l+1)\phi} + K_{l-1}(qr)e^{i(l-1)\phi}] \exp[i(\omega t - \beta z)] \\
H_x &\cong 0 \\
H_y &= \frac{\beta}{\omega\mu} BK_l(qr)e^{il\phi} \exp[i(\omega t - \beta z)] \\
H_z &= \frac{q}{\omega\mu}\frac{B}{2}[K_{l+1}(qr)e^{i(l+1)\phi} - K_{l-1}(qr)e^{i(l-1)\phi}] \exp[i(\omega t - \beta z)]
\end{aligned}
\tag{3.3-25}
$$

In arriving at Equations (3.3-24) and (3.3-25), we again made the assumption that $\beta \cong n_1 k_0 \cong n_2 k_0$ because of Equation (3.3-1b). We note that E_x and H_y are again the dominant field components in this solution. Therefore the mode is again nearly transverse and linearly polarized along the x direction. The constant B is again given by Equation (3.3-20) to ensure the continuity of E_x ($E_\phi \propto E_x$) at the core boundary $r = a$.

We have obtained the field expressions for two types of guided modes whose transverse fields are polarized orthogonally to each other. Based on the circular symmetry of the fiber, we can be sure that these two guided modes must have the same propagation constant and the same intensity and power distribution. The field expressions for the linearly polarized modes are solutions of Maxwell's equations, provided the tangential components of the field vectors are continuous at the dielectric interface $r = a$. The continuity of E_ϕ at $r = a$ leads to Equation (3.3-20). The H_ϕ components are proportional to the E_ϕ components, according to the field expressions (3.3-18), (3.3-19), (3.3-24), and (3.3-25) in this approximation. Therefore the continuity of E_ϕ results in the continuity of H_ϕ. We now consider the continuity of E_z at $r = a$. Since the continuity condition must hold for all azimuth angles ϕ, we must equate the coefficients of $\exp[i(l+1)\phi]$ and $\exp[i(l-1)\phi]$ separately. Using the field expressions (3.3-18) and (3.3-19) and (3.3-20), we obtain the following mode conditions:

$$
ha\frac{J_{l+1}(ha)}{J_l(ha)} = qa\frac{K_{l+1}(qa)}{K_l(qa)} \tag{3.3-26}
$$

and

$$
ha\frac{J_{l-1}(ha)}{J_l(ha)} = -qa\frac{K_{l-1}(qa)}{K_l(qa)} \tag{3.3-27}
$$

The same equations result from the continuity of H_z. In addition, if we use the field expressions (3.3-24) and (3.3-25) for the x-polarized mode, we will arrive at the same mode conditions (3.3-26) and (3.3-27). This means that these two transversely orthogonal modes are degenerate in the propagation constant β. The mode condition (3.3-27) is mathematically equivalent to condition (3.3-26) if we use the recurrence relation of the Bessel functions (3.3-17).

The mode condition (3.3-26) obtained in this approximation is much simpler than the exact expression for *HE* and *EH* modes in Appendix B. The exact mode condition (B-11) has twice as many solutions as the simple one (3.3-26) because condition (B-11) is quadratic $J_l'(ha)/J_l(ha)$. This indicates that each solution of Equation (3.3-26) is really twofold degenerate. In fact, the propagation constants of the exact $HE_{l+1,m}$ and $EH_{l-1,m}$ modes are nearly degenerate [6]. They become exactly the same in the limit $n_2 \to n_1$. This can also be seen from the expressions of the field components E_z and H_z in Equations (3.3-18), (3.3-19), (3.3-24), and (3.3-25). Comparison of the linearly polarized mode expressions with the exact modes (B-6)–(B-9) shows that the linearly polarized modes are actually a superposition of $HE_{l+1,m}$ and $EH_{l-1,m}$ modes [6]. Two independent linear superpositions lead to the x-polarized and y-polarized modes. The total number of modes is the same in both theories. The eigenvalues obtained from Equation (3.3-26) are labeled β_{lm} with $l = 0, 1, 2, 3, \ldots$ and $m = 1, 2, 3, \ldots$, where the subscript m indicates the mth root of the transcendental Equation (3.3-26). The modes are designated LP_{lm}. The lowest order mode is LP_{01} with a propagation constant labeled β_{01}. This mode corresponds to the HE_{11} mode of the exact solutions. It is important to note that LP_{x01} and LP_{y01} are degenerate and have the same propagation constant β_{01}. This is consistent with the circular symmetry of the waveguide. In practical fibers, the cores are not exactly circular due to production imperfection. Such an asymmetry breaks the degeneracy and leads to $\beta_{x01} \neq \beta_{y01}$. In other words, the x-polarized LP mode and the y-polarized LP mode are propagating at a slightly different propagation constant and thus a slightly different group velocity. This leads to a phenomenon known as polarization mode dispersion (PMD), which will be described in Chapter 7.

The mode conditions for those linearly polarized waves (3.3-26) or (3.3-27) can also be solved numerically. Here we examine the case of $l = 0$. For the purpose of discussion we define an important parameter

$$V = k_0 a(n_1^2 - n_2^2)^{1/2} = \frac{2\pi a}{\lambda}(n_1^2 - n_2^2)^{1/2} = \sqrt{(ha)^2 + (qa)^2} \tag{3.3-28}$$

where a is the core radius. As we will see later, this parameter determines how many confined modes can be supported by the step-index circular fiber. If we replace qa in Equation (3.3-26) with $\sqrt{V^2 - (ha)^2}$, we can then plot both sides of Equation (3.3-26), for a given V, as functions of ha. The intersections are the solutions. Figure 3.11 shows such a graphical

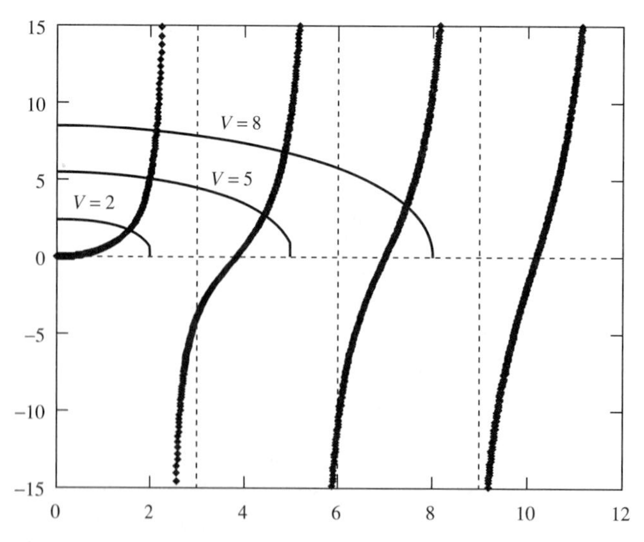

Figure 3.11 Graphical method of determining ha of LP modes with $l = 0$. The left side of Equation (3.3-26) is plotted as the dotted lines. The right side is plotted for $V = 2$, 5, and 8. For $V = 2$, only one solution exists (LP_{01} mode). For $V = 8$, there are three solutions: LP_{01}, LP_{02}, and LP_{03} modes.

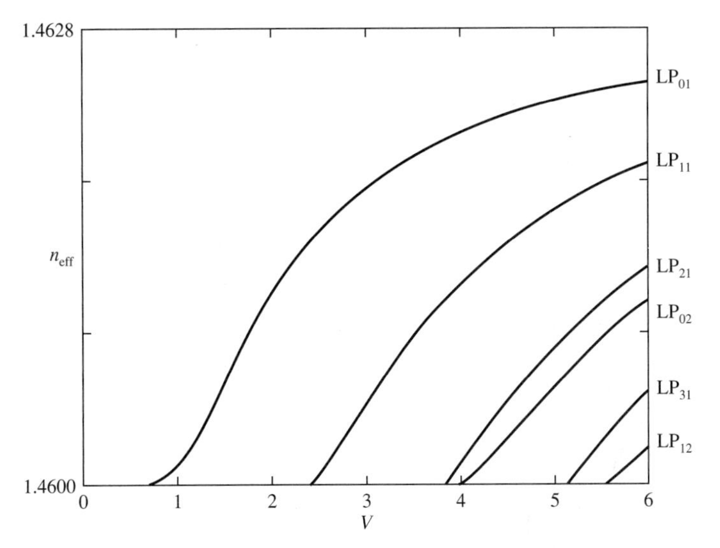

Figure 3.12 Normalized propagation constant n_{eff} as function of normalized frequency V for some of the guided modes of the optical fiber, $n_{eff} = \beta/k_0$. The fiber parameters are $n_1 = 1.4628$, $n_2 = 1.4600$, and $a = 4.7$ μm.

method of determining the solution for $l = 0$ for the cases of $V = 2$, 5, and 8. For the left side, the denominator vanishes at roots of $J_0(ha)$. Thus discontinuity at $\pm\infty$ occurs at $ha = 2.405$, 5.520, 8.654, ... The numerator vanishes at roots of $J_1(ha)$, which occur at $ha = 0$, 3.832, 7.016, ... So, near $ha = 0$, the left side is a positive increasing function of ha starting from zero and reaching $+\infty$ at $ha = 2.405$. For the right side, we note that both $K_0(qa)$ and $K_1(qa)$ are positive functions. Near $ha = 0$, the right side starts from a positive value of $VK_1(V)/K_0(V)$ and decreases as a function of ha, reaching zero at $ha = V$. As a result, there is at least one intersection even if V is very small. As V increases, the number of intersections increases.

Figure 3.12 shows the normalized propagation constant (or effective index) of several modes as a function of the normalized frequency V. We note that for $V < 2.405$, there is only one mode (LP_{01}). For $V = 6$, there are six modes: LP_{01}, LP_{02}, LP_{11}, LP_{12}, LP_{21}, and LP_{31}.

We note that the LP_{01} mode always exists regardless of the fiber parameter V. As V increases beyond $V = 2.405$, the LP_{11} mode starts to appear. Thus we call $V = 2.405$ the cutoff value for the LP_{11} mode. As V increases beyond $V = 3.832$, LP_{21} and LP_{02} modes start to appear. Thus we call $V = 3.832$ the cutoff value for LP_{21} and LP_{02} modes. In general, the mode cutoff corresponds to the condition $q = 0$, which, according to Equation (3.3-27), leads to the condition

$$J_{l-1}(V) = 0 \tag{3.3-29}$$

where $V = k_0 a(n_1^2 - n_2^2)^{1/2} = 2\pi a(n_1^2 - n_2^2)^{1/2}/\lambda$ is the fiber parameter defined in Equation (3.3-28). It follows that the lowest order mode, characterized by $l = 0$, has a cutoff given by the lowest root of the equation

$$J_{-1}(V) = -J_1(V) = 0 \tag{3.3-30}$$

Hence $V = 0$. In other words, the lowest order mode does not have a cutoff. This is the HE_{11} mode and is now labeled LP_{01}. The next mode of the type $l = 0$ cuts off when $J_1(V)$ next equals zero, that is, when $V \sim 3.832$. This mode is labeled LP_{02}. The cutoff values of V for some low-order LP_{lm} modes are given in Table 3.1.

TABLE 3.1 Cutoff Values of V for Some Low-Order LP Modes

V_{cutoff}	$m = 1$	$m = 2$	$m = 3$	$m = 4$
$l = 0$	0	3.832	7.016	10.173
$l = 1$	2.405	5.520	8.654	11.792
$l = 2$	3.832	7.016	10.173	13.323
$l = 3$	5.136	8.417	11.620	14.796
$l = 4$	6.379	9.760	13.017	16.224

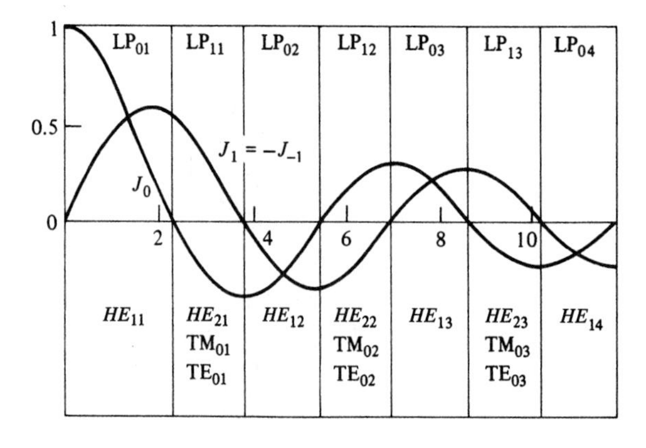

Figure 3.13 The regions of the parameter V for modes of order $l = 0$, 1.

All these values are zeros of the Bessel function. For high-order modes, the cutoff value of V is given approximately according to Equations (3.3-29) and (A-13)

$$V(\mathrm{LP}_{lm}) \cong m\pi + \left(l - \frac{3}{2}\right)\frac{\pi}{2} \tag{3.3-31}$$

Figure 3.13 shows the regions in which a given mode is the highest one allowed for a given l value group, labeled in LP mode designation. Also shown in the figure are the associated HE, EH, TE, and TM mode notations that are the exact modes. Figure 3.14 shows the field distribution of the LP_{11} modes [6]. We note that the intensity distribution consists of two lobes inside the core. This is distinctly different from that of the fundamental mode (the LP_{01} mode), which has a radially symmetric field distribution $J_0(hr)$ in the core.

One of the most important advantages of using the linearly polarized mode is that the modes are almost transversely polarized and are dominated by one transverse electric field component (E_x or E_y) and one transverse magnetic field component (H_y or H_x). The **E** vector can be chosen to be along any arbitrary radial direction with the **H** vector along a perpendicular radial direction. Once this mode is chosen, there exists another independent mode with **E** and **H** orthogonal to the first pair.

Power Flow and Power Density

We now derive expressions for the Poynting vector and the power flow in the core and cladding. The time-averaged Poynting vector along the waveguide is, acccording to Equation (1.3-17)

$$S_z = \tfrac{1}{2}\mathrm{Re}[E_x H_y^* - E_y H_x^*] \tag{3.3-32}$$

Substituting the field components from Equations (3.3-18) and (3.3-19) or (3.3-24) and (3.3-25) into Equation (3.3-32), we obtain

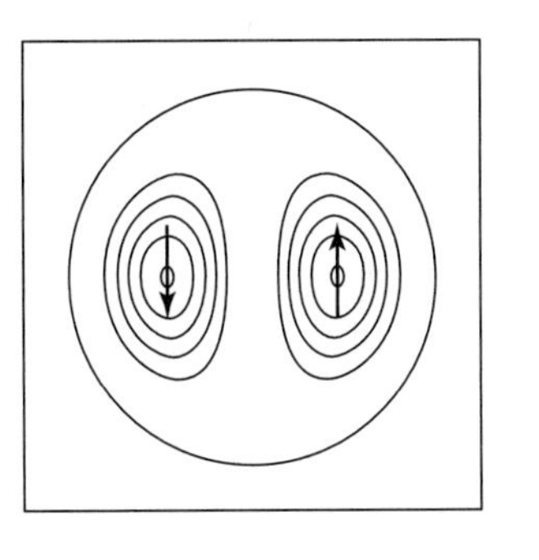
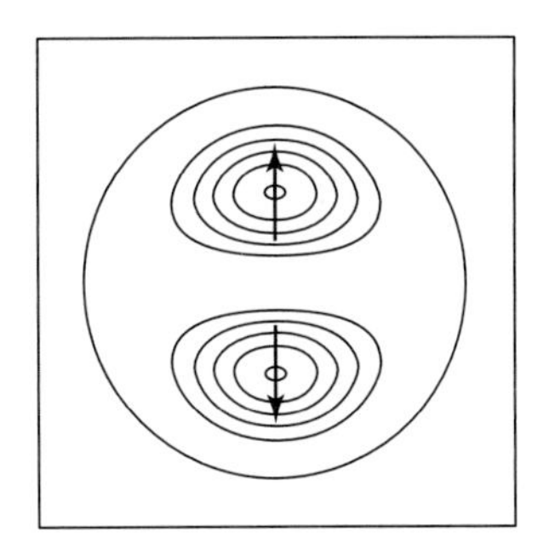
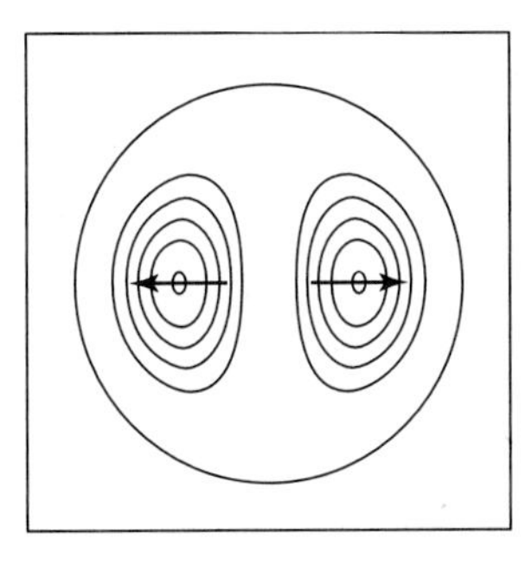
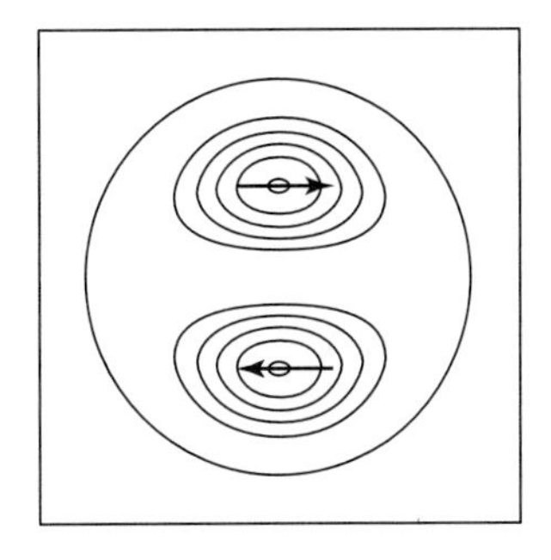

Figure 3.14 Sketch of the four possible field distributions of LP_{11} modes inside the core of the fiber. The arrows indicate the polarization of the electric field vector.

$$S_z = \begin{cases} \dfrac{\beta}{2\omega\mu}|A|^2 J_l^2(hr), & r < a \\ \dfrac{\beta}{2\omega\mu}|B|^2 K_l^2(qr), & r > a \end{cases} \tag{3.3-33}$$

Note that the intensity distribution is cylindrically symmetric (i.e., no ϕ dependence). It is important to note that a combination of the degenerate modes can lead to intensity distribution with azimuth (ϕ) dependence. For example, a linear combination of LP_{11x} mode with $\exp(i\phi)$ and $\exp(-i\phi)$ can lead to an intensity distribution of $\cos^2\phi$ or $\sin^2\phi$ (see Figure 3.14).

The amount of power that is contained in the core and the cladding is given, respectively, by

$$P_{\text{core}} = \int_0^{2\pi}\int_0^a S_z r \, dr \, d\phi \tag{3.3-34}$$

$$P_{\text{clad}} = \int_0^{2\pi}\int_a^\infty S_z r \, dr \, d\phi \tag{3.3-35}$$

Using the following integrals of Bessel functions [7] (see also Problem 3.15),

$$\int_0^a r J_l^2(hr)\, dr = \frac{a^2}{2}[J_l^2(ha) - J_{l-1}(ha)J_{l+1}(ha)] \tag{3.3-36a}$$

$$\int_a^\infty r K_l^2(hr)\, dr = \frac{a^2}{2}[-K_l^2(qa) + K_{l-1}(qa)K_{l+1}(qa)] \tag{3.3-36b}$$

the powers P_{core} and P_{clad} can be written, respectively,

$$P_{\text{core}} = \frac{\beta}{2\omega\mu}\pi a^2|A|^2[J_l^2(ha) - J_{l-1}(ha)J_{l+1}(ha)] \tag{3.3-37}$$

$$P_{\text{clad}} = \frac{\beta}{2\omega\mu}\pi a^2|B|^2[-K_l^2(qa) + K_{l-1}(qa)K_{l+1}(qa)] \tag{3.3-38}$$

By using Equation (3.3-20) for B and the mode conditions (3.3-26) and (3.3-27), the power P_{clad} can be written

$$P_{\text{clad}} = \frac{\beta}{2\omega\mu}\pi a^2|A|^2\left[-J_l^2(ha) - \left(\frac{h}{q}\right)^2 J_{l-1}(ha)J_{l+1}(ha)\right] \tag{3.3-39}$$

For those ha values that are allowed by the mode condition (3.3-26) or (3.3-27), $J_{l-1}(ha)J_{l+1}(ha)$ is always negative, so that P_{core} is always positive. The negativeness of $J_{l-1}(ha)J_{l+1}(ha)$ can be seen from Equations (3.3-26) and (3.3-27), since the $K_l(qa)$ terms are always positive. According to Equations (3.3-37) and (3.3-39), the total power flow is thus given by

$$P = \frac{\beta}{2\omega\mu}\pi a^2|A|^2\left(1 + \frac{h^2}{q^2}\right)[-J_{l-1}(ha)J_{l+1}(ha)] \tag{3.3-40}$$

The ratio of cladding power to the total power, $\Gamma_2 = (P_{\text{clad}}/P)$, which measures the fraction of mode power flowing in the cladding layer, is given, according to Equations (3.3-39) and (3.3-40), by

$$\Gamma_2 = \frac{P_{\text{clad}}}{P} = \frac{1}{V^2}\left((ha)^2 + \frac{(qa)^2 J_l^2(ha)}{J_{l-1}(ha)J_{l+1}(ha)}\right) \tag{3.3-41}$$

where we used $(ha)^2 + (qa)^2 = k_0^2a^2(n_1^2 - n_2^2) = V^2$. The confinement factor Γ_1 is given by

$$\Gamma_1 = \frac{P_{\text{core}}}{P} = 1 - \Gamma_2 = \frac{1}{V^2}\left((qa)^2 - \frac{(qa)^2 J_l^2(ha)}{J_{l-1}(ha)J_{l+1}(ha)}\right) \tag{3.3-42}$$

Figure 3.15 shows the confinement factor $\Gamma_1 = P_{\text{core}}/P$ for several modes as a function of the normalized frequency V. Note that the fundamental mode LP_{01} is best confined. Generally, the confinement factor for a given mode $\Gamma_1 = P_{\text{core}}/P$ increases with V.

Generally, the confinement factor for confined modes is zero at cutoff ($qa = 0$). We have seen this in previous sections for dielectric slab waveguides. This is also true for LP modes with $l = 0$ and 1. For high-order LP modes with $l > 1$, the confinement factor at cutoff is obtained by taking the limit $qa \to 0$. This leads to the following limiting value of confinement factor at cutoff:

$$\Gamma_1(qa \to 0) = 1 - \frac{1}{l} \quad (\text{for } l > 1 \text{ at cutoff}) \tag{3.3-43}$$

For example, the confinement factor of the LP_{31} mode at cutoff is $\frac{2}{3}$. Further discussion of the confinement factor for high-order LP modes can be found in Problem 3.11.

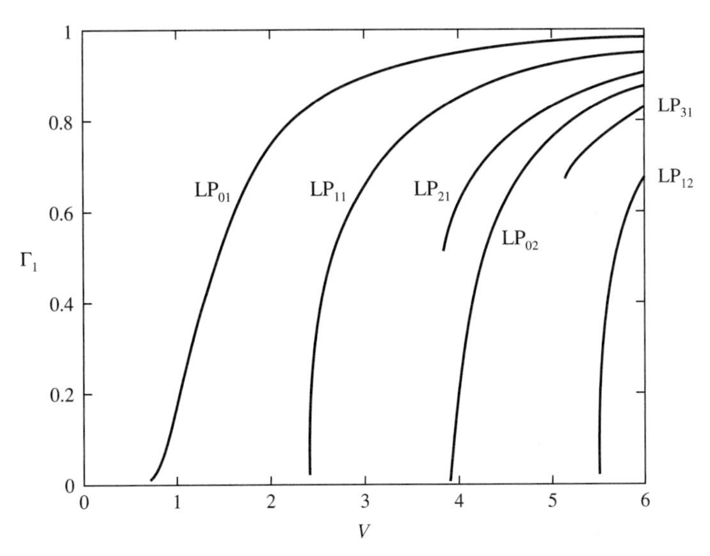

Figure 3.15 Confinement factor Γ_1 versus V. Note that, at cutoff, the confinement factor is zero for LP modes with $l = 0, 1$; whereas the confinement factor is $\frac{1}{2}$ for $l = 2$; $\frac{2}{3}$ for $l = 3$; $\frac{3}{4}$ for $l = 4$; and so on.

3.4 EFFECTIVE INDEX THEORY

Although simple layered structures such as dielectric slabs can be used for waveguiding purposes, the confinement of energy is only limited to one dimension. In practice, more complicated waveguide structures are used. For example, the waveguides used in integrated optics or guided-wave optics (also known as planar light circuits (PLCs)) are usually two-dimensional waveguides (e.g., channel waveguides, ridge waveguides). Figure 3.16 shows examples of such two-dimensional waveguides. Exact analytical treatment of these waveguide structures is not possible, except for some special cases. Although numerical solutions can be obtained by various methods, there are several approximate analytical approaches. Here, we will introduce one of the simplest approaches, the *effective index theory*.

Referring to Figure 3.16c, we consider the guiding of electromagnetic radiation in a ridge waveguide. The rectangular waveguide shown in Figure 3.16b can be considered as a special case of this ridge waveguide by taking $d = 0$, where d is the thickness of the guiding layer on both sides of the ridge. The thickness of the guiding layer at the ridge is t, which is chosen to be greater than d because we are interested in the confinement of electromagnetic radiation at

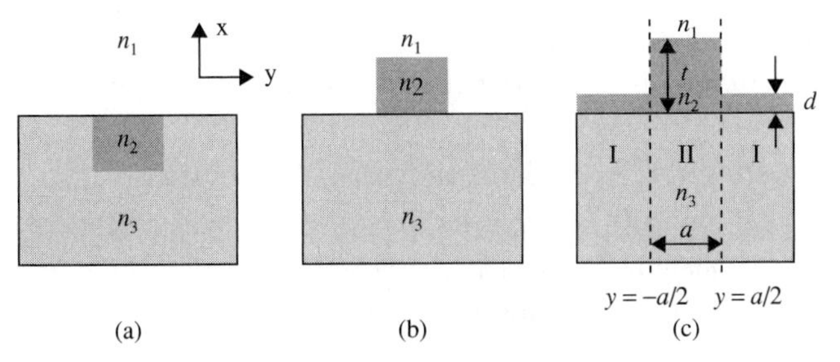

Figure 3.16 Schematic drawing of several two-dimensional waveguides: (a) a rectangular strip of dielectric medium embedded in another dielectric medium of lower index of refraction; (b) ridge waveguide structure; and (c) another ridge structure.

the ridge. The width of the ridge is a. For confined propagation, the index of refraction n_2 is greater than either n_1 or n_3.

We now divide the structure into three regions for the purpose of introducing the concept of effective index theory. These regions are $y < -\frac{1}{2}a$, $-\frac{1}{2}a < y < \frac{1}{2}a$, and $\frac{1}{2}a < y$. In each of the three regions, the confinement along the vertical direction (x) is exactly identical to that of the asymmetric slab waveguide. For simplicity, we assume that only one TE mode is supported by all three regions of the dielectric structure. The propagation constant of the confined mode inside the ridge is, however, different from those on both sides of the ridge because of the difference in layer thickness. The effective index of refraction (or the normalized propagation constant) defined in Equation (3.1-20) for these three segments is

$$n_{\text{eff}}(y) = \begin{cases} n_{\text{I}}, & y < -\frac{1}{2}a \\ n_{\text{II}}, & -\frac{1}{2}a < y < \frac{1}{2}a \\ n_{\text{I}}, & \frac{1}{2}a < y \end{cases} \tag{3.4-1}$$

where n_{I} and n_{II} are the effective indices of refraction of the asymmetric slab waveguides, respectively.

In the theory of effective index, the lateral waveguiding is treated by taking Equation (3.4-1) as a "slab" waveguide structure along the y direction. Thus confinement of electromagnetic radiation requires that $n_{\text{II}} > n_{\text{I}}$. This is always true when the thickness t is greater than d because, according to Figure 3.7, the effective index of refraction of any confined mode is an increasing function of thickness. To illustrate further the lateral guiding, we will consider the following example.

GaAs Ridge Waveguide

We consider a ridge waveguide made of GaAs layers and AlGaAs substrate. Let the indices of refraction be $n_1 = 1$, $n_2 = 3.5$, and $n_3 = 3.2$. The thicknesses are $t = 0.40\lambda$ and $d = 0.25\lambda$. According to a numerical solution of Equation (3.2-5), the effective indices are

$$n_{\text{I}} = 3.301 \quad \text{and} \quad n_{\text{II}} = 3.388$$

We note that a step height of 0.15λ at the ridge gives rise to an increase in the effective index of 0.087. Such an index difference is sufficient to provide lateral waveguiding. In fact, if we take $a = 0.5\lambda$ as the width of the ridge and solve for the confined TE modes of such a symmetric waveguide, we obtain a single TE mode with a normalized propagation constant of 3.348. The wavefunction of such a mode is similar to the fundamental mode shown in Figure 3.3. This wavefunction shows the lateral confinement.

It is important to note that the effective index theory is a good approximation, provided the index difference between the core and cladding is small so that the scalar wave approximation is valid. In the scalar approximation, we ignore the vector nature of the electromagnetic waves.

Ridge Waveguides or Two-Dimensional Waveguides

The ridge waveguide structures described in Figure 3.16 can be obtained by several different approaches, including etching and diffusion. In the following, we describe some examples of channel waveguides that are important in optical communications.

Thermal Indiffusion

A channel waveguide in a $LiNbO_3$ crystal can be obtained by using a conventional thermal indiffusion process. Prior to the diffusion process, a thin strip of metallic Ti layer about 100 nm

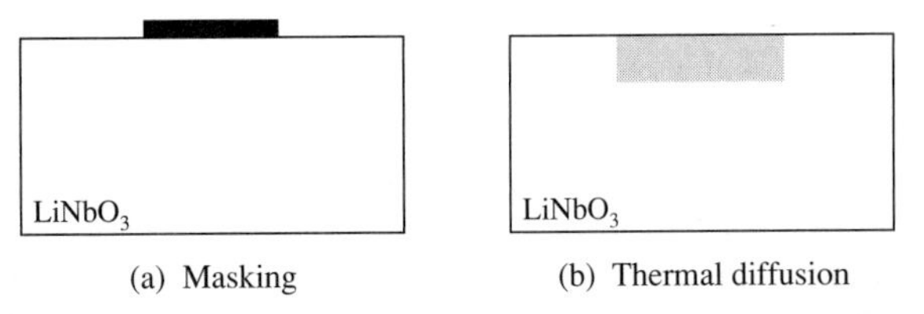

Figure 3.17 Fabrication steps in Ti-diffused $LiNbO_3$ channel waveguide. (a) First, a thin strip of Ti metal is deposited on the optically polished surface of the $LiNbO_3$ crystal wafer. Such a strip or a more complicated planar optical circuit pattern can be obtained by a photolithographic method that defines the channel waveguide circuit. (b) The wafer is then heat treated at an elevated temperature. As a result of thermal diffusion, Ti atoms migrate into the crystal wafer. This leads to an optical circuits with a higher refractive index at the core of the waveguide circuit.

in thickness is deposited on the surface of a $LiNbO_3$ wafer by a vacuum evaporation technique. A planar optical circuit can be patterned by a conventional photolithography technique. The wafer is then heat treated at around 1000 °C for a period of time (e.g., 20 hours). As a result of the heating-assisted diffusion, the Ti atoms migrate into the crystal. The presence of Ti atoms in the $LiNbO_3$ crystal increases both the ordinary and extraordinary index of refraction. This creates a channel waveguide in the region of Ti-atom concentration. Figure 3.17 illustrates the steps in a Ti-diffused $LiNbO_3$ channel waveguide.

A channel waveguide in silica (SiO_2) wafer can be obtained by a similar thermal indiffusion process using Pb or other metals. The presence of Pb atoms in silica increases the index of refraction and thus creates a channel waveguide.

Proton Exchange in Lithium Niobate

A channel waveguide in a lithium niobate wafer can also be obtained by using the proton exchange process (see Figure 3.18). First, a patterned material is applied to the optically polished surface of the $LiNbO_3$ crystal. The pattern can be obtained by a photolithographic method that defines the channel waveguide optical circuit. The wafer is then immersed in a bath of liquid acid that provides a source of protons that exchanges with lithium ions, creating a thin layer of protons at the surface. The protons at the $LiNbO_3$ surface can penetrate further into the crystal via diffusion during an anneal process at high temperature. Once cooled, the channel waveguide is extremely stable. As a result, the extraordinary index of refraction is increased in the proton-exchanged regions and the ordinary index is reduced. This leads to a channel waveguide that supports confined propagation with the **E** field polarized along the c axis of the crystal. There are other techniques that can be employed to increase the index of refraction and to produce optical waveguides. These include ion implantation and the sol-gel process.

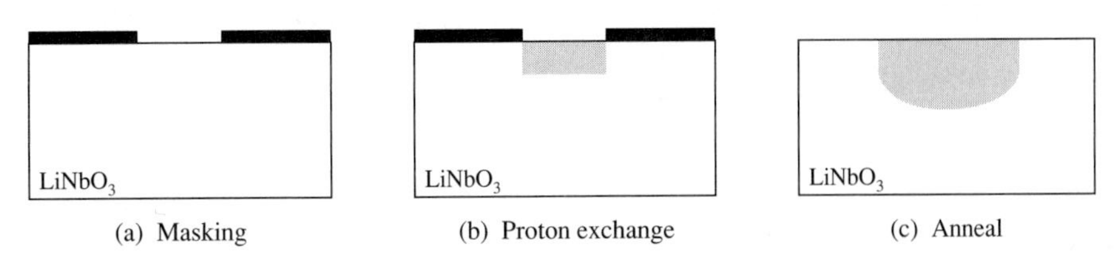

Figure 3.18 Fabrication steps in proton-exchange process in a $LiNbO_3$ channel waveguide.

3.5 WAVEGUIDE DISPERSION IN OPTICAL FIBERS

So far in this chapter, we have discussed various optical waveguides, including dielectric slab waveguides and step-index optical fibers. These waveguides can be described in terms of the index of refraction as a function of space. Confined propagation in these waveguides is governed by the wave equation, which provides the wavefunction and the propagation constant once the index distribution of the waveguide is given. The wave equation is often written

$$\left\{\nabla_{\perp}^{2}+\frac{\omega^{2}}{c^{2}}n^{2}(x, y)\right\}E_{m}(x, y)=\beta_{m}^{2}E_{m}(x, y) \tag{3.5-1}$$

where $\nabla_{\perp}^{2}$ is the transverse Laplacian, $E_m(x, y)$ is the wavefunction of the mth mode, and β_m is the propagation constant. Equation (3.5-1) may be viewed as an eigenvalue problem with β_m^2 as the eigenvalue. Consider a general mode of propagation

$$E(x, y, z, t)=E_{m}(x, y)\exp[i(\omega t-\beta_{m}z)] \tag{3.5-2}$$

In multimode waveguides, a general wave of propagation can be written as a linear combination of all the modes.

For the convenience of discussion, we define an effective index of a mode as

$$n_{\text{eff}}=\frac{\beta_{m}}{\omega/c} \tag{3.5-3}$$

It is important to note that both the wavefunction and the propagation constant β_m depend on the waveguide geometry $n^2(x, y)$ as well as the angular frequency ω. For confined modes, the effective index is always between the core index and the clad index. In other words,

$$n_{\text{clad}}<n_{\text{eff}}<n_{\text{core}} \tag{3.5-4}$$

For multimode waveguides, each confined mode has a distinct effective index. The fundamental mode has the largest effective index. Generally, each mode has its own phase velocity and group velocity. As a result, modal dispersion occurs in multimode waveguides. This was discussed in Section 2.9.

In a multimode waveguide, if an optical wave (e.g., a sequence of pulses) is represented by a superposition of a number of modes, then the pulses will spread as they propagate in the waveguide. The spread can be due to (a) modal dispersion, (b) waveguide dispersion, and (c) material dispersion. In single-mode waveguides, the pulse spreading is due to (a) waveguide dispersion and (b) material dispersion. In this section, we discuss waveguide dispersion and the pulse spreading in single-mode optical fibers.

The effective index of a step-index fiber (or slab waveguide) depends on the frequency of light ω, even if the indices of refraction of the fiber are independent of the frequency of light ω. This is a result of the different wavefunctions and ray paths inside the waveguide at different frequencies. The dispersion figures in Sections 3.1–3.3 (e.g., Figure 3.12) show the dependence of n_{eff} on frequency ω. Thus we may write

$$n_{\text{eff}}=n_{\text{eff}}[n_{1}(\omega), n_{2}(\omega), \omega] \tag{3.5-5}$$

where the effective index depends on the frequency directly due to waveguide dispersion and indirectly due to material dispersion through $n_1(\omega)$ and $n_2(\omega)$.

We now examine these two contributions of the dispersion. Using the method of variation in perturbation theory, we apply a small variation of the frequency $\delta\omega$ to the wave equation (3.5-1). Let $\delta\beta_m^2$, δn_1^2, and δn_2^2 be the corresponding variations in the propagation constant of the mth mode, and the refractive indices, respectively. We obtain from the wave equation (3.5-1)

$$\delta\beta_m^2 = \Gamma_1 \frac{\omega^2}{c^2}\delta n_1^2 + \Gamma_2 \frac{\omega^2}{c^2}\delta n_2^2 + \Gamma_1 \frac{2\omega\,\delta\omega}{c^2} n_1^2 + \Gamma_2 \frac{2\omega\,\delta\omega}{c^2} n_2^2 \tag{3.5-6}$$

where the first two terms on the right side are due to the changes of the index of refraction δn_1^2 and δn_2^2, while the last two terms are due to a change of the frequency $\delta\omega$; Γ_1 and Γ_2 are confinement factors given by

$$\Gamma_\alpha = \frac{\displaystyle\iint_{n_\alpha} \mathbf{E}_m^* \cdot \mathbf{E}_m \, dx\, dy}{\displaystyle\iint_{\infty} \mathbf{E}_m^* \cdot \mathbf{E}_m \, dx\, dy} \qquad (\alpha = 1, 2) \tag{3.5-7}$$

where the integral in the numerator is over the region where the index of refraction is n_α.

Equation (3.5-6) can be rewritten as

$$\frac{d\beta}{d\omega} = \Gamma_1 \frac{\omega}{c}\frac{n_1}{n_{\text{eff}}}\frac{\partial n_1}{\partial\omega} + \Gamma_2 \frac{\omega}{c}\frac{n_2}{n_{\text{eff}}}\frac{\partial n_2}{\partial\omega} + \frac{1}{cn_{\text{eff}}}(\Gamma_1 n_1^2 + \Gamma_2 n_2^2) \tag{3.5-8}$$

where we have dropped the subscript m for simplicity. Formula (3.5-8) can be employed for each mode. Using Equation (3.5-3) for β, the above equation can further be written

$$\frac{d}{d\omega} n_{\text{eff}} = \Gamma_1 \frac{n_1}{n_{\text{eff}}}\frac{\partial n_1}{\partial\omega} + \Gamma_2 \frac{n_2}{n_{\text{eff}}}\frac{\partial n_2}{\partial\omega} + \frac{1}{\omega n_{\text{eff}}}(\Gamma_1 n_1^2 + \Gamma_2 n_2^2 - n_{\text{eff}}^2) \tag{3.5-9}$$

where the first two terms on the right side are due to material dispersion, and the last term is due to waveguide dispersion, which represents the variation of the effective index of the guided mode with respect to the frequency by keeping both the core and cladding index constant. In other words, the dispersion of a guided wave consists of the sum of an intrinsic material dispersion and the waveguide dispersion. Thus we can write

$$\frac{d}{d\omega} n_{\text{eff}} = \left(\frac{\partial n_{\text{eff}}}{\partial\omega}\right)_{\text{material}} + \left(\frac{\partial n_{\text{eff}}}{\partial\omega}\right)_{\text{waveguide}} \tag{3.5-10}$$

where the material dispersion is given by

$$\left(\frac{\partial n_{\text{eff}}}{\partial\omega}\right)_{\text{material}} = \Gamma_1 \frac{n_1}{n_{\text{eff}}}\frac{\partial n_1}{\partial\omega} + \Gamma_2 \frac{n_2}{n_{\text{eff}}}\frac{\partial n_2}{\partial\omega} \tag{3.5-11}$$

and the waveguide dispersion is given by

$$\left(\frac{\partial n_{\text{eff}}}{\partial\omega}\right)_{\text{waveguide}} = \frac{1}{\omega n_{\text{eff}}}(\Gamma_1 n_1^2 + \Gamma_2 n_2^2 - n_{\text{eff}}^2) \tag{3.5-12}$$

For single-mode silica fibers used in modern optical communications, the clad and core are both made of silica with $n_1 \approx n_2$, so Equations (3.5-10)–(3.5-12) can be written approximately

$$\left(\frac{\partial n_{\text{eff}}}{\partial\omega}\right)_{\text{material}} = \frac{\partial n}{\partial\omega} \tag{3.5-13}$$

$$\left(\frac{\partial n_{\text{eff}}}{\partial \omega}\right)_{\text{waveguide}} = \frac{1}{\omega n_{\text{eff}}}(n_2^2 + \Gamma_1[n_1^2 - n_2^2] - n_{\text{eff}}^2) \tag{3.5-14}$$

where n is the index of refraction of the fiber material (silica), and we have used $n_2 \approx n_{\text{eff}} \approx n_1$ and $\Gamma_2 = 1 - \Gamma_1$. We note that the term on the right side of Equation (3.5-13) is purely material dispersion, while Equation (3.5-14) addresses the waveguide dispersion.

Using the confinement factor expression in Section 3.3, the waveguide dispersion can be written (see Problem 3.12)

$$\left(\frac{\partial n_{\text{eff}}}{\partial \omega}\right)_{\text{waveguide}} = \frac{1}{\omega n_{\text{eff}}}(n_{\text{eff}}^2 - n_2^2)\left(\frac{-J_m^2(ha)}{J_{m-1}(ha)J_{m+1}(ha)}\right) \tag{3.5-15}$$

Since $J_{m-1}(ha)J_{m+1}(ha)$ is always negative for the confined-modes that satisfy the mode conditions (3.3-26) and (3.3-27), we note that the right side of Equation (3.5-15) is always positive. In other words, the effective index is an increasing function of frequency. This is consistent with the dispersion curves shown in Figure 3.12.

In optical networks, the dispersion of a mode of propagation is often written

$$\frac{dn_{\text{eff}}}{d\lambda} = \left(\frac{\partial n}{\partial \lambda}\right)_{\text{material}} + \left(\frac{\partial n_{\text{eff}}}{\partial \lambda}\right)_{\text{waveguide}} \tag{3.5-16}$$

and

$$\frac{d^2 n_{\text{eff}}}{d\lambda^2} = \left(\frac{\partial^2 n}{\partial \lambda^2}\right)_{\text{material}} + \left(\frac{\partial^2 n_{\text{eff}}}{\partial \lambda^2}\right)_{\text{waveguide}} \tag{3.5-17}$$

Consider the propagation of an optical pulse in a single-mode waveguide of length L. Let n_{eff} be the effective index of the mode of propagation. The phase shift of the beam at output due to propagation is $\phi = n_{\text{eff}}(\omega/c)L$. It can be shown that the group delay (flight time) is

$$\tau = \frac{d\phi}{d\omega} = L\left(\frac{n_{\text{eff}}}{c} + \frac{\omega}{c}\frac{dn_{\text{eff}}}{d\omega}\right) = L\left(\frac{n_{\text{eff}}}{c} - \frac{\lambda}{c}\frac{dn_{\text{eff}}}{d\lambda}\right) \tag{3.5-18}$$

The group velocity dispersion (in units of picaseconds per nanometer of bandwidth per kilometer of fiber) for optical fibers is defined as

$$D = \frac{d}{d\lambda}\left(\frac{\tau}{L}\right) \quad \text{(ps/nm-km)} \tag{3.5-19}$$

Using Equations (3.5-18) and (3.5-19), we obtain

$$D = -\frac{1}{c\lambda}\left(\lambda^2 \frac{d^2 n_{\text{eff}}}{d\lambda^2}\right) \tag{3.5-20}$$

If the waveguide material is dispersive, the group velocity dispersion parameter D can be written, according to Equation (3.5-17),

$$D = -\frac{1}{c\lambda}\left(\lambda^2 \frac{\partial^2 n}{\partial \lambda^2}\right)_{\text{material}} - \frac{1}{c\lambda}\left(\lambda^2 \frac{\partial^2 n_{\text{eff}}}{\partial \lambda^2}\right)_{\text{waveguide}} \tag{3.5-21}$$

Note that the terms inside the parentheses are dimensionless. The parameter D has a dimension of s/m-m. For practical application, D is often expressed in units of ps/nm-km.

EXAMPLE: PULSE SPREADING

Consider the transmission of 10 Gb/s signals at $\lambda = 1500$ nm in a single-mode fiber of 100 km, with a group velocity dispersion of $D = 17$ ps/nm-km. The pulse spreading after a transmission of distance L can be written

$$\Delta\tau = DL\,\Delta\lambda \tag{3.5-22}$$

where $\Delta\lambda$ is the spectral bandwidth of the signals. For 10 Gb/s, the pulse width is $\tau_0 = 100$ ps. The spectral bandwidth is approximately $\Delta\nu = 1/\tau_0 = 10$ GHz. In terms of nanometers, the bandwidth is

$$\Delta\lambda = \frac{\lambda^2}{c}\Delta\nu = 0.075 \text{ nm}$$

The pulse spreading is thus given by, using Equation (3.5-22), $\Delta\tau = 128$ ps

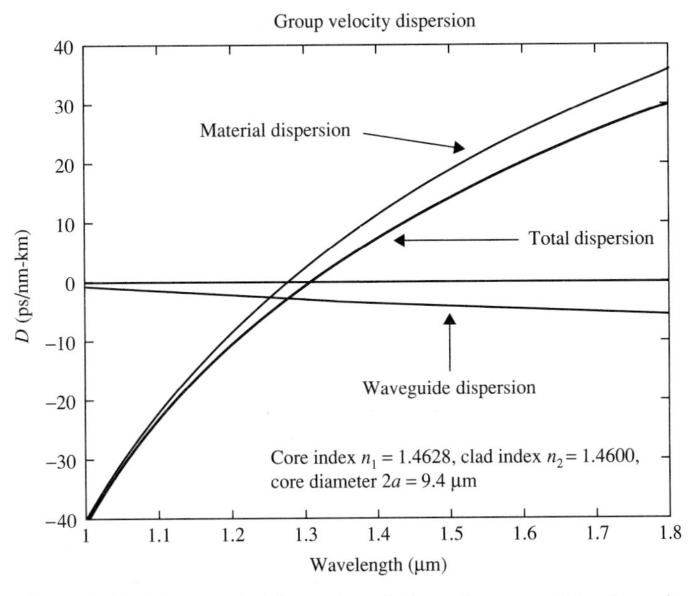

Figure 3.19 The material dispersion of silica, the waveguide dispersion of a single-mode GeO_2-doped silica fiber, and the resultant total dispersion in the spectral region between 1 and 1.8 μm.

Figure 3.19 shows the material dispersion of silica, the waveguide dispersion of a single-mode GeO_2-doped silica fiber with a core diameter of 9.4 μm, and the resultant total dispersion. Not that the total dispersion is zero at $\lambda = 1.3$ μm [8, 9]. It is important to note that the waveguide dispersion depends on core diameter a as well as the core and clad indices n_1 and n_2. It is possible to tailor the zero dispersion wavelength by balancing the positive material dispersion against the negative waveguide dispersion [10]. Thus, by choosing a proper core diameter a between 4 and 5 μm, and a relative refractive index difference of $(n_1 - n_2)/n_1 > 0.004$, the wavelength of zero dispersion can be shifted to the 1.5–1.6 μm region, where the propagation loss is minimum [11–16].

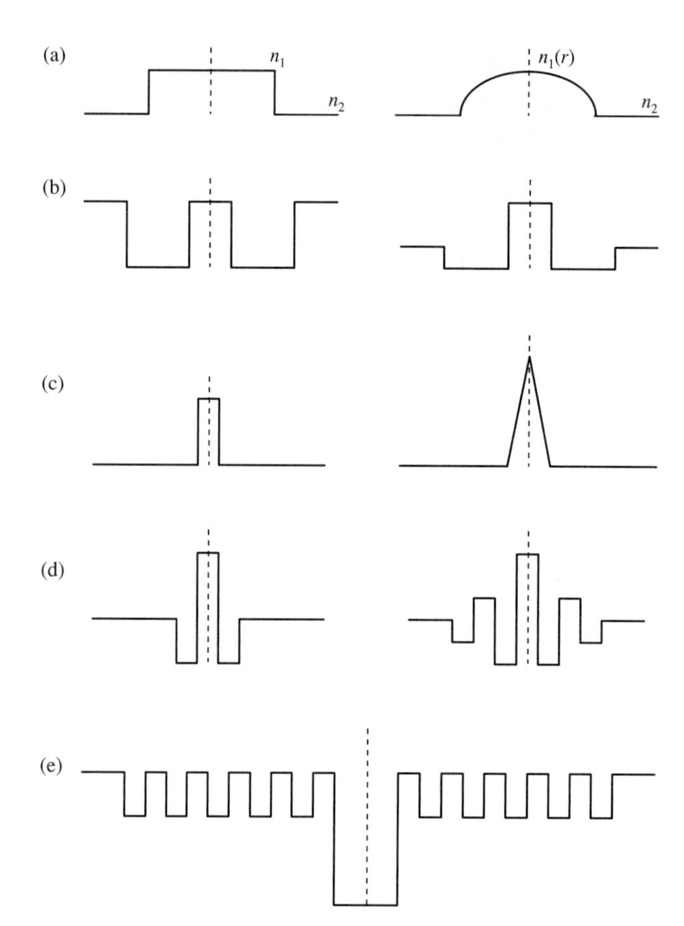

Figure 3.20 Various index profiles of fibers. (a) Multimode fibers with a core diameter of about 50 μm, and a clad diameter of 120 μm. (b) Single-mode fibers with a core diameter of about 10 μm. (c) Dispersion-shifted fibers with a core diameter of about 5 μm. (d) Dispersion-flattened fibers with double clad or triple clad and a core diameter of about 6 μm. (e) Bragg fibers with a core of low index of refraction, or a hollow core.

We have so far discussed the step-index circular fibers. Closed-form solutions for the modes are available. There are optical fibers with different refractive index profiles. These fibers are designed to provide minimum loss or minimum dispersion in the spectral region of interest. Some of the fibers can provide near-zero dispersion over a broad spectral region of interest. Figure 3.20 shows some of the index profiles. Some of the profiles involve additional cladding layers of different refractive indices [17–21]. These index profiles offer additional degrees of freedom in the design of fibers for specific applications.

According to Figure 3.19, we note that at $\lambda = 1550$ nm the chromatic dispersion is mostly due to material dispersion. If a fiber can be fabricated with an air core, the material dispersion can be virtually eliminated. To support a confined propagation in a low-index region requires a multilayer cladding with a periodic variation of the clad index in the radial direction. High reflection via Bragg scattering in the radial direction provides the confinement (see Figure 3.20e). The discussion of such fiber is beyond the scope of this book. Interested readers are referred to Reference [22].

3.6 ATTENUATION IN SILICA FIBERS

Propagation attenuation exists in virtually all waveguides. As a result of attenuation, signal power $P(z)$ will decay exponentially according to the following equation:

$$P(L) = P(0)\exp(-\alpha L) \tag{3.6-1}$$

where $P(0)$ is the power at the input, L is the distance of propagation, and α is the linear attenuation coefficient. In optical communications, attenuation is often measured in units of dB, and the attenuation coefficient is measured in units of dB/km. Thus signal attenuation (dB) per unit length is defined as

$$\alpha = \frac{10}{L}\log_{10}\left(\frac{P(L)}{P(0)}\right) \tag{3.6-2}$$

There are many sources of propagation attenuation in optical fibers. In the early days of silica fiber development, propagation attenuation was mainly due to the presence of impurities that absorb optical energy. Trace metal impurities, such as iron, nickel, and chromium, are introduced into the fiber during the fabrication. The electronic energy levels of these metal ions are broadened by the random atomic environment in the silica. Some of the electronic transitions fall in the spectral regime of the propagating optical beam. Photons in the optical beam are absorbed by these electronic transitions. The absorption also occurs when a small amount of water is present in the silica glass. Water in silica glass forms a silicon-hydroxyl (Si—OH) bond, which can cause absorption in the spectral regime around $\lambda = 1.4$ μm.

Although the intrinsic absorption due to the vibration of the SiO_2 is relatively insignificant in the spectral regime of optical communications, propagation attenuation is still present in pure silica fibers. The attenuation is mainly due to manufacturing imperfection of the fiber structure, as well as the fundamental process of Rayleigh scattering. Probably the single most important factor responsible for the emergence of silica glass optical fiber as the premium information transmission medium is the low optical propagation losses in such fibers. Figure 3.21 shows the measured losses as a function of wavelength of a high-quality, GeO_2-doped single-mode fiber. The loss peak at around 1.4 μm is due to residual OH contamination of the glass. A low value of loss ~ 0.2 dB/km obtains near $\lambda = 1.55$ μm. In 1986, a low transmission attenuation of 0.154 dB/km was realized [23]. Recently, a pure silica core fiber, with the index profile shown in Figure 3.20d, has been fabricated with a low transmission loss of 0.1484 dB/km at $\lambda = 1570$ nm [20, 21]. Consequently, this region of the spectrum is now favored for long-distance optical communications. Recent experiments have taken advantage of the small pulse spreading near the zero group velocity dispersion wavelength and the low losses to demonstrate high-data-rate transmission (data rate exceeding 400 Mb/s) over a propagation path exceeding 100 km [24, 25] at $\lambda \sim 1.55$ μm. Longer distance and higher data rates are possible provided optical amplifiers are employed to boost the energy attenuation due to propagation as well as chromatic dispersion compensation to eliminate the pulse distortion. For a more detailed discussion of propagation effects in optical fibers, the student can consult Reference [26].

Another source of attenuation in single-mode fibers is the bending loss. This is usually not a problem in the transmission optical networks where the curvature of bending is very small. In optical fiber sensors such as fiberoptic gyros, a long segment of fibers (kilometers) must be coiled in a small box of a few centimeters. The radius of curvature in this case is in the range of centimeters. In Dense Wavelength Division Multiplexing(DWDM) optical networks, there are situations when many passive or active components must be connected by fibers inside a

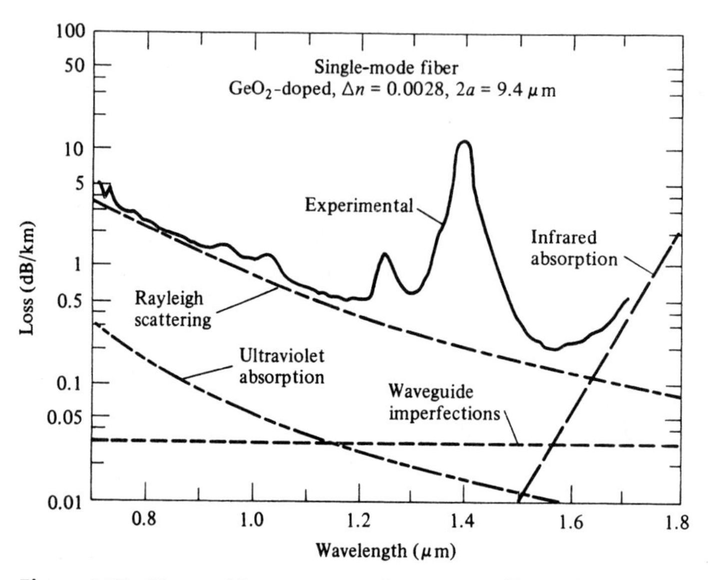

Figure 3.21 Observed loss spectrum of a germanosilicate single-mode fiber. Estimated loss spectra for various intrinsic material effects and waveguide imperfections are also shown. (From Reference [24].)

small box. Bending of fibers is unavoidable in these situations. To minimize the loss due to bending, the radius of curvature must be kept as large as possible inside the box. In this section we briefly discuss the loss due to bending of fibers.

Referring to Figure 3.22, we consider the propagation of a LP_{01} mode inside a single-mode fiber. In the straight fiber, the field distribution will remain unchanged as it propagates along the fiber. When the fiber is bent with a radius of curvature of R, we may assume that the fiber mode pattern remains the same, provided the radius of curvature of bending is much larger than the fiber core radius a (i.e., $a \ll R$). It is known that bent waveguides are intrinsically leaky. The source of leak can be explained as follows. To maintain the same mode pattern in the bent waveguide, the planar wavefront of the mode must be pivoted around the center of curvature O. This leads to a velocity mismatch problem. To keep up with the mode, the wavefront on the outer evanescent tail must travel faster than the wavefront at the center of the

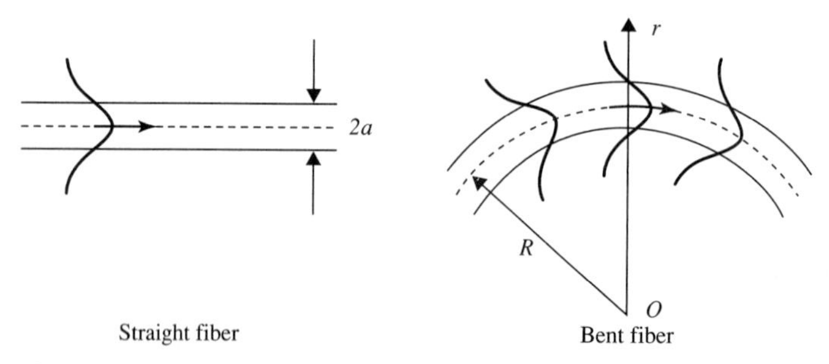

Figure 3.22 Schematic drawing of a straight fiber and a bent fiber. To maintain the mode pattern in the bent fiber, the wavefront of the outer tail of the mode must travel faster than the wavefront of the mode at the center of the fiber core. R is the radius of curvature of the bending and r is the radial distance measured from the center of curvature O.

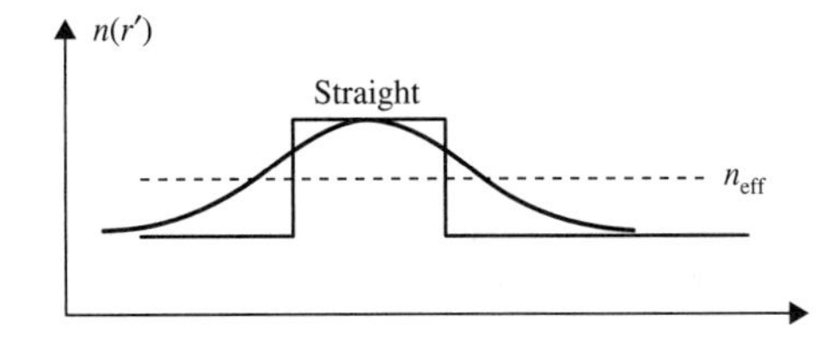

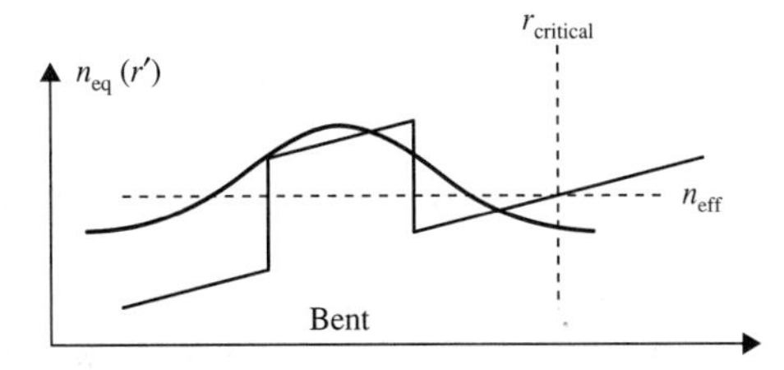

Figure 3.23 Index profile and mode wavefunction of the straight waveguide and the equivalent index profile of the bent waveguide. The horizontal dashed line indicates the effective index of the fiber mode, n_{eff}.

core. In fact, the speed of the wavefront at location r' from the center of the fiber core must be $(1 + r'/R)$ times c/n_{eff}. At some critical distance from the core of the fiber, this speed can be greater than c/n_2. Since this is impossible, the field beyond this critical distance is coupled to the radiation modes. Thus part of the mode energy will break away from the guide. This leads to propagation attenuation.

Based on the above argument and the schematic drawing shown in Figure 3.23, the wave propagation in a bent waveguide can be described by the wave propagation in an equivalent waveguide. Since the outer side of the mode must travel more distance and experience more phase shift for a given propagation constant β, the index profile of the equivalent waveguide can be written

$$n_{eq}(r') = n(r')\left(1 + \frac{r'}{R}\cos\phi'\right) \tag{3.6-3}$$

where r' is the radial distance measured from the center of the fiber core, and ϕ' is the azimuthal angle measured in the cross-sectional plane of the fiber. Note that the radial coordinate r is measured from the center of curvature of the bending O. Figure 3.23 is a sketch of the equivalent index profile. The horizontal dashed line indicates the effective index of the fiber mode, n_{eff}. Based on the discussion earlier in the chapter, the effective index of a confined mode must be greater than the index of the clad. Beyond the critical radius ($r_{critical}$), the equivalent index in the clad is higher than the effective index n_{eff}. This is the region where the guide mode is coupled to the radiation mode. A guided mode in the fiber can thus tunnel through the triangular barrier between the outer edge of the core and the critical radius.

As a result of the equivalent index profile shown in Figure 3.23, all modes of propagation in the bent waveguide are leaky. Solving the wave equation with the equivalent index profile given by Equation (3.6-3) would require extensive numerical techniques. The attenuation coefficient due to the bending of the fibers is usually obtained by using various different approaches [27–41]. Here we describe a result computed with the help of scalar diffraction theory [27]. In this approach, we consider the propagation of light inside the core of a circular ring of fiber. The radius of the ring R is much greater than the core radius of the fiber

a ($a \ll R$). A guided mode of the fiber is propagating in the ring. We can regard the field of the guided mode at the surface of the ring of fiber as the source of a field whose values at all points in space can be obtained by the well-known scalar diffraction theory. This approach is particularly useful in this case, as we are interested in the radiation of power into the far field. Furthermore, the scalar diffraction integral is reduced to the simple Fraunhofer diffraction integral. Based on scalar diffraction theory, the field $\psi(\mathbf{r})$ at an arbitrary point of observation $\mathbf{r}$ in the far field can be written

$$\psi(\mathbf{r}) = \frac{e^{-ikr}}{4\pi r} \int_S e^{i\mathbf{k}\cdot\mathbf{r}'} [i\mathbf{k} \cdot \mathbf{n}\psi(\mathbf{r}') - \mathbf{n} \cdot \nabla'\psi(\mathbf{r}')] \, da' \tag{3.6-4}$$

where $\mathbf{k}$ is the wavevector in the direction of observation, $\mathbf{r}'$ is the coordinate of an arbitrary point on the surface of the ring of fiber S "seen" by an observer at point $\mathbf{r}$, $\mathbf{n}$ is an outward normal to the surface S, and $\psi(\mathbf{r}')$ is the field of the mode of propagation at point $\mathbf{r}'$ on the surface S [42]. For optical fibers with, $n_{\text{core}} - n_{\text{clad}} \ll 1$, $\psi(\mathbf{r}')$ is simply the LP modes derived in Section 3.3. For planar surfaces in the xy plane, Equation (3.6-4) reduces to the Fraunhofer approximation of Equation (2.12-32).

The power flow of the radiated energy can be written in terms of the field (3.6-4):

$$\Delta S_r = \frac{1}{2} \text{Re}\left(\psi^* \frac{i}{\omega\mu} \frac{\partial}{\partial r} \psi \right) = \frac{n_2 k_0}{2\omega\mu} |\psi(\mathbf{r})|^2 \tag{3.6-5}$$

where we assume that the cladding of the fiber is infinite (i.e., $b = \infty$ in Figure 3.10), $n_2 = n_{\text{clad}}$ is the clad index, and the position $\mathbf{r}$ is very far from the ring of fiber (i.e., $R \ll r$). The attenuation coefficient is then obtained from the following equation:

$$\alpha = \frac{1}{2\pi RP} \int_{r\to\infty} \Delta S_r r^2 \, d\Omega \tag{3.6-6}$$

where $d\Omega$ is an element of solid angle, r is the distance of observation measured from the center of the ring, R is the radius of the ring, and P is the mode power in the fiber as given by Equation (3.3-40). It can easily be shown that the Poynting power flow ΔS_r at the far field drops like $1/r$. So the integral converges as $r \to \infty$. Using the method of steepest descent (the saddle point method) as well as the assumption of $1 \ll qa \ll qR$, an approximate expression for the attenuation coefficient is obtained [27]:

$$\alpha = \frac{e^{2qa}}{\sqrt{\pi qa}} \left(\frac{ha}{n_2 V} \right)^2 \frac{1}{\sqrt{Ra}} \exp\left(-\frac{2(qa)^3}{3(\beta a)^2} \frac{R}{a} \right) \tag{3.6-7}$$

where R is the radius of curvature of the bending, a is the core radius of the fiber, and q, h, V, and β are the fiber and mode parameters defined in Section 3.3. We note that qa, ha, V, and βa are all dimensionless parameters. According to Equation (3.6-7), the attenuation coefficient due to bending decays exponentially as a function of the radius of curvature of the bending. As an example, Figure 3.24 is a plot of Equation (3.6-7) as a function of the radius of curvature of bending of a single-mode fiber. We note that the bending loss is significant only when the radius of bending is less than a few centimeters.

According to Equation (3.6-7), fiber modes with a larger qa will have a smaller bending loss. This can be achieved by designing fibers with a larger index step ($n_1 - n_2$). In fiberoptic gyros or planar light circuits (PLCs), where the radius of curvature of bending is in the range of centimeters, it is important to use waveguides with a larger index step to minimize the bending loss.

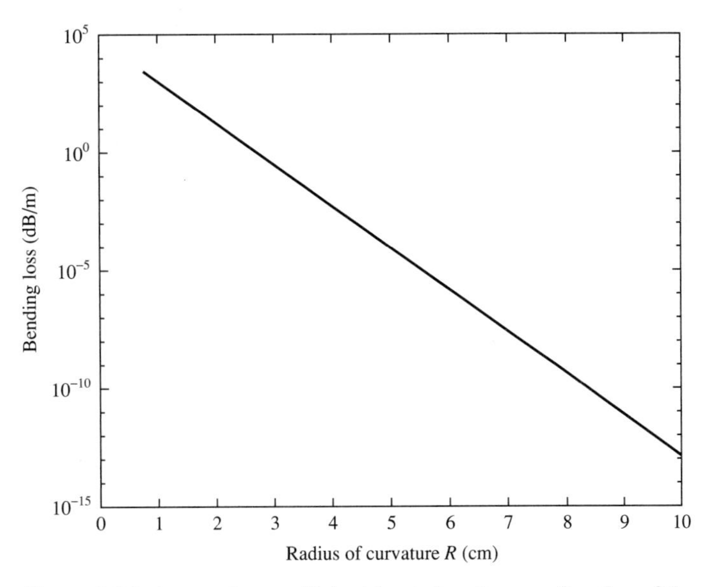

Figure 3.24 Attenuation coefficient due to bending as a function of the radius of curvature of bending, calculated using Equation (3.6-7). The fiber parameters used are: core index $n_1 = 1.4628$, clad index $n_2 = 1.4600$, core radius $a = 5.49\ \mu\text{m}$, $\lambda = 1.30\ \mu\text{m}$. These parameters lead to $V = 2.400$, $n_{\text{eff}} = 1.4614$, $ha = 1.6978$, $qa = 1.6969$, and $\beta a = 38.777$.

PROBLEMS

3.1

(a) Show that in the layers where $\beta^2 > n^2(\omega/c)^2$, the wavefunction $E(x)$ can have, at most, one zero crossing. [Hint: Use the continuous nature of $E(x)$ and $(1/E)(\partial^2 E/\partial x^2) > 0$.]

(b) Show that local maxima of $|E(x)|$ can only occur in the layers where $\beta^2 < n^2(\omega/c)^2$.

3.2 Let Γ_i be the fraction of power flowing in medium i ($i = 1, 2, 3$) of a slab waveguide. In particular, Γ_2 is the fraction of mode power flowing in the guiding layer n_2 and is often called the mode confinement factor. If the mode is normalized to a power of 1 W, the Γ values are defined as

$$\Gamma_i = \frac{1}{2}\text{Re}\int_i (\mathbf{E} \times \mathbf{H}^*) \cdot \hat{\mathbf{z}}\, dx, \quad i = 1, 2, 3$$

where the integral is over the region occupied by medium i where the index of refraction is n_i.

(a) Show that, for TE modes,

$$\Gamma_1 = \frac{1/q}{t + 1/q + 1/p}\frac{h^2}{h^2 + q^2}$$

$$\Gamma_2 = \frac{t}{t + 1/q + 1/p} + \frac{1/p}{t + 1/q + 1/p}\frac{p^2}{h^2 + p^2} + \frac{1/q}{t + 1/q + 1/p}\frac{q^2}{h^2 + q^2}$$

$$\Gamma_3 = \frac{1/p}{t + 1/q + 1/p}\frac{h^2}{h^2 + p^2}$$

Note that $\Gamma_1 + \Gamma_2 + \Gamma_3 = 1$. This proves Equation (3.2-7).

(b) Show that, for TM modes,

$$\Gamma_1 = \frac{1/q'}{t' + 1/q' + 1/p'}\frac{h^2}{h^2 + q^2}$$

$$\Gamma_2 = \frac{t'}{t' + 1/q' + 1/p'} + \frac{1/p'}{t' + 1/q' + 1/p'}\frac{p^2}{h^2 + p^2} + \frac{1/q'}{t' + 1/q' + 1/p'}\frac{q^2}{h^2 + q^2}$$

$$\Gamma_3 = \frac{1/p'}{t' + 1/q' + 1/p'}\frac{h^2}{h^2 + p^2}$$

where

$$t' = \frac{t}{n_2^2}, \quad \frac{1}{q'} = \frac{q^2 + h^2}{\bar{q}^2 + h^2}\frac{1}{n_1^2 q}, \quad \frac{1}{p'} = \frac{p^2 + h^2}{\bar{p}^2 + h^2}\frac{1}{n_3^2 p}$$

(c) Show that Γ_2 increases as β increases, and

$$\Gamma_2 \to 1, \quad \text{as} \quad \beta \to \frac{n_2 \omega}{c}$$

Thus lower-order modes are more confined than higher-order modes because lower-order modes have a larger propagation constant. [Hint: Use $h^2 + p^2 = (n_2^2 - n_3^2)(\omega/c)^2$ and $h^2 + q^2 = (n_2^2 - n_1^2)(\omega/c)^2$.]

(d) Compare the mode confinement factor Γ_2 for TE_m and TM_m modes.

(e) Show that in the limit of well confinement (i.e., $h \to 0$),

$$\frac{\Gamma_3^{TM}}{\Gamma_3^{TE}} = \frac{1 + p/q + pt}{1 + (n_1^2/n_3^2)(p/q) + (n_2^2/n_3^2)pt}$$

where $p = (\omega/c)(n_2^2 - n_3^2)^{1/2}$ and $q = (\omega/c)(n_2^2 - n_1^2)^{1/2}$.

3.3 The interface between a metal and a dielectric can support the propagation of a surface wave, known as a surface plasmon, provided the metal has a negative dielectric constant.

(a) If the metal is viewed as a semi-infinite free electron gas with a dielectric constant given by

$$\varepsilon = \varepsilon_0 \left(1 - \frac{\omega_p^2}{\omega^2}\right)$$

where ω_p is the *plasma frequency*, show that *surface plasmon* modes exist only when $\omega^2 < \frac{1}{2}\omega_p^2$. Here, we assume that $n_1^2 = 1$ for the dielectric. For an electron gas with an electron number density of N, the plasma frequency is $\omega_p^2 = Ne^2/m\varepsilon_0$, where m is the electron mass.

(b) Show that the propagation constant β is given by

$$\beta = \left(\frac{n_1^2 n_3^2}{n_1^2 + n_3^2}\right)^{1/2} \frac{\omega}{c}$$

[Hint: Use Equation (3.2-11). Show that β is always greater than $n_1 k_0$ (i.e., $\beta > n_1 k_0$), provided $n_1^2 > 0$ and $n_1^2 + n_3^2 < 0$. Here, $k_0 = \omega/c$.]

(c) Show that $\beta^2 = pq$ and that the polarization states of the **E** vector in the two media are mutually orthogonal.

(d) Obtain an expression for the z component of the Poynting vector. Show that the Poynting power flow in the positive and negative dielectric media are opposite in direction. Thus a surface plasmon wave propagating along the silver surface along the $+z$ direction will have a negative Poynting power flow in the silver and a positive Poynting power flow in the air.

(e) The complex refractive index of gold at $\lambda = 10\ \mu\text{m}$ is $n - i\kappa = 7.4 - i53.4$. Find the propagation constant $\beta^{(0)}$ and the attenuation coefficient α of the surface plasmon wave.

3.4 The number of confined modes that can be supported by a circular dielectric waveguide depends on the refractive index profile and the wavelength.

(a) Using the cutoff value for the LP_{lm} mode, show that the mode subscripts (l, m) for a step-index fiber must satisfy the condition

$$m\pi + \left(l - \frac{3}{2}\right)\frac{\pi}{2} \le V$$

where $V = k_0 a(n_1^2 - n_2^2)^{1/2}$. Show that each LP_{lm} mode is fourfold degenerate.

(b) By counting the allowed mode subscripts (l, m), show that the total number of confined modes that can be supported by a step-index fiber is

$$N \cong \frac{4}{\pi^2}V^2 \cong \frac{1}{2}V^2$$

(c) Using Equation (2.9-15), show that for a truncated quadratic-index fiber ($n(r) = n_2$, for $r > a$)

$$N \cong \tfrac{1}{4}V^2$$

Note that the total number of modes in a truncated quadratic-index fiber is one-half that of a step-index fiber.

(d) Estimate the number of confined modes in a multimode step-index fiber with $a = 50\ \mu\text{m}$, $n_1 = 1.52$, and $n_2 = 1.50$ at a carrier wavelength of $\lambda = 1\ \mu\text{m}$.

(e) In a general, truncated graded-index fiber with a core radius a and a cladding index n_2, it is convenient to define an effective V number such that

$$V_{\text{eff}}^2 = 2k_0^2 \int_0^a [n^2(r) - n_2^2] r\, dr$$

and the number of confined modes is approximately given by

$$N \cong \tfrac{1}{2}V_{\text{eff}}^2$$

Show that this approximation agrees with (b) and (c) for step-index and quadratic-index fibers, respectively.

(f) Show that, according to (e), the number of confined modes in a power-law (power p) graded-index fiber is given by

$$N = \frac{1}{2(1+2/p)} V^2 = \frac{k_0^2 a^2}{2(1+2/p)} (n_1^2 - n_2^2)$$

Show that this expression again agrees with the results obtained in (b) for step-index fibers ($p = \infty$) and in (c) for quadratic-index fibers ($p = 2$).

3.5 The numerical aperture (NA) is a measure of the light-gathering capability of a fiber. It is defined as the sine of the maximum external angle of the entrance ray (measured with respect to the axis of the fiber) that is trapped in the core by total internal reflection.

(a) Show that

$$NA = n_1 \sin \theta_1 = (n_1^2 - n_2^2)^{1/2}$$

(b) Show that the solid acceptance angle in air is

$$\Omega = n\pi(n_1^2 - n_2^2)^{1/2} = \pi(NA)^2$$

(c) Show that the solid angle (in air) for a single electromagnetic radiation mode leaving or entering the core aperture is

$$\Omega_{\text{mode}} = \frac{\lambda^2}{\pi a^2}$$

(d) The total number of modes that the fiber can support, couple to, and radiate into air is therefore

$$N = 2\frac{\Omega}{\Omega_{\text{mode}}}$$

where the factor of 2 accounts for the two independent polarizations in air. Show that this estimate agrees with Problem 3.4.

(e) Find the numerical aperture of a multimode fiber with $n_1 = 1.52$ and $n_2 = 1.50$.

3.6 A single-mode step-index fiber must have a V number less than 2.405; that is,

$$V = k_0 a(n_1^2 - n_2^2)^{1/2} < 2.405$$

(a) Show that the expression derived in Problem 3.4(b) ($N \cong 4V^2/\pi^2$) still applies, provided we realize that a single-mode fiber supports two independently polarized HE_{11} modes (or LP_{01} modes).

(b) With $a = 5\ \mu\text{m}$, $n_2 = 1.50$, and $\lambda = 1\ \mu\text{m}$, find the maximum core index for a single-mode fiber. [Answer: $n_1 = 1.50195$.]

(c) With $n_1 = 1.501$, $n_2 = 1.500$, and $\lambda = 1\ \mu\text{m}$, find the maximum core radius for a single-mode fiber. [Answer: $a = 7\ \mu\text{m}$.]

(d) Show that the confinement factor for a single-mode fiber is

$$\Gamma_1 = \frac{P_{\text{core}}}{P} = \frac{(qa)^2}{V^2}\left(1 + \frac{J_0^2(ha)}{J_1(ha)}\right)$$

where ha satisfies the mode condition (3.3-26)

$$ha\frac{J_1(ha)}{J_0(ha)} = qa\frac{K_1(qa)}{K_0(qa)}$$

(e) Show that, by using the table of Bessel functions, $ha = 1.647$ is an approximate solution to the mode condition for $V = 2.405$. Evaluate the confinement factor Γ_1 for the LP_{01} mode of this single-mode fiber [Answer: $\Gamma_1 = 83\%$.] Note that this is the maximum confinement factor for a single-mode fiber. Compare this value with the curves in Figure 3.12.

3.7

(a) Derive the mode condition for step-index fibers (Equation (B-11)).

(b) Derive the expressions for the constants B, C, and D in terms of A (Equation (B-12)).

(c) Derive the mode condition for TE and TM modes (Equations (B-17a) and (B-17b)).

(d) Show that $E_z = E_r = 0$ for TE modes and $H_z = H_r = 0$ for TM modes.

(e) Show that in the limit $n_1 - n_2 \ll n_1$, TE and TM modes become identical.

3.8

(a) Derive Equations (3.3-6) and (3.3-7).

(b) Derive Equations (3.3-18) and (3.3-19).

(c) Derive the field components (3.3-23).

(d) Derive Equations (3.3-24) and (3.3-25).

3.9 There are several approximations for an analytic form of the effective index of the fundamental mode of a symmetric slab waveguide. Using the definition $u = hd/2$ and $v = qd/2$, the mode condition for the fundamental mode can be written $\tan u = v/u$.

(a) Show that the mode condition can also be written

$$u = V\cos u, \quad \text{or} \quad v = V\sin u, \quad \text{or} \quad \tan u = \frac{\sin u}{u}V$$

where $V = \sqrt{n_2^2 - n_1^2}\,(\pi d/\lambda)$ for the slab waveguide.

(b) If $u \ll 1$, then $u = \tan^{-1}V$ is an approximation. Show that the following is also an approximation:

$$u = \tan^{-1}\left(\frac{V}{\tan^{-1}V}\frac{V}{\sqrt{1+V^2}}\right)$$

(c) The following is also a useful approximation [43]:

$$u^2 = \ln\sqrt{1+2V^2}$$

Compare the approximations in (b) and (c) with the exact mode condition by plotting u versus V. Show that for $V < \pi/2$, the approximation in (c) yields more accurate results.

3.10 The propagation β of the fundamental mode (LP_{01}) satisfies the following mode condition:

$$ha\frac{J_1(ha)}{J_0(ha)} = qa\frac{K_1(qa)}{K_0(qa)}$$

In the following, we find asymptotic solutions for β in the regions $V \ll 1$ and $V \gg 1$.

(a) Using the asymptotic forms of the Bessel functions, show that, for $V \ll 1$, the mode condition reduces to

$$qa = \frac{2}{\Gamma}e^{-2/V^2}$$

or equivalently,

$$n_{\text{eff}}^2 = \left(\frac{\beta}{k_0}\right)^2 = n_2^2 + \frac{4(n_1^2 - n_2^2)}{V^2}e^{-2(2/V^2+\gamma)}$$

where γ is Euler's constant $\gamma = 0.5772$, and $\Gamma = \exp(\gamma) = 1.782$. We note that n_{eff} approaches the clad index n_2 very quickly as V approaches zero (see Figure 3.12).

(b) Using the asymptotic forms of the Bessel functions, show that, for $V \gg 1$, the mode condition reduces to

$$ha = x_1(1 - 1/V)$$

$$n_{\text{eff}}^2 = \left(\frac{\beta}{k_0}\right)^2 = n_1^2 - \frac{x_1^2}{V^2}(n_1^2 - n_2^2)\left(1 - \frac{1}{V}\right)^2$$

where $x_1 = 2.405$ is the first root of $J_0(x) = 0$. [Hint: See Figure 3.11. ha approaches x_1 for LP_{01} mode at large V.]

3.11 The confinement factor Γ_1 is defined as $\Gamma_1 = 1 - \Gamma_2$, where

$$\Gamma_2 = \frac{P_{\text{clad}}}{P} = \frac{1}{V^2}\left((ha)^2 + \frac{(qa)^2 J_l^2(ha)}{J_{l-1}(ha)J_{l+1}(ha)}\right)$$

(a) Using the asymptotic forms of the Bessel functions, show that, at cutoff ($qa \to 0$), the confinement factor is

$$\Gamma_1 = \frac{l-1}{l} = 1 - \frac{1}{l} \quad \text{for} \quad 1 \le l$$

$$\Gamma_1 = 0 \quad \text{for} \quad l = 0, 1$$

(b) At cutoff ($qa = 0$), the solutions of the wave equation in the cladding region are no longer Bessel functions. According to Equation (A-7), the wave equation for the field in the cladding region at cutoff ($qa = 0$ and $\beta = n_2 k_0$) becomes

$$\frac{\partial^2\psi}{\partial r^2} + \frac{1}{r}\frac{\partial\psi}{\partial r} + \left(-\frac{l^2}{r^2}\right)\psi = 0$$

Show that the solution for the y-polarized LP mode can be written

$$E_y = \begin{cases} AJ_l(hr)e^{il\phi}\exp[i(\omega t - \beta z)], & r < a \\ Br^{-l}e^{il\phi}\exp[i(\omega t - \beta z)], & r > a \end{cases}$$

where $B = a^l AJ_l(ha) = a^l AJ_l(V)$ due to continuity at the core boundary.

(c) Using $S_z = (\beta/2\omega\mu)|B|^2 r^{-2l}$ for the cladding region, evaluate the following integral:

$$P_{\text{clad}} = \int_0^{2\pi}\int_a^{\infty} S_z r\, dr\, d\phi$$

Show that the integral diverges for the case when $l = 0$ and 1. This is consistent with the zero confinement factor at cutoff. For $l > 1$, the integral converges to a finite value, leading to a finite confinement factor at cutoff.

3.12 The mode condition for the LP modes is given by Equation (3.3-26),

$$ha\frac{J_{l+1}(ha)}{J_l(ha)} = qa\frac{K_{l+1}(qa)}{K_l(qa)}$$

(a) Differentiating both sides of the mode condition, show that

$$ha\left(\frac{J_{l+1}(ha)J_{l-1}(ha)}{J_l^2(ha)} - 1\right)\frac{d(ha)}{dV} + qa\left(\frac{K_{l+1}(qa)K_{l-1}(qa)}{K_l^2(qa)} - 1\right)\frac{d(qa)}{dV} = 0$$

[Note: V is proportional to frequency ω.]

(b) Using $(qa)^2 + (ha)^2 = V^2$ and the results in (a), show that

$$qa\frac{d(qa)}{dV} = \Gamma_1 V \quad \text{and} \quad ha\frac{d(ha)}{dV} = \Gamma_2 V$$

where Γ_1 is the confinement factor and $\Gamma_2 = 1 - \Gamma_1$:

$$\Gamma_1 = \frac{(qa)^2}{V^2}\left(1 - \frac{J_l^2(ha)}{J_{l-1}(ha)J_{l+1}(ha)}\right),$$

$$\Gamma_2 = \frac{1}{V^2}\left((ha)^2 + \frac{(qa)^2 J_l^2(ha)}{J_{l-1}(ha)J_{l+1}(ha)}\right)$$

(c) Using the results in (a) and (b) and the definition of the effective index $n_{\text{eff}} = \beta/(\omega/c)$, show that

$$\left(\frac{\partial n_{\text{eff}}}{\partial \omega}\right)_{\text{waveguide}} = \frac{1}{\omega n_{\text{eff}}}(\Gamma_1 n_1^2 + \Gamma_2 n_2^2 - n_{\text{eff}}^2)$$

It is important to note that the right side is always positive.

(d) Show that

$$\left(\frac{\partial n_{\text{eff}}}{\partial \omega}\right)_{\text{waveguide}} = \frac{1}{\omega n_{\text{eff}}}(n_{\text{eff}}^2 - n_2^2)\left(-\frac{J_m^2(ha)}{J_{m-1}(ha)J_{m+1}(ha)}\right)$$

Show that the right side is always positive for the modes of propagation.

3.13

(a) Solve for the mode of propagation along the z axis by assuming $n = 1$ (vacuum). The partial differential equation is separable. You may assume a solution of the form

$$E_z(r, t) = AJ_l(hr) \exp[i(\omega t + l\phi - \beta z)]$$

$$H_z(r, t) = BJ_l(hr) \exp[i(\omega t + l\phi - \beta z)]$$

and use

$$E_r = \frac{-i\beta}{\omega^2\mu\varepsilon - \beta^2}\left(\frac{\partial}{\partial r}E_z + \frac{\omega\mu}{\beta}\frac{\partial}{r\,\partial\phi}H_z\right),$$

$$E_\phi = \frac{-i\beta}{\omega^2\mu\varepsilon - \beta^2}\left(\frac{\partial}{r\partial\phi}E_z - \frac{\omega\mu}{\beta}\frac{\partial}{\partial r}H_z\right)$$

$$H_r = \frac{-i\beta}{\omega^2\mu\varepsilon - \beta^2}\left(\frac{\partial}{\partial r}H_z - \frac{\omega\varepsilon}{\beta}\frac{\partial}{r\partial\phi}E_z\right),$$

$$H_\phi = \frac{-i\beta}{\omega^2\mu\varepsilon - \beta^2}\left(\frac{\partial}{r\partial\phi}H_z + \frac{\omega\varepsilon}{\beta}\frac{\partial}{\partial r}E_z\right)$$

(b) Obtain expressions for the energy density and Poynting vector. Consider TE, TM, and hybrid modes.

(c) Sketch the energy density and the Poynting vector components (S_ϕ, S_z) as functions of r, for $l = 0, \pm 1, \pm 2$. Show that for $l \neq 0$, the Poynting power flows spirally around the z axis, with a handedness that depends on the sign of l.

(d) Using the angular momentum defined as $\mathbf{L} = \mathbf{r} \times \mathbf{P}$, where $\mathbf{P}$ is the linear momentum defined in Problem 1.5, calculate the time-averaged component of the angular momentum along the direction of propagation ($+z$). Show that this component of the angular momentum is $l\hbar$ provided the energy of the electromagnetic wave is normalized to $\hbar\omega$.

(e) What are the mode orthogonal relationships?

3.14

Using Equations (3.1-14) and (3.1-15), the mode condition can be written

$$[h \tan(\tfrac{1}{2}hd) - q][h \cot(\tfrac{1}{2}hd) + q] = 0$$

Show that the above equation reduces to $\tan(hd) = 2hq/(h^2 - q^2)$. [Hint: $\tan 2\theta = 2 \tan\theta/(1 - \tan^2\theta)$.]

3.15

(a) Using Equations (3.3-16) and (3.3-17), show that

$$\int rJ_l^2(\alpha r)\, dr = \frac{r^2}{2}[J_l^2(\alpha r) - J_{l-1}(\alpha r)J_{l+1}(\alpha r)]$$

(b) Similarly, show that

$$\int rK_l^2(\alpha r)\, dr = \frac{r^2}{2}[K_l^2(\alpha r) - K_{l-1}(\alpha r)K_{l+1}(\alpha r)]$$

where α is an arbitrary constant. [Hint: Differentiate both sides with respect to r.]

REFERENCES

1. Yariv, A., and P. Yeh, *Optical Waves in Crystals*. Wiley, New York, 1984.
2. Yeh, P., *Optical Waves in Layered Media*. Wiley, New York, 1988.
3. Kao, C. K., and T. W. Davies, Spectroscopic studies of ultra low loss optical glasses. *J. Sci. Instrum.* **1**(2):1063 (1968).
4. Kapron, F. P., D. B. Keck, and R. D. Maurer, Radiation losses in glass optical waveguides. *Appl. Phys. Lett.* **17**:423 (1970).
5. Gloge, D., Weakly guiding fibers. *Appl. Opt.* **10**.2252 (1971).
6. See, for example, D. Marcuse, *Theory of Dielectric Optical Waveguides*. Academic Press, New York, 1974.
7. See, for example, I. S. Gradshteyn and I. M. Ryzhik, *Table of Integrals, Series, and Products*. Academic Press, New York, 1965, p. 634, Equation (5.54-2).

8. Payne, D. N., and W. A. Gambling, Zero material dispersion in optical fibers. *Electron. Lett.* **11**:176 (1975).
9. Cohen, L. G., and C. Lin, Pulse delay measurements in the zero material dispersion wave length region for optical fibers. *Appl. Opt.* **12**:3136 (1977).
10. Li, T., Structures, parameters, and transmission properties of optical fibers. *Proc. IEEE* **68**:1175 (1980).
11. Cohen, L. G., C. L. Mammel, and H. M. Presby, Correlation between numerical predications and measurements of single-mode fiber dispersion characteristics. *Appl. Opt.* **19**:2007 (1980).
12. Tsuchiya, H., and N. Imoto, Dispersion-free single mode fibers in 1.5 μm wavelength region. *Electron. Lett.* **15**:476 (1979).
13. Cohen, L. G., C. Lin, and W. G. French, Tailoring zero chromatic dispersion into the 1.5–1.6 μm low-loss spectral region of single-mode fibers. *Electron. Lett.* **15**:334 (1979).
14. White, K. I., and B. P. Nelson, Zero total dispersion in step-index monomode fibers at 1.30 and 1.55 μm. *Electron. Lett.* **15**:396 (1979).
15. Gambling, W. A., H. Matsumara, and C. M. Ragdale, Zero total dispersion in graded-index single-mode fibers. *Electron. Lett.* **15**:474 (1979).
16. Bloom, D. M., L. F. Mollenauer, Chinlon Lin, and A. M. Del Gaudio, Demonstration of pulse propagation in km-length fibers. *Opt. Lett.* **4**:297 (1979).
17. Monroe, M., Propagation in doubly clad single mode optical fibres. *IEEE J. Quantum Electron.* **18**(4):535 (1992).
18. Li, Y., C. D. Hussey, and T. A. Birks, Triple-clad single-mode fibers for dispersion shifting. *IEEE J. Lightwave Technol.* **11**:1812 (1993).
19. Li, Y. and C. D. Hussey, Triple-clad single-mode fibers for dispersion flattening. *Opt. Eng.* **33**:3999 (1994).
20. Nagayama, K., T. Saitoh, M. Kakui, K. Kawasaki, M. Matsui, H. Takamizawa, H. Miyaki, Y. Ooga, I. Tsuchiya, and Y. Chigusa, Ultra low loss pure silica core fiber and its impact on submarine transmission systems. *Tech. Digest of OFC2002*, PD FA10 (2002).
21. Nagayama, K., M. Kakui, M. Matsui, T. Saitoh, and Y. Chigusa, Ultra low loss pure silica core fiber and extension of transmission distance. *Electron. Lett.* **38**:1168 (2002).
22. Yeh, P., A. Yariv, and E. Marom, Theory of Bragg fibers. *J. Opt. Soc. Am.* **68**(9):1196 (1978).
23. Kanamori, H., H. Yokota, G. Tanaka, M. Watanabe, Y. Ishiguro, I. Yoshida, T. Kakii, S. Itou, Y. Asano, and S. Tanaka, Transmission characteristics and reliability of pure silica core single-mode fibers. *J. Lightwave Technol.* **4**(8):1144 (1986).
24. Miya, T., Y. Terunuma, T. Hosaka, and T. Miyashita, Ultimate low-loss single-mode fiber at 1.55 μm. *Electron. Lett.* **15**:106 (1979).
25. Suematsu, Y., Long wavelength optical fiber communication. *Proc. IEEE* **71**:692 (1983).
26. Miller, S., and I. P. Kaminow (eds.), *Optical Fiber Telecommunication II*. Academic Press, San Diego, 1988, Chapters 2 and 3.
27. Marcuse, D., Bend loss of slab and fiber modes computed with diffraction theory. *IEEE J. Quantum Electron.* **29**:2957 (1993).
28. Arnaud, J. A., Transverse coupling in fiber optics, Part III: Bending loss. *Bell Syst. Tech. J.* **53**:1379 (1974).
29. Marcatili, E. A. J., Bends in optical dielectric guides. *Bell Syst. Tech. J.* **48**:2103 (1969).
30. Marcuse, D., Curvature loss formula for optical fibers. *J. Opt. Soc. Am.* **66**:216 (1976).
31. Tsao, C., *Optical Fiber Waveguide Analysis*. Oxford University Press, Oxford, UK, 1992, Chapter 11.
32. Lewin, L., Radiation from curved dielectric slabs and fibers. *IEEE Trans. Microwave Theory Tech.* **22**:718 (1974).

33. Gambling, W. A., D. N. Payne, and H. Matsumura, Radiation from curved single-mode fibers. *Electron. Lett.* **12**:567 (1976).
34. Sakai, J., and T. Kimura, Bending loss of propagation modes in arbitrary-index profile optical fibers. *Appl. Opt.* **17**:1499 (1978).
35. Snyder, A. W., and J. D. Love, *Optical Waveguide Theory*. Chapman & Hall, New York, 1983.
36. Kawakami, S., M. Miyagi, and S. Nishida, Bending losses of of dielectric slab optical waveguide with double or multiple cladding: theory. *Appl. Opt.* **14**:2588 (1975).
37. Vassallo, C., Scalar-field theory and 2-D ray theory for bent single-mode weakly guiding optical fibers. *J. Lightwave Technol.* **3**:416 (1985).
38. Murakami, Y., and H. Tsuchiya, Bending losses of coated single-mode optical fibers. *IEEE J. Quantum Electron.* **14**:495 (1978).
39. Morgan, R., J. S. Barton, P. G. Harper, and J. D. C. Jones, Wavelength dependence of bending loss in monomode fibers: effect of the buffer coating. *Opt. Lett.* **15**:947 (1990).
40. Valiente, I., and C. Vassallo, New formalism for bending losses in coated single-mode optical fibers. *Electron. Lett.* **25**:1544 (1989).
41. Renner, H., Bending losses of coated single-mode fibers: a simple approach. *J. Lightwave Technol.* **10**:544 (1992).
42. See, for example, J. D. Jackson, *Classical Electrodynamics*, 3rd ed. Wiley, New York, 1999, p. 491, Equation (10-108).
43. Chen, K.-L., and S. Wang, An approximate expression for the effective index in symmetric DH lasers. *IEEE J. Quantum Electron.* **19**:1354 (1983).

ADDITIONAL READING

Collin, R. E., *Field Theory of Guided Waves*. McGraw-Hill, New York, 1960.

CHAPTER 4

OPTICAL RESONATORS

4.0 INTRODUCTION

Optical resonators, like their microwave counterparts, are used primarily for the purpose of generating monochromatic optical beams with high intensities (e.g., lasers). These resonators, also known as optical cavities, consist in most cases of two or more mirrors that are aligned to provide multiple reflections and refocusing of light waves. Under appropriate conditions, the energy of an optical beam can be "trapped" inside the resonator. In this case, the optical beam becomes a mode of the resonator. A universal measure of this property is the quality factor Q of the resonator. Q is defined by the relation [1, 2]

$$Q = \omega \times \frac{\text{field energy stored by resonator}}{\text{power dissipated by resonator}} \tag{4.0-1}$$

where ω is the frequency of the wave. As an example, consider the case of a simple resonator formed by bouncing a plane wave between two perfectly reflecting planes of separation L so that the field inside is

$$E(z, t) = E_0 \sin \omega t \sin kz \tag{4.0-2}$$

where E_0 is a constant, the z axis is perpendicular to the reflecting planes, ω is the angular frequency, and k is the wavenumber. According to the discussion in Chapter 1, the average electric energy stored in the resonator is

$$U_{\text{electric}} = \frac{A\varepsilon}{2T} \int_0^L \int_0^T E^2(z, t)\, dz\, dt \tag{4.0-3}$$

where A is the cross-sectional area, ε is the dielectric constant, and $T = 2\pi/\omega$ is the period. Using Equation (4.0-2) we obtain

$$U_{\text{electric}} = \tfrac{1}{8}\varepsilon E_0^2 V \tag{4.0-4}$$

where $V = AL$ is the resonator volume. Since the average magnetic energy stored in a resonator is equal to the electric energy [1], the total stored energy is

$$U = \tfrac{1}{4}\varepsilon E_0^2 V \tag{4.0-5}$$

Thus, recognizing that in the steady state the input power is equal to the dissipated power, and designating the power input to the resonator by P, we obtain from Equation (4.0-1)

$$Q = \frac{\omega \varepsilon E_0^2 V}{4P}$$

The peak electric field is given by

$$E_0 = \sqrt{\frac{4QP}{\omega \varepsilon V}} \tag{4.0-6}$$

We note that the electric field amplitude is proportional to the square root of Q. At a given power input, a high-Q resonator can provide a high electric field amplitude, or a high field intensity.

Mode Density in Optical Resonators

For the purpose of frequency definition, typical dimensions of resonators (or cavities) are of the order of the wavelength of the electromagnetic waves. This is important to ensure that the resonators possess a very small number (ideally only one) of high-Q modes in a given spectral region. For microwaves, the wavelengths are in the ranges of centimeters. Resonators of these dimensions can easily be obtained. In optical regimes, the resonator's dimensions need to be of order of micrometers. These micrometer optical resonators can be a manufacturing challenge.

EXAMPLE: ONE-DIMENSIONAL RESONATOR

We consider the case of a one-dimensional resonator consisting of two mirrors (also known as the Fabry–Perot resonator). The Fabry–Perot resonator is widely used as a multiple-beam interferometer, an instrument first constructed in the early 1800s by Charles Fabry and Alfred Perot [3]. The Fabry–Perot interferometer has an extremely high resolving power—about 10 times better than a grating spectrometer (which is already at least an order of magnitude better than a prism spectrometer). When the mirrors are aligned perfectly parallel to each other, the reflections of the light waves between the two mirrors can interfere constructively, giving rise to a standing wave pattern as described by Equation (4.0-2) between the mirror surfaces, just like standing waves on a string. This occurs when the mirror separation (also known as the cavity spacing or resonator length L) is an exact integer multiple of half a wavelength of light. If the resonator is made of dielectric mirrors of partial reflection, then an incident beam traveling along the axis can be "trapped" by the multiple reflections inside the resonator, provided the resonator length is an integral number of the half-wavelength of the beam. The "trapping" leads to a build up of a large field intensity inside the resonator. This phenomenon is often called resonance. The resonant frequencies are determined by requiring that the field vanishes at $z = 0$ and at the location $z = L$ of the second reflector. This happens when

$$kL = m\pi, \quad \text{with } m = 1, 2, 3, \ldots$$

Using

$$k = \frac{2\pi\nu}{c} n = \frac{2\pi}{\lambda} n$$

where n is the index of refraction and λ is the wavelength of light in vacuum, the resonance condition can be written

$$L = m\frac{\lambda}{2n}$$

where m is an integer ($m = 1, 2, 3, \ldots$). In terms of the frequency of light (ν), the resonance condition is often written

$$\nu = m\frac{c}{2nL}$$

The resonance condition ensures that the multiple reflections inside the cavity add up constructively. The frequency spacing between adjacent resonances of a one-dimensional resonator is $c/2nL$. Thus the total number of modes between 0 and ν is $4nL\nu/c$ (a factor of 2 is added to account for 2 orthogonal polarization states). For single-mode operation (actually two polarization modes), we obtain $L = \lambda/2n$. In other words, the *linear dimension needs to be comparable to the wavelength* (in the medium).

Mode control in the optical regime would thus seem to require that we construct resonators with volume ~ λ^3(~10^{-12} cm^3 at $\lambda = 1$ μm). This is not easily achievable. An alternative is to build large ($L \gg \lambda$) resonators but to use a geometry that endows only a small fraction of these modes with low losses (a high Q). In our two-mirror example, any mode that does not travel normally to the mirror will "walk off" after a few bounces and thus will possess a low Q factor. We will show later that when the resonator contains an amplifying (inverted population) medium, oscillation will occur preferentially at high-Q modes, so that the strategy of modal discrimination by controlling Q is sensible. We shall also find that further modal discrimination is due to the fact that the atomic medium is capable of amplifying radiation only within a limited frequency region so that modes outside this region, even if possessing high Q, do not oscillate.

One question asked often is the following: Given a large optical resonator ($L \gg \lambda$), how many of its modes will have their resonant frequencies in a given frequency interval, say, between ν and $\nu + \Delta\nu$? To answer this question, consider a large, perfectly reflecting box resonator with sides a, b, c along the x, y, z directions. Resonators of different shapes will differ in detail, but for large resonators ($L \gg \lambda$), the results are similar. Without going into modal details, it is sufficient for our purpose to take the amplitude field solution in the form [2]

$$E(x, y, z) = E_0 \sin k_x x \sin k_y y \sin k_z z \tag{4.0-7}$$

where the components of the wavevector must satisfy the following condition:

$$k_x^2 + k_y^2 + k_z^2 = \left(\frac{\omega}{c}n\right)^2 \tag{4.0-8a}$$

where n is the index of refraction of the cavity medium. For the field to vanish at the boundaries, we thus need to satisfy

$$k_x = \frac{r\pi}{a}, \quad k_y = \frac{s\pi}{b}, \quad k_z = \frac{t\pi}{c}, \quad r, s, t = \text{any integers} \tag{4.0-8b}$$

With each such mode, we may thus associate a propagation vector $k = \hat{\mathbf{x}}k_x + \hat{\mathbf{y}}k_y + \hat{\mathbf{z}}k_z$. The triplet r, s, t defines a mode. Since replacing any integer with its negative does not, according to Equation (4.0-7), generate an independent mode, we will restrict, without loss of generality, r, s, t to positive integers. It is convenient to describe the modal distribution in **k** space, as in

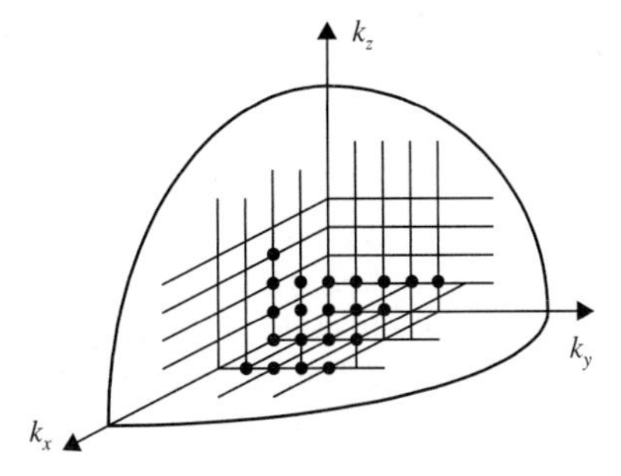

Figure 4.1 The **k**-space description of modes. Every positive triplet of integers r, s, t defines a unique mode. We can thus associate a primitive volume π^3/abc in **k** space with each mode.

Figure 4-1. Since each lattice point (r, s, t) in **k** space represents an independent mode, we can associate with each mode an elemental volume in **k** space.

$$V_{\text{mode}} = \frac{\pi^3}{abc} = \frac{\pi^3}{V} \tag{4.0-9}$$

where V is the physical volume of the resonator. We recall that the length of the vector **k** satisfies Equation (4.0-8a), rewritten here as

$$k(r, s, t) = \frac{2\pi\nu(r, s, t)}{c} n \tag{4.0-10}$$

To find the total number of modes with wavenumber between 0 and k, we divide the corresponding volume in **k** space by the volume per mode:

$$N(k) = \frac{\left(\frac{1}{8}\right)\frac{4\pi}{3}k^3}{\pi^3/V} = \frac{k^3 V}{6\pi^2}$$

where the factor $\frac{1}{8}$ is due to the restriction of $r, s, t > 0$.

We next use Equation (4.0-10) to obtain the number of modes with resonant frequencies between 0 and ν:

$$N(k) = \frac{4\pi\nu^3 n^3 V}{3c^3}$$

The mode density, that is, the number of modes per unit ν near ν in a resonator with volume V ($\gg\lambda^3$), is thus

$$p(\nu) = \frac{dN(\nu)}{d\nu} = \frac{8\pi\nu^2 n^3 V}{c^3} \tag{4.0-11}$$

where we multiplied the final result by 2 to account for the two independent orthogonally polarized modes that are associated with each r, s, t triplet.

The number of modes that fall within the interval $d\nu$ centered on ν is thus

$$N \cong \frac{8\pi\nu^2 n^3 V}{c^3} d\nu \tag{4.0-12}$$

where V is the volume of the resonator. For the case of $V = 1\ \text{cm}^3$, $\nu = 3 \times 10^{14}$ Hz, and $d\nu = 3 \times 10^{10}$, as an example, Equation (4.0-12) yields $N \sim 2 \times 10^9$ modes. If the resonator were closed, all these modes would have similar values of Q. This situation is to be avoided in the case of lasers, since it will cause the atoms to emit power (thus causing oscillation) into a large number of modes, which may differ in their frequencies as well as in their spatial characteristics.

This objection is overcome to a large extent by the use of open resonators, which consist essentially of a pair of opposing flat or curved reflectors. In such resonators the energy of the vast majority of the modes does not travel at right angles to the mirrors and will thus be lost in essentially a single traversal. These modes will consequently possess a very low Q. If the mirrors are curved, the transverse confinement is obtained automatically by the Gaussian modes: thus the diffraction losses caused by the open sides can be made small compared with other loss mechanisms such as mirror transmission. In arriving at Equation (4.0-12), we ignored the dependence of the index of refraction n on the optical frequency. As discussed in previous chapters, the index of refraction depends on the frequency ν. When the dispersion cannot be ignored, the density of optical modes becomes

$$p(\nu) = \frac{dN(\nu)}{d\nu} = \frac{8\pi\nu^2 n^2 n_g V}{c^3} \tag{4.0-13}$$

where n_g is the group index

$$n_g = n + \nu\frac{\partial n}{\partial \nu} \tag{4.0-14}$$

In this chapter, we discuss some important properties of optical resonators that are related to optical electronics. Starting with a simple plane wave analysis of the optical properties of Fabry–Perot resonators, we discuss the reflection and transmission of light through a Fabry–Perot etalon. We then discuss the property of asymmetric Fabry–Perot etalons, including Gires–Tournois etalons and ring resonators. We also discuss optical properties of coupled ring resonators, as well as multicavity etalons. To take diffraction loss into account, we consider Gaussian beams inside optical resonators consisting of spherical mirrors. The mode stability condition is then derived. Such a mode stability condition is essential in the design of stable optical resonators for laser applications. At the end of this chapter, we discuss an important concept of mode coupling in resonators.

4.1 FABRY–PEROT ETALON

The Fabry–Perot etalon, or interferometer, named after its inventors [3], can be considered as the best example of an optical resonator. It consists of a plane-parallel plate of thickness l and refractive index n that is immersed in a medium of index n'. Generally, an etalon can be obtained by spacing two partially reflecting mirrors at a distance l apart so that $n = n' = 1$. Another common form of etalon is produced by grinding two plane-parallel (or curved) faces on a transparent solid and then evaporating a metallic or dielectric layer (or layers) on the surfaces. The term etalon is often reserved for a plane-parallel plate of solid transparent material with reflecting surfaces on both sides, while interferometers (or cavities) are reserved for structures that consist of two parallel mirrors with an empty space in between.

Let a plane wave be incident on the etalon at an angle θ' to the normal, as shown in Figure 4.2. We can treat the problem of transmission (and reflection) of the plane wave through the etalon by considering the infinite number of partial waves produced by multiple reflections at the two end surfaces. The phase delay between two partial waves—which is attributable to one additional round trip—is given, according to Figure 4.3, by

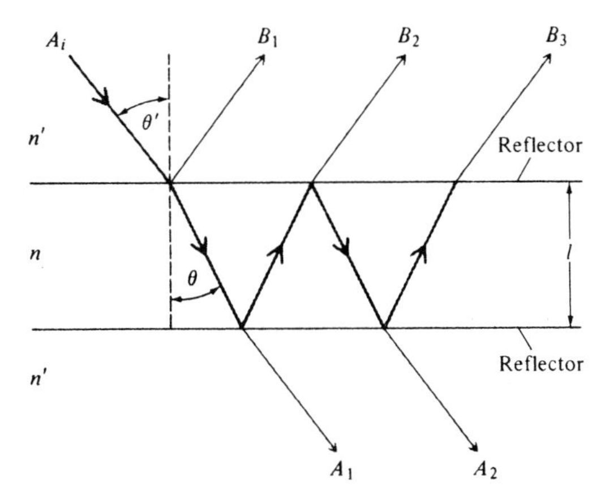

Figure 4.2 Multiple-reflection model for analyzing the Fabry–Perot etalon.

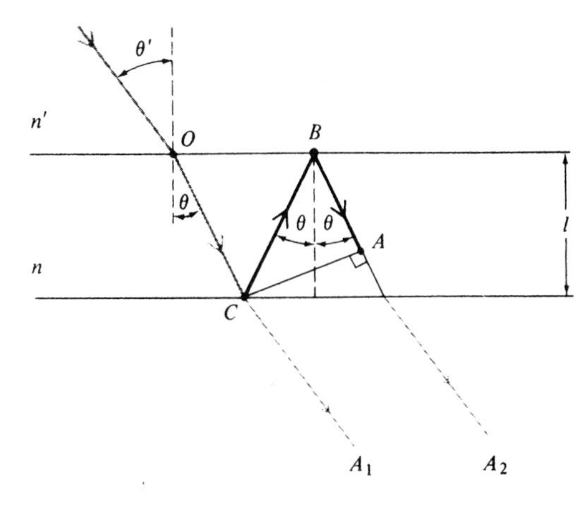

Figure 4.3 Path difference between successive reflections inside the Fabry–Perot etalon. Their path difference ΔL is given by $\Delta L = AB + BC = l(\cos 2\theta/\cos\theta) + l/\cos\theta = 2l\cos\theta$, $\rightarrow \delta = (2\pi(\Delta L)n)/\lambda = (4\pi nl\cos\theta)/\lambda$

$$\delta = \frac{4\pi nl\cos\theta}{\lambda} = 2k_x l \tag{4.1-1}$$

where λ is the vacuum wavelength of the incident wave, θ is the internal angle of incidence, and k_x is the x component of the wavevector (x axis is perpendicular to the mirrors). At normal incidence, this phase shift is simply $\delta = 2kl$, where k is the wavenumber of propagation in the medium. Thus the phase shift is often called the round-trip phase shift. If the complex amplitude of the incident wave is taken as A_i, then the partial reflection amplitudes, B_1, B_2, and so forth, are given by

$$B_1 = rA_i, \quad B_2 = tt'r'A_i e^{-i\delta}, \quad B_3 = tt'r'^3A_i e^{-i2\delta}, \quad \ldots$$

where r is the amplitude reflection coefficient of the interface (ratio of reflected to incident amplitude), t is the amplitude transmission coefficient for waves incident from n' toward n, and r' and t' are the corresponding quantities for waves traveling from n toward n'. The complex amplitude of the (total) reflected wave is $A_r = B_1 + B_2 + B_3 + \cdots$, or

$$A_r = [r + tt'r'e^{-i\delta}(1 + r'^2e^{-i\delta} + r'^4e^{-2i\delta} + \cdots)]A_i \tag{4.1-2}$$

For the transmitted wave, the amplitudes of the partial waves are

$$A_1 = tt'A_ie^{-i\delta/2}, \quad A_2 = tt'r'^2e^{-i\delta}A_ie^{-i\delta/2}, \quad A_3 = tt'r'^4e^{-2i\delta}A_ie^{-i\delta/2}, \quad \ldots$$

where the phase factor $e^{-i\delta/2}$ corresponds to a single traversal of the plate and is common to all the terms. Adding up the A terms, we obtain

$$A_t = A_i tt'(1 + r'^2 e^{-i\delta} + r'^4 e^{-2i\delta} + \cdots)e^{-i\delta/2} \tag{4.1-3}$$

for the complex amplitude of the total transmitted wave. We notice that the terms within the parentheses in Equations (4.1-2) and (4.1-3) form an infinite geometric series; adding them, we get

$$A_r = \frac{(1 - e^{-i\delta})\sqrt{R}}{1 - Re^{-i\delta}} A_i \tag{4.1-4}$$

and

$$A_t = \frac{Te^{-i\delta/2}}{1 - Re^{-i\delta}} A_i \tag{4.1-5}$$

where we used the fact that $r' = -r$ for the dielectric interface, the conservation-of-energy relation that applies to lossless mirrors,

$$r^2 + tt' = 1$$

as well as the definitions

$$R \equiv r^2 = r'^2 \quad \text{and} \quad T \equiv tt'$$

R and T are, respectively, the fraction of the intensity reflected and transmitted at each interface and will be referred to in the following discussion as the mirrors' reflectance and transmittance. It is important to note $r' = -r$ and $r^2 + tt' = 1$ are valid only for the interface between two dielectric media. For mirrors with dielectric coatings, similar relationships exist. They will be discussed in Section 4.9.

If the incident intensity (watts per square meter) is taken as $A_i A_i^*$, we obtain from Equation (4.1-4) the following expression for the fraction of the incident intensity that is reflected by the etalon:

$$\frac{I_r}{I_i} = \frac{A_r A_r^*}{A_i A_i^*} = \frac{4R \sin^2(\delta/2)}{(1 - R)^2 + 4R \sin^2(\delta/2)} \tag{4.1-6}$$

Moreover, from Equation (4.1-5),

$$\frac{I_t}{I_i} = \frac{A_t A_t^*}{A_i A_i^*} = \frac{(1 - R)^2}{(1 - R)^2 + 4R \sin^2(\delta/2)} \tag{4.1-7}$$

for the transmitted fraction. Our basic model contains no loss mechanisms, so conservation of energy requires that $I_r + I_t$ be equal to I_i, as is indeed the case.

Let us consider the transmission characteristics of a Fabry–Perot etalon. According to Equation (4.1-7) the transmission is unity whenever

$$\delta = \frac{4\pi nl \cos\theta}{\lambda} = 2m\pi, \quad m = \text{any integer} \tag{4.1-8}$$

The condition (4.1-8) for maximum transmission can be written

$$\nu_m = m\frac{c}{2nl\cos\theta}, \quad m = \text{any integer} \tag{4.1-9}$$

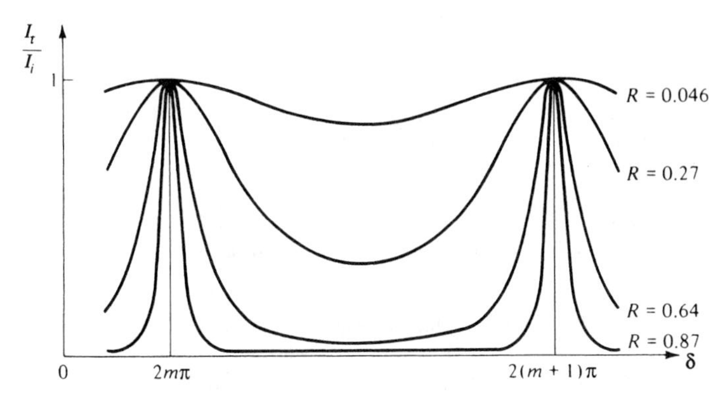

Figure 4.4 Transmission characteristics (theoretical) of a Fabry–Perot etalon. (After Reference [4].)

where $c = \nu\lambda$ is the velocity of light in vacuum and ν is the optical frequency. For a fixed l and θ, Equation (4.1-9) defines the unity transmission (resonance) frequencies of the etalon. These are separated by the *free spectral range*,

$$\Delta\nu \equiv \nu_{m+1} - \nu_m = \frac{c}{2nl\cos\theta} \tag{4.1-10}$$

Equation (4.1-10) is obtained by assuming that the index of refraction n is a constant. In the case when the spectral dispersion cannot be neglected, the free spectral range becomes

$$\Delta\nu \equiv \nu_{m+1} - \nu_m = \frac{c}{2n_g l\cos\theta} = \frac{v_g}{2l\cos\theta} \tag{4.1-11}$$

where v_g is the group velocity of light in the medium and n_g is the group index defined in Equation (4.0-14).

Theoretical transmission plots of a Fabry–Perot etalon are shown in Figure 4.4. The maximum transmission is unity, as stated previously. The minimum transmission, on the other hand, approaches zero as the mirror reflectivity R approaches unity.

If we allow for the existence of loss or gain in the etalon medium, we find that the peak transmission is no longer unity. Let the intensity gain (or loss) per pass be G, defined as $G = I_{\text{output}}/I_{\text{input}}$. We find that the intensity transmission becomes

$$\frac{I_t}{I_i} = \frac{A_t A_t^*}{A_i A_i^*} = \frac{(1-R)^2 G}{(1-GR)^2 + 4GR\sin^2(\delta/2)} \tag{4.1-12}$$

The proof of Equation (4.1-12) is left as an exercise (Problem 4.2). The maximum transmission can be greater than unity when gain is present. In fact, the maximum transmission can be much greater than unity when the product GR is approaching unity. According to Equation (4.1-12), the maximum transmission can be written

$$\left(\frac{I_t}{I_i}\right)_{\max} = \frac{(1-R)^2 G}{(1-GR)^2} \tag{4.1-13}$$

In the case of loss, G is less than 1, and the maximum transmission is less than unity.

If the loss is due to linear absorption with an absorption coefficient α in the medium, then G can be written

$$G = \exp(-\alpha l) \tag{4.1-14}$$

An experimental transmission plot of a Fabry–Perot etalon is shown in Figure 4.5.

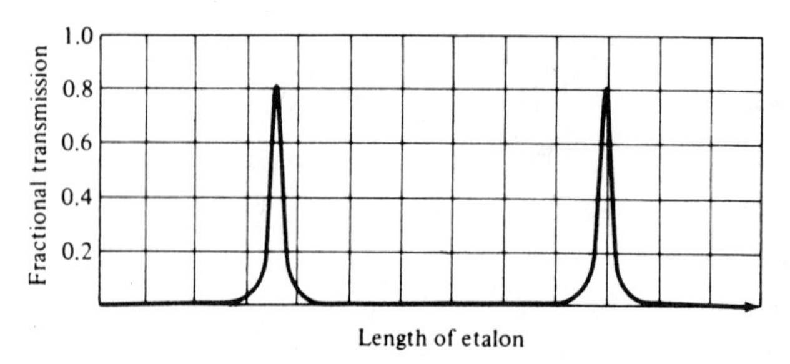

Figure 4.5 Experimental transmission characteristics of a Fabry–Perot etalon at $\lambda = 632.8$ nm as a function of the etalon optical length with $R = 0.9$ and $G = 0.98$. The two peaks shown correspond to a change in the optical length $\Delta(nl) = \lambda/2$. (After Reference [5].)

Intracavity field

We mentioned earlier that the field intensity inside the cavity can be built up significantly at resonance. We now examine the field inside the cavity. Let I_i be the intensity of the incident beam, and I_o be the intensity of the transmitted beam. Inside the Fabry–Perot etalon, the electric field can be written as the sum of a right-traveling wave with an intensity I_1 (traveling along the $+x$ direction) and a left-traveling wave with an intensity I_2 (traveling along the $-x$ direction). If the cavity medium is lossless, the Poynting vector inside the cavity can be written as the algebraic sum of the Poynting vector of these two traveling waves. Let I_1 and I_2 be the magnitudes of the Poynting vectors of these two waves, respectively. If we further assume that the mirrors are lossless, then these intensities are related by

$$I_o = \frac{(1-R)^2}{(1-R)^2 + 4R\sin^2(\delta/2)} I_i \tag{4.1-15}$$

$$I_o = TI_1 = (1-R)I_1 \tag{4.1-16}$$

$$I_2 = RI_1 \tag{4.1-17}$$

where we recall that R is the mirror reflectivity, and $T = 1 - R$ is the mirror transmittance. According to these equations, the intracavity intensities at resonance ($\delta = 2\pi, 4\pi, \ldots$) are given by

$$I_1 = \frac{1}{(1-R)} I_i, \quad I_2 = \frac{R}{(1-R)} I_i \tag{4.1-18}$$

The right-traveling wave and the left-traveling wave form an interference (standing wave) pattern inside the cavity (see Figure 4.6). The intensity pattern in the cavity can be written

$$I(x) = I_1 + I_2 + 2\sqrt{I_1 I_2}\cos(2kx + \alpha_0) \tag{4.1-19}$$

where k is the wavenumber in the cavity, and α_0 is a constant phase due to possible phase shifts upon reflection from the cavity mirrors. It can easily be shown that the average intensity inside the cavity can be written, according to Equations (4.1-18) and (4.1-19),

$$I_{\text{ave}} = \frac{1+R}{1-R} I_i \tag{4.1-20}$$

For a mirror reflectivity of $R = 0.99$, the average intracavity intensity in such a Fabry–Perot etalon is 199 times that of the incident beam at resonance when maximum transmission occurs.

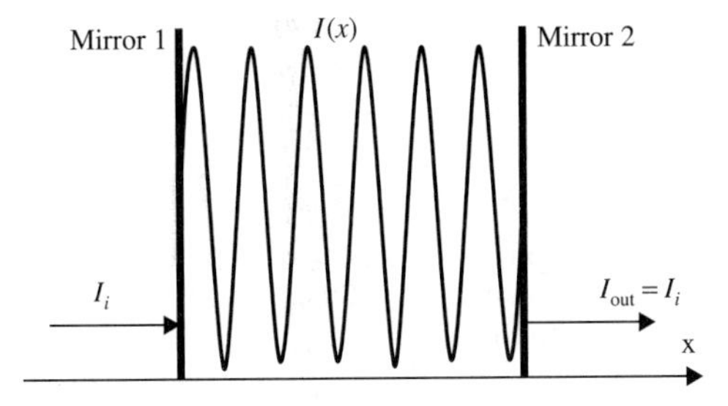

Figure 4.6 Intensity distribution inside the cavity at resonance.

As shown in Figure 4.4, the transmitted beam intensity is spectrally modulated. The transmission is maximum at resonance when the cavity length is an integral number of half-wavelengths. In addition to spectral modulation, the phase shift of the transmitted beam is also being spectrally modulated. The phase dispersion is especially strong at resonance. This is related to the build up of field intensity inside the cavity. We define ϕ as the one-pass phase shift inside the cavity (i.e., $\phi = \delta/2$). The etalon transmission coefficient can be written, according to Equation (4.1-5),

$$t_e = \frac{Te^{-i\phi}}{1 - Re^{-2i\phi}} = |t_e|e^{-i\psi} \tag{4.1-21}$$

where ψ is the phase shift of the transmitted beam. It can be shown that

$$\psi = \phi + \tan^{-1}\left(\frac{R \sin 2\phi}{1 - R \cos 2\phi}\right) \tag{4.1-22}$$

We now examine the derivative of ψ with respect to the angular frequency ω:

$$\frac{d\psi}{d\omega} = \frac{d\psi}{d\phi}\frac{d\phi}{d\omega} \tag{4.1-23}$$

Physically, the frequency derivative of the phase shift is the flight time (or group delay). If we write

$$\tau_e = \frac{d\psi}{d\omega} \quad \text{and} \quad \tau = \frac{d\phi}{d\omega} \tag{4.1-24}$$

where τ is the one-pass flight time inside the cavity medium and τ_e is the flight time through the etalon, then we obtain

$$\tau_e = \frac{d\psi}{d\phi}\tau \tag{4.1-25}$$

Thus $d\psi/d\phi$ may be viewed as an enhancement factor of the time of flight due to the etalon. By taking the derivative, we obtain

$$\frac{d\psi}{d\phi} = \left(\frac{1 + R}{1 - R}\right) \quad \text{when } \phi = \pi, 2\pi, 3\pi, \ldots \tag{4.1-26}$$

Combining the above two equations, we obtain

$$\tau_e = \left(\frac{1 + R}{1 - R}\right)\tau \quad \text{when } \phi = \pi, 2\pi, 3\pi, \ldots \tag{4.1-27}$$

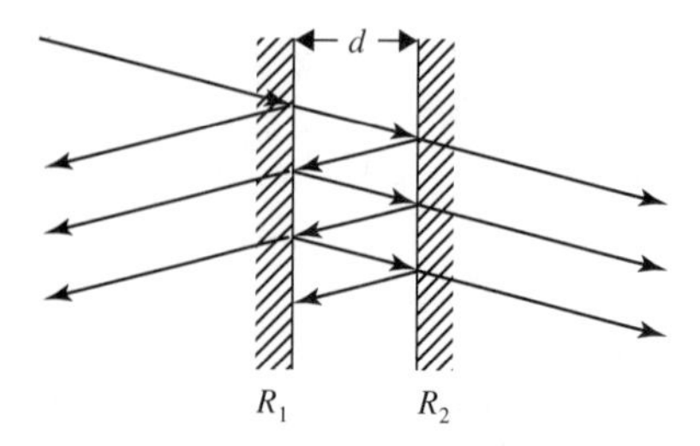

Figure 4.7 Schematic drawing of an asymmetric Fabry–Perot etalon with mirror reflectivities R_1 and R_2, and a cavity spacing of d.

We note that the time of flight (group delay) is enhanced by a factor of $(1 + R)/(1 - R)$ at resonance. For a mirror reflectivity of $R = 0.99$, the group delay at resonance is enhanced by a factor of 199. This is in agreement with the increase in the intracavity energy density at resonance. In a lossless system, the Poynting vector is a constant. If we write the Poynting vector as the product of energy density and the speed of light, then the effective speed of light must be low in the region where the energy density is high.

Asymmetric Fabry–Perot Etalons

Most Fabry–Perot etalons employed in spectral analysis are made of two identical dielectric mirrors facing each other. These mirrors are usually made of multilayer dielectric coatings on optically flat dielectric substrates. The symmetric etalons can provide a high finesse as well as a maximum transmission of 100% at specific wavelengths. Generally, an etalon can be made of two mirrors of different mirror reflectivities. Figure 4.7 shows a schematic drawing of an asymmetric Fabry–Perot etalon, with mirror reflectivities R_1 and R_2. An extreme case of the asymmetric Fabry–Perot etalon is the Gires–Tournois etalon, where $R_2 = 1$.

Using a similar analysis leading to Equations (4.1-2) and (4.1-3), we obtain

$$A_r = \{r_1 + t_1 t_1' r_2' e^{-i\delta}[1 + r_1' r_2' e^{-i\delta} + (r_1' r_2' e^{-i\delta})^2 + \cdots]\}A_i \tag{4.1-28}$$

$$A_t = t_1 t_2' e^{-i\delta/2}[1 + r_1' r_2' e^{-i\delta} + (r_1' r_2' e^{-i\delta})^2 + \cdots]A_i \tag{4.1-29}$$

with

$$\delta = \frac{4\pi n d \cos\theta}{\lambda} = 2k_x d \tag{4.1-30}$$

where n is the index of refraction of the cavity medium, k_x is the x component of the wavevector, d is the cavity spacing, and r_1, r_1', r_2, r_2', t_1, t_1', t_2 and t_2' are reflection and transmission coefficients of the mirrors. The prime indicates that the mirror reflection and transmission coefficients are defined for the case when the incident beam is launched from inside the etalon. The four coefficients associated with each mirror are, in general, complex numbers. They are related by the fundamental principles of physics, including reciprocity and time-reversal symmetry (see Problem 4.16). Physically, δ is the round-trip propagation phase shift inside the etalon.

By summing up the infinite geometric series, we obtain the etalon reflection coefficient and transmission coefficient, respectively:

$$r \equiv \frac{A_r}{A_i} = r_1 + \frac{t_1 t_1' r_2' e^{-i\delta}}{1 - r_1' r_2' e^{-i\delta}} = \frac{r_1 + (t_1 t_1' - r_1 r_1')r_2' e^{-i\delta}}{1 - r_1' r_2' e^{-i\delta}} \tag{4.1-31}$$

$$t \equiv \frac{A_t}{A_i} = \frac{t_1 t_2' e^{-i\delta/2}}{1 - r_1' r_2' e^{-i\delta}} \tag{4.1-32}$$

For dielectric mirrors made of lossless materials, the reflection coefficients and transmission coefficients are related by the Stokes relationships [6] (see Problems 4.16 and 4.18). Using the Stokes relationships and Equations (4.1-31) and (4.1-32), the intensity transmission and reflection coefficient of the etalon can be written

$$T_{\text{etalon}} = \frac{|A_t|^2}{|A_i|^2} = \frac{(1-R_1)(1-R_2)}{(1-\sqrt{R_1}\sqrt{R_2})^2 + 4\sqrt{R_1}\sqrt{R_2}\sin^2(\phi)} \qquad (4.1\text{-}33)$$

$$R_{\text{etalon}} = \frac{|A_r|^2}{|A_i|^2} = \frac{(\sqrt{R_1}-\sqrt{R_2})^2 + 4\sqrt{R_1}\sqrt{R_2}\sin^2(\phi)}{(1-\sqrt{R_1}\sqrt{R_2})^2 + 4\sqrt{R_1}\sqrt{R_2}\sin^2(\phi)} \qquad (4.1\text{-}34)$$

where R_1 and R_2 are the mirror reflectivities, and the phase ϕ is given by

$$2\phi = \delta + \rho_1' + \rho_2' \qquad (4.1\text{-}35)$$

where ρ_1' and ρ_2' are the phase shifts of reflections from mirrors 1 and 2, respectively (i.e., $r_1' = |r_1'| \exp(-i\rho_1')$ and $r_2' = |r_2'| \exp(-i\rho_2')$). We note that both the etalon reflectivity and transmittance are periodic functions of ϕ, which is a linear function of the frequency ω. For lossless mirrors with $R_1 + T_1 = 1$ and $R_2 + T_2 = 1$, it can easily be shown that

$$T_{\text{etalon}} + R_{\text{etalon}} = 1 \qquad (4.1\text{-}36)$$

which is in agreement with the conservation of energy. Note that zero etalon reflection is possible only when $R_1 = R_2$, according to Equation (4.1-34). In other words, zero reflection (or unity maximum transmission) occurs only in symmetric Fabry–Perot etalons at resonance.

Absorption in Fabry–Perot Etalons

Unity maximum transmission is possible only when the mirrors as well as the cavity medium are lossless. In practice, the transmission of light through an etalon can be reduced due to material absorption in the cavity, scattering loss in the cavity or in the mirror, and material absorption in the mirror. Let α be the intensity attenuation coefficient (or gain if $\alpha < 0$) in the cavity medium. The etalon transmission and reflection coefficient become

$$T_{\text{etalon}} = \frac{|A_t|^2}{|A_i|^2} = \frac{(1-R_1)(1-R_2)e^{-\alpha L}}{(1-\sqrt{R_1}\sqrt{R_2}e^{-\alpha L})^2 + 4\sqrt{R_1}\sqrt{R_2}e^{-\alpha L}\sin^2(\phi)} \qquad (4.1\text{-}37)$$

$$R_{\text{etalon}} = \frac{|A_r|^2}{|A_i|^2} = \frac{(\sqrt{R_1}-\sqrt{R_2}e^{-\alpha L})^2 + 4\sqrt{R_1}\sqrt{R_2}e^{-\alpha L}\sin^2(\phi)}{(1-\sqrt{R_1}\sqrt{R_2}e^{-\alpha L})^2 + 4\sqrt{R_1}\sqrt{R_2}e^{-\alpha L}\sin^2(\phi)} \qquad (4.1\text{-}38)$$

where L is the length of the medium in which the intensity attenuation is present. Equations (4.1-37) and (4.1-38) are general forms of intensity transmission and reflection of an asymmetric Fabry–Perot etalon. They can be used to describe some special cases of interest.

If we define $1 - A$ as the one-way passage loss in the cavity, then $A = \exp(-\alpha L)$. The transmission at resonance is given by, according to Equation (4.1-37),

$$(T_{\text{etalon}})_{\text{resonance}} = \frac{|A_t|^2}{|A_i|^2} = \frac{(1-R_1)(1-R_2)A}{(1-\sqrt{R_1}\sqrt{R_2}A)^2} \qquad (4.1\text{-}39)$$

which is in agreement with Equation (4.1-13).

Gires–Tournois Etalons

A very special class of asymmetric Fabry–Perot etalons is known as the Gires–Tournois etalons (GTEs). In these etalons, the rear mirror has a mirror reflectivity of 100% ($R_2 = 1$).

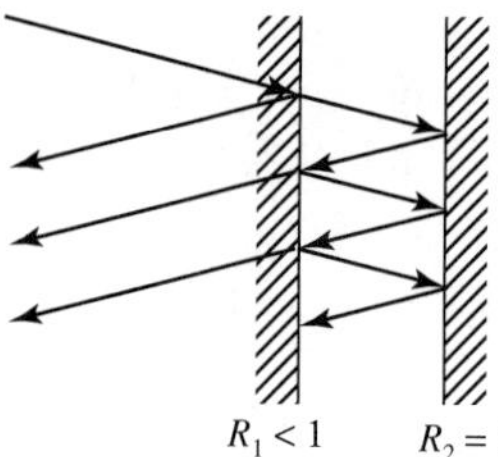

Figure 4.8 Schematic drawing of a Gires–Tournois etalon (GTE).

In this case, all wavelengths of light are reflected, as nothing can transmit through the rear mirror, provided the structure is lossless ($\alpha = 0$). Figure 4.8 shows a schematic drawing of a Gires–Tournois etalon.

In the case of lossless structures, a Gires–Tournois etalon is an all-pass reflector, as nothing can be transmitted through the rear mirror. Without loss of generality, we let $r_1' = -r_1 = \sqrt{R}$, $r_2' = 1$, and $t_1 t_1' - r_1 r_1' = 1$ in Equation (4.1-31) and obtain

$$r_{\text{etalon}} = \exp(-i\Phi) = \frac{-\sqrt{R} + e^{-2i\phi}}{1 - \sqrt{R}e^{-2i\phi}} \tag{4.1-40}$$

where the phase shift ϕ is given by Equation (4.1-35). In cases where the mirror phase shifts can be ignored, the phase can be written (for the case of normal incidence)

$$\phi = \frac{2\pi}{\lambda} nd = \frac{\omega}{c} nd \tag{4.1-41}$$

where d is the space between the mirrors and n is the index of refraction of the medium. It can be shown that the phase shift of a Gires–Tournois etalon can be written, according to Equation (4.1-40),

$$\Phi = 2 \tan^{-1}\left(\frac{1 + \sqrt{R}}{1 - \sqrt{R}} \tan \phi\right) \equiv 2 \tan^{-1}(\sigma \tan \phi) \tag{4.1-42}$$

where

$$\sigma = \frac{1 + \sqrt{R}}{1 - \sqrt{R}} \tag{4.1-43}$$

Figure 4.9 shows the etalon phase shift as a function of ϕ, which is proportional to the frequency. We note that strong phase dispersion occurs at resonance when $\phi = \pi, 2\pi, 3\pi, \ldots$.

By taking the derivative with respect to ω, we obtain

$$\frac{d\Phi}{d\omega} = \frac{d\Phi}{d\phi}\frac{d\phi}{d\omega} = \frac{2\sigma}{1 + (\sigma^2 - 1)\sin^2\phi}\frac{d\phi}{d\omega} = \frac{\sigma}{1 + (\sigma^2 - 1)\sin^2\phi}\tau_0 \tag{4.1-44}$$

where $\tau_0 = 2\, d\phi/d\omega$ is the round-trip group delay of the etalon when the front mirror reflectivity is zero. If we ignore the phase shift dispersion of the rear reflector, then $\tau_0 = 2d/v_g = 2n_g d/c$, where v_g is the group velocity of light in the medium between the mirrors, and n_g is the corresponding group index. Equation (4.1-44) is an expression of the time delay (group delay) due to reflection from the etalon when the front mirror reflectivity is finite. The dimensionless factor $\sigma/[1 + (\sigma^2 - 1)\sin^2\phi]$ in Equation (4.1-44) is exactly one-half of the slope of the etalon phase shift as shown in Figure 4.9. We note that the slope achieves a maximum of 2σ at resonance when $\phi = \pi, 2\pi, 3\pi, \ldots$ and reaches a minimum of $2/\sigma$ when $\phi = m\pi + \pi/2$, m = integer. According to Equation (4.1-43), σ is an increasing function of the front mirror

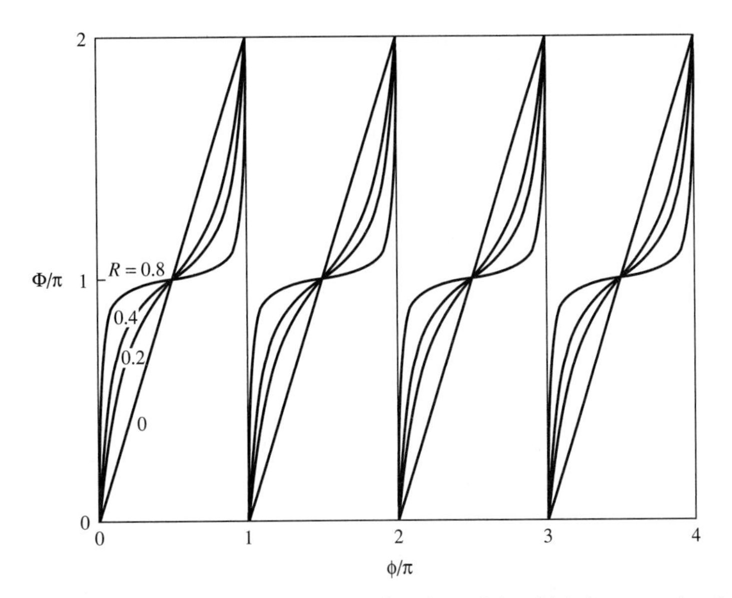

Figure 4.9 Etalon phase shift as a function of ϕ, which is proportional to the frequency. The front mirror reflectivities are $R = 0, 0.2, 0.4, 0.8$.

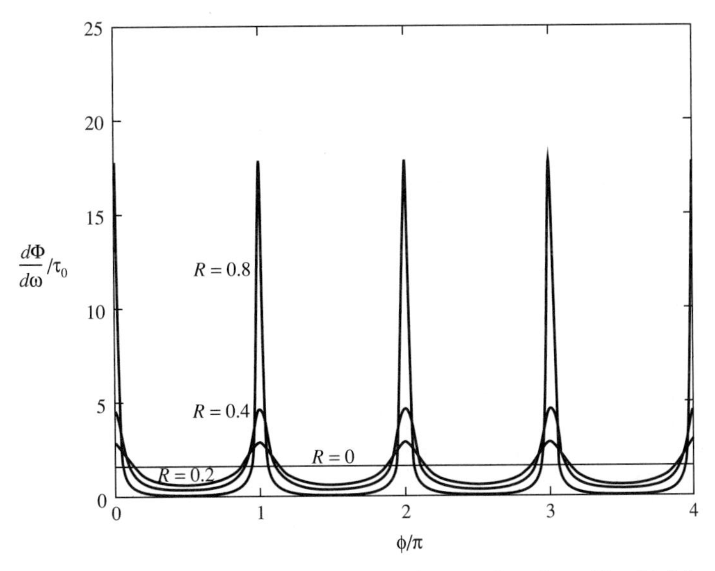

Figure 4.10 Etalon group delay in units of τ_0 as a function of ϕ, which is proportional to the frequency. The front mirror reflectivities are $R = 0, 0.2, 0.4, 0.8$.

reflectivity R. At resonance, the group delay of the etalon becomes $\sigma\tau_0$ and the field energy builds up inside the cavity. Figure 4.10 shows the group delay of the etalon as a function of ϕ, which is proportional to the frequency. For a front mirror reflectivity of $R = 0.8$, the group delay becomes about $18\tau_0$.

When $\phi = m\pi + \pi/2$, $m =$ integer, the etalon group delay becomes τ_0/σ. This is the situation of antiresonance when destructive interference occurs in the cavity.

4.2 FABRY–PEROT ETALONS AS OPTICAL SPECTRUM ANALYZERS

According to Equation (4.1-8), the maximum transmission of a Fabry–Perot etalon occurs when

$$\frac{2nl\cos\theta}{\lambda} = m \tag{4.2-1}$$

Taking, for simplicity, the case of normal incidence ($\theta = 0°$), we obtain the following expression for the change $d\nu$ in the resonance frequency of a given transmission peak due to a length variation dl:

$$\frac{d\nu}{\Delta\nu} = -\frac{dl}{(\lambda/2n)} \tag{4.2-2}$$

where $\Delta\nu$ is the intermode frequency separation as given by Equation (4.1-10). According to Equation (4.2-2), we can tune the peak transmission frequency of the etalon by $\Delta\nu$ by changing its length by half a wavelength. This property is utilized in operating the etalon as a scanning interferometer. The optical signal to be analyzed passes through the etalon as its length is being swept. If the width of the transmission peaks is small compared to that of the spectral detail in the incident optical beam signal, the output of the etalon will constitute a replica of the spectral profile of the signal. In this application it is important that the spectral width of the signal beam be smaller than the intermode spacing of the etalon ($c/2nl$) so that the ambiguity due to simultaneous transmission through more than one transmission peak is avoided. For the same reason, the total length scan is limited to $dl < \lambda/2n$. Figure 4.11 demonstrates the operation of a scanning Fabry–Perot etalon; Figure 4.12 shows intensity versus frequency data obtained by analyzing the output of a multimode He–Ne laser oscillating near 6328 Å. The peaks shown correspond to longitudinal laser modes, which will be discussed in Section 4.5.

It is clear from the foregoing that when operating as a spectrum analyzer the etalon resolution—that is, its ability to distinguish details in the spectrum—is limited by the finite width of its transmission peaks. If we take, somewhat arbitrarily,[1] the limiting resolution of

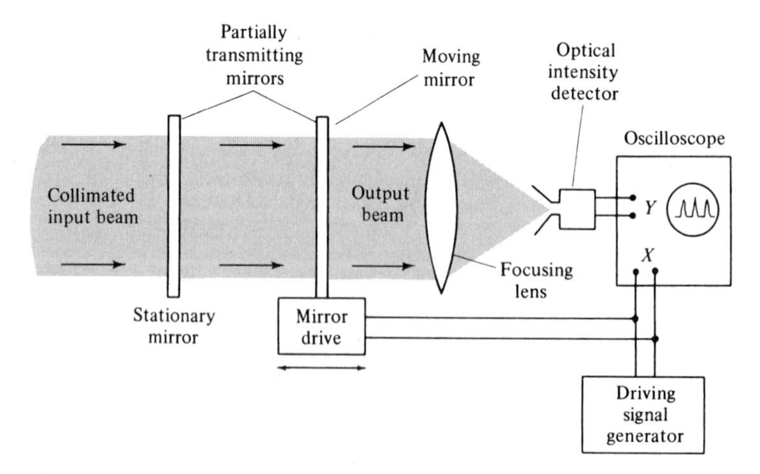

Figure 4.11 Typical scanning Fabry–Perot interferometer experimental arrangement.

[1] For a more complete discussion concerning the definition of resolution, see Reference [4].

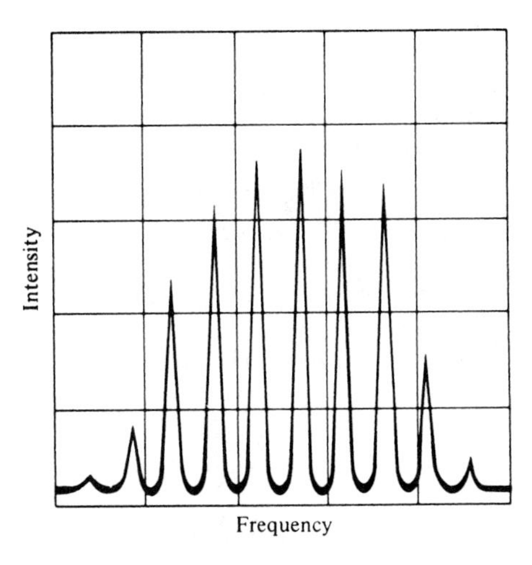

Figure 4.12 Intensity versus frequency analysis of the output of a He–Ne 6328 Å laser obtained with a scanning Fabry–Perot etalon. The horizontal scale is 250 MHz per division.

the etalon as the separation $\Delta\nu_{1/2}$ between the two frequencies at which the transmission is down to half of its peak value, then from Equation (4.1-7) we obtain

$$\sin^2\left(\frac{\delta_{1/2} - 2m\pi}{2}\right) = \frac{(1-R)^2}{4R} \tag{4.2-3}$$

where $\delta_{1/2}$ is the value of δ corresponding to the two half-power points—that is, the value of δ at which the denominator of Equation (4.1-7) is equal to $2(1-R)^2$. If we assume $(\delta_{1/2} - 2m\pi) \ll \pi$, so that the width of the high-transmission regions in Figure 4.4 is small compared to the separation between the peaks, we obtain

$$\Delta\nu_{1/2} = \frac{c}{2\pi nl \cos\theta}(\delta_{1/2} - 2m\pi) \simeq \frac{c}{2\pi nl \cos\theta}\frac{1-R}{\sqrt{R}} \tag{4.2-4}$$

or using Equation (4.1-10) and defining the etalon *finesse* as

$$F \equiv \frac{\pi\sqrt{R}}{1-R} \tag{4.2-5}$$

we obtain

$$\Delta\nu_{1/2} = \frac{\Delta\nu}{F} = \frac{c}{2nl\cos\theta F} \tag{4.2-6}$$

for the limiting resolution. The finesse F (which is used as a measure of the resolution of a Fabry–Perot etalon) is, according to Equation (4.2-6), the ratio of the separation between peaks to the width of a transmission bandpass. This ratio can be read directly from the transmission characteristics such as those of Figure 4.5, for which we obtain $F \simeq 26$.

EXAMPLE: DESIGN OF A FABRY–PEROT ETALON

Consider the problem of designing a scanning Fabry–Perot etalon to be used in studying the mode structure of a He–Ne laser with the following characteristics: $l_{\text{laser}} = 100$ cm and the region of oscillation $= \Delta\nu_{\text{gain}} \simeq 1.5 \times 10^9$ Hz.

The free spectral range of the etalon (i.e., its intermode spacing) must exceed the spectral region of interest, so from Equation (4.1-10) we obtain

$$\frac{c}{2nl_{\text{etal}}} \geq 1.5 \times 10^9 \text{ Hz} \quad \text{or} \quad 2nl_{\text{etal}} \leq 20 \text{ cm} \tag{4.2-7}$$

The separation between longitudinal modes of the laser oscillation is $c/2nl_{\text{laser}} = 1.5 \times 10^8$ Hz (here we assume $n = 1$). We choose the resolution of the etalon to be a tenth of this value, so spectral details as narrow as 1.5×10^7 Hz can be resolved. According to Equation (4.2-6), this resolution can be achieved if

$$\Delta\nu_{1/2} = \frac{c}{2nl_{\text{etal}}F} \leq 1.5 \times 10^7 \text{ Hz} \quad \text{or} \quad 2nl_{\text{etal}}F \geq 2 \times 10^3 \text{ cm} \tag{4.2-8}$$

To satisfy condition (4.2-7), we choose $2nl_{\text{etal}} = 20$ cm; thus condition (4.2-8) is satisfied when

$$F \geq 100 \tag{4.2-9}$$

A finesse of 100 requires, according to Equation (4.2-5), a mirror reflectivity of approximately 97%.

As a practical note we may add that the finesse, as defined by the first equality in Equation (4.2-6), depends not only on R but also on the mirror flatness and the beam angular spread. These points are taken up in Problems 4.3 and 4.4.

Another important mode of optical spectrum analysis performed with Fabry–Perot etalons involves the fact that a noncollimated monochromatic beam incident on the etalon will emerge simultaneously, according to Equation (4.1-8), along many directions θ,[2] which corresponds to the various orders m. If the output is then focused by a lens, each such direction θ will give rise to a circle in the focal plane on the lens, and, therefore, each frequency component present in the beam leads to a family of circles. This mode of spectrum analysis is especially useful under transient conditions where scanning etalons cannot be employed. Further discussion of this topic is included in Problem 4.6.

4.3 OPTICAL RESONATORS WITH SPHERICAL MIRRORS

In this section we study the properties of optical resonators formed by two opposing spherical mirrors; see References [7] and [8]. We will show that the field solutions inside the resonators are those of the propagating Gaussian beams, which were considered in Chapter 2. Consequently, it is useful to start by reviewing the properties of these beams.

The field distribution corresponding to the (l, m) transverse mode is given, according to Equation (2.8-1), by

$$E_{l,m}(\mathbf{r}) = E_0 \frac{\omega_0}{\omega(z)} H_l\left(\sqrt{2}\frac{x}{\omega(z)}\right) H_m\left(\sqrt{2}\frac{y}{\omega(z)}\right) \times \exp\left(-\frac{x^2+y^2}{\omega^2(z)} - ik\frac{x^2+y^2}{2R(z)} - ikz + i(l+m+1)\eta\right) \tag{4.3-1}$$

where the spot size $\omega(z)$ is

$$\omega(z) = \omega_0\left[1 + \left(\frac{z}{z_0}\right)^2\right]^{1/2}, \quad z_0 = \frac{\pi\omega_0^2 n}{\lambda} \tag{4.3-2}$$

[2] Each direction θ corresponds in three dimensions to the surface of a cone with a half-apex angle θ.

and where ω_0, the minimum spot size, is a parameter characterizing the beam. The radius of curvature of the wavefront is

$$R(z) = z\left[1 + \left(\frac{\pi\omega_0^2 n}{\lambda z}\right)^2\right] = \frac{1}{z}[z^2 + z_0^2] \tag{4.3-3}$$

and the phase factor η is as follows:

$$\eta = \tan^{-1}\left(\frac{\lambda z}{\pi\omega_0^2 n}\right) \tag{4.3-4}$$

The sign of $R(z)$ is taken as positive when the center of curvature is to the left of the wavefront, and vice versa. According to Equations (4.3-1) and (4.3-2), the loci of the points at which the beam intensity (watts per square meter) is a given fraction of its intensity on the axis are the hyperboloids

$$x^2 + y^2 = \text{const.} \times \omega^2(z) \tag{4.3-5}$$

The hyperbolas generated by the intersection of these surfaces with planes that include the z axis are shown in Figure 4.13. These hyperbolas are normal to the phase fronts and thus correspond to the local direction of energy flow. The hyperboloid $x^2 + y^2 = \omega^2(z)$ is, according to Equation (4.3-1), the locus of the points where the exponential factor in the field amplitude is down to e^{-1} from its value on the axis. The quantity $\omega(z)$ is thus defined as the *mode spot size* at the plane z.

Given a beam of the type described by Equation (4.3-1), we can form an optical resonator merely by inserting at points z_1 and z_2 two reflectors with radii of curvature that match those of the propagating beam spherical phase fronts at these points. Since the surfaces are normal to the direction of energy propagation as shown in Figure 4.13, the reflected beam retraces itself; thus, if the phase shift between the mirrors is some multiple of 2π radians, a *self-reproducing stable field* configuration results.

Alternatively, given two mirrors with spherical radii of curvature R_1 and R_2 and some distance of separation l, we can, under certain conditions to be derived later, adjust the position $z = 0$ and the parameter ω_0 so that the mirrors coincide with two spherical wavefronts of the propagating beam defined by the position of the waist ($z = 0$) and ω_0. If, in addition, the mirrors can be made large enough to intercept the majority (99%, say) of the incident beam energy in the fundamental ($l = m = 0$) transverse mode, we may expect this mode to have a larger Q than higher-order transverse modes, which, according to Figure 2.7, have fields extending farther from the axis and consequently lose a larger fraction of their energy by "spilling" over the mirror edges (diffraction losses).

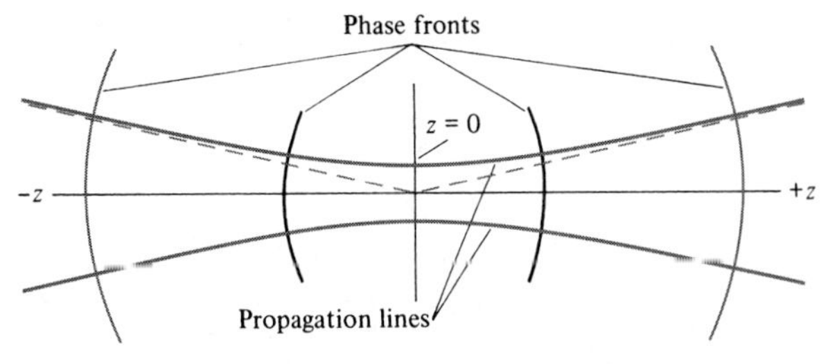

Figure 4.13 Hyperbolic curves corresponding to the local directions of propagation. The nearly spherical phase fronts represent possible positions for reflectors. Any two reflectors form a resonator with a transverse field distribution given by Equation (4.3-1).

Optical Resonator Algebra

As mentioned in the preceding paragraphs, we can form an optical resonator by using two reflectors, one at z_1 and the other at z_2, chosen so that their radii of curvature are the same as those of the beam wavefronts at the two locations. The propagating beam mode (4.3-1) is then reflected back and forth between the reflectors without a change in the transverse profile. The requisite radii of curvature, determined by Equation (4.3-3), are

$$R_1 = z_1 + \frac{z_0^2}{z_1}$$

$$R_2 = z_2 + \frac{z_0^2}{z_2}$$

from which we get

$$\begin{aligned} z_1 &= \frac{R_1}{2} \pm \frac{1}{2}\sqrt{R_1^2 - 4z_0^2} \\ z_2 &= \frac{R_2}{2} \pm \frac{1}{2}\sqrt{R_2^2 - 4z_0^2} \end{aligned} \tag{4.3-6}$$

For a given minimum spot size $\omega_0 = (\lambda z_0/\pi n)^{1/2}$, we can use (4.3-6) to find the positions z_1 and z_2 at which to place mirrors with curvatures R_1 and R_2, respectively.

In practice, we often start with given mirror curvatures R_1 and R_2 and a mirror separation l. The problem is then to find the minimum spot size ω_0, its location with respect to the reflectors, and the mirror spot sizes ω_1 and ω_2. Taking the mirror spacing as $l = z_2 - z_1$, we can solve (4.3-6) for z_0^2, obtaining

$$z_0^2 = \frac{l(-R_1 - l)(R_2 - l)(R_2 - R_1 - l)}{(R_2 - R_1 - 2l)^2} \tag{4.3-7}$$

where z_2 is to the right of z_1 (so that $l = z_2 - z_1 > 0$) and the mirror curvature is taken as positive when the center of curvature is to the left of the mirror.

The minimum spot size $\omega_0 = (\lambda z_0/\pi n)^{1/2}$ and its position is next determined from (4.3-6). The mirror spot sizes are then calculated by the use of Equation (4.3-2).

The Symmetrical Mirror Resonator

The special case of a resonator with symmetrically (about $z = 0$) placed mirrors merits a few comments. The planar phase front at which the minimum spot size occurs is, by symmetry, at $z = 0$. Putting $R_2 = -R_1 = R$ in Equation (4.3-7) gives

$$z_0^2 = \frac{(2R - l)l}{4} \tag{4.3-8}$$

and

$$\omega_0 = \left(\frac{\lambda z_0}{\pi n}\right)^{1/2} = \left(\frac{\lambda}{\pi n}\right)^{1/2}\left(\frac{l}{2}\right)^{1/4}\left(R - \frac{l}{2}\right)^{1/4} \tag{4.3-9}$$

which, when substituted in Equation (4.3-2) with $z = l/2$, yields the following expression for the spot size at the mirrors:

$$\omega_{1,2} = \left(\frac{\lambda l}{2\pi n}\right)^{1/2}\left(\frac{2R^2}{l(R - l/2)}\right)^{1/4} \tag{4.3-10}$$

A comparison with Equation (4.3-9) shows that, for $R \gg l$, $\omega_{1,2} \simeq \omega_0$ and the beam spread inside the resonator is small.

The value of R (for a given l) for which the mirror spot size is a minimum, is readily found from Equation (4.3-10) to be $R = l$. When this condition is fulfilled, we have what is called a symmetrical *confocal resonator*, since the two foci, occurring at a distance of $R/2$ from the mirrors, coincide. From Equation (4.3-9) we obtain

$$(\omega_0)_{\text{conf}} = \left(\frac{\lambda l}{2\pi n}\right)^{1/2} \tag{4.3-11}$$

whereas from Equation (4.3-10) we get

$$(\omega_{1,2})_{\text{conf}} = (\omega_0)_{\text{conf}}\sqrt{2} \tag{4.3-12}$$

so the beam spot size increases by $\sqrt{2}$ between the center and the mirrors.

EXAMPLE: DESIGN OF A SYMMETRICAL RESONATOR

Consider the problem of designing a symmetrical resonator for $\lambda = 10^{-4}$ cm with a mirror separation $l = 2m$. If we were to choose the confocal geometry with $R = l = 2$m, the minimum spot size (at the resonator center) would be, from Equation (4.3-11) and for $n = 1$,

$$(\omega_0)_{\text{conf}} = \left(\frac{\lambda l}{2\pi n}\right)^{1/2} = 0.0564 \text{ cm}$$

whereas, using Equation (4.3-12), the spot size at the mirrors would have the value

$$(\omega_{1,2})_{\text{conf}} = \omega_0\sqrt{2} \approx 0.0798 \text{ cm}$$

Assume next that a mirror spot size $\omega_{1,2} = 0.3$ cm is desired. Using this value in Equation (4.3-10) and assuming $R \gg l$, we get

$$\frac{\omega_{1,2}}{(\lambda l/2\pi n)^{1/2}} = \frac{0.3}{0.056} = \left(\frac{2R}{l}\right)^{1/4}$$

Thus

$$R \simeq 400l \simeq 800 \text{ m}$$

so that the assumption $R \gg l$ is valid. The minimum beam spot size ω_0 is found, through Equations (4.3-2) and (4.3-8), to be

$$\omega_0 = 0.994\omega_{1/2} \simeq 0.3 \text{ cm}$$

Thus to increase the mirror spot size from its minimum (confocal) value of 0.0798 cm to 0.3 cm, we must use exceedingly plane mirrors ($R = 800$ m). This also shows that even small mirror curvatures (i.e., large R) give rise to "narrow" beams.

The numerical example we have worked out applies equally well to the case in which a plane mirror is placed at $z = 0$. The beam pattern is equal to that existing in the corresponding half of the symmetric resonator in the example, so the spot size on the planar reflector is ω_0.

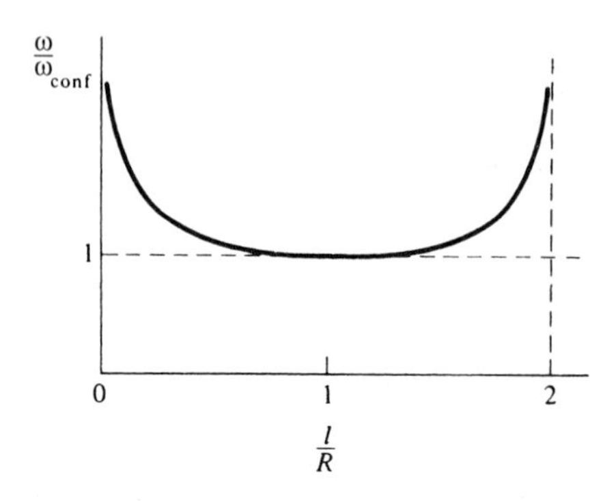

Figure 4.14 Ratio of beam spot size at the mirrors of a symmetrical resonator to its confocal ($l/R = 1$) value.

4.4 MODE STABILITY CRITERIA

The ability of an optical resonator to support low (diffraction) loss[3] modes depends on the mirrors' separation l and their radii of curvature R_1 and R_2. To illustrate this point, consider first the symmetric resonator with $R_2 = -R_1 = R$.

The ratio of the mirror spot size at a given l/R to its minimum confocal ($l/R = 1$) value, given by the ratio of Equation (4.3-10) to (4.3-12), is

$$\frac{\omega_{1,2}}{(\omega_{1,2})_{\text{conf}}} = \left(\frac{1}{(l/R)[2-(l/R)]}\right)^{1/4} \tag{4.4-1}$$

This ratio is plotted in Figure 4.14. For $l/R = 0$ (plane-parallel mirrors) and for $l/R = 2$ (two concentric mirrors), the spot size becomes infinite. It is clear that the diffraction losses from these cases are very high, since most of the beam energy "spills over" the reflector edges. Since, according to Table 2.1, the reflection of a Gaussian beam from a mirror with a radius of curvature R is formally equivalent to its transmission through a lens with a focal length $f = R/2$, the problem of the existence of stable confined optical modes in a resonator is formally the same as that of the existence of stable solutions for the propagation of a Gaussian beam in a biperiodic lens sequence, as shown in Figure 4.15. This problem was considered in Section 2.1 and led to the stability condition (2.1-16).

If, in Equation (2.1-16), we replace f_1 by $R_1/2$ and f_2 by $R_2/2$, we obtain the stability condition for optical resonators[4]:

$$0 \le \left(1 - \frac{l}{R_1}\right)\left(1 - \frac{l}{R_2}\right) \le 1 \tag{4.4-2}$$

A convenient representation of the stability condition (4.4-2) is by means of the diagram [8] shown in Figure 4.16. From this diagram, for example, it can be seen that the symmetric concentric ($R_1 = R_2 = l/2$), the confocal ($R_1 = R_2 = l$), and the plane-parallel ($R_1 = R_2 = \infty$) resonators are all on the verge of instability and thus may become extremely lossy by small deviations of the parameters in the direction of instability.

---

[3] By diffraction loss we refer to the fact that due to the beam spread (see Equation (2.5-18)), a fraction of the Gaussian beam energy 'misses" the mirror and is not reflected and is thus lost.

[4] This causes the sign convention of R_1 and R_2 to be different from that used in preceding sections. The sign of R is the same as that of the focal length of the equivalent lens. This makes R_1 (or R_2) positive when the center of curvature of mirror 1 (or 2) is in the direction of mirror 2 (or 1), and negative otherwise.

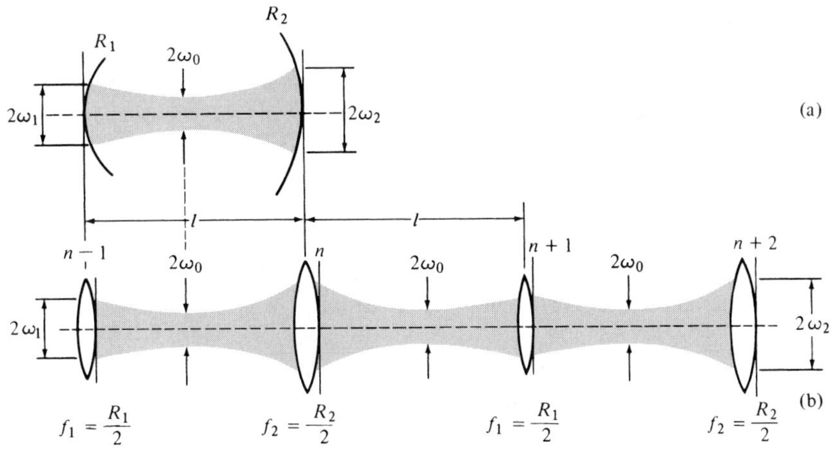

Figure 4.15 (a) Asymmetric resonator ($R_1 \neq R_2$) with mirror curvatures R_1 and R_2. (b) Biperiodic lens system (lens waveguide) equivalent to resonator shown in (a).

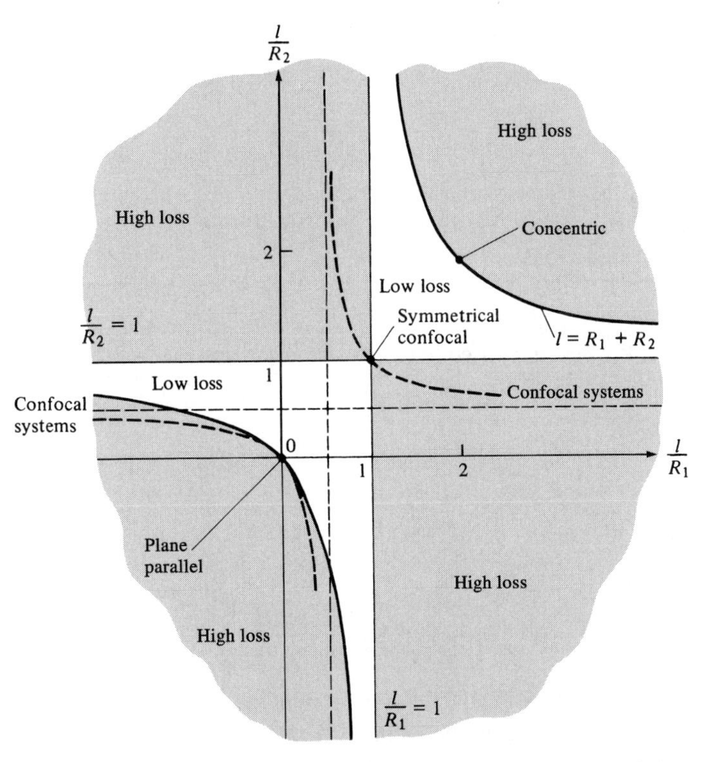

Figure 4.16 Stability diagram of optical resonator. Shaded (high-loss) areas are those in which the stability condition $0 \leq (1 - l/R_1)(1 - l/R_2) \leq 1$ is violated and the clear (low-loss) areas are those in which it is fulfilled. The sign convention for R_1 and R_2 is discussed in footnote 4. (After Reference [8].)

4.5 MODES IN A GENERALIZED RESONATOR—SELF-CONSISTENT METHOD

Up to this point we have treated resonators consisting of two opposing spherical mirrors. We may sometimes wish to consider the properties of more complex resonators made up of an arbitrary number of lenslike elements such as those shown in Table 2.1. A simple case of such a resonator may involve placing a lens between two spherical reflectors or constructing an off-axis three-reflector resonator. Yet another case is that of a traveling wave resonator in which the beam propagates in one sense only.

In either of these cases, we need to find if low-loss (i.e., stable) modes exist in the complex resonator and, if so, to solve for the spot size $\omega(z)$ and the radius of curvature $R(z)$ everywhere.

We apply the self-consistency condition and require that a stable eigenmode of the resonator is one that *reproduces itself after one round trip*. We choose an *arbitrary* reference plane in the resonator, denote the steady-state complex beam parameter at this plane as q_s, and, using the *ABCD* law (2.6-9), require that

$$q_s = \frac{Aq_s + B}{Cq_s + D} \tag{4.5-1}$$

where A, B, C, D are the "ray" matrix elements for one complete round trip—starting and ending at the chosen reference plane.

Solving Equation (4.5-1) for $1/q_s$ gives

$$\frac{1}{q_s} = \frac{(D - A) \pm \sqrt{(D - A)^2 + 4BC}}{2B} \tag{4.5-2}$$

Since the individual elements in the resonator are described by unimodular matrices, that is, $A_iD_i - B_iC_i = 1$ (see Table 2.1), it follows that the matrix A, B, C, D, which is the product of individual matrices, satisfies

$$AD - BC = 1$$

and Equation (4.5-2) can, consequently, be written

$$\frac{1}{q_s} = \frac{D - A}{2B} \pm i\frac{\sqrt{1 - [(D + A)/2]^2}}{B} = \frac{D - A}{2B} + \frac{i \sin\theta}{B} \tag{4.5-3}$$

where

$$\cos\theta = \frac{D + A}{2}$$

$$\theta = \pm\left|\cos^{-1}\left(\frac{D + A}{2}\right)\right| \tag{4.5-4}$$

According to Equation (2.5-14) the condition for a confined Gaussian beam is that the square of the beam spot size ω^2 be a finite positive number. Recalling that q is related to the spot size ω and the radius of curvature R as

$$\frac{1}{q} = \frac{1}{R} - i\frac{\lambda}{\pi\omega^2 n}$$

we find by comparing the last expression to (4.5-3) that the condition for a confined beam is satisfied by choosing θ in Equation (4.5-4), so that $\sin\theta/B < 0$ provided

$$\left|\frac{D+A}{2}\right| < 1 \tag{4.5-5}$$

and the steady-state beam parameter is

$$\frac{1}{q_s} = \frac{D-A}{2B} - i\frac{\sqrt{1-[(D+A)/2]^2}}{|B|} = \frac{D-A}{2B} + \frac{i\sin\theta}{|B|}, \quad \theta < 0 \tag{4.5-6}$$

Equation (4.5-5) can thus be viewed as the generalization of the stability condition (4.4-2) to the case of an arbitrary resonator. When applied to a resonator composed of two spherical reflectors, it reduces to Equation (4.4-2).

The radius of curvature R and the spot size ω at the reference plane are obtained from Equation (4.5-6) by using Equation (2.6-7)

$$\begin{aligned} R &= \frac{2B}{D-A} \\ \omega &= \left(\frac{\lambda}{\pi n}\right)^{1/2} \frac{(|B|)^{1/2}}{[1-[(D+A)/2]^2]^{1/4}} \end{aligned} \tag{4.5-7}$$

The complex beam parameter q, and hence ω and R, at any other plane can be obtained by applying the $ABCD$ law (2.6-9) to q_s.

Stability of the Resonator Modes

The treatment just concluded dealt with the existence of steady-state (self-reproducing) resonator modes. Having found that such modes do exist, we need to inquire whether the modes are stable. This can be done by perturbing the steady-state solution $1/q_s$ as given by Equation (4.5-6) and following the evolution of the perturbation with propagation [9].

We start with Equation (2.6-9), which relates the beam parameter q_{out} to the beam parameter q_{in}, after one round trip:

$$q_{\text{out}} = \frac{Aq_{\text{in}} + B}{Cq_{\text{in}} + D}$$

where A, B, C, D are the ray matrix elements for one complete round trip inside the optical resonator. Rewriting the last expression as

$$q_{\text{out}}^{-1} = \frac{C + Dq_{\text{in}}^{-1}}{A + Bq_{\text{in}}^{-1}} \tag{4.5-8}$$

we obtain by differentiation

$$\begin{aligned} \frac{dq_{\text{out}}^{-1}}{dq_{\text{in}}^{-1}} &= \frac{D - [(C + Dq_{\text{in}}^{-1})/(A + Bq_{\text{in}}^{-1})]B}{A + Bq_{\text{in}}^{-1}} \\ &= \frac{D - Bq_{\text{out}}^{-1}}{A + Bq_{\text{in}}^{-1}} \end{aligned} \tag{4.5-9}$$

At steady state, $q_{\text{out}} = q_{\text{in}} \equiv q_s$:

$$\left.\frac{dq_{\text{out}}^{-1}}{dq_{\text{in}}^{-1}}\right|_{q_{\text{in}}=q_s} = \frac{D - Bq_s^{-1}}{A + Bq_s^{-1}} \tag{4.5-10}$$

Using Equation (4.5-6) we obtain

$$D - Bq_s^{-1} = \frac{D + A}{2} - i \sin \theta = e^{-i\theta}$$

$$A + Bq_s^{-1} = \frac{D + A}{2} + i \sin \theta = e^{i\theta}$$

so that

$$\left. \frac{dq_{\text{out}}^{-1}}{dq_{\text{in}}^{-1}} \right|_{q_{\text{in}}=q_s} = e^{-2i\theta} \tag{4.5-11}$$

Because confined modes require, according to Equations (4.5-4) and (4.5-5), that θ be real, it follows from Equation (4.5-11) that a small perturbation $\Delta q_{\text{in}}^{-1}$ of the beam parameter q^{-1} from the steady-state value q_s^{-1} does not decay, since the perturbation after one round trip $(\Delta q_{\text{out}}^{-1})$ satisfies

$$|\Delta q_{\text{out}}^{-1}| = |\Delta q_{\text{in}}^{-1}| \tag{4.5-12}$$

We thus find that the theory predicts that mode perturbations in Gaussian mode resonators do not decay. This does not agree with experience, which shows that the mode characteristics of laser oscillators are highly stable, thus implying a strong perturbational decay, that is, $|\Delta q_{\text{out}}^{-1}| < |\Delta q_{\text{in}}^{-1}|$. The discrepancy is resolved if we include in the analysis leading to Equation (4.5-11) the fact that the resonator mirrors are of finite extent.

4.6 RESONANCE FREQUENCIES OF OPTICAL RESONATORS

Up to this point we have considered only the dependence of the spatial mode characteristics on the resonator mirrors (their radii of curvature and separation). Another important consideration is that of determining the resonance frequency of a given spatial mode.

The frequencies are determined by the condition that the complete round-trip phase delay of a resonant mode be some multiple of 2π. This requirement is equivalent to that in microwave waveguide resonators where the resonator length must be equal to an integral number of half-guide wavelengths [1]. This requirement makes it possible for a stable standing wave pattern to establish itself along the axis with a transverse field distribution equal to that of the propagating mode.

If we consider a spherical mirror resonator with mirrors at z_2 and z_1, the resonance condition for the l, m mode can be written[5]

$$\theta_{l,m}(z_2) - \theta_{l,m}(z_1) = q\pi \tag{4.6-1}$$

where q is some integer and $\theta_{l,m}(z)$, the phase shift, is given according to Equation (2.8-2) by

$$\theta_{l,m}(z) = kz - (l + m + 1) \tan^{-1} \frac{z}{z_0} \quad (z_0 = \pi\omega_0^2 n/\lambda) \tag{4.6-2}$$

---

[5] In obtaining Equation (4.6-1) we did not allow for the phase shift upon reflection. This correction does not affect any of the results of this section, since these shifts cancel out in the subtraction of Equation (4.6-3).

The resonance condition (4.6-1) is thus

$$k_q d - (l + m + 1)\left(\tan^{-1}\frac{z_2}{z_0} - \tan^{-1}\frac{z_1}{z_0}\right) = q\pi$$

where $d = z_2 - z_1$ is the resonator length. It follows that

$$k_{q+1} - k_q = \frac{\pi}{d}$$

or, using $k = 2\pi\nu n/c$,

$$\nu_{q+1} - \nu_q = \frac{c}{2nd} \tag{4.6-3}$$

for the intermode frequency spacing.

Let us consider, next, the effect of varying the transverse mode indices l and m in a mode with a fixed q. We notice from Equation (4.6-3) that the resonant frequencies depend on the sum $l + m$ and not on l and m separately, so for a given q all the modes with the same value of $l + m$ are degenerate (i.e., they have the same resonance frequencies). Considering Equation (4.6-3) at two different values of $l + m$ gives

$$\begin{aligned} k_1 d - (l + m + 1)_1\left(\tan^{-1}\frac{z_2}{z_0} - \tan^{-1}\frac{z_1}{z_0}\right) &= q\pi \\ k_2 d - (l + m + 1)_2\left(\tan^{-1}\frac{z_2}{z_0} - \tan^{-1}\frac{z_1}{z_0}\right) &= q\pi \end{aligned} \tag{4.6-4}$$

and, by subtraction.

$$(k_1 - k_2)d = [(l + m + 1)_1 - (l + m + 1)_2]\left(\tan^{-1}\frac{z_2}{z_0} - \tan^{-1}\frac{z_1}{z_0}\right) \tag{4.6-5}$$

and

$$\Delta\nu = \frac{c}{2\pi nd}\Delta(l + m)\left(\tan^{-1}\frac{z_2}{z_0} - \tan^{-1}\frac{z_1}{z_0}\right) \tag{4.6-6}$$

for the change $\Delta\nu$ in the resonance frequency caused by a change $\Delta(l + m)$ in the sum $l + m$. As an example, in the case of a Symmetrical confocal resonator ($R = d$) we have, according to Equation (4.3-6), $z_2 = -z_1 = z_0$; therefore $\tan^{-1}(z_2/z_0) = -\tan^{-1}(z_1/z_0) = \pi/4$, and Equation (4.6-6) becomes

$$\Delta\nu_{\text{conf}} = \frac{1}{2}[\Delta(l + m)]\frac{c}{2nd} \tag{4.6-7}$$

Comparing Equation (4.6-7) to (4.6-4), we find that in the confocal resonator the resonance frequencies of the transverse modes, resulting from changing l and m, either coincide or fall halfway between those resulting from a change of the longitudinal mode index q. This situation is depicted in Figure 4.17.

To see what happens to the transverse resonance frequencies (i.e., those due to a variation of l and m) in a non-confocal resonator, we may consider the nearly planar resonator in which $|z_1|$ and z_2 are small compared to z_0 (i.e., $d \ll |R_1|$ and R_2). In this case, Equation (4.6-6) becomes

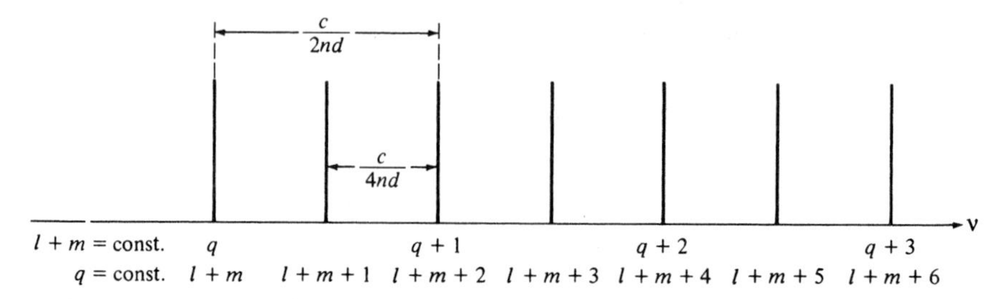

Figure 4.17 Position of resonance frequencies of a confocal ($d = R$) optical resonator as a function of the mode indices l, m, and q.

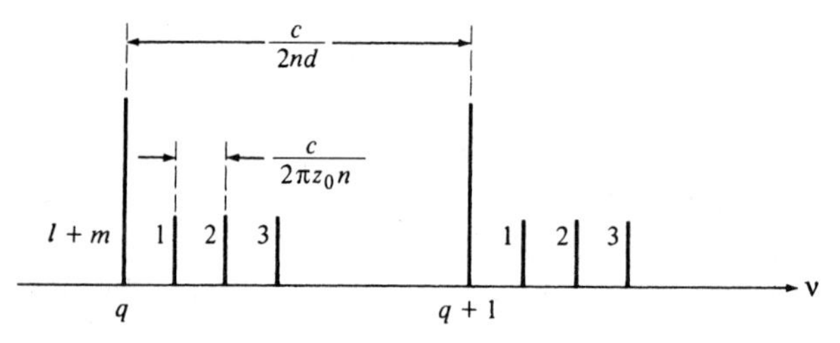

Figure 4.18 Resonant frequencies of a near-planar ($R \gg d$) optical resonator as a function of the mode indices l, m, and q.

$$\Delta\nu \simeq \frac{c}{2\pi n z_0} \Delta(l + m) \tag{4.6-8}$$

The mode grouping for this case is illustrated in Figure 4.18.

In the general case where $|z_1|$ and z_2 are comparable to z_0, the approximation used to derive Equation (4.6-8) does not hold. In this case, it is possible to show using a lengthy, but straightforward, algebra that

$$\tan^{-1}\frac{z_2}{z_0} - \tan^{-1}\frac{z_1}{z_0} = \cos^{-1}\left[\pm\sqrt{\left(1 - \frac{d}{R_1}\right)\left(1 - \frac{d}{R_2}\right)}\right] \tag{4.6-9}$$

The plus (+) sign is used when both $(1 - d/R)$ factors are positive while the minus (−) sign applies when both are negative. (The other options correspond to unstable resonators.) We can then solve either of Equations (4.6-4) for

$$\nu_{q,l,m} = \frac{c}{2nd}\left[q + (l + m + 1)\frac{\cos^{-1}\left[\pm\sqrt{\left(1 - \frac{d}{R_2}\right)\left(1 - \frac{d}{R_1}\right)}\right]}{\pi}\right] \tag{4.6-10}$$

the resonant frequency of mode q, l, m.

The situation depicted in Figure 4.18 is highly objectionable if the resonator is to be used as a scanning interferometer. The reason is that in reconstructing the spectral profile of the unknown signal, an ambiguity is caused by the simultaneous transmission of more than one frequency. This ambiguity is resolved by using a confocal etalon whose mode spacing is as shown in Figure 4.17 and by choosing d to be small enough that the intermode spacing $c/4nd$ exceeds the width of the spectral region that is scanned.

4.7 LOSSES IN OPTICAL RESONATORS

An understanding of the mechanisms by which electromagnetic energy is dissipated in optical resonators and the ability to control them are of major importance in understanding and operating a variety of optical devices. For historical reasons as well as for reasons of convenience, these losses are often characterized by a number of different parameters. This book uses, in different places, the concepts of loss per pass, photon lifetime, and quality factor Q to describe losses in resonators. Let us see how these quantities are related to each other.

The decay lifetime (photon lifetime) t_c of a cavity mode is defined by means of the equation

$$\frac{d\mathscr{E}}{dt} = -\frac{\mathscr{E}}{t_c} \tag{4.7-1}$$

where $\mathscr{E}$ is the energy stored in the mode so that in a passive resonator $\mathscr{E}(t) = \mathscr{E}(0)\exp(-t/t_c) = \mathscr{E}(0)\exp(-\omega t/Q)$. If the fractional (intensity) loss per pass is L and the length of the resonator is l, then the fractional loss per unit time is cL/nl; therefore

$$\frac{d\mathscr{E}}{dt} = -\frac{cL}{nl}\mathscr{E}$$

and, from Equation (4.7-1),

$$t_c = \frac{nl}{cL} \tag{4.7-2}$$

for the case of a resonator with mirrors' reflectivities R_1 and R_2 and an average distributed loss constant α, the average loss per pass is for small losses $L = \alpha l - \ln\sqrt{R_1R_2}$ so that

$$t_c = \frac{n}{c[\alpha - (1/l)\ln\sqrt{R_1R_2}\,]} \approx \frac{nl}{c[\alpha l + (1 - \sqrt{R_1R_2}\,)]} \tag{4.7-3}$$

where the approximate equality applies when $R_1R_2 \approx 1$.

The quality factor of the resonator is defined universally as

$$Q = \frac{\omega\mathscr{E}}{P} = -\frac{\omega\mathscr{E}}{d\mathscr{E}/dt} \tag{4.7-4}$$

where $\mathscr{E}$ is the stored energy, ω is the resonant frequency, and $P = -d\mathscr{E}/dt$ is the power dissipated. By comparing Equations (4.7-4) and (4.7-1) we obtain

$$Q = \omega t_c \tag{4.7-5}$$

The Q factor is related to the full width $\Delta\nu_{1/2}$ (at the half-power points) of the resonator's Lorentzian response curve as

$$\Delta\nu_{1/2} = \frac{\nu}{Q} = \frac{1}{2\pi t_c} \tag{4.7-6}$$

([4] and Section 5.5) so that, according to (4.7-3)

$$\Delta\nu_{1/2} = \frac{c[\alpha - (1/l)\ln\sqrt{R_1R_2}\,]}{2\pi n} \tag{4.7-7}$$

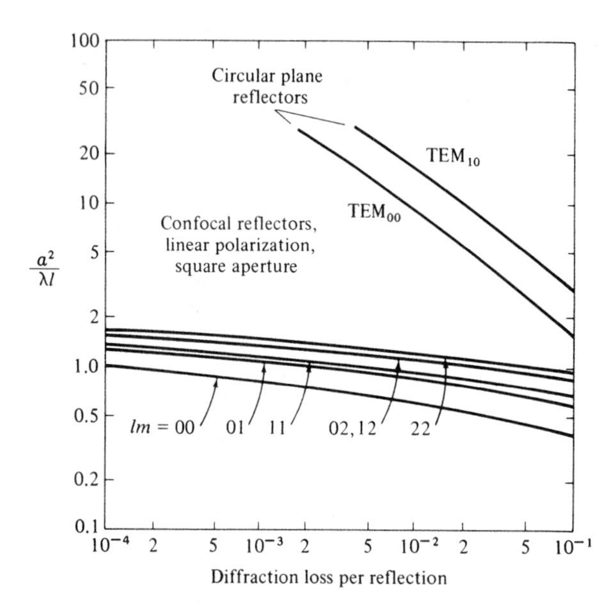

Figure 4.19 Diffraction losses for a plane-parallel resonator and several low-order confocal resonators; a is the mirror radius and l is their spacing. The pairs of numbers under the arrows refer to the transverse-mode indices l, m. (After Reference [7].)

The most common loss mechanisms in optical resonators are the following.

1. *Loss resulting from nonperfect reflection.* Reflection loss is unavoidable, since without some transmission no power output is possible. In addition, no mirror is ideal; and even when mirrors are made to yield the highest possible reflectivities, some residual absorption and scattering reduce the reflectivity to somewhat less than 100%.
2. *Absorption and scattering in the laser medium.* Transitions from some of the atomic levels, which are populated in the process of pumping, to higher-lying levels constitute a loss mechanism in optical resonators when they are used as laser oscillators. Scattering from inhomogeneities and imperfections is especially serious in solid-state laser media.
3. *Diffraction losses.* From Equation (4.3-1) or from Figure 2.7, we find that the energy of propagating-beam modes extends to considerable distances from the axis. When a resonator is formed by "trapping" a propagating beam between two reflectors, it is clear that for finite-dimension reflectors some of the beam energy will not be intercepted by the mirrors and will therefore be lost. For a given set of mirrors this loss will be greater, the higher the transverse mode indices l, m, since in this case the energy extends farther. This fact is used to prevent the oscillation of higher-order modes by inserting apertures into the laser resonator whose opening is large enough to allow most of the fundamental $(0, 0, q)$ mode energy through, but small enough to increase substantially the losses of the higher-order modes. Figure 4.19 shows the diffraction losses of a number of low-order confocal resonators. Of special interest is the dramatic decrease of the diffraction losses that results from the use of spherical reflectors instead of the plane-parallel ones.

4.8 RING RESONATORS

Etalons, resonators, or interferometers can also be made with a ring configuration. These resonators (or interferometers) are employed in ring laser gyros, microresonators, and microdisk

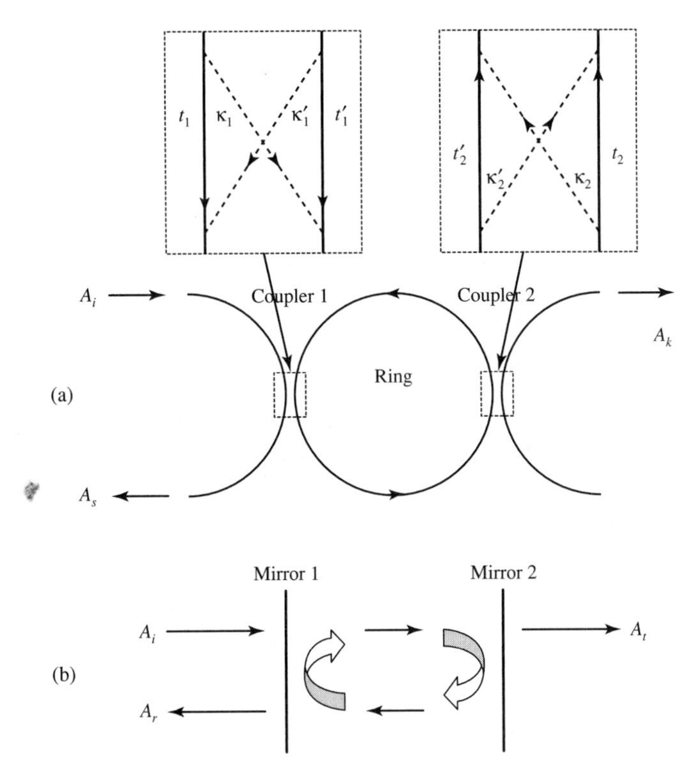

Figure 4.20 Schematic drawings of (a) a ring resonator consisting of a ring of fiber with two couplers and (b) a Fabry–Perot etalon made of two partial reflecting mirrors.

lasers [10–12]. Under appropriate conditions, optical electronic devices can be designed to provide the possibility of switching or modulations. In free space, a ring resonator can be formed by using three mirrors oriented in such a way that a beam of light can circulate along the triangle defined by the three mirrors. In guided waves, a ring resonator can easily be obtained by bending a fiber into a ring, or fabricating a channel waveguide with a circular path. For the guided wave versions, couplers are employed to couple the ring with external waveguides. Referring to Figure 4.20a, we consider a schematic drawing of a ring resonator that is made of a ring of fiber with two couplers. The couplers allow the possibility of energy exchange between the two waveguides (or fibers) with the ring resonator at the location of the couplers. We consider an input beam of amplitude A_i. At the first coupler, a fraction of the input beam will be propagating straight through, while a fraction of the input beam will be coupled into the ring. As the beam circulates inside the ring, multiple couplings occur at the couplers. The situation is similar to the case of multiple reflections in a Fabry–Perot etalon, as shown in Figure 4.20b.

We define the cross-coupling coefficients for the couplers as κ_1, κ_1', κ_2, and κ_2' and the straight-through coupling coefficients as t_1, t_1', t_2, and t_2' (see Figure 4.20a). The multiple circulations inside the ring resonator are physically identical to the multiple reflections inside a Fabry–Perot etalon. Thus we may view Figure 4.20b as an equivalent optical circuit of the ring resonator. Using a similar analysis leading to Equations (4.1-2) and (4.1-3), we obtain the straight-through mode amplitude A_s and the crossover mode amplitude A_k:

$$A_s = \{t_1 + \kappa_1\kappa_1' t_2' e^{-i\delta}[1 + t_1' t_2' e^{-i\delta} + (t_1' t_2' e^{-i\delta})^2 + \cdots]\}A_i \tag{4.8-1}$$

$$A_k = \kappa_1\kappa_2' e^{-i\delta/2}[1 + t_1' t_2' e^{-i\delta} + (t_1' t_2' e^{-i\delta})^2 + \cdots]A_i \tag{4.8-2}$$

with

$$\delta = 2\pi n d/\lambda \tag{4.8-3}$$

where d is the circumference of the ring and n is the effective index of the mode of propagation in the fiber. We note that these two expressions are identical to those of the Fabry–Perot etalon, Equations (4.1-28) and (4.1-29), provided we associate the straight-through coupling coefficients (t_1, t_1', t_2, t_2') with the amplitude reflection coefficients of the mirrors, and the cross-coupling coefficients (κ_1, κ_1', κ_2, κ_2') with the amplitude transmission coefficients of the mirrors. Here, without loss of generality, we assume that the couplers are situated symmetrically in the ring. Physically, δ is the round-trip propagation phase shift inside the resonator.

By summing up the infinite geometric series, we obtain the straight-through (transmission through the fiber) coefficient and crossover (coupled through the ring and finally to the other fiber) coefficient, respectively:

$$\sigma \equiv \frac{A_s}{A_i} = t_1 + \frac{\kappa_1\kappa_1' t_2' e^{-i\delta}}{1 - t_1' t_2' e^{-i\delta}} = \frac{t_1 + (\kappa_1\kappa_1' - t_1 t_1')t_2' e^{-i\delta}}{1 - t_1' t_2' e^{-i\delta}} \tag{4.8-4}$$

$$\chi \equiv \frac{A_k}{A_i} = \frac{\kappa_1\kappa_2' e^{-i\delta/2}}{1 - t_1' t_2' e^{-i\delta}} \tag{4.8-5}$$

Again, these two expressions are formally identical to those of the Fabry–Perot etalons, Equations (4.1-31) and (4.1-32), provided we associate the straight-through coupling coefficients (t_1, t_1', t_2, t_2') with the amplitude reflection coefficients of the mirrors, and the cross-coupling coefficients (κ_1, κ_1', κ_2, κ_2') with the amplitude transmission coefficients of the mirrors. These eight coupling coefficients for the couplers are not independent. They are related by the fundamental principles of reciprocity, conservation of energy, and the time-reversal symmetry. For simplicity, we assume that the couplers are lossless, and the coupling is limited to waves traveling in one sense; that is, no reflection takes place inside the waveguides (or fibers). In this case, these coupling coefficients reduce to two independent constants [13]. The coupling can conveniently be described in terms of a unitary coupling matrix. Before discussing the property of the ring resonators, we must have a clear understanding of the couplers.

Referring to Figure 4.21, we consider the coupling between two single-mode waveguides (or fibers). Let A_1 and A_2 be the mode amplitudes of the input, and B_1 and B_2 be the mode

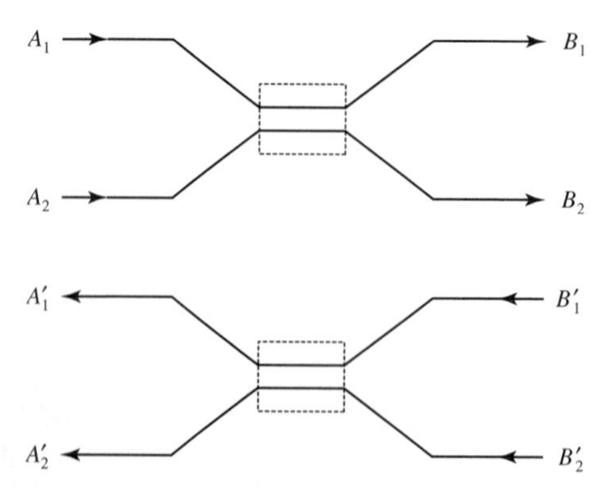

Figure 4.21 Schematic drawings of the couplers.

amplitudes of the output. The modes of propagation in the waveguides are normalized so that the absolute squares of the mode amplitudes represent the modal power flow. The linear input–output relationship of the coupler can be written

$$\begin{aligned} B_1 &= A_1\kappa_{11} + A_2\kappa_{21} \\ B_2 &= A_1\kappa_{12} + A_2\kappa_{22} \end{aligned} \tag{4.8-6}$$

or in matrix form

$$\begin{bmatrix} B_1 \\ B_2 \end{bmatrix} = \begin{bmatrix} \kappa_{11} & \kappa_{21} \\ \kappa_{12} & \kappa_{22} \end{bmatrix} \begin{bmatrix} A_1 \\ A_2 \end{bmatrix} \equiv X \begin{bmatrix} A_1 \\ A_2 \end{bmatrix} \tag{4.8-7}$$

where κ_{12} and κ_{21} are the cross-coupling coefficients, and κ_{11} and κ_{22} are the straight-through coupling coefficients. If we reverse the direction of propagation (see Figure 4.21), we can define another matrix X' such that

$$\begin{bmatrix} A'_1 \\ A'_2 \end{bmatrix} = \begin{bmatrix} \kappa'_{11} & \kappa'_{21} \\ \kappa'_{12} & \kappa'_{22} \end{bmatrix} \begin{bmatrix} B'_1 \\ B'_2 \end{bmatrix} \equiv X' \begin{bmatrix} B'_1 \\ B'_2 \end{bmatrix} \tag{4.8-8}$$

By virtue of time-reversal symmetry, these two matrices must satisfy the following relationship:

$$X'X^* = 1 \tag{4.8-9}$$

provided the coupler is lossless. Furthermore, the principle of reciprocity requires that

$$\kappa'_{11} = \kappa_{11}, \quad \kappa'_{12} = \kappa_{21}, \quad \kappa'_{22} = \kappa_{22}, \quad \kappa'_{12} = \kappa_{21} \tag{4.8-10}$$

or equivalently,

$$X' = \tilde{X} \tag{4.8-11}$$

In other words, X' is the transpose of X. As a result, according to Equations (4.8-9)–(4.8-11), we recognize that the matrix X is unitary; that is,

$$X^\dagger X = 1 \tag{4.8-12}$$

where the dagger † stands for Hermitian conjugate. By virtue of the unitary property, the matrix elements of X satisfy the following relations, according to Equation (4.8-12):

$$\begin{aligned} &\kappa_{22} = |X|\kappa_{11}^*, \qquad \kappa_{21} = -|X|\kappa_{12}^* \\ &|\kappa_{11}|^2 + |\kappa_{12}|^2 = |\kappa_{22}|^2 + |\kappa_{21}|^2 = 1 \\ &\kappa_{11}^*\kappa_{21} + \kappa_{12}^*\kappa_{22} = \kappa_{21}^*\kappa_{11} + \kappa_{22}^*\kappa_{12} = 0 \end{aligned} \tag{4.8-13}$$

where $|X|$ is the determinant of the matrix X,

$$|X| = \kappa_{11}\kappa_{22} - \kappa_{12}\kappa_{21} \tag{4.8-14}$$

Equation (4.8-13) is consistent with the conservation of energy.

It can easily be shown that the determinant $|X|$ is, in general, a complex number of a unit modulus (i.e., $|X| = \exp(i\theta)$, with θ being a real number). Without loss of generality, we can take

$$|X| = \kappa_{11}\kappa_{22} - \kappa_{12}\kappa_{21} = -1 \tag{4.8-15}$$

The choice leads to a convenience in the matrix analysis of coupled ring resonators. With this choice of the determinant ($|X| = -1$), the linear input–output relationship of the coupler can be written

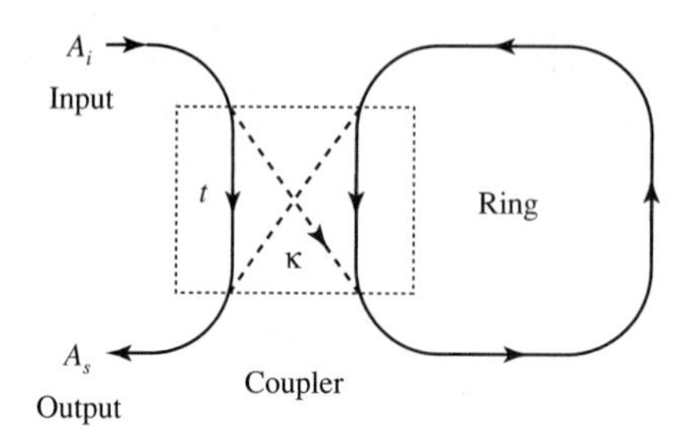

Figure 4.22 Schematic drawing of a ring resonator coupled with a waveguide. (t, κ) are the coupling coefficients of the coupler.

$$\begin{bmatrix} B_1 \\ B_2 \end{bmatrix} = X \begin{bmatrix} A_1 \\ A_2 \end{bmatrix} = \begin{bmatrix} t & \kappa^* \\ \kappa & -t^* \end{bmatrix} \begin{bmatrix} A_1 \\ A_2 \end{bmatrix} \tag{4.8-16}$$

where $t = \kappa_{11}$ is the straight-through coupling coefficient and $\kappa = \kappa_{12}$ is the cross-coupling coefficient of the coupler. Note that the coupling matrix X is Hermitian ($X^\dagger = X$), as a result of the choice of $|X| = -1$. These two coupling coefficients satisfy the following relationship, according to Equation (4.8-13):

$$|t|^2 + |\kappa|^2 = 1 \tag{4.8-17}$$

Matrix formulation can conveniently be employed to investigate the optical properties of coupled resonators when several ring resonators are involved. Using the unitary conditions (4.8-13) and (4.8-15), the straight-through transmission coefficient of the ring resonator of Figure 4.20a can be written, according to Equation (4.8-4),

$$\sigma \equiv \frac{A_s}{A_i} = \frac{t_1 + t_2' a e^{-i\delta}}{1 - t_1' t_2' a e^{-i\delta}} = \frac{t_1 + t_2' e^{-i\delta} e^{-\alpha d}}{1 - t_1' t_2' e^{-i\delta} e^{-\alpha d}} \tag{4.8-18}$$

where $a = \exp(-\alpha d)$, with α being an amplitude attenuation coefficient to account for loss due to bending or scattering. We now consider a special case of a ring resonator where $\kappa_2 = 0$ and $t_2 = -1$. This resonator is illustrated in Figure 4.22. For this case, the straight-through transmission coefficient becomes

$$\sigma \equiv \frac{A_s}{A_i} = \frac{t_1 - a e^{-i\delta}}{1 + t_1' a e^{-i\delta}} = \frac{t - a e^{-i\delta}}{1 - t^* a e^{-i\delta}} \tag{4.8-19}$$

where we recall δ is the round-trip phase shift inside the ring resonator, and $t_1' = -t_1^* \equiv -t^*$. Here, we drop the subscript 1, as there is only one coupler. By taking the absolute square of Equation (4.8-19), we obtain the intensity transmission (straight-through)

$$|\sigma|^2 \equiv \left| \frac{A_s}{A_i} \right|^2 = \frac{a^2 + t^2 - 2at\cos\delta}{1 + a^2 t^2 - 2at\cos\delta} \tag{4.8-20}$$

where, without loss of generality, we have taken t as a real number. This is legitimate, as the constant phase of t can always be lumped with the phase δ.

At resonance $\delta = m2\pi$ (m = integer), the straight-through intensity transmission becomes

$$|\sigma|^2 \equiv \left| \frac{A_s}{A_i} \right|^2 = \frac{(a - |t|)^2}{(1 - a|t|)^2} = \frac{(a - t)^2}{(1 - at)^2} \tag{4.8-21}$$

This simple relation, plotted in Figure 4.23 as a function of the cross-coupling efficiency $|\kappa|^2$, has two important features that are the key for some potential applications. (1) The straight-through transmission $|\sigma|^2$ is zero at a value of coupling $a = t = \sqrt{1 - |\kappa|^2}$, known as the

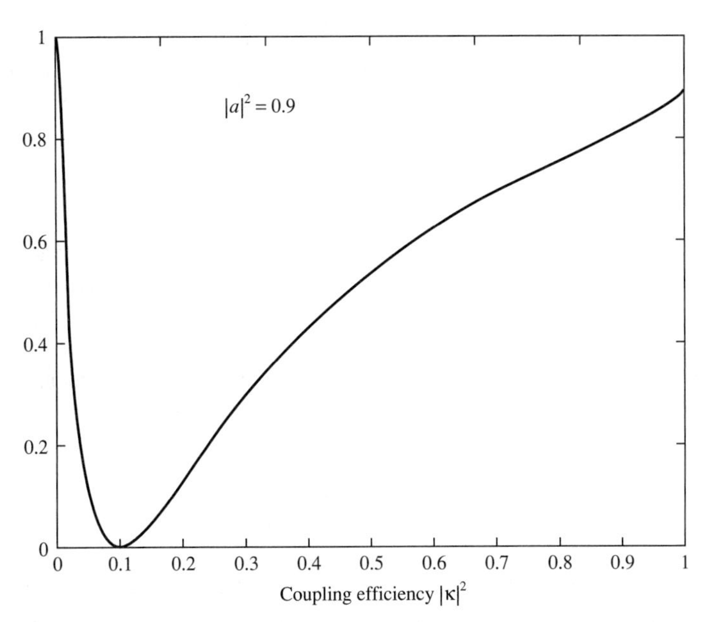

Figure 4.23 Straight-through transmission of a ring resonator is plotted as a function of the coupling efficiency $|\kappa|^2$. Critical coupling occurs when the coupling efficiency equals the ring resonator internal loss $(1 - |a|^2)$.

"critical coupling." (2) For high-finesse ring resonators with a low internal loss (round-trip loss $= 1 - |a|^2$, a is near unity), the portion of the curve to the left of the critical coupling point is extremely steep. "Small changes" in a for a given coupling efficiency $|\kappa|^2$, or vice versa, can control the transmission between zero and unity. Such a feature is desirable in the design of switching of modulation devices with a high modulation depth [13]. Electroabsorption or optical amplifiers can be employed to control the loss [14].

Coupled Ring Resonators

We next consider the situation when several ring resonators are coupled serially. The optical property of such coupled ring resonators can conveniently be described by using the unitary matrix relationship, Equation (4.8-15), discussed earlier. Referring to Figure 4.24, we consider serially coupled ring resonators. Fiber 1 is for the input and straight-through output, fibers 2,

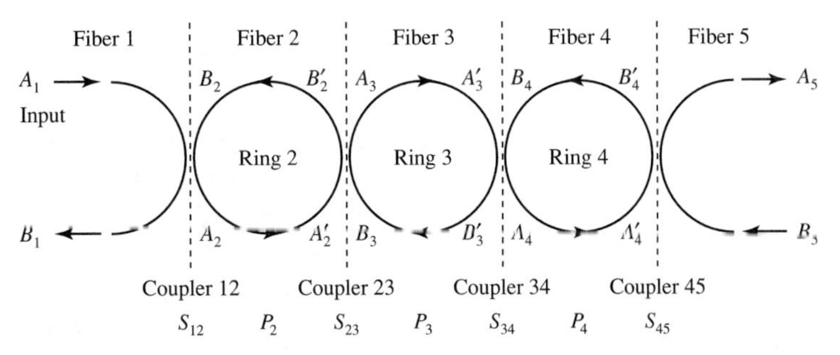

Figure 4.24 Schematic drawing of a system of serially coupled ring resonators, consisting of three ring resonators. These three ring resonators are labeled as ring 2, ring 3, and ring 4.

3, and 4 are ring resonators, and fiber 5 is for the cross output. For the first coupler (12) between fibers 1 and 2 in Figure 4.24, the input–output matrix relationship can be written, according to Equation (4.8-15),

$$\begin{bmatrix} B_1 \\ A_2 \end{bmatrix} = X_{12} \begin{bmatrix} A_1 \\ B_2 \end{bmatrix} = \begin{bmatrix} \kappa_{11} & \kappa_{21} \\ \kappa_{12} & \kappa_{22} \end{bmatrix} \begin{bmatrix} A_1 \\ B_2 \end{bmatrix} \tag{4.8-22}$$

where κ_{11}, κ_{22}, κ_{12}, and κ_{21} are the coupling coefficients of the coupler. These coupling coefficients are related by the unitary conditions, according to Equations (4.8-13) and (4.8-15),

$$\kappa_{22} = -\kappa_{11}^*, \qquad \kappa_{21} = \kappa_{12}^* \tag{4.8-23}$$

Note that for this coupler (12), A_1 and B_2 are the input mode amplitudes, whereas B_1 and A_2 are the output mode amplitudes. For the purpose of analyzing the serially coupled ring resonators, we rewrite Equation (4.8-22) as

$$\begin{bmatrix} A_1 \\ B_1 \end{bmatrix} = \frac{1}{\kappa_{12}^2} \begin{bmatrix} 1 & \kappa_{11}^* \\ \kappa_{11} & 1 \end{bmatrix} \begin{bmatrix} A_2 \\ B_2 \end{bmatrix} \equiv S_{12} \begin{bmatrix} A_2 \\ B_2 \end{bmatrix} \tag{4.8-24}$$

where S_{12} is defined as the scattering matrix of the coupler, A_1 is the incident mode amplitude, B_1 is the straight-through mode amplitude in the same fiber, and A_2 and B_2 are mode amplitudes inside the first ring resonator (see Figure 4.24). It is important to distinguish the difference between matrices X and S. Coupling matrix X is a linear relationship between the two input mode amplitudes and the two output mode amplitudes to a coupler, whereas matrix S is a linear relationship between the two mode amplitudes of one waveguide to the two mode amplitudes of the other waveguides. The determinant of the scattering matrix is given by, according to Equations (4.8-13), (4.8-23), and (4.8-24),

$$|S_{12}| = \frac{1}{\kappa_{12}^2}(1 - |\kappa_{11}|^2) = \frac{\kappa_{21}}{\kappa_{12}} = \frac{\kappa_{12}^*}{\kappa_{12}} \tag{4.8-25}$$

We note that the determinant is, in general, a complex number with a unit modulus. It becomes 1 when the cross-coupling constant κ_{12} is real. In this case, the scattering matrix S is Hermitian and unimodular.

We define A_2' and B_2' as the mode amplitudes in the first ring resonator (fiber 2, or ring 2) at the location of the next coupler (23). Similarly, (A_m, B_m) and (A_m', B_m') are the mode amplitudes in ring resonator m (fiber m). The amplitudes in the resonators are related by the propagation in the ring of fibers. Without loss of generality, we assume that the couplers are situated symmetrically. The relationship can be written

$$\begin{bmatrix} A_m \\ B_m \end{bmatrix} = \begin{bmatrix} e^{i\phi_m/2} & 0 \\ 0 & e^{-i\phi_m/2} \end{bmatrix} \begin{bmatrix} A_m' \\ B_m' \end{bmatrix} \equiv P_m \begin{bmatrix} A_m' \\ B_m' \end{bmatrix} \tag{4.8-26}$$

where ϕ_m is the round-trip phase shift in ring resonator m, and P_m is defined as the propagation matrix. A complex phase shift ϕ_m should be employed to account for propagation or bending loss. At the coupler between the mth fiber and the $(m+1)$st fiber, the mode amplitudes are related by, according to Equation (4.8-24),

$$\begin{bmatrix} A_m' \\ B_m' \end{bmatrix} = \frac{1}{\kappa_{m,m+1}} \begin{bmatrix} 1 & \kappa_{mm}^* \\ \kappa_{mm} & 1 \end{bmatrix} \begin{bmatrix} A_{m+1} \\ B_{m+1} \end{bmatrix} \equiv S_{m,m+1} \begin{bmatrix} A_{m+1} \\ B_{m+1} \end{bmatrix}, \quad m \geq 1 \tag{4.8-27}$$

where $S_{m,m+1}$ is the scattering matrix for the coupler $(m, m+1)$, for $m > 0$.

Using Equations (4.8-24), (4.8-26), and (4.8-27), the mode amplitudes of fiber 1 and fiber 5 can be written

$$\begin{bmatrix} A_1 \\ B_1 \end{bmatrix} = S_{12}\, P_2\, S_{23}\, P_3\, S_{34}\, P_4\, S_{45} \begin{bmatrix} A_5 \\ B_5 \end{bmatrix} \tag{4.8-28}$$

where S_{12}, S_{23}, S_{34}, and S_{45} are the scattering matrices of the couplers, and P_2, P_3, and P_4 are the propagation matrices of the rings. The same equation can easily be extended for the case of N serially coupled ring resonators (with $N > 3$),

$$\begin{bmatrix} A_0 \\ B_0 \end{bmatrix} = S_{01}\, P_1\, S_{12}\, P_2\, S_{23} \cdots P_N\, S_{N,N+1} \begin{bmatrix} A_{N+1} \\ B_{N+1} \end{bmatrix} \equiv M \begin{bmatrix} A_{N+1} \\ B_{N+1} \end{bmatrix} \equiv \begin{bmatrix} M_{11} & M_{12} \\ M_{21} & M_{22} \end{bmatrix} \begin{bmatrix} A_{N+1} \\ B_{N+1} \end{bmatrix} \tag{4.8-29}$$

where A_0 and B_0 are the mode amplitudes in fiber 0, and A_{N+1} and B_{N+1} are the mode amplitudes in fiber $N + 1$; fibers 1 through N are ring resonators. The matrix M is the product of all the matrices in sequence, similar to Equation (4.8-28). The straight-through transmission coefficient in fiber 0 is thus given by, according to Equation (4.8-29),

$$\sigma \equiv \left(\frac{B_0}{A_0} \right)_{B_{N+1}=0} = \frac{M_{21}}{M_{11}} \tag{4.8-30}$$

The crossover transmission coefficient from fiber 0 to fiber $N + 1$ is given by

$$\chi \equiv \left(\frac{A_{N+1}}{A_0} \right)_{B_{N+1}=0} = \frac{1}{M_{11}} \tag{4.8-31}$$

The matrix elements M_{11}, M_{12}, M_{21}, and M_{22} satisfy some relationship similar to that of S_{12} (see Equation (4.8-24)). Based on the conservation of power flow in a lossless system, the mode amplitudes must satisfy the following condition:

$$|A_0|^2 - |B_0|^2 = |A_{N+1}|^2 - |B_{N+1}|^2 \tag{4.8-32}$$

Using Equations (4.8-29) and (4.8-32), we obtain

$$\begin{gathered} |M_{11}|^2 - |M_{21}|^2 = |M_{22}|^2 - |M_{12}|^2 = 1 \\ M_{11}M_{12}^* - M_{21}M_{22}^* = 0 \end{gathered} \tag{4.8-33}$$

These equations are consistent with time-reversal symmetry and the principle of reciprocity. According to Equation (4.8-33), we have

$$|\sigma|^2 + |\chi|^2 = 1 \tag{4.8-34}$$

which is exactly conservation of energy. It follows directly from Equation (4.8-33) that the matrix elements satisfy the following relationships:

$$|M_{22}| = |M_{11}|, \quad |M_{21}| = |M_{12}| \tag{4.8-35}$$

For the case when all cross-coupling coefficients are real, the relationships become

$$M_{22} = M_{11}^*, \quad M_{21} = M_{12}^* \tag{4.8-36}$$

In the event when all couplers are identical, and all rings are identical, the matrix equation (4.8-29) can be written

$$\begin{bmatrix} A_0 \\ B_0 \end{bmatrix} = [SP]^N\, S \begin{bmatrix} A_{N+1} \\ B_{N+1} \end{bmatrix} \tag{4.8-37}$$

where S is the scattering matrix of the couplers, and P is the propagation matrix for the rings. Identical microrings can be fabricated using lithography replication techniques [15, 16]. Equations (4.8-31) and (4.8-32) can be employed to calculate the straight-through or cross-over transmission coefficients. Both intensity as well as the phase shift can then be obtained. The frequency derivative of the phase shift will yield the group delay. For the case of N identical ring resonators, the straight-through transmission efficiency can be written (see Problem 4.19)

$$|\sigma|^2 \equiv \left|\frac{B_0}{A_0}\right|^2_{B_{N+1}=0} = \left|\frac{M_{21}}{M_{11}}\right|^2 = \frac{(1-|\kappa|^2)}{(1-|\kappa|^2)+|\kappa|^2 \dfrac{\sin^2\theta}{\sin^2(N+1)\theta}} \tag{4.8-38}$$

where θ is given by

$$\cos\theta = \frac{1}{|\kappa|}\cos(\phi/2) \tag{4.8-39}$$

where $\phi = 2\pi n_{\text{eff}} d/\lambda$ is the round-trip phase shift due to propagation in the ring resonators. In the spectral regimes where $|\cos\theta|$ is greater than unity, θ is a complex number. The second term in the denominator of Equation (4.8-38) decays exponentially as a function of N. These regimes are the so-called photonic bandgaps. The straight-through transmission (or "reflection" from the rings) approaches unity. Figure 4.25 is a plot of the straight-through transmission $|\sigma|^2$ of a series of six rings ($N = 6$) as a function of the phase shift ϕ. We notice that $|\sigma|^2$ is a periodic function of ϕ. The spectral response consists of a periodic array of stopbands, separated by ($N - 1$) sidelobes. Further discussion of photonic bandgaps will be presented later in this book. Figure 4.26 is a photograph of a portion of a linear array of identical ring resonators [15, 16].

For an infinite chain of identical ring resonators with identical coupling constants, the fields are periodic at twice the lattice constant Λ. Here we define the lattice constant as a

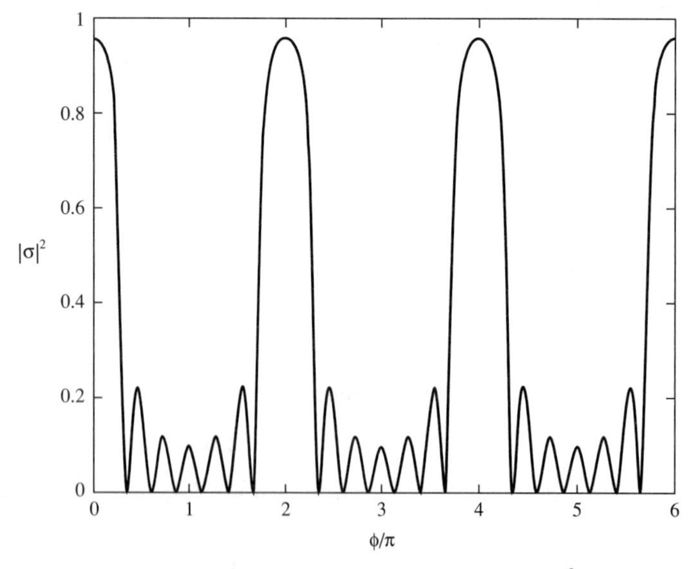

Figure 4.25 Straight-through transmission efficiency $|\sigma|^2$ as a function of phase shift ϕ. The parameters are $N = 6$ and $|\kappa| = 0.95$.

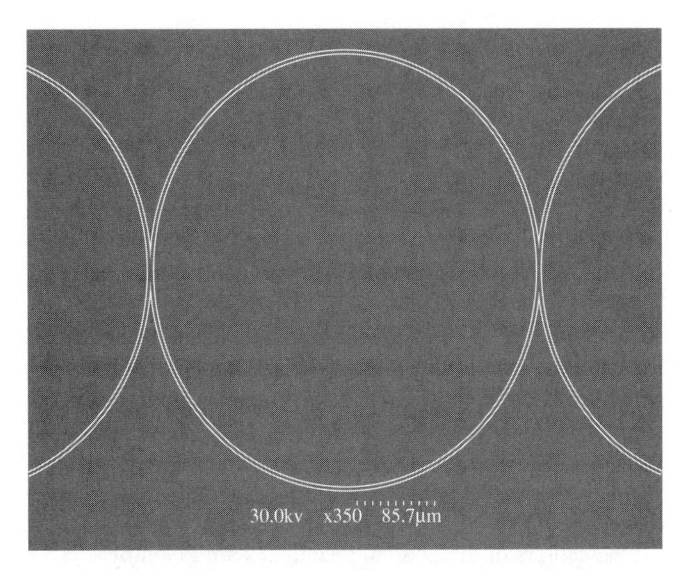

Figure 4.26 Photograph of a portion of a linear array of coupled ring resonators. The ring resonators have a radius of 50 μm, with a circumference of about 314 μm. The waveguides are made of a polymer (SU-8) with a refractive index of 1.57, on top of a silica substrate with a refractive index of 1.44 at $\lambda =$ 1550 nm. The waveguide dimensions are 1.6 μm thick and 2 μm wide. At the coupling region, the rings are separated by approximately 0.5 μm. (Courtesy of George T. Paloczi, References [15] and [16].)

characteristic length associated with a single ring. Using Bloch's theorem, the field amplitudes at the n-th ring are related to those of the (n + 2)-th ring by the following relationship,

$$\begin{pmatrix} A_n \\ B_n \end{pmatrix} = P_n\, S_{n,n+1}\, P_{n+1}\, S_{n+1,n+2} \begin{pmatrix} A_{n+2} \\ B_{n+2} \end{pmatrix} = (PS)^2 \begin{pmatrix} A_{n+2} \\ B_{n+2} \end{pmatrix} = e^{i2K\Lambda} \begin{pmatrix} A_{n+2} \\ B_{n+2} \end{pmatrix} \tag{4.8-40}$$

where P is the propagation matrix of a single ring ($P_n = P$ for all n), S is the scattering matrix of a single coupler ($S_{n,n+1} = S$ for all n), K is the Bloch wavenumber. The amplitudes of the normal modes of propagation thus satisfy the following eigenvalue equation

$$(PS)^2 \begin{pmatrix} A_0 \\ B_0 \end{pmatrix} = e^{i2K\Lambda} \begin{pmatrix} A_0 \\ B_0 \end{pmatrix} \tag{4.8-41}$$

and

$$\begin{pmatrix} A_{2n} \\ B_{2n} \end{pmatrix} = e^{-i2nK\Lambda} \begin{pmatrix} A_0 \\ B_0 \end{pmatrix} \tag{4.8-42}$$

The amplitudes A_0 and B_0 of the normal modes as well as the Block wavenumber can be obtained by solving Equation (4.8-41). The dispersion relation (K as a function of ω) is thus given by

$$\mathrm{Det}|(PS)^2 - e^{i2K\Lambda} I| = 0 \tag{4.8-43}$$

where I is the unit matrix. Using Equations (4.8-26) and (4.8-27) for P and S, we obtain

$$\cos K\Lambda = \frac{1}{|\kappa_{12}|} \cos \frac{1}{2}\phi \tag{4.8-44}$$

where we recall that ϕ is the roundtrip phase shift in a single ring resonator. For ring resonators made of single mode fibers, this phase shift can be written

$$\phi = \frac{\omega}{c} n_{eff} d = \frac{2\pi}{\lambda} n_{eff} d \tag{4.8-45}$$

where d is the circumference of the ring resonator and n_{eff} is the effective index of the mode of propagation in the fiber.

According to Equation (4.8-13), the magnitude of the coupling constant $|\kappa_{12}|$ is always less than (or equal to) unity ($0 \le |\kappa_{12}| \le 1$). As a result, the Bloch wavenumber K can be complex in some spectral regimes according to Equation (4.8-44). A Bloch wave with a complex wavenumber cannot propagate in the periodic medium. According to Equation (4.8-42), the field amplitudes decay exponentially in the spectral regimes where the Bloch wavenumber is complex. These spectral regimes are known as the photonic bandgaps. If we plot the dispersion relationship (ω vs K) using Equation (4.8-44), we find that photonic bandgaps occur in the spectral regime where the roundtrip phase shift ϕ is an integral number of 2π. The centers of the photonic bandgaps occur at exactly $\phi = 2m\pi$ ($m = 1, 2, 3 \ldots$). The Bloch wavenumbers at the centers of the photonic bandgaps are given by

$$K\Lambda = m\pi \pm i \cosh^{-1}\left(\frac{1}{|\kappa_{12}|}\right) \tag{4.8-46}$$

We note that the magnitude of the imaginary part of the Bloch wavenumber depends on the cross-coupling constant of the waveguide coupler. In Chapter 12, we will discuss more periodic structures that are also employed in photonics and optical electronics.

4.9 MULTICAVITY ETALONS

In the previous section, we described the transmission properties of single ring resonators as well as coupled ring resonators. Resonance occurs when the round-trip propagation phase shift inside the resonator is an integral multiple of 2π. For ring resonators made of single-mode fibers or other single-mode waveguides, propagation attenuation due to bending loss can be significant, particularly for ring resonators with circumferences in the range of micrometers. In this section, we discuss coupled etalons made of thin films or multilayer structures. Propagation attenuation is negligible for thin films with thicknesses in the range of micrometers. There are situations in optical electronics when multicavity etalons are needed for spectral filtering or dispersion compensation applications. The transmission and reflection properties of these multicavity etalons can be analyzed by using the results obtained in Sections 4.1 and 4.8. Particularly, the amplitude transmission and reflection coefficients of Equations (4.1-31) and (4.1-32), or equivalently Equations (4.8-4) and (4.8-5) can be very useful. Referring to Figure 4.27, we consider the transmission of light through an etalon consisting of two mirrors separated by a distance d. The transmission and reflection coefficients are written in a more general form as, according to Equations (4.1-31) and (4.1-32),

$$r = r_{12} + \frac{t_{12}t_{21}r_{23}e^{-i2\phi}}{1 - r_{21}r_{23}e^{-i2\phi}} = \frac{r_{12} + (t_{12}t_{21} - r_{12}r_{21})r_{23}e^{-i2\phi}}{1 - r_{21}r_{23}e^{-i2\phi}} \tag{4.9-1}$$

$$t = \frac{t_{12}t_{23}e^{-i\phi}}{1 - r_{21}r_{23}e^{-i2\phi}} \tag{4.9-2}$$

with

$$2\phi = 2k_x d = 2nd \cos\theta(\omega/c) \tag{4.9-3}$$

where d is the thickness of the cavity medium, n is the refractive index of the cavity medium, and k_x is the component of the wavevector along the normal of the mirror surface. The transmission and the reflection coefficients of the mirrors are defined as:

t_{12} = transmission coefficient of mirror 1 with incidence from medium 1
t_{21} = transmission coefficient of mirror 1 with incidence from medium 2
t_{23} = transmission coefficient of mirror 2 with incidence from medium 2
t_{32} = transmission coefficient of mirror 2 with incidence from medium 3
r_{12} = reflection coefficient of mirror 1 with incidence from medium 1
r_{21} = reflection coefficient of mirror 1 with incidence from medium 2
r_{23} = reflection coefficient of mirror 2 with incidence from medium 2
r_{32} = reflection coefficient of mirror 2 with incidence from medium 3

The above definitions are also illustrated in Figure 4.27.

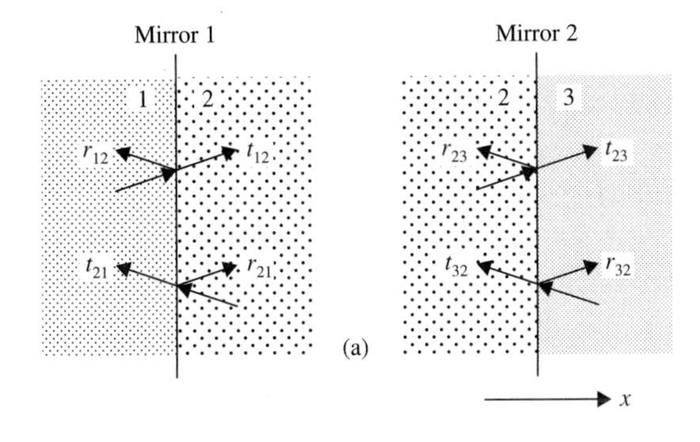

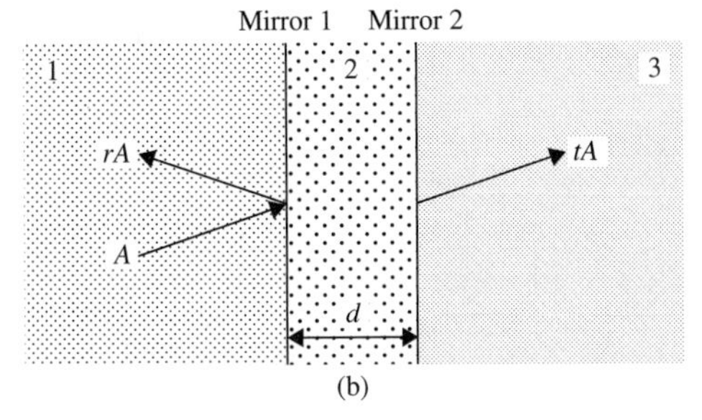

Figure 4.27 Definition of transmission and reflection coefficients. (a) Definition of the mirror transmission and reflection coefficients. (b) Transmission and reflection coefficients of the etalon formed by the two mirrors.

The four coefficients associated with each mirror are in general complex numbers. As discussed earlier, they are related by fundamental principles of physics, including reciprocity and time-reversal symmetry. Specifically, we have (see Problem 4.16)

$$\begin{aligned} t_{12}t_{21}^* + r_{21}r_{21}^* = 1, \quad t_{23}t_{32}^* + r_{32}r_{32}^* = 1 \\ t_{12}r_{12}^* + r_{21}t_{12}^* = 0, \quad t_{23}r_{23}^* + r_{32}t_{23}^* = 0 \end{aligned} \tag{4.9-4}$$

These equations are known as the Stokes relationships [6].

It is important to note that t_{12} and t_{21} are identical only when the refractive indices of media 1 and 2 are the same. For lossless media with different refractive indices, these two sets of transmission coefficients are related by

$$\frac{t_{12}}{k_{1x}} = \frac{t_{21}}{k_{2x}}, \quad \frac{t_{23}}{k_{2x}} = \frac{t_{32}}{k_{3x}} \tag{4.9-5}$$

where k_{1x}, k_{2x}, and k_{3x} are x components of the wavevectors in media 1, 2, and 3, respectively,

$$\begin{aligned} k_{1x} &= n_1 \cos\theta_1(\omega/c) \\ k_{2x} &= n_2 \cos\theta_2(\omega/c) \\ k_{3x} &= n_3 \cos\theta_3(\omega/c) \end{aligned} \tag{4.9-6}$$

where ω is the frequency of the incident beam of light, n_1, n_2, and n_3 are the refractive indices, and θ_1, θ_2, and θ_3 are the ray angles in the media, respectively. Equation (4.9-5) is consistent with the reciprocity of transmission of energy. The factors k_{1x}, k_{2x}, and k_{3x} are essential to account for the different speeds of light in the media.

To illustrate the use of the transmission and reflection formulas, we consider the following example.

EXAMPLE: THIN FILM SANDWICHED BETWEEN TWO DIFFERENT MEDIA

Referring to Figure 4.28, we consider a thin film of thickness d_2 and refractive index n_2. The film is sandwiched between two media with refractive indices n_1 and n_3.

The interface reflection and transmission coefficients are simply the Fresnel reflection and transmission coefficients. They are given by [6]

$$r_{12} = \begin{cases} \dfrac{k_{1x} - k_{2x}}{k_{1x} + k_{2x}} \\[2ex] \dfrac{n_1^2 k_{2x} - n_2^2 k_{1x}}{n_1^2 k_{2x} + n_2^2 k_{1x}} \end{cases} = \begin{cases} \dfrac{n_1 \cos\theta_1 - n_2 \cos\theta_2}{n_1 \cos\theta_1 + n_2 \cos\theta_2} & (s\text{-wave}) \\[2ex] \dfrac{n_1 \cos\theta_2 - n_2 \cos\theta_1}{n_1 \cos\theta_2 + n_2 \cos\theta_1} & (p\text{-wave}) \end{cases} \tag{4.9-7}$$

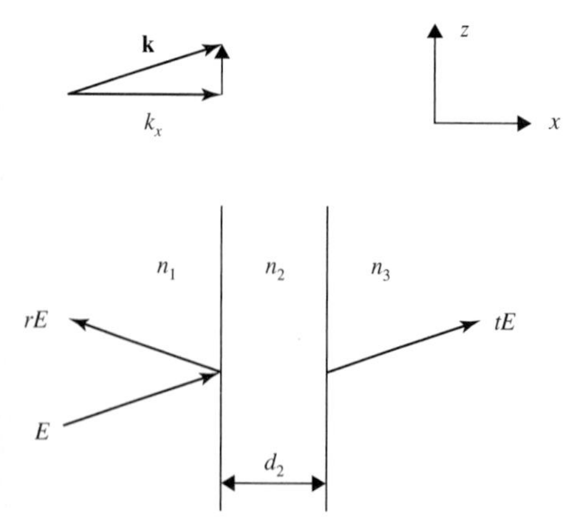

Figure 4.28 The transmission and reflection of light through and from a thin film. The x axis is perpendicular to the film surface.

$$t_{12} = \begin{cases} \dfrac{2k_{1x}}{k_{1x} + k_{2x}} = \dfrac{2n_1 \cos\theta_1}{n_1 \cos\theta_1 + n_2 \cos\theta_2} & (s\text{-wave}) \\[2ex] \dfrac{2n_1 n_2 k_{1x}}{n_1^2 k_{2x} + n_2^2 k_{1x}} = \dfrac{2n_1 \cos\theta_1}{n_1 \cos\theta_2 + n_2 \cos\theta_1} & (p\text{-wave}) \end{cases} \tag{4.9-8}$$

where k_{1x} and k_{2x} are the x components of the wavevectors in media 1 and 2, respectively. θ_1 and θ_2 are ray angles in media 1 and 2, respectively. The ray angles are related to the x components of the wavevectors by the relationships (4.9-6). For an s-wave, the electric field vector is perpendicular to the plane of incidence (xz plane), whereas for a p-wave, the electric field vector is parallel to the plane of incidence. A similar set of transmission and reflection coefficients for the interface between media 2 and 3 can be obtained from the above by replacing $2 \rightarrow 3$ and $1 \rightarrow 2$. We note that the Fresnel reflection and transmission coefficients satisfy the following relationships:

$$\begin{aligned} t_{12}t_{21} - r_{12}r_{21} &= 1 \\ r_{21} &= -r_{12} \end{aligned} \tag{4.9-9}$$

For the purpose of illustrating the concept, we consider the special case of normal incidence. At normal incidence, $\theta_1 = \theta_2 = 0$, and the s-wave and p-wave are identical. The Fresnel transmission and reflection become

$$\begin{aligned} r_{12} &= \frac{n_1 - n_2}{n_1 + n_2}, \quad r_{21} = -r_{12} \\ r_{23} &= \frac{n_2 - n_3}{n_2 + n_3}, \quad r_{32} = -r_{23} \\ t_{12} &= \frac{2n_1}{n_1 + n_2}, \quad t_{23} = \frac{2n_2}{n_2 + n_3} \\ t_{21} &= \frac{2n_2}{n_1 + n_2}, \quad t_{32} = \frac{2n_3}{n_2 + n_3} \end{aligned} \tag{4.9-10}$$

For lossless media, these coefficients are all real. The reflection and transmission coefficients for the thin film shown in Figure 4.28 can now be written, according to Equations (4.9-1), (4.9-2), and (4.9-10),

$$r = \frac{r_{12} + r_{23}e^{-i2\phi}}{1 - r_{21}r_{23}e^{-i2\phi}} = \frac{\left(\dfrac{n_1 - n_2}{n_1 + n_2}\right) + \left(\dfrac{n_2 - n_3}{n_2 + n_3}\right)e^{-i2\phi}}{1 + \left(\dfrac{n_1 - n_2}{n_1 + n_2}\right)\left(\dfrac{n_2 - n_3}{n_2 + n_3}\right)e^{-i2\phi}} \tag{4.9-11}$$

$$t = \frac{t_{12}t_{23}e^{-i\phi}}{1 - r_{21}r_{23}e^{-i2\phi}} = \frac{\left(\dfrac{2n_1}{n_1 + n_2}\right)\left(\dfrac{2n_2}{n_2 + n_3}\right)e^{-i\phi}}{1 + \left(\dfrac{n_1 - n_2}{n_1 + n_2}\right)\left(\dfrac{n_2 - n_3}{n_2 + n_3}\right)e^{i2\phi}} \tag{4.9-12}$$

where $\phi = k_2 d_2 = n_2 d_2(\omega/c) = 2\pi n_2 d_2/\lambda$.

EXAMPLE: ANTIREFLECTION (AR) COATING

We consider an example of a thin film coating to eliminate the Fresnel reflection at $\lambda = 1.5\ \mu m$. Let medium 1 be air with a refractive index of $n_1 = 1$, and medium 3 be a semiconductor with a refractive index of $n_3 = 3.5$. Equation (4.9-11) can be employed to investigate the reflection as a function of the film thickness and its index of refraction. According to Equation (4.9-11), the reflection coefficient vanishes when

$$r_{12} + r_{23}e^{-i2\phi} = \left(\frac{n_1 - n_2}{n_1 + n_2}\right) + \left(\frac{n_2 - n_3}{n_2 + n_3}\right)e^{-i2\phi} = 0 \tag{4.9-13}$$

The above equation can easily be satisfied by choosing

$$n_2 = \sqrt{n_1 n_3} \quad \text{and} \quad \phi = \pi/2, 3\pi/2, 5\pi/2, 7\pi/2, \ldots \tag{4.9-14}$$

In other words, a thin film with a quarter-wave thickness and a refractive index equal to the geometric average of the bounding media can lead to zero reflectance. Since $\phi = k_2 d_2 = n_2 d_2(\omega/c) = 2\pi n_2 d_2/\lambda$, the quarter thickness is

$$d_2 = \frac{\lambda}{4n_2} = \frac{\lambda}{4\sqrt{n_1 n_3}} \tag{4.9-15}$$

Using $n_1 = 1$ and $n_3 = 3.5$, the refractive index of the thin film must be

$$n_2 = \sqrt{n_1 n_3} = 1.87$$

and the thickness can be

$$d_2 = \frac{\lambda}{4n_2} = \frac{\lambda}{4\sqrt{n_1 n_3}} = 0.2\ \mu\text{m}$$

or an odd integral multiple of it. It is important to note that the antireflection coating described above is only limited to the wavelength at $\lambda = 1.5\ \mu m$. Equation (4.9-9) can be employed to investigate the reflectivity at wavelength around $\lambda_0 = 1.5\ \mu m$. Figure 4.29 shows the reflectance ($|r|^2$) as a function of the phase ϕ for various indices of refraction of the thin film coating. The reflectance is a periodic function of the phase with minima occurring at $\phi = \pi/2, 3\pi/2, 5\pi/2, 7\pi/2, \ldots$ for $n_2 < n_3$. When the film index is greater than n_3, the reflectance is higher and the minima occur at $\phi = \pi, 2\pi, 3\pi, \ldots$

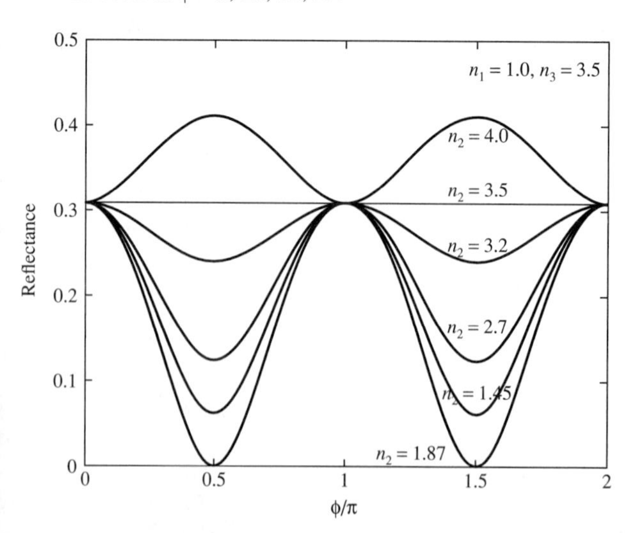

Figure 4.29 Reflectance versus ϕ for $n_1 = 1, n_3 = 3.5$, and various values of n_2.

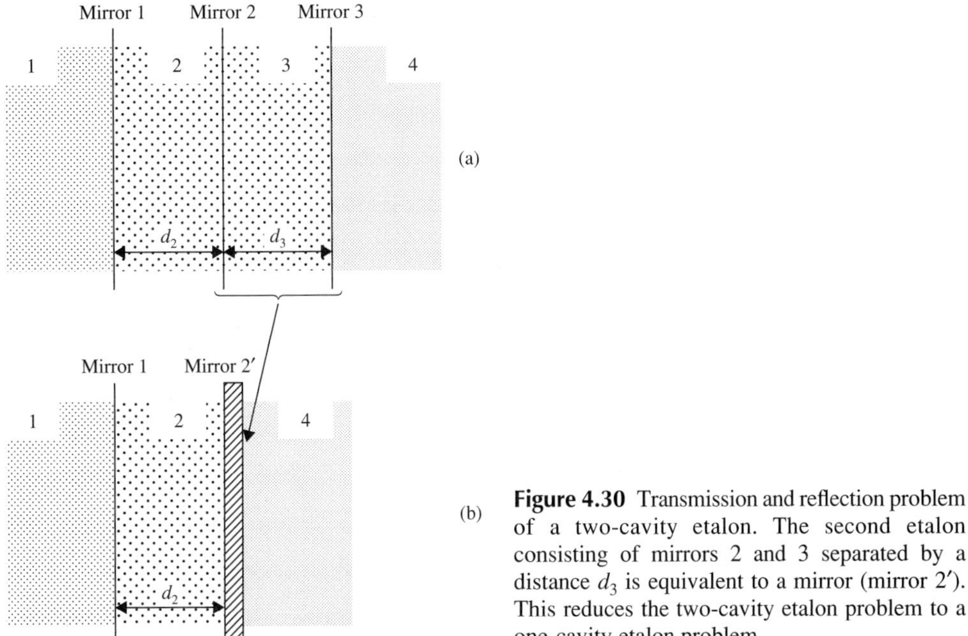

Figure 4.30 Transmission and reflection problem of a two-cavity etalon. The second etalon consisting of mirrors 2 and 3 separated by a distance d_3 is equivalent to a mirror (mirror 2′). This reduces the two-cavity etalon problem to a one-cavity etalon problem.

The formulas for transmission and reflection coefficients of the etalon, Equations (4.9-1) and (4.9-2), can be generalized to treat multicavity etalons. To illustrate this, we consider a two-cavity etalon as shown in Figure 4.30. To find the transmission and reflection coefficients, we first find the transmission and reflection coefficients of the second etalon formed by medium 3 sandwiched between mirrors 2 and 3. This is done by removing mirror 1 and medium 1. The formula in Equations (4.9-1) and (4.9-2) can be employed for this purpose. We then consider a one-cavity problem formed by mirror 1 and an equivalent mirror (mirror 2′). The equivalent mirror (mirror 2′) consists of the second etalon formed by mirrors 2 and 3, sandwiched between media 2 and 4. The equivalent one-cavity etalon is illustrated in Figure 4.30b. The transmission and reflection coefficients can then be obtained by employing again the general formulas of Equations (4.9-1) and (4.9-2). The method described here can be generalized to treat the transmission and reflection problem of multicavity etalons.

2 × 2 Matrix Method

In addition to the use of an equivalent mirror, the problem of transmission and reflection properties of multicavity etalons can be treated by using a 2 × 2 matrix method. The matrix method is so general that it can be applied to treat the transmission and reflection properties of a general multilayer structure. To introduce the matrix method, we consider a general case of plane wave propagation on both sides of a dielectric interface (see Figure 4.31). We can write the electric field as

$$E = \begin{cases} (A_1 e^{-ik_{1x}x} + B_1 e^{+ik_{1x}x}) e^{-i\beta z} & \text{medium 1} \\ (A_2 e^{-ik_{2x}x} + B_2 e^{+ik_{2x}x}) e^{-i\beta z} & \text{medium 2} \end{cases} \tag{4.9-16}$$

where A_1 and A_2 are the amplitudes of the right-traveling wave (along $+x$ direction), B_1 and B_2 are the amplitudes of the left-traveling wave (along $-x$ direction), β is the z component of

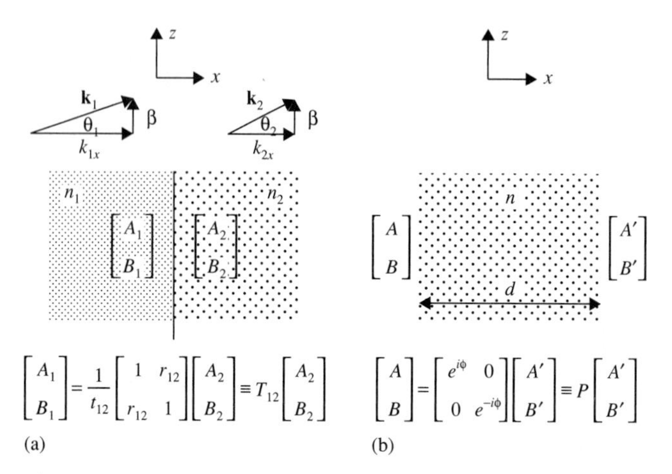

Figure 4.31 (a) Transition matrix for a dielectric interface between two media. (b) Propagation matrix for a homogeneous medium.

the wavevector, and k_{1x} and k_{2x} are x components of the wavevector. Note that β is the same in both media as the structure is homogeneous in the z direction. These components of the wavevectors are related by the following equations:

$$
\begin{aligned}
k_{1x}^2 + \beta^2 &= (n_1\omega/c)^2 \\
k_{2x}^2 + \beta^2 &= (n_2\omega/c)^2
\end{aligned}
\tag{4.9-17}
$$

Based on the definition of the transmission and reflection coefficients associated with the interface, the amplitudes are related by the following equations:

$$
\begin{aligned}
B_1 &= A_1 r_{12} + B_2 t_{21} \\
A_2 &= A_1 t_{12} + B_2 r_{21}
\end{aligned}
\tag{4.9-18}
$$

where r_{12}, r_{21}, t_{12}, and t_{21} are the Fresnel reflection and transmission coefficients discussed earlier. Solving for A_1 and B_1 in terms of A_2 and B_2, we obtain

$$
\begin{aligned}
A_1 &= \frac{1}{t_{12}} A_2 - \frac{r_{21}}{t_{12}} B_2 \\
B_1 &= \frac{r_{12}}{t_{12}} A_2 + \left(t_{21} - \frac{r_{12} r_{21}}{t_{12}} \right) B_2
\end{aligned}
\tag{4.9-19}
$$

Using Equation (4.10-9), the above can be rewritten

$$
\begin{aligned}
A_1 &= \frac{1}{t_{12}} A_2 + \frac{r_{12}}{t_{12}} B_2 \\
B_1 &= \frac{r_{12}}{t_{12}} A_2 + \frac{1}{t_{12}} B_2
\end{aligned}
\tag{4.9-20}
$$

or, equivalently, in a matrix form

$$
\begin{bmatrix} A_1 \\ B_1 \end{bmatrix} = \frac{1}{t_{12}} \begin{bmatrix} 1 & r_{12} \\ r_{12} & 1 \end{bmatrix} \begin{bmatrix} A_2 \\ B_2 \end{bmatrix} \equiv T_{12} \begin{bmatrix} A_2 \\ B_2 \end{bmatrix}
\tag{4.9-21}
$$

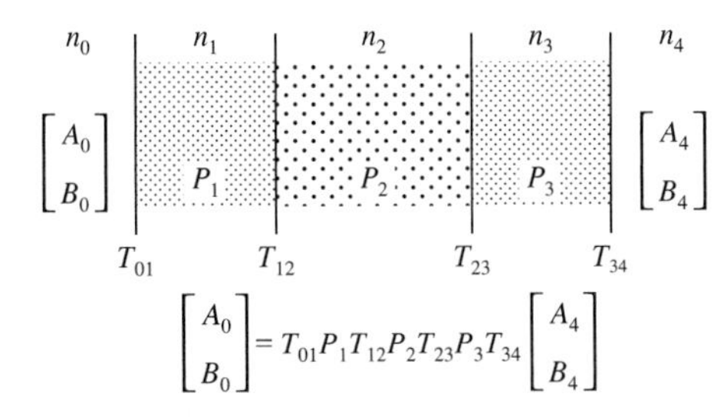

Figure 4.32 Schematic drawing showing the matrix method of treating an example of multilayer structures (with $N = 3$).

where T_{12} is defined as the transition matrix between layers 1 and 2. We notice the strong resemblance between the transition matrix T_{12} and the scattering matrix S_{12} of Equation (4.8-24) for ring resonators. Physically, t_{12} is equivalent to the cross-coupling coefficient κ_{12}, and r_{12} is equivalent to the straight-through coupling coefficient κ_{11}. It is important to note that the reflection and transmission coefficients of the interface, r_{12} and t_{12}, depend on the state of polarization (either s or p). Thus the problem of transmission and reflection is often treated separately for each of the polarization states. We note the transition matrix T_{12} is a symmetric matrix.

The transition matrix relates the field column vector on both sides of a dielectric interface. We now need a matrix for propagation from one end of a homogeneous layer to the other end of the layer. Referring to Figure 4.31b, we consider propagation of plane waves in a homogeneous medium with thickness d and refractive index n. Using the column vector for the field representation, we obtain

$$\begin{bmatrix} A \\ B \end{bmatrix} = \begin{bmatrix} e^{i\phi} & 0 \\ 0 & e^{-i\phi} \end{bmatrix} \begin{bmatrix} A' \\ B' \end{bmatrix} \equiv P \begin{bmatrix} A' \\ B' \end{bmatrix} \tag{4.9-22}$$

where A and B are the field amplitudes on the left end of the medium, whereas A' and B' are the field amplitudes on the right end of the medium. The phase shift is given by

$$\phi = k_x d = nd \cos\theta(\omega/c) \tag{4.9-23}$$

where k_x is the x component of the wavevector in the medium and θ is the ray angle (angle between the wavevector and the x axis). The diagonal matrix P is defined as the propagation matrix. For lossless media, the propagation matrix is unitary. The transition matrix T and the propagation matrix P can now be employed to treat a general multilayer structure. This is illustrated in Figure 4.32 for the case of a stack of three layers ($N = 3$) sandwiched between semi-infinite media of refractive indices n_0 and n_4. The column vector of the plane wave amplitudes (A_0, B_0) in medium 0 is related to the column vector of the plane wave amplitudes (A_{N+1}, B_{N+1}) by the following relationship:

$$\begin{bmatrix} A_0 \\ B_0 \end{bmatrix} = \begin{bmatrix} M_{11} & M_{12} \\ M_{21} & M_{22} \end{bmatrix} \begin{bmatrix} A_{N+1} \\ B_{N+1} \end{bmatrix} \equiv M \begin{bmatrix} A_{N+1} \\ B_{N+1} \end{bmatrix} \tag{4.9-24}$$

where M is the 2×2 matrix that can be obtained by multiplying the transition matrices and the propagation in sequence:

$$M = T_{01}P_1T_{12}P_2T_{23}P_3 \ldots P_{N-1}T_{N-1,N}P_NT_{N,N+1} \tag{4.9-25}$$

We note the propagation matrix is independent of the state of polarization, whereas the transition matrix depends on the state of polarization. Using the symmetric property of the transition matrix T_{12} and the unitary property of the propagation matrix, it can easily be shown that

$$M_{21} = M_{12}^*, \qquad M_{22} = M_{11}^* \tag{4.9-26}$$

for lossless multilayer structures. Using Equations (4.9-21) and (4.9-22) as well as the explicit forms of the Fresnel reflection and transmission coefficients (4.9-7) and (4.9-8), it can be shown that the determinant is given by

$$|M| = \frac{k_{N+1x}}{k_{0x}} \tag{4.9-27}$$

where k_{N+1x} and k_{0x} are the x components of the wavevectors in media 0 and $N + 1$, respectively. Equation (4.9-27) is valid even if loss is present in the layered structure. If the two media are identical (i.e., $n_0 = n_{N+1}$), then the matrix M is unimodular.

The transition matrices for both the s-wave and p-wave are given as, according to Equations (4.9-7), (4.9-8), and (4.9-21),

$$T_{12} = \begin{cases} \dfrac{1}{2k_{1x}} \begin{bmatrix} k_{1x} + k_{2x} & k_{1x} - k_{2x} \\ k_{1x} - k_{2x} & k_{1x} + k_{2x} \end{bmatrix} & (s\text{-wave}) \\[2ex] \dfrac{1}{2n_1 n_2 k_{1x}} \begin{bmatrix} n_1^2 k_{2x} + n_2^2 k_{1x} & n_1^2 k_{2x} - n_2^2 k_{1x} \\ n_1^2 k_{2x} - n_2^2 k_{1x} & n_1^2 k_{2x} + n_2^2 k_{1x} \end{bmatrix} & (p\text{-wave}) \end{cases} \tag{4.9-28}$$

or equivalently, in terms of ray angles

$$T_{12} = \begin{cases} \dfrac{1}{2n_1 \cos\theta_1} \begin{bmatrix} n_1\cos\theta_1 + n_2\cos\theta_2 & n_1\cos\theta_1 - n_2\cos\theta_2 \\ n_1\cos\theta_1 - n_2\cos\theta_2 & n_1\cos\theta_1 + n_2\cos\theta_2 \end{bmatrix} & (s\text{-wave}) \\[2ex] \dfrac{1}{2n_1 \cos\theta_1} \begin{bmatrix} n_1\cos\theta_2 + n_2\cos\theta_1 & n_1\cos\theta_2 - n_2\cos\theta_1 \\ n_1\cos\theta_2 - n_2\cos\theta_1 & n_1\cos\theta_2 + n_2\cos\theta_1 \end{bmatrix} & (p\text{-wave}) \end{cases} \tag{4.9-29}$$

The propagation matrix is given by

$$P = \begin{bmatrix} e^{ik_x d} & 0 \\ 0 & e^{-ik_x d} \end{bmatrix} \tag{4.9-30}$$

where d is the thickness of the medium, and k_x is the x component of the wavevector:

$$k_x = \sqrt{(n\omega/c)^2 - \beta^2} = (n\omega/c)\cos\theta \tag{4.9-31}$$

Table 4.1 is a summary of the 2×2 matrices for both the s-wave and p-wave. At normal incidence, the s-wave and p-wave are identical; the transition matrix becomes

$$T_{12} = \frac{1}{2n_1} \begin{bmatrix} n_1 + n_2 & n_1 - n_2 \\ n_1 - n_2 & n_1 + n_2 \end{bmatrix} \quad \text{(normal incidence)} \tag{4.9-32}$$

Once the matrix is obtained, the transmission and reflection coefficients of the multilayer structure can be calculated as follows. Let A_0 be the amplitude of the incident beam, B_0 be that of the reflected beam, and A_{N+1} be that of the transmitted beam. We set $B_{N+1} = 0$, as the beam is launched from the left side. These amplitudes are related by, according to Equation (4.9-24),

$$\begin{aligned} A_0 &= M_{11} A_{N+1} \\ B_0 &= M_{21} A_{N+1} \end{aligned} \tag{4.9-33}$$

TABLE 4.1 2 × 2 Matrices for Dielectric Interface and Homogeneous Layer

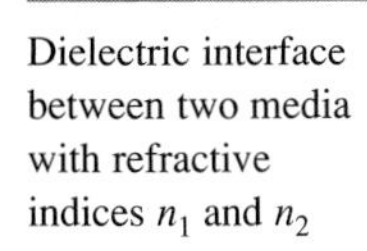 Dielectric interface between two media with refractive indices n_1 and n_2	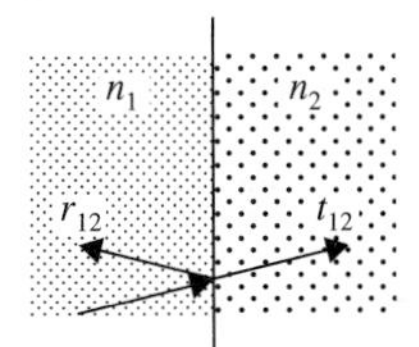	$\frac{1}{t_{12}}\begin{bmatrix} 1 & r_{12} \\ r_{12} & 1 \end{bmatrix}$
Dielectric interface between two media with refractive indices n_1 and n_2	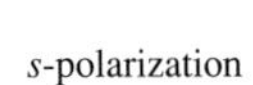 *s*-polarization ($\mathbf{E} \perp xz$ plane) 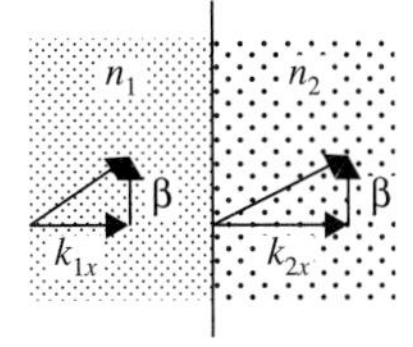	$\frac{1}{2k_{1x}}\begin{bmatrix} k_{1x}+k_{2x} & k_{1x}-k_{2x} \\ k_{1x}-k_{2x} & k_{1x}+k_{2x} \end{bmatrix}$
Dielectric interface between two media with refractive indices n_1 and n_2	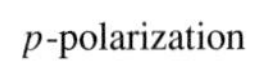 *p*-polarization ($\mathbf{E} \parallel xz$ plane) 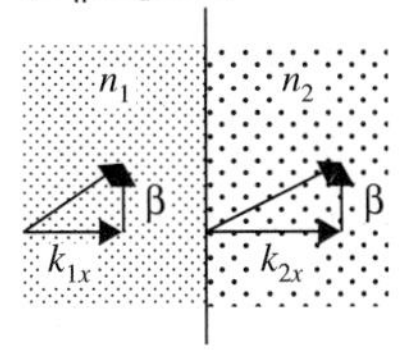	$\frac{1}{2n_1n_2k_{1x}}\begin{bmatrix} n_1^2k_{2x}+n_2^2k_{1x} & n_1^2k_{2x}-n_2^2k_{1x} \\ n_1^2k_{2x}-n_2^2k_{1x} & n_1^2k_{2x}+n_2^2k_{1x} \end{bmatrix}$
Homogeneous medium of thickness d	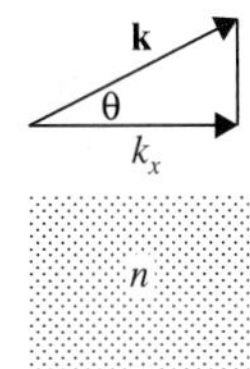	$\begin{bmatrix} e^{ik_xd} & 0 \\ 0 & e^{-ik_xd} \end{bmatrix}$

The transmission and reflection coefficients are thus given by

$$
\begin{aligned}
t &= \left(\frac{A_{N+1}}{A_0}\right)_{B_{N+1}=0} = \frac{1}{M_{11}} \\
r &= \left(\frac{B_0}{A_0}\right)_{B_{N+1}=0} = \frac{M_{21}}{M_{11}}
\end{aligned}
\tag{4.9-34}
$$

When the incident beam is launched from the right side, the transmission and reflection coefficients are given by

$$
\begin{aligned}
t' &= \left(\frac{B_0}{B_{N+1}}\right)_{A_0=0} = \frac{|M|}{M_{11}} \\
r' &= \left(\frac{A_{N+1}}{B_{N+1}}\right)_{A_0=0} = -\frac{M_{12}}{M_{11}}
\end{aligned}
\tag{4.9-35}
$$

where $|M|$ is the determinant of the matrix. Let T and T' be the transmittances of the layered structure when light is incident from the left medium (medium 0) and right medium (medium $N+1$), respectively. These two transmittances are given by

$$T = \frac{k_{N+1x}}{k_{0x}}|t|^2, \quad T' = \frac{k_{0x}}{k_{N+1x}}|t'|^2 \tag{4.9-36}$$

respectively. Using Equations (4.9-34) and (4.9-35) and the expression for $|M|$ in Equation (4.9-27), we obtain

$$T' = T \tag{4.9-37}$$

and

$$t' = |M|t = \frac{k_{N+1x}}{k_{0x}}t \tag{4.9-38}$$

Based on Equation (4.9-37), we note that the transmittances T' and T of a layered structure are identical, regardless of whether the beam is incident from the left side or the right side. This is true even if loss is present in the layered structure. In the event when media 0 and $N+1$ are identical (i.e., $n_0 = n_{N+1}$), we have

$$t' = t \tag{4.9-39}$$

This is consistent with the principle of reciprocity.

Using Equations (4.9-34) and (4.9-35) for the general expressions of the transmission and reflection coefficients, we obtain

$$tt'^* + rr^* = 1, \quad tr'^* + rt^* = 0 \tag{4.9-40}$$

which are identical to the results obtained from the symmetry argument (time-reversal symmetry) [6]. They are known as the Stokes relationships. It can also be shown that, for a lossless layered structure,

$$R + T = 1 \tag{4.9-41}$$

which is consistent with the conservation of energy.

4.10 MODE MATCHING AND COUPLING LOSS

A basic problem of both theoretical and practical interest is how to couple efficiently an incident beam of light with complex amplitude $E_{\text{in}}(x, y)$ to a given mode of an optical resonator or an optical fiber, and also derive a measure of the residual, undesirable, excitation of other resonator modes. In the case of an optical fiber, this process is known as end-fire coupling. The problem arises frequently in optical communications when the light is coupled in and out of a single-mode fiber for wavelength selection, power amplification, or dispersion management purposes, and when the light is coupled in and out of a laser resonator (or amplifier). In optical fibers, a poor mode matching will lead to a significant insertion loss.

Referring to Figure 4.33, we designate the electric field of an incident beam of light at the "input" plane z_1 as $E_{\text{in}}(x, y)$ and the wavefunction of modes of the fiber or resonator as $E_{mn}(x, y)$, where m, n are the transverse mode integers of the Gaussian beam of an optical resonator, or the LP mode of an optical fiber. The set $E_{mn}(x, y)$ of modes of a fiber or a resonator constitutes a complete orthogonal set of wavefunctions. They satisfy

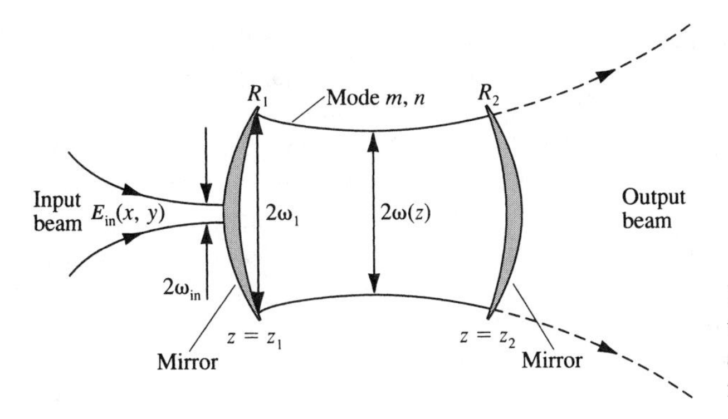

Figure 4.33 Excitation of a transverse mode m, n of a resonator (or an optical fiber) by an incident monochromatic beam of light.

$$\iint E_{mn}(x, y)E^*_{m'n'}(x, y)\, dx\, dy = 0, \quad \text{unless } m = m' \text{ and } n = n' \tag{4.10-1}$$

By taking advantage of the completeness of the set E_{mn} and expanding the incident field as a linear superposition of the modes, we can write

$$E_{in}(x, y) = \sum_{mn} a_{mn} E_{mn}(x, y) \tag{4.10-2}$$

where a_{mn} are constants. Multiplying both sides by E^*_{mn}, integrating over the entire xy plane, and making use of the orthogonal relationships (4.10-1), we obtain

$$a_{mn} = \frac{\iint E_{in}(x, y)E^*_{mn}(x, y)\, dx\, dy}{\iint E_{mn}(x, y)E^*_{mn}(x, y)\, dx\, dy} \tag{4.10-3}$$

The coupling efficiency of the incident field into a given spatial mode, say, E_{mn}, is defined as

$$\eta_{mn} = \frac{\text{Power coupled into mode } mn}{\text{Total incident power}} = \frac{\iint |a_{mn}E_{mn}(x, y)|^2\, dx\, dy}{\iint |E_{in}(x, y)|^2\, dx\, dy} \tag{4.10-4}$$

Substituting Equation (4.10-3) for the constant a_{mn} in Equation (4.10-4), we obtain

$$\eta_{mn} = \frac{\left| \iint E_{in}(x, y)\, E^*_{mn}(x, y)\, dx\, dy \right|^2}{\iint |E_{in}(x, y)|^2\, dx\, dy \cdot \iint |E_{mn}(x, y)|^2\, dx\, dy} \tag{4.10-5}$$

This is the main result of this section. We note η_{mn} is always less than or equal to unity, according to Equation (4.10-5) and the Schwarz inequality. This is in agreement with the conservation of energy.

It follows from Equation (4.10-5) that if the input beam at the input plane has the *same* spatial dependence as that of the mode to be excited, that is, when

$$E_{in}(x, y) \propto E_{mn}(x, y) \tag{4.10-6}$$

then $\eta_{mn} = 1$, and all other $\eta_{m'n'}$ are zero, and all the incident power goes into exciting the single E_{mn} mode. In practice, mode matching requires that, at the input facet of the fiber or the resonator mirror, the incident beam possesses the same transverse mode numbers m and n as the mode to be excited, as well as possessing the same spot size and radius of curvature. In arriving at the result (4.10-5), we assume that no waves are reflected at the input facet. An additional condition for the case of an optical resonator is that the mode m, n thus excited fulfill the longitudinal Fabry–Perot resonance condition of the resonator. Otherwise, most of the incident beam power will be reflected. In the situation depicted in Figure 4.33, where $\omega_{in} \ll \omega_1$, a large number of modes will be excited. It will be left as an exercise to show that the number of modes that are excited "substantially" is $\sim(\omega_1/\omega_{in})^2$. By "mode matching" we can avoid the excitation of higher-order modes when we use a scanning Fabry–Perot etalon (made of spherical mirrors) to analyze the spectrum of an incident fundamental Gaussian beam.

EXAMPLE: INSERTION LOSS

An obvious condition of an efficient coupling requires that the incident beam be propagating along the axis (z axis) of the resonator or optical fiber. Any lateral shift (x, y) of the incident beam can lead to a reduced coupling efficiency into a resonator or a single-mode fiber. The insertion loss can easily be calculated as follows. We consider the lateral shift only along the x axis. The y dependence can be ignored in the analysis. Let the wavefunction of the resonator or the fiber possess a Gaussian transverse field profile of the form $E_{00} = e^{-x^2/\omega^2}$, where ω is the spot size of the Gaussian beam at the input facet. Let the wavefunction of the laterally shifted beam be written $E_{in} = e^{-(x-a)^2/\omega^2}$, where a is the lateral displacement along the x axis. Substituting these wavefunctions into Equation (4.10-5) and carrying out the integration, we obtain

$$\eta = e^{-a^2/\omega^2}$$

For a Gaussian beam of spot size $\omega = 10\ \mu\text{m}$, and a lateral beam displacement of $a = 2\ \mu\text{m}$, the above formula yields $\eta = e^{-1/25} \approx 0.96$, which corresponds to a 4% insertion loss (−0.13 dB). Similarly, a displacement of $a = 10\ \mu\text{m}$ will lead to a coupling efficiency of $\eta = e^{-1} \approx 0.368$, which corresponds to a 63.2% insertion loss (−4.4 dB).

PROBLEMS

4.1 Plot I_r/I_i versus δ of a Fabry–Perot etalon with $R = 0.9$.

4.2 Show that if a Fabry–Perot etalon has an intensity gain per pass of G, its peak transmission is given as $(1 - R)^2G/(1 - RG)^2$.

4.3 Starting with the definition

$$F \equiv \frac{\nu_{m+1} - \nu_m}{\Delta\nu_{1/2}}$$

for the finesse of a Fabry–Perot etalon and using semiquantitative arguments, show why in the case where the root-mean-square surface deviation from perfect flatness is approximately λ/N, the finesse cannot exceed $F \cong N/2$. [Hint: Consider the spreading of the transmission peak due to a small number of etalons of nearly equal length transmitting in parallel.]

4.4 Show that the angular spread of a beam that is incident normally on a plane-parallel Fabry–Perot etalon must not exceed

$$\theta_{1/2} = \sqrt{\frac{2n\lambda}{lF}}$$

if its peak transmission is not to deviate substantially from unity.

4.5 Complete the derivation of Equations (4.1-4), (4.1-5), (4.1-6), and (4.1-7).

4.6 Consider a diverging monochromatic beam that is incident on a plane-parallel Fabry–Perot etalon.

(a) Obtain an expression for the various angles along which the output energy is propagating. [Hint: These correspond to the different values of θ in Equation (4.1-8) that result from changing m.]
(b) Let the output beam in (a) be incident on a lens with a focal length f. Show that the energy distribution in the focal plane consists of a series of circles, each corresponding to a different value of m. Obtain an expression for the radii of the circles.
(c) Consider the effect in (b) of having simultaneously two frequencies ν_1 and ν_2 present in the input beam. Derive an expression for the separation of the respective circles in the focal plane. Show that the smallest separation $\nu_1 - \nu_2$ that can be resolved by this technique is given by $(\Delta\nu) \sim c/2nlF$.

4.7 Calculate the fraction of the power of a fundamental ($l = m = 0$) Gaussian beam that passes through an aperture with a radius equal to the beam spot size.

4.8 Show that in the case of a conventional two-reflector resonator the stability condition, Equation (4.5-5), reduces to Equation (4.4-2).

4.9 Consider a spherical mirror with a radius of curvature R whose reflectivity varies as

$$\rho(r) = \rho_0 \exp(-r^2/a^2)$$

where r is the radial distance from the center. Show that the (A, B, C, D) matrix of this mirror is given by

$$\begin{bmatrix} A & B \\ C & D \end{bmatrix} = \begin{bmatrix} 1 & 0 \\ -\dfrac{2}{R} - i\dfrac{\lambda}{\pi a^2} & 1 \end{bmatrix}$$

4.10 Consider an optical resonator with values as shown in Figure P4.10.

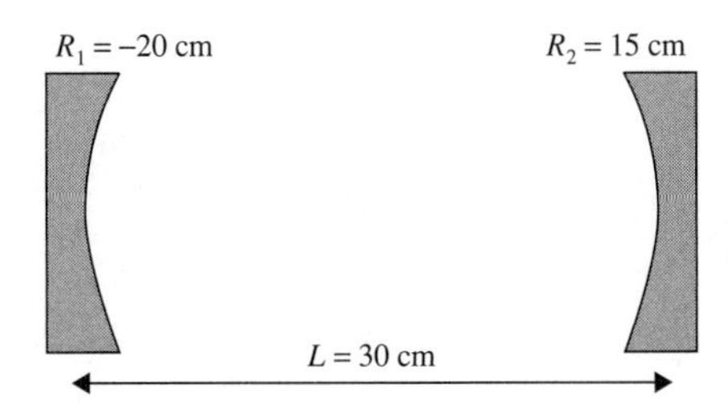

Figure P4.10

(a) Calculate the position of the waist of the mode at $\lambda = 1$ µm.
(b) Calculate the diameter of the waist.

4.11 Derive the following expression for the group delay of the Gires–Tournois etalon:

$$\tau = \frac{\sigma}{1 + (\sigma^2 - 1)\sin^2\phi}\tau_0$$

where $\tau_0 = 2d\phi/d\omega$ is the group delay of the etalon when the front mirror reflectivity is zero. Show that the group delay (time delay) reaches a maximum of $\sigma\tau_0$ at resonsnce ($\phi = m\pi$) and a minimum of τ_0/σ occurs at $\phi = m\pi + \pi/2$, $m =$ integer.

4.12 Show by simple arguments that in the situation depicted in Figure 4.33 the number of Hermite–Gaussian resonator modes that are excited substantially (say, within an order of magnitude of the maximal value) by a fundamental (incident) Gaussian beam is $\sim(\omega_1/\omega_{in})^2$. Ignore longitudinal resonance and consider only transverse (x, y) modes. [Hint: Consider the integral in the numerator of expression (4.10-5) for η_{mn}. Reason why the main contribution to the integral comes from a circle of radius ω_{in} or ω_1/l (l, m are the Hermite polynomial integers) whichever is smaller.]

4.13 Obtain an expression for the coupling efficiency of an incident Gaussian fundamental beam into an optical resonator whose mirrors have a radius of curvature R and a spot radius at the mirrors of ω. The incident beam has a radius R_b and spot size radius ω_b at the mirror.

4.14 The free spectral range (FSR $= \Delta\nu = c/2nd$)) of a Fabry–Perot etalon is determined by the resonance condition: $2kd = 2(2\pi\nu/c)nd = 2m\pi$, where n is the index of refraction of the cavity medium and d is the cavity spacing.

(a) Show that if material dispersion is present, the free spectral range becomes

$$\Delta\nu = \frac{v_g}{2d} = \frac{c}{2n_g d}$$

where v_g is the group velocity and n_g is the group index defined as $n_g = c/v_g$.
(b) In the above, we ignore the phase shifts upon reflection from the mirrors. These phase shifts can be a function of frequency. The resonance condition can be written $2kd + \phi_1 + \phi_2 = 2m\pi$. Derive an expression for the free spectral range by taking into account the frequency dependence of the phase shifts.

4.15 Consider a symmetric Fabry–Perot etalon with mirror reflectivity R.

(a) Derive an analytic expression for the phase delay of the transmitted beam (phase of E_t/E_i).
(b) Derive an expression for the group delay (derivative of phase with respect to angular frequency). Find the maximum and minimum of the group delay, in units of the cavity transit time. (Assume $n = 1$.)

4.16 As indicated in this book, the reflection and transmission coefficients r, r', t, and t' are related by the fundamental reciprocity and time-reversal symmetry. These coefficients are defined in Figure P4.16. When time is reversed, $(tA)^*$ and $(rA)^*$ should recombine and produce A^*. Show that

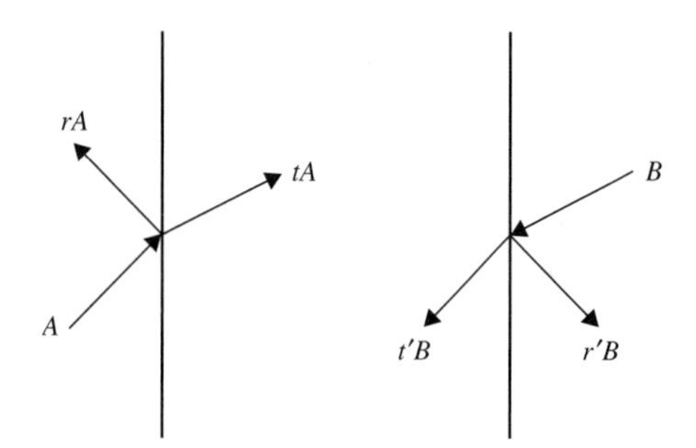

Figure P4.16

$$tt'^* + rr^* = 1$$

$$tr'^* + rt^* = 0$$

These are the Stokes relationships.

4.17 Optical interleavers are passive devices that can separate the spectral content of an incoming beam into two interleaving sets. In other words, if the incoming beam contains frequencies (ν_1, ν_2, ν_3, ν_4, ν_5, ν_6, ν_7, ν_8), the first set of output consists of (ν_1, ν_3, ν_5, ν_7) and the second set of output consists of (ν_2, ν_4, ν_6, ν_8). Interferometers like the Michelson (or Mach–Zehnder) can be employed for this purpose (see Figure P4.17). Let $\Delta L = L_2 - L_1$ be the path difference between the two arms of interferance; the intensity at one of the output ports is given by

$$I = \frac{1}{2}\left[1 + \cos\left(\frac{2\pi}{c}\nu 2\,\Delta L + \phi_2 - \phi_1\right)\right]$$

where ϕ_1 and ϕ_2 are phase shifts of the mirrors. Note that the round-trip path difference is $2\,\Delta L$.

(a) Assuming $\phi_1 = \phi_2 = 0$, find the frequencies where the output intensity is zero, and the frequencies where the intensity is maximum. Show that the maxima (or minima) are separated by

$$\Delta\nu = \frac{c}{2\Delta L}$$

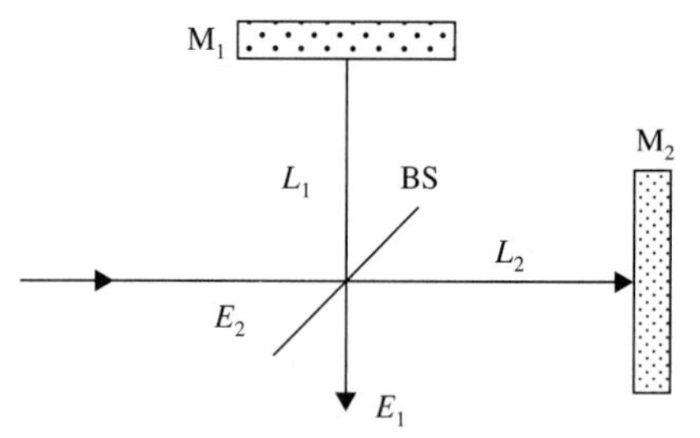

Figure P4.17

(b) A Gires–Tournois etalon can function like a mirror with a strong phase dispersion (i.e., phase shift depends on frequency). Several etalons can be used as mirrors in the interleavers, provided the period of the phase dispersion matches the period of the intensity output. Find the cavity spacing of the etalons.
(c) If the front mirror reflectivities of the etalons are properly chosen, a flat-top passband can be obtained. This is desirable in engineering application. Plot the output intensity for the case of $R_1 = 0\%$, $R_2 = 20\%$.
(d) Repeat (c) for the case when $R_1 = 3\%$, $R_2 = 40\%$.

4.18 Let the reflection and transmission coefficients for the beam splitter (BS) in Problem 4.17 be r, r', t, and t'. Assuming a lossless beam splitter, the conservation of energy requires that

$$|t|^2 + |r|^2 = 1 \quad \text{and} \quad |t'|^2 + |r'|^2 = 1$$

The principle of reciprocity requires that $t' = t$, assuming that the media on both sides of the beam splitter are identical.

(a) Show that $|r'| = |r|$.
(b) Show that the output fields (E_1, E_2) can be written

$$E_1 = rtAe^{-2ikL_1} + tr'Ae^{-2ikL_2} \quad \text{and}$$
$$E_2 = r^2Ae^{-2ikL_1} + tt'Ae^{-2ikL_2}$$

(c) Using the conservation of energy $|E_1|^2 + |E_2|^2 = |A|^2$, show that

$$tt'^* + rr^* = 1 \quad \text{and} \quad tr'^* + rt^* = 0$$

4.19 Consider the case of N identical rings with $(N + 1)$ identical couplers. The mode amplitudes are given by Equation (4.8-37).

(a) Show that Equation (4.8-37) can be rewritten

$$\begin{bmatrix} A_0 \\ B_0 \end{bmatrix} = [SP]^{N+1} P^{-1} \begin{bmatrix} A_{N+1} \\ B_{N+1} \end{bmatrix} = [SP]^{N+1} \begin{bmatrix} A_{N+1} \exp(-i\phi/2) \\ B_{N+1} \exp(+i\phi/2) \end{bmatrix}$$

(b) Assuming a real cross-coupling coefficient κ, show that

$$[SP] = \frac{1}{\kappa} \begin{bmatrix} \exp(i\phi/2) & t^*\exp(-i\phi/2) \\ t\exp(i\phi/2) & \exp(-i\phi/2) \end{bmatrix} \equiv \begin{bmatrix} A & B \\ C & D \end{bmatrix}$$

and

$$AD - BC = 1$$

(c) Using Chebyshev's identity (2.1-11), show that

$$|\chi|^2 \equiv \left| \frac{A_{N+1}}{A_0} \right|^2_{B_{N+1}=0} = \left| \frac{1}{M_{11}} \right|^2 = \frac{1}{1 + |C|^2 \dfrac{\sin^2(N+1)\theta}{\sin^2\theta}}$$

(d) Derive Equation (4.8-38).

REFERENCES

1. Ramo, S., J. R. Whinnery, and T. Van Duzer, *Fields and Waves in Communication Electronics*. Wiley, New York, 1965.
2. Yariv, A., *Quantum Electronics*, 3rd ed. Wiley, New York, 1989, p. 99.
3. Fabry, C., and A. Perot, Theorie et applications d'une nouvelle methode de spectroscopie interferentielle. *Ann. Chim. Phys.* **16**:115 (1899).
4. Born, M., and E. Wolf, *Principles of Optics*, 3rd ed. Pergamon, New York, 1965, Chapter 7.
5. Peterson, D. G., and A. Yariv, Interferometry and laser control with Fabry–Perot etalons. *Appl. Opt.* **5**:985 (1966).
6. See, for example, Pochi Yeh, *Optical Waves in Layered Media*. Wiley, New York, 1988.
7. Boyd, G. D., and J. P. Gordon, Confocal multimode resonator for millimeter through optical wavelength masers. *Bell Syst. Tech. J.* **40**:489 (1961).
8. Boyd, G. D., and H. Kogelnik, Generalized confocal resonator theory. *Bell Syst. Tech. J.* **41**:1347 (1962).
9. Casperson, L., Gaussian light beams in inhomogeneous media. *Appl. Opt.* **12**:2434 (1973).
10. Levi, A. F., R. E. Slusher, S. L. McCall, J. L. Glass, S. J. Pearson, and Ra. A. Logan, Directional light coupling from micro-disk lasers. *Appl. Phys. Lett.*, **62**:561 (1993).
11. Little, B. E., Ultra compact Si–SiO_2 micro-ring resonator optical channel dropping filters. *Opt. Lett.* **23**:1570 (1998).
12. Cai, M., O. Painter, and K. J. Vahala, Observation of critical coupling in a fiber taper to a silica-microsphere whispering gallery mode system. *Phys. Rev. Lett.* **85**:74 (2000).
13. Yariv, A., Universal relations for coupling of optical power between micro-resonators and dielectric waveguides. *Electron. Lett.* **36**:32 (2000).
14. Choi, J. M., R. K. Lee, and A. Yariv, Ring fiber resonators based on fused silica-fiber grating add-drop filters: application to resonator coupling, *Opt. Lett.* **27**:1598 (2002).
15. Huang, Y., G. T. Paloczi, J. Scheuer, and A. Yariv, Soft lithography replication of polymetric microring optical resonators. *Optics Express* **11**:2452 (2003).
16. Paloczi, G., Y. Huang, and A. Yariv, Polymeric Mach–Zehnder interferometer using serially coupled microring resonators. *Optics Express* **11**:2666 (2003).

ADDITIONAL READING AND REFERENCES

J. K. S. Poon, J. Scheuer, S. Mookherjea, G. T. Paloczi, Y. Huang, and A. Yariv, "Matrix analysis of microring coupled-resonator optical waveguides," Optics Express, 12(1), 90–103, 2004.

J. K. S. Poon, J. Scheuer, Y. Xu, and A. Yariv, "Designing coupled-resonator optical waveguide delay lines," J. Opt. Soc. Am., B, 21(9), 1665–1673, 2004.

J. E. Heebner and R.W. Boyd, "Slow and fast light in resonator-coupled waveguides," J. Modern Optics, 49, 2629–2636, 2002.

Y. Xu, Y. Li, R. K. Lee, and A. Yariv, "Scattering-theory analysis of waveguide-resonator coupling," Physical Review E. 62(5), 7389–7404, 2000.

CHAPTER 5

INTERACTION OF RADIATION AND ATOMIC SYSTEMS

5.0 INTRODUCTION

In optical electronics, especially optical communications, the index of refraction of a homogeneous medium is the key physical parameter that determines the transmission property of an optical beam. In the previous chapter we defined the index of refraction of a medium in terms of its dielectric constant. In this chapter we investigate the physical origins of the index of refraction, dispersion, and absorption. Specifically, we consider what happens to an electromagnetic wave propagating in an atomic medium (e.g., atomic vapors). Transitions between quantum states of an atomic system may involve the absorption or emission of photons. As a result of the interaction, the atomic system may behave like an oscillating electric dipole. A collection of these individual dipole moments leads to a macroscopic electric polarization of the medium. This is the origin of the electric susceptibility, dielectric constant, and index of refraction. We are also particularly concerned with the possibility of growth (or attenuation) of the radiation resulting from its interaction with the atoms, as well as the changes in the velocity of propagation of light due to such interaction. The concepts derived in this chapter will be used later in the text in treating the laser oscillator. We first discuss the connection between atomic transitions and the absorption and emission of electromagnetic radiation. We then discuss the classical electron model and derive the atomic polarizability. The index of refraction and its dependence on frequency (dispersion) of an atomic system can then be obtained. We then discuss induced transition, absorption, and amplification of an atomic system. Finally, we take up the subject of gain saturation in optical amplifiers.

5.1 ATOMIC TRANSITIONS AND ELECTROMAGNETIC WAVES

One of the fundamental results of the theory of quantum mechanics is that each physical system can be found, upon measurement, in only one of a predetermined set of energy states—the *eigenstates* of the system. With each of these states we associate an energy that corresponds to the total energy of the system when occupying the state. Some of the simpler systems, which are treated in any basic text on quantum mechanics, include the free electron, the hydrogen atom, and the harmonic oscillator. Examples of more complicated systems include the hydrogen molecule and the semiconducting crystal. With each state, the state m of the electron in a hydrogen atom, say, we associate an eigenfunction

$$\psi_m(\mathbf{r}, t) = u_m(\mathbf{r}) \exp(-iE_m t/\hbar), \quad m = 1, 2, 3, \ldots \tag{5.1-1}$$

where $|u_m(\mathbf{r})|^2\, dx\, dy\, dz$ gives the probability of finding the electron, once it is known to be in the state m, within the volume element $dx\, dy\, dz$, which is centered on the point $\mathbf{r}$. E_m is the state energy described above and $\hbar \equiv h/2\pi$, where $h = 6.626 \times 10^{-34}$ joule-second is Planck's constant.

One of the main tasks of quantum mechanics is the determination of the eigenfunctions $u_m(\mathbf{r})$ and the corresponding energy E_m of various physical systems. In this book, however, we will accept the existence of these states, their energy levels, as well as a number of other related results whose justification is provided by the experimentally proved formalism of quantum mechanics. Some of these results are discussed in the following.

Radiative Process and Atomic Transitions

There are, in general, many eigenstates in most atomic systems. Let us concentrate on two of these levels—1 and 2, for example. In the case of a hydrogen atom, these can be the 1s and 2p state of the electron. Such a two-level system greatly simplifies the discussion and introduction of the basic concepts. Figure 5.1 shows the schematic diagrams of the energy levels and the possible radiative processes.

If an atomic system is initially in its ground state (E_1 in Figure 5.1), and the atomic system is subject to the radiation of an incident photon, the photon may be absorbed by the atomic system provided the photon energy satisfies the follow condition:

$$h\nu = E_2 - E_1 \tag{5.1-2}$$

and certain selection rules (e.g., conservation of angular momentum) are also satisfied. As a result of the absorption of the incident photon, the atomic system is elevated to the upper state (with energy E_2). If the atomic system is known to be in state 2 at $t = 0$, there is a finite probability per unit time that it will undergo a transition to state 1, emitting in the process a photon of energy $h\nu = E_2 - E_1$. This process, occurring as it does without the inducement of a radiation field, is referred to as *spontaneous emission*. If the atomic system is initially in state 2 while at the same time under the action of an incident photon, the external action may also induce emission of a photon in addition to the spontaneous emission. This is called stimulated emission and is very important in many cases, including lasers and optical amplifiers.

The interaction of an atomic system with external electromagnetic radiation can be treated with the use of quantum mechanics, or a simple classical electron model. During the irradiation, the atomic system may behave as an oscillating electric dipole. In the next section,

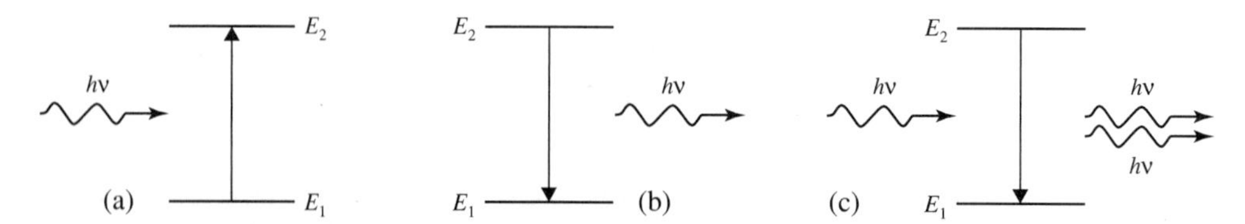

Figure 5.1 Radiative processes: (a) absorption—an incident photon is absorbed while the atomic state is elevated from eigenstate 1 to eigenstate 2; (b) spontaneous emission—a photon is emitted while the atomic system descends from eigenstate 2 to eigenstate 1; and (c) stimulated emission—an additional photon is emitted when an atomic system is under the action of an incident photon.

we will discuss the induced dipole moment of the atomic system due to the presence of an electromagnetic field.

5.2 ATOMIC POLARIZABILITY AND DIELECTRIC CONSTANT

Consider the transmission of a beam of light through a transparent isotropic medium. Under the action of the electric field of the optical beam, the elementary charge particles (mostly electrons) inside each atom are displaced from their equilibrium positions. In most dielectrics, this charge separation is directly proportional to and in the same direction as the electric field of the light beam. Thus the induced dipole moment $\mathbf{p}$ can be written

$$\mathbf{p} = \alpha \mathbf{E} \tag{5.2-1}$$

where the constant α is known as the atomic polarizability (or molecular polarizability) and $\mathbf{E}$ is the electric field. Generally, the molecular polarizability is a tensor. For atomic systems with spherical symmetry, the polarizability reduces to a scalar. This atomic polarizability α can be derived by using a simple classical electron model (see next section). The dielectric constant of a medium will depend on the manner in which the atoms are assembled. Consider a simple case of a gas medium. Let N be the number of atoms per unit volume. Then the polarization can be written approximately as

$$\mathbf{P} = N\,\mathbf{p} = N\alpha\mathbf{E} \equiv \varepsilon_0 \chi \mathbf{E} \tag{5.2-2}$$

where χ is the electric susceptibility of the medium. In Equation (5.2-2), we ignore the difference between the local field and the macroscopic field. This is legitimate for gas media provided $N\alpha/\varepsilon_0 \ll 1$. The dielectric constant ε of the medium is thus given by, according to Equations (1.1-5), and (5.2-2)

$$\varepsilon = \varepsilon_0(1 + \chi) = \varepsilon_0(1 + N\alpha/\varepsilon_0) \tag{5.2-3}$$

If the medium is nonmagnetic, the index of refraction of the medium is written, according to Equation (1.4-15),

$$n^2 = 1 + \chi = 1 + N\alpha/\varepsilon_0 \tag{5.2-4}$$

It is important to note that the index of refraction as well as the dielectric constant are dependent on the frequency of light. We mentioned this for silica in Chapter 3. This is known as the dispersion, which will be discussed further in the next section when we describe the classical electron model.

In the case of solids and liquids, the density N is about 1000 times higher. Thus the electric susceptibility is significantly higher. Under the circumstances, the electric field experienced by an individual atom (or molecule) is not the average electric field $\mathbf{E}$ in the medium. The field experienced by the atom (or molecule) is known as the local field. To find the local field, we consider a uniform medium in the presence of a mean electric field $\mathbf{E}$, which is uniform in the medium (on macroscopic length scales). Consider one of the atoms (or molecules) that constitutes the dielectric. We draw a sphere of radius r about this particular atom. This is intended to represent the boundary between the microscopic and the macroscopic range of phenomena affecting the atom. We shall treat the dielectric outside the sphere as a continuous medium and the dielectric inside the sphere as a collection of polarized atoms. As a result of uniform polarization, there are surface charges on the inner surface of the sphere. It can easily be shown that the field at the atom due to the surface charges on the sphere is

$$\mathbf{E}_{\text{surface charge}} = \frac{\mathbf{P}}{3\varepsilon_0} \tag{5.2-5}$$

Thus the net electric field (the local field) seen by an individual molecule is

$$\mathbf{E}_{\text{local}} = \mathbf{E} + \frac{\mathbf{P}}{3\varepsilon_0} \tag{5.2-6}$$

This is *larger* than the average electric field in the dielectric. The above analysis indicates that the local field is due to the long-range interactions of the atom (or molecule) with the other atoms (or molecules) in the medium. Substituting Equation (5.2-6) as the electric field in (5.2-2), we obtain

$$\mathbf{P} = N\alpha\mathbf{E}_{\text{local}} = N\alpha\left(\mathbf{E} + \frac{\mathbf{P}}{3\varepsilon_0}\right) = \varepsilon_0\chi\mathbf{E} \tag{5.2-7}$$

By eliminating **P**, we obtain

$$\chi = \frac{\dfrac{N\alpha}{\varepsilon_0}}{1 - \dfrac{N\alpha}{3\varepsilon_0}} \tag{5.2-8}$$

This is the dielectric susceptibility constant, by taking into account of the local field effect. The above equation is often written, according to Equation (5.2-3),

$$\frac{\varepsilon - \varepsilon_0}{\varepsilon + 2\varepsilon_0} = \frac{N\alpha}{3\varepsilon_0} \tag{5.2-9}$$

This is known as the *Clausius–Mossotti* relation. We note that the dielectric susceptibility reduces to $\chi = N\alpha/\varepsilon_0$ in the case of dilute systems where $\chi = N\alpha/\varepsilon_0 \ll 1$.

5.3 CLASSICAL ELECTRON MODEL

In order to find the atomic polarizability, we need to solve for the dipole moment of an atom induced by the electric field of an incident optical wave. Since the atoms are so small compared to the wavelength of light, we can assume that the electric field is uniform in each atom. Let the electric field of the optical wave in the atom be

$$E = E_0 \exp(i\omega t) \tag{5.3-1}$$

where E_0 is a constant amplitude and ω is the frequency. Each of the charged particles in the atom will feel this electric field and will be driven up and down (or back and forth) by the time-varying electric force. Because of the great difference in their masses, we expect that electrons contribute dominantly to the induced dipole moment. In classical physics, we further assume that the electrons are "fastened" elastically to the atoms and obey the following equation of motion:

$$m\frac{d^2}{dt^2}X + m\gamma\frac{d}{dt}X + m\omega_0^2 X = -eE \tag{5.3-2}$$

where X is the position of the electron relative to the atom, m is the electron mass, and $-e$ is the electron charge. The parameter ω_0 is the resonant frequency of the electron motion, and γ is the damping coefficient. For the case of a two-level system described in Figure 5.1,

$\omega_0 = (E_2 - E_1)/\hbar$. For a harmonic driving field, this equation has the following steady-state solution:

$$X = \frac{-eE_0}{m(\omega_0^2 - \omega^2 + i\gamma\omega)} \exp(i\omega t) \tag{5.3-3}$$

The induced dipole moment is thus given by

$$p = -eX = \frac{e^2}{m(\omega_0^2 - \omega^2 + i\gamma\omega)} E_0 \exp(i\omega t) \tag{5.3-4}$$

Using definition (5.2-2), we obtain the following expression for the atomic polarizability:

$$\alpha = \frac{e^2}{m(\omega_0^2 - \omega^2 + i\gamma\omega)} \tag{5.3-5}$$

If there are N atoms per unit volume, the index of refraction of such a medium is given by, according to Equations (5.3-5) and (5.2-4),

$$n^2 = 1 + \chi = 1 + \frac{Ne^2}{m\varepsilon_0(\omega_0^2 - \omega^2 + i\gamma\omega)} \tag{5.3-6}$$

where the electric susceptibility χ is often written in terms of its real and imaginary parts as

$$\chi = \chi' - i\chi'' = \frac{Ne^2}{m\varepsilon_0(\omega_0^2 - \omega^2 + i\gamma\omega)} \tag{5.3-7}$$

We note that the dielectric susceptibility χ is maximum when the driving frequency ω is close to the resonant frequency ω_0 of the atoms. At resonance, $\omega = \omega_0$, the susceptibility χ is a pure imaginary number. If the second term in Equation (5.3-6) is small compared to 1, the index of refraction can be written approximately as

$$n = 1 + \tfrac{1}{2}\chi = 1 + \frac{Ne^2}{2m\varepsilon_0(\omega_0^2 - \omega^2 + i\gamma\omega)} \quad (\text{when } \chi \ll 1) \tag{5.3-8}$$

In the spectral regime near resonance ($\omega \approx \omega_0$), these expressions can be written

$$n^2 = 1 + \chi = 1 + \frac{Ne^2}{2\omega_0 m\varepsilon_0(\omega_0 - \omega + i\gamma/2)} \tag{5.3-9}$$

$$\chi = \chi' - i\chi'' = \frac{Ne^2}{2\omega_0 m\varepsilon_0(\omega_0 - \omega + i\gamma/2)} \tag{5.3-10}$$

$$n = 1 + \tfrac{1}{2}\chi = 1 + \frac{Ne^2}{4\omega_0 m\varepsilon_0(\omega_0 - \omega + i\gamma/2)} \quad (\text{when } \chi \ll 1) \tag{5.3-11}$$

Equation (5.3-6) not only gives the index of refraction in terms of the basic atomic quantities, but also shows the variation with the frequency ω of light. However, it is not very practical to compute the index of refraction because both N and ω_0 vary from material to material. In addition, information about the nature of the resonance frequency ω_0 requires a calculation using quantum mechanics. Therefore we cannot expect to get a simple formula for the index of refraction that applies to all substances. However, Equations (5.3-9) and (5.3-11) are still very useful for many practical purposes. First, consider gas media such as air, helium, and hydrogen. The natural resonance frequencies of the electron oscillations correspond to ultraviolet light. These frequencies are higher than the frequencies of visible light; that is,

ω_0 is much larger than the ω of visible light. To a first approximation, we can neglect ω^2 and $i\gamma\omega$ in the denominator in comparison with ω_0^2. Then we find that the index of refraction is nearly constant. If we take $N = 2.69 \times 10^{25}$ m^{-3} for air at standard temperature (0 °C) and pressure (760 mmHg) and take the absorption band at $\lambda_0 = 80$ nm, Equation (5.3-7) yields $n = 1.0002$ for air. For most transparent substances such as glass and water, ω_0 is also in the ultraviolet region. However, for solids and liquid, N is three orders of magnitude larger than that of air; thus the electric susceptibility is significantly higher. Equations (5.3-7) and (5.3-10) are also useful when we study the gain or loss in laser media. Equations (5.3-6)–(5.3-11) are also useful for describing the gain profiles of the lasing transitions in gas lasers (e.g., He–Ne, argon lasers).

A similar expression can be obtained by using the time-dependent perturbation theory in quantum mechanics. Assuming the atomic system is initially in the lower state (state 1 in Figure 5.1a), the atomic polarizability is given by [1]

$$\alpha = \frac{f_{21}e^2}{2\omega_0 m(\omega_0 - \omega + i\gamma/2)} \tag{5.3-12}$$

where γ is the linewidth of the transition, f_{21} is the oscillator strength given by

$$f_{21} = \frac{2m\omega_0}{\hbar}|\langle u_2|x|u_1\rangle|^2 = \frac{2m\omega_0}{\hbar}\iiint u_2^* x u_1 \, dx\, dy\, dz \tag{5.3-13}$$

where u_1 and u_2 are the wavefunctions of the eigenstates, and we assume that the electric field is polarized along the x direction. The oscillator strength is a dimensionless parameter. It's a measure of the relative strength of the various transitions in the atomic system. It can be shown that

$$\sum_{m=2}^{\infty} f_{m1} = 1 \tag{5.3-14}$$

where the summation is over all states, except the initial state (state 1). Equation (5.3-14) is known as the *sum rule*. We note that the classical electron model provides an accurate result as long as the two levels are the "dominant levels" of the atomic system.

5.4 DISPERSION AND COMPLEX REFRACTIVE INDEX

If we examine closely the expression for the index of refraction, Equation (5.3-11), we notice that as ω rises and becomes closer to ω_0 the index of refraction also rises. So n rises slowly with ω_0. This is true for almost all transparent materials. Thus the index of refraction is higher for blue light than for red light. This is the reason a prism bends the light more in the blue than in the red. The phenomenon that the refractive index depends on frequency is called *chromatic dispersion*. Equation (5.3-7) is called the dispersion equation. As we will see later, dispersion affects the propagation of signals in optical communications.

So far, we have been ignoring the imaginary term $i\gamma\omega$ in the denominator of Equation (5.3-6). This term accounts for the damping of electron motion (for $\gamma > 0$) and gives rise to the phenomenon of *optical absorption*. Because of this damping term, $i\gamma\omega$, the index of refraction is a complex number according to Equation (5.3-7). The imaginary part of the index of refraction is significant only when ω is close to ω_0. By working out the real and imaginary parts of Equation (5.3-8), the complex refractive index can be written

$$n' = n - i\kappa = 1 + \frac{Ne^2(\omega_0^2 - \omega^2)}{2m\varepsilon_0[(\omega_0^2 - \omega^2)^2 + \gamma^2\omega^2]} - i\frac{Ne^2\gamma\omega}{2m\varepsilon_0[(\omega_0^2 - \omega^2)^2 + \gamma^2\omega^2]} \quad (5.4\text{-}1)$$

where we use n' to denote the complex refractive index and $-\kappa$ is the imaginary part of the refractive index. From now on, we use n to represent the real part of the index of refraction. The constant κ is referred to as the *extinction coefficient* because it represents an absorption (or attenuation) of the electromagnetic wave. To see the effect of κ on electromagnetic radiation, we consider the propagation of a monochromatic plane wave in a medium with a complex refractive index $(n - i\kappa)$. According to the discussion in Section 1.4, such a wave can be written

$$E = A \exp[i(\omega t - k'z)] \quad (5.4\text{-}2)$$

where k' denotes the wavenumber in the complex medium and is given by

$$k' = \omega\sqrt{\mu\varepsilon} = \frac{\omega}{c}n' = \frac{2\pi}{\lambda}n' \quad (5.4\text{-}3)$$

Using the complex refractive index (5.4-1), the wavenumber k' becomes complex and is

$$k' = \frac{2\pi}{\lambda}(n - i\kappa) \quad (5.4\text{-}4)$$

where n is now the real part of the complex refractive index and κ is the extinction coefficient. Substituting Equation (5.4-4) for k' in Equation (5.4-2), the electric field of the wave becomes

$$E = A \exp\left[i\left(\omega t - \frac{2\pi}{\lambda}nz\right)\right]\exp\left(-\frac{2\pi}{\lambda}\kappa z\right) \quad (5.4\text{-}5)$$

We notice that the imaginary part of the complex refractive index leads to an attenuation of electromagnetic radiation along its direction of propagation. The *attenuation coefficient* is usually defined as

$$\alpha \equiv -\frac{1}{I}\frac{dI}{dz} \quad (5.4\text{-}6)$$

where I is the intensity of electromagnetic radiation and, aside from a constant factor, is proportional to E^*E. According to Equation (5.4-6), the intensity of an attenuating beam can be written

$$I(z) = I(0)\exp(-\alpha z) \quad (5.4\text{-}7)$$

where $I(0)$ is the intensity at $z = 0$. The attenuation coefficient α defined in Equation (5.4-6) is related to the extinction coefficient κ by

$$\alpha = \frac{4\pi}{\lambda}\kappa \quad (5.4\text{-}8)$$

according to Equations (5.4-7) and (5.4-5). The attenuation coefficient is proportional to the imaginary part of the complex refractive index. According to Equation (5.4-8), even a small κ will lead to a strong attenuation for visible light. For example, with $\kappa = 0.0001$ and $\lambda = 0.5\ \mu$m, Equation (5.4-8) yields an attenuation coefficient of $\alpha = 25\ \text{cm}^{-1}$.

Both the real part n and the imaginary part κ of the complex index of refraction are functions of ω. The dependence on ω is significant, especially when ω is close to ω_0. To examine the dispersion at frequencies near resonance, we put $\omega \cong \omega_0$ in Equation (5.4-1) and obtain

$$n - i\kappa = 1 + \frac{Ne^2(\omega_0 - \omega)}{4m\omega_0\varepsilon_0[(\omega_0 - \omega)^2 + (\gamma/2)^2]} - i\frac{Ne^2(\gamma/2)}{4m\omega_0\varepsilon_0[(\omega_0 - \omega)^2 + (\gamma/2)^2]} \tag{5.4-9}$$

Equivalently, the electric susceptibility can be written

$$\chi = \chi' - i\chi'' = \frac{Ne^2(\omega_0 - \omega)}{2m\omega_0\varepsilon_0[(\omega_0 - \omega)^2 + (\gamma/2)^2]} - i\frac{Ne^2(\gamma/2)}{2m\omega_0\varepsilon_0[(\omega_0 - \omega)^2 + (\gamma/2)^2]} \tag{5.4-10a}$$

The electric susceptibility is often written in terms of the frequency ν as

$$\chi' = \frac{Ne^2(\nu_0 - \nu)}{4\pi m\omega_0\varepsilon_0[(\nu_0 - \nu)^2 + (\Delta\nu/2)^2]}$$
$$\chi'' = \frac{Ne^2(\Delta\nu/2)}{4\pi m\omega_0\varepsilon_0[(\nu_0 - \nu)^2 + (\Delta\nu/2)^2]} \tag{5.4-10b}$$

where $\Delta\nu$ is the FWHM bandwidth in units of Hz. Figure 5.2 shows a normalized plot of $n - 1$ and κ as functions of ω. Note that the extinction coefficient κ is maximum at $\omega = \omega_0$ and decreases like $|\omega - \omega_0|^{-2}$ as $|\omega - \omega_0|$ increases. In fact, $\kappa(\omega)$ in Equation (5.4-9) has a Lorentzian lineshape. On the other hand, the real part of the index of refraction approaches 1, like $|\omega - \omega_0|^{-1}$ as $|\omega - \omega_0|$ increases. This explains the fact that dispersion exists in most transparent materials at frequencies where κ is negligible. According to Equation (5.4-9), the index of refraction is greater than unity for low frequencies ($\omega < \omega_0$) and increases with ω as the resonant frequency ω_0 is approached. Such an increasing behavior is the case of *normal dispersion* and occurs in most transparent media. Very near the resonance frequency, there is a small spectral region where the slope is negative. Such a negative slope is often referred to as *anomalous dispersion*. As we can see from Figure 5.2, anomalous dispersion always occurs when absorption is significant.

In the classical electron model discussed in Section 5.3, we only assumed a single resonance frequency ω_0, which is for a two-level system and is somewhat simpler than in nature.

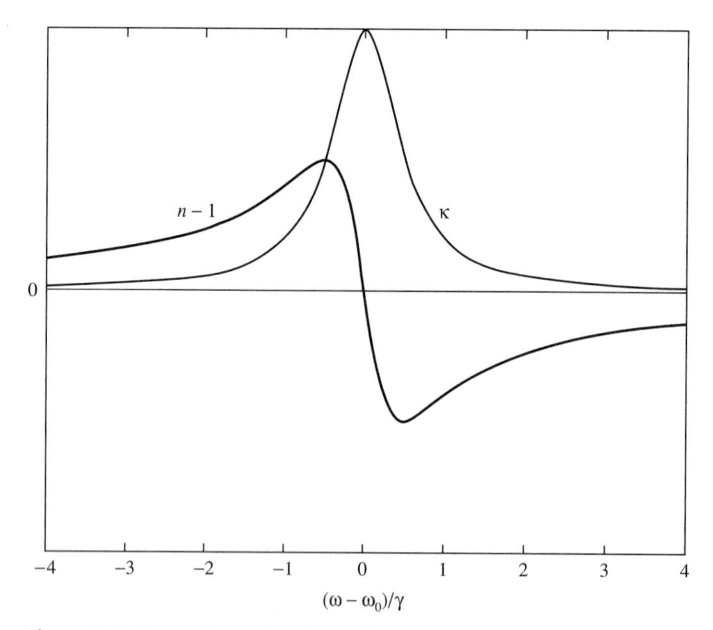

Figure 5.2 Normalized plot of n and κ versus ω.

In fact, there are several resonance frequencies for each atom. We can now modify the dispersion equation by assuming that not all the electrons are identically bound and that all the electrons oscillate independently. Let f_j be the fraction of electrons whose natural resonance frequency is ω_j and whose damping factor is γ_j. Then by adding the contribution from all the electron oscillators, we obtain

$$n^2 = 1 + \frac{Ne^2}{m\varepsilon_0} \sum_j \frac{f_j}{(\omega_j^2 - \omega^2 + i\gamma_j\omega)} \tag{5.4-11}$$

The fractions f_j are known as oscillator strengths, defined in Equation (5.3-13), and are dimensionless. By definition, the oscillator strengths obey the following sum rule:

$$\sum_j f_j = Z \tag{5.4-12}$$

where Z is the number of electrons per atom. Theoretical evaluation of oscillator strengths requires knowledge of the wavefunction of the atomic system. To illustrate the magnitude of the oscillator strengths and the rule (5.4-12), we note that the oscillator strength corresponding to the 1s $\rightarrow$ 2p transition ($\omega_{1s\text{-}2p} = 1.55 \times 10^{16}\ \text{s}^{-1}$) in the hydrogen atom is 0.4162. The sum of the oscillator strengths for transitions from the ground state 1s to all other states different from 2p is 0.5838. In the preceding discussion, we assumed that there is a damping coefficient γ_j associated with each resonance frequency. These coefficients are all positive, and electron damping leads to attenuation of electromagnetic radiation.

If some electrons are in their excited states, electromagnetic radiation will drive these electrons and induce the emission of radiation at the resonance frequency. For such an excited system (or inverted system), Equation (5.4-11) is still valid, provided we replace the parameter N by $N_g - N_j$, where N_g is the number of electrons in the ground state per unit volume and N_j is the number of electrons in the jth excited state per unit volume. In an inverted system, N_j is greater than N_g; thus we see that the complex refractive index of an excited system has a positive imaginary part in a small frequency range near the resonance frequency ω_j for which $N_g - N_j$ is negative. According to Equation (5.4-5), the propagation of electromagnetic radiation in a medium with $\kappa < 0$ will lead to amplification of the energy. Such a medium is called a *gain medium* and is described by a complex index of refraction with a positive imaginary part (i.e., $\kappa < 0$).

Figure 5.3 shows the dispersion of the index of refraction of several transparent optical materials in the spectral range $\lambda = 0.3$ to 30 μm. Traditionally, the dispersion of n is usually expressed in terms of wavelength λ.

Complex Refractive Index of Electron Gas (Metals)

In a metal, there are free electrons that do not oscillate around atoms but are free to move under the action of the applied electric field. For these electrons, there is no restoring force. The equation of motion (5.3-2) is still valid, provided we set $\omega_0 = 0$. Thus all the results obtained in Section 5.3 can still be used for metals. Putting $\omega_0 = 0$ in Equation (5.3-6), we obtain the complex refractive index

$$n^2 = 1 - \frac{Ne^2}{m\varepsilon_0(\omega^2 - i\gamma\omega)} \tag{5.4-13}$$

where N is electron density. In fact, Equation (5.4-13) becomes a good approximation to the actual complex refractive index, provided we can neglect the contribution from the deeply

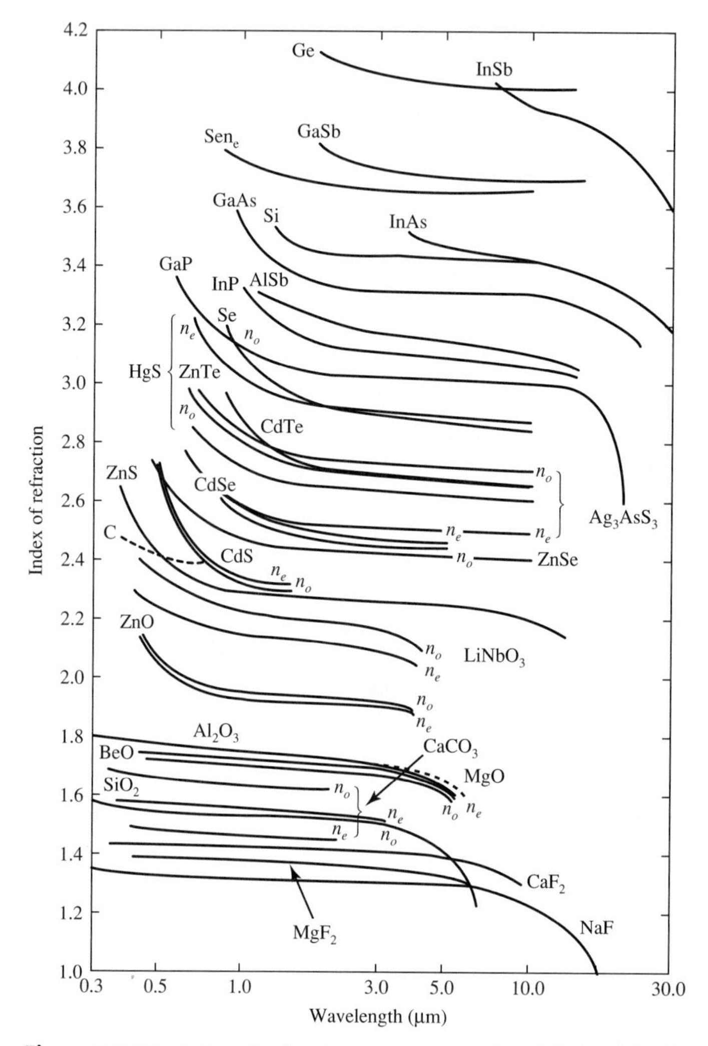

Figure 5.3 The index of refraction n versus wavelength λ. For birefringent crystals, n_o is the ordinary refractive index and n_e is the extraordinary refractive index. (Data adapted from Reference [2].)

bound electrons in metals and some semiconductors. The damping constant is some sort of average rate of collisions involving appreciable momentum transfer. If we further assume that $\gamma \ll \omega$, the index of refraction is given by

$$n^2 = 1 - \frac{Ne^2}{m\varepsilon_0\omega^2} = 1 - \frac{\omega_p^2}{\omega^2} \tag{5.4-14}$$

where ω_p is the *plasma angular frequency* given by

$$\omega_p^2 = \frac{Ne^2}{m\varepsilon_0} \tag{5.4-15}$$

For high-frequency radiation ($\omega > \omega_p$), the index of refraction is real and the waves propagate freely. For frequencies lower than the plasma frequency ω_p, n is purely imaginary. Thus, within the metal, the fields fall off exponentially with distance from the surface (see also Problem 5.4). Consequently, such electromagnetic radiation incident on a high-conductivity metal will be reflected from the surface. For metals such as aluminum, copper, gold, and silver, the density of the free electrons is on the order of $N = 10^{23}$ cm^{-3}. This means that $\omega_p \approx 2 \times 10^{16}$ s^{-1}, so that for visible and infrared radiation with $\omega < \omega_p$, the index of refraction is imaginary according to Equation (5.4-14). The index of refraction, in general, is complex when γ is finite.

5.5 LINESHAPE FUNCTION—HOMOGENEOUS AND INHOMOGENEOUS BROADENING

According to Equations (5.4-8) and (5.4-9), the absorption coefficient of a two-level atomic system can be written

$$\alpha(\omega) = \frac{4\pi}{\lambda} \frac{Ne^2\gamma}{8m\omega_0\varepsilon_0[(\omega_0 - \omega)^2 + (\gamma/2)^2]} \tag{5.5-1}$$

where, we recall, ω is the frequency of light, and ω_0 is the natural resonance frequency of the atomic system. By examining the above expression, we note that the absorption is not strictly monochromatic (i.e., of one frequency). In fact, absorption peaks at ω_0 and drops off rapidly as the frequency of light ω is detuned from ω_0. The lineshape described by Equation (5.5-1) is known as a Lorentzian lineshape. In the classical electron model, the absorption spectrum is the same as the spontaneous emission spectrum. So, if one performs a spectral analysis of the radiation emitted by spontaneous $2 \rightarrow 1$ transitions, one finds that the radiation is not strictly monochromatic but occupies a finite frequency bandwidth. The normalized lineshape function describing the spectral distribution of emitted radiation of a two-level atomic system in the vicinity of ω_0 can thus be written, according to Equation (5.5-1),

$$f(\omega) = \frac{\gamma}{[(\omega_0 - \omega)^2 + (\gamma/2)^2]} \tag{5.5-2}$$

The above lineshape is exactly the Fourier spectrum of a damped oscillating dipole (see Problem 5.7). In optical electronics, the lineshape function $g(\nu)$ is often expressed in terms of the frequency ν and is normalized according to

$$\int_{-\infty}^{+\infty} g(\nu)\, d\nu = 1 \tag{5.5-3}$$

We can consequently view $g(\nu)\, d\nu$ as the a priori probability that a given spontaneous emission from level 2 to level 1 will result in a photon whose frequency is between ν and $\nu + d\nu$.

From the experimental point of view, $g(\nu)$ can be obtained by spectrally analyzing the emitted radiation from the atomic system, or by performing a transmission measurement of a beam of light through a sample containing the atomic system as a function of the frequency. The fact that both the emission and the absorption are described by the same lineshape function $g(\nu)$ can be verified experimentally and follows from basic quantum mechanical considerations.

Homogeneous and Inhomogeneous Broadening

Using the result obtained from the classical electron model, the lineshape function of the atomic transition is written

$$g(\nu) = \frac{1}{2\pi} \frac{\gamma/2\pi}{(\nu - \nu_0)^2 + (\gamma/4\pi)^2} \tag{5.5-4}$$

where $\nu_0 = \omega_0/2\pi$, and γ is the damping factor. The separation $\Delta\nu$ between the two frequencies at which the Lorentzian lineshape function is down to half its peak value is referred to as the linewidth and is given by

$$\Delta\nu = \gamma/2\pi \tag{5.5-5}$$

In terms of the linewidth $\Delta\nu$, the normalized lineshape function can be written

$$g(\nu) = \frac{\Delta\nu}{2\pi[(\nu - \nu_0)^2 + (\Delta\nu/2)^2]} \tag{5.5-6}$$

One of the possible causes for the frequency spread of spontaneous emission is the finite lifetime τ of the emitting state. According to Equation (5.3-2), the oscillation amplitude of the electron will decay when the driving field is turned off. If we solve Equation (5.3-2) in the absence of the driving field, we obtain

$$|X(t)|^2 \propto \exp(-\gamma t)$$

Thus the damping factor is related to the spontaneous lifetime as

$$\gamma = 1/\tau \tag{5.5-7}$$

Using Equation (5.5-5), the linewidth can be written

$$\Delta\nu = \frac{1}{2\pi\tau} \tag{5.5-8}$$

Here we have an expression of the linewidth broadening due to the finite lifetime of the upper state (τ_u). In many laser systems, the lower state (τ_l) may also have a finite lifetime. In this case, the total linewidth broadening is a sum of the two individual broadenings. Further broadening can be due to collisions. In the process of coherent radiation, the radiating field from an atom can experience an abrupt change of phase due to elastic collisions from other atoms (see Problem 5.8). Let τ_{co} be the collision time. We thus generalize Equation (5.5-8) to read

$$\Delta\nu = \frac{1}{2\pi}\left(\tau_u^{-1} + \tau_l^{-1} + \tau_{\text{co}}^{-1}\right) \tag{5.5-9}$$

The type of broadening (i.e., the finite width of the emitted spectrum) described above is called *homogeneous broadening*. It is characterized by the fact that the spread of the response over a band $\Delta\nu$ is characteristic of *each* atom in the sample. The function $g(\nu)$ thus describes the response of any of the atoms, which are indistinguishable.

As mentioned earlier, homogeneous broadening is due most often to the finite interaction lifetime of the emitting or absorbing atoms. Some of the most common mechanisms are:

1. The spontaneous lifetime of the excited state.
2. Collision of an atom embedded in a crystal with a phonon. This may involve the emission or absorption of acoustic energy. Such a collision does not terminate the lifetime of the atom in its absorbing or emitting state. It does, however, interrupt the relative phase between the atomic oscillation and that of the field, thus causing a broadening of the radiation.

3. Pressure broadening of atoms in a gas. At sufficiently high atomic densities, the collisions between atoms become frequent enough that lifetime termination and phase interruption as in the preceding mechanism dominate the broadening mechanism.

There are, however, many physical situations in which the individual atoms are distinguishable, each having a slightly different transition frequency ν_0. If one observes, in this case, the spectrum of the spontaneous emission, its spectral distribution will reflect the spread in the individual transition frequencies and not the broadening due to the finite lifetime of the excited state. Two typical situations give rise to this type of broadening, referred to as *inhomogeneous broadening*.

First, the energy levels, hence the transition frequencies, of ions present as impurities in a host crystal depend on the immediate crystalline surroundings. The ever present random strain, as well as other types of crystal imperfections, cause the crystal surroundings to vary from one ion to the next, thus effecting a spread in the transition frequencies. Second, the transition frequency ν of a gaseous atom (or molecule) is Doppler-shifted due to the finite velocity of the atom along the direction of observation according to

$$\nu = \nu_0 + \frac{v_x}{c}\nu_0 \tag{5.5-10}$$

where v_x is the component of the velocity along the direction connecting the observer with the moving atom, c is the velocity of light in the medium, and ν_0 is the frequency corresponding to a stationary atom. The Maxwell velocity distribution function of a gas with atomic mass M that is at equilibrium at temperature T is [3]

$$f(v_x, v_y, v_z) = \left(\frac{M}{2\pi kT}\right)^{3/2} \exp\left(-\frac{M}{2kT}(v_x^2 + v_y^2 + v_z^2)\right) \tag{5.5-11}$$

where $k = 1.38 \times 10^{-23}$ J/K is the Boltzmann constant, and $f(v_x, v_y, v_z)\, dv_x\, dv_y\, dv_z$ is thus the fraction of all the atoms whose x component of velocity is contained in the interval v_x to $v_x + dv_x$ while, simultaneously, their y and z components lie between v_y and $v_y + dv_y$, v_z and $v_z + dv_z$, respectively. Alternatively, we may view $f(v_x, v_y, v_z)\, dv_x\, dv_y\, dv_z$ as the a priori probability that the velocity vector $\mathbf{v}$ of any given atom terminates within the differential volume $dv_x\, dv_y\, dv_z$ centered on $\mathbf{v}$ in velocity space so that

$$\iiint_{-\infty}^{\infty} f(v_x, v_y, v_z)\, dv_x\, dv_y\, dv_z = 1 \tag{5.5-12}$$

According to Equation (5.5-10) the probability $g(\nu)\, d\nu$ that the transition frequency is between ν and $\nu + d\nu$ is equal to the probability that v_x will be found between $v_x = (\nu - \nu_o)(c/\nu_o)$ and $(\nu + d\nu - \nu_o)(c/\nu_o)$ irrespective of the values of v_y and v_z (since if $v_x = (\nu - \nu_o)(c/\nu_o)$, the Doppler-shifted frequency will be equal to ν regardless of v_y and v_z). This probability is thus obtained by substituting $v_x = (\nu - \nu_o)(c/\nu_o)$ in $f(v_x, v_y, v_z)\, dv_x\, dv_y\, dv_z$ and then integrating over all values of v_y and v_z. The result is

$$g(\nu)\, d\nu = \left(\frac{Mc^2}{2\pi kT}\right)^{3/2} \int_{-\infty}^{\infty}\int_{-\infty}^{\infty} e^{-(M/2kT)(v_y^2+v_z^2)} dv_y\, dv_z \times e^{-(Mc^2/2kT)(\nu-\nu_0)^2/\nu_0^2} \left(\frac{1}{\nu_0}\right) d\nu \tag{5.5-13}$$

Using the definite integral

$$\int_{-\infty}^{\infty} e^{-(M/2kT)v_z^2}\, dv_z = \left(\frac{2\pi kT}{M}\right)^{1/2}$$

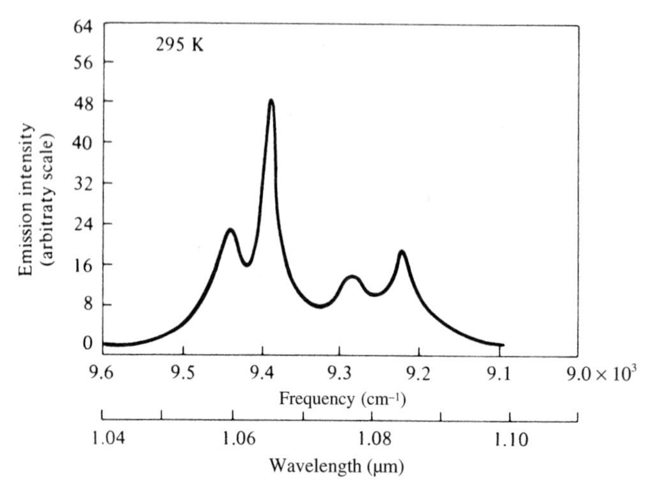

Figure 5.4 Emission spectrum of Nd^{3+}:$CaWO_4$ in the vicinity of the 1.06 μm laser transition. The main peak corresponds to the laser transition. (After Reference [4].) The unit 1 cm^{-1} corresponds to a frequency of $\nu = 30$ GHz, or a photon energy of $h\nu = 1.24 \times 10^{-4}$ eV.

we obtain, from Equation (5.5-13),

$$g(\nu)\, d\nu = \frac{1}{\nu_0}\left(\frac{Mc^2}{2\pi kT}\right)^{1/2} e^{-(Mc^2/2kT)(\nu-\nu_0)^2/\nu_0^2}\, d\nu \tag{5.5-14}$$

for the *normalized Doppler-broadened lineshape*. The functional dependence of $g(\nu)$ in Equation (5.5-14) is referred to as Gaussian. The width of $g(\nu)$ in this case is taken as the frequency separation between the points where $g(\nu)$ is down to half its peak value. It is obtained from Equation (5.5-14) as

$$\Delta\nu_D = 2\nu_0\sqrt{\frac{2kT}{Mc^2}\ln 2} \tag{5.5-15}$$

where the subscript D stands for *Doppler*. We can reexpress $g(\nu)$ in terms of $\Delta\nu_D$, obtaining

$$g(\nu) = \frac{2(\ln 2)^{1/2}}{\pi^{1/2}\Delta\nu_D} e^{-[4(\ln 2)(\nu-\nu_0)^2/\Delta\nu_D^2]} \tag{5.5-16}$$

In Figure 5.4 we show, as an example of a lineshape function, the spontaneous emission spectrum of Nd^{3+} when present as an impurity ion in a $CaWO_4$ lattice. The spectrum consists of a number of transitions, which are partially overlapping.

EXAMPLE: THE DOPPLER LINEWIDTH OF Ne

Consider the 6328 Å transition in Ne, which is used in the popular He–Ne lasers. Using atomic mass 20 for neon in Equation (5.5-15) and taking $T = 300$ K, we obtain

$$\Delta\nu_D \approx 1.5 \times 10^9 \text{ Hz}$$

for the Doppler linewidth. The 10.6 μm transition in the CO_2 laser has, according to Equation (5.5-15), a linewidth $\Delta\nu_D \approx 6 \times 10^7$ Hz.

5.6 INDUCED TRANSITIONS—ABSORPTION AND AMPLIFICATION

Consider a two-level atomic system as shown in Figure 5.1. In the presence of an electromagnetic field of frequency $\nu \sim (E_2 - E_1)/h$, an atom can undergo a transition from state 1 to 2, *absorbing* in the process a quantum of excitation (photon) with energy $h\nu$ from the field. If the atom happens to be in state 2 at the moment when it is first subjected to the electromagnetic field, it may make a downward transition to state 1, *emitting* a photon of energy $h\nu$. What distinguishes the process of induced transition from the spontaneous one described in Section 5.1 is the fact that the induced rate for 2→1 and 1→2 transitions is *equal*, whereas the spontaneous 1→2 (i.e., the one in which the atomic energy increases) transition rate is zero. Another fundamental difference—one that again follows from quantum mechanical considerations—is that the induced rate is *proportional* to the *intensity* of the electromagnetic field, whereas the spontaneous rate is independent of it. The relationship between the induced transition rate and the (inducing) field intensity is of fundamental importance in treating the interaction of atomic systems with electromagnetic fields. Its derivation follows.

Consider first the interaction of an assembly of identical atoms with a radiation field whose energy density is distributed uniformly in frequency in the vicinity of the transition frequency (within the atomic lineshape function). Let the energy density per unit frequency be $\rho(\nu)$. We assume that the induced transition rates per atom from 2→1 and 1→2 are both proportional to $\rho(\nu)$ and take them as

$$
\begin{aligned}
(W'_{21})_{\text{induced}} &= B_{21}\rho(\nu) \\
(W'_{12})_{\text{induced}} &= B_{12}\rho(\nu)
\end{aligned}
\tag{5.6-1}
$$

where B_{21} and B_{12} are constants to be determined. The total downward (2→1) transition rate is the sum of the induced and spontaneous contributions

$$
W'_{21} = B_{21}\rho(\nu) + A_{21} \tag{5.6-2}
$$

where A_{21} is the spontaneous transition rate. The total upward (1→2) transition rate is

$$
W'_{12} = (W'_{12})_{\text{induced}} = B_{12}\rho(\nu) \tag{5.6-3}
$$

Our first task is to obtain an expression for B_{12} and B_{21}. Since the magnitudes of the coefficients B_{21} and B_{12} depend on the atoms and not on the radiation field, we consider, without loss of generality, the case where the atoms are in a thermal equilibrium with a blackbody (thermal) radiation field at temperature T. In this case the radiation density is given by [5]

$$
\rho(\nu) = \frac{8\pi n^3 h\nu^3}{c^3}\frac{1}{e^{h\nu/kT} - 1} \tag{5.6-4}
$$

Since at thermal equilibrium the average populations of levels 2 and 1 are constant with time, it follows that the number of 2→1 transitions in a given time interval is equal to the number of 1→2 transitions; that is,

$$
N_2 W'_{21} = N_1 W'_{12} \tag{5.6-5}
$$

where N_1 and N_2 are the equilibrium population densities of levels 1 and 2, respectively. Using Equations (5.6-2) and (5.6-3) in Equation (5.6-5), we obtain

$$N_2[B_{21}\rho(\nu) + A_{21}] = N_1 B_{12}\rho(\nu)$$

and, substituting for $\rho(\nu)$ from Equation (5.6-4),

$$N_2\left(B_{21}\frac{8\pi n^3 h\nu^3}{c^3(e^{h\nu/kT}-1)} + A_{21}\right) = N_1\left(B_{12}\frac{8\pi n^3 h\nu^3}{c^3(e^{h\nu/kT}-1)}\right) \tag{5.6-6}$$

Since the atoms are in thermal equilibrium, the ratio N_2/N_1 is given by the Boltzmann factor [5] as

$$\frac{N_2}{N_1} = e^{-h\nu/kT} \tag{5.6-7}$$

Equating N_2/N_1 as given by Equation (5.6-6) to that in Equation (5.6-7) gives

$$\frac{8\pi n^3 h\nu^3}{c^3(e^{h\nu/kT}-1)} = \frac{A_{21}}{B_{12}e^{h\nu/kT} - B_{21}} \tag{5.6-8}$$

The last equality can be satisfied only when

$$B_{12} = B_{21} \tag{5.6-9}$$

and simultaneously

$$\frac{A_{21}}{B_{21}} = \frac{8\pi n^3 h\nu^3}{c^3} \tag{5.6-10}$$

The last two equations were first derived by Einstein [6]. We can, using Equation (5.6-10), rewrite the induced transition rate (5.6-1) as

$$W_i' = \frac{A_{21}c^3}{8\pi n^3 h\nu^3}\rho(\nu) = \frac{c^3}{8\pi n^3 h\nu^3 t_{\text{spont}}}\rho(\nu) \tag{5.6-11}$$

where $t_{\text{spont}} \equiv 1/A_{21}$ is the spontaneous lifetime of the atom. By virtue of Equation (5.6-9), the distinction between $2\rightarrow1$ and $1\rightarrow2$ induced transition rates is superfluous.

Equation (5.6-11) gives the transition rate per atom due to a field with a uniform (white) spectrum with energy density per unit frequency $\rho(\nu)$. In the event when the energy density per unit frequency $\rho(\nu)$ is not exactly uniform within the atomic lineshape function $g(\nu)$, the induced transition rate should be written

$$W_i = \int \frac{c^3}{8\pi n^3 h\nu^3 t_{\text{spont}}}\rho(\nu)g(\nu)\,d\nu \tag{5.6-12}$$

In optical electronics our main concern is the transition rates induced by a monochromatic (i.e., single-frequency) field of frequency ν. Let us denote this transition rate as $W_i(\nu)$. For a monochromatic field at frequency ν, the energy density per unit frequency $\rho(\nu')$ is a delta function,

$$\rho(\nu') = U_0\delta(\nu' - \nu) \tag{5.6-13}$$

where U_0 is the energy density (joules per unit volume) of the monochromatic field. Substituting Equation (5.6-13) for the energy density per unit frequency in Equation (5.6-12), we obtain

$$W_i = \frac{c^3 U_0}{8\pi n^3 h\nu^3 t_{\text{spont}}}g(\nu) \tag{5.6-14}$$

where U_0 is the energy density (joules per cubic meter) of the electromagnetic field inducing the transitions. This is the induced transition rate due to the presence of a monochromatic field with an energy density U_0 and a frequency ν.

Returning to our central result, Equation (5.6-14), we can rewrite it in terms of the beam intensity $I_0 = cU_0/n$ (watts per square meter) of the optical wave as

$$W_i = \frac{A_{21}c^2 I_0}{8\pi n^2 h\nu^3} g(\nu) = \frac{\lambda^2 I_0}{8\pi n^2 h\nu t_{\text{spont}}} g(\nu) \tag{5.6-15}$$

where I_0 is the beam intensity at frequency ν, c is the velocity of propagation of light in vacuum, λ is the vacuum wavelength, and $t_{\text{spont}} \equiv 1/A_{21}$. We note that the induced transition rate is proportional to the intensity of the incoming optical beam.

Absorption and Amplification

Consider the case of a monochromatic plane wave of frequency ν and intensity I_0 propagating through an atomic medium with N_2 atoms per unit volume in level 2 and N_1 in level 1. According to Equation (5.6-15) there will occur $N_2 W_i$ induced transitions per unit time per unit volume from level 2 to level 1 and $N_1 W_i$ transitions from level 1 to level 2. The net power generated within a unit volume is thus

$$\frac{P}{\text{Volume}} = (N_2 - N_1) W_i h\nu \tag{5.6-16}$$

This radiation is added coherently (i.e., with a definite phase relationship) to that of the traveling wave so that it is equal, in the absence of any dissipation mechanisms, to the increase in the intensity per unit length. Consider the increase of power in a differential distance of dz along the direction of propagation; using Equation (5.6-15), we obtain

$$\frac{dI_0}{dz} = (N_2 - N_1) \frac{c^2 g(\nu)}{8\pi n^2 \nu^2 t_{\text{spont}}} I_0 \tag{5.6-17}$$

The solution of Equation (5.6-17) is

$$I_0(z) = I_0(0) e^{\gamma(\nu) z} \tag{5.6-18}$$

where $\gamma(\nu)$ is the gain coefficient

$$\gamma(\nu) = (N_2 - N_1) \frac{c^2}{8\pi n^2 \nu^2 t_{\text{spont}}} g(\nu) \tag{5.6-19}$$

We note that the intensity grows exponentially when the population is inverted ($N_2 > N_1$) or is attenuated when $N_2 < N_1$. The first case corresponds to laser-type amplification, whereas the second case is the one encountered in atomic systems at thermal equilibrium. The two situations are depicted in Figure 5.5. We recall that at thermal equilibrium

$$\frac{N_2}{N_1} = e^{-h\nu/kT} \tag{5.6-20}$$

so that systems at thermal equilibrium are always absorbing. The inversion condition $N_2 > N_1$ can still be represented by Equation (5.6-20), provided we take T as negative. As a matter of fact, the condition $N_2 > N_1$ is often referred to as one of "negative temperature"—the

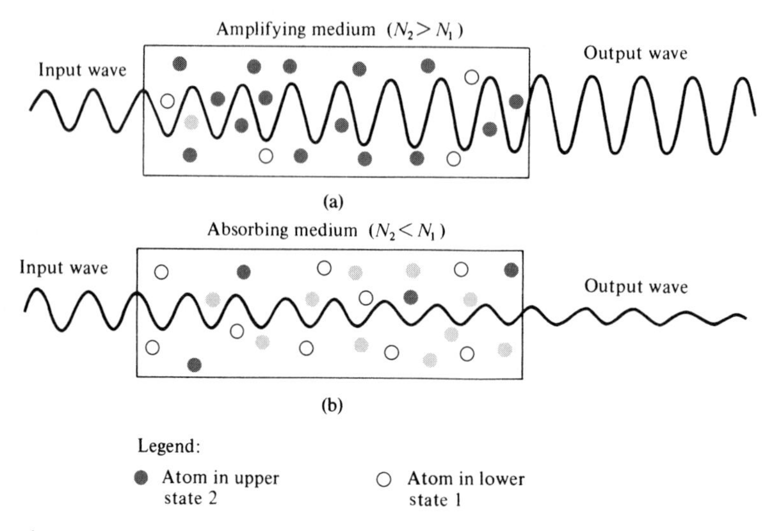

Figure 5.5 Amplification of a traveling electromagnetic wave in (a) an inverted population ($N_2 > N_1$) and (b) attenuation in an absorbing ($N_2 < N_1$) medium.

"temperature" in this case serving as an indicator of the population ratio, in accordance with Equation (5.6-20). The absorption, or amplification, of electromagnetic radiation by an atomic transition can be described not only by means of the exponential gain constant $\gamma(\nu)$, but also alternatively, in terms of the imaginary part of the electric susceptibility $\chi''(\nu)$ of the propagation medium. According to Equation (1.3-24), the density of absorbed power is

$$\overline{\frac{\text{Power}}{\text{Volume}}} = \frac{\omega \varepsilon_0 \chi''(\nu)}{2} |E|^2 \tag{5.6-21}$$

where $\chi''(\nu)$ is the imaginary part of the electric susceptibilities. This last result must agree with a derivation using the concept of the induced transition rate $W_i(\nu)$ according to which

$$\overline{\frac{\text{Power}}{\text{Volume}}} = (N_1 - N_2) W_i(\nu) h\nu \tag{5.6-22}$$

Equating (5.6-21) and (5.6-22), substituting (5.6-15) for $W_i(\nu)$, and using the relation $I_0 = (c/n)\varepsilon |E|^2/2$ (see Equation (1.4-21)), we obtain

$$\chi''(\nu) = \frac{(N_1 - N_2)\lambda^3}{16\pi^2 n t_{\text{spont}}} g(\nu) \tag{5.6-23}$$

where $n^2 \equiv \varepsilon/\varepsilon_0$ and λ is the wavelength in vacuum. In the case of a Lorentzian lineshape function $g(\nu)$, the last result can be written

$$\chi''(\nu) = \frac{(N_1 - N_2)\lambda^3}{8\pi^3 n t_{\text{spont}} \, \Delta\nu} \frac{1}{1 + [4(\nu - \nu_0)^2]/(\Delta\nu)^2} \tag{5.6-24}$$

This is a key result and it will be used on numerous occasions.

EXAMPLE: EXPONENTIAL GAIN CONSTANT IN A RUBY LASER

Let us estimate the exponential gain constant at line center of a ruby (Al_2O_3 doped with Cr^{3+} ions) crystal having the following characteristics:

$$N_2 - N_1 = 5 \times 10^{17}/\text{cm}^3$$

$$\Delta\nu \cong \frac{1}{g(\nu_0)} = 2 \times 10^{11} \text{ Hz at 300 K}$$

$$t_{\text{spont}} = 3 \times 10^{-3} \text{ s}$$

$$\nu = 4.326 \times 10^{14} \text{ Hz}$$

$$\frac{c}{n} (\text{in ruby}) \cong 1.69 \times 10^{10} \text{ cm/s}$$

Using these values in Equation (5.6-19) gives

$$\gamma(\nu) = 5 \times 10^{-2} \text{ cm}^{-1}$$

Thus the intensity of a wave with a frequency corresponding to the center of the transition is amplified by approximately 5% per centimeter in its passage through a ruby rod with the foregoing characteristics.

By comparing Equations (5.6-19) and (5.6-23), we obtain the following relationship between the gain coefficient and χ'':

$$\gamma(\nu) = -\frac{2\pi}{\lambda n}\chi'' = -\frac{k}{n^2}\chi'' \tag{5.6-25}$$

In a passive medium with $\chi'' > 0$, the gain coefficient is negative. Many laser media consist of a host medium and an assembly of active atoms whose atomic transitions are responsible for the emission of laser radiation. For example, chromium atoms are the active ingredient in ruby lasers, whereas an Al_2O_3 crystal is the host medium. The complex index of refraction of the laser medium can be written

$$n'^2 = n^2 + \chi' - i\chi'' \tag{5.6-26}$$

where n is the index of refraction of the host medium, and $\chi' - i\chi''$ is the dielectric susceptibility of the active atoms. The complex wavenumber of propagation in such a medium can thus be written

$$k' = k_0 n' = nk_0\left(1 + \frac{\chi'}{n^2} - i\frac{\chi''}{n^2}\right)^{1/2} \cong nk_0\left(1 + \frac{\chi'}{2n^2} - i\frac{\chi''}{2n^2}\right) \tag{5.6-27}$$

where k_0 is the wavenumber of light in vacuum. If we write $k = nk_0$, then the complex wavenumber can be written

$$k' = k\left(1 + \frac{\chi'}{2n^2} - i\frac{\chi''}{2n^2}\right) = k + k\frac{\chi'}{2n^2} - ik\frac{\chi''}{2n^2} \tag{5.6-28}$$

As we know, the intensity of a plane wave with a complex propagation constant will grow exponentially with distance of propagation. Using Equation (5.6-28), we can obtain a gain coefficient that is consistent with Equation (5.6-25).

Combining Equations (5.6-24) and (5.6-25), we can write the exponential gain coefficient of an active medium with a Lorentzian lineshape as

$$\gamma(\nu) = \frac{(N_2 - N_1)\lambda^2}{4\pi^2 n^2 t_{\text{spont}}\,\Delta\nu} \frac{1}{1 + [4(\nu - \nu_0)^2]/(\Delta\nu)^2} \tag{5.6-29}$$

with a gain bandwidth (FWHM) of $\Delta\nu$. The gain at the line center is, according to Equation (5.6-29),

$$\gamma_0 = \gamma(\nu_0) = \frac{(N_2 - N_1)\lambda^2}{4\pi^2 n^2 t_{\text{spont}}\,\Delta\nu} \tag{5.6-30}$$

where we note that the peak gain coefficient is inversely proportional to the linewidth of the gain profile. We can also write, according to Equations (5.6-29) and (5.6-30),

$$\gamma(\nu) = \frac{\gamma_0}{1 + [4(\nu - \nu_0)^2]/(\Delta\nu)^2} \tag{5.6-31}$$

where $\gamma_0 = \gamma(\nu_0)$ is the peak gain coefficient at the line center.

5.7 GAIN SATURATION IN HOMOGENEOUS LASER MEDIA

In the last section we derived an expression (5.6-19) for the exponential gain constant due to a population inversion. It is given by

$$\gamma(\nu) = (N_2 - N_1)\frac{c^2}{8\pi n^2 \nu^2 t_{\text{spont}}} g(\nu) \tag{5.7-1}$$

where N_2 and N_1 are the population densities of the two atomic levels involved in the induced transition. There is nothing in Equation (5.7-1) to indicate what causes the inversion $(N_2 - N_1)$, and this quantity can be considered as a parameter of the system. In practice, the inversion is caused by a "pumping" agent, hereafter referred to as the pump, that can take various forms such as the electric current in injection lasers, the flashlamp light in pulsed ruby lasers, or the energetic electrons in plasma-discharge gas lasers.

Consider next the situation prevailing at some point *inside* a laser medium in the presence of an optical wave. The pump establishes a population inversion, which in the absence of any optical field has a value ΔN^0. The presence of the optical field induces $2 \to 1$ and $1 \to 2$ transitions. Since $N_2 > N_1$ and the induced rates for $2 \to 1$ and $1 \to 2$ transitions are equal, it follows that more atoms are induced to undergo a transition from level 2 to level 1 than in the opposite direction. This can lead to a net decrease of the population difference ΔN^0.

The reduction in the population inversion and hence of the gain constant brought about by the presence of an electromagnetic field is called *gain saturation*. Its understanding is of fundamental importance in optical electronics. As an example, which will be treated in the next chapter, we may point out that gain saturation is the mechanism that reduces the gain inside laser oscillators to a point where it just balances the losses so that steady oscillation can result. The same mechanism also reduces the gain inside optical amplifiers in optical networks.

In Figure 5.6 we show the ground state 0 as well as the two laser levels 2 and 1 of a four-level laser system. The density of atoms pumped per unit time into level 2 is taken as R_2, and that pumped into level 1 is R_1. Pumping into level 1 is, of course, undesirable since it leads to a reduction of the inversion. In many practical situations it cannot be avoided. The actual "decay" lifetime of atoms in level 2 in the absence of any radiation field is taken as t_2. This

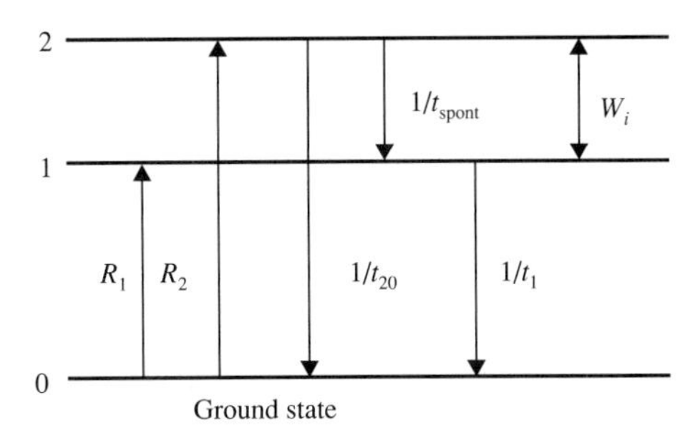

Figure 5.6 Energy levels and transition rates of a four-level laser system. (The fourth level, which is involved in the original excitation by the pump, is not shown and the pumping is shown as proceeding directly into levels 1 and 2.) The total lifetime of level 2 is t_2, where $1/t_2 = 1/t_{spont} + 1/t_{20}$, where $1/t_{20}$ is the decay rate to the ground state (0).

decay rate has a contribution t_{spont}^{-1} that is due to spontaneous (photon emitting) 2→1 transitions as well as to additional nonradiative relaxation from 2 to 1. The lifetime of atoms in level 1 is t_1. The induced rate for 2→1 and 1→2 transitions due to a radiation field at frequency ν is denoted by $W_i(\nu)$ and, according to Equation (5.6-15), is given by

$$W_i(\nu) = \frac{\lambda^2 g(\nu)}{8\pi n^2 h\nu t_{spont}} I_0 \tag{5.7-2}$$

where $g(\nu)$ is the normalized lineshape of the transition and I_0 is the intensity (watts per square meter) of the optical field.

The equations describing the populations of levels 2 and 1 in the combined presence of a radiation field at ν and a pump are

$$\frac{dN_2}{dt} = R_2 - \frac{N_2}{t_2} - (N_2 - N_1)W_i(\nu) \tag{5.7-3}$$

$$\frac{dN_1}{dt} = R_1 - \frac{N_1}{t_1} + \frac{N_2}{t_{spont}} + (N_2 - N_1)W_i(\nu) \tag{5.7-4}$$

where N_2 and N_1 are the population densities (m^{-3}) of levels 2 and 1, respectively. R_2 and R_1 are the pumping rates (m^{-3}-s^{-1}) into these levels. N_2/t_2 is the change per unit time in the population of level 2 due to decay out of level 2 to all levels. This includes spontaneous transitions to level 1 but *not* induced transitions. The rate for the latter is $N_2W_i(\nu)$ so that the net change in N_2 due to induced transitions is given by the last term of Equation (5.7-3). At steady state the populations are constant with time, so putting $d/dt = 0$ in the two preceding equations, we can solve for N_1 and N_2 and obtain

$$N_2 - N_1 = \frac{R_2t_2 - (R_1 + \delta R_2)t_1}{1 + [t_2 + (1 - \delta)t_1]W_i(\nu)} \tag{5.7-5}$$

where $\delta = t_2/t_{spont}$. (Note: $0 < \delta < 1$.) Here we assume that levels 1 and 2 are high enough (in energy) so that the role of thermal processes in populating them can be neglected. If the optical field is absent, $W_i(\nu) = 0$, and the population inversion density is given by, according to Equation (5.7-5),

$$\Delta N^0 = R_2t_2 - (R_1 + \delta R_2)t_1 \tag{5.7-6}$$

We can use Equation (5.7-6) to rewrite (5.7-5) as

$$N_2 - N_1 = \frac{\Delta N^0}{1 + \phi t_{spont} W_i(\nu)} \tag{5.7-7}$$

where the dimensionless parameter ϕ is defined by

$$\phi = \delta\left(1 + (1 - \delta)\frac{t_1}{t_2}\right)$$

We note that ϕ is always positive since $0 < \delta < 1$. We also note that in efficient laser systems $t_2 \sim t_{\text{spont}}$, so $\delta \approx 1$, and that $t_1 \ll t_2$, so $\phi \approx 1$. Substituting Equation (5.7-2) for $W_i(\nu)$, the last equation becomes

$$N_2 - N_1 = \frac{\Delta N^0}{1 + [\phi\lambda^2 g(\nu)/8\pi n^2 h\nu]I_0} = \frac{\Delta N^0}{1 + I_0/I_s(\nu)} \tag{5.7-8}$$

where $I_s(\nu)$, the saturation intensity, is given by

$$I_s(\nu) = \frac{8\pi n^2 h\nu}{\phi\lambda^2 g(\nu)} = \frac{8\pi n^2 h\nu}{(t_2/t_{\text{spont}})\lambda^2 g(\nu)} = \frac{8\pi n^2 h\nu\,\Delta\nu}{(t_2/t_{\text{spont}})\lambda^2} \tag{5.7-9}$$

and corresponds to the intensity level (watts per square meter) that causes the inversion to drop to one-half of its nonsaturated value (ΔN^0). In Equation (5.7-9), the last equality is obtained by putting $1/g(\nu) = \Delta\nu$, which is valid for frequencies near the peak of the lineshape function. By using Equation (5.7-8) in the gain expression (5.7-1), we obtain our final result:

$$\gamma(\nu) = \frac{1}{1 + I_0/I_s(\nu)}\left(\frac{\Delta N^0 \lambda^2}{8\pi n^2 t_{\text{spont}}}\right)g(\nu) = \frac{\gamma_0(\nu)}{1 + I_0/I_s(\nu)} \tag{5.7-10}$$

where $\gamma_0(\nu)$ is the gain coefficient at low intensities ($I_0 \ll I_s$). Equation (5.7-10) shows the dependence of the gain constant on the optical intensity. In closing, we recall that Equation (5.7-10) applies to a homogeneous laser system. This is due to the fact that, in the rate equations (5.7-3) and (5.7-4), we considered all the atoms as equivalent and, consequently, experiencing the same transition rates. This assumption is no longer valid in inhomogeneous laser systems. This case is treated in the next section.

5.8 GAIN SATURATION IN INHOMOGENEOUS LASER MEDIA

In Section 5.7 we considered the reduction in optical gain—that is, saturation—due to the optical field in a homogeneous laser medium. In this section we treat the problem of gain saturation in inhomogeneous systems. According to the discussion of Section 5.5, in an inhomogeneous atomic system the individual atoms are distinguishable, with each atom having a unique transition frequency $(E_2 - E_1)/h$. We can thus imagine the inhomogeneous medium as made up of classes of atoms each designated by a continuous variable ξ. The variable ξ can, as an example, correspond to the center frequency of the lineshape function $g^{\xi}(\nu)$ of atoms in group ξ. Furthermore, we define a function $p(\xi)$ so that the a priori probability that an atom has its ξ parameter between ξ and $\xi + d\xi$ is $p(\xi)\,d\xi$. It follows that

$$\int_{-\infty}^{\infty} p(\xi)\,d\xi = 1 \tag{5.8-1}$$

since any atom has a unit probability of having its ξ value between $-\infty$ and ∞.

The atoms within a given class ξ are considered as homogeneously broadened, having a lineshape function $g^{\xi}(\nu)$ that is normalized so that

$$\int_{-\infty}^{\infty} g^{\xi}(\nu)\,d\nu = 1 \tag{5.8-2}$$

In Section 5.5 we defined the transition lineshape $g(\nu)$ by taking $g(\nu)\,d\nu$ to represent the a priori probability that a spontaneous emission will result in a photon whose frequency is between ν and $\nu + d\nu$. Using this definition, we obtain

$$g(\nu)\,d\nu = \left[\int_{-\infty}^{\infty} p(\xi)g^{\xi}(\nu)\,d\xi\right]d\nu \tag{5.8-3}$$

which is a statement of the fact that the probability of emitting a photon of frequency between ν and $\nu + d\nu$ is equal to the probability $g^{\xi}(\nu)\,d\nu$ of this occurrence, given that the atom belongs to class ξ, summed up over all the classes.

Next, we proceed to find the contribution to the inversion that is due to a single class ξ. The rate equations are

$$\begin{aligned}\frac{dN_2^{\xi}}{dt} &= R_2 p(\xi) - \frac{N_2^{\xi}}{t_2} - (N_2^{\xi} - N_1^{\xi})W_i^{\xi}(\nu)\\ \frac{dN_1^{\xi}}{dt} &= R_1 p(\xi) - \frac{N_1^{\xi}}{t_1} + \frac{N_2^{\xi}}{t_{\text{spont}}} + (N_2^{\xi} - N_1^{\xi})W_i^{\xi}(\nu)\end{aligned} \tag{5.8-4}$$

and are similar to Equations (5.7-3) and (5.7-4), except that N_2^{ξ} and N_1^{ξ} refer to the upper and lower level densities of atoms in class ξ only. The pumping rate (atoms/m^3-s) into levels 2 and 1 is taken to be proportional to the probability of finding an atom in class ξ and is given by $R_1 p(\xi)$ and $R_2 p(\xi)$ respectively. The total pumping rate into level 2 is, as in Section 5.7, R_2 since

$$\int_{-\infty}^{\infty} R_2 p(\xi)\,d\xi = R_2 \int_{-\infty}^{\infty} p(\xi)\,d\xi = R_2$$

where we made use of Equation (5.8-1). The induced transition rate N_2^{ξ} is given, according to Equation (5.6-15), by

$$W_i^{\xi}(\nu) = \frac{\lambda^2}{8\pi n^2 h\nu t_{\text{spont}}} g^{\xi}(\nu) I_0 \tag{5.8-5}$$

which is of a form identical to (5.7-2) except that $g^{\xi}(\nu)$ refers to the lineshape function of atoms in class ξ. The steady-state $d/dt = 0$ solution of Equation (5.8-4) yields

$$N_2^{\xi} - N_1^{\xi} = \frac{\Delta N^0 p(\xi)}{1 + \phi t_{\text{spont}} W_i^{\xi}(\nu)} \tag{5.8-6}$$

where ΔN^0 and ϕ have the same significance as in Section 5.7. The total power emitted by induced transitions per unit volume by atoms in class ξ is thus

$$\frac{P^{\xi}(\nu)}{V} = (N_2^{\xi} - N_1^{\xi})h\nu W_i^{\xi}(\nu) = \frac{\Delta N^0 p(\xi) h\nu}{1/W_i^{\xi}(\nu) + \phi t_{\text{spont}}} \tag{5.8-7}$$

where the spontaneous lifetime is assumed the same for all the groups ξ.

Summing Equation (5.8-7) over all the classes, we obtain an expression for the total power at ν per unit volume emitted by the atoms

$$\frac{P(\nu)}{V} = \frac{\Delta N^0 h\nu}{t_{\text{spont}}} \int_{-\infty}^{\infty} \frac{p(\xi)\,d\xi}{1/(W_i^{\xi}(\nu)t_{\text{spont}}) + \phi} \tag{5.8-8}$$

which, by the use of Equation (5.8-5), can be rewritten

$$\frac{P(\nu)}{V} = \frac{\Delta N^0\, h\nu}{t_{\text{spont}}} \int_{-\infty}^{\infty} \frac{p(\xi)\, d\xi}{8\pi n^2 h\nu/[\lambda^2 I_0 g^{\xi}(\nu)] + \phi} \tag{5.8-9}$$

The stimulated emission of power causes the intensity of the traveling optical wave to increase with distance z according to $I_0 = I_0(0)\exp[\gamma(\nu)z]$, where

$$\begin{aligned} \gamma(\nu) &= \frac{dI_0/dz}{I_0} = \frac{P(\nu)/V}{I_0} \\ &= \frac{\Delta N^0\, \lambda^2}{8\pi n^2 t_{\text{spont}}} \int_{-\infty}^{\infty} \frac{p(\nu_\xi)\, d\nu_\xi}{[1/g^{\xi}(\nu)] + (\phi\lambda^2 I_0/8\pi n^2 h\nu)} \end{aligned} \tag{5.8-10}$$

where we replaced $p(\xi)\, d\xi$ by $p(\nu_\xi)\, d\nu_\xi$. Here we may view ν_ξ as the center frequency of the lineshape function $g^{\xi}(\nu)$ of atoms in group ξ. This is our basic result.

As a first check on Equation (5.8-10), we shall consider the case in which $I_0 \ll 8\pi n^2 h\nu/\phi\lambda^2 g^{\xi}(\nu)$ and therefore the effects of saturation can be ignored. Using Equation (5.8-3) in (5.8-10), we obtain

$$\gamma(\nu) = \frac{\Delta N^0\, \lambda^2}{8\pi n^2 t_{\text{spont}}}\, g(\nu)$$

which is the same as Equation (5.6-19). This shows that in the absence of saturation the expressions for the gain of a homogeneous and an inhomogeneous atomic system are identical.

Our main interest in this treatment is in deriving the saturated gain constant for an inhomogeneously broadened atomic transition. If we assume that in each class ξ all the atoms are identical (homogeneous broadening), we can use Equation (5.5-4) for the lineshape function $g^{\xi}(\nu)$, and therefore

$$g^{\xi}(\nu) = \frac{\Delta\nu}{2\pi[(\nu - \nu_\xi)^2 + (\Delta\nu/2)^2]} \tag{5.8-11}$$

where $\Delta\nu$ is called the homogeneous linewidth of the inhomogeneous line. Atoms with transition frequencies that are clustered within $\Delta\nu$ from each other can be considered as indistinguishable. The term "homogeneous packet" is often used to describe them. Using Equation (5.8-11) in (5.8-10) leads to

$$\gamma(\nu) = \frac{\Delta N^0\, \lambda^2\, \Delta\nu}{16\pi^2 n^2 t_{\text{spont}}} \int_{-\infty}^{\infty} \frac{p(\nu_\xi)\, d\nu_\xi}{(\nu - \nu_\xi)^2 + (\Delta\nu/2)^2 + (\phi\lambda^2 I_0\, \Delta\nu/16\pi^2 n^2 h\nu)} \tag{5.8-12}$$

In extreme inhomogeneous cases, the width of $p(\nu_\xi)$ is by definition very much larger than the remainder of the integrand in Equation (5.8-12) and thus it is essentially a constant over the region in which the integrand is appreciable. In this case, we can put $p(\nu_\xi)_{\nu_\xi=\nu} = p(\nu)$ outside the integral sign in Equation (5.8-12), obtaining

$$\gamma(\nu) = \frac{\Delta N^0\, \lambda^2\, \Delta\nu}{16\pi^2 n^2 t_{\text{spont}}}\, p(\nu) \times \int_{-\infty}^{\infty} \frac{d\nu_\xi}{(\nu - \nu_\xi)^2 + (\Delta\nu/2)^2 + (\phi\lambda^2 I_0\, \Delta\nu/16\pi^2 n^2 h\nu)} \tag{5.8-13}$$

Using the definite integral

$$\int_{-\infty}^{\infty} \frac{dx}{x^2 + a^2} = \frac{\pi}{a}$$

to evaluate Equation (5.8-13), we obtain

$$\gamma(\nu) = \frac{\Delta N^0 \lambda^2 p(\nu)}{8\pi n^2 t_{\text{spont}}} \frac{1}{\sqrt{1 + \phi\lambda^2 I_0 / 4\pi n^2 h\nu\,\Delta\nu}} \tag{5.8-14}$$

$$= \gamma_0(\nu) \frac{1}{\sqrt{1 + I_0/I_s}} \tag{5.8-15}$$

where

$$I_s = \frac{4\pi^2 n^2 h\nu\,\Delta\nu}{\phi\lambda^2} \tag{5.8-16}$$

is the saturation intensity. In Equation (5.8-16), we recall that the parameter ϕ is defined in Section 5.7, and $\Delta\nu$ is the width of the lineshape function $g^{\xi}(\nu)$. A comparison of Equations (5.8-15) and (5.7-10) shows that, because of the square root, the saturation—that is, decrease in gain—sets in more slowly as the intensity I_0 is increased in the case of inhomogeneous broadening.

PROBLEMS

5.1 Use Equations (5.4-8) and (5.4-9) for the following.

(a) Show that the absorption coefficient can be written

$$\alpha(\omega) = \alpha_0 \frac{(\gamma/2)^2}{(\omega - \omega_0)^2 + (\gamma/2)^2} \quad \text{with} \quad \alpha_0 = \frac{Ne^2}{mc\varepsilon_0\gamma}$$

(b) Show that

$$I = \int_{-\infty}^{\infty} \alpha(\omega)\, d\omega = \pi\alpha_0\gamma/4$$

(c) Show that the integrated absorption is independent of γ (even when $\gamma \to 0$).

5.2 Consider a two-level atomic gas with an absorption line center at $\nu_0 = 2.0 \times 10^{15}$ Hz and a Lorentzian lineshape. The atomic gas has a refractive index of 1.0003 (real part) at $\nu = 2.0 \times 10^{14}$ Hz. Assume a Lorentzian linewidth of $\Delta\nu = 1$ GHz.

(a) Find the absorption coefficient in units of m^{-1} at the line center and at $\nu = 2.0 \times 10^{14}$ Hz.

(b) Find the index of refraction (real part) at the line center, and at a frequency detuning of 1 GHz from the line center.

5.3 Show that if $\chi(\nu)$ possesses a pole in the lower half of the complex plane, a step excitation $e(t)$ would lead to a response $p(t)$ that grows exponentially in time. [Hint: Use the Fourier integral relation

$$p(t) = \int_{-\infty}^{\infty} \tilde{p}(\nu) \exp(i2\pi\nu t)\, d\nu$$

where

$$\tilde{p}(\nu) = \varepsilon_0 \chi(\nu) \tilde{e}(\nu)$$

5.4 Show that for $p(t)$ to be a real function in Problem 5.3

$$\chi(-\nu) = \chi^*(\nu) \tag{5.P-1}$$

or equivalently,

$$\chi'(\nu) = \chi'(-\nu)$$
$$\chi''(\nu) = -\chi''(-\nu) \tag{5.P-2}$$

5.5

(a) Show that, according to Equation (5.3-7), the poles of χ occur at

$$\omega = i\frac{\gamma}{2} \pm \sqrt{\omega_0^2 - \gamma^2/4}$$

Note that these poles are all in the upper half of the complex plane.

(b) Show that the following integral is zero for $\tau < 0$:

$$F(\tau) = \frac{1}{2\pi} \int_{-\infty}^{\infty} \left[\varepsilon(\omega) - \varepsilon_0\right] e^{i\omega\tau}\, d\omega$$

According to Equation (C-9), the displacement field is given by

$$D(t) = \varepsilon_0 E(t) + \int_{-\infty}^{\infty} F(\tau) E(t - \tau)\, d\tau$$

$$= \varepsilon_0 E(t) + \int_0^{\infty} F(\tau) E(t - \tau)\, d\tau$$

The displacement field $D(t)$ at time t depends on the value of electric field E at all earlier times.

5.6 Show that, for a Lorenztian lineshape centered at ν_0, the peak gain can be written

$$\gamma(\nu_0) = (N_2 - N_1)\frac{c^2}{8\pi n^2 \nu^2 t_{spont}} \frac{2}{\pi\,\Delta\nu}$$

where $\Delta\nu$ is the full width at half-maximum of the lineshape function.

5.7

(a) Show that in the absence of a driving field, the solution of the classical electron model can be written

$$X(t) = A\exp(-\gamma t/2)\cos(\omega' t + \alpha)$$

where A and α are constants, and ω' is given by

$$\omega'^2 = \omega_0^2 - (\gamma/2)^2$$

(b) Given a damped oscillating dipole $X(t) = A\exp(-\gamma t/2)\cos(\omega_0 t)$, show that the Fourier transform is approximately given by

$$f(\omega) = \frac{A}{4\pi}\frac{i}{(\omega_0 - \omega) + i\gamma/2}$$

5.8 Consider two atoms (or molecules) approaching each other in an elastic collision. The quantum states, strictly speaking, depend on the distance between these two atoms as well as on their orientations, which are functions of time.

(a) If the atoms are radiating with a photon frequency of $\omega = (E_2 - E_1)/\hbar$, show that the net result of an elastic collision is to add a phase to the radiating field:

$$\phi = \int \frac{[\Delta E_2(t) - \Delta E_1(t)]}{\hbar}\,dt$$

If $\phi \gg 1$, one collision is enough to completely dephase the oscillation of the radiating field. For a sequence of collisions, we may treat ϕ as a random number.

(b) Using the model in (a) and examining the Fourier spectrum of the radiating field, show that the broadening of the linewidth is proportional to $1/\tau_{co}$.

REFERENCES

1. See, for example, A. Yariv, *Quantum Electronics*, 3rd ed. Wiley, New York, 1989.
2. Yeh, P., *Optical Waves in Layered Media*. Wiley, New York, 1988.
3. See, for example, R. Kubo, *Statistical Mechanics*. North Holland, Amsterdam, 1964, p. 31.
4. Johnson, L. F., Optically pumped pulsed crystal lasers other than ruby. In *Lasers*, Vol. I, A. K. Levine (ed.). Marcel Dekker, New York, 1966, p. 137.
5. Kittel, C., *Elementary Statistical Physics*. Wiley, New York, 1958, p. 197.
6. Einstein, A., Zur Quantentheorie der Strahlung, *Phys. Z.* **18**:121 (1917).

CHAPTER 6

THEORY OF LASER OSCILLATION AND SOME SPECIFIC LASER SYSTEMS

6.0 INTRODUCTION

In Chapter 5 we found that an atomic medium (e.g., a two-level atomic system) with an inverted population ($N_2 > N_1$) is capable of amplifying an electromagnetic wave in the spectral regime within the transition lineshape (or gain profile). In this chapter we consider the situation in which the inverted atomic medium (also called laser medium) is placed inside an optical resonator. As the electromagnetic wave bounces back and forth between the two reflectors, it passes through the inverted atomic medium and is amplified. If the amplification exceeds the losses caused by transmission (or absorption) loss in the mirrors and scattering loss in the laser medium, the field energy stored in the resonator will increase with time under the appropriate conditions. The increase in the field energy causes the gain coefficient of the gain medium to decrease as a result of gain saturation discussed in the last chapter. The field energy in the resonator will keep increasing until the saturated gain per pass just equals the losses. At this point in time, the net gain per pass is unity and no further increase in the radiation intensity is possible—that is, steady-state oscillation is reached. In this chapter we will derive the threshold population inversion of the atomic medium needed to sustain laser oscillation, beginning with the theory of the Fabry–Perot laser. We will also obtain an expression for the oscillation frequency of the laser oscillator and show how it is affected by the dispersion of the atomic medium. We then consider the problem of optimum output coupling. Multimode laser oscillation and mode locking will be addressed later in the chapter. Several laser systems will be described, including ruby lasers, Nd:YAG lasers, and Er:silica lasers. An important class of lasers, the semiconductor lasers, will be discussed in Chapters 15 and 16.

6.1 FABRY–PEROT LASER

Referring to Figure 6.1, we consider an optical resonator consisting of two mirrors. This is basically a Fabry–Perot etalon. Let the space between the mirrors be filled with an amplifying medium with an inverted atomic population. As a result of the presence of the amplifying medium, the complex index of refraction n' can be written

$$n'^2 = n^2 + \chi' - i\chi'' \tag{6.1-1}$$

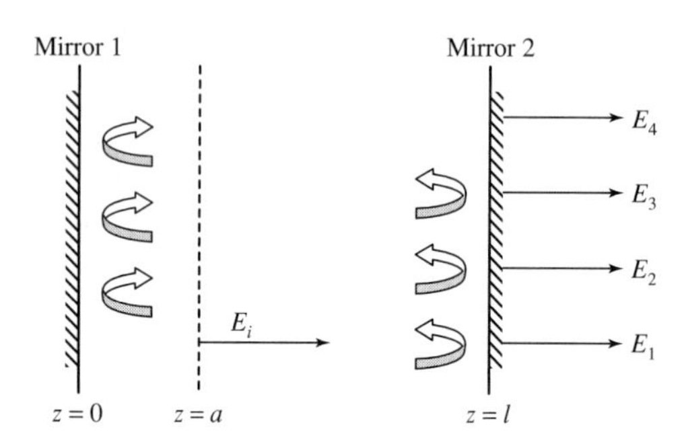

Figure 6.1 Schematic drawing of multiple reflections inside a Fabry–Perot cavity and the corresponding partially transmitted beams. E_1 is the first transmitted beam, E_2 is the transmitted beam after one round-trip inside the cavity, E_3 is the transmitted beam after two round-trips inside the cavity, and so on.

where n is the index of refraction of the host medium, and $\chi' - i\chi''$ is the complex dielectric susceptibility due to the presence of the amplifying materials (the active atoms that are responsible for the gain). In the case of ruby lasers, $n = 1.77$ is the index of refraction of the host material Al_2O_3, and $\chi' - i\chi''$ is due to the chromium atoms (actually Cr^{3+} ions). In most laser systems, $|\chi' - i\chi''| \ll n^2$. The complex propagation constant of plane waves in the gain medium can thus be written

$$k'(\omega) = k_0 n' = k_0\sqrt{n^2 + \chi' - i\chi''} - i\frac{\alpha}{2} \cong k + k\frac{\chi'(\omega)}{2n^2} - ik\frac{\chi''(\omega)}{2n^2} - i\frac{\alpha}{2} \qquad (6.1\text{-}2)$$

where $k_0 = \omega/c$, $k = n\omega/c$, and α is the linear attenuation coefficient to account for a residual loss due to a variety of mechanisms, such as scattering at imperfections, absorption by excited atomic levels, and others. The attenuation resulting from all of these mechanisms is lumped into the distributed loss constant α. We may view $k - i\alpha/2$ as the propagation constant of the medium at frequencies well removed from that of the laser transition. Since α accounts for the distributed passive losses of the medium, the intensity loss factor per pass is $\exp(-\alpha l)$, where l is the length of the cavity.

Referring again to Figure 6.1, we consider now a white noise with a uniform spectral distribution inside the cavity. The noise can be originating from blackbody radiation or spontaneous emission. Let E_i be the amplitude of the plane wave component (due to the noise) propagating along the positive axis of the cavity at frequency ω.

The electromagnetic wave E_i will undergo multiple reflections inside the resonator, as shown in Figure 6.1. Let us examine the partially transmitted waves at mirror 2 located at $z = l$. We follow the same procedure employed in Chapter 4 to obtain the output wave at mirror 2. The output wave can be written as a sum of all partially transmitted waves,

$$E_{\text{out}} = E_1 + E_2 + E_3 + E_4 + \cdots \qquad (6.1\text{-}3)$$

where

$$\begin{aligned} E_1 &= t_2 e^{-ik'(l-a)} E_i \\ E_2 &= (r_1 r_2 e^{-i2k'l}) t_2 e^{-ik'(l-a)} E_i \\ E_3 &= (r_1 r_2 e^{-i2k'l})^2 t_2 e^{-ik'(l-a)} E_i \\ E_4 &= (r_1 r_2 e^{-i2k'l})^3 t_2 e^{-ik'(l-a)} E_i \end{aligned} \qquad (6.1\text{-}4)$$

where a is the location of the noise source inside the resonator, t_2 is the amplitude transmission coefficient of the second mirror (at $z = l$), and r_1 and r_2 are the amplitude reflection coefficients (looking from inside the cavity) of the mirrors, respectively. We note that as a result of the multiple reflections inside the resonator, the sum of the transmitted waves forms a geometric series. Using Equations (6.1-3) and (6.1-4), we obtain

$$E_{\text{out}} = t_2 e^{-ik'(l-a)} E_i \{1 + (r_1 r_2 e^{-i2k'l})^1 + (r_1 r_2 e^{-i2k'l})^2 + (r_1 r_2 e^{-i2k'l})^3 + \cdots \}$$
$$= \frac{t_2 e^{-ik'(l-a)}}{1 - r_1 r_2 e^{-i2k'l}} E_i \tag{6.1-5}$$

where k' is the complex propagation constant given by Equation (6.1-1), and l is the length of the etalon. Since χ'' is related to the gain coefficient, we can rewrite the complex propagation constant as

$$k' = k + \Delta k + i(\gamma - \alpha)/2 \tag{6.1-6}$$

with

$$\Delta k = k \frac{\chi'(\omega)}{2n^2} \tag{6.1-7}$$

$$\gamma = -k \frac{\chi''(\omega)}{n^2} = (N_2 - N_1) \frac{\lambda^2}{8\pi n^2 t_{\text{spont}}} g(\nu) \tag{6.1-8}$$

Physically, Δk is a correction to the propagation constant due to the presence of the active atoms (e.g., Cr atoms in ruby lasers). γ is a gain coefficient due to the active atoms. The output wave can thus be written

$$E_{\text{out}} = \frac{t_2 e^{-i[(k+\Delta k)(l-a)]} e^{(\gamma-\alpha)(l-a)/2}}{1 - r_1 r_2 e^{-i2(k+\Delta k)l} e^{(\gamma-\alpha)l}} E_i \tag{6.1-9}$$

If the noise source is located outside the resonator (say, at $z < 0$), then E_i can be viewed as an incident wave from the mirror at $z = 0$ (mirror 1 in Figure 6.1). In this case, the output wave can be written

$$E_{\text{out}} = \frac{t_1 t_2 e^{-i(k+\Delta k)l} e^{(\gamma-\alpha)l/2}}{1 - r_1 r_2 e^{-i2(k+\Delta k)l} e^{(\gamma-\alpha)l}} (E_i)_{\text{outside}} \tag{6.1-10}$$

where t_1 is the amplitude transmission coefficient of mirror 1 and $(E_i)_{\text{outside}}$ is the amplitude of an incident wave arriving at mirror 1. We notice that the denominators in the above two equations are identical. If the atomic transition is inverted ($N_2 > N_1$), then $\gamma > 0$ and the denominator of Equation (6.1-9) or (6.1-10) can become very small. The output wave E_{out} can become much larger than the source wave E_i. The Fabry–Perot etalon (with the laser medium) in this case acts as an amplifier with a power gain $|E_{\text{out}}/E_i|^2$.

If the denominator of Equation (6.1-9) or (6.1-10) becomes zero, which happens when

$$r_1 r_2 e^{-i2(k+\Delta k)l} e^{(\gamma-\alpha)l} = 1 \tag{6.1-11}$$

then the ratio E_{out}/E_i becomes infinite. This corresponds to a finite output wave E_{out}, with an infinitesimal source wave ($E_i \cong 0$)—that is, to *oscillation*. Physically, condition (6.1-11) represents the case in which a wave making a complete round trip inside the resonator returns to the starting plane with the *same amplitude* and, except for some integral multiple of 2π, with the *same phase*. Separating the oscillation condition (6.1-11) into the amplitude and phase requirements gives

$$r_1 r_2 e^{[\gamma(\omega)-\alpha]l} = 1 \tag{6.1-12}$$

for the threshold gain constant $\gamma_t(\omega)$ and

$$2[k + \Delta k(\omega)]l = 2m\pi, \quad m = 1, 2, 3, \ldots \tag{6.1-13}$$

for the phase condition. Here, without loss of generality, we assume real reflection coefficients r_1 and r_2. The amplitude condition (6.1-12) can be written

$$\gamma_t(\omega) = \alpha - \frac{1}{l} \ln r_1 r_2 \tag{6.1-14}$$

where $\gamma_t(\omega)$ is known as the threshold gain coefficient. The threshold gain condition can be written in terms of the threshold population inversion, according to Equation (6.1-8),

$$N_t \equiv (N_2 - N_1)_t = \frac{8\pi n^2 t_{\text{spont}}}{g(\nu)\lambda^2}\left(\alpha - \frac{1}{l} \ln r_1 r_2\right) \tag{6.1-15}$$

This is the population inversion density at threshold. It was derived originally by Schawlow and Townes in their classic paper on the feasibility of lasers; see Reference [1]. The population inversion density threshold is often written in terms of the photon cavity lifetime, which is defined as follows. Consider the case in which the mirror losses and the distributed losses are all small, and therefore $r_1^2 \approx 1$, $r_2^2 \approx 1$, and $\exp(-\alpha l) \approx 1$. An electromagnetic wave starting with a unit intensity will return after one round trip with an intensity $R_1 R_2 \exp(-2\alpha l)$, where $R_1 \equiv r_1^2$ and $R_2 \equiv r_2^2$ are the mirrors' reflectivities. The fractional intensity loss per round trip is thus $1 - R_1 R_2 \exp(-2\alpha l)$. Since this loss occurs in a round-trip flight time of $2ln/c$. The intensity of the electromagnetic wave will decay exponentially with decay time constant t_c given by (see Problem 6.15)

$$\frac{1}{t_c} = \frac{(1 - R_1 R_2 e^{-2\alpha l})c}{2nl} \tag{6.1-16}$$

This is known as the photon cavity lifetime.

With this definition, the energy U stored in the passive resonator decays as $dU/dt = -U/t_c$. Since $R_1 R_2 e^{-2\alpha l} \cdot e^{2\gamma_t l} = 1$ at oscillation, the photon cavity lifetime can also be written

$$\frac{1}{t_c} = \frac{(1 - e^{-2\gamma_t l})c}{2nl} = \frac{c}{n}\gamma_t = \frac{c}{n}\left(\alpha - \frac{1}{l} \ln r_1 r_2\right) \tag{6.1-17}$$

where we assume a small threshold gain, so that $e^{-2\gamma_t l} \approx 1 - 2\gamma_t l$. Using this expression of the photon cavity lifetime, the threshold condition (6.1-14) becomes

$$N_t \equiv (N_2 - N_1)_t = \frac{8\pi n^3 \nu^2 t_{\text{spont}}}{c^3 t_c g(\nu)} \tag{6.1-18}$$

where $N \equiv N_2 - N_1$, and the subscript t signifies threshold.

EXAMPLE: POPULATION INVERSION

To get an order of magnitude estimate of the critical population inversion $(N_2 - N_1)$, we use data typical of a 6328 Å He–Ne laser. The appropriate constants are

$$\lambda = 6.328 \times 10^{-5} \text{ cm}$$

$$t_{\text{spont}} = 10^{-7} \text{ s}$$

$$l = 12 \text{ cm}$$

$$\frac{1}{g(\nu_0)} \approx \Delta\nu \approx 10^9 \text{ Hz}$$

The last figure is the Doppler-broadened width of the laser transition.

The cavity decay time t_c is calculated from Equation (6.1-17) assuming $\alpha = 0$ and $R_1 = R_2 = 0.98$. Since $R_1 = R_2 \sim 1$, we can use the approximation $\ln r_1 r_2 = \ln R \approx R - 1$ to write

$$t_c \approx \frac{nl}{c(1-R)} = 2 \times 10^{-8} \text{ s}$$

Using the foregoing data in Equation (6.1-18), we obtain

$$N_t \approx 10^9 \text{ cm}^{-3}$$

Figure 6.2 consists of a plot of the output wave intensity, at a given noise source, as a function of the round-trip phase shift. In Figure 6.2a, a uniform gain is assumed. Each curve in Figure 6.2a is for a different value of the distributed gain constant γ. We note that when $\exp[(\gamma - \alpha)l] > 1$—that is, when the net gain per pass exceeds unity—the output can exceed the input source and the Fabry–Perot etalon functions as an amplifier. Figure 6.2b shows the output intensity by assuming a Lorentzian gain profile. We notice that the output intensities of the modes are different. This is a result of the difference in the gain for each of the modes. As a result of the gain profile, only one mode can satisfy the oscillation condition (6.1-11). At oscillation, the gain exactly balances the losses, and a steady-state field energy is built up

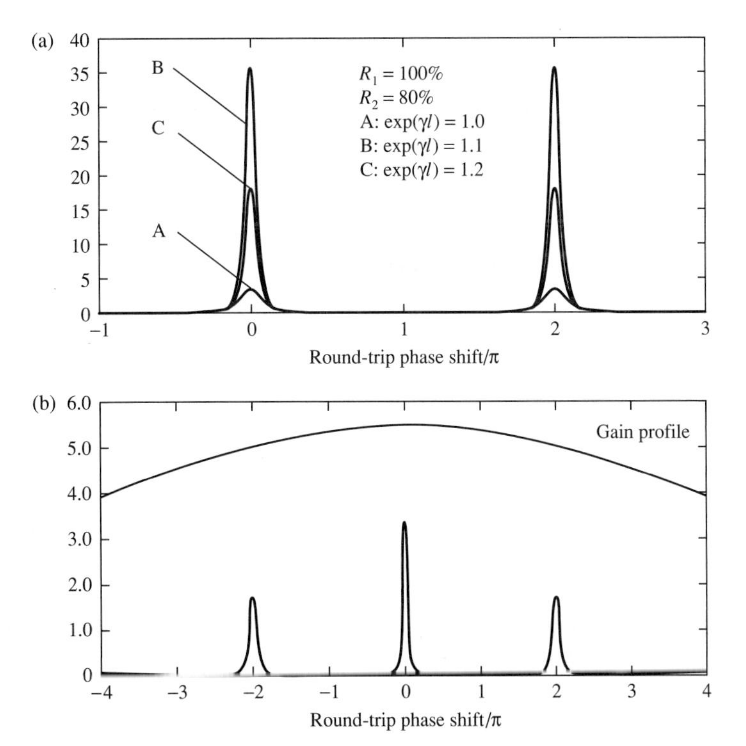

Figure 6.2 Output intensity as a function of the round-trip phase shift ($2kl - 2m\pi$) in a Fabry–Perot resonator: (a) uniform gain and (b) a Lorentzian gain profile centered at a resonance frequency of the Fabry–Perot etalon.

inside the resonator. Thus, in the case of a homogeneously broadened gain medium, only one mode will oscillate.

It is especially interesting to note the narrowing of the peaks as the oscillation condition $\exp(2\gamma l)R_1R_2 = 1$ is approached. At oscillation, the peak transmission becomes infinity, and the width of the transmission peak becomes zero. So the spectral bandwidth of an ideal laser is zero. In practice, the exact laser frequency is determined by the cavity length l, which can change due to external perturbations (e.g., thermal, acoustic). This can lead to a change of the oscillation frequency with time. Furthermore, the presence of spontaneous emission also leads to a broadening of the spectral distribution of the laser output. This problem is explored further in Section 10.6.

6.2 OSCILLATION FREQUENCY

The phase part of the oscillation condition as given by Equation (6.1-13) is satisfied at an infinite set of frequencies, which correspond to the different values of the integer m. If, in addition, the gain condition (6.1-12) is satisfied at one or more of these frequencies, the laser will oscillate at this frequency. To solve for the oscillation frequency, we rewrite Equation (6.1-13) as

$$kl\left(1 + \frac{\chi'(\nu)}{2n^2}\right) = m\pi \tag{6.2-1}$$

We introduce

$$\nu_m = \frac{mc}{2nl} \tag{6.2-2}$$

so that it corresponds to the mth resonance frequency of the passive $(N_2 - N_1 = 0)$ resonator. Using Equations (5.4-10a) and (5.4-10b), we can write

$$\chi'(\nu) = \frac{2(\nu_0 - \nu)}{\Delta\nu}\chi''(\nu) \tag{6.2-3}$$

where $\Delta\nu$ is the FWHM bandwidth of the gain profile, and

$$\gamma(\nu) = -\frac{k\chi''(\nu)}{n^2} \tag{6.2-4}$$

We obtain from Equation (6.2-1) the oscillation frequency of the laser,

$$\nu\left[1 - \left(\frac{\nu_0 - \nu}{\Delta\nu}\right)\frac{\gamma(\nu)}{k}\right] = \nu_m \tag{6.2-5}$$

where ν_0 is the center frequency of the atomic lineshape function (gain profile). Let us assume that the laser resonator length l is adjusted so that one of its resonance frequencies ν_m is very near ν_0. We anticipate that the oscillation frequency ν will also be close to ν_m and take advantage of the fact that when $\nu \sim \nu_0$, the gain constant $\gamma(\nu)$ is a slowly varying function of ν; see Figure 5.2 for κ or $\chi''(\nu)$, which is proportional to $\gamma(\nu)$. We can consequently replace $\gamma(\nu)$ in Equation (6.2-5) by $\gamma(\nu_m)$, and $(\nu_0 - \nu)$ by $(\nu_0 - \nu_m)$, obtaining

$$\nu = \nu_m - (\nu_m - \nu_0)\frac{\gamma(\nu_m)c}{2\pi n\,\Delta\nu} \tag{6.2-6}$$

as the solution for the oscillation frequency ν.

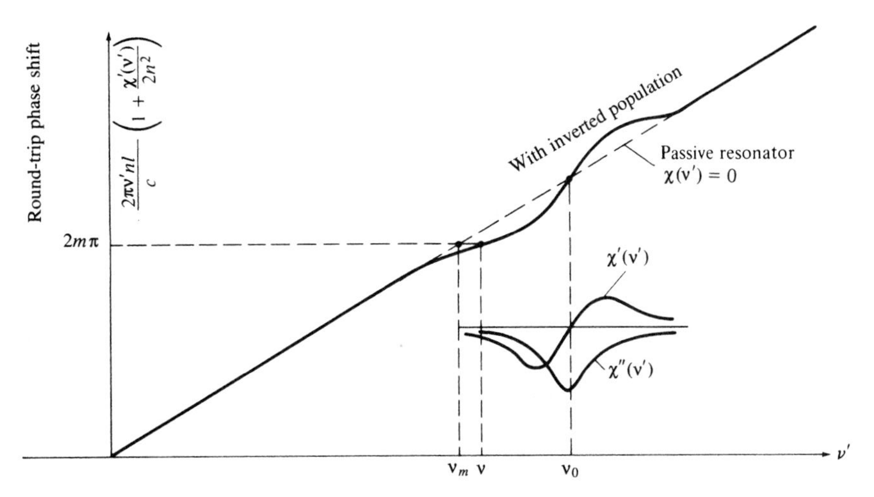

Figure 6.3 Graphical illustration of the laser frequency condition (Equation (6.2-1)) showing how the atomic dispersion $\chi'(\nu)$ "pulls" the laser oscillation frequency, ν, from the passive resonator value, ν_m, toward that of the atomic resonance at ν_0.

We can recast Equation (6.2-6) in a slightly different, and easier to use, form by starting with the gain threshold condition (6.1-12). Taking, for simplicity, $r_1 = r_2 = \sqrt{R}$ and assuming that $R \sim 1$ and $\alpha = 0$, we can write Equation (6.1-14) as

$$\gamma_t(\nu) \approx \frac{1-R}{l}$$

We also take advantage of the relation, according to Equation (4.2-4),

$$\Delta\nu_{1/2} \approx \frac{c(1-R)}{2\pi nl}$$

which relates the passive resonator linewidth $\Delta\nu_{1/2}$ to mirror reflectivity (with $R \sim 1$) and write Equation (6.2-6) as

$$\nu = \nu_m - (\nu_m - \nu_0)\frac{\Delta\nu_{1/2}}{\Delta\nu} \tag{6.2-7}$$

A study of Equation (6.2-7) shows that if the passive cavity resonance ν_m coincides with the atomic line center—that is, $\nu_m = \nu_0$—oscillation takes place at $\nu = \nu_0$. If $\nu_m \neq \nu_0$, oscillation takes place near ν_m but is shifted slightly toward ν_0. This phenomenon is referred to as *frequency pulling* and is demonstrated by Figure 6.3.

The frequency pulling is a result of dispersion in the spectral regime near the line center. Based on the discussion in Chapter 4, we can write the free spectral range (FSR) as

$$(\Delta\nu)_{\mathrm{FSR}} = \frac{c}{2n_g l} \tag{6.2-8}$$

where n_g is the group index given by

$$n_g = n_a + \omega\frac{dn_a}{d\omega} = n_a + \nu\frac{dn_a}{d\nu} \tag{6.2-9}$$

where n_a is the index of refraction including the contribution from the active medium,

$$n_a = n + \tfrac{1}{2}\chi' \tag{6.2-10}$$

In the spectral regime near the line center of an inverted system, $\nu(dn_a/d\nu)$ is a positive number. This is reflected in the increase in the slope near the line center as shown in Figure 6.3. As a result of the positive slope, the group index is larger than the refractive index. Thus the free spectral range is expected to be smaller in the spectral regime near the line center where laser oscillation occurs. The smaller free spectral range is consistent with the pulling of the oscillation frequency toward the line center.

6.3 THREE- AND FOUR-LEVEL LASERS

Lasers are commonly classified into "three-level" or "four-level" lasers. An idealized model of a four-level laser is shown in Figure 6.4. The feature characterizing this laser is that the separation E_1 of the terminal laser level from the ground state is large enough that at the temperature T at which the laser is operated, $E_1 \gg kT$. This guarantees that the thermal equilibrium population of level 1 can be neglected. If, in addition, the lifetime t_1 of atoms in level 1 is short compared to t_2, we can neglect N_1 compared to N_2 and the threshold condition (6.1-18) is satisfied when

$$N_2 \simeq N_t \tag{6.3-1}$$

Therefore laser oscillation begins when the upper laser level acquires, by pumping, a population density equal to the threshold value N_t.

A three-level laser is one in which the lower laser level is either the ground state or a level whose separation E_1 from the ground state is small compared to kT, so that at thermal equilibrium a substantial fraction of the total population occupies this level. An idealized three-level laser system is shown in Figure 6.5.

At a pumping level that is strong enough to create a population $N_2 = N_1 = N_0/2$ in the upper laser level,[1] the optical gain γ is zero, since $\gamma \propto N_2 - N_1 = 0$. To satisfy the oscillation condition, the pumping rate has to be further increased until

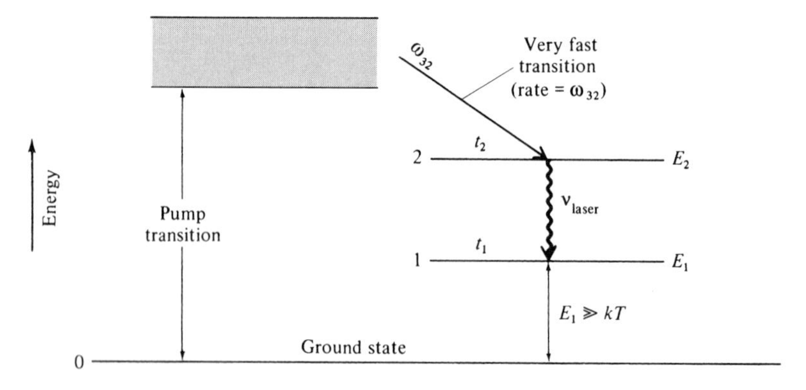

Figure 6.4 Energy-level diagram of an idealized four-level laser.

[1] Here we assume that because of the very fast transition rate ω_{32} out of level 3, the population of this level is negligible and $N_1 + N_2 = N_0$, where N_0 is the density of the active atoms.

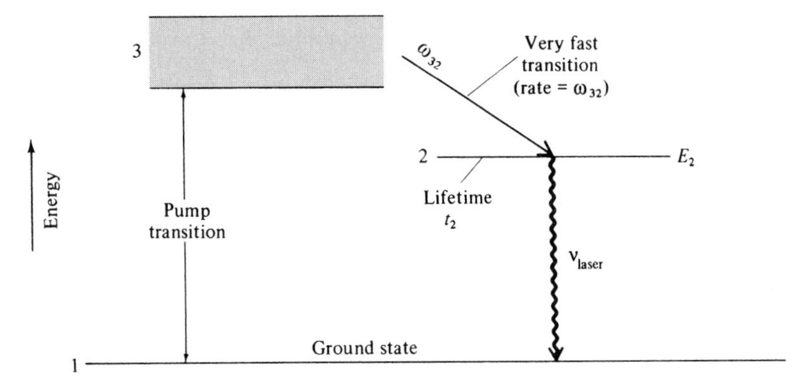

Figure 6.5 Energy-level diagram of an idealized three-level laser.

$$N_2 = \frac{N_0}{2} + \frac{N_t}{2}$$

and

$$N_1 = \frac{N_0}{2} - \frac{N_t}{2} \tag{6.3-2}$$

so $N_2 - N_1 = N_t$. Since in most laser systems $N_0 \gg N_t$, we find by comparing Equation (6.3-1) to (6.3-2) that the pump rate at threshold in a three-level laser must exceed that of a four-level laser—all other factors being equal—by

$$\frac{(N_2)_{\text{3-level}}}{(N_2)_{\text{4-level}}} \sim \frac{N_0}{2N_t}$$

In the example given in the next chapter, we will find that in the case of the ruby laser this factor is ~100.

The need to maintain about $N_0/2$ atoms in the upper level of a three-level laser calls for a *minimum* expenditure of power of

$$(P_s)_{\text{3-level}} = \frac{N_0 h\nu V}{2t_2} \tag{6.3-3}$$

and of

$$(P_s)_{\text{4-level}} = \frac{N_t h\nu V}{t_2} \tag{6.3-4}$$

in a four-level laser. V is the volume. The last two expressions are derived by multiplying the decay rate (atoms per second) from the upper level at threshold, which is $N_0V/2t_2$ and N_tV/t_2 in the two cases, by the energy $h\nu$ per transition. If the decay rate per atom t_2^{-1} (s^{-1}) from the upper level is due to spontaneous emission only, we can replace t_2 by t_{spont}. P_s is then equal to the power emitted through fluorescence by atoms within the (mode) volume V at threshold. We will refer to it as the *critical fluorescence* power. In the case of the four-level laser we use Equation (6.1-18) for N_t and obtain

$$(P_s)_{\text{4-level}} = \frac{N_t h\nu V}{t_2} = \frac{8\pi n^3 h\, \Delta\nu\, V}{\lambda^3 t_c} \frac{t_{\text{spont}}}{t_2} \tag{6.3-5}$$

where $\Delta\nu \cong 1/g(\nu_0)$ is the width of the laser transition lineshape.

EXAMPLE: CRITICAL FLUORESCENCE POWER OF AN ND^{3+}:GLASS LASER

The critical fluorescence power of an Nd^{3+}:glass laser is calculated using the following data:

$$l = 10 \text{ cm}$$

$$V = 10 \text{ cm}^3$$

$$\lambda = 1.06 \times 10^{-6} \text{ m}$$

$$R = (\text{mirror reflectivity}) = 0.95$$

$$n \simeq 1.5$$

$$t_c \simeq \frac{nl}{(1-R)c} = 10^{-8} \text{ s}$$

$$\Delta\nu = 3 \times 10^{12} \text{ Hz}$$

The Nd^{3+} : glass is a four-level laser system (see Figure 6.49), since level 1 is about 2000 cm^{-1} above the ground state so that at room temperature $E_1 \approx 10kT$. We can thus use Equation (6.3-5), obtaining $N_t = 8.5 \times 10^{15}$ cm^{-3} and

$$P_s \simeq 150 \text{ W}$$

6.4 POWER IN LASER OSCILLATORS

In Section 6.1 we derived an expression for the threshold population inversion N_t at which the laser gain becomes equal to the losses. We would expect that as the pumping intensity is increased beyond the point at which $N_2 - N_1 = N_t$ the laser will break into oscillation and emit power. In this section, we obtain the expression relating the laser power output to the pumping intensity. We also treat the problem of optimum coupling—that is, of the mirror transmission that results in the maximum power output.

Rate Equations

Consider an ideal four-level laser such as the one shown in Figure 6.4. We take $E_1 \gg kT$ so that the thermal population of the lower laser level 1 can be neglected. We assume that the critical inversion density N_t is very small compared to the ground-state population, so during oscillation the latter is hardly affected. We can consequently characterize the pumping intensity by R_2 and R_1, the density of atoms pumped per second into levels 2 and 1, respectively. Process R_1, which populates the lower level 1, causes a reduction of the gain and is thus detrimental to the laser operation. In many laser systems, such as discharge gas lasers, considerable pumping into the lower laser level is unavoidable, and therefore a realistic analysis of such systems must take R_1 into consideration.

The rate equations that describe the populations of levels 1 and 2 become

$$\frac{dN_2}{dt} = -N_2\omega_{21} - W_i(N_2 - N_1) + R_2 \tag{6.4-1}$$

$$\frac{dN_1}{dt} = -N_1\omega_{10} + N_2\omega_{21} + W_i(N_2 - N_1) + R_1 \tag{6.4-2}$$

ω_{ij} is the decay rate per atom from level i to j; thus the density of atoms per second undergoing decay from i to j is $N_i\omega_{ij}$. If the decay rate is due entirely to spontaneous transitions, then ω_{ij} is equal to the Einstein A_{ij} coefficient introduced in Section 5.6. W_i is the probability per unit time that an atom in level 2 will undergo an *induced* (stimulated) transition to level 1 (or vice versa). W_i, given by Equation (5.6-15), is proportional to the energy density of the radiation field inside the cavity.

Implied in the foregoing rate equations is the fact that we are dealing with a homogeneously broadened system. In an inhomogeneously broadened atomic transition, atoms with different transition frequencies $(E_2 - E_1)/h$ experience different induced transition rates and a single parameter W_i is not sufficient to characterize them.

In a steady-state situation, we have $\dot{N}_1 = \dot{N}_2 = 0$. In this case, we can solve Equations (6.4-1) and (6.4-2) for N_1 and N_2, obtaining

$$N_2 - N_1 = \frac{R_2[1 - (\omega_{21}/\omega_{10})(1 + R_1/R_2)]}{W_i + \omega_{21}} \tag{6.4-3}$$

A necessary condition for population inversion in our model is thus $\omega_{21} < \omega_{10}$, which is equivalent to requiring that the lifetime of the upper laser level ω_{21}^{-1} exceed that of the lower one. The effectiveness of the pumping is, according to Equation (6.4-3), reduced by the finite pumping rate R_1 and lifetime ω_{10}^{-1} of level 1 to an effective value

$$R = R_2\left[1 - \frac{\omega_{21}}{\omega_{10}}\left(1 + \frac{R_1}{R_2}\right)\right] \tag{6.4-4}$$

so Equation (6.4-3) can be written

$$N_2 - N_1 = \frac{R}{W_i + \omega_{21}} \tag{6.4-5}$$

Below the oscillation threshold the induced transition rate W_i is zero (since the oscillation energy density is zero) and $N_2 - N_1$ is, according to Equation (6.4-5), proportional to the pumping rate R. This state of affairs continues until $R = N_t\omega_{21}$, at which point $N_2 - N_1$ reaches the threshold value (see Equation (6.1-18))

$$N_t = \frac{8\pi n^3\nu^2 t_{\text{spont}}}{c^3 t_c g(\nu_0)} = \frac{8\pi n^3\nu^2 t_{\text{spont}}\,\Delta\nu}{c^3 t_c} \tag{6.4-6}$$

This is the point at which the gain at ν_0 due to the inversion is large enough to make up *exactly* for the cavity losses (the criterion that was used to derive N_t). Further increase of $N_2 - N_1$ with pumping is impossible in a *steady-state situation*, since it would result in a rate of induced (energy) emission that exceeds the losses so that the field energy stored in the resonator will increase with time in violation of the steady-state assumption.

This argument suggests that, under steady-state conditions, $N_2 - N_1$ must remain equal to N_t regardless of the amount by which the threshold pumping rate is exceeded. An examination of Equation (6.4-5) shows that this is possible, provided W_i is allowed to increase once R exceeds its threshold value $\omega_{21}N_t$, so that the equality

$$N_t = \frac{R}{W_i + \omega_{21}} \tag{6.4-7}$$

is satisfied. Since, according to Equation (5.6-15), W_i is proportional to the energy density in the resonator, Equation (6.4-7) relates the electromagnetic energy stored in the resonator to the pumping rate R. To derive this relationship we first solve Equation (6.4-7) for W_i, obtaining

$$W_i = \frac{R}{N_t} - \omega_{21}, \quad R \geq N_t\omega_{21} \tag{6.4-8}$$

The total power generated by stimulated emission is

$$P_e = (N_t V)W_i h\nu \tag{6.4-9}$$

where V is the volume of the oscillating mode. Using Equation (6.4-8) in (6.4-9) gives

$$\frac{P_e}{Vh\nu} = N_t\omega_{21}\left(\frac{R}{N_t\omega_{21}} - 1\right), \quad R \geq N_t\omega_{21} \tag{6.4-10}$$

This expression may be recast in a slightly different form, which we will find useful later on. We use expression (6.4-6) for N_t and, recalling that in our idealized model $\omega_{21}^{-1} = t_{\text{spont}}$, obtain

$$\frac{P_e}{Vh\nu} = N_t\omega_{21}\left(\frac{R}{p/t_c} - 1\right), \quad R \geq \frac{p}{t_c} \tag{6.4-11}$$

where

$$p = \frac{8\pi n^3\nu^2}{c^3 g(\nu_0)} = \frac{8\pi n^3\nu^2\,\Delta\nu}{c^3} \tag{6.4-12}$$

According to Equation (4.0-12), p corresponds to the density (m^{-3}) of radiation modes whose resonance frequencies fall within the atomic transition linewidth $\Delta\nu$—that is, the density of radiation modes that are capable of interacting with the transition.

Returning to the expression for the power output of a laser oscillator (6.4-11), we find that the term $R/(p/t_c)$ is the factor by which the pumping rate R exceeds its threshold value p/t_c. In addition, in an ideal laser system, $\omega_{21} = t_{\text{spont}}^{-1}$, so we can identify $N_t\omega_{21}h\nu V$ with the power P_s going into spontaneous emission at threshold, which is defined by Equation (6.3-5). We can consequently rewrite Equation (6.4-11) as

$$P_e = P_s\left(\frac{R}{R_t} - 1\right) \tag{6.4-13}$$

The main attraction of Equation (6.4-13) is in the fact that, in addition to providing an extremely simple expression for the power emitted by the laser atoms, it shows that for each unity increment of pumping, measured relative to the threshold value, the power increases by P_s. An experimental plot showing the linear relation predicted by Equation (6.4-13) is shown in Figure 6.6.

In the example of Section 6.3, which was based on an Nd^{3+}:glass laser, we obtained $P_s = 150$ W. We may expect on this basis that the power from this laser for, say, $(R/R_t) \simeq 2$ (i.e., twice above threshold) will be on the order of 150 W.

6.5 OPTIMUM OUTPUT COUPLING IN LASER OSCILLATORS

The total loss encountered by the oscillating laser mode can conveniently be attributed to two different sources: (a) the inevitable residual loss due to absorption and scattering in the laser material and in the mirrors, as well as diffraction losses in the finite diameter reflectors;

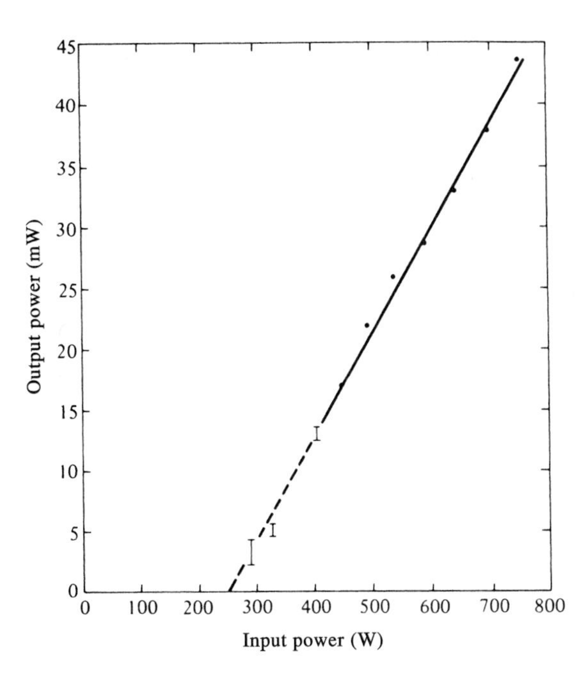

Figure 6.6 Plot of output power versus electric power input to a xenon lamp in a CW $CaF_2 : U^{3+}$ laser. Mirror transmittance at 2.61 μm is 0.2 percent, $T = 77$ K. (After Reference [2].)

(b) the (useful) loss due to coupling of output power through the partially transmissive reflector. It is obvious that loss (a) should be made as small as possible since it raises the oscillation threshold without contributing to the output power. The problem of the coupling loss (b), however, is more subtle. At zero coupling (i.e., both mirrors have zero transmission) the threshold will be at its minimum value and the power P_e emitted by the atoms will be maximum. But since none of this power is available as output, this is not a useful state of affairs. If, on the other hand, we keep increasing the coupling loss, the increasing threshold pumping will at some point exceed the actual pumping level. When this happens, oscillation will cease and the power output will again be zero. Between these two extremes there exists an optimum value of coupling (i.e., mirror transmission) at which the power output is a maximum.

The expression for the population inversion was shown in Equation (6.4-5) to have the form

$$N_2 - N_1 = \frac{R/\omega_{21}}{1 + W_i/\omega_{21}} \tag{6.5-1}$$

Since the exponential gain constant $\gamma(\nu)$ is, according to Equation (5.6-19), proportional to $N_2 - N_1$, we can use Equation (6.5-1) to write it as

$$\gamma = \frac{\gamma_0}{1 + W_i/\omega_{21}} \tag{6.5-2}$$

where γ_0 is the unsaturated ($W_i = 0$) gain constant (i.e., the gain exercised by a very weak field, so that $W_i \ll \omega_{21}$). We can use Equation (6.4-9) to express W_i in (6.5-2) in terms of the total emitted power P_e and then, in the resulting expression, replace $N_t V h\nu\omega_{21}$ by P_s. The result is

$$\gamma = \frac{\gamma_0}{1 + P_e/P_s} \tag{6.5-3}$$

where P_s, the saturation power, is given by Equation (6.3-4). The oscillation condition (6.1-12) can be written

$$e^{\gamma_t l}(1 - L) = 1 \tag{6.5-4}$$

where $L = 1 - r_1 r_2 \exp(-\alpha l)$ is the fraction of the intensity lost per pass. In the case of small losses ($L \ll 1$), Equation (6.5-4) can be written

$$\gamma_t l = L \tag{6.5-5}$$

According to the discussion in the introduction to this chapter, once the oscillation threshold is exceeded, the actual gain γ exercised by the laser oscillation is clamped at the threshold value γ_t regardless of the pumping. We can thus replace γ by γ_t in Equation (6.5-3) and, solving for P_e, obtain

$$P_e = P_s\left(\frac{g_0}{L} - 1\right) \tag{6.5-6}$$

where $g_0 = \gamma_0 l$ (i.e., the unsaturated gain per pass in nepers). P_e, we recall, is the *total* power given off by the atoms due to stimulated emission. The total loss per pass L can be expressed as the sum of the residual (unavoidable) loss L_i and the useful mirror transmission[2] T, so

$$L = L_i + T \tag{6.5-7}$$

The fraction of the total power P_e that is coupled out of the laser as useful output is thus $T/(T + L_i)$. Therefore, using Equation (6.5-6), we can write the (useful) power output as

$$P_o = P_s\left(\frac{g_0}{L_i + T} - 1\right)\frac{T}{T + L_i} \tag{6.5-8}$$

Replacing P_s in Equation (6.5-8) by the right side of Equation (6.3-5), and recalling from Equation (4.7-2) that for small losses

$$t_c = \frac{nl}{(L_i + T)c} = \frac{nl}{Lc} \tag{6.5-9}$$

Equation (6.5-8) becomes

$$P_o = \frac{8\pi n^2 h\nu\, \Delta\nu\, A}{\lambda^2 (t_2/t_{\text{spont}})} T\left(\frac{g_0}{L_i + T} - 1\right) = I_s AT\left(\frac{g_0}{L_i + T} - 1\right) \tag{6.5-10}$$

where $A = V/l$ is the cross-sectional area of the mode (assumed constant) and I_s is the saturation intensity as given in Equation (5.7-9). Maximizing P_0 with respect to T by setting $\partial P_0/\partial T = 0$ yields

$$T_{\text{opt}} = -L_i + \sqrt{g_0 L_i} \tag{6.5-11}$$

as the condition for the mirror transmission that yields the maximum power output.

The expression for the power output at optimum coupling is obtained by substituting Equation (6.5-11) for T in (6.5-10). The result, using Equation (5.7-9), is

$$(P_o)_{\text{opt}} = \frac{8\pi n^2 h\nu\, \Delta\nu\, A}{(t_2/t_{\text{spont}})\lambda^2}(\sqrt{g_0} - \sqrt{L_i})^2 = I_s A(\sqrt{g_0} - \sqrt{L_i})^2 \equiv S(\sqrt{g_0} - \sqrt{L_i})^2 \tag{6.5-12}$$

---

[2] For the sake of simplicity we can imagine one mirror as being perfectly reflecting, whereas the second (output) mirror has a transmittance T.

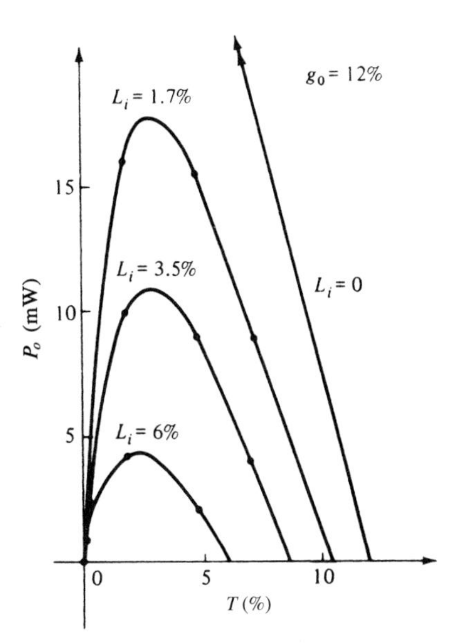

Figure 6.7 Useful power output (P_o) versus mirror transmission T for various values of internal loss L_i in an He–Ne 6328 Å laser. (After Reference [3].)

where the parameter $S = I_s A$ is defined by Equation (6.5-12) and is independent of the excitation level (pumping) or losses.

Theoretical plots of Equation (6.5-10) with L_i as a parameter are shown in Figure 6.7. Also shown are experimental data points obtained in a He–Ne 6328 Å laser. Note that the value of g_0 is given by the intercept of the $L_i = 0$ curve and is equal to 12%. The existence of an optimum coupling resulting in a maximum power output for each L_i is evident.

It is instructive to consider what happens to the energy $\mathscr{E}$ stored in the laser resonator as the coupling T is varied. A little thinking will convince us that $\mathscr{E}$ is proportional to P_o/T.[3] A plot of P_o (taken from Figure 6.7) and $\mathscr{E} \propto P_o/T$ as a function of the coupling T is shown in Figure 6.8. As we may expect, $\mathscr{E}$ is a monotonically decreasing function of T.

6.6 MULTIMODE LASER OSCILLATION AND MODE LOCKING

In this section, we discuss the effect of homogeneous or inhomogeneous broadening of the gain profile on the laser oscillation. As a result of the broadening, several modes can achieve enough gain to reach oscillation. We discuss both single-mode oscillation and multimode oscillation. In addition, we will discuss the locking of phases of multiple longitudinal modes to achieve short laser pulses. We start by reminding ourselves of some basic results pertinent to this discussion:

1. The actual gain constant prevailing inside a laser oscillator *at the oscillation frequency* ν is clamped, at steady state, at a value that is equal to the losses, according to Equation (6.1-14),

$$\gamma_t(\omega) = \alpha - \frac{1}{l} \ln r_1 r_2 \tag{6.6-1}$$

[3] The internal one-way power P_i incident on the mirrors is related, by definition, to P_o by $P_o = P_i T$. The total energy $\mathscr{E}$ is proportional to P_i.

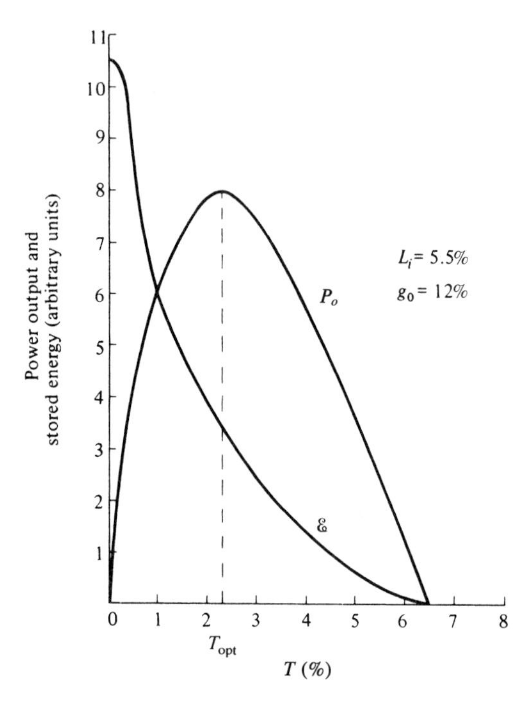

Figure 6.8 Power output P_o and stored energy $\mathscr{E}$ plotted against mirror transmission T.

where l is the length of the gain medium as well as the distance between the mirrors, which are taken here to be the same, α is the linear attenuation coefficient in the resonator (or cavity), and r_1 and r_2 are amplitude reflection coefficients of the mirror.

2. The gain constant of a distributed laser medium is given, according to Equation (5.6-19), by

$$\gamma(\nu) = (N_2 - N_1)\frac{c^2}{8\pi n^2 \nu^2 t_{spont}} g(\nu) \tag{6.6-2}$$

3. The optical resonator can support oscillations, provided sufficient gain is present to overcome losses, at resonance frequencies ν_q of the resonator separated according to Equation (4.6-3) by

$$\nu_{q+1} - \nu_q = \frac{c}{2nl} \tag{6.6-3}$$

where q is an integer, and we ignore the high-order transverse modes and the dispersion. Now consider what happens to the gain constant $\gamma(\nu)$ inside a laser oscillator as the pumping is increased from some value below threshold. Operationally, we can imagine an extremely weak wave of frequency ν launched into the laser medium and then measuring the gain constant $\gamma(\nu)$ as "seen" by this signal as ν is varied.

We treat first the case of a laser with a homogeneous broadening of the gain medium. Below threshold the inversion $(N_2 - N_1)$ is proportional to the pumping rate and $\gamma(\nu)$, which is given by Equation (6.6-2), is proportional to $g(\nu)$. This situation is illustrated by curve A in Figure 6.9a. The resonance frequencies as given by Equation (6.6-3) of the passive resonator are shown in Figure 6.9b. As the pumping rate is increased, the point is reached at which the gain per pass at the center resonance frequency ν_0 is equal to the average loss per

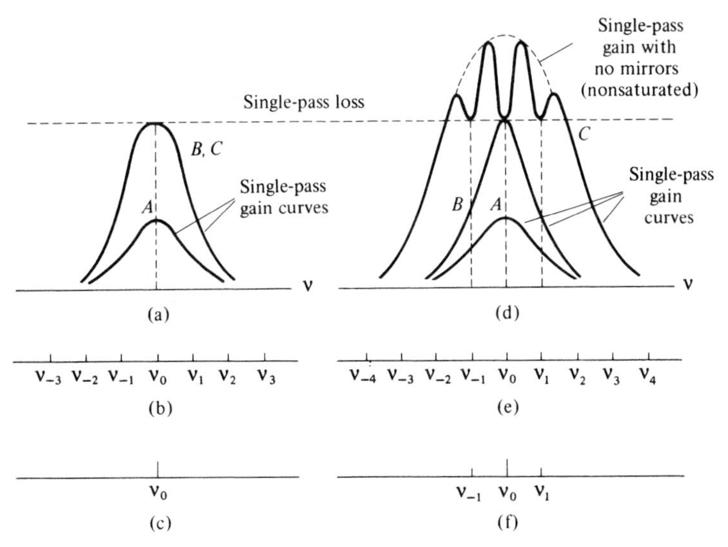

Figure 6.9 (a) Single-pass gain curves for a homogeneous atomic system (*A*—below threshold; *B*—at threshold; *C*—well above threshold). (b) Mode spectrum of optical resonator. (c) Oscillation spectrum (only one mode oscillates). (d) Single-pass gain curves for an inhomogeneous atomic system (*A*—below threshold; *B*—at threshold; *C*—well above threshold). (e) Mode spectrum of optical resonator. (f) Oscillation spectrum for pumping level *C*, showing three oscillating modes.

pass. This is shown in curve *B*. At this point, oscillation at ν_0 starts. An increase in the pumping cannot increase the inversion since this will cause $\gamma(\nu_0)$ to increase beyond its clamped value as given by Equation (6.6-1). Since the spectral lineshape function $g(\nu)$ describes the response of each individual atom, all the atoms being identical, it follows that the gain profile $\gamma(\nu)$ above threshold as in curve *C* is identical to that at threshold curve *B*. The result of further pumping only leads to an increase of the oscillation intensity, while the gain is clamped at the threshold gain value of Equation (6.6-1). Further increase in pumping, and the resulting increase in intracavity optical intensity, will eventually cause an additional broadening of $\gamma(\nu)$ due to the shortening of the lifetime by induced emission. The gain at other frequencies—such as ν_{-1}, ν_1, ν_{-2}, ν_2, and so forth—remains below the threshold value so that the ideal homogeneously broadened laser can oscillate only at a single mode.

In the extreme inhomogeneous case, the individual atoms can be considered as being all different from one another and acting independently. The lineshape function $g(\nu)$ reflects the distribution of the transition frequencies of the individual atoms. The gain profile $\gamma(\nu)$ below threshold is proportional to $g(\nu)$, and its behavior is similar to that of the homogeneous case. Once the threshold is reached as in curve *B*, the gain at ν_0 remains clamped at the threshold value. There is no reason, however, why the gain at other frequencies should not increase with further pumping. This gain is due to atoms that do not communicate with those contributing to the gain at ν_0. Further pumping will thus lead to oscillation at additional longitudinal mode frequencies as shown in curve *C*. Since the gain at each oscillating frequency is clamped, the gain profile curve acquires depressions at the oscillation frequencies. This phenomenon is referred to as "hole burning".[4] More discussion on hole burning will be given in Section 6.10.

A plot of the output frequency spectrum showing the multimode oscillation of a He–Ne 6328 Å laser is shown in Figure 6.10.

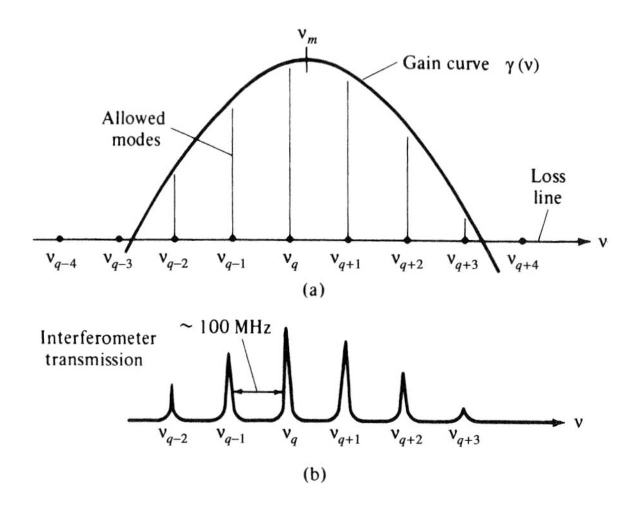

Figure 6.10 (a) Inhomogeneously broadened Doppler gain curve of the 6328 Å Ne transition and position of allowed longitudinal-mode frequencies. (b) Intensity versus frequency profile of an oscillating He–Ne laser. Six modes have sufficient gain to oscillate (After Reference [5].)

Mode Locking

We have argued above that in an inhomogeneously broadened laser, oscillation can take place at a number of frequencies, which are separated by (assuming a refractive index of $n = 1$)

$$\omega_q - \omega_{q-1} = \frac{\pi c}{l} \equiv \Omega \quad (q = \text{integer}) \tag{6.6-4}$$

where Ω is the free spectral range (or the mode spacing) in angular frequency. Now consider the total optical electric field resulting from such multimode oscillation at some arbitrary point, say, next to one of the mirrors, in the optical resonator. It can be taken, using complex notation, as

$$E(t) = \sum_m C_m e^{i[(\omega_0 + m\Omega)t + \phi_m]} \tag{6.6-5}$$

where C_m is the amplitude of the mth mode, and ϕ_m is the phase of the mth mode. The summation is extended over all the oscillating modes and ω_0 is chosen, arbitrarily, as oscillation frequency of one of the modes (usually chosen as the one closest to the line center). One property of Equation (6.6-5) is that $|E(t)|$ is periodic in time with a period of $\tau \equiv 2\pi/\Omega = 2l/c$, which is the round-trip transit time inside the resonator. Using Equation (6.6-5), the field at $t + \tau$ can be written

$$\begin{aligned} E(t+\tau) &= \sum_m C_m \exp\left\{ i\left[(\omega_0 + m\Omega)\left(t + \frac{2\pi}{\Omega} \right) + \phi_m \right] \right\} \\ &= \sum_m C_m \exp\{ i[(\omega_0 + m\Omega)t + \phi_m] \} \exp\left\{ i\left[2\pi\left(\frac{\omega_0}{\Omega} + m \right) \right] \right\} \\ &= E(t)\exp(i2\pi\omega_0/\Omega) \end{aligned} \tag{6.6-6}$$

Not that $E(t + \tau)$ is identical to $E(t)$, except for a constant phase factor.

Note that the periodic property of $E(t)$ depends on the fact that the modes are equally spaced and the phases ϕ_m are fixed. In typical lasers the phases ϕ_m are likely to vary randomly with time. This causes the intensity of the laser output to fluctuate randomly and greatly reduces its usefulness for many applications where temporal coherence is important. It should be noted that this fluctuation takes place because of random interference between modes and not because of intensity fluctuations of individual modes.

There are two ways in which the laser can be made coherent. First, make it possible for the laser to oscillate at a single frequency only so that mode interference is eliminated. This can be achieved in a variety of ways, including shortening the resonator length l, thus increasing the mode spacing ($\Omega = \pi c/l$) to a point where only one mode has sufficient gain to oscillate. The second approach is to force the modes' phases ϕ_m to maintain their relative values (ideally zero, so that they all oscillate in phase). This is the "mode locking" technique proposed and demonstrated in the early history of the laser [6–8]. This mode locking causes the oscillation intensity to consist of a periodic train with a period of $\tau = 2l/c = 2\pi/\Omega$.

One of the most useful forms of mode locking results when the phases ϕ_m are made equal to zero. To simplify the analysis of this case, assume that there are N oscillating modes with equal amplitudes. Taking $C_m = 1/\sqrt{N}$ and $\phi_m = 0$ in Equation (6.6-5), we obtain

$$E(t) = \frac{1}{\sqrt{N}} \sum_{m=1}^{N} e^{i(\omega_0 + m\Omega)t} = \frac{1}{\sqrt{N}} e^{i[(\omega_0 + (N+1)\Omega/2]t} \frac{\sin(N\Omega t/2)}{\sin(\Omega t/2)} \tag{6.6-7}$$

where the field is normalized to have a constant energy (independent of N). The last equality in the above equation is obtained by summing up the geometric series. The average laser power output is proportional to $E(t)E^*(t)$ and is given by

$$P(t) \propto \frac{1}{N} \frac{\sin^2(N\Omega t/2)}{\sin^2(\Omega t/2)} \tag{6.6-8}$$

where the averaging is performed over a time that is long compared with the optical period $2\pi/\omega_0$ but short compared with the modulation period $2\pi/\Omega$.

Some of the analytic properties of $P(t)$ are immediately apparent.

1. The power is emitted in the form of a train of pulses with a period $\tau = 2l/c$, that is, the round-trip transit time.
2. The peak power, $P(s\tau)$ (for $s = 0, 1, 2, 3, \ldots$), is equal to N times the average power, where N is the number of modes locked together.
3. The peak field amplitude is equal to N times the amplitude of a single mode.
4. The pulse width of the main peaks, defined as the time from the peak to the first zero, is $\tau_0 = \tau/N$. This is approximately the FWHM of the main peaks of $P(t)$ (for $N \gg 1$) (see Problem 6.18). There are $(N - 2)$ sidelobes between the neighboring main peaks.

The number of oscillating modes can be estimated by $N \cong \Delta\omega/\Omega$—that is, the ratio of the transition lineshape width $\Delta\omega$ (or gain bandwidth) to the frequency spacing Ω between the modes. Using this relation, as well as $\tau = 2\pi/\Omega$ in $\tau_0 = \tau/N$, we obtain

$$\tau_0 \sim \frac{2\pi}{\Delta\omega} = \frac{1}{\Delta\nu} = \frac{\tau}{N} \tag{6.6-9}$$

where $\Delta\nu$ is the gain bandwidth. Thus the temporal length of the mode-locked laser pulses is approximately the inverse of the gain linewidth.

A theoretical plot of $\sqrt{P(t)}$ as given by Equation (6.6-8) for the case of eight modes ($N = 8$) is shown in Figure 6.11. The ordinate may also be considered as being proportional to the instantaneous field amplitude. The foregoing discussion was limited to consideration of mode locking as a function of time. It is clear, however, that since the solution of Maxwell's equations in the cavity involves traveling waves (a standing wave can be considered as the

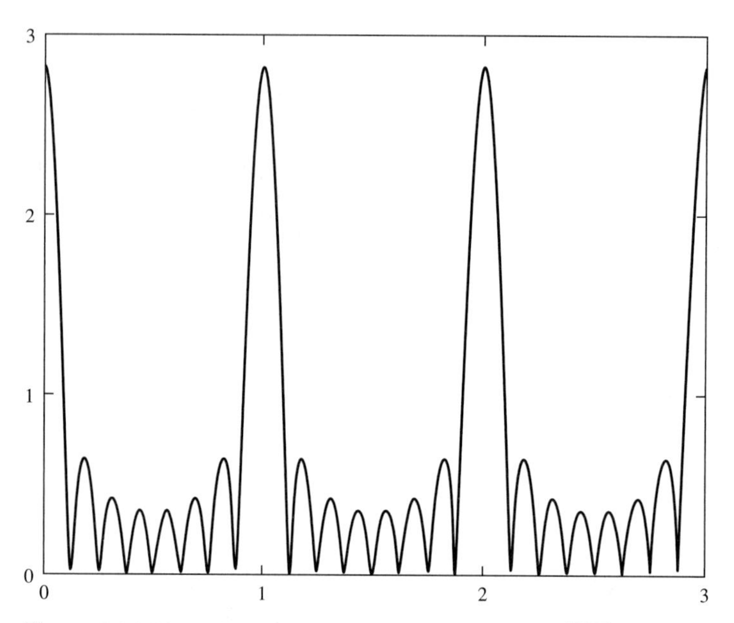

Figure 6.11 Theoretical plot of optical field amplitude $\sqrt{P(t)}$: $|\sin(N\Omega t/2)/\sin(\Omega t/2)|$ resulting from phase locking of eight ($N = 8$) equal-amplitude modes separated from each other by a frequency interval $\Omega = 2\pi/\tau$.

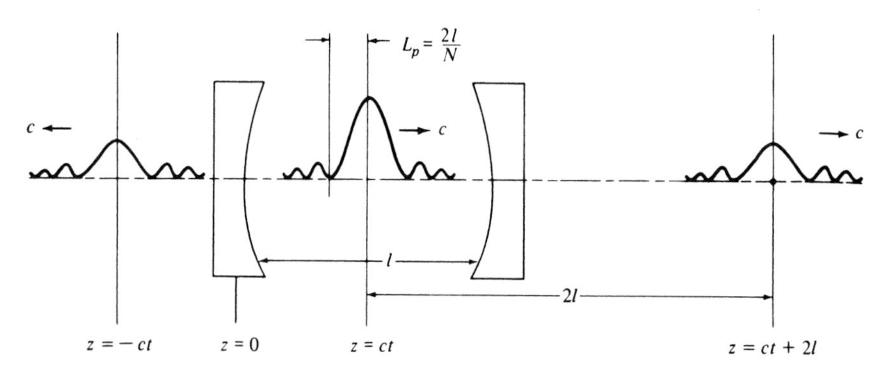

Figure 6.12 Traveling pulse of energy resulting from the mode locking of N laser modes.

sum of two waves traveling in opposite directions), mode locking causes the oscillation energy of the laser to be condensed into a packet that travels back and forth between the mirrors with the velocity of light c (see Figure 6.12). The pulsation period $\tau = 2l/c$ corresponds simply to the time interval between two successive arrivals of the pulse at the mirror. The spatial length of the pulse L_p must correspond to its time duration multiplied by its group velocity c. Using $\tau_0 = \tau/N$, we obtain

$$L_p = c\tau_0 = \frac{c\tau}{N} = \frac{2l}{N} \tag{6.6-10}$$

Figure 6.12 shows a schematic drawing of the spatial profile of the mode-locked pulses at a given instant in time.

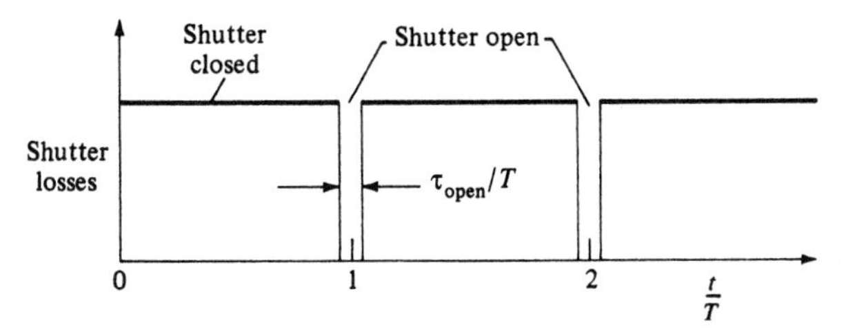

Figure 6.13 Periodic losses introduced by a shutter to induce mode locking. The presence of these losses favors the choice of mode phases that result in a pulse passing through the shutter during open intervals—that is, mode locking.

Methods of Mode Locking

In the preceding discussion we considered the consequences of fixing the phases of the longitudinal modes of a laser mode locking. Mode locking can be achieved by modulating the losses (or gain) of the laser at an angular frequency $\Omega = \pi c/l$, which is equal to the intermode frequency spacing. The theoretical proof of mode locking by loss modulation [2, 6, 7] is rather formal, but a good plausibility argument can be made as follows: As a form of loss modulation consider a thin shutter inserted inside the laser resonator. Let the shutter be closed (high optical loss) most of the time except for brief periodic openings for a duration of τ_{open} every $\tau = 2l/c$ s. This situation is illustrated in Figure 6.13.

A single laser mode will not oscillate in this case because of the high losses (we assume that τ_{open} is too short to allow the oscillation to build up during each opening). The same applies to multimode oscillation with arbitrary phases. There is one exception, however. If the phases were "locked" as in Equation (6.6-7), the energy distribution inside the resonator would correspond to that shown in Figure 6.12 and would consist of a narrow ($L_p = 2l/N$) traveling pulse. If this pulse should arrive at the shutter's position when it is open and if the pulse (temporal) length τ_0 is short compared to the opening time τ_{open}, the mode-locked pulse will be "unaware" of the shutter's existence and, consequently, *will not be attenuated by it.* We may thus reach the conclusion that loss modulation causes mode locking through some kind of "survival of the fittest" mechanism. In reality, the periodic shutter chops off any intensity tails acquired by the mode-locked pulses due to a "wandering" of the phases from their ideal ($\phi_n = 0$) values. This has the effect of continuously restoring the phases.

An experimental setup used to mode lock a He–Ne laser is shown in Figure 6.14; the periodic loss [8] is introduced by Bragg diffraction of a portion of the laser intensity from a standing acoustic wave. The standing-wave nature of the acoustic oscillation causes the strain to have a form $S(z, t) = S_0 \cos(\omega_a t)\cos(k_a z)$, where the acoustic velocity is $v_a = \omega_a/k_a$, and S_0 is a constant amplitude. Since the change in the index of refraction is, to first order, proportional to the strain $S(z, t)$, a phase diffraction grating is created with a spatial period $2\pi/k_a$, which is equal to the acoustic wavelength. The diffraction loss of the incident laser beam due to the grating reaches its peak twice in each acoustic period when $S(z, t)$ has its maximum and minimum values. The loss modulation frequency is thus $2\omega_a$, and mode locking occurs when $2\omega_a = \Omega$, where Ω is the angular frequency separation between two longitudinal laser modes.

Figure 6.15 shows the pulses resulting from mode locking a rhodamine 6G dye laser. Mode locking occurs spontaneously in some lasers if the optical path contains a saturable absorber (an absorber whose opacity decreases with increasing optical intensity). This method is used to induce mode locking in the high-power pulsed solid-state lasers [10, 11]

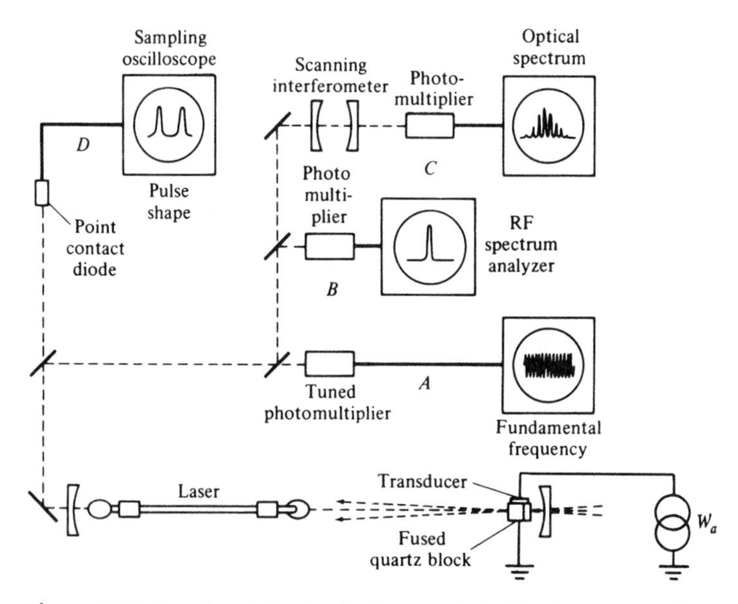

Figure 6.14 Experimental setup for laser mode locking by acoustic (Bragg) loss modulation. The loss is due to Bragg diffraction of the main laser beam by a standing acoustic wave. Parts *A*, *B*, *C*, and *D* of the experimental setup are designed to display the fundamental component of the intensity modulation, the power spectrum of the intensity modulation, the power spectrum of the optical field $E(t)$, and the optical intensity, respectively. (After Reference [9].)

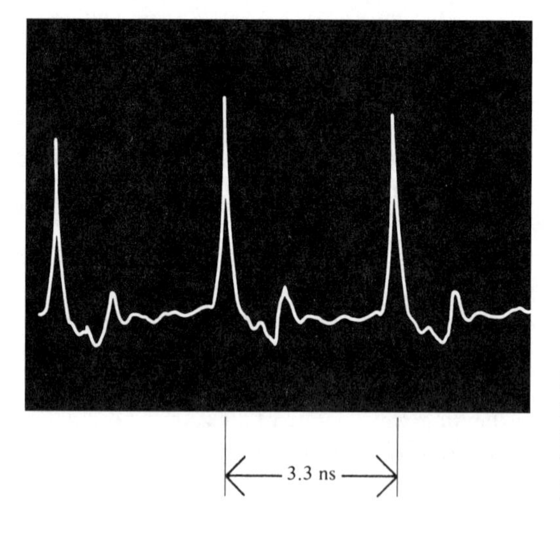

Figure 6.15 Power output as a function of time of a mode-locked dye laser, using rhodamine 6G. The oscillation is at $\lambda = 0.61\ \mu\text{m}$. The pulse width is detector limited. (After Reference [9].)

and in continuous dye lasers. This is due to the fact that such a dye will absorb less power from a mode-locked train of pulses than from a random phase oscillation of many modes [2], since the first form of oscillation leads to the highest possible peak intensities, for a given average power from the laser, and is consequently attenuated less severely. From arguments identical to those advanced in connection with the periodic shutter (Figure 6.13), it follows that the presence of a saturable absorber in the laser cavity will "force" the laser, by a "survival of the fittest" mechanism, to lock its modes' phases as in Equation (6.6-7).

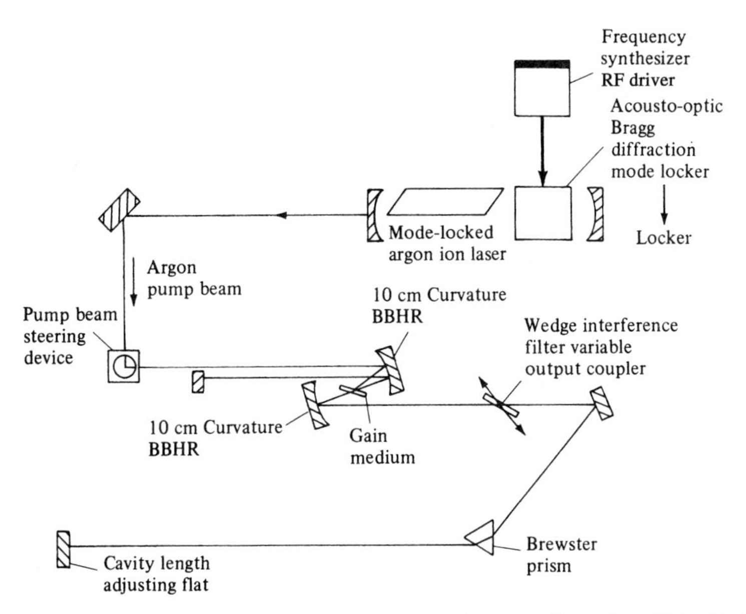

Figure 6.16 Synchronously mode-locked dye laser configuration. (After Reference [19].)

Some of the shortest mode-locked pulses to date were obtained from dye lasers employing rhodamine 6G as the gain medium. The mode locking is caused by synchronous gain modulation that is due to the fact that the pumping (blue-green) argon gas laser is itself mode locked. The pump pulses are synchronized exactly to the pulse repetition rate of the dye laser. (This requires that both lasers have precisely the same optical resonator length.) When this is done, the dye laser gain medium will be pumped once in each round-trip period so that the pumping pulse and the mode-locked pulse overlap spatially and temporally in the dye cell.

Additional sharpening of the mode-locked pulses can result from inclusion of a saturable (dye) absorber in the resonator. A sketch of a synchronously mode-locked dye laser configuration is shown in Figure 6.16. Additional amplification of the output pulses of the dye laser by a sequence of three to four dye laser amplifier cells (consisting of rhodamine 6G pumped by the pulsed second harmonic of Q-switched Nd^{3+}:YAG lasers) has yielded subpicosecond pulses with peak power exceeding 10^9 W. Mode-locked optical pulses with pulse duration of $\sim 30 \times 10^{-15}$ s have been obtained [12]. These pulses have been narrowed down further to $\sim 6 \times 10^{-15}$ s by the use of nonlinear optical techniques that are described in Section 6.8.

Ultrashort mode-locked pulses are now used in an ever-widening circle of applications involving the measurement and study of short-lived molecular and electronic phenomena. The use of ultrashort optical pulses has led to an improvement of the temporal resolution of such experiments by more than three orders of magnitude. For a description of many of these applications as well as of the many methods used to measure the pulse duration, the student should consult References [13–15].

Mode locking in semiconductor lasers is of particular interest owing to the very large gain bandwidth in these media. These lasers offer potential operation in the 10–20 fs range although present results are far from this goal [16, 17]. Of special interest is the possibility of controlling the gain and loss by means of multiple electrodes [18]. Table 6.1 lists some of the lasers commonly used in mode locking, along with the achievable pulse durations.

TABLE 6.1 Some Laser Systems, Their Gain Linewidth $\Delta\nu$, and the Length of Their Pulses in Mode-Locked Operation

Laser Medium	$\Delta\nu$ (Hz)	$(\Delta\nu)^{-1}$ (second)	Observed τ_0 (second)
He–Ne (0.6328 μm)	$\sim 1.5 \times 10^9$	6.66×10^{-10}	$\sim 6 \times 10^{-10}$
Nd:YAG (1.06 μm)	$\sim 1.2 \times 10^{10}$	8.34×10^{-11}	$\sim 7.6 \times 10^{-11}$
Ruby (0.6934 μm)	6×10^{10}	1.66×10^{-11}	$\sim 1.2 \times 10^{-11}$
Nd^{3+}:glass	3×10^{12}	3.33×10^{-13}	$\sim 3 \times 10^{-13}$
Rhodamine 6G dye laser (0.6 μm)	10^{13}	10^{-13}	3×10^{-14}
Diode laser	10^{13}	10^{-13}	4×10^{-13}

Theory of Mode Locking

We have argued earlier that the introduction of a time-periodic loss, or gain, into a laser oscillator will cause phase locking between the otherwise independent, longitudinal modes of the resonator. A necessary condition is that the frequency of this modulation be equal to an integral multiple of the intermode frequency separation (free spectral range) of the resonator. This follows from the gating argument leading to Figure 6.13 or, equivalently, from the requirement that optical sidebands at $\omega_0 \pm \omega_m$, which arise from loss modulation of optical mode ω_0, coincide in frequency with the resonance frequency of the modes ω_q. The theoretical treatment of this problem consists of solving Maxwell's equations for oscillation of an optical resonator with gain (due to inversion) and a periodic loss modulation. The periodic loss modulation leads to a coupling among the modes, and eventually a locking of the phases of the modes. A detailed derivation is presented, for example, in References [7, 20]. The key result is that when the modulation frequency ω_m is equal to the intermode frequency separation, $\Omega = \pi c/l$, we obtain a solution wherein all the modes have the *same* phase. This corresponds to the ultra short pulse form of mode locking discussed at the beginning of this section.

In what follows, we describe an elementary theory of mode locking via a periodic loss modulation. We represent the loss element as a transmission element with a time-varying amplitude modulation. Let the round-trip amplitude transmission coefficient of the loss modulator be written

$$T(t) = \sigma_0 + 2\sigma_1 \cos(\Omega t) = \sigma_0 + 2\sigma_1 \cos\left(\frac{2\pi}{\tau} t\right) \tag{6.6-11}$$

where σ_0 and σ_1 are constants, τ is the round-trip flight time inside the resonator, and $\Omega = 2\pi/\tau$ is the angular frequency of the longitudinal mode spacing. For simplicity in the discussion, we assume that $T(t)$ is a round-trip transmission modulation with real constants σ_0 and σ_1. Let the field amplitude at the input of the loss modulator be written

$$E_{\text{in}}(t) = \sum_{n=1}^{N} C_n e^{in\Omega t} \tag{6.6-12}$$

where C_n (the mode amplitudes) are arbitrary constants, and N is the number of modes involved. For simplicity, we ignore the constant reference frequency ω_0. Physically, N is the total number of modes that are oscillating in the resonator. The field amplitude of the output after the transmission modulation element can be written

$$E_{\text{out}}(t) = T(t)E_{\text{in}}(t) = \sum_{n=1}^{N} C'_n e^{in\Omega t} \tag{6.6-13}$$

where C'_n are constants. Using Equation (6.6-11) for $T(t)$ in Equation (6.6-12), we obtain

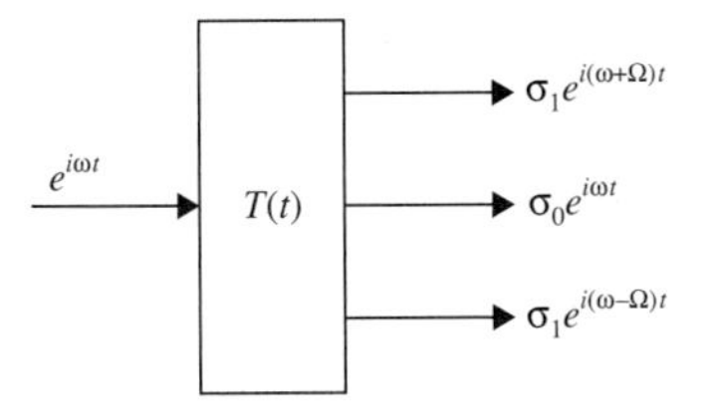

Figure 6.17 A single frequency (ω) input at the modulator leads to an output beam consisting of three frequency components (ω, $\omega + \Omega$, $\omega - \Omega$).

$$E_{\text{out}}(t) = T(t)E_{\text{in}}(t) = \sum_{n=1}^{N} (\sigma_0 C_n + \sigma_1 C_{n+1} + \sigma_1 C_{n-1})\, e^{in\Omega t} \tag{6.6-14}$$

We notice that each mode is coupled with its neighboring modes, when the modulation frequency is equal to the longitudinal mode spacing of the resonator. This coupling mechanism is illustrated in Figure 6.17. For each mode transmitting through the periodic loss modulator, two side bands ($\omega \pm \Omega$) are generated.

As the wave bounces back and forth inside the resonator, all the modes are coupled eventually. Thus the time modulation creates a mechanism that couples all the modes. We will now show that the coupling leads to a constant phase relationship between the modes. This is the origin of mode locking.

If we represent the field amplitude in terms of a column vector of the constants C_n, then the input–output relationship (Equations (6.6-13) and (6.6-14)) can be written

$$\begin{bmatrix} \sigma_0 & \sigma_1 & 0 & \cdots & 0 \\ \sigma_1 & \sigma_0 & \sigma_1 & \cdots & 0 \\ 0 & \sigma_1 & \sigma_0 & \cdots & 0 \\ \vdots & \vdots & \vdots & \vdots & \vdots \\ 0 & 0 & 0 & \cdots & \sigma_0 \end{bmatrix} \begin{bmatrix} C_1 \\ C_2 \\ C_3 \\ \vdots \\ C_N \end{bmatrix} = \begin{bmatrix} C_1' \\ C_2' \\ C_3' \\ \vdots \\ C_N' \end{bmatrix} \tag{6.6-15}$$

We note that all the elements of the matrix are zero except the diagonal elements and those immediately next to the diagonal elements. This is a result of the particular modulation function that creates only one upper and one lower sideband, leading to nearest-neighbor mode coupling. In an inhomogeneously broadened laser system, we may assume an equal gain for the modes that are oscillating. Let g be the round-trip net gain coefficient for the field amplitude; then the round-trip input–output relationship becomes

$$\begin{bmatrix} \sigma_0 & \sigma_1 & 0 & \cdots & 0 \\ \sigma_1 & \sigma_0 & \sigma_1 & \cdots & 0 \\ 0 & \sigma_1 & \sigma_0 & \cdots & 0 \\ \vdots & \vdots & \vdots & \vdots & \vdots \\ 0 & 0 & 0 & \cdots & \sigma_0 \end{bmatrix} \begin{bmatrix} g & 0 & 0 & \cdots & 0 \\ 0 & g & 0 & \cdots & 0 \\ 0 & 0 & g & \cdots & 0 \\ \vdots & \vdots & \vdots & \vdots & \vdots \\ 0 & 0 & 0 & \cdots & g \end{bmatrix} \begin{bmatrix} C_1 \\ C_2 \\ C_3 \\ \vdots \\ C_N \end{bmatrix} = \begin{bmatrix} C_1' \\ C_2' \\ C_3' \\ \vdots \\ C_N' \end{bmatrix}$$

or equivalently,

$$\begin{bmatrix} g\sigma_0 & g\sigma_1 & 0 & \cdots & 0 \\ g\sigma_1 & g\sigma_0 & g\sigma_1 & \cdots & 0 \\ 0 & g\sigma_1 & g\sigma_0 & \cdots & 0 \\ \vdots & \vdots & \vdots & \vdots & \vdots \\ 0 & 0 & 0 & \cdots & g\sigma_0 \end{bmatrix} \begin{bmatrix} C_1 \\ C_2 \\ C_3 \\ \vdots \\ C_N \end{bmatrix} = \begin{bmatrix} C_1' \\ C_2' \\ C_3' \\ \vdots \\ C_N' \end{bmatrix} \tag{6.6-16}$$

At steady state, the field must reproduce itself after one round-trip. In other words, the output field is proportional to the input field. We can write

$$E_{\text{out}}(t) = \eta E_{\text{in}}(t) \tag{6.6-17}$$

where η is a constant. Physically, η is a round-trip amplitude transmission coefficient of the field inside the resonator. Thus the matrix equation can now be written

$$\begin{bmatrix} g\sigma_0 & g\sigma_1 & 0 & \cdots & 0 \\ g\sigma_1 & g\sigma_0 & g\sigma_1 & \cdots & 0 \\ 0 & g\sigma_1 & g\sigma_0 & \cdots & 0 \\ \vdots & \vdots & \vdots & \vdots & \vdots \\ 0 & 0 & 0 & \cdots & g\sigma_0 \end{bmatrix} \begin{bmatrix} C_1 \\ C_2 \\ C_3 \\ \vdots \\ C_N \end{bmatrix} = \eta \begin{bmatrix} C_1 \\ C_2 \\ C_3 \\ \vdots \\ C_N \end{bmatrix} \tag{6.6-18}$$

We note that the steady-state oscillation is an eigenvector of the matrix equation and η is an eigenvalue of the matrix equation. The eigenvectors represent a superposition of the N modes in the resonator and can be viewed as the "supermodes." This matrix problem can be solved by rewriting the equation as the following recurrence equation,

$$g\sigma_0 C_n + g\sigma_1 C_{n+1} + g\sigma_1 C_{n-1} = \eta C_n \tag{6.6-19}$$

or equivalently,

$$g\sigma_1 C_{n+1} + (g\sigma_0 - \eta)C_n + g\sigma_1 C_{n-1} = 0 \tag{6.6-20}$$

subject to the boundary condition

$$C_0 = C_{N+1} = 0 \tag{6.6-21}$$

The solution for the mode amplitudes subject to the preceding boundary condition can be written

$$C_n = \sqrt{\frac{2}{N+1}} \sin\left(\frac{ns\pi}{N+1}\right) \tag{6.6-22}$$

where s is an integer ($s = 1, 2, 3, \ldots, N$) and the square root $\sqrt{2/(N+1)}$ is a normalization factor. The eigenvalues are given by

$$\eta = g\sigma_0 + 2g\sigma_1 \cos\left(\frac{s\pi}{N+1}\right), \quad s = 1, 2, 3, \ldots, N \tag{6.6-23}$$

It is interesting to note that the eigenvectors as given by Equation (6.6-22) are independent of the absolute magnitude of (g, σ_0, σ_1). Physically, this means mode locking can be achieved with an arbitrary depth of amplitude modulation (σ_1). With η being the effective round-trip amplitude transmission coefficient, we note that the fundamental solution (with $s = 1$) has the maximum effective transmission (assuming positive σ_0 and σ_1). For this solution, the mode amplitudes are given by

$$C_n = \sqrt{\frac{2}{N+1}} \sin\left(\frac{n\pi}{N+1}\right), \quad n = 1, 2, 3, \ldots, N \tag{6.6-24}$$

Not that the amplitudes are all real. In other words, all phases of the mode amplitudes are locked at zero. We can also say that all modes are phase locked.

We can now construct the temporal profile of the steady-state field inside the resonator for this solution. The field can be written, according to Equation (6.6-24),

$$E(t) = \sum_{n=1}^{N} C_n e^{in\Omega t} = \sqrt{\frac{2}{N+1}} \sum_{n=1}^{N} \sin\left(\frac{n\pi}{N+1}\right) e^{in\Omega t} \tag{6.6-25}$$

After a few steps of algebra, we obtain

$$E(t) = \frac{1}{2}\sqrt{\frac{2}{N+1}} \left(\frac{\sin\frac{N}{2}(\Omega t + \alpha)}{\sin\frac{1}{2}(\Omega t + \alpha)} + \frac{\sin\frac{N}{2}(\Omega t - \alpha)}{\sin\frac{1}{2}(\Omega t - \alpha)} \right) e^{i[(N+1)/2]\Omega t} \tag{6.6-26}$$

where α is given by

$$\alpha = \frac{\pi}{N+1} \tag{6.6-27}$$

Equation (6.6-26) is slightly different from the previous expression, Equation (6.6-7), where constant mode amplitudes were assumed. The field consists of two pulses, each with a width of τ/N, centered at $\Omega t = \alpha, -\alpha$, respectively. As a result, the mode-locked pulse is broader and the sidelobes are lower. The field (6.6-26) is periodic with a period of τ. Peak intensity occurs at $t = 0, \tau, 2\tau, 3\tau, 4\tau, \ldots$. If we define the pulse width of the main peaks as the time from the peak to the first zero, it can easily be shown that, for large N, the width of the mode-locked pulse (6.6-26) is

$$\tau_0 = \frac{3}{2}\frac{\tau}{N} \tag{6.6-28}$$

where we recall that τ is the round-trip flight time in the cavity. This is approximately the full width at half maximum of the pulse (see Problem 6.18). We note that the pulse width is approximately 50% broader due to the tapering of the mode amplitudes as given by Equation (6.6-24). Figure 6.18 is a plot of the amplitude profile of the steady-state field inside the resonator.

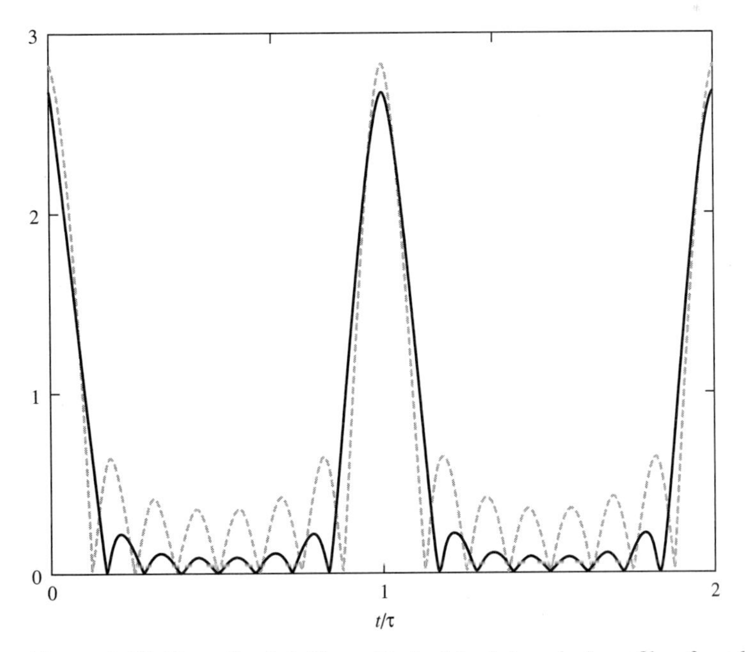

Figure 6.18 Normalized field amplitude (absolute value) profile of mode-locked laser beam with $N = 8$. The solid curve is from Equation (6.6-25), and the dotted curve is from Equation (6.6-7). The field amplitudes are normalized so that $\sum|C_n|^2 = 1$.

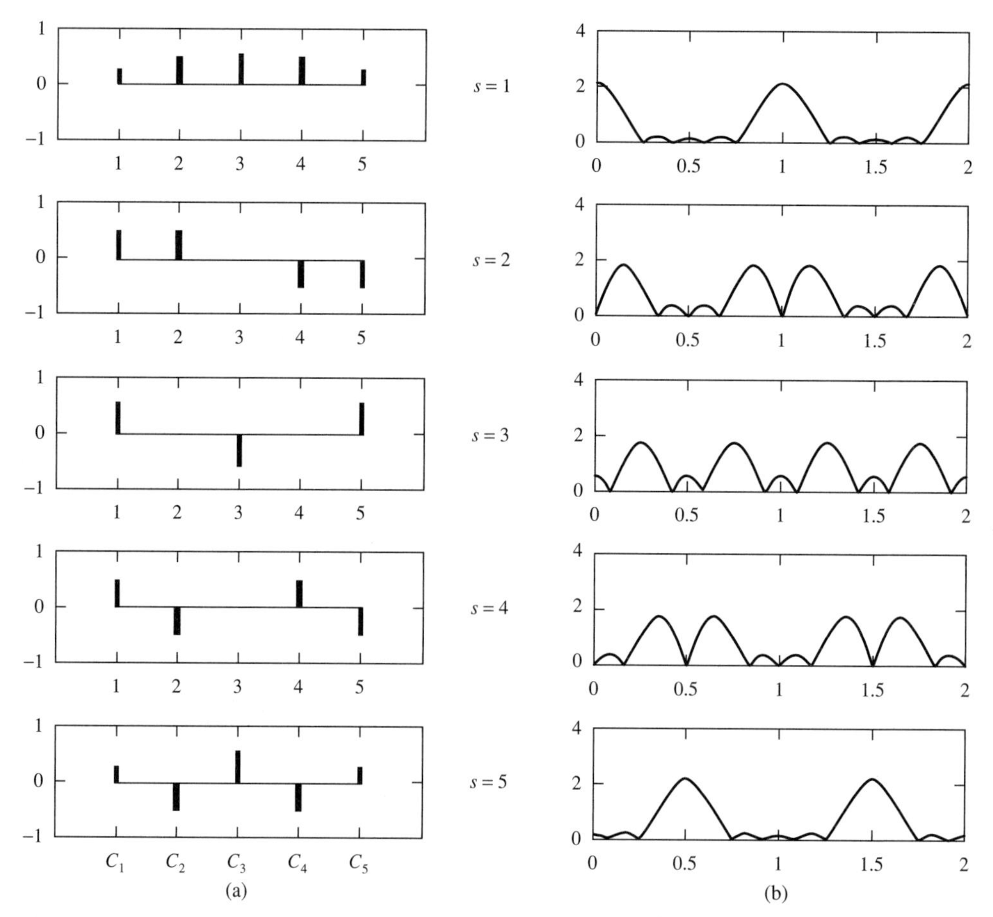

Figure 6.19 Normalized field amplitude (absolute value) profile of (a) the mode coefficients C_n with $N = 5$ and (b) the supermodes. The field amplitudes are normalized so that $\Sigma|C_n|^2 = 1$. The top row is for the supermode with $s = 1$, and the bottom row is for the supermode with $s = 5$.

Strictly speaking, only the supermode with $\eta = 1$ can sustain oscillation. In general, there is only one supermode whose round-trip amplitude transmission coefficient equals one. Depending on the pumping level and the gain coefficient, any one of the supermodes can oscillate. Figure 6.19 shows the normalized field amplitude (absolute value) profile of the supermodes with $N = 5$. The field amplitude of the supermodes can be written

$$E(t) = \frac{1}{2}\sqrt{\frac{2}{N+1}}\left(\frac{\sin\frac{N}{2}(\Omega t + s\alpha)}{\sin\frac{1}{2}(\Omega t + s\alpha)} + \frac{\sin\frac{N}{2}(\Omega t - s\alpha)}{\sin\frac{1}{2}(\Omega t - s\alpha)}\right)e^{i[(N+1)/2]\Omega t} \tag{6.6-29}$$

where $s = 1, 2, 3, \ldots, N$.

We note that the main peaks of the supermode with $s = 1$ coincide with the maximum transmission windows of the amplitude modulator. This is consistent with the maximum averaged transmission. The main peaks of the supermode with $s = 5$ coincide with the minimum transmission windows of the amplitude modulator. Figure 6.20 illustrates the absolute value of the field amplitude for the supermodes with $s = 1$ and $s = 5$. Note the temporal locations

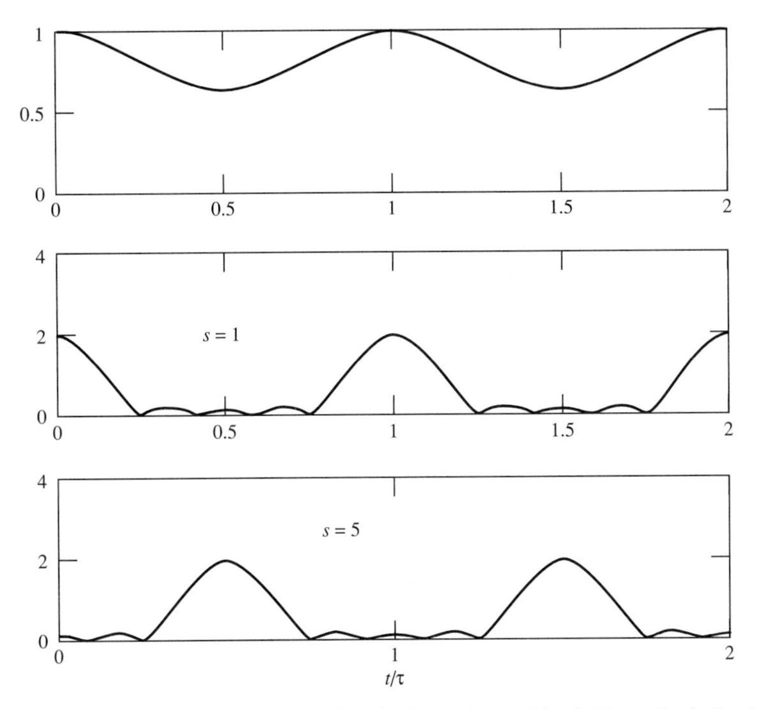

Figure 6.20 Theoretical plots of the absolute values of the field amplitude for the supermodes with $s = 1$ and $s = 5$. The top plot is the amplitude transmission as a function of time (in units of τ). The lower two plots are the absolute values of the field of supermodes $s = 1$ and $s = 5$, respectively.

of the main peaks of these modes relative to the maximum transmission windows of the amplitude modulator.

6.7 MODE LOCKING IN HOMOGENEOUSLY BROADENED LASER SYSTEMS

The analysis of mode locking in inhomogeneous laser systems in Section 6.6 assumed that the role of internal modulation was that of locking together the phases of modes that, in the absence of modulation, oscillate with random phases. In the case of homogeneous broadening, only one mode can normally oscillate. Experiments, however, reveal that mode locking leads to short pulses in a manner quite similar to that described in Section 6.6 and therefore must involve multimode oscillation. One way to reconcile the two points of view and the experiments is to realize that *in the presence of internal modulation*, power is transferred continuously from the high-gain mode to those of lower gain (i.e., those that would not normally oscillate). This power can be viewed simply as that of the sidebands at $(\omega_0 \pm n\Omega)$ of the mode at ω_0 created by a modulation at frequency Ω. Armed with this understanding, we see that the physical phenomenon is not only one of mode locking but also one of mode generation. The net result, however, is that of a large number of oscillating modes with equal frequency spacing and fixed phases, as in the inhomogeneous case, leading to ultrashort pulses.

The analytical solution to this case [21–23] follows an approach used originally to analyze short pulses in traveling wave microwave oscillators [24]. Referring to Figure 6.21, we consider an optical resonator with mirror reflectivity R_1 and R_2 that contains, in addition to the gain medium, a periodically modulated loss cell. The method of solution is to follow

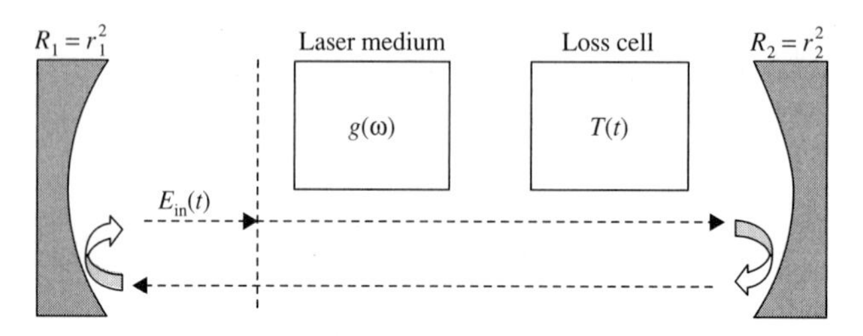

Figure 6.21 The experimental arrangement assumed in the theoretical analysis of mode locking in homogeneously broadened lasers. An input field $E_{in}(t)$ at an arbitrary reference point will experience an amplitude gain of $g(\omega)$ and an amplitude loss modulation of $T(t)$ after one round-trip and becomes $E_{out}(t) = g(\omega)T(t)\,E_{in}(t)$.

one pulse through a complete round trip through the resonator and to require that the pulse reproduce itself. For simplicity in the discussion, we assume a round-trip gain profile of $g(\omega)$ for the field amplitude and a round-trip amplitude transmission modulation function of $T(t)$ for the loss cell.

The elementary theory of mode locking described in the previous section must be modified to account for the different gain in a homogeneously broadened gain profile. Here, again, we assume a round-trip amplitude transmission function of

$$T(t) = \sigma_0 + 2\sigma_1 \cos(\Omega t) = \sigma_0 + 2\sigma_1 \cos\left(\frac{2\pi}{\tau} t\right) \tag{6.7-1}$$

for the loss cell. Where we recall that σ_0 and σ_1 are constants, τ is the round-trip flight time inside the resonator, and Ω is the mode spacing in angular frequency. For simplicity, we assume that σ_0 and σ_1 are real constants. The constant σ_1 is the so-called modulation depth. Let the round-trip amplitude gain be written

$$g_n = g(\omega_0 + n\Omega) \tag{6.7-2}$$

for the nth mode, where ω_0 is a constant reference frequency. Let the field amplitude at an arbitrary reference point inside the resonator be written

$$E_{in}(t) = \sum_{n=1}^{N} C_n e^{in\Omega t} \tag{6.7-3}$$

where C_n is the amplitude of the nth mode and N is the total number of modes involved. Here again, we ignore the constant reference frequency ω_0. The field amplitude after one round-trip inside the resonator can be written

$$E_{out}(t) = g(\omega)T(t)E_{in}(t) = \sum_{n=1}^{N} C'_n e^{in\Omega t} \tag{6.7-4}$$

where we have lumped the mirror reflectivity into the net gain $g(\omega)$, and C_n are constants.

Using Equations (6.7-1) and (6.7-2), we obtain

$$E_{out}(t) = g(\omega)T(t)E_{in}(t) = \sum_{n=1}^{N} (g_n\sigma_0 C_n + \sigma_1 g_{n+1} C_{n+1} + \sigma_1 g_{n-1} C_{n-1})\, e^{in\Omega t} \tag{6.7-5}$$

We notice that, again, each mode is coupled with its neighboring modes, when the modulation frequency is equal to the mode spacing of the resonator. As the wave bounces back and forth inside the resonator, all the modes are coupled eventually. Thus the time modulation creates a mechanism that couples all the modes. As a result of the coupling, energy exchange occurs among the modes. We will now show that the coupling leads to a constant phase relationship between the modes. This is the origin of mode locking.

If we represent the field amplitude in terms of a column vector of the constants C_n, then the input–output relationship can be written

$$\begin{bmatrix} \sigma_0 & \sigma_1 & 0 & \cdots & 0 \\ \sigma_1 & \sigma_0 & \sigma_1 & \cdots & 0 \\ 0 & \sigma_1 & \sigma_0 & \cdots & 0 \\ \vdots & \vdots & \vdots & \vdots & \vdots \\ 0 & 0 & 0 & \cdots & \sigma_0 \end{bmatrix} \begin{bmatrix} g_1 & 0 & 0 & \cdots & 0 \\ 0 & g_2 & 0 & \cdots & 0 \\ 0 & 0 & g_2 & \cdots & 0 \\ \vdots & \vdots & \vdots & \vdots & \vdots \\ 0 & 0 & 0 & \cdots & g_N \end{bmatrix} \begin{bmatrix} C_1 \\ C_2 \\ C_3 \\ \vdots \\ C_N \end{bmatrix} = \begin{bmatrix} C_1' \\ C_2' \\ C_3' \\ \vdots \\ C_N' \end{bmatrix} \tag{6.7-6}$$

Here the second diagonal matrix accounts for the different gains for each of the modes in a homogeneously broadened medium. Multiplying the two matrices, we obtain

$$\begin{bmatrix} g_1\sigma_0 & g_2\sigma_1 & 0 & \cdots & 0 \\ g_1\sigma_1 & g_2\sigma_0 & g_3\sigma_1 & \cdots & 0 \\ 0 & g_2\sigma_1 & g_3\sigma_0 & \cdots & 0 \\ \vdots & \vdots & \vdots & \vdots & \vdots \\ 0 & 0 & 0 & \cdots & g_N\sigma_0 \end{bmatrix} \begin{bmatrix} C_1 \\ C_2 \\ C_3 \\ \vdots \\ C_N \end{bmatrix} = \begin{bmatrix} C_1' \\ C_2' \\ C_3' \\ \vdots \\ C_N' \end{bmatrix} \tag{6.7-7}$$

At steady state, the field must reproduce itself after one round-trip. Thus the output field is proportional to the input field. We can write

$$E_{\text{out}}(t) = \eta E_{\text{in}}(t) \tag{6.7-8}$$

where η is a constant. Physically, η is the net round-trip amplitude transmission coefficient of the combination of the gain medium, mirrors, and the loss cell in the resonator. Thus the matrix equation can now be written

$$\begin{bmatrix} g_1\sigma_0 & g_2\sigma_1 & 0 & \cdots & 0 \\ g_1\sigma_1 & g_2\sigma_0 & g_3\sigma_1 & \cdots & 0 \\ 0 & g_2\sigma_1 & g_3\sigma_0 & \cdots & 0 \\ \vdots & \vdots & \vdots & \vdots & \vdots \\ 0 & 0 & 0 & \cdots & g_N\sigma_0 \end{bmatrix} \begin{bmatrix} C_1 \\ C_2 \\ C_3 \\ \vdots \\ C_N \end{bmatrix} = \eta \begin{bmatrix} C_1 \\ C_2 \\ C_3 \\ \vdots \\ C_N \end{bmatrix} \tag{6.7-9}$$

We note that the steady-state oscillation is an eigenvector of the matrix equation and η is an eigenvalue of the matrix equation. The eigenvectors represent a superposition of the N modes in the resonator and can be viewed as the supermodes. This matrix problem is equivalent to the following recurrence equation

$$g_n\sigma_0 C_n + g_{n+1}\sigma_1 C_{n+1} + g_{n-1}\sigma_1 C_{n-1} = \eta C_n \tag{6.7-10}$$

or equivalently,

$$g_{n+1}\sigma_1 C_{n+1} + (g_n\sigma_0 - \eta)C_n + g_{n-1}\sigma_1 C_{n-1} = 0 \tag{6.7-11}$$

subject to the boundary condition

$$C_0 = C_{N+1} = 0 \tag{6.7-12}$$

Closed-form solutions of Equation (6.7-10) subject to the preceding boundary condition are, in general, not easy to obtain. Since the matrix in Equation (6.7-9) is a product of a real symmetric matrix and a real diagonal matrix, all eigenvalues and eigenvectors are real. It can

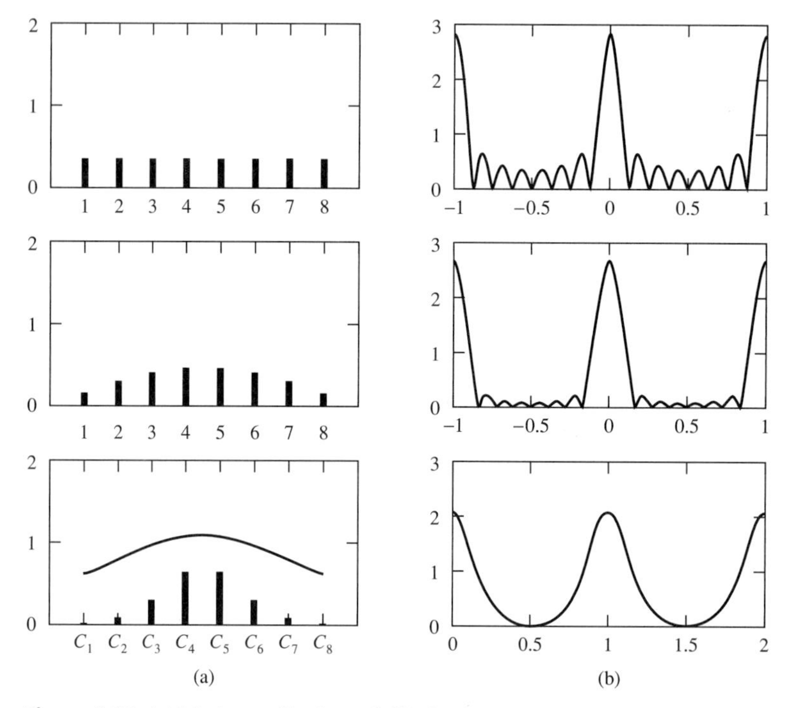

Figure 6.22 (a) Mode amplitudes and (b) absolute values of the field of the fundamental supermode of mode-locked laser pulses. The top row if for the case when all modes have the same amplitude. The middle row is for the case of inhomogeneous broadening with $C_n \propto \sin[n\pi/(N+1)]$. The bottom row is for the case of homogeneous broadening. The curve in the lower left box shows the gain profile of the homogeneous broadening.

be shown that the components of the eigenvector corresponding to the largest eigenvalue (the fundamental supermode) are all phase locked with the same phase. A simple computer program can be employed to obtain all the supermodes and their corresponding eigenvalues. As a result of the homogeneous broadening of the gain profile, the mode amplitudes C_n are further tapered (see Figure 6.22). This leads to a broadening of the main peak and the suppression of the sidelobes. Figure 6.22 shows a comparison of the three different mode-locked pulses and the corresponding mode amplitudes for $N = 8$.

The pulse width of the mode-locked laser with a homogeneous broadening can be estimated as follows. According to Equation (6.7-5), we can write the nth Fourier component of the pulse as

$$C'_n = g_n\sigma_0 C_n + g_{n+1}\sigma_1 C_{n+1} + g_{n-1}\sigma_1 C_{n-1} \tag{6.7-13}$$

The round-trip amplitude change can thus be written

$$\Delta C_n = C'_n - C_n = (g_n\sigma_0 + 2g_n\sigma_1 - 1)C_n + \sigma_1(g_{n+1}C_{n+1} - 2g_nC_n + g_{n-1}C_{n-1}) \tag{6.7-14}$$

where we recall that g_n is the round-trip gain for the nth frequency component. We consider the case of a Lorentzian gain profile,

$$g_n = 1 + \frac{g_0}{1 + \left(\dfrac{n\Omega}{\Omega_g}\right)^2} - L = 1 + g_0 - L - g_0\left(\frac{n\Omega}{\Omega_g}\right)^2 \tag{6.7-15}$$

where g_0 is a constant (the peak round-trip fractional gain), L is the round-trip fractional loss, and Ω_g is the gain bandwidth. Equation (6.7-14) can be transformed into a differential equation by introducing three approximations [25]: (1) the frequency dependence of the gain profile can be expanded to second order in $(\omega - \omega_0) = n\Omega$; (2) the discrete frequency spectrum with Fourier component at $n\Omega$ is replaced with a continuum spectrum, a function of $(\omega - \omega_0) = n\Omega$; and (3) the sum $(g_{n+1}C_{n+1} - 2g_nC_n + g_{n-1}C_{n-1})$ can be replaced by a second-order derivative with respective to frequency if the spectrum is very dense (usually thousands of modes are involved in mode locking). Thus we obtain

$$\Delta C(\omega) = (g_0 - L)C - g_0\left(\frac{\omega - \omega_0}{\Omega_g}\right)^2 C + \sigma_1\Omega^2 \frac{d^2 gC}{d\omega^2} \tag{6.7-16}$$

where we assume that $\sigma_0 + 2\sigma_1 = 1$ (which corresponds to a peak transmission of 100% for the loss cell, according to Equation (6.7-1)), and Ω is the modulation frequency as well as the intermode frequency spacing. In the steady state, the round-trip change of the amplitude is zero. Hence the mode-locked pulse must be a solution of the differential equation

$$(g_0 - L)C - g_0\left(\frac{\omega - \omega_0}{\Omega_g}\right)^2 C + \sigma_1\Omega^2 \frac{d^2 gC}{d\omega^2} = 0 \tag{6.7-17}$$

where C is now the Fourier transform of the mode-locked field $E_{in}(t)$ (see Equation (6.7-3)).

The solution of Equation (6.7-17) is a Gaussian pulse (see Problem 6.16)

$$C(\omega - \omega_0) = C_0 \exp[-(\omega - \omega_0)^2\tau^2] \tag{6.7-18}$$

where the constant τ (pulse width) must satisfy the following equations:

$$\left(1 + \frac{2g_0}{\Omega_g^2\tau^2}\right)\tau^4 = \frac{g_0}{4\sigma_1(\Omega\Omega_g)^2} \quad \text{and} \quad g_0 - L = 2\sigma_1\Omega^2\tau^2\left(1 + \frac{g_0}{\Omega_g^2\tau^2}\right) \tag{6.7-19}$$

For mode-locked lasers with $g_0 \ll \Omega_g^2\tau^2$, the above equation becomes

$$\tau^4 = \frac{g_0}{4\sigma_1(\Omega\Omega_g)^2} \quad \text{and} \quad g_0 - L = 2\sigma_1\Omega^2\tau^2 \tag{6.7-20}$$

The first equation in (6.7-20) is known as the Kuizenga–Siegman formula [26] for the pulse width, which is inversely proportional to the geometric mean of the gain bandwidth and the modulation frequency. The second equation gives an expression for the round-trip net gain

$$g_0 - L = 2\sigma_1\Omega^2\tau^2 \tag{6.7-21}$$

The peak gain is greater than the loss. This is legitimate and does not cause instabilities, since the modulation leads to an increase of loss in the pulse wings. The temporal pulse profile $E(t)$ is thus given by

$$E(t) = \int_{-\infty}^{\infty} C(\omega)e^{i\omega t}d\omega = \frac{C_0}{\tau}\sqrt{\pi}\,e^{i\omega_0 t}\exp\left(-\frac{t^2}{4\tau^2}\right) \tag{6.7-22}$$

The full width at half-maximum of the intensity profile is thus given by

$$\tau_p = 2\tau\sqrt{2\ln 2} = 2\sqrt{\ln 2}\left(\frac{g_0}{\sigma_1}\right)^{1/4}\frac{1}{\sqrt{\Omega\Omega_g}} \tag{6.7-23}$$

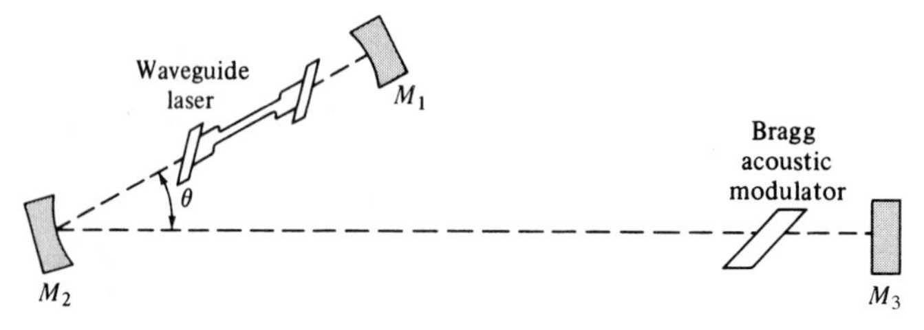

Figure 6.23 Schematic drawing of the mode-locking experiment in a high-pressure CO_2 laser. (After Reference [27].)

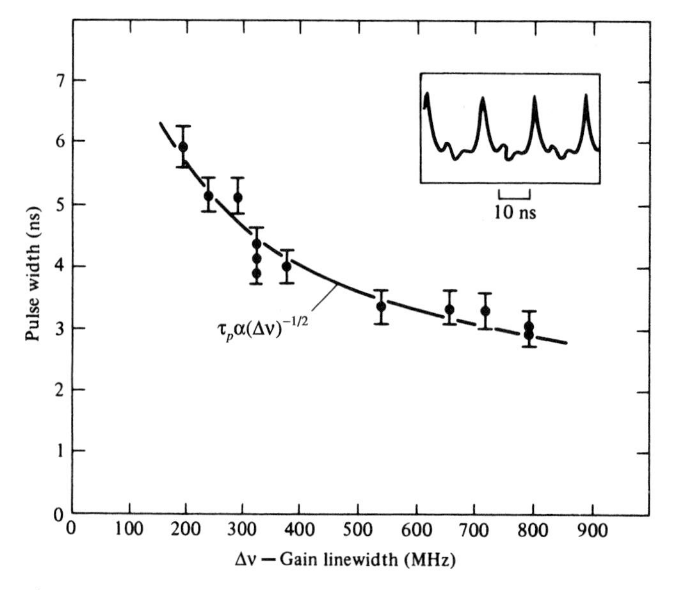

Figure 6.24 The dependence of the pulse width on the gain linewidth, $\Delta\nu = \Omega_g/2\pi$, that is controlled by varying the pressure ($\Delta\nu = 8 \times 10^8$ at 150 torr). (After Reference [27].)

where we recall that Ω is the angular frequency spacing between the longitudinal modes and Ω_g is the gain bandwidth.

An experimental setup demonstrating mode locking in a pressure-broadened CO_2 laser is sketched in Figure 6.23. The inverse square root dependence of τ_p on $\Delta\nu$ is displayed by the data of Figure 6.24, while the dependence on the modulation parameter σ_1 is shown in Figure 6.25.

Mode Locking by Phase Modulation

Mode locking can be induced by internal phase, rather than loss, modulation. This is usually done by using an electro-optic crystal inside the resonator oriented in the appropriate orientation such that the passing wave undergoes a phase delay proportional to the instantaneous electric field in the crystal. The frequency of the modulating signal is equal, as in the loss modulation case, to the inverse of the round-trip group delay time, that is, to the longitudinal intermode frequency separation.

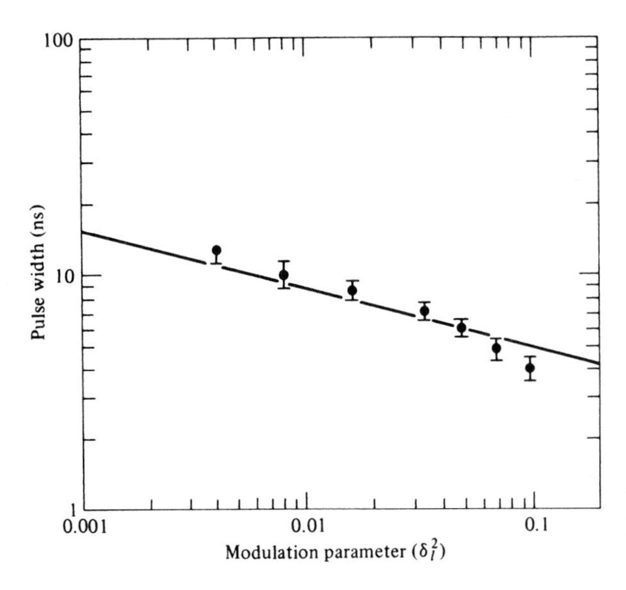

Figure 6.25 The mode-locked pulse width as a function of the modulation parameter, $\delta_l^2 = 2\sigma_1$. (After Reference [27].)

We consider a transmission function of a phase modulator as

$$T(t) = \exp(-i\delta \sin \Omega t) = \exp(-i\delta \sin 2\pi\, \Delta\nu_{\text{axial}}\, t) \tag{6.7-24}$$

where δ is a constant, known as the depth of modulation. Such an amplitude transmission function can be written in terms of its Fourier components as

$$T(t) = \exp(-i\delta \sin \Omega t) = \sum_{n=-\infty}^{+\infty} J_n(\delta)\, e^{-in\Omega t} \tag{6.7-25}$$

where $J_n(\delta)$ is the Bessel function. We note that a phase modulator will generate an infinite series of sidebands. These sidebands coincide with the modes of the resonator. Thus a periodically modulated phase modulator will couple all the modes of the resonator provided that the modulation frequency equals the longitudinal mode spacing.

Following the matrix analysis described earlier, we can obtain the supermodes of oscillation of the mode-locked laser by solving the following eigenvalue problem:

$$g\begin{bmatrix} J_0(\delta) & J_1(\delta) & J_2(\delta) & \cdots & J_{N-1}(\delta) \\ J_{-1}(\delta) & J_0(\delta) & J_1(\delta) & \cdots & J_{N-2}(\delta) \\ J_{-2}(\delta) & J_{-1}(\delta) & J_0(\delta) & \cdots & J_{N-3}(\delta) \\ \vdots & \vdots & \vdots & \vdots & \vdots \\ J_{-N+1}(\delta) & J_{-N+2}(\delta) & J_{-N+3}(\delta) & \dots & J_0(\delta) \end{bmatrix}\begin{bmatrix} C_1 \\ C_2 \\ C_3 \\ \vdots \\ C_N \end{bmatrix} = \eta \begin{bmatrix} C_1 \\ C_2 \\ C_3 \\ \vdots \\ C_N \end{bmatrix} \tag{6.7-26}$$

where g is the round-trip net gain. The mode amplitudes C_n can be obtained by solving the eigenvalue problem. As a result of the antisymmetric property of the matrix, the eigenvectors and eigenvalues are in general complex. Although the mode amplitudes are locked, the complex amplitudes lead to "fixed" phase delays among the modes. The phase delays lead to a mode-locked pulse with a frequency chirping.

Passive Mode Locking Via Saturable Absorber

As mentioned in Section 6.6, a saturable absorber exhibits a low absorption for optical beams with high intensities, and a high absorption for optical beams with low intensities. As a result, such an absorber will favor the transmission of short optical pulses with high peak intensities in the laser resonator. In what follows, we describe an elementary theory of passive mode locking by assuming a fast response of the saturable absorber [25]. In passive mode locking, the loss modulator is replaced by a saturable absorber. The loss modulation of the saturable absorber $s(t)$ in transmission through the absorber is given by

$$s(t) = \frac{s_0}{1 + I(t)/I_{sat}} \tag{6.7-27}$$

where s_0 is the unsaturated loss (i.e., loss at low intensities), $I(t)$ is the instantaneous intensity of the pulse, and I_{sat} is the saturation intensity. If the saturation is relatively weak, Equation (6.7-27) can be written

$$s(t) = s_0 - s_0 I(t)/I_{sat} \tag{6.7-28}$$

If we normalize the mode amplitude so that $|E(t)|^2 = \text{Power}$, then the transmission through the absorber can be written

$$T(t) = 1 - s(t) = 1 - s_0 + s_0 \frac{|E(t)|^2}{I_{sat} A_{eff}} = 1 - s_0 + \gamma |E(t)|^2 \tag{6.7-29}$$

where A_{eff} is the effective area of the mode and γ is the self amplitude modulation (SAM) coefficient. The self amplitude modulation is a direct result of the saturation behavior of the absorption. According to Equation (6.7-29), the transmission is an increasing function of the field intensity. The round-trip amplitude change in the case of passive mode locking can now be obtained by modifying Equation (6.7-16). Using the Fourier transform relationship (6.7-22), we can transform Equation (6.7-16) to the time domain. This leads to

$$\Delta E(t) = (g_0 - L)E(t) + \frac{g_0}{\Omega_g^2} \frac{\partial^2}{\partial t^2} E(t) - \sigma_1 \Omega^2 t^2 \int gCe^{i\omega t}\, d\omega \tag{6.7-30}$$

We replace the active mode-locking term $\sigma_1 \Omega^2 t^2 \int gCe^{i\omega t}\, d\omega$ with the self amplitude modulation (there's no active modulation). This leads to

$$\Delta E(t) = (g_0 - L)E(t) + \frac{g_0}{\Omega_g^2} \frac{\partial^2}{\partial t^2} E(t) + \gamma |E(t)|^2 E(t) \tag{6.7-31}$$

At steady state, $\Delta E(t) = 0$, the solution is a simple hyperbolic secant (see Problem 6.17):

$$E(t) = E_0 \operatorname{sech}(t/\tau)$$

where E_0 and τ are constants, with

$$\tau^2 = \frac{2g_0}{\gamma E_0^2 \Omega_g^2} \tag{6.7-32}$$

and

$$L - g_0 = \frac{g_0}{\Omega_g^2 \tau^2} \tag{6.7-33}$$

The full width at half-maximum of the intensity profile is thus given by

$$\tau_p = 1.67\tau = 1.67 \sqrt{\frac{2g_0}{\gamma}} \frac{1}{E_0 \Omega_g} \tag{6.7-34}$$

It is interesting to note that expression (6.7-32) for the pulse width of passive mode locking is related to that of active mode locking (6.7-19). In active mode locking, $\sigma_1\Omega^2$ is proportional to the curvature of the amplitude modulation (second-order derivative of transmission). In the case of self amplitude modulation in passive mode locking, the curvature is proportional to $\gamma E_0^2/\tau^2$. Thus Equations (6.7-32) and (6.7-19) are indeed related. This comparison also explains why passive mode locking can result in much shorter pulse widths for the same gain bandwidth Ω_g. As the pulse gets shorter, the curvature of self-modulation increases as $\gamma E_0^2/\tau^2$, whereas it remains unchanged ($\sigma_1\Omega^2$) for active mode locking. In the resonator, the net gain is negative preceding and following the pulse (i.e., at low intensities, according to Equation (6.7-33)). At the peak of the pulse, the net gain is positive due to bleaching of the saturable absorber.

6.8 PULSE LENGTH MEASUREMENT AND NARROWING OF CHIRPED PULSES

The problem of measuring the duration of mode-locked ultrashort pulses is of great practical and theoretical interest. Since the fastest conventional optical detectors possess response times of ~2×10^{-11} s, it is impossible to use these optical detectors to measure directly the short ($\tau < 10^{-11}$ s) mode-locked pulses. A number of techniques invented for this purpose all take advantage of some nonlinear process to obtain a spatial autocorrelation trace of the optical intensity pulse. The measurement of a pulse of duration, say, $\tau_0 = 10^{-12}$ s is thus replaced with measuring the spatial extent of an autocorrelation trace of length $c\tau_0 = 0.3$ mm, which is a relatively simple task.

In what follows we will describe one such method, the one most widely used, that is based on the phenomenon of optical second harmonic generation. The process of second harmonic generation is developed in detail in Chapter 8. It will suffice for the purpose of the present discussion to state that when an optical pulse

$$e_1(t) = \text{Re}[\mathscr{E}_1(t)e^{i\omega t}] \tag{6.8-1}$$

is incident on a nonlinear optical crystal, it generates an output optical pulse $e_2(t)$ at twice the frequency with

$$e_2(t) = \text{Re}[\mathscr{E}_2(t)e^{2i\omega t}] \propto \text{Re}[\mathscr{E}_1^2(t)e^{2i\omega t}] \tag{6.8-2}$$

A sketch of a second harmonic system for measuring the pulse length is shown in Figure 6.26. The laser emits a continuous stream of mode-locked pulses. Each individual pulse $\mathscr{E}(t)e^{i\omega t}$ is divided by a beam splitter into two equal intensity pulses. One of these pulses is advanced (or delayed) by τ seconds relative to the other. The two pulses recombine again in a nonlinear optical crystal. The second harmonic (2ω) pulse generated by the crystal is incident on a "slow" detector whose output current is integrated over a time long compared to the optical pulse duration.

The total optical field incident on the nonlinear crystal is the sum of the direct and retarded fields:

$$\begin{aligned} e_{\text{tot}}(t) &= \text{Re}\{[\mathscr{E}_1(t) + \mathscr{E}_1(t-\tau)e^{-i\omega\tau}]e^{i\omega t}\} = \text{Re}[\mathscr{E}(t)e^{i\omega t}] \\ \mathscr{E}(t) &= \mathscr{E}_1(t) + \mathscr{E}_1(t-\tau)e^{-i\omega\tau} \end{aligned} \tag{6.8-3}$$

According to Equation (6.8-2), the second harmonic field radiated by the crystal has a complex amplitude that is proportional to the square of the complex amplitude $\mathscr{E}(t)$ of the incident fundamental field:

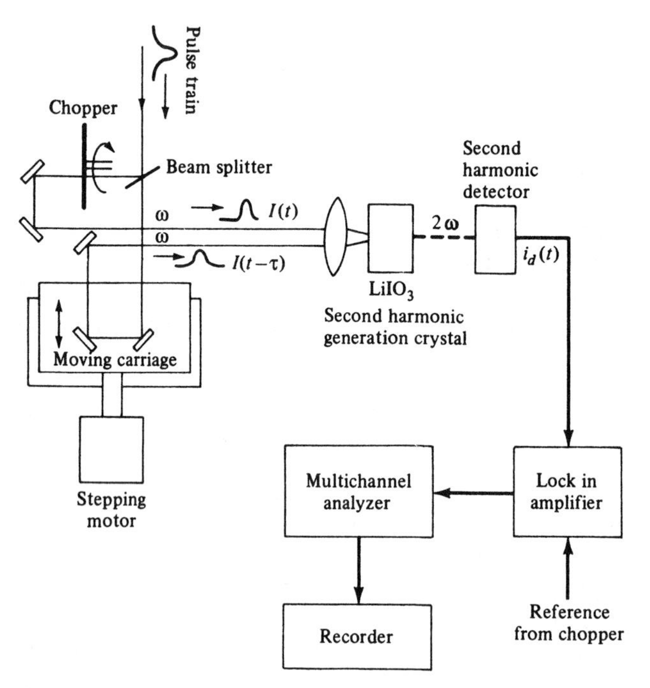

Figure 6.26 The second harmonic generation autocorrelation setup for measuring the width of mode-locked ultrashort pulses.

$$
\begin{aligned}
\mathscr{E}_2(t) &\propto [\mathscr{E}_1(t) + \mathscr{E}_1(t-\tau)e^{-i\omega\tau}]^2 \\
&= \mathscr{E}_1^2(t) + \mathscr{E}_1^2(t-\tau)e^{-2i\omega\tau} + 2\mathscr{E}_1(t)\mathscr{E}_1(t-\tau)e^{-i\omega\tau}
\end{aligned}
\tag{6.8-4}
$$

The second harmonic field, $e_2(t) = \mathrm{Re}[\mathscr{E}_2(t)\exp(i2\omega t)]$, is incident next on the optical detector (photomultiplier, diode, etc.) whose output current i_d (see Section 11.1) is proportional to the incident intensity. Using Equation (6.8-4), we can obtain

$$
\begin{aligned}
i_d(t) \propto \mathscr{E}_2(t)\mathscr{E}_2^*(t) &= [\mathscr{E}_1(t)\mathscr{E}_1^*(t)]^2 + [\mathscr{E}_1(t-\tau)\mathscr{E}_1^*(t-\tau)]^2 \\
&\quad + 4\mathscr{E}_1(t)\mathscr{E}_1^*(t)\mathscr{E}_1(t-\tau)\mathscr{E}_1^*(t-\tau) + s(\tau)
\end{aligned}
\tag{6.8-5}
$$

where $s(\tau)$ is composed of terms with $\cos\omega\tau$ and $\cos 2\omega\tau$ dependence. Since these terms fluctuate with a delay period $\Delta\tau \sim 10^{-15}$ s, a small unintentional or deliberate integration (smearing) over the delay τ averages them out to near zero. The term $s(\tau)$ is consequently left out.

Since the temporal (t) variation of the first three terms in Equation (6.8-5) is on the scale of picoseconds (or less), the much slower optical detector inevitably integrates the current $i_d(t)$, with the result that the actual output from the optical detector is a function of the delay (τ) only,

$$
i_d(\tau) \propto \langle I^2(t)\rangle + \langle I^2(t-\tau)\rangle + 4\langle I(t)I(t-\tau)\rangle \tag{6.8-6}
$$

where the angle brackets signify time averaging and the *intensity* $I(t)$ is defined[4] as $I(t) = \mathscr{E}_1(t)\mathscr{E}_1^*(t)$. By dividing both sides of Equation (6.8-6) by $\langle I^2(t)\rangle$ and recognizing that $\langle I^2(t)\rangle = \langle I^2(t-\tau)\rangle$, the normalized detector output becomes

[4] A proportionality constant involved in this definition is left out since it cancels out in the subsequent division of Equation (6.8-9).

$$i_d(\tau) = 1 + 2G^{(2)}(\tau) \tag{6.8-7}$$

where $G^{(2)}(\tau)$, the second-order autocorrelation function of the intensity pulse, is defined by

$$G^{(2)}(\tau) \equiv \frac{\langle I(t)I(t-\tau)\rangle}{\langle I^2(t)\rangle} \tag{6.8-8}$$

In the case of a well-behaved ultrashort coherent light pulse of duration τ_0, we have

$$i_d(0) = 3, \quad i_d(\tau \gg \tau_0) = 1$$

since $G^{(2)}(0) = 1$ and $G^{(2)}(\tau \gg \tau_0) = 0$.

A plot of $i_d(\tau)$ versus τ will consist of a peak of (normalized) height of 3 atop a background of unity height. The central peak will have a width $\sim\tau_0$.

It is important in practice to be able to distinguish between the case just discussed and that of incoherent light (such as light due to a laser oscillating in a large number of independent modes). In this case, we have $i_d(0) = 3$ (since even incoherent light is correlated with itself at zero delay). For $\tau > 0$, or more precisely for τ longer than the coherence time of the light, we have

$$G^{(2)}(\tau) = \frac{\langle I(t)I(t-\tau)\rangle}{\langle I^2(t)\rangle} = \frac{\langle I(t)\rangle^2}{\langle I^2(t)\rangle} \tag{6.8-9}$$

since $I(t)$ and $I(t-\tau)$ are completely uncorrelated. For truly incoherent light of the type we are considering here, the time averaging indicated in Equation (6.8-9) can be replaced by ensemble averaging so that

$$\langle I^2(t)\rangle = \int_0^\infty p(I)I^2\, dI \tag{6.8-10}$$

where $p(I)$ is the intensity distribution function so that $p(I)\, dI$ is the probability that a measurement of I will result in a value between I and $I + dI$. For incoherent light[5]

$$p(I) = \frac{1}{\langle I\rangle} e^{-I/\langle I\rangle}$$

which when used in Equation (6.8-10) gives

$$\langle I^2(t)\rangle = 2\langle I\rangle^2$$

and, returning to Equation (6.8-7),

$$i_d(\tau > 0) \propto 1 + 2\frac{\langle I(t)\rangle^2}{\langle I^2(t)\rangle} = 2$$

A plot of $i_d(\tau)$ versus τ in the case of incoherent light should thus consist of a very narrow peak of height 3 on a background of height 2. The general features of the coherent mode-locked pulses and the incoherent light are depicted in Figure 6.27.

The determination of the original pulse width from the width of the second harmonic correlation trace is somewhat ambiguous. We can show by performing the integration indicated by Equation (6.8-8) that the width (at half-maximum) t_0 of $G^{(2)}(\tau)$ and τ_0 of $I(t)$ are related as in the case of the "popular" waveforms tabulated in the left column of Table 6.2.

---

[5] This follows directly from the fact that the optical field distribution $p(e)$ for the field of an incoherent beam function is Gaussian.

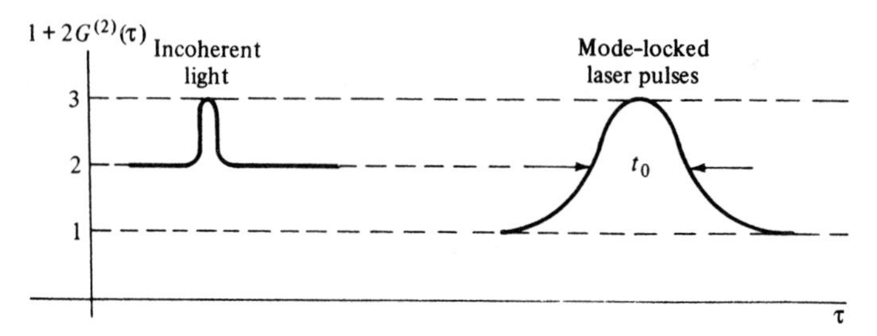

Figure 6.27 The second harmonic (averaged) integrated intensity due to two optical pulses as a function of time delay between them for the case (left) of an incoherent source and (right) coherent mode-locked optical pulses.

TABLE 6.2 Some Simple Pulse Widths

$I(t)$	t_0/τ_0
$1\ (0 \le t \le t_0)$, zero otherwise	1
$\exp\left(-\frac{(4\ln 2)t^2}{\tau_0^2}\right)$	$\sqrt{2}$
$\operatorname{sech}^2\left(\frac{1.76t}{\tau_0^2}\right)$	1.55
$\exp\left(-\frac{(\ln 2)t}{\tau_0}\right)\quad (t \ge 0)$	2

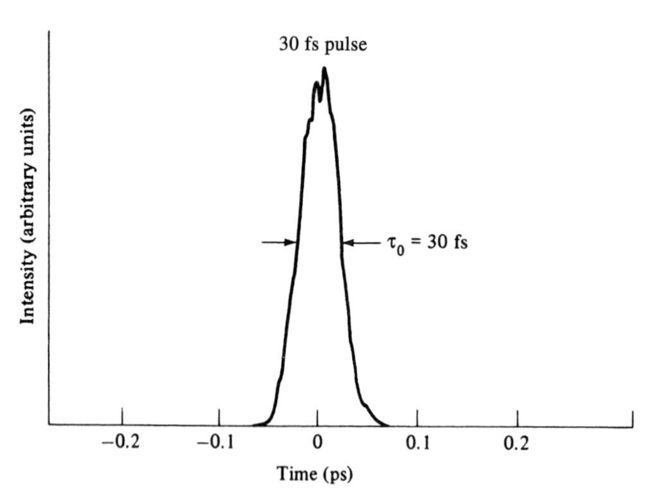

Figure 6.28 The autocorrelation trace of a mode-locked dye laser ($\lambda \sim 6100$ Å) pulse.

We conclude this section by showing in Figure 6.28 the autocorrelation trace of one of the shortest optical pulses ($\tau_0 \sim 30 \times 10^{-15}$ s) produced to date. It is interesting to note that within such a pulse the light rises and falls a mere 15(!) times.

With just one order of magnitude improvement we should thus be able to isolate a single optical cycle. There may, however, be some compelling reasons (and the student is encouraged to think of some) why this development may not take place anytime soon.

Pulse Narrowing by Chirping and Compression

Mode locking of a laser oscillator was shown in Section 6.6 to lead to a mode of oscillation consisting of a continuous train of ultrashort pulses. The width of each pulse is $\tau_p \sim 1/(\Delta\omega)_{\text{gain}}$, where $(\Delta\omega)_{\text{gain}}$ is the spectral width over which the gain is capable of sustaining oscillation. It is, however, possible to further reduce the pulse width beyond the limit $(\Delta\omega)^{-1}_{\text{gain}}$ by employing nonlinear optical techniques. To lend some plausibility to the new ideas that we will soon introduce, we remind ourselves of the exact formal analogy that exists between the diffractive propagation of an optical beam in free homogeneous space and that of an optical pulse propagation in a dispersive channel (fiber).

The propagation of an optical pulse, $f_1(t)\exp(i\omega_0 t)$, through a length L of a dispersive fiber can be described (see Equation (7.2-5)) by

$$\tilde{f}_2(\Omega) = \tilde{f}_1(\Omega)\exp\left(-i\frac{\beta''}{2}L\Omega^2\right) \tag{6.8-11}$$

where $\tilde{f}_1(\Omega)$ and $\tilde{f}_2(\Omega)$ are the Fourier transforms of the input and output pulse envelopes $f_1(t)$ and $f_2(t)$, respectively, and $\beta'' \equiv d^2\beta/d\omega^2$.

The propagation of a confined optical beam with a transverse profile $u_1(x)$ was found in Section 2.12 (see Equation (2.12-8)) to obey

$$\tilde{u}_2(K) = \tilde{u}_1(K)\exp\left(i\frac{LK^2}{2k}\right) \tag{6.8-12}$$

The exact formal analogy of Equations (6.8-11) and (6.8-12) suggests that is should be possible to construct a temporal analog of the familiar spatial optical demagnification by a lens as illustrated in Figure 6.29. Such a system can be used to narrow optical pulses. All that is needed is a time lens, a device that multiplies an incoming optical field by $\exp(iat^2)$, which is the exact analog of the factor $\exp(ikx^2/2f)$ by which the spatial lens multiplies an incoming beam.

Our temporal pulse narrower is illustrated in Figure 6.29a. The input pulse envelope, before narrowing, is taken as a Gaussian

$$f_1(t) = \exp(-t^2/\tau_p^2) \tag{6.8-13}$$

with a (FWHM) intensity width $\Delta\tau = \sqrt{2\ln 2}\,\tau_p$. Passage through the time lens multiplies $f_1(t)$ by a phase factor $\exp(iAt^2/\tau_p^2)$:

$$f_2(t) = \exp[-(1-iA)(t/\tau_p)^2] \tag{6.8-14}$$

The total field at plane 2 is

$$E_2(t) = f_2(t)\exp(i\omega_0 t) = \exp[-(t/\tau_p)^2]\exp i\left[\omega_0 t + A\left(\frac{t}{\tau_p}\right)^2\right] \tag{6.8-15}$$

If we write $E_2(t) = \exp[-(t/\tau_p)^2 + i\phi(t)]$, the instantaneous frequency is $\omega(t) = d\phi/dt$

$$\omega(t) = \frac{d}{dt}\left[\omega_0 t + A\left(\frac{t}{\tau_p}\right)^2\right] = \omega_0 + 2A\frac{t}{\tau_p^2} \tag{6.8-16}$$

corresponding to a linear "chirp." The pulse width, however, is unaffected by the lens. The lens, however, modifies the spectrum of the pulse. To find the spectrum we take the Fourier transform of $f_2(t)$:

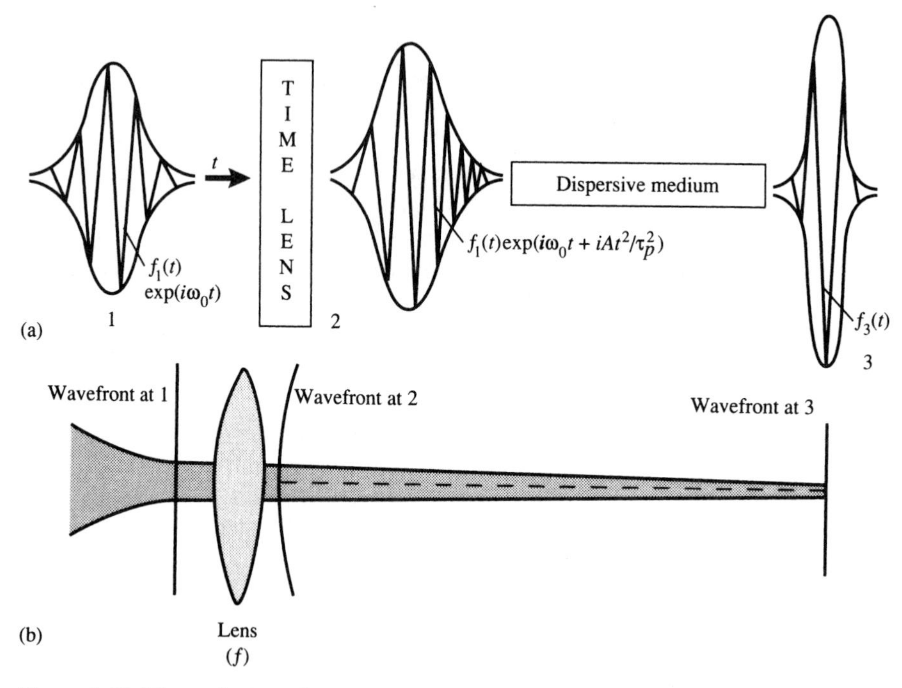

Figure 6.29 The equivalence between temporal focusing (pulse narrowing) in (a) and spatial focusing by a conventional lens in (b). Note that a chirped optical pulse corresponds to a Gaussian spatial beam with a curved wavefront.

$$\tilde{f}_2(\Omega) = \sqrt{\frac{\tau_p^2}{4\pi(1 - iA)}} \exp\left(-\frac{1 + iA}{4(1 + A^2)}(\Omega\tau_p)^2\right) \tag{6.8-17}$$

The effect of the chirp is thus to increase the spectral width (FWHM) by the factor $\sqrt{1 + A^2}$

TIME LENSES

The fact that the time lens has increased the spectral width while preserving the total pulse energy implies redistribution of power among the spectral components. This, by definition, cannot be accomplished by a linear passive device. We will, indeed, find that our practical candidates for time lenses all employ nonlinear techniques. One such method employs self phase modulation in a medium with a large Kerr coefficient, that is, a medium where the index of refraction depends on the light intensity, $n = n_0 + n_2 I$ (Kerr effect). I = intensity (W/m^2), and n_2 a constant of the material. Consider, for example, a Gaussian intensity pulse,

$$I = I_0 \exp(-2\alpha t^2)$$

entering a nonlinear medium characterized by n_2 so that the index of refraction is $n = n_0 + n_2 I(t)$

$$E_{\text{in}} = E_0 \exp(-\alpha t^2)$$

$$I_{\text{in}} = I_0 \exp(-2\alpha t^2)$$

The pulse will emerge at $z = L$ with a phase delay factor,

$$\exp\left[i\frac{\omega}{c}(n_0 + n_2 I(t))L\right]$$

If we expand the Gaussian pulse as $I(t) = I_0(1 - 2\alpha t^2 + \ldots)$ and keep only the first two terms, we find that the delay factor becomes

$$E_{\text{out}} = E_{\text{in}} \times \text{delay factor}$$
$$= E_0 \exp(-\alpha t^2) \exp\left(i\frac{2\omega n_2 \alpha I_0 L}{c} t^2\right) \tag{6.8-18}$$

and has, according to Equation (6.8-15), the requisite chirp to serve as a time lens. We left out an inconsequential fixed phase and group delay.

The pulse $f_2(t)$ is thus no longer transform-limited since its time–bandwidth product

$$\Delta\nu\,\Delta\tau = \frac{2\ln 2}{\pi}(1 + A^2)^{1/2} = 0.4413(1 + A^2)^{1/2} \tag{6.8-19}$$

exceeds the minimum value for a Gaussian pulse by $(1 + A^2)^{1/2}$. This excess bandwidth should make it possible, in principle, to reduce the pulse width, employing passive means, by the factor $(1 + A^2)^{1/2}$, which will result, again, in a compressed transform-limited pulse.

In the spatial example of Figure 6.29a, the narrowing of the spatially chirped pulse is accomplished by propagating it through the appropriate distance L in space. The temporal equivalent is to propagate the chirped pulse through a dispersive device with transfer characteristics of the form (6.8-11). This results in

$$\tilde{f}_3(\Omega) = \tilde{f}_2(\Omega)\exp\left(-i\frac{\beta''}{2}L\Omega^2\right)$$
$$= \sqrt{\frac{\tau_p^2}{4\pi(1 - iA)}}\exp\left[-\frac{\Omega^2\tau_p^2}{4(1 + A^2)} - i\left(\frac{A\tau_p^2}{4(1 + A^2)} + \frac{\beta'' L}{2}\right)\Omega^2\right] \tag{6.8-20}$$

where L is the length of the dispersive channel. If the condition

$$\beta'' L = -\frac{A\tau_p^2}{2(1 + A^2)} \tag{6.8-21}$$

is satisfied, the output is given by

$$\tilde{f}_3(\Omega) = \sqrt{\frac{\tau_p^2}{4\pi(1 - iA)}}\exp\left(-\frac{(\Omega\tau_p)^2}{4(1 + A^2)}\right) \tag{6.8-22}$$

which corresponds in the time domain to

$$f_3(t) = F^{-1}\{\tilde{f}_3(\Omega)\} = \sqrt{1 + iA}\exp\left(-(1 + A^2)\frac{t^2}{\tau_p^2}\right) \tag{6.8-23}$$

The width of the output is thus compressed to

$$(\Delta\tau)_{\text{comp}} = \frac{\Delta\tau}{\sqrt{1 + A^2}} \tag{6.8-24}$$

For $A \gg 1$, the compression ratio is $\sim A$. This compression ratio is thus exactly the factor by which the spectral width is increased by the chirping. A simple integration shows that

$$\int_{-\infty}^{\infty} |f_1(t)|^2\,dt = \int_{-\infty}^{\infty} |f_2(t)|^2\,dt = \int_{-\infty}^{\infty} |f_3(t)|^2\,dt \tag{6.8-25}$$

so that the pulse energy is conserved.

The Grating Pair Compressor

In the above example, a fiber with group velocity dispersion ($\beta'' \neq 0$) was employed in order to compress the chirped pulse. The essential feature of the fiber was its transfer function, given by Equation (6.8-20) as

$$\text{Transfer function of fiber (length } L) = \exp\left(-\frac{i\beta''}{2} L\Omega^2\right)$$

Any other device with a transfer function of the form $\exp(ib\Omega^2)$, where b is some real constant, can thus serve as compressor provided b can be adjusted in magnitude as well as sign. A commonly used pulse-narrowing configuration is the dual grating telescope compressor [16, 28, 29] illustrated in Figure 6.30. It is based on the fact that different Fourier components (Ω) of the incident optical beam are diffracted by the gratings along different directions, thus following paths of different length between the input and output, accumulating in the process a differential phase $b\Omega^2$. (See Problem 6.19)

Figure 6.31 shows intensity profiles, obtained by second harmonic autocorrelation, of a chirped and a compressed pulse in the experiment depicted by Figure 6.30. The input pulse is compressed from an initial width of 5.2 ps to 0.32 ps. This corresponds, according to Equation (6.8-24), to a compression ratio $A \sim 16$. The $\Delta\nu(\Delta\tau)_{\text{comp}}$ product is $\sim$0.45, close to the theoretical limit of 0.44, which indicates that the pulse is as narrow as allowed by its spectral content; that is, it is transform-limited.

A semiconductor laser excited by current pulses or pulsating due to mode locking is naturally chirped. This is due to carrier density—hence, index of refraction transients in the active region. This makes the output pulses of such lasers natural candidates for compression. Chirping in semiconductor lasers is treated in Chapter 15.

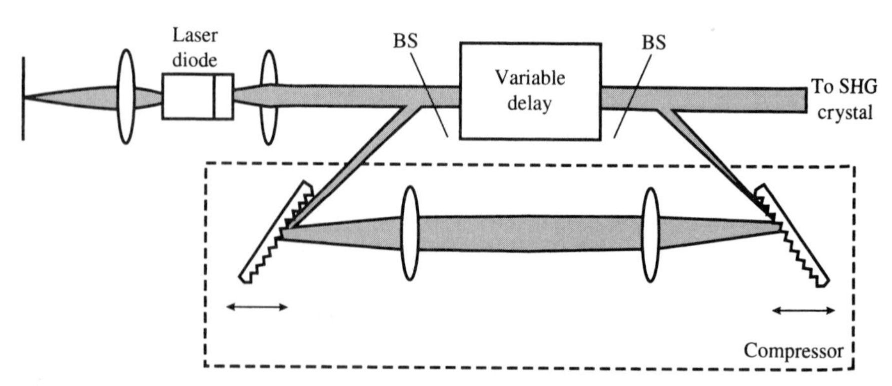

Figure 6.30 A dual grating telescope pulse compressor. (After Reference [43].)

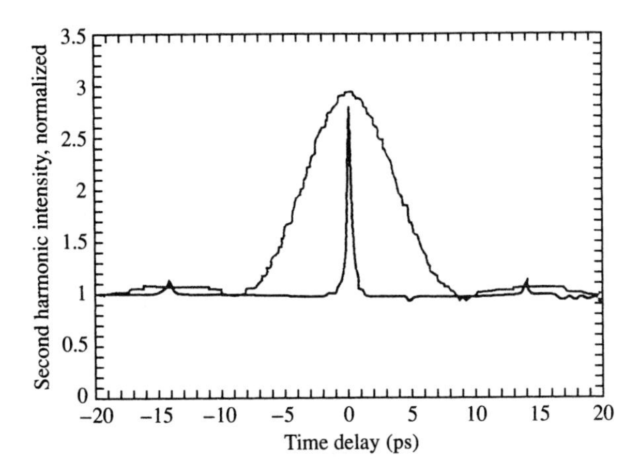

Figure 6.31 Intensity autocorrelation of uncompressed and compressed pulses. The compression ratio is ~A = 16. (After Reference [16].)

6.9 GIANT PULSE (*Q*-SWITCHED) LASERS

The "*Q*-switching" technique [30] is used to obtain intense and short bursts of oscillation from lasers; see References [31–33]. The quality factor Q of the optical resonator is degraded (lowered) by some means during the pumping so that the gain (i.e., inversion $N_2 - N_1$) can build up to a very high value without oscillation. (The spoiling of the Q raises the threshold inversion to a value higher than that obtained by pumping.) When the inversion reaches its peak, the Q is restored abruptly to its (ordinary) high value. The gain (per pass) in the laser medium is now well above threshold. This causes an extremely rapid buildup of the oscillation and a simultaneous exhaustion of the inversion by stimulated 2→1 transitions. This process converts most of the energy that was stored by atoms pumped into the upper laser level into photons, which are now inside the optical resonator. These proceed to bounce back and forth between the reflectors with a fraction $(1 - R)$ "escaping" from the resonator each time. This causes a decay of the pulse with a characteristic time constant (the "photon lifetime") given in Equation (4.7-3) as

$$t_c \simeq \frac{n_0 l}{c(1-R)}$$

where n_0 is the index of refraction of the medium.

Both experiment and theory indicate that the total evolution of a giant laser pulse as described above is typically completed in ~2×10^{-8} s. We will consequently neglect the effects of population relaxation and pumping that take place during the pulse. We will also assume that the switching of the Q from the low to the high value is accomplished instantaneously.

The laser is characterized by the following variables: ϕ—the total number of photons in the optical resonator; $n \equiv (N_2 - N_1)V$—the total inversion; and t_c—the decay time constant for photons in the *passive* resonator. The exponential gain constant γ is proportional to n. The radiation intensity I thus grows with distance as $I(z) = I_0 \exp(\gamma z)$ and $dI/dz = \gamma I$. An observer traveling with the wave velocity will see it grow at a rate

$$\frac{dI}{dt} = \frac{dI}{dz}\frac{dz}{dt} = \gamma\left(\frac{c}{n_0}\right)I$$

and thus the temporal exponential growth constant is $\gamma(c/n_0)$. If the laser rod is of length L while the resonator length is l, then only a fraction L/l of the photons are undergoing amplification at any one time and the average growth constant is $\gamma c(L/n_0 l)$. We can thus write

$$\frac{d\phi}{dt} = \phi\left(\frac{\gamma c L}{n_0 l} - \frac{1}{t_c}\right) \tag{6.9-1}$$

where $-\phi/t_c$ is the decrease in the number of resonator photons per unit time due to incidental resonator losses and to the output coupling. Defining a dimensionless time by $\tau = t/t_c$, we obtain, upon multiplying Equation (6.9-1) by t_c,

$$\frac{d\phi}{d\tau} = \phi\left[\left(\frac{\gamma}{n_0 l/cLt_c}\right) - 1\right] = \phi\left(\frac{\gamma}{\gamma_t} - 1\right)$$

where $\gamma_t = (n_0 l/cLt_c)$ is the minimum value of the gain constant at which oscillation (i.e., $d\phi/d\tau = 0$) can be sustained. Since, according to Equation (5.6-19), γ is proportional to the inversion n, the last equation can also be written

$$\frac{d\phi}{d\tau} = \phi\left(\frac{n}{n_t} - 1\right) \tag{6.9-2}$$

where $n_t = N_t V$ is the total inversion at threshold as given by Equation (6.1-15).

The term $\phi(n/n_t)$ in Equation (6.9-2) gives the number of photons generated by induced emission per unit of normalized time. Since each generated photon results from a single transition, it corresponds to a decrease of $\Delta n = -2$ in the total inversion. We can thus write directly

$$\frac{dn}{d\tau} = -2\phi\frac{n}{n_t} \tag{6.9-3}$$

The coupled pair of equations, (6.9-2) and (6.9-3), describes the evolution of ϕ and n. It can be solved easily by numerical techniques. Before we proceed to give the results of such calculation, we will consider some of the consequences that can be deduced analytically.

Dividing Equation (6.9-2) by (6.9-3) results in

$$\frac{d\phi}{dn} = \frac{n_t}{2n} - \frac{1}{2}$$

and, by integration,

$$\phi - \phi_i = \frac{1}{2}\left[n_t \ln\frac{n}{n_i} - (n - n_i)\right]$$

where n_i is the initial population inversion and ϕ_i is the initial photon number. Assuming that ϕ_i, the initial number of photons in the cavity, is negligible, we obtain

$$\phi = \frac{1}{2}\left(n_t \ln\frac{n}{n_i} - (n - n_i)\right) \tag{6.9-4}$$

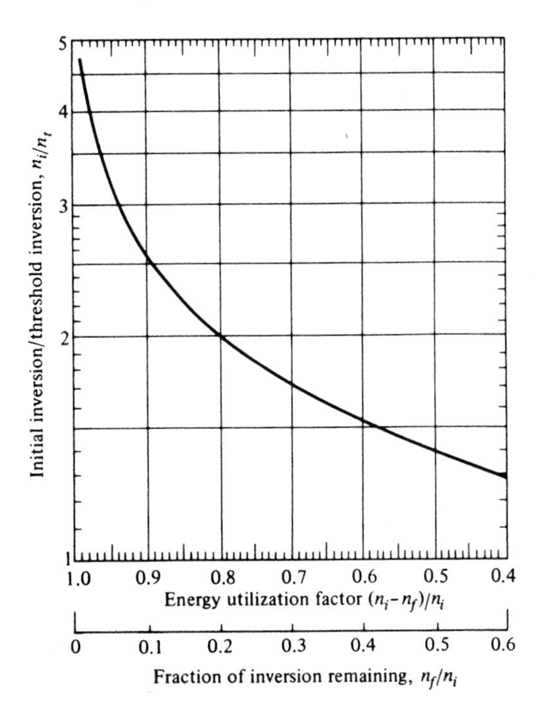

Figure 6.32 Energy utilization factor $(n_i - n_f)/n_i$ and inversion remaining after the giant pulse. (After Reference [30].)

for the relation between the number of photons ϕ and the inversion n at any moment. At $t \gg t_c$, the photon density ϕ will be zero so that setting $\phi = 0$ in Equation (6.9-4) results in the following expression for the final inversion n_f:

$$\frac{n_f}{n_i} = \exp\left(\frac{n_f - n_i}{n_t}\right) \tag{6.9-5}$$

This equation is of the form $(x/a) = \exp(x - a)$, where $x = n_f/n_t$ and $a = n_i/n_t$, so that it can be solved graphically (or numerically) for n_f/n_i as a function of n_i/n_t.[6] The result is shown in Figure 6.32. We notice that the fraction of the energy originally stored in the inversion that is converted into laser oscillation energy is $(n_i - n_f)/n_i$ and that it tends to unity as n_i/n_t increases.

The instantaneous power output of the laser is given by $P = \phi h\nu/t_c$ or, using Equation (6.9-4), by

$$P = \frac{h\nu}{2t_c}\left(n_t \ln\frac{n}{n_i} - (n - n_i)\right) \tag{6.9-6}$$

Of special interest to us is the peak power output. Setting $\partial P/\partial n = 0$, we find that maximum power occurs when $n = n_t$. Putting $n = n_t$ in Equation (6.9-6) gives

$$P_p = \frac{h\nu}{2t_c}\left(n_t \ln\frac{n_t}{n_i} - (n_t - n_i)\right) \tag{6.9-7}$$

---

[6] This can be done by assuming a value of a and finding the corresponding x at which the plots of x/a and $\exp(x - a)$ intersect.

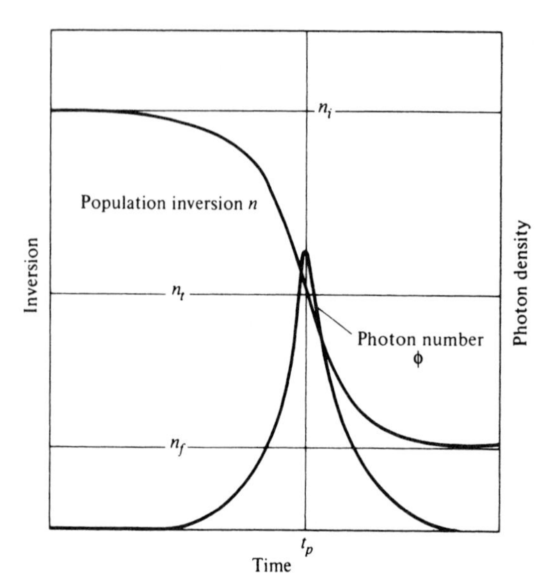

Figure 6.33 Inversion and photon density during a giant pulse. (After Reference [30].)

for the peak power. If the initial inversion is well in excess of the (high-Q) threshold value (i.e., $n_i \gg n_t$), we obtain from Equation (6.9-7)

$$(P_p)_{n_i \gg n_t} \simeq \frac{n_i h\nu}{2t_c} \tag{6.9-8}$$

Since the power P at any moment is related to the number of photons ϕ by $P = \phi h\nu/t_c$, it follows from Equation (6.9-8) that the maximum number of stored photons inside the resonator is $n_i/2$. This can be explained by the fact that if $n_i \gg n_t$, the buildup of the pulse to its peak value occurs in a time short compared to t_c so that at the peak of the pulse, when $n = n_t$, most of the photons that were generated by stimulated emission are still present in the resonator. Moreover, since $n_i \gg n_t$, the number of these photons $(n_i - n_t)/2$ is very nearly $n_i/2$.

A typical numerical solution of Equations (6.9-2) and (6.9-3) is given in Figure 6.33.

To initiate the pulse we need, according to Equations (6.9-2) and (6.9-3), to have $\phi_i \neq 0$. Otherwise the solution is trivial ($\phi = 0$, $n = n_i$). The appropriate value of ϕ_i is usually estimated on the basis of the number of spontaneously emitted photons within the acceptance solid angle of the laser mode at $t = 0$. We also notice, as discussed earlier, that the photon density—hence the power—reaches a peak when $n = n_t$. The energy stored in the cavity ($\propto \phi$) at this point is maximum, so stimulated transitions from the upper to the lower laser levels continue to reduce the inversion to a final value $n_f < n_t$.

Numerical solutions of Equations (6.9-2) and (6.9-3) corresponding to different initial inversions n_i/n_t are shown in Figure 6.34. We notice that for $n_i \gg n_t$ the rise time becomes short compared to t_c but the fall time approaches a value nearly equal to t_c. The reason is that the process of stimulated emission is essentially over at the peak of the pulse ($\tau = 0$) and the observed output is due to the free decay of the photons in the resonator.

In Figure 6.35, we show an actual oscilloscope trace of a giant pulse. Giant laser pulses are used extensively in applications that require high peak powers and short duration. These applications include experiments in nonlinear optics, ranging, material machining and drilling, initiation of chemical reactions, and plasma diagnostics.

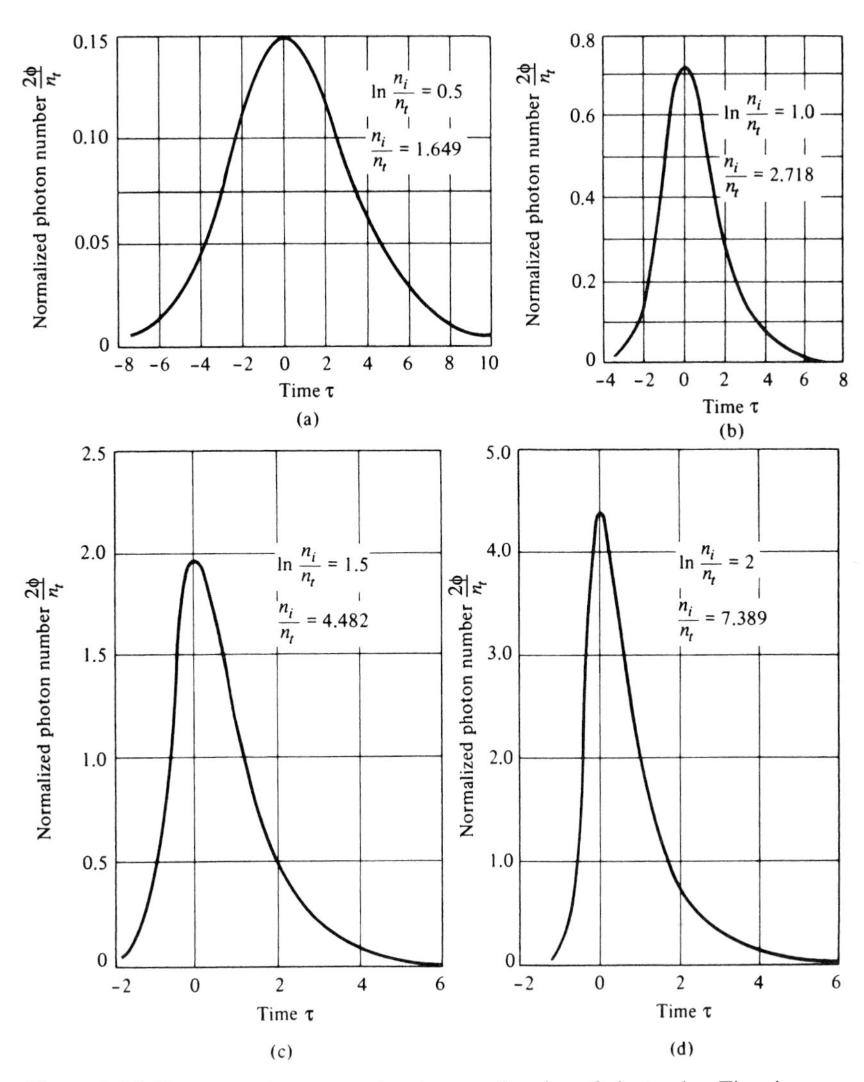

Figure 6.34 Photon number versus time in central region of giant pulse. Time is measured in units of photon lifetime. (After Reference [30].)

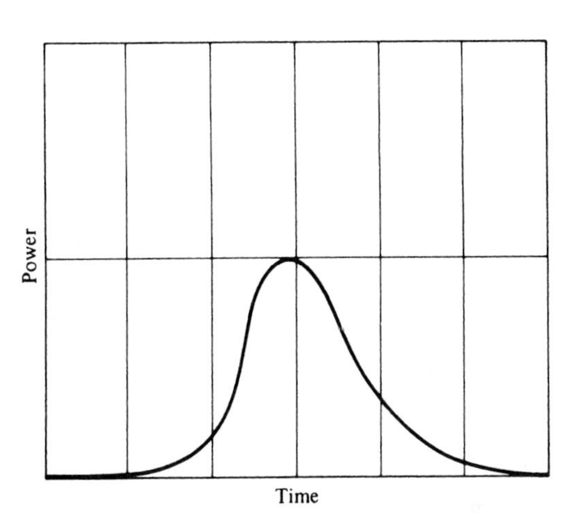

Figure 6.35 An oscilloscope trace of the intensity of a giant pulse. Time scale is 20 ns per division.

EXAMPLE: GIANT PULSE RUBY LASER

Consider the case of pink ruby with a chromium ion density of $N = 1.58 \times 10^{19}$ cm^{-3}. Its absorption coefficient is taken from Figure 6.42, where it corresponds to that of the R_1 line at 6943 Å, and is $\alpha \simeq 0.2$ cm^{-1} (at 300 K). Other assumed characteristics are:

$$l = \text{length of ruby rod} = 10 \text{ cm}$$

$$A = \text{cross-sectional area of mode} = 1 \text{ cm}^2$$

$$(1 - R) = \text{fractional intensity loss per pass} = 20\%$$

$$n_0 = 1.78$$

Since, according to Equation (5.6-19), the exponential loss coefficient is proportional to $N_1 - N_2$, we have

$$\alpha(\text{cm}^{-1}) = 0.2 \frac{N_1 - N_2}{1.58 \times 10^{19}} \tag{6.9-9}$$

Thus, at room temperature, where $N_2 \ll N_1$ when $N_1 - N_2 \cong 1.58 \times 10^{19}$ cm^{-3}, Equation (6.9-9) yields $\alpha = 0.2$ cm^{-1} as observed. The expression for the gain coefficient follows directly from Equation (6.9-9):

$$\gamma(\text{cm}^{-1}) = 0.2 \frac{N_2 - N_1}{1.58 \times 10^9}$$
$$= 0.2 \frac{n}{1.58 \times 10^9 V} \tag{6.9-10}$$

where n is the total inversion and $V = AL$ is the crystal volume in cm^3.

Threshold is achieved when the net gain per pass is unity. This happens when

$$e^{\gamma_t l} R = 1 \quad \text{or} \quad \gamma_t = -\frac{1}{l} \ln R \tag{6.9-11}$$

where the subscript t indicates the threshold condition.

Using Equation (6.9-10) in the threshold condition (6.9-11) plus the appropriate data from above gives

$$n_t = 1.8 \times 10^{19} \tag{6.9-12}$$

Assuming that the initial inversion is $n_i = 5n_t = 9 \times 10^{19}$, we find from Equation (6.9-8) that the peak power is approximately

$$P_p = \frac{n_i h\nu}{2t_c} = 5.1 \times 10^9 \text{ W} \tag{6.9-13}$$

where $t_c = n_0 l / c(1 - R) \simeq 2.5 \times 10^{-9}$ s.

The total pulse energy is

$$\mathscr{E} \sim \frac{n_i h\nu}{2} \sim 13 \text{ J}$$

while the pulse duration (see Figure 6.34) $\simeq 3t_c \simeq 7.5 \times 10^{-9}$ s.

Methods of *Q*-Switching

Some of the schemes used in Q-switching are:

1. Mounting one of the two end reflectors on a rotating shaft so that the optical losses are extremely high except for the brief interval in each rotation cycle in which the mirrors are nearly parallel.
2. The inclusion of a saturable absorber (bleachable dye) in the optical resonator; see References [10, 11, 34]. The absorber whose opacity decreases (saturates) with increasing optical intensity prevents rapid inversion depletion due to buildup of oscillation by presenting a high loss to the early stages of oscillation during which the slowly increasing intensity is not high enough to saturate the absorption. As the intensity increases the loss decreases, and the effect is similar, but not as abrupt, as that of a sudden increase of Q.
3. The use of an electro-optic crystal (or liquid Kerr cell) as a voltage-controlled gate inside the optical resonator. It provides a more precise control over the losses (Q) than schemes 1 and 2. Its operation is illustrated by Figure 6.36 and is discussed in some detail in the following. The control of the phase delay in the electro-optic crystal by the applied voltage is discussed in detail in Chapter 9.

During the pumping of the laser by the light from a flashlamp, a voltage is applied to the electro-optic crystal of such magnitude as to introduce a $\pi/2$ relative phase shift (retardation) between the two mutually orthogonal components (x' and y') that make up the linearly polarized (x) laser field. On exiting from the electro-optic crystal at point f, the light traveling to the right is circularly polarized. After reflection from the right mirror, the light passes once more through the crystal. The additional retardation of $\pi/2$ adds to the earlier one to give a total retardation of π, thus causing the emerging beam at d to be linearly polarized a long y and consequently to be blocked by the polarizer.

It follows that with the voltage on, the losses are high, so oscillation is prevented. The Q-switching is timed to coincide with the point at which the inversion reaches its peak and is achieved by a removal of the voltage applied to the electro-optic crystal. This reduces the retardation to zero so that the state of polarization of the wave passing through the crystal is unaffected and the Q regains its high value associated with the ordinary losses of the system.

6.10 HOLE BURNING AND THE LAMB DIP IN DOPPLER-BROADENED GAS LASERS

In this section we concern ourselves with some of the consequences of Doppler broadening in low-pressure gas lasers.

Consider an atom with a transition frequency $\nu_0 = (E_2 - E_1)/h$, where 2 and 1 refer to the upper and lower laser levels, respectively. Let the component of the velocity of the a tom parallel to the wave propagation direction be v. This component, thus, has the value

$$v = \frac{\mathbf{v}_{\text{atom}} \cdot \mathbf{k}}{k} \tag{6.10-1}$$

where the electromagnetic wave is described by

$$\mathbf{E} = \mathbf{E}e^{i(2\pi\nu t - \mathbf{k}\cdot\mathbf{r})} \tag{6.10-2}$$

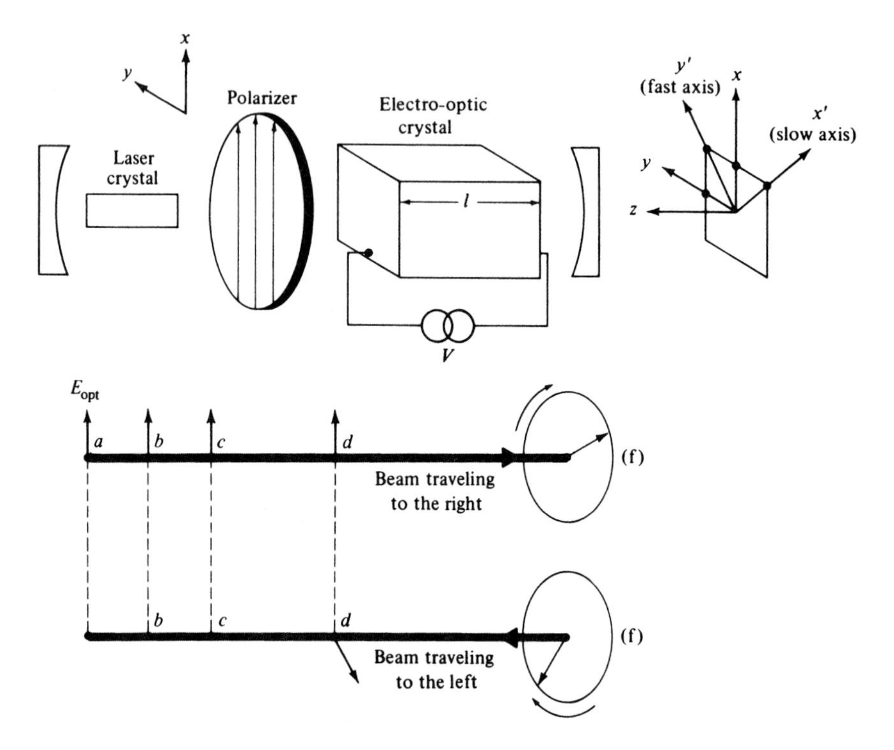

For beam traveling to right:

At point d,

$$E_{x'} = \frac{E}{\sqrt{2}} \cos \omega t$$
$$E_{y'} = \frac{E}{\sqrt{2}} \cos \omega t$$

The optical field is linearly polarized with its electric field vector parallel to x

At point f,

$$E_{x'} = \frac{E}{\sqrt{2}} \cos \left(\omega t - kl - \frac{\pi}{2}\right)$$
$$E_{y'} = \frac{E}{\sqrt{2}} \cos (\omega t - kl)$$

Circularly polarized

For beam traveling to left:

At point f,

$$E_{x'} = -\frac{E}{\sqrt{2}} \cos \left(\omega t - kl - \frac{\pi}{2}\right)$$
$$E_{y'} = -\frac{E}{\sqrt{2}} \cos (\omega t - kl)$$

Circularly polarized

At point d,

$$E_{x'} = -\frac{E}{\sqrt{2}} \cos (\omega t - 2kl - \pi)$$
$$E_{y'} = -\frac{E}{\sqrt{2}} \cos (\omega t - 2kl)$$

Linearly polarized along y

Figure 6.36 Electro-optic crystal used as voltage-controlled gate in Q-switching a laser.

An atom moving with a constant velocity $\mathbf{v}$, so that $\mathbf{r} = \mathbf{v}t + \mathbf{r}_0$, will experience a field

$$\begin{aligned} \mathbf{E}_{\text{atom}} &= \mathbf{E}e^{i[2\pi\nu t - \mathbf{k}\cdot(\mathbf{r}_0 + \mathbf{v}t)]} \\ &= \mathbf{E}e^{i[(2\pi\nu - \mathbf{v}\cdot\mathbf{k})t - \mathbf{k}\cdot\mathbf{r}_0]} \end{aligned} \qquad (6.10\text{-}3)$$

and will thus "see" a Doppler-shifted frequency

$$\nu_D = \nu - \frac{\mathbf{v}\cdot\mathbf{k}}{2\pi} = \nu - \frac{v}{c}\nu \qquad (6.10\text{-}4)$$

where in the second equality we took $n = 1$ so that $k = 2\pi\nu/c$ and used Equation (6.10-1).

The condition for the maximum strength of interaction (i.e., emission or absorption) between the moving atom and the wave is that the apparent (Doppler) frequency ν_D "seen" by the atom be equal to the atomic resonant frequency ν_0

$$\nu_0 = \nu - \frac{v}{c}\nu \tag{6.10-5}$$

or reversing the argument, a wave of frequency ν moving through an ensemble of atoms will "seek out" and interact most strongly with those atoms whose velocity component v satisfies

$$\nu = \frac{\nu_0}{1 - \dfrac{v}{c}} \approx \nu_0\left(1 + \frac{v}{c}\right) \tag{6.10-6}$$

where the approximation is valid for $v \ll c$.

Now consider a gas laser oscillating at a single frequency ν, where, for the sake of definiteness, we take $\nu > \nu_0$. The standing-wave electromagnetic field at ν inside the laser resonator consists of two waves traveling in opposite directions. Consider, first, the wave traveling in the positive x direction (the resonator axis is taken parallel to the axis). Since $\nu > \nu_0$, the wave interacts, according to Equation (6.10-6), with atoms having $v > 0$, that is, atoms with

$$v_x = +\frac{c}{\nu}(\nu - \nu_0) \tag{6.10-7}$$

The wave traveling in the opposite direction ($-x$) must also interact with atoms moving in the same direction so that the Doppler-shifted frequency is reduced from ν to ν_0. These are atoms with

$$v_x = -\frac{c}{\nu}(\nu - \nu_0) \tag{6.10-8}$$

We conclude that due to the standing-wave nature of the field inside a conventional two-mirror laser oscillator, a given frequency of oscillation interacts with two velocity classes of atoms.

Consider, next, a four-level gas laser oscillating at a frequency $\nu > \nu_0$. At negligibly low levels of oscillation and at low gas pressure, the velocity distribution function of atoms in the upper laser level is given, according to Equation (5.5-11), by

$$f(v_x) \propto e^{-Mv_x^2/2kT} \tag{6.10-9}$$

where $f(v_x)\,dv_x$ is proportional to the number of atoms (in the upper laser level) with the x component of velocity between v_x and $v_x + dv_x$. As the oscillation level is increased, say, by reducing the laser losses, we expect the number of atoms in the upper laser level, with x velocities near $v_x = \pm(c/\nu)(\nu - \nu_0)$, to decrease from their equilibrium value as given by Equation (6.10-9). This is due to the fact that these atoms undergo stimulated downward transitions from level 2 to 1, thus reducing the number of atoms in level 2. The velocity distribution function under conditions of oscillation consequently has two depressions as shown schematically in Figure 6.37.

If the oscillation frequency ν is equal to ν_0, only a single "hole" exists in the velocity distribution function of the inverted atoms. This hole is centered on $v_x = 0$. We may thus expect the power output of a laser oscillating at $\nu = \nu_0$ to be less than that of a laser in which ν is tuned slightly to one side or the other of ν_0 (this tuning can be achieved by moving one of the laser mirrors). This power dip first predicted by Lamb [35] is indeed observed in gas lasers [36]. An experimental plot of the power versus frequency in a He–Ne 1.15 μm laser

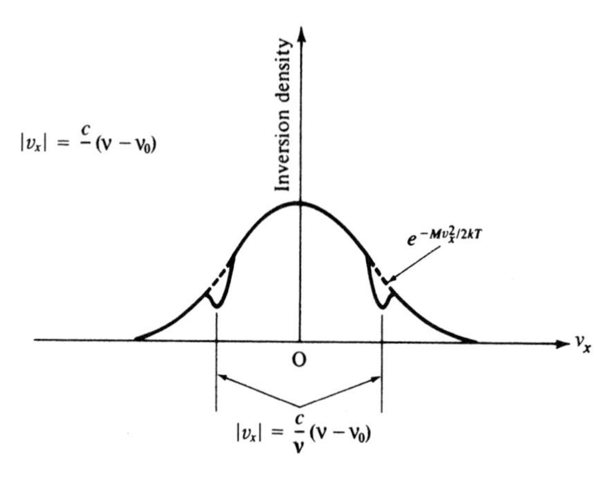

Figure 6.37 The distribution of inverted atoms as a function of v_x. The dashed curve that is proportional to $\exp(-Mv_x^2/2kT)$ corresponds to the case of zero field intensity. The solid curve corresponds to a standing-wave field at $\nu = \nu_0/(1 - v_x/c)$ or one at $\nu = \nu_0/(1 + v_x/c)$.

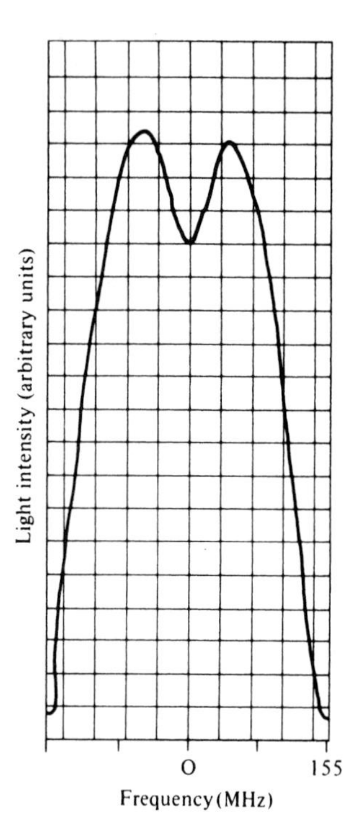

Figure 6.38 The power output as a function of the frequency of a single-mode 1.15 μm He–Ne laser using the ^{20}Ne isotope. (After Reference [36].)

is shown in Figure 6.38. The phenomenon is referred to as the "Lamb dip" and is used in frequency stabilization schemes of gas lasers [37].

6.11 SOME SPECIFIC LASER SYSTEMS

In this section, we discuss some specific laser systems. The pumping of the atoms into the upper laser level is accomplished in a variety of ways, depending on the type of laser. Here we will review some of the more common laser systems and in the process describe their pumping mechanisms. The laser systems described include $Cr^{3+}:Al_2O_3$ (ruby), Nd^{3+}:YAG, Nd^{3+}:glass, and Er^{3+}:silica lasers. The semiconductor current-pumped (or carrier injection) laser, because

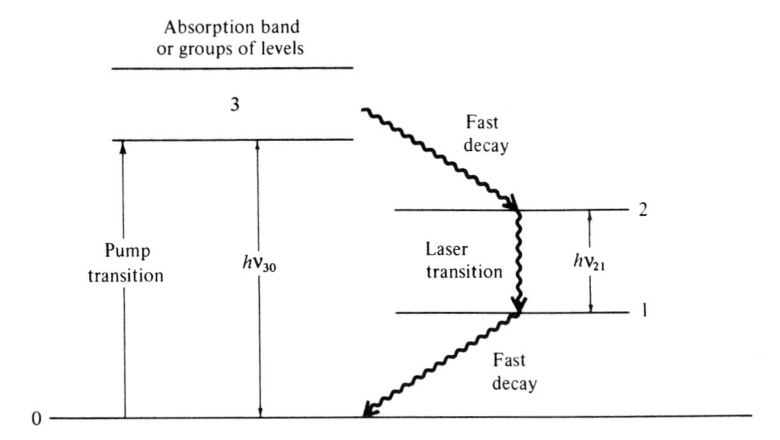

Figure 6.39 Pumping–oscillation cycle of a typical laser.

of its unique technological importance, receives special billing in Chapters 15 and 16. Figure 6.39 shows the pumping–oscillation cycle of a (hypothetical) representative laser. The pumping agent elevates the atoms into some excited state 3 from which they relax into the upper laser level 2. The stimulated laser transition takes place between levels 2 and 1 and results in the emission of a photon of frequency ν_{21}. It is evident from this figure that the minimum energy input per output photon is $h\nu_{30}$, so the power efficiency of the laser cannot exceed

$$\eta_{\text{atomic}} = \frac{\nu_{21}}{\nu_{30}} \tag{6.11-1}$$

referred to as the "atomic quantum efficiency." The overall laser efficiency depends on the fraction of the total pump power that is effective in transferring atoms into level 3 and on the pumping quantum efficiency defined as the fraction of the atoms that once in level 3, make a transition to level 2. The product of the last two factors, which constitutes an upper limit on the efficiency of optically pumped lasers, ranges from about 1% for solid-state lasers such as Nd^{3+}:YAG, to about 30% in the CO_2 laser, and to near unity in the GaAs junction laser. We shall discuss these factors when we get down to some specific laser systems. We may note, however, that according to Equation (6.11-1), in an efficient laser system ν_{21} and ν_{30} must be of the same order of magnitude, so the laser transition should involve low-lying levels.

The laser examples discussed in this section consist of an optically transparent host material such as glass (SiO_2) or sapphire crystal (Al_2O_3) doped with a small amount of Cr, Nd, or Er ions. In a glass or crystal host, the quantum states of these active atoms (ionized) are modified by the local electric field that causes Stark splitting and shifts. For the purpose of discussion, we briefly review the spectroscopic notation used for the quantum states. For atoms with many electrons, the quantum state is characterized by a set L, S, J of quantum numbers. They are total orbital angular momentum, total spin angular momentum, and total angular momentum of the electrons, respectively. The state is designated as

$$^{2S+1}L_J$$

where $2S + 1$, the multiplicity, denotes the total number of different J-values associated with a given set of (L, S), assuming $S \leq L$, which is normally the case. The total angular momentum is the vector sum of the total orbital angular momentum and the total spin angular momentum. Thus the quantum number J is often limited within $L - S$ and $L + S$. The letter

TABLE 6.3 Electronic State Designations

Letter	$L =$
S	0
P	1
D	2
F	3
G	4
H	5
I	6
⋮	⋮

symbol L, denoting the orbital angular momentum, is designated as shown in Table 6.3. For each state $^{2S+1}L_J$, there are $2J + 1$ mutually orthogonal sublevels with the same energy. An external magnetic field can cause the splitting of the state $^{2S+1}L_J$ into $2J + 1$ energy levels. For example, the ground state of Er^{3+} ion is designated as $^4I_{15/2}$, which means $S = 3/2$, $L = 6$, and $J = 15/2$. The multiplicity of $2S + 1 = 4$ indicates that the multiplet consists of four states with the same L and S. These four states are $^4I_{9/2}$, $^4I_{11/2}$, $^4I_{13/2}$, and $^4I_{15/2}$. These states within the same multiplet have different total angular momentum J and have different energies due to the so-called spin–orbit interaction.

Ruby Laser

The first material in which laser action was demonstrated [17] and still one of the most useful laser materials is ruby, whose output is at $\lambda_0 = 0.6943\ \mu m$. The active laser particles are Cr^{3+} ions present as impurities in Al_2O_3 crystal. Typical Cr^{3+} concentrations are ~0.05% by weight. The triply ionized Cr ion has 3 electrons in the 3d state. These unpaired d electrons form a high-spin ground state referred to as ^4F ($S = 3/2$). The first excited state of the free ion is a ^2G state ($S = 1/2$) about 2 eV higher in energy than the ground state. Both of the F and G levels have high orbital degeneracy. There are seven orbital sublevels of the F level ($L = 3$) and nine sublevels of the G level ($L = 4$). Under the action of crystalline electric fields in the sapphire crystal (Al_2O_3), the ^4F state splits into three states ($^4A_2$, $^4T_2$, $^4T_1$), and the ^2G state splits into four levels (^2E, $^2T_1$, $^2T_2$, $^2A_1$). As a result, the ground state becomes $^4A_2$. In these notations, the A level is singly degenerate, the E level is doubly degenerate, and the T level is triply degenerate. The pertinent energy level diagram is shown in Figure 6.40.

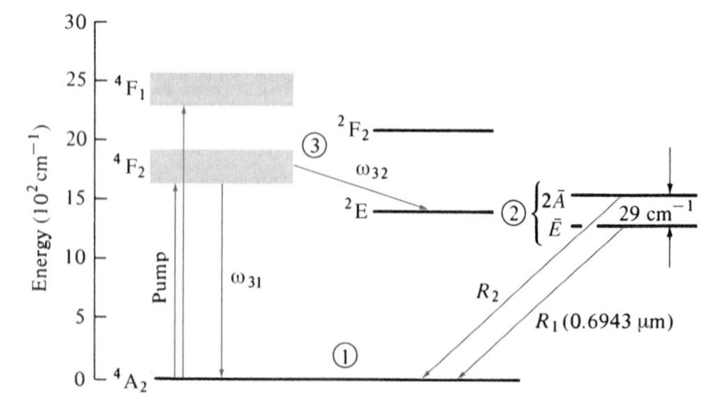

Figure 6.40 Energy levels pertinent to the operation of a ruby laser. The unit 1 cm^{-1} corresponds to $\nu = 30$ GHz, or $h\nu = 1.24 \times 10^{-4}$ eV. (After Reference [38].)

The pumping of ruby is usually performed by subjecting it to the light of intense flashlamps (quite similar to the types used in flash photography). A portion of this light that corresponds in frequency to the two absorption bands $^4F_2$ ($^4T_2$) and $^4F_1$ ($^4T_1$) is absorbed, thereby causing Cr^{3+} ions to be transferred into these levels. The ions proceed to decay, within an average time of $\omega_{32}^{-1} \cong 5 \times 10^{-8}$ s [38], into the upper laser level ^2E. Level ^2E is composed of two separate levels $2\bar{A}$ and $\bar{E}$ separated by 29 cm^{-1} (due to spin–orbit interaction, etc.) The unit 1 cm^{-1} (one wavenumber) is equivalent to $\nu = 30$ GHz, or $h\nu = 1.24 \times 10^{-4}$ eV. It is also used as a measure of energy where 1 cm^{-1} corresponds to the energy $h\nu$ of a photon with $\nu = 30$ GHz. The lower of these two, $\bar{E}$, is the upper laser level. The lower laser level is the ground state and thus, according to the discussion earlier in this section, a ruby laser is a three-level laser. The lifetime of atoms in the upper laser level $\bar{E}$ is $t_2 = 3 \times 10^{-3}$ s. Each decay results in the (spontaneous) emission of a photon, so $t_2 = t_{\text{spont}}$.

An absorption spectrum of a typical ruby with two orientations of the optical field relative to the c (optic) axis is shown in Figure 6.41. The two main peaks correspond to absorption into the useful $^4F_1$ and $^4F_2$ bands, which are responsible for the characteristic (ruby) color. The ordinate is labeled in terms of the absorption coefficient and in terms of the transition cross section σ, which may be defined as the absorption coefficient per unit inversion per unit volume and has consequently the dimension of area. According to this definition, the linear absorption coefficient $\alpha(\nu)$ is given by

$$\alpha(\nu) = (N_1 - N_2)\sigma(\nu) \tag{6.11-2}$$

A more detailed plot of the absorption near the laser emission wavelength is shown in Figure 6.42. The width $\Delta\nu$ of the laser transition as a function of temperature is shown in Figure 6.43. At room temperature, $\Delta\nu = 11$ cm^{-1} (or 3.3×10^{11} Hz). We can use ruby to illustrate some of the considerations involved in optical pumping of solid-state lasers. Figure 6.44 shows a typical setup of an optically pumped laser, such as ruby. The helical flashlamp

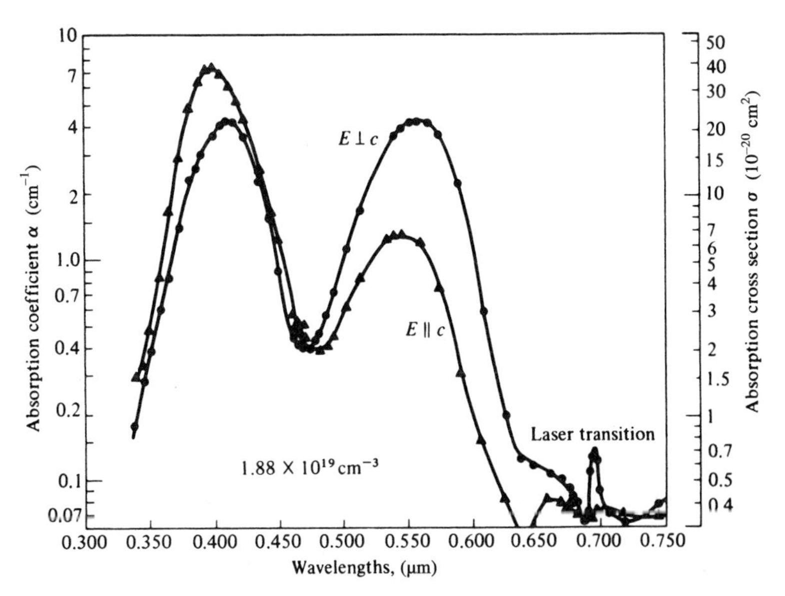

Figure 6.41 Absorption coefficient and absorption cross section as functions of wavelength for $E \parallel c$ and $E \perp c$. The 300 K data were derived from transmittance measurements on pink ruby with an average Cr ion concentration of 1.88×10^{19} cm^{-3}. (After Reference [39].)

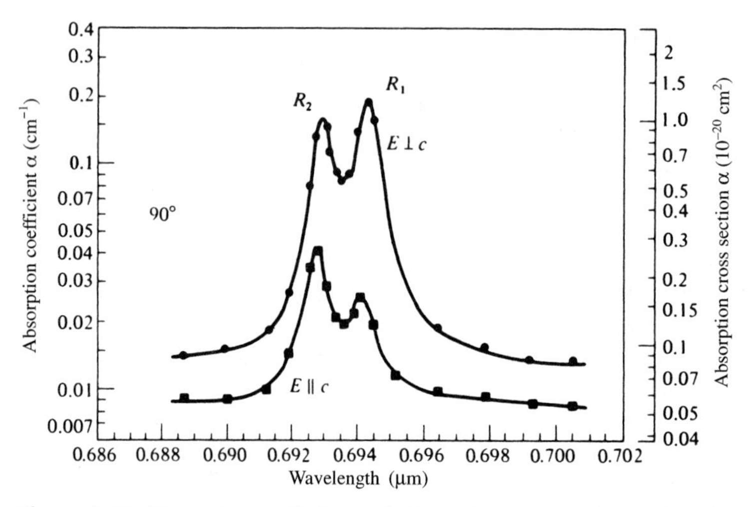

Figure 6.42 Absorption coefficient and absorption cross section as functions of wavelength for $E \parallel c$ and $E \perp c$. Sample was a pink ruby laser rod having a 90° c-axis orientation with respect to the rod axis and a Cr ion concentration of 1.58×10^{19} cm^{-3}. (After Reference [39].)

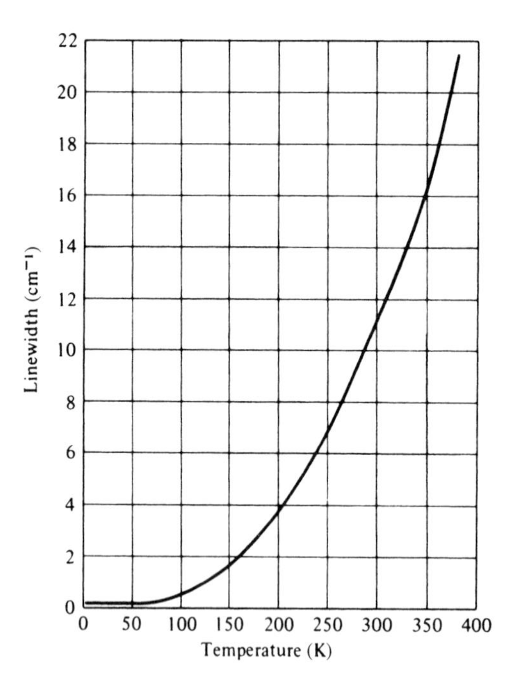

Figure 6.43 Linewidth of the R_1 line of ruby as a function of temperature. (After Reference [40].)

surrounds the ruby rod. The flash excitation is provided by the discharge of the charge stored in a capacitor bank across the lamp.

The typical flash output consists of a pulse of light of duration $t_{\text{flash}} = 5 \times 10^{-4}$ s. Let us, for the sake of simplicity, assume that the flash pulse is rectangular in time and of duration t_{flash} and that it results in an optical flux at the crystal surface having $s(\nu)$ watts per unit area per unit frequency at the frequency ν. If the absorption coefficient of the crystal is $\alpha(\nu)$, then the amount of energy absorbed by the crystal per unit volume is

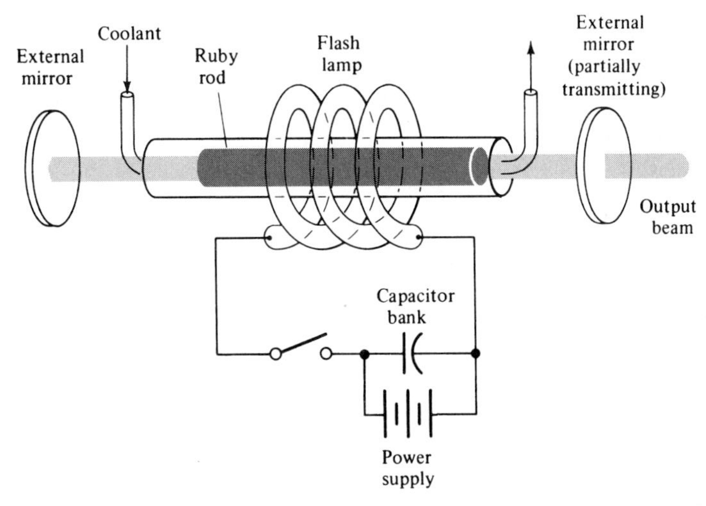

Figure 6.44 Typical setup of a pulsed ruby laser using flashlamp pumping and external mirrors.

$$t_{\text{flash}} \int_0^\infty S(\nu)\alpha(\nu)\, d\nu$$

where we assume that the total absorption in passing the crystal is small, so $s(\nu)$ is taken to be independent of the distance through the crystal.

If the absorption quantum efficiency (the probability that the absorption of a pump photon at ν results in transferring one atom into the upper laser level) is $\eta(\nu)$, the number of atoms pumped into level 2 per unit volume is

$$N_2 = t_{\text{flash}} \int_0^\infty \frac{s(\nu)\alpha(\nu)\eta(\nu)}{h\nu}\, d\nu \tag{6.11-3}$$

Since the lifetime $t_2 = 3 \times 10^{-3}$ s of atoms in level 2 is considerably longer than the flash duration (~5×10^{-4} s), we may neglect the spontaneous decay out of level 2 during the time of the flash pulse, so N_2 represents the population of level 2 after the flash.

EXAMPLE: FLASH PUMPING OF A PULSED RUBY LASER

Consider the case of a ruby laser with the following parameters:

$$N_0 = 2 \times 10^{19} \text{ atoms/cm}^3 \text{ (Cr}^{3+}\text{ concentration)}$$

$$t_2 = t_{\text{spont}} = 3 \times 10^{-3} \text{ s}$$

$$t_{\text{flash}} = 5 \times 10^{-4} \text{ s}$$

If the useful absorption is limited to relatively narrow spectral regions, we may approximate Equation (6.11-3) by

$$N_2 = \frac{t_{\text{flash}}\, \overline{s(\nu)}\, \overline{\alpha(\nu)}\, \overline{\eta(\nu)}}{h\bar{\nu}} \overline{\Delta\nu} \tag{6.11-4}$$

where the bars represent average values over the useful absorption region whose width is $\overline{\Delta\nu}$.

From Figure 6.41, we deduce an average absorption coefficient of $\overline{\alpha(\nu)} \cong 2\ \text{cm}^{-1}$ over the two central peaks. Since ruby is a three-level laser, the upper level population is, according to Equation (6.3-2), $N_2 \sim N_0/2 = 10^{19}\ \text{cm}^{-3}$. Using $\bar{\nu} \cong 5 \times 10^{14}$ Hz and $\overline{\eta(\nu)} \cong 1$, Equation (6.11-4) yields

$$\bar{s}\ \overline{\Delta\nu}\ t_{\text{flash}} \cong 1.5\ \text{J/cm}^2$$

for the pump energy in the useful absorption region that must fall on each square centimeter of crystal surface in order to obtain threshold inversion. To calculate the total lamp energy that is incident on the crystal, we need to know the spectral characteristics of the lamp output. Typical data of this sort are shown in Figure 6.45. The mercury-discharge lamp is seen to contain considerable output in the useful absorption regions (near 4000 and 5500 Å) of ruby. If we estimate the useful fraction of the lamp output at 10%, the fraction of the lamp light actually incident on the crystal as 20%, and the conversion of electrical-to-optical energy as 50%, we find that the threshold electric energy input to the flashlamp per square centimeter of laser surface is

$$\frac{1.5}{0.1 \times 0.2 \times 0.5} = 150\ \text{J/cm}^2$$

These are, admittedly, extremely crude calculations. They are included not only to illustrate the order of magnitude numbers involved in laser pumping, but also as an example of the quick and rough estimates needed to discriminate between feasible ideas and "pie-in-the-sky" schemes.

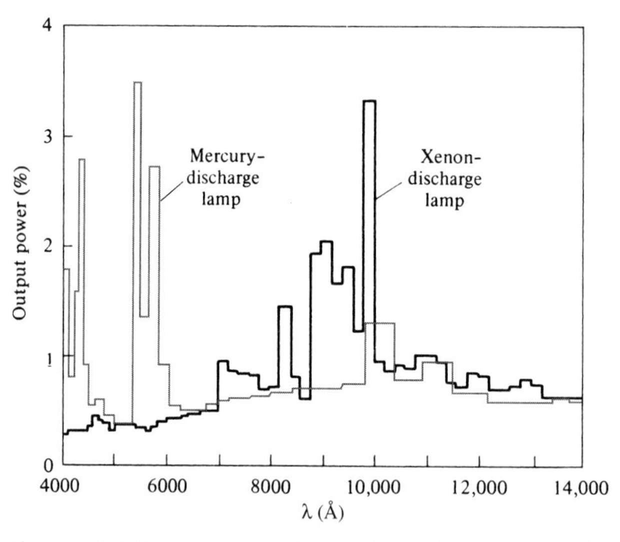

Figure 6.45 Spectral output characteristics of two commercial high-pressure lamps. Output is plotted as a fraction of electrical input to lamp over certain wavelength intervals (mostly 200 Å) between 0.4 and 1.4 μm. (After Reference [41].)

Nd^{3+}:YAG Laser

One of the most important laser systems is that using trivalent neodymium ions (Nd^{3+}), which are present as impurities in yttrium aluminum garnet (YAG = $Y_3Al_5O_{12}$); see References [42, 43]. The laser emission occurs at $\lambda_0 = 1.0641\ \mu\text{m}$ at room temperature. Figure 6.46 illustrates the relevant energy levels of the neodymium ions (Nd^{3+}) in the host material. The lower

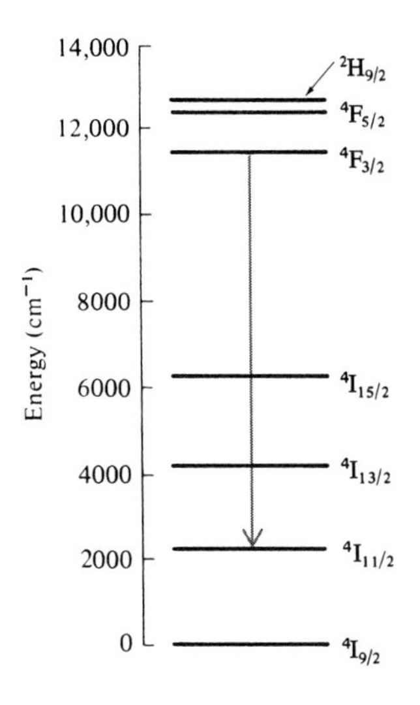

Figure 6.46 Energy-level diagram of Nd^{3+} in YAG. (After Reference [42].) The unit 1 cm^{-1} corresponds to $\nu = 30$ GHz, or $h\nu = 1.24 \times 10^{-4}$ eV.

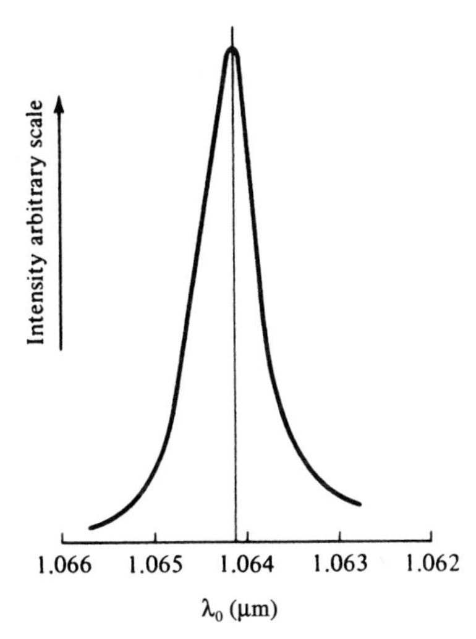

Figure 6.47 Spontaneous emission spectrum of Nd^{3+} in YAG near the laser transition at $\lambda_0 = 1.064$ μm. (After Reference [43].)

laser level is at $E_2 = 2111$ cm^{-1} (0.26 eV) from the ground state so that at room temperature its population is down by a factor of $\exp(-E_2/kT) = e^{-10}$ from that of the ground state and can be neglected. The Nd^{3+}:YAG thus fits our definition of a four-level laser.

The spontaneous emission spectrum of the laser transition is shown in Figure 6.47. The width of the gain linewidth at room temperature is $\Delta\nu = 6$ cm^{-1} (or 180 GHz). The spontaneous lifetime for the laser transition has been measured [43] as $t_{spont} = 5.5 \times 10^{-4}$ s. The room-temperature cross section at the center of the laser transition is $\sigma = 9 \times 10^{-19}$ cm^2. If we compare this number to $\sigma = 1.22 \times 10^{-20}$ cm^2 in ruby (see Figure 6.42), we expect that at a given inversion the optical gain constant γ in Nd^{3+}:YAG is approximately 75 times that of ruby. This causes the oscillation threshold to be very low and explains the easy continuous (CW) operation of this laser compared to the ruby laser.

The absorption responsible for populating the upper laser level takes place in a number of bands between 13,000 and 25,000 cm^{-1}. Here, we recall the unit 1 cm^{-1} corresponds to $\nu = 30$ GHz, or $h\nu = 1.24 \times 10^{-4}$ eV.

EXAMPLE: THRESHOLD OF AN Nd^{3+}:YAG LASER

Pulsed Threshold

First we estimate the energy needed to excite a typical Nd^{3+}:YAG laser on a pulse basis so that we can compare it with that of the ruby laser. We use the following data:

$$\left.\begin{aligned} l &= 20 \text{ cm (length optical resonator)} \\ \text{L} &= 4\% \\ n &= 1.8 \end{aligned}\right\} t_c = \frac{nl}{Lc} = 2.5 \times 10^{-8} \text{ s}$$

$$\Delta\nu = 6 \text{ cm}^{-1} \ (= 6 \times 3 \times 10^{10} \text{ Hz}) \quad (\text{Note: } 1 \text{ cm}^{-1} = 30 \text{ GHz})$$

$$t_{\text{spont}} = 5.5 \times 10^{-4} \text{ s}$$

$$\lambda = 1.06 \ \mu\text{m}$$

Using the foregoing data in Equation (6.1-18) gives

$$N_t = \frac{8\pi n^3 t_{\text{spont}} \Delta\nu}{c t_c \lambda^2} \cong 1.7 \times 10^{15} \text{ cm}^{-3}$$

Assuming that 5% of the exciting light energy falls within the useful absorption bands, that 5% of this light is actually absorbed by the crystal, that the average ratio of laser frequency to the pump frequency is 0.5, and that the lamp efficiency (optical output/electrical input) is 0.5, we obtain

$$E_{\text{lamp}} = \frac{N_t h\nu_{\text{laser}}}{5 \times 10^{-2} \times 5 \times 10^{-2} \times 0.5 \times 0.5} \cong 0.3 \text{ J/cm}^3$$

for the energy input to the lamp at threshold. It is interesting to compare this last number to the figure of 150 joules per square centimeter of surface area obtained in the ruby example. For reasonable crystal dimensions (say, length = 5 cm, r = 2 mm), we obtain $E_{\text{lamp}} = 0.19$ J. We expect the ruby threshold to exceed that of Nd^{3+}:YAG by three orders of magnitude, which is indeed the case.

Continuous Operation

The critical fluorescence power—that is, the actual power given off by spontaneous emission just below threshold—is given by Equation (6.3-4) as

$$\left(\frac{P_s}{V}\right) = \frac{N_t h\nu}{t_{\text{spont}}} \cong 0.34 \text{ W/cm}^3$$

Taking the crystal diameter as 0.25 cm and its length as 3 cm and using the same efficiency factors assumed in the first part of this example, we can estimate the power input to the lamp at threshold as

$$P_{\text{lamp}} = \frac{0.34 \times (\pi/4) \times (0.25)^2 \times 3}{5 \times 10^{-2} \times 5 \times 10^{-2} \ 0.5 \times 0.5} \cong 81 \text{ W}$$

which is in reasonable agreement with experimental values [42]. A typical arrangement used in continuous solid-state lasers is shown in Figure 6.48. The highly polished elliptic cylinder is used to concentrate the light from the lamp, which is placed along one focal axis, onto the laser rod, which occupies the other axis. This configuration guarantees that most of the light emitted by the lamp passes through the laser rod. The reflecting mirrors (for the laser resonator) are placed outside the cylinder.

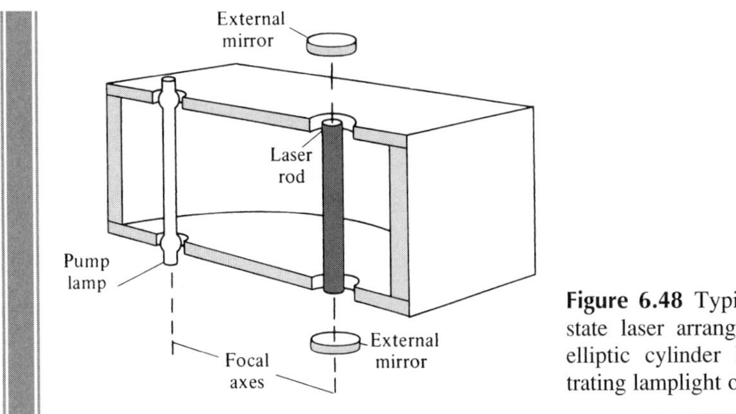

Figure 6.48 Typical continuous solid-state laser arrangement employing an elliptic cylinder housing for concentrating lamplight onto laser.

Neodymium–Glass Laser

One of the most useful laser systems is that which results when the Nd^{3+} ion is present as an impurity atom in glass [44]. The energy levels involved in the laser transition in a typical glass are shown in Figure 6.49. The laser emission wavelength is at $\lambda = 1.059$ µm and the lower level is approximately 1950 cm^{-1} (0.24 eV) above the ground state. As in the case of Nd^{3+}:YAG described above, we have here a four-level laser, since the thermal population of the lower laser level is negligible. The fluorescent emission near $\lambda_0 = 1.06$ µm is shown in Figure 6.50. The fluorescent linewidth can be measured off directly and ranges, for the glasses shown, around 300 cm^{-1} (9 THz). This width is approximately a factor of 50 larger than that of Nd^{3+} in YAG. This is due to the amorphous structure of glass, which causes different Nd^{3+} ions to "see" slightly different surroundings. This causes their energy splittings to vary slightly. Different ions consequently radiate at slightly different frequencies, causing a broadening of the spontaneous emission spectrum. This is a good example of inhomogeneous broadening. The absorption bands responsible for pumping the laser level are shown in Figure 6.51. The probability that the absorption of a photon in any of these bands will result in pumping an atom to the upper laser level (i.e., the absorption quantum efficiency) has been estimated [44] at about 0.4.

The lifetime t_2 of the upper laser level depends on the host glass and on the Nd^{3+} concentration. This variation in two glass series is shown in Figure 6.52.

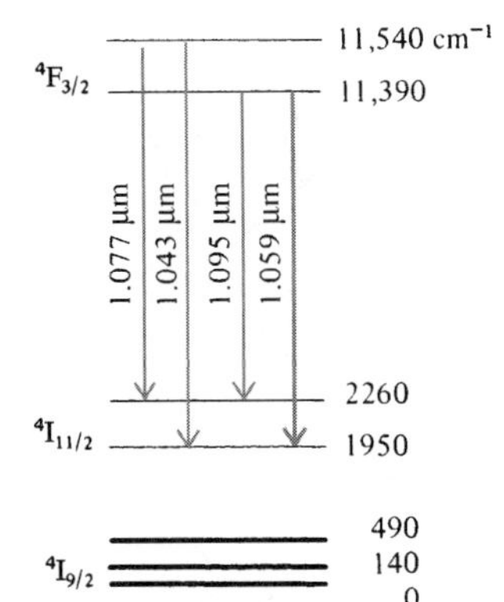

Figure 6.49 Energy-level diagram for the ground state and the states involved in laser emission at 1.059 µm for Nd^{3+} in a rubidium potassium barium silicate glass. The unit 1 cm^{-1} corresponds to $\nu = 30$ GHz, or $h\nu = 1.24 \times 10^{-4}$ eV. (After Reference [44].)

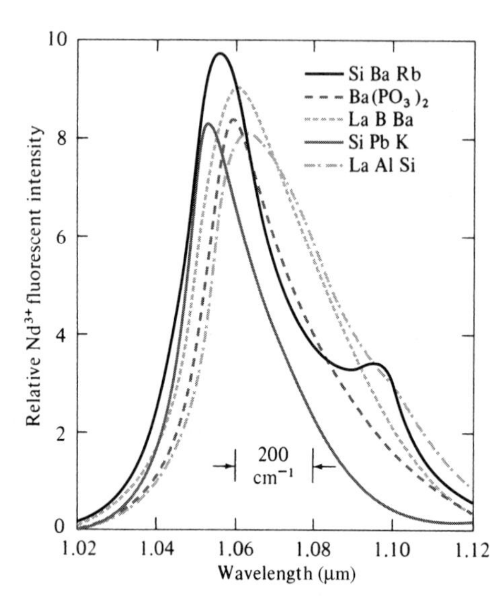

Figure 6.50 Fluorescent emission of the 1.06 μm line of Nd^{3+} at 300 K in various glass bases. (After Reference [44].)

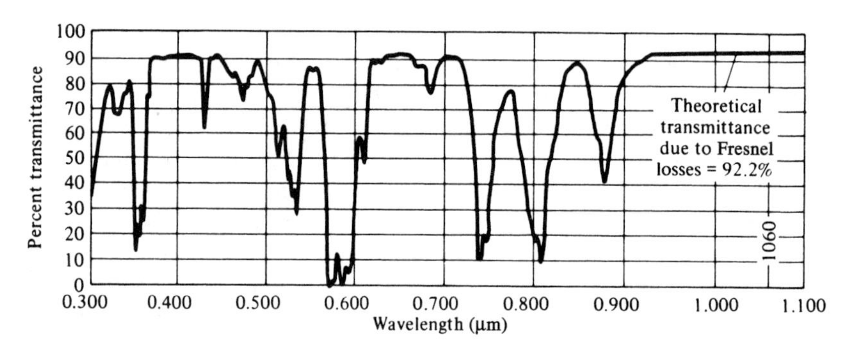

Figure 6.51 Nd^{3+} absorption spectrum for a sample of glass 6.4 mm thick with the composition 66 wt.% SiO_2, 5 wt.% Nd_2O_3, 16 wt.% Na_2O, 5 wt.% BaO, 2 wt.% Al_2O_3, and 1 wt.% Sb_2O_3. (After Reference [44].)

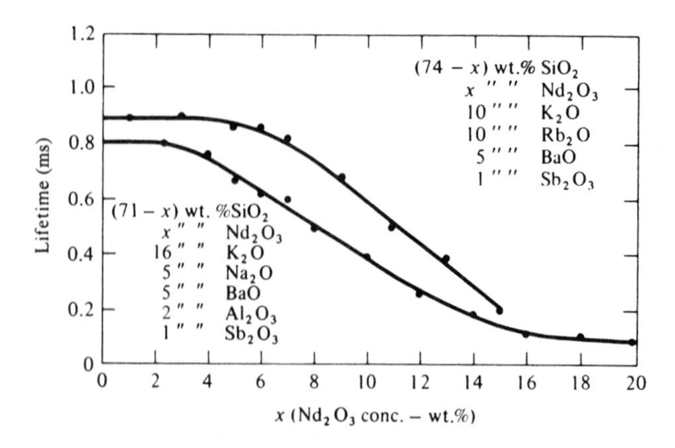

Figure 6.52 Lifetime as a function of concentration for two glass series. (After Reference [44].)

EXAMPLE: THRESHOLDS FOR CW AND PULSED OPERATION OF Nd^{3+}:GLASS LASERS

Let us estimate first the threshold for continuous (CW) laser action in a Nd^{3+}:glass laser using the following data:

$$\Delta\nu = 200\ \text{cm}^{-1}\ (\text{or}\ \Delta\nu = 6\ \text{THz, see Figure 6.50})$$

$$n = 1.5$$

$$t_{\text{spont}} \sim t_2 = 3 \times 10^{-4}\ \text{s}$$

$$\left.\begin{aligned} &l = \text{length optical resonator} = 20\ \text{cm} \\ &\text{L} = \text{loss per pass} = 4\% \end{aligned}\right\} t_c \cong \frac{nl}{Lc} = 5 \times 10^{-8}\ \text{s}$$

Using Equation (6.1-18), we obtain

$$N_t = \frac{8\pi t_{\text{spont}} n^3\, \Delta\nu}{c t_c \lambda^2} = 9.05 \times 10^{15}\ \text{atoms/cm}^3$$

for the critical inversion. The fluorescence power at threshold P_s is thus (see Equation (6.3-5))

$$P_s = \frac{N_t h\nu V}{t_{\text{spont}}} = 5.65\ \text{watts}$$

in a crystal volume $V = 1\ \text{cm}^3$.

We assume (a) that only 10% of the pump light lies within the useful absorption bands, (b) that because of the optical coupling inefficiency and the relative transparency of the crystal only 10% of the energy leaving the lamp within the absorption bands is actually absorbed, (c) that the absorption quantum efficiency is 40%, and (d) that the average pumping frequency is twice that of the emitted radiation. The lamp output at threshold is thus

$$\frac{2 \times 5.65}{0.1 \times 0.1 \times 0.4} = 2825\ \text{W}$$

If the efficiency of the lamp in converting electrical to optical energy is about 50%, we find that continuous operation of the laser requires about 5 kW of power. This number is to be contrasted with a threshold of approximately 100 watts for the Nd:YAG laser, which helps explain why Nd:glass lasers are not operated continuously.

If we consider the pulsed operation of a Nd:glass laser by flash excitation, we have to estimate the minimum energy needed to pump the laser at threshold. Let us assume here that the losses (attributable mostly to the output mirror transmittance) are $L = 20\%$. (Because of the higher pumping rate available with flash pumping, optimum coupling in this case calls for larger mirror transmittances compared to the CW case.) A recalculation of N_t gives

$$N_t = 9.05 \times 10^{16}\ \text{atoms/cm}^3$$

The minimum energy needed to pump N_t atoms into level 2 is then

$$\frac{E_{\min}}{V} = N_t(h\nu) = 1.7 \times 10^{-2}\ \text{J/cm}^3$$

Assuming a crystal volume $V = 10\ \text{cm}^3$ and the same efficiency factors used in the previous CW example, we find that the input energy to the flashlamp at thresholds $\sim 2 \times 1.7 \times 10^{-2} \times$

$10/(0.1 \times 0.1 \times 0.4) = 85$ J. Typical Nd^{3+}:glass lasers with characteristics similar to those used in this example are found to require an input of about 150–300 joules at threshold.

Er^{3+}:Silica Laser

In addition to neodymium atoms, many other rare-earth atoms can be employed as dopants in glass or crystals. These active elements can be pumped with external light sources to reach population inversion. As a result of their intrinsic atomic energy states, these rare-earth atoms cover various spectral regimes of interest. In the area of fiberoptic communications, we are particularly interested in the spectral regime around $\lambda = 1550$ nm. One of the most important laser systems is that of *Er-doped* silica fibers [45–50] at $\lambda = 1.54–1.55$ μm. Such fibers pumped at $\lambda = 0.98$ μm or $\lambda = 1.48$ μm are used as inline optical amplifiers in optical communications fiber systems and have major system implications [47]. Oscillation can be obtained if the Er:silica fiber is placed inside an optical resonator. This is often achieved by using fiber Bragg gratings (FBGs) at both ends of a section of these fibers. The fiber Bragg gratings in this case act as the mirrors of the laser resonator. Optical properties of fiber Bragg gratings will be discussed in Chapter 12.

The energy levels involved in the laser transition in a typical Er:silica fiber laser are shown in Figure 6.53. Er^{3+} is a rare-earth ion from the lanthanide family. The erbium ion Er^{3+} has 11 electrons in the 4f orbital. It is an element of choice for lasing and amplification in the 1.5 μm spectral regime, due to its extremely wide gain bandwidth resulting from the transition between the states ${}^4I_{15/2}$ and ${}^4I_{13/2}$. In the host medium, all $2J + 1$ components of both multiplets of the erbium ion are split and homogeneously broadened by the environmental perturbations (Stark effect, etc.) due to the neighboring oxygen ions in the silica glass (SiO_2). As a result of the random environments for each of the erbium ions, all these sublevels are further broadened due to inhomogeneous broadening. This leads to the extremely wide gain bandwidth, which is desirable for broadband optical communications applications. There are two strong peaks in absorption cross section of the erbium-doped fiber, at 1480 nm and 980 nm corresponding to the transitions ${}^4I_{15/2} \rightarrow {}^4I_{11/2}$ and ${}^4I_{15/2} \rightarrow {}^4I_{13/2}$, respectively. The lifetime of atoms in the upper laser level ${}^4I_{13/2}$ is $t_2 = 10.2 \times 10^{-3}$ s. Each decay results in the (spontaneous) emission of a photon, so $t_2 = t_{spont}$. Since the laser emission involves the ground state as the lower level, an Er:silica fiber laser is a three-level laser (similar to the ruby laser).

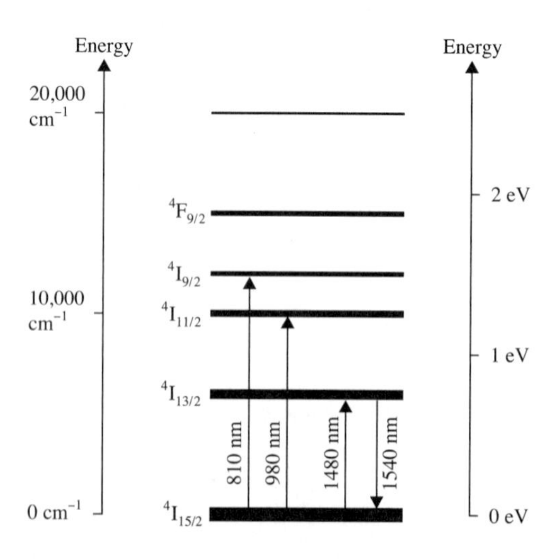

Figure 6.53 Energy levels of Er^{3+} ion and some dominant transitions. The unit 1 cm^{-1} corresponds to $\nu = 30$ GHz, or $h\nu = 1.24 \times 10^{-4}$ eV.

As a result, the fiber must be pumped relatively hard to achieve gain. This problem is eased by the fiber geometry, where the pumped beam can be copropagating with the lasing beam inside the fiber core.

An Er:silica fiber without the fiber Bragg gratings can be employed as an inline optical amplifier in telecommunications networks. More discussion on erbium-doped fiber amplifiers (EDFAs) can be found in Chapter 17.

6.12 FREQUENCY COMB AND OPTICAL FREQUENCY METROLOGY

The mode-locked lasers discussed in Section 6.6 provide a train of ultrashort optical pulses. These optical pulses are identical in shape and equally spaced in time (or space). The periodic pulses are a result of the coupling of the modes of oscillation of the laser resonator via either active or passive mode locking. We showed that the output beam of a mode-locked laser can be written

$$E(t) = \sum_m C_m e^{i[(\Omega_0 + m\Omega)t + \phi_m]} \tag{6.12-1}$$

where Ω is the longitudinal mode spacing in angular frequency, C_m is the amplitude of the mth mode, and ϕ_m is the phase of the mth mode (m = integer). The summation is extended over all the oscillating modes and Ω_0 is a reference frequency. In Equation (6.12-1), we assume that the oscillation frequencies of the resonator are

$$\omega_m = \Omega_0 + m\Omega \tag{6.12-2}$$

where m is an integer and the longitudinal mode spacing Ω is given by

$$\Omega = 2\pi \frac{c}{2n_g l} \tag{6.12-3}$$

where l is the length of the resonator and n_g is the group index. In active mode locking, the modulation frequency is chosen to be exactly Ω. The modulation at this frequency leads to coupling among the modes of oscillation.

In cases where the group velocity dispersion (n_g depends on ω) is present, the mode spacing varies with ω slightly over the spectrum. In these cases the modulation frequency is chosen to be the average of the mode spacing over the entire gain spectrum. When mode locking occurs, coupling among the modes of oscillation causes the "actual frequency" of oscillation to be locked with the frequency given by Equation (6.12-2). In other words, they are exactly equally spaced, even though the modes of the cold resonator may not be exactly equally spaced. As a result, the field envelope is exactly periodic in time with a period of $\tau \equiv 2\pi/\Omega = 2l/v_g = 2n_g l/c$, which is the average round-trip transit time of the mode-locked pulses inside the resonator. According to Equation (6.6-5), we can write

$$\begin{aligned} E(t+\tau) &= \sum_m C_m \exp\left\{ i\left[(\Omega_0 + m\Omega)\left(t + \frac{2\pi}{\Omega} \right) + \phi_m \right] \right\} \\ &= E(t)\exp(i2\pi\Omega_0/\Omega) \end{aligned} \tag{6.12-4}$$

Note that $E(t + \tau)$ is identical to $E(t)$, except for a constant phase factor $\Delta\phi = 2\pi\Omega_0/\Omega$. Since Ω_0 is, in general, not an integral multiple of Ω, this phase shift at $t = \tau$ is not an integral multiple of 2π. As a result, the peak of the actual electric field may not coincide with

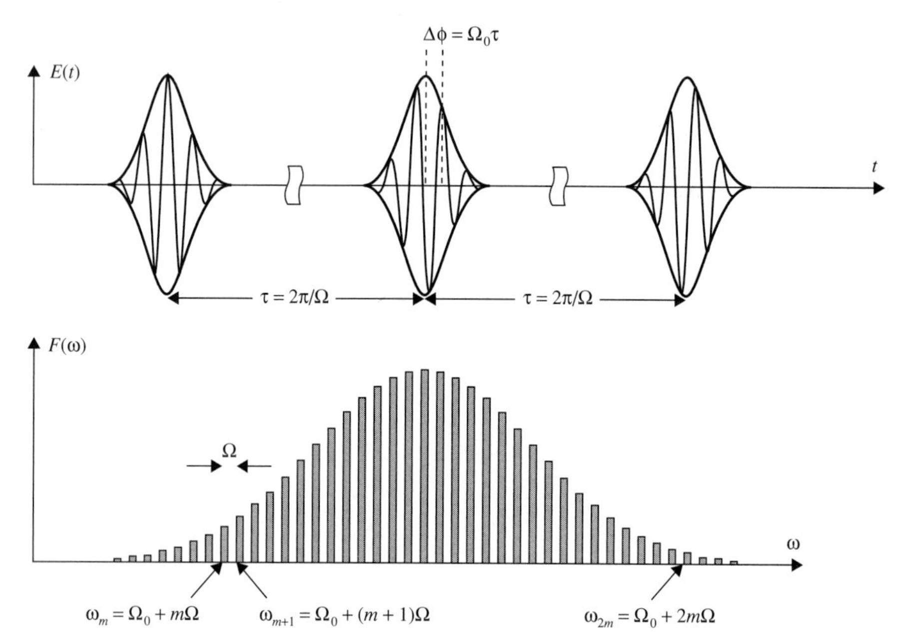

Figure 6.54 A periodic pulse train and its frequency spectrum. The frequency spectrum consists of a discrete set of frequencies that are exactly equally spaced, provided the pulse train is exactly periodic in time.

the peak of the envelope. Figure 6.54 shows the periodic pulse train and the corresponding frequency comb.

As discussed earlier in this chapter, the pulse width is inversely proportional to the gain bandwidth. In the optical spectral regime, the carrier frequency is in the range of a few hundred terahertz (e.g., green light at $\lambda = 500$ nm corresponds to a frequency of $\nu = 600$ THz). A pulse with a temporal duration of few femtoseconds would require a gain bandwidth of approximately a few hundred terahertz. These ultrashort pulses consist of a few optical cycles within the pulse envelope (see Figure 6.54). Mode-locked pulses in the femtosecond regime can be employed for application in optical frequency metrology and, equivalently, the standard of time unit. The discussion here follows closely that of Reference [51]. In the microwave regime, we are able to count the exact number of cycles of the electromagnetic waves. One second in SI units is defined as the time duration that is exactly 9,192,631,770 cycles of oscillation of the cesium clock. The cesium clock is tuned to resonance with the ground-state hyperfine splitting of the cesium atoms, which are kept at an extremely low temperature. In the optical regime, most spectroscopic approaches rely on the measurement of optical wavelengths rather than frequencies. The spectral resolution is limited by several practical issues, including the transverse aperture of the optical beam as well as the physical size of the instrument. For example, the spectral resolution of a grating is of the order of $1/N$, where N is the number of periods. A resolution of $\Delta\lambda/\lambda \approx 10^{-12}$ would require a grating with 10^{12} periods, which is virtually impossible. If we are able to count the number of cycles of oscillation of an optical beam, then the spectral resolution can be at least $\Delta\nu/\nu \approx 10^{-15}$.

By virtue of its periodic nature, the frequency comb provided by a mode-locked laser,

$$\omega_m = \Omega_0 + m\Omega \tag{6.12-5}$$

can be employed as an "optical ruler" for the measurement of optical frequencies, provided both Ω_0 and Ω are kept constant. The integer m (in the range of 10^5–10^6) in the above equation can be defined in such a way that the offset frequency Ω_0 is less than Ω. In this case, Ω_0 may not be the frequency of a mode of oscillation. Equation (6.12-5) maps two radio frequencies Ω_0 and Ω onto the optical frequency ω. The modulation frequency Ω can be kept constant, for example, by using a modulator that is locked to the cesium clock, operating at 9,192,631,770 cycles of oscillation per second. In the case of passive mode locking (e.g., via saturable absorbers), the repetition frequency Ω, which lies between a few tens of megahertz and a few gigahertz, can easily be measured with a photodetector. The offset frequency Ω_0, however, is not easily measurable unless the frequency comb contains more than an optical octave. The measurement of the offset frequency in this case involves the beating of a selected frequency from the comb and the second harmonic of a different frequency of the same comb. This is described as follows.

Referring to the frequency comb in Figure 6.54, we consider a mode of oscillation in the low-frequency portion of the comb whose frequency is given according to Equation (6.12-5) by $\omega_m = \Omega_0 + m\Omega$. Using second-harmonic generation in a nonlinear crystal, we obtain a beam of light at a frequency of $2\omega_m = 2\Omega_0 + 2m\Omega$. If the frequency comb covers a full optical octave, then a mode with a mode number $2m$ should oscillate simultaneously at $\omega_{2m} = \Omega_0 + 2m\Omega$. The mixing of the frequency-doubled mode (m) and the mode at $2m$ yields a beating (difference) frequency of

$$2\omega_m - \omega_{2m} = (2\Omega_0 + 2m\Omega) - (\Omega_0 + 2m\Omega) = \Omega_0 \tag{6.12-6}$$

which is exactly the offset frequency. With this approach, both the repetition frequency Ω and the offset frequency Ω_0 are determined. As discussed in Chapter 4, both frequencies Ω and ω_0 depend on the cavity length. Frequency stabilization is needed to ensure a stable comb. This can be achieved by phase locking both Ω and Ω_0 to a precise radio frequency reference (e.g., Cs clock). Once Ω and Ω_0 are precisely determined, each frequency of the comb is precisely determined. It is even possible to lock the offset frequency at exactly zero ($\Omega_0 = 0$). In this case, the frequency comb is an exact integral multiple of Ω and the electric field is a periodic function with a period of $2\pi/\Omega$.

For the purpose of optical frequency metrology, a mode-locked laser with a gain spectrum spanning a full optical octave is needed. In the following we discuss the Ti:sapphire laser, which can provide the broad bandwidth needed for optical frequency metrology. The discussion here follows closely that of Reference [52]. The active laser particles in the Ti:sapphire laser are Ti^{3+} ions present in Al_2O_3 crystals. Typical concentrations are $\sim 10^{19}$–10^{20} cm^{-3} [53]. The energy level diagram of Ti^{3+} ions in sapphire crystal is shown in Figure 6.55.

The electronic structure of the Ti^{3+} ion is a closed shell (argon) plus a single 3d electron. The energy state (3d) of the free Ti^{3+} ion has a degeneracy of 5 due to the spherical symmetry of free space. In the host crystal (Al_2O_3), the Ti^{3+} ion is surrounded by six oxygen atoms situated at the corners of a slightly distorted octahedron. These oxygen atoms produce a nonspherical electric potential at the Ti^{3+} ion. The lower symmetry of the crystal field leads to removal of the degeneracy. As a result, the energy levels are split by the crystal field, which can be viewed as a combination of a cubic field and a trigonal field (sapphire crystal has a point group symmetry of $\bar{3}\,m$). The cubic field dominates and splits the energy state into a triply degenerate $^2T_{2g}$ ground state and a doubly degenerate excited state ^2E_g [54]. The trigonal field further splits the ground state $^2T_{2g}$ into two levels, ^2E and $^2A_1$, and the lower one is further split into two levels by the spin–orbit interaction. Figure 6.56 shows the absorption cross section of Ti^{3+} ion due to the transition $^2T_2 \rightarrow {}^2E$. Pumping of the Ti:sapphire laser can

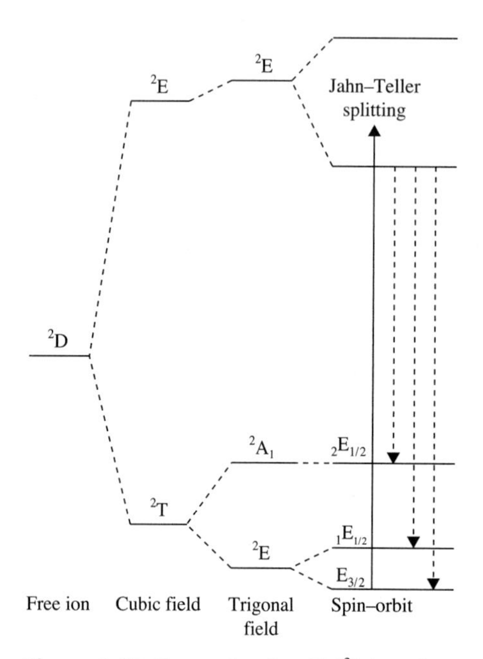

Figure 6.55 Energy levels of Ti^{3+} ions in sapphire crystal showing the effect due to crystal–field, spin–orbit, and Jahn–Teller splitting.

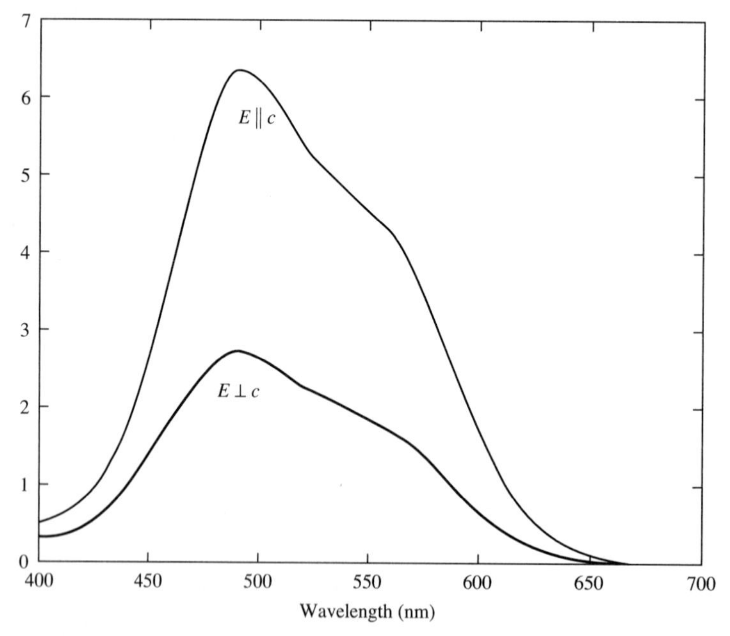

Figure 6.56 Absorption cross section (in units of 10^{-20} cm^2) for the transition $^2T_2 \rightarrow {}^2E$ in the Ti:sapphire laser as a function of wavelength (in units of nm). E = polarization of electric field; c = axis of symmetry of crystal. (Data from Reference [52].)

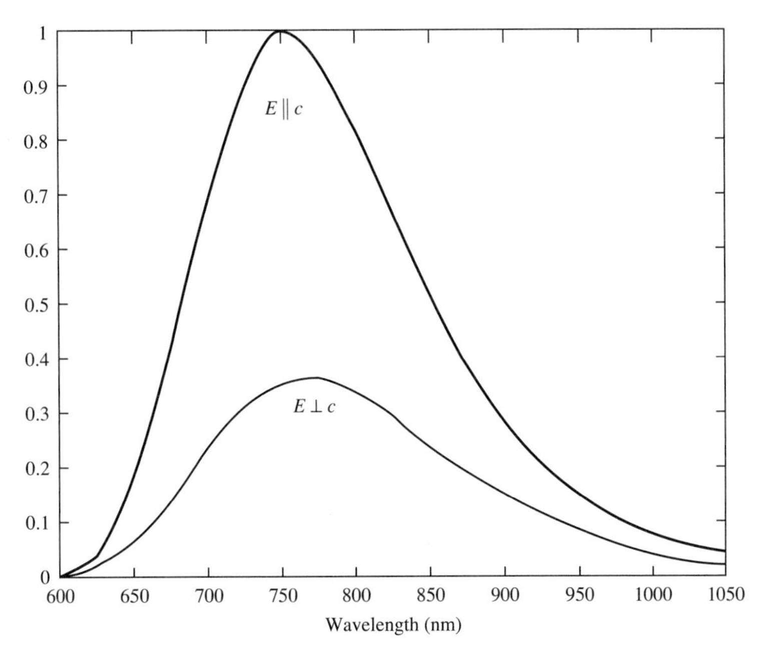

Figure 6.57 Emission spectra of the Ti:sapphire laser. E = polarization of electric field; c = axis of symmetry of crystal. (Data from Reference [52].)

be performed by subjecting it to the light of intense flashlamps (see Figure 6.44). The main peak of absorption spectrum occurs at around λ = 500 nm. The broad absorption peak at around 500 nm is a good match to the 514.5 nm line of the argon ion laser or the doubled YAG wavelength at 532 nm for the purpose of laser optical pumping. When the Ti^{3+} ion is excited to the ^2E_g manifold, it quickly (~10^{-12} s) decays to the minimum energy states within the manifold via the emission of phonons. The system can then decay to the ground state with the emission of a photon and one or more phonons, with a decay lifetime of 3.15 μs at room temperature. Phonons and the nature of the electronic transition are responsible for the broad gain bandwidth of the Ti:sapphire laser.

A close examination of the absorption spectrum reveals actually a dual peak, which is due to the Jahn–Teller splitting of the ^2E state [54]. In addition to the effect of the crystal field, the electronic states are further affected by the interaction between the electron and the vibration of the crystal lattice (phonon). As a result, the doubly degenerate excited state is split into two levels (known as Jahn–Teller splitting [55]). At finite temperatures, the vibration of the Ti ion and the vibration of the crystal lattice lead to a multitude of vibrational states. As a result of homogeneous broadening and inhomogeneous broadening, these states overlap and form a continuum. Depending on the temperature of the crystal and the phonon density in the crystal, these two upper levels can span an energy range of several thousand cm^{-1}. (The unit 1 cm^{-1} corresponds to ν = 30 GHz, or an energy of $h\nu = 1.24 \times 10^{-4}$ eV). As a result, the Ti:sapphire laser exhibits a broad emission spectrum that spans from 600 to 1100 nm (see Figure 6.57). As discussed earlier, a broad gain bandwidth is a prerequisite for ultrashort pulses. To ensure the whole frequency comb spans an optical octave, pulse compression schemes described earlier in this chapter can be employed. This can lead to a frequency comb that is broader than that provided by the gain bandwidth of the laser medium.

PROBLEMS

6.1 Show that the effect of frequency pulling by the atomic medium is to reduce the intermode frequency separation from $c/2l$ to

$$\frac{c}{2l}\left(1-\frac{\gamma c}{2\pi\Delta\nu}\right)$$

where the symbols are defined in Section 6.2. Calculate the reduction for the case of a laser with $\Delta\nu = 10^9$ Hz, $\gamma = 4 \times 10^{-2}$ m^{-1}, and $l = 100$ cm.

6.2 Derive Equation (6.4-3).

6.3 Derive the optimum coupling condition (Equation (6.5-11)).

6.4 Calculate the critical fluorescence power P_s of the He–Ne laser operating at 6328 Å. Assume $V = 2$ cm^3, $L = 1\%$ per pass, $l = 30$ cm, and $\Delta\nu = 1.5 \times 10^9$ Hz.

6.5 Calculate the critical inversion density N_t of the He–Ne laser described in Problem 6.4.

6.6 Derive an expression for the finesse of a Fabry–Perot etalon containing an inverted population medium. Assume that $r_1^2 = r_2^2 \cong 1$ and that the inversion is insufficient to result in oscillation. Compare the finesse to that of a passive Fabry–Perot etalon.

6.7 Derive an expression for the maximum gain–bandwidth product of a Fabry–Perot regenerative amplifier, that is, with an amplifying medium. Define the bandwidth as the frequency region in which the intensity gain $(E_t E_t^*)/(E_i E_i^*)$ exceeds half its peak value. Assume that $\nu_0 = \nu_m$.

6.8

(a) Derive Equation (6.6-7) and show that the intensity is periodic in time.

(b) Derive Equation (6.6-26)

6.9

(a) Describe qualitatively what one may expect to see in parts A, B, C, and D of the mode-locking experiment sketched in Figure 6.14. (The reader may find it useful to read first the section on photomultipliers in Chapter 11.)

(b) What is the effect of mode locking on the intensity of the beat signal (at $\Omega = \pi c/l$) displayed by the RF spectrum analyzer in B? Assume N equal amplitude modes spaced by Ω whose phases before mode locking are random. [Answer: Mode locking increases the beat signal power by N.]

(c) Show that a standing wave at $\nu_0 + \delta$ (ν_0 is the center frequency of the Doppler-broadened lineshape function) in a gas laser will burn the same two holes in the velocity distribution function (see Figure 6.37) as a field at $\nu_0 - \delta$.

(d) Can two traveling waves, one at $\nu_0 + \delta$ the other at $\nu_0 - \delta$, interact with the same class of atoms? If the answer is yes, under what conditions?

6.10 Design a frequency stabilization scheme for gas lasers based on the Lamb dip (see Figure 6.38). [Hint: You may invent a new scheme, but, failing that, consider what happens to the phase of the modulation in the power output when the cavity length is modulated sinusoidally near the bottom of the Lamb dip. Can you derive an error correction signal from this phase that will control the cavity length?]

6.11 Verify the relations of Table 6.2.

6.12 A helium–neon laser ($\lambda = 0.63$ μm) operating in the fundamental transverse mode has mirrors separated by $l = 30$ cm. The Doppler width is $\Delta\nu_D = 1.5$ GHz, and the effective refractive index is $n = 1$. The output mirror is flat, and the other mirror is spherical with a radius of curvature 16 m.

(a) What is the frequency difference between longitudinal modes in the resonator?

(b) Show that the resonator is stable.

(c) What would the Doppler width become if the temperature of the laser medium were doubled?

(d) What is the spot size at the flat mirror?

(e) If the output is taken from the flat mirror, what is the spot size 16 km away?

(f) Given that the internal cavity loss is $L_i = 10^{-1}$, the small signal gain coefficient is $\gamma_0 = 10^{-3}$ cm^{-1}, and the reflection coefficient of the spherical mirror is 1.0, what is the reflection coefficient (R) of the output mirror that will give the maximum output power?

(g) A thin lens of focal length f is placed against the output mirror. Find the radius of curvature of the Gaussian beam at a distance $d = f$ from the lens.

6.13 Derive the expression relating the absorption cross section at ν in a given $a{\to}b$ transition to the spontaneous $b{\to}a$ lifetime.

6.14

$$E(t) = \sum_{n=1}^{N} C_n \exp(i\omega_0 t + in\Omega t)$$

For the following, take $N = 5$ and plot in the range Ωt $(-4\pi, 4\pi)$.

(a) Generate a set of N random phases (ϕ_n), and then plot $|E(t)|$ versus Ωt, with $C_n = \exp(i\phi_n)/\sqrt{N}$.
(b) Let $C_n = 1/\sqrt{N}$. Plot $|E(t)|$ versus Ωt.
(c) Let $C_n = \sin\{[s\pi/(N+1)]n\}/\sqrt{(N+1)/2}$. Plot $|E(t)|$ versus Ωt for $s = 1, 2, 3, 4, 5$.
(d) For (b) and (c), derive analytic expressions for the electric field. [Hint: For (c), use $\sin x = (e^{ix} - e^{-ix})/2i$, and sum up the geometric series.]

6.15 Consider an asymmetric Fabry–Perot etalon, with mirror reflectivities R_1 and R_2. The linear attenuation coefficient of the cavity medium is α.

(a) Find the etalon transmission as a function of frequency for normal incidence.
(b) Find the full width at half-maximum $(\Delta\nu_{1/2})_e$ of the transmission peaks.
(c) An external laser beam is in resonance with the etalon. Find the exponential decay constant of the energy stored in the etalon when the laser beam is turned off. This is the photon lifetime t_c in the etalon.
(d) Write down an expression for the E field of electromagnetic radiation emitted from the etalon after the laser beam is turned off. Find the Fourier transform of the E field. Find the full width at half-maximum $(\Delta\nu_{1/2})_s$ of the Fourier spectrum.
(e) Assume $|1 - \sqrt{R_1R_2}\exp(-\alpha l)| \ll 1$. Compare $(\Delta\nu_{1/2})_e$ and $(\Delta\nu_{1/2})_s$ and find their relationship with t_c.
(f) Assume $|1 - \sqrt{R_1R_2}\exp(-\alpha l)| \ll 1$. Find the relationship between cavity Q-factor and the finesse F.

6.16

(a) Substitute Equation (6.7-18) into Equation (6.7-17) and show that

$$\left(1 + \frac{2g_0}{\Omega_g^2\tau^2}\right)\tau^4 = \frac{g_0}{4\sigma_1(\Omega\Omega_g)^2} \quad \text{and}$$

$$g_0 - L = 2\sigma_1\Omega^2\tau^2\left(1 + \frac{g_0}{\Omega_g^2\tau^2}\right)$$

(b) For mode-locked lasers with $g_0 \ll \Omega_g^2\tau^2$, the above equations become

$$\tau^4 = \frac{g_0}{4\sigma_1(\Omega\Omega_g)^2} \quad \text{and} \quad g_0 - L = 2\sigma_1\Omega^2\tau^2$$

Show that the condition $g_0 \ll \Omega_g^2\tau^2$ is equivalent to $4\sigma_1\Omega^2\tau^2 \ll 1$.

6.17 Substitute $E(t) = E_0\,\mathrm{sech}(t/\tau)$ into Equation (6.7-31). Show that $\Delta E(t) = 0$ requires

$$\tau^2 = \frac{2g_0}{\gamma E_0^2\Omega_g^2} \quad \text{and} \quad L - g_0 = \frac{g_0}{\Omega_g^2\tau^2}$$

6.18

(a) Show that for large N $(N \to \infty)$, the full width at half-maximum (FWHM) of the pulse given by Equation (6.6-8) is

$$\Delta t_{1/2} = 0.886\frac{\tau}{N}$$

where $\tau = 2\pi/\Omega$.

(b) Show that for large $N(N \to \infty)$, the FWHM of the intensity of the pulse given by Equation (6.6-26) is

$$\Delta t_{1/2} = 1.19\frac{\tau}{N}$$

We note that the pulse width is about 34% broader.

6.19 Consider a pair of identical diffraction gratings with a separation d and a period Λ. The two gratings are parallel. For simplicity, we assume that they are transmission gratings. Let τ be the group delay, find the group velocity dispersion $d\tau/d\lambda$ as a function of the angle of incidence. Group delay dispersion in double gratings can be employed for pulse compression.

REFERENCES

1. Schawlow, A. L., and C. H. Townes, Infrared and optical masers. *Phys. Rev.* **112**:1940 (1958).
2. Yariv, A., *Quantum Electronics*, 2nd ed. Wiley, New York, 1975.
3. Laures, P., Variation of the 6328 Å gas laser output power with mirror transmission. *Phys. Lett.* **10**:61 (1964).
4. Bennett, W. R., Jr., Gaseous optical masers. *Appl. Opt. Suppl. 1 Optical Masers*, 24 (1962).
5. Fork, R. L., D. R. Herriott, and H. Kogelnik, A scanning spherical mirror interferometer for spectral analysis of laser radiation. *Appl. Opt.* **3**:1471 (1964).

6. DiDomenico, M., Jr., Small signal analysis of internal modulation of lasers. *J. Appl. Phys.* **35**:2870 (1964).
7. Yariv, A., Internal modulation in multimode laser oscillators. *J. Appl. Phys.* **36**:388 (1965).
8. Hargrove, L. E., R. L. Fork, and M. A. Pollack, Locking of He–Ne laser modes induced by synchronous intracavity modulation. *Appl. Phys. Lett.* **5**:4 (1964).
9. DiDomenico, M., Jr., J. E. Geusic, H. M. Marcos, and R. G. Smith, Generation of ultrashort optical pulses by mode locking the Nd^{3+}:YAG laser. *Appl. Phys. Lett.* **8**:180 (1966).
10. Mocker, H., and R. J. Collins, Mode competition and self-locking effects in a *Q*-switched ruby laser. *Appl. Phys. Lett.* **7**:270 (1965).
11. DeMaria, A. J., Mode locking. *Electronics* Sept. 16:112 (1968).
12. Valdmanis, J. A., R. L. Ford, and J. P. Gordon, *Opt. Lett.* **10**:131 (1985).
13. Shapiro, S. L. (ed.), *Ultrashort Light Pulses* (Topics in Applied Physics, Vol. 18). Springer-Verlag, New York, 1977.
14. Hochstrasser, R. M., W. Kaiser, and C. V. Shank (eds.), *Picosecond Phenomena II* (Series in Chemical Physics, Vol. 14). Springer-Verlag, New York, 1980.
15. Eisenthal, K. B., R. M. Hochstrasser, W. Kaiser, and A. Laubereau (eds.), *Picosecond Phenomena III* (Series in Chemical Physics, Vol. 23). Springer-Verlag, New York, 1982.
16. Schrans, T., *Part I: Longitudinal Static and Dynamic Effects in Semiconductor Lasers; Part II: Spectral Characteristics of Passively Mode-Locked Quantum Well Lasers*. Ph.D. thesis, California Institute of Technology, 1994.
17. Maiman, T. H., Stimulated optical radiation in ruby masers. *Nature* **187**:493 (1960).
18. Sanders, S., L. Eng, J. Paslaski, and A. Yariv, 108 GHz passive mode locking of a multiple quantum well semiconductor laser with an intracavity absorber. *Appl. Phys. Lett.* **56**:310 (1990).
19. Koch, T. L., *Gigawatt Picosecond Dye Lasers and Ultrafast Processes in Semiconductor Lasers*. Ph.D. thesis, California Institute of Technology, 1982.
20. Yariv, A., *Quantum Electronics*, 3rd ed. Wiley, New York, 1989, p. 550.
21. Siegman, A. E., and D. J. Kuizenga, Simple analytic expressions for AM and FM mode locked pulses in homogeneous lasers. *Appl. Phys. Lett.* **14**:181 (1969).
22. Kuizenga, D. J., and A. E. Siegman, FM and AM mode locking of the homogeneous laser: Part I, Theory; Part II, Experiment. *J. Quantum Electron.* **QE- 6**:694 (1970).
23. Siegman, A. E., *Lasers*. University Science Books, Mill Valley, CA, 1986.
24. Tracy, E. B., *IEEE J. Quantum Electron.* **5**:454 (1969).
25. Haus, H. A., Mode-locking of lasers. *IEEE J. Selected Topics Quantum Electron.* **6**(6):1173 (2000).
26. Kuizenga, D. I., and A. E. Siegman, Modulator frequency detuning effects in the FM mode-locked lasers. *IEEE J. Quantum Electron.* **6**:803 (1970).
27. Smith, P. J., T. J. Bridges, and E. J. Burkhardt, Mode locked high pressure CO_2 laser. *Appl. Phys. Lett.* **21**:470 (1972).
28. Martinez, O. E., J. P. Gordon, and R. L. Fork, *J. Opt. Soc. Am. A* **1**:1003 (1984).
29. Martinez, O. E., *IEEE J. Quantum Electron.* **23**:59 (1987).
30. Wagner, W. G., and B. A. Lengyel, Evolution of the giant pulse in a laser. *J. Appl. Phys.* **34**:2042 (1963).
31. Hellwarth, R. W., Control of fluorescent pulsations. In: *Advances in Quantum Electronics*, J. R. Singer (ed.). Columbia University Press, New York, 1961, p. 334.
32. McClung, F. J., and R. W. Hellwarth, Giant Optical Pulsations from Ruby. *J. Appl. Phys.* **33**:828 (1962).

33. Hellwarth, R. W., *Q* modulation of lasers. In: *Lasers*, Vol. 1, A. K. Levine (ed.). Marcel Dekker, New York, 1966, p. 253.
34. DeMaria, A. J., Picosecond laser pulses. *Proc. IEEE* **57**:3 (1969).
35. Lamb, W. E., Jr., Theory of an optical maser. *Phys. Rev. A* **134**:1429 (1964).
36. Szöke, A., and A. Javan, Isotope shift and saturation behavior of the *1.15jL* transition of neon. *Phys. Rev. Lett.* **10**:512 (1963).
37. Bloom, A., *Gas Lasers*. Wiley, New York, 1963, p. 93.
38. Maiman, T. H., Optical and microwave-optical experiments in ruby. *Phys. Rev. Lett.* **4**:564 (1960).
39. Cronemeyer, D. C., Optical absorption characteristics of pink ruby. *J. Opt. Soc. Am.* **56**:1703 (1966).
40. Schawlow, A. L., Fine structure and properties of chromium fluorescence. In: *Advances in Quantum Electronics*, J. R. Singer (ed.). Columbia University Press, New York, 1961, p. 53.
41. Yariv, A., Energy and power considerations in injection and optically pumped lasers. *Proc. IEEE* **51**:1723 (1963).
42. Geusic, J. E., H. M. Marcos, and L. G. Van Uitert, Laser oscillations in Nd-doped yttrium aluminum, yttrium gallium and gadolinium garnets. *Appl. Phys. Lett.* **4**:182 (1964).
43. Kushida, T., H. M. Marcos, and J. E. Geusic, Laser transition cross section and fluorescence branching ratio for Nd^{3+} in yttrium aluminum garnet. *Phys. Rev.* **167**:1289 (1968).
44. Snitzer, E., and C. G. Young, Glass lasers. In: *Lasers*, Vol. 2, A. K. Levine (ed.). Marcel Dekker, New York, 1968, p. 191.
45. Simon, J. C., Semiconductor laser amplifier for single mode optical fiber communications. *J. Opt. Commun.* **4**:51 (1983).
46. Mears, R. J., L. Reekie, I. M. Jauncey, and D. N. Payne, Low noise erbium-doped fiber amplifier operating at 1.54 mm. *Electron. Lett.* **23**:1026 (1987).
47. Hagimoto, K., et al., A 212 km non-repeatered transmission experiment at 1.8 Gb/s using LD pumped Er^{+}-doped fiber amplifiers in an 1m/direct-detection repeater system. In: *Proceedings of the Optical Fiber Conference, Houston, IX*. Postdeadline Paper PD15, 1989.
48. Olshansky, R., Noise figure for Er-doped optical fibre amplifiers. *Electron. Lett.* **24**:1363 (1988).
49. Payne, D. N., Tutorial session abstracts. In: *Optical Fiber Communication (OFC 1990) Conference*, San Francisco, 1990.
50. See, for example, G. Eisenstein, U. Koren, G. Raybon, T. L. Koch, M. Wiesenfeld, M. Wegener, R. S. Tucker, and B. I. Miller, Large-signal and small-signal gain characteristics of 1.5 μm quantum well optical amplifiers. *Appl. Phys. Lett.* **56**:201 (1990).
51. Udem, Th., R. Holzwarth, and T. W. Hänsch, Optical frequency metrology. *Nature* **416**:233 (14 Mar. 2002).
52. Moulton, P. F., Spectroscopic and laser characteristics of $Ti{:}Al_2O_3$. *J. Opt. Soc. Am. B* **3**(1):125 (1986).
53. McKinnie, I. T., A. L. Oien, D. M. Warrington, P. N. Tonga, L. A. W. Gloster, and T. A. King, Ti ion concentration and Ti:sapphire laser performance. *IEEE J. Quantum Electron.* **33**(7):1221 (1997).
54. In group theory, A_1 and A_2 stand for one-dimensional irreducible representations, E stands for a two-dimensional irreducible representation, and T_1 and T_2 stand for three-dimensional irreducible representation of point group O. The subscripts *g* and *u*, for gerade and ungerade, indicate even and odd representations under inversion. See, for example, M. Tinkham, *Group Theory and Quantum Mechanics*. McGraw-Hill, New York, 1964.
55. Jahn, H. A., and E. Teller, *Proc. Roy. Soc. London A* **161**:220 (1937).

ADDITIONAL READING

Becker, P. C., N. A. Olsson, and J. R. Simpson, *Erbium-Doped Fiber Amplifiers: Fundamentals and Technology*, Academic Press, San Diego, 1999.

Desurvire, E., *Erbium-Doped Fiber Amplifiers: Principles and Applications*. Wiley, New York, 2002.

Digonnet, M. J. F., *Rare-Earth Doped Fiber Lasers and Amplifiers*. Marcel Dekker, New York, 2001.

CHAPTER 7

CHROMATIC DISPERSION AND POLARIZATION MODE DISPERSION IN FIBERS

7.0 INTRODUCTION

Chromatic dispersion is an intrinsic property of virtually all optical materials. The phenomenon is manifested as a dependence of the phase velocity (and group velocity) of a beam of light in a transparent material on the frequency (or wavelength) of light. Similarly, chromatic dispersion occurs in optical fibers that are made of transparent materials. As a result of chromatic dispersion, an optical pulse spreads as it propagates along the fiber. The spreading leads to signal fading when neighboring pulses overlap. The problem becomes progressively more severe as the propagation distance and the data rate increase. In addition to chromatic dispersion, optical fibers also exhibit birefringent properties in which the phase and (group) velocity depend on the beam polarization. The birefringence is a result of fiber imperfection and/or external perturbations. As a result, the two LP modes (e.g., LP_{01x} and LP_{01y}) are no longer degenerate. In other words, a time delay occurs between the two principal modes of propagation—the slow mode and the fast mode. The time delay leads to a broadening of the optical pulse and to signal fading. In this chapter we investigate quantitatively the effect of chromatic dispersion and birefringence on the propagation of pulses in optical fibers.

7.1 CHROMATIC DISPERSION IN OPTICAL TRANSMISSION SYSTEMS

In Chapter 1 we describe chromatic dispersion in terms of the dependence of the index of refraction on the frequency (or wavelength) of light. This is important for studying the propagation of optical pulses in a homogeneous medium. As a result of chromatic dispersion, the group velocity of light depends on the frequency (or wavelength). This is known as the group velocity dispersion (GVD). In optical electronics, we often deal with the transmission of optical waves in various optical systems, including fibers, modulators, and amplifiers. The group velocity dispersion in a general optical system can be described by examining the phase shift as a function of the frequency.

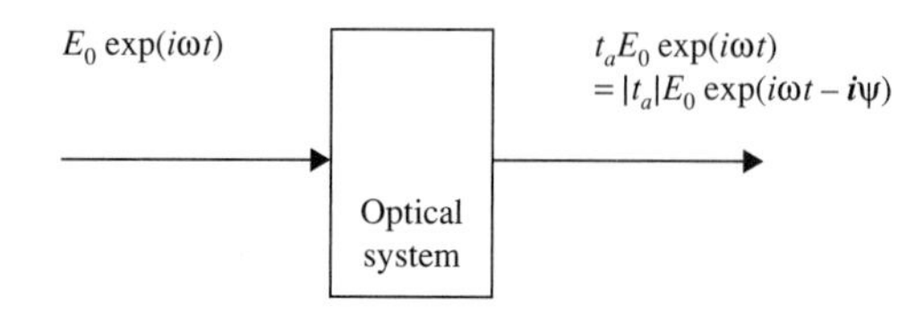

Figure 7.1 Schematic drawing of a general optical transmission system, where t_a is the amplitude transmission coefficient. The box can be a spool of fiber, a Fabry–Perot etalon, a spectral filter, and so on.

Referring to Figure 7.1, we consider the transmission of a monochromatic beam of light through a general optical system.

The amplitude transmission coefficient t_a can be written

$$t_a = \sqrt{T} \exp(-i\psi) \tag{7.1-1}$$

where T is the intensity transmission coefficient and ψ is the phase shift. In a homogeneous medium, the phase shift is simply

$$\psi = kL = \frac{\omega}{c} nL \tag{7.1-2}$$

where ω is the angular frequency, n is the index of refraction, and L is the length of propagation.

We consider the transmission of an optical pulse. Let the input optical pulse be written as a product of the optical carrier $\exp(i\omega_0 t)$ and an envelope function $s(t)$:

$$E_{\text{in}}(t) = s(t)e^{i\omega_0 t} = e^{i\omega_0 t} \int A(\Omega) e^{i\Omega t} d\Omega \tag{7.1-3}$$

where ω_0 is the carrier frequency of the pulse and $A(\Omega)$ is the Fourier spectrum of the envelope pulse. The envelope function $s(t)$ can be viewed as a time-varying amplitude. The output pulse after transmitting through the optical system in Figure 7.1 can be written

$$E_{\text{out}}(t) = s'(t)e^{i\omega_0 t} = e^{i\omega_0 t} \int A(\Omega) e^{i\Omega t} \sqrt{T} e^{-i\psi} d\Omega \tag{7.1-4}$$

We consider the situation where the intensity transmission is uniform over the optical spectral regime of the input pulse. In this case, the output pulse can be written

$$E_{\text{out}}(t) = s'(t)e^{i\omega_0 t} = e^{i\omega_0 t} \sqrt{T} \int A(\Omega) e^{i\Omega t} e^{-i\psi} d\Omega \tag{7.1-5}$$

Generally, the output pulse shape $s'(t)$ is different from the input pulse shape $s(t)$. This is a consequence of the dependence of the phase shift $\psi(\Omega)$ on frequency.

In optical communications, especially when the carrier is a laser beam with a well-defined carrier frequency ω_0, the Fourier spectrum of the envelope pulse $A(\Omega)$ is often sharply peaked around $\Omega = 0$. To study the pulse shape of the output pulse, we expand $\psi(\Omega)$ around $\Omega = 0$ in terms of a Taylor series:

$$\psi(\Omega) = \psi_0 + \left(\frac{d\psi}{d\Omega}\right)_0 \Omega + \cdots \tag{7.1-6}$$

which when substituted in Equation (7.1-5) results in

$$E_{\text{out}}(t) = s'(t)e^{i\omega_0 t} = e^{i\omega_0 t} \sqrt{T} e^{-i\psi_0} \int A(\Omega) e^{i\Omega t} e^{-i(d\psi/d\Omega)_0 \Omega} d\Omega \tag{7.1-7}$$

where we have neglected the higher-order terms in Ω. Comparing the last equation with Equation (7.1-3), we can rewrite it as

$$E_{out}(t) = s'(t)e^{i\omega_0 t} = e^{i\omega_0 t}\sqrt{T}\,e^{-i\psi_0}s(t-\tau_0) \tag{7.1-8}$$

where

$$\tau_0 = \left(\frac{d\psi}{d\Omega}\right)_0 \tag{7.1-9}$$

is the time of flight (or propagation time) of the pulse envelope.

This shows that apart from an overall phase factor and an amplitude factor, the output pulse remains identical to that of the input pulse, with a time delay of $(d\psi/d\Omega)_0$. This approximation is legitimate provided the spectral distribution $A(\Omega)$ is sharply peaked around $\Omega = 0$, and the phase shift is a smoothly varying function of Ω. Thus the propagation time delay, referred to as the group delay, due to propagation through the optical system described in Figure 7.1 is given by

$$\tau = \frac{d\psi}{d\omega} \tag{7.1-10}$$

which is, in general, a function of frequency. In a homogeneous medium, $\psi = kL$, Equation (7.1-10) becomes

$$\tau = \frac{dk}{d\omega}L = \frac{L}{v_g} \tag{7.1-11}$$

where v_g is the group velocity.

If the flight time τ is independent of the frequency, then the group delay has no dispersion. This is the situation when all frequency components of an optical pulse travel at exactly the same group velocity, and thus the same flight time τ. In this case, the pulse shape remains the same. In the event when the group delay τ is a function of frequency, then the output pulse may spread. The pulse spread can be estimated as follows, provided the pulse is not severely distorted:

$$\Delta\tau = \left(\frac{d\tau}{d\omega}\right)_0 \Delta\omega = \left(\frac{d^2\psi}{d\omega^2}\right)_0 \Delta\omega \tag{7.1-12}$$

where $\Delta\omega$ is the spectral spread of the input pulse. We note that the pulse spread is proportional to the product of the second-order derivative of the phase shift with respect to the frequency, and the spectral bandwidth of the input pulse.

In optical communications, the pulse spread is often written in terms of the wavelength spread $\Delta\lambda$:

$$\Delta\tau = \left(\frac{d\tau}{d\lambda}\right)_0 \Delta\lambda \tag{7.1-13}$$

where we can define

$$\tau_\lambda = \left(\frac{d\tau}{d\lambda}\right)_0 \quad \text{(ps/nm)} \tag{7.1-14}$$

as the group delay dispersion, which is often expressed in units of ps/nm. For propagation in a single-mode fiber, the phase shift is simply the product of the propagation constant and the length of the fiber,

$$\psi = \beta L \tag{7.1-15}$$

where β is the propagation constant of the mode. Using Equations (7.1-12), (7.1-13), (7.1-14), and (7.1-15), we obtain

$$\tau_\lambda = -\frac{2\pi c}{\lambda^2}\left(\frac{d^2\psi}{d\omega^2}\right)_0 = -\frac{2\pi c}{\lambda^2}\left(\frac{d^2\beta}{d\omega^2}\right)_0 L \tag{7.1-16}$$

We note that τ_λ is proportional to the length of propagation. For optical engineering, it is convenient to define the group velocity dispersion D in units of ps/km-nm as

$$D = -\frac{2\pi c}{\lambda^2}\left(\frac{d^2\beta}{d\omega^2}\right)_0 \tag{7.1-17}$$

so that τ_λ is given by

$$\tau_\lambda = DL \tag{7.1-18}$$

where L is the length of the fiber. For example, consider a single-mode fiber with a length of $L = 100$ km and a dispersion of $D = 17$ ps/nm-km; then the group delay dispersion is $\tau_\lambda = 1700$ ps/nm.

In the event when the group delay τ is dependent on the frequency, the output shape can be obtained by using the following expression of the phase shift and Equation (7.1-5):

$$\begin{aligned}\psi(\Omega) &= \psi_0 + \tau_0\Omega + \frac{1}{2}\left(\frac{d^2\psi}{d\omega^2}\right)_0 \Omega^2 + \cdots \\ &= \psi_0 + \tau_0\Omega - \frac{\pi c}{\omega_0^2}\tau_\lambda\Omega^2 + \cdots\end{aligned} \tag{7.1-19}$$

In the next section, we consider the transmission of an optical pulse with a Gaussian pulse shape through a dispersive system.

In optical communications networks, signal pulses may have to travel through various optical elements or subsystems. The total group delay is the sum of the group delays in each transmission system. Consider, for example, the transmission through two optical systems in series. The total phase shift can be written

$$\psi = \psi_1 + \psi_2 \tag{7.1-20}$$

where ψ_1 and ψ_2 are the phase shifts of the optical systems, respectively. The total group delay is given by

$$\tau = \frac{d\psi}{d\omega} = \frac{d\psi_1}{d\omega} + \frac{d\psi_2}{d\omega} = \tau_1 + \tau_2 \tag{7.1-21}$$

where τ_1 and τ_2 are the individual group delays. We note that τ_1 and τ_2 are both positive. The pulse broadening due to group velocity dispersion (GVD) is given by

$$\Delta\tau = \left(\frac{d\tau_1}{d\lambda}\right)_0 \Delta\lambda + \left(\frac{d\tau_2}{d\lambda}\right)_0 \Delta\lambda \tag{7.1-22}$$

where $\Delta\lambda$ is the spectral bandwidth (in units of nm) of the pulse.

If the two optical systems have group velocity dispersion of opposite signs, then the pulse broadening can vanish provided

$$\left(\frac{d\tau_1}{d\lambda}\right)_0 + \left(\frac{d\tau_2}{d\lambda}\right)_0 = 0 \tag{7.1-23}$$

This is the basic principle of dispersion compensation. In the case of transmission through two optical fibers, Equation (7.1-23) is written, according to Equation (7.1-18),

$$D_1 L_1 + D_2 L_2 = 0 \tag{7.1-24}$$

where D_1 and D_2 are the dispersions of the fibers, and L_1 and L_2 are the fiber lengths. Most single-mode fibers for optical communications exhibit a positive dispersion ($D_1 > 0$). Thus a specially designed optical fiber with negative dispersion ($D_2 < 0$) can be employed for the restoration of pulse shapes.

7.2 OPTICAL PULSE SPREADING IN DISPERSIVE MEDIA

Virtually all the traffic carried by optical fibers in modern communications is in the form of digital pulses, each representing one bit of information. It thus follows that the narrower the pulses, the more of them can be packed into a given (transmission) time slot, and thus more data (bits) can be transmitted during that time. As a matter of fact, modern communications systems are designed with pulse widths as narrow as 3×10^{-11} s and with data rates exceeding 10^{10} bits/s. In a 10 Gb/s system, there are 10 billion bits per second. The trend to ever narrower pulses and higher rates continues unabated. What limits our ability to reduce the pulse width even further is the basic phenomenon of pulse broadening due to the dependence of its group velocity on the frequency. This phenomenon is termed *group velocity dispersion* (*GVD*). To be specific, we consider the realistic scenario of a fiber supporting a single spatial mode, usually the lowest-order fundamental mode, which is excited at $z = 0$ by a temporal Gaussian pulse

$$E(x, y, z = 0, t) = u_0(x, y) \exp(-\alpha t^2 + i\omega_0 t) \tag{7.2-1}$$

where $u_0(x, y)$ is the wavefunction of a confined mode, α is a constant, and ω_0 is the optical carrier frequency. We consider the case of a slowly varying envelope so that there are many optical oscillations (cycles) within the envelope. This corresponds to the situation where $\alpha^{1/2} \ll \omega_0$. We may express the input pulse $E(x, y, 0, t)$ as a Fourier integral

$$E(z = 0, t) = \exp(i\omega_0 t) \int F(\Omega) e^{i\Omega t}\, d\Omega \tag{7.2-2}$$

where $F(\Omega)$ is the Fourier transform of the Gaussian envelope $\exp(-\alpha t^2)$

$$F(\Omega) = \sqrt{\frac{1}{4\pi\alpha}} \exp\left(-\frac{\Omega^2}{4\alpha}\right) \tag{7.2-3}$$

In the above equations, we ignore the wavefunction $u_0(x, y)$. This is legitimate provided the wavefunction remains the same within the spectral band of the signal. Note that the frequency spectrum of a Gaussian pulse is also a Gaussian. We may view Equation (7.2-2) as an assembly of harmonic fields, each with its unique frequency ($\omega_0 + \Omega$) and amplitude $F(\Omega)\, d\Omega$. To obtain the field at an output plane z, we need to multiply each frequency component $F(\Omega)\, d\Omega \exp[i(\omega_0 + \Omega)t]$ in Equation (7.2-2) by its propagation phase delay factor,

$$\exp[-i\beta(\omega_0 + \Omega)z] \tag{7.2-4}$$

The result is

$$E(z, t) = \int F(\Omega) \exp\{i[(\omega_0 + \Omega)t - \beta(\omega_0 + \Omega)z]\} \, d\Omega \tag{7.2-5}$$

We can expand $\beta(\omega_0 + \Omega)$ near the center (optical) frequency ω_0 in a Taylor series

$$\beta(\omega_0 + \Omega) = \beta(\omega_0) + \left.\frac{d\beta}{d\omega}\right|_{\omega_0} \Omega + \frac{1}{2} \left.\frac{d^2\beta}{d\omega^2}\right|_{\omega_0} \Omega^2 + \cdots \tag{7.2-6}$$

and obtain

$$\begin{aligned} E(z, t) &= \exp[i(\omega_0 t - \beta_0 z)] \int_{-\infty}^{\infty} d\Omega \, F(\Omega) \exp\left\{i\left[\Omega t - \frac{\Omega z}{v_g} - \frac{1}{2}\frac{d}{d\omega}\left(\frac{1}{v_g}\right)\Omega^2 z\right]\right\} \\ &\equiv \exp[i(\omega_0 t - \beta_0 z)]\mathscr{E}(z, t) \end{aligned} \tag{7.2-7}$$

where

$$\beta_0 \equiv \beta(\omega_0), \quad \left.\frac{d\beta}{d\omega}\right|_{\omega_0} = \frac{1}{v_g} = \frac{1}{\text{Group velocity}} \tag{7.2-8}$$

The field envelope is given by the integral (7.2-7)

$$\begin{aligned} \mathscr{E}(z, t) &= \int_{-\infty}^{\infty} d\Omega \, F(\Omega) \exp\left\{i\Omega\left[\left(t - \frac{z}{v_g}\right) - \frac{1}{2}\frac{d}{d\omega}\left(\frac{1}{v_g}\right)\Omega z\right]\right\} \\ &= \int_{-\infty}^{\infty} d\Omega \, F(\Omega) \exp\left\{i\Omega\left[\left(t - \frac{z}{v_g}\right) - a\Omega z\right]\right\} \end{aligned} \tag{7.2-9}$$

where a is a constant defined as $a = (d^2\beta/d\omega^2)/2$. Here we neglect the high-order terms (Ω^3 and up) in the Taylor's series expansion in Equation (7.2-6). We note that in the absence of group velocity dispersion when $a = 0$, the envelope can be written

$$\mathscr{E}(z, t) = \int_{-\infty}^{\infty} d\Omega \, F(\Omega) \exp\left\{i\Omega\left(t - \frac{z}{v_g}\right)\right\} = \mathscr{E}\left[0, \left(t - \frac{z}{v_g}\right)\right] \tag{7.2-10}$$

In other words, the pulse envelope remains unchanged and propagates at the group velocity, v_g. This is in agreement with the general discussion in the last section (Equation (7.1-8)). The pulse spreading is caused by the group velocity dispersion characterized by the parameter

$$a \equiv \frac{1}{2}\left.\frac{d^2\beta}{d\omega^2}\right|_{\omega=\omega_0} = \frac{1}{2}\frac{d}{d\omega}\left(\frac{1}{v_g}\right) = -\frac{1}{2v_g^2}\frac{dv_g}{d\omega} \tag{7.2-11}$$

After substituting $F(\Omega)$ from (7.2-3), Equation (7.2-9) becomes

$$\mathscr{E}(z, t) = \sqrt{\frac{1}{4\pi\alpha}} \int_{-\infty}^{\infty} \exp\left\{-\left[\Omega^2\left(\frac{1}{4\alpha} + iaz\right) - i\left(t - \frac{z}{v_g}\right)\Omega\right]\right\} d\Omega \tag{7.2-12}$$

Carrying out the integration, we obtain

$$\mathscr{E}(z,t) = \frac{1}{\sqrt{1+i4a\alpha z}} \exp\left(-\frac{(t-z/v_g)^2}{1/\alpha + 16a^2z^2\alpha}\right) \exp\left(i\frac{4az(t-z/v_g)^2}{1/\alpha^2 + 16a^2z^2}\right) \tag{7.2-13}$$

The pulse duration τ at z can be taken as the separation between the two times when the pulse envelope squared is reduced by a factor of $\frac{1}{2}$ from its peak value (so-called FWHM), that is,

$$\tau(z) = \sqrt{2\ln 2}\sqrt{\frac{1}{\alpha} + 16a^2z^2\alpha} \tag{7.2-14}$$

The initial pulse width is

$$\tau_0 = \left(\frac{2\ln 2}{\alpha}\right)^{1/2} \tag{7.2-15}$$

The pulse width after propagating a distance L can thus be expressed as

$$\tau(L) = \tau_0\sqrt{1 + \left(\frac{8aL\ln 2}{\tau_0^2}\right)^2} \tag{7.2-16}$$

At large distances such that $|aL| \gg \tau_0^2$, we obtain

$$\tau(L) \sim \frac{(8\ln 2)aL}{\tau_0} \tag{7.2-17}$$

If we use the definition (7.2-11) of the factor a, the last expression becomes

$$\tau(L) = \frac{4\ln 2}{v_g^2}\left|\frac{dv_g}{d\omega}\right|\frac{L}{\tau_0} \tag{7.2-18}$$

The group velocity dispersion is often characterized by $D \equiv L^{-1}(dT/d\lambda)$, where T is the pulse transmission time (group delay) through length L of the fiber. This definition is related to the second-order derivative of β with respect to ω as, according to Equation (7.1-17),

$$D = -\frac{2\pi c}{\lambda^2}\left(\frac{d^2\beta}{d\omega^2}\right) \tag{7.2-19}$$

and is related to the parameter a used above by

$$D = -\frac{4\pi c}{\lambda^2}a \tag{7.2-20}$$

With this new definition, the pulse width expression (7.2-9) can be written

$$\tau(L) = \tau_0\sqrt{1 + \left(\frac{2\ln 2}{\pi c}\frac{DL\lambda^2}{\tau_0^2}\right)^2} \tag{7.2-21}$$

If DL is in units of picoseconds per nanometer, λ is in units of micrometers, and τ_0 is in units of picoseconds, the pulse width in units of picoseconds can be written

$$\tau(L) = \tau_0\sqrt{1 + \left(\frac{1.47DL\lambda^2}{\tau_0^2}\right)^2} \tag{7.2-22}$$

At large distances such that $|DL\lambda^2| \gg \tau_0^2$, Equation (7.2-21) becomes

$$\tau(L) = \frac{2 \ln 2}{\pi c} \frac{DL\lambda^2}{\tau_0} \tag{7.2-23a}$$

or equivalently (when DL is in units of ps/nm, λ is in units of μm, and τ_0 is in units of ps), according to Equation (7.2-22),

$$\tau(L) = \frac{1.47DL\lambda^2}{\tau_0} \tag{7.2-23b}$$

The expression (7.2-23a) is consistent with

$$\Delta\tau = DL\,\Delta\lambda \tag{7.2-24}$$

provided we take the bandwidth of the pulse as

$$\Delta\lambda = \frac{2 \ln 2}{\pi c} \frac{\lambda^2}{\tau_0} \tag{7.2-25}$$

This bandwidth $\Delta\lambda$ is consistent with the FWHM of the square of the Fourier spectrum (7.2-3) (see Problem 7.2).

The group velocity dispersion, that is, the dependence of v_g on ω, in optical fibers is due to the sum of material dispersion and waveguide dispersion (see Section 3.5). As we discussed in the last section, it is possible to design optical fibers such that the group velocity dispersion is either at a minimum or at a desired value in the spectral regime of interest.

Frequency Chirp

In addition to pulse broadening, the effect of pulse propagation in a dispersive fiber ($d^2\beta/d\omega^2 \neq 0$) is to modify the optical frequency. In the case of a Gaussian pulse after propagating a distance z, the result, according to Equations (7.2-7) and (7.2-13), is

$$E(z,t) = \frac{1}{\sqrt{1 + i4a\alpha z}} \exp\left(-\frac{(t - z/v_g)^2}{1/\alpha + 16a^2z^2\alpha}\right) \exp\left(i(\omega_0 t - \beta_0 z) + i\frac{4az(t - z/v_g)^2}{1/\alpha^2 + 16a^2z^2}\right)$$

$$a = \frac{1}{2}\frac{d^2\beta}{d\omega^2} = -\frac{1}{2v_g^2}\frac{dv_g}{d\omega} \tag{7.2-26}$$

The overall phase of the optical field is

$$\phi(z,t) = \omega_0 t - \beta_0 z + \frac{4az(t - z/v_g)^2}{1/\alpha^2 + 16a^2z^2} \tag{7.2-27}$$

The optical frequency $\omega(z, t)$ is fundamentally the number of (optical) oscillations per second as measured by a stationary observer at z. It is thus given by

$$\omega(z,t) = \frac{\partial}{\partial t}\phi(z,t) = \omega_0 + 8\frac{az}{(1/\alpha^2 + 16a^2z^2)}(t - z/v_g) \tag{7.2-28}$$

The frequency is not a constant but is "chirped." The linear chirp is a consequence of the group velocity dispersion causing different "groups" of frequencies to travel at different velocities and thus spread themselves along the pulse.

A chirped optical pulse resulting from propagation in a fiber with $d^2\beta/d\omega^2 < 0$ ($dv_g/d\omega > 0$) is shown in Figure 7.2, where the horizontal axis is the position. The "blue" (high-frequency)

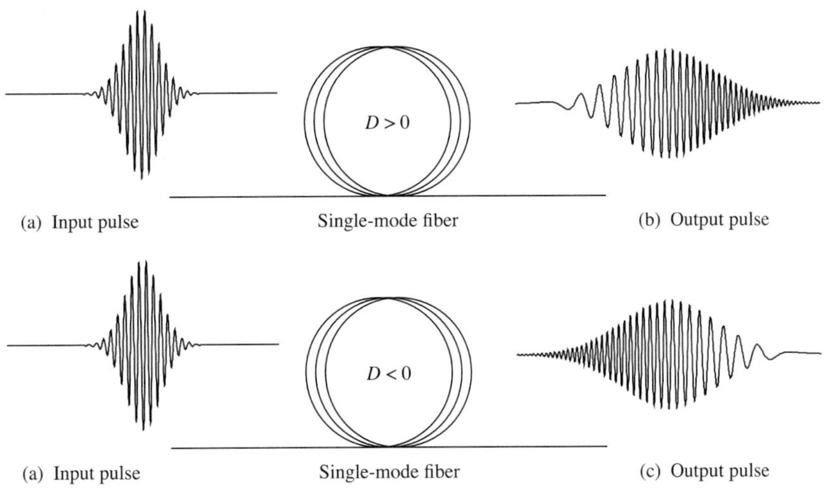

Figure 7.2 (a) An unchirped optical pulse with a Gaussian envelope (spatial profile). (b) The spatial profile of the same pulse after propagation in a single-mode fiber with positive dispersion ($D > 0$). The pulse is broadened and chirped. Note that the leading edge of the pulse consists of high-frequency components that travel faster in positive dispersion fibers. (c) The spatial profile of the same pulse after propagation in a single-mode fiber with negative dispersion ($D < 0$). The pulse is broadened and chirped. Note that the leading edge of the pulse consists of low-frequency components that travel faster in negative dispersion fibers.

portion of the pulse spectrum travels faster and arrives at z before the "red" portion, so that the frequency decreases (linearly) with time.

Pulses with Other Profiles

For optical pulses with a Gaussian profile, the Fourier transforms are also Gaussian. In other words, the spectrum follows a Gaussian distribution around the carrier frequency. In practice, the pulses in optical communications are not exactly Gaussian in shape. The variation in shape can lead to a change of the spectral distribution and hence can affect the broadening of the pulse after propagation through a dispersive medium. Figure 7.3 shows the pulse broadening of three different pulses. They are a trapezoid pulse, a Gaussian pulse, and a cosine pulse. Note that they have different spectral distributions and different pulse broadenings. The pulse with a trapezoid shape has the widest spectral bandwidth. This is a result of the sharp edges (steep rise and fall). The broader spectral distribution leads to a larger broadening and a smaller "eye" (see Figure 7.3). Generally, a Gaussian pulse shape has the best pulse width–spectral width product. This is consistent with the concept of minimum wave packet in physics.

The "eye" diagram is a method used by transmission and network system engineers to depict the degradation of digital binary pulses due to pulse spreading, distortion, and noise. The stream of received pulses is displayed on a storage oscilloscope whose horizontal scan is triggered in synchronism with the bit rate. The storage oscilloscope thus records multiple sequences of bits of zeros and/or ones (with spreading and possible distortion), adding up to an "eye" diagram. An eye diagram with large open eyes indicates a clear transmission with a low bit error rate.

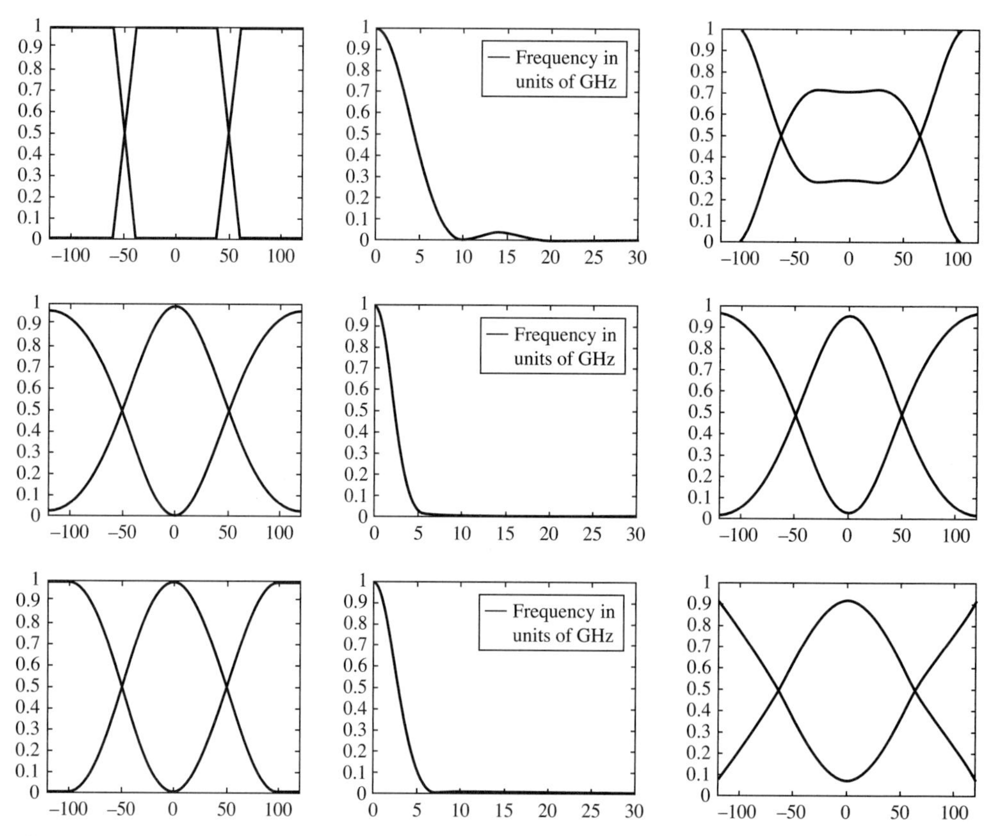

Figure 7.3 Eye diagrams of three different pulses. Initial pulse shape and eye diagram, spectral distributions, and final pulse shape and eye diagram after propagating through a dispersive medium of 800 ps/nm. Column 1—initial pulse shape; column 2—spectral distribution; column 3—final pulse shape after propagation. Row 1 is for an initial pulse shape of a trapezoid; row 2 is for an initial pulse of Gaussian shape; and row 3 is for an initial pulse of cosine shape. The time scale in columns 1 and 3 is in units of picosecond.

7.3 POLARIZATION EFFECTS IN OPTICAL FIBERS

We have demonstrated that chromatic dispersion can lead to the spreading of optical pulses. In this section, we show that polarization effects in fibers can also lead to the spreading of optical pulses. In Chapter 3, we discussed the modes of propagation in circular dielectric fibers. Specifically, we discussed linearly polarized modes in circular dielectric single-mode fibers. Despite their name, these "single-mode" fibers actually support two modes of propagation distinguished by their polarization states (LP_{01x} and LP_{01y}). As a result of the circular symmetry, these modes of propagation (LP_{01x} and LP_{01y}) are degenerate. In other words, at any given frequency they have exactly the same propagation constant and thus travel at exactly the same phase and group velocities.

In practice, all circular dielectric fibers are not exactly circular. The fiber cross section can be slightly flattened in one direction and slightly elongated in the other direction. This leads to a slightly elliptical cross section of the core. Figure 7.4 shows the polarization states of LP

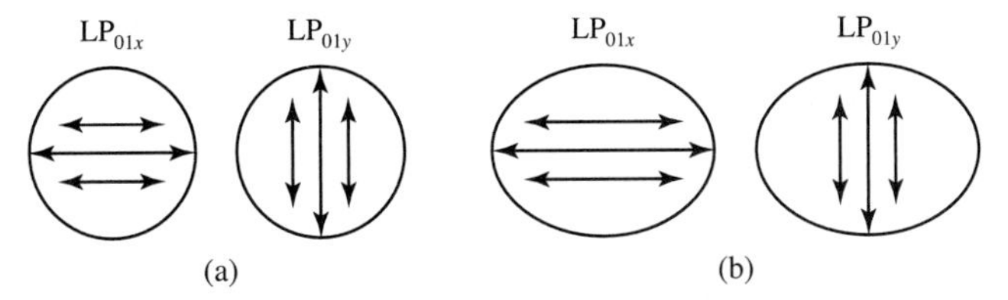

Figure 7.4 Polarization states of LP modes in optical fiber with an elliptical core. (a) In a circular fiber, the two modes are degenerate, that is, $\beta_{LP_{01x}} = \beta_{LP_{01y}}$ (b) In an elliptical fiber, $\beta_{LP_{01x}} \neq \beta_{LP_{01y}}$.

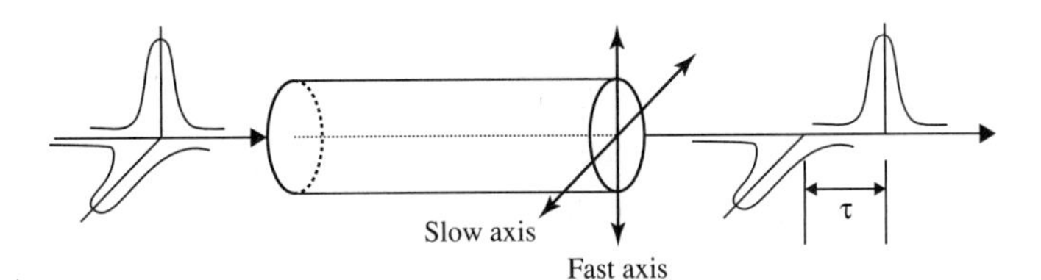

Figure 7.5 Pulse propagation through a uniform birefringent fiber. A time delay occurs between the slow and fast components of the pulse.

modes in an optical fiber with an elliptical core. The deviation from circular symmetry leads to a small difference in the propagation constant of the modes of propagation. In other words, the elliptical core leads to an optical birefringence in single-mode fibers. If the x and y axes coincide with the long and short axes of the elliptical core, then these two modes (LP_{01x} and LP_{01y}) travel with different group velocities. In practical fibers, the orientation of the long axis of the elliptical core may change with position along the fiber, as the fiberoptic cable twists and bends. Furthermore, the photoelastic effect due to stress along the fiber can cause a coupling between the modes of propagation. So, as a result of manufacturing imperfections and environmental perturbations (bending, twisting, etc.), an optical fiber actually behaves like a birefringent medium with the orientation of the slow axis that changes with position along the fiber. We will show that such a birefringence can lead to pulse distortion, broadening, and system impairments that limit the transmission capacity of the fiber.

Referring to Figure 7.5, we consider the simplest case of a single segment of fiber with a uniform birefringence. For a beam of unpolarized light, 50% of the energy is propagating along the fiber with polarization parallel to the slow axis of the fiber, and the other 50% is parallel to the fast axis of the fiber. As a result of the different speeds of propagation, a time delay occurs between these two components. Let n_s and n_f be the effective indices of the slow mode and the fast mode, respectively. It can easily be shown that the (group) time delay τ between these two components can be written (see Problem 7.6)

$$\tau = \left(\frac{d}{c} \Delta n + \frac{\omega d}{c} \frac{\partial \Delta n}{\partial \omega} \right) \tag{7.3-1}$$

where d is the length of the fiber segment and Δn is the birefringence given by

$$\Delta n = n_s - n_f \tag{7.3-2}$$

The spreading of optical pulses due to birefringence in fibers is known as polarization mode dispersion (PMD).

Mean Signal Delay and Broadening Due to Birefringence

Let the normalized input signal intensity be $I(t)$; then the output intensity can be written

$$I_{out}(t) = \gamma I(t-\tau_1) + (1-\gamma)I(t-\tau_2) \tag{7.3-3}$$

where γ is the fraction of energy of the input wave along the direction of the slow axis and times τ_1 and τ_2 are the flight times of these modes, respectively. For simplicity, we assume $I(t)$ has a centroid at $t = 0$. In other words,

$$\langle tI(t)\rangle = 0 \tag{7.3-4}$$

In terms of the normalized Jones vector of the polarization state of the input wave **J**, the fraction γ is given by

$$\gamma = |\mathbf{J}^* \cdot \mathbf{J}_{slow}|^2 \tag{7.3-5}$$

where $\mathbf{J}_{slow}$ is the normalized Jones vector of the polarization state of the slow mode. In investigating PMD, we often use the Poincaré sphere to represent the polarization states of a beam of polarized light. For polarized light, we can ignore the S_0 component. Thus the Stokes vector of a beam of polarized light is often written $\mathbf{s} = (S_1, S_2, S_3)$ as a real unit vector in Poincaré space. In terms of the Stokes vector of the input wave **s**, the fraction γ is given by, according to Equation (1.6-41),

$$\gamma = \tfrac{1}{2}(1 + \mathbf{p}\cdot\mathbf{s}) \tag{7.3-6}$$

where **s** is the three-component Stokes vector of the polarization state of the input wave, and **p** is the three-component Stokes vector of the polarization state of the slow mode.

The mean signal delay due to birefringence is given by

$$\tau_g = \langle t\rangle = \int tI_{out}(t)\,dt = \gamma\tau_1 + (1-\gamma)\tau_2 = (\gamma - \tfrac{1}{2})\tau \tag{7.3-7}$$

where we have taken $\tau_1 = \tau/2$ (slow mode) and $\tau_2 = -\tau/2$ (fast mode) with τ being the time delay between these two modes. For a beam of unpolarized light, $\gamma = \frac{1}{2}$, the mean delay is zero.

The mean delay can also be written, according to Equations (7.3-6) and (7.3-7),

$$\tau_g = \tfrac{1}{2}\boldsymbol{\tau}\cdot\mathbf{s} \tag{7.3-8}$$

where $\boldsymbol{\tau} = \tau\mathbf{p}$ is the so-called PMD vector. We will discuss the PMD vector in more detail in Section 7.5.

The pulse broadening can be evaluated by calculating the variance of the output intensity profile. Using Equation (7.3-3), we obtain

$$(\Delta\tau_b)^2 = \langle(t-\tau_g)^2\rangle - b^2 = \int (t-\tau_g)^2 I_{out}(t)\,dt - b^2 = \gamma(1-\gamma)(\tau)^2 \tag{7.3-9}$$

where b is a measure of the original pulse width of the input signal $I(t)$,

$$b^2 = \langle t^2 I(t)\rangle = \int t^2 I(t)\,dt \tag{7.3-10}$$

In terms of the PMD vector $\boldsymbol{\tau} = \tau\mathbf{p}$ and the Stokes vector **s** of the input wave, the PMD broadening can be written, according to Equations (7.3-6) and (7.3-9),

$$(\Delta\tau_b)^2 = \tfrac{1}{4}[\tau^2 - (\boldsymbol{\tau}\cdot\mathbf{s})^2] \tag{7.3-11}$$

where **s** is the Stokes vector of the input wave. We note that when the input Stokes vector **s** is aligned along the PMD vector $\boldsymbol{\tau} = \tau\mathbf{p}$, the PMD broadening is zero. For an unpolarized wave $\gamma = 0.5$ and $\boldsymbol{\tau}\cdot\mathbf{s} = 0$, the PMD broadening is, according to Equation (7.3-9) or (7.3-11),

$$(\Delta\tau_b)^2 = \tfrac{1}{4}(\tau)^2 \tag{7.3-12}$$

We note that the pulse broadening due to birefringence in a short segment of fiber with uniform birefringence is proportional to the length of the fiber d. Although the result (7.3-11) is obtained by assuming a linear birefringence of the medium, it is actually valid for a general birefringent medium with elliptically polarized modes of propagation.

In real fiberoptic telecommunications systems, the fibers are much more complicated than a single segment of uniformly birefringent medium. We can try to understand the pulse broadening by dividing a long segment of fiber into a large number of small segments of uniformly birefringent fibers. Each segment is assumed to be uniformly birefringent with a fixed polarization state of the slow and fast modes. Based on the analysis above, each segment will introduce a broadening of the pulse. The total of the broadening is a result of the contributions from all the segments. In the simplest case, when the birefringence is completely random, the total broadening is a random walk process and can be written

$$(\Delta\tau_b)^2 = \sum_{i=1}^{N} \frac{1}{4}(\tau)_i^2 \tag{7.3-13}$$

$$= \sum_{i=1}^{N} \frac{1}{4}\left(\frac{\Delta n}{c} + \frac{\omega}{c}\frac{\partial \Delta n}{\partial \omega}\right)_i^2 d_i^2 \tag{7.3-14}$$

where d_i is the length of the ith segment. If we assume that all segments are of equal broadening (i.e., $\tau_i = \tau_1$ for all i), then it is clear from Equation (7.3-14), that the total broadening $\Delta\tau_b$ is proportional to the square root of the length of the fiber (see Problem 7.7). It is important to note that Equation (7.3-14) is valid only in the oversimplified case when the birefringence is completely random (see Problem 7.8). In practical fibers, although the birefringence may be random, the relative orientations of the slow modes (or fast modes) between neighboring elements are not completely random. A more rigorous analysis of PMD in multisegment uniform birefringent fibers will be discussed in Section 7.5. In the next section, we introduce the important concept of the principal states of polarization (PSP).

7.4 PRINCIPAL STATES OF POLARIZATION

For the purpose of analyzing the evolution of the polarization state, we may divide a long segment of fiber into a large number of small segments of uniformly birefringent fibers. Each segment is assumed to be uniaxially birefringent with a fixed orientation of the slow and fast axes. In such a birefringent network, the input and output polarization states can be obtained by using the Jones matrix method, which was discussed in detail in Chapter 1. Figure 7.6 shows

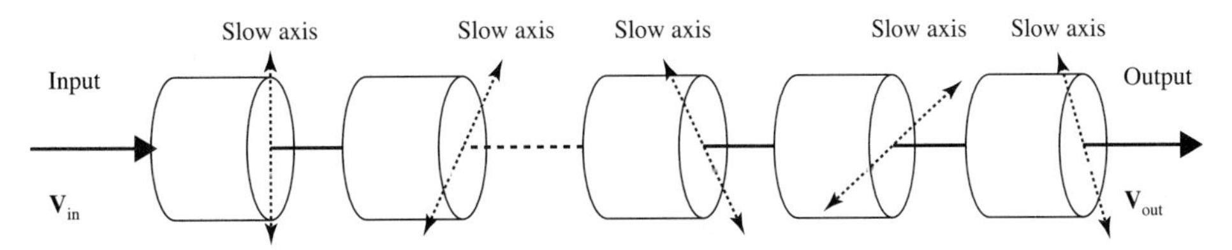

Figure 7.6 A long segment of fiber can be represented by an equivalent circuit that consists of a series of birefringent elements. The slow axis of each of the birefringent elements can be in an arbitrary orientation. The fast axis is perpendicular to the slow axis in each birefringent element.

a schematic drawing of the birefringent optical circuit representing a long segment of real fibers. In real fibers, the loss of circular symmetry can be a result of an elliptical core or cladding of the fibers, or a result of stress-induced anisotropy due to bending, twisting, and so on.

Given a fixed input state of polarization, the output state of polarization is in general a function of the frequency of the light. In what follows, we describe a set of particular states of input polarization, known as the principal states of polarization (PSP), such that the output states of polarization are insensitive to the variation of the frequency of the light. The concept of PSP was first introduced by Poole and Wagner in 1986 [1]. For the purpose of describing the concept of PSP, we assume that the fiber is lossless. Let $\mathbf{V}_{\text{in}}$ be the Jones vector of the input optical wave. The output polarization state $\mathbf{V}_{\text{out}}$ is given by

$$\mathbf{V}_{\text{out}} = M_N M_{N-1} \cdots M_3 M_2 M_1 \mathbf{V}_{\text{in}} = U\mathbf{V}_{\text{in}} \tag{7.4-1}$$

where M_m ($m = 1, 2, 3, \ldots, N$) is the Jones matrix of the mth birefringent element. The overall Jones matrix U is the product of all the individual Jones matrices in sequence. For lossless fibers, all these Jones matrices are unitary. Thus we can write

$$\mathbf{V}_{\text{out}} = \begin{bmatrix} a & b \\ -b^* & a^* \end{bmatrix} \mathbf{V}_{\text{in}} \tag{7.4-2}$$

where a and b are elements of the overall Jones matrix U of the whole birefringent network. In practice, these matrix elements are obtained by multiplying all the individual Jones matrices in sequence. These elements (a and b) depend on the frequency of light, as well as the birefringence distribution of the fiber. The Jones matrix U in Equation (7.4-2) is unitary and thus

$$|a|^2 + |b|^2 = 1 \tag{7.4-3}$$

Generally, if the input polarization state $\mathbf{V}_{\text{in}}$ is fixed, the output polarization state $\mathbf{V}_{\text{out}}$ may depend on the frequency of light. The dependence can be investigated by examining the derivative of $\mathbf{V}_{\text{out}}$ in Equation (7.4-2) with respect to ω; in other words,

$$\frac{\partial \mathbf{V}_{\text{out}}}{\partial \omega} = \begin{bmatrix} a' & b' \\ -b'^* & a'^* \end{bmatrix} \mathbf{V}_{\text{in}} \tag{7.4-4}$$

where a prime indicates differentiation with respective to ω. In arriving at Equation (7.4-4), we assume a fixed input polarization state $\mathbf{V}_{\text{in}}$ over the spectral regime of interest.

If the output polarization state satisfies the following equation,

$$\frac{\partial \mathbf{V}_{\text{out}}}{\partial \omega} = -i\delta \mathbf{V}_{\text{out}} \tag{7.4-5}$$

where δ is a constant, then the Jones vector of the output beam $\mathbf{V}_{\text{out}}$ represents a polarization state that is independent of the frequency (to the first order in frequency). This can be seen by integrating the above equation with respect to the frequency ω. We obtain

$$\mathbf{V}_{\text{out}}(\omega) = \mathbf{V}_{\text{out}}(\omega_0) \exp\left(-i \int_{\omega_0}^{\omega} \delta \, d\omega\right) \tag{7.4-6}$$

where ω_0 is a constant, usually the carrier frequency. Note that the exponential multiplicative factor of a Jones vector does not change the polarization state. By examining Equation (7.4-6), we note that δ is actually the propagation time delay.

We now find the input polarization states $\mathbf{V}_{\text{in}}$ such that the output polarization state satisfies Equation (7.4-5). Substituting Equations (7.4-2) and (7.4-4) into Equation (7.4-5), we obtain after few steps of algebra

$$\begin{bmatrix} A & B \\ C & D \end{bmatrix} \mathbf{V}_{\text{in}} = -i\delta \mathbf{V}_{\text{in}} \tag{7.4-7}$$

where

$$\begin{aligned} A &= a^*a' + bb'^* \\ B &= a^*b' - a'^*b \\ C &= a'b^* - ab'^* = -B^* \\ D &= aa'^* + b'b^* = A^* \end{aligned} \tag{7.4-8}$$

In other words, the input principal polarization states must be eigenvectors of the above matrix equation (7.4-7). When input polarization states are chosen so that Equation (7.4-7) is satisfied, the output polarization states will be independent of frequency to the first order in ω. Using Equations (7.4-3) and (7.4-8), it can be shown that $A + D = 0$. In other words, the trace of the matrix in Equation (7.4-7) is zero. This is a direct consequence of the unitary property of the Jones matrix and can easily be obtained by differentiating Equation (7.4-3) with respect to ω. Thus the matrix element A in Equation (7.4-7) is purely imaginary.

The eigenvectors and eigenvalues of Equation (7.4-7) are

$$\mathbf{V}_{\text{in}} = \begin{bmatrix} -B \\ A + i\delta \end{bmatrix} \tag{7.4-9}$$

and

$$\delta = \pm\sqrt{AD - BC} = \pm\sqrt{|a'|^2 + |b'|^2} \tag{7.4-10}$$

respectively. Let these two input principal states be designated as $\mathbf{V}_1$ and $\mathbf{V}_2$ with eigenvalues $\delta = \tau/2, -\tau/2$. Let τ_0 be the time of flight in the absence of the birefringence; then the time of flight of these two modes are, according to Equation (7.4-6),

$$\tau_0 + \tau/2 \quad \text{(slow mode)}$$

$$\tau_0 - \tau/2 \quad \text{(fast mode)}$$

respectively. These two eigenvectors are mutually orthogonal (i.e., $\mathbf{V}_1^* \cdot \mathbf{V}_2 = 0$). Using these two eigenvectors as the basis, a general input polarization state can be written

$$\mathbf{V}_{\text{in}} = c_1\mathbf{V}_1 + c_2\mathbf{V}_2 \tag{7.4-11}$$

where c_1 and c_2 are constants. The output polarization state can be written

$$\mathbf{V}_{\text{out}} = c_1\mathbf{V}_1' + c_2\mathbf{V}_2' \tag{7.4-12}$$

where $\mathbf{V}_1'$ and $\mathbf{V}_2'$ are the corresponding output principal polarization states (i.e., $\mathbf{V}_1' = U\mathbf{V}_1$ and $\mathbf{V}_2' = U\mathbf{V}_2$). It can also be written in terms of the frequency dependence, according to Equation (7.4-6),

$$\mathbf{V}_{\text{out}} = c_1\mathbf{V}_1'(\omega_0)\exp\left(-i\int_{\omega_0}^{\omega}\frac{\tau}{2}\,d\omega\right) + c_2\mathbf{V}_2'(\omega_0)\exp\left(i\int_{\omega_0}^{\omega}\frac{\tau}{2}\,d\omega\right) \tag{7.4-13}$$

The phase difference between these two polarization components can be written

$$\Delta\phi = -\int_{\omega_0}^{\omega}\tau\,d\omega \tag{7.4-14}$$

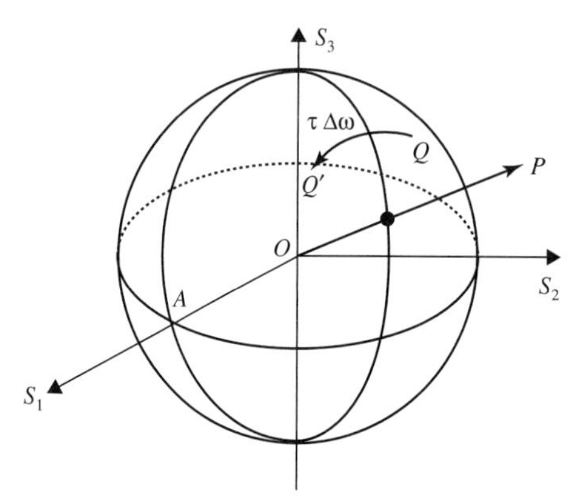

Figure 7.7 Schematic drawing showing the output state of polarization at various frequencies. Q is a point of the Poincaré sphere representing the output state $\mathbf{V}_{out}(\omega_0)$. Q' is a point of the Poincaré sphere representing the output state $\mathbf{V}_{out}(\omega_0 + \Delta\omega)$. OP is a unit vector along the PMD vector. Equivalently, OP is the Stokes vector of the slow PSP. Note that the output state Q' at $\omega_0 + \Delta\omega$ is obtained from the output state Q at ω_0 by rotating an angle of $\tau\,\Delta\omega$ around the PMD vector.

The group delay between these two components is given by, according to Equation (7.4-10),

$$\Delta\tau = -\frac{\partial}{\partial\omega}\Delta\phi = \tau = 2\sqrt{|a'|^2 + |b'|^2} \tag{7.4-15}$$

The group delay between these two polarization components is equivalent to a broadening of the pulse, according to the discussion in Section 7.3.

If the input state of polarization is aligned along one of the principal states of polarization, the broadening due to the birefringence disappears (to first order in frequency). We now examine the output state of polarization over a small frequency spread $\Delta\omega$ around ω_0. Using Equation (7.4-13), we can write the output state as

$$\mathbf{V}_{out}(\omega_0 + \Delta\omega) = c_1\mathbf{V}_1'(\omega_0)\exp(-i\tau\,\Delta\omega/2) + c_2\mathbf{V}_2'(\omega_0)\exp(i\tau\,\Delta\omega/2) \tag{7.4-16}$$

We note that the output polarization state at $\omega_0 + \Delta\omega$ can be obtained from the output polarization state at ω_0, by a rotation around the PMD vector on the Poincaré sphere with an angle of $\tau\,\Delta\omega$. This is illustrated in Figure 7.7.

The above treatment of PSP is based on the Jones matrix formulation. As we know, the polarization states of a beam of light can also be represented in terms of points on the Poincaré sphere. In this case, the output polarization state is obtained by a three-dimensional rotation of the input polarization state in Stokes space. Thus the same subject of PSP can be treated equivalently by using the Stokes formulation. This treatment will be described in Section 7.5.

EXAMPLE: SINGLE UNIFORM BIREFRINGENT FIBER

As an example to illustrate the concept of PSP, we consider the simplest case of a single segment of a birefringent medium with a uniform birefringence. For this case, the Jones matrix is given by

$$M = \begin{bmatrix} e^{-i\Gamma/2} & 0 \\ 0 & e^{i\Gamma/2} \end{bmatrix}$$

where

$$\Gamma = \frac{\omega}{c}(n_s - n_f)d = \frac{\omega}{c}\Delta n\, d$$

So $a = e^{-i\Gamma/2}$, $b = 0$, and

$$a' = -ia\frac{1}{2}\left(\frac{d}{c}\Delta n + \frac{\omega d}{c}\frac{d\Delta n}{d\omega}\right)$$

$$b' = 0$$

$$A = a^*a' + bb'^* = -i\frac{1}{2}\left(\frac{d}{c}\Delta n + \frac{\omega d}{c}\frac{d\Delta n}{d\omega}\right)$$

$$B = a^*b' - a'^*b = 0$$

The eigenvalues are, according to Equation (7.4-10),

$$\delta = +\tau/2 = \frac{1}{2}\left(\frac{d}{c}\Delta n + \frac{\omega d}{c}\frac{d\Delta n}{d\omega}\right) \quad \text{(slow mode)}$$

$$\delta = -\tau/2 = -\frac{1}{2}\left(\frac{d}{c}\Delta n + \frac{\omega d}{c}\frac{d\Delta n}{d\omega}\right) \quad \text{(fast mode)}$$

The principal states are simply the slow mode or the fast mode.

In other words,

$$\mathbf{V}_{\text{in1}} = \begin{bmatrix} 1 \\ 0 \end{bmatrix}, \quad \text{with time of flight } \tau_1 = \tau_0 + \tau/2 \quad \text{(slow mode)}$$

$$\mathbf{V}_{\text{in2}} = \begin{bmatrix} 0 \\ 1 \end{bmatrix}, \quad \text{with time of flight } \tau_2 = \tau_0 - \tau/2 \quad \text{(fast mode)}$$

And the PMD is

$$\Delta\tau = \left(\frac{d}{c}\Delta n + \frac{\omega d}{c}\frac{d\Delta n}{d\omega}\right)$$

Problem 7.9 deals with the PMD of a two-element birefringent fiber.

7.5 VECTOR ANALYSIS OF POLARIZATION MODE DISPERSION

In this section, we describe the polarization mode dispersion (PMD) using both the Jones vector and Stokes vector representations. Pauli spin matrices are employed to provide a convenient relationship between these two representations. The Stokes vectors are particularly useful for graphic representation of the variation of the polarization state of the output beam on the Poincaré sphere. We first introduce the definitions and notations of some of the important parameters used in polarization mode dispersion analysis. We then describe the relationship between the Jones vector representation and the Stokes vector representation. In the last section, we describe briefly the concept of principal states of polarization (PSP) by using the Jones matrix formulation. In this section, we introduce the Stokes vector treatment of the same. In addition, we also introduce important concepts, including the PMD vector and the birefringence vector, which are the main tools for analyzing PMD in fiberoptic networks. Near the end, we describe a dynamical equation for the evolution of the PMD vector in arbitrary birefringent fibers, as well as a concatenation rule for PMD vectors. The discussion in this section follows closely that of Reference [2].

Definitions and Notations

Coordinate

z	Direction of propagation along the axis of the fiber
x, y	Transverse coordinates
$\psi(x, y) \exp[i(\omega t - kz]$	Continuous wave with frequency ω and wavenumber k traveling along the fiber

Polarization State Representations and Vectors

$\lvert s\rangle$	Dirac notation of 2-D complex Jones vector, defined in Chapter 1; using complex numbers, $\lvert s\rangle$ can be written as a column vector of (s_x, s_y)
$\mathbf{s}$	3-D real Stokes vector $\mathbf{s} = (s_1, s_2, s_3)$, defined in Chapter 1; in terms of the Jones vector elements, the Stokes vector $\mathbf{s}$ can be written $s_1 = s_x s_x^* - s_y s_y^*, \quad s_2 = s_x s_y^* + s_y s_x^*, \quad s_3 = i(s_x s_y^* - s_y s_x^*)$ The definition here is consistent with that of Chapter 1
$\lvert p\rangle$	Unit Jones vector of slow output PSP
$\lvert p_-\rangle$	Unit Jones vector of fast output PSP
$\mathbf{p}$	Unit Stokes vector along the direction of slow output PSP
$\mathbf{p}_-$	Unit Stokes vector along the direction of fast output PSP, $\mathbf{p}_- = -\mathbf{p}$
$\mathbf{r}$	Axis of rotation of a rotation matrix $\mathbf{R}$ (3×3) in Stokes space, $\mathbf{r}_- = -\mathbf{r}$
$\boldsymbol{\tau}$	Output PMD vector in Stokes space: its length is $\Delta\tau$, the differential group delay (DGD) between the two PSP, and its direction is that of the Stokes vector of the slow output PSP, that is, $\boldsymbol{\tau} = \Delta\tau\, \mathbf{p}$.
$\boldsymbol{\beta}$	3-D birefringence vector in Stokes space describing local birefringence properties; its length is the angle of rotation on a Poincaré sphere per unit length of fiber, while its direction is the axis of rotation of R

Matrices

I	Identity matrix
U	2×2 Jones matrix Input–output relationship: $\lvert v\rangle = U\lvert s\rangle$ U is unitary and can be written $U = \begin{bmatrix} a & b \\ -b^* & a^* \end{bmatrix}$ with $\lvert a\rvert^2 + \lvert b\rvert = 1$
R	3×3 Rotation matrix in Stokes space isomorphic to U Input–output relationship: $\mathbf{v} = R\mathbf{s}$

$\sigma_1, \sigma_2, \sigma_3$ — 2×2 Pauli spin matrix

$$\sigma_1 = \begin{bmatrix} 1 & 0 \\ 0 & -1 \end{bmatrix}, \quad \sigma_2 = \begin{bmatrix} 0 & 1 \\ 1 & 0 \end{bmatrix}, \quad \sigma_3 = \begin{bmatrix} 0 & -i \\ i & 0 \end{bmatrix}$$

$\boldsymbol{\sigma}$ — Pauli spin vector in Stokes space

Using Pauli spin matrices, we can write

$$s_1 = \langle s|\sigma_1|s\rangle, \quad s_2 = \langle s|\sigma_2|s\rangle, \quad s_3 = \langle s|\sigma_3|s\rangle,$$

or symbolically

$$\mathbf{s} = \langle s|\boldsymbol{\sigma}|s\rangle$$

$|s\rangle\langle s|$ — Projection operator (matrix) in Jones space

$$|s\rangle\langle s| = \begin{bmatrix} s_x s_x^* & s_x s_y^* \\ s_y s_x^* & s_y s_y^* \end{bmatrix} = \tfrac{1}{2}(\mathrm{I} + \mathbf{s} \cdot \boldsymbol{\sigma})$$

$\mathbf{ss}$ — Projection operator (matrix) in Stokes space

$$\mathbf{ss} = \begin{bmatrix} s_1 s_1 & s_1 s_2 & s_1 s_3 \\ s_2 s_1 & s_2 s_2 & s_2 s_3 \\ s_3 s_1 & s_3 s_2 & s_3 s_3 \end{bmatrix}$$

In addition to the above notations, we also introduce the matrix representation of vector cross multiplication. This is purely a matter of mathematical convenience. Consider two vectors in Stokes space and their cross product:

$$\mathbf{A} = \begin{bmatrix} A_1 \\ A_2 \\ A_3 \end{bmatrix}, \quad \mathbf{B} = \begin{bmatrix} B_1 \\ B_2 \\ B_3 \end{bmatrix}, \quad \text{and} \quad \mathbf{A} \times \mathbf{B} = \begin{bmatrix} A_2B_3 - A_3B_2 \\ A_3B_1 - A_1B_3 \\ A_1B_2 - A_2B_1 \end{bmatrix}$$

Note that the product can be conveniently written as a matrix–vector multiplication.

$$\mathbf{A} \times \mathbf{B} = \begin{bmatrix} A_2B_3 - A_3B_2 \\ A_3B_1 - A_1B_3 \\ A_1B_2 - A_2B_1 \end{bmatrix} = \begin{bmatrix} 0 & -A_3 & A_2 \\ A_3 & 0 & -A_1 \\ -A_2 & A_1 & 0 \end{bmatrix} \begin{bmatrix} B_1 \\ B_2 \\ B_3 \end{bmatrix}$$

So it is convenient to define a matrix representation of the cross multiplication as

$$\mathbf{A}\times = \begin{bmatrix} 0 & -A_3 & A_2 \\ A_3 & 0 & -A_1 \\ -A_2 & A_1 & 0 \end{bmatrix}$$

Jones Vectors and Stokes Vectors

We now describe the relationship between the Jones vector representation and the Stokes vector representation. The Jones matrix is a 2×2 matrix that relates the output states of polarization to the input states of polarization when a beam of polarized light propagates through a birefringent network. The linear relationship can be written

$$|v\rangle = U|s\rangle \tag{7.5-1}$$

where $|s\rangle$ is an input polarization state, $|v\rangle$ is the output polarization state, and U is the 2×2 Jones matrix. We note that the vectors and matrix are complex. Equation (7.5-1) can be viewed as a transformation of the state of polarization by the birefringent network. The Jones matrix operation is equivalent to a "rotation" in the two-dimensional complex Jones space.

If we represent the states of polarization as unit vectors in Stokes space, then the transformation of the state of polarization can be depicted as a rotation in the three-dimensional Stokes space. Mathematically, this is written

$$\mathbf{v} = R\mathbf{s} \tag{7.5-2}$$

where $\mathbf{s}$ is the three-component column vector representing an input state of polarization and $\mathbf{v}$ is the three-component column vector representing the output state of polarization. R is a 3×3 matrix. Note that the Stokes vectors and matrix are all real. We recall that for a beam of polarized light, the Stokes parameter $s_0 = 1$. So we only need the three components (s_1, s_2, s_3) to describe the polarization state. Based on the definition of the Stokes vector components (defined in Chapter 1), they can be written conveniently in terms of the Jones vector and the Pauli spin matrices:

$$s_1 = \langle s|\sigma_1|s\rangle, \quad s_2 = \langle s|\sigma_2|s\rangle, \quad s_3 = \langle s|\sigma_3|s\rangle \tag{7.5-3}$$

or symbolically,

$$\mathbf{s} = \langle s|\boldsymbol{\sigma}|s\rangle \tag{7.5-4}$$

These two equations are the relationships between Jones vectors and Stokes vectors.

Obviously, the Jones matrix U and the rotation matrix R are also related. In fact, there is a unique one-to-one relationship between them. Given a 2×2 Jones matrix, we should be able to find the 3×3 rotation matrix R, or equivalently the axis of rotation and the angle of rotation in Stokes space. Let $\mathbf{r}$ be the axis of rotation of R in Stokes space (see Figure 7.8); then it's obvious that any input Stokes vector collinear with the rotation axis $\mathbf{r}$ will remain invariant upon transmission through the birefringent optical system (the fiber). Let $|r\rangle$ and $|r_-\rangle$ be the Jones vectors corresponding to Stokes vectors $\mathbf{r}$ and $-\mathbf{r}$, then the corresponding Jones vectors must be eigenvectors of the 2×2 Jones matrix U. Thus we can write

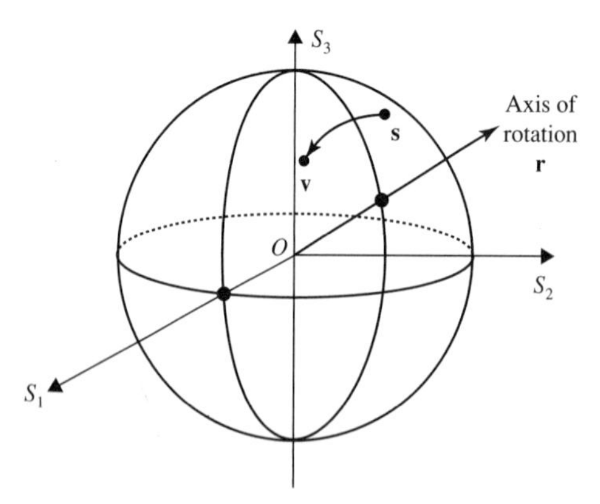

Figure 7.8 The transformation of polarization states by a birefringent network is equivalent to a rotation in Stokes space. The output Stokes vector **v** is obtained by a rotation of the input Stokes vector **s** around the axis **r**. The axis of rotation and the angle of rotation are determined by the birefringent network.

$$U|r\rangle = e^{-i\gamma/2}|r\rangle \tag{7.5-5}$$

$$U|r\_\rangle = e^{+i\gamma/2}|r\_\rangle \tag{7.5-6}$$

where $\exp(-i\gamma/2)$ and $\exp(+i\gamma/2)$ are eigenvalues. Since U is both unitary and unimodular, the eigenvalues are unimodular and their product must be unity. According to Equations (7.5-5) and (7.5-6), the Stokes vector **r** corresponds to the polarization state of the slow mode of the birefringent network, whereas −**r** corresponds to the polarization state of the fast mode of the birefringent network.

Let $|s\rangle$ be an arbitrary input Jones vector. Since $|r\rangle$ and $|r\_\rangle$ constitute a complete orthogonal set of Jones vectors, we can write

$$|s\rangle = c_1|r\rangle + c_2|r\_\rangle \tag{7.5-7}$$

where c_1 and c_2 are constants. Upon transmission through the birefringent optical system, the output Jones vector can be obtained according to Equation (7.5-1). Using Equations (7.5-5) and (7.5-6), we obtain the output state of polarization

$$|v\rangle = c_1|r\rangle e^{-i\gamma/2} + c_2|r\_\rangle e^{+i\gamma/2} \tag{7.5-8}$$

Based on the discussion in Chapter 1, we recognize that the output Stokes vector **v** can be obtained by a right-handed rotation of the input Stokes vector **s** by an angle γ around the axis **r**. Thus we are able to obtain the 3×3 rotation matrix R by solving a simple eigenvector problem in Jones space. Explicit expressions of the axis of rotation as well as the angle of rotation can be obtained in terms of the matrix elements of the Jones matrix. After solving the eigenvector problem (7.5-5) and (7.5-6), we obtain the following eigenvalues and eigenvectors:

$$\cos(\gamma/2) = \frac{a + a^*}{2} \tag{7.5-9}$$

$$|r\rangle = N\begin{bmatrix} -b \\ a - e^{-i\gamma/2} \end{bmatrix} \tag{7.5-10}$$

$$|r\_\rangle = N\_\begin{bmatrix} -b \\ a - e^{+i\gamma/2} \end{bmatrix} \tag{7.5-11}$$

where a and b are matrix elements of the Jones matrix, and N and $N\_$ are normalization constants. Note that $|r\rangle$ is a two-component vector in Jones space. These two Jones vectors are mutually orthogonal. The corresponding vector $\mathbf{r} = (r_1, r_2, r_3)$ in Stokes space can be obtained by using Equation (7.5-3). This leads to

$$\begin{aligned} r_1 &= \frac{-2|a|^2 + ae^{+i\gamma/2} + a^*e^{-i\gamma/2}}{2 - ae^{+i\gamma/2} - a^*e^{-i\gamma/2}} \\ r_2 &= \frac{-(ab^* + a^*b) + be^{+i\gamma/2} + b^*e^{-i\gamma/2}}{2 - ae^{+i\gamma/2} - a^*e^{-i\gamma/2}} \\ r_3 &= \frac{-(ia^*b - iab^*) + ibe^{+i\gamma/2} - ib^*e^{-i\gamma/2}}{2 - ae^{+i\gamma/2} - a^*e^{-i\gamma/2}} \end{aligned} \tag{7.5-12}$$

These are explicit expressions of rotation angle γ and rotation axis **r** in terms of the matrix elements a and b of the Jones matrix.

EXAMPLE: AZIMUTH ANGLE OF $\psi = 0$

Consider a wave plate of uniaxial crystal with its c axis (or slow axis) oriented at an azimuth angle of $\psi = 0$. According to Chapter 1, the matrix elements are

$$a = e^{-i\Gamma/2}, \quad b = 0$$

Using Equation (7.5-12), we obtain

$$\gamma = \Gamma, \quad r_1 = 1, \quad r_2 = 0, \quad r_3 = 0$$

Note that the angle of rotation is exactly the phase retardation Γ, and the rotation axis is simply the s_1 axis.

EXAMPLE: AZIMUTH ANGLE OF $\psi \neq 0$

Consider now a wave plate of uniaxial crystal with its c axis (or slow axis) oriented at an azimuth angle of $\psi \neq 0$. According to Chapter 1, the matrix elements are

$$a = e^{-i\Gamma/2} \cos^2\Psi + e^{+i\Gamma/2} \sin^2\Psi, \quad b = -i \sin(\Gamma/2) \sin(2\Psi)$$

Using Equation (7.5-12), we obtain

$$\gamma = \Gamma, \quad r_1 = \cos(2\psi), \quad r_2 = \sin(2\psi), \quad r_3 = 0$$

Note that the angle of rotation is exactly the phase retardation Γ, and the rotation axis is on the equatorial plane, at an angle of 2ψ from the s_1 axis. Let R be the point on the Poincaré sphere representing the polarization state of the slow mode; then the output polarization state P' is obtained by rotating P an angle Γ around the axis OR, where O is the origin of the sphere. Figure 7.9 illustrates the polarization transformation on the Poincaré sphere.

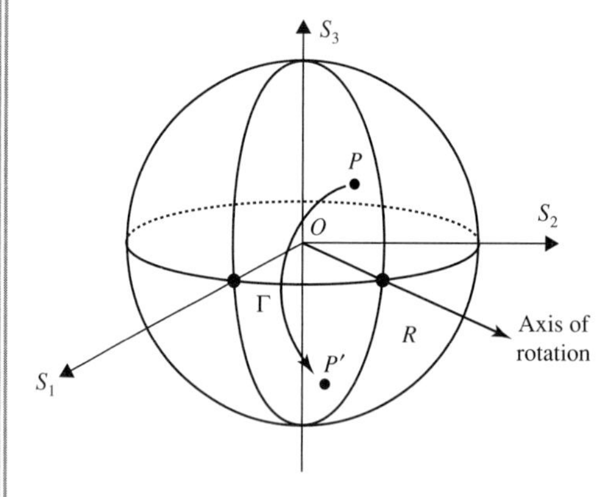

Figure 7.9 Polarization transformation on the Poincaré sphere. P is the input polarization state, P' is the output polarization state, and R is the polarization state of the slow mode of the wave plate.

Furthermore, we are able to obtain the 2×2 Jones matrix from the rotation angle γ and the axis of rotation $\mathbf{r}$. This is done as follows. We can write the Jones matrix in terms of the projection operators and eigenvalues as

$$U = e^{-i\gamma/2}|r\rangle\langle r| + e^{+i\gamma/2}|r_{\perp}\rangle\langle r_{\perp}| \tag{7.5-13}$$

Using the Pauli representation of the projection operators described in the beginning of this section, the Jones matrix (7.5-13) becomes

$$U = \cos(\gamma/2)I - i\sin(\gamma/2)\mathbf{r}\cdot\boldsymbol{\sigma} \tag{7.5-14}$$

where I is the identity matrix. Thus we can write the Jones matrix by using the explicit form of the Pauli matrices and the components of **r**,

$$U = \begin{bmatrix} \cos(\gamma/2) - ir_1\sin(\gamma/2) & -(r_3 + ir_1)\sin(\gamma/2) \\ (r_3 - ir_1)\sin(\gamma/2) & \cos(\gamma/2) + ir_1\sin(\gamma/2) \end{bmatrix} \tag{7.5-15}$$

where r_1, r_2, r_3 are the components of **r** (the axis of rotation). This is an explicit expression of the Jones matrix in terms of the axis of rotation **r** and angle of rotation γ in Stokes space.

Conversely, we can obtain an explicit expression of the 3×3 rotation matrix R in terms of the Jones matrix. This is done as follows. Using Equations (7.5-1), (7.5-2), and (7.5-4), we can write

$$\mathbf{v} = R\mathbf{s} = R\langle s|\boldsymbol{\sigma}|s\rangle = \langle s|R\boldsymbol{\sigma}|s\rangle = \langle v|\boldsymbol{\sigma}|v\rangle = \langle s|U^{\dagger}\boldsymbol{\sigma}U|s\rangle \tag{7.5-16}$$

where the dagger (†) stands for Hermitian conjugate (transpose and then complex conjugate). Since this is valid for an arbitrary Jones vector $|s\rangle$, we obtain

$$R\boldsymbol{\sigma} = U^{\dagger}\boldsymbol{\sigma}U \tag{7.5-17}$$

This is an expression of the 3×3 rotation matrix R in terms of the 2×2 Jones matrix U.

Substituting Equation (7.5-14) for U in Equation (7.5-17), using properties of the Pauli matrices, and after several algebraic steps (see Problem 7.11), we obtain

$$R = \mathbf{rr} + \sin(\gamma)\mathbf{r}\times - \cos(\gamma)(\mathbf{r}\times)(\mathbf{r}\times) \tag{7.5-18}$$

or equivalently,

$$R = \begin{bmatrix} r_1r_1 & r_1r_2 & r_1r_3 \\ r_2r_1 & r_2r_2 & r_2r_3 \\ r_3r_1 & r_3r_2 & r_3r_3 \end{bmatrix} + \sin(\gamma)\begin{bmatrix} 0 & -r_3 & r_2 \\ r_3 & 0 & -r_1 \\ -r_2 & r_1 & 0 \end{bmatrix} - \cos(\gamma)\begin{bmatrix} r_1r_1 - 1 & r_1r_2 & r_1r_3 \\ r_2r_1 & r_2r_2 - 1 & r_2r_3 \\ r_3r_1 & r_3r_2 & r_3r_3 - 1 \end{bmatrix} \tag{7.5-19}$$

This is an expression of the 3×3 rotation matrix in terms of the axis of rotation **r**, and the angle of rotation γ in Stokes space. Table 7.1 lists some examples of Jones matrices and the corresponding 3×3 rotation matrices and the angles of rotation.

PSP (Principal States of Polarization)

We now describe the principal states of polarization in both the Jones space and the Stokes space. Consider the input of an optical wave with a fixed polarization state over a spectral regime of interest. We investigate the variation of the output polarization state as a function of frequency ω. This is done by examining the derivative of the output state with respect to frequency. From Equation (7.5-1), we obtain

$$\frac{\partial}{\partial\omega}|v\rangle = \left(\frac{\partial}{\partial\omega}U\right)|s\rangle \tag{7.5-20}$$

where $|s\rangle$ is a fixed input state of polarization and U is the Jones matrix of the birefringent network. For a general birefringent network, the Jones matrix U is a function of frequency. In general, the output state of polarization will vary with the frequency. We are interested

TABLE 7.1 Jones Matrices and Their Corresponding Rotation Matrices in Stokes Space

	Jones Space	Stokes Space		
Physical Object	Jones Matrix	Axis of Rotation	Angle of Rotation	3 × 3 Rotation Matrix
Wave plate with phase retardation γ and azimuth angle[a] $\psi = 0$	$U = \begin{bmatrix} e^{-i\gamma/2} & 0 \\ 0 & e^{+i\gamma/2} \end{bmatrix}$	s_1	γ	$R = \begin{bmatrix} 1 & 0 & 0 \\ 0 & \cos\gamma & -\sin\gamma \\ 0 & \sin\gamma & \cos\gamma \end{bmatrix}$
Wave plate with phase retardation γ and azimuth angle[a] $\psi = 45°$	$U = \begin{bmatrix} \cos(\gamma/2) & -i\sin(\gamma/2) \\ -i\sin(\gamma/2) & \cos(\gamma/2) \end{bmatrix}$	s_2	γ	$R = \begin{bmatrix} \cos\gamma & 0 & \sin\gamma \\ 0 & 1 & 0 \\ -\sin\gamma & 0 & \cos\gamma \end{bmatrix}$
Polarization rotator of angle $\gamma/2$	$U = \begin{bmatrix} \cos(\gamma/2) & -\sin(\gamma/2) \\ \sin(\gamma/2) & \cos(\gamma/2) \end{bmatrix}$	s_3	γ	$R = \begin{bmatrix} \cos\gamma & -\sin\gamma & 0 \\ \sin\gamma & \cos\gamma & 0 \\ 0 & 0 & 1 \end{bmatrix}$

[a] The azimuth angle ψ is defined as the angle of the slow axis measured from the x axis.

in a set of particular input polarization states such that the output states of polarization are insensitive to the variation of frequency (to the first order). Mathematically, this is written

$$\frac{\partial}{\partial\omega}|v\rangle = -i\delta\,|v\rangle \tag{7.5-21}$$

where δ is a constant. Physically, δ is the group delay due to propagation. According to this equation, the variation in frequency only leads to a phase factor that does not change the state of polarization of an output Jones vector. The states that satisfy the above equation are known as the principal states of polarization (PSP). These states must satisfy the following equation, according to Equations (7.5-1), (7.5-20), and (7.5-21):

$$\left(\frac{\partial}{\partial\omega}U\right)U^{-1}|v\rangle = -i\delta|v\rangle \tag{7.5-22}$$

or equivalently,

$$U'U^{-1}|v\rangle = \begin{bmatrix} a'a^* + b'b^* & ab' - a'b \\ a'^*b^* - a^*b'^* & a'^*a + b'^*b \end{bmatrix}|v\rangle = -i\delta|v\rangle \tag{7.5-23}$$

where a prime indicates differentiation with respect to frequency ω.

As a result of the unitary property of U, the matrix $U'U^{-1}$ has a zero trace. Furthermore, this matrix $U'U^{-1}$ is anti-Hermitian with pure imaginary eigenvalues (see Problem 7.12). So the eigenvalues can be written $-i\tau/2$ and $i\tau/2$, with real τ. The determinant of a square matrix is the product of the eigenvalues. So

$$\det(U'U^{-1}) = \frac{\tau^2}{4} \tag{7.5-24}$$

Since U is unitary, we have

$$\det(U') = \frac{\tau^2}{4} = |a'|^2 + |b'|^2 \tag{7.5-25}$$

The time delay between these two PSP can thus be written, according to Equations (7.5-24) and (7.5-25),

$$\tau = 2\sqrt{\det(U')} = 2\sqrt{|a'|^2 + |b'|^2} \tag{7.5-26}$$

The Jones vectors of the output PSP can be written, according to Equation (7.5-23),

$$|p\rangle = N_- \begin{bmatrix} a'b - b'a \\ a'a^* + b'b^* + i\dfrac{\tau}{2} \end{bmatrix} \quad \text{(slow PSP with eigenvalue } \delta = +\tau/2\text{)}$$
$$|p_-\rangle = N \begin{bmatrix} a'b - b'a \\ a'a^* + b'b^* - i\dfrac{\tau}{2} \end{bmatrix} \quad \text{(fast PSP with eigenvalue } \delta = -\tau/2\text{)} \tag{7.5-27}$$

where N and N_- are normalization constants. These two output PSP are mutually orthogonal. The Jones vectors of the input PSP can be obtained by multiplying Equation (7.5-27) with U^{-1}.

Using these output PSP as the basis, any output state can be written

$$|v(\omega_0)\rangle = c_1|p\rangle + c_2|p_-\rangle$$
$$|v(\omega)\rangle = c_1|p\rangle e^{-i\Delta\omega\tau/2} + c_2|p_-\rangle e^{+i\Delta\omega\tau/2} \tag{7.5-28}$$

where ω_0 and ω are frequencies, and $\Delta\omega = \omega - \omega_0$. By examining Equation (7.5-28), we note that $\mathbf{v}(\omega)$ is obtained by rotating $\mathbf{v}(\omega_0)$ around the axis $\mathbf{p}$ with an angle $\tau\,\Delta\omega$ in the Stokes space (see Figure 7.7).

It is important to note that this is only valid for a very small frequency range around ω_0. In most fibers, the principal states of polarization depend on the frequency. So this picture is valid only in a small frequency regime around ω_0 where the variation of the PSP with the frequency can be neglected.

Output PMD Vector $\boldsymbol{\tau}$

The output PMD vector is defined as a vector in the Stokes space, which is parallel to the Stokes vector of the slow output PSP mode, with a length of τ. For a single birefringent network, the group delay τ is enough to describe the broadening of the signal. When several birefringent networks are present in series, the total group delay due to PMD is not a simple sum of the individual group delays. The total group delay is a complicated vector summation of the PMD vectors of the individual birefringent elements. This will be discussed near the end of this section. For that purpose, we need to introduce the concept of PMD vector. The PMD vector can be obtained by using the Jones vectors of the output PSP as follows:

$$\boldsymbol{\tau} = \tau\mathbf{p} = \tau\langle p|\boldsymbol{\sigma}|p\rangle \tag{7.5-29}$$

where the group delay τ is given by Equation (7.5-26), and the Jones vector of the slow output PSP $|p\rangle$ is given by Equation (7.5-27). For a uniform birefringent fiber or a wave plate, the PMD vector is parallel to the polarization direction of the slow mode, with a magnitude given by Equation (7.3-1).

We can also obtain the PMD vector by using the projection operators and Pauli matrix representation. The operator (or matrix) can be represented by the projection operators of the eigenvectors as follows:

$$U'U^{-1} = -i\frac{\tau}{2}|p\rangle\langle p| + i\frac{\tau}{2}|p_-\rangle\langle p_-| \tag{7.5-30}$$

where $-i\tau/2$ and $+i\tau/2$ are eigenvalues of $U'U^{-1}$, and $|p\rangle$ and $|p_-\rangle$ are the corresponding eigenvectors. Furthermore, the projection operators can be written, according to definitions in this section,

$$\begin{aligned} |p\rangle\langle p| &= \tfrac{1}{2}(I + \mathbf{p}\cdot\boldsymbol{\sigma}) \\ |p_-\rangle\langle p_-| &= \tfrac{1}{2}(I + \mathbf{p}_-\cdot\boldsymbol{\sigma}) = \tfrac{1}{2}(I - \mathbf{p}\cdot\boldsymbol{\sigma}) \end{aligned} \tag{7.5-31}$$

Thus we obtain from Equations (7.5-30) and (7.5-31)

$$U'U^{-1} = -\frac{i}{2}\boldsymbol{\tau}\cdot\boldsymbol{\sigma} \tag{7.5-32}$$

or equivalently,

$$2iU'U^{-1} = \boldsymbol{\tau}\cdot\boldsymbol{\sigma} = \begin{bmatrix} \tau_1 & \tau_2 - i\tau_3 \\ \tau_2 + i\tau_3 & -\tau_1 \end{bmatrix} \tag{7.5-33}$$

The components of the PMD vector can easily be obtained by examining the matrix elements of the operator $2iU'U^{-1}$. The results are, according to Equation (7.5-23),

$$\begin{aligned} \tau_1 &= 2i(a'a^* + b'b^*) \\ \tau_2 &= 2\,\mathrm{Im}(a'b - ab') \\ \tau_3 &= 2\,\mathrm{Re}(a'b - ab') \end{aligned} \tag{7.5-34}$$

These are explicit expressions of the PMD vector in terms of the matrix elements (a, b) of the Jones matrix U. Once the output PMD vector is obtained, the input PMD vector can be obtained by $\boldsymbol{\tau}_{\text{in}} = R^{-1}\boldsymbol{\tau}$, where R is the rotation matrix representing the polarization transformation due to the birefringent network in the Stokes space. Note that $|\boldsymbol{\tau}_{\text{in}}| = |\boldsymbol{\tau}|$.

EXAMPLE: UNIFORM BIREFRINGENT FIBER

As an example to illustrate the concept of the PMD vector, we consider the simplest case of a single segment of a birefringent medium with a uniform birefringence. For this case, the Jones matrix is given by

$$M = \begin{bmatrix} e^{-i\Gamma/2} & 0 \\ 0 & e^{i\Gamma/2} \end{bmatrix}$$

where

$$\Gamma = \frac{\omega}{c}(n_s - n_f)d = \frac{\omega}{c}\Delta n\, d$$

So $a = e^{-i\Gamma/2}$, $b = 0$, and

$$a' = -ia\frac{1}{2}\left(\frac{d}{c}\Delta n + \frac{\omega d}{c}\frac{d\Delta n}{d\omega}\right)$$

$$b' = 0$$

We obtain, according to Equation (7.5-34),

$$\tau_1 = 2i\,(a'a^* + b'b^*) = \left(\frac{d}{c}\Delta n + \frac{\omega d}{c}\frac{\partial \Delta n}{\partial \omega}\right)$$

$$\tau_2 = 2\,\mathrm{Im}(a'b - ab') = 0$$

$$\tau_3 = 2\,\mathrm{Re}(a'b - ab') = 0$$

We note that the PMD vector is parallel to the S_1 axis (or the slow axis of the birefringent medium), with a magnitude equal to the group delay between the polarization modes.

Infinitesimal Rotations and Differential Equations

It is known from the mechanics of rigid bodies that the law of rotations becomes particularly simple when the rotations are infinitesimally small. We consider an infinitesimally small segment of birefringent fiber. It is reasonable to believe that the angle of rotation in the Stokes space is proportional to the length of the fiber dz. In other words, we can write

$$\gamma = \beta\, dz = k(n_s - n_f)\, dz = k\,\Delta n\, dz \tag{7.5-35}$$

where $\beta = k\,\Delta n$ is a measure of the birefringence, and n_s and n_f are the refractive indices of the slow mode and fast mode, respectively. The Jones matrix responsible for the input–output relationship can be written, according to Equation (7.5-14),

$$U = I - \frac{i}{2}\beta\, dz\, \mathbf{r}\cdot\boldsymbol{\sigma} \tag{7.5-36}$$

where $\mathbf{r}$ is the axis of rotation in Stokes space. The 3×3 rotation matrix can be written explicitly as, according to Equation (7.5-19),

$$R = I + \beta\, dz \begin{bmatrix} 0 & -r_3 & r_2 \\ r_3 & 0 & -r_1 \\ -r_2 & r_1 & 0 \end{bmatrix} = I + \beta\, dz\, \mathbf{r}\times \tag{7.5-37}$$

Let the output Jones vector after the infinitesimal rotation be written $|v\rangle = |s\rangle + d|s\rangle$, where $d|s\rangle$ is a differential of the Jones vector, representing the change of the polarization state after transmission through the birefringent element. Using these matrices, the input–output relationship can be written

$$|s\rangle + d|s\rangle = U|s\rangle = |s\rangle - \frac{i}{2}\beta\, dz\, \mathbf{r}\cdot\boldsymbol{\sigma}|s\rangle \tag{7.5-38}$$

or equivalently,

$$\frac{d}{dz}|s\rangle = -\frac{i}{2}\boldsymbol{\beta}\cdot\boldsymbol{\sigma}|s\rangle \tag{7.5-39}$$

where

$$\boldsymbol{\beta} = \beta\mathbf{r} = k\,\Delta n\,\mathbf{r} \tag{7.5-40}$$

is known as the birefringence vector. The birefringence vector, a real vector in Stokes space, is defined at each location of a fiber. For each location, a differential element dz of the fiber between z and $z + dz$ is considered as a birefringent element with an infinitesimal rotation in Stokes space. The birefringence vector is parallel to the axis of rotation in the Stokes space corresponding to the polarization transformation by the differential element, while the magnitude of the vector is the product of the wavenumber with the local birefringence, $k\,\Delta n$. For a uniform birefringent fiber (or a wave plate), the birefringence vector is parallel to the polarization direction (in Stokes space) of the slow mode. With this definition, a birefringent property of a fiber is completely determined if $\boldsymbol{\beta} = \boldsymbol{\beta}(z)$ is given for all z.

Using the rotation matrix R, the input–output relationship in Stokes space can be written, according to Equation (7.5-37),

$$\mathbf{s} + d\mathbf{s} = R\mathbf{s} = \mathbf{s} + \beta\, dz\, \mathbf{r} \times \mathbf{s} \tag{7.5-41}$$

or equivalently,

$$\frac{\partial}{\partial z}\mathbf{s} = \boldsymbol{\beta} \times \mathbf{s} \tag{7.5-42}$$

This equation is particularly useful. If the birefringence vector $\boldsymbol{\beta}(z)$ of a fiber is known, this equation can be integrated to obtain the output polarization state at any location z in the fiber.

We now consider the case of an input beam of light with a fixed state of polarization, over a small spectral regime of interest. As we have discussed earlier, the output state of polarization depends on the frequency. We now investigate the change of the state of polarization of the output beam due to an infinitesimal increase of frequency $d\omega$. The change in the state of output polarization, denoted as $d|s\rangle$, can be described as an infinitesimal rotation in both the Jones space and the Stokes space. In the Jones space, the change of output polarization state $d|s\rangle$ can be written, according to Equations (7.5-22) and (7.5-32),

$$d|s\rangle = \left(\frac{\partial}{\partial \omega}U\right)U^{-1}|s\rangle\, d\omega = -\frac{i}{2}\boldsymbol{\tau}\, d\omega \cdot \boldsymbol{\sigma}|s\rangle \tag{7.5-43}$$

or equivalently,

$$\frac{\partial}{\partial \omega}|s\rangle = -\frac{i}{2}\boldsymbol{\tau} \cdot \boldsymbol{\sigma}|s\rangle \tag{7.5-44}$$

where $\boldsymbol{\tau}$ is the PMD vector Here we are interested in the description of the output polarization state $|v\rangle$ as a continuous function of both z and ω. It suffices to use the notation $|s\rangle \equiv |s(z, \omega)\rangle$ as the Jones vector of the polarization state at any point in the fiber. Following a similar mathematical argument leading from Equation (7.5-39) to Equation (7.5-42), we obtain

$$\frac{\partial}{\partial \omega}\mathbf{s} = \boldsymbol{\tau} \times \mathbf{s} \tag{7.5-45}$$

This differential equation describes the change of the state of polarization of the output beam in Stokes space as a function of frequency given an arbitrary input state of polarization. If $\boldsymbol{\tau}$ is a constant, Equation (7.5-45) can easily be integrated. The solution $\mathbf{s}(\omega)$ is obtained by rotating $\mathbf{s}(\omega_0)$ around $\boldsymbol{\tau}$ by an angle $\tau(\omega - \omega_0)$. This is consistent with the description in Figure 7.7.

The Dynamical PMD Equation

In analyzing the pulse broadening in birefringent fibers, it is desirable to obtain a differential equation for the evolution of the PMD vector along the fibers. As we recall, the magnitude of the PMD vector gives information about the broadening. We now take a partial derivative with respect to ω on both sides of Equation (7.5-42), and a partial derivative with respect to z on both sides of Equation (7.5-45). By eliminating $\partial^2\mathbf{s}/\partial z\, \partial\omega$, we obtain

$$\frac{\partial}{\partial z}(\boldsymbol{\tau} \times \mathbf{s}) = \frac{\partial}{\partial \omega}(\boldsymbol{\beta} \times \mathbf{s}) \tag{7.5-46}$$

Using a vector identity $\mathbf{a} \times (\mathbf{b} \times \mathbf{c}) = \mathbf{b}(\mathbf{c} \cdot \mathbf{a}) - \mathbf{c}(\mathbf{a} \cdot \mathbf{b})$, we obtain, after few mathematical steps,

$$\frac{\partial}{\partial z}\boldsymbol{\tau} = \frac{\partial}{\partial \omega}\boldsymbol{\beta} + \boldsymbol{\beta} \times \boldsymbol{\tau} \tag{7.5-47}$$

This is the differential equation describing the evolution of the PMD vector along the fiber. The same equation has been employed as a basis for the statistical theory of PMD. Given a birefringence vector $\boldsymbol{\beta}(z, \omega)$, Equation (7.6-47) can be employed to obtain the output PMD vector.

PMD Concatenation

The PMD vector defined earlier is an important parameter of a birefringent network. It contains information about the output PSP as well as the group delay between the two PSP. The PMD vector of a sequence of birefringent networks is related to the PMD vectors of each element of the networks. Here we describe a method of obtaining the PMD vector of a sequence of birefringent networks.

Referring to Figure 7.10, we consider a single section of a birefringent fiber. The input–output relationship can be written

$$\mathbf{v} = \mathbf{s}(z) = R\mathbf{s}(0) \tag{7.5-48}$$

where R is the 3×3 rotation matrix. The PMD vector $\boldsymbol{\tau}$ can be obtained in two ways:

1. The Jones matrix approach.
2. The PSP approach.

The Jones Matrix Approach

We start from the Jones equation,

$$|v\rangle = |s(z)\rangle = U|s\rangle \tag{7.5-49}$$

where U is the Jones matrix. Once the Jones matrix is obtained, the PMD vector can be obtained by using Equation (7.5-34).

Alternately, we can find the eigenvectors of U. The Stokes vectors of these two eigenvectors determine the axis of rotation of R, while the phases of the eigenvalues determine the angle of rotation. These eigenvectors and eigenvalues uniquely determine the 3×3 rotation matrix R. Once R is obtained by using Equation (7.5-19), the PMD vector can be obtained as follows. Using Equation (7.5-45), we have

$$\frac{\partial \mathbf{s}(z)}{\partial \omega} = \boldsymbol{\tau} \times \mathbf{s}(z) = \boldsymbol{\tau} \times R\mathbf{s}(0) = \frac{\partial R}{\partial \omega}\mathbf{s}(0) \tag{7.5-50}$$

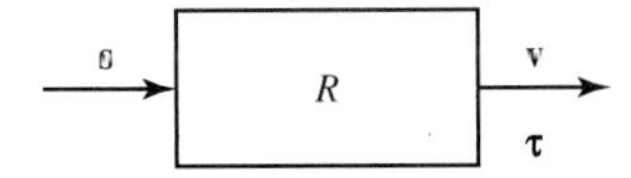

Figure 7.10 Linear input–output relationship of a birefringent fiber system. In the figure, R is the rotation matrix (3×3) that transforms the input Stokes vector $\mathbf{s}$ at $z = 0$ into the output Stokes vector $\mathbf{v} = \mathbf{s}(z)$ at z, and $\boldsymbol{\tau}$ is the PMD vector of the birefringent fiber system.

Since $\mathbf{s}(0)$ is arbitrary, we obtain

$$\frac{\partial R}{\partial \omega} = \boldsymbol{\tau} \times R$$

or equivalently,

$$\boldsymbol{\tau}\times = \frac{\partial R}{\partial \omega} R^{-1} \tag{7.5-51}$$

This equation can be employed to obtain all three components of the PMD vector in terms of the rotation matrix.

The PSP Approach

In the PSP approach, we first find the principal states of polarization by solving the eigenvector problem (7.5-23) and obtaining the eigenvectors and the eigenvalues. The magnitude of the PMD vector is the group delay between the two PSP modes, while the direction of the PMD vector is parallel to the polarization direction of the slow PSP mode. Or alternatively, the PMD vector can be obtained by using Equation (7.5-29), where $|p\rangle$ is the Jones vector of the slow PSP mode.

Once the PMD vector of a birefringent network is obtained, we can find the PMD vector of a series of birefringent networks. We start with the discussion of two concatenated birefringent networks.

PMD of Two Concatenated Sections

Referring to Figure 7.11, we consider two sections of birefringent elements in sequence. Let $\boldsymbol{\tau}_1$ and $\boldsymbol{\tau}_2$ be the PMD vectors of the individual sections.

Let $\boldsymbol{\tau}$ be the overall PMD vector for the whole system shown in Figure 7.9. We need to obtain the PMD vector $\boldsymbol{\tau}$ in terms of $\boldsymbol{\tau}_1$ and $\boldsymbol{\tau}_2$. Let R be the rotation matrix of the overall system, $R = R_2R_1$. The PMD vector $\boldsymbol{\tau}$ can be written, according to Equation (7.5-51),

$$\boldsymbol{\tau}\times = \frac{\partial R}{\partial \omega} R^{-1}$$

Using $R = R_2R_1$, where R_1 and R_2 are the 3×3 rotation matrices of the individual sections, we obtain

$$\boldsymbol{\tau}\times = \frac{\partial R}{\partial \omega} R^{-1} = \frac{\partial (R_2R_1)}{\partial \omega}(R_2R_1)^{-1} = \frac{\partial R_2}{\partial \omega} R_1R_1^{-1}R_2^{-1} + R_2\frac{\partial R_1}{\partial \omega} R_1^{-1}R_2^{-1} \tag{7.5-52}$$

Using $R_1R_1^{-1} = 1$ and $\boldsymbol{\tau}_1\times = (\partial R_1/\partial\omega)R_1^{-1}$, the above equation can be simplified to

$$\boldsymbol{\tau}\times = \frac{\partial R}{\partial \omega} R^{-1} = \frac{\partial R_2}{\partial \omega} R_2^{-1} + R_2\boldsymbol{\tau}_1 \times R_2^{-1} = \boldsymbol{\tau}_2\times + (R_2\boldsymbol{\tau}_1)\times \tag{7.5-53}$$

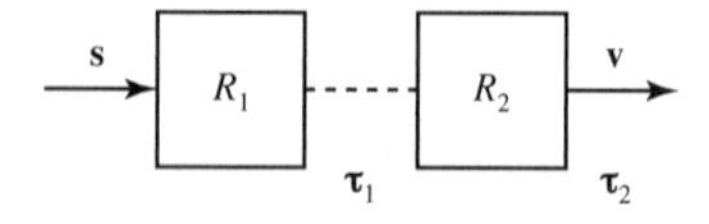

Figure 7.11 Concatenation of two sections of birefringent media.

where the last equality is obtained by using the fact that $R_2\boldsymbol{\tau}_1 \times R_2^{-1}$ is a transformation of the matrix $\boldsymbol{\tau}_1\times$ via a coordinate rotation R_2 and thus can be written $(R_2\boldsymbol{\tau}_1)\times$. So we obtain

$$\boldsymbol{\tau} = \boldsymbol{\tau}_2 + R_2\boldsymbol{\tau}_1 \tag{7.5-54}$$

This is the basic concatenation rule of PMD vectors. According to this equation, the total PMD vector is the sum of the PMD vector of the last element, plus the PMD vector of the previous element after being rotated by the 3×3 rotation matrix of the last element.

EXAMPLE: PMD VECTOR FOR *N* CONCATENATED SECTIONS

Consider the concatenation of three sections of birefringent fibers. Let $\boldsymbol{\tau}_1$, $\boldsymbol{\tau}_2$, and $\boldsymbol{\tau}_3$ be the PMD vectors of the individual sections. The overall PMD vector is obtained as follows:

$$\boldsymbol{\tau} = \boldsymbol{\tau}_3 + R_3(\boldsymbol{\tau}_2 + R_2\boldsymbol{\tau}_1) = \boldsymbol{\tau}_3 + R_3\boldsymbol{\tau}_2 + R_3R_2\,\boldsymbol{\tau}_1 \tag{7.5-55}$$

The basic concatenation rule can easily be extended to N sections of birefringent fibers.

$$\boldsymbol{\tau} = \boldsymbol{\tau}_N + R_N\boldsymbol{\tau}_{N-1} + R_NR_{N-1}\boldsymbol{\tau}_{N-2} + R_NR_{N-1}R_{N-2}\boldsymbol{\tau}_{N-3} + \cdots + R_NR_{N-1}\cdots R_2\boldsymbol{\tau}_1 \tag{7.5-56}$$

If we define

$$R(N, n) = R_NR_{N-1}R_{N-2}\cdots R_{n+1}R_n \tag{7.5-57}$$

then the concatenation rule for N sections can be written

$$\boldsymbol{\tau} = \sum_{n=1}^{N} R(N, n+1)\boldsymbol{\tau}_n \tag{7.5-58}$$

where $\boldsymbol{\tau}_n$ is the PMD vector of the nth section.

PMD Vector of an Infinitesimal Section

The PMD vector of an infinitesimal section of birefringent fiber is particularly simple.

According to Equation (7.5-37), the 3×3 rotation matrix of an infinitesimal section of birefringent fiber can be written

$$R = I + dz\,\boldsymbol{\beta}\times \tag{7.5-59}$$

where $\boldsymbol{\beta}$ is the birefringence vector of the element and dz is the length of the infinitesimal element. Taking the derivative with respect to frequency, we obtain

$$\frac{\partial}{\partial\omega}R = dz\frac{\partial\boldsymbol{\beta}}{\partial\omega}\times \tag{7.5-60}$$

The inverse rotation matrix can be written

$$R^{-1} = I - dz\,\boldsymbol{\beta}\times \tag{7.5-61}$$

Substituting Equations (7.5-60) and (7.5-61) into Equation (7.5-51), we obtain

$$\boldsymbol{\tau} = \frac{\partial\boldsymbol{\beta}}{\partial\omega}dz \tag{7.5-62}$$

This is the expression of the PMD vector of an infinitesimal element dz. The same result can also be easily obtained by a direct integration of the dynamical equation, and by neglecting

TABLE 7.2 Birefringence Vectors and PMD Vectors of Some Birefringent Elements

Physical Object	Birefringence Vector	PMD Vector
Wave plate with birefringence Δn, thickness d, and azimuth angle $\psi = 0$ (angle between slow axis and x axis)	$\boldsymbol{\beta} = \begin{bmatrix} 1 \\ 0 \\ 0 \end{bmatrix} k\,\Delta n$	$\boldsymbol{\tau} = \tau \begin{bmatrix} 1 \\ 0 \\ 0 \end{bmatrix}$
Wave plate with birefringence Δn, thickness d, and azimuth angle ψ (angle between slow axis and x axis)	$\boldsymbol{\beta} = \begin{bmatrix} \cos(2\psi) \\ \sin(2\psi) \\ 0 \end{bmatrix} k\,\Delta n$	$\boldsymbol{\tau} = \tau \begin{bmatrix} \cos(2\psi) \\ \sin(2\psi) \\ 0 \end{bmatrix}$
Polarization rotator of angle $\gamma/2$ (circular birefringence Δn and thickness d)	$\boldsymbol{\beta} = \begin{bmatrix} 0 \\ 0 \\ 1 \end{bmatrix} k\,\Delta n$	$\boldsymbol{\tau} = \tau \begin{bmatrix} 0 \\ 0 \\ 1 \end{bmatrix}$

Note: Here $\tau = d\partial(k\,\Delta n)/\partial\omega$ is the group delay between the two polarization modes.

the second-order term $\boldsymbol{\beta} \times \boldsymbol{\tau}\, dz$. For a uniform birefringent fiber, the PMD vector and the birefringence vector are both parallel to the polarization direction of the slow mode. Table 7.2 is a list of PMD vectors and birefringence vectors of some simple birefringent elements.

Using Equation (7.5-62) for $\boldsymbol{\tau}_n$ in Equation (7.5-58), we obtain the following integral form of the concatenation rule:

$$\boldsymbol{\tau}(L) = \int_0^L R(L, z) \frac{\partial \boldsymbol{\beta}(z)}{\partial \omega} dz \tag{7.5-63}$$

where L is the length of the birefringent fiber, and $R(L, z)$ is the 3×3 rotation matrix corresponding to the segment between (z, L).

PMD Vector and the Broadening of Pulses

The PMD vector discussed earlier is a measure of the pulse broadening due to PMD in birefringent networks. As discussed in Section 7.3, the pulse broadening of a beam of light due to PMD in a birefringent network can be written in terms of the input PMD vector as, according to Equation (7.3-11),

$$(\Delta\tau_b)^2 = \tfrac{1}{4}[|\boldsymbol{\tau}_{\text{in}}|^2 - (\boldsymbol{\tau}_{\text{in}} \cdot s)^2] = \tfrac{1}{4}[|\boldsymbol{\tau}|^2 - (\boldsymbol{\tau}_{\text{in}} \cdot s)^2] \tag{7.5-64}$$

where $\boldsymbol{\tau}_{\text{in}}$ is the input PMD vector, $\boldsymbol{\tau} = R\boldsymbol{\tau}_{\text{in}}$ is the output PMD vector, $\mathbf{s}$ and is the Stokes vector of the input wave (note that $|\boldsymbol{\tau}_{\text{in}}| = |\boldsymbol{\tau}|$). We note that when the input Stokes vector $\mathbf{s}$ is aligned along the input PMD vector $\boldsymbol{\tau}_{\text{in}}$ (or the input PSP), the PMD broadening (first order) is zero. For a beam of unpolarized wave $\boldsymbol{\tau}_{\text{in}} \cdot \mathbf{s} = 0$, the PMD broadening is, according to Equation (7.5-64) or (7.3-11),

$$(\Delta\tau_b)^2 = \tfrac{1}{4}|\boldsymbol{\tau}|^2 \tag{7.5-65}$$

where $\boldsymbol{\tau}$ is the PMD vector of the birefringent network. If the birefringent network consists of a sequence of N birefringent elements (fibers), the PMD vector of the whole network can be obtained from the PMD vectors ($\boldsymbol{\tau}_n$, $n = 1, 2, 3 \ldots, N$) of the individual elements by using the concatenation rule (7.5-56) described earlier. To illustrate the concatenation rule of PMD vectors, we consider the following example.

EXAMPLE: TWO UNIFORM BIREFRINGENT ELEMENTS IN SEQUENCE

Referring to Figure 7.12, we consider the transmission of a beam of light through two uniform birefringent elements in sequence. Let the orientation of the slow axes be ψ_1, ψ_2 measured relative to the x axis. According to the discussion above and Table 7.2, the PMD vectors in Stokes space for each of the elements are

$$\boldsymbol{\tau}_1 = \tau_1 \begin{bmatrix} \cos(2\psi_1) \\ \sin(2\psi_1) \\ 0 \end{bmatrix} \quad \text{and} \quad \boldsymbol{\tau}_2 = \tau_2 \begin{bmatrix} \cos(2\psi_2) \\ \sin(2\psi_2) \\ 0 \end{bmatrix}$$

with $\tau_1 = d_1\partial(k\,\Delta n_1)/\partial\omega$ and $\tau_2 = d_2\partial(k\,\Delta n_2)/\partial\omega$, where Δn_1 and Δn_2 are birefringence, and d_1 and d_2 are lengths of the birefringent elements. These two PMD vectors lie in the equatorial plane of the Poincaré sphere. They are oriented at $2\psi_1$ and $2\psi_2$ from the s_1 axis (see Figure 7.12).

The overall PMD vector of the whole system is obtained by using the concatenation rule (7.5-54). We first rotate PMD vector $\boldsymbol{\tau}_1$ by an angle Γ_2 around the PMD vector $\boldsymbol{\tau}_2$ of the second element. The overall PMD vector is then obtained by a vector summation of $\boldsymbol{\tau}_2$ and $R_2\boldsymbol{\tau}_1$. This is illustrated in Figure 7.12. Following a simple algebra, we obtain the length of the overall PMD vector:

$$\tau^2 = \tau_1^2 + \tau_2^2 + 2\tau_1\tau_2 \cos[2(\psi_2 - \psi_1)] \tag{7.5-66}$$

The length of the overall PMD vector is a measure of the pulse broadening for a beam of unpolarized light, according to Equation (7.5-65). Depending on the relative orientation between the slow axes of the two birefringent elements, the length of the overall PMD vector can vary from $(\tau_1 + \tau_2)$ to $|\tau_1 - \tau_2|$, depending on the angle $(\psi_2 - \psi_1)$. If the slow axes of the second element is completely random relative to the orientation of the rotated PMD vector

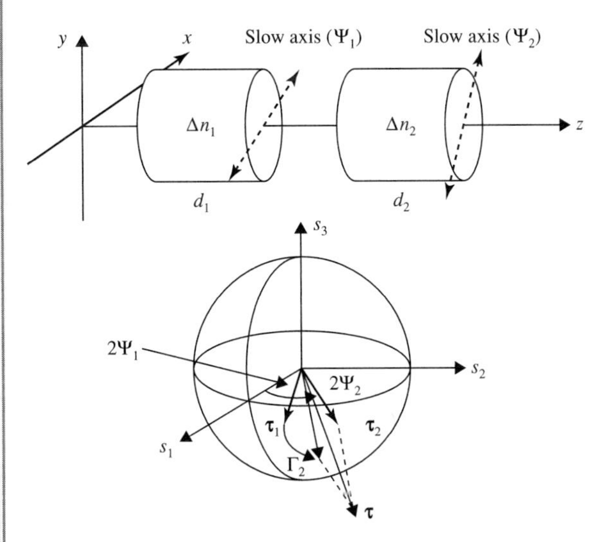

Figure 7.12 Schematic drawing of transmission of a beam of light through a series of two birefringent elements, with their slow axes oriented at azimuth angles ψ_1 and ψ_2, respectively. The lower figure is the Poincaré sphere representation of the PMD vectors, and the vector addition of $\boldsymbol{\tau}_2$ and $R_2\boldsymbol{\tau}_1$. R_2 is a rotation of angle Γ_2 around the vector $\boldsymbol{\tau}_2$, where $\Gamma_2 = d_2k\,\Delta n_2$.

$R_2\boldsymbol{\tau}_1$, then the third term on the right side of Equation (7.5-66) will average out to zero. In this case, the pulse broadening is $(\tau_1^2 + \tau_2^2)/4$, according to Equation (7.6-65). In other words, the last element with a PMD vector $\boldsymbol{\tau}_2$ contributes an additional broadening of $\tau_2^2/4$. Based on this argument, the pulse broadening of a sequence of N birefringent elements with completely random orientation of the slow axes can be written

$$(\Delta\tau_b)^2 = \frac{1}{4}\sum_{n=1}^{N} |\boldsymbol{\tau}_n|^2 \tag{7.5-67}$$

where $\boldsymbol{\tau}_n$ is the PMD vector of the nth element. It is important to note that this is a purely hypothetical case. In real fiberoptic networks, the orientation of the slow axes of each of the birefringent elements may not be completely random.

7.6 HIGH-ORDER PMD AND COMPENSATORS

The principal states of polarization (PSP) described earlier are useful in the so-called first-order PMD compensator (PMDC). Figure 7.13 shows a schematic drawing of such a compensator. An input pulse of a beam of unpolarized light is split into two smaller pulses. The two smaller pulses are polarized along the fast PSP mode and the slow PSP mode, respectively. These two modes are mutually orthogonal in their polarization states that are generally elliptical. At the output of the birefringent network, these two pulses are separated by a polarization group delay. They are polarized along the output PSP directions. Using a quarter-wave plate at the appropriate orientation, these two polarized modes can be converted into linearly polarized states. A birefringent wave plate with the proper retardance can then be employed to compensate for the time delay between the two pulses. This leads to a single pulse at output and thus removes the pulse spreading (or splitting) due to PMD.

Polarization mode dispersion (PMD) is currently one of the primary obstacles to ultrahigh-speed optical communications systems. As discussed earlier in this chapter, the effects of PMD were traditionally described in terms of a pair of principal states of polarization (PSP) and the differential group delay (DGD) between them. At low communication speeds, the PSP and the DGD at the center of the signal spectrum (this is usually the carrier frequency) are sufficient to fully describe the effects of PMD over the spectrum of the transmitted signal. As the communication speed increases, the spectral bandwidth of the signals increases accordingly. The PSP and DGD at the center frequency are no longer a good representation

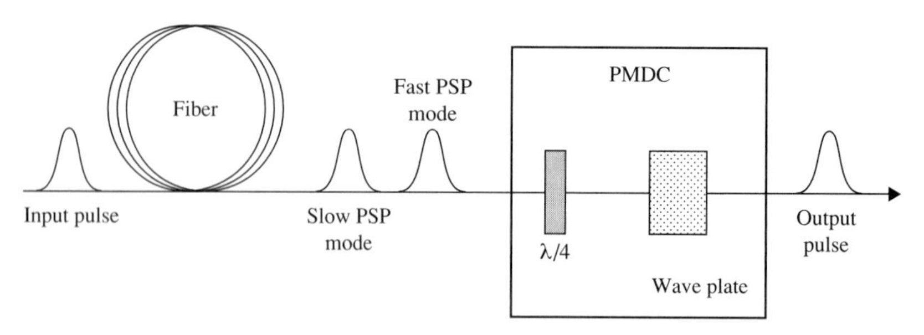

Figure 7.13 Schematic drawing of a first-order PMDC.

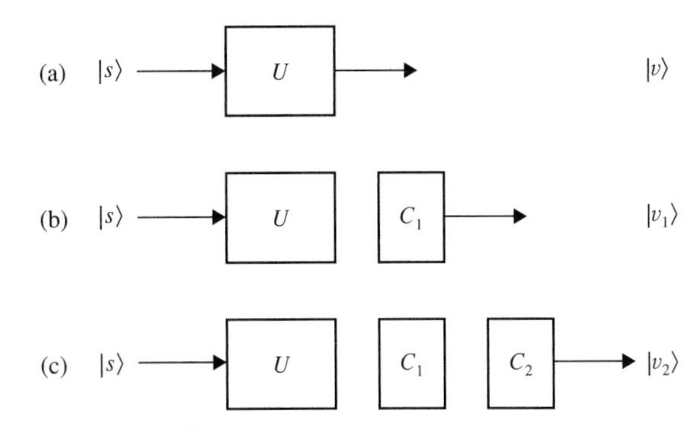

Figure 7.14 Schematic drawing of PMD compensators and the input–output Jones vectors. U is the Jones matrix of the fiber; C_1 and C_2 are the Jones matrices of PMD compensators.

of the PSP and DGD within the whole signal spectral bandwidth. Under the circumstances, we need to consider the second-order variation of the PSP and DGDs over the bandwidth [3–5]. Furthermore, we need to find compensators that can undo the PMD over the bandwidth for these high-speed signals.

Commonly termed as high-order PMD, the variation of first-order PSP and DGD over the bandwidth of high-speed signals necessitates the development of analytical tools for its description. Several approaches appear in the literature for the description of high-order PMD. The most common approach is based on the output PMD vector, $\boldsymbol{\tau}$, which in the absence of polarization-dependent loss (PDL) is a real, three-dimensional vector whose magnitude is equal to the DGD and whose orientation is parallel to the Poincaré sphere representation of the slow output PSP. Using this approach, the high-order PMD can be described in terms of the derivatives of the PMD vector with respect to frequency [2, 4].

Here we provide a general description of high-order PMD and discuss optical means to compensate for the high-order PMD. Using the Jones matrix method, we generalize the phenomenological approach of Poole and Wagner [1], by considering the high-order derivatives of the Jones matrix equation [6–8]. Referring to Figure 7.14, we consider a fiber whose birefringent property is described by a Jones matrix U. The input–output relationship is written

$$|v\rangle = U\,|s\rangle \tag{7.6-1}$$

where $|s\rangle$ is an arbitrary input Jones vector and $|v\rangle$ is the corresponding output Jones vector.

The first-order PSP satisfy the following vector equation:

$$\frac{\partial}{\partial\omega}|v\rangle = \left(\frac{\partial}{\partial\omega}U\right)|s\rangle = -i\delta_1\,|v\rangle \tag{7.6-2}$$

where $|s\rangle$ is a fixed input state of polarization and δ_1 is an eigenvalue. The DGD is the difference between the two eigenvalues $(\delta_{1+} - \delta_{1-})$. For the purpose of discussion, we use the following notations for the PSP and their eigenvalues:

$|p_{1+}\rangle$ Unit Jones vector of slow first-order output PSP with eigenvalue δ_{1+}

$|p_{1-}\rangle$ Unit Jones vector of fast first-order output PSP with eigenvalue δ_1

Let C_1 represent the Jones matrix of a first-order PMD compensator. Then, by definition, C_1 must satisfy the following equation:

$$\frac{\partial}{\partial\omega}|v_1\rangle = \left(\frac{\partial}{\partial\omega}C_1U\right)|s\rangle = 0 \tag{7.6-3}$$

where $|s\rangle$ is a fixed input state of polarization and C_1U is the Jones matrix of the fiber plus the compensator (see Figure 7.14b). With the presence of the compensator C_1, the DGD of the overall system becomes zero. Carrying out the differentiation in Equation (7.6-3), we obtain

$$(C_1'U + C_1U')|s\rangle = 0 \tag{7.6-4}$$

where a prime indicates differentiation with respect to ω. To solve for the Jones matrix C_1, we let $|s\rangle$ be the input PSP. According to Equations (7.6-1) and (7.6-2), we have $U|s\rangle = |p_1\rangle$ and $U'|s\rangle = -i\delta_1|p_1\rangle$, where $|p_1\rangle$ is a first-order output PSP. The above equation becomes

$$[C_1' + C_1(-i\delta_1)]|p_1\rangle = 0 \tag{7.6-5}$$

The Jones matrix of the first-order PMD compensator C_1 can thus be written, using the first-order output PSP as the basis,

$$C_1 = \begin{bmatrix} \exp(-i\delta_{1+}\omega) & 0 \\ 0 & \exp(-i\delta_{1-}\omega) \end{bmatrix}_{\text{PSP}} \tag{7.6-6}$$

In xy coordinates, the Jones matrix C_1 can be obtained by using the following formula:

$$\langle i|C_1|j\rangle = \sum_{\alpha\beta} \langle i|\alpha\rangle\langle\alpha|C_1|\beta\rangle\langle\beta|j\rangle \quad \text{with } i, j = x, y \quad \text{and} \quad \alpha, \beta = |p_{1+}\rangle, |p_{1-}\rangle \tag{7.6-7}$$

The Jones matrix C_1 in xy coordinates can also be written

$$C_1 = \begin{bmatrix} p_{1+x} & p_{1-x} \\ p_{1+y} & p_{1-y} \end{bmatrix} \begin{bmatrix} \exp(-i\delta_{1+}\omega) & 0 \\ 0 & \exp(-i\delta_{1-}\omega) \end{bmatrix} \begin{bmatrix} p_{1+x} & p_{1+y} \\ p_{1-x} & p_{1-y} \end{bmatrix} \tag{7.6-8}$$

where p_{1+x}, p_{1+y}, p_{1-x}, p_{1+y} are xy components of the Jones vectors of the first-order output PSP. We note that the first matrix in the above equation consists of two column vectors of the first-order output PSP, and the third matrix consists of two row vectors of the first-order output PSP. The above equation is an explicit expression of the Jones matrix of the first-order PMD compensator.

We now consider the second-order effect. Referring to Figure 7.14b, the input–output Jones vector relationship is now written

$$|v_1\rangle = C_1U|s\rangle \tag{7.6-9}$$

With the presence of the first-order PMD compensator, all input states are first-order principal states and the DGD is zero to the first order. Since the DGD is zero for the first order, we consider the second-order effect. This is particularly important in high-speed transmission where the signal bandwidth is quite broad. The second-order PMD is described as the second-order differential variation of the output polarization states and the associated DGD. The second-order PSP satisfy the following vector equation:

$$\frac{\partial^2}{\partial\omega^2}|v_1\rangle = \left(\frac{\partial^2}{\partial\omega^2}C_1U\right)|s\rangle = -i\delta_2|v_1\rangle \tag{7.6-10}$$

where $|s\rangle$ is a fixed input state of polarization and δ_2 is an eigenvalue. Let the second-order output PSP and their eigenvalues be written

$|p_{2+}\rangle$ Unit Jones vector of slow second-order output PSP with eigenvalue δ_{2+}

$|p_{2-}\rangle$ Unit Jones vector of fast second-order output PSP with eigenvalue δ_{2-}

The DGD between the two eigenmodes is

$$\text{DGD} = (\delta_{2+} - \delta_{2-})(\omega - \omega_0) \tag{7.6-11}$$

where ω_0 is the center of the signal spectrum, and δ_{2+} and δ_{2-} are the eigenvalues.

Let C_2 represent the Jones matrix of the second-order PMD compensator (see Figure 7.14c). Then, by definition, C_2 must satisfy the following equation:

$$\frac{\partial^2}{\partial\omega^2}|v_2\rangle = \left(\frac{\partial^2}{\partial\omega^2}C_2 M_1\right)|s\rangle = 0 \tag{7.6-12}$$

where $|s\rangle$ is a fixed input state of polarization and M_1 is the Jones matrix of the fiber plus the compensator (see Figure 7.14b),

$$M_1 = C_1 U \tag{7.6-13}$$

With the presence of the compensator C_2, the second-order DGD of the overall system (Figure 7.14c) becomes zero.

Carrying out the differentiation in Equation (7.6-12), we obtain

$$(C_2'' M_1 + 2C_2' M_1' + C_2 M_1'')|s\rangle = 0 \tag{7.6-14}$$

To solve for the Jones matrix C_2, we let $|s\rangle$ be the input second-order PSP. According to Equations (7.6-3), (7.6-9), and (7.6-10), we have $M_1|s\rangle = |p_2\rangle$, $M_1'|s\rangle = 0$, and $M_1''|s\rangle = -i\delta_2|p_2\rangle$, where $|p_2\rangle$ is a second-order output PSP. Equation (7.6-14) becomes

$$(C_2'' + C_2(-i\delta_2))|p_2\rangle = 0 \tag{7.6-15}$$

The Jones matrix of the second-order PMD compensator C_2 can thus be written, using the second-order PSP as the basis,

$$C_2 = \begin{bmatrix} \exp(-i\frac{1}{2}\delta_{2+}\omega^2) & 0 \\ 0 & \exp(-i\frac{1}{2}\delta_{2-}\omega^2) \end{bmatrix}_{\text{PSP}} \tag{7.6-16}$$

This is the matrix in the coordinate system where the second-order output PSP are the axes.

Using Equation (7.6-7), we can write the Jones matrix C_2 in xy coordinates as

$$C_2 = \begin{bmatrix} p_{2+x} & p_{2-x} \\ p_{2+y} & p_{2-y} \end{bmatrix} \begin{bmatrix} \exp(-i\frac{1}{2}\delta_{2+}\omega^2) & 0 \\ 0 & \exp(-i\frac{1}{2}\delta_{2-}\omega^2) \end{bmatrix} \begin{bmatrix} p_{2+x} & p_{2+y} \\ p_{2-x} & p_{2-y} \end{bmatrix} \tag{7.6-17}$$

where p_{2+x}, p_{2+y}, p_{2-x}, p_{2+y} are the xy components of the Jones vectors of the second-order output PSP.

We note that the first matrix in the above equation consists of two column vectors of the second-order PSP, and the third matrix consists of two row vectors of the second-order PSP. The above equation is an explicit expression of the Jones matrix of the second-order PMD compensator. With the presence of both the first-order and second-order PMD compensator, the overall DGD is zero to the second order of variation in frequency. The second matrix in Equation (7.6-17) represents a wave plate with a DGD given by Equation (7.6-11). Such a wave plate would require a birefringent material with birefringence dispersion.

Once the second-order PMD is eliminated with the presence of the second-order PMD compensator, we can now move on to consider the third-order PMD effect. Similar analysis can be carried out, and the Jones matrix of the third-order PMD compensator can be obtained. The process can continue indefinitely until the effect of PMD is eliminated completely within the bandwidth of interest.

PROBLEMS

7.1 Show that in a quadratic index fiber in which g^2 is independent of ω, the spreading of a pulse of spectral width Δω can be described by

$$\Delta\tau \cong \frac{L}{c}\left|\frac{n^2g^2}{ck^3}(l+m+1)^2 - \frac{dn}{d\omega}\right|\Delta\omega$$

where l and m are the transverse mode indices. In the expression, the term proportional to $(l+m+1)^2$ is due to modal dispersion, whereas the term proportional to $dn/d\omega$ is due to material dispersion.

7.2 Show that the FWHM bandwidth of the square of the Fourier spectrum (7.2-3) is given by

$$\Delta\Omega_{\mathrm{FWHM}} = \frac{4\ln 2}{\tau_0} = 2\pi\,\Delta\upsilon_{\mathrm{FWHM}}$$

where $\tau_0 = \sqrt{2\ln 2/\alpha}$ is the full width at half-maximum. Show that the same bandwidth can be written

$$\Delta\lambda = \frac{2\ln 2}{\pi c}\frac{\lambda^2}{\tau_0}$$

7.3 In optical communications, signal pulses are often in the shape of a trapezoid, as shown in Figure P7.3.

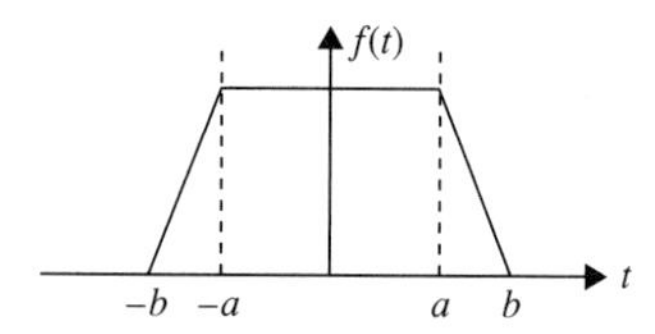

Figure P7.3

(a) Show that the pulse can be written analytically as

$$f(t) = \begin{cases} 1, & 0<|t|<a \\ \dfrac{b-|t|}{b-a}, & a<|t|<b \\ 0, & \text{elsewhere} \end{cases}$$

(b) Show that the Fourier transform can be written

$$F(\Omega) = a\frac{\sin\Omega a}{\pi\Omega a} + b\frac{\sin\Omega b - \sin\Omega a}{\pi\Omega(b-a)}$$
$$-\frac{\cos\Omega b - \cos\Omega a + \Omega b\sin\Omega b - \Omega a\sin\Omega a}{\pi\Omega^2(b-a)}$$
$$= \frac{\cos\Omega a - \cos\Omega b}{\pi\Omega^2(b-a)}$$

(c) Show that, in the limit when $b = a$, the Fourier transform becomes

$$F(\Omega) = a\frac{\sin\Omega a}{\pi\Omega a}$$

(d) Assume the propagation of a trapezoid pulse with $a = 40$ ps and $b = 60$ ps in a single-mode fiber with $D = 17$ ps/nm-km. Estimate the pulse width at the end of a 100 km fiber. Try to find the pulse shape.

7.4 Consider a pulse given by

$$f(t) = \begin{cases} \cos\left(\dfrac{\pi t}{2b}\right), & -b<t<b \\ 0, & \text{elsewhere} \end{cases}$$

(a) Find the full width at half-maximum of the pulse intensity. [Answer: b.]
(b) Show that the Fourier transform is given by

$$F(\Omega) = \frac{1}{2\pi}\left(\frac{\sin(\Omega-\alpha)b}{(\Omega-\alpha)} + \frac{\sin(\Omega+\alpha)b}{(\Omega+\alpha)}\right)$$

where $\alpha = \pi/2b$.
(c) Assume propagation of a cosine pulse with $b = 100$ ps in a single-mode fiber with $D = 17$ ps/nm-km. Estimate the pulse width at the end of a 100 km fiber. Try to find the pulse shape.

7.5 Consider a general pulse $f(t)$ with a Fourier transform $F(\Omega)$:

$$f(t) = \int F(\Omega)e^{i\Omega t}\,d\Omega$$

Assume both the pulse function and its Fourier transform have a mean at zero. Let $(\Delta t)^2$ be the variance of $f(t)$ and $(\Delta\Omega)^2$ be the variance of $F(\Omega)$. They are defined as

$$(\Delta t)^2 = \frac{\int f^*(t)t^2 f(t)\,dt}{\int f^*(t)f(t)\,dt} \equiv \frac{\langle f|t^2|f\rangle}{\langle f|f\rangle},$$

$$(\Delta\Omega)^2 = \frac{\int F^*(\Omega)\Omega^2 F(\Omega)\,d\Omega}{\int F^*(\Omega)F(\Omega)\,d\Omega} \equiv \frac{\langle F|\Omega^2|F\rangle}{\langle F|F\rangle}$$

(a) Show that

$$\frac{\partial}{\partial t}f(t) = \int i\Omega F(\Omega)e^{i\Omega t}\,d\Omega$$

and

$$(\Delta\Omega)^2 = \frac{1}{\int f^*(t)f(t)\,dt} \int f^*(t)\left(-\frac{\partial^2}{\partial t^2}\right)f(t)\,dt$$

(b) Define the following pair of Hermitian operators:

$$A = \lambda t + i\Omega/\lambda, \quad A^\dagger = \lambda t - i\Omega/\lambda$$

where Ω is replaced with its operator $-i\,\partial/\partial t$. Show that

$$\langle f | A^\dagger A | f\rangle = \langle f | \lambda^2 t^2 - \frac{\partial^2}{\lambda^2\,\partial t^2} - 1 | f\rangle$$

The above integral is positive definite for all real λ.

(c) Using (b), show that

$$\Delta t\,\Delta\Omega \geq \tfrac{1}{2}$$

This is the uncertainty principle.

(d) Show that the equality holds for Gaussian pulses.

(e) The standard deviations Δt and $\Delta\Omega$ are different from the FWHM values discussed in this chapter and Problem 7.2. Show that

$$\Delta t_{\mathrm{FWHM}}\Delta\Omega_{\mathrm{FWHM}} = 4\ln 2 = 2.77$$

7.6 Let n_s and n_f be the indices of refraction for the slow mode and the fast mode, respectively. The PMD is defined as

$$\Delta\tau = \frac{d}{v_{gs}} - \frac{d}{v_{gf}}$$

where v_{gs} and v_{gf} are the group velocities of the modes. Show that

$$\Delta\tau = \left(\frac{d}{c}\Delta n + \frac{\omega d}{c}\frac{\partial \Delta n}{\partial\omega}\right)$$

7.7 PMD in a short segment of fiber is proportional to the length. In a long segment of fiber with random birefringence, the PMD is proportional to the square root of the length. Consider a long fiber consisting of N segments of the following parameters:

$$d_i = L/N$$

$$\tau_i = \tau_1$$

(a) Show that

$$(\Delta\tau_b)^2 = \frac{1}{4}N\tau_1^2 = \frac{1}{4}N\left(\frac{\Delta n}{c} + \frac{\omega}{c}\frac{\partial\Delta n}{\partial\omega}\right)^2 d_1^2$$

(b) Show that the broadening can be written

$$(\Delta\tau_b) = \frac{1}{2}\sqrt{N}\left(\frac{\Delta n}{c} + \frac{\omega}{c}\frac{\partial\Delta n}{\partial\omega}\right)d_1$$

$$= \frac{1}{2}\left(\frac{\Delta n}{c} + \frac{\omega}{c}\frac{\partial\Delta n}{\partial\omega}\right)\sqrt{Ld_1}$$

7.8 Consider a birefringent network consisting of two segments of uniform birefringent fibers. Let θ be the angle between the polarization directions of the slow modes of the two fibers. Define τ_1 and τ_2 as

$$\tau_1 = \left(\frac{d_1}{c}\Delta n_1 + \frac{\omega d_1}{c}\frac{\partial\Delta n_1}{\partial\omega}\right), \quad \tau_2 = \left(\frac{d_2}{c}\Delta n_2 + \frac{\omega d_2}{c}\frac{\partial\Delta n_2}{\partial\omega}\right)$$

$$t_1 = \left(\frac{d_1}{c}\bar{n}_1 + \frac{\omega d_1}{c}\frac{\partial\bar{n}_1}{\partial\omega}\right), \quad t_2 = \left(\frac{d_2}{c}\bar{n}_2 + \frac{\omega d_2}{c}\frac{\partial\bar{n}_2}{\partial\omega}\right)$$

where d_1 and d_2 are lengths of the fibers, Δn_1 and Δn_2 are the birefringence values of the fibers, and $\bar{n}_1$ and $\bar{n}_2$ are the average indices. A beam of unpolarized light is launched into the fiber system with an input pulse of intensity $I_0(t)$. For simplicity, we assume $I_0(t)$ has a centroid at $t = 0$.

(a) Show that, at the end of the first segment of fibers, the intensity is

$$I_1(t) = \tfrac{1}{2}I_0(t - t_1 - \tau_1/2) + \tfrac{1}{2}I_0(t - t_1 + \tau_1/2)$$
$$\equiv I_{1s}(t) + I_{1f}(t)$$

where t_1 is the average group delay of the two polarization modes, I_{1s} is the intensity of the slow mode, and I_{1f} is the intensity of the fast mode.

(b) Show that, at the end of the second element, the intensity is

$$I_2(t) = I_{2s}(t) + I_{2f}(t)$$

where

$$I_{2s}(t) = \gamma I_{1s}(t - t_2 - \tau_2/2) + (1-\gamma)I_{1f}(t - t_2 - \tau_2/2)$$
$$I_{2f}(t) = \gamma I_{1f}(t - t_2 + \tau_2/2) + (1-\gamma)I_{1s}(t - t_2 + \tau_2/2)$$

where $\gamma = \cos^2\theta$.

(c) Show that

$$\tau_{g1} \equiv \int tI_1(t)\,dt = t_1 \quad \text{and} \quad \tau_{g2} \equiv \int tI_2(t)\,dt = t_1 + t_2$$

(d) Show that

$$b_1^2 \equiv \langle (t - \tau_{g1})^2\rangle \equiv \int (t - \tau_{g1})^2 I_1(t)\,dt = b_0^2 + \tfrac{1}{4}\tau_1^2$$

$$b_2^2 \equiv \langle (t - \tau_{g2})^2\rangle \equiv \int (t - \tau_{g2})^2 I_2(t)\,dt$$
$$= b_0^2 + \tfrac{1}{4}(\tau_1 - \tau_2)^2 + \gamma\tau_1\tau_2$$

where

$$b_0^2 \equiv \int t^2 I_0(t)\,dt$$

Note that $b_2^2 = b_0^2 + \frac{1}{4}\tau_1^2 + \frac{1}{4}\tau_2^2$ is valid only when $\gamma = \frac{1}{2}$.

7.9 Consider the same birefringent network of Problem 7.8. Let the azimuth angle of the slow axis of the first element be zero ($\psi_1 = 0$) and let θ be the azimuth angle of the slow axis of the second element ($\psi_2 = \theta$).

(a) The Jones matrix of the system can be written

$$M = \begin{bmatrix} e^{-i\Gamma_2/2}\cos^2\theta + e^{i\Gamma_2/2}\sin^2\theta & -i\sin(\Gamma_2/2)\sin(2\theta) \\ -i\sin(\Gamma_2/2)\sin(2\theta) & e^{-i\Gamma_2/2}\sin^2\theta + e^{i\Gamma_2/2}\cos^2\theta \end{bmatrix} \begin{bmatrix} e^{-i\Gamma_1/2} & 0 \\ 0 & e^{i\Gamma_1/2} \end{bmatrix}$$

where Γ_1 and Γ_2 are

$$\Gamma_1 = \Delta n_1 d_1 \omega / c \quad \text{and} \quad \Gamma_2 = \Delta n_2 d_2 \omega / c$$

Multiply the matrices and find the matrix elements. Write down expressions for a and b.

(b) Show that τ_1 and τ_2 defined in Problem 7.8 can be written

$$\tau_1 = \frac{\partial \Gamma_1}{\partial \omega} \quad \text{and} \quad \tau_2 = \frac{\partial \Gamma_2}{\partial \omega}$$

(c) Show that

$$|a'|^2 + |b'|^2 = \tfrac{1}{4}(\tau_2 - \tau_1)^2 + \tau_2\tau_1\cos^2\theta$$

The group delay between the two principal states of polarization is thus

$$\Delta\tau = 2\sqrt{|a'|^2 + |b'|^2} = \sqrt{(\tau_2 - \tau_1)^2 + 4\tau_2\tau_1\cos^2\theta}$$

Note that the PMD is minimum when $\theta = 90°$ when the slow axes of the two elements are mutually orthogonal, and maximum when the slow axes are parallel ($\theta = 0$).

(d) Find the Jones vectors of the principal states of polarization.

(e) For an input beam of unpolarized light, the broadening is given by, according to Equations (7.3-9) and (7.3-11),

$$(\Delta\tau_b)^2 = \tfrac{1}{4}(\Delta\tau)^2$$

Show that this broadening is consistent with the result of Problem 7.8(d).

7.10

(a) Using the matrix elements obtained in Problem 7.9(a), show that

$$A = a^*a' + bb'^* = -\frac{i}{2}[\tau_1 + \tau_2(\cos^2\theta - \sin^2\theta)]$$

$$B = a^*b' - a'^*b = -\frac{i}{2}\tau_2\sin 2\theta\exp(i\Gamma_1)$$

(b) Show that the group delay obtained from

$$\tau = \pm\sqrt{AD - BC} = \pm\sqrt{|A|^2 + |B|}$$

agrees with the result of Problem 7.9(c).

(c) The principal states of polarization are given by Equation (7.4-9):

$$\mathbf{V}_{\text{in}} = \begin{bmatrix} -B \\ A + i\delta \end{bmatrix}$$

where $\delta = \tau/2, -\tau/2$. Investigate the frequency dependence of the PSP by plotting the polarization ellipse as a function of frequency. You may plot the ellipticity and the inclination angle of the ellipse as functions of the frequency.

7.11

(a) Show that the eigenvectors of the Jones matrix in Equations (7.5-10) and (7.5-11) are mutually orgthogonal.

(b) Derive Equation (7.5-12).

(c) Using Equation (7.5-14) for U, show that

$$U^\dagger\boldsymbol{\sigma}U = \cos^2(\gamma/2)\boldsymbol{\sigma} + \sin^2(\gamma/2)[(\mathbf{r}\cdot\boldsymbol{\sigma})\boldsymbol{\sigma}(\mathbf{r}\cdot\boldsymbol{\sigma})] + i\sin(\gamma/2)\cos(\gamma/2)[(\mathbf{r}\cdot\boldsymbol{\sigma})\boldsymbol{\sigma} - \boldsymbol{\sigma}(\mathbf{r}\cdot\boldsymbol{\sigma})]$$

(d) Using $\sigma_1\sigma_2 = -\sigma_2\sigma_1 = i\sigma_3$, $\sigma_2\sigma_3 = -\sigma_3\sigma_2 = i\sigma_1$, $\sigma_3\sigma_1 = -\sigma_1\sigma_3 = i\sigma_2$, and $\sigma_1^2 = \sigma_2^2 = \sigma_3^2 = 1$, show that

$$(\mathbf{r}\cdot\boldsymbol{\sigma})\boldsymbol{\sigma} - \boldsymbol{\sigma}(\mathbf{r}\cdot\boldsymbol{\sigma}) = -2i(\mathbf{r}\times\boldsymbol{\sigma})$$

$$(\mathbf{r}\cdot\boldsymbol{\sigma})\boldsymbol{\sigma}(\mathbf{r}\cdot\boldsymbol{\sigma}) = 2\mathbf{r}(\mathbf{r}\cdot\boldsymbol{\sigma}) - \boldsymbol{\sigma}$$

(e) Show that $U^\dagger\boldsymbol{\sigma}U = \cos(\gamma)\boldsymbol{\sigma} + 2\sin^2(\gamma/2)[\mathbf{r}(\mathbf{r}\cdot\boldsymbol{\sigma})] + \sin(\gamma)(\mathbf{r}\times\boldsymbol{\sigma})$.

(f) Using $\mathbf{r}\times(\mathbf{r}\times\boldsymbol{\sigma}) = \mathbf{r}(\mathbf{r}\cdot\boldsymbol{\sigma}) - \boldsymbol{\sigma} = (\mathbf{rr} - 1)\boldsymbol{\sigma}$, derive Equations (7.5-18) and (7.5-19).

7.12 By definition an anti-Hermitian matrix A satisfies $A^\dagger = -A$.

(a) Using $|a|^2 + |b|^2 = 1$, show that the trace of the matrix in Equation (7.5-23) is zero.

(b) Show that the matrix is anti-Hermitian.

(c) Show that the eigenvalues of an anti-Hermitian matrix are pure imaginary. [Hint: An anti-Hermitian matrix can be written $A = iH$, where H is a Hermitian matrix.]

REFERENCES

1. Poole, C. D., and R. E. Wagner, Phenomenological approach to polarization dispersion in long single-mode fibers. *Electron. Lett.* **22**(19):1029 (1986).
2. Gordon, J. P., and H. Kogelnik, *Proc. Natl. Acad. Sci. USA* **97**(9):4541 (2000).
3. Ciprut, P., B. Gisin, N. Gisin, R. Passy, J. P. Von der Weid, F. Prieto, and C. W. Zimmer, Second-order polarization mode dispersion: impact on analog and digital transmissions. *J. Lightwave Techno.* **16**(5):757 (1998).
4. Poole, C. D., and C. R. Giles, Polarization-dependent pulse compression and broadening due to polarization dispersion in dispersion shifted fibers. *Opt. Lett.* **13**(2):155 (1988).
5. Francia, C., F. Bruyere, D. Penninckx, and M. Chbat, PMD second-order effects on pulse propagation in single-mode optical fibers. *IEEE Photonics Technol. Lett.* **10**(12):1739 (1998).
6. Eyal, A., W. K. Marshall, M. Tur, and A. Yariv, Representation of second-order polarization mode dispersion. *Electron Lett.* **35**(19):1658 (1999).
7. Eyal, A., Y. Li, W. K. Marshall, M. Tur, and A. Yariv, Statistical determination of the length dependence of high-order PMD. *Opt. Lett.* **25**:875 (2000).
8. Li, Y., A. Eyal, P.-O. Hedekvist, and A. Yariv, Measurement of high-order polarization mode dispersion. *IEEE Photonics Techno. Lett.* **12**:861 (2000).

CHAPTER 8

NONLINEAR OPTICS

8.0 INTRODUCTION

In Chapter 1 we considered the propagation of electromagnetic radiation in linear media in which the induced polarization is proportional to the electric field. In this chapter we consider some of the consequences of the nonlinear dielectric properties of media in which, in addition to the linear response, an electric field produces a polarization that is a nonlinear function of the field. The nonlinear response can give rise to an exchange of energy between a number of electromagnetic fields of different frequencies. Some of the most important applications of this phenomenon include: (1) second-harmonic generation (SHG) in which part of the energy of an optical wave of frequency ω propagating through a crystal is converted to that of a wave at 2ω; (2) parametric oscillation in which a strong pump wave at ω_3 causes the simultaneous generation in a nonlinear crystal of radiation at lower frequencies ω_1 and ω_2, where $\omega_3 = \omega_1 + \omega_2$; and (3) stimulated Raman scattering (SRS), stimulated Brillouin scattering (SBS), and so on. Starting from the physical origin of the dielectric nonlinearity, we first describe a general methodology of nonlinear interaction of optical beams. We then provide an electromagnetic formulation of the coupling between optical beams in nonlinear media. Second-harmonic generation, parametric amplification, and third-order phenomena including the Kerr effect and Raman scattering will be addressed in detail in this chapter.

8.1 ON THE PHYSICAL ORIGIN OF NONLINEAR POLARIZATION

In any real atomic system, the polarization induced in the medium by the presence of an electric field is not exactly proportional to the electric field, but can be expressed in a Taylor series expansion, which in a lossless medium takes the form

$$P_i = \varepsilon_0 \chi_{ij} E_j + 2d_{ijk} E_j E_k + 4\chi_{ijkl} E_j E_k E_l + \cdots \tag{8.1-1}$$

where P_i is the ith component of the instantaneous polarization and E_i is the ith component of the instantaneous electric field of the optical beams ($i, j, k, l = x, y, z$). Summations over repeated indices are assumed. χ_{ij} is the linear susceptibility, while d_{ijk} and χ_{ijkl} are the second-order and third-order nonlinear optical susceptibilities, respectively. Since the order of E_j, E_k, E_l are immaterial in the above definition, the nonlinear optical susceptibilities observe the following symmetry:

$$d_{ijk} = d_{ikj}$$
$$\chi_{ijkl} = \chi_{i[jkl]} \tag{8.1-2}$$

where $[jkl]$ is any permutation of (jkl). It is known that the linear dielectric susceptibility tensor χ_{ij} is symmetric (i.e., $\chi_{ij} = \chi_{ji}$) in lossless media (see Problem 1.37). In the extreme case of a dc field (or low frequencies), the dispersion of the nonlinear coefficients d_{ijk} and χ_{ijkl} can be neglected. In this regime, all the relevant frequencies are far away from the absorption lines (or resonance frequencies), the system is lossless, and the material response is instan-taneous. In this case, the high-order terms can be viewed as a correction to the dielectric tensor. As a result, the nonlinear coefficients d_{ijk} and χ_{ijkl} are invariant under any reshuffling of their subscripts; for example, $\chi_{1231} = \chi_{2113}$ (see Problem 8.15). For the second-order nonlinear optical coefficient d_{ijk}, this symmetry reduces the 27 tensor elements to 10 independent elements. In lossy and consequently dispersive systems, d_{ijk} will in general depend on the frequencies of the waves involved in the nonlinear processes (e.g., sum or difference mixing, $\omega_3 = \omega_1 + \omega_2$ or $\omega_3 = \omega_1 - \omega_2$). In this case, the symmetry properties of the nonlinear coefficients d_{ijk} and χ_{ijkl} are limited to Equation (8.1-2). Throughout this book, we will use (1, 2, 3) and (x, y, z) interchangeably for the subscript indices of the nonlinear optical coefficients. For example, we write $\chi_{1111} = \chi_{xxxx}$, $\chi_{1122} = \chi_{xxyy}$.

The nonlinear optical response characterized by the parameters d_{ijk} and χ_{ijkl} give rise to numerous interesting phenomena and useful applications. The second-order nonlinearity $P_i = 2d_{ijk}E_jE_k$ is responsible for second-harmonic generation (SHG, or frequency doubling), for sum and difference frequency generation, and for parametric amplification and oscillation. The third-order term $P_i = 4\chi_{ijkl}E_jE_kE_l$ figures in such diverse phenomena as third-harmonic generation (THG), Raman and Brillouin scattering, self-focusing, the Kerr effect, the optical soliton, four-wave mixing, and phase conjugation. In this chapter, we will not concern ourselves with the physical origin of the nonlinear coefficients d_{ijk} and χ_{ijkl}. We will take them as material parameters and merely explore the electromagnetic phenomena and applications made possible by the nonlinearities.

8.2 SECOND-ORDER NONLINEAR PHENOMENA—GENERAL METHODOLOGY

Consider the nonlinear coupling of two optical fields. The first field, described in terms of its electric-field components, is given by

$$E_j^{\omega_1}(t) = \mathrm{Re}(E_{0j}^{\omega_1} e^{i\omega_1 t}) = \tfrac{1}{2}(E_{0j}^{\omega_1} e^{i\omega_1 t} + \text{c.c.}) \quad (j = x, y, z) \tag{8.2-1}$$

while the second field at ω_2 is

$$E_k^{\omega_2}(t) = \mathrm{Re}(E_{0k}^{\omega_2} e^{i\omega_2 t}) = \tfrac{1}{2}(E_{0k}^{\omega_2} e^{i\omega_2 t} + \text{c.c.}) \quad (k = x, y, z) \tag{8.2-2}$$

where c.c. stands for complex conjugate, and $E_{0j}^{\omega_1}$ and $E_{0k}^{\omega_1}$ are constant amplitudes of the two optical fields, respectively. If the medium is nonlinear, the presence of these field components can give rise to polarizations at frequencies $(n\omega_1 + m\omega_2)$, where n and m are any integers. To illustrate the general methodology, we consider the ith Cartesian component of the nonlinear polarization at $\omega_3 = \omega_1 + \omega_2$ as

$$P_i^{\omega_3=\omega_1+\omega_2}(t) = \mathrm{Re}(P_{0i}^{\omega_3} e^{i\omega_3 t}) = \tfrac{1}{2}(P_{0i}^{\omega_3} e^{i\omega_3 t} + \text{c.c.}) \quad (i = x, y, z) \tag{8.2-3}$$

where $P_{0i}^{\omega_3}$ is a constant amplitude of the polarization. Limiting our attention to the second-order terms in Equation (8.1-1)—that is,

$$P_i = 2d_{ijk}E_jE_k \tag{8.2-4}$$

we obtain

$$P_i(t) = 2d_{ijk}\tfrac{1}{2}(E_{0j}^{\omega_1}e^{i\omega_1 t} + E_{0j}^{\omega_2}e^{i\omega_2 t} + \text{c.c.}) \times \tfrac{1}{2}(E_{0k}^{\omega_1}e^{i\omega_1 t} + E_{0k}^{\omega_2}e^{i\omega_2 t} + \text{c.c.}) \tag{8.2-5}$$

Consider only the sum frequency term

$$P_i^{\omega_1+\omega_2}(t) = (\tfrac{1}{2}d_{ijk}E_{0j}^{\omega_1}E_{0k}^{\omega_2}e^{i(\omega_1+\omega_2)t} + \tfrac{1}{2}d_{ikj}E_{0k}^{\omega_2}E_{0j}^{\omega_1}e^{i(\omega_2+\omega_1)t} + \text{c.c.}) \tag{8.2-6}$$

where we recall the convention of summation over repeated indices. In a lossless (instantaneous response) system, $d_{ijk} = d_{ikj}$, so that

$$P_i^{\omega_1+\omega_2}(t) = \tfrac{1}{2}P_{0i}^{\omega_1+\omega_2}e^{i(\omega_1+\omega_2)t} + \text{c.c.} = d_{ijk}E_{0j}^{\omega_1}E_{0k}^{\omega_2}e^{i(\omega_1+\omega_2)t} + \text{c.c.} \tag{8.2-7a}$$

or in terms of the field amplitudes,

$$P_{0i}^{\omega_1+\omega_2} = 2d_{ijk}E_{0j}^{\omega_1}E_{0k}^{\omega_2} \tag{8.2-7b}$$

It is important to note that d_{ijk} will in general depend on ω_1 and ω_2 and whether the sum or their difference is considered. Our primary interest will be in the case of transparent lossless media where the coefficients d_{ijk} do not depend on the frequencies involved or on whether their sum or difference is generated.

Only noncentrosymmetric crystals can possess a nonvanishing d_{ijk} tensor. This follows from the requirement that in a centrosymmetric crystal a reversal of the signs of $E_j^{\omega_1}$ and $E_k^{\omega_2}$ must also cause a reversal in the sign of $P_i^{\omega_1+\omega_2}$ and will not affect the amplitude. Using Equation (8.2-7b), we get

$$-2d_{ijk}E_{0j}^{\omega_1}E_{0k}^{\omega_2} = 2d_{ijk}(-E_{0j}^{\omega_1})(-E_{0k}^{\omega_2}) \tag{8.2-8}$$

so that $d_{ijk} = 0$. Lack of inversion symmetry is also the prerequisite for a linear electro-optic effect (Pockels effect) and piezoelectricity, so that all electro-optic and piezoelectric crystals can be expected to display second-order ($P \propto E^2$) nonlinear optical properties. The same argument can also be used to show that all crystals as well as liquids and gases can display third-order optical nonlinearities.

The nonlinear coefficients d_{ijk} are measured most often in second-harmonic generation experiments, where $\omega_1 = \omega_2 = \omega$. In this case, we have, according to Equation (8.2-5),

$$P_i(t) = (\tfrac{1}{2}P_{0i}^{2\omega}e^{i2\omega t} + \text{c.c.}) = 2d_{ijk}(\tfrac{1}{2}E_{0j}^{\omega}e^{i\omega t} + \text{c.c.}) \times (\tfrac{1}{2}E_{0k}^{\omega}e^{i\omega t} + \text{c.c.}) \tag{8.2-9}$$

or equivalently,

$$P_{0i}^{2\omega} = d_{ijk}E_{0j}^{\omega}E_{0k}^{\omega} \tag{8.2-10}$$

where we observe the convention of summation over repeated indices. Note the factor of 2 difference between Equations (8.2-10) and (8.2-7b). It is important to note that we cannot obtain Equation (8.2-10) by setting $\omega_1 = \omega_2$ in Equation (8.2-7b). This is a result of the field normalization in Equation (8.2-5) when ω_2 approaches ω_1.

Since no physical significance can be attached to the interchange of j and k in Equation (8.2-10), we can replace the subscripts kj and jk by the contracted indices as follows:

$$xx = 1, \quad yy = 2, \quad zz = 3$$

$$yz = zy = 4, \quad xz = zx = 5, \quad xy = yx = 6$$

The resulting d_{iK} components ($i = 1, 2, 3$ and $K = 1, 2, 3, 4, 5, 6$) form a 3×6 matrix that operates on the E^2 column tensor to yield the amplitude of the second-harmonic nonlinear polarization $\mathbf{P}^{2\omega}$. The second-harmonic polarization amplitude can thus be written, according to Equation (8.2-10),

$$
\begin{bmatrix} P_{0x}^{2\omega} \\ P_{0y}^{2\omega} \\ P_{0z}^{2\omega} \end{bmatrix} = \begin{bmatrix} d_{11} & d_{12} & d_{13} & d_{14} & d_{15} & d_{16} \\ d_{21} & d_{22} & d_{23} & d_{24} & d_{25} & d_{26} \\ d_{31} & d_{32} & d_{33} & d_{34} & d_{35} & d_{36} \end{bmatrix} \begin{bmatrix} E_{0x}^{\omega} E_{0x}^{\omega} \\ E_{0y}^{\omega} E_{0y}^{\omega} \\ E_{0z}^{\omega} E_{0z}^{\omega} \\ 2E_{0y}^{\omega} E_{0z}^{\omega} \\ 2E_{0z}^{\omega} E_{0x}^{\omega} \\ 2E_{0x}^{\omega} E_{0y}^{\omega} \end{bmatrix} \tag{8.2-11}
$$

The contracted d_{iK} tensor obeys the same symmetry restrictions as the piezoelectric tensor and the electro-optic tensor (see Chapter 9), and in crystals of a given point-group symmetry it has the same form. The tensor forms are given in Table 8.1.

TABLE 8.1 The Form of the Second-Order Nonlinear Optical Tensor in Contracted Notation for All Crystal Classes

Centrosymmetric: ($\bar{1}$, $2/m.mmm.4/m.4/mmm$, $\bar{3}$, $\bar{3}m$, $6/m$, $6/mmm$, $m3$, $m3m$)

$$
\begin{bmatrix} 0 & 0 & 0 & 0 & 0 & 0 \\ 0 & 0 & 0 & 0 & 0 & 0 \\ 0 & 0 & 0 & 0 & 0 & 0 \end{bmatrix}
$$

Triclinic: (1)

$$
\begin{bmatrix} d_{11} & d_{12} & d_{13} & d_{14} & d_{15} & d_{16} \\ d_{21} & d_{22} & d_{23} & d_{24} & d_{25} & d_{26} \\ d_{31} & d_{32} & d_{33} & d_{34} & d_{35} & d_{36} \end{bmatrix}
$$

Monoclinic: 2 ($2 \parallel x_2$)

$$
\begin{bmatrix} 0 & 0 & 0 & d_{14} & 0 & d_{16} \\ d_{21} & d_{22} & d_{23} & 0 & d_{25} & 0 \\ 0 & 0 & 0 & d_{34} & 0 & d_{36} \end{bmatrix}
$$

2 ($2 \parallel x_3$)

$$
\begin{bmatrix} 0 & 0 & 0 & d_{14} & d_{15} & 0 \\ 0 & 0 & 0 & d_{24} & d_{25} & 0 \\ d_{31} & d_{32} & d_{33} & 0 & 0 & d_{36} \end{bmatrix}
$$

m ($m \perp x_2$)

$$
\begin{bmatrix} d_{11} & d_{12} & d_{13} & 0 & d_{15} & 0 \\ 0 & 0 & 0 & d_{24} & 0 & d_{26} \\ d_{31} & d_{32} & d_{33} & 0 & d_{35} & 0 \end{bmatrix}
$$

m ($m \perp x_3$)

$$
\begin{bmatrix} d_{11} & d_{12} & d_{13} & 0 & 0 & d_{16} \\ d_{21} & d_{22} & d_{23} & 0 & 0 & d_{26} \\ 0 & 0 & 0 & d_{34} & d_{35} & 0 \end{bmatrix}
$$

Orthorhombic: 222

$$
\begin{bmatrix} 0 & 0 & 0 & d_{14} & 0 & 0 \\ 0 & 0 & 0 & 0 & d_{25} & 0 \\ 0 & 0 & 0 & 0 & 0 & d_{36} \end{bmatrix}
$$

$2mm$

$$
\begin{bmatrix} 0 & 0 & 0 & 0 & d_{15} & 0 \\ 0 & 0 & 0 & d_{24} & 0 & 0 \\ d_{31} & d_{32} & d_{33} & 0 & 0 & 0 \end{bmatrix}
$$

Tetragonal: 4

$$
\begin{bmatrix} 0 & 0 & 0 & d_{14} & d_{15} & 0 \\ 0 & 0 & 0 & d_{15} & -d_{14} & 0 \\ d_{31} & d_{31} & d_{33} & 0 & 0 & 0 \end{bmatrix}
$$

$\bar{4}$

$$
\begin{bmatrix} 0 & 0 & 0 & d_{14} & d_{15} & 0 \\ 0 & 0 & 0 & -d_{15} & d_{14} & 0 \\ d_{31} & -d_{31} & 0 & 0 & 0 & d_{36} \end{bmatrix}
$$

422

$$
\begin{bmatrix} 0 & 0 & 0 & d_{14} & 0 & 0 \\ 0 & 0 & 0 & 0 & -d_{14} & 0 \\ 0 & 0 & 0 & 0 & 0 & 0 \end{bmatrix}
$$

$4mm$

$$
\begin{bmatrix} 0 & 0 & 0 & 0 & d_{15} & 0 \\ 0 & 0 & 0 & d_{15} & 0 & 0 \\ d_{31} & d_{31} & d_{33} & 0 & 0 & 0 \end{bmatrix}
$$

$\bar{4}2m$ ($2 \parallel x_1$)

$$
\begin{bmatrix} 0 & 0 & 0 & d_{14} & 0 & 0 \\ 0 & 0 & 0 & 0 & d_{14} & 0 \\ 0 & 0 & 0 & 0 & 0 & d_{36} \end{bmatrix}
$$

Trigonal: 3

$$
\begin{bmatrix} d_{11} & -d_{11} & 0 & d_{14} & d_{15} & -d_{22} \\ -d_{22} & d_{22} & 0 & d_{15} & -d_{14} & -d_{11} \\ d_{31} & d_{31} & d_{33} & 0 & 0 & 0 \end{bmatrix}
$$

32

$$
\begin{bmatrix} d_{11} & -d_{11} & 0 & d_{14} & 0 & 0 \\ 0 & 0 & 0 & 0 & -d_{14} & -d_{11} \\ 0 & 0 & 0 & 0 & 0 & 0 \end{bmatrix}
$$

TABLE 8.1 (cont'd)

System	Class	Matrix
	$3m\ (m \perp x_1)$	$\begin{bmatrix} 0 & 0 & 0 & 0 & d_{15} & -d_{22} \\ -d_{22} & d_{22} & 0 & d_{15} & 0 & 0 \\ d_{31} & d_{31} & d_{33} & 0 & 0 & 0 \end{bmatrix}$
	$3m\ (m \perp x_2)$	$\begin{bmatrix} d_{11} & -d_{11} & 0 & 0 & d_{15} & 0 \\ 0 & 0 & 0 & d_{15} & 0 & -d_{11} \\ d_{31} & d_{31} & d_{33} & 0 & 0 & 0 \end{bmatrix}$
Hexagonal:	6	$\begin{bmatrix} 0 & 0 & 0 & d_{14} & d_{15} & 0 \\ 0 & 0 & 0 & d_{15} & -d_{14} & 0 \\ d_{31} & d_{31} & d_{33} & 0 & 0 & 0 \end{bmatrix}$
	$6mm$	$\begin{bmatrix} 0 & 0 & 0 & 0 & d_{15} & 0 \\ 0 & 0 & 0 & d_{15} & 0 & 0 \\ d_{31} & d_{31} & d_{33} & 0 & 0 & 0 \end{bmatrix}$
	622	$\begin{bmatrix} 0 & 0 & 0 & d_{14} & 0 & 0 \\ 0 & 0 & 0 & 0 & -d_{14} & 0 \\ 0 & 0 & 0 & 0 & 0 & 0 \end{bmatrix}$
	$\bar{6}$	$\begin{bmatrix} d_{11} & -d_{11} & 0 & 0 & 0 & -d_{22} \\ -d_{22} & d_{22} & 0 & 0 & 0 & -d_{11} \\ 0 & 0 & 0 & 0 & 0 & 0 \end{bmatrix}$
	$\bar{6}m2\ (m \perp x_1)$	$\begin{bmatrix} 0 & 0 & 0 & 0 & 0 & -d_{22} \\ -d_{22} & d_{22} & 0 & 0 & 0 & 0 \\ 0 & 0 & 0 & 0 & 0 & 0 \end{bmatrix}$
	$\bar{6}m2\ (m \perp x_2)$	$\begin{bmatrix} d_{11} & -d_{11} & 0 & 0 & 0 & 0 \\ 0 & 0 & 0 & 0 & 0 & -d_{11} \\ 0 & 0 & 0 & 0 & 0 & 0 \end{bmatrix}$
Cubic:	$\bar{4}3m$, 23	$\begin{bmatrix} 0 & 0 & 0 & d_{14} & 0 & 0 \\ 0 & 0 & 0 & 0 & d_{14} & 0 \\ 0 & 0 & 0 & 0 & 0 & d_{14} \end{bmatrix}$
	432	$\begin{bmatrix} 0 & 0 & 0 & 0 & 0 & 0 \\ 0 & 0 & 0 & 0 & 0 & 0 \\ 0 & 0 & 0 & 0 & 0 & 0 \end{bmatrix}$

EXAMPLE: SHG IN KDP (KH_2PO_4)

As a result of the crystal symmetry $\bar{4}2m$, the nonvanishing nonlinear coefficients are, according to Table 8.1, d_{14}, $d_{25} = d_{14}$, and d_{36}. So the second-order nonlinear polarization amplitude can be written, according to Equation (8.2-11),

$$P_{0x} = 2d_{xzy}E_{0z}E_{0y} = 2d_{14}E_{0z}E_{0y}$$

$$P_{0y} = 2d_{yzx}E_{0z}E_{0x} = 2d_{25}E_{0z}E_{0y} = 2d_{14}E_{0z}E_{0y}$$

$$P_{0z} = 2d_{zxy}E_{0x}E_{0y} = 2d_{36}E_{0x}E_{0y}$$

where E_{0x}, E_{0y}, E_{0z} are the amplitudes of the electric field of the fundamental optical wave at frequency ω. Table 8.2 lists the nonlinear optical coefficients for a number of crystals.

8.3 ELECTROMAGNETIC FORMULATION AND OPTICAL SECOND-HARMONIC GENERATION

We consider the interaction of three waves with $\omega_3 = \omega_1 + \omega_2$ via the second-order optical nonlinearity. Let the field be written

$$E_i(t) = E_i^{\omega_1}(t) + E_i^{\omega_2}(t) + E_i^{\omega_3}(t) \quad (i = x', y', z') \tag{8.3-1}$$

with

$$E_i^{\omega_1}(t) = \tfrac{1}{2}(E_{0i}^{\omega_1}e^{i(\omega_1 t - k_1 z)} + \text{c.c.}) = \tfrac{1}{2}(a_{1i}E_1 e^{i(\omega_1 t - k_1 z)} + \text{c.c.}) \qquad (i = x', y', z')$$

$$E_i^{\omega_2}(t) = \tfrac{1}{2}(E_{0i}^{\omega_2}e^{i(\omega_2 t - k_2 z)} + \text{c.c.}) = \tfrac{1}{2}(a_{2i}E_2 e^{i(\omega_2 t - k_2 z)} + \text{c.c.}) \qquad (i = x', y', z')$$

$$E_i^{\omega_3}(t) = \tfrac{1}{2}(E_{0i}^{\omega_3}e^{i(\omega_3 t - k_3 z)} + \text{c.c.}) = \tfrac{1}{2}(a_{3i}E_3 e^{i(\omega_3 t - k_3 z)} + \text{c.c.}) \qquad (i = x', y', z')$$

where we assume that all three fields are normal modes of propagation along the z direction in the nonlinear medium, with unique wavenumbers k_1, k_2, and k_3 and amplitudes E_1, E_2, and

TABLE 8.2 Second-Order Nonlinear Optical Coefficients for a Number of Crystals

Crystal	$d_{ijk}^{(2\omega)}$ in Units of $\frac{1}{9} \times 10^{-22}$ MKS[a]
Ag_3AsS_3	$d_{22} = 22.5$
(Proustite)	$d_{36} = 13.5$
$AgGaSe_2$	$d_{36} = 27 \pm 3$
$AgSbS_3$	$d_{36} = 9.5$
$AlPO_4$	$d_{11} = 0.38 \pm 0.03$
β-BaB_2O_4 (BBO)	$d_{11} = 5.8 \times d_{36}$ (KDP)
	$d_{31} = 0.05 \times d_{11}$
	$d_{22} < 0.05 \times d_{11}$
$Ba_2NaNb_5O_{15}$	$d_{33} = 10.4 \pm 0.7$
	$d_{32} = 7.4 \pm 0.7$
$BaTiO_3$	$d_{33} = 6.4 \pm 0.5$
	$d_{31} = 18 \pm 2$
	$d_{15} = 17 \pm 2$
CdS	$d_{33} = 28.6 \pm 2$
	$d_{31} = 30 \pm 10$
	$d_{36} = 33$
CdSe	$d_{31} = 22.5 \pm 3$
$CdGeAs_2$	$d_{36} = 363 \pm 70$
GaP	$d_{14} = 80 \pm 14$
GaAs	$d_{14} = 72$
KH_2PO_4	$d_{36} = 0.45 \pm 0.03$
(KDP)	$d_{14} = 0.35$
KD_2PO_4	$d_{36} = 0.42 \pm 0.02$
	$d_{14} = 0.42 \pm 0.02$
KH_2AsO_4	$d_{36} = 0.48 \pm 0.03$
	$d_{14} = 0.51 \pm 0.03$
LiB_3O_5 (LBO)	$d_{31} = 0.84$
	$d_{32} = -0.78$
$LiIO_3$	$d_{15} = 4.4$
$LiNbO_3$	$d_{15} = 4.4$
	$d_{22} = 2.3 \pm 1.0$
$NH_4H_2PO_4$	$d_{36} = 0.45$
(ADP)	$d_{14} = 0.50 \pm 0.02$
Quartz (SiO_2)	$d_{11} = 0.37 \pm 0.02$
Se	$d_{11} = 130 \pm 30$
Te	$d_{11} = 517$
ZnO	$d_{33} = 6.5 \pm 0.2$
	$d_{31} = 1.95 \pm 0.2$
	$d_{15} = 2.1 \pm 0.2$
ZnS	$d_{36} = 13$

[a] Some authors define the nonlinear coefficient d by $P = \varepsilon_0 dE^2$ (where d is in units of pm/V) rather than by the relation $P = dE^2$ used in this book. To convert: 1 pm/V = 0.7969 × ($\frac{1}{9} \times 10^{-22}$ MKS) and ($\frac{1}{9} \times 10^{-22}$ MKS) = 1.255 pm/V.

E_3. In the above equations, $\mathbf{a}_1$, $\mathbf{a}_2$, and $\mathbf{a}_3$ are unit vectors representing the polarization direction of the normal modes of propagation of these three fields in the nonlinear medium. In formulating the interaction we adopt the z axis as the common direction of propagation, and (x', y', z') as the Cartesian components in the principal coordinates of the nonlinear medium. These three fields are propagating in the same direction (z axis) but may assume different polarization states. Generally, the direction of propagation (z) may not be parallel to one of the principal axes. The collinear propagation in the nonlinear medium is to ensure maximum physical overlap. In the absence of a nonlinear dielectric response, these three fields are solutions of the wave equation and are propagating independently in the medium.

We now start with the wave equation in a form that includes the polarization explicitly:

$$\nabla^2 \mathbf{E} = \mu_0 \frac{\partial^2}{\partial t^2}(\varepsilon_0 \mathbf{E} + \mathbf{P}) = \mu_0 \varepsilon \frac{\partial^2 \mathbf{E}}{\partial t^2} + \mu_0 \frac{\partial^2}{\partial t^2} \mathbf{P}_{\mathrm{NL}} \tag{8.3-2}$$

where $\mathbf{P}_{\mathrm{NL}}$ stands for the nonlinear polarization and can be written

$$(P_{\mathrm{NL}})_i = 2 d_{ijk} E_j E_k \quad (i, j, k = x', y', z') \tag{8.3-3}$$

It is important to note that $\mathbf{E}$ in the above equation is the sum of the three fields given by Equation (8.3-1). As a result of the nonlinear polarization, the three fields are coupled. To obtain the coupled equations for the field amplitudes, we start with the nonlinear polarization at $\omega_1 = \omega_3 - \omega_2$. According to Equation (8.2-7a), we obtain

$$[P_{\mathrm{NL}}^{\omega_3-\omega_2}(z, t)]_i = d_{ijk} a_{3j} a_{2k} E_3 E_2^* e^{i[(\omega_3-\omega_2)t-(k_3-k_2)z]} + \text{c.c.} \tag{8.3-4}$$

Similarly, we obtain the other two nonlinear polarizations

$$[P_{\mathrm{NL}}^{\omega_3-\omega_1}(z, t)]_i = d_{ijk} a_{3j} a_{1k} E_3 E_1^* e^{i[(\omega_3-\omega_1)t-(k_3-k_1)z]} + \text{c.c.} \tag{8.3-5}$$

$$[P_{\mathrm{NL}}^{\omega_1+\omega_2}(z, t)]_i = d_{ijk} a_{1j} a_{2k} E_1 E_2 e^{i[(\omega_1+\omega_2)t-(k_1+k_2)z]} + \text{c.c.} \tag{8.3-6}$$

where we observe the convention of summation over repeated indices ($i, j, k = x', y', z'$). These nonlinear polarizations can be viewed as distributed dipole sources, which can radiate and generate waves at the oscillating frequencies. We now substitute these nonlinear polarizations into the wave equation. By carrying out the indicated differentiation and assuming the following slowly varying amplitude approximation,

$$\frac{d^2}{dz^2} E_s \ll k_s \frac{d}{dz} E_s \quad (s = 1, 2, 3) \tag{8.3-7}$$

we obtain, after a few steps of algebra,

$$\begin{aligned} \frac{d}{dz} E_1 &= -i\omega_1 \sqrt{\frac{\mu_0}{\varepsilon_1}}\, d E_3 E_2^* e^{-i(k_3-k_2-k_1)z} \\ \frac{d}{dz} E_2^* &= +i\omega_2 \sqrt{\frac{\mu_0}{\varepsilon_2}}\, d E_1 E_3^* e^{+i(k_3-k_2-k_1)z} \\ \frac{d}{dz} E_3 &= -i\omega_3 \sqrt{\frac{\mu_0}{\varepsilon_3}}\, d E_1 E_2 e^{+i(k_3-k_2-k_1)z} \end{aligned} \tag{8.3-8}$$

where d is the effective second-order nonlinear coefficient

$$d = \sum_{ijk} d_{ijk} a_{1i} a_{2j} a_{3k} \tag{8.3-9}$$

The coupled equations (8.3-8) constitute the main result of this section. In arriving at the coupled equations, we have employed the cyclic symmetry of the nonlinear coefficients, that is, $d_{ijk} = d_{jik} = d_{ikj}$. We will apply them in the following sections to some specific cases. We note the coupled equations are in agreement with the conservation of energy. It can be shown that

$$\frac{d}{dz}\left(\sqrt{\varepsilon_1}\,|E_1|^2 + \sqrt{\varepsilon_2}\,|E_2|^2 + \sqrt{\varepsilon_3}\,|E_3|^2\right) = 0 \tag{8.3-10}$$

provided $\omega_3 = \omega_1 + \omega_2$.

We define the following new field variables A_1, A_2, and A_3:

$$A_m = \sqrt{\frac{n_m}{\omega_m}}\, E_m, \quad m = 1, 2, 3 \tag{8.3-11}$$

where n_m is the index of refraction associated with wave E_m and ω_m is the corresponding frequency. The beam intensity can be written

$$I_m = \frac{1}{2}\sqrt{\frac{\varepsilon_0}{\mu_0}}\, n_m |E_m|^2 = \frac{1}{2}\sqrt{\frac{\varepsilon_0}{\mu_0}}\, \omega_m |A_m|^2 \tag{8.3-12}$$

Since a photon's energy is $\hbar\omega$, it follows that $|A_m|^2$ is proportional to the photon flux of the beam at frequency ω_m, the proportional constant being independent of frequency. The coupled equations (8.3-8) can now be written

$$\begin{aligned} \frac{d}{dz} A_1 &= -i\kappa A_3 A_2^* e^{-i\Delta k z} \\ \frac{d}{dz} A_2^* &= +i\kappa A_1 A_3^* e^{+i\Delta k z} \\ \frac{d}{dz} A_3 &= -i\kappa A_1 A_2 e^{+i\Delta k z} \end{aligned} \tag{8.3-13}$$

where the momentum mismatch Δk (or wavenumber mismatch) and the coupling constant κ are given by

$$\Delta k = k_3 - (k_1 + k_2) \tag{8.3-14}$$

$$\kappa = d\sqrt{\frac{\mu_0\omega_1\omega_2\omega_3}{\varepsilon_0 n_1 n_2 n_3}} = \left(\sum_{ijk} d_{ijk} a_{1i} a_{2j} a_{3k}\right)\sqrt{\frac{\mu_0\omega_1\omega_2\omega_3}{\varepsilon_0 n_1 n_2 n_3}} \tag{8.3-15}$$

where the summations are over all the components of the polarization unit vectors. Using Equation (8.3-13), the conservation of energy becomes

$$\frac{d}{dz}(\omega_1|A_1|^2 + \omega_2|A_2|^2 + \omega_3|A_3|^2) = 0 \tag{8.3-16}$$

Second-Harmonic Generation

For the case of second-harmonic generation (SHG), we consider the interaction of two waves with frequencies ω_1 and ω_3 (with $\omega_3 = 2\omega_1$) via the second-order optical nonlinearity. Let the fields be written

$$\begin{aligned} E_i^{\omega_1}(t) &= \tfrac{1}{2}(E_{0i}^{\omega_1} e^{i(\omega_1 t - k_1 z)} + \text{c.c.}) = \tfrac{1}{2}(a_{1i} E_1 e^{i(\omega_1 t - k_1 z)} + \text{c.c.}) \quad (i = x', y', z') \\ E_i^{\omega_3}(t) &= \tfrac{1}{2}(E_{0i}^{\omega_3} e^{i(\omega_3 t - k_3 z)} + \text{c.c.}) = \tfrac{1}{2}(a_{3i} E_3 e^{i(\omega_3 t - k_3 z)} + \text{c.c.}) \quad (i = x', y', z') \end{aligned} \tag{8.3-17}$$

where a_1 and a_3 are unit vectors representing the polarization states of the waves.

The nonlinear polarizations for SHG can be written

$$[P_{\text{NL}}^{\omega_3-\omega_1}(z, t)]_i = d_{ijk} a_{3j} a_{1k} E_3 E_1^* e^{i[(\omega_3-\omega_1)t-(k_3-k_1)z]} + \text{c.c.} \tag{8.3-18}$$

$$[P_{\text{NL}}^{2\omega_1}(z, t)]_i = \tfrac{1}{2} d_{ijk} a_{1j} a_{1k} E_1 E_1 e^{i(2\omega_1 t - 2k_1 z)} + \text{c.c.} \tag{8.3-19}$$

where summations over repeated indices are assumed. The coupled equations for SHG can now be written

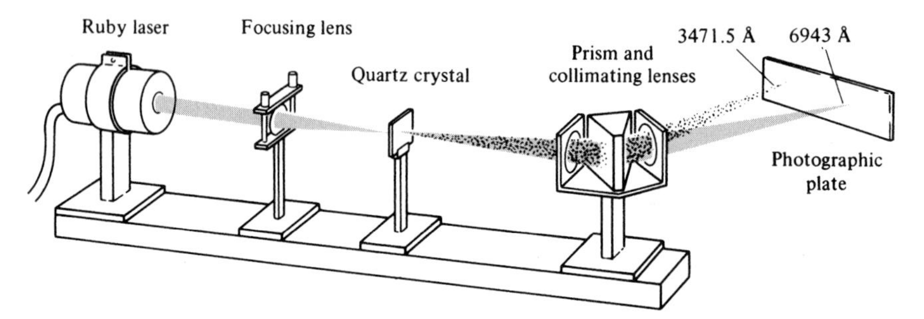

Figure 8.1 Arrangement used in the first experimental demonstration of second-harmonic generation [1]. A ruby laser beam at $\lambda_0 = 0.694$ μm is focused on a quartz crystal, causing generation of a (weak) beam at $\lambda_0/2 = 0.347$ μm. The two beams are then separated by a prism and detected on a photographic plate.

$$\frac{d}{dz}E_1 = -i\omega_1\sqrt{\frac{\mu_0}{\varepsilon_1}}\,dE_3E_1^*e^{-i(k_3-2k_1)z}$$

$$\frac{d}{dz}E_3 = -i\omega_3\sqrt{\frac{\mu_0}{\varepsilon_3}}\,\frac{1}{2}dE_1E_1e^{+i(k_3-2k_1)z} \tag{8.3-20}$$

where d is the effective second-order nonlinear coefficient

$$d = \sum_{ijk} d_{ijk}a_{3i}a_{1j}a_{1k} \tag{8.3-21}$$

The coupled equations (8.3-20) can now be applied for investigating the energy coupling between the fundamental and the second-harmonic wave.

It can be shown that

$$\frac{d}{dz}(\sqrt{\varepsilon_1}\,|E_1|^2 + \sqrt{\varepsilon_3}\,|E_3|^2) = 0 \tag{8.3-22}$$

provided $\omega_3 = 2\omega_1$.

The second-harmonic generation experiment that ushered in the field of nonlinear optics was performed by Franken, Hill, Peters, and Weinreich [1] in 1961. In their experiment (see Figure 8.1), a ruby laser beam at 694.3 nm was focused on the front surface of a crystalline quartz plate. The emergent radiation was examined with a spectrometer and was found to contain radiation at twice the input frequency (i.e., at λ = 347.15 nm). The conversion efficiency in this first experiment was $\sim 10^{-8}$. The utilization of more efficient materials, higher-intensity lasers, and phase-matching techniques has resulted, in the last few years, in conversion efficiencies approaching 100%. These factors will be discussed later in the section.

Phase Matching and Coherence Length

To illustrate the concept of phase matching, we consider a case of nondepletion approximation in which the amount of power lost from the input (ω_1) beam (by conversion to $2\omega_1$) is negligible, so that we may treat E_1 as a constant in the coupled equations (8.3-20) and we need consider only the second of the equations (8.3-20).

$$\frac{d}{dz}E_3 = -i\omega\sqrt{\frac{\mu_0}{\varepsilon}}\,dE_1^2e^{+i\Delta kz} \tag{8.3-23}$$

where

$$\omega = \omega_1 = \omega_3/2, \quad \varepsilon = \varepsilon_3$$

and

$$\Delta k = k_3 - 2k_1 \tag{8.3-24}$$

We now integrate Equation (8.3-23), assuming $E_3(0) = 0$ (i.e., no second-harmonic input at $z = 0$) and for a crystal of length L. We obtain

$$E_3(L) = -i\omega\sqrt{\frac{\mu_0}{\varepsilon}}\, dE_1^2\, \frac{e^{+i(\Delta k)L} - 1}{i(\Delta k)L} L \tag{8.3-25}$$

and the beam intensity $I^{(2\omega)}$ of the second-harmonic output, or the conversion efficiency η_{SHG}, can be written

$$\eta_{SHG} \equiv \frac{I^{(2\omega)}}{I^{(\omega)}} = \frac{2\omega^2 d^2 L^2}{n^3}\left(\frac{\mu_0}{\varepsilon_0}\right)^{3/2} \frac{\sin^2[(\Delta k)L/2]}{[(\Delta k)L/2]^2} I^{(\omega)} \tag{8.3-26}$$

where $I^{(\omega)}$ is the beam intensity of the fundamental wave, $n = n^{(\omega)} \approx n^{(2\omega)}$, and Δk is the wavenumber mismatch given by Equation (8.3-24). According to this equation, a prerequisite for efficient second-harmonic generation is that $\Delta k = 0$, or

$$k^{(2\omega)} = 2k^{(\omega)} \tag{8.3-27}$$

If $\Delta k \neq 0$, the second-harmonic wave generated at some plane (e.g., z_1), having propagated to some other plane (e.g., z_2), is not in phase with the second-harmonic wave generated at z_2. This results in the interference described by the sinusoidal factor on the right side of Equation (8.3-26). Two adjacent peaks of this spatial interference pattern are separated by the "coherence length,"

$$l_c = \frac{2\pi}{\Delta k} = \frac{2\pi}{k^{(2\omega)} - 2k^{(\omega)}} \tag{8.3-28}$$

The coherence length l_c is thus a measure of the maximum crystal length that is useful in producing the second-harmonic power. In terms of the indices of refraction, the wavenumber mismatch can be written

$$\Delta k = k^{(2\omega)} - 2k^{(\omega)} = \frac{2\omega}{c} n^{(2\omega)} - 2\frac{\omega}{c} n^{(\omega)} = \frac{2\omega}{c}\left(n^{(2\omega)} - n^{(\omega)}\right) \tag{8.3-29}$$

and the coherence length can be written

$$l_c = \frac{2\pi}{\Delta k} = \frac{2\pi}{k^{(2\omega)} - 2k^{(\omega)}} = \frac{2c\pi}{2\omega\left(n^{(2\omega)} - n^{(\omega)}\right)} = \frac{\lambda}{2\left(n^{(2\omega)} - n^{(\omega)}\right)} \tag{8.3-30}$$

where λ is the wavelength of the fundamental wave. Under ordinary circumstances, it may be no longer than 10^{-2} cm. This is because the index of refraction normally increases with ω. To illustrate this, we consider the following example.

EXAMPLE: COHERENCE LENGTH OF FREQUENCY DOUBLING IN KDP

Let $\lambda = 1\ \mu$m, and assume ordinary waves in KDP. The refractive indices (from Table 8.3) are

$$n(2\omega) = 1.514928$$

$$n(\omega) = 1.496044$$

Using Equation (8.3-30), we obtain

$$l_c = \frac{\lambda}{2\left(n^{(2\omega)} - n^{(\omega)}\right)} = \frac{10^{-6}}{2(1.514928 - 1.496044)}\ \text{m} = 26.5\ \mu\text{m}$$

TABLE 8.3 Index of Refraction Dispersion Data of KH_2PO_4

Wavelength (μm)	n_o (Ordinary Ray)	n_e (Extraordinary Ray)
0.2000	1.622630	1.563913
0.3000	1.545570	1.498153
0.4000	1.524481	1.480244
0.5000	1.514928	1.472486
0.6000	1.509274	1.468267
0.7000	1.505235	1.465601
0.8000	1.501924	1.463708
0.9000	1.498930	1.462234
1.0000	1.496044	1.460993
1.1000	1.493147	1.459884
1.2000	1.490169	1.458845
1.3000	1.487064	1.457838
1.4000	1.483803	1.456838
1.5000	1.480363	1.455829
1.6000	1.476729	1.454797
1.7000	1.472890	1.453735
1.8000	1.468834	1.452636
1.9000	1.464555	1.451495
2.0000	1.460044	1.450308

After Reference [2].

The technique that is used widely (see References [3, 4]) to satisfy the *phase-matching* requirement, $\Delta k = 0$, takes advantage of the natural birefringence of anisotropic crystals, which was discussed in Chapter 1. Using the relation $k^{(\omega)} = \omega\sqrt{\mu\varepsilon_0}n^{\omega}$, the phase-matching condition (8.3-27) becomes

$$n^{2\omega} = n^{\omega} \tag{8.3-31}$$

so the indices of refraction at the fundamental and second-harmonic frequencies must be equal. In normally dispersive materials, the index of the ordinary wave or the extraordinary wave along a given direction increases with ω, as can be seen from Table 8.3. This makes it impossible to satisfy (8.3-31) when both the ω and 2ω beams are of the same polarization type—that is, when both waves are extraordinary or ordinary. We can, however, under certain circumstances, satisfy (8.3-31) by making the two waves be of different types (different polarization states). To illustrate the point, consider the dependence of the index of refraction of the extraordinary wave in a uniaxial crystal on the angle θ between the propagation direction and the crystal optic (z') axis. It is given by

$$\frac{1}{n_e^2(\theta)} = \frac{\cos^2\theta}{n_o^2} + \frac{\sin^2\theta}{n_e^2} \tag{8.3-32}$$

According to Equation (8.3-32), we note that the index of refraction associated with the propagation of the extraordinary wave can vary from n_o at $\theta = 0$ to n_e at $\theta = 90°$. In a negative birefringent crystal (e.g., KDP), if $n_e^{2\omega} < n_o^{\omega}$, there exists an angle θ_m at which $n_e^{2\omega}(\theta_m) = n_o^{\omega}$; so if the fundamental beam (at ω) is launched along θ_m as an ordinary ray, the second-harmonic beam will be generated along the *same direction* as an extraordinary ray. The situation is illustrated by Figure 8.2. The angle θ_m is determined by the intersection between the sphere (shown as a circle in the figure) corresponding to the index surface of the ordinary beam at ω, and the index surface of the extraordinary ray $n_e^{2\omega}(\theta_m)$. The angle θ_m, which defines a

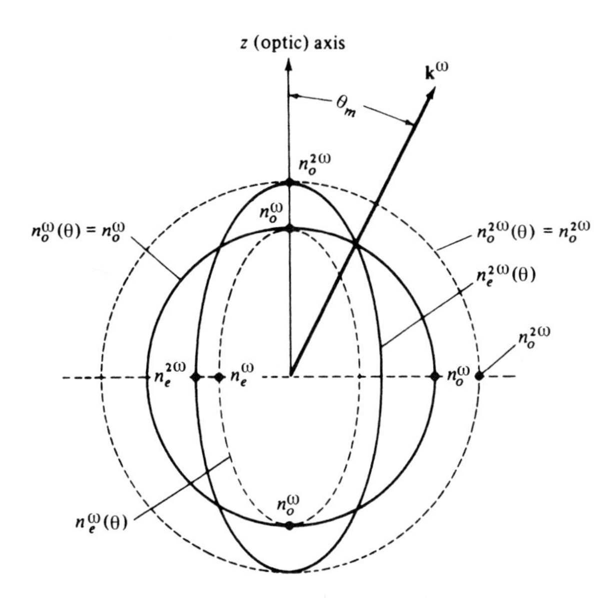

Figure 8.2 Normal (index) surfaces for the ordinary and extraordinary rays in a negative ($n_e < n_o$) uniaxial crystal. If $n_e^{2\omega} - n_o^{\omega}$, the condition $n_e^{2\omega}(\theta) = n_o^{\omega}$ is satisfied at $\theta = \theta_m$. The eccentricities shown are vastly exaggerated.

cone, for negative uniaxial crystals—that is, crystals in which $n_e^{2\omega} < n_o^{\omega}$ —is that satisfying $n_e^{2\omega}(\theta_m) = n_o^{\omega}$ or, using Equation (8.3-32),

$$\frac{\cos^2\theta_m}{(n_o^{2\omega})^2} + \frac{\sin^2\theta_m}{(n_e^{2\omega})^2} = \frac{1}{(n_o^{\omega})^2} \tag{8.3-33}$$

and, solving for θ_m,

$$\sin^2\theta_m = \frac{(n_o^{\omega})^{-2} - (n_o^{2\omega})^{-2}}{(n_e^{2\omega})^{-2} - (n_o^{2\omega})^{-2}} \tag{8.3-34}$$

The method of phase matching depicted in Figure 8.2 is known as type I phase matching. In type I phase matching, the polarization states of the fundamental and the second harmonic are mutually orthogonal. In a uniaxial crystal, if the fundamental is an ordinary wave, then the second harmonic is an extraordinary, or vice versa. Another mode of phase matching is possible in which the fundamental wave is a mixture of ordinary wave and extraordinary wave, and the second-harmonic wave can be either the ordinary or the extraordinary wave. The latter is known as type II phase matching. Generally, type I phase matching in uniaxial crystals can be obtained provided the birefringence ($|\Delta n| = |n_e - n_o|$) is larger than the index change ($n^{(2\omega)} - n^{(\omega)}$) due to chromatic dispersion of the crystal. If the birefringence is larger than twice the index change due to chromatic dispersion, then type II phase matching is possible. Table 8.4 is a summary of phase matching in uniaxial crystals.

TABLE 8.4 Summary of Phase Matching in Uniaxial Crystals

Birefringence	Type	Fundamental ω	Second Harmonic 2 ω	Condition
Negative ($n_e < n_o$)	I	O	E	$n_e^{(2\omega)} - n_e^{(\omega)} < n_o^{(\omega)} - n_e^{(\omega)}$
	II	O + E	E	$2(n_e^{(2\omega)} - n_e^{(\omega)}) < (n_o^{(\omega)} - n_e^{(\omega)})$
Positive ($n_o < n_e$)	I	E	O	$n_o^{(2\omega)} - n_o^{(\omega)} < n_e^{(\omega)} - n_o^{(\omega)}$
	II	O + E	O	$2(n_o^{(2\omega)} - n_o^{(\omega)}) < (n_e^{(\omega)} - n_o^{(\omega)})$

Note: E = extraordinary wave; O = ordinary wave.

EXAMPLE: SECOND-HARMONIC GENERATION

Consider the problem of second-harmonic generation using the output of a pulsed ruby laser ($\lambda_0 = 0.6943\ \mu$m) in a KH_2PO_4 crystal (KDP). The appropriate d coefficient can be obtained by using Equation (8.3-21). The unit vectors representing the polarization states of the waves are given by

$$\mathbf{a}_3 = \begin{bmatrix} \cos\theta_m/\sqrt{2} \\ -\cos\theta_m/\sqrt{2} \\ \sin\theta_m \end{bmatrix} \quad \text{and} \quad \mathbf{a}_1 = \begin{bmatrix} 1/\sqrt{2} \\ 1/\sqrt{2} \\ 0 \end{bmatrix}$$

where $\mathbf{a}_3$ is the polarization unit vector of the second-harmonic wave, while $\mathbf{a}_1$ is that of the fundamental wave. Using Equation (8.3-21) and carrying out the summation, we obtain

$$d = d_{312}\sin\theta_m = d_{36}\sin\theta_m$$

where d_{36} is listed in Table 8.2. The angle θ_m between the optic axis (c axis) and the direction of propagation for which $\Delta k = 0$ is given by Equation (8.3-34). The appropriate indices are taken from Table 8.2 and are

$$n_e(\lambda = 0.694\ \mu\text{m}) = 1.466, \quad n_e(\lambda = 0.347\ \mu\text{m}) = 1.490$$

$$n_o(\lambda = 0.694\ \mu\text{m}) = 1.506, \quad n_o(\lambda = 0.347\ \mu\text{m}) = 1.534$$

Substituting the foregoing data into Equation (8.3-34) gives

$$\theta_m = 52^\circ$$

To obtain phase matching along this direction, the fundamental beam in the crystal must be polarized as appropriate to an ordinary ray in accordance with the discussion following Equation (8.3-32).

Experimental Verification of Phase Matching

According to Equation (8.3-26), if the phase-matching condition $\Delta k = 0$ is violated, the output power is reduced by a factor

$$F = \frac{\sin^2[(\Delta k)L/2]}{[(\Delta k)L/2]^2} \tag{8.3-35}$$

from its (maximum) phase-matched value. The phase mismatch $(\Delta k)L/2$ is given, according to Equation (8.3-29), by

$$\frac{(\Delta k)L}{2} = \frac{\omega L}{c}[n_e^{2\omega}(\theta) - n_o^{\omega}] \tag{8.3-36}$$

and is thus a function of θ. If we use Equation (8.3-32) to expand $n_e^{2\omega}(\theta)$ as a Taylor series near $\theta \cong \theta_m$, retain the first two terms only, and assume perfect phase matching at $\theta = \theta_m$ so $n_e^{2\omega}(\theta_m) = n_o^{\omega}$, we obtain

$$\Delta k(\theta)L = -\frac{2\omega L}{c}\sin(2\theta_m)\frac{(n_e^{2\omega})^{-2} - (n_o^{2\omega})^{-2}}{2(n_o^{\omega})^{-3}}(\theta - \theta_m) \equiv 2\beta(\theta - \theta_m) \tag{8.3-37}$$

where β, as defined by Equation (8.3-37), is a constant depending on $n_e^{2\omega}$, $n_o^{2\omega}$, n_o^{ω}, ω, and L. If we plot the output power at 2ω as a function of θ, we would expect, according to Equations (8.3-26) and (8.3-37), to find it varying as

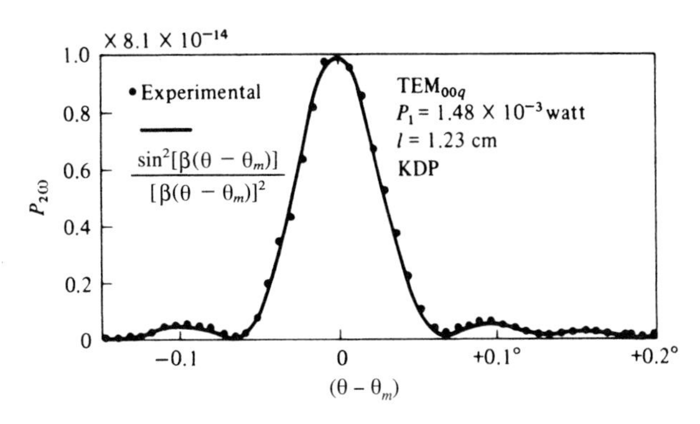

Figure 8.3 Variation of the second-harmonic power $P_{2\omega}$ with the angular departure $(\theta - \theta_m)$ from the phase-matching angle. (After Reference [5].)

$$P_{2\omega}(\theta) \propto \frac{\sin^2[\beta(\theta - \theta_m)]}{[\beta(\theta - \theta_m)]^2} \tag{8.3-38}$$

Figure 8.3 shows an experimental plot of $P_{2\omega}(\theta)$ as well as a theoretical plot of Equation (8.3-38). Another phase-matching technique involves the introduction of an artificial spatial periodicity $\Lambda = 2\pi/\Delta k$ into the beams' path. This method is discussed later in this chapter.

Second-Harmonic Generation with a Depleted Input

Expression (8.3-26) for the conversion efficiency in second-harmonic generation was derived assuming negligible depletion of the fundamental beam at ω. It is therefore valid only for cases where the conversion efficiency is small, that is, $\eta_{SHG} \ll 1$. Here we consider the general case of high-efficiency second-harmonic generation. In this case, we need to consider the depletion of the fundamental beam. Using the photon flux amplitude A_m defined in Equation (8.3-11), the coupled equations for second-harmonic generation can be written

$$\begin{aligned} \frac{dA_1}{dz} &= -i\kappa A_3 A_1^* e^{-i(\Delta k)z} \\ \frac{dA_3}{dz} &= -i\frac{\kappa}{2} A_1^2 e^{+i(\Delta k)z} \end{aligned} \tag{8.3-39}$$

where A_1 is the flux amplitude of the fundamental beam and A_3 is that of the second harmonic beam, and the coupling constant and the phase mismatch are given by

$$\kappa \equiv \sqrt{\left(\frac{\mu}{\varepsilon_0}\right)\frac{\omega_1\omega_1\omega_3}{n_1 n_1 n_3}}\, d \approx \left(\frac{\omega}{n}\right)^{3/2}\sqrt{2\left(\frac{\mu_0}{\varepsilon_0}\right)}\, d \tag{8.3-40}$$

$$\Delta k \equiv k_3 - (k_1 + k_2) = k^{(2\omega)} - 2k^{(\omega)} \tag{8.3-41}$$

In Equation (8.3-40), we used $\omega_3 = 2\omega_1 \equiv 2\omega$ and assumed $n = n(\omega) \approx n(2\omega)$. For the case of SHG with phase matching ($\Delta k = 0$), the coupled equations become

$$\begin{aligned} \frac{dA_1}{dz} &= -i\kappa A_3 A_1^* \\ \frac{dA_3}{dz} &= -i\frac{\kappa}{2} A_1^2 \end{aligned} \tag{8.3-42}$$

It follows from Equation (8.3-42) that if we choose, without loss of generality, $A_1(0)$ as a real number, then $A_1(z)$ is real and Equation (8.3-42) can be rewritten in the form

$$\frac{dA_1}{dz} = -\kappa A_3' A_1$$
$$\frac{dA_3'}{dz} = \frac{1}{2}\kappa A_1^2 \tag{8.3-43}$$

where $A_3 \equiv -iA_3'$. It follows from Equation (8.3-43) that

$$\frac{d}{dz}(A_1^2 + 2A_3'^2) = 0$$

(i.e., for every two photons "removed" from beam 1 (the fundamental beam at ω), one photon is added to beam 3 (the second harmonic at 2ω)).

Assuming no input at ω_3 (i.e., $A_3(0) = 0$), we have $A_1^2 + 2A_3'^2 = A_1^2(0)$, and the second of (8.3-43) becomes

$$\frac{dA_3'}{dz} = \frac{1}{2}\kappa\left(A_1^2(0) - 2A_3'^2\right) \tag{8.3-44}$$

A direct integration of the above equation leads to a solution

$$A_3'(z) = \frac{1}{\sqrt{2}}A_1(0)\tanh\left(\frac{1}{\sqrt{2}}\kappa A_1(0)z\right) \tag{8.3-45}$$

for the photon flux amplitude of the second-harmonic beam.

We note that as $\kappa A_1(0)z \to \infty$, $A_3'(z) \to A_1(0)/\sqrt{2}$ so that all the input photons at ω are converted into half as many photons at 2ω and the power conversion efficiency approaches unity. In the general case,

$$\eta_{SHG} \equiv \frac{P^{(2\omega)}}{P^{(\omega)}} = \frac{\hbar\omega_3|A_3(z)|^2}{\hbar\omega_1|A_1(z)|^2} = \frac{2|A_3(z)|^2}{|A_1(z)|^2} = \tanh^2[\tfrac{1}{2}\kappa A_1(0)z] \tag{8.3-46}$$

A plot of the theoretically predicted relation (8.3-46) as well as of experimental data obtained in converting from $\lambda = 1.06$ to $\lambda = 0.53$ μm is shown in Figure 8.4.

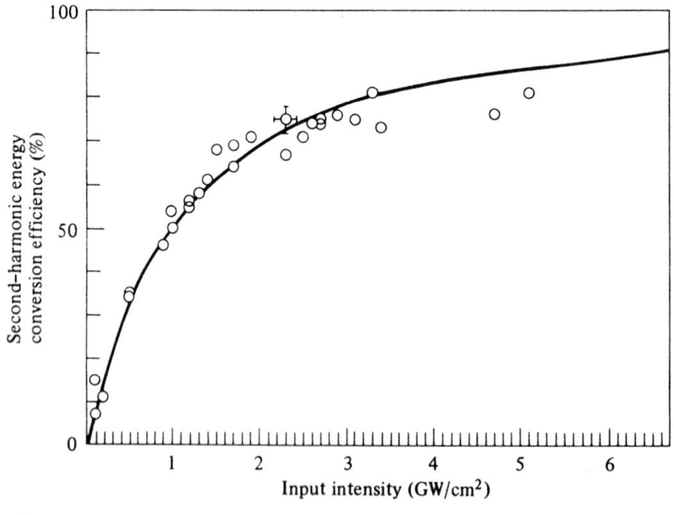

Figure 8.4 Frequency doubling energy conversion efficiency. Solid curve: theoretical prediction of Equation (8.3-46) (recall that $A_1(0) \propto \sqrt{I_{in}}$). The circles correspond to experimental points. (After supplementary Reference [6].)

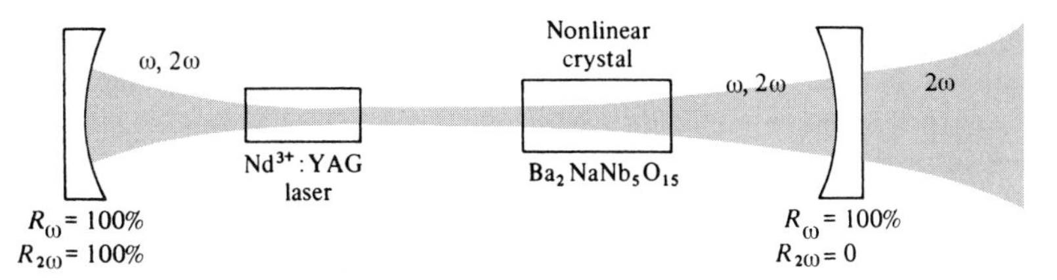

Figure 8.5 Typical setup for second-harmonic conversion inside a laser resonator. (After Reference [8].)

As a nonlinear optical process, an efficient second-harmonic generation usually requires a fundamental beam with a very high intensity. In the example shown in Figure 8.4, an input intensity of GW/cm^2 is needed to achieve a significant conversion efficiency. For such a high intensity, the nonlinear optical crystal must have a high damage threshold. To achieve a high conversion efficiency with a relatively low input intensity would require the development of new materials with a large nonlinear optical coefficient d. Another approach to achieve high conversion efficiency is to take advantage of the high intensity in an optical resonator and to place the nonlinear crystal inside the laser resonator where the energy flux can be made very large. This approach has been used successfully [7]. Figure 8.5 shows a schematic drawing of the experiment. In this example, the fundamental beam is a Nd:YAG laser beam oscillating at $\lambda = 1.064$ μm. The resonator mirrors are designed so that the Nd:YAG laser beam is confined inside the resonator. A crystal of $Ba_2NaNb_5O_{15}$ is placed inside the laser resonator as the nonlinear optical medium for SHG.

8.4 OTHER SECOND-ORDER NONLINEAR PROCESSES

In addition to the second-harmonic generation (SHG), there are other second-order nonlinear processes, including optical rectification, optical parametric amplification (OPA), optical parametric oscillation (OPO), and frequency up-conversion. In this section, we will discuss some of the most important processes. Optical parametric amplification can be employed to produce squeezed states of optical beams. This subject will be discussed in Chapter 18. Optical parametric oscillation is widely used in tunable solid-state lasers (e.g., Ti: sapphire lasers).

Optical Parametric Amplification

Optical parametric amplification (OPA) in its simplest form involves the transfer of power from a "pump" wave at ω_3 to two waves at lower frequencies ω_1 and ω_2, with $\omega_3 = \omega_1 + \omega_2$. It is fundamentally similar to the case of second-harmonic generation treated in Section 8.3. The only difference is in the direction of power flow. In SHG, power is fed from the low-frequency optical field at ω to the field at 2ω. In OPA, the power flow is from the high-frequency field (ω_3) to the low-frequency fields at ω_1 and ω_2. In the special case where $\omega_1 = \omega_2$, we have the exact reverse of SHG. These processes are the result of the nonlinear optical mixing of three beams. The equations governing the interaction among the three waves are identical to those described in Section 8.2. The only differences are the boundary conditions. Let the field of the three waves be written

$$\mathbf{E} = \sum_{m=1,2,3} \mathbf{a}_m E_m \exp[i(\omega_m t - k_m z)] \tag{8.4-1}$$

where $\mathbf{a}_m$ is the unit vector representing the polarization state of the mth beam in the nonlinear crystal, and E_m is the amplitude of the mth beam with frequency ω_m. Each of the waves is a normal mode of propagation with a well-defined state of polarization in the medium. The field amplitudes $E_1(z)$, $E_2(z)$, and $E_3(z)$ are functions of z, to reflect the coupling. We assume that these three beams are copropagating along the same direction (z axis), with $\omega_3 = \omega_1 + \omega_2$. As a result of the second-order optical nonlinearity, these three waves are coupled. The coupled equations can be written, according to Equation (8.3-8),

$$\frac{dE_1}{dz} = -\frac{i\omega_1}{2}\sqrt{\frac{\mu}{\varepsilon_1}}\, dE_3E_2^* e^{-i(k_3-k_2-k_1)z}$$

$$\frac{dE_2^*}{dz} = +\frac{i\omega_2}{2}\sqrt{\frac{\mu}{\varepsilon_2}}\, dE_1E_3^* e^{-i(k_1-k_3+k_2)z} \tag{8.4-2}$$

$$\frac{dE_3}{dz} = -\frac{i\omega_3}{2}\sqrt{\frac{\mu}{\varepsilon_3}}\, dE_1E_2 e^{-i(k_1+k_2-k_3)z}$$

where the effective nonlinear coefficient d is given by

$$d = \sum_{ijk} d_{ijk} a_{1i} a_{2j} a_{3k} \tag{8.4-3}$$

In arriving at Equations (8.4-2), we have employed the permutation symmetry of the nonlinear coefficients (i.e., $d_{ijk} = d_{jik} = d_{ikj}$). We now transform to a new set of field variables A_l defined by

$$A_m \equiv \sqrt{\frac{n_m}{\omega_m}}\, E_m, \quad m = 1, 2, 3 \tag{8.4-4}$$

where $n_m^2 = \varepsilon_m/\varepsilon_0$; that is, n_m is the index of refraction of wave m, so that the beam intensity (power flow per unit area) at ω_m is given by

$$I_m = \frac{P_m}{A} = \frac{1}{2}\sqrt{\frac{\varepsilon_0}{\mu}}\, n_m |E_m|^2 = \frac{1}{2}\sqrt{\frac{\varepsilon_0}{\mu}}\, \omega_m |A_m|^2 \tag{8.4-5}$$

The power flow P_m/A per unit area is related to the flux N_m (photons per square meter per second) by

$$\frac{P_m}{A} = N_m \hbar\omega_m = \frac{1}{2}\sqrt{\frac{\varepsilon_0}{\mu}}\, |A_m|^2 \omega_m \tag{8.4-6}$$

so that $|A_m|^2$ is proportional to the photon flux at ω_m. The coupled equations (8.4-2) for the photon flux amplitudes A_m become

$$\frac{dA_1}{dz} = -\frac{\alpha_1}{2}A_1 - \frac{i}{2}\kappa A_2^* A_3 e^{-i(\Delta k)z}$$

$$\frac{dA_2^*}{dz} = -\frac{\alpha_2}{2}A_2^* + \frac{i}{2}\kappa A_1 A_3^* e^{i(\Delta k)z} \tag{8.4-7}$$

$$\frac{dA_3}{dz} = -\frac{\alpha_3}{2}A_3 - \frac{i}{2}\kappa A_1 A_2 e^{i(\Delta k)z}$$

where

$$\kappa \equiv d\sqrt{\left(\frac{\mu}{\varepsilon_0}\right)\frac{\omega_1\omega_2\omega_3}{n_1n_2n_3}} \tag{8.4-8}$$

$$\Delta k \equiv k_3 - (k_1 + k_2) \tag{8.4-9}$$

and α_m is the linear attenuation coefficient for the mth beam. The advantage of using A_m instead of E_m is now apparent since, unlike (8.4-2), relations (8.4-7) involve a single coupling parameter κ.

We will now use (8.4-7) to obtain the field amplitudes $A_1(z)$, $A_2(z)$, and $A_3(z)$ for the case in which three waves with amplitudes $A_1(0)$, $A_2(0)$, and $A_3(0)$ at frequencies ω_1, ω_2, and ω_3, respectively, are incident on a nonlinear crystal at $z = 0$. We take $\omega_3 = \omega_1 + \omega_2$, $\alpha_1 = \alpha_2 = \alpha_3 = 0$ (no losses), and $\Delta k = k_3 - k_1 - k_2 = 0$. In addition, we assume that $\omega_1|A_1(z)|^2$ and $\omega_2|A_2(z)|^2$ remain small compared to $\omega_3|A_3(z)|^2$ throughout the interaction region. This last condition, in view of (8.4-7), is equivalent to assuming that the power drained off the "pump" (at ω_3) by the "signal" (ω_1) and idler (ω_2) is negligible compared to the input power at ω_3. This enables us to view $A_3(z)$ as a constant. With the assumptions stated above, Equations (8.4-7) become

$$\frac{dA_1}{dz} = -\frac{ig}{2}A_2^* \quad \text{and} \quad \frac{dA_2^*}{dz} = \frac{ig^*}{2}A_1 \tag{8.4-10}$$

where g is a constant given by, according to Equations (8.4-4) and (8.4-8),

$$g \equiv \kappa A_3(0) = \sqrt{\left(\frac{\mu}{\varepsilon_0}\right)\frac{\omega_1\omega_2}{n_1n_2}}\,E_3(0)d \tag{8.4-11}$$

We note that the coupling constant g is proportional to the (pump) field amplitude $E_3(0)$. Generally, g is a complex number, depending on the phase of the pump field amplitude $E_3(0)$.

The solutions of the coupled equations (8.4-10), subject to the boundary conditions $A_1(z = 0) \equiv A_1(0)$, $A_2(z = 0) \equiv A_2(0)$, are

$$\begin{aligned} A_1(z) &= A_1(0)\cosh\gamma z - i\frac{g}{|g|}A_2^*(0)\sinh\gamma z \\ A_2^*(z) &= A_2^*(0)\cosh\gamma z + i\frac{g}{|g|}A_1(0)\sinh\gamma z \end{aligned} \tag{8.4-12}$$

where the gain coefficient γ is given by

$$\gamma = \left|\frac{g}{2}\right| = \left|\frac{\kappa A_3(0)}{2}\right| = \frac{1}{2}\left|\sqrt{\left(\frac{\mu}{\varepsilon_0}\right)\frac{\omega_1\omega_2}{n_1n_2}}\,E_3(0)d\right| \tag{8.4-13}$$

where we recall that d is the effective nonlinear optical coefficient.

Equations (8.4-12) describe the growth of the signal and idler waves under phase-matching conditions. In the case of parametric amplification, the input will consist of the pump (ω_3) wave and one of the other two fields, say, ω_1. In this case $A_2(0) = 0$, and using the relation $N_i \propto A_iA_i^*$ for the photon flux, we obtain from (8.4-12)

$$\begin{aligned} N_1(z) &\propto A_1^*(z)A_1(z) = |A_1(0)|^2\cosh^2\gamma z \xrightarrow[\gamma z \gg 1]{} \frac{|A_1(0)|^2}{4}e^{2\gamma z} \\ N_2(z) &\propto A_2^*(z)A_2(z) = |A_1(0)|^2\sinh^2\gamma z \xrightarrow[\gamma z \gg 1]{} \frac{|A_1(0)|^2}{4}e^{2\gamma z} \end{aligned} \tag{8.4-14a}$$

Thus, for $\gamma z \gg 1$, the photon fluxes at ω_1 and ω_2 grow exponentially. If we limit our attention to the signal wave at ω_1, we note that the signal power (or signal photon flux) undergoes an amplification by a factor

$$\frac{A_1^*(z)A_1(z)}{A_1^*(0)A_1(0)} \underset{\gamma z \gg 1}{=} \frac{1}{4}e^{2\gamma z} \tag{8.4-14b}$$

EXAMPLE: PARAMETRIC AMPLIFICATION

The magnitude of the gain coefficient γ available in a traveling-wave parametric interaction is estimated for the following case involving the use of a $LiNbO_3$ crystal:

$$d_{15} = d_{113} = 5 \times 10^{-23} \text{ MKS (see Table 8.2)}$$

$$\nu_1 \cong \nu_2 = 3 \times 10^{14} \text{ Hz}$$

$P_3/\text{Area} = \text{Pump power} = 5 \times 10^6 \text{ W/cm}^2$

$$n_1 \cong n_3 = 2.2$$

Converting P_3 to $|E_3|^2$ with the use of Equation (8.4-5) and then substituting into Equation (8.4-11) or (8.4-13) yields

$$\gamma = 0.35 \text{ cm}^{-1}$$

This shows that traveling-wave parametric amplification is not expected to lead to large values of gain except for extremely large pump-power densities. The main attraction of the parametric amplification just described is probably in giving rise to optical parametric oscillation (OPO), which will be described later in this section.

In the preceding section the analysis of parametric amplification assumed that the phase-matching condition

$$k_3 = k_1 + k_2 \tag{8.4-15}$$

is satisfied. It is important to determine the consequences of violating this condition.

We start with Equations (8.4-10) by adding the exponential terms accounting for the phase mismatch, and taking the loss coefficients $\alpha_1 = \alpha_2 = 0$:

$$\begin{aligned} \frac{dA_1}{dz} &= -i\frac{g}{2}A_2^* e^{-i(\Delta k)z} \\ \frac{dA_2^*}{dz} &= +i\frac{g^*}{2}A_1 e^{i(\Delta k)z} \end{aligned} \tag{8.4-16}$$

The above coupled equations can be solved as follows. We multiply the first equation of (8.4-16) by $\exp[+i(\Delta k)z]$ and then differentiate with respect to z. This leads to

$$\frac{d}{dz}\left(e^{+i(\Delta k)z}\frac{dA_1}{dz}\right) = -i\frac{g}{2}\frac{d}{dz}A_2^* \tag{8.4-17}$$

We can now eliminate A_2^* by using the second equation of (8.4-16). After a few steps of algebra, we obtain

$$\frac{d^2A_1}{dz^2} + i(\Delta k)\frac{dA_1}{dz} = \left|\frac{g}{2}\right|^2 A_1 \tag{8.4-18}$$

This is an ordinary second-order differential equation. The general solution can be written

$$A_1(z) = e^{-i(\Delta k/2)z}[C_1 \cosh(sz) + C_2 \sinh(sz)] \tag{8.4-19}$$

where C_1 and C_2 are constants, and

$$s = \sqrt{|g/2|^2 - (\Delta k/2)^2} \tag{8.4-20}$$

The general solutions of (8.4-16) subject to the boundary condition of $A_1(0)$ and $A_2^*(0)$ can be written

$$\begin{aligned} A_1(z)e^{i(\Delta k/2)z} &= A_1(0)\left[\cosh(sz) + \frac{i(\Delta k)}{2s}\sinh(sz)\right] - i\frac{g}{2s}A_2^*(0)\sinh(sz) \\ A_2^*(z)e^{-i(\Delta k/2)z} &= A_2^*(0)\left[\cosh(sz) - \frac{i(\Delta k)}{2s}\sinh(sz)\right] + i\frac{g}{2s}A_1(0)\sinh(sz) \end{aligned} \tag{8.4-21}$$

The last result reduces, as it should, to (8.4-12) if we put $\Delta k = 0$.

By examining Equations (8.4-21), we note that exponential gain is possible only when the gain coefficient s is a real number. This is possible provided

$$|g| = 2\gamma \geq \Delta k \tag{8.4-22}$$

according to Equation (8.4-20). When $|g| < \Delta k$, no sustained growth of the signal (A_1) and idler (A_2) waves is possible, since in this case the sinh and cosh functions in (8.4-21) become

$$\begin{aligned} &i\sin\{\tfrac{1}{2}[(\Delta k)^2 - |g|^2]^{1/2} z\} \\ &i\cos\{\tfrac{1}{2}[(\Delta k)^2 - |g|^2]^{1/2} z\} \end{aligned} \tag{8.4-23}$$

respectively, and the energies of the waves at ω_1 and ω_2 oscillate as functions of the distance z.

The problem of phase matching in parametric amplification is fundamentally the same as that in second-harmonic generation. Instead of satisfying the condition $k^{2\omega} = 2k^{\omega}$, we have, according to Equation (8.4-15), to satisfy the condition

$$k_3 = k_1 + k_2$$

This is done, as in second-harmonic generation, by using the dependence of the phase velocity of the extraordinary wave in uniaxial crystals on the direction of propagation. In a negative uniaxial crystal ($n_e < n_o$), we can, as an example, choose the signal and idler waves as ordinary while the pump at ω_3 is applied as an extraordinary wave. This is type I phase matching. Using Equation (8.3-32) and the relation $k^{\omega} = (\omega/c)n^{\omega}$, the phase-matching condition (8.4-15) is satisfied when all three waves propagate at an angle θ_m to the z (optic) axis, where

$$n_e^{\omega_3}(\theta_m) = \left[\left(\frac{\cos\theta_m}{n_o^{\omega_3}}\right)^2 + \left(\frac{\sin\theta_m}{n_e^{\omega_3}}\right)^2\right]^{-1/2} = \frac{\omega_1}{\omega_3}n_o^{\omega_1} + \frac{\omega_2}{\omega_3}n_o^{\omega_2} \tag{8.4-24}$$

Optical Parametric Oscillation (OPO)

In the previous discussion, we have demonstrated that a pump wave at ω_3 can cause a simultaneous amplification in a nonlinear medium of signal and idler waves at frequencies ω_1 and ω_2, respectively, where $\omega_3 = \omega_1 + \omega_2$. If the nonlinear crystal is placed within an optical resonator (as shown in Figure 8.6) that provides resonance for the signal or idler waves (or

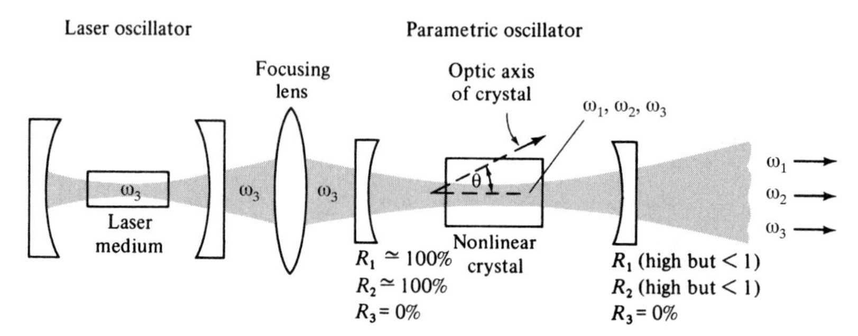

Figure 8.6 Schematic diagram of an optical parametric oscillator in which the laser output at ω_3 is used as the pump, giving rise to oscillations at ω_1 and ω_2 (where $\omega_3 = \omega_1 + \omega_2$) in an optical cavity that contains the nonlinear crystal and resonates at ω_1 and ω_2.

both), the parametric gain will, at some threshold pumping intensity, cause a simultaneous oscillation at the signal and idler frequencies. The threshold pumping corresponds to the point at which the parametric gain just balances the losses of the signal and idler waves. This is the physical basis of the optical parametric oscillator. Its practical importance derives from its ability to convert the power output of the pump laser to power at the signal and idler frequencies that, as will be shown later, can be tuned continuously over large spectral ranges.

To analyze this situation, we return to (8.4-7). We take $\Delta k = 0$ and neglect the depletion of the pump waves, so $A_3(z) = A_3(0)$. The result is

$$\begin{aligned} \frac{dA_1}{dz} &= -\frac{1}{2}\alpha_1 A_1 - i\frac{g}{2}A_2^* \\ \frac{dA_2^*}{dz} &= -\frac{1}{2}\alpha_2 A_2^* + i\frac{g^*}{2}A_1 \end{aligned} \tag{8.4-25}$$

where, as in Equation (8.4-11), the complex constant g is given by

$$g \equiv \sqrt{\left(\frac{\mu}{\varepsilon_0}\right)\frac{\omega_1\omega_2}{n_1 n_2}}\,E_3(0)d \tag{8.4-26}$$

Equations (8.4-25) describe traveling-wave parametric interactions. We will use them to describe the interaction inside a resonator. This procedure seems plausible if we think of propagation inside an optical resonator as a folded optical path. The magnitudes of the spatial distributed loss constants α_1 and α_2 must then be chosen so that they account for the actual losses in the resonator. The latter will include losses caused by the less than perfect reflection at the mirrors, as well as distributed loss in the nonlinear crystal and that due to diffraction. The effective loss constant α_m is chosen so that $\exp(-\alpha_m l)$ is the total attenuation in intensity per resonator pass at ω_m, where l is the crystal length.

If the parametric gain is sufficiently high to overcome the losses, steady-state oscillation results. When this is the case,

$$\frac{dA_1}{dz} = \frac{dA_2^*}{dz} = 0 \tag{8.4-27}$$

and thus the power gained via the parametric interaction just balances the losses. Putting $d/dz = 0$ in (8.4-25) gives

$$-\frac{1}{2}\alpha_1 A_1 - i\frac{g}{2}A_2^* = 0$$
$$i\frac{g^*}{2}A_1 - \frac{1}{2}\alpha_2 A_2^* = 0 \tag{8.4-28}$$

The condition for nontrivial solutions for A_1 and A_2^* is that the determinant of (8.4-28) vanish; that is,

$$\det\begin{vmatrix} -\frac{\alpha_1}{2} & -i\frac{g}{2} \\ i\frac{g^*}{2} & -\frac{\alpha_2}{2} \end{vmatrix} = 0$$

and, therefore,

$$|g|^2 = \alpha_1\alpha_2 \tag{8.4-29}$$

This is the *threshold condition* for *parametric oscillation.*

We have shown above that the pair of signals (ω_1) and idler frequencies that are caused to oscillate by parametric pumping at ω_3 satisfy the condition $k_3 = k_1 + k_2$. Using $k_i = \omega_i n_i/c$, we can write it as

$$\omega_3 n_3 = \omega_1 n_1 + \omega_2 n_2 \tag{8.4-30}$$

In a crystal the indices of refraction generally depend, as shown in Section 8.3, on the frequency, crystal orientation (if the wave is extraordinary), electric field (in electro-optic crystals), and the temperature. If, as an example, we change the crystal orientation in the oscillator, the oscillation frequencies ω_1 and ω_2 will change so as to compensate for the change in indices, and thus condition (8.4-30) will be satisfied at the new frequencies. Such an angular dependence can be employed as a tuning mechanism for tunable OPO. We will discuss the tuning of OPO in the last section of this chapter.

Frequency Up-Conversion

Parametric interactions in a crystal can be used to convert a signal from a "low" frequency ω_1 to a "high" frequency ω_3 by mixing it with a strong laser beam at ω_2, where

$$\omega_1 + \omega_2 = \omega_3 \tag{8.4-31}$$

This is particularly useful in situations where the detection system is more efficient in the spectral regime around ω_3. Using the quantum mechanical photon picture, we can consider the basic process taking place in frequency up-conversion as one in which a signal (ω_1) photon and a pump (ω_2) photon are annihilated while, simultaneously, one photon at ω_3 is generated. Since a photon energy is $\hbar\omega$; conservation of energy dictates that $\omega_3 = \omega_1 + \omega_2$ and, in a manner similar to Equation (8.4-15), the conservation of momentum leads to the relationship

$$\mathbf{k}_3 = \mathbf{k}_1 + \mathbf{k}_2 \tag{8.4-32}$$

between the wavevectors at the three frequencies. This point of view also suggests that the number of output photons at ω_3 cannot exceed the input number of photons at ω_1.

The analysis of frequency up-conversion starts with Equations (8.4-7). Assuming negligible depletion of the pump wave A_2, no losses ($\alpha = 0$) at ω_1 and ω_3, and taking $\Delta k = 0$, we can write the first and third of these equations as

$$\frac{dA_1}{dz} = -i\frac{g}{2}A_3$$
$$\frac{dA_3}{dz} = -i\frac{g}{2}A_1 \qquad (8.4\text{-}33)$$

where, choosing without loss of generality the pump phase as zero so that $A_2(0) = A_2^*(0)$,

$$g \equiv \sqrt{\left(\frac{\mu}{\varepsilon_0}\right)\frac{\omega_1\omega_3}{n_1 n_3}}\, E_2 d \qquad (8.4\text{-}34)$$

where E_2 is the amplitude of the electric field of the pump laser. Taking the input waves with initial (complex) amplitudes $A_1(0)$ and $A_3(0)$, the general solutions of (8.4-33) are

$$A_1(z) = A_1(0)\cos\left(\frac{g}{2}z\right) - iA_3(0)\sin\left(\frac{g}{2}z\right)$$
$$A_3(z) = A_3(0)\cos\left(\frac{g}{2}z\right) - iA_1(0)\sin\left(\frac{g}{2}z\right) \qquad (8.4\text{-}35)$$

In the case of a single (low) frequency input at ω_1, we have $A_3(0) = 0$. In this case,

$$|A_1(z)|^2 = |A_1(0)|^2\cos^2\left(\frac{g}{2}z\right)$$
$$|A_3(z)|^2 = |A_3(0)|^2\sin^2\left(\frac{g}{2}z\right) \qquad (8.4\text{-}36)$$

therefore,

$$|A_1(z)|^2 + |A_3(z)|^2 = |A_1(0)|^2 \qquad (8.4\text{-}37)$$

In the discussion following Equation (8.4-5), we pointed out that $|A_l(z)|^2$ is proportional to the photon flux (photons per square meter per second) at ω_l. Using this fact, we may interpret (8.4-35) as stating that the photon flux at ω_1 plus that at ω_3 at any plane z is a constant equal to the input ($z = 0$) flux at ω_1. If we rewrite (8.4-36) in terms of powers, we obtain

$$P_1(z) = P_1(0)\cos^2\left(\frac{g}{2}z\right)$$
$$P_3(z) = \frac{\omega_3}{\omega_1}P_1(0)\sin^2\left(\frac{g}{2}z\right) \qquad (8.4\text{-}38)$$

In a crystal of length l, the conversion efficiency is thus

$$\frac{P_3(l)}{P_1(0)} = \frac{\omega_3}{\omega_1}\sin^2\left(\frac{g}{2}l\right) \qquad (8.4\text{-}39)$$

and can have a maximum value of ω_3/ω_1, corresponding to the case in which all the input (ω_1) photons are converted to ω_3 photons.

In most practical situations, the conversion efficiency is small (see the following example) so using $\sin x \sim x$ for $x \ll 1$, we get

$$\frac{P_3(l)}{P_1(0)} \approx \frac{\omega_3}{\omega_1}\left(\frac{g^2 l^2}{4}\right)$$

which, by the use of Equations (8.4-34) and (8.4-5), can be written

$$\frac{P_3(l)}{P_1(0)} \approx \frac{\omega_3^2 l^2 d^2}{2n_1 n_2 n_3}\left(\frac{\mu}{\varepsilon_0}\right)^{3/2}\left(\frac{P_2}{A}\right) \tag{8.4-40}$$

where A is the cross-sectional area of the interaction region.

EXAMPLE: FREQUENCY UP-CONVERSION

The main practical interest in parametric frequency up-conversion stems from the fact that it offers a means of detecting infrared radiation (a region where detectors are either inefficient, very slow, or require cooling to cryogenic temperatures) by converting the frequency into the visible or near-visible part of the spectrum. The radiation can then be detected by means of efficient and fast detectors such as photomultipliers or photodiodes; see References [9–12].

As an example of this application, consider the problem of up-converting a 10.6 μm signal, originating in a CO_2 laser, to 0.96 μm by mixing it with the 1.06 μm output of a Nd^{3+}:YAG laser. The nonlinear crystal chosen for this application has to have low losses at 1.06 μm and 10.6 μm, as well as at 0.96 μm. In addition, its birefringence has to be such as to make phase matching possible. The crystal proustite (Ag_3AsS_3) meets these requirements [12].

Using the data

$$\frac{P_{1.06\mu\text{m}}}{A} = 10^4 \text{ W/cm}^2 = 10^8 \text{ W/cm}^2$$

$$l = 10^{-2} \text{ m}$$

$n_1 \sim n_2 \sim n_3 = 2.6$ (an average number based on the data of Reference [12])

$d_{\text{eff}} = 1.1 \times 10^{-22}$ (MKS) (taken conservatively as a little less than half the value for d_{22})

we obtain, from Equation (8.4-40),

$$\frac{P_{\lambda=0.96\mu\text{m}}(l = 1 \text{ cm})}{P_{\lambda=10.6\mu\text{m}}(l = 0)} = 7.1 \times 10^{-4}$$

indicating a useful amount of conversion efficiency.

8.5 QUASI PHASE MATCHING

An alternative technique for achieving phase matching in crystals is referred to as *quasi phase matching* [13], a reference to a crystal fashioned in such a way that the direction of one of its principal axes, say, z, is reversed periodically. This, in properly chosen crystal orientation and polarization directions of the participating optical fields, results in a periodic modulation of the nonlinear coefficient tensor element d_{ijk} responsible for the interaction. The coupled wave equations (8.4-2) remain unchanged, except that d is replaced by $d(z)$, which, being periodic, can be expanded in a Fourier series

$$d(z) = d_{\text{bulk}}\left(\sum_{m=-\infty}^{\infty} c_m \exp\left(im\frac{2\pi}{\Lambda}z\right)\right) \tag{8.5-1}$$

where c_m are the Fourier expansion coefficients and Λ is the period of $d(z)$. The effect on the first of Equations (8.4-2), as an example, is to transform it to

$$\frac{dE_1}{dz} = -\frac{i\omega_1}{2}\sqrt{\frac{\mu}{\varepsilon_1}}\, d_{\text{bulk}} E_3 E_2^* \sum_{m=-\infty}^{\infty} c_m \exp\left[i\left(m\frac{2\pi}{\Lambda} - k_3 + k_2 + k_1\right)z\right] \tag{8.5-2}$$

Phase matching can be obtained if, for some integer m, the condition

$$m\frac{2\pi}{\Lambda} = k_3 - k_2 - k_1 \tag{8.5-3}$$

is satisfied. Ignoring non-phase-matched terms in Equation (8.5-2) (their contribution averages out to zero over distances that are large compared to the coherence length), we rewrite (8.5-2) as

$$\frac{dE_1}{dz} = -i\frac{\omega_1}{2}\sqrt{\frac{\mu}{\varepsilon_1}}\, d_{\text{bulk}} c_m \exp\left[i\left(m\frac{2\pi}{\Lambda} - k_3 + k_2 + k_1\right)z\right] \tag{8.5-4}$$

$$c_m = \frac{1}{\Lambda}\int_0^{\Lambda} \frac{d(z)}{d_{\text{bulk}}} \exp\left(-im\frac{2\pi}{\Lambda}z\right) dz \tag{8.5-5}$$

where m is an integer that satisfies Equation (8.5-3).

The simplest case of a spatially periodic $d(z)$ is one in which $d(z)$ switches from d_{bulk} to $-d_{\text{bulk}}$ every $\Lambda/2$. In this case,

$$c_m = \frac{1 - \cos m\pi}{m\pi} \quad \text{for } m \neq 0 \tag{8.5-6}$$

so that, choosing $m = 1$, the effective nonlinear constant is

$$d_{\text{eff}} = c_m d_{\text{bulk}} = \frac{2}{\pi} d_{\text{bulk}} \tag{8.5-7}$$

For $m = 1$ quasi phase matching, the spatial period is given by $\Lambda = 2\pi/\Delta k = 2L_c$, according to Equation (8.5-3). It is clear from Equation (8.5-4) that, in principle, quasi-phase-matched configurations can give rise to the same conversion efficiency as in the ideal, $\Delta k = 0$, phase-matched case, except that we require a longer interaction path to achieve it. The length penalty factor is $d_{\text{bulk}}/d_{\text{eff}} = c_m^{-1}$.

To appreciate quasi phase matching on an intuitive basis, we note that it involves reversing the sign of the nonlinear interaction at

$$z = L_c, 2L_c, \ldots \quad \text{with} \quad L_c = \frac{\pi}{\Delta k} \tag{8.5-8}$$

These are the locations where the amplitude of the newly generated second-harmonic (SH) wave is opposite in sign relative to the already generated second-harmonic wave in the non-phase-matched case. Reversing the sign of the newly generated SH wave leads to constructive buildup of $E^{2\omega}(z)$. This keeps the power flowing from ω to 2ω along the length of the crystal and leads to cumulative buildup of $E^{2\omega}(z)$. The buildup of $E^{2\omega}(z)$ via quasi phase matching can best be visualized using a phasor plot of the interaction. Taking the specific case of second-harmonic generation as an example, we can divide the interaction path L into sufficiently short segments, each of length δz, such that $(\Delta k)\,\delta z \ll \pi$ and obtain from Equation (8.3-23)

$$\Delta E^{(2\omega)}(z) = -i\omega\sqrt{\frac{\mu}{\varepsilon}}\, d(E^{(\omega)})^2 e^{i(\Delta k)z}\delta z \tag{8.5-9}$$

where $\Delta E^{(2\omega)}$ is the complex increment to the phasor $E^{(2\omega)}(z)$ due to the segment of length δz centered at z. By adding the increments $\Delta E^{(2\omega)}$ vectorially, or rather phasorially, we obtain the phasor diagram shown in Figure 8.7. In the non-phase-matched case (Figure 8.7a), the generated second-harmonic field keeps growing, reaching a maximum, usually insignificantly small, at

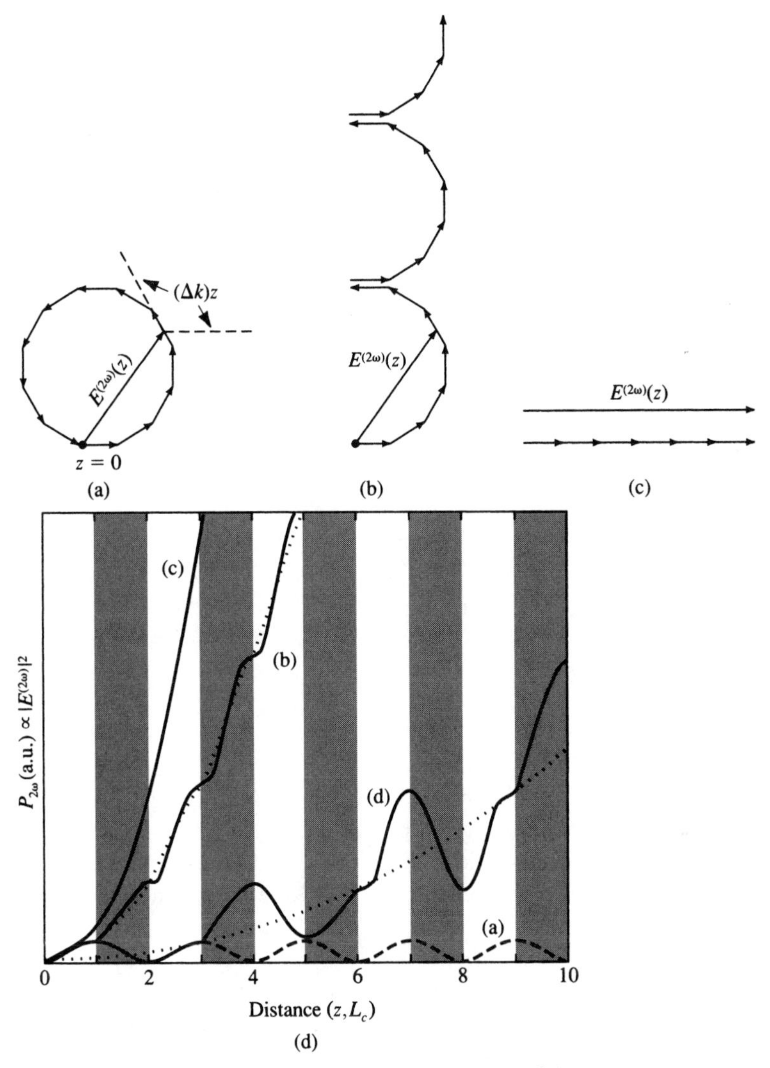

Figure 8.7 Evolution of the second-harmonic phasor $E^{(2\omega)}(z)$ in (a) a non-phase-matched case, (b) quasi-phase-matched case, and (c) bulk birefringent phase matching case ($\Delta k = 0$). (d) The second-harmonic power field $E^{(2\omega)}$ in a crystal for cases (a), (b), and (c), as well as curve (d) for quasi-phase-matched operation using the third Fourier coefficient $m = 3$, $L_c \equiv \pi/\Delta k$. Part (d) is reproduced from Reference [14].

$z = \pi/\Delta k$. At longer distances, $E^{(2\omega)}$ begins to shrink, returning to zero at $z = 2\pi/\Delta k$. In the quasi-phase-matched case (Figure 8.7b), the sign of the interaction reverses every $\Delta z = \pi/\Delta k$. This is done by reversing the sign of $d(z)$. The resultant $E^{(2\omega)}(z)$ thus keeps growing monotonically, albeit at a (spatial) rate smaller than in the case of ideal bulk phase matched (Figure 8.7c).

Quasi Phase Matching in Crystal Dielectric Waveguides

Quasi phase matching has been practiced almost exclusively in configurations involving dielectric waveguiding. The main reason is that in a single-mode waveguide, the small cross-sectional area ($\sim 10^{-8}$ cm^2) leads to very large intensities even at modest input power levels. This enables efficient conversion efficiencies with reasonable, say, <1 cm, crystal length. In addition, the lack of diffraction makes it possible to maintain the high intensity throughout the crystal length. Most of the work, to date, involves optical waveguides fabricated in $LiNbO_3$ [15]. The spatial modulation of the nonlinear coefficient is achieved by reversing locally the direction of the c axis of the crystal by means of periodically applied electric fields [16, 17] or diffusion of Ti through openings in a mask with a period $\Lambda = 2L_c$ [17]. This periodic reversal is caused by a periodic reversal of the direction of the crystal's permanent electric polarization by a periodically (spatially) reversing electric field or impurity diffusion.

8.6 THIRD-ORDER NONLINEAR OPTICAL PROCESSES

The third-order nonlinear polarization

$$P_i^{(3)} = 4\chi_{ijkl} E_j E_k E_l \tag{8.6-1}$$

is responsible for third-harmonic generation (THG), stimulated Raman scattering (SRS), stimulated Brillouin scattering (SBS), four-wave mixing (FWM), optical Kerr effect, and so on. In this section, we will treat the coupling of waves due to the third-order nonlinearity. Particularly, we will treat some of the important third-order processes that are relevant to optical communications.

Optical Kerr Effect

It was first discovered by J. Kerr in 1875 that a transparent liquid becomes doubly refracting (birefringent) when placed in a strong electric field. Because the effect is quadratic with respect to the applied electric field E, it is sometimes known as the quadratic electro-optical effect. The amount of birefringence due to the Kerr effect can be written

$$\Delta n = n_{\parallel} - n_{\perp} = \lambda K E^2 \tag{8.6-2}$$

where K is the Kerr constant and λ is the vacuum wavelength. In Equation (8.6-2), $n_{\parallel}$ is the refractive index for the optical wave polarized along the direction of the applied electric field, while $n_{\perp}$ is for the optical wave polarized perpendicular to the applied electric field.

According to Equation (8.6-1), the presence of an optical beam in a nonlinear medium can cause a change of the index of refraction. Let the field of an optical beam be written

$$E_j(t) = \mathrm{Re}(E_{0j}^{\omega} e^{i\omega t}) = \tfrac{1}{2}(E_{0j}^{\omega} e^{i\omega t} + \text{c.c.}) \quad (j = x, y, z) \tag{8.6-3}$$

where E_{0j}^{ω} is a constant. The third-order nonlinear polarization can be written

$$P_i^{(3)}(t) = 4\chi_{ijkl} \tfrac{1}{2}(E_{0j}^{\omega} e^{i\omega t} + \text{c.c.}) \tfrac{1}{2}(E_{0k}^{\omega} e^{i\omega t} + \text{c.c.}) \tfrac{1}{2}(E_{0l}^{\omega} e^{i\omega t} + \text{c.c.}) \tag{8.6-4}$$

where summations over repeated indices (j, k, l) are assumed.

We focus our attention on the amplitude of the nonlinear term that oscillates at the same frequency as the input optical beam ω. Ignoring all other terms, we obtain the amplitude of the nonlinear polarization that oscillates at frequency ω:

$$P_{\mathrm{NL}i}^{(\omega)} = \chi_{ijkl}\{E_{0j}^{\omega} E_{0k}^{\omega} E_{0l}^{\omega *} + E_{0j}^{\omega *} E_{0k}^{\omega} E_{0l}^{\omega} + E_{0j}^{\omega} E_{0k}^{\omega *} E_{0l}^{\omega}\} \tag{8.6-5}$$

where summations over repeated indices (j, k, l) are assumed. This nonlinear polarization oscillating at frequency ω is equivalent to a self-induced change in the dielectric susceptibility.

Consider a simple case of an optical beam polarized along the x axis in an isotropic medium with an amplitude E_{0x}^{ω}. The amplitude of nonlinear polarization at ω is written

$$P_{NLx}^{(\omega)} = 3\chi_{xxxx} E_{0x}^{\omega}(E_{0x}^{\omega} E_{0x}^{\omega *}) \tag{8.6-6}$$

This is equivalent to a change of the linear susceptibility $\Delta\chi$. We recall that $n^2 = 1 + \chi$. Thus the third-order nonlinear polarization leads to a change of the index of refraction of

$$\Delta n = n_2 I = \frac{\Delta\chi}{2n} = \frac{3\chi_{xxxx}}{2n\varepsilon_0} |E_{0x}|^2 = \frac{3c\mu_0\chi_{xxxx}}{n^2\varepsilon_0} I \tag{8.6-7}$$

where I is the optical beam intensity and n_2 is known as the Kerr coefficient. For silica fibers doped with germanium oxide, the Kerr coefficient is around (depending on the density of impurities)

$$n_2 \approx 3.0 \times 10^{-20}\ \mathrm{m^2/W} \tag{8.6-8}$$

Like most of the third-order nonlinear processes, the Kerr effect is relatively weak in most media. In optical fibers, the nonlinear Kerr effect can be significant due to the high optical intensity in the core and the long interaction lengths afforded by these waveguides. The third-order nonlinearity in optical fibers can lead to four-wave mixing, self-phase modulation, cross-phase modulation, and so on.

According to Equation (8.6-7), the Kerr coefficient for a beam of linearly polarized light is given by

$$n_2 = \frac{3c\mu_0\chi_{xxxx}}{n^2\varepsilon_0} = \frac{3}{cn^2\varepsilon_0^2}\chi_{xxxx} \tag{8.6-9}$$

For a general polarization state, the Kerr coefficient is slightly different from (8.6-9). Let the electric field of the optical wave with an arbitrary polarization state be written

$$\mathbf{E}(t) = \tfrac{1}{2}(\hat{\mathbf{x}} E_{0x}^{\omega} e^{i\omega t} + \hat{\mathbf{y}} E_{0y}^{\omega} e^{i\omega t} + \text{c.c.}) \tag{8.6-10}$$

where E_{0x} and E_{0y} are the components. A similar derivation leads to the following amplitudes of the nonlinear polarization at frequency ω (see also Section 14.2):

$$\begin{aligned} P_{NLx}^{(\omega)} &= \chi_{xxxx}\{(3|E_{0x}^{\omega}|^2 + 2|E_{0y}^{\omega}|^2)E_{0x}^{\omega} + E_{0y}^{\omega 2} E_{0x}^{\omega *}\} \\ P_{NLy}^{(\omega)} &= \chi_{xxxx}\{(3|E_{0y}^{\omega}|^2 + 2|E_{0x}^{\omega}|^2)E_{0y}^{\omega} + E_{0x}^{\omega 2} E_{0y}^{\omega *}\} \end{aligned} \tag{8.6-11}$$

where we have used the following symmetry relationships (see Section 14.2 and Problem 14.8):

$$\chi_{1111} = \chi_{1122} + \chi_{1212} + \chi_{1221} \quad \text{and} \quad \chi_{1111} = \chi_{2222} \tag{8.6-12}$$

We note that cross-phase modulation exists when both polarization components are present. In other words, the amplitude of the x component can induce a change of the refractive index for y-polarized light, and vice versa. Kerr coefficients for circularly polarized light and unpolarized light can still be defined. Using Equation (8.6-10), we can show that the Kerr coefficients are given by (see Problem 8.14)

$$n_{2,\text{unpolarized}} = \frac{5}{6} n_2 = \frac{5}{2cn^2\varepsilon_0^2}\chi_{xxxx} \tag{8.6-13}$$

and

$$n_{2,\text{circular}} = \frac{2}{3} n_2 = \frac{2}{cn^2\varepsilon_0^2}\chi_{xxxx} \tag{8.6-14}$$

TABLE 8.5 Kerr Coefficients of Some Selected Substances

Material	Wavelength (μm)	Index of Refraction	χ_{1111} (esu) (10^{-13})	χ_{1111} (MKS) (10^{-33})	n_2 (MKS) (10^{-18} m^2/W)
CCl_4	0.694	1.454	0.05	0.62	0.04
CS_2	1.06	1.594	4.65	57.50	2.89
	0.694	1.612	3.57	44.10	2.16
Lucite	1.06	1.49	0.11	1.34	0.08
Ruby	1.06	1.76	0.07	0.85	0.04
SF_6 glass	1.06	1.77	0.42	5.22	0.21
SiO_2	0.694	1.455	0.01	0.15	0.01
YAG	1.06	1.83	0.15	1.90	0.07
	0.694	1.829	0.06	0.74	0.03

Data adapted from Reference [18].

Table 8.5 lists the Kerr coefficients of selected nonlinear materials. Many of the early publications on third-order susceptibility are in units of esu. We have converted them into MKS units for convenience (see Problem 8.13). According to (8.6-11), a beam of elliptically polarized light may undergo a change of the state of polarization as the beam propagates in a Kerr medium.

Stimulated Raman Scattering

The Raman effect, discovered in 1928 by Sir Chandrasekhera Venkata Raman, is an inelastic scattering of light by elementary excitations of matter. When a photon interacts with an atomic system (e.g., molecule), it can be scattered in one of the three ways: (1) it can be elastically scattered and thus retain its incident photon energy (Rayleigh scattering), (2) it can be inelastically scattered by quasi-particle excitations (e.g., molecular vibration) of the atomic system, thereby either giving energy to the medium (Stokes scattering), or (3) it can be scattered by removing energy from it (anti-Stokes scattering). Thus the Raman effect involves a coupling between incident photons and the quasi-particle excitations such as phonons (lattice vibrations). In terms of energy-level diagrams, the Raman effect is often described as in Figure 8.8.

The Raman frequency shifts ($\omega_L - \omega_S$), in the range of 10^{12}–10^{13} Hz, are determined by different vibration frequencies of the molecules. Raman spectroscopy is an excellent probe to

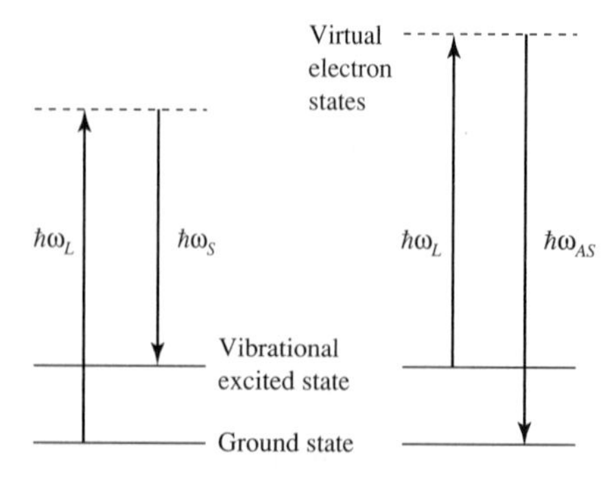

Figure 8.8 Schematic drawing of the energy diagram of Raman scattering. In Stokes scattering, the atomic system is excited from the ground state to an excited state (vibrational excited state). In anti-Stokes scattering, the atomic system is donating energy to the optical beam by decaying from an excited state to the ground state.

investigate the low-energy elementary excitations in materials as well as to characterize their structural, electronic, vibrational, and magnetic properties. As depicted in Figure 8.8, for each molecular vibration we observe two Raman lines. The Stokes and anti-Stokes lines are connected with transitions from the ground state to the first excited vibrational state and vice versa. Since the excited states are populated only slightly at room temperature, the anti-Stokes intensities are relatively small compared with the Stokes intensities. The lifetime of the molecular vibration is often deduced from the width of the Raman line.

Like the Kerr effect, Raman scattering also is relatively weak in most media. In optical fibers, however, the nonlinear effect due to Raman scattering can be significant due to the high optical intensity of a laser beam in the core and the long interaction lengths (in the range of 100 km) afforded by these waveguides. Both stimulated Raman scattering (SRS) and stimulated Brillouin scattering (SBS) occur when the light launched into the fiber exceeds a threshold power level for each process. Under the conditions of stimulated scattering, optical power is more efficiently converted from the input pump wave to the scattered Stokes wave.

The scattered wave is frequency shifted from the pump and in the case of SBS propagates in the opposite direction. This means that the amount of optical power leaving the far end of the fiber no longer increases linearly with the input power. The maximum launch power becomes clamped and the excess is simply reflected back out of the fiber. For long distance or highly branched fiber links, it is important that as much power as possible can be launched into the fiber to compensate for attenuation and power splitting. Limits on the maximum output power due to SBS must therefore be avoided.

In SRS, the Stokes wave can be shifted from the pump wave by typically 10–100 nm and continues to propagate forward along the fiber with the pump wave. If the pump is actually one channel of a multiwavelength WDM communications system, then its Stokes wave may overlap with other channels at longer wavelengths—leading to crosstalk and Raman amplification. Raman amplification causes shorter wavelength channels to experience power depletion and act as a pump for the amplification of longer wavelength channels. This can skew the power distribution among the WDM channels—reducing the signal-to-noise ratio of the lowest frequency channels and introducing crosstalk on the high-frequency channels. Both of these effects can lower the information-carrying capacity of the optical system. On the other hand, with a proper spacing of the channels, SRS can be employed for the amplification of signal waves. This subject is taken up in Chapter 17.

We now consider the energy coupling in SRS. For simplicity, we assume that both the laser and the Stokes scattering are polarized along the x axis. The electric field can be written

$$E_x(t) = \mathrm{Re}(E_L e^{i(\omega_L t - k_L z)} + E_S e^{i(\omega_S t - k_S z)}) = \tfrac{1}{2}(E_L e^{i(\omega_L t - k_L z)} + E_S e^{i(\omega_S t - k_S z)} + \mathrm{c.c.}) \tag{8.6-15}$$

where E_L and E_S are the field amplitudes, ω_L and ω_S are frequencies, and k_L and k_S are wavenumbers. Using Equation (8.6-1), the amplitudes of the nonlinear polarization oscillating at frequencies ω_L and ω_S can be written

$$P_{\mathrm{NL}}^{(\omega_L)} = E_L(6\chi_{R1}^{(3)} E_S E_S^* + 3\chi_1^{(3)} E_L E_L^*) \tag{8.6-16}$$

$$P_{\mathrm{NL}}^{(\omega_S)} = E_S(3\chi_2^{(3)} E_S E_S^* + 6\chi_{R2}^{(3)} E_L E_L^*) \tag{8.6-17}$$

where we use different effective third-order nonlinear susceptibilities to account for the possible dependence on the frequencies. By examining the above equations, we note that the $\chi_1^{(3)}$ and $\chi_2^{(3)}$ terms in the nonlinear polarization only act to modify the dielectric constant (or index of refraction). They are responsible for the optical Kerr effect discussed earlier. These terms have no direct effect on stimulated Raman scattering and thus will be neglected in the following discussion. The $\chi_{R1}^{(3)}$ and $\chi_{R2}^{(3)}$ terms in the nonlinear polarization, on the other hand, effectively couple E_L and E_S and can cause energy exchange between these two waves. These two nonlinear susceptibilities are responsible for the stimulated Raman process and are often called Raman susceptibilities. Using quantum mechanical theory, it is possible to show that [19]

$$\chi_{R1}^{(3)} = \chi_{R2}^{*(3)} \tag{8.6-18}$$

In other words, these two Raman susceptibilities are complex conjugates of each other.

Following the procedure leading to the coupled equations discussed in Section 8.3, we obtain

$$\frac{dE_L}{dz} = -\frac{1}{2}\alpha_L E_L + i\frac{6\mu_0\omega_L^2}{k_L}\chi_{R1}^{(3)}|E_S|^2 E_L \tag{8.6-19}$$

$$\frac{dE_S}{dz} = -\frac{1}{2}\alpha_S E_S + i\frac{6\mu_0\omega_S^2}{k_S}\chi_{R2}^{(3)}|E_L|^2 E_S \tag{8.6-20}$$

where we have added terms (α_L and α_S) to account for the linear attenuation due to absorption and scattering. These are the equations that govern the energy coupling between the laser (pump) beam and the Stokes beam.

To understand the Raman gain, first we consider a simplified problem in which the laser amplitude E_L is assumed constant with z. This applies when the energy depletion from the laser field (at ω_L) by the nonlinear interaction is small compared to the laser beam energy. This makes it possible to integrate Equation (8.6-20) for the Stokes field (at ω_S). We obtain, according to Equation (8.6-20),

$$|E_S|^2(z) = |E_S|^2(0)\exp(G_R z - \alpha_S z) \tag{8.6-21}$$

where the Raman gain coefficient G_R is given by

$$G_R = \frac{12\mu_0\omega_S^2}{k_S}|E_L|^2\,\mathrm{Im}(\chi_{R1}^{(3)}) = -\frac{12\mu_0\omega_S^2}{k_S}|E_L|^2\,\mathrm{Im}(\chi_{R2}^{(3)}) \tag{8.6-22}$$

We note that the Stokes beam is amplified with an exponential gain coefficient G_R that is proportional to the laser beam intensity. Thus a Raman medium pumped by a laser beam can be employed as an optical amplifier for the spectral regime around the Stokes line.

In stimulated Raman scattering (SRS), there's no input Stokes beam. The laser beam is focused onto a Raman medium in which the laser beam is trapped in the form of a thin filament with a diameter of a few microns. This self-induced trapping is due to a combination of the optical Kerr effect described earlier and the Gaussian intensity profile of the laser beam. This effect gives rise to an extremely large laser beam intensity and hence a very large Raman gain coefficient at the Stokes line. This gain is so large that any electromagnetic radiation at the Stokes line is amplified significantly in a single pass, leading to stimulated Raman scattering.

As we indicated, the Raman scattering involves the interaction of the input laser beam with the molecular vibration. The scattering process creates the Stokes wave at a lower frequency. Once the Stokes wave is generated, it beats with the input laser wave at a beat frequency, $\omega_L - \omega_S$, that is automatically equal to that of the molecular vibration frequency. Therefore a positive feedback loop is created as the beating term acts as a driving force that further increases the amplitude of the molecular vibration, which in turn increases the amplitude of the Raman scattered wave.

In nonlinear optical interaction, where photons of different frequencies are involved, it is convenient to describe the energy exchange in terms of photon flux, which is a measure of the number of photons per unit area per unit time. The photon flux for a beam of light with an electric field amplitude E is defined as

$$\phi = \frac{I}{\hbar\omega} = \frac{I}{\hbar\omega}\frac{k}{2\mu_0\omega}|E|^2 \tag{8.6-23}$$

where I is the beam intensity. Using photon fluxes, the coupled equations can be written

$$\frac{d}{dz}\phi_L = -\alpha_L\phi_L - G_R'\phi_L\phi_S \tag{8.6-24}$$

$$\frac{d}{dz}\phi_S = -\alpha_S\phi_S + G_R'\phi_L\phi_S \tag{8.6-25}$$

where the Raman gain coefficient for photon fluxes is given by

$$G_R' = \frac{12\hbar\mu_0^2\omega_L^2\omega_S^2}{k_L k_S}\,\mathrm{Im}(\chi_{R1}^{(3)}) = -\frac{12\hbar\mu_0^2\omega_L^2\omega_S^2}{k_L k_S}\,\mathrm{Im}(\chi_{R2}^{(3)}) \tag{8.6-26}$$

We note that Equations (8.6-24) and (8.6-25) are consistent with the conservation of photon flux in the event when the linear attenuation coefficients vanish ($\alpha_L = \alpha_S = 0$). In other words, one laser photon (at ω_L) is converted into one Stokes photon (at ω_S).

In the event when the difference between the two linear attenuation coefficients can be neglected (i.e., $\alpha_L = \alpha_S = \alpha$), exact solutions of the coupled equations (8.6-24) and (8.6-25) are available. They are written

$$\phi_L(z) = \phi_L(0)\frac{1 + m^{-1}}{1 + m^{-1}\exp\left(\frac{\gamma}{\alpha}(1 - e^{-\alpha z})\right)}\exp(-\alpha z) \tag{8.6-27}$$

$$\phi_S(z) = \phi_S(0)\frac{1 + m}{1 + m\exp\left(-\frac{\gamma}{\alpha}(1 - e^{-\alpha z})\right)}\exp(-\alpha z) \tag{8.6-28}$$

where the exponential gain coefficient γ is given by

$$\gamma = G_R'[\phi_S(0) + \phi_L(0)] \tag{8.6-29}$$

and m is the photon flux ratio at input ($z = 0$)

$$m = \phi_L(0)/\phi_S(0) \tag{8.6-30}$$

In the event when the linear attenuation coefficient becomes zero ($\alpha = 0$), the solutions reduce to

$$\phi_L(z) = \phi_L(0)\frac{1 + m^{-1}}{1 + m^{-1}\exp(\gamma z)} \tag{8.6-31}$$

$$\phi_S(z) = \phi_S(0)\frac{1 + m}{1 + m\exp(-\gamma z)} \tag{8.6-32}$$

In typical stimulated Raman scattering, the initial Stokes photon flux is very small, $\phi_S(0) \ll \phi_L(0)$ (or $1 \ll m$). According to Equation (8.6-32), we note that the Stokes photon flux increases exponential initially and then reaches saturation as the laser beam is depleted.

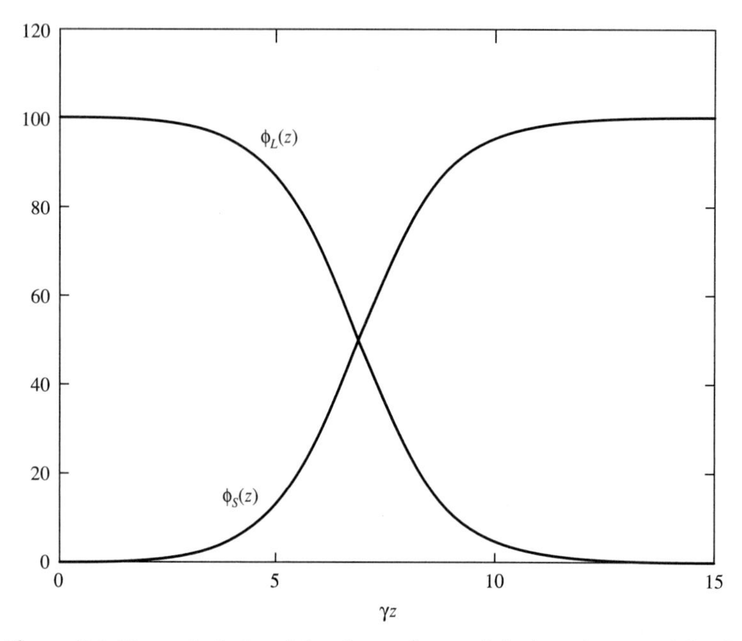

Figure 8.9 Theoretical plot of the photon fluxes of the laser beam and the Stokes beam as a function of interaction distance γz for the case when $\alpha = 0$.

Figure 8.9 shows the energy exchange in terms of the photon flux as a function of interaction distance.

Based on the quantum mechanical energy diagrams shown in Figure 8.8, the peak of Raman emission occurs when the frequency shift equals the molecular vibration frequency. In other words,

$$\omega_L - \omega_S = \omega_v \tag{8.6-33}$$

where ω_v is the molecular vibration frequency.

We conclude by noting that the nonlinear polarization (8.6-17) at the Stokes line (with frequency ω_S) has a space–time dependence of $\exp(i\omega_S t - ik_S z)$, which is exactly the same as the Stokes field. This explains why the "phase-matching" conditions that had to be satisfied in parametric interactions are automatically satisfied in stimulated Raman scattering. Table 8.6 provides a list of Raman properties for selected materials.

TABLE 8.6 Raman Properties for Selected Materials

Material	Raman Shift (THz)	Linewidth (GHz)	Raman Gain (10^{-3} cm/MW)
Liquid oxygen, O_2	46.6	3.5	14.5 ± 4
Liquid nitrogen, N_2	69.8	2.01	17 ± 5
Benzene, C_6H_6	29.8	64.5	2.8
CS_2	19.7	15	24
Nitrobenzene	40.4	198	2.1
$LiNbO_3$	7.74	210	28.7
$LiTaO_3$	6.45	360	10
SiO_2	14		0.8

From Reference [19].

EXAMPLE: RAMAN GAIN IN SiO_2

Consider a laser beam power of 10 mW in a single-mode fiber with an effective core area of 100 μm^2. The intensity is thus $I = 10^8$ W/m^2. Using the gain coefficient of 0.8×10^{-3} cm/MW, we obtain a Raman gain coefficient of 8×10^{-4} m^{-1}.

8.7 STIMULATED BRILLOUIN SCATTERING

In Raman scattering, incoming photons are interacting with the molecules, causing molecular vibrations. The scattered photons are downshifted in frequency by an amount that equals the internal vibration frequency of the molecules. The frequency shift is in the range of 10^{12}–10^{13} Hz. Similar inelastic scatterings can occur in which the frequency shift is caused by the physical motion of the atoms or molecules. For example, spontaneous Brillouin scattering involves the interaction with thermally excited acoustic waves. Such a scattering of light was considered as early as 1922 by L. Brillouin. The frequency shift in Brillouin scattering is in the range of 10^{10} Hz.

Similar to SRS, stimulated Brillouin scattering (SBS) occurs when the pump laser beam is intense enough. The theory of the nonlinear interaction in SBS follows the formalism of SRS given in Section 8.6. In general, stimulated Brillouin scattering can occur in all directions. Here we consider the most important case of contradirectional Brillouin emission. Let the electric field (assume x-polarized) be written

$$E_x(t) = \mathrm{Re}(E_L e^{i(\omega_L t - k_L z)} + E_B e^{i(\omega_B t + k_B z)}) = \tfrac{1}{2}(E_L e^{i(\omega_L t - k_L z)} + E_B e^{i(\omega_B t + k_B z)} + \text{c.c.}) \quad (8.7\text{-}1)$$

where E_L and E_B are the field amplitudes, ω_L and ω_B are frequencies, and k_L and k_B are wavenumbers. Note that the Brillouin beam is propagating backward. Using Equation (8.6-1), the amplitudes of the nonlinear polarization oscillating at frequencies ω_L and ω_B can be written

$$P_{\mathrm{NL}}^{(\omega_L)} = E_L(6\chi_B^{(3)} E_B E_B^* + 3\chi_1^{(3)} E_L E_L^*) \quad (8.7\text{-}2)$$

$$P_{\mathrm{NL}}^{(\omega_B)} = E_B(3\chi_2^{(3)} E_B E_B^* + 6\chi_B^{(3)} E_L E_L^*) \quad (8.7\text{-}3)$$

where we use different effective third-order nonlinear susceptibilities to account for the possible dependence on the frequencies. By examining the above equations, we note that the $\chi_1^{(3)}$ and $\chi_2^{(3)}$ terms in the nonlinear polarization only act to modify the dielectric constant (or index of refraction). They are responsible for the optical Kerr effect discussed earlier. These terms have no direct effect on stimulated Brillouin scattering and thus will be neglected in the following discussion. The $\chi_B^{(3)}$ terms in the nonlinear polarization, on the other hand, effectively couple E_L and E_B and can cause energy exchange between these two waves. These two nonlinear susceptibilities are responsible for the stimulated Brillouin process and are often called Brillouin susceptibilities.

Following the procedure leading to the coupled equations of SRS, we obtain

$$\frac{dE_L}{dz} = -\frac{1}{2}\alpha E_L + i\frac{6\mu_0\omega_L^2}{k_L}\chi_B^{(3)}|E_B|^2 E_L \quad (8.7\text{-}4)$$

$$\frac{dE_B^*}{dz} = +\frac{1}{2}\alpha E_B^* + i\frac{6\mu_0\omega_B^2}{k_B}\chi_B^{(3)}|E_L|^2 E_B^* \quad (8.7\text{-}5)$$

where α accounts for the linear attenuation due to absorption and other scattering loss. These are the coupled equations governing the energy exchange between the laser beam and the Brillouin beam.

To understand the Brillouin gain, first we consider a simplified problem in which the laser amplitude E_L is assumed constant with z. This applies when the energy depletion from the laser field (at ω_L) by the nonlinear interaction is small compared to the laser beam energy. This makes it possible to integrate Equation (8.7-5) for the Brillouin field (at ω_B). We obtain, according to Equation (8.7-5),

$$|E_B|^2(z) = |E_B|^2(L)\exp[(G_B - \alpha)(L - z)] \tag{8.7-6}$$

or

$$\frac{|E_B|^2(0)}{|E_B|^2(L)} = \exp[(G_B L - \alpha L)]$$

where the Brillouin gain coefficient G_B is given by

$$G_B = \frac{12\mu_0\omega_B^2}{k_B}|E_L|^2\,\mathrm{Im}(\chi_B^{(3)}) \tag{8.7-7}$$

We note that the Brillouin beam is amplified with an exponential gain coefficient G_B, which is proportional to the laser beam intensity. Thus a Brillouin medium pumped by a laser beam can be employed as an optical amplifier for the spectral regime around the Brillouin line. The gain bandwidth, however, is relatively narrow.

In practice, it is legitimate to ignore the small difference between ω_L and ω_B. From the coupled equations for the field amplitudes, we obtain

$$\frac{d}{dz}|E_L|^2 = -\alpha|E_L|^2 - G_B|E_L|^2|E_B|^2 \tag{8.7-8}$$

$$\frac{d}{dz}|E_B|^2 = +\alpha|E_B|^2 - G_B|E_L|^2|E_B|^2 \tag{8.7-9}$$

or equivalently, in terms of beam intensities,

$$\frac{d}{dz}I_L = -\alpha I_L - gI_L I_B \tag{8.7-10}$$

$$\frac{d}{dz}I_B = +\alpha I_B - gI_L I_B \tag{8.7-11}$$

where g is the Brillouin gain coefficient in units of cm/MW.

The coupled equations for the beam intensities can be solved analytically in the absence of linear attenuation ($\alpha = 0$). The solutions can be written

$$I_L(z) = \frac{c_0}{1 - \rho\exp(-gc_0 z)} \tag{8.7-12}$$

$$I_B(z) = \frac{c_0}{1 - \rho\exp(-gc_0 z)}\rho\exp(-gc_0 z) \tag{8.7-13}$$

where c_0 and ρ are constants related to the incident laser beam intensity $I_L(0)$ at $z = 0$, and the noise Brillouin seed intensity $I_B(0)$ at $z = L$,

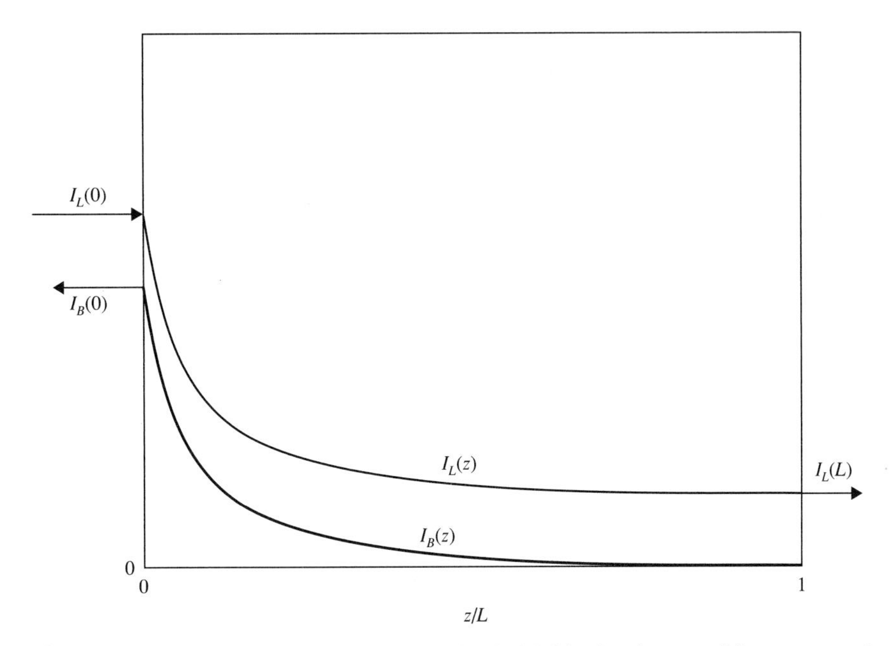

Figure 8.10 Theoretical plot of beam intensity in SBS in the absence of linear attenuation. Note that the SBS beam intensity grows from a very low noise intensity $I_B(L)$ at $z = L$ to a significant value of $I_B(0)$ at $z = 0$.

$$\frac{I_B(L)}{c_0 + I_B(L)} = \frac{I_L(0) - c_0}{I_L(0)} \exp(-gc_0L) \tag{8.7-14}$$

$$\rho = \frac{I_L(0) - c_0}{I_L(0)} = \frac{I_B(0)}{I_L(0)} \tag{8.7-15}$$

The parameter ρ is actually the fraction of input beam power converted into the SBS beam. Figure 8.10 shows the beam intensities as a function of interaction distance z.

For transmission in optical networks, we are particularly interested in the relationship between the incident intensity (or power) $I_L(0)$ and the transmitted intensity (or power) $I_L(L)$. Using Equations (8.7-12)–(8.7-15), we obtain

$$I_L(0) = I_L(L) \frac{I_L(L) - I_B(L)}{I_L(L) - I_B(L) \exp\{g[I_L(L) - I_B(L)]L\}} \tag{8.7-16}$$

where $I_B(L)$ is an infinitesimal SBS seed intensity due to noise at $z = L$.

Although it may be difficult to invert this equation to obtain an expression of the output intensity as a function of the input intensity, it is still quite useful. Figures 8.11 and 8.12 are plots of the input–output relationship. We note that as the input intensity increases, the nonlinear SBS interaction also increases. The nonlinear interaction leads to a stronger SBS beam, and thus the transmitted beam is a nonlinear function of the input intensity. At low intensities, the transmitted beam initially increases linearly with the input intensity. At high intensities (e.g., $I_L(0) > 15$ kW/cm^2), the transmitted intensity increases only logarithmically (almost flat). In other words, the transmitted intensity does not increase proportionally with the input intensity. SBS is a nonlinear optical process. It is interesting to examine the transmitted laser beam intensity $I_L(L)$ as a function of the input laser beam intensity $I_L(0)$. This is particularly important in single-mode fibers, where the interaction length can be many kilometers, and the cumulative nonlinear optical effect can be quite strong.

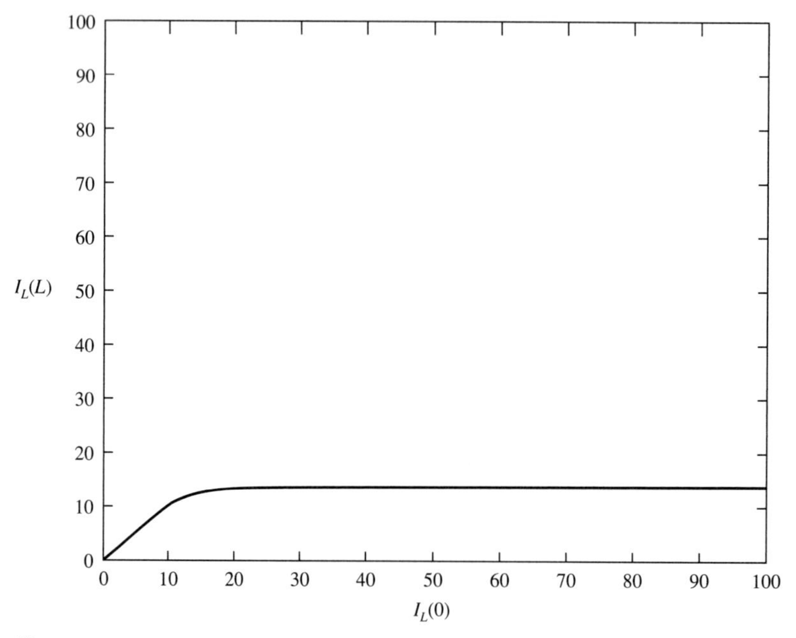

Figure 8.11 Theoretical plot of transmitted intensity versus input intensity in the SBS process (in units of kW/cm^2). We assume $gL = 0.00135$ cm^2/W and a seed intensity of $I_B(L) = 0.0001$ kW/cm^2.

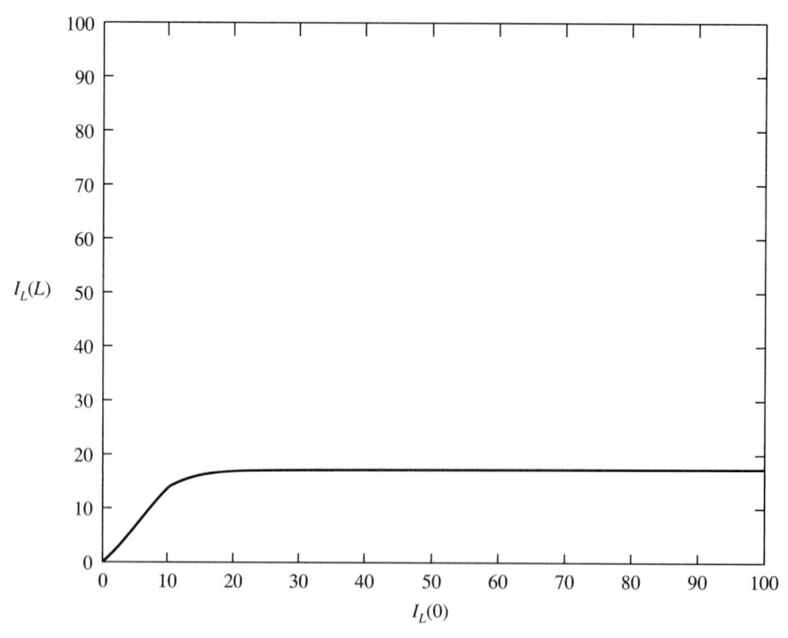

Figure 8.12 Transmitted intensity $I_L(L)$ versus input intensity $I_L(0)$ in SBS process (in units of kW/cm^2). We assume $gL = 0.00135$ cm^2/W and a seed intensity of $I_B(L) = 0.000001$ kW/cm^2.

As we indicated, Brillouin scattering involves interaction with an acoustic wave in the nonlinear medium. The scattering process must conserve both energy and momentum, so the frequencies and wavevectors of the three waves are related by

$$\begin{aligned} \Omega_A &= \omega_L - \omega_B \\ K_A &= k_L - (-k_B) \end{aligned} \tag{8.7-17}$$

where we note that the Brillouin wave is propagating backward, ω_L and ω_B are frequencies, and k_L and k_B are the wavenumbers of the laser and Brillouin waves, respectively. An electrostrictive force $\propto \chi_B^{(3)} E_L E_B^*$ is created by the simultaneous presence of the laser beam and the Brillouin wave. This force drives the acoustic wave at frequency ($\Omega_A = \omega_L - \omega_B$). The acoustic wave modulates the dielectric constant ε and creates a sideband (the Brillouin wave) at $\omega_L - (\omega_L - \omega_B) = \omega_B$. A positive feedback loop is created as the electrostrictive force increases the amplitude of the acoustic wave, which in turn increases the amplitude of the Brillouin wave. The frequency Ω_A and wavevector K_A of the acoustic wave satisfy the dispersion relation

$$\Omega_A = v_A K_A \tag{8.7-18}$$

where v_A is the velocity of acoustic wave in the medium. Since k_L and k_B are almost equal, the Brillouin frequency shift can be written

$$\Omega_A = 2 v_A k_L = \frac{4\pi n}{\lambda_L} v_A \tag{8.7-19}$$

or equivalently,

$$f_A = \frac{2n}{\lambda_L} v_A \quad \text{(in Hz)} \tag{8.7-20}$$

Table 8.7 provides a list of Brillouin properties for selected materials, including gain, frequency shift, and bandwidth.

EXAMPLE: SBS GAIN IN SiO_2

According to Table 8.7, the Brillouin gain coefficient is 0.0045 cm/MW. Assuming 10 mW of power in a core of 100 μm², the beam intensity is 10^8 W/m². The Brillouin gain is thus given by $g_B I_L = 0.0045$ cm/MW $\times 10^8$ W/m² $= 4.5 \times 10^{-3}$ m^{-1}.

TABLE 8.7 Brillouin Gain, Frequency Shift, and Gain Bandwidth for Selected Materials

Material	Acoustic Velocity (km/s)	Frequency Shift (GHz)	Bandwidth (MHz)	Gain (cm/MW)
CS_2		5.85	52.3	0.13
Acetone		4.6	224	0.02
CCl_4		4.39	520	0.006
Benzene		6.47	289	0.018
Water	1.5	5.69	317	0.0048
Optical glasses		11–16	10–106	0.004–0.025
SiO_2		23	26	0.0045
Silica	5.95	11.2		
GaAs	5.15	23.3		
$LiNbO_3$	6.57	18.7		
$LiTaO_3$	6.19	17.6		
α-Al_2O_3	11.15	25.3		

8.8 FOUR-WAVE MIXING AND PHASE CONJUGATION

We have so far discussed several special cases of third-order nonlinear optical phenomena, including the Kerr effect, SRS, and SBS. Generally, a third-order nonlinear process involves the participation of four optical waves. Such a general third-order process can be viewed as the generation of a fourth optical wave with an input of three optical waves. Let the frequencies and wavevectors of the three input waves be $(\omega_1, \mathbf{k}_1)$, $(\omega_2, \mathbf{k}_2)$, and $(\omega_3, \mathbf{k}_3)$; then the frequencies and wavevectors of the fourth optical wave $(\omega_4, \mathbf{k}_4)$ can be written

$$\omega_4 = \omega_1 \pm \omega_2 \pm \omega_3 \tag{8.8-1}$$

$$\mathbf{k}_4 = \mathbf{k}_1 \pm \mathbf{k}_2 \pm \mathbf{k}_3 \tag{8.8-2}$$

These two equations can be viewed as the conservation of energy and momentum. Since the wavevector $(\mathbf{k}_4)$ is related to the frequency (ω_4) by the dispersion relationship, these two equations may not be satisfied simultaneously. Thus nonlinear optical four-wave mixing occurs only in some particular input configurations when the above two equations are satisfied simultaneously. In this section, we will discuss a very interesting and useful scheme of nonlinear optical four-wave mixings in which the phase-matching condition is satisfied. In particular, we will be discussing the degenerate (or near-degenerate) four-wave mixing process, which has important applications, including phase conjugation and spectral reversal.

We now consider a special case of four-wave mixing in which three input optical waves with frequencies ω_1, ω_2, and ω_3 mix and produce a fourth optical wave at $\omega_1 + \omega_2 - \omega_3$. To be specific, we assume that a nonlinear optical medium is irradiated simultaneously by three input optical fields:

$$\begin{aligned} \mathbf{E}_1(\mathbf{r}, t) &= \tfrac{1}{2}\mathbf{A}_1(\mathbf{r})e^{i(\omega_1 t - \mathbf{k}_1\cdot\mathbf{r})} + \text{c.c.} \\ \mathbf{E}_2(\mathbf{r}, t) &= \tfrac{1}{2}\mathbf{A}_2(\mathbf{r})e^{i(\omega_2 t - \mathbf{k}_2\cdot\mathbf{r})} + \text{c.c.} \\ \mathbf{E}_3(\mathbf{r}, t) &= \tfrac{1}{2}\mathbf{A}_3(\mathbf{r})e^{i(\omega_3 t - \mathbf{k}_3\cdot\mathbf{r})} + \text{c.c.} \end{aligned} \tag{8.8-3}$$

where $\mathbf{A}_1$, $\mathbf{A}_2$, and $\mathbf{A}_3$ are amplitudes of the waves. According to the third-order term in Equation (8.1-1), a nonlinear (NL) polarization oscillating at $\omega_1 + \omega_2 - \omega_3$ is generated:

$$\begin{aligned} P_i^{(\text{NL})}(\mathbf{r}, t) &= 3\chi_{ijkl}^{(3)} A_{1j} A_{2k} A_{3l}^*(\mathbf{r}) e^{i[(\omega_1+\omega_2-\omega_3)t - (\mathbf{k}_1+\mathbf{k}_2-\mathbf{k}_3)\cdot\mathbf{r}]} + \text{c.c.} \\ &= \text{Re}[6\chi_{ijkl}^{(3)} A_{1j} A_{2k} A_{3l}^*(\mathbf{r}) e^{i[(\omega_1+\omega_2-\omega_3)t - (\mathbf{k}_1+\mathbf{k}_2-\mathbf{k}_3)\cdot\mathbf{r}]}] \end{aligned} \tag{8.8-4}$$

where i, j, k, l refer to Cartesian coordinates, and summations over repeated indices are assumed. $\chi_{ijkl}^{(3)}$ is a fourth-rank tensor characteristic of the medium [18] that depends on the input frequencies ω_1, ω_2, ω_3. If we apply to (8.8-4) the argument leading to Equation (8.2-8), we can convince ourselves that, unlike the phenomenon of second-harmonic generation (SHG), the third-order optical effects considered here exist in all media, including centrosymmetric crystals and isotropic media. The form of χ_{ijkl} but not its magnitude is determined by the symmetry properties of the media that are discussed and tabulated in References [18, 19]. The nonlinear polarization is actually a volume distribution of oscillating dipoles, which can radiate a new wave at the frequency $\omega_4 = \omega_1 + \omega_2 - \omega_3$, provided the wavevector of the new wave satisfies the momentum conservation condition $\mathbf{k}_4 = \mathbf{k}_1 + \mathbf{k}_2 - \mathbf{k}_3$.

Here we consider a special case of interest. In this case, all frequencies of the three input waves are identical. In other words, $\omega_1 = \omega_2 = \omega_3 \equiv \omega_0$ and $\omega_4 = \omega_0$. Assuming the medium

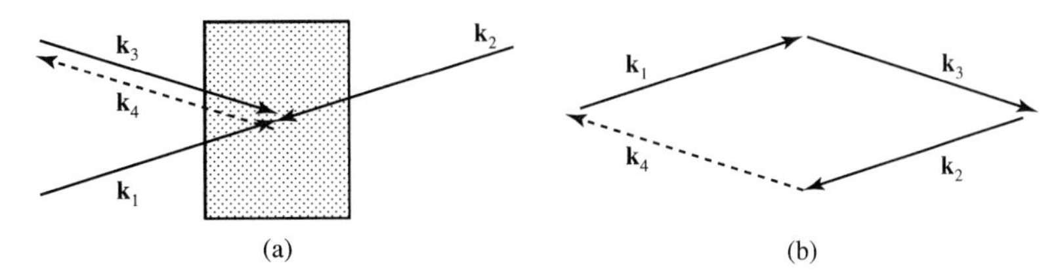

Figure 8.13 (a) Schematic drawing of degenerate four-wave mixing in an isotropic nonlinear medium. (b) Vector diagram of the conservation of momentum.

is optically isotropic, then all four wavevectors are of equal length. Thus conservation of momentum requires that the wavevectors form an equilateral parallelogram (see Figure 8.13b). In a typical degenerate four-wave mixing (shown in Figure 8.13a), two of the input beams ($\mathbf{k}_1$, $\mathbf{k}_2$) are counterpropagating, so that $\mathbf{k}_1 + \mathbf{k}_2 = 0$. A third input wave arrives at the interaction region. The fourth wave is emitted in a direction opposite to the third one. If we view the third wave as a signal wave, then the newly generated wave (the fourth wave) is always counterpropagating with respect to the signal wave. In other words, the generated wave (ω_4, $\mathbf{k}_4$) is retracing the path of the third input wave (ω_3, $\mathbf{k}_3$). The retroreflection nature of the generated wave relative to the third input wave is responsible for the phenomenon of phase conjugation.

A list of the nonlinear coefficients of some optical materials is included in Table 8.8. We also include in the table a listing of the Kerr constant n_2 [18]. This constant, as discussed in Section 8.6, describes the dependence of the index of refraction of an isotropic medium on the optical intensity I according to

$$n = n_0 + n_2 I \tag{8.8-5}$$

TABLE 8.8 Nonlinear Optical Coefficient $\chi^{(3)}_{ijkl}(-\omega, \omega, \omega, -\omega)$ of Selected Materials

Material	λ (μm)	Index	$\chi^{(3)}_{ijkl}$ (esu)	$\chi^{(3)}_{ijkl}$ (MKS)	n_2 (m²/W)
CS_2	1.06	1.594	$\chi_{1111} = 4.65 \times 10^{-13}$	$\chi_{1111} = 5.7 \times 10^{-32}$	2.9×10^{-18}
	0.694	1.612	$\chi_{1221} = 3.6 \times 10^{-13}$	$\chi_{1221} = 4.4 \times 10^{-32}$	
YAG	1.06	1.83	$\chi_{1111} = 0.15 \times 10^{-13}$	$\chi_{1111} = 19 \times 10^{-34}$	7.2×10^{-20}
	0.694	1.829	$\chi_{1221} = 0.06 \times 10^{-13}$	$\chi_{1221} = 7.4 \times 10^{-34}$	
CCl_4	0.694	1.454	$\chi_{1221} = 0.05 \times 10^{-13}$	$\chi_{1221} = 6.2 \times 10^{-34}$	
Fused silica	0.694	1.455	$\chi_{1111} = 4 \times 10^{-15}$	$\chi_{1111} = 5 \times 10^{-34}$	3×10^{-20}
			$\chi_{1221} = 1.2 \times 10^{-15}$	$\chi_{1221} = 1.5 \times 10^{-34}$	
Ruby	1.06	1.76	$\chi_{1111} = 6.9 \times 10^{-15}$	$\chi_{1111} = 8.5 \times 10^{-34}$	3.5×10^{-20}
SF_6 glass	1.06	1.77	$\chi_{1111} = 4.2 \times 10^{-14}$	$\chi_{1111} = 5.2 \times 10^{-33}$	2.1×10^{-19}
Lucite	1.06	1.49	$\chi_{1111} = 1.1 \times 10^{-14}$	$\chi_{1111} = 1.3 \times 10^{-33}$	7.7×10^{-20}
2-Methyl-4-nitroaniline (MNA)		1.8	$\chi_{1111} = 1.19 \times 10^{-11}$	$\chi_{1111} = 1.5 \times 10^{-30}$	5.8×10^{-17}
PTS polydiacetylene		1.88	$\chi_{1111} = 4.0 \times 10^{-11}$	$\chi_{1111} = 4.9 \times 10^{-30}$	1.8×10^{-16}

From Reference [18].

It follows directly from Equation (8.6-1) that (see also Equation (8.6-7))

$$n_2 = \frac{3}{n_0^2 \varepsilon_0} \sqrt{\frac{\mu}{\varepsilon_0}} \chi_{1111}(\text{MKS}) = \frac{3}{c n_0^2 \varepsilon_0^2} \chi_{1111}(\text{MKS}) \tag{8.8-6}$$

where the last equality holds for nonmagnetic materials. The relatively large value of χ_{1111} in MNA is due to a large charge separation (on the order of 30 Å) and hence large induced dipoles that can obtain in certain organic molecules.

Many of the published results on third-order nonlinear coefficients are expressed in units of esu. Based on the definition, these two units can be converted as follows:

$$\chi_{ijkl}^{(3)}(\text{MKS}) = \frac{4\pi\varepsilon_0}{9 \times 10^8} \chi_{ijkl}^{(3)}(\text{esu}) \quad (\text{based on } P_i = \chi_{ijkl}^{(3)} E_j E_k E_l, \text{ in this book}) \tag{8.8-7}$$

Some authors define $P_i = \varepsilon_0 \chi_{ijkl}^{(3)} E_j E_k E_l$. In this case, the nonlinear coefficients $\chi_{ijkl}^{(3)}$ are in units of m^2/V^2 and the conversion formula is as follows:

$$\varepsilon_0 \chi_{ijkl}^{(3)}(\text{m}^2/\text{V}^2) = \frac{4\pi\varepsilon_0}{9 \times 10^8} \chi_{ijkl}^{(3)}(\text{esu}) \quad (\text{based on the definition of } P_i = \varepsilon_0 \chi_{ijkl}^{(3)} E_j E_k E_l) \tag{8.8-8}$$

Coupled-Mode Formulation of Degenerate Four-Wave Mixing for Phase Conjugation

Referring to Figure 8.14, we consider the four-wave mixing process in a nonlinear medium between $z = 0$ and $z = L$. We will find that the result is a new wave generated in and radiated by the nonlinear medium. For traditional reasons, waves 1 and 2 are called pump beams, and wave 4 is an input beam. The new wave generated by the nonlinear polarization is labeled as wave 3. Let the electric fields of the waves be written

$$\begin{aligned}
\mathbf{E}_1(\mathbf{r}, t) &= \tfrac{1}{2}\mathbf{A}_1(\mathbf{r})e^{i(\omega t - \mathbf{k}_1 \cdot \mathbf{r})} + \text{c.c.} \\
\mathbf{E}_2(\mathbf{r}, t) &= \tfrac{1}{2}\mathbf{A}_2(\mathbf{r})e^{i(\omega t - \mathbf{k}_2 \cdot \mathbf{r})} + \text{c.c.} \\
\mathbf{E}_3(\mathbf{r}, t) &= \tfrac{1}{2}\mathbf{A}_3(z)e^{i(\omega t + kz)} + \text{c.c.} \\
\mathbf{E}_4(\mathbf{r}, t) &= \tfrac{1}{2}\mathbf{A}_4(z)e^{i(\omega t - kz)} + \text{c.c.} \\
k^2 &\equiv \omega^2 \mu \varepsilon
\end{aligned} \tag{8.8-9}$$

Waves 1 and 2 propagate along the directions $\mathbf{k}_1$ and $\mathbf{k}_2$, respectively. In the analysis that follows, these two waves correspond to the pump beams and their amplitudes $|A_1|$ and $|A_2|$

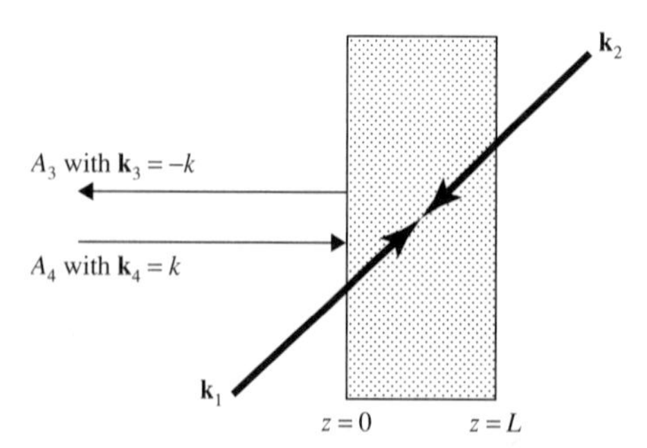

Figure 8.14 The "canonical" geometry of phase conjugation by four-wave mixing.

will be taken as much larger than $|A_3|$ and $|A_4|$ and thus will be scarcely affected by the interaction so that they will be taken as constants throughout the interaction volume. We will further take $|A_1| = |A_2|$. This causes the effect of each one of these two waves on the phase velocity of the other to be the same so that $k_2 = k_1$ [20]. Furthermore, the pump waves 1 and 2 are made to propagate through the nonlinear medium in opposite directions so that

$$\mathbf{k}_1 + \mathbf{k}_2 = 0 \tag{8.8-10}$$

As we will see later, this condition is essential for the generation of phase-conjugated waves.

Wave 4 corresponds to the input beam and in what follows wave 3, which is generated by the interaction of beams 1, 2, and 4, is the (desired) phase-conjugate replica of wave 4.

We now examine the nonlinear polarizations that are relevant to the interaction between waves 3 and 4. The third-order nonlinear polarization is given by Equation (8.6-1), where E is the sum of the electric fields of the four waves (8.8-9). There are a total of 512 terms. The choice of the relevant terms in $P_i^{(\mathrm{NL})}$ requires some judicious reasoning. First, since we assumed that the amplitudes A_3 and A_4 are not time dependent, $P_i^{(\mathrm{NL})}$ must contain the exponential time factor $\exp(i\omega t)$ to radiate efficiently into these two waves. Second, in order that $A_3(z)$ and $A_4(z)$ not vary significantly on the scale of one optical wavelength (i.e., "slow" variation), $P_i^{(\mathrm{NL})}$ must also include the factors $\exp(-ikz)$ and $\exp(+ikz)$ so that we must look for nonlinear polarization terms that contain the wave factors $\exp[i(\omega t - kz)]$ and $\exp[i(\omega t + kz)]$.

Recalling that $\mathbf{k}_1 + \mathbf{k}_2 = 0$, it follows from Equation (8.6-1) that the only third-order products of the fields $\mathbf{E}_1$, $\mathbf{E}_2$, $\mathbf{E}_3$, and $\mathbf{E}_4$ that contain the factors $\exp[\pm i(\omega t - kz)]$ and $\exp[\pm i(\omega t + kz)]$ are (see Problem 8.12)

$$\begin{aligned} P_i^{(\mathrm{NL})}(\mathbf{r}, t) = {} & \tfrac{1}{2}\chi_{ijkl}^{(3)} \times \{6A_{1j}A_{2k}A_{3l}^* + 6A_{1j}A_{1k}^*A_{4l} + 6A_{2j}A_{2k}^*A_{4l} + 6A_{3j}A_{3k}^*A_{4l} \\ & + 3A_{4j}A_{4k}^*A_{4l}\} \exp[i(\omega t - kz)] + \tfrac{1}{2}\chi_{ijkl}^{(3)} \times \{6A_{1j}A_{2k}A_{4l}^* + 6A_{1j}A_{1k}^*A_{3l} + 6A_{2j}A_{2k}^*A_{3l} \\ & + 6A_{4j}A_{4k}^*A_{3l} + 3A_{3j}A_{3k}^*A_{3l}\} \exp[i(\omega t + kz)] + \text{c.c.} \end{aligned} \tag{8.8-11}$$

where summations over repeated indices are assumed, and we have used the cyclic symmetry of the third-order nonlinear coefficients,

$$\chi_{ijkl}^{(3)} = \chi_{[ijkl]}^{(3)} \tag{8.8-12}$$

with $[ijkl]$ being any permutation of $(ijkl)$.

For the purpose of clarity in introducing the concept, we will consider a simple case of importance. This simple case allows us to drop the tensorial notation and limit ourselves to cases where a single χ_{ijkl} (χ_{1111}) is involved. In this case, we assume that all waves are polarized as s-waves (perpendicular to the plane of incidence). By taking 1 as the axis perpendicular to the plane of incidence, the nonlinear polarization can be written

$$\begin{aligned} P_i^{(\mathrm{NL})}(\mathbf{r}, t) = {} & \tfrac{1}{2}\chi_{1111}^{(3)} \times \{6A_1A_2A_3^* + 6A_1A_1^*A_4 + 6A_2A_2^*A_4 + 6A_3A_3^*A_4 + 3A_4A_4^*A_4\} \exp[i(\omega t - kz)] \\ & + \tfrac{1}{2}\chi_{1111}^{(3)} \times \{6A_1A_2A_4^* + 6A_1A_1^*A_3 + 6A_2A_2^*A_3 + 6A_4A_4^*A_3 + 3A_3A_3^*A_3\} \exp[i(\omega t + kz)] + \text{c.c.} \end{aligned} \tag{8.8-13}$$

Furthermore, we will neglect the small terms on the right side of Equation (8.8-13). This leads to

$$\begin{aligned} P_i^{(\mathrm{NL})}(\mathbf{r}, t) = {} & 3\chi_{1111}^{(3)} \times \{A_1A_2A_3^* + A_1A_1^*A_4 + A_2A_2^*A_4\} \exp[i(\omega t - kz)] \\ & + 3\chi_{1111}^{(3)} \times \{A_1A_2A_4^* + A_1A_1^*A_3 + A_2A_2^*A_3\} \exp[i(\omega t + kz)] + \text{c.c.} \end{aligned} \tag{8.8-14}$$

since $|A_3|, |A_4| \ll |A_1|, |A_2|$. Using Equation (8.8-14) in the wave equation (8.3-2) results in

$$\frac{dA_4}{dz} = -i\frac{\omega}{2}\sqrt{\frac{\mu}{\varepsilon}}\,\chi^{(3)}(|A_1|^2 + |A_2|^2)A_4 - i\frac{\omega}{2}\sqrt{\frac{\mu}{\varepsilon}}\,\chi^{(3)}A_1A_2A_3^* \tag{8.8-15}$$

$$\frac{dA_3}{dz} = +i\frac{\omega}{2}\sqrt{\frac{\mu}{\varepsilon}}\,\chi^{(3)}(|A_1|^2 + |A_2|^2)A_3 + i\frac{\omega}{2}\sqrt{\frac{\mu}{\varepsilon}}\,\chi^{(3)}A_1A_2A_4^* \tag{8.8-16}$$

where $\chi^{(3)} = 6\chi_{1111}$.

We note that the first term on the right side of Equations (8.8-15) and (8.8-16), acting alone, merely modifies the propagation constant of waves 3 and 4 from k to $k + (\omega/2)\sqrt{\mu/\varepsilon}\,\chi^{(3)}(|A_1|^2 + |A_2|^2)$. This is the optical Kerr effect discussed earlier. Using the modified wavenumbers, we can simplify the analysis by removing the first term on the right side of Equations (8.8-15) and (8.8-16). This leads to the following coupled equations:

$$\frac{dA_4}{dz} = -i\frac{\omega}{2}\sqrt{\frac{\mu}{\varepsilon}}\,\chi^{(3)}A_1A_2A_3^* \tag{8.8-17}$$

$$\frac{dA_3^*}{dz} = -i\frac{\omega}{2}\sqrt{\frac{\mu}{\varepsilon}}\,\chi^{(3)*}A_1^*A_2^*A_4 \tag{8.8-18}$$

Defining

$$\kappa^* \equiv \frac{\omega}{2}\sqrt{\frac{\mu}{\varepsilon}}\,\chi^{(3)}A_1A_2 \tag{8.8-19}$$

and taking the complex conjugate of Equations (8.8-17) and (8.8-18) results in our final form of the coupled-mode equations for phase-conjugate optics [21]:

$$\begin{aligned}\frac{dA_4^*}{dz} &= i\kappa A_3\\ \frac{dA_3}{dz} &= i\kappa^* A_4^*\end{aligned} \tag{8.8-20}$$

The student is urged to ponder at this point how a relatively complex physical experiment involving four optical beams interacting through the nonlinear optical response of a material medium can be described by equations as simple as (8.8-20). This is possible through a "ruthless," but justifiable, elimination of mathematical terms whose effects are physically negligible but whose inclusion will have rendered the analysis intractable. This is an *essential* difference between mathematics and physics.

Since wave 4 propagates in the $+z$ direction, while wave 3 propagates in the $-z$ direction, we can specify their complex amplitudes at their respective input planes $z = 0$ and $z = L$ (see Figure 8.14). These are taken as $A_4(0)$ and $A_3(L)$. Subject to these boundary conditions, the solutions of (8.8-20) are

$$\begin{aligned}A_3(z) &= \frac{\cos|\kappa|z}{\cos|\kappa|L}A_3(L) + i\frac{\kappa^*\sin|\kappa|(z-L)}{|\kappa|\cos|\kappa|L}A_4^*(0)\\ A_4(z) &= -i\frac{|\kappa|\sin|\kappa|z}{\kappa\cos|\kappa|L}A_3^*(L) + \frac{\cos|\kappa|(z-L)}{\cos|\kappa|L}A_4(0)\end{aligned} \tag{8.8-21}$$

In the basic phase-conjugate experiments, there is but a single input, $A_4(0)$ (the "pump" beams A_1 and A_2 are considered here as part of the apparatus and are lumped in our analysis into the coupling constant κ). Putting $A_3(L) = 0$, we obtain, from (8.8-21) for the reflected wave at the input,

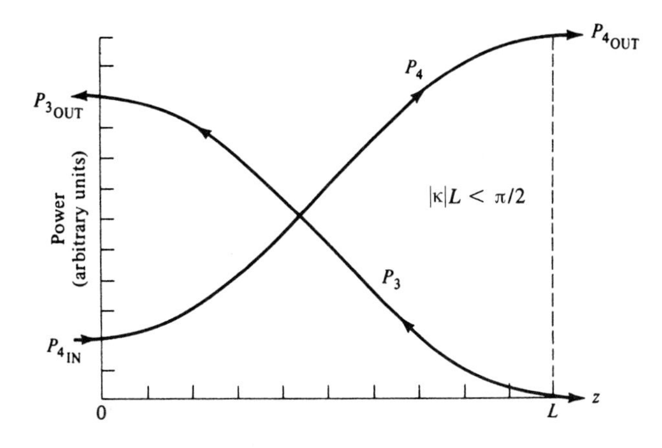

Figure 8.15 Intensity distribution inside the interaction region corresponding to the amplifier case $\pi/4 < |\kappa|L < \pi/2$.

$$A_3(0) = -i\left(\frac{\kappa^*}{|\kappa|}\tan|\kappa|L\right)A_4^*(0) \tag{8.8-22}$$

while at the output ($z = L$) we have

$$A_4(L) = \frac{A_4(0)}{\cos|\kappa|L} \tag{8.8-23}$$

We note that $|A_4(L)| > |A_4(0)|$; that is, the device acts as a phase *coherent optical amplifier* with a gain of $|(\cos|\kappa|L)^{-1}|$. When

$$\frac{\pi}{4} \le |\kappa|L \le \frac{3\pi}{4} \tag{8.8-24}$$

the result $|A_3(0)| > |A_4(0)|$ obtains, so that *the reflectivity of the phase conjugate mirror exceeds unity*. The intensity distribution of the two waves inside the nonlinear medium for a value of $|\kappa|L$ satisfying Equation (8.8-24) is shown in Figure 8.15.

Of particular interest is the condition $|\kappa|L = \pi/2$. In this case,

$$\frac{A_3(0)}{A_4(0)} = \infty \qquad \frac{A_4(L)}{A_4(0)} = \infty \tag{8.8-25}$$

that is, both the transmission gain $[A_4(L)/A_4(0)]$ and the reflection gain $[A_3(0)/A_4(0)]$ become infinite so that finite outputs $A_3(0)$ and $A_4(L)$ can result even when the input $A_4(0)$ is zero. This corresponds to *oscillation*. This oscillation takes place *without the benefit of mirror feedback*. The feedback process that is essential to oscillation is provided by the fact that waves 3 and 4 propagate in opposite directions, so that $A_4(z_1)$, for example, is influenced by $A_4(z_2)$ even when $z_2 > z_1$, the information being carried from z_2 to z_1 by the backward-going wave 3. The intensity distribution corresponding to the oscillation condition is shown in Figure 8.16.

Another point of physical interest is that of the source of the power. Since energy is conserved, it follows that the increase in the output powers of beams 3 and 4 relative to their input values must come at the expense of the pump beams 1 and 2. A more exact analysis that does not neglect the spatial dependence of beams 1 and 2 shows that this indeed is the case. A quantum mechanical description of this process [20] shows that on the atomic scale the basic process is one where, simultaneously, two photons, one from beam 1 and one from beam 2,

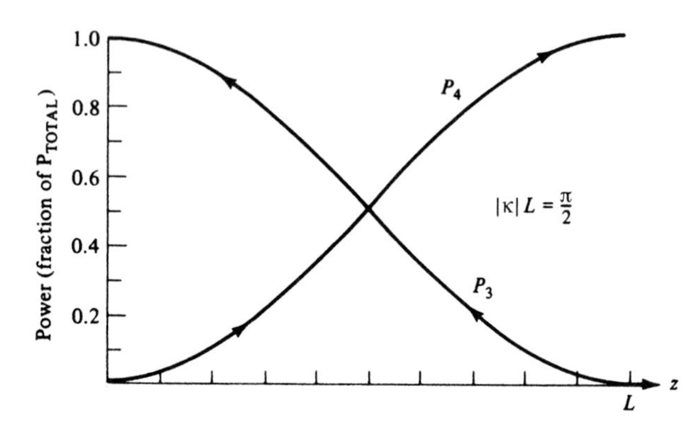

Figure 8.16 Intensity distribution inside the interaction region when the oscillation condition $|\kappa|L = \pi/2$ is satisfied.

are annihilated while two photons are created—one of these photons is added to beam 3 and the other to beam 4.

We have strayed somewhat from our main purpose, which is to show that the four-wave mixing geometry of Figure 8.14 is capable of rendering in real time the phase-conjugate (reflected) replica of an input beam. Returning to our basic plane wave result (8.8-22),

$$A_3(0) = -i\left(\frac{\kappa^*}{|\kappa|}\tan|\kappa|L\right)A_4^*(0) \tag{8.8-26}$$

It follows that the amplitude A_3 of the backward wave 3 at the input plane $z = 0$ is proportional to the complex amplitude $A_4^*(0)$ of the input wave 4 at the same plane. It follows directly that since an input wave with an arbitrarily complex wavefront $A_4(x, y, z)$ can be expanded in terms of plane wave components (which in the paraxial limit adopted here will span a small solid angle centered about the $+z$ axis), we can extend Equation (8.8-26) to each plane wave component individually [21] to obtain

$$A_3(x, y, z < 0) = -i\left(\frac{\kappa^*}{|\kappa|}\tan|\kappa|L\right)A_4^*(x, y, z < 0) \tag{8.8-27}$$

This is the basic result of phase conjugation by four-wave mixing. It shows that the reflected beam $A_3(\mathbf{r})$ to the left of the nonlinear medium ($z < 0$) is the phase conjugate of the input beam $A_4(\mathbf{r})$ (see Figure 8.17).

As in the rest of this book, our analysis of phase conjugation employs the MKS system of units. Much of the research literature, unfortunately, uses the esu system. The relations described in Equations (8.8-7) and (8.8-8) earlier in this section should facilitate the translation from one system to another.

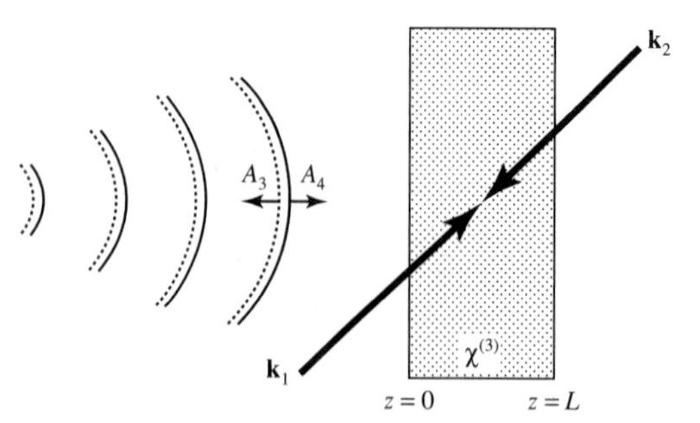

Figure 8.17 Optical phase conjugation by four-wave mixing. Solid curves represent wavefronts of incident wave A_4, and dotted curves are for the phase-conjugated wave A_3.

EXAMPLE: PHASE CONJUGATION IN CS_2

Many of the first experiments in phase-conjugate optics were performed in carbon disulfide, CS_2. In the case when all are polarized along the same direction that we will call x, the relevant nonlinear coefficient is χ_{1111}, which in MKS units has the value (Table 8.8)

$$\chi_{1111} = 5.7 \times 10^{-32}$$

The coefficient $\chi^{(3)}$ used in our analysis, we recall, is according to Equation (8.8-16) equal to $6\chi_{1111}$. We will further assume that the experiment is carried out with waves at $\lambda = 1.06\ \mu\text{m}$ and that the two pump beams are of equal intensities with $I_1 = I_2 = 5 \times 10^{10}\ \text{W/m}^2$ (i.e., $5 \times 10^6\ \text{W/cm}^2$). The index of refraction of CS_2 is $n = 1.594$. Using these data, we obtain

$$A_1 = A_2 = \sqrt{\frac{2I_1}{\varepsilon_0 cn}} = 4.86 \times 10^6\ \text{V/m}$$

and from Equation (8.8-22), $|\kappa| = 1.1\ \text{m}^{-1}$.

It follows that in CS_2 we need to use pump beams with intensities on the order of megawatts per square centimeter with path lengths on the order of 1 m in order to satisfy the condition $|\kappa|L \sim 1$ needed according to Equation (8.8-22) for appreciable (~ 1) phase-conjugate reflectivities.

8.9 FREQUENCY TUNING IN PARAMETRIC OSCILLATION

We have shown earlier in this chapter that the pair of signal (ω_1) and idler (ω_2) frequencies that are caused to oscillate by parametric pumping at ω_3 satisfy the condition $k_3 = k_1 + k_2$. Using $k_i = \omega_i n_i/c$, we can write it as

$$\omega_3 n_3 = \omega_1 n_1 + \omega_2 n_2 \tag{8.9-1}$$

where $n_1 = n(\omega_1)$, $n_2 = n(\omega_2)$, and $n_3 = n(\omega_3)$ are the indices of refraction. This is the phase-matching condition. In a crystal the indices of refraction generally depend, as shown in Section 8.3, on the frequency, crystal orientation (if the wave is extraordinary), electric field (in electro-optic crystals), and temperature. If, as an example, we change the crystal orientation in the oscillator shown in Figure 8.6, the oscillation frequencies ω_1 and ω_2 will change so as to compensate for the change in indices, and thus condition (8.9-1) will be satisfied at the new frequencies. To be specific, we consider the case of a parametric oscillator pumped by an extraordinary beam at a fixed frequency ω_3. The signal (ω_1) and the idler (ω_2) are ordinary waves. At some crystal orientation θ_0 (angle between the c axis and the direction of propagation), the oscillation takes place at frequencies ω_{10} and ω_{20}. Let the indices of refraction at ω_{10}, ω_{20}, and ω_3 under those conditions be n_{10}, n_{20}, and n_{30}, respectively. We want to find the change in ω_1 and ω_2 due to a small change $\Delta\theta$ in the crystal orientation.

From Equation (8.9-1) we have, at $\theta = \theta_0$,

$$\omega_3 n_{30} = \omega_{10} n_{10} + \omega_{20} n_{20} \tag{8.9-2}$$

After the crystal orientation has been changed from θ_0 to $\theta_0 + \Delta\theta$, the following changes occur:

$$n_{30} \rightarrow n_{30} + \Delta n_3$$
$$n_{10} \rightarrow n_{10} + \Delta n_1$$
$$n_{20} \rightarrow n_{20} + \Delta n_2$$
$$\omega_{10} \rightarrow \omega_{10} + \Delta\omega_1$$

Since $\omega_1 + \omega_2 = \omega_3 = \text{constant}$,

$$\omega_{20} \to \omega_{20} + \Delta\omega_2 = \omega_{20} - \Delta\omega_1$$

that is, $\Delta\omega_2 = -\Delta\omega_2$. Furthermore, since Equation (8.9-1) must be satisfied at $\theta_0 = \theta_0 + \Delta\theta$, we have

$$\omega_3(n_{30} + \Delta n_3) = (\omega_{10} + \Delta\omega_1)(n_{10} + \Delta n_1) + (\omega_{20} - \Delta\omega_1)(n_{20} + \Delta n_2)$$

Neglecting the second-order terms $\Delta n_1\,\Delta\omega_1$ and $\Delta n_2\,\Delta\omega_1$ and using Equation (8.9-2), we obtain

$$\Delta\omega_1\Big|_{\substack{\omega_1\approx\omega_{10}\\ \omega_2\approx\omega_{20}}} = \frac{\omega_3\,\Delta n_3 - \omega_{10}\,\Delta n_1 - \omega_{20}\,\Delta n_2}{n_{10} - n_{20}} \tag{8.9-3}$$

According to our starting hypotheses the pump is an extraordinary ray; therefore, according to Equation (8.3-32), its index depends on the orientation θ, giving

$$\Delta n_3 = \frac{\partial n_3}{\partial\theta}\Big|_{\theta_0}\Delta\theta \tag{8.9-4}$$

The signal and idler are ordinary rays, so their indices depend on the frequencies but not on the direction. It follows that

$$\Delta n_1 = \frac{\partial n_1}{\partial\omega_1}\Big|_{\omega_{10}}\Delta\omega_1$$
$$\Delta n_2 = \frac{\partial n_2}{\partial\omega_2}\Big|_{\omega_{20}}\Delta\omega_2 \tag{8.9-5}$$

Using the last two equations in (8.9-3) results in

$$\frac{\partial\omega_1}{\partial\theta} = \frac{\omega_3(\partial n_3/\partial\theta)}{(n_{10} - n_{20}) + [\omega_{10}(\partial n_1/\partial\omega_1) - \omega_{20}(\partial n_2/\partial\omega_2)]} \tag{8.9-6}$$

for the rate of change of the oscillation frequency with respect to the crystal orientation. Using Equation (8.3-32) and the relation $d(1/x^2) = -(2/x^3)dx$, we obtain

$$\frac{\partial n_3}{\partial\theta} = -\frac{n_3^3}{2}\sin(2\theta)\left[\left(\frac{1}{n_e^{\omega_3}}\right)^2 - \left(\frac{1}{n_o^{\omega_3}}\right)^2\right]$$

which, when substituted in Equation (8.9-6), gives

$$\frac{\partial\omega_1}{\partial\theta} = \frac{-\frac{1}{2}\omega_3 n_{30}^3\left[\left(\frac{1}{n_e^{\omega_3}}\right)^2 - \left(\frac{1}{n_o^{\omega_3}}\right)^2\right]\sin(2\theta_0)}{(n_{10} - n_{20}) + \left(\omega_{10}\frac{\partial n_1}{\partial\omega_1} - \omega_{20}\frac{\partial n_2}{\partial\omega_2}\right)} \tag{8.9-7}$$

An experimental curve showing the dependence of the signal and idler frequencies on θ in $NH_4H_2PO_4$ (ADP) is shown in Figure 8.18. Also shown is a theoretical curve based on a quadratic approximation of Equation (8.9-7), which was plotted using the dispersion (i.e., n versus ω) data of ADP; see Reference [22]. Reasoning similar to that used to derive the angle-tuning expression (8.9-7) can be applied to determine the dependence of the oscillation frequency on temperature. Here we need to know the dependence of the various indices on temperature. This is discussed further in Problem 8.6. An experimental temperature-tuning curve is shown in Figure 8.19.

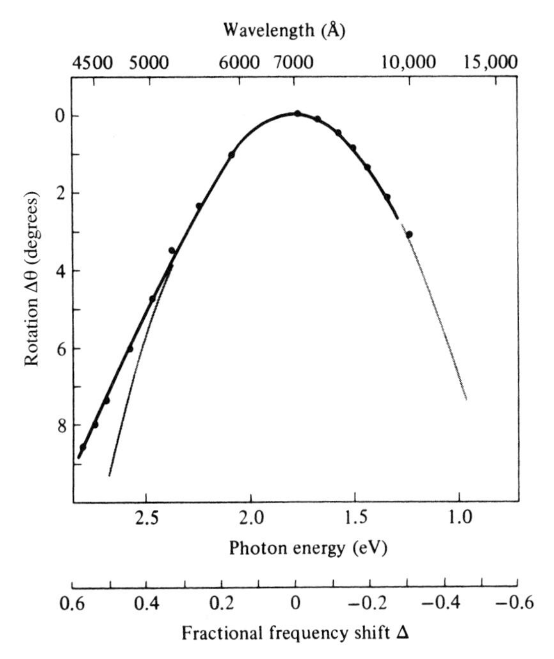

Figure 8.18 Dependence of the signal (ω_1) frequency on the angle between the pump propagation direction and the optic axis of the ADP crystal. The angle θ is measured with respect to the angle for which $\omega_1 = \omega_1/2$. $\Delta \equiv (\omega_1 - \omega_3/2)/(\omega_3/2)$. (After Reference [22].)

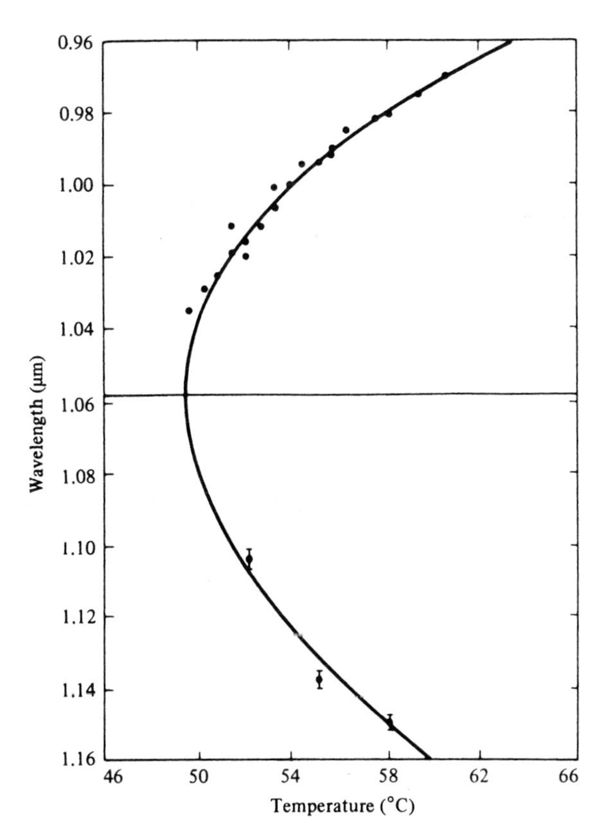

Figure 8.19 Signal and idler wavelength as a function of the temperature of the oscillator crystal. (After Reference [23].)

PROBLEMS

8.1 Show that if θ_m is the phase-matching angle for an ordinary wave at ω and an extraordinary wave at 2ω, then

$$\Delta k(\theta)l\,\big|_{\theta\approx\theta_m} = -\frac{2\omega l}{c_0}\sin(2\theta_m)\frac{(n_e^{2\omega})^{-2} - (n_o^{2\omega})^{-2}}{2(n_o^{\omega})^{-3}}(\theta - \theta_m)$$

8.2 Derive the expression for the phase-matching angle of a parametric amplifier using KDP, in which two of the waves are extraordinary while the third is ordinary. Which of the three waves (i.e., signal, idler, or pump) would you choose as ordinary? Can this type of phase matching be accomplished with $\omega_3 = 10{,}000\ \text{cm}^{-1}$ and $\omega_1 = \omega_2 = 5000\ \text{cm}^{-1}$? If so, what is θ_m?

8.3 Show that Equations (8.4-12) are consistent with the fact that the increases in the photon flux at ω_1 and ω_2 are identical—that is, $A_1^*(z)A_1(z) - A_1^*(0)A_1(0) = A_2^*(z)A_2(z) - A_2^*(0)A_2(0)$. Derive the same result directly from the coupled equations (8.4-10).

8.4 Complete the missing steps in the derivation of Equation (8.4-21).

8.5 Show, using the arguments of Section 8.1, that third-order optical effects as defined by Equation (8.8-4) can exist in all homogeneous media.

8.6 Consider a parametric oscillator setup such as that shown in Figure 8.19. The crystal orientation angle is θ, its temperature is T, and the signal and idler frequencies are ω_{10} and ω_{20}, respectively, with $\omega_{10} + \omega_{20} = \omega_3$. Show that a small temperature change ΔT causes the signal frequency to change by

$$\Delta\omega_1 = \Delta T \times \left\{\omega_3\left[\cos^2\theta\left(\frac{n_e^{\omega_3}(\theta)}{n_o^{\omega_3}}\right)^3\frac{\partial n_o^{\omega_3}}{\partial T} + \sin^2\theta\left(\frac{n_e^{\omega_3}(\theta)}{n_e^{\omega_3}}\right)^3\frac{\partial n_e^{\omega_3}}{\partial T}\right] - \omega_{10}\frac{\partial n_o^{\omega_1}}{\partial T} - \omega_{20}\frac{\partial n_o^{\omega_2}}{\partial T}\right\}\times\frac{1}{n_{10} - n_{20}}$$

The pump is taken as an extraordinary ray, whereas the signal and idler are ordinary. [Hint: The starting point is Equation (8.9-3), which is valid regardless of the nature of the perturbation.]

8.7 Using the published dispersion data of proustite (Reference [12]), calculate the maximum angular deviation of the input beam at ν_1 (from parallelism with the pump beam at ν_2) that results in a reduction by a factor of 2 in the up-conversion efficiency. Take $\lambda_1 = 10.6\ \mu\text{m}$, $\lambda_2 = 1.06\ \mu\text{m}$, and $\lambda_3 = 0.964\ \mu\text{m}$. [Hint: A proper choice must be made for the polarizations at ω_1, ω_2, and ω_3 so that phase matching can be achieved along some angle.] The maximum angular deviation is that for which

$$\frac{\sin^2[\Delta k(\theta)l/2]}{[\Delta k(\theta)l/2]^2} = \frac{1}{2}$$

where, at the phase-matching angle θ_m, $\Delta k(\theta_m) = 0$. Approximate the dispersion data by a Taylor-series expansion about the nominal ($\Delta k = 0$) frequencies.

8.8 Using the dispersion data of Reference [12], discuss what happens to phase matching in an up-conversion experiment due to a deviation of the input frequency from the nominal ($\Delta k = 0$) ν_{10} value. Derive an expression for the spectral width of the output in the case where the input spectral density (power per unit frequency) in the vicinity of ν_{10} is uniform. [Hint: Use a Taylor-series expansion of the dispersion data about the phase-matching ($\Delta k = 0$) frequencies to obtain an expression for $\Delta k(\nu_3)$.] Define the output spectral width as twice the frequency deviation at which the output is one-half its maximum ($\Delta k = 0$) value.

8.9 Derive the coupled-mode equations in a manner similar to that leading to (8.8-15) and (8.8-16) for the case where the frequency of the incident wave ω_4 is detuned from that of the pump beams (ω) by

$$\omega_4 = \omega - \delta$$

(a) Show that the reflected wave frequency is $\omega_3 = \omega + \delta$.
(b) Solve the coupled-mode equations for the reflection coefficient $|A_3(0)/A_4(0)|^2$. Plot it as a function of the frequency offset δ.

8.10

(a) Solve the degenerate ($\omega_1 = \omega_2 = \omega_3 = \omega_4$) coupled-mode equations (8.8-15) and (8.8-16) as modified for (ordinary) optical losses. The new equations are

$$\frac{dA_4^*}{dz} = i\kappa e^{-\alpha L}A_3 - \alpha A_4^*$$

$$\frac{dA_3}{dz} = i\kappa^* e^{-\alpha L}A_4^* + \alpha A_3$$

$$\kappa^* = \frac{\omega}{2}\sqrt{\frac{\mu}{\varepsilon}}\,\chi^{(3)}A_1(L)A_2(0)$$

where α is the optical amplitude loss coefficient (assumed the same for all four beams).

(b) Plot the reflection coefficient $|A_3(0)/A_4(0)|^2$ as a function of κL for $\alpha L = 0.1, 0.5, 1, 2$. Discuss qualitatively the effect of the losses.

8.11

(a) Derive Equations (8.3-8) and (8.3-9).

(b) Show that for phase-matched SHG in KDP, $d = d_{36} \sin \theta_m$.

8.12 Starting from the definition of third order and the presence of four waves given by Equation (8.8-9), show that the nonlinear polarization with spatial–temporal dependence of $\exp[i(\omega t - kz)]$ is given by

$$\begin{aligned}P_i^{(\mathrm{NL})}(\mathbf{r}, t) = \tfrac{1}{2}\chi_{ijkl}^{(3)} \times \{&(A_{1j}A_{2k}A_{3l}^* + A_{1j}A_{2l}A_{3k}^* + A_{1k}A_{2j}A_{3l}^* \\ &+ A_{1k}A_{2l}A_{3j}^* + A_{1l}A_{2k}A_{3j}^* + A_{1l}A_{2j}A_{3k}^*) \\ &+ (A_{1j}A_{1k}^*A_{4l} + A_{1k}A_{1j}^*A_{4l} + A_{1j}A_{1l}^*A_{4k} + A_{1l}A_{1j}^*A_{4k} \\ &+ A_{1l}A_{1k}^*A_{4j} + A_{1k}A_{1l}^*A_{4j}) + (A_{2j}A_{2k}^*A_{4l} + A_{2k}A_{2j}^*A_{4l} \\ &+ A_{2j}A_{2l}^*A_{4k} + A_{2l}A_{2j}^*A_{4k} + A_{2l}A_{2k}^*A_{4j} + A_{2k}A_{2l}^*A_{4j}) \\ &+ (A_{3j}A_{3k}^*A_{4l} + A_{3k}A_{3j}^*A_{4l} + A_{3j}A_{3l}^*A_{4k} + A_{3l}A_{3j}^*A_{4k} \\ &+ A_{3l}A_{3k}^*A_{4j} + A_{3k}A_{3l}^*A_{4j}) + (A_{4j}A_{4k}^*A_{4l} + A_{4j}A_{4l}^*A_{4k} \\ &+ A_{4l}A_{4j}^*A_{4k})\} \exp[i(\omega t - kz)]\end{aligned}$$

[Hint: Use the symmetric relation $\chi_{ijkl}^{(3)} = \chi_{[ijkl]}^{(3)}$, where $[ijkl]$ is any permutation of $(ijkl)$.] Derive a similar expression with $\exp[i(\omega t + kz)]$.

8.13 The polarization including linear and third-order electric susceptibility as well as the refractive index can be written as in Table P8.13.

TABLE P8.13

MKS	esu (CGS)
$P = \varepsilon_0 \chi E + \chi^{(3)} E^3$	$P = \chi E + \chi^{(3)} E^3$
$n^2 = 1 + \chi$	$n^2 = 1 + 4\pi\chi$

(a) Treat the contribution from third-order susceptibility as a correction to Δn^2. Show that

$$\frac{1}{\varepsilon_0}\chi_{\mathrm{MKS}}^{(3)} E_{\mathrm{MKS}}^2 = 4\pi\chi_{\mathrm{esu}}^{(3)} E_{\mathrm{esu}}^2$$

(b) $E_{\mathrm{esu}} = 1$ corresponds to $E_{\mathrm{MKS}} = 3 \times 10^4$ V/m. Show that

$$\chi_{\mathrm{MKS}}^{(3)} = \frac{4\pi\varepsilon_0}{9 \times 10^8}\chi_{\mathrm{esu}}^{(3)}$$

8.14

(a) Start from Equations (8.6-11). Show that for unpolarized light, the third terms average out to zero. Derive Equation (8.6-13).

(b) For circularly polarized light, let $E_{0x}^{\omega} = A$ and $E_{0y}^{\omega} = Ae^{i\delta}$, where A and $\delta = \pi/2$ are real numbers. Show that

$$P_{\mathrm{NL}x}^{(\omega)} = 4\chi_{xxxx}|A|^2 E_{0x}^{(\omega)} \quad \text{and} \quad P_{\mathrm{NL}y}^{(\omega)} = 4\chi_{xxxx}|A|^2 E_{0y}^{(\omega)}$$

We note that the third-order nonlinear polarization is parallel to the electric field vector. This ensures that the polarization state will remain the same. Derive Equation (8.6-14).

(c) For a general linear polarized light, $E_{0x}^{\omega} = A$ and $E_{0y}^{\omega} = B$, where A and B are real numbers. Show that the third-order nonlinear polarization is parallel to the electric field vector. This is consistent with the isotropic nature of the medium.

(d) Consider the situation of a beam of elliptically polarized light. Without loss of generality, we can choose the x axis parallel to the long axis of the polarization ellipse at input. Let $E_{0x}^{\omega} = A$ and $E_{0y}^{\omega} = Be^{i\delta}$, where A, B, and $\delta = \pi/2$ are real numbers. Show that

$$P_{\mathrm{NL}x}^{(\omega)} = \chi_{xxxx}(3A^2 + B^2)E_{0x}^{(\omega)}$$

$$P_{\mathrm{NL}y}^{(\omega)} = \chi_{xxxx}(3B^2 + A^2)E_{0y}^{(\omega)}$$

Note that the third-order nonlinear polarization is not parallel to the electric field vector. As a result, a phase retardation exists between the two components as the wave propagates in the Kerr medium.

(e) Show, with the aid of a Poincaré sphere, that the polarization ellipse will rotate while maintaining the ellipticity as the wave propagates in the Kerr medium. Show that the rate of rotation per unit distance of propagation is given by

$$\frac{d\psi}{dz} = -\frac{1}{3}\frac{2AB}{A^2 + B^2}\frac{2\pi}{\lambda}n_2 I$$

where I is the intensity of the beam. The sense of rotation is the same as the handedness of the elliptical polarization state. The rate of rotation is zero for the special case of linearly polarized light (A or $B = 0$).

8.15 Consider a nonlinear medium with $P_i = \varepsilon_0\chi_{ij}E_j + 2d_{ijk}E_jE_k$. According to Problem 1.6, the work done per unit volume by applying an electric field in a dielectric medium is $W = \int \mathbf{E} \cdot d\mathbf{P}$, where $\mathbf{P}$ is the polarization.

Carry out the integration from (0, 0) to (E_{10}, E_{20}) in a two-dimensional case by using two different paths:

Path A: First integrate from (0, 0) to $(E_{10}, 0)$, and then from $(E_{10}, 0)$ to (E_{10}, E_{20}).
Path B: First integrate from (0, 0) to $(0, E_{20})$, and then from $(0, E_{20})$ to (E_{10}, E_{20}).

(a) Show that for path A, the work done due to the nonlinear polarization is

$$W = \tfrac{4}{3}d_{111}E_{10}^3 + \tfrac{4}{3}d_{222}E_{20}^3 + 4d_{112}E_{10}^2E_{20} + 2d_{122}E_{10}E_{20}^2 + 2d_{212}E_{10}E_{20}^2$$

(b) Show that for path B, the work done due to the nonlinear polarization is

$$W = \tfrac{4}{3}d_{111}E_{10}^3 + \tfrac{4}{3}d_{222}E_{20}^3 + 2d_{112}E_{10}^2E_{20} + 2d_{211}E_{20}E_{10}^2 + 4d_{212}E_{10}E_{20}^2$$

In a lossless medium, the integration should be independent of the path. This leads to $d_{122} = d_{212}$ and $d_{121} = d_{211}$. A similar analysis in the three-dimensional case will yield $d_{ijk} = d_{jik}$. Combining the result here with Equations (8.1-2), we obtain $d_{ijk} = d_{[ijk]}$, where $[ijk]$ is any permutation of the subscript indices.

REFERENCES

1. Franken, P. A., A. E. Hill, C. W. Peters, and G. Weinreich, Generation of optical harmonics. *Phys. Rev. Lett.* **7**:118 (1961).
2. Zemike, F., Jr., Refractive indices of ammonium dihydrogen phosphate and potassium dihydrogen phosphate between 2000 Å and 1.5 μ. *J. Opt. Soc. Am.* **54**:1215 (1964).
3. Maker, P. D., R. W. Terhune, M. Nisenoff, and C. M. Savage, Effects of dispersion and focusing on the production of optical harmonics. *Phys. Rev. Lett.* **8**:21 (1962).
4. Giordmaine, J. A., Mixing of light beams in crystals. *Phys. Rev. Lett.* **8**:19 (1962).
5. Ashkin, A., G. D. Boyd, and J. M. Dziedzic, Observation of continuous second harmonic generation with gas lasers. *Phys. Rev. Lett.* **11**:14 (1963).
6. Seka, W., S. D. Jacobs, J. E. Rizzo, R. Boni, and R. S. Craxton, Demonstration of high efficiency third harmonic conversion of high power Nd–glass laser radiation. *Opt. Commun.* **34**:469 (1980). Also see R. S. Craxton, High efficiency frequency tripling schemes for high power Nd:glass lasers. *IEEE J. Quantum Electron.* **17**:177 (1981). (Additional articles on doubling and frequency conversion are to be found in the same issue.)
7. Geusic, J. E., H. J. Levinstein, S. Singh, R. G. Smith, and L. G. Van Uitert, Continuous 0.53 μm solid-state source using $Ba_2NaNb_5O_{15}$. *IEEE J. Quantum Electron.* **4**:352 (1968).
8. Thorsos, E. I. (ed.), *Laboratory for Laser Analytics Review*, Vol. II, p. 7, 1979–1980.
9. Johnson, F. M., and J. A. Durado, Frequency up-conversion. *Laser Focus* **3**:31 (1967).
10. Midwinter, J. E., and J. Warner, Up-conversion of near infrared to visible radiation in lithium–meta-niobate. *J. Appl. Phys.* **38**:519 (1967).
11. Warner, J., Photomultiplier detection of 10.6 μm radiation using optical up-conversion in proustite. *Appl. Phys. Lett.* **12**:222 (1968).
12. Hulme, K. F., O. Jones, P. H. Davies, and M. V. Hobden, Synthetic proustite (Ag_3AsS_3): a new material for optical mixing. *Appl. Phys. Lett.* **10**:133 (1967).
13. Somekh, S., and A. Yariv, Phase matching by periodic modulation of the nonlinear optical properties. *Opt. Commun.* **6**(3):301 (1972).
14. Bonz, M. L., *Quasi Phase Matched Optical Frequency Conversion in Lithium Niobate Waveguides*. Ph.D. thesis, Stanford University, 1994.
15. Jackel, J. L., C. E. Rice, and J. J. Vesseka, Proton exchange for high index waveguides in $LiNbO_3$. *Appl. Phys. Lett.* **41**:607 (1982).
16. Fiesst, A., and P. Koidl, Current induced periodic ferroelectric domain reversal in $LiNbO_3$ for efficient second harmonic generation. *Appl. Phys. Lett.* **47**:1125 (1985).

17. Yamada, M., N. Nada, M. Saito, and K. Watanabe, First order quasi phase-matched $LiNbO_3$ waveguide periodically poled by applying an external field for efficient blue second harmonic generation. *Appl. Phys. Lett.* **62**:435 (1993).
18. Hellwarth, R. W., Third order susceptibilities of liquids and gases. *Progr. Quantum Electron.* **5**:1 (1977).
19. See, for example, Y. R. Shen, *Principles of Nonlinear Optics*. Wiley, New York, 1984, Chapter 10.
20. Fisher, R. A. (ed.), *Optical Phase Conjugation*. Academic Press, New York, 1983. A comprehensive collection of material covering many aspects of phase conjugation.
21. Yariv, A., and D. M. Pepper, Amplified reflection, phase conjugation, and oscillation in degenerate four-wave mixing. *Opt. Lett.* **1**:16 (1977).
22. Magde, D., and H. Mahr, Study in ammonium dihydrogen phosphate of spontaneous parametric interaction tunable from 4400 to 16000 Å. *Phys. Rev. Lett.* **18**:905 (1967).
23. The first demonstration of optical parametric oscillation is that of J. A. Giordmaine and R. C. Miller Tunable optical parametric oscillation in $LiNbO_3$ at optical frequencies. *Phys. Rev. Lett.* **14**:973 (1965).

CHAPTER 9

ELECTRO-OPTIC MODULATION OF LASER BEAMS

9.0 INTRODUCTION

In Chapter 1 we treated the propagation of electromagnetic waves in anisotropic media. It was shown how the properties of the propagating wave can be determined from the dielectric tensor ε_{ij}, or more conveniently from the index ellipsoid. In this chapter we consider the problem of propagation of optical radiation in crystals in the presence of an applied electric field. We find that in certain types of crystals it is possible to effect a change in the index of refraction that is proportional to the applied electric field. This is the linear electro-optic effect (also known as the Pockels effect, named after F. Pockels who studied the effect in 1893). It affords a convenient and widely used means of controlling the intensity or phase of the propagating radiation. This modulation is used in an ever expanding number of applications, including the impression of information onto optical beams, active mode locking of lasers for generation of ultrashort optical pulses, and optical beam deflection. Some of these applications will be discussed further in this chapter. Modulation and deflection of laser beams by acoustic beams are also considered later in this chapter.

9.1 LINEAR ELECTRO-OPTIC EFFECT

The propagation of optical radiation in a crystal can be described completely in terms of the dielectric tensor. In Chapter 1 we found that, given a direction of propagation in a crystal, in general two possible linearly polarized modes exist—the so-called normal modes of propagation. Each mode possesses a unique direction of polarization (i.e., direction of **D**) and a corresponding index of refraction (i.e., a velocity of propagation). The mutually orthogonal polarization directions (**D** vectors) and the indices of the two modes are found most conveniently by using the index ellipsoid, which assumes its simplest form in the principal coordinate system:

$$\frac{x^2}{n_x^2} + \frac{y^2}{n_y^2} + \frac{z^2}{n_z^2} = 1 \tag{9.1-1}$$

where the directions x, y, and z are the principal dielectric axes—that is, the directions in the crystal along which the **D** and **E** field vectors are parallel. We note that $(1/n_x^2)$, $(1/n_y^2)$, and $(1/n_z^2)$ are the principal values of the impermeability tensor, which is defined as

$$\eta = \frac{\varepsilon_0}{\varepsilon} \tag{9.1-2}$$

where ε is the dielectric tensor.

According to the quantum theory of solids, the optical dielectric tensor depends on the distribution of charges in the crystal. The application of an applied electric field can cause a redistribution of the charges and possibly a slight deformation of the crystal lattice. The net result is a change in the optical impermeability tensor. This known as the electo-optic effect. The linear electro-optic coefficients are defined traditionally as

$$\Delta\eta_{ij} = \eta(\mathbf{E}) - \eta(0) \equiv \Delta\left(\frac{1}{n^2}\right)_{ij} \equiv r_{ijk}E_k \tag{9.1-3}$$

where **E** is the applied electric field, E_k is the k component ($k = x, y, z$) of the electric field, and the summation over repeated indices is assumed. In the above equation, we neglected the high-order terms. The constants r_{ijk} are the linear electro-optic coefficients. For convenience, we use the convention $1 = x$, $2 = y$, $3 = z$.

Thus the index ellipsoid of a crystal in the presence of an applied electric field is given by

$$\eta_{ij}(\mathbf{E})x_i x_j = 1 \tag{9.1-4}$$

or equivalently,

$$\left(\frac{1}{n_x^2} + r_{1 1k}E_k\right)x^2 + \left(\frac{1}{n_y^2} + r_{22k}E_k\right)y^2 + \left(\frac{1}{n_z^2} + r_{33k}E_k\right)z^2 + 2xyr_{12k}E_k + 2yzr_{23k}E_k + 2zxr_{13k}E_k = 1 \tag{9.1-5}$$

where E_k is the k component ($k = x, y, z$) of the electric field and the summation over repeated indices is assumed. In Equation (9.1-5), we have used the symmetry property of the impermeability tensor ($\eta_{ij} = \eta_{ji}$). As a result of the "mixed terms" in Equation (9.1-5), the principal axes (x, y, z) of the crystal are no longer the principal axes of the index ellipsoid.

The impermeability tensor is symmetric provided that the medium is lossless and optically inactive (i.e., no optical activity or optical rotatory power). In this case the linear electro-optic coefficients satisfy the following symmetric relationship:

$$r_{ijk} = r_{jik} \tag{9.1-6}$$

Because of the symmetry, it is convenient to introduce contracted indices to abbreviate the notation. They are defined as

$$\begin{aligned} 1 &= (11) = (xx) \\ 2 &= (22) = (yy) \\ 3 &= (33) = (zz) \\ 4 &= (23) = (32) = (yz) = (zy) \\ 5 &= (13) = (31) = (xz) = (zx) \\ 6 &= (12) = (21) = (xy) = (yx) \end{aligned} \tag{9.1-7}$$

Using these contracted indices, we can write

$$\begin{aligned} r_{1k} &= r_{11k} \\ r_{2k} &= r_{22k} \\ r_{3k} &= r_{33k} \\ r_{4k} &= r_{23k} = r_{32k} \\ r_{5k} &= r_{13k} = r_{31k} \\ r_{6k} &= r_{12k} = r_{21k} \end{aligned} \qquad (k = 1, 2, 3) \tag{9.1-8}$$

It is important to remember that the contraction of the indices is just a matter of convenience. These matrix elements (6×3) do not have the usual tensor transformation or multiplication properties. The permutation symmetry reduces the number of independent elements of r_{ijk} from 27 to 18.

According to Equation (9.1-5), the normal modes of propagation and the corresponding refractive indices associated with the modes in the presence of an applied electric field will depend on the magnitude as well as the direction of the applied field. Using the method of the index ellipsoid, the refractive indices of the normal modes of propagation can be obtained. The linear electro-optic effect is the change in the refractive indices of the normal modes of propagation (e.g., ordinary and extraordinary modes in uniaxial crystals) that is caused by and is proportional to an applied electric field. This effect exists only in crystals that do not possess inversion symmetry (centrosymmetric crystals). The structure of these crystals remains invariant under the inversion operation (a coordinate transformation by replacing **r** with −**r**.) This statement can be justified as follows: Assume that in a crystal possessing an inversion symmetry, the application of an electric field E along some direction causes a change $\Delta n_1 = sE$ in the index, where s is a constant characterizing the linear electro-optic effect. If the direction of the field is reversed, the change in the index is given by $\Delta n_2 = s(-E)$, but because of the inversion symmetry the two directions are physically equivalent, so $\Delta n_1 = \Delta n_2$. This requires that $s = -s$, which is possible only for $s = 0$, so no linear electro-optic effect can exist in centrosymmetric crystals. The division of all crystal classes into those that do and those that do not possess an inversion symmetry is an elementary consideration in crystallography and this information is widely tabulated [1].

Using the contracted notation, the index ellipsoid in the presence of the applied electric field can be written

$$\left[\frac{1}{n_x^2} + \Delta\left(\frac{1}{n^2}\right)_1\right]x^2 + \left[\frac{1}{n_y^2} + \Delta\left(\frac{1}{n^2}\right)_2\right]y^2 + \left[\frac{1}{n_z^2} + \Delta\left(\frac{1}{n^2}\right)_3\right]z^2$$
$$+ 2yz\,\Delta\left(\frac{1}{n^2}\right)_4 + 2zx\,\Delta\left(\frac{1}{n^2}\right)_5 + 2xy\,\Delta\left(\frac{1}{n^2}\right)_6 = 1$$

or equivalently,

$$\left(\frac{1}{n_x^2} + r_{1k}E_k\right)x^2 + \left(\frac{1}{n_y^2} + r_{2k}E_k\right)y^2 + \left(\frac{1}{n_z^2} + r_{3k}E_k\right)z^2$$
$$+ 2yzr_{4k}E_k + 2zxr_{5k}E_k + 2xyr_{6k}E_k = 1 \tag{9.1-9}$$

where, again, summations over repeated indices are assumed.

Furthermore, the linear change in the coefficients

$$\left(\frac{1}{n^2}\right)_i, \quad i = 1, \ldots, 6$$

due to an arbitrary dc electric field $\mathbf{E}(E_x, E_y, E_z)$ can be expressed in a matrix form as

$$\begin{bmatrix} \Delta(1/n^2)_1 \\ \Delta(1/n^2)_2 \\ \Delta(1/n^2)_3 \\ \Delta(1/n^2)_4 \\ \Delta(1/n^2)_5 \\ \Delta(1/n^2)_6 \end{bmatrix} = \begin{bmatrix} r_{11} & r_{12} & r_{13} \\ r_{21} & r_{22} & r_{23} \\ r_{31} & r_{32} & r_{33} \\ r_{41} & r_{42} & r_{43} \\ r_{51} & r_{52} & r_{53} \\ r_{61} & r_{62} & r_{63} \end{bmatrix} \begin{bmatrix} E_x \\ E_y \\ E_z \end{bmatrix} \tag{9.1-10}$$

where, using the rules for matrix multiplication, we have, for example,

$$\Delta\left(\frac{1}{n^2}\right)_6 = r_{61}E_x + r_{62}E_y + r_{63}E_z \tag{9.1-11}$$

The 6×3 matrix with elements r_{ij} is called the electro-optic tensor. We have shown above that in crystals possessing an inversion symmetry (centrosymmetric), $r_{ij} = 0$. The form, but not the magnitude, of the tensor r_{ij} can be derived from symmetry considerations [1], which dictate which of the 18 r_{ij} coefficients are zero, as well as the relationships that exist between the remaining coefficients. In Table 9.1 we give the form of the electro-optic tensor for all the noncentrosymmetric crystal classes. The electro-optic coefficients of some crystals are given in Table 9.2.

TABLE 9.1 Electro-optic Coefficients in Contracted Notation for all Crystal Symmetry Classes[a]

Centrosymmetric: ($\bar{1}$,2/*m*.*mmm*.4/*m*.4/*mmm*, $\bar{3}$, $\bar{3}m$, 6/*m*, 6/*mmm*, *m*3, *m*3*m*)

$$\begin{bmatrix} 0 & 0 & 0 \\ 0 & 0 & 0 \\ 0 & 0 & 0 \\ 0 & 0 & 0 \\ 0 & 0 & 0 \\ 0 & 0 & 0 \end{bmatrix}$$

Triclinic: (1)

$$\begin{bmatrix} r_{11} & r_{12} & r_{13} \\ r_{21} & r_{22} & r_{23} \\ r_{31} & r_{32} & r_{33} \\ r_{41} & r_{42} & r_{43} \\ r_{51} & r_{52} & r_{53} \\ r_{61} & r_{62} & r_{63} \end{bmatrix}$$

Monoclinic: $2\ (2 \parallel x_2)$ and $2\ (2 \parallel x_3)$

$$\begin{bmatrix} 0 & r_{12} & 0 \\ 0 & r_{22} & 0 \\ 0 & r_{32} & 0 \\ r_{41} & 0 & r_{43} \\ 0 & r_{52} & 0 \\ r_{61} & 0 & r_{63} \end{bmatrix} \quad \begin{bmatrix} 0 & 0 & r_{13} \\ 0 & 0 & r_{23} \\ 0 & 0 & r_{33} \\ r_{41} & r_{42} & 0 \\ r_{51} & r_{52} & 0 \\ 0 & 0 & r_{63} \end{bmatrix}$$

TABLE 9.1 *(cont'd)*

$m\ (m \perp x_2)$

$$\begin{bmatrix} r_{11} & 0 & r_{13} \\ r_{21} & 0 & r_{23} \\ r_{31} & 0 & r_{33} \\ 0 & r_{42} & 0 \\ r_{51} & 0 & r_{53} \\ 0 & r_{62} & 0 \end{bmatrix}$$

$m\ (m \perp x_3)$

$$\begin{bmatrix} r_{11} & r_{12} & 0 \\ r_{21} & r_{22} & 0 \\ r_{31} & r_{32} & 0 \\ 0 & 0 & r_{43} \\ 0 & 0 & r_{53} \\ r_{61} & r_{62} & 0 \end{bmatrix}$$

Orthorhombic:

222

$$\begin{bmatrix} 0 & 0 & 0 \\ 0 & 0 & 0 \\ 0 & 0 & 0 \\ r_{41} & 0 & 0 \\ 0 & r_{52} & 0 \\ 0 & 0 & r_{63} \end{bmatrix}$$

$2mm$

$$\begin{bmatrix} 0 & 0 & r_{13} \\ 0 & 0 & r_{23} \\ 0 & 0 & r_{33} \\ 0 & r_{42} & 0 \\ r_{51} & 0 & 0 \\ 0 & 0 & 0 \end{bmatrix}$$

Tetragonal:

4

$$\begin{bmatrix} 0 & 0 & r_{13} \\ 0 & 0 & r_{13} \\ 0 & 0 & r_{33} \\ r_{41} & r_{51} & 0 \\ r_{51} & -r_{41} & 0 \\ 0 & 0 & 0 \end{bmatrix}$$

$\bar{4}$

$$\begin{bmatrix} 0 & 0 & r_{13} \\ 0 & 0 & -r_{13} \\ 0 & 0 & 0 \\ r_{41} & -r_{51} & 0 \\ r_{51} & r_{41} & 0 \\ 0 & 0 & r_{63} \end{bmatrix}$$

422

$$\begin{bmatrix} 0 & 0 & 0 \\ 0 & 0 & 0 \\ 0 & 0 & 0 \\ r_{41} & 0 & 0 \\ 0 & -r_{41} & 0 \\ 0 & 0 & 0 \end{bmatrix}$$

$4mm$

$$\begin{bmatrix} 0 & 0 & r_{13} \\ 0 & 0 & r_{13} \\ 0 & 0 & r_{33} \\ 0 & r_{51} & 0 \\ r_{51} & 0 & 0 \\ 0 & 0 & 0 \end{bmatrix}$$

$\bar{4}2m\ (2 \parallel x_1)$

$$\begin{bmatrix} 0 & 0 & 0 \\ 0 & 0 & 0 \\ 0 & 0 & 0 \\ r_{41} & 0 & 0 \\ 0 & r_{41} & 0 \\ 0 & 0 & r_{63} \end{bmatrix}$$

Trigonal:

3

$$\begin{bmatrix} r_{11} & -r_{22} & r_{13} \\ -r_{11} & r_{22} & r_{13} \\ 0 & 0 & r_{33} \\ r_{41} & r_{51} & 0 \\ r_{51} & -r_{41} & 0 \\ -r_{22} & -r_{11} & 0 \end{bmatrix}$$

32

$$\begin{bmatrix} r_{11} & 0 & 0 \\ -r_{11} & 0 & 0 \\ 0 & 0 & 0 \\ r_{41} & 0 & 0 \\ 0 & -r_{41} & 0 \\ 0 & -r_{11} & 0 \end{bmatrix}$$

$3m\ (m \perp x_1)$

$$\begin{bmatrix} 0 & -r_{22} & r_{13} \\ 0 & r_{22} & r_{13} \\ 0 & 0 & r_{33} \\ 0 & r_{51} & 0 \\ r_{51} & 0 & 0 \\ -r_{22} & 0 & 0 \end{bmatrix}$$

$3m\ (m \perp x_2)$

$$\begin{bmatrix} r_{11} & 0 & r_{13} \\ -r_{11} & 0 & r_{13} \\ 0 & 0 & r_{33} \\ 0 & r_{51} & 0 \\ r_{51} & 0 & 0 \\ 0 & -r_{11} & 0 \end{bmatrix}$$

TABLE 9.1 *(cont'd)*

Hexagonal:

6
$$\begin{bmatrix} 0 & 0 & r_{13} \\ 0 & 0 & r_{13} \\ 0 & 0 & r_{33} \\ r_{41} & r_{51} & 0 \\ r_{51} & -r_{41} & 0 \\ 0 & 0 & 0 \end{bmatrix}$$

$6mm$
$$\begin{bmatrix} 0 & 0 & r_{13} \\ 0 & 0 & r_{13} \\ 0 & 0 & r_{33} \\ 0 & r_{51} & 0 \\ r_{51} & 0 & 0 \\ 0 & 0 & 0 \end{bmatrix}$$

622
$$\begin{bmatrix} 0 & 0 & 0 \\ 0 & 0 & 0 \\ 0 & 0 & 0 \\ r_{41} & 0 & 0 \\ 0 & -r_{41} & 0 \\ 0 & 0 & 0 \end{bmatrix}$$

$\bar{6}$
$$\begin{bmatrix} r_{11} & -r_{22} & 0 \\ -r_{11} & r_{22} & 0 \\ 0 & 0 & 0 \\ 0 & 0 & 0 \\ 0 & 0 & 0 \\ -r_{22} & -r_{11} & 0 \end{bmatrix}$$

$\bar{6}m2\ (m \perp x_1)$
$$\begin{bmatrix} 0 & -r_{22} & 0 \\ 0 & r_{22} & 0 \\ 0 & 0 & 0 \\ 0 & 0 & 0 \\ 0 & 0 & 0 \\ -r_{22} & 0 & 0 \end{bmatrix}$$

$\bar{6}m2\ (m \perp x_2)$
$$\begin{bmatrix} r_{11} & 0 & 0 \\ -r_{11} & 0 & 0 \\ 0 & 0 & 0 \\ 0 & 0 & 0 \\ 0 & 0 & 0 \\ 0 & -r_{11} & 0 \end{bmatrix}$$

Cubic:

$\bar{4}3m$, 23
$$\begin{bmatrix} 0 & 0 & 0 \\ 0 & 0 & 0 \\ 0 & 0 & 0 \\ r_{41} & 0 & 0 \\ 0 & r_{41} & 0 \\ 0 & 0 & r_{41} \end{bmatrix}$$

432
$$\begin{bmatrix} 0 & 0 & 0 \\ 0 & 0 & 0 \\ 0 & 0 & 0 \\ 0 & 0 & 0 \\ 0 & 0 & 0 \\ 0 & 0 & 0 \end{bmatrix}$$

[a] The symbol over each matrix is the conventional symmetry-group designation.

TABLE 9.2 Linear Electro-optic Coefficients of Some Commonly Used Crystals

Substance	Symmetry	Wavelength λ (μm)	Electro-optic Coefficients r_{lk} (10^{-12} m/V)	Index of Refraction n_i	n^3r (10^{-12} m/V)	Dielectric Constant[a] $\varepsilon_i(\varepsilon_0)$
CdTe	$\bar{4}3m$	1.0	$(T)\ r_{41} = 4.5$	$n = 2.84$	103	$(S)\ \varepsilon = 9.4$
		3.39	$(T)\ r_{41} = 6.8$			
		10.6	$(T)\ r_{41} = 6.8$	$n = 2.60$	120	
		23.35	$(T)\ r_{41} = 5.47$	$n = 2.58$	94	
		27.95	$(T)\ r_{41} = 5.04$	$n = 2.53$	82	
GaAs	$\bar{4}3m$	0.9	$r_{41} = 1.1$	$n = 3.60$	51	$(S)\ \varepsilon = 13.2$
		1.15	$(T)\ r_{41} = 1.43$	$n = 3.43$	58	$(T)\ \varepsilon = 12.3$
		3.39	$(T)\ r_{41} = 1.24$	$n = 3.3$	45	
		10.6	$(T)\ r_{41} = 1.51$	$n = 3.3$	54	
GaP	$\bar{4}3m$	0.55–1.3	$(T)\ r_{41} = -1.0$	$n = 3.66$–3.08		$(S)\ \varepsilon = 10$
		0.633	$(S)\ r_{41} = -0.97$	$n = 3.32$	35	
		1.15	$(S)\ r_{41} = -1.10$	$n = 3.10$	33	
		3.39	$(S)\ r_{41} = -0.97$	$n = 3.02$	27	
β-ZnS (sphalerite)	$\bar{4}3m$	0.4	$(T)\ r_{41} = 1.1$	$n = 2.52$	18	$(T)\ \varepsilon = 16$
		0.5	$(T)\ r_{41} = 1.81$	$n = 2.42$		$(S)\ \varepsilon = 12.5$
		0.6	$(T)\ r_{41} = 2.1$	$n = 2.36$		
		0.633	$(S)\ r_{41} = -1.6$	$n = 2.35$		
		3.39	$(S)\ r_{41} = -1.4$			

TABLE 9.2 *(cont'd)*

Substance	Symmetry	Wavelength λ (μm)	Electro-optic Coefficients r_{lk} (10^{-12} m/V)		Index of Refraction n_i	n^3r (10^{-12} m/V)	Dielectric Constant[a] $\varepsilon_i(\varepsilon_0)$
ZnSe	$\bar{4}3m$	0.548	$(T)\ r_{41} = 2.0$		$n = 2.66$		$(T)\ \varepsilon = 9.1$
		0.633	$(S)\ r_{41} = 2.0$		$n = 2.60$	35	$(S)\ \varepsilon = 9.1$
		10.6	$(T)\ r_{41} = 2.2$		$n = 2.39$		
ZnTe	$\bar{4}3m$	0.589	$(T)\ r_{41} = 4.51$		$n = 3.06$		$(T)\ \varepsilon = 10.1$
		0.616	$(T)\ r_{41} = 4.27$		$n = 3.01$		$(S)\ \varepsilon = 10.1$
		0.633	$(T)\ r_{41} = 4.04$		$n = 2.99$	108	
			$(S)\ r_{41} = 4.3$				
		0.690	$(T)\ r_{41} = 3.97$		$n = 2.93$		
		3.41	$(T)\ r_{41} = 4.2$		$n = 2.70$	83	
		10.6	$(T)\ r_{41} = 3.9$		$n = 2.70$	77	
$Bi_{12}SiO_{20}$	23	0.633	$r_{41} = 5.0$		$n = 2.54$	82	
CdSe	$6mm$	3.39	$(S)\ r_{13} = 1.8$		$n_o = 2.452$		$(T)\ \varepsilon_1 = 9.70$
							$(T)\ \varepsilon_3 = 10.65$
			$(T)\ r_{33} = 4.3$		$n_e = 2.471$		$(S)\ \varepsilon_1 = 9.33$
							$(S)\ \varepsilon_3 = 10.20$
α-ZnS (wurtzite)	$6mm$	0.633	$(S)\ r_{13} = 0.9$		$n_o = 2.347$		$(T)\ \varepsilon_1 = \varepsilon_2 = 8.7$
			$(S)\ r_{33} = 1.8$		$n_e = 2.360$		$(S)\ \varepsilon_1 = 8.7$
$Pb_{0.814}La_{0.214}(Ti_{0.6}Zr_{0.4})O_3$ (PLZT)	∞m	0.546	$n_e^3 r_{33} - n_o^3 r_{13} = 2320$		$n_o = 2.55$		
$LiIO_3$	6	0.633	$(S)\ r_{13} = 4.1$	$(S)\ r_{33} = 6.4$	$n_o = 1.8830$		
			$(S)\ r_{41} = 1.4$	$(S)\ r_{51} = 3.3$	$n_o = 1.7367$		
Ag_3AsS_3	$3m$	0.633	$(S)\ n_o^3 r_e = 70$		$n_o = 3.019$		
			$(S)\ n_e^3 r_{22} = 29$		$n_e = 2.739$		
$LiNbO_3$ ($T_c = 1230$ °C)	$3m$	0.633	$(T_4)\ r_{13} = 9.6$	$(S)\ r_{13} = 8.6$	$n_o = 2.286$		$(T)\ \varepsilon_1 = \varepsilon_2 = 78$
			$(T)\ r_{22} = 6.8$	$(S)\ r_{22} = 3.4$	$n_e = 2.200$		$(T)\ \varepsilon_2 = 32$
			$(T)\ r_{33} = 30.9$	$(S)\ r_{33} = 30.8$			$(S)\ \varepsilon_1 = \varepsilon_2 = 43$
			$(T)\ r_{51} = 32.6$	$(T)\ r_{51} = 28$			$(S)\ \varepsilon_3 = 28$
			$(T)\ r_c = 21.1$				
		1.15	$(T)\ r_{22} = 5.4$		$n_o = 2.229$		
			$(T)\ r_c = 19$		$n_e = 2.150$		
		3.39	$(T)\ r_{22} = 3.1$	$(S)\ r_{33} = 28$	$n_o = 2.136$		
			$(T)\ r_c = 18$	$(S)\ r_{22} = 3.1$	$n_e = 2.073$		
				$(S)\ r_{13} = 6.5$			
				$(S)\ r_{51} = 23$			
$LiTaO_3$	$3m$	0.633	$(T)\ r_{13} = 8.4$	$(S)\ r_{13} = 7.5$	$n_o = 2.176$		$(T)\ \varepsilon_1 = \varepsilon_2 = 51$
			$(T)\ r_{33} = 30.5$	$(S)\ r_{33} = 33$	$n_e = 2.180$		$(T)\ \varepsilon_3 = 45$
			$(T)\ r_{22} = -0.2$	$(S)\ r_{51} = 20$			$(S)\ \varepsilon_1 = \varepsilon_2 = 41$
			$(T)\ r_c = 22$	$(S)\ r_{22} = 1$			$(S)\ \varepsilon_3 = 43$
		3.39	$(S)\ r_{33} = 27$		$n_o = 2.060$		
			$(S)\ r_{13} = 4.5$		$n_e = 2.065$		
			$(S)\ r_{51} = 15$				
			$(S)\ r_{22} = 0.3$				
$AgGaS_2$	$\bar{4}2m$	0.633	$(T)\ r_{41} = 4.0$		$n_o = 2.553$		
			$(T)\ r_{63} = 3.0$		$n_e = 2.507$		
CsH_2AsO_4 (CDA)	$\bar{4}2m$	0.55	$(T)\ r_{41} = 14.8$		$n_o = 1.572$		
			$(T)\ r_{63} = 18.2$		$n_e = 1.550$		
KH_2PO_4 (KDP)	$\bar{4}2m$	0.546	$(T)\ r_{41} = 8.77$		$n_o = 1.5115$		$(T)\ \varepsilon_1 = \varepsilon_2 = 42$
			$(T)\ r_{63} = 10.3$		$n_e = 1.4698$		$(T)\ \varepsilon_3 = 21$
		0.633	$(T)\ r_{41} = 8$		$n_o = 1.5074$		$(S)\ \varepsilon_1 = \varepsilon_2 = 44$
			$(T)\ r_{63} = 11$		$n_e = 1.4669$		$(S)\ \varepsilon_3 = 21$
		3.39	$(T)\ r_{63} = 9.7$				
			$(T)\ n_o^3 r_{63} = 33$				

TABLE 9.2 *(cont'd)*

Substance	Symmetry	Wavelength λ (μm)	Electro-optic Coefficients r_{lk} (10^{-12} m/V)		Index of Refraction n_i	$n^3 r$ (10^{-12} m/V)	Dielectric Constant[a] $\varepsilon_i(\varepsilon_0)$
KD_2PO_4 (KD^*P)	$\bar{4}2m$	0.546 0.633	$(T)\ r_{63} = 26.8$ $(T)\ r_{41} = 8.8$ $(T)\ r_{63} = 24.1$		$n_o = 1.5079$ $n_e = 1.4683$ $n_o = 1.502$ $n_e = 1.462$		$(T)\ \varepsilon_3 = 50$ $(S)\ \varepsilon_1 = \varepsilon_2 = 58$ $(S)\ \varepsilon_3 = 48$
$(NH_4)H_2PO_4$ (ADP)	$\bar{4}2m$	0.546 0.633	$(T)\ r_{41} = 23.76$ $(T)\ r_{63} = 8.56$ $(T)\ r_{41} = 23.41$ $(T)\ n_o^3 r_{63} = 27.6$		$n_o = 1.5266$ $n_e = 1.4808$ $n_o = 1.5220$ $n_e = 1.4773$		$(T)\ \varepsilon_1 = \varepsilon_2 = 56$ $(T)\ \varepsilon_3 = 15$ $(S)\ \varepsilon_1 = \varepsilon_2 = 58$ $(S)\ \varepsilon_3 = 14$
$(NH_4)D_2PO_4$ (AD^*P)	$\bar{4}2m$	0.633	$(T)\ r_{41} = 40$ $(T)\ r_{63} = 10$		$n_o = 1.516$ $n_e = 1.475$		
$BaTiO_3$ ($T_c = 395$ K)	$4mm$	0.546	$(T)\ r_{51} = 1640$ $(T)\ r_c = 108$	$(S)\ r_{51} = 820$ $(S)\ r_c = 23$	$n_o = 2.437$ $n_e = 2.365$		$(T)\ \varepsilon_1 = \varepsilon_2 = 3600$ $(T)\ \varepsilon_3 = 135$
$(KTa_xNb_{1-x}O_3)$ (KTN), $x = 0.35$ ($T_c = 40$–60 °C)		0.633	$(T)\ r_{51} = 8000(T_c - 28)$ $(T)\ r_c = 500(T_c - 28)$ $(T)\ r_{51} = 3000(T_c - 16)$ $(T)\ r_c = 700(T_c - 16)$		$n_o = 2.318$ $n_e = 2.277$ $n_o = 2.318$ $n_e = 2.281$		
$Ba_{0.25}Sr_{0.75}Nb_2O_6$ ($T_c = 395$ K)	$4mm$	0.633	$(T)\ r_{13} = 67$ $(T)\ r_{33} = 1340$	$(T)\ r_{51} = 42$ $(S)\ r_c = 1090$	$n_o = 2.3117$ $n_e = 2.2987$		$\varepsilon_3 = 3400$ (15 MHz)
α-HIO_3	222	0.633	$(T)\ r_{41} = 6.6$ $(T)\ r_{52} = 7.0$ $(T)\ r_{63} = 6.0$	$(S)\ r_{41} = 2.3$ $(S)\ r_{52} = 2.6$ $(S)\ r_{63} = 4.3$	$n_1 = 1.8365$ $n_2 = 1.984$ $n_3 = 1.960$		
$KNbO_3$	$2mm$	0.633	$(T)\ r_{13} = 28$ $(T)\ r_{42} = 380$ $(T)\ r_{51} = 105$	$(T)\ r_{23} = 1.3$ $(T)\ r_{33} = 64$ $(S)\ r_{42} = 270$	$n_1 = 2.280$ $n_2 = 2.329$ $n_3 = 2.169$		
KIO_3	1	0.500	$r_{62} = 90$		$n_1 = 1.700$ $n_2 = 1.828$ (5893 Å) $n_3 = 1.832$		

[a] (T) = low frequency from dc through audio range; (S) = high frequency.

EXAMPLE: ELECTRO-OPTIC EFFECT IN KH_2PO_4

Consider the specific example of a crystal of potassium dihydrogen phosphate (KH_2PO_4), also known as KDP. The crystal has a fourfold axis of symmetry (C_4), which by strict convention is taken as the z (optic) axis, as well as two mutually orthogonal twofold axes of symmetry that lie in the plane normal to z. These are designated as the x and y axes. The symmetry group of this crystal is $\bar{4}2m$. A crystal with an n-fold axis of symmetry (C_n) is invariant under a rotation of $2\pi/n$ around the axis. Using Table 9.1, we take the electro-optic tensor in the form of

$$r_{ij} = \begin{bmatrix} 0 & 0 & 0 \\ 0 & 0 & 0 \\ 0 & 0 & 0 \\ r_{41} & 0 & 0 \\ 0 & r_{41} & 0 \\ 0 & 0 & r_{63} \end{bmatrix} \tag{9.1-12}$$

so the only nonvanishing elements are $r_{41} = r_{52}$ and r_{63}. Using Equation (9.1-9), we obtain the equation of the index ellipsoid in the presence of a field $\mathbf{E}(E_x, E_y, E_z)$ as

$$\frac{x^2}{n_o^2} + \frac{y^2}{n_o^2} + \frac{z^2}{n_e^2} + 2r_{41}E_x yz + 2r_{41}E_y xz + 2r_{63}E_z xy = 1 \tag{9.1-13}$$

where the constants involved in the first three terms do not depend on the field and, since the crystal is uniaxial, are taken as $n_x = n_y = n_o$ and $n_z = n_e$. We thus find that the application of an electric field causes the appearance of "mixed" terms in the equation of the index ellipsoid. These are the terms with xy, xz, and yz. This means that the major axes of the ellipsoid, with a field applied, are no longer parallel to the x, y, and z axes. It becomes necessary, then, to find the directions and magnitudes of the new axes, in the presence of $\mathbf{E}$, so that we may determine the effect of the field on the propagation. To be specific, we choose the direction of the applied field parallel to the z axis, so Equation (9.1-13) becomes

$$\frac{x^2 + y^2}{n_o^2} + \frac{z^2}{n_e^2} + 2r_{63}E_z xy = 1 \tag{9.1-14}$$

The problem is one of finding a new coordinate system—x', y', z'—in which the equation of the ellipsoid (9.1-14) contains no mixed terms; that is, it is of the form

$$\frac{x'^2}{n_{x'}^2} + \frac{y'^2}{n_{y'}^2} + \frac{z'^2}{n_{z'}^2} = 1 \tag{9.1-15}$$

x', y', and z' are then the directions of the major axes of the ellipsoid in the presence of an external field applied parallel to z. The lengths of the major axes of the ellipsoid are, according to Equation (9.1-15), $2n_{x'}$, $2n_{y'}$, and $2n_{z'}$ and these will, in general, depend on the applied field.

In the case of (9.1-14) it is clear from inspection that in order to put the equation in diagonal form, we need to choose a coordinate system x', y', z' where z' is parallel to z, and because of the symmetry of (9.1-14) in x and y, x' and y' are related to x and y by a 45° rotation, as shown in Figure 9.1. The transformation relations from x, y to x', y' are thus

$$x = x' \cos 45° + y' \sin 45°$$

$$y = -x' \sin 45° + y' \cos 45°$$

which, upon substitution in Equation (9.1-14), yield

$$\left(\frac{1}{n_o^2} - r_{63}E_z\right)x'^2 + \left(\frac{1}{n_o^2} + r_{63}E_z\right)y'^2 + \frac{z^2}{n_e^2} = 1 \tag{9.1-16}$$

Equation (9.1-16) shows that x', y', and z are indeed the principal axes of the ellipsoid when a field is applied along the z direction. According to Equation (9.1-16), the length of the x' axis of the ellipsoid is $2n_{x'}$, where

$$\frac{1}{n_{x'}^2} = \frac{1}{n_o^2} - r_{63}E_z$$

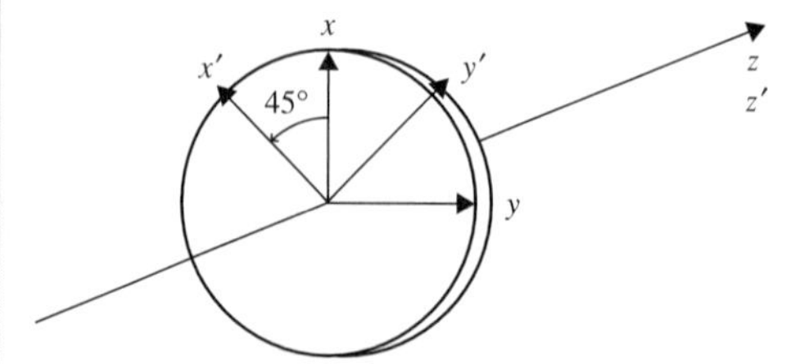

Figure 9.1 Coordinate rotation to transform the quadratic equation into a diagonal form. In this case, the z-axis is the fourfold axis of symmetry, and x and y are the twofold axes of symmetry in the crystals with $\bar{4}2m$ symmetry.

which, assuming $r_{63}E_z \ll n_o^{-2}$ and using the differential relation $dn = -(n^3/2)d(1/n^2)$, gives for the change in $n_{x'}$, $dn_{x'} = (n_o^3/2)r_{63}E_z$ so that

$$n_{x'} = n_o + \frac{n_o^3}{2} r_{63}E_z \tag{9.1-17}$$

and, similarly,

$$n_{y'} = n_o - \frac{n_o^3}{2} r_{63}E_z \tag{9.1-18}$$

$$n_{z'} = n_z = n_e \tag{9.1-19}$$

Let us now consider the case when the applied electric field is parallel to the x axis, so that Equation (9.1-13) becomes

$$\frac{x^2}{n_o^2} + \frac{y^2}{n_o^2} + \frac{z^2}{n_e^2} + 2r_{41}E_x yz = 1 \tag{9.1-20}$$

In this case it is clear from inspection of Equation (9.1-20) that the new principal axis x' will coincide with the x axis, because the "mixed" term involves only y and z. A rotation in the yz plane is therefore required to put it in diagonal form. Let θ be the angle between the new coordinate $y'z'$ and the old coordinate yz. The transformation from x, y, z to x', y', z' is given by

$$\begin{aligned} x &= x' \\ y &= y' \cos\theta - z' \sin\theta \\ z &= y' \sin\theta + z' \cos\theta \end{aligned} \tag{9.1-21}$$

We now substitute Equation (9.1-21) for y and z in Equation (9.1-20) and require that the coefficient for the $y'z'$ term vanishes. This yields

$$\frac{x'^2}{n_o^2} + \left(\frac{1}{n_o^2} + r_{41}E_x \tan\theta\right)y'^2 + \left(\frac{1}{n_e^2} - r_{41}E_x \tan\theta\right)z'^2 = 1 \tag{9.1-22}$$

with θ given by

$$\tan\theta = \frac{2n_e^2 n_o^2}{n_e^2 - n_o^2} r_{41}E_x \tag{9.1-23}$$

The new index ellipsoid (9.1-22) has its principal axes rotated at an angle θ about the x axis with respect to the principal axes of the crystal when a field E_x is applied along the x axis. This angle is very small, even for a moderately high field. For KDP with an applied field of $E_x = 10^6$ V/m, this angle is only 0.04°. According to Equation (9.1-23), this angle is only significant for]materials with $n_o \approx n_e$. In particular, $\theta = 45°$ when $n_o = n_e$. The new principal refractive indices, according to Equation (9.1-22), are given by

$$\begin{aligned} n_{x'} &= n_o \\ n_{y'} &= n_o - \tfrac{1}{2} n_o^3 r_{41}E_x \tan\theta \\ n_{z'} &= n_e + \tfrac{1}{2} n_e^3 r_{41}E_x \tan\theta \end{aligned} \tag{9.1-24}$$

For KDP with a moderate field E_x, θ is small and is almost linearly proportional to $r_{41}E_x$ according to Equation (9.1-23). Therefore the change in the refractive indices is of second order in E_x.

EXAMPLE: ELECTRO-OPTIC EFFECT IN $LiNbO_3$

The $LiNbO_3$ crystal has a crystal symmetry of $3m$. The electro-optic coefficients are in the form, according to Table 9.1,

$$r_{ij} = \begin{bmatrix} 0 & -r_{22} & r_{13} \\ 0 & r_{22} & r_{13} \\ 0 & 0 & r_{33} \\ 0 & r_{51} & 0 \\ r_{51} & 0 & 0 \\ -r_{22} & 0 & 0 \end{bmatrix} \tag{9.1-25}$$

We now consider the case when the applied electric field is along the c axis (z axis) of the crystal so that the equation of the index ellipsoid can be written, according to Equation (9.1-9),

$$\left(\frac{1}{n_o^2} + r_{13}E\right)x^2 + \left(\frac{1}{n_o^2} + r_{13}E\right)y^2 + \left(\frac{1}{n_e^2} + r_{33}E\right)z^2 = 1 \tag{9.1-26}$$

where n_o and n_e are the ordinary and extraordinary refractive indices, respectively. Since no mixed terms appear in Equation (9.1-26), the principal axes of the new index ellipsoid remain unchanged. The lengths of the new semiaxes are

$$\begin{aligned} n_x &= n_o - \tfrac{1}{2}n_o^3 r_{13}E \\ n_y &= n_o - \tfrac{1}{2}n_o^3 r_{13}E \\ n_z &= n_e - \tfrac{1}{2}n_e^3 r_{33}E \end{aligned} \tag{9.1-27}$$

Note that under the influence of an applied electric field in the direction of the c axis, the crystal remains uniaxially anisotropic. If a beam of light is propagating along the x axis, the birefringence seen by it is

$$n_z - n_y = (n_e - n_o) - \tfrac{1}{2}(n_e^3 r_{33} - n_o^3 r_{13})E \tag{9.1-28}$$

Note that the birefringence can be tuned electrically.

The electro-optic effect in the important $\bar{4}3m$ crystal class (GaAs, InP, ZnS) is treated in detail in Appendix F.

The General Solution

We now consider the problem of optical propagation in a crystal in the presence of an external dc field along an arbitrary direction. The index ellipsoid with the dc field on is given by Equation (9.1-4), which is written in the quadratic form

$$\eta_{ij}x_i x_j = 1 \tag{9.1-29}$$

where η_{ij} is the impermeability tensor in the presence of the applied electric field. We also use the convention of summation over repeated indices. Our problem consists of finding the directions and magnitudes of the principal axes of the ellipsoid (9.1-29). This is often achieved by a rotation of the coordinate system. In the new coordinate system, the quadratic from (9.1-29) becomes

$$\eta'_{11}x_1'^2 + \eta'_{22}x_2'^2 + \eta'_{33}x_3'^2 = 1 \tag{9.1-30}$$

The axes of the new coordinate system are the principal axes of the ellipsoid, and the lengths of the principal axes are $2\sqrt{1/\eta'_{11}}$, $2\sqrt{1/\eta'_{22}}$, $2\sqrt{1/\eta'_{33}}$.

The mathematical process of transforming a quadratic form (9.1-29) into its principal form, where all mixed terms disappear, is equivalent to the diagonalization of the matrix η_{ij}:

$$\eta_{ij} = \begin{bmatrix} \eta_{11} & \eta_{12} & \eta_{13} \\ \eta_{21} & \eta_{22} & \eta_{23} \\ \eta_{31} & \eta_{32} & \eta_{33} \end{bmatrix} = \begin{bmatrix} \dfrac{1}{n_x^2} + r_{1k}E_k & r_{6k}E_k & r_{5k}E_k \\ r_{6k}E_k & \dfrac{1}{n_y^2} + r_{2k}E_k & r_{4k}E_k \\ r_{5k}E_k & r_{4k}E_k & \dfrac{1}{n_z^2} + r_{3k}E_k \end{bmatrix} \tag{9.1-31}$$

where summations over repeated indices k are assumed. The diagonalization of the impermeability tensor (or matrix) is often done by solving the following eigenvalue problem:

$$\begin{bmatrix} \eta_{11} & \eta_{12} & \eta_{13} \\ \eta_{21} & \eta_{22} & \eta_{23} \\ \eta_{31} & \eta_{32} & \eta_{33} \end{bmatrix} \mathbf{V} = \eta \mathbf{V} \tag{9.1-32}$$

For a real symmetric matrix (or tensor), the above equation yields three real eigenvectors and three real eigenvalues. The eigenvalues are exactly η'_{11}, η'_{22}, and η'_{33}, and the eigenvectors are parallel to the principal axes of the ellipsoid.

EXAMPLE: ELECTRO-OPTIC EFFECT IN KH_2PO_4

To illustrate the method of matrix diagonalization, we use the example of KH_2PO_4 (KDP) with a dc field along the crystal z axis, which was solved earlier in a somewhat less formal fashion.

The index ellipsoid is given by Equation (9.1-14) as

$$\frac{x^2}{n_o^2} + \frac{y^2}{n_o^2} + \frac{z^2}{n_e^2} + 2r_{63}E_z xy = 1 \tag{9.1-33}$$

The impermeability matrix is thus

$$\eta_{ij} = \begin{bmatrix} \dfrac{1}{n_o^2} & r_{63}E_z & 0 \\ r_{63}E_z & \dfrac{1}{n_o^2} & 0 \\ 0 & 0 & \dfrac{1}{n_e^2} \end{bmatrix} \tag{9.1-34}$$

The eigenvalues are given according to Equation (9.1-32) as the roots of the equation

$$\det \begin{vmatrix} \dfrac{1}{n_o^2} - \eta & r_{63}E_z & 0 \\ r_{63}E_z & \dfrac{1}{n_o^2} - \eta & 0 \\ 0 & 0 & \dfrac{1}{n_e^2} - \eta \end{vmatrix} \tag{9.1-35}$$

which upon evaluation is

$$\left(\frac{1}{n_e^2}-\eta\right)\left[\left(\frac{1}{n_o^2}-\eta\right)^2-(r_{63}E_z)^2\right]=0 \tag{9.1-36}$$

The roots are

$$\eta'_{11}=\frac{1}{n_e^2}$$
$$\eta'_{22}=\frac{1}{n_o^2}+r_{63}E_z \tag{9.1-37}$$
$$\eta'_{33}=\frac{1}{n_o^2}-r_{63}E_z$$

in agreement with Equation (9.1-16). These roots are used, one at a time, in Equation (9.1-32) to obtain the eigenvectors (or the principal axes of the ellipsoid).

The linear electro-optic effect (Pockels effect) can also be written in terms of the change in the dielectric constant. Starting from the definition of the dielectric impermeability tensor of Equation (9.1-2), $\eta\varepsilon=\varepsilon_0$, we take a differentiation and obtain

$$(\Delta\eta)\varepsilon+\eta(\Delta\varepsilon)=0 \tag{9.1-38}$$

Multiplying both sides of Equation (9.1-38) by ε from the left, we obtain

$$\Delta\varepsilon=-\frac{\varepsilon(\Delta\eta)\varepsilon}{\varepsilon_0} \tag{9.1-39}$$

where $\Delta\eta$ is the change of the dielectric impermeability tensor due to an applied electric field. In terms of tensor elements, the above equation can be written

$$\Delta\varepsilon_{ij}=\varepsilon_{ij}(\mathbf{E})-\varepsilon_{ij}(0)=-\sum_k\varepsilon_0 n_i^2 n_j^2 r_{ijk}E_k \tag{9.1-40}$$

where n_i and n_j are the principal refractive indices for polarization along the principal axes i and j, respectively.

9.2 ELECTRO-OPTIC MODULATION—PHASE, AMPLITUDE

We have demonstrated in the previous section that an applied electric field can change the index ellipsoid of certain crystals. We also know that the propagation characteristics are governed by the index ellipsoid. Consequently, we can employ the electro-optic effect of these crystals to manipulate the propagation of optical waves, fundamentally their phase and polarization state. This modulation can, if desired, be converted into amplitude modulation. As an example, let us consider the propagation of a beam of polarized light in a KDP crystal under the influence of an applied electric field. Figure 9.2 shows a schematic drawing of a *c*-cut KDP crystal plate (crystal cut with surface perpendicular to the *c* axis) with an electric field **E** applied parallel to the *z* axis (*z* axis is parallel to *c* axis).

The new principal axes (x', y', z) for KDP with **E** applied parallel to z are shown in Figure 9.2. If we consider propagation along the z direction, then according to the procedure described in Chapter 1, we need to determine the ellipse formed by the intersection of the

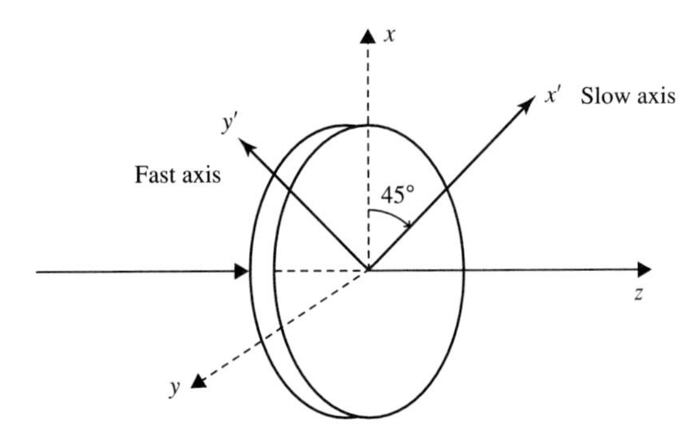

Figure 9.2 A plate of c-cut KDP crystal, showing the principal dielectric axes x', y', and z due to an electric field applied along the z axis. Assuming a positive $r_{63}E_z$, the x' axis is the slow axis, and the y' axis is the fast axis.

plane $z = 0$ (in general, the plane that contains the origin and is normal to the propagation direction) and the index ellipsoid. The equation of this ellipse is obtained from Equation (9.1-16) by putting $z = 0$ and is

$$\left(\frac{1}{n_o^2} - r_{63}E_z\right)x'^2 + \left(\frac{1}{n_o^2} + r_{63}E_z\right)y'^2 = 1 \tag{9.2-1}$$

It follows from the index ellipsoid method that the two normal modes of propagation are polarized along x' and y' and that their indices of refraction are $n_{x'}$ and $n_{y'}$, which are given by Equations (9.1-17) and (9.1-18).

We are now in a position to take up the concept of electro-optic phase retardation. We consider an optical field that is incident normally on the $x'y'$ plane. For this propagation, the birefringence is given by

$$n_{x'} - n_{y'} = n_o^3 r_{63} E \tag{9.2-2}$$

Let l be the thickness of the KDP plate. The phase retardation of this plate is then given by

$$\Gamma = \frac{\omega}{c}(n_{x'} - n_{y'})l = \frac{2\pi}{\lambda}(n_{x'} - n_{y'})l = \frac{2\pi}{\lambda}n_o^3 r_{63} El = \frac{2\pi}{\lambda}n_o^3 r_{63} V \tag{9.2-3}$$

where V is the voltage applied and is given by $V = El$. We know from our discussion in Chapter 1 that phase retardation plates are polarization-state converters. Here we have a plate with a phase retardation proportional to the applied voltage. Consequently, we are able to convert the polarization state of the incident beam of light into a desired polarization state electrically. To illustrate this, let us assume that the incident beam of light is linearly polarized with its $\mathbf{E}_{\text{in}}$ vector along the x direction. The polarization state at the input plane ($z = 0$) can be represented by a Jones vector (in the principal coordinate axes x', y'),

$$\mathbf{E}_{\text{in}} = \frac{1}{\sqrt{2}}\begin{bmatrix} 1 \\ 1 \end{bmatrix} \tag{9.2-4}$$

The polarization state (in the principal coordinate axes x', y') of the emerging beam of light at the output plane ($z = d$) is given by

$$\mathbf{E}_{\text{out}} = \frac{1}{\sqrt{2}}\begin{bmatrix} e^{-i\Gamma/2} \\ e^{+i\Gamma/2} \end{bmatrix} \tag{9.2-5}$$

where Γ is given by Equation (9.2-3). Figure 9.3 shows the polarization ellipse of the output beam of light at various values of the phase retardation Γ. The voltage that yields a phase retardation of π is known as the *half-wave voltage* and is given in this case by

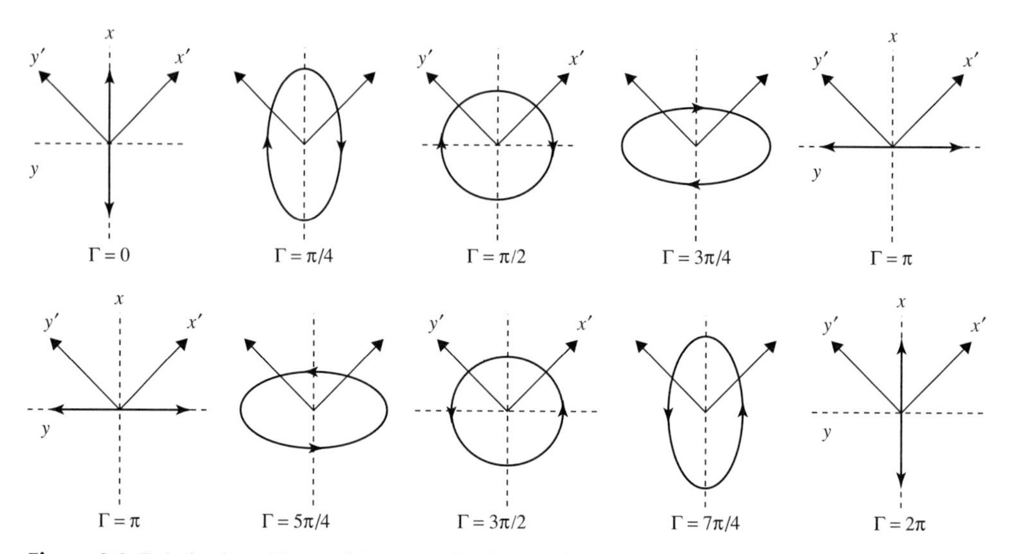

Figure 9.3 Polarization ellipses of the emerging beam with various values of phase retardation.

$$V_\pi = \frac{\lambda}{2n_o^3 r_{63}} \tag{9.2-6}$$

where λ is the wavelength of light. The half-wave voltage for a z-cut KDP plate at $\lambda = 633$ nm is about 9.3 kV. Note that the half-wave voltage is proportional to the wavelength and is inversely proportional to the relevant electro-optic coefficient. A phase retardation of π is needed to transform a polarization state into its orthogonal polarization state.

Amplitude Modulation

An examination of Figure 9.3 reveals that the electrically induced birefringence causes a wave launched at $z = 0$ with its polarization along the x axis to acquire a polarization component along the y axis, which grows with distance at the expense of the x component until, at a position at which $\Gamma = \pi$, the polarization state becomes parallel to the y axis. If this position corresponds to the output plane of the crystal and if one inserts at this point a polarizer at right angles to the input polarization—that is, one that allows only E_y to pass—then with the field on, the optical beam passes through unattenuated, whereas with the field off ($\Gamma = 0$), the output beam is blocked off completely by the crossed output polarizer. This control of the optical energy flow serves as the basis of the electro-optic amplitude modulation of light.

A typical arrangement of an electro-optic amplitude modulator is shown in Figure 9.4. It consists of an electro-optic crystal placed between two crossed polarizers, which, in turn, are at an angle of 45° with respect to the new (electrically induced) principal axes x' and y'. To be specific, we show how this arrangement is achieved using a KDP crystal. Also included in the optical path is a naturally birefringent crystal that introduces a fixed retardation, so the total retardation Γ is the sum of the retardation due to this crystal and the electrically induced one.

The polarization state of the beam emerging from the crystal is given by Equation (9.2-5). With the presence of a polarizer along the y axis, the amplitude of the transmitted beam can be obtained by a simple geometric projection:

$$E'_{\text{out}} = \mathbf{E}_{\text{out}} \cdot \hat{\mathbf{y}} \tag{9.2-7}$$

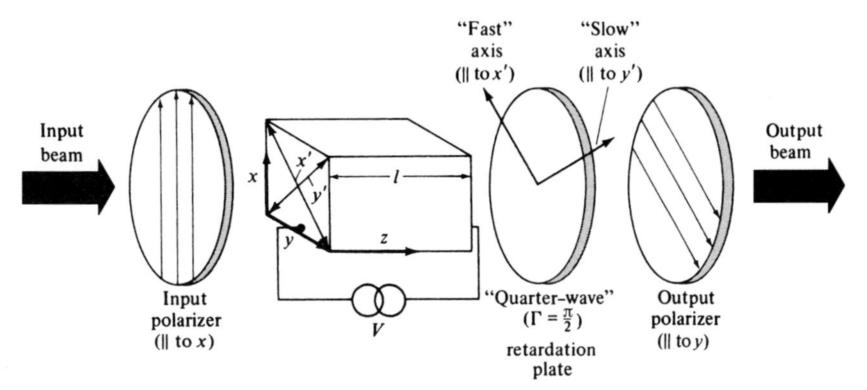

Figure 9.4 A typical electro-optic amplitude modulator. The total retardation Γ is the sum of the fixed retardation bias ($\Gamma_B = \pi/2$) introduced by the quarter-wave plate and that attributable to the electro-optic crystal.

Using

$$\hat{\mathbf{y}} = \frac{1}{\sqrt{2}}\begin{bmatrix} 1 \\ -1 \end{bmatrix} \tag{9.2-8}$$

we obtain

$$E'_{\text{out}} = -i \sin(\Gamma/2) \tag{9.2-9}$$

The ratio of the output intensity to the input is thus

$$\frac{I_o}{I_i} = \sin^2\left(\frac{\Gamma}{2}\right) = \sin^2\left[\left(\frac{\pi}{2}\right)\frac{V}{V_\pi}\right] \tag{9.2-10}$$

The second equality in Equation (9.2-10) was obtained from Equation (9.2-6). The transmission factor (I_o/I_i) is plotted in Figure 9.5 against the applied voltage.

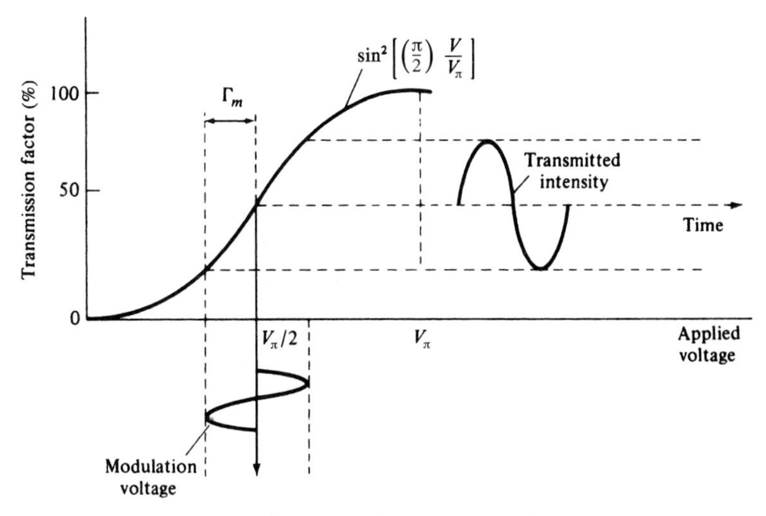

Figure 9.5 Transmission factor of a cross-polarized electro-optic modulator as a function of an applied voltage. The modulator is biased to the point $\Gamma = \pi/2$, which results in a 50% intensity transmission. A small applied sinusoidal voltage modulates the transmitted intensity about the bias point.

The process of amplitude modulation of an optical signal is also illustrated in Figure 9.5. The modulator is usually biased with a fixed retardation $\Gamma = \pi/2$ to the 50% transmission point where the slope is maximum. This bias can be achieved by applying a voltage $V = V_\pi/2$ or, more conveniently, by using a naturally birefringent crystal as in Figure 9.4 to introduce a phase difference (retardation) of $\pi/2$ between the x' and y' components. A small sinusoidal modulation voltage would then cause a nearly sinusoidal modulation of the transmitted intensity as shown.

To treat the situation depicted by Figure 9.5 mathematically, we take

$$\Gamma = \frac{\pi}{2} + \Gamma_m \sin \omega_m t \tag{9.2-11}$$

where the retardation bias is taken as $\pi/2$, and Γ_m is related to the amplitude V_m of the modulation voltage $V_m \sin \omega_m t$ by Equation (9.2-6); thus $\Gamma_m = \pi(V_m/V_\pi)$.

Using Equation (9.2-10), we obtain

$$\frac{I_o}{I_i} = \sin^2\left(\frac{\pi}{4} + \frac{\Gamma_m}{2} \sin \omega_m t\right) \tag{9.2-12}$$

$$= \tfrac{1}{2}[1 + \sin(\Gamma_m \sin \omega_m t)] \tag{9.2-13}$$

which, for $\Gamma_m \ll 1$, becomes

$$\frac{I_o}{I_i} \approx \tfrac{1}{2}(1 + \Gamma_m \sin \omega_m t) \tag{9.2-14}$$

so that the intensity modulation is a linear replica of the modulating voltage $V_m \sin \omega_m t$. If the condition $\Gamma_m \ll 1$ is not fulfilled, it follows from Figure 9.4 or from Equation (9.2-12) that the intensity variation is distorted and will contain an appreciable amount of the higher (odd) harmonics. The dependence of the distortion on Γ_m is discussed further in Problem 9.3.

In Figure 9.6 we show how some information signal $f(t)$ (the electric output of a phonograph stylus in this case) can be impressed electro-optically as an amplitude modulation on a

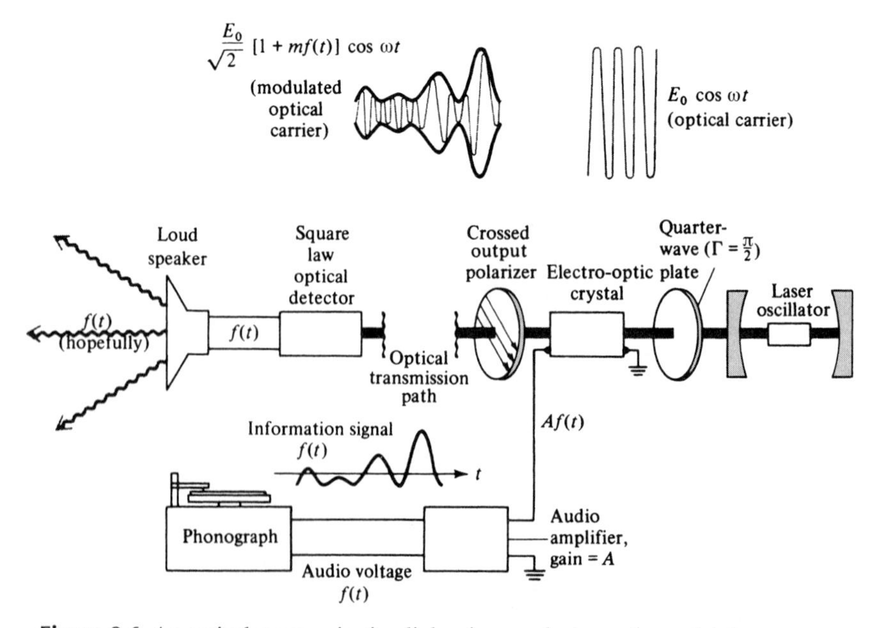

Figure 9.6 An optical communication link using an electro-optic modulator.

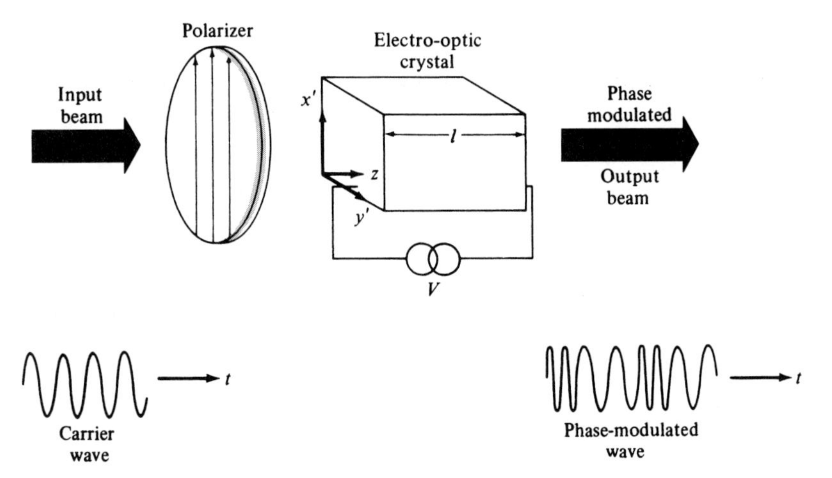

Figure 9.7 An electro-optic phase modulator. The crystal orientation and applied directions are appropriate to KDP. The optical polarization is parallel to an electrically induced principal dielectric axis (x').

laser beam and subsequently be recovered by an optical detector. The details of the optical detection are considered in Chapter 11.

Phase Modulation of Light

In the preceding section we saw how the modulation of the state of polarization, from linear to elliptic, of an optical beam by means of the electro-optic effect can be converted, using polarizers, to intensity modulation. Here we consider the situation depicted by Figure 9.7, in which, instead of there being equal components along the induced birefringent axes (x' and y' in Figure 9.2), the incident beam is polarized parallel to one of them, x' say. In this case the application of the electric field does not change the state of polarization, but merely changes the output phase by

$$\Delta\phi_{x'} = \frac{\omega}{c}\Delta n_{x'}\, l = \frac{2\pi}{\lambda}\Delta n_{x'}\, l \tag{9.2-15}$$

where l is the length of the crystal. From Equation (9.1-17),

$$\Delta\phi_{x'} = \frac{\omega n_o^3 r_{63}}{2c} E_z l = \frac{\pi n_o^3 r_{63}}{\lambda} V = \pi \frac{V}{V_\pi} \tag{9.2-16}$$

If the applied voltage is sinusoidal and is taken as

$$V = V_m \sin \omega_m t \tag{9.2-17}$$

then an incident optical field, which at the input ($z = 0$) face of the crystal is given by $E_{\text{in}} = A \exp(i\omega t)$, will emerge as

$$E_{\text{out}} = A \exp\left[i\left(\omega t - \frac{\omega}{c} n_o l - \frac{V_m}{V_\pi}\pi \sin \omega_m t\right)\right] \tag{9.2-18}$$

where l is the length of the crystal. Dropping the constant phase factor, which is of no consequence here, we rewrite the last equation as

$$E_{\text{out}} = A \exp[i(\omega t - \delta \sin \omega_m t)] \tag{9.2-19}$$

where

$$\delta = \frac{\omega n_o^3 r_{63} E_m l}{2c} = \frac{\pi n_o^3 r_{63} V_m}{\lambda} = \frac{V_m}{V_\pi}\pi \tag{9.2-20}$$

is referred to as the phase modulation index. The optical field is thus phase modulated with a modulation index δ. If we use the Bessel function identity

$$\exp(-i\delta \sin \omega_m t) = \sum_{n=-\infty}^{\infty} J_n(\delta) \exp(-in\omega_m t) \tag{9.2-21}$$

we can rewrite Equation (9.2-20) as

$$E_{\text{out}} = A \sum_{n=-\infty}^{\infty} J_n(\delta) e^{i(\omega - n\omega_m)t} \tag{9.2-22}$$

which form gives the distribution of energy in the sidebands as a function of the modulation index δ. We note that, for $\delta = 0$, $J_0(0) = 1$ and $J_n(\delta) = 0$, $n \neq 0$. Another point of interest is that the phase modulation index δ as given by Equation (9.2-20) is one-half the phase retardation Γ as given by Equation (9.2-3).

Transverse Electro-optic Modulators

In the examples of electro-optic phase retardation discussed earlier in this section, the electric field was applied along the direction of light propagation. This is the so-called longitudinal mode of modulation. A more desirable mode of operation is the transverse one, in which the field is applied perpendicular to the direction of propagation. The reason is that in this case the field electrodes do not interfere with the optical beam, and the retardation, being proportional to the product of the field times the crystal length, can be increased by the use of longer crystals. In the longitudinal case the retardation, according to Equation (9.2-3), is proportional to $E_z l = V$ and is independent of the crystal length l. Figures 9.1 and 9.2 suggest how transverse retardation can be obtained using a KDP crystal with the actual arrangement shown in Figure 9.8. The light propagates along y' and its polarization is in the $x'z$ plane at 45° from the z axis. The retardation, with a field applied along z, is, from Equations (9.1-10) and (9.1-12),

$$\Gamma = \phi_{x'} - \phi_z = \frac{\omega l}{c}\left[(n_o - n_e) + \frac{n_o^3}{2} r_{63}\left(\frac{V}{d}\right)\right] \tag{9.2-23}$$

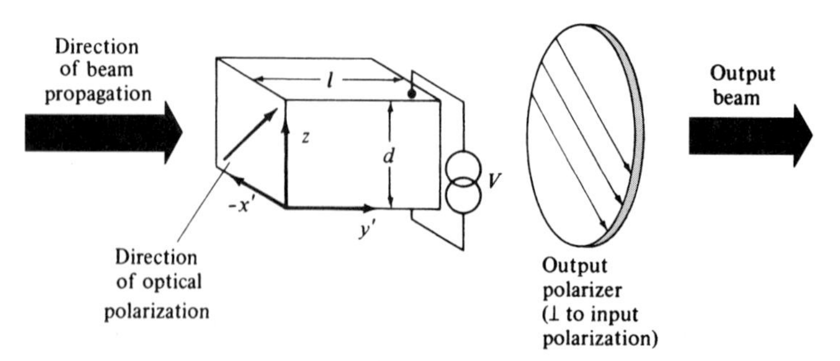

Figure 9.8 A transverse electro-optic amplitude modulator using a KH_2PO_4 (KDP) crystal in which the field is applied normal to the direction of propagation.

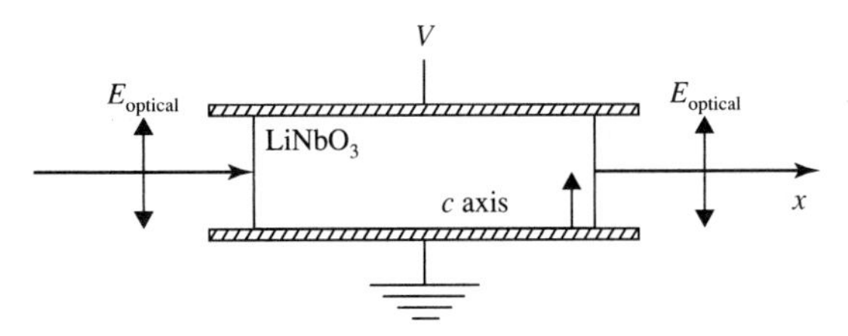

Figure 9.9 Schematic drawing of a transverse electro-optic phase modulator based on $LiNbO_3$ crystal. The double-headed arrows show the direction of polarization of the optical beam.

where d is the crystal dimension along the direction of the applied field. We note that the electro-optic phase retardation is proportional to the length of the crystal, at a given applied voltage. The total phase retardation Γ contains a term that does not depend on the applied voltage. This point will be discussed in Problem 9.2.

We consider next a transverse electro-optic phase modulator made of $LiNbO_3$ crystal (see Figure 9.9). The crystal is cut in a configuration so that the applied electric field is along the direction of the c axis (z axis). An optical beam is propagating along the x axis, with its direction of polarization parallel to the z axis. This choice takes advantage of the largest electro-optic coefficient r_{33} in $LiNbO_3$ crystals.

According to Equations (9.1-27), the principal refractive indices are given by

$$\begin{aligned} n_x &= n_o - \tfrac{1}{2}n_o^3 r_{13}E \\ n_y &= n_o - \tfrac{1}{2}n_o^3 r_{13}E \\ n_z &= n_e - \tfrac{1}{2}n_e^3 r_{33}E \end{aligned} \tag{9.2-24}$$

where E is the applied electric field along the z axis. For a beam of light polarized along the z axis, the phase shift upon emerging from the output face is given by

$$\phi = \frac{2\pi}{\lambda} n_z l = \frac{2\pi}{\lambda}\left(n_e - \frac{1}{2} n_e^3 r_{33} \frac{V}{d}\right) l \tag{9.2-25}$$

We note that the electro-optic phase shift is proportional to the length of the crystal, at a given applied voltage. The half-wave voltage that provides a phase modulation index of π in this case is given by

$$V_\pi = \frac{\lambda}{n_e^3 r_{33}} \frac{d}{l} \tag{9.2-26}$$

where l is the length of interaction. The half-wave voltage can be made smaller by choosing a smaller gap d between the electrodes, and a longer interaction length l. In practice, the gap between the electrodes is limited by the beam size and the problem of diffraction. This is the situation where guided waves can play an important role in the electro-optical modulation. In a dielectric waveguide, the transverse dimension can be on the order of a wavelength, with no diffraction problem. In what follows, we discuss a waveguide version of transverse modulators by using Mach–Zehnder interferometry.

Mach–Zehnder Interferometer Modulators

In optical communications, most external electro-optic modulations are achieved by using a Mach–Zehnder interferometer in an optical circuit. Referring to Figure 9.10, we consider a waveguide version of a Mach–Zehnder interferometer. The particular waveguide circuit can

(a)

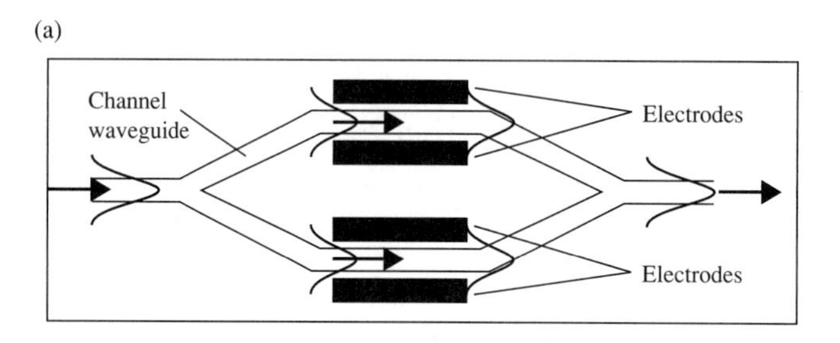

(b)

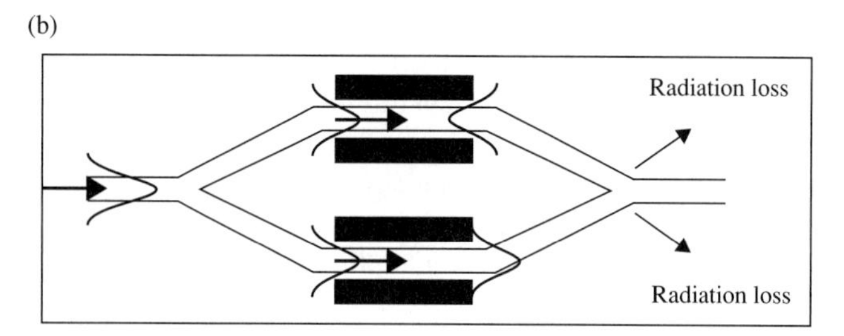

Figure 9.10 Schematic drawing of the top view of a Mach–Zehnder interferometer modulator. (a) Recombination with in-phase beams. (b) Recombination with out-of-phase beams.

be fabricated using various techniques discussed in Chapter 3. For example, a Ti-diffused channel waveguide circuit can be produced on a lithium niobate crystal plate. A channel waveguide is split into two via a Y-junction. Electrodes are deposited on both sides of the channel waveguides on the surface of the crystal plate. When voltages are applied, electric fields are created parallel to the surface and across the waveguides. The phase shifts are proportional to the voltages applied. Following the electrodes, the optical beams in the waveguides are recombined via a second Y-junction. The output intensity is proportional to the square of the combined field,

$$I_{\text{output}} \propto \left| \frac{1}{\sqrt{2}} E e^{-i\phi_1} + \frac{1}{\sqrt{2}} E e^{-i\phi_2} \right|^2 = \tfrac{1}{2}(1 + \cos \Delta\phi) I_{\text{input}} \tag{9.2-27}$$

where I_{input} and I_{output} are input and output beam intensities, and $\Delta\phi = \phi_2 - \phi_1$ is the relative phase shift between the two interfering arms. In the operation, a set of the electrodes astride one of the waveguides can be employed to provide a fixed bias voltage, while the other set of electrodes can be employed to provide the signal modulating voltage.

We note that at the Y-junction where the two beams inside the waveguides recombine, the output beam intensity depends on the phase difference between the two optical beams. When the two optical beams are in-phase ($\Delta\phi = 0$), the recombined field profile matches well that of the mode of the waveguide. As a result, the recombined field continues to propagate past the junction having suffered a small loss only. However, if the two beams are out of phase ($\Delta\phi = \pi$), the recombined field will possess an odd symmetry about the midplane of the waveguide and thus will be orthogonal to the fundamental mode of the waveguide at the junction and will not excite it. It will, however, excite nonconfined radiation modes. This leads to a zero output in the waveguide past the junction. Thus, by controlling the voltage across the electrodes, such an interferometer can be employed for the external electro-optical modulation of laser beams in channel waveguides.

EXAMPLE: HALF-WAVE VOLTAGE IN A WAVEGUIDE MODULATOR

Consider an example of a channel waveguide with a transverse dimension of $d = 10$ μm and an interaction length of $l = 5$ cm. From Table 9.2, we take $n_e = 2.2$ and $r_{33} = 30$ pm/V. The half-wave voltage at $\lambda = 1.5$ μm is

$$V_\pi = \frac{\lambda}{n_e^3 r_{33}} \frac{d}{l} = 0.94 \text{ V}$$

9.3 HIGH-FREQUENCY MODULATION CONSIDERATIONS

In the examples considered in the preceding sections, we derived expressions for the phase retardation caused by electric fields of low frequencies. In many practical situations the modulation signal is often at very high frequencies and, in order to utilize the wide frequency spectrum available with lasers, may occupy a large bandwidth. In this section we consider some of the basic factors limiting the highest usable modulation frequencies in a number of typical experimental situations. Consider first the situation described by Figure 9.11.

The electro-optic crystal is placed between two electrodes with a modulation field at frequency $\omega_0/2\pi$ applied to it. Let R_s be the internal resistance of the modulation source and let C represent the parallel-plate capacitance due to the electro-optic crystal. If $R_s > (\omega_0 C)^{-1}$, most of the modulation voltage drop is across R_s and is thus wasted, since it does not contribute to the retardation. This can be remedied by resonating the crystal capacitance with an inductance L, where $\omega_0^2 = (LC)^{-1}$, as shown in Figure 9.11. In addition, a shunting resistance R_L is used so that at $\omega = \omega_0$ the impedance of the parallel RLC circuit is R_L, which is chosen to be larger than R_s so most of the modulation voltage appears across the crystal. The resonant circuit has a finite bandwidth—that is, its impedance is high only over a frequency interval $\Delta\omega/2\pi = \frac{1}{2}\pi R_L C$ (centered on ω_0). Therefore the maximum modulation bandwidth (the frequency spectrum occupied by the modulation signal) must be less than

$$\Delta\nu = \frac{\Delta\omega}{2\pi} \approx \frac{1}{2\pi R_L C} \tag{9.3-1}$$

if the modulation field is to be a faithful replica of the modulation signal.

In practice, the size of the modulation bandwidth $\Delta\omega/2\pi$ is dictated by the specific application. In addition, one requires a certain peak retardation (or phase shift) Γ_m. Using Equation (9.2-3) to relate Γ_m to the peak modulation voltage $V_m = (E_z)_m L$, we can show, with the aid of Equation (9.3-1), that the power needed in KDP-type crystals to obtain a peak retardation Γ_m is related to the modulation bandwidth $\Delta\nu = \Delta\omega/2\pi$ as

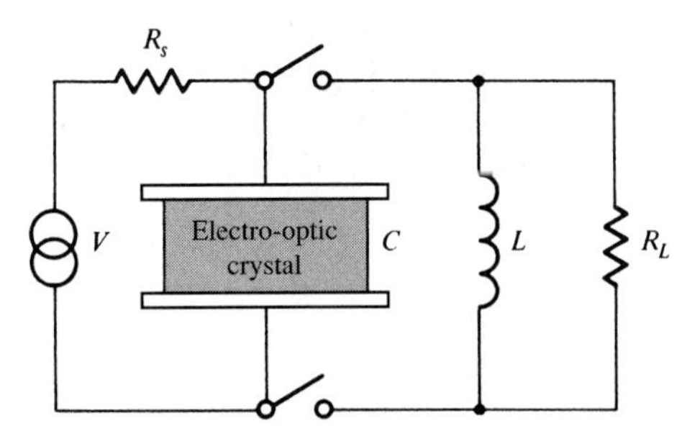

Figure 9.11 Equivalent circuit of an electro-optic modulation crystal in a parallel-plate configuration.

$$P = \frac{\Gamma_m^2 \lambda^2 A \varepsilon \, \Delta\nu}{4\pi L n^6 r_{63}^2} \tag{9.3-2}$$

where n is the effective index of refraction, L is the length of the optical path in the crystal, A is the cross-sectional area of the crystal normal to L, and ε is the dielectric constant at the modulation frequency ω_0.

Transit-Time Limitations to High-Frequency Electro-optic Modulation

According to Equation (9.2-3), the electro-optic retardation (or phase shift) due to an applied electric field E can be written

$$\Gamma = \frac{\omega}{c} n^3 r_{63} E L \equiv \alpha E L \tag{9.3-3}$$

with $\alpha = \omega n^3 r_{\text{eff}}/c$, where ω is the frequency of the optical beam, r_{eff} is the effective electro-optic coefficient, and L is the length of the optical path in the crystal. If the field E changes appreciably during the transit time $\tau = nL/c$ of light through the crystal, we must replace Equation (9.3-3) by

$$\Gamma = \alpha \int_0^L E(z)\, dz = \alpha \frac{c}{n} \int_{t-\tau}^{t} E(t')\, dt' \tag{9.3-4}$$

where c/n is the velocity of light in the modulator medium and $E(t')$ is the instantaneous electric field. In the second integral we replace integration over z by integration over time, recognizing that the portion of the wave that reaches the output face $z = L$ at time $t + \tau$ entered the crystal at time t. We also assumed that at any given moment the field $E(t)$ has the same value throughout the crystal. Taking $E(t')$ as a sinusoid,

$$E(t') = E_m \cos(\omega_m t')$$

where ω_m is the modulation frequency. We obtain from Equation (9.3-4)

$$\Gamma = \alpha \frac{c}{n} E_m \int_t^{t+\tau} \cos(\omega_m t')\, dt' = \Gamma_0 \left(\frac{\sin(\omega_m \tau/2)}{\omega_m \tau/2} \right) \cos(\omega_m t + \omega_m \tau/2) \tag{9.3-5}$$

where $\Gamma_0 = \alpha(c/n)\tau E_m = \alpha L E_m$ is the phase shift (or retardation) when $\omega_m \tau \ll 1$. The reduction factor

$$\eta = \frac{\sin(\omega_m \tau/2)}{\omega_m \tau/2} \tag{9.3-6}$$

gives the decrease in phase shift (or retardation) resulting from the finite transit time τ. For $\eta \sim 1$ (i.e., no reduction), the condition $\omega_m \tau \ll 1$ must be satisfied, so the transit time must be small compared to the shortest modulation period. The factor η is plotted in Figure 9.12.

If, somewhat arbitrarily, we take the highest useful modulation frequency as that for which $\omega_m \tau = \pi/2$ (at this point, according to Figure 9.12, $|\eta| = 0.9$) and we use the relation $\tau = nL/c$, we obtain

$$(\nu_m)_{\max} = \frac{c}{4nL} \tag{9.3-7}$$

which, using a KDP crystal ($n = 1.5$) and a length $L = 1$ cm, yields $(\nu_m)_{\max} = 5 \times 10^9$ Hz.

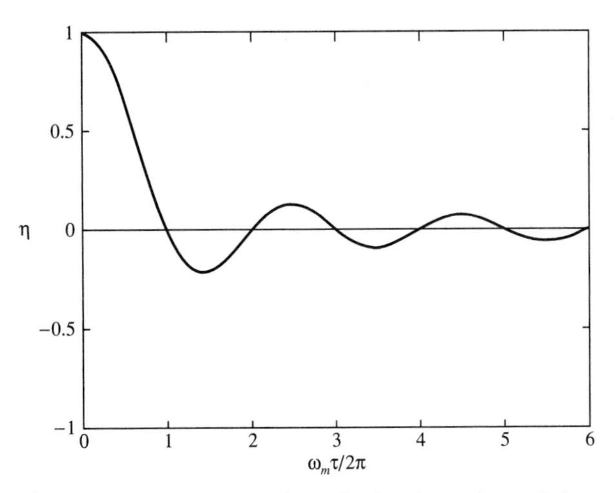

Figure 9.12 Phase retardation reduction factor due to finite transit time τ_o.

Traveling-Wave Modulators

One method that can, in principle, overcome the transit-time limitation involves applying the modulation signal in the form of a traveling wave [2], as shown in Figure 9.13. If the phase velocities of both the optical and modulation fields are equal to each other, any given optical wavefront will exercise the same instantaneous electric field, which corresponds to the field it encounters at the entrance face, as it propagates through the crystal and the transit-time problem discussed earlier is eliminated. This form of modulation can be used only in the transverse geometry that was discussed in the preceding section, since the RF field in most propagating structures is predominantly transverse.

Consider an element of the optical wavefront that *enters* the crystal at $z = 0$ at time t. The position z of this element of optical wavefront at some later time t' is

$$z(t') = \frac{c}{n}(t' - t) \tag{9.3-8}$$

where c/n is the phase velocity of the optical beam. The phase retardation (or phase shift) exercised by this element is given similarly to Equation (9.3-4) by

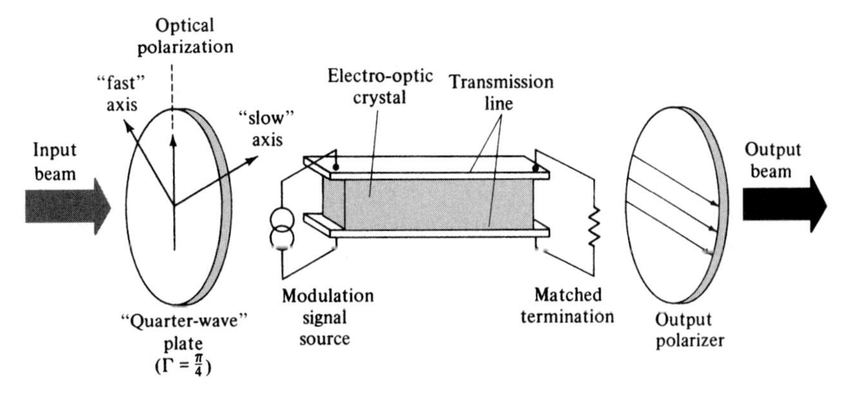

Figure 9.13 A traveling-wave electro-optic modulator.

$$\Gamma = \alpha \frac{c}{n} \int_{t}^{t+\tau} E[t', z(t')]dt' \tag{9.3-9}$$

where $E[t', z(t')]$ is the instantaneous modulation field as seen by an observer traveling with the optical phase front. Take the traveling modulation field as

$$E(t', z) = E_m \cos(\omega_m t' - k_m z) \tag{9.3-10}$$

where E_m is a constant, ω_m is the modulation frequency, and k_m is the wavenumber of the traveling modulation field. Using Equation (9.3-8), we obtain

$$E(t', z) = E_m \cos[\omega_m t' - k_m(c/n)(t' - t)] \tag{9.3-11}$$

Recalling that $k_m = \omega_m/c_m$, where c_m is the phase velocity of the traveling modulation field, we substitute Equation (9.3-11) in (9.3-9) and, carrying out the simple integration, obtain

$$\Gamma = \Gamma_0 \left(\frac{\sin[\omega_m \tau(1 - c/nc_m)/2]}{\omega_m \tau(1 - c/nc_m)/2} \right) \cos[\omega_m t + \omega_m \tau(1 - c/nc_m)/2] \tag{9.3-12}$$

where $\Gamma_0 = \alpha L E_m = \alpha(c/n)\tau E_m$ is the retardation that would result from a dc field equal to E_m.

The reduction factor for the traveling-wave modulator is now written

$$\eta = \frac{\sin[\omega_m \tau(1 - c/nc_m)/2]}{\omega_m \tau(1 - c/nc_m)/2} = \frac{\sin[\omega_m \tau(\Delta n/n)/2]}{\omega_m \tau(\Delta n/n)/2} \tag{9.3-13}$$

where $n_m = c/c_m$ is the "index of refraction" of the modulating field, $\Delta n = (n - n_m)$. We note that the reduction factor is of the same form as that of the lumped-constant modulator (9.3-6) except that τ is replaced by $\tau\, \Delta n/n$. If the two phase velocities are made equal so that $\Delta n = 0$, then $\eta = 1$ and maximum retardation is obtained *regardless* of the crystal length L (or the transit time τ). This is the situation when the modulation field (microwave) and the optical beam are propagating at the same phase velocity in the modulator.

To find the maximum useful modulation frequency and the maximum modulation depth, we rewrite Equation (9.3-12) as

$$\Gamma = \delta \cos\left(\omega_m t + \omega_m \Delta n \frac{L}{2c} \right) \tag{9.3-14}$$

where $\Delta n = (n - n_m)$, and δ is the modulation depth given by

$$\delta = \frac{2c\alpha E_m}{\omega_m\, \Delta n} \sin\left(\omega_m\, \Delta n \frac{L}{2c} \right) \equiv \delta_{\max} \sin\left(\omega_m\, \Delta n \frac{L}{2c} \right) \tag{9.3-15}$$

where $\delta_{\max}$ is the maximum modulation depth

$$\delta_{\max} = \frac{2c\alpha E_m}{\omega_m\, \Delta n} = \frac{\omega}{\omega_m\, \Delta n} n^3 r_{\text{eff}} E_m \tag{9.3-16}$$

We note that the depth of modulation δ is a sinusoidal function of the crystal length L. The maximum modulation occurs at $\omega_m\, \Delta n L/2c = \pi/2$, according to Equation (9.3-15), with a crystal length of

$$L = \frac{\pi c}{\omega_m\, \Delta n} = \frac{\lambda_m}{2\Delta n} \tag{9.3-17}$$

where λ_m is the wavelength of the modulation field, $\Delta n = (n - n_m)$.

The maximum useful modulation frequency is taken, as in the treatment leading to Equation (9.3-7), as that for which $\omega_m \tau(1 - c/nc_m) = \pi/2$, yielding

$$(\nu_m)_{\max} = \frac{c}{4L(n - n_m)} = \frac{c}{4L\,\Delta n} \tag{9.3-18}$$

which, upon comparison with Equation (9.3-7), shows an increase in the frequency limit or useful crystal length by a factor of $n_0/\Delta n$. According to Equations (9.3-16) and (9.3-18), high modulation depth at high-frequency operation requires velocity matching in traveling-wave modulators. The problem of designing traveling-wave electro-optic modulators is considered in References [3–5]. For a more detailed treatment of electro-optic modulation including the traveling-wave and high-frequency cases, the student should consult Reference [6].

9.4 ELECTROABSORPTION AND ELECTROABSORPTION MODULATORS

Electroabsorption is a phenomenon in semiconductors in which the application of an electric field can cause a change in the linear absorption coefficient. Particularly, the Franz–Keldysh effect (FKE) is a phenomenon in which the absorption edge in semiconductors shifts toward longer wavelengths (often called red shift) in the presence of an applied electric field. When an electric field is applied to the semiconductor, the energy bands of electrons are tilted. The slope of the tilt is exactly the electric field (V/cm). As a result of the tilt, the electron wavefunctions are no longer sinusoidal. In fact, the electron wavefunction becomes an Airy function as shown in Figure 9.14. There exists a finite probability that the electron may tunnel into the classically forbidden region where its energy is less than the potential energy (see Figure 9.14). This reduces the effective bandgap $(E_g)_{\text{eff}}$. As the degree of tilt is related to the amount of the applied electric field, the effective bandgap becomes smaller as a stronger electric field is applied.

The decrease in the bandgap can be estimated as follows. Referring to Figure 9.14, we consider the electronic transition from the valence band to the conduction band. As a result of the tilt, the wavefunctions become Airy functions. These wavefunctions have exponentially decaying tails representing the tunneling into the gap. Thus the transition rate will decrease exponentially as a function of $(E_{\text{gap}} - h\nu)^{3/2}/qE$, where E is the applied electric field and $h\nu$ is the photon energy. The effective bandgap is taken as the photon energy where the transition rate drops to $1/e$ of its peak value. Thus we obtain

$$\Delta E_g = \frac{(3)^{2/3}}{(m^*)^{1/3}} (q\hbar E)^{2/3} \tag{9.4-1}$$

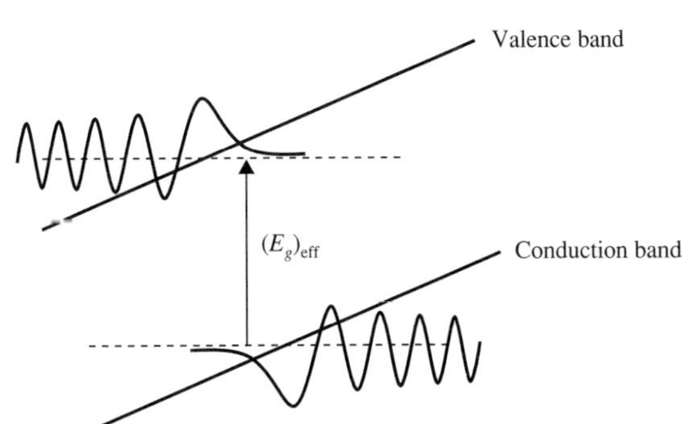

Figure 9.14 Schematic drawing of the Franz–Keldysh effect (FKE) in semiconductors. The wavefunctions of carriers become Airy functions that tunnel into the forbidden gap.

where E is the applied electric field, m^* is the effective mass of the electrons, and q is the electronic charge. For photon energy near the bandedge, the Franz–Keldysh effect leads to an increase in the absorption coefficient. The amount of absorption coefficient change is as large as a few hundred inverse centimeters as the electric field is changed from zero to hundreds of kV/cm in the FKE.

Optical modulators can be built based on the FKE (also known as the electroabsorption effect). These electroabsorption modulators (EAMs) control the light intensity by changing the absorption coefficient via the application of an electric field. Because the absorption coefficient can be changed quite drastically by applying the electric field in some semiconductors, the EAMs are very effective and the size can be quite small. EAMs are typically around 200 μm long while the electro-optic modulators are centimeters long. Another advantage of EAMs is the ease of integration with the light source, as the laser diode and the EAMs are based on very similar material and structure (e.g., InP-based).

Referring to Figure 9.15, we consider the propagation of an optical beam at the operation wavelength λ_0. In the absence of the applied electric field, the linear absorption coefficient is α_0 (typically around 10 cm^{-1}). If an electric field is applied (typically 10^3 V/cm), the absorption coefficient will increase to α_1 (typically around 10^3 cm^{-1}). The actual change in the transmission intensity will depend on the path length through the modulator. Since the transmitted intensity is an exponential function of the path length, the modulation ratio can be written

$$\frac{I_{\text{max}}}{I_{\text{min}}} = \frac{e^{-\alpha_0 L}}{e^{-\alpha_1 L}} \tag{9.4-2}$$

where L is the optical path in the modulator.

To have an efficient optical modulator, we need a minimum absorption ($\alpha_0 L \ll 1$) when no field is applied and a large absorption ($\alpha_1 L \gg 1$) when we apply the electric field. To achieve this, we select a material whose bandgap energy (E_g) is larger than the photon energy ($h\nu$) of the carrier beam. For 1.3 or 1.55 μm, the most commonly used material system with a bandgap in this range is that of InP-based semiconductors.

Electroabsorption modulators can also be based on quantum-confined Stark effect (QCSE). The QCSE is a phenomenon in quantum wells whose energy levels may shift as a result of the applied electric field. The shift of the energy levels as well as the change of the wavefunctions lead to a change in the absorption coefficient, as well as the index of refraction. A quantum well, which we will discuss in detail in Chapter 16 (on quantum-well semiconductor lasers), can easily be made in semiconductor material systems by sandwiching, during epitaxial crystal growth, a thin (10–100 Å thickness is typical) semiconductor layer

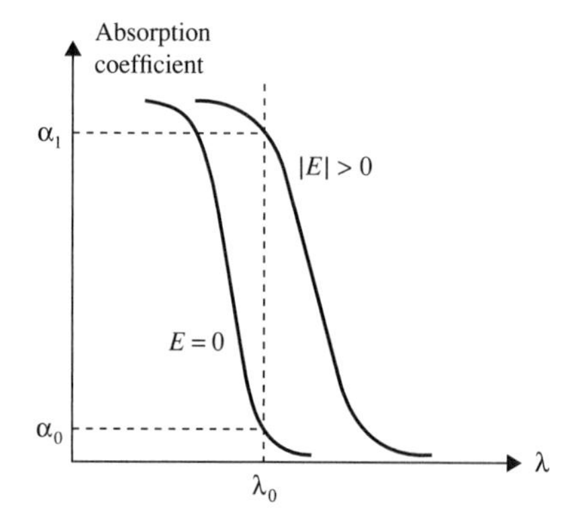

Figure 9.15 The Franz–Keldysh effect in semiconductors leads to a shift (red shift) in the absorption bandedge. For an operation wavelength (λ_0) near the bandedge, the application of an electric field leads to an increase in the absorption coefficient.

between two semiconductors whose energy gaps are larger than that of the thin layer. The conduction or valance band discontinuity (sometimes called band offset) at the interfaces forms a potential barrier such that the carriers (electrons in the conduction band and holes in the valence band) are trapped in an effective quantum potential well. Examples are AlAs–GaAs–AlAs, or InP–InGaAs–InP. So if AlAs is first grown followed by a thin layer of GaAs and then by another layer of AlAs, the sandwich AlAs–GaAs–AlAs forms a quantum well. For the QCSE to be useful as an optical modulator, the quantum potential wells in the conduction band and in the valence band must spatially coincide as in the case of AlAs–GaAs–AlAs or InP–InGaAs–InP (this is called type I heterostructure). To enhance the effect, often several quantum wells are used. When more than one quantum well is grown, the barrier material (AlAs or InP in the above examples) between two wells should be thick enough so that coupling, via quantum mechanical tunneling, between two neighboring wells is negligible. Such coupling would lead to broadening of the quantized energy levels (or cause the splitting of the energy levels and the formation of mini-bands) inside each well. When the barrier layers are thick enough so that the coupling between adjacent wells is negligible, this group of quantum wells are called multiple quantum wells (MQWs).

Most EAMs are based on the reversed bias of an electric voltage in either Schottky diodes or PIN structures. When a PIN diode is reverse-biased, almost all the voltage drop occurs at the intrinsic layer as its conductivity is lowest (thus possessing the highest resistance). As a result, the Franz–Keldysh effect is most significant in the intrinsic layer. If a laser beam is transmitted through the instrinsic layer, then an EAM can be obtained by modulating the voltages. Figure 9.16 shows schematic drawings of EAMs based on Schottky diodes or PIN structures. The thickness of the intrinsic layer is usually very thin so that a strong electric field is obtained at a small applying voltage. This limits the spot size of the optical beam. The small spot size leads to a spreading of the beam due to diffraction. The problem is solved by using a waveguide structure to confine the propagation of the optical beam, and at the same time to attain a strong electric field.

As a diode, the EAM has a characteristic capacitance determined by the device dimensions and the intrinsic layer thickness when it is reverse-biased. The capacitance can be calculated approximately by using the formula for the parallel-plate capacitor. As the capacitor stores charges, it delays the response of the circuit driven by the voltage source. The time constant of the circuit with series resistance R and the capacitor C is given by RC. Hence the bigger the capacitance, the slower the response of the circuit to the driving signal. Thus, for high-speed operation, the device dimensions should be small and the intrinsic layer thickness

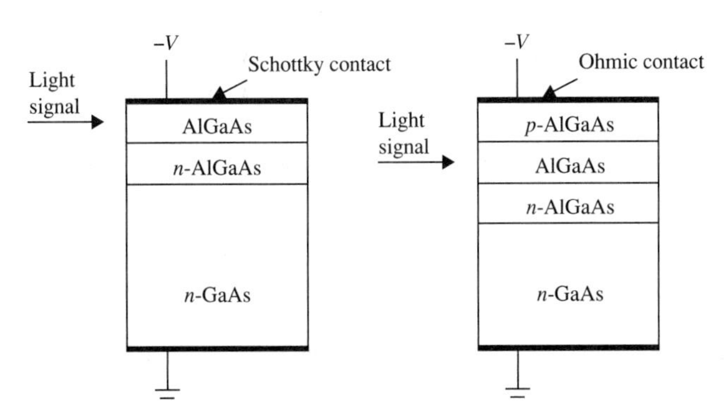

Figure 9.16 Schematic drawings of EAMs based on Schottky diodes or PIN structures.

should not be too thin. In addition to the issue of the device speed, we must also consider the issue of modulation efficiency of the EAM. For most applications, the modulation efficiency is measured in terms of output intensity versus voltage modulation. In digital applications (e.g., for telecommunications), if the modulator efficiency is good, it requires less voltage to change the light intensity from its on-state to off-state, and vice versa. In practical EAM design, a trade-off is needed to determine the thickness of the intrinsic layer so that both the *RC* time constant and the modulation efficiency are acceptable.

9.5 ELECTRO-OPTICAL EFFECT IN LIQUID CRYSTALS

The liquid crystal is a state of matter that is intermediate between that of a crystalline solid and amorphous liquid. It may also be viewed as a liquid in which an ordered arrangement of molecules exists. Liquid crystals arise under certain conditions in organic substances having sharply anisotropic molecules, that is, highly elongated (rodlike) molecules or flat (disklike) molecules. A direct consequence of the ordering of the anisotropic molecules is the anisotropy of mechanical, electric, magnetic, and optical properties. A good example of such a substance is hexyl-cyanobiphenyl (also known as 6CB), which shows a liquid crystalline phase of the simplest type (nematic) in the temperature range 15 °C $< T <$ 29 °C. A glass flask of 6CB in this temperature range often appears milky. The milky appearance is a result of the scattering of light due to the presence of multiple domains in the medium. When 6CB is heated at room temperature, a nematic–isotropic transition occurs at $T = 29$ °C. The material "then becomes suddenly completely clear." 6CB has the following rodlike molecular structure:

C_6H_{13}—(benzene ring)—(benzene ring)—CN

which consists of two benzene rings and two terminal groups. In the nematic phase, only the long axes of the molecules have a statistical preferential orientation.

Generally, there are three phases of liquid crystals, known as the smectic phase, the nematic phase, and the cholesteric phase. For the sake of clarity, we assume that the liquid crystals are made of rodlike molecules. Figure 9.17a illustrates the smectic phase in which one-dimensional translational order as well as orientational order exist. The local average of the orientation of the molecular axis is known as the *director*. Figure 9.17b illustrates the nematic phase in which only a long-range orientational order of the molecular axes exists. The cholesteric phase is also a nematic type of liquid crystal except that it is composed of chiral molecules. As a consequence, the structure acquires a spontaneous twist about a helical axis normal to the directors. The twist may be right-handed or left-handed depending on the molecular chirality. Figure 9.17c illustrates the cholesteric phase of a liquid crystal by viewing the distribution of molecules at several planes that are perpendicular to the helical axis. Note that the liquid crystal is in a nematic phase in each of the perpendicular planes.

A smectic liquid crystal is closest in structure to a solid crystal. It is interesting to note that in substances that form both a nematic phase and a smectic phase, the sequence of phase changes on rising temperature is

Solid crystal → smectic liquid crystal → nematic liquid crystal → isotropic liquid

Although the smectic phase possesses the highest degree of order, it is the nematic and cholesteric phases that have the greatest number of electro-optical applications. In the nematic

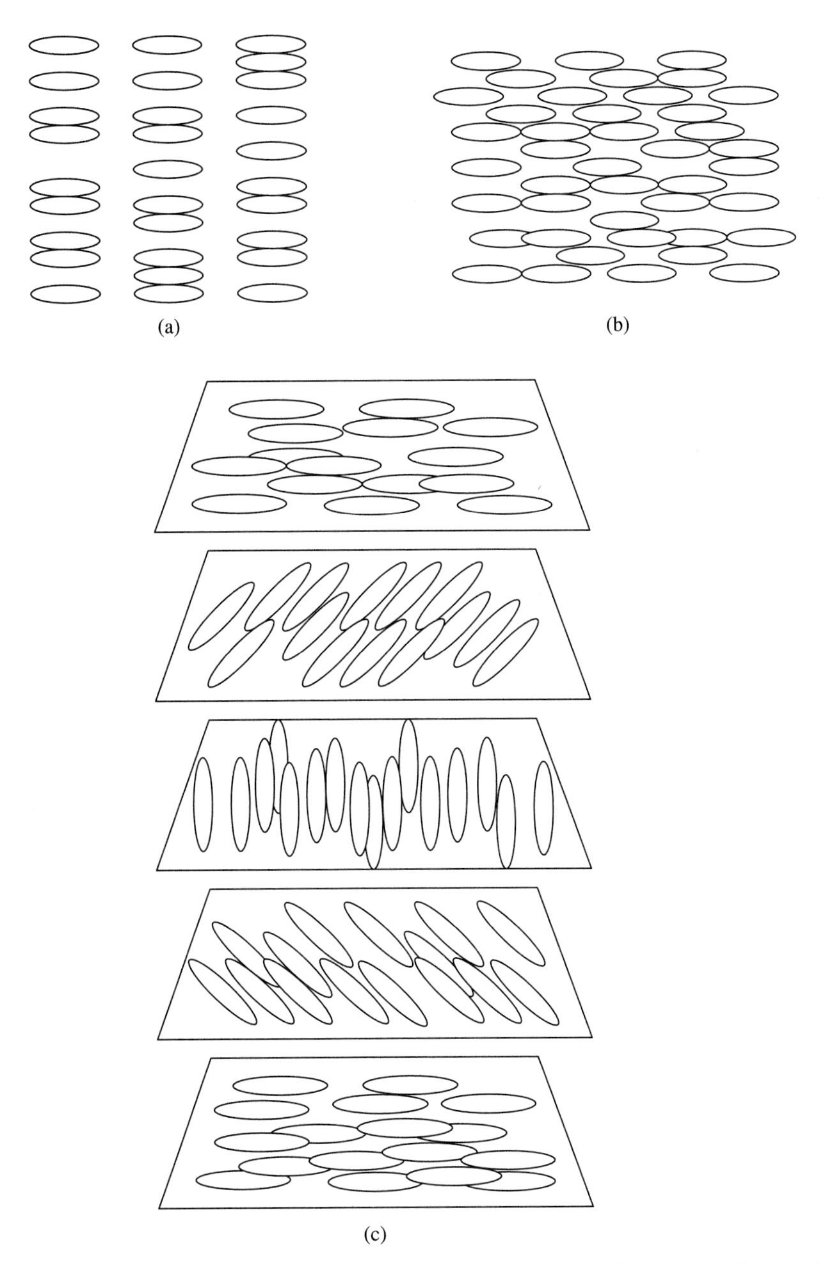

Figure 9.17 Phases of liquid crystals: (a) smectic phase, (b) nematic phase, (c) cholesteric phase. The director in the cholesteric phase follows a helical twist as illustrated.

phase, the medium may appear milky if the orientation order exists in many different domains. The nematic liquid crystal is clear only when a long-range order exists in the whole medium. At the nematic–isotropic transition temperature, the medium becomes isotropic and looks clear and transparent. Thus the temperature is also known as the clearing point. In the following, we discuss various important physical properties that are the results of the orientational order.

Orientational Order Parameter

In the nematic phase, the molecules are rodlike with their long axes aligned approximately parallel to one another. Thus at any point in the medium, we can define a vector **n** to represent the averaged orientation in the immediate neighborhood of the point. This vector is known as the *director*. In a homogeneous nematic liquid crystal, the director is a constant throughout the medium. In an inhomogeneous nematic liquid crystal, the director **n** can change from point to point and is, in general, a function of space (x, y, z). If we define a unit vector to represent the long axis of each molecule, then the director **n** is the statistical average of the unit vectors over a small volume element around the point.

The order parameter S of a liquid crystal is defined as

$$S = \tfrac{1}{2}\langle 3\cos^2\theta_m - 1\rangle \tag{9.5-1}$$

where θ_m is the angle between the long axis of an individual molecule and the director **n**, and the angular brackets denote a statistical average. For perfectly parallel alignment, $S = 1$, while for totally random orientations, $S = 0$. In the nematic phase, the order parameter S has an intermediate value that is strongly temperature dependent. It is evident that $S = 0$ at the clearing point. Typical values of the order parameter S are in the range between 0.6 and 0.4 at low temperatures. As the clearing point is approached, the order parameter S drops abruptly to zero. The value of the order parameter S also depends on the structure of the molecules. Experimental observation indicates that liquid crystals based on cyclohexane rings exhibit higher values of S than aromatic systems. A nematic liquid crystal medium with $S > 0$ appears optically birefringent with uniaxial symmetry.

Dielectric Constants

Because of the orientational ordering of the rodlike molecules, the smectic and nematic liquid crystals are uniaxially symmetric, with the axis of symmetry parallel to the axes of the molecules (director **n**). As a result of the uniaxial symmetry, the dielectric constants differ in value along the preferred axis ($\varepsilon_\parallel$) and perpendicular to this axis ($\varepsilon_\perp$). The dielectric anisotropy is defined as

$$\Delta\varepsilon = \varepsilon_\parallel - \varepsilon_\perp \tag{9.5-2}$$

The sign and magnitude of the dielectric anisotropy $\Delta\varepsilon$ are of the utmost importance in electro-optical applications. To illustrate this, we consider an applied electric field along the z axis in a homogeneous nematic liquid crystal. As a result of the anisotropy, the induced dipole moment of the molecules is not parallel to the applied electric field, except when the molecular axis is parallel or perpendicular to the electric field. This creates a net torque that tends to align the molecules along the direction of the electric field for most rodlike molecules. The macroscopic electrostatic energy can be written

$$U = \tfrac{1}{2}\mathbf{D}\cdot\mathbf{E} \tag{9.5-3}$$

where **E** is the electric field vector and **D** is the displacement field vector. In a homogeneous medium, the displacement field vector **D** is independent of the orientation of the liquid crystal. Let θ be the angle between the director and the z axis. The z component of the displacement field vector can be written

$$D_z = (\varepsilon_\parallel \cos^2\theta + \varepsilon_\perp \sin^2\theta)E \tag{9.5-4}$$

Thus the electrostatic energy can be written

$$U = \frac{1}{2} \frac{D_z^2}{\varepsilon_\parallel \cos^2\theta + \varepsilon_\perp \sin^2\theta} \tag{9.5-5}$$

For liquid crystals with positive dielectric anisotropy ($\varepsilon_\perp < \varepsilon_\parallel$), the lowest electrostatic energy occurs at $\theta = 0$ when the director is parallel to the applied electric field.

In classical dielectric theory, the macroscopic dielectric constant is often proportional to the molecular polarizability. For rodlike molecules, the longitudinal polarizability (parallel to molecular axis) is often greater than the transverse polarizability (perpendicular to axis). So even in nonpolar liquid crystals with rodlike molecules, the dielectric anisotropy is positive ($0 < \Delta\varepsilon$). In the case of polar liquid crystal compounds, there is an additional contribution to the dielectric constant due to the permanent dipole moment. Depending on the angle between the dipole moment and the molecular axis, the dipole contribution can cause an increase or a decrease of $\Delta\varepsilon$, leading eventually to a negative value of $\Delta\varepsilon$ (e.g., MBBA). In practice, dielectric anisotropy varies between $-2\varepsilon_0$ and $+15\varepsilon_0$. Liquid crystal compounds with a large anisotropy can be synthesized by substitution of a strongly polar group (e.g., cyanide group) in specific positions. The dielectric anisotropy also depends on the temperature and approaches zero abruptly at the clearing point. Beyond the clearing point, the dielectric constant becomes the mean dielectric constant

$$\bar{\varepsilon} = \frac{\varepsilon_\parallel + 2\varepsilon_\perp}{3} \tag{9.5-6}$$

Refractive Index and Birefringence

In a glass flask, nematic liquid crystals often appear as an opaque milky fluid. The scattering of light is due to the random fluctuation of the refractive index of the sample. With no proper boundaries to define the preferred orientation, the sample consists of many domains of nematic liquid crystal. The discontinuity of the refractive index at the domain boundaries is the main cause of the scattering leading to the milky appearance. Under proper treatment (e.g., rubbing the alignment layer on the glass substrate), a slab of nematic liquid crystal can be obtained with a uniform alignment of the director. Such a sample exhibits uniaxial optical symmetry with two principal refractive indices n_o and n_e. The ordinary refractive index n_o is for light with electric field polarization perpendicular to the director and the extraordinary refractive index n_e is for light with electric field polarization parallel to the director. The birefringence (or optical anisotropy) is defined as

$$\Delta n = n_e - n_o \tag{9.5-7}$$

If $n_o < n_e$, the liquid crystal is said to be positive birefringent, whereas if $n_e < n_o$, it is said to be negative birefringent. In classical dielectric theory, the macroscopic refractive index is related to the molecular polarizability at optical frequencies. The existence of the optical anisotropy is due mainly to the anisotropic molecular structures. Most liquid crystals with rodlike molecules exhibit positive birefringence ranging from 0.05 to 0.45. Optical anisotropy plays an essential role in changing the polarization state of light in liquid crystals.

There are many electro-optical effects in liquid crystals. Most of them involve the switching of their optical properties by an external electric field. As a result of their dielectric anisotropy, the molecules in the liquid crystal with an applied electrostatic field experience a torque, which tends to reorient them in such a way as to minimize the electrostatic energy. In the case when the liquid crystal molecules at the boundary are anchored along the direction determined by surface treatment (e.g., via rubbing), a distribution of the liquid crystal orientation is obtained in such a way that the electrostatic force (or torque) is balanced with the restoring elastic force (or torque). Since the optical property of a liquid crystal is very anisotropic, any change in the orientation of the liquid crystal molecules is easily observed optically. The time constant involved in the reorientation of molecules in the liquid crystal is on the order of 10^{-3} s, depending on the viscosity of the material.

Let θ be the angle between the director **n** (or the optic axis of the liquid crystal) and the direction of propagation of the optical beam inside the liquid crystal. The extraordinary index of refraction seen by the optical beam is given by, according to Equation (1.8-3),

$$\frac{1}{n_e^2(\theta)} = \frac{\cos^2\theta}{n_o^2} + \frac{\sin^2\theta}{n_e^2} \tag{9.5-8}$$

where n_o and n_e are the ordinary index and extraordinary index of refraction of the liquid crystal, respectively. In this case the phase retardation between the two modes of propagation is given by

$$\Gamma = \frac{2\pi}{\lambda}[n_e(\theta) - n_o]d \tag{9.5-9}$$

where d is the interaction length inside the liquid crystal cell.

In most cases, the director orientation θ is a function of position (z) inside the liquid crystal cell. This is a result of the boundary conditions due to alignment treatment at the surfaces. In this case, the phase retardation is given by

$$\Gamma = \frac{2\pi}{\lambda}\int_0^d [n_e(\theta) - n_o]\, dz \tag{9.5-10}$$

In most liquid crystal cells, the director orientation θ is a function of position inside the cell as well as the strength of the applied electric field. In the case of strong electric fields, the director is nearly parallel to the direction of the applied electric field. When the electric field is removed, the director orientation θ is restored to its original direction. The restoration is due to the elastic restoring force of the liquid crystal medium. In most liquid crystal cells, the directors at the cell surfaces are anchored to the surface due to the presence of an alignment layer. The directors away from the surfaces are subject to reorientation by the applied electrostatic field. The actual director distribution $\theta(z)$ is determined by the minimum of total free energy of the liquid crystal cell. In most liquid crystal cells the free energy is the sum of the elastic potential energy and the electrostatic energy. In some cases, the director orientation θ can be electrically tuned from zero to $\pi/2$. As a result, the phase retardation of a thin layer of properly oriented liquid crystal can be tuned from zero to $\Gamma = 2\pi(n_e - n_o)d/\lambda$ with the application of an electric field. In most cases, the phase retardation can be tuned continuously. The electric tuning of the phase retardation in liquid crystal cells can be employed in many optical electronic devices, including modulators and tunable filters for optical network applications. Figure 9.18 shows a schematic drawing of such an electro-optical phase retardation cell.

There are many different liquid crystal cell configurations that can be employed for electro-optical applications. In some cases, the applied electric field can be perpendicular to the direction of beam propagation. This leads to the possibility of electrically tuning the azimuth angle ψ of the direction relative to the direction of propagation of the optical beam.

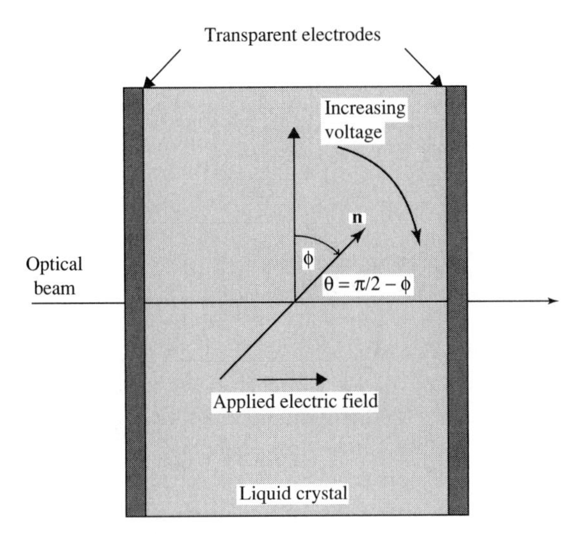

Figure 9.18 Schematic drawing of a liquid crystal cell with an applied electric field parallel to the direction of propagation of the optical beam. In the field-off state, the director is perpendicular to the direction of propagation of the optical beam. As the field strength increases, the director at the center of the cell tilts toward the direction of the applied field. This leads to a change of the phase retardation. The angle ϕ is an increasing function of the applied voltage.

Twisted Nematic Liquid Crystal Cell

A thin layer of nematic liquid crystal layer with its director parallel to the surface may be twisted in such a way that the director orientation varies as a function of z. This is often done by alignment treatment (e.g., rubbing) of the boundary surfaces. In a twisted nematic liquid crystal cell, the alignment directions of the two surfaces are different. The director of the liquid crystal is often anchored in a direction parallel to the alignment direction. As a result, a twisted nematic liquid crystal cell is obtained. In a linearly twisted nematic crystal, the azimuth angle of the director is a linear function of z:

$$\psi = \alpha z \tag{9.5-11}$$

where α is a constant.

The propagation of optical beams in a linearly twisted anisotropic medium has been treated by many workers. It can be shown that linearly polarized light with its plane of polarization parallel or perpendicular to the local director will remain "locked in" with the local director as it propagates in the medium, provided that the twisting rate α is small (i.e., $\alpha \ll 2\pi(n_e - n_o)/\lambda$). In other words, the plane of polarization of such linearly polarized light will rotate in the same way as the director does. This is known as "waveguiding" in liquid crystal displays. The same phenomenon has often been misinterpreted as optical rotatory power.

For electro-optical applications, a nematic liquid crystal with a twist angle of 90° is often involved. The cell is placed between a pair of parallel polarizers with the transmission axis parallel to the director at the entrance plane. This structure transmits no light unless a field is applied (known as normally black). The electric field (longitudinal) tends to set the local director along the field direction (z axis), giving rise to the transmission of light due to the vanishing of birefringence. Using the Jones matrix method, the transmission through a 90° twisted nematic cell sandwiched between parallel polarizers can be written [7]

$$T = \frac{\sin^2\left(\pi\sqrt{1+u^2}/2\right)}{1+u^2} \tag{9.5-12}$$

with

$$u = \frac{\Gamma}{\pi} = \frac{2}{\lambda}(n_e - n_o)d \tag{9.5-13}$$

where d is the thickness of the cell. For most twisted nematic cells, $1 \ll u$ in the field-off state, and the transmission is near zero for all wavelengths in the visible spectrum, according to Equation (9.5-12). When the electric field is turned on, the birefringence disappears ($u = 0$), and the transmission is unity according to the same equation.

If the cell is placed between a pair of crossed polarizers with the transmission axes parallel to the director at the boundary of the cell, this structure transmits light due to the phenomenon of waveguiding in the field-off state (known as normally white). The electric field (longitudinal) tends to set the local director along the field direction (z axis), giving rise to the extinction of light due to the vanishing of birefringence and the pair of crossed polarizers.

There are other electro-optical effects, such as dynamic light scattering, cholesteric–nematic phase transition, and the guest–host effect. Discussion of these effects is beyond the scope of this book. Interested readers are referred to Reference [7]. Due to their strong electro-optical effect, liquid crystals are widely used in many optical electronic instruments, mostly for the purpose of displaying information. In the area of optical communications, liquid crystals are employed as electro-optical media for switching and modulation as well as tunable filter applications.

9.6 ACOUSTO-OPTIC EFFECT (PHOTOELASTIC EFFECT)

We consider the effect of mechanical strain on the optical properties of materials. As we know, the index of refraction of a gas medium changes with the density (or pressure) of the gas medium. Similarly, the refractive index may change as a solid or liquid undergoes elastic deformation. The photoelastic effect in materials couples the mechanical strain to the optical index of refraction. Mechanical strain in materials due to elastic deformation is often defined as

$$S_{ij} = \frac{1}{2}\left(\frac{\partial u_i}{\partial x_j} + \frac{\partial u_j}{\partial x_i}\right) \tag{9.6-1}$$

where x_i ($i = 1, 2, 3$) is the coordinate of a particle (or a small element of the material) in the material medium, and u_i ($i = 1, 2, 3$) is the physical displacement of the particle from its equilibrium position. A general deformation can be described by $u_i(x_1, x_2, x_3)$. Note that this tensor (9.6-1) is symmetric by definition. The mechanical deformation of an elastic material is traditionally described by the strain tensor (9.6-1). Physically, S_{11} represents a linear expansion (or contraction, if negative) along the x direction. The off-diagonal elements (e.g., S_{12}) represent a shear strain. The best example of a shear strain is the deformation of a rectangular-shaped solid, turning into a parallelogram-shaped solid. We note the strain tensor element S_{ij} is dimensionless.

As an elastic material is deformed mechanically, the index of refraction may change. This effect, known as the photoelastic effect, occurs in all states of matter and is traditionally described by

$$\Delta\eta_{ij} = \Delta\left(\frac{1}{n^2}\right)_{ij} = p_{ijkl}S_{kl} \tag{9.6-2}$$

where $\Delta\eta_{ij}$ (or $\Delta(1/n^2)_{ij}$) is the change in the optical impermeability tensor and S_{kl} is the strain tensor. The dimensionless coefficients p_{ijkl} form the strain-optic tensor. In the above equation, we have neglected high-order terms involving powers of S_{kl}. These high-order terms are usually small compared with the linear term (S_{kl} is in the range of 10^{-5}). The index ellipsoid of an optical medium in the presence of an applied strain field is thus given by

$$(\eta_{ij} + p_{ijkl}S_{kl})x_i x_j = 1 \tag{9.6-3}$$

In the event when the strain field vanishes ($S_{kl} = 0$), the index ellipsoid reduces to Equation (9.1-4) in the principal coordinate system. Since both η_{ij} and S_{kl} are symmetric tensors, the indices i and j as well as k and l in Equation (9.6-2) can be permuted. The permutation symmetry of the strain-optic tensor can thus be written

$$p_{ijkl} = p_{jikl} = p_{ijlk} = p_{jilk} \tag{9.6-4}$$

As a result of this symmetry, it is convenient to use the contracted indices (9.1-7) to abbreviate the notation. Equation (9.6-2) can thus be written

$$\Delta\left(\frac{1}{n^2}\right)_I = p_{IJ}S_J \quad (I, J = 1, 2, 3, 4, 5, 6) \tag{9.6-5}$$

where S_J are the strain components. The equation of the index ellipsoid in the presence of a strain field can now be written

$$\begin{aligned}
&\left(\frac{1}{n_x^2} + p_{11}S_1 + p_{12}S_2 + p_{13}S_3 + p_{14}S_4 + p_{15}S_5 + p_{16}S_6\right)x^2 \\
&+ \left(\frac{1}{n_y^2} + p_{21}S_1 + p_{22}S_2 + p_{23}S_3 + p_{24}S_4 + p_{25}S_5 + p_{26}S_6\right)y^2 \\
&+ \left(\frac{1}{n_z^2} + p_{31}S_1 + p_{32}S_2 + p_{33}S_3 + p_{34}S_4 + p_{35}S_5 + p_{36}S_6\right)z^2 \\
&+ 2yz(p_{41}S_1 + p_{42}S_2 + p_{43}S_3 + p_{44}S_4 + p_{45}S_5 + p_{46}S_6) \\
&+ 2zx(p_{51}S_1 + p_{52}S_2 + p_{53}S_3 + p_{54}S_4 + p_{55}S_5 + p_{56}S_6) \\
&+ 2xy(p_{61}S_1 + p_{62}S_2 + p_{63}S_3 + p_{64}S_4 + p_{65}S_5 + p_{66}S_6) = 1
\end{aligned} \tag{9.6-6}$$

where n_x, n_y, n_z are the principal indices of refraction. The strain-optic coefficients p_{IJ} are usually defined in the principal coordinate system. There are in general 36 independent strain-optic coefficients. The point-group symmetry of the material dictates which of the 36 coefficients are zero, as well as relationships that may exist between the nonvanishing coefficients. The new index ellipsoid in the presence of the strain field is in general different from the zero-field index ellipsoid. For specific forms of the strain-optic tensors and data on the photoelastic coefficients p_{IJ}, students are referred to Reference [6].

EXAMPLE: ACOUSTO-OPTIC EFFECT IN ISOTROPIC SOLID

Consider the propagation of a sound wave in an isotropic solid (e.g., fused silica). The photoelastic coefficients in the contracted notation can be written [6]

$$\begin{bmatrix} p_{11} & p_{12} & p_{12} & 0 & 0 & 0 \\ p_{12} & p_{11} & p_{12} & 0 & 0 & 0 \\ p_{12} & p_{12} & p_{11} & 0 & 0 & 0 \\ 0 & 0 & 0 & p_{44} & 0 & 0 \\ 0 & 0 & 0 & 0 & p_{44} & 0 \\ 0 & 0 & 0 & 0 & 0 & p_{44} \end{bmatrix}, \quad \text{with } p_{44} = \tfrac{1}{2}(p_{11} - p_{12}) \tag{9.6-7}$$

Let the sound wave be a longitudinal wave propagating along the z axis with the particle displacement given by

$$u(z, t) = A\hat{\mathbf{z}} \cos(\Omega t - Kz) \tag{9.6-8}$$

where A is the amplitude of the oscillation, Ω is the angular frequency of the sound, and K is the wavevector. The strain field associated with this sound wave is given by

$$S_3(z, t) = KA \sin(\Omega t - Kz) = S_0 \sin(\Omega t - Kz) \tag{9.6-9}$$

where S_0 is a constant defined as KA. We note that Equation (9.6-9) represents a compressional wave traveling along the z axis. The change in the impermeability tensor is given by, according to Equation (9.6-5),

$$\begin{bmatrix} \Delta\eta_{11} \\ \Delta\eta_{22} \\ \Delta\eta_{33} \\ \Delta\eta_{23} \\ \Delta\eta_{31} \\ \Delta\eta_{12} \end{bmatrix} = \begin{bmatrix} p_{11} & p_{12} & p_{12} & 0 & 0 & 0 \\ p_{12} & p_{11} & p_{12} & 0 & 0 & 0 \\ p_{12} & p_{12} & p_{11} & 0 & 0 & 0 \\ 0 & 0 & 0 & p_{44} & 0 & 0 \\ 0 & 0 & 0 & 0 & p_{44} & 0 \\ 0 & 0 & 0 & 0 & 0 & p_{44} \end{bmatrix} \begin{bmatrix} 0 \\ 0 \\ S_3 \\ 0 \\ 0 \\ 0 \end{bmatrix} \tag{9.6-10}$$

or equivalently,

$$\begin{aligned} \Delta\eta_{11} &= \Delta\left(\frac{1}{n^2}\right)_1 = p_{12} S_0 \sin(\Omega t - Kz) \\ \Delta\eta_{22} &= \Delta\left(\frac{1}{n^2}\right)_2 = p_{12} S_0 \sin(\Omega t - Kz) \\ \Delta\eta_{33} &= \Delta\left(\frac{1}{n^2}\right)_3 = p_{11} S_0 \sin(\Omega t - Kz) \\ \Delta\eta_{ij} &= 0, \quad \text{for } i \neq j \end{aligned} \tag{9.6-11}$$

The new index ellipsoid is given by Equation (9.6-3) or (9.6-6) and can now be written

$$\begin{aligned} &\left(\frac{1}{n^2} + p_{12} S_0 \sin(\Omega t - Kz)\right) x^2 + \left(\frac{1}{n^2} + p_{12} S_0 \sin(\Omega t - Kz)\right) y^2 \\ &+ \left(\frac{1}{n^2} + p_{11} S_0 \sin(\Omega t - Kz)\right) z^2 = 1 \end{aligned} \tag{9.6-12}$$

Since no mixed terms are involved, the principal axes remain unchanged. The new principal indices of refraction are thus given by

$$\begin{aligned} n_x &= n - \tfrac{1}{2}n^3 p_{12} S_0 \sin(\Omega t - Kz) \\ n_y &= n - \tfrac{1}{2}n^3 p_{12} S_0 \sin(\Omega t - Kz) \\ n_z &= n - \tfrac{1}{2}n^3 p_{11} S_0 \sin(\Omega t - Kz) \end{aligned} \tag{9.6-13}$$

where p_{11} and p_{12} are the strain-optic coefficients and n is the index of refraction of the isotropic solid. We notice that in the presence of the sound wave, the solid becomes a periodic medium that is equivalent to a volume index grating with a grating period of $\Lambda = 2\pi/K$.

We next consider a transverse sound wave propagating along the x axis with the particle displacement given by

$$u(x, t) = A\hat{\mathbf{y}} \cos(\Omega t - Kx) \tag{9.6-14}$$

where A is the amplitude of the oscillation, Ω is the angular frequency of the sound, and K is the wavevector. The strain field associated with this sound wave is given by

$$S_6(x, t) = S_{12}(x, t) = \tfrac{1}{2}KA \sin(\Omega t - Kx) = S_0 \sin(\Omega t - Kx) \tag{9.6-15}$$

where S_0 is a constant defined as $KA/2$. This is a shear wave traveling along the x axis. The change in the impermeability tensor is given by, according to Equation (9.6-5),

$$\begin{bmatrix} \Delta\eta_{11} \\ \Delta\eta_{22} \\ \Delta\eta_{33} \\ \Delta\eta_{23} \\ \Delta\eta_{31} \\ \Delta\eta_{12} \end{bmatrix} = \begin{bmatrix} p_{11} & p_{12} & p_{12} & 0 & 0 & 0 \\ p_{12} & p_{11} & p_{12} & 0 & 0 & 0 \\ p_{12} & p_{12} & p_{11} & 0 & 0 & 0 \\ 0 & 0 & 0 & p_{44} & 0 & 0 \\ 0 & 0 & 0 & 0 & p_{44} & 0 \\ 0 & 0 & 0 & 0 & 0 & p_{44} \end{bmatrix} \begin{bmatrix} 0 \\ 0 \\ 0 \\ 0 \\ 0 \\ S_6 \end{bmatrix} \tag{9.6-16}$$

or equivalently,

$$\begin{aligned} \Delta\eta_{12} &= \Delta\left(\frac{1}{n^2}\right)_{12} = p_{44} S_0 \sin(\Omega t - Kz) \\ \Delta\eta_{11} &= \Delta\eta_{22} = \Delta\eta_{33} = \Delta\eta_{31} = \Delta\eta_{23} = 0 \end{aligned} \tag{9.6-17}$$

The new index ellipsoid is given by Equation (9.6-3) or (9.6-6) and can now be written

$$\frac{x^2 + y^2 + z^2}{n^2} + 2xy p_{44} S_0 \sin(\Omega t - Kz) = 1 \tag{9.6-18}$$

Since the mixed term involves x and y only, the index ellipsoid can easily be diagonalized in the following form:

$$\left(\frac{1}{n^2} + p_{44} S_0 \sin(\Omega t - Kz)\right) x'^2 + \left(\frac{1}{n^2} - p_{44} S_0 \sin(\Omega t - Kz)\right) y'^2 + \frac{z^2}{n^2} = 1 \tag{9.6-19}$$

where the axes x' and y' are at 45° from the x and y axes. The new principal indices of refraction are thus given by

$$\begin{aligned} n_x &= n - \tfrac{1}{2}n^3 p_{44} S_0 \sin(\Omega t - Kz) \\ n_y &= n + \tfrac{1}{2}n^3 p_{44} S_0 \sin(\Omega t - Kz) \\ n_z &= n \end{aligned} \tag{9.6-20}$$

where we recall that $p_{44} = (p_{11} - p_{12})/2$. For fused silica ($SiO_2$), the strain-optic coefficients [6] are $p_{11} = 0.12$ and $p_{12} = 0.27$.

EXAMPLE: ACOUSTO-OPTIC EFFECT IN CRYSTALS OF TRIGONAL SYMMETRY (E.G., $LiNbO_3$)

Consider the propagation of a sound wave in a crystalline medium of trigonal symmetry with point groups of $3m$, 32, $\bar{3}m$ (e.g., $LiNbO_3$, quartz (SiO_2), sapphire (Al_2O_3)). The photoelastic coefficients for these crystals in the contracted notation can be written

$$\begin{bmatrix} p_{11} & p_{12} & p_{13} & p_{14} & 0 & 0 \\ p_{12} & p_{11} & p_{13} & -p_{14} & 0 & 0 \\ p_{31} & p_{31} & p_{33} & 0 & 0 & 0 \\ p_{41} & -p_{41} & 0 & p_{44} & 0 & 0 \\ 0 & 0 & 0 & 0 & p_{44} & p_{41} \\ 0 & 0 & 0 & 0 & p_{14} & p_{66} \end{bmatrix}, \quad \text{with } p_{66} = \tfrac{1}{2}(p_{11} - p_{12}) \tag{9.6-21}$$

Let the sound wave be a longitudinal wave propagating along the z axis with the particle displacement given by

$$u(z, t) = A\hat{\mathbf{z}} \cos(\Omega t - Kz) \tag{9.6-22}$$

where A is the amplitude of the oscillation, Ω is the angular frequency of the sound, and K is the wavevector. The strain field associated with this sound wave is given by

$$S_3(z, t) = KA \sin(\Omega t - Kz) = S_0 \sin(\Omega t - Kz) \tag{9.6-23}$$

where S_0 is a constant defined as KA. The change in the impermeability tensor is given by, according to Equation (9.6-5),

$$\begin{bmatrix} \Delta\eta_{11} \\ \Delta\eta_{22} \\ \Delta\eta_{33} \\ \Delta\eta_{23} \\ \Delta\eta_{31} \\ \Delta\eta_{12} \end{bmatrix} = \begin{bmatrix} p_{11} & p_{12} & p_{13} & p_{14} & 0 & 0 \\ p_{12} & p_{11} & p_{13} & -p_{14} & 0 & 0 \\ p_{31} & p_{31} & p_{33} & 0 & 0 & 0 \\ p_{41} & -p_{41} & 0 & p_{44} & 0 & 0 \\ 0 & 0 & 0 & 0 & p_{44} & p_{41} \\ 0 & 0 & 0 & 0 & p_{14} & p_{66} \end{bmatrix} \begin{bmatrix} 0 \\ 0 \\ S_3 \\ 0 \\ 0 \\ 0 \end{bmatrix} \tag{9.6-24}$$

or equivalently,

$$\begin{aligned} \Delta\eta_{11} &= \Delta\left(\frac{1}{n^2}\right)_1 = p_{13} S_0 \sin(\Omega t - Kz) \\ \Delta\eta_{22} &= \Delta\left(\frac{1}{n^2}\right)_2 = p_{13} S_0 \sin(\Omega t - Kz) \\ \Delta\eta_{33} &= \Delta\left(\frac{1}{n^2}\right)_3 = p_{33} S_0 \sin(\Omega t - Kz) \\ \Delta\eta_{ij} &= 0, \quad \text{for } i \neq j \end{aligned} \tag{9.6-25}$$

The new index ellipsoid is given by Equation (9.6-3) or (9.6-6) and can now be written

$$\left(\frac{1}{n_o^2} + p_{13} S_0 \sin(\Omega t - Kz)\right) x^2 + \left(\frac{1}{n_o^2} + p_{13} S_0 \sin(\Omega t - Kz)\right) y^2 + \left(\frac{1}{n_e^2} + p_{33} S_0 \sin(\Omega t - Kz)\right) z^2 = 1 \tag{9.6-26}$$

Since no mixed terms are involved, the principal axes remain unchanged. The new principal indices of refraction are thus given by

$$\begin{aligned} n_x &= n_o - \tfrac{1}{2} n_o^3 p_{13} S_0 \sin(\Omega t - Kz) \\ n_y &= n_o - \tfrac{1}{2} n_o^3 p_{13} S_0 \sin(\Omega t - Kz) \\ n_z &= n_e - \tfrac{1}{2} n_e^3 p_{33} S_0 \sin(\Omega t - Kz) \end{aligned} \tag{9.6-27}$$

where p_{13} and p_{33} are the strain-optic coefficients, and n_e and n_o are the extraordinary and ordinary indices of refraction of the crystal. We note that in the presence of the sound wave, the crystal becomes a periodic medium that is equivalent to a volume index grating with a grating period of $\Lambda = 2\pi/K$.

We next consider a transverse sound wave propagating along the x axis with the particle displacement given by

$$u(x, t) = A\hat{\mathbf{y}} \cos(\Omega t - Kx) \tag{9.6-28}$$

where A is the amplitude of the oscillation, Ω is the angular frequency of the sound, and K is the wavevector. The strain field associated with this sound wave is given by

$$S_6(x, t) = S_{12}(x, t) = \tfrac{1}{2} KA \sin(\Omega t - Kx) = S_0 \sin(\Omega t - Kx) \tag{9.6-29}$$

where S_0 is a constant defined as $KA/2$. The change in the impermeability tensor is given by, according to Equation (9.6-5),

$$\begin{bmatrix} \Delta\eta_{11} \\ \Delta\eta_{22} \\ \Delta\eta_{33} \\ \Delta\eta_{23} \\ \Delta\eta_{31} \\ \Delta\eta_{12} \end{bmatrix} = \begin{bmatrix} p_{11} & p_{12} & p_{13} & p_{14} & 0 & 0 \\ p_{12} & p_{11} & p_{13} & -p_{14} & 0 & 0 \\ p_{31} & p_{31} & p_{33} & 0 & 0 & 0 \\ p_{41} & -p_{41} & 0 & p_{44} & 0 & 0 \\ 0 & 0 & 0 & 0 & p_{44} & p_{41} \\ 0 & 0 & 0 & 0 & p_{14} & p_{66} \end{bmatrix} \begin{bmatrix} 0 \\ 0 \\ 0 \\ 0 \\ 0 \\ S_6 \end{bmatrix} \tag{9.6-30}$$

or equivalently,

$$\begin{aligned} \Delta\eta_{31} &= \Delta\left(\frac{1}{n^2}\right)_{31} = p_{41} S_0 \sin(\Omega t - Kz) \\ \Delta\eta_{12} &= \Delta\left(\frac{1}{n^2}\right)_{12} = p_{66} S_0 \sin(\Omega t - Kz) = \tfrac{1}{2}(p_{11} - p_{12}) S_0 \sin(\Omega t - Kz) \\ \Delta\eta_{11} &= \Delta\eta_{22} = \Delta\eta_{33} = \Delta\eta_{23} = 0 \end{aligned} \tag{9.6-31}$$

The new index ellipsoid is given by Equation (9.6-3) or (9.6-6) and can now be written

$$\frac{x^2 + y^2 + z^2}{n^2} + 2zxp_{41} S_0 \sin(\Omega t - Kz) + 2xyp_{66} S_0 \sin(\Omega t - Kz) = 1 \tag{9.6-32}$$

The index ellipsoid can be transformed into a diagonal form by using a coordinate rotation. For $LiNbO_3$ crystal, the strain-optic coefficients [6] are $p_{11} = -0.026$, $p_{12} = 0.090$, $p_{13} = 0.133$, $p_{14} = -0.075$, $p_{31} = 0.179$, $p_{33} = 0.071$, $p_{41} = -0.151$, $p_{44} = 0.146$.

The photoelastic effect can also be expressed in terms of the change in the dielectric tensor. Using the definition of the impermeability tensor,

$$\eta = \frac{\varepsilon_0}{\varepsilon} \tag{9.6-33}$$

we obtain

$$\Delta\varepsilon = \varepsilon\left(\frac{\Delta\eta}{\varepsilon_0}\right)\varepsilon \tag{9.6-34}$$

where the change in the dielecric tensor is expressed in a matrix form. The right side of Equation (9.6-34) is a product of three matrices. Thus the photoelastic effect can also be written

$$\Delta\varepsilon = \varepsilon\left(\frac{pS}{\varepsilon_0}\right)\varepsilon \tag{9.6-35}$$

where pS is the matrix formed by the elements $p_{ijkl}S_{kl}$. In an isotropic medium where $\varepsilon = n^2\varepsilon_0$, the photoelastic effect can be further simplified as

$$\Delta\varepsilon = \varepsilon_0 n^2 pS \tag{9.6-36}$$

For more discussions on the photoelastic effect, students are referred to Reference [6].

9.7 SCATTERING OF LIGHT BY SOUND

In the last section, we described the creation of an index grating due to the presence of an acoustic wave in solids. When an acoustic wave propagates in a medium, there is an associated strain field. The strain results in a change of the index of refraction. This is referred to as the photoelastic effect. The strain field is a periodic function of position for a plane acoustic wave. As a result of the photoelastic effect, the index of refraction of the medium becomes periodically modulated. And thus an index grating is created.

We now consider the interaction of an optical beam with the index gratings produced by acoustic waves in material media. Such an interaction (acousto-optic interaction) was first predicted by Brillouin in 1922 [8] and demonstrated experimentally in 1932 by Debye and Sears in the United States and by Lucas and Biguard in France [9]. Recent developments in high-frequency acoustics [10] and in lasers created a renewed interest in this field because the scattering of light from sound (Bragg scattering) affords a convenient means of controlling the frequency, intensity, and direction of an optical beam. This type of control enables a large number of applications involving the transmission, display, and processing of information [11].

Bragg Scattering

A sound wave consists of a sinusoidal perturbation of the density (or strain) of the material that travels at the sound velocity v_s, as shown in Figure 9.19. According to the examples described in the last section, the index grating induced by the sound wave is of the form

$$\Delta n(z, t) = \Delta n_0 \sin(\Omega t - Kz) \tag{9.7-1}$$

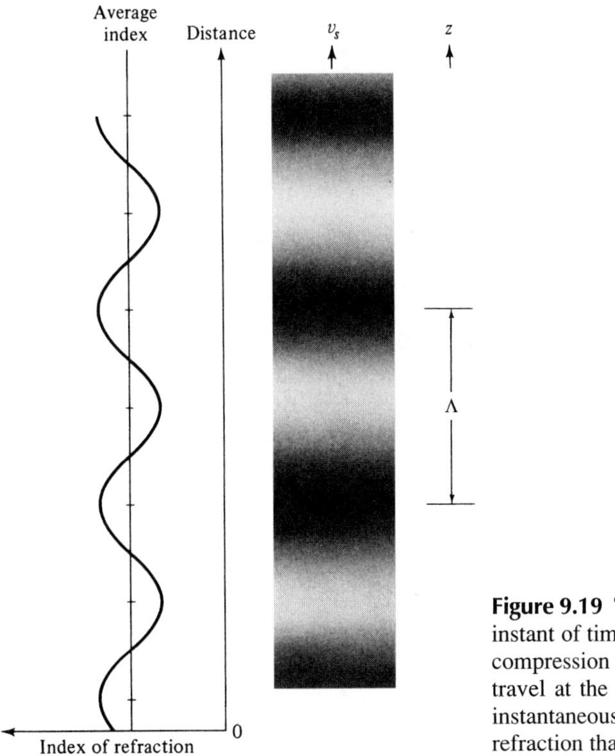

Figure 9.19 Traveling sound wave "frozen" at some instant of time. It consists of alternating regions of compression (dark) and rarefaction (white), which travel at the sound velocity v_s. Also shown is the instantaneous spatial variation of the index of refraction that accompanies the sound wave.

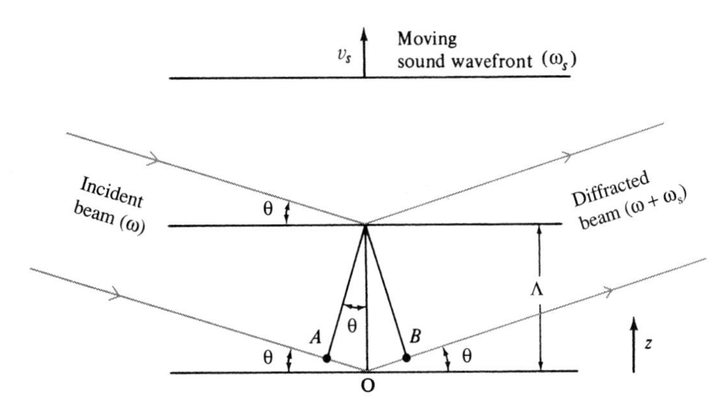

Figure 9.20 Reflections from two equivalent planes in the sound beam (i.e., planes separated by the sound wavelength Λ), which add up in phase along the direction θ if the optical path difference $AO + OB$ is equal to an integral multiple of optical wavelength.

where Δn_0 is a constant, Ω is the angular frequency of the sound wave, and $\Omega/K = v_s$. In this chapter we use the word *sound* to describe acoustic waves with frequencies that in practice may range from MHz to 10 GHz.

Next, consider an optical beam incident on a sound wave at an angle θ_i as shown in Figure 9.20. For the purpose of the immediate discussion, we can characterize the sound wave as a series of partially reflecting mirrors, separated by the sound wavelength Λ, that are moving at a velocity v_s. This is an oversimplified picture. However, it is useful in describing the concept of acousto-optic interaction. According to Equation (9.7-1), an acoustic wave consists of a traveling strain wave, which can be viewed as an array of alternating layers of high and low index of refraction. Since a change in index causes reflection, the mirrors analogy follows.

Ignoring, for the moment, the motion of the mirrors, let us consider the diffracted wave (or scattered wave). Using the mirror analogy and the specular nature of mirror reflection, we conclude that the diffraction angle θ_r must be the same as the incident angle θ_i.

Let

$$\theta = \theta_i = \theta_r \tag{9.7-2}$$

This is similar to Snell's law of reflection applied to a flat mirror. In addition to the condition on the diffraction angle (9.7-2), we require that diffraction from any two acoustic phase fronts add up in phase along the direction of the reflected beam (or diffracted wave). The path difference, $AO + OB$ shown in Figure 9.20, of a given optical wavefront resulting from reflection from two equivalent acoustic wavefronts (i.e., planes separated by Λ) must thus be equal to an integral multiple of optical wavelength λ. Using Figure 9.20, we find that this condition can be written

$$2n\Lambda \sin\theta = m\lambda, \quad m = 1 \text{ for sinusoidal sound wave} \tag{9.7-3}$$

The diffraction of light that satisfies Equation (9.7-3) is known as Bragg diffraction after a similar law for X-ray diffraction from crystals ($2d\Lambda \sin\theta = m\lambda$, with d = atomic lattice spacing, $m = 1, 2, 3, \ldots$) [12]. We note that Equation (9.7-3) is derived based on constructive interference. It will be shown later in this section that Bragg diffraction occurs with $m = 1$ only, when a sinusoidal sound wave is involved. This is a result of the physical nature of the coupling between the optical waves and a monochromatic sound wave. If the sound wave is a nonsinusoidal (but periodic) wave (and thus polychromatic), then diffraction occurs with $m = 2, 3, \ldots$. To get an idea of the order of magnitude of the angle θ, consider the case of diffraction of light with $\lambda/n = 0.5$ μm from a 500 MHz sound wave ($f_s = 500$ MHz). Taking the sound velocity as $v_s = 3 \times 10^5$ cm/s, we have $\Lambda = v_s/f_s = 6 \times 10^{-4}$ cm and, from Equation (9.7-3), $\theta \approx 4 \times 10^{-2}$ rad $\approx 3.5°$.

Particle Picture of Bragg Diffraction of Light by Sound

Many of the features of Bragg diffraction of light by sound can be deduced if we take advantage of the dual particle–wave nature of light and of sound. According to this picture, an optical monochromatic plane wave of the form $\cos(\omega t - \mathbf{k} \cdot \mathbf{r})$, with a propagation vector $\mathbf{k}$ and frequency ω, can be considered to consist of a stream of particles (photons) with momentum $\hbar\mathbf{k}$ and energy $\hbar\omega$. The sound wave, likewise, can be thought of as made up of particles (phonons) with momentum $\hbar\mathbf{K}$ and energy $\hbar\Omega$. The diffraction of light by an approaching sound beam illustrated in Figure 9.20 can be described as a series of collisions, each of which involves an annihilation of *one* incident photon at ω_i and *one* phonon and a simultaneous creation of a new (diffracted) photon at a frequency ω_d, which propagates along the direction of the scattered beam. The conservation of momentum requires that the momentum $\hbar(\mathbf{K} + \mathbf{k}_i)$ of the colliding particles be equal to the momentum $\hbar\mathbf{k}_d$ of the scattered photon, so

$$\mathbf{k}_d = \mathbf{K} + \mathbf{k}_i \tag{9.7-4}$$

The conservation of energy takes the form

$$\omega_d = \Omega + \omega_i \tag{9.7-5}$$

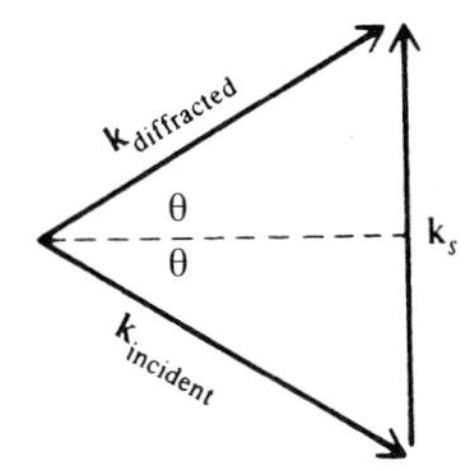

Figure 9.21 The momentum conservation relation, Equation (9.7-4), used to derive the Bragg condition $2\Lambda \sin\theta = \lambda$, for an optical beam that is diffracted by an approaching sound wave. θ is the angle between the incident or diffracted beam and the acoustic wavefront.

From Equation (9.7-5) we learn that the diffracted beam is shifted in frequency by an amount equal to the sound frequency. Since the interaction involves the annihilation of a phonon, conservation of energy decrees that the shift in frequency is such that $\omega_d > \omega_i$ and the phonon energy is *added* to that of the annihilated photon to form a new photon. Using this argument, it follows that if the direction of the sound beam in Figure 9.20 were reversed, so that it was receding from the incident optical wave, the scattering process could be considered as one in which a new photon (diffracted photon) and a *new* phonon are generated while the incident photon is annihilated. In this case, the conservation of energy principle yields

$$\omega_d = \omega_i - \Omega$$

The relation between the sign of the frequency change and the sound propagation direction will become clearer using Doppler-shift arguments, as is done at the end of this section.

The conservation of momentum condition (9.7-4) is equivalent to the Bragg condition (9.7-3) with $m = 1$. To show why this is true, consider Figure 9.21. Since the sound frequencies of interest are below 10^{10} Hz and those of the optical beams are usually above 10^{13} Hz, we have

$$\omega_d = \omega_i + \Omega \approx \omega_i, \quad \text{so} \quad k_d \cong k_i$$

and the magnitude of the two optical wavevectors is taken as k (see also Problem 9.4). The magnitude of the sound wavevector is thus

$$k_s = 2k \sin\theta \tag{9.7-6}$$

Using $K = 2\pi/\Lambda$, this equation becomes

$$2\Lambda \sin\theta = \lambda \tag{9.7-7}$$

which is the same as the Bragg-diffraction condition (9.7-3) with $m = 1$.

In the case of nonsinusoidal (but periodic) sound waves, the sound waves consist of a superposition of plane waves of the form $\cos(m\Omega t - m\mathbf{K}\cdot\mathbf{r})$, with $m = 1, 2, 3, \ldots$. In this case, the conservation of momentum with the mth harmonic becomes

$$\mathbf{k}_d = m\mathbf{K} + \mathbf{k}_i \tag{9.7-8}$$

or equivalently,

$$mK = 2k \sin\theta \tag{9.7-9}$$

Using $K = 2\pi/\Lambda$, this equation becomes

$$2n\Lambda \sin\theta = m\lambda \tag{9.7-10}$$

which is exactly the same as the Bragg-diffraction condition (9.7-3) with $m = 1$.

Doppler Derivation of the Frequency Shift

The frequency-shift condition (9.7-5) can also be derived by considering the Doppler shift experienced by an optical beam incident on a mirror moving at the sound velocity v_s, at an angle satisfying the Bragg condition (9.7-7). The formula for the Doppler frequency shift of a wave reflected from a moving object is

$$\Delta\omega = 2\omega \frac{v}{c/n} \tag{9.7-11}$$

where ω is the optical frequency and v is the component of the object velocity that is parallel to the wave propagation direction. From Figure 9.20 we have $v = v_s \sin\theta$, and thus

$$\Delta\omega = 2\omega \frac{v_s \sin\theta}{c/n} \tag{9.7-12}$$

Using Equation (9.7-10) for $\sin\theta$, we obtain

$$\Delta\omega \equiv \frac{2\pi v_s}{\Lambda} = \Omega \tag{9.7-13}$$

and, therefore, $\omega_d = \omega + \Omega$.

If the direction of propagation of the sound beam is reversed so that, in Figure 9.20, the sound recedes from the optical beam, the Doppler shift changes sign and the diffracted beam has a frequency $\omega - \Omega$.

9.8 BRAGG DIFFRACTION—COUPLED-WAVE ANALYSIS

In the particle picture of the acousto-optic interaction, we obtained the relationship involving the frequencies and the directions between the incident beam and the scattered beam. The discussion was based on the conservation of energy and momentum only. To find out the efficiency of the interaction, particularly the diffraction efficiency, we need to solve for the amplitude of the interacting beams. This requires solving the wave equation. To do so we will use a coupled-wave analysis. In treating the diffraction of light by acoustic waves, we assume a long interaction path so that higher diffraction orders [13] are missing and the only two waves coupled by the sound are the incident wave at ω_i and a diffracted wave at $\omega_d = \omega_i + \Omega$ or at $\omega_i - \Omega$, depending on the direction of the Doppler shift as discussed in the last section.

Referring to Figure 9.22, we consider the interaction of an optical beam with a sound wave. According to the discussion in Section 9.6, the sound wave causes a traveling modulation of the index of refraction (an index grating).

For simplicity, we consider an index grating in an isotropic medium given by

$$\Delta n(z, t) = \Delta n_0 \cos(\Omega t - Kz) \tag{9.8-1}$$

or equivalently, in terms of the dielectric constant,

$$\Delta\varepsilon(z, t) = \Delta\varepsilon_0 \cos(\Omega t - Kz) = 2\varepsilon_0 n\, \Delta n_0 \cos(\Omega t - Kz) \tag{9.8-2}$$

where Δn_0 is the depth of the index modulation, and $\Delta\varepsilon_0 = 2\varepsilon_0 n\, \Delta n_0$ is the depth of modulation in terms of the dielectric constant. Let the electric fields of the incident beam and the diffracted beam be written

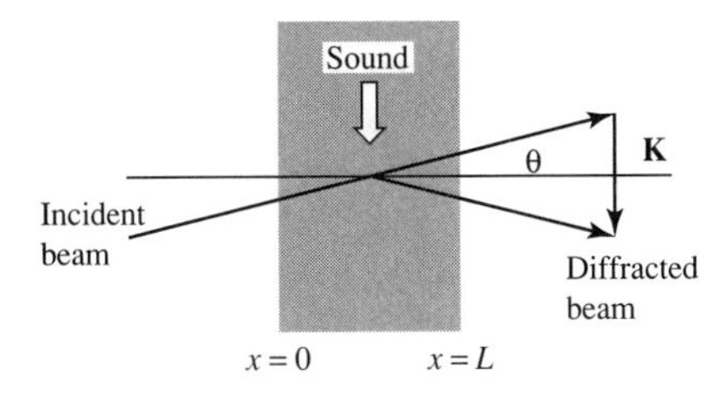

Figure 9.22 Schematic drawing of an interaction configuration of Bragg scattering. K is the wavenumber of the sound wave.

$$\mathbf{E}(x, z, t) = A_1\mathbf{E}_1 e^{i(\omega_1 t - \mathbf{k}_1\cdot\mathbf{r})} + A_2\mathbf{E}_2 e^{i(\omega_2 t - \mathbf{k}_2\cdot\mathbf{r})} \tag{9.8-3}$$

where $\mathbf{k}_1$ and $\mathbf{k}_2$ are the propagation wavevectors of the incident and diffracted beams, respectively. ω_1 and ω_2 are the corresponding frequencies of the optical beams. $\mathbf{E}_1$ and $\mathbf{E}_2$ are the normal modes of propagation, which are constant vectors in a homogeneous medium. A_1 and A_2 are the mode amplitudes of the optical beams. In the presence of the index grating $\Delta\varepsilon(x, z, t)$, the two modes are coupled and the mode amplitudes are a function of position. For continuous waves (both optical and acoustic), we can ignore the time dependence of the mode amplitudes, provided the frequencies satisfy the conservation of energy ($\omega_2 = \omega_1 \pm \Omega$). Let the plane of incidence (i.e., plane formed by $\mathbf{k}_1$ and $\mathbf{K}$) be the xz plane. The conservation of momentum requires that $\mathbf{k}_2$ must also be in this plane.

The electric field can thus be written

$$\mathbf{E}(x, z, t) = A_1(x)\mathbf{E}_1 e^{i(\omega_1 t - \alpha_1 x - \beta_1 z)} + A_2(x)\mathbf{E}_2 e^{i(\omega_2 t - \alpha_2 x - \beta_2 z)} \tag{9.8-4}$$

where β_1 and β_2 are the z components of the wavevectors $\mathbf{k}_1$ and $\mathbf{k}_2$, respectively, and α_1 and α_2 are the x components of these wavevectors parallel to the wavefront of the acoustic wave. The mode amplitudes A_1 and A_2 are in general functions of both x and z. However, there are a number of cases in which the interaction configuration requires that the mode amplitudes A_1 and A_2 be functions of either x or z only. In the case shown in Figure 9.22, where the optical waves are of infinite extent along the z axis, the mode amplitudes are functions of x only. This is a result of the boundary conditions on the continuity of the field amplitudes.

The electric field (9.8-4) must satisfy the following wave equation:

$$(\nabla^2 + \omega^2\mu\varepsilon + \omega^2\mu\Delta\varepsilon)\mathbf{E} = 0 \tag{9.8-5}$$

where ε is the dielectric constant of the medium in the absence of the sound wave and $\Delta\varepsilon$ is the dielectric perturbation (the index grating) due to the sound. Both $\mathbf{E}_1 e^{i(\omega_1 t - \alpha_1 x - \beta_1 z)}$ and $\mathbf{E}_2 e^{i(\omega_2 t - \alpha_2 x - \beta_2 z)}$ are solutions of the wave equation when $\Delta\varepsilon = 0$. Substitution of Equation (9.8-4) for $\mathbf{E}$ in Equation (9.8-5) leads to

$$\sum_{m=1,2}\left(\frac{\partial^2 A_m}{\partial x^2} + \frac{\partial^2 A_m}{\partial z^2} - 2i\alpha_m\frac{\partial}{\partial x}A_m - 2i\beta_m\frac{\partial}{\partial z}A_m\right)\mathbf{E}_m e^{i(\omega_m t - \alpha_m x - \beta_m z)}$$
$$= -\omega^2\mu\sum_{l=1,2}\Delta\varepsilon A_l\mathbf{E}_l e^{i(\omega_l t - \alpha_l x - \beta_l z)} \tag{9.8-6}$$

The second-order derivatives $\partial^2 A/\partial x^2$ and $\partial^2 A/\partial z^2$ are often neglected in the slowly varying amplitude (SVA) approximation. For acousto-optic interaction this is legitimate as the index perturbation is usually very small ($\Delta\varepsilon/\varepsilon \approx 10^{-5}$). The differential equation is thus dominated by the first-order derivatives.

We now limit ourselves to the interaction configuration shown in Figure 9.22 so that the mode amplitudes are functions of x only. The wave equation becomes

$$-2i\alpha_1\left(\frac{\partial}{\partial x}A_1\right)\mathbf{E}_1e^{i(\omega_1t-\alpha_1x-\beta_1z)} - 2i\alpha_2\left(\frac{\partial}{\partial x}A_2\right)\mathbf{E}_2e^{i(\omega_2t-\alpha_2x-\beta_2z)}$$
$$= -\omega^2\mu\Delta\varepsilon(A_1\mathbf{E}_1e^{i(\omega_1t-\alpha_1x-\beta_1z)} + A_2\mathbf{E}_2e^{i(\omega_2t-\alpha_2x-\beta_2z)}) \tag{9.8-7}$$

or equivalently,

$$-2i\alpha_1\left(\frac{\partial}{\partial x}A_1\right)\mathbf{E}_1e^{i(\omega_1t-\alpha_1x-\beta_1z)} - 2i\alpha_2\left(\frac{\partial}{\partial x}A_2\right)\mathbf{E}_2e^{i(\omega_2t-\alpha_2x-\beta_2z)}$$
$$= -\omega^2\mu\varepsilon_0 n\Delta n_0(e^{i(\Omega t-Kz)} + e^{-i(\Omega t-Kz)})(A_1\mathbf{E}_1e^{i(\omega_1t-\alpha_1x-\beta_1z)} + A_2\mathbf{E}_2e^{i(\omega_2t-\alpha_2x-\beta_2z)}) \tag{9.8-8}$$

Before we derive the coupled equations for the field amplitude, we will first normalize the field vectors $\mathbf{E}_1$ and $\mathbf{E}_2$ in such a way that

$$\mathbf{E}_m = \sqrt{\frac{2\mu\omega_m}{\alpha_m}}\,\mathbf{p}_m, \quad m = 1, 2 \tag{9.8-9}$$

where $\mathbf{p}_m$ are the unit vectors representing the polarization states of the beams, and α_m are the x components of the wavevectors. With this normalization, a unity mode amplitude represents a power flow of 1 W/m^2 in the x direction for the case of an isotropic medium.

By scalar-multiplying Equation (9.8-8) with $\mathbf{E}_l^* e^{-i(\omega_l t-\alpha_l x-\beta_l z)}$, $l = 1, 2$, one at a time and integrating over z and t, we obtain the following set of coupled equations:

$$\begin{aligned}\frac{d}{dx}A_1 &= -i\kappa_{12}A_2e^{i\Delta\alpha x}\\ \frac{d}{dx}A_2 &= -i\kappa_{12}^*A_1e^{-i\Delta\alpha x}\end{aligned} \tag{9.8-10}$$

where κ_{12} is the coupling constant and is given by

$$\kappa_{12} = \frac{\omega^2\mu\varepsilon_0}{2\sqrt{\alpha_1\alpha_2}}\,n\,\Delta n_0\,\mathbf{p}_1^*\cdot\mathbf{p}_2 \tag{9.8-11}$$

provided the following conditions are satisfied:

$$\beta_2 = \beta_1 \pm K \tag{9.8-12}$$

and

$$\omega_2 = \omega_1 \pm \Omega \tag{9.8-13}$$

The coupled equations are not valid when either one of the above conditions is not satisfied. This is a result of the assumption of continuous wave (cw) interaction as well as an infinite dimension in the z direction. The quantity $\Delta\alpha$ is the momentum mismatch in the x direction and is given by

$$\Delta\alpha = \alpha_1 - \alpha_2 \tag{9.8-14}$$

These two components, α_1 and α_2, are assumed to be positive. For the case of an isotropic medium, the Bragg condition (9.8-12) can be satisfied when $\beta_1 = -\beta_2 = -K/2 = -k\sin\theta_B$, in which case a phonon is absorbed, or $\beta_1 = -\beta_2 = K/2 = k\sin\theta_B$ when the emission of a phonon takes place. At this Bragg angle of incidence,

$$\theta_B = \sin^{-1}\left(\frac{K}{2k}\right) = \sin^{-1}\left(\frac{\lambda}{2\Lambda}\right) \tag{9.8-15}$$

the momentum mismatch $\Delta\alpha$ vanishes ($\Delta\alpha = 0$), the x axis bisects the angle between $\mathbf{k}_1$ and $\mathbf{k}_2$, and the coupled equations become

$$\frac{d}{dx}A_1 = -i\kappa_{12}A_2$$
$$\frac{d}{dx}A_2 = -i\kappa_{12}^* A_1 \qquad (9.8\text{-}16)$$

The general solutions to (9.8-16) are

$$A_1(x) = A_1(0)\cos\kappa x - i\frac{\kappa_{12}}{\kappa}A_2(0)\sin\kappa x$$
$$A_2(x) = A_2(0)\cos\kappa x - i\frac{\kappa_{12}^*}{\kappa}A_1(0)\sin\kappa x \qquad (9.8\text{-}17)$$

where $A_1(0)$ and $A_2(0)$ are the amplitudes of the beams at $x = 0$, and κ is the magnitude of κ_{12},

$$\kappa = |\kappa_{12}| \qquad (9.8\text{-}18)$$

In the practical case of a single beam incidence at $x = 0$ (i.e., $A_2(0) = 0$), the solutions become

$$A_1(x) = A_1(0)\cos\kappa x$$
$$A_2(x) = -i\frac{\kappa_{12}^*}{\kappa}A_1(0)\sin\kappa x \qquad (9.8\text{-}19)$$

We note that

$$|A_1(x)|^2 + |A_1(x)|^2 = |A_1(0)|^2 \qquad (9.8\text{-}20)$$

so that the total power carried by both beams of light is conserved. The conservation of energy is true in the general case when both beam amplitudes are finite at $x = 0$. Figure 9.23 shows the power of the beams as a function of interaction distance κL. We note that the optical energy is transferred back and forth as the beams propagate in the medium.

If the interaction distance L between the two beams is such that $\kappa L = \pi/2$, the total power of the incident beam is transferred into the diffracted beam. Since this process is used in a large number of technological and scientific applications, it may be worthwhile to gain some appreciation for the diffraction efficiencies using data of some known acoustic media and conveniently available acoustic power levels. The coupling constant derived above is for a scalar index grating. We note that the coupling constant is maximum when the scattered beam

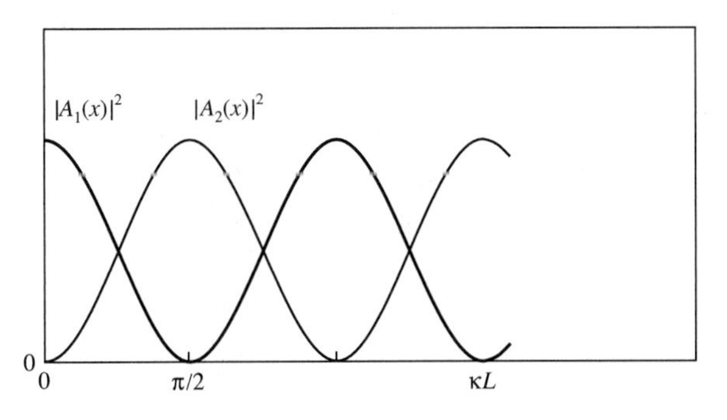

Figure 9.23 Power of beams as a function of interaction distance in acousto-optic diffraction.

has the same polarization state as that of the incident beam. In this case, the coupling is written

$$\kappa_{12} = \frac{\omega^2 \mu \varepsilon_0}{2\sqrt{\alpha_1 \alpha_2}} n \, \Delta n_0 \tag{9.8-21}$$

where Δn_0 is the amplitude of the index grating. In the case of a birefringent index grating, the coupling constant can be written

$$\kappa_{12} = \frac{\omega^2 \mu}{4\sqrt{\alpha_1 \alpha_2}} \mathbf{p}_1^* \cdot \Delta\varepsilon_0 \, \mathbf{p}_2 \tag{9.8-22}$$

The fraction of the power of the incident beam transferred in a distance L into the diffracted beam is given, using (9.8-19), by

$$\frac{I_{\text{diffracted}}}{I_{\text{incident}}} = \frac{|A_2(L)|^2}{|A_1(0)|^2} = \sin^2 \kappa L \tag{9.8-23}$$

where κ is given, according to Equations (9.8-11), (9.8-22), and (9.8-18), by

$$\kappa = |\kappa_{12}| = \frac{\omega^2 \mu}{4\sqrt{\alpha_1 \alpha_2}} |\mathbf{p}_1^* \cdot \Delta\varepsilon_0 \, \mathbf{p}_2| \tag{9.8-24}$$

At the Bragg condition (9.8-12), we have $|\beta_1| = |\beta_2| = k \sin\theta = K/2$ and $\alpha_1 = \alpha_2 = k \cos\theta_B = (\omega/c) n \cos\theta_B$, and the coupling constant can be written

$$\kappa = \frac{\omega \mu c}{4n \cos\theta_B} |\mathbf{p}_1^* \cdot \Delta\varepsilon_0 \, \mathbf{p}_2| \tag{9.8-25}$$

where we recall that $\Delta\varepsilon_0$ is the amplitude of the modulation in the dielectric constant due to the presence of the acoustic wave, and $\mathbf{p}_1$ and $\mathbf{p}_2$ are the unit vectors representing the polarization states of the incident and diffracted beams, respectively. By assuming an isotropic medium and using Equation (9.6-36), the coupling constant can also be expressed in terms of the strain components and the photoelastic constants as

$$\kappa = \frac{n^3 \omega \mu \varepsilon_0 c}{4 \cos\theta_B} |\mathbf{p}_1^* \cdot pS_0 \mathbf{p}_2| = \frac{\omega n^3}{4n \cos\theta_B} |\mathbf{p}_1^* \cdot pS_0 \mathbf{p}_2| \tag{9.8-26}$$

where pS_0 is a matrix representing the change of the impermeability tensor due to the sound wave, with S_0 being the amplitude of the strain wave.

Since the strain amplitude S_0 is directly related to the acoustic power flow or the intensity of the sound, it is convenient to express the diffraction efficiency (9.8-23) in terms of the acoustic intensity I_{acoustic} (W/m^2). For this purpose, the coupling constant is written

$$\kappa = \frac{\omega n^3}{4c} pS_0 \tag{9.8-27}$$

where p is the effective photoelastic constant of the medium, and S_0 is the effective strain amplitude. According to elastic theory, the strain S_0 is related to the acoustic intensity I_{acoustic} by

$$I_{\text{acoustic}} = \tfrac{1}{2} \rho v_s^3 S_0^2 \tag{9.8-28}$$

or equivalently,

$$S_0 = \sqrt{\frac{2I_{\text{acoustic}}}{\rho v_s^3}} \tag{9.8-29}$$

where v_s is the velocity of sound in the medium and ρ is the mass density (kg/m^3). Assuming a scalar index grating and same polarization states (i.e., $\mathbf{p}_1 = \mathbf{p}_2$), we obtain after combining Equations (9.8-27), (9.8-28), (9.8-26), and (9.6-23),

$$\frac{I_{\text{diffracted}}}{I_{\text{incident}}} = \sin^2\left(\frac{\pi L}{\sqrt{2}\lambda}\sqrt{\frac{n^6 p^2}{\rho v_s^3}\, I_{\text{acoustic}}}\right) \tag{9.8-30}$$

Using the following definition for the acousto-optic diffraction figure of merit,

$$S_0 = \sqrt{\frac{2I_{\text{acoustic}}}{\rho v_s^3}} \tag{9.8-31}$$

Equation (9.8-30) becomes

$$\frac{I_{\text{diffracted}}}{I_{\text{incident}}} = \sin^2\left(\frac{\pi L}{\sqrt{2}\lambda}\sqrt{MI_{\text{acoustic}}}\right) \tag{9.8-32}$$

Taking water as an example, an optical wavelength of $\lambda = 0.6328$ μm, and the constants (taken from Table 9.3)

$$n = 1.33$$

$$p = 0.31$$

$$v_s = 1.5 \times 10^3 \text{ m/s}$$

$$\rho = 1000 \text{ kg/m}^3$$

Equation (9.8-32) gives

$$\left(\frac{I_{\text{diffracted}}}{I_{\text{incident}}}\right)_{\substack{H_2O \\ \text{at } \lambda=0.6328\ \mu\text{m}}} = \sin^2\left(1.4L\sqrt{I_{\text{acoustic}}}\right) \tag{9.8-33}$$

TABLE 9.3 A List of Some Materials Commonly Used in the Diffraction of Light by Sound and Some of Their Relevant Properties

Material	ρ (mg/m^3)	v_s (km/s)	n	p	M_ω
Water	1.0	1.5	1.33	0.31	1.0
Extra-dense flint glass	6.3	3.1	1.92	0.25	0.12
Fused quartz (SiO_2)	2.2	5.97	1.46	0.20	0.006
Polystyrene	1.06	2.35	1.59	0.31	0.8
KRS-5	7.4	2.11	2.60	0.21	1.6
Lithium niobate ($LiNbO_3$)	4.7	7.40	2.25	0.15	0.012
Lithium fluoride (LiF)	2.6	6.00	1.39	0.13	0.001
Rutile (TiO_2)	4.26	10.30	2.60	0.05	0.001
Sapphire (Al_2O_3)	4.0	11.00	1.76	0.17	0.001
Lead molybdate ($PbMoO_4$)	6.95	3.75	2.30	0.28	0.22
Alpha iodic acid (HIO_3)	4.63	2.44	1.90	0.41	0.5
Tellurium dioxide (TeO_2) (slow shear wave)	5.99	0.617	2.35	0.09	5.0

Note: ρ is the Density, v_s the velocity of sound, n the index of refraction, p the photoelastic constant as defined by Equation (9.6-36), and M_ω is the relative diffraction constant defined on the next page. (After Reference [11].)

For other materials and at other wavelengths we can combine the last two equations to obtain a convenient working formula:

$$\frac{I_{\text{diffracted}}}{I_{\text{incident}}} = \sin^2\left(1.4\frac{0.6328}{\lambda(\mu\text{m})} L\sqrt{M_\omega I_{\text{acoustic}}}\right) \tag{9.8-34}$$

where $M_\omega = M_{\text{material}}/M_{H_2O}$ is the diffraction figure of merit of the material relative to water. It is important to note that Equation (9.8-34) is valid provided that the interaction configurations (angle of incidence, acoustic frequency, polarization states, etc.) are assumed to be the same. Values of M and M_ω for some common materials are listed in Tables 9.3 and 9.4.

According to Equation (9.8-30), at small diffraction efficiencies, the diffracted light intensity is proportional to the acoustic intensity. This fact is used in acoustic modulation of optical radiation. The information signal is used to modulate the intensity of the acoustic beam. This modulation is then transferred, according to Equation (9.8-30), as intensity modulation onto the diffracted optical beam.

TABLE 9.4 A List of Materials Commonly Used in Acousto-optic Interactions and Some of Their Relevant Properties

Material	λ (μm)	n	ρ (g/cm^3)	Acoustic Wave Polarization and Direction	v_s (10^5 cm/s)	Optical Wave Polarization and Direction[a]	$M = n^6p^2/\rho v_s^3$
Fused quartz	0.63	1.46	2.2	long.	5.95	⊥	1.51×10^{-15}
Fused quartz	0.63			trans.	3.76	‖ or ⊥	0.467
GaP	0.63	3.31	4.13	long. in [110]	6.32	‖	44.6
GaP	0.63			trans. in [100]	4.13	‖ or ⊥ in [010]	24.1
GaAs	1.15	3.37	5.34	long. in [110]	5.15	‖	104
GaAs	1.15			trans. in [100]	3.32	‖ or ⊥ in [010]	46.3
TiO_2	0.63	2.58	4.6	long. in [11–20]	7.86	⊥ in [001]	3.93
$LiNbO_3$	0.63	2.20	4.7	long. in [11–20]	6.57		6.99
YAG	0.63	1.83	4.2	long. in [100]	8.53	‖	0.073
YAG	0.63			long. in [110]	8.60	⊥	0.012
YIG	1.15	2.22	5.17	long. in [100]	7.21	⊥	0.33
$LiTaO_3$	0.63	2.18	7.45	long. in [001]	6.19	‖	1.37
As_2S_3	0.63	2.61	3.20	long.	2.6	⊥	433
As_2S_3	1.15	2.46		long.		‖	347
SF-4	0.63	1.616	3.59	long.	3.63	⊥	4.51
β-ZnS	0.63	2.35	4.10	long. in [110]	5.51	‖ in [001]	3.41
β-ZnS	0.63			trans. in [110]	2.165	‖ or ⊥ in [001]	0.57
α-Al_2O_3	0.63	1.76	4.0	long. in [001]	11.15	‖ in [11–20]	0.34
CdS	0.63	2.44	4.82	long. in [11–20]	4.17	‖	12.1
ADP	0.63	1.58	1.803	long. in [100]	6.15	‖ in [010]	2.78
ADP	0.63			trans. in [100]	1.83	‖ or ⊥ in [001]	6.43
KDP	0.63	1.51	2.34	long. in [100]	5.50	‖ in [010]	1.91
KDP	0.63			trans. in [100]		‖ or ⊥ in [001]	3.83
H_2O	0.63	1.33	1.0	long.	1.5		160
Te	10.6	4.8	6.24	long. in [11–20]	2.2	‖ in [001]	4400
$PbMoO_4$ [15]	0.63	2.4		long. ‖ c axis	3.75	‖ or ⊥	73

Note: $M = n^6p^2/\rho v_s^3$ is the figure of merit, defined by Equation (9.8-31), and is given in MKS units. (After Reference [14].)

[a] The optical beam direction actually differs from that indicated by the magnitude of the Bragg angle. The polarization is defined as parallel or perpendicular to the scattering plane formed by the acoustic and optical **k** vectors.

EXAMPLE: SCATTERING IN $PbMoO_4$

Calculate the fraction of 0.633 μm light that is diffracted under Bragg conditions from a sound wave in $PbMoO_4$ with the following characteristics:

Acoustic power = 1 W

Acoustic beam cross section = 1 mm × 1 mm

l = optical path in acoustic beam = 1 mm

M_ω (from Table 9.3) = 0.22

Substituting these data into Equation (9.8-23) yields

$$\frac{I_{\text{diffracted}}}{I_{\text{incident}}} \approx 37\%$$

Diffraction with Bragg Mismatch

Let us now consider the situation when the angle of incidence is slightly deviated from the Bragg angle:

$$\theta_B = \sin^{-1}\left(\frac{K}{2k}\right) = \sin^{-1}\left(\frac{\lambda}{2\Lambda}\right) \tag{9.8-35}$$

Let the incident angle be written

$$\theta_1 = \theta_B + \Delta\theta \tag{9.8-36}$$

where $\Delta\theta$ is a small angle. The Bragg condition (9.8-12) requires that the angle of the diffracted beam be written

$$\theta_2 = \theta_B - \Delta\theta \tag{9.8-37}$$

Therefore, $\Delta\alpha$ (9.8-14) becomes

$$\Delta\alpha = 2k\,\Delta\theta \sin\theta_B = K\,\Delta\theta \tag{9.8-38}$$

Figure 9.24 illustrates the Bragg mismatch $\Delta\alpha$ when the incident beam is deviated from the Bragg angle θ_B.

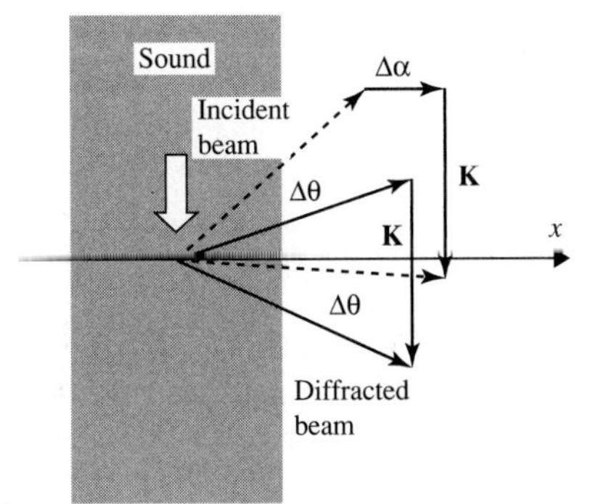

Figure 9.24 Acousto-optic scattering with a Bragg mismatch $\Delta\alpha$. Dashed arrows are for optical beams with Bragg mismatch. **K** is the wavevector of the acoustic wave.

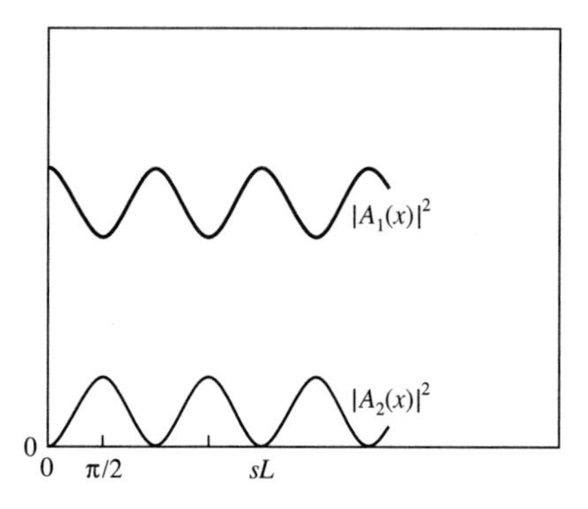

Figure 9.25 Power of beams as a function of interaction distance in acousto-optic diffraction.

Since $\Delta\alpha$ is not zero, the solutions to (9.8-10), subject to the boundary condition $A_2(0) = 0$, are given by

$$A_1(x) = e^{i\Delta\alpha x/2} A_1(0)\left(\cos sx - i\frac{\Delta\alpha}{2s}\sin sx\right)$$
$$A_2(x) = -ie^{-i\Delta\alpha x/2} A_1(0)\frac{\kappa_{12}^*}{s}\sin sx \tag{9.8-39}$$

where

$$s^2 = \kappa^2 + (\Delta\alpha/2)^2 = \kappa^2 + (K\,\Delta\theta/2)^2 \tag{9.8-40}$$

An inspection of (9.8-39) reveals that the power transfer, or diffraction efficiency, can never be complete and is given by

$$\frac{I_{\text{diffracted}}}{I_{\text{incident}}} = \frac{|A_2(L)|^2}{|A_1(0)|^2} = \frac{\kappa^2}{s^2}\sin^2 sL = \frac{\kappa^2}{\kappa^2 + (K\,\Delta\theta/2)^2}\sin^2\kappa L\sqrt{1 + \left(\frac{K\,\Delta\theta}{2\kappa}\right)^2} \tag{9.8-41}$$

Figure 9.25 shows the power transfer as a function of the interaction distance sL.

The maximum diffraction efficiency $\kappa^2/[\kappa^2 + (K\,\Delta\theta/2)^2]$, which occurs at $sL = \pi/2$, becomes small when $K\,\Delta\theta \gg \kappa$. The angular deviation can be due to either misalignment of the incident laser beam or the distortion of the acoustic wavefront due to the finite size of the transducer. The latter connection can be used to describe the angular plane wave spectrum of a transducer with width L. Thus, if the power of the diffraction light beam is measured and plotted as a function of $\Delta\theta$, the radiation pattern of the transducer will be traced out.

9.9 BRAGG CELLS AND BEAM DEFLECTORS

One of the most important applications of acousto-optic interactions is in the deflection of optical beams. Basically, acousto-optic deflectors operate in the same way as Bragg diffraction modulators, the only difference being that the frequency rather than the amplitude of the sound wave is varied. So acousto-optic deflectors are obtained by changing the sound frequency while operating near the Bragg matching condition. The use of acousto-optic interaction offers the possibility of high-resolution beam deflection. Both random access and continuously scanned deflectors can be obtained. The basic principle is depicted in Figure 9.26 and can be understood using Figure 9.27.

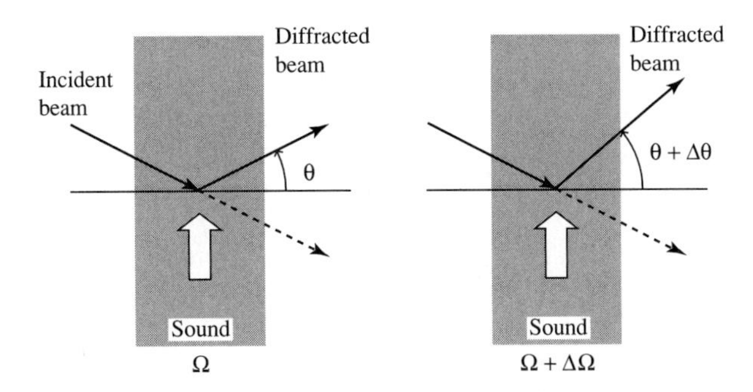

Figure 9.26 A change of frequency of the sound wave from Ω to $\Omega + \Delta\Omega$ causes a change $\Delta\theta$ in the direction of the diffracted beam.

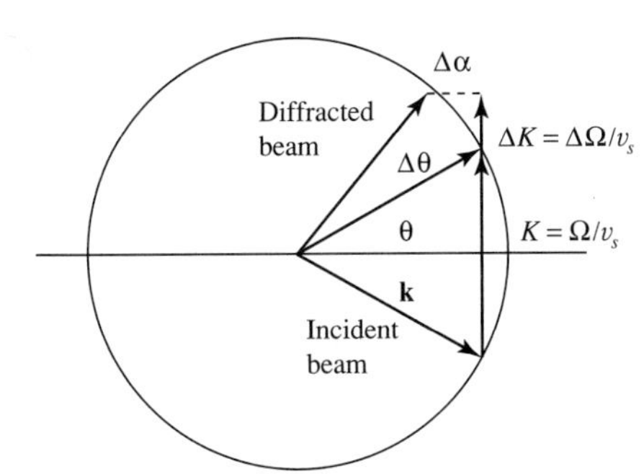

Figure 9.27 Momentum diagram, illustrating how the change in sound frequency from Ω to $\Omega + \Delta\Omega$ deflects the diffracted light beam from θ to $\theta + \Delta\theta$. $\Delta\alpha$ is the momentum mismatch. K is the wavenumber of the sound wave and v_s is the sound velocity.

Let us assume first that the Bragg condition (9.8-12) is satisfied. The momentum vector diagram shown Figure 9.27 is closed, and the beam is diffracted along the direction θ_B as given by Equation (9.8-15). We rewrite the Bragg condition as

$$2k \sin \theta_B = K = \frac{\Omega}{v_s} = \frac{2\pi}{v_s} f \tag{9.9-1}$$

where k is the wavenumber of the optical beam, K is the wavenumber of the acoustic beam, v_s is the sound velocity, and f is the frequency of the sound wave. Now let the sound frequency change from f_0 to $f_0 + \Delta f$. Since $K = 2\pi f/v_s$, this causes a change of $\Delta K = 2\pi\, \Delta f/v_s$ in the magnitude of the sound wavevector as shown. Now the angle of incidence remains the same (θ_B), and the magnitude of the wavevector of the diffracted beam is virtually unchanged, so its tip is constrained to the circle locus shown in Figure 9.27. Thus we can no longer close the momentum diagram, and the momentum is no longer strictly conserved. The beam will be diffracted along the direction shown in the figure. A momentum mismatch $\Delta\alpha$ exists. Diffraction under such circumstances is possible due to the finite interaction length L along the x axis. Recalling that the angle θ and $\Delta\theta$ are both small and that $k = K \sin \theta$, we obtain

$$\Delta\theta = \frac{\Delta K}{k \cos \theta_B} = \frac{\lambda}{n v_s \cos \theta_B} \Delta f \tag{9.9-2}$$

where θ_B is the Bragg angle at the center frequency f_0. The angle of deflection is thus proportional to the change in the sound frequency.

In acousto-optic beam deflection, the angle of deflection is often very small. For practical applications, we are interested in the number of resolvable spots—that is, the factor by which $\Delta\theta$ exceeds the beam divergence angle. If we take the beam divergence angle as

$$\delta\theta = 2\theta_{\text{beam}} = \frac{2\lambda}{\pi n \omega_0} \tag{9.9-3}$$

where θ_{beam} is given by Equation (2.5-22), ω_0 is the Gaussian beam spot size, and the number of resolvable spots is

$$N = \frac{\Delta\theta}{\delta\theta} = \frac{\pi\omega_0}{2v_s \cos\theta_B}\Delta f = \tau\,\Delta f \tag{9.9-4}$$

where $\tau = \pi\omega_0/2v_s \cos\theta_B$ is approximately the time it takes the sound to cross the Gaussian optical beam spot and is usually called the access time of the deflector.

A common figure of merit for a beam deflector is the ratio of the total number of resolvable spots to the access time, expressed as a speed–capacity product, which, according to Equation (9.9-4), is

$$F = \frac{N}{\tau} = \Delta f \tag{9.9-5}$$

Thus a high speed–capacity product is only available when the bandwidth (or frequency tuning range) Δf is large. According to Figure 9.27, the Bragg mismatch $\Delta\alpha$ increases as Δf increases. The diffraction efficiency drops as the Bragg mismatch increases. Thus the frequency tuning range Δf is usually a fraction of the center frequency f_0. As a result, it is desirable to use high-frequency sound waves for beam deflectors, because large bandwidth Δf is only possible when the center frequency is high. In practice, the bandwidth Δf is limited by the availability of a broadband transducer and by the Bragg angle tolerance. The latter refers to the fact that a proper angle of incidence of the incident beam for Bragg-matched diffraction is a function of the sound frequency ($\theta_B = \sin^{-1}(\lambda f/2nv_s)$).

EXAMPLE: BEAM DEFLECTION

Consider a deflection system using flint glass and a sound beam that can be varied in frequency from 80 MHz to 120 MHz; thus $\Delta f = 40$ MHz. Let the optical beam diameter be 1 cm. From Table 9.3 we obtain $v_s = 3.1 \times 10^5$ cm/s; therefore the access time $\tau = 3.23 \times 10^{-6}$ s and the number of resolvable spots is $N = \Delta f\,\tau \sim 130$.

Bragg interactions have been demonstrated experimentally [16] between surface acoustic waves and optical modes confined in thin film dielectric waveguides. Since the modulation efficiency depends, according to Equation (9.8-30), on the acoustic intensity, the confinement of the acoustic power near the surface (to a distance $\sim\Lambda$) leads to low modulation or switching power. Figure 9.28 shows an experimental setup in which both the acoustic surface wave and the optical wave are guided in a single crystal of $LiNbO_3$. The dielectric waveguide is produced by out-diffusion of Ti from a layer of ~10 μm near the surface, which raises the index of refraction. For a more advanced treatment of the subject of light and sound interaction in crystals and for some new devices that are based on it, the student should consult Reference [6].

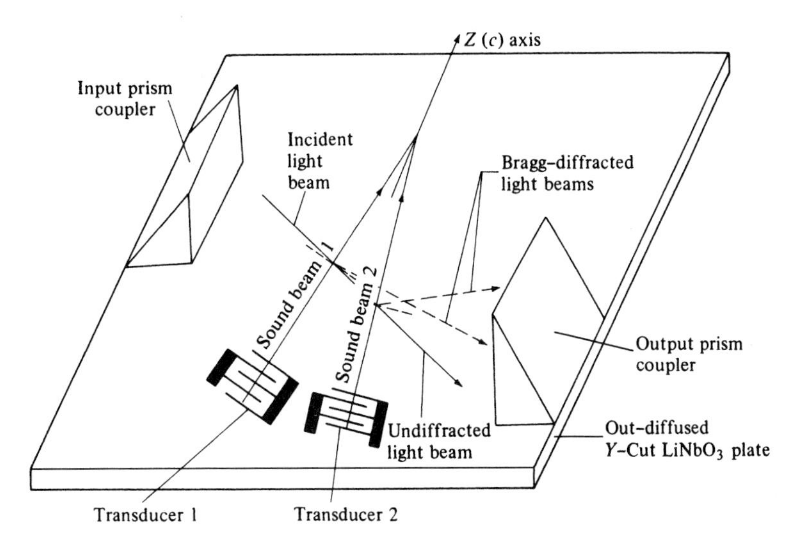

Figure 9.28 Guided-wave acousto-optic Bragg diffraction from two tilted surface acoustic waves. (From Reference [17].)

PROBLEMS

9.1 Derive the equations of the ellipses traced during one period by the optical field vector as shown in Figure 9.3 for $\Gamma = 0, \pi/4, \pi/2, 3\pi/4, \pi, 5\pi/4$.

9.2 Discuss the consequence of the field-independent retardation $(\omega l/c_c)(n_0 - n_e)$ in Equation (9.2-23) on an amplitude modulator such as that shown in Figure 9.4.

9.3 Use the Bessel function expansion of $\sin(a \sin x)$ to express Equation (9.2-13) in terms of the harmonics of the modulation frequency ω_m. Plot the ratio of the third harmonic $(3\omega_m)$ of the output intensity to the fundamental as a function of Γ_m. What is the maximum allowed Γ_m if this ratio is not to exceed 10^{-2}? [Answer: $\Gamma_m < 0.5$.]

9.4 Show that, if a phase-modulated optical wave is incident on a square-law detector, the output contains no alternating currents.

9.5 Using References [3] and [4], design a partially loaded KDP traveling-wave phase modulator that operates at $\nu_m = 10^9$ Hz and yields a peak phase excursion of $\delta = \pi/3$. What is the modulation power?

9.6 Derive the expression (similar to Equation (9.3-2)) for the modulation power of a transverse $\bar{4}3m$ crystal electro-optic modulator of the type described in the Appendix F.

9.7 Derive an expression for the modulation power requirement (corresponding to Equation (9.3-2)) for a GaAs transverse modulator.

9.8 Show that if a ray propagates at an angle θ ($\ll 1$) to the z axis in the arrangement of Figure 9.4, it exercises a birefringent contribution to the retardation:

$$\Delta\Gamma_{\text{birefringent}} = \frac{\omega l}{2c} n_o \left(\frac{n_o^2}{n_e^2} - 1\right)\theta^2$$

which corresponds to a change in index

$$n_o - n_e = \frac{n_o\theta^2}{2}\left(\frac{n_o^2}{n_e^2} - 1\right)$$

9.9 Derive an approximate expression for the maximum allowable beam-spreading angle in Problem 9.8 for which $\Delta\Gamma_{\text{birefringent}}$ does not interfere with the operation of the modulator. [Answer: $\theta < [\lambda/4n_o l(n_o^2/n_e^2 - 1)]^{1/2}$.]

9.10 Consider the index ellipsoid S defined by

$$\eta_{ij}x_i x_j = 1$$

Show that the vector $\mathbf{N}$ defined by

$$N_i = \eta_{ij}x_j$$

is perpendicular to S at the point (x_1, x_2, x_3) on S.

9.11 Consider the case of a KH_2PO_4 (KDP) crystal with an applied field along the x axis. Show that in the new principal dielectric axes coordinate system (x', y', z'), x' coincides with x while y' and z' are in the yz plane, but rotated from their original positions by θ, where

$$\tan 2\theta = \frac{2r_{41}E_z}{1/n_o^2 - 1/n_e^2}$$

Show that in the x, y', z' system the equation for the index ellipsoid is

$$\frac{x^2}{n_o^2} + \left(\frac{1}{n_o^2} + r_{41}E_x \tan\theta\right)y'^2 + \left(\frac{1}{n_e^2} - r_{41}E_x \tan\theta\right)z'^2 = 1$$

9.12 An optical beam with amplitude E_0 and frequency $\omega/2\pi$ is split, equally, in two. One of the beams is left as is, while the other is phase modulated according to

$$\Delta\phi = a + \delta \cos \omega_m t \quad (\omega_m \ll \omega)$$

The two beams are then recombined coherently (the whole procedure can be accomplished by a Michelson–Morley or a Mach–Zehnder interferometer with a phase modulator placed in one arm).

(a) Express the recombined field in the form

$$E_{\text{rec}} = f(t)e^{i(\alpha + \beta\cos\omega_m t)}$$

(b) Show that for $a = \pi/2$, $\delta \ll 1$,

$$E_{\text{rec}} \approx E_0\left(1 - \frac{\delta}{2}\cos\omega_m t\right)e^{i[\pi/4 + (\delta/2)\cos\omega_m t]}$$

(c) Obtain the (approximate) optical spectrum of the output beam.
(d) Derive the intensity modulation characteristics $|E_{\text{rec}}|^2/|E_0|^2$ for the general case.
(e) Using the results from (d), determine how we can obtain a nearly linear modulation response in which the detected photocurrent, $I_{\text{det}} \propto |E_{\text{rec}}|^2$, is proportional to the modulation signal $\delta \cos \omega_m t$.

9.13 Show that the three principal vectors X', X'', and X''' of the index ellipsoid are perpendicular to each other.

9.14 Let x, y, z be the principal dielectric axes of a crystal with dielectric tensor elements $\varepsilon_{xx}, \varepsilon_{yy}, \varepsilon_{zz}$. Consider a new coordinate system ξ, η, z, where ξ and η are rotated at an angle θ about the z axis (the z axis is the same in both systems). Show that the dielectric tensor in the new system is

$$\varepsilon = \begin{bmatrix} \varepsilon_{xx} + \delta\sin^2\theta & \delta\sin(2\theta)/2 & 0 \\ \delta\sin(2\theta)/2 & \varepsilon_{xx} + \delta\cos^2\theta & 0 \\ 0 & 0 & \varepsilon_{zz} \end{bmatrix}$$

where $\delta \equiv \varepsilon_{yy} - \varepsilon_{xx}$.

9.15 Consider a crystal with principal dielectric axes x, y, z and corresponding $\varepsilon_{xx}, \varepsilon_{yy}, \varepsilon_{zz}$. Let the application of an electric field (or strain) cause an off-diagonal element ε_{zy} to appear.

(a) Show the new principal dielectric axes are rotated about the x axis by an angle

$$\theta \cong \frac{\varepsilon_{zy}}{\varepsilon_{yy} - \varepsilon_{zz}} \quad (\varepsilon_{zy} \ll \varepsilon_{yy}, \varepsilon_{zz})$$

(b) Show that in KDP the application of a dc field $\mathbf{E} = \hat{\mathbf{e}}_x E_x$ causes a rotation β of the z and y principal axes about the x axis, where

$$\beta = -\frac{n_e^2 n_o^2 r_{41} E_z}{n_o^2 - n_e^2}$$

9.16

(a) Design an electro-optic waveguide modulator in a $LiTaO_3$ crystal as shown in Figure P9.16. Show that the phase retardation $\Gamma \equiv \theta_{\text{TE}} - \theta_{\text{TM}}$ is given by

$$\Gamma = \frac{\omega l}{c}(n_o^3 r_{13} - n_e^3 r_{33})E_z$$

(b) Describe how you will use the waveguide as (1) an amplitude modulator and (2) a phase modulator. Calculate the requisite modulation voltage assuming a width in the z direction of 5 μm and $\lambda = 0.6328$ μm.

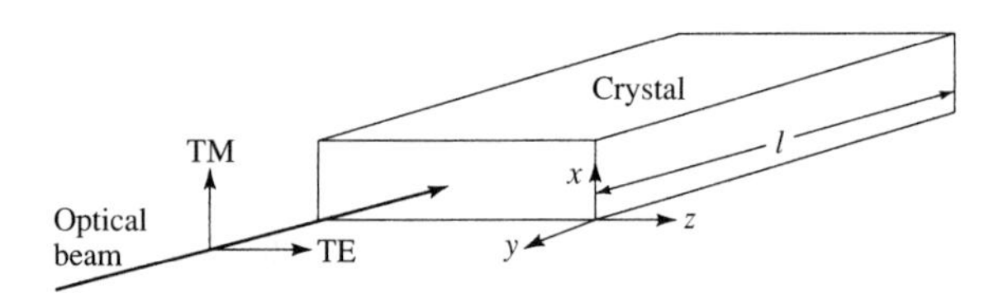

Figure P9.16

9.17

(a) Design a polarization switch (mode coupler) TE–TM using $LiNbO_3$, the crystal geometry of Problem 9.15, and a dc field parallel to the x axis. Show how you can overcome the velocity mismatch problem $(n_o \neq n_e)$ by using a spatially periodic dc field

$$E_x = E_0 \cos\frac{2\pi}{\Lambda}y$$

with a proper choice of the period Λ.

(b) What is the value of Λ at $\lambda = 1.15$ μm? (See Table 9.2 for dispersion data.)

9.18 Derive the expression of the frequency shift, under Bragg conditions, from a receding sound wave.

9.19 Design an acoustic modulation system for transferring the output of a magnetic-cartridge phonograph onto an optical beam with $\lambda_0 = 0.6328\ \mu m$ and $I_{incident} = 10^{-3}$ W. Specify the power levels involved and the essential characteristics of all the key components. [Hint: Use the audio output of the cartridge to modulate a high-frequency (100 MHz, say) carrier, which is then used to transduce an acoustic beam.]

9.20 What happens in the Bragg diffraction of light from a standing sound wave? Describe the frequency shifts and direction of diffraction.

9.21 Using Figure 9.21, show that under Bragg conditions the change in wavevector of the diffracted wave is

$$\frac{k_{diffracted} - k}{k} = 2 \sin\theta \frac{v_s}{c}$$

9.22 Consult the literature (see, e.g., References [11, 13]) and describe the difference between Bragg diffraction and Debye–Sears diffraction. Under what conditions is each observed?

9.23 Bragg's law for diffraction of X-rays in crystals is [12]

$$2d \sin\theta = m\frac{\lambda}{n}, \quad m = 1, 2, 3, \ldots$$

where d is the distance between equivalent atomic planes, θ is the angle of incidence, and λ/n is the wavelength of the diffracted radiation. Bragg diffraction of light from sound (see Equation (9.7-3)) takes place when

$$2n\Lambda \sin\theta = m\lambda$$

Thus, if we compare it to the X-ray result and take $\Lambda = d$, only the case of $m = 1$ is allowed. Explain the difference. Why don't we get light diffracted along directions θ corresponding to $m = 2, 3, \ldots$? [Hint: The diffraction of X-rays takes place at discrete atomic planes, which can be idealized as infinitely thin sheets, whereas the sound wave is continuous in z; see Figure 9.20.]

9.24 Design an acoustic deflection system using $LiTaO_3$ to be used in scanning an optical beam in a manner compatible with that of commercial television receivers.

REFERENCES

1. See, for example, J. F. Nye, *Physical Properties of Crystals.* Oxford University Press, New York, 1957, p. 123.
2. Peters, L. C., Gigacycle bandwidth coherent light traveling-wave phase modulators. *Proc. IEEE* **51**:147 (1963).
3. Rigrod, W. W., and I. P. Kaminow, Wideband microwave light modulation. *Proc. IEEE* **51**:137 (1963).
4. Kaminow, I. P., and J. Lin, Propagation characteristics of partially loaded two-conductor transmission lines for broadband light modulators. *Proc. IEEE* **51**:132 (1963).
5. White, R. M., and C. E. Enderby, Electro-optical modulators employing intermittent interaction. *Proc. IEEE* **51**:214 (1963).
6. Yariv, A., and P. Yeh, *Optical Waves in Crystals.* Wiley-Interscience, New York, 1983.
7. Yeh, P., and C. Gu, *Optics of Liquid Crystal Displays.* Wiley, New York, 1999.
8. Brillouin, L., Diffusion de la lumiere et des rayons X par un corps transparent homogene. *Ann. Phys.* **17**:88 (1922).
9. Debye, P., and F. W. Sears, On the scattering of light by supersonic waves. *Proc. Natl. Acad. Sci. U.S.A.* **18**:409 (1932).
10. Dransfeld, K., Kilomegacycle ultrasonics. *Sci. Am.* **208**:60 (1963).
11. See, for example, R. Adler, Interaction between light and sound. *IEEE Spectrum* **4**(May):42 (1967).
12. Kittel, C., *Introduction to Solid State Physics*, 3rd ed. Wiley, New York, 1967, p.38.
13. Born, M., and E. Wolf, *Principles of Optics.* Pergamon, New York, 1965, Chapter 12.

14. Dixon, R. W., Photoelastic properties of selected materials and their relevance for applications to acoustic light modulators and scanners. *J. Appl. Phys.* **38**:5149 (1967).
15. Pinnow, D. A., L. G. Van Uitert, A. W. Warner, and W. A. Bonner, $PbMO_4$: A melt grown crystal with a high figure of merit for acoustooptic device applications. *Appl. Phys. Lett.* **15**:83 (1969).
16. Kuhn, L. M., L. Dakss, F. P. Heidrich, and B. A. Scott, Deflection of optical guided waves by a surface acoustic wave. *Appl. Phys. Lett.* **17**:265 (1970).
17. Tsai, C. S., Le T. Nguyen, S. K. Yao, and M. H. Alhaider, High performance acousto-optic guided light beam device using two tilting surface acoustic waves. *Appl. Phys. Lett.* **26**:140 (1975).

CHAPTER 10

NOISE IN OPTICAL DETECTION AND GENERATION

10.0 INTRODUCTION

In this chapter we study the effect of noise on a number of important physical processes. We will take the term noise to represent random electromagnetic fields occupying the same spectral region as that occupied by some "signal." The effect of noise will be considered in the following cases.

1. *Measurement of optical power*. In this case the noise causes fluctuations in the measurement, thus placing a lower limit on the smallest amount of power that can be measured.
2. *Linewidth of laser oscillators*. The presence of incoherent spontaneous emission power will be found to be the cause for a finite amount of spectral line broadening in the output of single-mode laser oscillators. This broadening manifests itself as a limited coherence time.
3. *Optical communications system*. We will consider the case of an optical communications system using a binary pulse code modulation in which the information is carried by means of a string of 1 and 0 pulses. The presence of noise will be shown to lead to a certain probability that any given pulse in the reconstructed train pulse is in error.

In this chapter we consider optical detectors utilizing light-generated charge carriers. These include the photomultiplier, the photoconductive detector, the *p-n* junction photodiode, and the avalanche photodiode (APD). These detectors are the main ones used in the field of quantum electronics and optical communications, because they combine high sensitivity with very short response times. Other types of detectors, such as bolometers, Golay cells, and thermocouples, whose operation depends on temperature changes induced by the absorbed radiation, will not be discussed. The interested reader will find a good description of these devices in Reference [1].

Two types of noise will be discussed in detail. The first type is thermal (Johnson) noise, which represents noise power generated by thermally agitated charge carriers. The expression for this noise will be derived by using the conventional thermodynamic treatment as well as by a statistical analysis of a particular model in which the physical origin of the noise is more apparent. The second type, shot noise (or generation-recombination noise in photoconductive detectors), is attributable to the random way in which electrons are emitted or generated in the process of interacting with a radiation field. This noise exists even at zero temperature, where thermal agitation or generation of carriers can be neglected. In this case it results from the randomness with which carriers are generated by the *very signal that is measured.* Detection in the limit of signal-generated shot noise is called quantum-limited detection, since the corresponding sensitivity is that allowed by the uncertainty principle in quantum mechanics. This point will be brought out in the next chapter. Further discussions on noise due to uncertainty principles, vacuum fluctuation, and the squeezing of field fluctuation [2–4] are given in Chapter 18.

10.1 LIMITATIONS DUE TO NOISE POWER

Measurement of Optical Power

Consider the problem of measuring an optical signal field

$$v_S(t) = V_S \cos \omega t \tag{10.1-1}$$

in the presence of a noise field, where ω is the frequency and V_S is a real constant. The instantaneous noise field that adds to that of the signal can be taken as the sum of an in-phase component and a quadrature component according to

$$v_N(t) = V_{NC}(t)\cos \omega t + V_{NS}(t)\sin \omega t \tag{10.1-2}$$

where $V_{NC}(t)$ and $V_{NS}(t)$ are slowly (compared to $\exp(i\omega t)$) varying random uncorrelated quantities with a zero mean. Both $V_{NC}(t)$ and $V_{NS}(t)$ are real. The total field at the detector $v(t) = v_S(t) + v_N(t)$ can be written

$$v(t) = \text{Re}\{[V_S + V_{NC}(t) - iV_{NS}(t)]e^{i\omega t}\} \tag{10.1-3}$$

$$\equiv \text{Re}[V(t)e^{i\omega t}] \tag{10.1-4}$$

where $V(t)$ is the complex analytic representation of the total field. The total (signal plus noise) field phasor $V(t)$ is shown in Figure 10.1. In most situations of interest to optical detection,

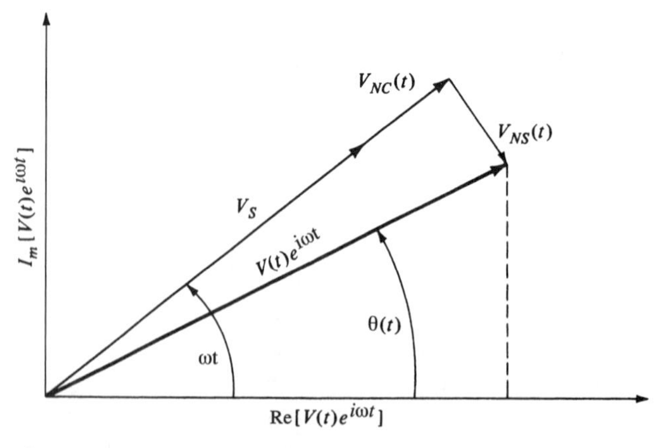

Figure 10.1 A phasor diagram showing the total (signal plus noise) field phasor $V(t)$ at time t. The instantaneous field is given by the horizontal projection of $V(t)\exp(i\omega t)$.

the sources of noise are due to the concerted action of a large number of independent agents. In this case the central limit theorem of statistics [5] tells us that the probability distribution function for finding $V_{NC}(t)$ at time t between V_{NC} and $V_{NC} + dV_{NC}$ is described by a Gaussian

$$p(V_{NC})\, dV_{NC} = \frac{1}{\sqrt{2\pi}\,\sigma} e^{-V_{NC}^2/2\sigma^2} dV_{NC} \tag{10.1-5}$$

where σ is the standard deviation of the distribution. The quadrature component V_{NS} satisfies a similar distribution in which V_{NS} replaces V_{NC} for $p(V_{NS})$. Since $V_{NC}(t)$ has a unity probability of having some value between $-\infty$ and ∞, it follows that

$$\int_{-\infty}^{\infty} p(V_{NC})\, dV_{NC} = 1 \tag{10.1-6}$$

It follows from Equation (10.1-5) that $\overline{V_{NC}}$, the ensemble average (denoted by a horizontal bar) of V_{NC}, is zero, whereas the mean square value is

$$\overline{V_{NC}^2} = \overline{V_{NS}^2} = \int_{-\infty}^{\infty} V_{NC}^2\, p(V_{NC})\, dV_{NC} = \sigma^2 \tag{10.1-7}$$

The ensemble average $\overline{A(t)}$ of a quantity $A(t)$ is obtained by measuring A simultaneously at time t in a very large number of systems that, *to the best of our knowledge*, are identical. Mathematically,

$$\bar{A} = \int_{-\infty}^{\infty} Ap(A)\, dA$$

where $p(A)$ is the probability distribution function, in the sense of Equation (10.1-5), of a random variable A. For a time-invariant phenomenon, the ensemble average can be replaced by temporal average

$$\bar{A} = \frac{1}{T} \int_{-T/2}^{T/2} A(t)\, dt$$

where T, which must approach infinity, is the temporal period for averaging (the integration time).

The reason for $\overline{V_{NC}}\,(t) = 0$ can be appreciated from Figure 10.1. $V_{NC}(t)$ has an equal probability of being in phase with V_S as being out phase, thus averaging out to zero.

The "power" corresponding to the field $v(t)$ is obtained using Equation (10.1-2) as

$$\begin{aligned} P(t) &\equiv \tfrac{1}{2}(V(t)e^{i\omega t})(V^*(t)e^{-i\omega t}) \\ &= \tfrac{1}{2}(V_S^2 + 2V_S V_{NC} + V_{NC}^2 + V_{NS}^2) \end{aligned} \tag{10.1-8a}$$

The ensemble average (or *long*-time average) of $P(t)$ is

$$\bar{P} \equiv \overline{P(t)} = \tfrac{1}{2}(\overline{V_S^2} + \overline{V_{NC}^2} + \overline{V_{NS}^2}) = \tfrac{1}{2}(\overline{V_S^2} + 2\sigma^2) \tag{10.1-8b}$$

where use has been made of the fact that $\overline{V_{NC}} = 0$ and of Equation (10.1-7). The physical significance of the time-varying power $P(t)$ and its long-time (or ensemble) average P is illustrated by Figure 10.2.

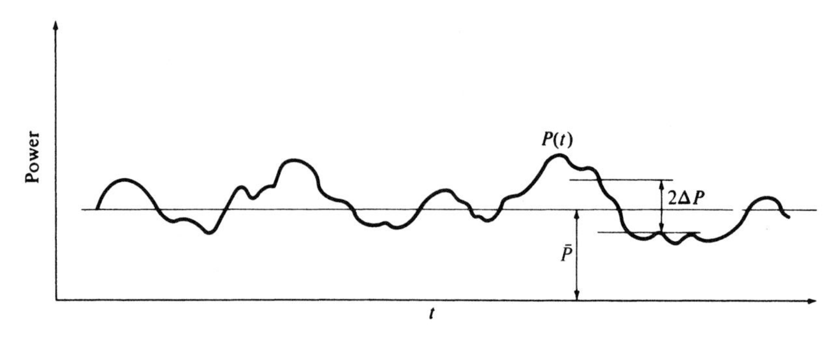

Figure 10.2 The intermingling of noise power with that of a signal causes the total power to fluctuate. The rms fluctuation ΔP limits the accuracy of power measurements.

It is clear from the fluctuating nature of $P(t)$ that any measurement of this power is subject to an uncertainty due to the random nature of V_{NC} and V_{NS} in Equation (10.1-8a). As a measure of the uncertainty in power measurement, we may reasonably take the root mean square (rms) power deviation

$$\Delta P \equiv [\overline{(P(t) - \bar{P})^2}]^{1/2} \tag{10.1-9}$$

Substituting Equations (10.1-8a) and (10.1-8b) into (10.1-9) and using

$$\overline{V_{NC}^4} = \int_{-\infty}^{\infty} V_{NC}^4 p(V_{NC})\, dV_{NC} = 3\sigma^4 \tag{10.1-10}$$

and $\overline{V_{NC}^2} = \overline{V_{NS}^2} = \sigma^2$, we obtain after some algebra

$$\Delta P = \sigma(\overline{V_S^2} + \sigma^2)^{1/2} = \sigma(2P_S + \sigma^2)^{1/2} \tag{10.1-11}$$

where according to Equation (10.1-8a) we may associate $P_S = \frac{1}{2}\overline{V_S^2}$ with the signal power, that is, the power that would be measured if V_{NC} and V_{NS} were, hypothetically, rendered zero.

A question of practical importance involves the minimum signal power that can be measured in the presence of noise. We may, somewhat arbitrarily, take this power P_{limit} to be that at which the uncertainty ΔP becomes equal to the signal power P_S.

At this point we have from Equation (10.1-11)

$$P_{\text{limit}} = \sigma(2P_{\text{limit}} + \sigma^2)^{1/2}$$

or, after solving for P_{limit},

$$P_{\text{limit}} = \sigma^2(1 + \sqrt{2}) = P_N(1 + \sqrt{2}) \tag{10.1-12}$$

where $P_N = \sigma^2 = \frac{1}{2}(\overline{V_{NC}^2} + \overline{V_{NS}^2})$ is the noise power. Widespread convention chooses to define the minimum detectable signal power as equal to P_N instead of $2.414P_N$, as obtained above. This simplification is understandable, since our choice of the limit of detectability $\Delta P = P_S$ was somewhat arbitrary. In any case, the main conclusion to remember is that near the limit of detectivity, the rms power fluctuation is comparable to the signal power. The next task, which will be taken up in this chapter and in Chapter 11, is to find out the main sources of noise power and consequently ways to minimize them. Before tackling this task, however, we need to develop some mathematical tools for dealing with random processes.

10.2 NOISE—BASIC DEFINITIONS AND THEOREMS

A real function $v(t)$ and its Fourier transform $V(\omega)$ are related by

$$V(\omega) = \frac{1}{2\pi} \int_{-\infty}^{\infty} v(t) e^{-i\omega t} dt \tag{10.2-1}$$

and

$$v(t) = \int_{-\infty}^{\infty} V(\omega) e^{i\omega t} d\omega \tag{10.2-2}$$

In the process of measuring a signal $v(t)$, we are not in a position to use the infinite time interval needed, according to Equation (10.2-1), to evaluate $V(\omega)$. If the time duration of the measurement is T, we may consider the function $v(t)$ to be zero when $t \leq -T/2$ and $t \geq T/2$ and, instead of Equation (10.2-1), get

$$V_T(\omega) = \frac{1}{2\pi} \int_{-T/2}^{T/2} v(t) e^{-i\omega t} dt \tag{10.2-3}$$

Since $v(t)$ is real, it follows that

$$V_T(\omega) = V_T^*(-\omega) \tag{10.2-4}$$

T is usually called the resolution or integration time of the system. Let us evaluate the average power P associated with $v(t)$. Taking the instantaneous power as $v^2(t)$, we obtain

$$P = \frac{1}{T} \int_{-T/2}^{T/2} v^2(t)\, dt = \frac{1}{T} \int_{-T/2}^{T/2} \left\{ v(t) \left[\int_{-\infty}^{\infty} V_T(\omega) e^{i\omega t} d\omega \right] \right\} dt \tag{10.2-5}$$

It may be convenient for our purpose to think of $v(t)$ as the voltage across a 1 ohm resistance.

Using Equation (10.2-3) and (10.2-4) in the last equation and interchanging the order of integration, we obtain

$$P = \frac{2\pi}{T} \int_{-\infty}^{\infty} |V_T(\omega)|^2 d\omega \tag{10.2-6}$$

or

$$P = \frac{4\pi}{T} \int_{0}^{\infty} |V_T(\omega)|^2 d\omega \tag{10.2-7}$$

where we used

$$\lim_{T\to\infty} (2\pi)^{-1} \int_{-T/2}^{T/2} dt \exp[i(\omega + \omega')t] = \delta(\omega + \omega')$$

If we define the *spectral density function* $S_v(\omega)$ of $v(t)$ by

$$S_v(\omega) = \lim_{T\to\infty} \frac{4\pi |V_T(\omega)|^2}{T} \tag{10.2-8}$$

then, according to Equation (10.2-7), $S_v(\omega)\, d\omega$ is the portion of the average power of $v(t)$ that is due to frequency components between ω and $\omega + d\omega$. According to this physical interpretation, we may measure $S_v(\omega)$ by separating the spectrum of $v(t)$ into its various frequency classes as shown in Figure 10.3 and then measuring the power output $S_v(\omega_i)\, \Delta\omega_i$ of each of the filters [6].

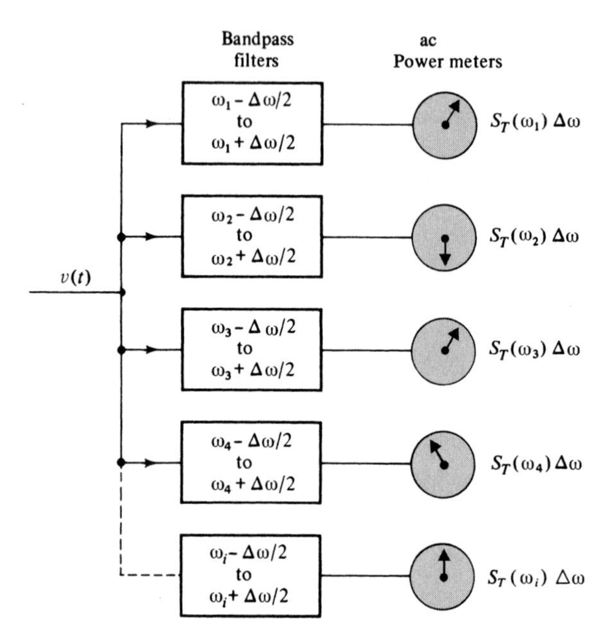

Figure 10.3 Diagram illustrating how the spectral density function $S_T(\omega)$ of a signal $v(t)$ can be obtained by measuring the power due to different frequency intervals.

Wiener–Khintchine Theorem

We will next derive another formal result involving the spectral density function. Consider the time average of the product of some field quantity $v(t)$ with its delayed version $v(t+T)$:

$$C_v(\tau) = \overline{v(t)v(t+\tau)} \tag{10.2-9}$$

The function $C_v(\tau)$ is termed the autocorrelation function of $v(t)$. We use Equation (10.2-2) to carry out the integration indicated in Equation (10.2-9):

$$\begin{aligned} C_v(\tau) &= \frac{1}{T}\int_{-T/2}^{T/2} v(t)v(t+\tau)\,dt \\ &= \frac{1}{T}\int_{-\infty}^{\infty}\int_{-\infty}^{\infty}\int_{-T/2}^{T/2} d\omega\, d\omega'\, dt\, V_T(\omega)V_T(\omega')e^{i(\omega+\omega')t}e^{i\omega\tau} \end{aligned} \tag{10.2-10}$$

In the limit $T \to \infty$,

$$\lim_{T\to\infty}\int_{-T/2}^{T/2} dt e^{i(\omega+\omega')t} = 2\pi\delta(\omega+\omega') \tag{10.2-11}$$

so that

$$\begin{aligned} C_v(\tau) &= \lim_{T\to\infty}\frac{2\pi}{T}\int_{-\infty}^{\infty}\int_{-\infty}^{\infty} V_T(\omega')V_T(\omega)\delta(\omega+\omega')e^{i\omega\tau}d\omega\, d\omega' \\ &= \lim_{T\to\infty}\frac{1}{2}\int_{-\infty}^{\infty}\frac{4\pi|V_T(\omega)|^2}{T}e^{i\omega\tau}d\omega \end{aligned} \tag{10.2-12}$$

The quantity $4\pi|V_T(\omega)|^2/T$ is, according to Equation (10.2-8), the spectral density function of $S_v(\omega)$ of $v(t)$, so that

$$C_v(\tau) = \frac{1}{2}\int_{-\infty}^{\infty} S_v(\omega)e^{i\omega\tau}d\omega \tag{10.2-13}$$

so that using Equation (10.2-1),

$$S_v(\tau) = \frac{1}{\pi}\int_{-\infty}^{\infty} C_v(\tau)e^{-i\omega\tau}d\tau \tag{10.2-14}$$

The last two equations state that the spectral density function $S_v(\omega)$ and the autocorrelation function $C_v(\tau)$ form a Fourier transform pair. This result is one of the more important theoretical and practical tools of information theory and of the mathematics of random processes, and it is known, after the American and Russian mathematicians who, independently, formulated it, as the Wiener–Khintchine theorem. Its main importance for our purposes lies in the fact that it is often easier to obtain, experimentally or theoretically, $C_v(\tau)$ rather than $S_v(\omega)$, so that $S_v(\omega)$ is derived by a Fourier transformation of $C_v(\tau)$.

10.3 SPECTRAL DENSITY FUNCTION OF A TRAIN OF RANDOMLY OCCURRING EVENTS

Consider a time-dependent random variable $i(t)$ made up of a very large number of individual events $f(t - t_i)$ that occur at random times t_i. For example, the events can be the arrival of a photon or the generation of an electron due to the arrival of a photon. An observation of $i(t)$ during a period T will yield

$$i_T(t) = \sum_{i=1}^{N_T} f(t - t_i), \quad 0 \le t \le T \tag{10.3-1}$$

where N_T is the total number of events occurring in T. Here we assume that the a priori probability that a given event will occur in any time interval is distributed uniformly over the interval T. In this case, the probability $p(n)$ for n events to occur in an observation period T is given by the Poisson distribution function [6]

$$p(n) = \frac{(\bar{n})^n e^{-\bar{n}}}{n!}$$

where $\bar{n}$ is the average number of events occurring in an observation time period T. Typical examples of a random function $i(t)$ are provided by the thermionic emission current from a hot cathode (under temperature-limited conditions), or the electron current caused by photoemission from a surface. In these cases, $f(t - t_i)$ represents the current resulting from a single electron emission occurring at t_i.

The Fourier transform of $i_T(t)$ is given according to Equation (10.2-3) by

$$I_T(\omega) = \sum_{i=1}^{N_T} F_i(\omega) \tag{10.3-2}$$

where $F_i(\omega)$ is the Fourier transform of $f(t - t_i)$:

$$F_i(\omega) = \frac{1}{2\pi}\int_{-\infty}^{\infty} f(t - t_i)e^{-i\omega t}dt = \frac{e^{-i\omega t_i}}{2\pi}\int_{-\infty}^{\infty} f(t)e^{-i\omega t}dt = e^{-i\omega t_i}F(\omega) \tag{10.3-3}$$

Here we assume that the individual event $f(t - t_i)$ is over in a short time compared to the observation period T, so the integration limits can be taken as $-\infty$ to ∞ instead of 0 to T.

From Equations (10.3-2) and (10.3-3), we obtain

$$|I_T(\omega)|^2 = |F(\omega)|^2 \sum_{i=1}^{N_T} \sum_{j=1}^{N_T} e^{-i\omega(t_i - t_j)} = |F(\omega)|^2 \left(N_T + \sum_{i \neq j}^{N_T} \sum_{j}^{N_T} e^{i\omega(t_j - t_i)} \right) \tag{10.3-4}$$

If we take the average of Equation (10.3-4) over an ensemble of a very large number of physically identical systems, the second term on the right side of (10.3-4) can be neglected in comparison to N_T, since the times t_i are random. This results in

$$\overline{|I_T(\omega)|^2} = \bar{N}_T \,|\, F(\omega) \,|^2 \equiv \bar{N} T \,|\, F(\omega) \,|^2 \tag{10.3-5}$$

where the horizontal bar denotes ensemble averaging and where $\bar{N}$ is the average rate at which the events occur so that $\bar{N}_T = \bar{N}T$. The spectral density function $S_i(\omega)$ of the function $i_T(t)$ is given according to Equations (10.2-8) and (10.3-5) as

$$S_i(\omega) = 4\pi\bar{N}\,|F(\omega)|^2 \tag{10.3-6}$$

In practice, one uses more often the spectral density function $S(\nu)$ defined so that the average power due to frequencies between ν and $\nu + d\nu$ is equal to $S(\nu)\, d\nu$. It follows, then, that $S(\nu)\, d\nu = S(\omega)\, d\omega$; thus, since $\omega = 2\pi\nu$,

$$S_i(\nu) = 8\pi^2\bar{N}\,|F(2\pi\nu)|^2 \tag{10.3-7}$$

The last result is known as Carson's theorem and its usefulness will be demonstrated in the following sections, where we employ it in deriving the spectral density function associated with a number of different physical processes related to optical detection.

Equation (10.3-7) was derived for the case in which the individual events $f(t - t_i)$ were displaced in time but were otherwise identical. There are physical situations in which the individual events may depend on one or more additional parameters. Denoting the parameter (or group of parameters) as α, we can clearly single out the subclass of events $f_\alpha(t - t_i)$ whose α is nearly the same and use Equation (10.3-7) to obtain directly

$$S_\alpha(\nu) = 8\pi^2\bar{N}(\alpha)\,|F_\alpha(2\pi\nu)|^2\Delta\alpha \tag{10.3-8}$$

for the contribution of this subclass of events to $S(\nu)$. $F_\alpha(\omega)$ is the Fourier transform of $f_\alpha(t)$, and thus $\bar{N}(\alpha)\,\Delta\alpha$ is the average number of events per second whose parameter falls between α and $\alpha + \Delta\alpha$:

$$\int_{-\infty}^{\infty} \bar{N}(\alpha)\, d\alpha = \bar{N}$$

The probability distribution function for α is $p(\alpha) = \bar{N}(\alpha)/\bar{N}$; therefore

$$\int_{-\infty}^{\infty} p(\alpha)\, d\alpha = \frac{1}{\bar{N}} \int_{-\infty}^{\infty} \bar{N}(\alpha)\, d\alpha = 1 \tag{10.3-9}$$

Summing Equation (10.3-8) over all classes α and weighting each class by the probability $p(\alpha)\,\Delta\alpha$ of its occurrence, we obtain

$$\begin{aligned} S_i(\nu) &= \sum_\alpha S_\alpha(\nu) = 8\pi^2 \sum_\alpha \bar{N}(\alpha)\,|F_\alpha(2\pi\nu)|^2\Delta\alpha \\ &= 8\pi^2\,\bar{N} \sum_\alpha |F_\alpha(2\pi\nu)|^2\, p(\alpha)\,\Delta\alpha \\ &= 8\pi^2\,\bar{N} \int_{-\infty}^{\infty} |F_\alpha(2\pi\nu)|^2\, p(\alpha)\, d\alpha = 8\pi^2\bar{N}\,\overline{|F_\alpha(2\pi\nu)|^2} \end{aligned} \tag{10.3-10}$$

where the bar denotes averaging over α. Equation (10.3-10) is thus the extension of (10.3-7) to the case of events whose characterization involves, in addition to their time t_i, some added parameters. We will use it later in this chapter to derive the noise spectrum of photoconductive detectors, in which case α is the lifetime of the excited photocarriers.

10.4 SHOT NOISE

Let us consider the spectral density function of current arising from random generation and flow of mobile charge carriers. This current is identified with "shot noise" [7]. To be specific, we consider the case illustrated in Figure 10.4, in which electrons are released at random into the vacuum from electrode A to be collected at electrode B, which is maintained at a slight positive potential relative to A. The average *rate* $\bar{N}$ of electron emission from A is $\bar{N} = \bar{I}/e$, where $\bar{I}$ is the average current and the electronic charge is taken as $-e$. The current pulse due to a single electron as observed in the external circuit is

$$i_e(t) = \frac{ev(t)}{d} \tag{10.4-1}$$

where $v(t)$ is the instantaneous velocity and d is the separation between A and B. To prove (10.4-1), consider the case in which the moving electron is replaced by a thin sheet of a very large area and of total charge $-e$ moving between the plates, as illustrated in Figure 10.5. It is a simple matter to show (see Problem 10.1), using the relation $\nabla \cdot \mathbf{E} = \rho/\varepsilon$, that the charge induced by the moving sheet on the left electrode is

$$Q_1 = \frac{e(d - x)}{d} \tag{10.4-2}$$

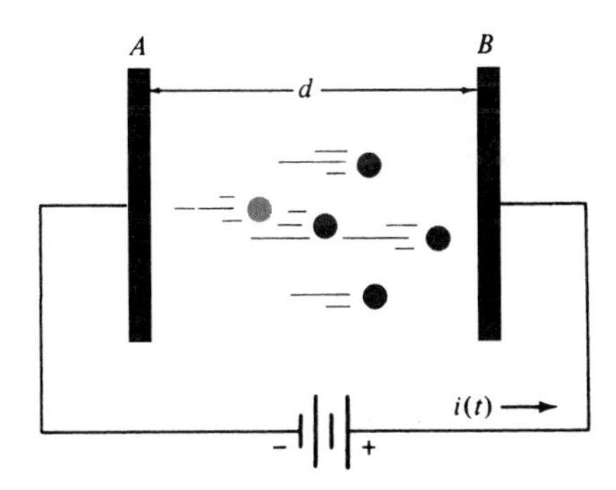

Figure 10.4 Random electron flow between two electrodes. This basic configuration is used in the derivation of shot noise.

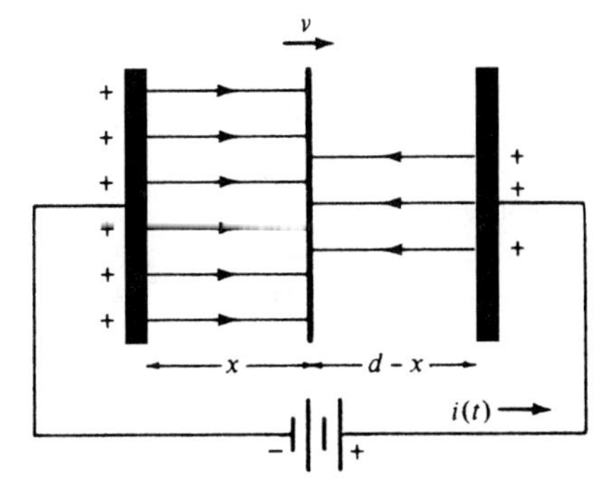

Figure 10.5 Induced charges and field lines due to a thin charge layer between the electrodes.

and that on the right electrode is

$$Q_2 = \frac{ex}{d} \tag{10.4-3}$$

where x is the position of the charged sheet measured from the left electrode. The current in the external circuit due to a single electron is thus

$$i_e(t) = \frac{dQ_2}{dt} = \frac{e}{d}\frac{dx}{dt} = \frac{e}{d}v(t) \tag{10.4-4}$$

as claimed.

The Fourier transform of a single current pulse is

$$F(\omega) = \frac{e}{2\pi d}\int_0^{t_a} v(t)e^{-i\omega t}dt \tag{10.4-5}$$

where t_a is the arrival time of an electron emitted at $t = 0$. If the transit time of an electron is sufficiently small that, at the frequency of interest ω,

$$\omega t_a \ll 1 \tag{10.4-6}$$

that is, $i_e(t) \sim \delta(t)$, we can replace $\exp(-i\omega t)$ in Equation (10.4-5) by unity and obtain

$$F(\omega) = \frac{e}{2\pi d}\int_0^{t_a} \frac{dx}{dt}dt = \frac{e}{2\pi} \tag{10.4-7}$$

since $x(t_a)$ is, by definition, equal to d. Using Equation (10.4-7) in (10.3-7) and recalling that $\bar{I} = e\bar{N}$, we have

$$S(\nu) = 8\pi^2\bar{N}\left(\frac{e}{2\pi}\right)^2 = 2e\bar{I} \tag{10.4-8}$$

The power (in the sense of Equation (10.2-5) in the frequency interval ν to $\nu + \Delta\nu$ associated with the current is, according to the discussion following Equation (10.2-8), given by $S(\nu)\,\Delta\nu$. It is convenient to represent this power by an *equivalent noise generator* at ν with a mean-square current amplitude

$$\overline{i_N^2}(\nu) \equiv S(\nu)\,\Delta\nu = 2e\bar{I}\,\Delta\nu \tag{10.4-9}$$

The noise mechanism described above is referred to as *shot noise.*

It is interesting to note that e in Equation (10.4-9) is the charge of the particle responsible for the current flow. If, hypothetically, these carriers had a charge of $2e$, then at the *same average current* $\bar{I}$ the shot-noise power would double. Conversely, shot noise would disappear if the magnitude of an individual charge tended to zero. This is a reflection of the fact that shot noise is caused by fluctuations in the current that are due to the discreteness of the charge carriers and to the random electronic emission (for which the number of electrons emitted per unit time obey Poisson statistics [5–7]).

The ratio of the fluctuations to the average current decreases with increasing number of events. More precisely, for events obeying Poisson statistics we have (Reference [5] or derivable directly from the Poisson distribution in Section 10.3) $(\Delta n)^2 \equiv (n - \bar{n})^2 = \bar{n}$. Thus the ratio of the fluctuation to the average is

$$\frac{\Delta n}{\bar{n}} = \frac{1}{\sqrt{\bar{n}}}$$

where n is the number of events in an observation time and $\bar{n}$ is the average value of n. Another point to remember is that, in spite of the appearance of $\bar{I}$ on the right side of Equation (10.4-9), $\overline{i_N^2}(\nu)$ represents an alternating current with frequencies near ν.

10.5 JOHNSON NOISE

Johnson noise, or *Nyquist noise*, describes the fluctuations in the voltage across a dissipative circuit element; see References [8, 9]. These fluctuations are most often caused by the thermal motion *of* the charge carriers. Here we use the word "carriers" rather than "electrons" to include cases of ionic conduction or conduction by holes in semiconductors. The charge neutrality of an electrical resistance is satisfied when we consider the whole volume, but locally the random thermal motion of the carriers sets up fluctuating charge gradients and, correspondingly, a fluctuating (ac) voltage. If we now connect a second resistance across the first one, the thermally induced voltage described above will give rise to a current and hence to a power transfer to the second resistor. The same argument applies to the second resistor, so at thermal equilibrium the net power leaving each resistor is zero. This is *Johnson noise*, whose derivation follows.

Consider the case illustrated in Figure 10.6 of a transmission line connected between two similar resistances R, which are maintained at the same temperature T.

We choose the resistance R to be equal to the characteristic impedance Z_0 of the line, so that no reflection can take place at the ends. The transmission line can support traveling voltage waves of the form

$$v(t) = A\cos(\omega t \pm kz) \tag{10.5-1}$$

where A is a constant, ω is the frequency, $k = 2\pi/\lambda$ is the wavenumber, and the phase velocity is $c = \omega/k$. For simplicity, we require that the allowed solutions be periodic in the distance L, so if we extend the solution outside the limits $0 \le z \le L$, we obtain

$$v(t) = A\cos[\omega t \pm k(z + L)] = A\cos(\omega t \pm kz)$$

This is the so-called periodic boundary condition that is used extensively in similar situations in thermodynamics to derive the blackbody radiation density, or in solid-state physics to derive the density of electronic states in crystals. The periodic boundary condition is fulfilled when

$$kL = 2m\pi, \quad m = 1, 2, 3, \ldots \tag{10.5-2}$$

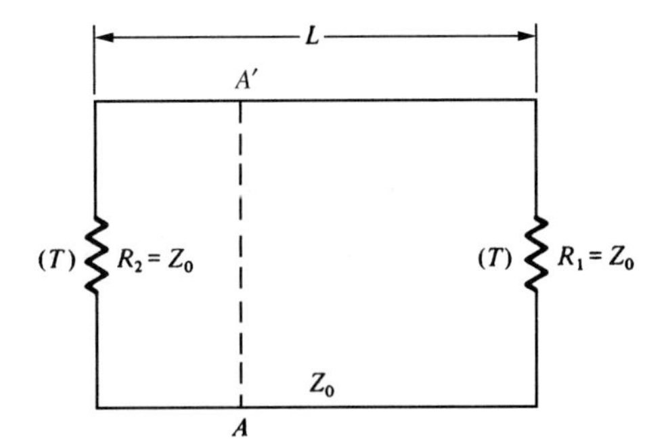

Figure 10.6 Lossless transmission line of characteristic impedance Z_0 connected between two matched loads ($R = Z_0$) at temperature T.

Therefore two adjacent modes differ in their value of k by

$$\Delta k = \frac{2\pi}{L} \tag{10.5-3}$$

and the number of modes having their k values somewhere between zero and $+k$ is

$$N_k = \frac{k}{\Delta k} = \frac{kL}{2\pi} \tag{10.5-4}$$

or, using $k = 2\pi\nu/c$, we obtain

$$N(\nu) = \frac{\nu L}{c}$$

for the number of positively traveling modes with frequencies between zero and ν. Negative k values correspond, according to Equation (10.5-1), to waves traveling in the $-z$ direction. Our bookkeeping is thus limited to modes carrying power in the $+z$ direction.

The number of modes per unit frequency interval is

$$p(\nu) = \frac{dN(\nu)}{d\nu} = \frac{L}{c} \tag{10.5-5}$$

Consider the power flowing in the $+z$ direction across some arbitrary plane, A–A' say. It is clear that due to the lack of reflection this power must originate in R_2. Since the power is carried by the electromagnetic modes of the system, we have

$$\text{Power} = \frac{\text{Energy}}{\text{Distance}}(\text{Velocity of energy})$$

We find, taking the velocity of light as c, that the power P due to frequencies between ν and $\nu + \Delta\nu$ is given by

$$\begin{aligned} P &= \left(\frac{1}{L}\right)\begin{pmatrix}\text{Number of modes between}\\ \nu \text{ and } \nu + \Delta\nu\end{pmatrix}(\text{Energy per mode})(c) \\ &= \left(\frac{1}{L}\right)\left(\frac{L}{c}\Delta\nu\right)\left(\frac{h\nu}{e^{h\nu/kT} - 1}\right)(c) \end{aligned}$$

or

$$P = \frac{h\nu\,\Delta\nu}{e^{h\nu/kT} - 1} \approx kT\,\Delta\nu \quad (kT \gg h\nu) \tag{10.5-6}$$

where we used the fact that in thermal equilibrium the energy of a mode is given by [10]

$$E = \frac{h\nu}{e^{h\nu/kT} - 1} \tag{10.5-7}$$

This result is also obtained in Section 17.3 (see Equation (17.3-21)) from a different point of view. An equal amount of noise power is, of course, generated in the right resistor and is dissipated in the left one, so in thermal equilibrium the net power crossing any plane is zero.

The power given by Equation (10.5-6) represents the maximum noise power available from the resistance, since it is delivered to a matched load. If the load connected across R has a resistance different from R, the noise power delivered is less than that given by Equation (10.5-6).

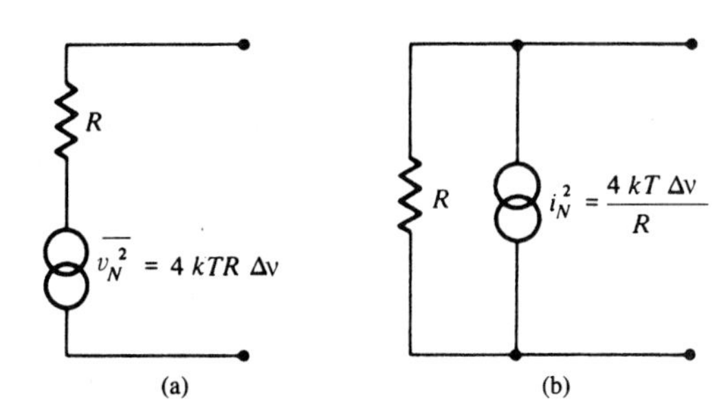

Figure 10.7 (a) Voltage and (b) current noise equivalent circuits of a resistance.

The noise-power bookkeeping is done correctly if the resistance R appearing in a circuit is replaced by either one of the following two equivalent circuits: a noise generator in series with R with mean square voltage amplitude

$$\overline{v_N^2}(\nu) = \frac{4h\nu R\,\Delta\nu}{e^{h\nu/kT}-1} \underset{kT \gg h\nu}{\approx} 4kTR\,\Delta\nu \tag{10.5-8}$$

or a noise current generator of mean square value

$$\overline{i_N^2}(\nu) = \frac{4h\nu\,\Delta\nu}{R(e^{h\nu/kT}-1)} \underset{kT \gg h\nu}{\approx} \frac{4kT\,\Delta\nu}{R} \tag{10.5-9}$$

in parallel with R. The noise representations of the resistor are shown in Figure 10.7. There are numerous other derivations of the formula for Johnson noise. For derivations using lumped-circuit concepts and an antenna example, the reader is referred to References [1, 10], respectively.

Statistical Derivation of Johnson Noise

The derivation of Johnson noise leading to Equation (10.5-6) leans heavily on thermodynamic and statistical mechanics considerations. It may be instructive to obtain this result using a physical model for a resistance and applying the mathematical tools developed in this chapter. The model used is shown in Figure 10.8. The resistor consists of a medium of volume $V = Ad$, which contains N_e free electrons per unit volume. In addition, there are N_e positively charged ions, which preserve the (average) charge neutrality. The electrons move about randomly with an average kinetic energy per electron of

$$\bar{E} = \tfrac{3}{2}kT = \tfrac{1}{2}m(\overline{v_x^2} + \overline{v_y^2} + \overline{v_z^2}) \tag{10.5-10}$$

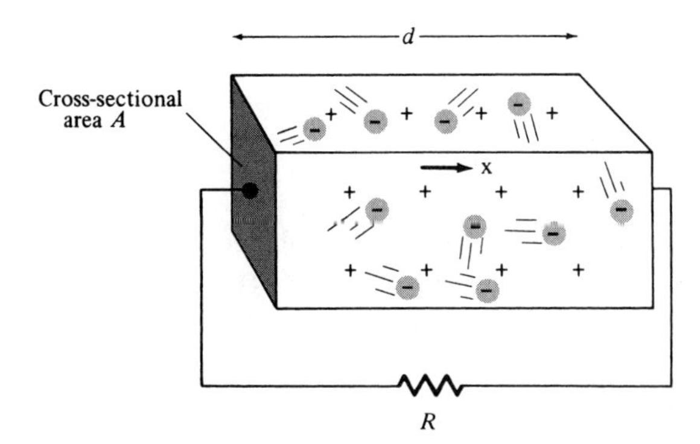

Figure 10.8 Model of a resistance used in deriving the Johnson-noise formula.

where m is the mass of the electron, and $(\overline{v_x^2} + \overline{v_y^2} + \overline{v_z^2})$ refer to thermal averages. A variety of scattering mechanisms including electron–electron, electron–ion, and electron–phonon collisions act to interrupt the electron motion at an average rate of τ_0^{-1} times per second. τ_0 is thus the mean scattering time. These scattering mechanisms are responsible for the electrical resistance and give rise to a dc conductivity

$$\sigma = \frac{N_e e^2 \tau_0}{m} \tag{10.5-11}$$

where m is the mass of the electron. In a semiconductor we use the effective mass of the charge carrier (see Chapter 15). The derivation of Equation (10.5-11) can be found in any introductory book on solid-state physics. The sample dc resistance is thus

$$R = \frac{d}{\sigma A} = \frac{md}{N_e e^2 \tau_0 A} \tag{10.5-12}$$

while its ac resistance $R(\omega)$ is $md(1 + \omega^2\tau_0^2)/Ne^2\tau_0 A$.

We apply next the results of Section 10.3 to the problem and choose as our basic single event the current pulse $i_e(t)$ in the external circuit due to the motion of *one* electron between two successive scattering events. Using Equation (10.4-1), we write

$$i_e(t) = \begin{cases} \dfrac{ev_x}{d}, & 0 \le t \le \tau \\ 0, & \text{otherwise} \end{cases} \tag{10.5-13}$$

where v_x is the x component of the velocity (assumed constant) and where τ is the scattering time of the electron under observation. Taking the Fourier transform of $i_e(t)$, we have

$$I_e(\omega, \tau, v_x) = \frac{1}{2\pi}\int_0^\tau i_e(t)e^{-i\omega t}dt = \frac{(1/2\pi)ev_x}{-i\omega d}(e^{-i\omega\tau} - 1) \tag{10.5-14}$$

from which

$$|I_e(\omega, \tau, v_x)|^2 = \frac{e^2 v_x^2}{4\pi^2\omega^2 d^2}(2 - e^{i\omega\tau} - e^{-i\omega\tau}) \tag{10.5-15}$$

According to Equation (10.3-10), we need to average $|I_e(\omega, \tau, v_x)|^2$ over the parameters τ and v_x. We assume that τ and v_x are independent variables so that the probability function can be written

$$p(\tau, v_x) = g(\tau)f(v_x)$$

where $g(\tau)$ is the probability distribution of τ. If we assume the collision probability per carrier per unit time is $1/\tau_0$, then $g(\tau)$ can be written

$$g(\tau) = \frac{1}{\tau_0}e^{-\tau/\tau_0} \tag{10.5-16}$$

Physically, $g(\tau)\, d\tau$ is the probability that a collision will occur between $(\tau, \tau + d\tau)$, and τ_0 is the average time of the collision. In other words, $\int \tau g(\tau)\, d\tau = \tau_0$.

We now perform the averaging over τ. This leads to

$$\overline{|I_e(\omega, v_x)|^2} = \int_0^\infty g(\tau)|I_e(\omega, v_x, \tau)|^2 d\tau = \frac{2e^2 v_x^2 \tau_0^2}{4\pi^2 d^2(1 + \omega^2\tau_0^2)} \tag{10.5-17}$$

The second averaging over v_x^2 is particularly simple, since it results in the replacement of v_x^2 in Equation (10.5-17) by its average $\overline{v_x^2}$, which, for a sample at thermal equilibrium, is given according to Equation (10.5-10) by $\overline{v_x^2} = kT/m$. The final result is then

$$\overline{|I_e(\omega)|^2} = \frac{2e^2\tau_0^2 kT}{4\pi^2 md^2(1 + \omega^2\tau_0^2)} \tag{10.5-18}$$

The average number of scattering events per second $\bar{N}$ is equal to the total number of electrons $N_e V$ divided by the mean scattering time τ_0

$$\bar{N} = \frac{N_e V}{\tau_0} \tag{10.5-19}$$

Thus, from Equation (10.3-10), we obtain

$$S_i(\nu) = 8\pi^2 \bar{N}\, \overline{|I_e(\omega)|^2} = \frac{4N_e V e^2 \tau_0 kT}{md^2(1 + \omega^2\tau_0^2)}$$

and, after using Equation (10.5-12) and limiting ourselves as in Equation (10.4-6) to frequencies where $\omega\tau_0 \ll 1$, we get

$$\overline{i_N^2}(\nu) \equiv S_i(\nu)\,\Delta\nu = \frac{4kT\,\Delta\nu}{R(\nu)} \tag{10.5-20}$$

in agreement with Equation (10.5-9).

10.6 SPONTANEOUS EMISSION NOISE IN LASER OSCILLATORS

Another type of noise that plays an important role in quantum electronics and optical communications is that of spontaneous emission in laser oscillators and amplifiers. As shown in Chapter 5, a necessary condition for laser amplification is that the atomic population of a pair of levels 1 and 2 be inverted. If $E_2 > E_1$, gain occurs when the population is inverted, that is, $N_2 > N_1$. Assume that an optical wave with frequency $\nu \approx (E_2 - E_1)/h$ is propagating through an inverted population medium. This wave will grow coherently due to the effect of stimulated emission. In addition, its radiation will be contaminated by noise radiation caused by spontaneous emission from level 2 to level 1. Some of the radiation emitted by the spontaneous emission will propagate very nearly along the same direction as that of the stimulated emission and cannot be separated from it. This has two main consequences. First, the laser output will have a finite spectral width. This effect is described in this section. Second, the signal-to-noise ratio achievable at the output of laser amplifiers [10] is limited because of the intermingling of spontaneous emission noise power with that of the amplified signal (see Figure 10.9). The issue of amplified spontaneous emission (ASE) and its impact on the signal-to-noise ratio will be addressed in Section 17.3.

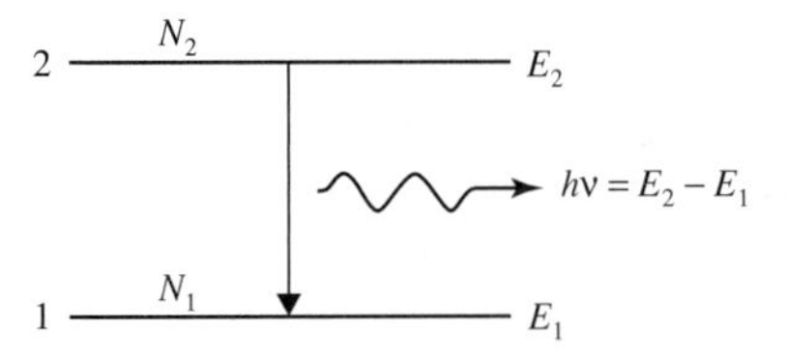

Figure 10.9 An atomic transition with $N_2 > N_1$ providing gain for laser oscillation or optical amplification. Spontaneous emission of photon with energy $h\nu = (E_2 - E_1)$ can contaminate the laser oscillation as well as the amplified radiation.

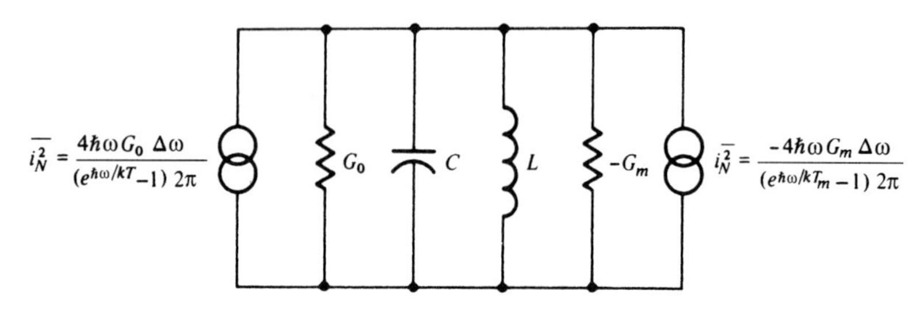

Figure 10.10 Equivalent circuit of a laser oscillator.

Returning to the case of a laser oscillator, we represent it by an *RLC* circuit, as shown in Figure 10.10. The presence of the laser medium with negative loss (i.e., gain) is accounted for by including a negative conductance $-G_m$ while the ordinary loss mechanisms described in Chapter 6 are represented by the positive conductance G_0. The noise generator associated with the losses G_0 is given according to Equation (10.5-9) as

$$\overline{i_N^2} = \frac{4\hbar\omega G_0(\Delta\omega/2\pi)}{e^{\hbar\omega/kT} - 1}$$

where $\omega = 2\pi\nu$, $\Delta\omega = 2\pi\,\Delta\nu$, and T is the actual temperature of the losses. This is the Johnson noise due to a resistor with $R = 1/G_0$. Noise due to spontaneous emission is represented by a similar expression for the noise current

$$\overline{(i_N^2)}_{\substack{\text{spont}\\\text{emission}}} = \frac{4\hbar\omega(-G_m)(\Delta\omega/2\pi)}{e^{\hbar\omega/kT_m} - 1} \tag{10.6-1a}$$

where the term $(-G_m)$ represents negative losses and T_m is a temperature determined by the population ratio according to

$$\frac{N_2}{N_1} = e^{-\hbar\omega/kT_m} \tag{10.6-1b}$$

Since $N_2 > N_1$, then $T_m < 0$, and $\overline{(i_N^2)}$ in Equation (10.6-1a) is positive definite. The 2π factor appearing in the denominators of $\overline{(i_N^2)}$ is due to the fact that here we use $\overline{i_N^2}(\omega)$ instead of $\overline{i_N^2}(\nu)$ with $\overline{i_N^2}(\omega)\,\Delta\omega = \overline{i_N^2}(\nu)\,\Delta\nu$. Using Equation (10.6-1b), the noise current due to spontaneous emission (10.6-1a) can be written

$$\overline{(i_N^2)}_{\substack{\text{spont}\\\text{emission}}} = \frac{4\hbar\omega(G_m)N_2(\Delta\omega/2\pi)}{N_2 - N_1} \tag{10.6-2}$$

Although a detailed justification of Equation (10.6-1a) is outside the scope of the present treatment, a strong case for its plausibility can be made by noting that since the gain is proportional to the population difference, that is, $G_m \propto N_2 - N_1$, $\overline{(i_N^2)}$ in Equation (10.6-1a) can be written, using Equation (10.6-2),

$$\overline{(i_N^2)}_{\substack{\text{spont}\\\text{emission}}} \propto \frac{-4\hbar\omega\,\Delta\omega(N_2 - N_1)}{(N_1/N_2) - 1} = 4\hbar\omega\,\Delta\omega\,N_2 \tag{10.6-3}$$

and is thus proportional to N_2. The proportionality of G_m to $N_2 - N_1$ can also be justified by noting that in the equivalent circuit (Figure 10.10) the stimulated emission power is given by $v^2 G_m$, where v is the voltage. Using the field approach, this power is proportional to $E^2(N_2 - N_1)$, where E is the field amplitude. Since v is proportional to E, G_m is proportional to $N_2 - N_1$. Equation (10.6-3) makes sense, since spontaneous emission power is due to $2 \to 1$ transitions and should consequently be proportional to N_2.

Returning to the equivalent circuit of Figure 10.10, its quality factor Q is given by

$$Q^{-1} = \frac{G_0 - G_m}{\omega_0 C} = \frac{1}{Q_0} - \frac{1}{Q_m} \tag{10.6-4}$$

where $\omega_0^2 = (LC)^{-1}$ is the resonance frequency of the LC circuit. The impedance of the circuit shown in Figure 10.10 is

$$\begin{aligned} Z(\omega) &= \frac{1}{(G_0 - G_m) + (1/i\omega L) + i\omega C} \\ &= \frac{i\omega}{C} \frac{1}{(i\omega\omega_0/Q) + (\omega_0^2 - \omega^2)} \end{aligned} \tag{10.6-5}$$

so the voltage across this impedance due to a current source with a complex amplitude $I(\omega)$ is

$$V(\omega) = \frac{i}{C} \frac{I(\omega)}{[(\omega_0^2 - \omega^2)/\omega] + (i\omega_0/Q)} \tag{10.6-6}$$

which, near $\omega = \omega_0$, becomes

$$\overline{|V(\omega)|^2} = \frac{1}{4C^2} \frac{\overline{|I(\omega)|^2}}{(\omega_0 - \omega)^2 + (\omega_0^2/4Q^2)} \tag{10.6-7}$$

The current sources driving the resonant circuit are those shown in Figure 10.10; since they are not correlated, we may take $|I(\omega)|^2$ as the sum of their mean-square values, according to Equation (10.6-2),

$$\overline{|I(\omega)|^2} = 4\hbar\omega \left(\frac{G_m N_2}{N_2 - N_1} + \frac{G_0}{e^{\hbar\omega/kT} - 1} \right) \frac{d\omega}{2\pi} \tag{10.6-8}$$

where the first term is due to the spontaneous emission, and the second term is due to the losses. In the optical region, $\lambda = 1$ μm say, and for $T = 300$ K we have $\hbar\omega/kT \approx 50$. Thus, since near oscillation $G_m \sim G_0$, we may neglect the thermal (Johnson) noise term in Equation (10.6-8), thereby obtaining

$$\overline{|V(\omega)|^2_{\omega \equiv \omega_0}} = \frac{\hbar G_m}{2\pi C^2} \left(\frac{N_2}{N_2 - N_1} \right) \frac{\omega \, d\omega}{(\omega_0 - \omega)^2 + (\omega_0^2/4Q^2)} \tag{10.6-9}$$

Equation (10.6-9) represents the spectral distribution of the laser output. If we subject the output to high-resolution spectral analysis, we should, according to (10.6-9), measure a linewidth

$$\Delta\omega = \frac{\omega_0}{Q} \tag{10.6-10}$$

between the half-intensity points. The trouble is that, though correct, Equation (10.6-10) is not of much use in practice. The reason is that, according to Equation (10.6-4), Q^{-1} is equal to the difference of two nearly equal quantities, neither of which is known with high enough accuracy. We can avoid this difficulty by showing that Q is related to the laser power output, and thus $\Delta\omega$ may be expressed in terms of the power.

The total optical oscillation power extracted from the atoms comprising the laser is

$$P = G_0 \int_0^\infty \frac{|V(\omega)|^2}{d\omega} d\omega$$

$$= \frac{\hbar G_m G_0}{2\pi C^2} \left(\frac{N_2}{N_2 - N_1} \right) \int_0^\infty \frac{\omega \, d\omega}{(\omega_0 - \omega)^2 + (\omega_0/2Q)^2} \tag{10.6-11}$$

Since the integrand peaks sharply near $\omega \approx \omega_0$, we may replace ω in the numerator of Equation (10.6-11) by ω_0 and, after integration, obtain

$$P = \frac{\hbar G_m G_0 Q}{C^2} \left(\frac{N_2}{N_2 - N_1} \right) \tag{10.6-12}$$

which is the desired result linking P to Q. In a laser oscillator, the gain very nearly equals the loss, or in our notation, $G_m \sim G_0$. Using this result in Equation (10.6-12), we obtain

$$Q = \frac{C^2}{\hbar G_0^2} \left(\frac{N_2 - N_1}{N_2} \right) P$$

which, when substituted in Equation (10.6-10), yields

$$\Delta\nu = \frac{2\pi h \nu_0 (\Delta\nu_{1/2})^2}{P} \left(\frac{N_2}{N_2 - N_1} \right) \tag{10.6-13}$$

where $\Delta\nu_{1/2}$ is the full width of the passive cavity resonance given in Equation (4.7-6) as $\Delta\nu_{1/2} = \nu_0/Q_0 = (1/2\pi)(G_0/C)$. It is worthwhile to recall here that $\Delta\nu$ represents, in the quantum limit, the laser field spectral width. The expression (10.6-13) is known as the Schawlow–Townes linewidth after the two American coinventors of the laser [11] who first derived it.

Equation (10.6-13) does not predict an inverse dependence of $\Delta\nu$ on P, as may be deduced at first glance, because of the dependence of N_2 on P. For very large powers, $P \to \infty$, N_2 is proportional to P, while $N_2 - N_1$ remains clamped at its threshold value $(N_2 - N_1)_{th}$. This leads to a residual power-independent value of $\Delta\nu$. To appreciate this argument qualitatively, we note that unless the lifetime t_1 of the lower laser level is zero, as P increases, N_1 must increase since the increased (net) induced transition rate into level 1 must equal, in steady state N_1/t_1, the rate of emptying of level 1. This causes the population N_2 to increase in order to keep $N_2 - N_1$, and thus the gain, a constant. At sufficiently high values of P, N_2 becomes and stays proportional to P and the ratio N_2/P in Equation (10.6-13) approaches a constant value, thus leading to a residual power independent linewidth.

To obtain the power dependence of the population inversion factor

$$\mu \equiv \frac{N_2}{(N_2 - N_1)_{th}}$$

we need to solve the rate equations for the atomic populations plus the equation for the photon number p (p = number of photons in the optical resonator). They are given by

$$\frac{dN_2}{dt} = R - \frac{N_2}{t_2} - (N_2 - N_1)W_i$$

$$\frac{dN_1}{dt} = -\frac{N_1}{t_1} + (N_2 - N_1)W_i + \frac{N_2}{t_2} \tag{10.6-14}$$

$$\frac{dp}{dt} = (N_2 - N_1)W_i - \frac{p}{t_c}$$

where we recall that R is the pump rate and W_i is the induced transition rate. The first two equations for the populations are similar to Equations (5.7-3) and (5.7-4) with $R_1 = 0$, $t_2 \rightarrow t_{\text{spont}}$, $R_2 \rightarrow R$, and N_2 and N_1 representing the total atomic populations of the laser transition levels 2 and 1, respectively. The third equation is a conservation equation for the total number of photons. The photon lifetime t_c is related to the cavity linewidth $\Delta\nu_{1/2}$ by $\Delta\nu_{1/2} = (2\pi t_c)^{-1}$. At equilibrium (or steady state), $d/dt = 0$, we can solve Equation (10.6-14) to obtain

$$N_2 - N_1 = \frac{R(t_2 - t_1)}{1 + W_i t_2}$$
$$N_2 = \frac{R t_2 (1 + W_i t_1)}{1 + W_i t_2} \qquad (10.6\text{-}15)$$

so that

$$\frac{N_2}{(N_2 - N_1)_{\text{th}}} = \frac{t_2}{t_2 - t_1}(1 + W_i t_1) \qquad (10.6\text{-}16)$$

where the subscript th indicates the value at threshold. The power output, including "wall losses" of the laser, is

$$P = (N_2 - N_1)_{\text{th}} W_i h\nu_0 \qquad (10.6\text{-}17)$$

which, when used together with Equation (10.6-16) in Equation (10.6-13) gives

$$\Delta\nu_{\text{laser}} = \frac{2\pi h\nu_0 (\Delta\nu_{1/2})^2}{P} \frac{t_2}{t_2 - t_1} + \frac{c\, \Delta\nu_{1/2}\, \lambda_0^2}{8\pi n^3\, \Delta\nu_{\text{gain}}\, V} \frac{t_1}{t_2 - t_1} \qquad (10.6\text{-}18)$$

where $\Delta\nu_{\text{gain}}$ is the linewidth of atomic transition responsible for the laser gain. V is the mode volume. In arriving at Equation (10.6-18), we used

$$(N_2 - N_1)_{\text{th}} = \frac{8\pi\nu_0^2 n^3\, \Delta\nu_{\text{gain}}\, V t_2}{c^3 t_c} \quad (t_2 = t_{\text{spont}}) \qquad (10.6\text{-}19)$$

which is obtained from Equation (6.1-18) if we put $\Delta\nu_{\text{gain}} = 1/g(\nu)$. The first term on the right-hand side of Equation (10.6-18) is the conventional Schawlow–Townes expression containing the inverse P dependence. The second term is power independent and corresponds to a residual linewidth as $P \rightarrow \infty$.

To get an idea of the magnitudes involved, we consider the case of a 0.6328 μm He–Ne laser with mirror reflectivities of $R = 0.99$, a resonator length of $l = 30$ cm, and $t_1/t_2 = 0.1$. We obtain

$$\Delta\nu_{1/2}(\text{Hz}) = \frac{(1 - R)c}{2\pi n l} = 1.6 \times 10^6$$

and

$$\Delta\nu_{\text{laser}}(\text{Hz}) \cong \frac{10^{-3}}{P(\text{mW})} + 3.8 \times 10^{-4}$$

The residual linewidth thus dominates at power levels exceeding a few milliwatts.

10.7 PHASOR DERIVATION OF LASER LINEWIDTH

The derivation of the laser linewidth in Section 10.6 takes advantage of the highly sophisticated and efficient concepts and phenomena represented by the seemingly simple circuit model of a laser oscillator. The price we pay when taking this approach is a certain loss of physical insight into the mechanisms whereby spontaneous emission affects the laser linewidth. In this section we will derive expression (10.6-13) for the laser linewidth using a different approach. This is done not only for pedagogic purposes, but because some of the interim results involving phase fluctuations are useful in their own right.

The Phase Noise

An ideal monochromatic radiation field can be written as

$$E(t) = \mathrm{Re}\,[E_0 e^{i(\omega_0 t+\theta)}] \tag{10.7-1}$$

where ω_0 is the radian frequency, E_0 is the field amplitude, and θ is a constant. A real field including that of lasers undergoes random phase and amplitude fluctuations that can be represented by writing

$$E(t) = \mathrm{Re}\,[E_0 e^{i(\omega_0 t+\theta(t))}] \tag{10.7-2}$$

where $E(t)$ and $\theta(t)$ vary only "slightly" during one optical period.

There are many reasons in a practical laser for the random fluctuation in amplitude and phase. Most of these can be reduced, in theory, to inconsequence by various improvements such as ultrastabilization of the laser cavity length and the near elimination of microphonic and temperature variations. There remains, however, a basic source of noise that is quantum mechanical in origin. This is due to spontaneous emission that continually causes new power to be added to the laser oscillation field. The electromagnetic field represented by this new power, not being coherent with the old field, causes phase as well as amplitude fluctuations. These are responsible ultimately for the deviation of the evolution of the laser field from that of an ideal monochromatic field, that is, for the quantum mechanical noise.

Let us consider the effect of one spontaneous emission event on the electromagnetic field of a single oscillating laser mode. A field such as (10.7-1) can be represented by a phasor of length E_0 rotating with an angular (radian) rate ω_0. In a frame rotating at ω_0, we would see a constant vector E_0. Since $E_0^2 \propto \bar{n}$, the average number of quanta in the mode, we shall represent the laser field phasor before a spontaneous emission event by a phasor of length $\sqrt{\bar{n}}$ as in Figure 10.11. The spontaneous emission adds *one* photon to the field, and this is

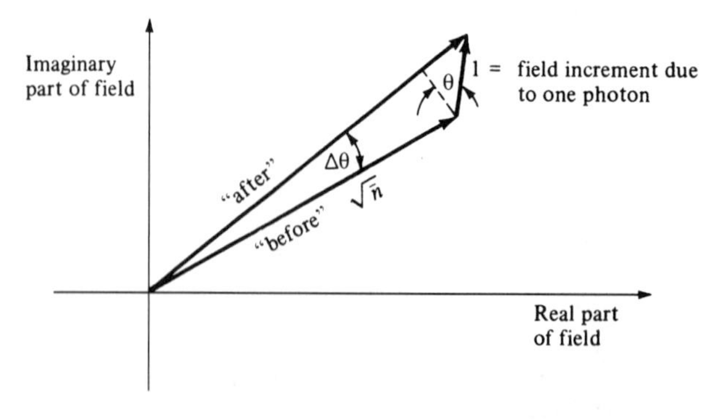

Figure 10.11 The phasor model for the effect of a single spontaneous emission event on the laser field phase.

represented, according to our conversion, by an incremental vector of unity length. Since this field increment is not correlated in phase with the original field, the angle ϕ is a random variable (i.e., it is distributed uniformly between zero and 2π). The resulting change $\Delta\theta$ of the field phase can be approximated for $\bar{n} \gg 1$ by

$$\Delta\theta_{\text{one emission}} = \frac{1}{\sqrt{\bar{n}}} \cos\phi \tag{10.7-3}$$

Next, consider the effect of N spontaneous emissions on the phase of the laser field. The problem is one of angular random walk, since ϕ may assume with equal probability any value between 0 and 2π. We can then write

$$\langle[\Delta\theta(N)]^2\rangle = \langle(\Delta\theta_{\text{one emission}})^2\rangle N \tag{10.7-4}$$

and from Equation (10.7-3),

$$\langle[\Delta\theta(N)]^2\rangle = \frac{1}{\bar{n}}\langle\cos^2\theta\rangle N$$

where $\langle\ \rangle$ denotes an ensemble average taken over a very large number of individual emission events.

Equation (10.7-4) is a statement of the fact that in a random walk problem the mean squared distance traversed after N steps is the square of the size of one step times N. The mean deviation $\langle\Delta\theta(N)\rangle$ after N spontaneous emissions is, of course, zero. Any one experiment, however, will yield a nonzero result. The mean squared deviation is thus nonzero and is a measure of the phase fluctuation. To obtain the root mean square (rms) phase deviation in a time t, we need to calculate the average number of spontaneous emission events $N(t)$ into a single laser mode in time t.

The total number of spontaneous emissions per second is VN_2/t_{spont}, where V is the volume, N_2 is the total number of atoms per unit volume in the upper laser level 2, and t_{spont} is the spontaneous lifetime of an atom in level 2. We assume a normalized lineshape function of $g(\nu)$, and the spontaneous emission in a spectral bandwidth of $\Delta\nu$. The number of spontaneous emissions per second within the bandwidth of $\Delta\nu$ is $(VN_2/t_{\text{spont}})g(\nu)\,\Delta\nu$. Note that these spontaneous emissions will spread over all optical modes in the volume V. The total number of transitions per second into one mode is thus

$$\frac{N_{\text{spont}}}{\text{mode-second}} = \frac{VN_2}{t_{\text{spont}}N_{\text{mode}}} g(\nu)\,\Delta\nu \tag{10.7-5}$$

where N_{mode} is the number of modes within the bandwidth $\Delta\nu$; according to Equation (4.0-12),

$$N_{\text{mode}} = \frac{8\pi\nu_0^2\,\Delta\nu\, Vn^3}{c^3} \tag{10.7-6}$$

Physically, N_{mode} is the number of modes interacting with the laser transition, that is, partaking in the spontaneous emission. We can rewrite Equation (10.7-5) as

$$\frac{N_{\text{spont}}}{\text{second-mode}} = \left(\frac{N_2}{\Delta N_t}\right)\frac{V\,\Delta N_t}{t_{\text{spont}}N_{\text{mode}}} g(\nu)\,\Delta\nu \tag{10.7-7}$$

where ΔN_t is the population inversion $(N_2 - N_1)$ at threshold. Using Equation (6.1-18) for the threshold population inversion and Equation (10.7-6), we obtain

$$\Delta N_t = \frac{N_{\text{mode}}}{Vg(\nu)\,\Delta\nu}\frac{t_{\text{spont}}}{t_c}$$

where t_c is the photon lifetime in the resonator. Thus Equation (10.7-7) can be written

$$\frac{N_{\text{spont}}}{\text{mode-second}} = \frac{\mu}{t_c}, \quad \mu \equiv \frac{N_2}{\Delta N_t} = \frac{N_2}{(N_2 - N_1)_t} \tag{10.7-8}$$

where μ is the population inversion factor. Equation (10.7-8) is an expression for the number of spontaneous emissions into a single mode of bandwidth $\Delta\nu$ per unit time. The number of spontaneous transitions into a single lasing mode in a time t is thus

$$N(t) = \frac{\mu}{t_c}t \tag{10.7-9}$$

We recall here that, in an ideal four-level laser, $N_1 = 0$ and $\Delta N_t = N_2$, that is, $\mu = 1$. In a three-level laser, on the other hand, μ can be appreciably larger than unity. In a ruby laser at room temperature, for example (see Section 6.11), $\mu \sim 50$. This reflects the fact that for a given gain the total excited population N_2 of a three-level laser must exceed that of a four-level laser by the factor μ, since gain is proportional to $N_2 - N_1$. Equation (10.7-8) is also equivalent to stating that above threshold there are μ spontaneously emitted photons present in a laser mode.

Using Equation (10.7-9) in (10.7-4), we obtain for the root mean square phase deviation after t seconds

$$\Delta\theta(t) = \langle[\Delta\theta(t)]^2\rangle^{1/2} = \sqrt{\frac{1}{2\bar{n}}\frac{\mu}{t_c}t}$$

The maximum time t available for such an experiment is the integration time T of the measuring apparatus so that

$$\Delta\theta(T) = \sqrt{\frac{1}{2\bar{n}}\frac{\mu}{t_c}T} \tag{10.7-10}$$

The rms frequency excursion caused by $\Delta\theta$ is

$$(\Delta\omega)_{\text{rms}} = \frac{\Delta\theta(T)}{T} = \sqrt{\frac{\mu}{2\bar{n}t_cT}} \tag{10.7-11a}$$

where $\bar{n}$ is the average number of photons in the lasing mode. We can cast the last result in a more familiar form by using the relations

$$P_e = \frac{\bar{n}\hbar\omega_0}{t_c}, \quad B = \frac{1}{2T} \tag{10.7-11b}$$

Here P_e is the power emitted by the atoms (i.e., the sum of the useful power output plus any power lost by scattering and absorption), and B is the bandwidth (in hertz) of the phase-measuring apparatus. The result is

$$(\Delta\omega)_{\text{rms}} = \sqrt{\frac{\mu\hbar\omega_0}{P_et_c^2}B} \tag{10.7-12}$$

From the experimental point of view $(\Delta\omega)_{\text{rms}}$ is the root mean square deviation of the reading of an instrument whose output is the frequency $\omega(t) \equiv d\theta/dt$. We will leave it as an exercise (Problem 10.11) for the student to design an experiment that measures $(\Delta\omega)_{\text{rms}}$.

Ring laser gyroscopes sense rotation by comparing the oscillation frequencies of two counterpropagating modes in a rotating ring resonator. Their sensitivity, that is, the smallest rotation rate that they can sense, is thus limited by any uncertainty $\Delta\omega$ in the laser frequency. Experiments have indeed demonstrated a rotation-measuring sensitivity approaching the quantum limit as given by Equation (10.7-12).

The Laser Field Spectrum

Next, we address the case where one measures directly the spectrum of the optical field

$$\mathscr{E}(t) = \mathrm{Re}\left[E(t)e^{i[\omega_0 t+\theta(t)]}\right] \tag{10.7-13}$$

using, say, a scanning Fabry–Perot etalon. If the etalon has a sufficiently high spectral resolution, the measurement should yield the spectral density function $S_{\mathscr{E}}(\omega)$ of the laser field. We will, consequently, proceed to obtain an expression for this quantity. We make use of the Wiener–Khintchine theorem (10.2-14) according to which $S_{\mathscr{E}}(\omega)$ is the Fourier integral transform of the field autocorrelation function $C_{\mathscr{E}}(\tau)$

$$\begin{aligned} S_{\mathscr{E}}(\omega) &= \frac{1}{\pi}\int_{-\infty}^{\infty} C_{\mathscr{E}}(\tau)e^{-i\omega\tau}\,d\tau \\ S_{\mathscr{E}}(\omega) &= \frac{4\pi}{T}\,|\mathscr{E}_T(\omega)|^2, \quad \mathscr{E}_T(\omega) = \frac{1}{2\pi}\int_{-T/2}^{T/2} \mathscr{E}(t)e^{-i\omega t}\,dt \end{aligned} \tag{10.7-14}$$

$$C_{\mathscr{E}}(\tau) \equiv \langle \mathscr{E}(t)\mathscr{E}(t+\tau)\rangle \tag{10.7-15}$$

where the symbol $\langle\ \rangle$ represents an ensemble, or time, average.

Using Equation (10.7-13), we obtain

$$\begin{aligned} C_{\mathscr{E}}(\tau) = \tfrac{1}{4}\langle &[E(t)e^{i[\omega_0 t+\theta(t)]} + E^*(t)e^{-i[\omega_0 t+\theta(t)]}] \\ &\times [E(t+\tau)e^{i[\omega_0(t+\tau)+\theta(t+\tau)]} + E^*(t+\tau)e^{-i[\omega_0(t+\tau)+\theta(t+\tau)]}]\rangle \end{aligned} \tag{10.7-16}$$

Now, for example,

$$\langle E(t)E(t+\tau)e^{i[2\omega_0 t+\theta(t)+\theta(t+\tau)]}\rangle = 0$$

since it corresponds to averaging a signal oscillating at twice the optical frequency over many periods. So if we keep only the slowly varying terms in $C_{\mathscr{E}}(\tau)$, we obtain

$$\begin{aligned} C_{\mathscr{E}}(\tau) &= \tfrac{1}{4}\langle E(t)E^*(t+\tau)e^{i[-\omega_0\tau+\theta(t)-\theta(t+\tau)]} + E^*(t)E(t+\tau)e^{i[\omega_0\tau-\theta(t)+\theta(t+\tau)]}\rangle \\ &= \tfrac{1}{4}[I(\tau) + I^*(\tau)] \end{aligned} \tag{10.7-17}$$

$$I(\tau) = \langle E^*(t)E(t+\tau)e^{i[\Delta\theta(t,\tau)+\omega_0\tau]}\rangle \tag{10.7-18}$$

$$\Delta\theta(t,\tau) \equiv \theta(t+\tau) - \theta(t) \tag{10.7-19}$$

The main contributions to the laser noise are due to fluctuations of the phase $\theta(t)$ and not the amplitude $E(t)$, since the amplitude fluctuations are kept negligibly small by gain saturation. Taking advantage of this fact, we write $\langle E^*(t)E(t+\tau)\rangle = \langle E^2\rangle \approx$ constant, so that

$$I(\tau) = \langle E^2\rangle e^{i\omega_0\tau}\langle e^{i\Delta\theta(t,\tau)}\rangle \tag{10.7-20}$$

Given a (normalized) probability distribution function for $\Delta\theta$, $g(\Delta\theta)$, the expectation value of $\exp[i\Delta\theta(t,\tau)]$ is obtained from

$$\langle e^{i\Delta\theta(t,\tau)}\rangle = \int_{-\infty}^{\infty} e^{i\Delta\theta(t,\tau)} g(\Delta\theta)\, d(\Delta\theta) \tag{10.7-21}$$

Since the total phase excursion $\Delta\theta$ is the net result of many small and statistically independent (spontaneous transitions) excursions, the central limit theorem of statistics applies, and $g(\Delta\theta)$ is a Gaussian, which we write as

$$g(\Delta\theta) = \frac{1}{\sqrt{2\pi\langle(\Delta\theta)^2\rangle}} e^{-(\Delta\theta)^2/2\langle(\Delta\theta)^2\rangle} \tag{10.7-22}$$

where

$$\langle(\Delta\theta)^2\rangle = \int_{-\infty}^{\infty} (\Delta\theta)^2 g(\Delta\theta)\, d(\Delta\theta) \tag{10.7-23}$$

Using Equation (10.7-22) in (10.7-21), we obtain

$$\langle e^{i\Delta\theta(t,\tau)}\rangle = e^{-\langle(\Delta\theta)^2\rangle/2} = e^{-\mu|\tau|/(4\bar{n}t_c)} \tag{10.7-24}$$

where, in order to obtain the last result, we used Equation (10.7-10) with $T = |\tau|$. Using Equation (10.7-24) in (10.7-20),

$$C_{\mathscr{E}}(\tau) = \tfrac{1}{4}\langle E^2\rangle e^{-\mu|\tau|/4\bar{n}t_c}(e^{i\omega_0\tau} + e^{-i\omega_0\tau}) \tag{10.7-25}$$

The spectral density function of the laser field $S_{\mathscr{E}}(\omega)$, the quantity observed by a spectral analysis of the field, is given according to Equations (10.7-14) and (10.7-25) by

$$S_{\mathscr{E}}(\omega) = \frac{\langle E^2\rangle}{4\pi}\int_{-\infty}^{\infty} e^{(-\mu|\tau|/4\bar{n}t_c)-i\omega\tau}\,(e^{i\omega_0\tau} + e^{-i\omega_0\tau}\, dt)\, d\tau \tag{10.7-26}$$

$$= \frac{\langle E^2\rangle}{2}\left(\frac{\mu/4\bar{n}t_c}{(\mu/4\bar{n}t_c)^2 + (\omega-\omega_0)^2} + \frac{\mu/4\bar{n}t_c}{(\mu/4\bar{n}t_c)^2 + (\omega+\omega_0)^2}\right) \tag{10.7-27}$$

We have defined in Equation (10.2-7) the spectral density function in such a way that only positive frequencies need to be considered. For $\omega > 0$, the second term on the right side of Equation (10.7-27) contributes negligibly so that

$$S_{\mathscr{E}}(\omega) = \frac{\langle E^2\rangle}{2\pi}\frac{\mu/4\bar{n}t_c}{(\mu/4\bar{n}t_c)^2 + (\omega-\omega_0)^2} \tag{10.7-28}$$

which corresponds to a Lorentzian-shaped function centered on the nominal laser frequency ω_0 with a full width at half-maximum of

$$(\Delta\omega)_{\text{laser}} = \frac{\mu}{2\bar{n}t_c} \tag{10.7-29}$$

Recalling that the total power emitted by the electrons is $P = \bar{n}\hbar\omega_0/t_c$ and defining the passive resonator linewidth $\Delta\nu_{1/2} = (2\pi t_c)^{-1}$, we can rewrite Equation (10.7-29) using (10.7-11b) as

$$(\Delta\nu)_{\text{laser}} = \frac{(\Delta\omega)_{\text{laser}}}{2\pi} = \frac{2\pi h\nu_0(\Delta\nu_{1/2})^2\mu}{P} \tag{10.7-30}$$

which, recalling the definition (10.7-8) of μ, is half the result of the circuit model (10.6-13).[1]

[1] The discrepancy by a factor of 2 should not be taken too seriously considering the very different mathematical approaches employed by the two derivations.

EXAMPLE: LINEWIDTH OF A HE–NE LASER AND A SEMICONDUCTOR DIODE LASER

To obtain an order of magnitude estimate of the linewidth $(\Delta\nu)_{laser}$ predicted by Equation (10.7-30), we will calculate it in the case of two largely different types of cw lasers: (1) a He–Ne laser and (2) a semiconductor GaInAsP laser.

1. *He–Ne laser.*

$$\nu = 4.741 \times 10^{14}\ \text{Hz}\ (\lambda = 6328\ \text{Å})$$

$$l\ \text{(distance between reflectors)} = 100\ \text{cm}$$

$$\text{Loss} = (1 - R) = 1\%\ \text{per pass}$$

From these numbers we get

$$(\Delta\nu_{1/2}) = \frac{1}{2\pi t_c} \approx \frac{(1-R)c}{2\pi n l} \approx 5 \times 10^5$$

(i.e., $t_c = 3.2 \times 10^{-7}$ s) and from Equation (10.7-30), assuming $\mu = 1$ (i.e., $N_1 \ll N_2$),

$$(\Delta\nu)_{laser} \cong 2 \times 10^{-3}\ \text{Hz}$$

at a power level $P = 1$ mW.

The predicted linewidth is thus so small as to be completely masked in almost all experimental situations by contributions due to extraneous causes, such as vibrations and temperature flucturations.

2. *Semiconductor laser.* We use as a typical example the case of a GaInAsP ($\lambda = 1.55\ \mu$m) laser with the following pertinent characteristics:

$$P = 3\ \text{mW}$$

$$\nu = 1.935 \times 10^{14}\ (\lambda_0 = 1.55\ \mu\text{m})$$

$$\Delta\nu_{1/2} = \frac{(1-R)c}{2\pi n l}$$

$$R\ \text{(reflectivity)} = 30\%$$

$$l = 300\ \mu\text{m}$$

$$n = 3.5$$

$$\mu = 3\ (\text{at}\ T = 300\ \text{K})$$

This results in $\Delta\nu_{1/2} \sim 3 \times 10^{10}$ (i.e., $t_c = 1/(2\pi\Delta\nu_{1/2}) = 5 \times 10^{-12}$ s) and

$$(\Delta\nu)_{laser} = 0.817 \times 10^6\ \text{Hz}$$

The experimental curve of Figure 10.12 shows the predicted (Equation (10.7-30)) P^{-1} dependence of $(\Delta\nu)_{laser}$, but the measured values of the linewidth are larger by a factor of ~70 than those predicted by the analysis. This discrepancy has been studied by a number of investigators [13–15], who have shown that the analysis leading to (10.7-30) ignores the modulation of the index of refraction of the laser medium, which is due to fluctuations of the

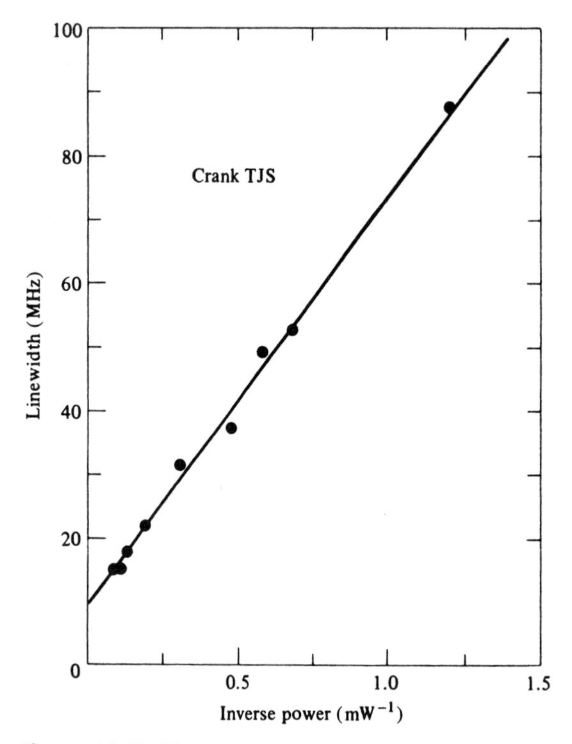

Figure 10.12 The measured dependence of the spectral linewidth of a semiconductor laser on the power output. (After Reference [12].)

electron density caused by spontaneous emission. When this effect is included, the result is to multiply Equation (10.7-30) by the factor

$$1+\left(\frac{\Delta n'}{\Delta n''}\right)^2 \equiv 1+\alpha^2 \tag{10.7-31}$$

where $\alpha = (\Delta n'/\Delta n'')$, and $\Delta n'$ and $\Delta n''$ are, respectively, the changes in the real and imaginary parts of the index of refraction "seen" by the laser field due to some change in the electron density. The factor $1 + (\Delta n'/\Delta n'')^2$ can be calculated from measured parameters of the laser or measured directly [1]. Its value is ~30 in typical cases, enough to reconcile the observed data of Figure 10.12 and the prediction of Equation (10.7-30). The same α factor is also responsible for the frequency chirp in the output beam of a current-modulated semiconductor laser (see Section 15.6).

The big difference, over nine orders of magnitude, between the limiting linewidth of conventional lasers, say, gas lasers and semiconductor lasers, is due mostly to the very short photon lifetime t_c in semiconductor laser resonators. At a given power output, we have from Equation (10.7-30) $(\Delta\nu)_{\text{laser}} \propto (\Delta\nu_{1/2})^2 \propto t_c^{-2})^2$. In the above examples we obtained $t_c \simeq 3\times10^{-8}$ s in the case of the He–Ne laser, and $t_c \simeq 5\times10^{-12}$ s in the semiconductor laser. Since $t_c \sim ln/c(1-R)$, the main hope for increasing t_c in a semiconductor laser, thus decreasing the linewidth $(\Delta\nu)_{\text{laser}}$, is to increase l by placing the laser in an external resonator and by using high reflectance mirrors $R \sim 1$. Semiconductor laser linewidths in the kilohertz regime are obtainable.

An actual (measured) GaAs/GaAlAs semiconductor laser, Lorentzian field spectrum is shown in Figure 10.13.

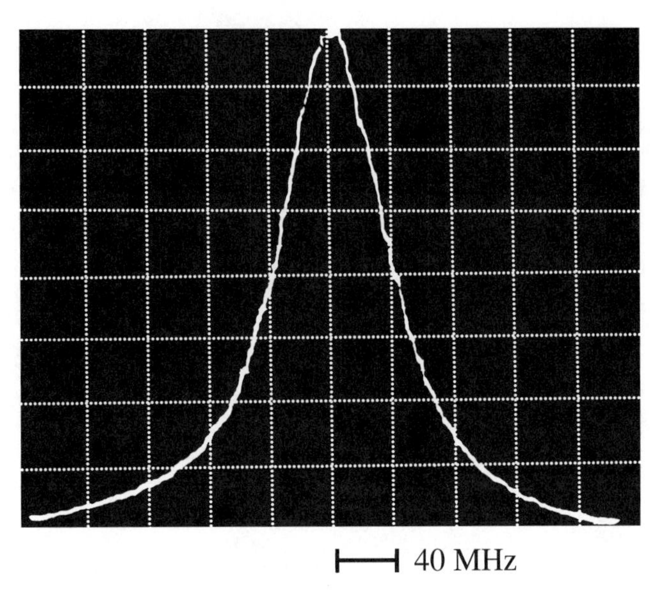

Figure 10.13 The measured Lorentzian field spectrum $S_{\mathscr{E}}(\omega)$ of a semiconductor laser. (After Reference [12].)

10.8 COHERENCE AND INTERFERENCE

In Section 10.7 (Equation (10.7-25)), we have derived the following expression for the autocorrelation function of the single-mode laser field:

$$C_{\mathscr{E}}(\tau) \equiv \langle \mathscr{E}(t)\mathscr{E}(t+\tau)\rangle \propto \cos \omega_0 \tau e^{-\mu|\tau|/4\bar{n}_c}$$
$$= \cos \omega_0 \tau e^{-|\tau|/\tau_c}$$
$$\tau_c = \frac{4\bar{n}t_c}{\mu} \tag{10.8-1}$$

where $\bar{n}$ is the number of photons inside the resonator, $\mu = N_2/(N_2 - N_1)$, and t_c is the photon lifetime (the decay time constant for the mode optical energy if the gain mechanism were turned off).

The parameter τ_c is called the coherence time of the laser field. According to Equation (10.7-29) it is equal to $2/(\Delta\omega)_{\text{laser}}$, where $(\Delta\omega)_{\text{laser}}$ is the laser output field linewidth. In practical terms, it is the time duration during which we can count on the laser to act as a well-behaved sinusoidal oscillator with a well-defined phase. If we try and correlate (by means to be discussed later) the laser field with itself using a time delay exceeding τ_c, the result approaches zero. One form of a field $\mathscr{E}(t)$ that will display this behavior is shown in Figure 10.14. The field undergoes a phase memory loss on the average every τ_c seconds. It is intuitively clear that performing the autocorrelation operation as defined by the first equality of (10.8-1) will yield a result whose rough features agree with the form $(\cos \omega_0 \tau)e^{-|\tau|/\tau_c}$.

Next, we will consider how the autocorrelation function $C_{\mathscr{E}}(\tau)$ is obtained in practice. The configuration used most often is the Michelson interferometer illustrated in Figure 10.15. An

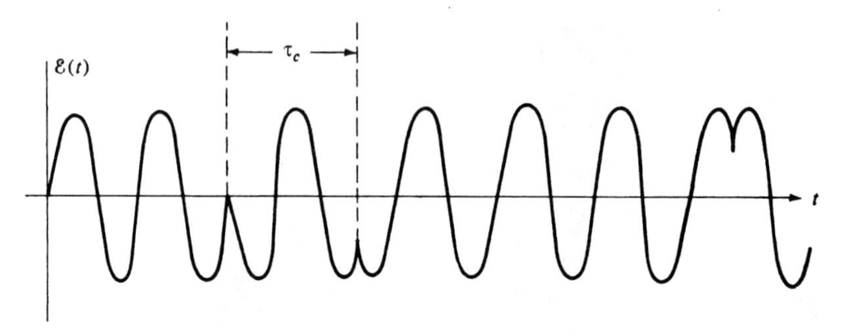

Figure 10.14 A sinusoidal field whose phase coherence is interrupted on the average every τ_c seconds.

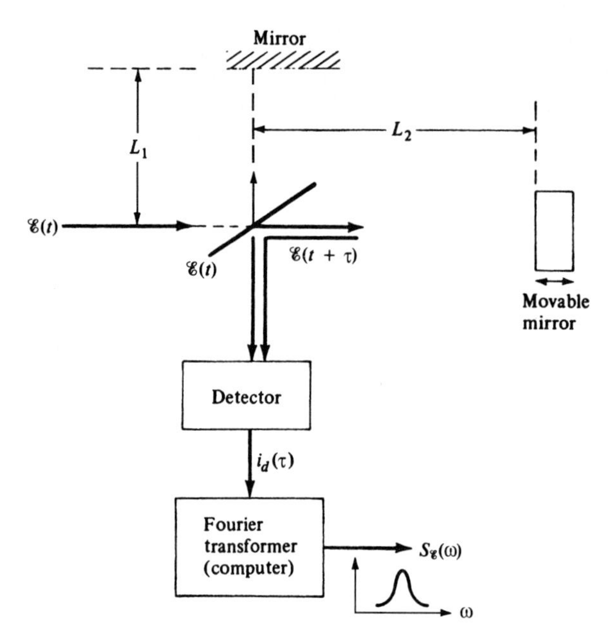

Figure 10.15 A Michelson interferometer "splits" an input beam into a two-component beam and then recombines them with a controlled time delay $\tau = 2(L_1 - L_2)/c$.

input field $\mathscr{E}_i(t)$ is split into two components. One of these fields is delayed relative to the second by a time delay

$$\tau = \frac{2(L_1 - L_2)}{c} \tag{10.8-2}$$

The two fields are then incident on a square-law detector whose current constitutes the useful output of the experiment.

Assuming equal division of power, the total optical field at the detector plane is

$$\mathscr{E}_d(t) = \mathscr{E}(t) + \mathscr{E}(t + \tau) \tag{10.8-3}$$

According to the discussion of Section 11.1, which the student is advised to preview at this point, the output current of the detector is

$$i_d = a\overline{\mathscr{E}_d^2(t)} \tag{10.8-4}$$

where a is some constant that is irrelevant in the present discussion, and the bar indicates, as it does throughout this book, time averaging. The duration of this averaging depends on the detector and its associated electrical circuitry and in the very fastest detectors may be as short as 10^{-11} s. It is thus *always* very long compared to the optical field period, which is $\sim 10^{-15}$ s.

The detector output is then

$$\begin{aligned} i_d &= a[\overline{\mathscr{E}^2(t)} + \overline{\mathscr{E}^2(t+\tau)} + \overline{2\mathscr{E}(t)\mathscr{E}(t+\tau)}] \\ &= 2a[\overline{\mathscr{E}^2} + \overline{\mathscr{E}(t)\mathscr{E}(t+\tau)}] \end{aligned} \tag{10.8-5}$$

since $\overline{\mathscr{E}^2(t)} = \overline{\mathscr{E}^2(t+\tau)} \equiv \overline{\mathscr{E}^2}$. The output current from the detector is thus made up of a dc component $2a\overline{\mathscr{E}^2}$ and a component $2a\overline{\mathscr{E}(t)\mathscr{E}(t+\tau)}$. The ratio of these two current components is, according to Equation (10.2-9), the (normalized) autocorrelation function of the optical field $\mathscr{E}(t)$.

$$\gamma(\tau) \equiv \frac{\tau_{\text{dependent part of } i_d}}{\tau_{\text{independent part}}} \propto C_{\mathscr{E}}(\tau) \tag{10.8-6}$$

The spectral density function $S_{\mathscr{E}}(\omega)$ is obtained, according to Equation (10.2-14), by a Fourier transformation

$$S_{\mathscr{E}}(\omega) = \frac{1}{\pi}\int_{-\infty}^{\infty} C_{\mathscr{E}}(\tau)e^{-i\omega\tau}\, d\tau \tag{10.8-7}$$

The above scheme for obtaining the spectrum (spectral density function) of optical fields is termed Fourier transform spectroscopy, and the configuration of Figure 10.15 is representative of commercial instruments designed for this purpose. These instruments are popular especially in the far infrared (say, $\lambda > 10\ \mu\text{m}$), since the relative inefficiency of detectors in this wavelength region can be compensated to some degree by a slow scanning rate (of τ) that allows for long integration times and better noise averaging.

A basic result of the Fourier integral transform relationships (10.2-13) and (10.2-14) between $C_{\mathscr{E}}(\tau)$ and $S_{\mathscr{E}}(\omega)$ is that in order to resolve $S_{\mathscr{E}}(\omega)$ to within, say, $\delta\omega$, that is, to discern structure in $S_{\mathscr{E}}(\omega)$ on the scale of $\delta\omega$, we need to employ time delays $\tau > \pi/\delta\omega$. If we were, as an example, to employ interference spectroscopy to measure the output spectrum of a commercial semiconductor laser with a linewidth of $(\Delta\omega)_{\text{laser}} = 2\pi \times 10^6$ Hz, we would need a delay time τ that could be varied from 0 to 5×10^{-7} s.

In the case of laser, the finite spectral width of the optical field is due predominantly to phase, rather than amplitude, fluctuations. In this case a rather simple technique that involves mixing (heterodyning) the laser field with a delayed version of itself is sufficient to obtain the laser spectrum. This method, which employs a fixed delay instead of the variable delay of the Fourier transform method, is described next.

Delayed Self-Heterodyning of Laser Fields

Consider the configuration of Figure 10.16. An optical field is split into two components that, after a relative path delay t_d, are recombined at a detector. The spectrum of the resulting photocurrent is displayed by a spectrum analyzer. This detection method is referred to as *delayed self-heterodyning* since it involves a "mixing" of the field with a delayed version of itself.

Since the main fluctuation of laser fields is that of the phase and not the amplitude (see comment following Equation (10.7-19)), we can approximate the field at the detector by the (complex) phasor

$$E_{\text{total}} = \tfrac{1}{2}E_0 e^{i\theta(t)} + \tfrac{1}{2}E_0 e^{i[\omega_0 t_d + \theta(t+t_d)]} \tag{10.8-8}$$

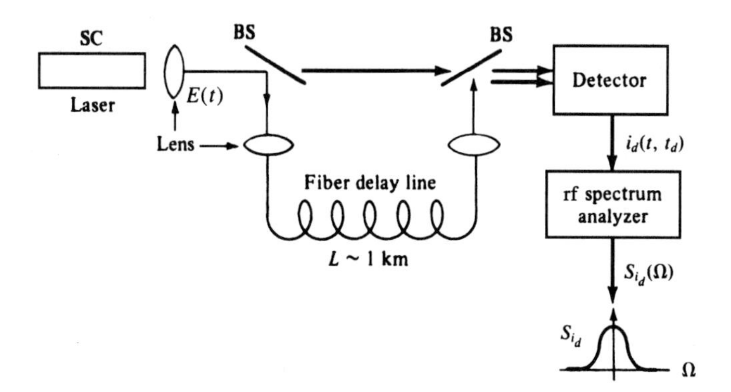

Figure 10.16 An interferometric arrangement employing a fiber delay for obtaining the spectrum $S_{\mathscr{E}}(\omega)$ of the laser field. (After Reference [16].)

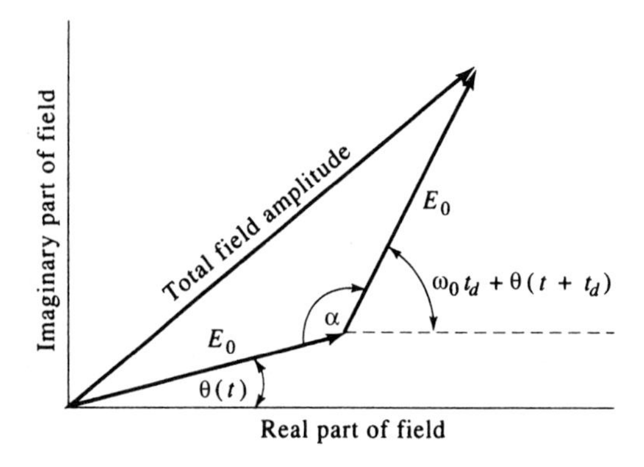

Figure 10.17 Construction showing the total optical field at the detector. For short delays, $t_d \ll \tau_c$, $\theta(t + t_d) \approx \theta(t)$ so that α and, consequently, the total field amplitude are constant.

This field is illustrated in Figure 10.17. For delays t_d that are considerably shorter than the phase coherence time τ_c of the laser field (defined by Equation (10.7-24)), $\theta(t + t_d) \simeq \theta(t)$ and the magnitude of the total field phasor is a constant as shown in Figure 10.17. Although the phase angle $\theta(t)$ varies randomly, the angle α that determines the magnitude of E_{total} depends only on the difference $\theta(t + t_d) - \theta(t)$ and, in the limit $t_d \ll \tau_c$, does not change with time. The output current from the detector is constant, and nothing can be learned from it about the laser field spectrum. It is clear that we need to consider the case of $t_d \gg \tau_c$. In what follows we will consider the general case of arbitrary t_d.

The output current i_d is proportional to the time average of the square of the total optical field incident on the detector. It is thus proportional (see Equation (1.3-16)) to the product of the complex amplitude of this field and its complex conjugate. Using Equation (10.8-8) leads to

$$i_d = SE_0^2\{e^{i\theta(t)} + e^{i[\omega_0 t_d + \theta(t+t_d)]}\} \times \{e^{-i\theta(t)} + e^{-i[\omega_0 t_d + \theta(t+t_d)]}\} \tag{10.8-9}$$

$$= SE_0^2\{2 + e^{i[\theta(t) - \omega_0 t_d - \theta(t+t_d)]} + e^{-i[\theta(t) - \omega_0 t_d - \theta(t+t_d)]}\} \tag{10.8-10}$$

where S is a constant depending on the detector. We will derive the spectrum of i_d by employing the Wiener–Khintchine theorem (Equation (10.2-14)) so that first we need to obtain the autocorrelation function of $C_{i_d}(\tau)$ of the current i_d. Define as in Equation (10.7-19)

$$\Delta\theta(t, \tau) \equiv \theta(t + \tau) - \theta(t) \tag{10.8-11}$$

We have reasoned in the last section (see discussion following Equation (10.7-21)) that $\Delta\theta(t, \tau)$ is a random Gaussian variable. It follows that the difference $\Delta\theta(t, \tau) - \Delta\theta(t + t_d, \tau)$ is also a

Gaussian variable so that, in a manner identical to that used to derive Equation (10.7-24), we obtain

$$\langle e^{i[\Delta\theta(t,\tau)-\Delta\theta(t+t_d,\tau)]}\rangle = e^{-\langle[\Delta\theta(t,\tau)-\Delta\theta(t+t_d,\tau)]^2\rangle/2} \tag{10.8-12}$$

Now

$$\langle[\Delta\theta(t,\tau)-\Delta\theta(t+t_d,\tau)]^2\rangle = 2\langle[\Delta\theta(\tau)]^2\rangle - 2\langle\Delta\theta(t,\tau)\Delta\theta(t+t_d,\tau)\rangle \tag{10.8-13}$$

where we used

$$\langle[\Delta\theta(t,\tau)]^2\rangle = \langle[\Delta\theta(t+t_d,\tau)]^2\rangle \equiv \langle[\Delta\theta(\tau)]^2\rangle$$

From the equation preceding (10.7-10) and putting $t = \tau$

$$\langle[\Delta\theta(\tau)]^2\rangle = \frac{\mu|\tau|}{2\bar{n}t_c} = \frac{2|\tau|}{\tau_c}, \quad \tau_c = 4\bar{n}t_c/\mu \tag{10.8-14}$$

Using Equations (10.8-13) and (10.8-14) in (10.8-12) and (10.8-10), we obtain

$$C_{i_d}(\tau) \equiv \langle i_d(t)i_d(t+\tau)\rangle = S^2E_0^4[4 + 2e^{-|\tau|/(\tau_c/2)}e^{\langle\Delta\theta(t,\tau)\Delta\theta(t+t_d,\tau)\rangle}] \tag{10.8-15}$$

Special Case $t_d \gg \tau_c$

In the special, but important, long delay case $t_d \gg \tau_c$, we have

$$\lim_{t_d\to\infty} \langle\Delta\theta(t,\tau)\Delta\theta(t+t_d,\tau)\rangle \to 0 \tag{10.8-16}$$

and

$$C_{i_d}(\tau)_{t_d\gg\tau_c} = S^2E_0^4(4 + e^{-|\tau|/(\tau_c/2)}) \tag{10.8-17}$$

Employing (10.7-14) or using directly the results of (10.7-28), we obtain the following expression for the spectral density of the current i_d:

$$S_{i_d}(\Omega)_{t_d\gg\tau_c} = \frac{2S^2E_0^4}{\pi}\left[\frac{(4/\tau_c)}{(2/\tau_c)^2 + \Omega^2} + 4\pi\delta(\Omega)\right] \tag{10.8-18}$$

The spectrum thus consists of a dc, $4\pi\delta(\Omega)$, term plus a Lorentzian distribution centered (if we count negative frequencies $\Omega < 0$) on $\Omega = 0$ with a full width at half maximum of

$$(\Delta\Omega)_{\text{FWHM}} = 4/\tau_c = 2(\Delta\omega)_{\text{laser}} \tag{10.8-19}$$

The last equality, derived from Equation (10.7-29), states that the width of the spectrum of the photodetected current in the limit $t_d \gg \tau_c$ is twice that of the laser field.

The rigorous treatment of the general case involving arbitrary values of the delay t_d is beyond the scope of this book, since it requires a knowledge of the function $\langle\Delta\theta(t,\tau)\Delta\theta(t+t_d,\tau)\rangle$. The derivation of this function involves the solution of the nonlinear, noise-driven laser equation. The result is (see Reference [15])

$$\langle\Delta\theta(t,\tau)\Delta\theta(t+t_d,\tau)\rangle = \frac{2|\tau|}{\tau_c} - \frac{2}{\tau_c}\min(|\tau|, t_d) \tag{10.8-20}$$

where $\min(\tau, t_d)$ signifies the smallest of τ and t_d. The last result together with Equations (10.8-13) and (10.8-20) when substituted in (10.8-15) give

$$C_{i_d}(\tau) \equiv \langle i_d(t)i_d(t+\tau)\rangle = S^2E_0^4[4 + 2e^{-(2/\tau_c)\min(|\tau|,t_d)}] \tag{10.8-21}$$

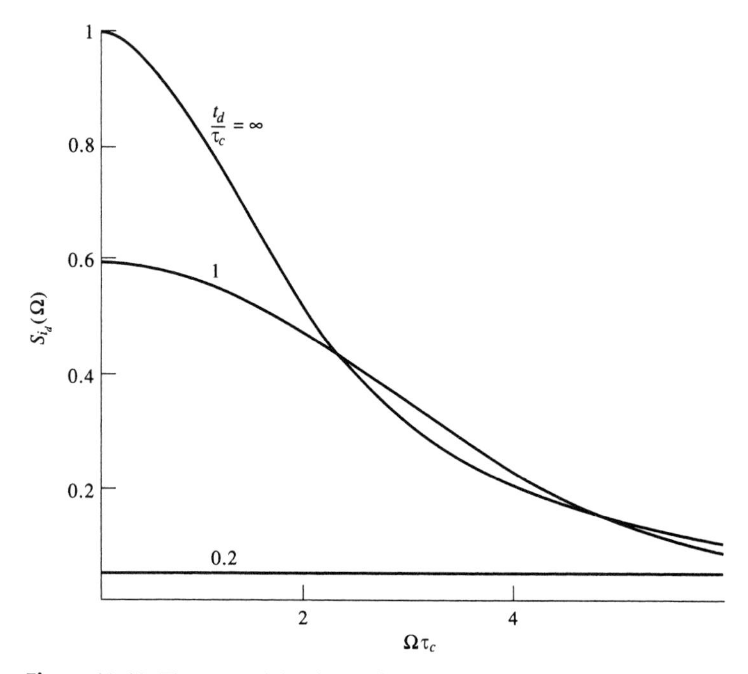

Figure 10.18 The spectral density $S_{i_d}(\Omega)$, as given by Equation (10.8-22), of the photocurrent in a delayed self-heterodyne detection of the output of a laser. The ratio of the delay time (t_d) to the laser field coherence time (τ_c) is a parameter. The frequency abscissa is in units of τ_c^{-1}, $\tau_c = (\Delta\nu)^{-1}_{\text{laser}}$.

$$\begin{aligned} S_{i_d}(\Omega) &= \frac{1}{\pi}\int_{-\infty}^{\infty} C_{i_d}(\tau)e^{-i\Omega\tau}\,d\tau \\ &= 8S^2E_0^4(1 + 0.5e^{-2t_d/\tau_c})\delta(\Omega) \\ &\quad + \left(\frac{8S^2E_0^4}{\pi\tau_c}\right)\frac{\left[1 - e^{-2t_d/\tau_c}\left(\cos\Omega t_d + \dfrac{2\sin\Omega t_d}{\Omega\tau_c}\right)\right]}{(2/\tau_c)^2 + \Omega^2} \end{aligned} \tag{10.8-22}$$

The integration leading to Equation (10.8-22) is long but straightforward. Equation (10.8-22) reduces, as it should, to (10.8-18) when $t_d/\tau_c \to \infty$. In summation, we recall that only in the case $t_d/\tau_c \gg 1$, that is, a long relative delay, is the spectrum $S_{i_d}(\Omega)$ a Lorentzian. A typical spectrum of a semiconductor laser obtained with a setup similar to that of Figure 10.16 is shown in Figure 10.13. A plot of the theoretical spectra of Equation (10.8-22) for the case $t_d/\tau_c = \infty$, 1, 0.2 is contained in Figure 10.18.

10.9 ERROR PROBABILITY IN A BINARY PULSE CODE MODULATION SYSTEM

The simplicity and reliability of digital processing by integrated electronic circuits has made it increasingly attractive to transmit information in the form of binary pulse trains. For optical communications systems, the analog data to be transmitted are coded into a train of 1 and 0 electrical pulses so that each pulse carries one bit of information. The electrical signal thus generated is impressed, say, by means of a modulator, on an optical beam, resulting in

A theoretical plot of BER as a function of the (peak) signal-to-noise ratio $i_S/\langle i_N \rangle$ is shown in Figure 10.20. We recall that $\overline{i_S^2}$ represents the electrical signal power at the detector output and not the optical power. It is interesting to note the extremely small error probabilities resulting from even moderate signal-to-noise power ratios. As an example, BER = 10^{-9} when $i_S/\langle i_N \rangle = 11.89$ (21.5 dB).

Experimental measurement of error probability in a detected optical pulse train is described by Reference [18]. Other pertinent discussions are to be found in References [19, 20].

A detailed example using the results of this section in designing a binary optical fiber communications system appears at the end of Chapter 11.

PROBLEMS

10.1 Derive Equations (10.4-2) and (10.4-3). [Hint: Apply the relation

$$\int_S \mathbf{D} \cdot \mathbf{n}\, ds = \int_v \rho\, dv$$

to a differential volume containing the charge sheet.]

10.2 Derive the shot-noise formula without making the restriction (Equation (10.4-6)) $\omega t_a \ll 1$. Assume the carriers move between the electrodes at a constant velocity.

10.3 Derive Equation (10.5-11).

10.4 Complete the missing steps in the derivation of Equation (10.5-20).

10.5 Estimate the scattering time τ_0 of carriers in copper at $T = 300$ K using a tabulated value for its conductivity. At what frequencies is the condition $\omega\tau_0 \ll 1$ violated?

10.6 Repeat Problem 10.5 for a material with a carrier density of 10^{22} cm^{-3} and $\sigma = 10^{-5}$ (ohm-cm)$^{-1}$.

10.7 What is the change $\Delta\nu$ in the resonant frequency of a laser whose cavity length changes by Δl?

10.8

(a) Estimate the frequency smearing $\Delta\nu$ of a laser in which fused-quartz rods are used to determine the length of the optical cavity in an environment where the temperature stability is ±0.5 K. [Caution: Do not forget the dependence of n on T.]

(b) What temperature stability is needed to reduce $\Delta\nu$ to less than 10^3 Hz?

10.9 Derive expression (10.5-9), $\overline{i_N^2}(\nu) = 4kT\,\Delta\nu/R$, for the Johnson noise by considering a high-Q parallel RLC circuit that is shunted by a current source of means quare amplitude $\overline{i_N^2}(\nu)$. The magnitude of $\overline{i_N^2}(\nu)$ is to be chosen so that the resulting excitation of the circuit corresponds to a stored electromagnetic energy of kT. [Hint: Since the magnetic and electric energies are equal, then

$$kT = C\overline{v^2(t)} = C\int_0^\infty \frac{\overline{V_N^2(\nu)}}{\Delta\nu}\, d\nu$$

where $\overline{V_N^2(\nu)} = \overline{i_N^2(\nu)}\,|Z(\nu)|^2$. Also assume that $\overline{i_N^2(\nu)}/\Delta\nu$ is independent of frequency.]

10.10 Derive and plot the error probability as a function of $i_S/\langle i_N \rangle$ for

(a) $k = 0.75$

(b) $k = 0.25$.

10.11 Design an experimental system for measuring the root mean square deviation of the laser frequency $(\Delta\omega)_{\text{rms}} \equiv \langle(\omega(t) - \omega_0)^2\rangle^{1/2}$.

10.12

(a) Please write a short report on the fundamentals of laser gyroscopes and of the Sagnac interferometer rotation sensor. You may, for example, look up the *Journal of Quantum Electronics* index section for listing of articles on "gyroscopes."

(b) What is the minimum rotation rate detectable by each of the two types of interferometers when the laser field spectral purity is limited by the quantum effects discussed in Section 10.7?

REFERENCES

1. Smith, R. A., F. A. Jones, and R. P. Chasmar, *The Detection and Measurement of Infrared Radiation*. Oxford University Press, New York, 1968.
2. Yuen, H. P., and J. H. Shapiro, Optical communication with two-photon coherent states, Part III. *IEEE Trans. Inf. Theory* **IT-26**:78 (1980).
3. Kimble, H. J., and D. F. Walls, Squeezed states of electromagnetic fields. (special issue) *J. Opt. Soc. B* **4**:1353 (1987).
4. Slusher, R. E., L. W. Hollberg, B. Yurke, D. C. Mertz, and J. F. Valley, Observation of squeezed states generated by four-wave mixing in an optical cavity. *Phys. Rev. Lett.* **55**:2409 (1985).
5. The basic concepts of noise theory used in this chapter can be found, for example, in W. B. Davenport and W. L. Root, *An Introduction to the Theory of Random Signals and Noise*. McGraw-Hill, New York, 1958.
6. Bennett, W. R., Methods of solving noise problems. *Proc. IRE* **44**:609 (1956).
7. The classic reference to this topic is S. O. Rice, Mathematical analysis of random noise. *Bell Syst. Tech. J.* **23**:282 (1944); **24**:46 (1945).
8. Johnson, J. B., Thermal agitation of electricity in conductors. *Phys. Rev.* **32**:97 (1928).
9. Nyquist, H., Thermal agitation of electric charge in conductors. *Phys. Rev.* **32**:110 (1928).
10. Yariv, A., *Quantum Electronics*, 2nd ed. Wiley, New York, 1975.
11. Schawlow, A. L., and C. H. Townes, *Phys. Rev.* **112**:1940 (1958).
12. Courtesy of Kerry Vahala and Chris Harder of the California Institute of Technology.
13. Henry, C. H., Theory of the linewidth of semiconductor lasers. *IEEE J. Quantum Electron* **18**(2):259 (1982).
14. Fleming, M., and A. Mooradian, Fundamental line broadening of single mode (GaAl)As diode lasers. *Appl. Phys. Lett.* **38**:511 (1981).
15. Vahala, K., and A. Yariv, Semiclassical theory of noise in semiconductor lasers. *IEEE J. Quantum Electron* **19**:1096 (1983).
16. Yamamoto, Y., T. Mukai, and S. Saito, Quantum phase noise and linewidth of a semiconductor laser. *Electron. Lett.* **17**:327 (1981). Also: Okoshi, T., K. Kikuchi, and A. Nakayma, Novel method for high resolution measurement of laser output spectrum. *Electron. Lett.* **6**:630 (1980).
17. Bennett, W. R., and J. R. Davey, *Data Transmission*. McGraw-Hill, New York, 1965, p. 100.
18. Goell, J. E., A 274 Mb/s optical repeater experiment employing a GaAs laser. *Proc. IEEE* **61**:1504 (1973).
19. Personick, S. D., Receiver design for digital fiber optic communication systems. *Bell Syst. Tech. J.* **52**:843 (1973).
20. Miller, S. E., T. Li, and E. A. J. Marcatili, Toward optic fiber transmission systems—devices and system considerations. *Proc. IEEE* **61**:1726 (1973).

CHAPTER 11

DETECTION OF OPTICAL RADIATION

11.0 INTRODUCTION

The detection of optical radiation is often accomplished by converting the radiant energy into an electric signal whose intensity is measured by conventional techniques. Some of the physical mechanisms that may be involved in this conversion include:

1. The generation of mobile charge carriers in solid-state photoconductive detectors.
2. Changing through absorption the temperature of thermocouples, thus causing a change in the junction voltage.
3. The release by the photoelectric effect of free electrons from photoemissive surfaces.

In this chapter we consider in some detail the operation of four of the most important detectors:

1. The photomultiplier.
2. The photoconductive detector.
3. The photodiode.
4. The avalanche photodiode.

The limiting sensitivity of each is discussed and compared to the theoretical limit. We will find that by use of the heterodyne mode of detection the theoretical limit of sensitivity may be approached.

11.1 OPTICALLY INDUCED TRANSITION RATES

A common feature of all the optical detection schemes discussed in this chapter is that the electric signal is proportional to the rate at which electrons are excited by the optical field. This excitation involves a transition of the electron from some initial bound state, say, a, to a final state (or a group of states) b in which it is free to move and contribute to the current flow. For example, in an n-type photoconductive detector, state a corresponds to electrons in the filled valence band or localized donor impurity atoms, while state b corresponds to electrons in the conduction band. The two levels involved are shown schematically in Figure 11.1.

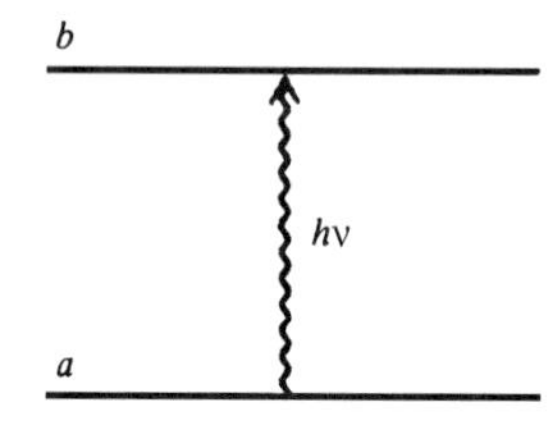

Figure 11.1 Most high-speed optical detectors depend on absorption of photons of energy $h\nu$ accompanied by a simultaneous transition of an electron (or hole) from a quantum state of low mobility (a) to one of higher mobility (b).

A photon of energy $h\nu$ is absorbed in the process of exciting an electron from a "bound" state a to a "free" state b in which the electron can contribute to the current flow.

An important point to understand before proceeding with the analysis of different detection schemes is the manner of relating the transition rate per electron from state a to b to the intensity of the optical field. This rate is derived by quantum mechanical considerations.[1] In our case, it can be stated in the following form: Given a nearly sinusoidal optical field[2]

$$e(t) = \tfrac{1}{2}[E(t)e^{i\omega_0 t} + E^*(t)e^{-i\omega_0 t}] \equiv \mathrm{Re}[V(t)] \tag{11.1-1}$$

where $V(t) = E(t)\exp(i\omega_0 t)$,[3] the transition rate per electron induced by this field is proportional to $V(t)V^*(t)$. Denoting the transition rate as $W_{a\to b}$, we have

$$W_{a\to b} \propto V(t)V^*(t) \tag{11.1-2}$$

We can easily show that $V(t)V^*(t)$ is equal to twice the average value of $e^2(t)$, where the averaging is performed over a few optical periods.

To illustrate the power of this seemingly simple result, consider the problem of determining the transition rate due to a field

$$e(t) = E_0 \cos(\omega_0 t + \phi_0) + E_1 \cos(\omega_1 t + \phi_1) \tag{11.1-3}$$

taking E_0 and E_1 real and $\omega_1 - \omega_0 \equiv \omega \ll \omega_0$. We can rewrite Equation (11.1-3) as

$$\begin{aligned} e(t) &= \mathrm{Re}\,(E_0\, e^{i(\omega_0 t+\phi_0)} + E_1\, e^{i(\omega_1 t+\phi_1)}) \\ &= \mathrm{Re}\,[(E_0\, e^{i\phi_0} + E_1\, e^{i(\omega t+\phi_1)})e^{i\omega_0 t}] \end{aligned} \tag{11.1-4}$$

and, using Equation (11.1-1), identify $V(t)$ as

$$V(t) = (E_0\, e^{i\phi_0} + E_1\, e^{i(\omega t+\phi_1)})e^{i\omega_0 t}$$

Thus, using Equation (11.1-2), we obtain

$$\begin{aligned} W_{a\to b} &\propto (E_0\, e^{i\phi_0} + E_1\, e^{i(\omega t+\phi_1)})(E_0\, e^{-i\phi_0} + E_1\, e^{-i(\omega t+\phi_1)}) \\ &= E_0^2 + E_1^2 + 2E_0E_1 \cos(\omega t + \phi_1 - \phi_0) \end{aligned} \tag{11.1-5}$$

This shows that the transition rate has, in addition to a constant term $E_0^2 + E_1^2$, a component oscillating at the difference frequency ω with a phase equal to the difference of the two original phases. This coherent "beating" effect forms the basis of the heterodyne detection scheme, which is discussed in detail in Section 11.4.

---

[1] More specifically, from first-order time-dependent perturbation theory; see, for example, Reference [1].

[2] By "nearly sinusoidal" we mean a field where $E(t)$ varies slowly compared to $\exp(i\omega_0 t)$ or, equivalently, where the Fourier spectrum of $E(t)$ occupies a bandwidth that is small compared to ω_0. Under these conditions the variation of the amplitude $E(t)$ during a few optical periods can be neglected.

[3] $V(t)$ is referred to as the "analytic signal" of $e(t)$. See Section 1.3.

Figure 11.2 Photocathode and focusing dynode configuration of a typical commercial photomultiplier. C = cathode; 1–8 = secondary-emission dynodes; A = collecting anode. (After Reference [3].)

11.2 PHOTOMULTIPLIER

The photomultiplier, one of the most common optical detectors, is used to measure radiation in the near ultraviolet, visible, and near infrared regions of the spectrum. Because of its inherent high current amplification and low noise, the photomultiplier is one of the most sensitive instruments devised and under optimal operation—which involves long integration time, cooling of the photocathode, and pulse-height discrimination—has been used to detect power levels as low as about 10^{-19} watt [2].

A schematic diagram of a conventional photomultiplier is shown in Figure 11.2. It consists of a photocathode (C) and a series of electrodes, called dynodes, that are labeled 1 through 8. The dynodes are kept at progressively higher potentials with respect to the cathode, with a typical potential difference between adjacent dynodes of 100 volts. The last electrode (A), the anode, is used to collect the electrons. The whole assembly is contained within a vacuum envelope in order to reduce the possibility of electronic collisions with gas molecules.

The photocathode is the most crucial part of the photomultiplier, since it converts the incident optical radiation to electronic current and thus determines the wavelength-response characteristics of the detector and, as will be seen, its limiting sensitivity. The photocathode consists of materials with low surface work functions. Compounds involving Ag–O–Cs and Sb–Cs are often used; see References [2, 3]. These compounds possess work functions as low as 1.5 eV, as compared to 4.5 eV in typical metals. As can be seen in Figure 11.3, this makes it possible to detect photons with longer wavelengths. It follows from the figure that the low-frequency detection limit corresponds to $h\nu = \phi$. At present, the lowest-work-function materials make possible photoemission at wavelengths as long as 1–1.1 μm.

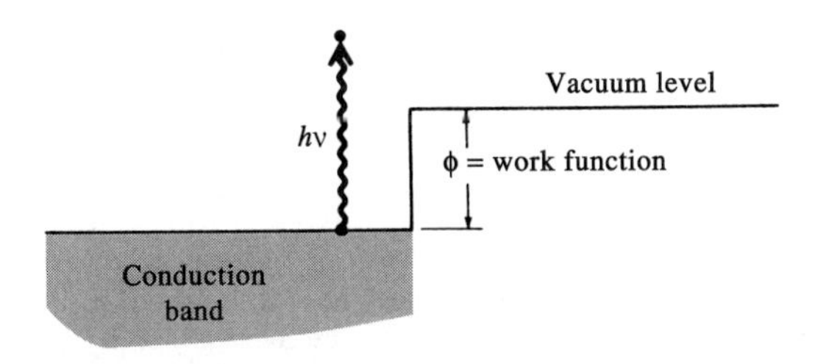

Figure 11.3 Photomultiplier photocathode. The vacuum level corresponds to the energy of an electron at rest at infinite distance from the cathode. The work function ϕ is the minimum energy required to lift an electron from the metal into the vacuum level, so only photons with $h\nu > \phi$ can be detected.

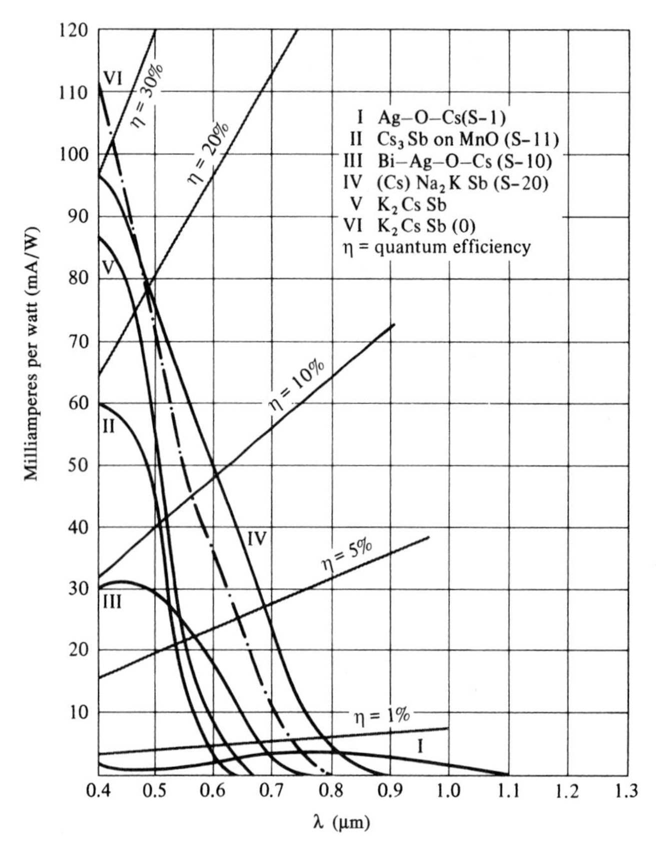

Figure 11.4 Photoresponse versus wavelength characteristics and quantum efficiency of a number of commercial photocathodes. (After Reference [3], p. 228.)

Spectral response curves of a number of commercial photocathodes are shown in Figure 11.4. The quantum efficiency (or quantum yield as it is often called) is defined as the number of electrons released per incident photon.

The electrons that are emitted from the photocathode are focused electrostatically and accelerated toward the first dynode, arriving with a kinetic energy of, typically, about 100 eV. Secondary emission from dynode surfaces causes a multiplication of the initial current. This process repeats itself at each dynode until the initial current emitted by the photocathode is amplified by a very large factor. If the average secondary emission multiplication at each dynode is δ (i.e., δ secondary electrons for each incident one) and the number of dynodes is N, the total current multiplication between the cathode and anode is

$$G = \delta^N$$

which, for typical values[4] of $\delta = 5$ and $N = 9$, gives $G \simeq 2 \times 10^6$.

[4] The value of δ depends on the voltage V between dynodes, and values of $\delta \simeq 10$ can be obtained (for $V \simeq 400$ volts). In commercial tubes, values of $\delta \simeq 5$, achievable with $V \simeq 100$ volts, are commonly used.

11.3 NOISE MECHANISMS IN PHOTOMULTIPLIERS

The random fluctuations observed in the photomultiplier output are due to the following:

1. Cathode shot noise, given according to Equation (10.4-9) by

$$\overline{(i_{N_1}^2)} = G^2 2\, e(\bar{i}_c + i_d)\Delta\nu \qquad (11.3\text{-}1)$$

where $\bar{i}_c$ is the average current emitted by the photocathode due to the signal power that is incident on it. The current i_d is the so-called dark current, which is due to random thermal excitation of electrons from the surface as well as to excitation by cosmic rays and radioactive bombardment.

2. Dynode shot noise, which is the shot noise due to the random nature of the secondary emission process at the dynodes. Since current originating at a dynode does not exercise the full gain of the tube, the contribution of all the dynodes to the total shot noise output is smaller by a factor of $\sim\delta^{-1}$ than that of the cathode; since $\delta \simeq 5$, it amounts to a small correction and will be ignored in the following.

3. Johnson noise, which is the thermal noise associated with the output resistance R connected across the anode. Its magnitude is given by Equation (10.5-9) as

$$\overline{(i_{N_2}^2)} = \frac{4kT\ \Delta\nu}{R} \qquad (11.3\text{-}2)$$

Minimum Detectable Power in Photomultipliers—Video Detection

Photomultipliers are used primarily in one of two ways. In the first, the optical wave to be detected is modulated at some low frequency ω_m before impinging on the photocathode. The signal consists of an output current oscillating at ω_m, which, as will be shown later, has an amplitude proportional to the optical intensity. This mode of operation is known as *video*, or straight, detection.

In the second mode of operation, the signal to be detected, whose optical frequency is ω_s, is combined at the photocathode with a much stronger optical wave of frequency $\omega_s + \omega$. The output signal is then a current at the offset frequency ω. This scheme, known as *heterodyne* detection, will be considered in detail in Section 11.4.

The optical signal in the case of video detection may be taken as

$$\begin{aligned} e_s(t) &= E_s(1 + m\cos\omega_m t)\cos\omega_s t \\ &= \mathrm{Re}\,[E_s(1 + m\cos\omega_m t)e^{i\omega_s t}] \end{aligned} \qquad (11.3\text{-}3)$$

where the factor $(1 + m\cos\omega_m t)$ represents amplitude modulation of the carrier.[5] The photocathode current is given, according to Equation (11.1-2), by

$$\begin{aligned} i_c(t) &\propto [E_s(1 + m\cos\omega_m t)]^2 \\ &= E_s^2\left[\left(1 + \frac{m^2}{2}\right) + 2m\cos\omega_m t + \frac{m^2}{2}\cos 2\omega_m t\right] \end{aligned} \qquad (11.3\text{-}4)$$

[5] The amplitude modulation can be due to the information carried by the optical wave or, as an example, to chopping before detection.

To determine the proportionality constant involved in Equation (11.3-4), consider the case of $m = 0$. The average photocathode current due to the signal is then[6]

$$\bar{i}_c = \frac{Pe\eta}{h\nu_s} \tag{11.3-5}$$

where $\nu_s = \omega_s/2\pi$, P is the average optical power, and η (the quantum efficiency) is the average number of electrons emitted from the photocathode per incident photon. This number depends on the photon frequency, the photocathode surface, and in practice (see Figure 11.4) is found to approach 0.3. Using Equation (11.3-5), we rewrite Equation (11.3-4) as

$$i_c(t) = \frac{Pe\eta}{h\nu_s}\left[\left(1 + \frac{m^2}{2}\right) + 2m\cos\omega_m t + \frac{m^2}{2}\cos 2\omega_m t\right] \tag{11.3-6}$$

The signal output current at ω_m is

$$i_s = \frac{GPe\eta}{h\nu_s}(2m)\cos\omega_m t \tag{11.3-7}$$

If the output of the detector is limited by filtering to a bandwidth $\Delta\nu$ centered on ω_m, it contains a shot-noise current, which, according to Equation (11.3-1), has a mean squared amplitude

$$\overline{(i_{N_1}^2)} = 2G^2 e(\bar{i}_c + i_d)\Delta\nu \tag{11.3-8}$$

where $\bar{i}_c$ is the average signal current and i_d is the dark current.

The noise and signal equivalent circuit is shown in Figure 11.5, where for the sake of definiteness we took the modulation index $m = 1$. R represents the output load of the photomultiplier. T_e is chosen so that the term $4kT_e\,\Delta\nu/R$ accounts for the thermal noise of R as well as for the noise generated by the amplifier that follows the photomultiplier.

The signal-to-noise power ratio at the output is thus

$$\frac{S}{N} = \frac{\overline{i_s^2}}{\overline{(i_{N_1}^2)} + \overline{(i_{N_2}^2)}} = \frac{2(Pe\eta/h\nu_s)^2 G^2}{2G^2 e(\bar{i}_c + i_d)\Delta\nu + (4kT_e\Delta\nu/R)} \tag{11.3-9}$$

Due to the large current gain ($G \simeq 10^6$), the first term in the denominator of Equation (11.3-9), which represents amplified cathode shot noise, is much larger than the thermal and amplifier

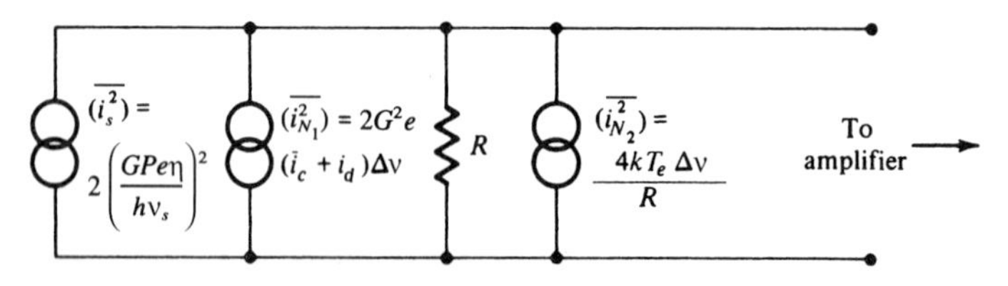

Figure 11.5 Equivalent circuit of a photomultiplier.

[6] $P/h\nu_s$ is the rate of photon incidence on the photocathode; thus, if it takes $1/\eta$ photons to generate one electron, the average current is given by Equation (11.3-5).

noise term $4kT_e\,\Delta\nu/R$. Neglecting the term $4kT_e\,\Delta\nu/R$, assuming $i_d \gg \bar{i}_c$, and setting $S/N = 1$, we can solve for the minimum detectable optical power as

$$P_{\min} = \frac{h\nu_s(i_d\,\Delta\nu)^{1/2}}{\eta e^{1/2}} \tag{11.3-10}$$

EXAMPLE: SENSITIVITY OF PHOTOMULTIPLIER

Consider a typical case of detecting an optical signal under the following conditions:

$\nu_s = 6 \times 10^{14}$ Hz ($\lambda = 0.5$ μm)

$\eta = 10\%$

$\Delta\nu = 1$ Hz

$i_d = 10^{-15}$ A (a typical value of the dark photocathode current)

Substitution in Equation (11.3-10) gives

$$P_{\min} = 3 \times 10^{-16}\ \text{W}$$

The corresponding cathode signal current is $\bar{i}_c \sim 2.4 \times 10^{-17}$ A, so the assumption $i_d \gg \bar{i}_c$ is justified.

Signal-Limited Shot Noise

If one could, somehow, eliminate the Johnson noise and the dark current altogether, so that the only contribution to the average photocathode current is $\bar{i}_c$, which is due to the optical signal, then, using Equations (11.3-5) and (11.3-9) to solve self-consistently for $P_{\min}$,

$$P_{\min} \simeq \frac{h\nu_s\,\Delta\nu}{\eta} \tag{11.3-11}$$

This corresponds to the quantum limit of optical detection. Its significance will be discussed in the next section. The practical achievement of this limit in video detection is nearly impossible since it depends on near total suppression of the dark current and other extraneous noise sources such as background radiation reaching the photocathode and causing shot noise.

The quantum detection limit (11.3-11) can, however, be achieved in the heterodyne mode of optical detection. This is discussed in the next section.

11.4 HETERODYNE DETECTION WITH PHOTOMULTIPLIERS

In the heterodyne mode of optical detection, the signal to be detected $E_s \cos \omega_s t$ is combined with a second optical field, referred to as the local-oscillator field, $E_L \cos(\omega_s + \omega)t$, shifted in frequency by ω ($\omega \ll \omega_s$). The total field incident on the photocathode is therefore given by

$$e(t) = \mathrm{Re}\,[E_L\, e^{i(\omega_s+\omega)t} + E_s\, e^{i\omega_s t}] \equiv \mathrm{Re}\,[V(t)] \tag{11.4-1}$$

The local-oscillator field originates usually at a laser at the receiving end, so that it can be made very large compared to the signal to be detected. In the following we will assume that

$$E_L \gg E_s \tag{11.4-2}$$

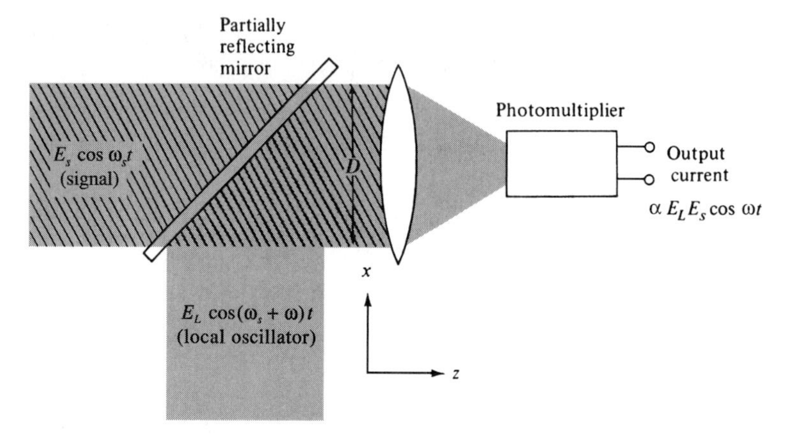

Figure 11.6 Schematic diagram of a heterodyne detector using a photomultiplier.

A schematic diagram of a heterodyne detection scheme is shown in Figure 11.6. The current emitted by the photocathode is given, according to Equations (11.1-2) and (11.4-1), by

$$i_c(t) \propto V(t)V^*(t) = E_L^2 + E_s^2 + 2E_L E_s \cos \omega t$$

which, using Equation (11.4-2) can be written

$$i_c(t) = aE_L^2\left(1 + \frac{2E_s}{E_L}\cos \omega t\right) = aE_L^2\left(1 + 2\sqrt{\frac{P_s}{P_L}}\cos \omega t\right) \tag{11.4-3}$$

where P_s and P_L are the signal and local-oscillator powers, respectively. The proportionality constant a in Equation (11.4-3) can be determined as in (11.3-6) by requiring that when $E_s = 0$ the direct current be related to the local-oscillator power P_L by $\bar{i}_c = P_L \eta e / h\nu_L$.[7] So taking $\nu \approx \nu_L \approx \nu_s$ and assuming $P_s \ll P_L$,

$$i_c(t) = \frac{P_L e\eta}{h\nu}\left(1 + 2\sqrt{\frac{P_s}{P_L}}\cos \omega t\right) \tag{11.4-4}$$

The total cathode shot noise is thus

$$\overline{(i_{N_1}^2)} = 2e\left(i_d + \frac{P_L e\eta}{h\nu}\right)\Delta\nu \tag{11.4-5}$$

where i_d is the average dark current while $P_L e\eta/h\nu$ is the dc cathode current due to the strong local-oscillator field. The shot-noise current is amplified by G, resulting in an output noise

$$\overline{(i_N^2)}_{\text{anode}} = G^2 2e\left(i_d + \frac{P_L e\eta}{h\nu}\right)\Delta\nu \tag{11.4-6}$$

The mean square signal current at the output is, according to Equation (11.4-4),

$$\overline{(i_s^2)}_{\text{anode}} = 2G^2\left(\frac{P_s}{P_L}\right)\left(\frac{P_L e\eta}{h\nu}\right)^2 \tag{11.4-7}$$

[7] This is just a statement of the fact that each incident photon has a probability η of releasing an electron.

The signal-to-noise power ratio at the output is given by

$$\frac{S}{N} = \frac{2G^2(P_sP_L)(e\eta/h\nu)^2}{[G^2 2e(i_d + P_L e\eta/h\nu) + 4kT_e/R]\Delta\nu} \tag{11.4-8}$$

where, as in Equation (11.3-9), the last term in the denominator represents the Johnson (thermal) noise generated in the output load, plus the effective input noise of the amplifier following the photomultiplier. The big advantage of the heterodyne detection scheme is now apparent. By increasing P_L, the S/N ratio increases until the denominator is dominated by the term $G^2 2eP_L e\eta/h\nu$. This corresponds to the point at which the *shot noise produced by the local-oscillator current dwarfs all the other noise contributions.* When this state of affairs prevails, we have, according to Equation (11.4-8),

$$\frac{S}{N} \simeq \frac{P_s}{h\nu\,\Delta\nu/\eta} \tag{11.4-9}$$

which corresponds to the quantum-limited detection limit. The minimum detectable signal—that is, the signal input power leading to an output signal-to-noise ratio of 1—is thus

$$(P_s)_{\min} = \frac{h\nu\,\Delta\nu}{\eta} \tag{11.4-10}$$

This power corresponds for $\eta = 1$ to a flux at a rate of one photon per $(\Delta\nu)^{-1}$ seconds—that is, one photon per resolution time of the system.[8]

EXAMPLE: MINIMUM DETECTABLE POWER WITH A HETERODYNE SYSTEM

It is interesting to compare the minimum detectable power for the heterodyne system as given by Equation (11.4-10) with that calculated in the example of Section 11.3 for the video system. Using the same data,

$$\nu = 6 \times 10^{14}\ \text{Hz}\ (\lambda = 0.5\ \mu\text{m})$$

$$\eta = 10\%$$

$$\Delta\nu = 1\ \text{Hz}$$

we obtain

$$(P_s)_{\min} \simeq 4 \times 10^{-18}\ \text{W}$$

to be compared with $P_{\min} \simeq 3 \times 10^{-16}$ W in the video case.

Limiting Sensitivity as a Result of the Particle Nature of Light

The quantum limit to optical detection sensitivity is given by Equation (11.4-10) as

$$(P_s)_{\min} = \frac{h\nu\,\Delta\nu}{\eta} \tag{11.4-11}$$

[8] A detection system that is limited in bandwidth to $\Delta\nu$ cannot resolve events in time that are separated by less than $\sim(\Delta\nu)^{-1}$ second. Thus $(\Delta\nu)^{-1}$ is the resolution time of the system.

This limit was shown to be due to the shot noise of the photoemitted current. We may alternatively attribute this noise to the granularity—that is, the particle nature—of light, according to which the minimum energy increment of an electromagnetic wave at frequency ν is $h\nu$. The average power P of an optical wave can be written

$$P = \bar{N}h\nu \tag{11.4-12}$$

where $\bar{N}$ is the average number of photons arriving at the photocathode per second. Next, assume a hypothetical noiseless photomultiplier in which *exactly* one electron is produced for each η^{-1} incident photons. The measurement of P is performed by counting the number of electrons produced during an observation period T and then averaging the result over a large number of similar observations.

The average number of electrons emitted per observation period T is

$$\bar{N}_e = \bar{N}T\eta \tag{11.4-13}$$

If the photons arrive in a perfectly random manner, then the number of photons arriving during the fixed observation period obeys Poissonian statistics.[9] Since in our ideal example, the electrons that are emitted mimic the arriving photons, they obey the same statistical distribution law. This leads to a fluctuation

$$\overline{(\Delta N_e)^2} \equiv \overline{(N_e - \bar{N}_e)^2} = \bar{N}_e = \bar{N}T\eta$$

Defining the minimum detectable number of quanta as that for which the rms fluctuation in the number of emitted photoelectrons equals the average value, we get

$$(\bar{N}_{\min}T\eta)^{1/2} = \bar{N}_{\min}T\eta$$

or

$$(\bar{N})_{\min} = \frac{1}{T\eta} \tag{11.4-14}$$

If we convert the last result to power by multiplying it by $h\nu$ and recall that $T^{-1} \simeq \Delta\nu$, where $\Delta\nu$ is the bandwidth of the system, we get

$$(P_s)_{\min} = \frac{h\nu\,\Delta\nu}{\eta} \tag{11.4-15}$$

in agreement with Equation (11.4-10).

---

[9] This follows from the assumption that the photon arrival is perfectly random, so the probability of having N photons arriving in a given time interval is given by the Poisson law

$$p(N) = \frac{(\bar{N})^N e^{-\bar{N}}}{N!}$$

The mean square fluctuation is given by

$$\overline{(\Delta N)^2} = \sum_{N=0}^{\infty} p(N)(N - \bar{N})^2 = \bar{N}$$

where

$$\bar{N} = \sum_{0}^{\infty} Np(N)$$

is the average N.

The above discussion points to the fact that the noise (fluctuation) in the photocurrent can be blamed on the physical process that introduces the randomness. In the case of Poissonian photon arrival statistics (as is the case with ordinary lasers) and perfect photon emission ($\eta = 1$), the fluctuations are due to the photons. The opposite, hypothetical, case of no photon fluctuations but random photoemission ($\eta < 1$) corresponds to pure shot noise. The electrical measurement of noise power will yield the same result in either case and one cannot distinguish between them.

11.5 PHOTOCONDUCTIVE DETECTORS

The operation of photoconductive detectors is illustrated in Figure 11.7. A semiconductor crystal is connected in series with a resistance R and a supply voltage V. The optical field to be detected is incident on and absorbed in the crystal, thereby exciting electrons into the conduction band (or, in p-type semiconductors, holes into the valence band). Such excitation results in a lowering of the resistance R_d of the semiconductor crystal and hence in an increase in the voltage drop across R, which, for $\Delta R_d/R_d \ll 1$, is proportional to the incident optical intensity.

To be specific, we show the energy levels involved in one of the more popular semiconductive detectors—mercury-doped germanium [4]. Mercury atoms enter germanium as acceptors with an ionization energy of 0.09 eV. It follows that it takes a photon energy of at least 0.09 eV (i.e., a photon with a wavelength shorter than 14 μm) to lift an electron from the top of the valence band and have it trapped by the Hg (acceptor) atom. Usually the germanium crystal contains a smaller density N_D of donor atoms, which at low temperatures find it energetically profitable to lose their valence electrons to one of the far more numerous Hg acceptor atoms, thereby becoming positively ionized and ionizing (negatively) an equal number of acceptors.

Since the acceptor density $N_A \gg N_D$, most of the acceptor atoms remain neutrally charged.

An incident photon is absorbed and lifts an electron from the valence band onto an acceptor atom, as shown in process A in Figure 11.8. The electronic deficiency (i.e., the hole) thus created is acted upon by the electric field, and its drift along the field direction gives rise to the signal current. The contribution of a given hole to the current ends when an electron drops from an ionized acceptor level back into the valence band, thus eliminating the hole as in B. This process is referred to as electron–hole recombination or trapping of a hole by an ionized acceptor atom.

By choosing impurities with lower ionization energies, even lower-energy photons can be detected, and, indeed, photoconductive detectors commonly operate at wavelengths up to $\lambda = 50$ μm. Cu, as an example, enters into Ge as an acceptor with an ionization energy of

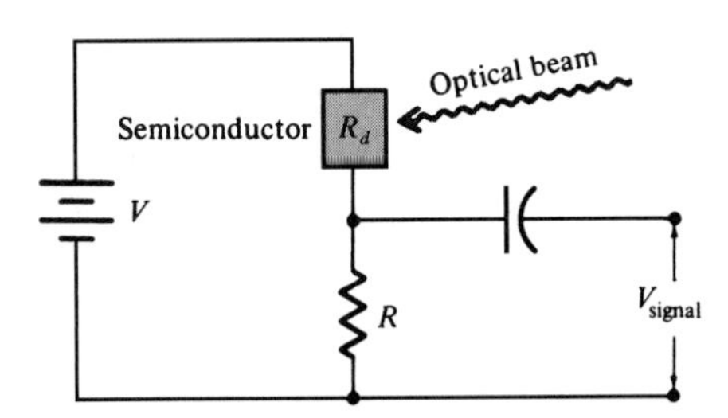

Figure 11.7 Typical biasing circuit of a photoconductive detector.

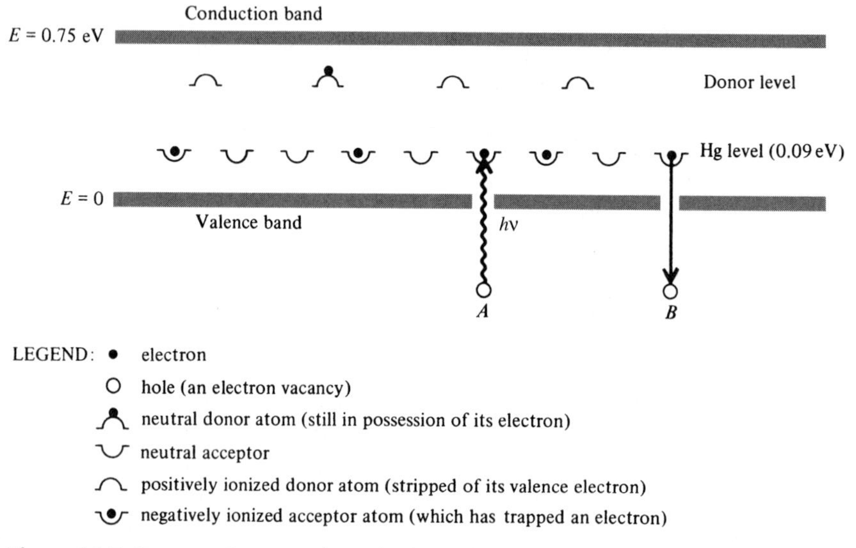

Figure 11.8 Donor and acceptor impurity levels involved in photoconductive semiconductors.

0.04 eV, which would correspond to long-wavelength detection cutoff of $\lambda \simeq 32\ \mu\text{m}$. The response of a number of commercial photoconductive detectors is shown in Figure 11.9.

It is clear from this discussion that the main advantage of photoconductors compared to photomultipliers is their ability to detect long-wavelength radiation, since the creation of mobile carriers does not involve overcoming the large surface potential barrier. On the debit side we find the lack of current multiplication and the need to cool the semiconductor so that photoexcitation of carriers will not be masked by thermal excitation.

Consider an optical beam, of power P and frequency ν, that is incident on a photoconductive detector. Taking the probability for excitation of a carrier by an incident photon—the so-called quantum efficiency—as η, the carrier generation rate is $G = P\eta/h\nu$. If the carriers last on the average τ_0 seconds before recombining, the average number of carriers N_c is found by equating the generation rate to the recombination rate (N_c/τ_0), so

$$N_c = G\tau_0 = \frac{P\eta\tau_0}{h\nu} \tag{11.5-1}$$

Each one of these carriers drifts under the electric field influence[10] at a velocity $\bar{v}$ giving rise, according to Equation (10.4-1), to a current in the external circuit of $i_e = e\bar{v}/d$, where d is the length (between electrodes) of the semiconductor crystal. The total current is thus the product of i_e and the number of carriers present, or, using Equation (11.5-1),

$$\bar{i} = N_c i_e = \frac{P\eta\tau_0 e\bar{v}}{h\nu d} = \frac{e\eta}{h\nu}\left(\frac{\tau_0}{\tau_d}\right)P \tag{11.5-2}$$

where $\tau_d = d/\bar{v}$ is the drift time for a carrier across the length d. The factor (τ_0/τ_d) is thus the fraction of the crystal length drifted by the average excited carrier before recombining.

---

[10] The drift velocity is equal to μE, where μ is the mobility and E is the electric field.

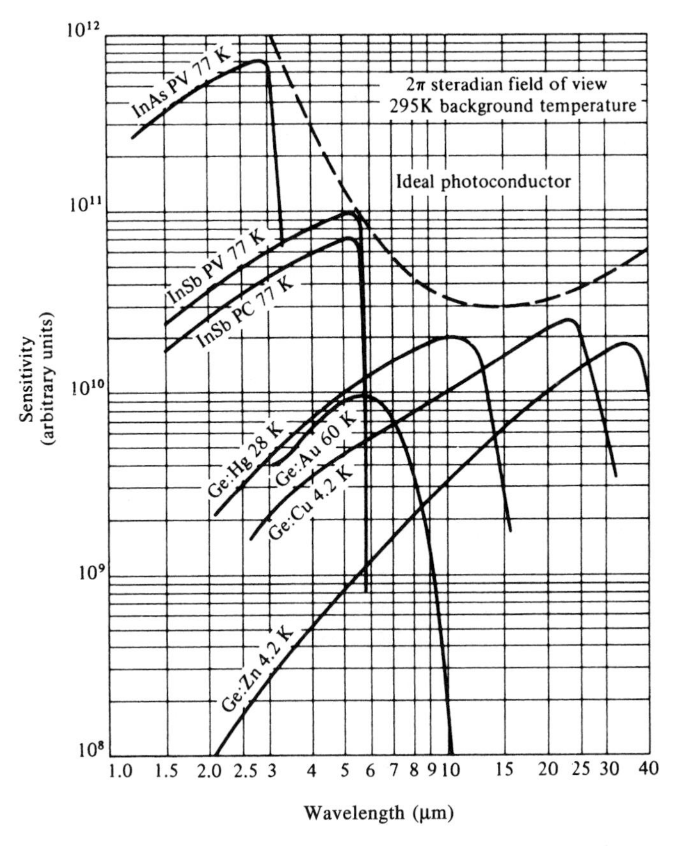

Figure 11.9 Relative sensitivity of a number of commercial photoconductors. (Courtesy Santa Barbara Research Corp.)

Equation (11.5-2) describes the response of a photoconductive detector to a constant optical flux. Our main interest, however, is in the heterodyne mode of photoconductive detection, which, as has been shown in Section 11.4, allows detection sensitivities approaching the quantum limit. In order to determine the limiting sensitivity of photoconductive detectors, we need first to understand the noise contribution in these devices.

Generation–Recombination Noise in Photoconductive Detectors

The principal noise mechanism in cooled photoconductive detectors reflects the randomness inherent in current flow. Even if the incident optical flux were constant in time, the generation of individual carriers by the flux would constitute a random process. This is exactly the type of randomness involved in photoemission, and we may export, likewise, that the resulting noise will be shot noise. This is almost true except for the fact that in a photoconductive detector a photoexcited carrier lasts τ seconds[11] (its recombination lifetime) before being captured by an ionized impurity. The contribution of the carrier to the charge flow in the external circuit is thus $e(\tau/\tau_d)$, as is evident from inspection of Equation (11.5-2). Since the

[11] The parameter τ_0 appearing in Equation (11.5-2) is the value of τ averaged over a large number of carriers.

lifetime τ is not a constant, but must be described statistically, another element of randomness is introduced into the current flow.

Consider a carrier excited by a photon absorption and lasting τ seconds. Its contribution to the external current is, according to Equation (10.4-1),

$$i_e(t) = \begin{cases} \dfrac{e\bar{v}}{d}, & 0 \le t \le \tau \\ 0, & \text{otherwise} \end{cases} \tag{11.5-3}$$

which has a Fourier transform

$$I_e(\omega, \tau) = \frac{e\bar{v}}{2\pi d}\int_0^\tau e^{-i\omega t}\,dt = \frac{-ie\bar{v}}{2\pi\omega d}(1 - e^{-i\omega\tau}) \tag{11.5-4}$$

so that

$$|I_e(\omega, \tau)|^2 = \frac{e^2\bar{v}^2}{4\pi^2\omega^2 d^2}(2 - e^{-i\omega\tau} - e^{i\omega\tau}) \tag{11.5-5}$$

According to Equation (10.3-10) we need to average $|I_e(\omega, \tau)|^2$ over τ. This is done in a manner similar to the procedure used in Section 10.5. Taking the probability function[12] $g(\tau) = \tau_0^{-1} \exp(-\tau/\tau_0)$, we average Equation (11.5-5) over all the possible values of τ according to

$$\overline{|I_e(\omega)|^2} = \int_0^\infty |I_e(\omega, \tau)|^2 g(\tau)\,d\tau = \frac{2e^2\bar{v}^2\tau_0^2}{4\pi^2 d^2(1 + \omega^2\tau_0^2)} \tag{11.5-6}$$

The spectral density function of the current fluctuations is obtained using Carson's theorem (10.3-10) as

$$S(\nu) = 2\bar{N}\,\frac{2e^2(\tau_0^2/\tau_d^2)}{1 + \omega^2\tau_0^2} \tag{11.5-7}$$

where we used $\tau_d = d/\bar{v}$ and where $\bar{N}$, the average number of carriers generated per second, can be expressed in terms of the average current $\bar{I}$ by use of the relation[13]

$$\bar{I} = \bar{N}\,\frac{\tau_0}{\tau_d}\,e \tag{11.5-8}$$

leading to

$$S(\nu) = \frac{4e\bar{I}(\tau_0/\tau_d)}{1 + 4\pi^2\nu^2\tau_0^2}$$

Therefore the mean square current representing the noise power in a frequency interval ν to $\nu + \Delta\nu$ is

$$\overline{i_N^2} \equiv S(\nu)\,\Delta\nu = \frac{4e\bar{I}(\tau_0/\tau_d)\,\Delta\nu}{1 + 4\pi^2\nu^2\tau_0^2} \tag{11.5-9}$$

which is the basic result for generation–recombination noise.

---

[12] $g(\tau)\,d\tau$ is the probability that a carrier lasts between τ and $\tau + d\tau$ seconds before recombining.

[13] This relation follows from the fact that the average charge per carrier flowing through the external circuit is $e(\tau_0/\tau_d)$, which, when multiplied by the generation rate $\bar{N}$, gives the current.

EXAMPLE: GENERATION–RECOMBINATION NOISE IN HG-DOPED GERMANIUM PHOTOCONDUCTIVE DETECTOR

To better appreciate the kind of numbers involved in the expression for $\overline{i_N^2}$, we may consider a typical mercury-doped germanium detector operating at 20 K with the following characteristics:

$$d = 10^{-1} \text{ cm}$$

$$\tau_0 = 10^{-9} \text{ s}$$

$$V \text{ (across the length } d) = 10 \text{ volts} \Rightarrow E = 10^2 \text{ V/cm}$$

$$\mu = 3 \times 10^4 \text{ cm}^2/\text{V-s}$$

The drift velocity is $\bar{v} = \mu E = 3 \times 10^6$ cm/s and $\tau_d = d/\bar{v} \simeq 3.3 \times 10^{-8}$ s, and therefore $\tau_0/\tau_d = 3 \times 10^{-2}$. Thus, on average, a carrier traverses only 3% of the length ($d = 1$ mm) of the sample before recombining. Comparing Equation (11.5-9) to the shot-noise result (10.4-9), we find that for a given average current $\bar{I}$ the generation–recombination noise is reduced from the shot-noise value by a factor

$$\frac{\overline{(i_N^2)}_{\text{generation–recombination}}}{\overline{(i_N^2)}_{\text{shot noise}}} \underset{\omega\tau_0 \ll 1}{=} 2\left(\frac{\tau_0}{\tau_d}\right) \tag{11.5-10}$$

which, in the foregoing example, has a value of about 1/15. Unfortunately, as will be shown subsequently, the reduced noise is accompanied by a reduction by a factor of τ_0/τ_d in the magnitude of the signal power, which wipes out the advantage of the lower noise.

Heterodyne Detection in Photoconductors

The situation here is similar to that described by Figure 11.6 in connection with heterodyne detection using photomultipliers. The signal field

$$e_s(t) = E_s \cos \omega_s t$$

is combined with a strong local-oscillator field

$$e_L(t) = E_L \cos(\omega + \omega_s)t, \quad E_L \gg E_s$$

so the total field incident on the photoconductor is

$$e(t) = \text{Re}(E_s e^{i\omega_s t} + E_L e^{i(\omega_s+\omega)t}) \equiv \text{Re}[V(t)] \tag{11.5-11}$$

The rate at which carriers are generated is taken, following Equation (11.1-2), as $aV(t)V^*(t)$, where a is a constant to be determined. The equation describing the number of excited carriers N_c is thus

$$\frac{dN_c}{dt} = aVV^* - \frac{N_c}{\tau_0} \tag{11.5-12}$$

where τ_0 is the average carrier lifetime, so N_c/τ_0 corresponds to the carrier's decay rate. We assume a solution for $N_c(t)$ that consists of the sum of dc and a simusoidal component in the form of

$$N_c(t) = N_0 + (N_1 e^{i\omega t} + \text{c.c.}) \tag{11.5-13}$$

where c.c. stands for complex conjugate.

Substitution in Equation (11.5-12) gives

$$N_c(t) = a\tau_0(E_s^2 + E_L^2) + a\tau_0\left(\frac{E_sE_Le^{i\omega t}}{1 + i\omega\tau_0} + \text{c.c.}\right) \tag{11.5-14}$$

where we took E_s and E_L as real. The current through the sample is given by the number of carriers per unit length N_c/d times $e\bar{v}$, where $\bar{v}$ is the drift velocity

$$i(t) = \frac{N_c(t)e\bar{v}}{d} \tag{11.5-15}$$

which, using Equation (11.5-14), gives

$$i(t) = \frac{e\bar{v}a\tau_0}{d}\left(E_s^2 + E_L^2 + \frac{2E_sE_L\cos(\omega t - \phi)}{\sqrt{1 + \omega^2\tau_0^2}}\right) \tag{11.5-16}$$

where $\phi = \tan^{-1}(\omega\tau_0)$.

The current is thus seen to contain a signal component that oscillates at ω and is proportional to E_s. The constant a in Equation (11.5-16) can be determined by requiring that, when $P_s = 0$, the expression for the direct current predicted by (11.5-16) agree with (11.5-2). This condition is satisfied if we rewrite Equation (11.5-16) as

$$i(t) = \frac{e\eta}{h\nu}\left(\frac{\tau_0}{\tau_d}\right)\left(P_s + P_L + \frac{2\sqrt{P_sP_L}}{\sqrt{1 + \omega^2\tau_0^2}}\cos(\omega t - \phi)\right) \tag{11.5-17}$$

where P_s and P_L refer, respectively, to the incident-signal and local-oscillator powers, $\nu = \nu_s = \omega_s/2\pi$, and η, the quantum efficiency, is the number of carriers excited per incident photon. The signal current is thus

$$i_s(t) = \frac{2e\eta}{h\nu}\left(\frac{\tau_0}{\tau_d}\right)\frac{\sqrt{P_sP_L}}{\sqrt{1 + \omega^2\tau_0^2}}\cos(\omega t - \phi) \tag{11.5-18}$$

while the dc (average) current is

$$\bar{I} = \frac{e\eta}{h\nu}\left(\frac{\tau_0}{\tau_d}\right)(P_s + P_L) \tag{11.5-19}$$

Since the average current $\bar{I}$ appearing in expression (11.5-9) for the generation–recombination noise is given in this case by

$$\bar{I} = \left(\frac{e\eta}{h\nu}\right)\left(\frac{\tau_0}{\tau_d}\right)P_L, \quad P_L \gg P_s$$

we can, by increasing P_L, increase the noise power $\overline{i_N^2}$ and at the same time, according to Equation (11.5-18), the signal $\overline{i_s^2}$ until the generation–recombination noise (11.5-9) is by far the largest contribution to the total output noise. When this condition is satisfied, the signal-to-noise ratio can be written, using Equations (11.5-9), (11.5-18), and (11.5-19) and taking $P_L \gg P_s$, as

$$\frac{S}{N} = \frac{\overline{i_s^2}}{\overline{i_N^2}} = \frac{2(e\eta\tau_0/h\nu\tau_d)^2P_sP_L/(1 + \omega^2\tau_0^2)}{4e^2\eta(\tau_0/\tau_d)^2P_L\,\Delta\nu/(1 + \omega^2\tau_0^2)h\nu} = \frac{P_s\eta}{2h\nu\,\Delta\nu} \tag{11.5-20}$$

The minimum detectable signal—that which leads to a signal-to-noise ratio of unity—is found by setting the left side of Equation (11.5-20) equal to unity and solving for P_s. It is

$$(P_s)_{\min} = \frac{2h\nu\,\Delta\nu}{\eta} \tag{11.5-21}$$

which, for the same η, is twice that of the photomultiplier heterodyne detection as given by Equation (11.4-10). In practice, however, η in photoconductive detectors can approach unity, whereas in the best photomultipliers $\eta \simeq 30\%$.

EXAMPLE: MINIMUM DETECTABLE POWER OF A HETERODYNE RECEIVER USING A PHOTOCONDUCTOR AT 10.6 μM

Assume the following:

$$\lambda = 10.6\ \mu\text{m}$$

$$\Delta\nu = 1\ \text{Hz}$$

$$\eta \simeq 1$$

Substitution in Equation (11.5-21) gives a minimum detectable power of

$$(P_s)_{\min} \simeq 10^{-19}\ \text{W}$$

Experiments [5, 6] have demonstrated that the theoretical signal-to-noise ratio as given by Equation (11.5-20) can be realized quite closely in practice; see Figure 11.10.

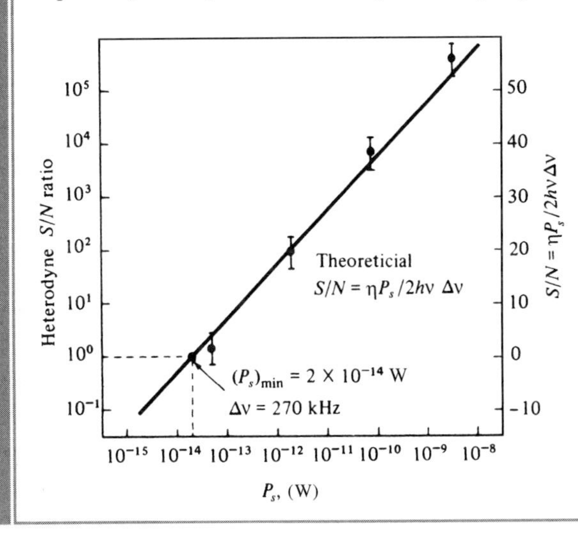

Figure 11.10 Signal-to-noise ratio of heterodyne signal to Ge:Cu detector at a heterodyne frequency of 70 MHz. Data points represent observed values. (After Reference [5].)

11.6 THE *p-n* JUNCTION

Before embarking on a description of the *p-n* diode detector, we need to understand the operation of the semiconductor *p-n* junction. Consider the junction illustrated in Figure 11.11. It consists of an abrupt transition from a donor-doped (i.e., *n*-type) region of a semiconductor, where the charge carriers are predominantly electrons, to an acceptor-doped (*p*-type) region, where the carriers are holes. The doping profile—that is, the density of excess donor (in the *n* region) atoms or acceptor atoms (in the *p* region)—is shown in Figure 11.11a. This abrupt transition results usually from diffusing suitable impurity atoms into a substrate of a

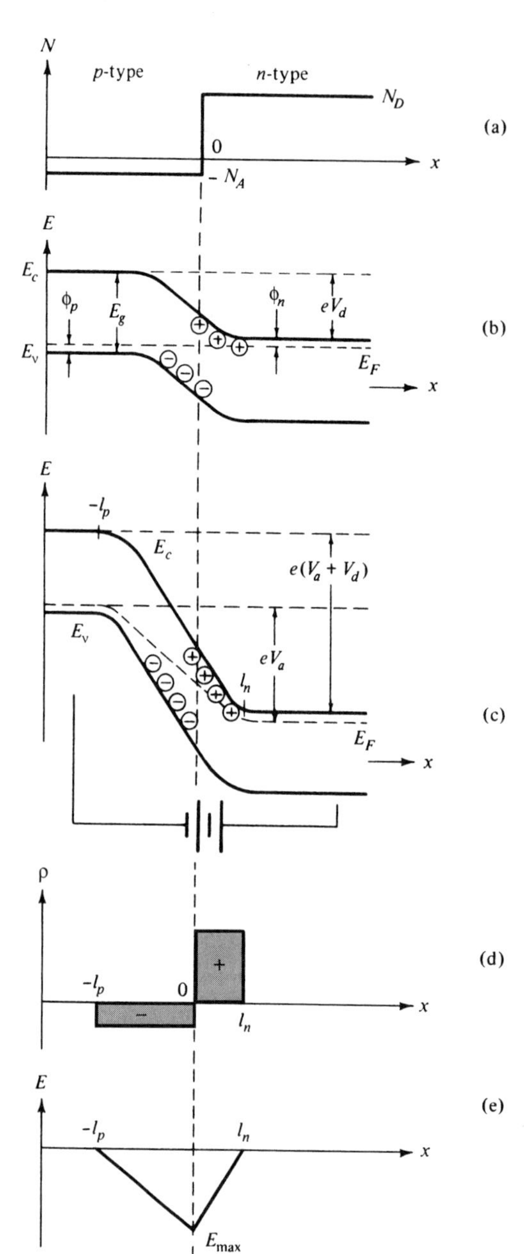

Figure 11.11 The abrupt p-n junction. (a) Impurity profile. (b) Energy-band diagram with zero applied bias. (c) Energy-band diagram with reverse applied bias. (d) Net charge density in the depletion layer. (e) The electric field. The circles in (b) and (c) represent ionized impurity atoms in the depletion layer.

semiconductor with the opposite type of conductivity. In our slightly idealized abrupt junction we assume that the n region ($x > 0$) has a constant (net) donor density N_D and the p region ($x < 0$) has a constant acceptor density N_A.

The energy-band diagram at zero applied bias is shown in Figure 11.11b. The top (or bottom) curve can be taken to represent the potential energy of an electron as a function of position x, so the minimum energy needed to take an electron from the n to the p side of the junction is eV_d. Taking the separations of the Fermi level from the respective bandedges as ϕ_n and ϕ_p as shown, we have

$$eV_d = E_g - (\phi_n + \phi_p)$$

V_d is referred to as the "built-in" junction potential.

Figure 11.11c shows the energy-band diagram in the junction with an applied reverse bias of magnitude V_a. This leads to a separation of eV_a between the Fermi levels in the p and n regions and causes the potential barrier across the junction to increase from eV_d to $e(V_d + V_a)$. The change of potential between the p and n regions is due to a sweeping of the mobile charge carriers from the region $-l_p < x < l_n$, giving rise to charged double layer of stationary (ionized) impurity atoms, as shown in Figure 11.11d.

In the analytical treatment of the problem, we assume that in the depletion layer ($-l_p < x < l_n$) the excess impurity atoms are fully ionized and, thus, using $\nabla \cdot \mathbf{E} = \rho/\varepsilon$ and $\mathbf{E} = -\nabla V$, where V is the potential, we have

$$\frac{d^2V}{dx^2} = \frac{eN_A}{\varepsilon}, \quad \text{for } -l_p < x < 0 \tag{11.6-1}$$

and

$$\frac{d^2V}{dx^2} = -\frac{eN_D}{\varepsilon}, \quad \text{for } 0 < x < l_n \tag{11.6-2}$$

where the charge of the electron is $-e$ and the permittivity is ε. The boundary conditions are

$$E = -\frac{dV}{dx} = 0 \quad \text{at } x = -l_p \text{ and } x = +l_n \tag{11.6-3}$$

$$V \text{ and } \frac{dV}{dx} \text{ are continuous at } x = 0 \tag{11.6-4}$$

$$V(l_n) - V(-l_p) = V_d + V_a \tag{11.6-5}$$

The solutions of Equations (11.6-1) and (11.6-2) conforming with the arbitrary choice of $V(0) = 0$ are

$$V = \frac{e}{2\varepsilon} N_A(x^2 + 2l_p x), \quad \text{for } -l_p < x < 0 \tag{11.6-6}$$

$$V = -\frac{e}{2\varepsilon} N_D(x^2 - 2l_n x), \quad \text{for } 0 < x < l_n \tag{11.6-7}$$

which, using Equation (11.6-4), gives

$$N_A l_p = N_D l_n \tag{11.6-8}$$

so the double layer contains an equal amount of positive and negative charge.

Condition (11.6-5) gives

$$V_d + V_a = \frac{e}{2\varepsilon}(N_D l_n^2 + N_A l_p^2) \tag{11.6-9}$$

which, together with Equation (11.6-8) leads to

$$l_p = (V_d + V_a)^{1/2}\left(\frac{2\varepsilon}{e}\right)^{1/2}\left(\frac{N_D}{N_A(N_A + N_D)}\right)^{1/2} \tag{11.6-10}$$

$$l_n = (V_d + V_a)^{1/2}\left(\frac{2\varepsilon}{e}\right)^{1/2}\left(\frac{N_A}{N_D(N_A + N_D)}\right)^{1/2} \tag{11.6-11}$$

and therefore, as before,

$$\frac{l_p}{l_n} = \frac{N_D}{N_A} \tag{11.6-12}$$

Differentiation of Equations (11.6-6) and (11.6-7) yields

$$E = -\frac{e}{\varepsilon} N_A(x + l_p), \quad \text{for } -l_p < x < 0$$

$$E = -\frac{e}{\varepsilon} N_D(l_n - x), \quad \text{for } 0 < x < l_n \tag{11.6-13}$$

The field distribution of (11.6-13) is shown in Figure 11.1e. The maximum field occurs at $x = 0$ and is given by

$$\begin{aligned} E_{\max} &= -2(V_d + V_a)^{1/2}\left(\frac{e}{2\varepsilon}\right)^{1/2}\left(\frac{N_D N_A}{N_A + N_D}\right)^{1/2} \\ &= -\frac{2(V_d + V_a)}{l_p + l_n} \end{aligned} \tag{11.6-14}$$

The presence of a charge $Q = -eN_A l_p$ per unit junction area on the p side and an equal and opposite charge on the n side leads to a junction capacitance. The reason is that l_p and l_n depend, according to Equations (11.6-10) and (11.6-11), on the applied voltage V_a, so a change in voltage leads to a change in the charge $eN_A l_p = eN_D l_n$ and hence to a differential capacitance per unit area,[14] given by

$$\begin{aligned} \frac{C_d}{\text{Area}} &\equiv \frac{dQ}{dV_a} = eN_A \frac{dl_p}{dV_a} \\ &= \left(\frac{\varepsilon e}{2}\right)^{1/2}\left(\frac{N_A N_D}{N_A + N_D}\right)^{1/2}\left(\frac{1}{V_a + V_d}\right)^{1/2} \end{aligned} \tag{11.6-15}$$

which, using Equations (11.6-10) and (11.6-11), can be shown to be equal to

$$\frac{C_d}{\text{Area}} = \frac{\varepsilon}{l_p + l_n} \tag{11.6-16}$$

as appropriate to a parallel-plate capacitance of separation $l = l_p + l_n$. The equivalent circuit of a p-n junction is shown in Figure 11.12. The capacitance C_d was discussed earlier. The

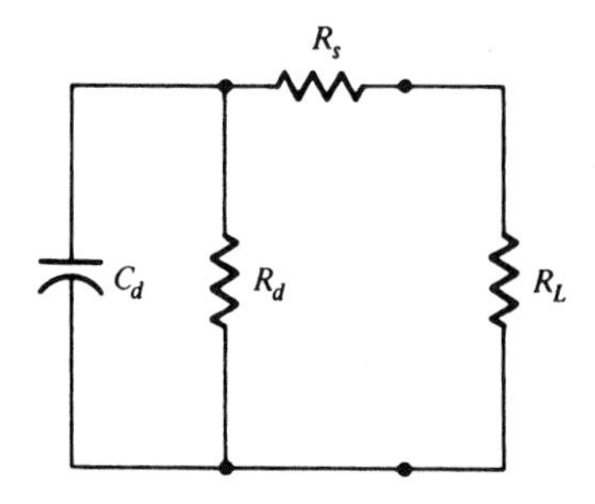

Figure 11.12 Equivalent circuit of a p-n junction. In typical back-biased diodes, $R_d \gg R_s$ and R_L, and $R_L \gg R_s$, so the resistance across the junction can be taken as equal to the load resistance R_L.

[14] The capacitance is defined by $C = Q/V_a$, whereas the differential capacitance $C_d = dQ/dV_a$ is the capacitance "seen" by a small ac voltage when the applied bias is V_a.

diode shunt resistance R_d in back-biased junctions is usually very large ($>10^6$ ohms) compared to the load impedance R_L and can be neglected. The resistance R_s represents ohmic losses in the bulk p and n regions adjacent to the junction.

11.7 SEMICONDUCTOR PHOTODIODES

Semiconductor *p-n* junctions are used widely for optical detection: see References [7–9]. In this role they are referred to as junction photodiodes. The main physical mechanisms involved in junction photodetection are illustrated in Figure 11.13. At *A*, an incoming photon is absorbed in the *p* side, creating a hole and a free electron. If this takes place within a diffusion length (the distance in which an excess minority concentration is reduced to e^{-1} of its peak value, or in physical terms, the average distance a minority carrier traverses before recombining with a carrier of the opposite type) of the depletion layer, the electron will, with high probability, reach the layer boundary and will drift under the field influence across it. An electron traversing the junction contributes a charge e to the current flow in the external circuit, as described in Section 10.4. If the photon is absorbed near the n side of the depletion layer, as shown at C, the resulting hole will diffuse to the junction and then drift across it again, giving rise to a flow of charge e in the external load. The photon may also be absorbed in the depletion layer as at B, in which case both the hole and electron that are created drift (in opposite directions) under the field until they reach the p and n sides, respectively. Since in this case each carrier traverses a distance that is less than the full junction width, the contribution of this process to charge flow in the external circuit is, according to Equations (10.4-1) and (10.4-7), e. In practice, this last process is the most desirable, since each absorption gives rise to a charge e, and delayed current response caused by finite diffusion time is avoided. As a result, photodiodes often use a *p-i-n* structure in which an intrinsic high-resistivity (i) layer is sandwiched between the p and n regions. The potential drop occurs mostly across this layer, which can be made long enough to ensure that most of the incident photons are absorbed within it. Typical construction of a *p-i-n* photodiode is shown in Figure 11.14.

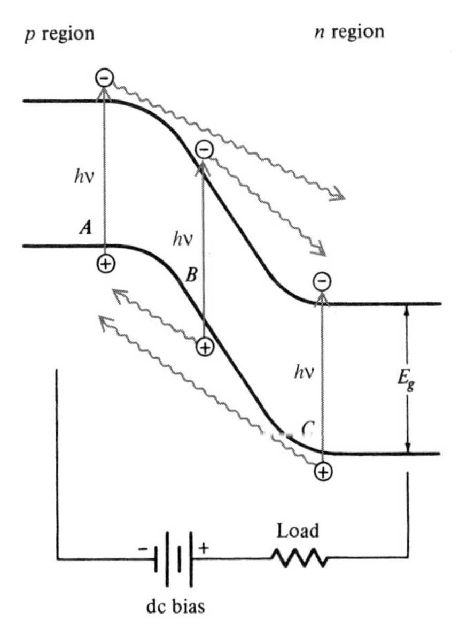

Figure 11.13 The three types of electron–hole pair creation by absorbed photons that contribute to current flow in a *p-n* photodiode.

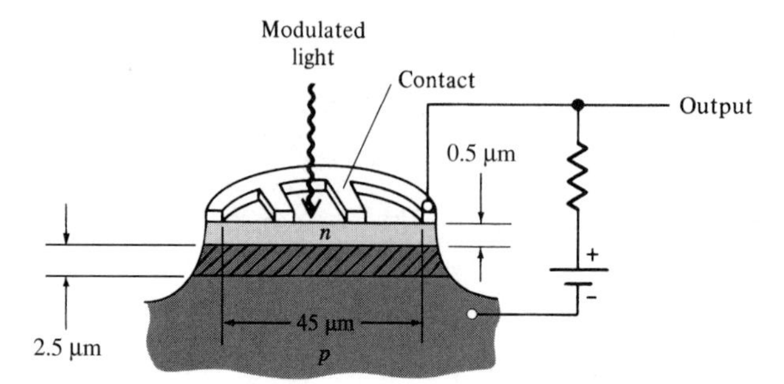

Figure 11.14 A *p-i-n* photodiode. (After Reference [10].)

It is clear from Figure 11.13 that a photodiode is capable of detecting only radiation with photon energy $h\nu > E_g$, where E_g is the energy gap of the semiconductor. If, on the other hand, $h\nu \gg E_g$, the absorption, which in a semiconductor increases strongly with frequency, will take place entirely near the input face (in the n region of Figure 11.14) and the minority carriers generated by absorbed photons will recombine with majority carriers before diffusing to the depletion layer. This event does not contribute to the current flow and, as far as the signal is concerned, is wasted. This is why the photoresponse of diodes drops off when $h\nu > E_g$. Typical frequency response curves of photodiodes are shown in Figure 11.15.

The quantum efficiency of a photodiode is defined as the number of carriers generated and flowing in the circuit per incoming photon. With a proper antireflection coating, the intrinsic quantum efficiency can approach 100%. Figure 11.15 shows the quantum efficiency of some photodiodes. In optical communications, the selection of photodiodes is based on several criteria, including speed, sensitivity, linearity, bias voltage, dark current, and cost. Generally, photodiodes made of semiconductors with indirect bandgap are relatively slow compared

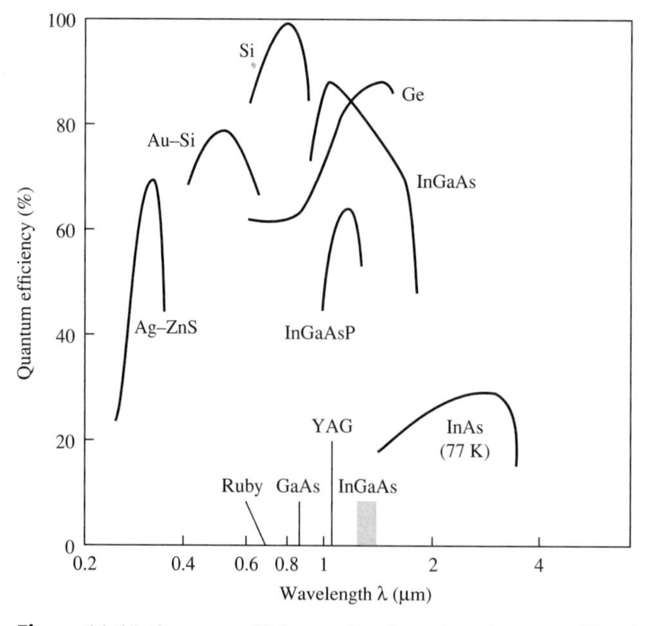

Figure 11.15 Quantum efficiency of various photodetectors. (The data are adapted from Reference [11].) Emission wavelengths for various lasers are also indicated.

with photodiodes made of direct bandgap materials. Stronger absorption in direct bandgap semiconductors is responsible for the high-speed feature of photodiodes such as InGaAs. According to Figure 11.15, InGaAs photodiodes also offer high quantum efficiency for detection in the spectral regime around 1300 and 1550 nm.

Frequency Response of Photodiodes

One of the major considerations in optical detectors is their frequency response—that is, the ability to respond to variations in the incident intensity such as those caused by high-frequency modulation. The three main mechanisms limiting the frequency response in photodiodes are:

1. The finite diffusion time of carriers produced in the p and n regions. This factor was described in the last section, and its effect can be minimized by a proper choice of the length of the depletion layer.
2. The shunting effect of the signal current by the junction capacitance C_d shown in Figure 11.12. This places an upper limit of

$$\omega_m \simeq \frac{1}{R_e C_d} \tag{11.7-1}$$

   on the intensity modulation frequency, where R_e is the equivalent resistance in parallel with the capacitance C_d.
3. The finite transit time of the carriers drifting across the depletion layer.

To analyze first the limitation due to transit time, we assume the slightly idealized case in which the carriers are generated in a *single* plane, say, point A in Figure 11.13, and then drift the full width of the depletion layer at a constant velocity v. For high enough electric fields, the drift velocity of carriers in semiconductors tends to saturate, so the constant-velocity assumption is not very far from reality even for a nonuniform field distribution, such as that shown in Figure 11.11e, provided the field exceeds its saturation value over most of the depletion layer length. The saturation of the whole velocity in germanium, as an example, is illustrated by the data of Figure 11.16.

The incident optical field is taken as

$$\begin{aligned} e(t) &= E_s(1 + m \cos \omega_m t)\cos \omega t \\ &\equiv \mathrm{Re}[V(t)] \end{aligned} \tag{11.7-2}$$

where

$$V(t) \equiv E_s(1 + m \cos \omega_m t)e^{i\omega t} \tag{11.7-3}$$

Thus the amplitude is modulated at a frequency $\omega_m/2\pi$. Following the discussion of Section 11.1, we take the generation rate $G(t)$—that is, the number of carriers generated per second—as proportional to the average of $e^2(t)$ over a time long compared to the optical period $2\pi/\omega$. This average is equal to $\frac{1}{2}V(t)V^*(t)$, so the generation rate is taken as

$$G(t) = aE_s^2\left[\left(1 + \frac{m^2}{2}\right) + 2m \cos \omega_m t + \frac{m^2}{2} \cos 2\omega_m t\right] \tag{11.7-4}$$

where a is a proportionality constant to be determined. Dropping the term involving $\cos 2\omega_m t$ and using complex notation, we rewrite $G(t)$ as

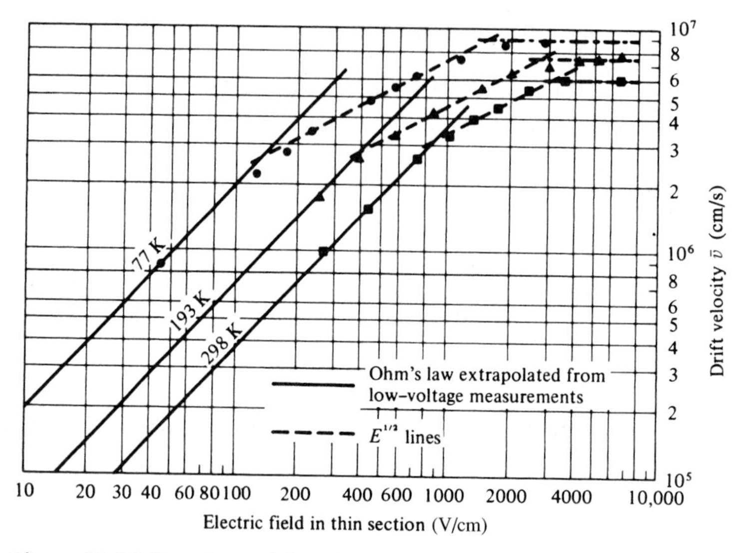

Figure 11.16 Experimental data showing the saturation of the drift velocity of holes in germanium at high electric fields. (After Reference [12].)

$$G(t) = aE_s^2\left(1 + \frac{m^2}{2} + 2me^{i\omega_m t}\right) \tag{11.7-5}$$

A single carrier drifting at a velocity $\bar{v}$ contributes, according to Equation (10.4-1), an instantaneous current

$$i = \frac{e\bar{v}}{d} \tag{11.7-6}$$

to the external circuit, where d is the width of the depletion layer. The current due to carriers generated between t' and $t' + dt'$ is $(e\bar{v}/d)G(t')dt'$ but, since each carrier spends a time $\tau_d = d/\bar{v}$ in transit, the instantaneous current at time t is the sum of contributions of carriers generated between t and $t - \tau_d$,

$$i(t) = \frac{e\bar{v}}{d}\int_{t-\tau_d}^{t} G(t')\,dt' = \frac{e\bar{v}aE_s^2}{d}\int_{t-\tau_d}^{t}\left(1 + \frac{m^2}{2} + 2me^{i\omega_m t'}\right)dt'$$

and, after integration,

$$i(t) = \left(1 + \frac{m^2}{2}\right)eaE_s^2 + 2meaE_s^2\left(\frac{1 - e^{-i\omega_m\tau_d}}{i\omega_m\tau_d}\right)e^{i\omega_m t} \tag{11.7-7}$$

The factor $(1 - e^{-i\omega_m\tau_d})/i\omega_m\tau_d$ represents the phase lag as well as the reduction in signal current due to the finite drift time τ_d. If the drift time is short compared to the modulation period, so $\omega_m\tau_d \ll 1$, it has its maximum value of unity, and the signal is maximum. This factor is plotted in Figure 11.17 as a function of the transit phase angle $\omega_m\tau_d$. We can determine the value of the constant a in Equation (11.7-7) by requiring that (11.7-7) agree with the experimental observation according to which in the absence of modulation, $m = 0$, each incident photon will create η carriers. Thus the dc (average) current is

$$\bar{I} = \frac{Pe\eta}{h\nu} \tag{11.7-8}$$

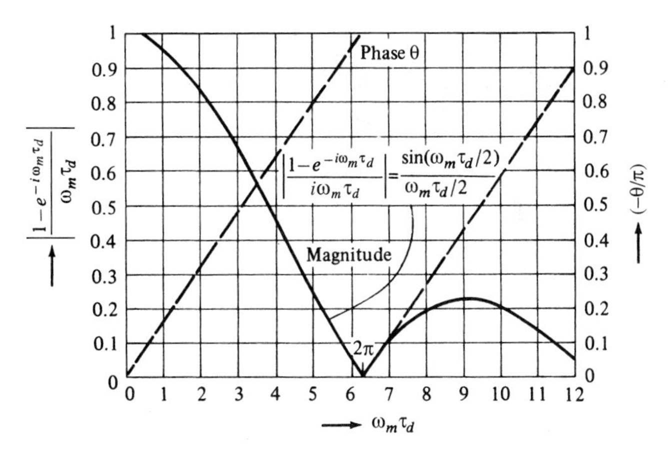

Figure 11.17 Phase and magnitude of the transit-time reduction factor $(1 - e^{-i\omega_m\tau_d})/\omega_m\tau_d$.

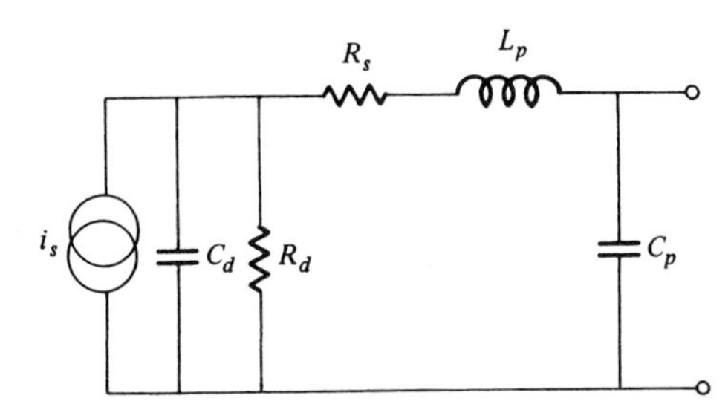

Figure 11.18 The equivalent high-frequency circuit of a semiconductor photodiode.

where P is the optical (signal) power when $m = 0$. Using Equation (11.7-8), we can rewrite Equation (11.7-7) as

$$i(t) = \frac{Pe\eta}{h\nu}\left(1 + \frac{m^2}{2}\right) + \frac{Pe\eta}{h\nu}2m\left(\frac{1 - e^{-i\omega_m\tau_d}}{i\omega_m\tau_d}\right)e^{i\omega_m t} \tag{11.7-9}$$

To evaluate the effect of the other limiting factors on the modulation frequency response of a photodiode, we refer to the diode equivalent ac circuit in Figure 11.18. Here R_d is the diode incremental (ac) resistance, C_d the junction capacitance, R_s represents the contact and series resistance, L_p the parasitic inductance associated mostly with the contact leads, and C_p the parasitic capacitance due to the contact leads and the contact pads.

Experimental studies [13–16] have resulted in metal–GaAs (Schottky) diodes with frequency response extending up to 10^{11} Hz. Figure 11.19 shows a schematic diagram of such a diode. This high-frequency limit was achieved by using a very small area (5 μm × 5 μm) that minimizes C_d, by using extremely short contact leads to reduce R_s and L_p, by fabricating the diode on semi-insulating GaAs substrate [13] to reduce C_p, and by using a thin (0.3 μm) n^-GaAs drift region to reduce the transit time. The resulting measured frequency response is shown in Figure 11.20. The measurement of the frequency response up to 100 GHz is by itself a considerable achievement. This was accomplished by first obtaining the impulse response of the photodiode by exciting it with picosecond pulses (which, for the range of frequencies of interest, may be considered as delta functions) from a mode-locked laser [14]. The diode response, which is only a few picoseconds long, is measured by an electro-optic sampling technique [16, 17]. The frequency response, as plotted in Figure 11.20, is obtained by taking the Fourier transform of the measured impulse response.

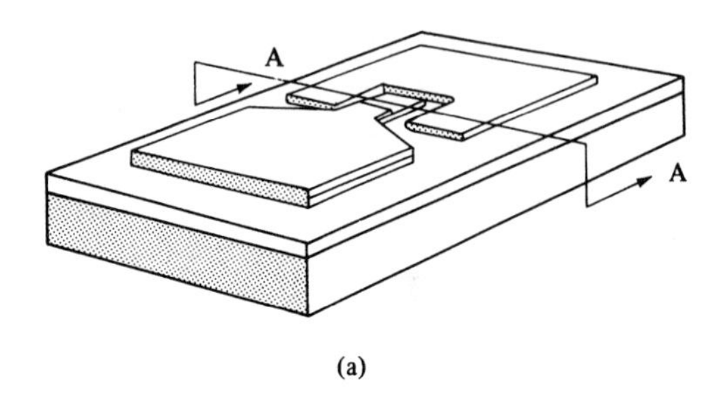

(a)

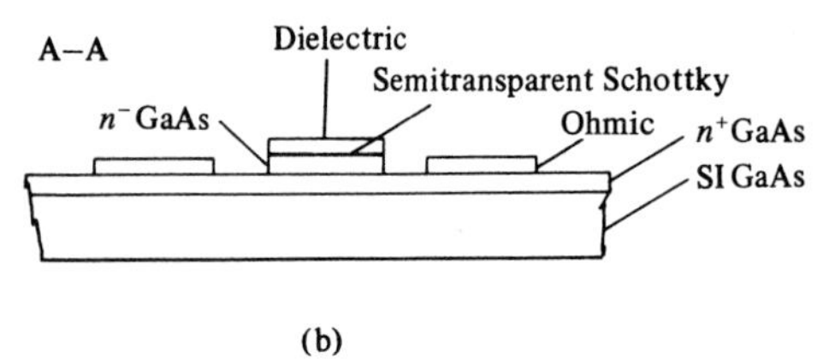

(b)

Figure 11.19 (a) Planar GaAs Schottky photodiode. (b) Cross section along A–A. The n^-GaAs layer (0.3 μm thick) and the n^+GaAs (0.4 μm thick) layer are grown by liquid-phase epitaxy on semi-insulating GaAs substrate. The semitransparent Schottky consists of 100 Å of Pt (After Reference [15].)

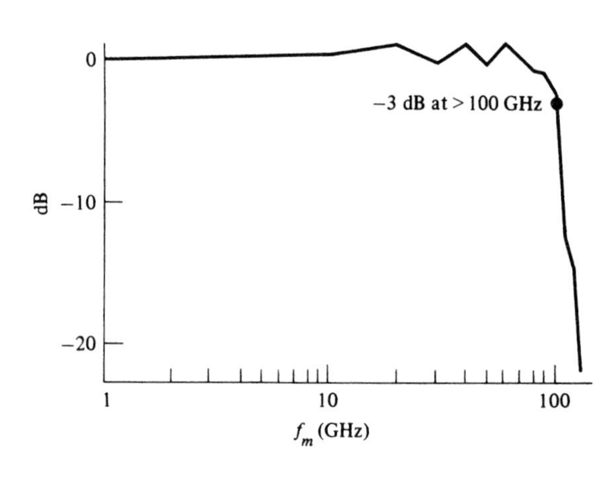

Figure 11.20 The modulation frequency response of the Schottky photodiode shown in Figure 11.19. (After Reference [15].)

EXAMPLE: MODULATION RESPONSE OF A GaAs *p-n* JUNCTION PHOTODIODE

Let us calculate the upper limit on the frequency response of the diode shown in Figure 11.19. The following data apply:

$$\text{Area} = 5\ \mu\text{m} \times 5\ \mu\text{m}$$

$$\varepsilon = 12.25\varepsilon_0$$

$$d = 0.3\ \mu\text{m}\ (= \text{thickness of drift region})$$

$$\bar{v} = 10^7\ \text{cm/s (saturation velocity of electron in GaAs)}$$

$$R_s \approx 10\ \Omega$$

The transit-time limit f_m is obtained from the condition $2\pi f_m \tau_d = 2$. This, according to Figure 11.17, is the frequency where the response is down to 84% of its maximum (zero-frequency) value. The result is

$$f_m \sim \frac{\bar{v}}{\pi d} \sim 1.06 \times 10^{11}\ \text{Hz}$$

The junction capacitance, based on the above data, is $\sim 10^{-14}$ F. The parasitic capacitance can be kept in this case to $\sim 10^{-13}$ F. Since the resistance R_d of the reverse-biased junction is very large, it is usually neglected.

The circuit limit to the frequency response is $f_m \sim 1/(2\pi R_s C_p) = 1.59 \times 10^{11}$ Hz. Since this value is larger than the transit-time limit, we conclude that the frequency response is transit-time limited to a value $\sim 10^{11}$ Hz, which is in agreement with the value obtained from Figure 11.20.

Detection Sensitivity of Photodiodes

We assume that the modulation frequency of the light to be detected is low enough that the transit-time factor is unity and that the condition

$$\omega_m \ll \frac{1}{R_e C_d} \tag{11.7-10}$$

is fulfilled and therefore, according to Equation (11.7-1), the shunting of signal current by the diode capacitance C_d can be neglected. The diode current is given by Equation (11.7-9) as

$$i(t) = \frac{Pe\eta}{h\nu}\left(1 + \frac{m^2}{2}\right) + \frac{Pe\eta}{h\nu} 2me^{i\omega_m t} \tag{11.7-11}$$

The noise equivalent circuit of a diode connected to a load resistance R_L is shown in Figure 11.21. The signal power is proportional to the mean square value of the sinusoidal current component, which, for $m = 1$, is

$$\overline{i_s^2} = 2\left(\frac{Pe\eta}{h\nu}\right)^2 \tag{11.7-12}$$

Two noise sources are shown. The first is the shot noise associated with the random generation of carriers. Using Equation (10.4-9), this is represented by a noise generator $\overline{i_{N_1}^2} = 2e\bar{I}\,\Delta\nu$, where $\bar{I}$ is the average current as given by the first term on the right side of Equation (11.7-11). Taking $m = 1$, we obtain

$$\overline{i_{N_1}^2} = \frac{3e^2(P + P_B)\eta\,\Delta\nu}{h\nu} + 2ei_d\,\Delta\nu \tag{11.7-13}$$

where P_B is the background optical power entering the detector (in addition to the signal power) and i_d is the "dark" direct current that exists even when $P_s = P_B = 0$. The second noise

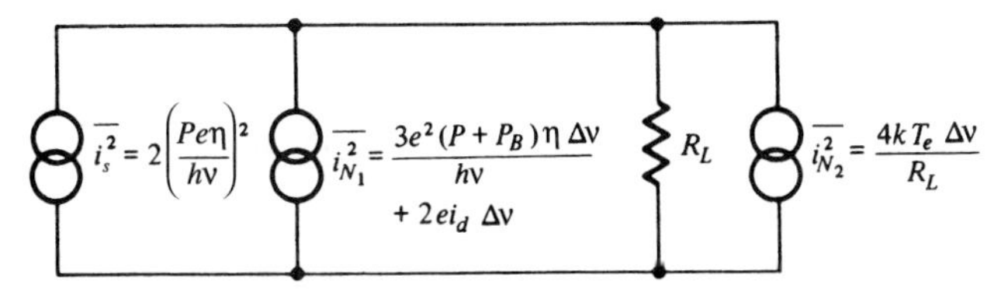

Figure 11.21 Noise equivalent circuit of a photodiode operating in the direct (video) mode. The modulation index m is taken as unity, and it is assumed that the modulation frequency is low enough that the junction capacitance and transit-time effects can be neglected. The resistance R_L is assumed to be much smaller than the shunt resistance R_d of the diode, so the latter is neglected. Also neglected is the series diode resistance, which is assumed small compared with R_L.

contribution is the thermal (Johnson) noise generated by the output load, which, using Equation (10.5-9), is given by

$$\overline{i_{N_2}^2} = \frac{4kT_e\,\Delta\nu}{R_L} \tag{11.7-14}$$

where T_e is chosen to include the equivalent input noise power of the amplifier following the diode.[15] The signal-to-noise power ratio at the amplifier output is thus

$$\frac{S}{N} = \frac{\overline{i_s^2}}{\overline{i_{N_1}^2} + \overline{i_{N_2}^2}} = \frac{2(Pe\eta/h\nu)^2}{3e^2(P + P_B)\eta\,\Delta\nu/h\nu + 2ei_d\,\Delta\nu + 4kT_e\,\Delta\nu/R_L} \tag{11.7-15}$$

In most practical systems the need to satisfy Equation (11.7-10) forces one to use small values of load resistance R_L. Under these conditions and for values of P that are near the detectability limit ($S/N = 1$), the noise term (11.7-14) is much larger than the shot noise (11.7-13) and the detector is consequently not operating near its quantum limit. Under these conditions we have

$$\frac{S}{N} \simeq \frac{2(Pe\eta/h\nu)^2}{4kT_e\,\Delta\nu/R_L} \tag{11.7-16}$$

The "minimum detectable optical power" is by definition that yielding $S/N = 1$ and is, from Equation (11.7-16),

$$(P)_{\min} = \frac{h\nu}{e\eta}\sqrt{\frac{2kT_e\,\Delta\nu}{R_L}} \tag{11.7-17}$$

which is to be compared to the theoretical limit of $h\nu\,\Delta\nu/\eta$, which, according to Equation (11.3-11), obtains when the signal shot-noise term predominates. In practice, the value of R_L is related to the desired modulation bandwidth $\Delta\nu$ and the junction capacitance C_d by

$$\Delta\nu \simeq \frac{1}{2\pi R_L C_d} \tag{11.7-18}$$

---

[15] In practice it is imperative that the signal-to-noise ratio take account of the noise power contributed by the amplifier. This is done by characterizing the "noisiness" of the amplifier by an effective input noise "temperature" T_A. The amplifier noise power measured at its output is taken as $GkT_A\Delta\nu$, where G is the power gain. (A hypothetical noiseless amplifier will thus be characterized by $T_A = 0$.) This power can be referred to the input by dividing by G, thus becoming $kT_A\Delta\nu$. The total effective noise power at the amplifier input is the sum of this power and the Johnson noise $kT\Delta\nu$ due to the diode load resistance; that is, $k(T + T_A)\Delta\nu \equiv kT_e\Delta\nu$. The amplifier noise temperature T_A is related to its "noise figure" F by the definition

$$F = 1 + \frac{T_A}{290}$$

It follows that the noise power generated within the amplifier and measured at its output is

$$N_A = GkT_A\Delta\nu = G(F-1)kT_0\Delta\nu$$

where $T_0 = 290$. The ratio of the signal-to-noise power ratio at the input of the amplifier to the same ratio at the output is thus

$$\frac{(S/N)_{\text{in}}}{(S/N)_{\text{out}}} = \frac{S_{\text{in}}[G(F-1)kT_0\,\Delta\nu + GkT\,\Delta\nu]}{kT\,\Delta\nu\,GS_{\text{in}}}$$

This ratio becomes equal to the "noise figure" F when the temperature T of the detector output load is equal to T_0. (Note that the choice $T_0 = 290$ is a matter of universal convention.)

which, when used in Equation (11.7-16), gives

$$P_{\min} \simeq 2\sqrt{\pi}\,\frac{h\nu\,\Delta\nu}{e\eta}\sqrt{kT_eC_d} \tag{11.7-19}$$

This shows that sensitive detection requires the use of small area junctions so that C_d will be at a minimum.

EXAMPLE: MINIMUM DETECTABLE POWER IN THE CASE OF AMPLIFIER-LIMITED DETECTION

Assume a typical Ge photodiode operating at $\lambda = 1.4\ \mu\text{m}$ with $C_d = 1$ pF, $\Delta\nu = 1$ GHz, and $\eta = 50\%$. Let the amplifier following the diode have an effective noise temperature $T_e = 1200 + 290 = 1490$ K (see footnote 15) [12, 18]. Substitution in Equation (11.7-19) gives

$$P_{\min} \simeq 3.34 \times 10^{-7}\ \text{W}$$

for the minimum detectable signal power.

11.8 AVALANCHE PHOTODIODE

By increasing the reverse bias across a *p-n* junction, the field in the depletion layer can increase to a point at which carriers (electrons or holes) that are accelerated across the depletion layer can gain enough kinetic energy to "kick" new electrons from the valence to the conduction band, while still traversing the layer. This process, illustrated in Figure 11.22, is referred to as avalanche multiplication. An absorbed photon (*A*) creates an electron–hole pair. The electron is accelerated until at point *C* it has gained sufficient energy to excite an electron from the valence to the conduction band, thus creating a new electron–hole pair. The newly generated carriers drift in turn in opposite directions. The hole (*F*) can also cause carrier multiplication as in *G*. The result is a dramatic increase (avalanche) in junction current that

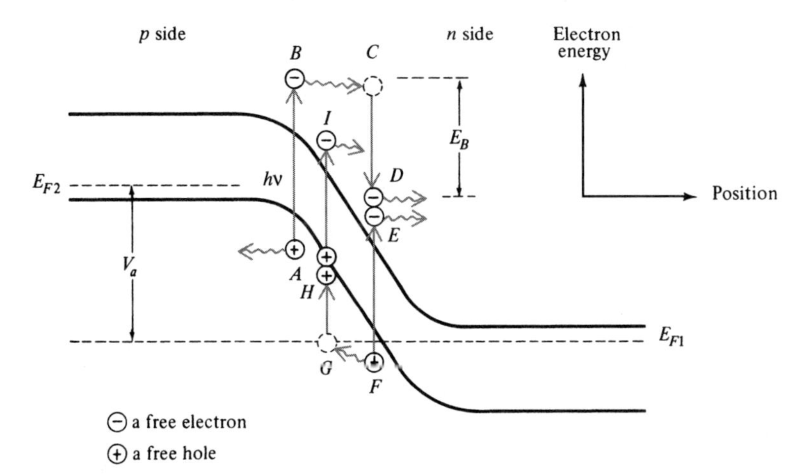

Figure 11.22 Energy-position diagram showing the carrier multiplication following a photon absorption in a reverse-biased avalanche photodiode.

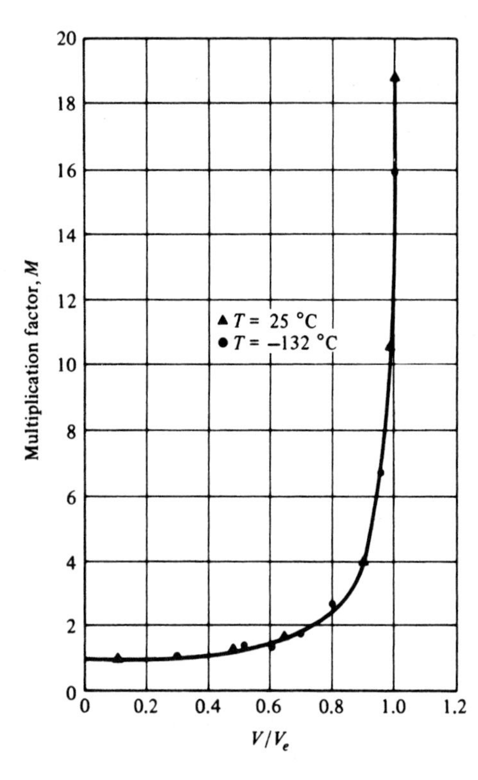

Figure 11.23 Current multiplication factor in an avalanche diode as a function of the electric field. (After Reference [19].)

sets in when the electric field becomes high enough. This effect, discovered first in gaseous plasmas and more recently in *p-n* junctions [18, 19], gives rise to a multiplication of the current over its value in an ordinary (nonavalanching) photodiode. An experimental plot of the current gain M as a function of the junction field is shown in Figure 11.23.[16]

Avalanche photodiodes are similar in their construction to ordinary photodiodes except that, because of the steep dependence of M on the applied field in the avalanche region, special care must be exercised to obtain very uniform junctions. A sketch of an avalanche photodiode is shown in Figure 11.24.

Since an avalanche photodiode is basically similar to a photodiode, its equivalent circuit elements are given by expressions similar to those given earlier for the photodiode. Its frequency response is similarly limited by diffusion, drift across the depletion layer, and capacitive loading, as discussed in Section 11.7.

A multiplication by a factor M of the photocurrent leads to an increase by M^2 of the signal power S over that which is available from a photodiode so that, using Equation (11.7-12), we get

$$S \propto \overline{i_s^2} = 2M^2 \left(\frac{Pe\eta}{h\nu} \right)^2 \tag{11.8-1}$$

[16] If the probability that a photoexcited electron–hole pair will create another pair during its drift is denoted by p, the current multiplication is

$$M = (1 + p + p^2 + p^3 + \cdots) = \frac{1}{1 - p}$$

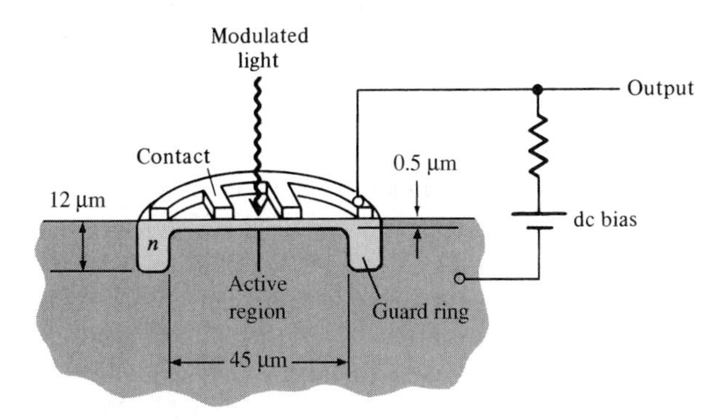

Figure 11.24 Planar avalanche photodiode. (After Reference [10].)

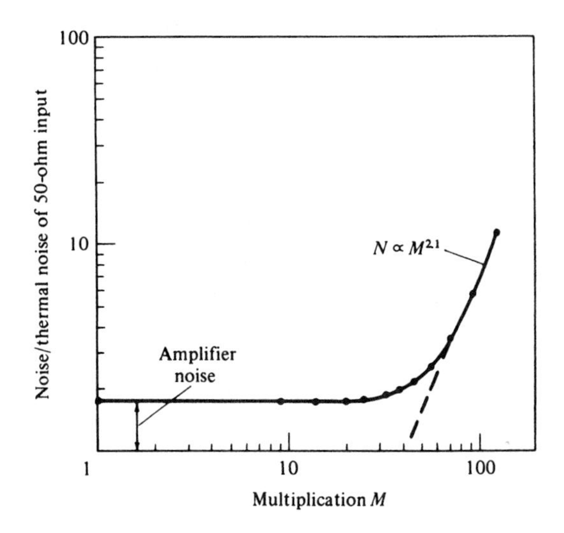

Figure 11.25 Noise power (measured at 30 MHz) as a function of photocurrent multiplication for an avalanche Schottky-barrier photodiode. (After Reference [21].)

where P is the optical power incident on the diode. This result is reminiscent of the signal power from a photomultiplier as given by the numerator of Equation (11.3-9), where the avalanche gain M plays the role of the secondary electron multiplication gain G. We may expect that, similarly, the shot-noise power will also increase by M^2. The shot noise, however, is observed to increase as M^n, where $2 < n < 3$.[17] Experimental observation of a near ideal $M^{2.1}$ behavior is shown in Figure 11.25.

The signal-to-noise power ratio at the output of the diode is thus given, following Equation (11.7-15), by

$$\frac{S}{N} = \frac{2M^2(Pe\eta/h\nu)^2}{[3e^2(P + P_B)\eta\ \Delta\nu/h\nu]M^n + 2ei_d\ \Delta\nu\ M^n + 4kT_e\ \Delta\nu/R_L} \tag{11.8-2}$$

The advantage of using an avalanche photodiode over an ordinary photodiode is now apparent. When $M = 1$, the situation is identical to that at the photodiode as described by Equation (11.7-15). Under these conditions the thermal term $4kT_e\ \Delta\nu/R_L$ in the denominator of Equation (11.8-2) is typically much larger than the shot-noise terms. This causes S/N to

[17] A theoretical study by McIntyre [20] predicts that if the multiplication is due to either holes or electrons, $n = 2$, whereas if both carriers are equally effective in producing electron–hole pairs, $n = 3$.

increase with M. This improvement continues until the shot-noise terms become comparable with $4kT_e\,\Delta\nu/R_L$. Further increases in M result in a reduction of S/N since $n > 2$, and the denominator of Equation (11.8-2) grows faster than the numerator. If we assume that M is adjusted optimally so that the denominator of Equation (11.8-2) is equal to twice the thermal term $4kT_e\,\Delta\nu/R_L$, we can solve for the minimum detectable power (i.e., the power input for which $S/N = 1$), obtaining

$$P_{\min} = \frac{2h\nu}{M'e\eta}\sqrt{\frac{kT_e\,\Delta\nu}{R_L}} \tag{11.8-3}$$

where M' is the optimum value of M as discussed previously. The improvement in sensitivity over the photodiode result (11.7-17) is thus approximately M'. Values of M' between 30 and 100 are commonly employed, so the use of avalanche photodiodes affords considerable improvement in sensitivity over that available from photodiodes.

11.9 POWER FLUCTUATION NOISE IN LASERS

The power output from lasers is ever fluctuating. This fluctuation may be due to temperature variations, acoustic vibrations, and other man-made causes. Even if all of these extraneous effects are eliminated, there remains a basic (quantum mechanical) contribution that is due to spontaneous emission of radiation into the laser mode by atoms dropping from the upper transition level into the lower level. The field due to this spontaneous emission is not coherent with that of the laser mode, thus causing phase and amplitude fluctuations [1]. Since these fluctuations are random, they are described and quantified in terms of the statistical noise tools developed earlier in this chapter and in Chapter 10.

Let the power output of the laser be

$$P(t) = P_0 + \Delta P(t) \tag{11.9-1}$$

where the time-averaging value of the fluctuation is zero.

$$\overline{\Delta P(t)} = 0$$

so that P_0 is the average optical power. Using Equations (10.2-6) and (10.2-8), we characterize the "power" of the fluctuation via the mean of the squared deviation[18]

$$\overline{(P(t) - P_0)^2} \equiv \overline{(\Delta P(t))^2} = \int_0^\infty S_{\Delta P}(f)\,df \tag{11.9-2}$$

$S_{\Delta P}(f)$ is related to the spectral density function $S_{\Delta P}(\omega)$, defined by Equations (10.2-8) and (10.2-14), by

$$S_{\Delta P}(f) = 2\pi S_{\Delta P}(\omega) \quad (\omega = 2\pi f) \tag{11.9-3}$$

If an optical field at frequency ν with a power $P(t)$ is incident on a detector whose quantum efficiency (electrons per photon) is η, the output current is

$$i(t) = \frac{e\eta P(t)}{h\nu}$$

[18] In this section, we will use f to denote "low" (RF) frequencies and ν for optical frequencies.

so that according to Equation (11.9-1) the optical power fluctuation $\Delta P(t)$ causes a fluctuating current component $\Delta i(t) = e\eta\Delta P(t)/h\nu$ with a mean square

$$\overline{i_{NL}^2}(t) \equiv \overline{[\Delta i(t)]^2} = \frac{e^2\eta^2}{(h\nu)^2}\overline{[\Delta P(t)]^2} = \frac{e^2\eta^2}{(h\nu)^2} S_{\Delta P}(f)\,\Delta f \tag{11.9-4}$$

where Δf is the bandwidth of the electronic detection circuit.

The relative intensity noise (RIN) is defined as the relative fluctuation "power" in a $\Delta f = 1$ Hz bandwidth

$$\text{RIN} \equiv \frac{S_{\Delta P}\,\Delta f(=1\text{ Hz})}{P_0^2} \tag{11.9-5}$$

A single-mode semiconductor laser might possess a value of RIN $\approx 10^{-16}$ (or -160 dB). Assuming that the detector circuit has a bandwidth of, say, $\Delta f = 10^9$ Hz, the relative mean squared fluctuation in the detected current is

$$\frac{\overline{(\Delta i_d)^2}}{i_{d0}^2} = \frac{\overline{(\Delta P)^2}}{P_0^2} = \frac{S_{\Delta P}(f)\,\Delta f}{P_0^2} = 10^{-16} \times 10^9 = 10^{-7}$$

The rms value of the power fluctuation is thus

$$\frac{\{\overline{[\Delta P(t)]^2}\}^{1/2}}{P_0} = 3.16 \times 10^{-4}$$

The mean squared noise current in the output of the detector due to these fluctuations is given by Equation (11.9-4):

$$\overline{i_{NL}^2}(t) = \frac{e^2\eta^2}{(h\nu)^2}(\text{RIN})P_0^2\,\Delta f \tag{11.9-6}$$

Assuming as an example that $\lambda = 1.3\ \mu\text{m}$, $P_0 = 3$ mW, RIN $= 10^{-16}\ \text{Hz}^{-1}$, $\Delta f = 10^9$ Hz, and $\eta = 0.6$, we obtain

$$(\overline{i_{NL}^2})^{1/2} = 5.95 \times 10^{-7}\ \text{A}$$

EXAMPLE: OPTICAL FIBER LINK DESIGN

Our task here is to determine the maximum allowed repeater spacing for an optical fiber communications link. We will assume that the optical source is a 1.3 μm GaInAsP laser ($\nu = c/\lambda = 2.31 \times 10^{14}$ Hz) and that the fiber possesses an attenuation of 0.3 dB/km (corresponding to an attenuation constant $\alpha = 0.3/4.343 = 0.0691\ \text{km}^{-1}$). The optical power launched into the fiber is $P_0 = 3$ mW. The channel is to transmit 10^9 bits/s so that the bandwidth of the detector circuit is taken as $\Delta f = 1/\text{period} = 10^9$ Hz. The system considerations dictate that the bit error probability at the detector output not exceed 10^{-10}. The detector output impedance is $R_L - 1000\ \Omega$, and the amplifier (following the detector) noise figure is 6 dB, that is, $F = 4$ (see footnote 15).

From Figure 10.20 we determine that the signal-to-noise power ratio at the amplifier output must exceed 22 dB to assure a bit error probability upon detection that is smaller than 10^{-10}. Our task is thus to calculate the signal power $\overline{i_s^2}$ and the total noise power $\overline{i_N^2}$ at the output of the detector as a function of the length L of the link.

The signal power is obtained from Equation (11.7-11), assuming a modulation index $m = 0.5$, as

$$\overline{i_s^2} = \frac{e^2\eta^2 P_0^2 e^{-2\alpha L}}{2(h\nu)^2} \tag{11.9-7}$$

The total noise power at the output of the amplifier referred to its input is

$$\overline{i_N^2} = \overline{i_{NL}^2} + \overline{i_{NS}^2} + \overline{i_{NA}^2} \underset{\text{(in order)}}{=} \frac{\eta^2 e^2}{(h\nu)^2}(\text{RIN})P_0^2 e^{-2\alpha L}\,\Delta f + \frac{2\eta e^2}{h\nu}P_0 e^{-\alpha L}\,\Delta f + \frac{4kT_e}{R_L}\Delta f \tag{11.9-8}$$

The first noise term is that due to power fluctuation (11.9-6); the second is the shot noise associated with the average current at the output of the detector $I_{d0} = \eta P_0 e \exp(-\alpha L)/(h\nu)$. The third term represents, as in Equation (11.7-15), both the Johnson noise of the output resistor R_L and the amplifier output noise power (referred to its input, see footnote 15). If the temperature of the output resistor R_L is $T = 290$ K, $T_E = T + (F - 1)290 = 1160$ K.

Figure 11.26a shows the main elements of an optical fiber link. Figure 11.26b shows a plot of $\overline{i_s^2}$, $\overline{i_{NL}^2}$, $\overline{i_{NS}^2}$, and $\overline{i_{NA}^2}$ as well as the total noise power as a function of the link length L. The important thing to note is the relative change of the various powers with distance. The distance L_0, where the detected signal-to-noise power ratio is down to 22 dB, is read off as $L_0 = 87$ km.[19] This distance is thus chosen as the link length. Note, as an example, that the dominant noise contribution at $L > 33$ km is the amplifier-detector noise $\overline{i_{NA}^2}$. If the latter were reduced by, say, 3 dB, the link length could be increased by 5 km, as indicated by the dashed line.

The signal-to-noise power ratio of a *p-n* diode detector is given by Equation (11.7-16) in the case where the dominant contributions to the noise power are the amplifier noise and the Johnson (thermal) noise of the load resistance R_L in the diode output circuit. The mean square noise current is then

$$\overline{i_N^2} \approx \frac{4kT_e\,\Delta f}{R_L} \tag{11.9-9}$$

The signal peak current is given by Equation (11.7-8) for the case $m = \frac{1}{2}$ is

$$i_s = \frac{P_s e\eta}{h\nu} \tag{11.9-10}$$

where P_s is the peak pulsed optical power incident on the detector. The signal-to-noise current ratio at the amplifier output is thus

$$\frac{i_s}{(\overline{i_N^2})^{1/2}} = \frac{P_s e\eta/h\nu}{(4kT_e\,\Delta f/R_L)^{1/2}} \tag{11.9-11}$$

Our next problem is that of finding the minimum value of the signal power P_s so that $i_s/(\overline{i_N^2})^{1/2}$ in Equation (11.9-11) exceeds the needed value of 12.59. We thus need to know T_e, R_L, and Δf. T_e is obtained from the given value of the amplifier noise figure ($F = 6$ dB).

[19] That is, $10\log(\overline{i_s^2}/\overline{i_N^2}) = 22$.

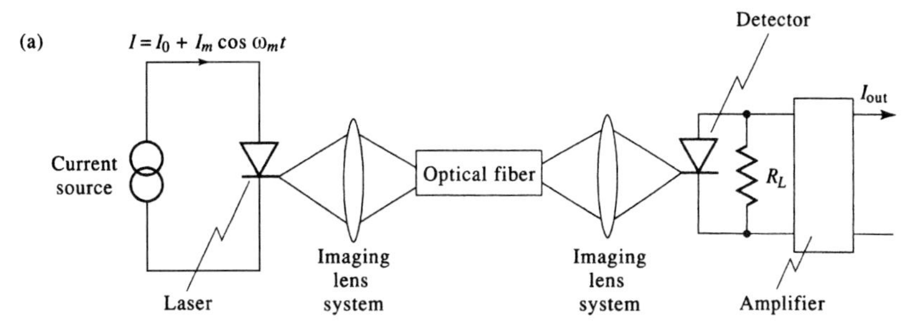

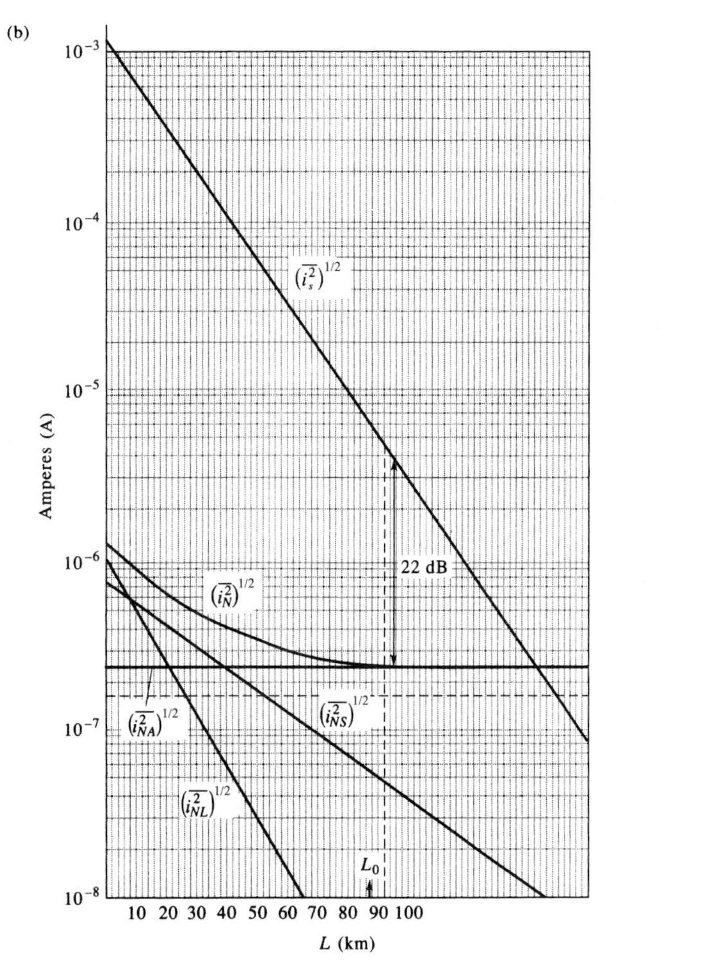

Figure 11.26 (a) An optical fiber communications link consisting of a laser, an optical coupling system c, a fiber L (km long), a detector D, an output resistance R_L, and an amplifier A with a current gain G and a noise figure F. (b) The signal $(\overline{i_s^2})$, laser fluctuation $(\overline{i_{NL}^2})$, detector shot noise $(\overline{i_{NS}^2})$, combined Johnson-amplifier noise $(\overline{i_{NA}^2})$, and the total noise

$$\overline{i_N^2} = \overline{i_{NL}^2} + \overline{i_{NS}^2} + \overline{i_{NA}^2}$$

currents as a function of the link length L. The currents are referred to the amplifier input plane S; that is, they correspond to output currents divided by the current gain G of the output amplifier.

Taking $T = 290$ K, we obtain, using footnote 15, $T_e = 290 + (4 - 1)290 = 1160$. In order to achieve this bandwidth, the load resistance R_L must not exceed (see Equation (11.7-18)) the value

$$R_L = \frac{1}{2\pi\, \Delta f\, C} \tag{11.9-12}$$

where C is the total output capacitance given as 3×10^{-12} F. Using the above value of Δf and C, we obtain

$$R_L \leq 53 \text{ ohms}$$

We return now to Equation (11.9-11), which, using $\eta = 0.5$, $\lambda = 1.35$ μm, and $i_s/(\overline{i_N^2})^{1/2} = 12.59$, yields

$$P_s \cong 2.55 \times 10^{-5} \text{ W}$$

for the minimum power input to the photodiode.

The total transmission loss in the 50 km fiber is 20 dB. We will assume that an additional 4 dB loss is caused by coupling the laser output to the fiber and at the fiber output so that the total loss is 24 dB (i.e., 251). The laser power output must thus exceed

$$P_{\text{laser}} = 6.4 \times 10^{-3} \text{ W}$$

which is a reasonable power level for cw diode lasers.

If the fiber had been substantially lossier than in the above example, we could still have met our design specifications by using an avalanche photodiode.

PROBLEMS

11.1 Show that the total output shot-noise power in a photomultiplier including that originating in the dynodes is given by

$$(\overline{i_N^2}) = G^2 2e(\bar{i}_c + i_d)\,\Delta f\,\frac{1 - \delta^{-N}}{1 - \delta^{-1}}$$

where δ is the secondary-emission multiplication factor and N is the number of stages.

11.2 Calculate the minimum power that can be detected by a photoconductor in the presence of a strong optical background power P_B. [Answer: $(P_s)_{\min} = 2(P_B h\nu\,\Delta f/\eta)^{1/2}$.]

11.3 Derive the expression for the minimum detectable power using a photoconductor in the video mode (i.e., no local-oscillator power) and assuming that the main noise contribution is the generation–recombination noise. The optical field is given by $e(t) = E(1 + \cos\omega_m t)\cos\omega t$, and the signal is taken as the component of the photocurrent at ω_m.

11.4 Derive the minimum detectable power of a Ge:Hg detector with characteristics similar to those described in Section 11.7 when the average current is due mostly to blackbody radiation incident on the photocathode. Assume $T = 295$ K, an acceptance solid angle $\Omega = \pi$, and a photocathode area of 1 mm^2. Assume that the quantum yield η for blackbody radiation at $\lambda < 14$ μm is unity and that for $\lambda > 14$ μm, $\eta = 0$. [Hint: Find the flux of photons with wavelengths 14 μm $> \lambda > 0$ using blackbody radiation formulas or, more easily, tables or a blackbody "slide rule."]

11.5 Find the minimum detectable power in Problem 11.4 when the input field of view is at $T = 4.2$ K.

11.6 Derive Equations (11.6-15) and (11.6-16).

11.7 Show that the transit-time reduction factor $(1 - e^{-i\omega_m\tau_d})/i\omega_m\tau_d$ in Equation (11.7-7) can be written

$$\alpha - i\beta$$

where

$$\alpha = \frac{\sin\omega_m\tau_d}{\omega_m\tau_d} \quad \text{and} \quad \beta = \frac{1 - \cos\omega_m\tau_d}{\omega_m\tau_d}$$

Plot α and β as functions of $\omega_m\tau_d$.

11.8 Derive the minimum detectable optical power for a photodiode operated in the heterodyne mode. [Answer: $P_{\min} = h\nu\,\Delta\nu/\eta$.]

11.9 Discuss the limiting sensitivity of an avalanche photodiode in which the noise increases as M^2. Compare it with that of a photomultiplier. What is the minimum detectable power in the limit of $M \gg 1$, and of zero background radiation and no dark current?

11.10 Derive an expression for the magnitude of the output current in a heterodyne detection scheme as a function of the angle θ between the signal and local-oscillator propagation directions. Taking the aperture diameter (see Figure 11.6) as D, show that if the output is to remain near its maximum ($\theta = 0°$) value, θ should not exceed λ/D. [Hint: You may replace the lens in Figure 11.6 by the photoemissive surface.] Show that instead of Equation (11.4-4) the current from an element $dx\,dy$ of the detector is

$$di(x,t) = \frac{P_L e\eta}{h\nu(\pi D^2/4)}\left(1 + 2\sqrt{\frac{P_s}{P_L}}\cos(\omega t + kx\sin\theta)\right)dx\,dy$$

The propagation directions lie in the zx plane. The contribution of $dx\,dy$ to the (complex) signal current is thus

$$dI_s(x,t) = \frac{2\sqrt{P_s P_L}}{h\nu(\pi D^2/4)}\,e^{ikx\sin\theta}\,dx\,dy$$

11.11 Show that for a Poisson distribution (see footnote 9) $(\Delta N)^2 = \bar{N}$.

11.12 Assume a fiber distribution network fed by a single semiconductor laser at $\lambda = 1.55$ μm with a power output $P_0 = 10$ mW. The power is divided into N branches, amplified by an optical fiber amplifier (in each branch) and then divided again into M branches. Determine the maximum number of "subscribers" NM that can be serviced by the system assuming: $\Delta f = 10^9$ Hz; R (receiver input impedance) is 10^3 ohms, $T_e = 1000$ K, and a minimum SNR at the subscriber of 42 dB. The maximum power level at the output of the amplifiers is 10 mW.

REFERENCES

1. Yariv, A., *Quantum Electronics*, 3rd ed. Wiley, New York, 1988, p. 54.
2. Engstrom, R. W., Multiplier phototube characteristics: application to low light levels. *J. Opt. Soc. Am.* **37**:420 (1947).
3. Sommer, A. H., *Photo-Emissive Materials*. Wiley, New York, 1968.
4. Chapman, R. A., and W. G. Hutchinson, Excitation spectra and photoionization of neutral mercury centers in germanium. *Phys. Rev.* **157**:615 (1967).
5. Teich, M. C., Infrared heterodyne detection. *Proc. IEEE* **56**:37 (1968).
6. Buczek, C., and G. Picus, Heterodyne performance of mercury doped germanium. *Appl. Phys. Lett.* **11**:125 (1967).
7. Lucovsky, G., M. E. Lasser, and R. B. Emmons, Coherent light detection in solid-state photodiodes. *Proc. IEEE* **51**:166 (1963).
8. Riesz, R. P., High speed semiconductor photodiodes. *Rev. Sci. Instrum.* **33**:994 (1962).
9. Anderson, L. K., and B. J. McMurtry, High speed photodetectors. *Appl. Opt.* **5**:1573 (1966).
10. D'Asaro, L. A., and L. K. Anderson, At the end of the laser beam, a more sensitive photodiode. *Electronics*, **May 30**:94 (1966).
11. Sze, S. M., *Physics of Semiconductor Devices*, 2nd ed. Wiley, New York, 1981.
12. Shockley, W., Hot electrons in germanium and Ohm's law. *Bell Syst. Tech. J.* **30**:990 (1951).
13. Bar-Chaim, N., K. Y. Lau, I. Ury, and A. Yariv, High speed GaAlAs/GaAs photodiode on a semi-insulating GaAs substrate. *Appl. Phys. Lett.* **43**:261 (1983).
14. Wang, S. Y., D. M. Bloom, and D. M. Collins, 20-GHz bandwidth GaAs photodiode. *Appl. Phys. Lett.* **42**:190 (1983).
15. Wang, S. Y., and D. M. Bloom, 100 GHz bandwidth planar GaAs Schottky photodiode. *Electron. Lett.* **19**:554 (1983).

16. Kolner, B. H., D. M. Bloom, and P. S. Cross, Characterization of high speed GaAs photodiodes using a 100-GHz electro-optic sampling system. 1983 Conference on Lasers and Electro-optics, paper ThGl.
17. Valdmanis, J. A., G. Mourou, and C. W. Gabel, Picosecond electro-optic sampling system. *Appl. Phys. Lett.* **41**:211 (1982).
18. McKay, K. G., and K. B. McAfee, Electron multiplication in silicon and germanium. *Phys. Rev.* **91**:1079 (1953).
19. McKay, K. G., Avalanche breakdown in silicon. *Phys. Rev.* **94**:877 (1954).
20. McIntyre, R., Multiplication noise in uniform avalanche diodes. *IEEE Trans. Electron. Devices* **ED-13**:164 (1966).
21. Lindley, W. T., R. J. Phelan, C. M. Wolfe, and A. J. Foyt, GaAs Schottky barrier avalanche photodiodes. *Appl. Phys. Lett.* **14**:197 (1969).

ADDITIONAL READING

Boyd, R. W., *Radiometry and the Detection of Optical Radiation*. Wiley, New York, 1983.

CHAPTER 12

WAVE PROPAGATION IN PERIODIC MEDIA

12.0 INTRODUCTION

The propagation of electromagnetic radiation in periodic media exhibits many interesting and useful phenomena. These include the diffraction of X-rays in crystals, the diffraction of light from the periodic strain variation accompanying a sound wave (Chapter 9), the photonic bandgap of light propagation in periodic layered media, and the fiber Bragg gratings. These phenomena are employed in many optical electronic devices, including diffraction gratings, high-reflectance Bragg mirrors, vertical cavity surface emitting lasers (VCSELs), distributed feedback lasers (DFB lasers), distributed Bragg reflection (DBR) lasers, fiber Bragg gratings (FBGs), and acousto-optic filters. In this chapter we will discuss some general properties of wave propagation in periodic media. We start from some basic concepts of Bragg scattering, including the conservation of energy and momentum. We then describe a matrix method for the special case of periodic layered media, as well as a coupled-mode theory for both transmission gratings and reflection gratings. The behavior of periodic media in optics bears a strong formal analogy to that of electrons in crystals and thus makes heavy use of the concepts of Bloch waves, forbidden gaps, and evanescent waves. In particular, we will discuss the transmission, reflection, and dispersion property of Bragg reflectors. Periodic structures also play an important role in guided-wave optics. We describe the propagation of guided waves in a waveguide with a periodic dielectric perturbation, and its application in fiber Bragg gratings for spectral filtering and dispersion compensation. Near the end, we briefly discuss the subject of two-dimensional and three-dimensional periodic media, and the so-called photonic crystal waveguides.

12.1 PERIODIC MEDIA

Generally, periodic media are any optical structures whose dielectric and permeability constants (tensors), reflecting the translational symmetry, are periodic functions of position:

$$\varepsilon(\mathbf{x}) = \varepsilon(\mathbf{x} + \mathbf{a}), \quad \mu(\mathbf{x}) = \mu(\mathbf{x} + \mathbf{a}) \tag{12.1 1}$$

where $\mathbf{a}$ is any arbitrary lattice vector. These equations merely state that the medium "looks" exactly the same to an observer at $\mathbf{x}$ as at $\mathbf{x} + \mathbf{a}$. The propagation of monochromatic optical waves in a periodic medium is described by Maxwell's equations,

$$\nabla \times \mathbf{H} = i\omega\varepsilon\mathbf{E} \tag{12.1-2}$$

$$\nabla \times \mathbf{E} = -i\omega\mu\mathbf{H} \tag{12.1-3}$$

where ω is the angular frequency of light. These equations must remain the same after we have substituted $\mathbf{x} + \mathbf{a}$ for $\mathbf{x}$ in the operator ∇ and ε, μ. The translational symmetry of the medium requires that we take its normal modes of propagation as

$$\mathbf{E} = \mathbf{E}_{\mathbf{K}}(\mathbf{x})e^{-i\mathbf{K}\cdot\mathbf{x}} \tag{12.1-4}$$

$$\mathbf{H} = \mathbf{H}_{\mathbf{K}}(\mathbf{x})e^{-i\mathbf{K}\cdot\mathbf{x}} \tag{12.1-5}$$

where both $\mathbf{E}_{\mathbf{K}}(\mathbf{x})$ and $\mathbf{H}_{\mathbf{K}}(\mathbf{x})$ are periodic functions of $\mathbf{x}$, that is,

$$\mathbf{E}_{\mathbf{K}}(\mathbf{x}) = \mathbf{E}_{\mathbf{K}}(\mathbf{x} + \mathbf{a}) \tag{12.1-6}$$

$$\mathbf{H}_{\mathbf{K}}(\mathbf{x}) = \mathbf{H}_{\mathbf{K}}(\mathbf{x} + \mathbf{a}) \tag{12.1-7}$$

This is known as the Floquet (or Bloch) theorem. A proof of the theorem is given at the end of this section. The subscript $\mathbf{K}$ indicates that the functions $\mathbf{E}_{\mathbf{K}}$ and $\mathbf{H}_{\mathbf{K}}$ depend on $\mathbf{K}$, which is known as the Bloch wavevector. A dispersion relation exists between ω and $\mathbf{K}$,

$$\omega = \omega(\mathbf{K}) \tag{12.1-8}$$

In the spectral regime where $\mathbf{K}$ is real, optical waves can propagate in the periodic medium without loss. They can be considered as the equivalent in periodic media of plane waves in the case of homogeneous media. The intensities of these waves are periodic functions of position in the medium. We will find out that there exist spectral regimes where $\mathbf{K}$ is a complex vector; the optical waves in this case become evanescent according to Equations (12.1-4) and (12.1-5). These spectral regimes are the so-called photonic bandgaps (or forbidden bands) where optical waves cannot propagate in an infinite periodic medium. Such waves (evanescent waves) can only exist in finite structures with energy concentrated near the surface or around "defects". Dielectric optical structures that exhibit forbidden bands are known as photonic bandgap structures.

In many applications in optical electronics, one often deals with a one-dimensional periodic medium such that

$$\varepsilon(z) = \varepsilon(z + \Lambda) \tag{12.1-9}$$

or equivalently,

$$n(z) = n(z + \Lambda) \tag{12.1-10}$$

where ε is the permittivity tensor, n is the index of refraction, and Λ is the period. In this chapter, we will limit ourselves to nonmagnetic materials so that $\mu = \mu_0$ throughout the periodic structure. In this case, the dielectric tensor alone is adequate for the description of the medium. Figure 12.1 shows schematic drawings of several periodic structures in optical electronics.

Bragg Law and Grating Equation

Before we carry out an electromagnetic analysis of the scattering (or propagation) of optical waves in periodic media, we consider some basic kinematic properties of the scattering. These basic kinematic properties are fundamental to all waves, including electron waves, optical waves, and acoustic waves. The study of such scattering benefits from early work on X-ray diffraction in crystalline solids. Consider an extreme situation when the index modulation (or

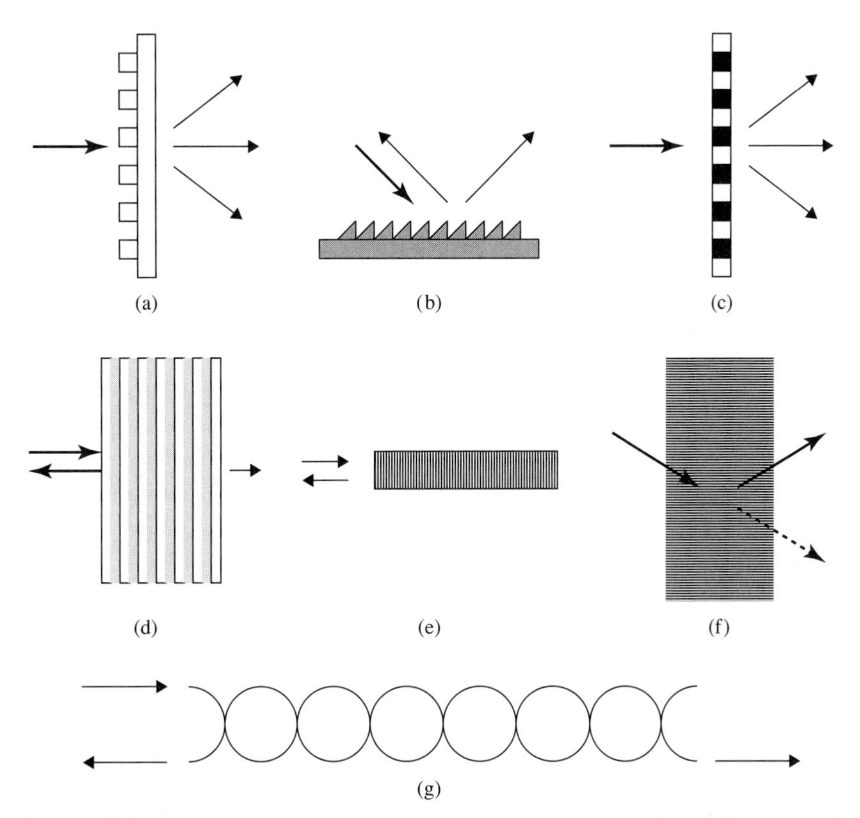

Figure 12.1 Schematic drawings of several periodic structures and the diffraction of light: (a) a transmission phase grating (e.g., surface relief on glass substrate), (b) a reflection phase grating (e.g., surface ruling on metallic surface), (c) an amplitude transmission grating (e.g., a multiple-slit grating), (d) a quarter-wave stack (e.g., multiple thin film deposition of alternating SiO_2 and TiO_2 layers), (e) a fiber Bragg grating (FBG), (f) a volume grating (e.g., Bragg cell), and (g) a periodic array of coupled microring resonators. The arrows indicate incident beam and diffracted beams.

dielectric modulation) of the periodic medium is concentrated in an array of equidistant planes as depicted in Figure 12.2. In addition, we assume that these planes are infinite so that reflection from these planes is specular (i.e., mirrorlike with angle of reflection equal to angle of incidence). Each of the planes reflects only a very small fraction of the incident plane wave. The total scattered wave consists of a linear superposition of all these partially reflected plane waves. The diffracted beams are found when all these reflected plane waves add up constructively.

Let Λ be the spacing between these planes (lattice planes in crystals). For a one-dimensional periodic medium, Λ is the period of the index variation in space. The path difference for rays reflected from two adjacent planes is $2\Lambda \sin\theta$, where θ is the angle of incidence measured from the planes. Constructive interference occurs when the path difference is an integral number of wavelengths λ/n in the medium, so that

$$2\Lambda \sin\theta = m(\lambda/n), \quad m = 1, 2, 3, \ldots \tag{12.1-11}$$

where n is the spatially averaged index of refraction of the periodic medium and m is an integer. For X-rays, the index of refraction n is one ($n = 1$). Equation (12.1-11) is known as the Bragg law. Although reflection from each plane is specular, beam diffraction occurs only

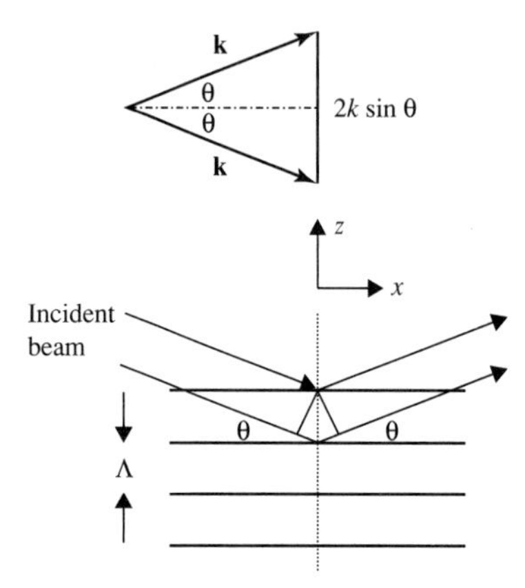

Figure 12.2 Scattering of monochromatic plane wave from a periodic medium.

for certain values of θ, which obey the Bragg law so that reflections from all planes add up in phase. Equation (12.1-11) can also be written

$$2k \sin \theta = m \frac{2\pi}{\Lambda}, \quad m = 1, 2, 3, \ldots \tag{12.1-12}$$

where k is the wavenumber of the light beam in the medium ($k = 2\pi n/\lambda$). The term $2\pi/\Lambda$ is known as the grating wavenumber and is often written

$$G = \frac{2\pi}{\Lambda} \tag{12.1-13}$$

The left side of Equation (12.1-12), $2k \sin \theta$, is the change of the wavevector upon diffraction from the periodic medium. Thus the Bragg law can be interpreted as simply the conservation of momentum. Upon diffraction from a periodic medium (grating), the change of wavevector is exactly an integral number of the grating wavevector G.

In fact, if we expand the periodic index function $n(z)$ (or dielectric function $\varepsilon(z)$) in a Fourier series, we obtain

$$n(z) = n_0 + n_1 \sum_m a_m \exp\left(-i \frac{2\pi}{\Lambda} mz\right) \tag{12.1-14}$$

where n_0 is the spatially averaged index of refraction, n_1 is a constant representing the amplitude of the periodic index modulation, and a_m is the mth Fourier component of the periodic index variation. We note that the mth Fourier component has a wavenumber of $m(2\pi/\Lambda)$. Each of the Fourier components, say, m, can be considered as an infinite set of partial reflectors with a spacing Λ/m responsible for a Bragg diffraction of order m according to Equation (12.1-12). Although the Bragg law is derived by assuming an infinite medium, the results (12.1-11) and (12.1-12) are valid provided the dimensions of these reflection planes are much larger than the beam size. Under these conditions, the periodic medium is called a thick grating (or volume grating). In the discussion of acousto-optic diffraction, only the diffraction order $m = 1$ occurs. This is due to the fact that the index perturbation due to most acoustic waves is sinusoidal, and the high-order Fourier components vanish; that is, $a_m = 0$ for $|m| > 1$.

Referring to Figure 12.3, we now consider the case of a thin grating. In a thin grating, the transverse dimension (thickness) of the periodic index variation is relatively small compared with the beam size and/or wavelength of light. When a plane wave is incident onto the periodic medium, diffraction from each of the planes occurs in addition to the specular

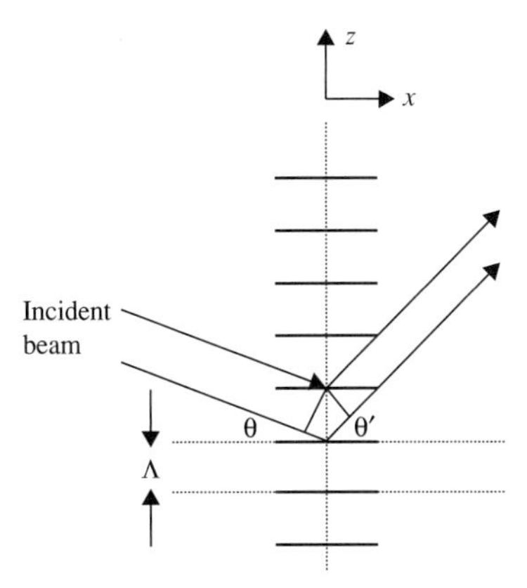

Figure 12.3 Diffraction from a thin grating (e.g., multiple slit).

reflection. The diffraction from each of these planes is a result of the finite size of the planes. This allows the scattered light to be directed along an angle θ', which can be different from the incident angle θ. Using a similar argument leading to Equation (12.1-11), we obtain the following condition for constructive interference:

$$\Lambda \sin \theta + \Lambda \sin \theta' = m(\lambda/n), \quad m = 1, 2, 3, \ldots \tag{12.1-15}$$

where m is an integer. This is known as the grating equation. According to this equation, there are a series of diffraction orders, each one corresponding to a different m, for any given angle of incidence. This is the main difference between a thin grating and a thick grating. In a thick grating, there is only one diffraction order at a given angle of incidence. In addition, diffraction occurs only when the incidence angle satisfies the Bragg law.

Group Velocity and Energy Velocity of Bloch Waves

We investigate the power flow associated with Bloch waves in periodic media. The time averaged flux of electromagnetic energy is given by

$$\mathbf{S} = \tfrac{1}{2}\mathrm{Re}[\mathbf{E} \times \mathbf{H}^*] \tag{12.1-16}$$

The time-averaged electromagnetic energy density is given by

$$U = \tfrac{1}{4}[\mathbf{E} \cdot \varepsilon \mathbf{E}^* + \mathbf{H} \cdot \mu \mathbf{H}^*] \tag{12.1-17}$$

where we assume a real dielectric tensor ε and a real permeability tensor μ. In the case of a Bloch wave in a periodic medium, the Poynting vector $\mathbf{S}$ and the energy density U are both periodic functions of space. We may thus define the energy velocity of the Bloch wave as

$$\mathbf{V}_e = \frac{\displaystyle\int \mathbf{S}\, d^3x}{\displaystyle\int U\, d^3x} = \frac{\langle \mathbf{S} \rangle}{\langle U \rangle} \tag{12.1-18}$$

where the integration is carried out over a single unit cell. The brackets $\langle\ \rangle$ represent the spatial average over a unit cell. By using the Bloch wavefunctions (12.1-4) and (12.1-5), we obtain

$$\mathbf{V}_e = \frac{\langle \frac{1}{2}\mathrm{Re}\,[\mathbf{E_K} \times \mathbf{H_K^*}] \rangle}{\langle \frac{1}{4}[\mathbf{E_K} \cdot \varepsilon \mathbf{E_K^*} + \mathbf{H_K} \cdot \mu \mathbf{H_K^*}] \rangle} \tag{12.1-19}$$

The group velocity associated with Bloch waves is defined as

$$\mathbf{V}_g = \nabla_{\mathbf{K}}\omega \tag{12.1-20}$$

which is a vector perpendicular to the normal surface (i.e., $\omega(\mathbf{K})$ = constant). If we now substitute the Bloch waves into Maxwell's equations (12.1-2) and (12.1-3), we obtain

$$\nabla \times \mathbf{H}_{\mathbf{K}} - i\mathbf{K} \times \mathbf{H}_{\mathbf{K}} = i\omega\varepsilon\mathbf{E}_{\mathbf{K}} \tag{12.1-21}$$

$$\nabla \times \mathbf{E}_{\mathbf{K}} - i\mathbf{K} \times \mathbf{E}_{\mathbf{K}} = -i\omega\mu\mathbf{H}_{\mathbf{K}} \tag{12.1-22}$$

Using the above equations, we now prove that the group velocity is equal to the energy velocity in a periodic medium. We start from Equations (12.1-21) and (12.1-22). Suppose now that $\mathbf{K}$ is changed by an infinitesimal amount $\delta\mathbf{K}$. If $\delta\omega$, $\delta\mathbf{E}_{\mathbf{K}}$, and $\delta\mathbf{H}_{\mathbf{K}}$ are the corresponding changes in ω, $\mathbf{E}_{\mathbf{K}}$, and $\mathbf{H}_{\mathbf{K}}$, respectively, we obtain from Equations (12.1-21) and (12.1-22), after a few steps of algebra,

$$\delta\omega = \mathbf{V}_e \cdot \delta\mathbf{K} \tag{12.1-23}$$

where we assume that ε and μ are independent of the frequency.

From the definition of the group velocity, we also have

$$\delta\omega = (\nabla_{\mathbf{K}}\omega) \cdot \delta\mathbf{K} = \mathbf{V}_g \cdot \delta\mathbf{K} \tag{12.1-24}$$

Since $\delta\mathbf{K}$ is an arbitrary vector, we conclude that

$$\mathbf{V}_e = \mathbf{V}_g \tag{12.1-25}$$

It is important to note that the above equation is valid when both ε and μ are independent of the frequency.

Mathematical Proof of the Bloch Theorem

Here we present a mathematical proof of the Bloch theorem for the case of a dielectric periodic medium with $\mu(\mathbf{x})$ = constant. Since the medium is periodic, we can expand the dielectric tensor $\varepsilon(\mathbf{x})$ in a Fourier series

$$\varepsilon(\mathbf{x}) = \sum_{\mathbf{G}} \varepsilon_{\mathbf{G}} e^{-i\mathbf{G}\cdot\mathbf{x}} \tag{12.1-26}$$

where $\varepsilon_{\mathbf{G}}$ is the Fourier expansion coefficient, and $\mathbf{G}$ runs over all the reciprocal lattice vectors, including $\mathbf{G} = 0$. For the case of one-dimensional periodic media (as described by Equation (12.1-14)),

$$\mathbf{G} = m\frac{2\pi}{\Lambda}\hat{\mathbf{z}}, \quad m = 0, \pm 1, \pm 2, \pm 3, \ldots \tag{12.1-27}$$

where $\hat{\mathbf{z}}$ is a unit vector along the direction of periodicity. The vector $\mathbf{G}$ is known as the reciprocal lattice vector in solid-state physics and plays a fundamental role in crystal physics and optics of periodic media.

The general solutions of Maxwell's equations in periodic media can be expressed as a Fourier integral

$$\mathbf{E}(\mathbf{x}) = \int d^3\mathbf{k}\, \mathbf{A}(\mathbf{k}) e^{-i\mathbf{k}\cdot\mathbf{x}} \tag{12.1-28}$$

where $\mathbf{A}(\mathbf{k})$ is the amplitude of the plane wave component with wavevector $\mathbf{k}$. Substituting Equations (12.1-26) and (12.1-28) into Equations (12.1-2) and (12.1-3) and eliminating the magnetic field vector $\mathbf{H}$, we obtain

$$\int d^3k\ \mathbf{k} \times [\mathbf{k} \times \mathbf{A}(\mathbf{k})]e^{-i\mathbf{k}\cdot\mathbf{x}} + \omega^2\mu \sum_{\mathbf{G}} \int d^3k\ \varepsilon_{\mathbf{G}}\mathbf{A}(\mathbf{k}-\mathbf{G})e^{-i\mathbf{k}\cdot\mathbf{x}} = 0 \qquad (12.1\text{-}29)$$

This equation is satisfied only when all the coefficients of $\exp(-i\mathbf{k}\cdot\mathbf{x})$ vanish. In other words,

$$\mathbf{k} \times [\mathbf{k} \times \mathbf{A}(\mathbf{k})] + \omega^2\mu \sum_{\mathbf{G}} \varepsilon_{\mathbf{G}}\mathbf{A}(\mathbf{k}-\mathbf{G}) = 0 \quad \text{for all } \mathbf{k} \qquad (12.1\text{-}30)$$

where the summation is over all the reciprocal lattice vectors. This is an infinite set of homogeneous equations for the unknown coefficients $\mathbf{A}(\mathbf{k})$. Each equation in this set has a different value of $\mathbf{k}$. The secular equation that results from setting the determinant of Equation (12.1-30) equal to zero can be solved in principle for the whole set. However, by inspecting Equation (12.1-30), we note that not all the coefficients $\mathbf{A}(\mathbf{k})$ are coupled. In fact, only the coefficients of the form $\mathbf{A}(\mathbf{k}-\mathbf{G})$ are coupled, for a given fixed wavevector $\mathbf{k}$. This makes it possible to divide the whole set (12.1-30) into many subsets, each labeled by a wavevector $\mathbf{K}$ and containing equations that involve $\mathbf{A}(\mathbf{K})$ and $\mathbf{A}(\mathbf{K}-\mathbf{G})$ with all possible $\mathbf{G}$'s. Each subset can be solved separately. Using this fact in Equation (12.1-30), the solution of a subset labeled by $\mathbf{K}$ can be written

$$\begin{aligned} \mathbf{E}_{\mathbf{K}} &= \sum_{\mathbf{G}} \mathbf{A}(\mathbf{K}-\mathbf{G})e^{-i(\mathbf{K}-\mathbf{G})\cdot\mathbf{x}} \\ &= e^{-i\mathbf{K}\cdot\mathbf{x}} \sum_{\mathbf{G}} \mathbf{A}(\mathbf{K}-\mathbf{G})e^{-i\mathbf{G}\cdot\mathbf{x}} \\ &= e^{-i\mathbf{K}\cdot\mathbf{x}}\mathbf{E}_{\mathbf{K}}(\mathbf{x}) \end{aligned} \qquad (12.1\text{-}31)$$

where $\mathbf{E}_{\mathbf{K}}(\mathbf{x})$ represents the sum of the Fourier series and is a periodic function of $\mathbf{x}$. In addition, a relationship between ω and $\mathbf{K}$ will be obtained by setting the determinant of this subset of homogeneous equations equal to zero. This is the dispersion relationship $\omega = \omega(\mathbf{K})$. The solution (12.1-31) labeled by $\mathbf{K}$ is a normal mode of propagation in the periodic medium. The label $\mathbf{K}$ is known as the Bloch wavevector. The most general solution (12.1-28) now becomes a linear superposition of these normal modes. This completes the proof of the Bloch theorem (12.1-4).

12.2 PERIODIC LAYERED MEDIA—BLOCH WAVES

We consider in this section the simplest periodic medium, which is made of alternating layers of transparent materials with different refractive indices. Figure 12.4 shows a schematic drawing of such a medium. Recent advances in thin film deposition make it possible to fabricate periodic layered media with well-controlled periodicities and layer thicknesses. In the area of epitaxial crystal growth (e.g., molecular beam epitaxy (MBE)), even atomic layer or molecular layers can be grown. Wave propagation in layered media has been studied by many workers. Exact solutions of the wave equation can be obtained in this case. The matrix method developed in Chapter 4 can be employed to study the transmission and reflection of optical waves in such a structure. In this chapter we will introduce a similar matrix method

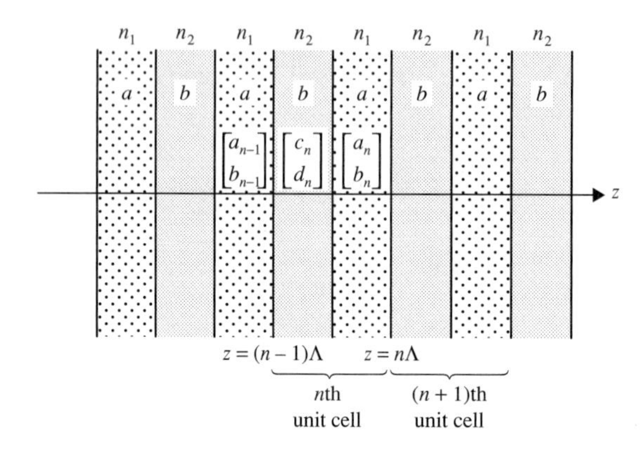

Figure 12.4 Schematic drawing of a periodic layered medium consisting of alternating layers of two different transparent materials with refractive indices n_1 and n_2 and thicknesses a and b, respectively.

to treat the periodic layered media. We will assume that the materials are isotropic and nonmagnetic. The simplest periodic layered medium consists of two different materials with a refractive index profile given by

$$n(z) = \begin{cases} n_2, & 0 < z < b \\ n_1, & b < z < \Lambda \end{cases} \tag{12.2-1}$$

with

$$n(z) = n(z + \Lambda) \tag{12.2-2}$$

where b is the thickness of the layer with refractive index n_2, $a = \Lambda - b$ is the thickness of the layer with refractive index n_1, the z axis is normal to the layer interfaces, and $\Lambda = a + b$ is the period. Without loss of generality, we consider the propagation of optical waves in the yz plane. Since the medium is homogeneous in the y direction, a general solution of the wave equation can be written

$$\mathbf{E}(y, z, t) = \mathbf{E}(z)e^{i(\omega t - k_y y)} \tag{12.2-3}$$

where k_y is the y component of the wavevector of propagation, which remains constant throughout the medium. In the following, we will treat s-polarization (**E** vector polarized perpendicular to the yz plane, the plane of propagation) and p-polarization (**E** vector polarized parallel to the yz plane) separately. The s-wave is also known as the TE wave, and the p-wave is also known as the TM wave. The electric field within each homogeneous layer can be expressed as a sum of a right-traveling ($+z$) and a left-traveling ($-z$) plane wave:

$$E(z) = \begin{cases} a_n e^{-ik_{1z}(z-n\Lambda)} + b_n e^{+ik_{1z}(z-n\Lambda)}, & n\Lambda - a < z < n\Lambda \\ c_n e^{-ik_{2z}(z-n\Lambda+a)} + d_n e^{+ik_{2z}(z-n\Lambda+a)}, & (n-1)\Lambda < z < n\Lambda - a \end{cases} \tag{12.2-4}$$

with

$$k_{1z} = \sqrt{\left(\frac{n_1\omega}{c}\right)^2 - k_y^2} \tag{12.2-5}$$

$$k_{2z} = \sqrt{\left(\frac{n_2\omega}{c}\right)^2 - k_y^2} \tag{12.2-6}$$

where n stands for the nth unit cell, and a_n, b_n, c_n, and d_n are constants. These constants are not independent. In fact, they are related by the continuity conditions at the interfaces.

In the case of TE waves, imposing the continuity of E_x and H_y (with $H_y \propto \partial E_x/\partial z$) at the interfaces $z = (n-1)\Lambda$ and $z = n\Lambda - a$, we obtain

$$\begin{aligned} a_{n-1} + b_{n-1} &= c_n e^{ik_{2z}b} + d_n e^{-ik_{2z}b} \\ ik_{1z}(a_{n-1} - b_{n-1}) &= ik_{2z}(c_n e^{ik_{2z}b} - d_n e^{-ik_{2z}b}) \\ c_n + d_n &= a_n e^{ik_{1z}a} + b_n e^{-ik_{1z}a} \\ ik_{2z}(c_n - d_n) &= ik_{1z}(a_n e^{ik_{1z}a} - b_n e^{-ik_{1z}a}) \end{aligned} \tag{12.2-7}$$

These four equations can be rewritten as two matrix equations:

$$\begin{bmatrix} 1 & 1 \\ ik_{1z} & -ik_{1z} \end{bmatrix} \begin{bmatrix} a_{n-1} \\ b_{n-1} \end{bmatrix} = \begin{bmatrix} e^{ik_{2z}b} & e^{-ik_{2z}b} \\ ik_{2z}e^{ik_{2z}b} & -ik_{2z}e^{-ik_{2z}b} \end{bmatrix} \begin{bmatrix} c_n \\ d_n \end{bmatrix} \tag{12.2-8}$$

$$\begin{bmatrix} 1 & 1 \\ ik_{2z} & -ik_{2z} \end{bmatrix} \begin{bmatrix} c_n \\ d_n \end{bmatrix} = \begin{bmatrix} e^{ik_{1z}a} & e^{-ik_{1z}a} \\ ik_{1z}e^{ik_{1z}a} & -ik_{1z}e^{-ik_{1z}a} \end{bmatrix} \begin{bmatrix} a_n \\ b_n \end{bmatrix} \tag{12.2-9}$$

We note that, in the matrix notation, the complex amplitudes of the two plane waves in each layer constitute the components of a two-component column vector. The electric field in each layer of a unit cell can thus be represented by a column vector. These column vectors are not independent. They are related by the continuity conditions at the interfaces. As a consequence, only one column vector (or two components of different column vectors) can be expressed arbitrarily.

By eliminating the column vector

$$\begin{bmatrix} c_n \\ d_n \end{bmatrix} \tag{12.2-10}$$

the matrix equation

$$\begin{bmatrix} a_{n-1} \\ b_{n-1} \end{bmatrix} = \begin{bmatrix} A & B \\ C & D \end{bmatrix} \begin{bmatrix} a_n \\ b_n \end{bmatrix} \tag{12.2-11}$$

is obtained. The matrix elements are

$$\begin{aligned} A &= e^{ik_{1z}a}\left[\cos k_{2z}b + \frac{i}{2}\left(\frac{k_{2z}}{k_{1z}} + \frac{k_{1z}}{k_{2z}}\right)\sin k_{2z}b\right] \\ B &= e^{-ik_{1z}a}\left[\frac{i}{2}\left(\frac{k_{2z}}{k_{1z}} - \frac{k_{1z}}{k_{2z}}\right)\sin k_{2z}b\right] \\ C &= e^{ik_{1z}a}\left[-\frac{i}{2}\left(\frac{k_{2z}}{k_{1z}} - \frac{k_{1z}}{k_{2z}}\right)\sin k_{2z}b\right] \\ D &= e^{-ik_{1z}a}\left[\cos k_{2z}b - \frac{i}{2}\left(\frac{k_{2z}}{k_{1z}} + \frac{k_{1z}}{k_{2z}}\right)\sin k_{2z}b\right] \end{aligned} \tag{12.2-12}$$

The matrix in Equation (12.2-11) is a unit-cell translation matrix, which relates the complex amplitudes of the plane waves in layer 1 (with index n_1) of a unit cell to those of the equivalent layer in the next unit cell. Because this matrix relates the field amplitudes of two equivalent layers with the same index of refraction, it is unimodular, that is,

$$\begin{vmatrix} A & B \\ C & D \end{vmatrix} = AD - BC = 1 \tag{12.2-13}$$

It is important to note that the unit-cell translation matrix that relates the field amplitudes in layer 2 (with index n_2) is different from the matrix in Equation (12.2-12). These matrices, however, do possess the same trace. It will be shown later that the trace of a unit-cell translation matrix is directly related to the photonic band structure $\omega = \omega(\mathbf{K})$ of the periodic medium.

The matrix elements (A, B, C, D) for TM waves (**H** vector perpendicular to yz plane) are slightly different from those of the TE waves. They are given by

$$\begin{aligned}
A_{\text{TM}} &= e^{ik_{1z}a}\left[\cos k_{2z}b + \frac{i}{2}\left(\frac{n_2^2k_{1z}}{n_1^2k_{2z}} + \frac{n_1^2k_{2z}}{n_2^2k_{1z}}\right)\sin k_{2z}b\right] \\
B_{\text{TM}} &= e^{-ik_{1z}a}\left[\frac{i}{2}\left(\frac{n_2^2k_{1z}}{n_1^2k_{2z}} - \frac{n_1^2k_{2z}}{n_2^2k_{1z}}\right)\sin k_{2z}b\right] \\
C_{\text{TM}} &= e^{ik_{1z}a}\left[-\frac{i}{2}\left(\frac{n_2^2k_{1z}}{n_1^2k_{2z}} - \frac{n_1^2k_{2z}}{n_2^2k_{1z}}\right)\sin k_{2z}b\right] \\
D_{\text{TM}} &= e^{-ik_{1z}a}\left[\cos k_{2z}b - \frac{i}{2}\left(\frac{n_2^2k_{1z}}{n_1^2k_{2z}} + \frac{n_1^2k_{2z}}{n_2^2k_{1z}}\right)\sin k_{2z}b\right]
\end{aligned} \tag{12.2-14}$$

As noted above, only one column vector is independent. We can choose it, for example, as the column vector of layer 1 in the zeroth unit cell. The remaining column vectors of the equivalent layers are related to that of the zeroth unit cell by

$$\begin{bmatrix} a_0 \\ b_0 \end{bmatrix} = \begin{bmatrix} A & B \\ C & D \end{bmatrix}^n \begin{bmatrix} a_n \\ b_n \end{bmatrix} \tag{12.2-15}$$

which can be inverted to yield

$$\begin{bmatrix} a_n \\ b_n \end{bmatrix} = \begin{bmatrix} A & B \\ C & D \end{bmatrix}^{-n} \begin{bmatrix} a_0 \\ b_0 \end{bmatrix} = \begin{bmatrix} D & -B \\ -C & A \end{bmatrix}^n \begin{bmatrix} a_0 \\ b_0 \end{bmatrix} \tag{12.2-16}$$

where the last equality is a result of the unimodular nature of the matrix. The column vector for layer 2 of the same unit cell can always be obtained by using Equation (12.2-9).

Bloch Waves and Photonic Band Structures

Wave propagation in periodic media is very similar to the motion of electrons in crystalline solids. In fact, the formulation of the Kronig–Penney model used in elementary energy band theory of solids is mathematically identical to that of the electromagnetic radiation (TE waves) in periodic layered media. Thus some of the physical concepts used in solid-state physics such as Bloch waves, Brillouin zones, and energy bands can also be used here. A periodic layered medium is equivalent to a one-dimensional crystal, which is invariant under lattice translations. In other words,

$$n^2(z) = n^2(z + \Lambda) \tag{12.2-17}$$

where Λ is the period.

According to the Bloch theorem, the electric field vector of a normal mode of propagation in a periodic medium is of the form

$$\mathbf{E} = \mathbf{E}_K(z)e^{-iKz}e^{i(\omega t - k_y y)} \tag{12.2-18}$$

where $\mathbf{E}_K(z)$ is a periodic function with period Λ, that is,

$$\mathbf{E}_K(z) = \mathbf{E}_K(z + \Lambda) \tag{12.2-19}$$

The subscript K indicates that the function $\mathbf{E}_K(z)$ depends on K. The constant K is known as the Bloch wavenumber. The problem at hand is thus that of determining K and $\mathbf{E}_K(z)$ as functions of ω and k_y. In terms of the column vector representation, and from Equation (12.2-4), the periodic condition (12.2-19) for the Bloch wave is simply

$$\begin{bmatrix} a_n \\ b_n \end{bmatrix} = e^{-iK\Lambda} \begin{bmatrix} a_{n-1} \\ b_{n-1} \end{bmatrix} \tag{12.2-20}$$

It follows from Equations (12.2-11) and (12.2-20) that the column vector of the Bloch wave satisfies the following eigenvalue problem:

$$\begin{bmatrix} A & B \\ C & D \end{bmatrix} \begin{bmatrix} a_n \\ b_n \end{bmatrix} = e^{iK\Lambda} \begin{bmatrix} a_n \\ b_n \end{bmatrix} \tag{12.2-21}$$

The phase factor $\exp(iK\Lambda)$ is thus the eigenvalue of the unit-cell translation matrix (A, B, C, D) and is given by

$$e^{iK\Lambda} = \tfrac{1}{2}(A + D) \pm \sqrt{\tfrac{1}{4}(A + D)^2 - 1} \tag{12.2-22}$$

or equivalently,

$$\cos K\Lambda = \tfrac{1}{2}(A + D) \tag{12.2-23}$$

The eigenvectors corresponding to the above eigenvalues are obtained from Equation (12.2-21) and are

$$\begin{bmatrix} a_n \\ b_n \end{bmatrix} = e^{-inK\Lambda} \begin{bmatrix} a_0 \\ b_0 \end{bmatrix} = e^{-inK\Lambda} \begin{bmatrix} B \\ e^{iK\Lambda} - A \end{bmatrix} \tag{12.2-24}$$

times any arbitrary constant. The Bloch waves that result from Equation (12.2-24) can be considered as the eigenvectors of the unit-cell translation matrix with eigenvalues $\exp(\pm iK\Lambda)$ given by Equation (12.2-22). The two eigenvalues are the inverse of each other since the translation matrix is unimodular. Equation (12.2-23) gives the dispersion relation between ω, k_y, and K for the Bloch wavefunction

$$K(\omega, k_y) = \frac{1}{\Lambda}\cos^{-1}[\tfrac{1}{2}(A + D)] \tag{12.2-25}$$

Regimes where $|(A + D)/2| < 1$ correspond to real K and thus to propagating Bloch waves; when $|(A + D)/2| > 1$, however, $K = m\pi/\Lambda + iK_i$, which has an imaginary part K_i so that the Bloch wave is evanescent. These are the so-called photonic bandgaps of the periodic medium. The photonic bandedges are the regimes where $|(A + D)/2| = 1$.

According to Equations (12.2-4) and (12.2-20), the final result for the Bloch wave in the n_1 layer of the nth unit cell is

$$E(x) = [(a_0 e^{-ik_{1z}(z-n\Lambda)} + b_0 e^{+ik_{1z}(z-n\Lambda)})e^{iK(z-n\Lambda)}]e^{-iKz} \tag{12.2-26}$$

where a_0 and b_0 are constants given by Equation (12.2-24). This completes the solution of the Bloch waves. Note that the expression inside the square brackets in Equation (12.2-26) is indeed periodic. The dispersion relationship (12.2-25) can be written

$$\cos K\Lambda = \begin{cases} \cos k_{1z}a \cos k_{2z}b - \dfrac{1}{2}\left(\dfrac{k_{2z}}{k_{1z}} + \dfrac{k_{1z}}{k_{2z}}\right)\sin k_{1z}a \sin k_{2z}b & \text{(TE)} \\ \cos k_{1z}a \cos k_{2z}b - \dfrac{1}{2}\left(\dfrac{n_2^2 k_{1z}}{n_1^2 k_{2z}} + \dfrac{n_1^2 k_{2z}}{n_2^2 k_{1z}}\right)\sin k_{1z}a \sin k_{2z}b & \text{(TM)} \end{cases} \qquad (12.2\text{-}27)$$

The photonic band structure for a typical periodic layered medium as obtained from Equation (12.2-25) is shown in Figures 12.5 and 12.6 for TE and TM waves, respectively. It is

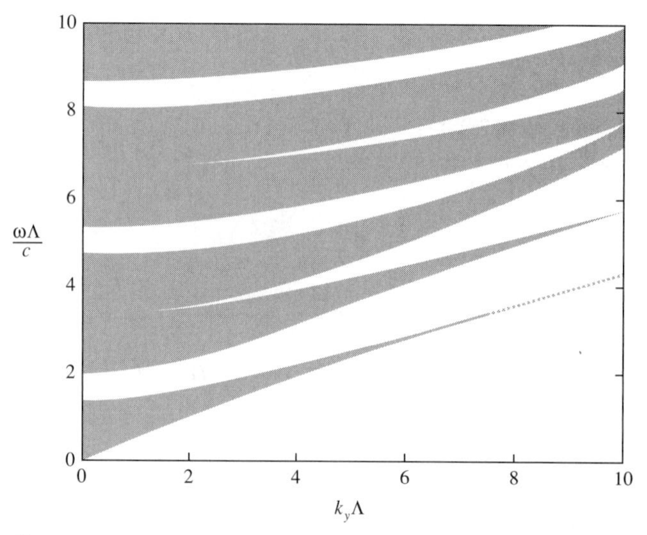

Figure 12.5 Photonic band structure of TE waves in a periodic medium. ($n_1 = 1.45$, $n_2 = 2.65$, $a = \lambda_0/4n_1$, $b = \lambda_0/4n_2$, with $\lambda_0 = 1.5$ μm.) The dark zones are allowed photonic bands in which $|\cos K\Lambda| < 1$.

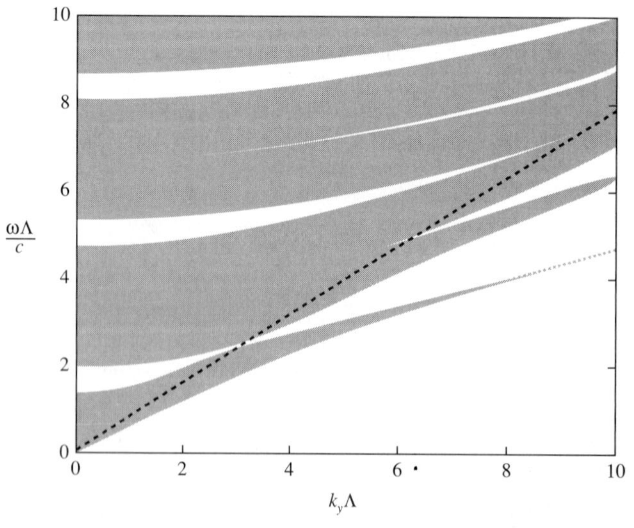

Figure 12.6 Photonic band structure of TM waves in a periodic medium. ($n_1 = 1.45$, $n_2 = 2.65$, $a = \lambda_0/4n_1$, $b = \lambda_0/4n_2$, with $\lambda_0 = 1.5$ μm.) The dark zones are allowed photonic bands in which $|\cos K\Lambda| < 1$. The dashed line is $k_y = (\omega/c)n_2 \sin\theta_B$. The photonic bandgaps vanish at the Brewster angle where the incident wave and the reflected wave are uncoupled.

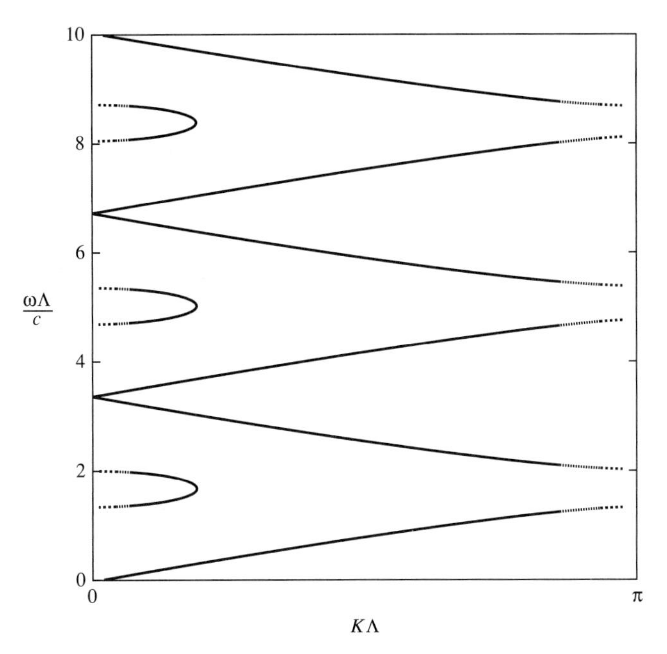

Figure 12.7 Photonic band structure of a periodic medium at normal incidence ($k_y = 0$). The dotted curve near the vertical axis is the imaginary part of the Bloch wavenumber. The periodic medium is a quarter-wave stack. ($n_1 = 1.45$, $n_2 = 2.65$, $a = \lambda_0/4n_1 = 259$ nm, $b = \lambda_0/4n_2 = 142$ nm, with $\lambda_0 = 1.5$ μm.)

interesting to note that the photonic bandgaps for TM waves vanish when $k_y = (\omega/c)n_2 \sin\theta_B$, where θ_B is the Brewster angle, since at this angle the incident and reflected waves are uncoupled.

The dispersion relation ω versus K for the special case $k_y = 0$ (i.e., normal incidence) is shown in Figure 12.7. The dispersion relation between ω and K in this case can be written

$$\cos K\Lambda = \cos k_1 a \cos k_2 b - \frac{1}{2}\left(\frac{n_2}{n_1} + \frac{n_1}{n_2}\right)\sin k_1 a \sin k_2 b \tag{12.2-28}$$

where $k_1 = (\omega/c)n_1$ and $k_2 = (\omega/c)n_2$.

Generally, forbidden gaps occur at $\text{Re}[K\Lambda] = m\pi$, with $m = 1, 2, 3, \ldots$. Using the concept of Brillouin zone borrowed from solid-state physics, the dispersion relationship (ω versus K) can be plotted in the region between 0 and π, as shown in Figure 12.7. Equation (12.2-27) can be solved approximately for K in the photonic bandgaps. The complex Bloch wavenumber of the first photonic bandgap can be written

$$K\Lambda = \pi + i\xi \tag{12.2-29}$$

Let ω_0 be the center of the photonic bandgap such that

$$k_1 a = k_2 b = \pi/2 \tag{12.2-30}$$

A structure with this condition is known as a quarter-wave stack (QWS). At frequency ω_0, Equation (12.2-28) becomes

$$\cos K\Lambda = -\frac{1}{2}\left(\frac{n_1}{n_2} + \frac{n_2}{n_1}\right) \tag{12.2-31}$$

By substituting Equation (12.2-29) into the above equation and solving for ξ, we obtain

$$\xi = \log\left|\frac{n_2}{n_1}\right| \cong \frac{2(n_2 - n_1)}{n_1 + n_2} \tag{12.2-32}$$

where the last equality holds for the case when $|n_2 - n_1| \ll n_{1,2}$. This is the imaginary part of $K\Lambda$ at the center of the first photonic bandgap. Within the photonic bandgap, the imaginary x varies from zero at the bandedges to this maximum value (12.3-32) at ω_0. Let ζ be the normalized frequency deviation from the center of the first photonic bandgap, ω_0:

$$\zeta = \frac{\omega - \omega_0}{c} n_1 a = \frac{\omega - \omega_0}{c} n_2 b \tag{12.2-33}$$

Substitution of Equations (12.2-29) and (12.2-33) into Equation (12.2-28) leads to

$$\cosh \xi = \frac{1}{2}\left(\frac{n_2}{n_1} + \frac{n_1}{n_2}\right)\cos^2\zeta - \sin^2\zeta \tag{12.2-34}$$

This is an expression for the imaginary part of $K\Lambda$ in the photonic bandgap as a function of frequency. The edges of the photonic bands are obtained by setting $\xi = 0$. The result is

$$\zeta_{\text{edge}} = \frac{\omega_{\text{edge}} - \omega_0}{c} n_1 a = \frac{\omega_{\text{edge}} - \omega_0}{c} n_2 b = \pm\sin^{-1}\frac{n_2 - n_1}{n_2 + n_1} \tag{12.2-35}$$

The size of the photonic bandgap is thus given by

$$\Delta\omega_{\text{gap}} = \omega_0 \frac{4}{\pi} \sin^{-1}\left|\frac{n_2 - n_1}{n_2 + n_1}\right| \tag{12.2-36}$$

whereas the imaginary part of $K\Lambda$ at the center of the photonic bandgap is

$$(K_i\Lambda)_{\max} = \log\left|\frac{n_2}{n_1}\right| \cong 2\left|\frac{n_2 - n_1}{n_2 + n_1}\right| \tag{12.2-37}$$

The above calculation is done for the first photonic bandgap. We notice from Figure 12.7 that the even photonic bandgaps ($K\Lambda = 2\pi, 4\pi, 6\pi, \ldots$) at normal incidence vanish. This is only valid for the special case of a quarter-wave stack. At the high-order photonic bandgaps, the quarter-wave stack actually becomes a half-wave stack, and a full-wave stack, and so on. According to Equation (12.2-28), $\cos K\Lambda = \pm 1$ when $k_1 a = k_2 b = m\pi$ with $m = 1, 2, 3, \ldots$. Since $|\cos K\Lambda| = 1$, the imaginary part vanishes.

As an exercise, the student may want to find the Fourier expansion coefficient of the dielectric constant for the periodic layered medium. The result is

$$\frac{\varepsilon_1}{\varepsilon_a} = \frac{4}{\pi}\left|\frac{n_2 - n_1}{n_2 + n_1}\right| \tag{12.2-38}$$

where ε_1 is the first-order Fourier expansion coefficient of the dielectric function $\varepsilon(z)$, and ε_a is the zeroth order Fourier expansion coefficient of the dielectric function $\varepsilon(z)$. We note that ε_a is also the spatial average of the dielectric function $\varepsilon(z)$. Note that the fundamental photonic bandgap of a periodic layered medium is directly proportional to the amplitude of the fundamental component of the Fourier expansion of the dielectric constant $\varepsilon(z)$. In a similar fashion, the higher order photonic bandgaps are proportional to the amplitudes of the corresponding spatial components of the Fourier expansion.

The dispersion relation ω versus K for the special case $k_y = 0$ (i.e., normal incidence) of a non-QWS is shown in Figure 12.8. We note that in this case all high-order photonic bandgaps

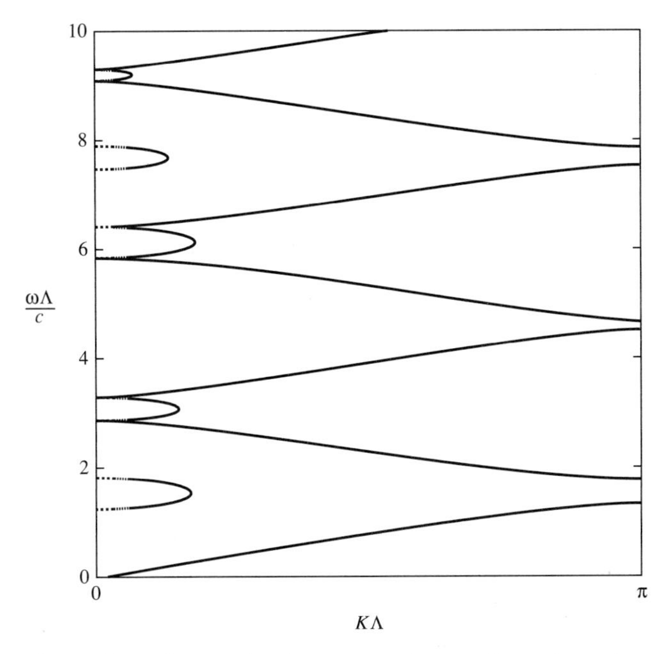

Figure 12.8 Photonic band structure of a periodic medium at normal incidence ($k_y = 0$). The dotted curve near the vertical axis is the imaginary part of the Bloch wavenumber. Note this is not a quarter-wave stack. ($n_1 = 1.45$, $n_2 = 2.65$, $a = b = 200$ nm.)

are finite. Figure 12.9 shows the electric field $E(z)$ of an evanescent Bloch wave with a frequency at the center of the first photonic bandgap of a periodic layered medium. We note that the field amplitude decays exponentially as a function of distance z.

General Properties of *ABCD* Matrix

The unit-cell translation matrix introduced earlier exhibits several interesting and useful properties, including unimodular and symmetric relationships between matrix elements. They are

$$\begin{aligned} &AD - BC = 1 \quad \text{(unimodular)} \\ &D = A^*, \quad C = B^* \end{aligned} \tag{12.2-39}$$

It is important to note that these properties are valid for a general periodic layered medium where each unit cell can consist of an arbitrary number of homogeneous layers. We derive these properties by using the conservation of energy and the time-reversal symmetry.

Time-Reversal Symmetry

The unit-cell translation matrix (A, B, C, D) provides a linear relationship between the plane wave amplitudes in one layer of one cell to those of the next cell. Referring to Figure 12.10, we consider a general unit cell consisting of an arbitrary number of layers.

The matrix relationship between the column vectors of field amplitudes is written

$$\begin{bmatrix} a_0 \\ b_0 \end{bmatrix} = \begin{bmatrix} A & B \\ C & D \end{bmatrix} \begin{bmatrix} a_1 \\ b_1 \end{bmatrix} \tag{12.2-40}$$

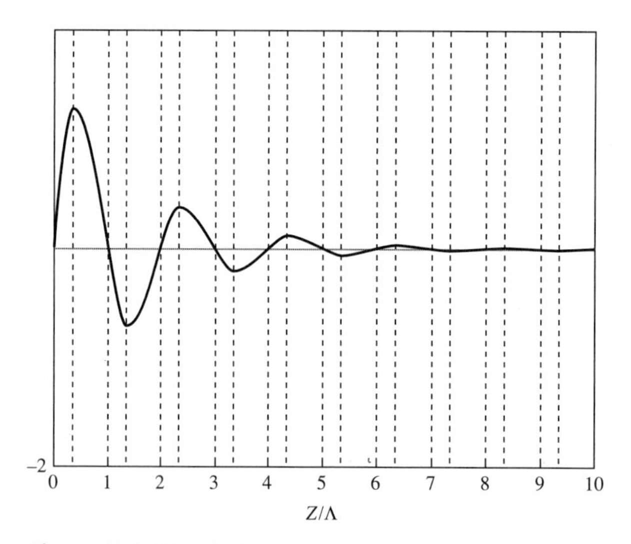

Figure 12.9 Electric field distribution $E(z)$ of an evanescent Bloch wave with a frequency at the center of the first photonic bandgap of a periodic layered medium. The periodic medium is a quarter-wave stack. ($n_1 = 1.45$, $n_2 = 2.65$, $a = \lambda_0/4n_1 = 259$ nm, $b = \lambda_0/4n_2 = 142$ nm, with $\lambda_0 = 1.5$ μm.)

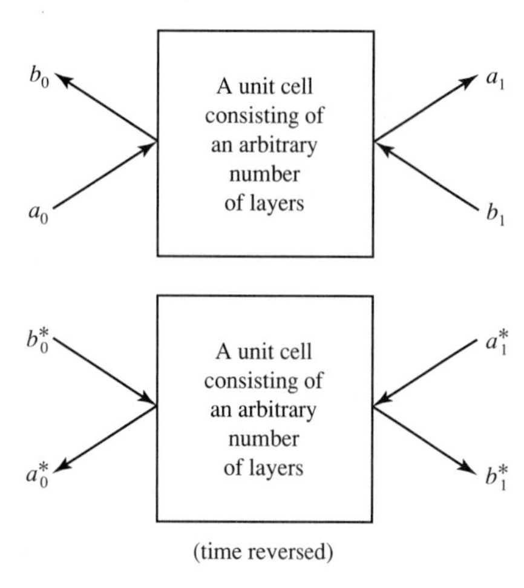

Figure 12.10 Schematic drawing of a unit cell consisting of an arbitrary number of layers.

where (a_0, b_0) are amplitudes of the plane wave components in the layer with refractive index n_1, and (a_1, b_1) are amplitudes of the plane wave components in the corresponding layer with the same refractive index n_1 in the next unit cell.

If the time is reversed, the directions of propagation of all the plane waves are also reversed; the matrix relationship should still hold. Thus we have

$$\begin{bmatrix} b_0^* \\ a_0^* \end{bmatrix} = \begin{bmatrix} A & B \\ C & D \end{bmatrix} \begin{bmatrix} b_1^* \\ a_1^* \end{bmatrix} \tag{12.2-41}$$

From these two matrix equations (12.2-40) and (12.2-41), we obtain

$$\begin{aligned} C &= B^* \\ D &= A^* \end{aligned} \tag{12.2-42}$$

Conservation of Energy

In a lossless periodic medium, the Poynting vector should be a constant throughout the medium. The z component of the Poynting vector for TE waves can be written

$$P_z = \tfrac{1}{2}\,\mathrm{Re}[E_x H_y^*] = \frac{k_{1z}}{2\omega\mu}(|a_0|^2 - |b_0|^2) = \frac{k_{1z}}{2\omega\mu}(|a_1|^2 - |b_1|^2) \tag{12.2-43}$$

The conservation of energy requires that

$$(|a_0|^2 - |b_0|^2) = (|a_1|^2 - |b_1|^2) \tag{12.2-44}$$

Using Equations (12.2-40), (12.2-42), and (12.2-44), we obtain the unimodular relationship,

$$AD - BC = 1 \tag{12.2-45}$$

A similar result can also be obtained for TM waves.

Using Equation (12.2-24) for the column vector of the Bloch wave, as well as Equation (12.2-43) for the z component of the Poynting vector, we can show that Poynting power flow along the z axis is zero in the photonic bandgaps where K is complex. In other words (see Problem 12.8),

$$P_z = \tfrac{1}{2}\,\mathrm{Re}[E_x H_y^*] = \frac{k_{1z}}{2\omega\mu}(|a_0|^2 - |b_0|^2) \propto (|B|^2 - |e^{iK\Lambda} - A|^2) = 0 \tag{12.2-46}$$

provided $K\Lambda = m\pi + iK_i\Lambda$. The zero power flow in the z direction is consistent with the exponentially decaying field amplitude shown in Figure 12.9.

12.3 BRAGG REFLECTORS

When a monochromatic plane wave is incident onto a periodic layered medium, a Bloch wave is generated in the periodic medium according to the discussion in the last section. If the frequency of the incident wave falls in the so-called photonic bandgap, the Bloch wave generated in the periodic medium is known as an evanescent wave, which cannot propagate in the medium. In fact, according to Figure 12.9, the electric field amplitude decays exponentially in the periodic medium. The power flow along the z axis is zero according to Equation (12.2-46). Thus the energy of the incident beam in this case is expected to be totally reflected, and the medium acts as a high-reflectance reflector for the incident wave. In the case of media with a finite number of periodic units, a residual amount of energy will be transmitted. By properly designing the periodic layered medium, it is possible to achieve near-unity reflectance for some selected spectral region.

In this section, we consider the reflection and transmission of electromagnetic radiation through a periodic layered medium. Such a structure exhibits resonance reflection very much like the diffraction of X-rays by crystal lattice planes. Therefore it is called a Bragg reflector. For the sake of illustrating the basic properties of Bragg reflectors, we consider a simple case of a periodic layered medium that consists of N unit cells (i.e., N pair of layers), which are bounded by homogeneous media of index n_1. The geometry of such a structure is sketched in Figure 12.11.

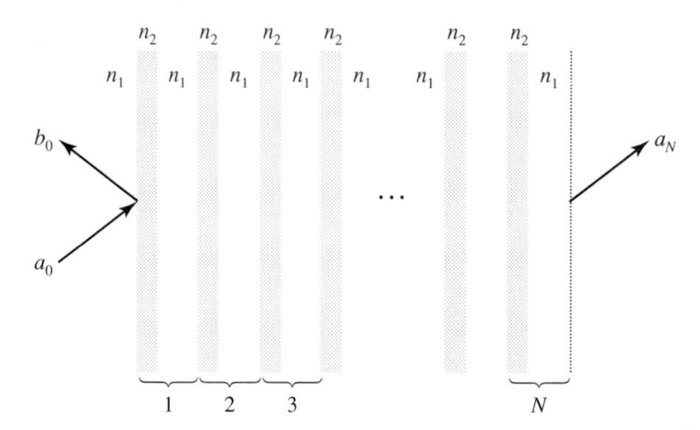

Figure 12.11 Schematic drawing of a Bragg reflector consisting of N pairs of alternating layers of two different materials.

The amplitude reflection coefficient is given by

$$r_N = \left(\frac{b_0}{a_0}\right)_{b_N=0} \tag{12.3-1}$$

From the discussion in the last section and Equation (12.2-15), we have

$$\begin{bmatrix} a_0 \\ b_0 \end{bmatrix} = \begin{bmatrix} A & B \\ C & D \end{bmatrix}^N \begin{bmatrix} a_N \\ b_N \end{bmatrix} \tag{12.3-2}$$

The Nth power of a unimodular matrix can be simplified by using the following matrix identity:

$$\begin{bmatrix} A & B \\ C & D \end{bmatrix}^N = \begin{bmatrix} AU_{N-1} - U_{N-2} & BU_{N-1} \\ CU_{N-1} & DU_{N-1} - U_{N-2} \end{bmatrix} \tag{12.3-3}$$

where

$$U_N = \frac{\sin(N+1)K\Lambda}{\sin K\Lambda} \tag{12.3-4}$$

with K given by Equation (12.2-25).

The reflection coefficient for the electric amplitude is immediately obtained from Equations (12.3-1)–(12.3-3) as

$$r_N = \frac{CU_{N-1}}{AU_{N-1} - U_{N-2}} \tag{12.3-5}$$

The reflectance (or intensity reflectivity) is obtained by taking the absolute square of r_N. After a few steps of elementary algebra, we obtain

$$|r_N|^2 = \frac{|C|^2}{|C|^2 + (\sin K\Lambda/\sin NK\Lambda)^2} \tag{12.3-6}$$

The above is an analytic expression of the reflectance of a Bragg reflector made of a periodic layered medium consisting of alternating layers of two different transparent materials. The special case of $N = 1$ is of particular interest:

$$|r_1|^2 = \frac{|C|^2}{|C|^2 + 1} \tag{12.3-7}$$

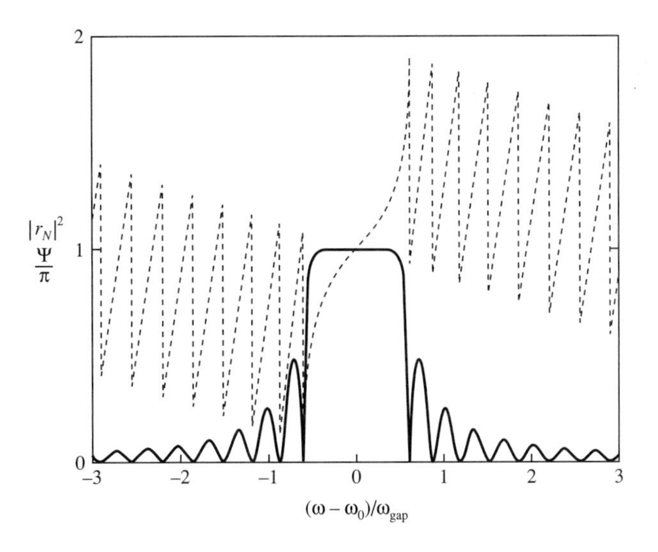

Figure 12.12 Reflectance spectrum (solid curve) as well as phase shift (dotted curve) of a 10-period Bragg reflector, with $N = 10$, $n_1 = 1.45$, $n_2 = 2.25$, $a = 259$ nm, and $b = 167$ nm. The fundamental photonic bandgap is centered at $\lambda_0 = 1500$ nm.

This is the reflectance due to one period of the structure. We note that $|C|^2$ is directly related to the reflectance of a single period of the structure. From the above equation, we obtain

$$|C|^2 = \frac{|r_1|^2}{1 - |r_1|^2} \tag{12.3-8}$$

For Bragg reflectors with $|r_1|^2$ much less that 1, $|C|^2$ is roughly equal to $|r_1|^2$. The second term in the denominator of Equation (12.3-6) is a fast varying function of K or, alternatively, of β and ω. Therefore it dominates the structure of the reflectance spectrum. Figure 12.12 shows the reflectance spectrum of a 10-period ($N = 10$) Bragg reflector centered around the fundamental photonic bandgap. We note that the peak of reflectance occurs at the center of the photonic bandgap. There are sidelobes outside the photonic bandgap. These sidelobes are under the envelope $|C|^2/(|C|^2 + \sin^2 K\Lambda)$. At the edges of photonic bands, $K\Lambda = m\pi$, the reflectance is given by

$$|r_N|^2 = \frac{|C|^2}{|C|^2 + (1/N)^2} \tag{12.3-9}$$

We note that the reflectance is approaching unity at the edges of the photonic bandgap for a Bragg reflector with a large number of periods ($N \gg 1$). In the photonic bandgap, $K\Lambda$ is a complex number

$$K\Lambda = m\pi + iK_i\Lambda \tag{12.3-10}$$

The reflectance formula of Equation (12.3-6) becomes

$$|r_N|^2 = \frac{|C|^2}{|C|^2 + (\sinh K_i\Lambda/\sinh NK_i\Lambda)^2} \tag{12.3-11}$$

For large N, the second term in the denominator approaches zero exponentially as $\exp[-2(N-1)K_i\Lambda]$. It follows that the reflectance in the photonic bandgap is near unity for a Bragg reflector with a substantial number of periods. At the center of each photonic bandgap, each period of the layered medium is approximately an integral multiple of half-waves. In the case of a quarter-wave stack, each period is an exact half-wave at the center of the photonic bandgap. The light waves will be highly reflected, since successive reflections from the

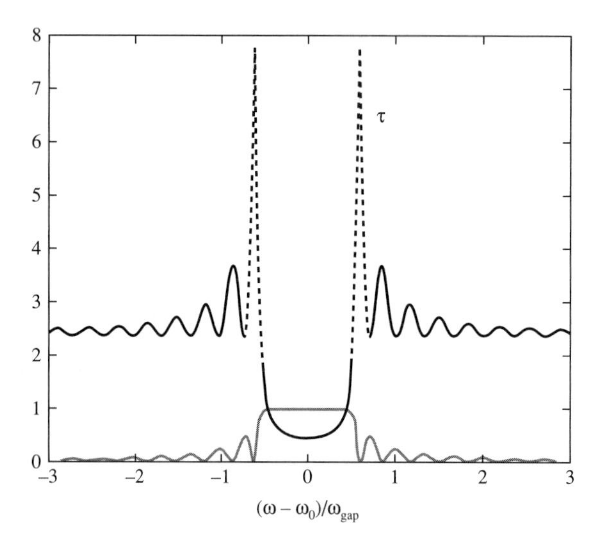

Figure 12.13 Group delay (dotted curve, in units of 10 fs) of a 10-period Bragg reflector as a function of frequency, with $N = 10$, $n_1 = 1.45$, $n_2 = 2.25$, $a = 259$ nm, and $b = 167$ nm. The solid curve is the reflectance spectrum. The fundamental photonic bandgap is centered at $\lambda_0 =$ 1500 nm.

neighboring periods will be in phase with one another and therefore will be constructively superimposed. This situation is analogous to the Bragg reflection of X-rays from crystal planes.

The phase shift Ψ of the optical beam upon reflection from a Bragg reflector is also of particular interest. It is defined as

$$r_N = |r_N| \exp(-i\Psi) \tag{12.3-12}$$

Figure 12.12 also shows the phase shift of a 10-period Bragg reflector as a function of frequency. We note that the phase shift exhibits a strong dispersion as a function of the frequency. As we recall, the group delay is related to the derivative of the phase shift,

$$\tau = \frac{d\Psi}{d\omega} \tag{12.3-13}$$

Figure 12.13 shows the group delay τ as a function of frequency of a 10-period Bragg reflector. We note that the group delay is minimum at the center of the photonic bandgap. This is a result of the minimum optical penetration into the periodic medium at the center of the photonic bandgap.

Transmission Filters

A Bragg reflector can act as a high-reflectance mirror in a selected spectral regime of the photonic bandgap. For a quarter-wave stack, the fundamental photonic bandgap is given, according to Equation (12.2-35), by

$$\Delta\omega_{gap} = \omega_0 \frac{4}{\pi} \sin^{-1} \left| \frac{n_2 - n_1}{n_2 + n_1} \right| \tag{12.3-14}$$

where ω_0 is the center of the photonic bandgap. We note that the stopband is proportional to the index step $|n_2 - n_1|$. Thus a narrowband reflection filter can be obtained by using a quarter-wave stack with small index difference $|n_2 - n_1|$. However, a high reflectance at the center of the band requires a large number of periods due to the small $|n_2 - n_1|$.

A narrowband spectral transmission filter can be obtained by using a half-wave layer sandwiched between two quarter-wave stacks. Figure 12.14 shows a schematic drawing of

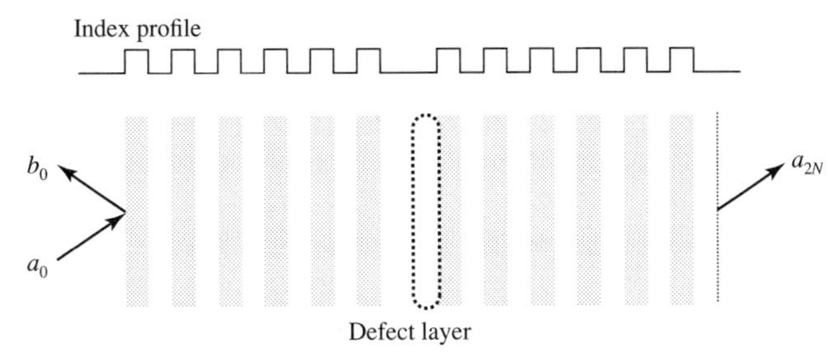

Figure 12.14 Schematic drawing of a transmission filter that consists of two quarter-wave stacks separated by a half-wave layer. The structure may also be viewed as a quarter-wave stack with a defect (i.e., an additional quarter-wave layer) at the center of the stack.

the filter structure. The structure may also be viewed as a quarter-wave stack with a defect (i.e., an additional quarter-wave layer) at the center of the stack. As a result of the presence of the additional layer, the structure acts like a narrowband transmission filter with a passband located at the center of the photonic bandgap of the quarter-wave section. The structure can also be viewed as a Fabry–Perot etalon with a thin cavity of only a quarter-wave layer. The front quarter-wave stack and the rear quarter-wave stack act as mirrors of the etalon. Since quarter-wave stacks are high-reflectance mirrors, the whole structure acts as a high finesse Fabry–Perot etalon with a sharp transmission peak.

Figure 12.15 shows the calculated reflectance spectrum of the filter that is made of a single quarter-wave layer sandwiched between two 10-period quarter-wave stacks. The reflectance spectrum can be obtained by using the matrix method described earlier in this chapter. Specifically, for the structure shown in Figure 12.14, the matrix equation relating the incident amplitude a_0 and the transmitted amplitude a_{2N} can be written

$$\begin{bmatrix} a_0 \\ b_0 \end{bmatrix} = \begin{bmatrix} A & B \\ C & D \end{bmatrix}^N \begin{bmatrix} e^{ik_1 a} & 0 \\ 0 & e^{-ik_1 a} \end{bmatrix} \begin{bmatrix} A & B \\ C & D \end{bmatrix}^N \begin{bmatrix} a_{2N} \\ 0 \end{bmatrix} \tag{12.3-15}$$

where a is the thickness of the defect layer, and $k_1 = n_1\omega/c$.

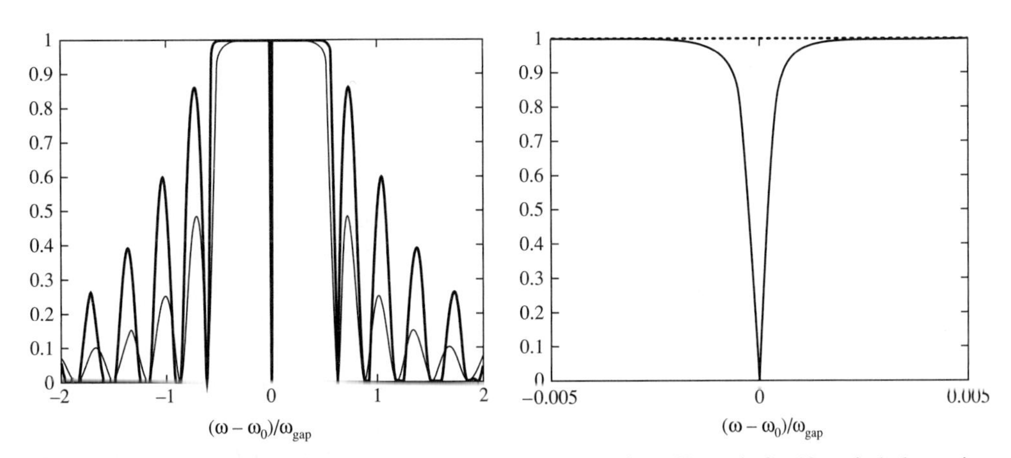

Figure 12.15 Reflectance spectrum of a $(N)Q(N)$ structure, where N stands for 10-period alternating layers of periodic layered stack, Q stands for a quarter-wave layer of n_1 layer, with $n_1 = 1.45$, $n_2 = 2.25$, $a = 259$ nm, and $b = 167$ nm. The fundamental photonic bandgap is centered at $\lambda_0 = 1500$ nm. This structure provides a narrow passband at the center of the photonic bandgap.

12.4 COUPLED-WAVE ANALYSIS

In Section 12.2 we obtained an exact solution for the propagation of electromagnetic radiation in a periodic layered medium. Closed-form Bloch waves of photonic band structures are obtained. There are, however, many periodic media in which only approximate solutions of Maxwell's equations can be obtained. In this section, we describe a coupled-wave analysis in which the periodic variation of the dielectric tensor is considered as a perturbation that couples the unperturbed normal modes of propagation of the structure. In other words, the dielectric tensor as a function of space is written

$$\varepsilon(x, y, z) = \varepsilon_a(x, y) + \Delta\varepsilon(x, y, z) \tag{12.4-1}$$

where $\varepsilon_a(x, y)$ is the unperturbed part of the dielectric tensor, and $\Delta\varepsilon(x, y, z)$, representing the dielectric perturbation, is periodic in the z direction and is the only periodically varying part of the dielectric tensor.

We assume that the normal modes of propagation in the unperturbed dielectric medium described by the dielectric tensor $\varepsilon_a(x, y)$ are known. Examples of the unperturbed structures include free space, slab waveguides, and circular dielectric waveguides (fibers). Since the unperturbed dielectric medium is homogeneous in the z direction, the normal modes of propagation of the unperturbed structure can be written in the form

$$\mathbf{E}(x, y, z, t) = \mathbf{E}_m(x, y)e^{i(\omega t - \beta_m z)} \tag{12.4-2}$$

where β_m is the propagation constant of the mth mode, m is a mode index (usually an integer for confined modes), and $\mathbf{E}_m(x, y)$ is the wavefunction of the normal mode. These normal modes satisfy

$$\left(\frac{\partial^2}{\partial x^2} + \frac{\partial^2}{\partial y^2} + \omega^2\mu\varepsilon_a(x, y) - \beta_m^2\right)\mathbf{E}_m(x, y) = 0 \tag{12.4-3}$$

where we assumed that $(\nabla \cdot \mathbf{E}) = 0$ and the above equation is an approximation of the wave equation. The LP modes in step-index circular fibers and TE modes in dielectric slabs described earlier in Chapter 3 satisfy the above equation. If an arbitrary field of frequency ω is excited at $z = 0$, the propagation of this field in the unperturbed medium can always be expressed in terms of a linear combination of the normal modes:

$$\mathbf{E} = \sum_m A_m \mathbf{E}_m(x, y)e^{i(\omega t - \beta_m z)} \tag{12.4-4}$$

where the A_m are constants and the summation is over all modes of propagation. Equation (12.4-4) is a legitimate expansion provided the normal modes form a complete set. For convenience, the normal modes are usually normalized to a power flow of 1 W in the z direction. Thus the orthogonal relation of the modes can be written

$$\frac{1}{2}\int (\mathbf{E}_m \times \mathbf{H}_n^*)_z \, dx\, dy = \delta_{mn} \tag{12.4-5}$$

where δ_{mn} is the Kronecker delta for discrete modes (confined modes) and the Dirac delta function for continuum modes (unbounded modes), $\mathbf{H}_n$ is the magnetic field associated with the nth mode, and $\mathbf{E}_m$ is the electric field of the mth mode. When $(\nabla \cdot \mathbf{E}) = 0$ and the modes satisfy Equation (12.4-3), this orthogonality relation becomes (see Problem 13.7)

$$\int \mathbf{E}_n^*(x, y) \cdot \mathbf{E}_m(x, y)\, dx\, dy = \frac{2\omega\mu}{|\beta_m|}\delta_{mn} \tag{12.4-6}$$

where δ_{mn} is the Kronecker delta for confined modes and the Dirac delta function for the unbounded modes (e.g., radiation modes). If a single mode is excited at $z = 0$, say, $\mathbf{E}_1(x, y)e^{i(\omega t-\beta_1 z)}$, then the electromagnetic wave will remain in this mode throughout the unperturbed medium.

Let's consider next the propagation of an unperturbed mode $\mathbf{E}_1(x, y)e^{i(\omega t-\beta_1 z)}$ that is excited at $z = 0$ in the periodically perturbed medium described by the dielectric tensor $\varepsilon(x, y, z) = \varepsilon_a(x, y) + \Delta\varepsilon(x, y, z)$. The presence of the dielectric perturbation $\Delta\varepsilon(x, y, z)$ gives rise to an additional polarization

$$\Delta\mathbf{P} = \Delta\varepsilon(x, y, z)\mathbf{E}_1(x, y)e^{i(\omega t-\beta_1 z)} \tag{12.4-7}$$

If this polarization wave, acting as a distributed radiating source, can feed energy into (or out of) some other mode $\mathbf{E}_2(x, y)e^{i(\omega t-\beta_2 z)}$, then we say that the dielectric perturbation $\Delta\varepsilon(x, y, z)$ couples (i.e., causes energy exchange) between modes $\mathbf{E}_1$ and $\mathbf{E}_2$. Let us find, next, under what conditions this coupling takes place.

The energy exchange between unperturbed modes due to the dielectric perturbation $\Delta\varepsilon(x, y, z)$ is analogous to the transition between the electronic eigenstates of an atom (e.g., hydrogen atom) under the influence of a time-dependent perturbation. An incoming photon can be absorbed, and an electron is excited from the ground state to an upper state. The mathematical formalism for the mode coupling is very similar to that of the time-dependent perturbation theory of quantum mechanics. The mathematical approach is sometimes called the method of variation of constants. The procedure consists of expressing the electric field vector of the electromagnetic wave as a linear combination of the normal modes of the unperturbed dielectric medium, where the expansion coefficients evidently depend on z, since for $\Delta\varepsilon(x, y, z) \neq 0$ the waves $\mathbf{E}_m(x, y)e^{i(\omega t-\beta_m z)}$ are no longer eigenmodes:

$$\mathbf{E} = \sum_m A_m(z)\mathbf{E}_m(x, y)e^{i(\omega t-\beta_m z)} \tag{12.4-8}$$

The z dependence of the amplitudes $A_m(z)$ is a mere reflection of the coupling of the normal modes. We now substitute Equation (12.4-8) into the wave equation

$$\{\nabla^2 + \omega^2\mu[\varepsilon_a(x, y) + \Delta\varepsilon(x, y, z)]\}\mathbf{E} = 0 \tag{12.4-9}$$

Using Equation (12.4-3), we obtain

$$\sum_m \left(\frac{d^2}{dz^2}A_m - 2i\beta_m\frac{d}{dz}A_m\right)\mathbf{E}_m(x, y)e^{i(\omega t-\beta_m z)} = -\omega^2\mu\sum_n \Delta\varepsilon(x, y, z)A_n\mathbf{E}_n(x, y)e^{i(\omega t-\beta_n z)} \tag{12.4-10}$$

We now assume further that the dielectric perturbation is "weak," so that the variation of the mode amplitudes is "slow" and satisfies the condition

$$\frac{d^2}{dz^2}A_m \ll \beta_m\frac{d}{dz}A_m \tag{12.4-11}$$

This is also called the slowly varying amplitude (SVA) approximation and is often used when the perturbation is small. Thus neglecting the second derivative in Equation (12.4-10) leads to

$$-2i\sum_m \beta_m\left(\frac{d}{dz}A_m\right)\mathbf{E}_m(x, y)e^{-i\beta_m z} = -\omega^2\mu\sum_n \Delta\varepsilon(x, y, z)A_n\mathbf{E}_n(x, y)e^{-i\beta_n z} \tag{12.4-12}$$

We next take the scalar product of Equation (12.4-12) with $\mathbf{E}_k^*(x, y)$ and integrate over all x and y. The result, using the orthogonal property (12.4-6) of the normal modes, is

$$\langle k|k\rangle \frac{d}{dz} A_k = \frac{\omega^2\mu}{2i\beta_k} \sum_n \langle k|\Delta\varepsilon|n\rangle A_n(z) e^{i(\beta_k-\beta_n)z} \tag{12.4-13}$$

where

$$\langle k|k\rangle = \int E_k^* \cdot E_k \, dx\, dy = \frac{2\omega\mu}{|\beta_k|} \tag{12.4-14}$$

$$\langle k|\Delta\varepsilon|n\rangle = \int \mathbf{E}_k^* \cdot \Delta\varepsilon(x, y, z)\mathbf{E}_n \, dx\, dy \tag{12.4-15}$$

Since the dielectric perturbation $\Delta\varepsilon(x, y, z)$ is periodic in z, we can expand it as a Fourier series

$$\Delta\varepsilon(x, y, z) = \sum_{m\neq 0} \varepsilon_m(x, y) \exp\left(-im\frac{2\pi}{\Lambda} z\right) \tag{12.4-16}$$

where the summation is over all integers m, except $m = 0$ because of the definition of $\Delta\varepsilon(x, y, z)$ in Equation (12.4-1).

Substitution of Equations (12.4-14), (12.4-15), and (12.4-16) in Equation (12.4-13) leads to

$$\frac{d}{dz} A_k = -i\frac{\beta_k}{|\beta_k|} \sum_m \sum_n C_{kn}^{(m)} A_n(z) e^{i(\beta_k-\beta_n-m2\pi/\Lambda)z} \tag{12.4-17}$$

where the coupling coefficient $C_{kn}^{(m)}$ is defined as

$$C_{kn}^{(m)} = \frac{\omega}{4}\langle k|\varepsilon_m(x, y)|n\rangle = \frac{\omega}{4}\int \mathbf{E}_k^* \cdot \varepsilon_m(x, y)\mathbf{E}_n \, dx\, dy \tag{12.4-18}$$

This coefficient $C_{kn}^{(m)}$ reflects the magnitude of the coupling between the kth and nth modes due to the mth Fourier component of the dielectric perturbation.

Equation (12.4-17) constitutes a set of coupled linear differential equations. In principle, an infinite number of mode amplitudes are involved. However, in practice, especially near the condition of resonant coupling, only two modes are strongly coupled, and Equation (12.4-17) reduces to a set of two equations for the two mode amplitudes. By resonant coupling, we mean a mode coupling between two modes (modes k and n) at the condition when

$$\beta_k - \beta_n - m\frac{2\pi}{\Lambda} = 0 \tag{12.4-19}$$

for some integer m. This condition is of fundamental importance, and we will refer to it as "longitudinal phase matching" or just as phase matching. This condition is the spatial analogy of the conservation of energy in time-dependent perturbation theory in quantum mechanics and therefore may be called the conservation of momentum. The resonant coupling can be explained as follows. According to Equation (12.4-17), we notice that the increment in the field amplitude of the kth mode dA_k, due to the mode coupling with the nth mode in the region between z and $z + dz$ via the mth Fourier component of the dielectric perturbation, is given by

$$dA_k = -i\frac{\beta_k}{|\beta_k|} C_{kn}^{(m)} A_n(z) e^{i(\beta_k-\beta_n-m2\pi/\Lambda)z} dz \tag{12.4-20}$$

Since the field amplitudes are slowly varying functions of space, we may integrate the above equation over a distance L which is much larger than the period Λ, yet is much smaller than the variation scale of the field amplitudes. This leads to an expression of the net increment of the field amplitude, ΔA_k, due to the mode coupling with the nth mode over the distance between z and $z + L$, mediated by the mth Fourier component of the dielectric perturbation:

$$\Delta A_k = -i\frac{\beta_k}{|\beta_k|} C_{kn}^{(m)} A_n \int_{L \gg \Lambda} e^{i(\beta_k - \beta_n - m2\pi/\Lambda)z} dz \tag{12.4-21}$$

From this equation we find that the mode coupling between the kth mode and the nth mode is insignificant when condition (12.4-19) is not satisfied for some integer m, because the integral (12.4-21) is nonvanishing only when the exponent is zero, which is exactly the phase-matching condition (12.4-19).

In summary, the propagation of electromagnetic radiation in a periodically perturbed medium can be described by the method of variation of constants. These mode amplitudes ("constants") are governed by the coupled equations (12.4-17). For significant mode coupling to take place, two conditions must be satisfied. The first is expressed by Equation (12.4-19), the kinematical condition. The second condition requires that the coupling constant $C_{kn}^{(m)}$ must not vanish. The latter is also called the dynamical condition, since it depends on characteristics of the waves such as their polarizations and the mode wavefunctions.

The general properties of mode coupling described above are of great importance in that they can be used to determine what process can take place and what kind of perturbation is needed to couple a given pair of normal modes. Several examples will be discussed in this chapter to demonstrate this point.

Equation (12.4-19) is also known as the Bragg condition because of its analogy to X-ray diffraction by crystals. In that case an incident wave, represented by a plane wave with an electric field proportional to $\exp(-ik_y y - i\beta z)$, is strongly coupled to a reflected wave with an electric field proportional to $\exp(-ik_y y + i\beta z)$. The constant β is the component of the wavevector perpendicular to the relevant crystal plane. It follows from Equation (12.4-19) that the spacing of the crystal planes Λ needs to satisfy

$$\beta - (-\beta) - m\frac{2\pi}{\Lambda} = 0 \quad (m = 1, 2, 3, \ldots) \tag{12.4-22}$$

or equivalently, since $\beta = k \sin\theta$, where θ is the angle of incidence measured from the crystal planes,

$$2\Lambda \sin\theta = m\lambda \quad (m = 1, 2, 3, \ldots) \tag{12.4-23}$$

The above equation is the well-known Bragg condition for X-ray diffraction. As noted above, this condition is necessary but not sufficient. The intensity of diffraction depends on the Fourier expansion coefficients of the periodic dielectric perturbation as well as on the polarization of the waves.

Coupled-Mode Equations

Equation (12.4-17) describes the most general case of mode coupling due to a periodic dielectric perturbation. In practice, often only the coupling between two modes is involved. Let the two coupled modes be designated as $\mathbf{E}_1(x, y)e^{i(\omega t - \beta_1 z)}$ and $\mathbf{E}_2(x, y)e^{i(\omega t - \beta_2 z)}$. Neglecting interaction with any other modes, the coupled equations in this case become

$$\frac{d}{dz}A_1 = -i\frac{\beta_1}{|\beta_1|}C_{12}^{(m)}A_2(z)e^{i\Delta\beta z}$$
$$\frac{d}{dz}A_2 = -i\frac{\beta_2}{|\beta_2|}C_{21}^{(-m)}A_1(z)e^{-i\Delta\beta z} \quad (12.4\text{-}24)$$

where

$$\Delta\beta = \beta_1 - \beta_2 - m\frac{2\pi}{\Lambda} = \beta_1 - \beta_2 - mK \quad (12.4\text{-}25)$$

where $K = 2\pi/\Lambda$, and $C_{12}^{(m)}$ and $C_{21}^{(-m)}$ are the coupling coefficients given by Equation (12.4-18). The constant K is also known as the grating wavenumber. It can be shown directly from the definition (12.4-18) that

$$C_{12}^{(m)} = [C_{21}^{(-m)}]^* \quad (12.4\text{-}26)$$

provided that the dielectric perturbation $\Delta\varepsilon(x, y, z)$ is a Hermitian dielectric tensor.

In the event that the dielectric tensor ε is a function of z only (i.e., no x, y dependence), the normal modes of the unperturbed medium are plane waves and the Fourier expansion coefficients ε_m of the periodic dielectric perturbation are constants. The coupling coefficients for this special case become

$$C_{kn}^{(m)} = \frac{\omega^2\mu}{2\sqrt{|\beta_k\beta_n|}}\mathbf{p}_k^* \cdot \varepsilon_m \mathbf{p}_n \quad (12.4\text{-}27)$$

where $\mathbf{p}_k$ and $\mathbf{p}_n$ are unit vectors representing the polarization states of the plane waves. We notice that the coupling coefficient depends on the polarization states of the coupled modes as well as the tensor property of the Fourier expansion coefficient ε_m of dielectric perturbation.

The signs of the factors $\beta_1/|\beta_1|$ and $\beta_2/|\beta_2|$ in the coupled equations (12.4-24) are very important and will determine the behavior of the coupling. These signs, of course, depend on the direction of propagation of the coupled modes. The coupling is therefore divided into two categories: codirectional coupling and contradirectional coupling.

Codirectional Coupling ($\beta_1\beta_2 > 0$)

When the coupled modes are propagating in the same direction, say, the $+z$ direction, the sign factors $\beta_1/|\beta_1|$ and $\beta_2/|\beta_2|$ are both equal to 1. The coupled equations (12.4-24) become

$$\frac{d}{dz}A_1 = -i\kappa A_2(z)e^{i\Delta\beta z}$$
$$\frac{d}{dz}A_2 = -i\kappa^* A_1(z)e^{-i\Delta\beta z} \quad (12.4\text{-}28)$$

where the coupling constant is written

$$\kappa = C_{12}^{(m)} \quad (12.4\text{-}29)$$

Remember that A_1 and A_2 are the complex amplitudes of the normalized modes. Therefore $|A_1|^2$ and $|A_2|^2$ represent the power flow in modes 1 and 2, respectively. The coupled equations (12.4-28) are consistent with the conservation of energy (see Problem 12.10), which requires that

$$\frac{d}{dz}(|A_1|^2 + |A_2|^2) = 0 \quad (12.4\text{-}30)$$

To solve the coupled equations, we first consider the simple case of perfect phase matching ($\Delta\beta = 0$).

The Case Where $\Delta\beta = 0$

Phase matching occurs when

$$\Delta\beta = \beta_1 - \beta_2 - m\frac{2\pi}{\Lambda} = 0 \tag{12.4-31}$$

for some integer m. Under the condition of phase matching, the coupled equations become

$$\begin{aligned} \frac{d}{dz} A_1 &= -i\kappa A_2 \\ \frac{d}{dz} A_2 &= -i\kappa^* A_1 \end{aligned} \tag{12.4-32}$$

Solutions of (12.4-32) can easily be obtained by first eliminating A_2 and then solving the resultant differential equation for A_1. This leads to

$$\begin{aligned} A_1(z) &= A_1(0)\cos|\kappa|z - i\frac{\kappa}{|\kappa|}A_2(0)\sin|\kappa|z \\ A_2(z) &= -i\frac{\kappa^*}{|\kappa|}A_1(0)\sin|\kappa|z + A_2(0)\cos|\kappa|z \end{aligned} \tag{12.4-33}$$

where $A_1(0)$ and $A_2(0)$ are the mode amplitudes at $z = 0$. In the special case of a single wave incident at $z = 0$ (i.e., $A_2(0) = 0$), the solutions become

$$\begin{aligned} A_1(z) &= A_1(0)\cos|\kappa|z \\ A_2(z) &= -i\frac{\kappa^*}{|\kappa|}A_1(0)\sin|\kappa|z \end{aligned} \tag{12.4-34}$$

We note that

$$|A_1(z)|^2 + |A_2(z)|^2 = |A_1(0)|^2 \tag{12.4-35}$$

so that the total power carried by both light waves is conserved. If the interaction distance between the two waves L is such that $|\kappa|L = \pi/2$, the total power of mode 1 is transferred completely into the power of mode 2. For a general interaction distance, we may define a coupling efficiency as

$$\eta = \left|\frac{A_2(L)}{A_1(0)}\right|^2 = \sin^2|\kappa|L \tag{12.4-36}$$

Figure 12.16a plots the mode powers $|A_1|^2$ and $|A_2|^2$ as functions of the interaction distance z. We notice that the power is transferring back and forth between the coupled modes, with a complete transfer occurring at $|\kappa|L = \pi/2, 3\pi/2, 5\pi/2, \ldots$. Next, we consider the case of coupling with a phase mismatch. We will find that in this case only a partial power transfer takes place.

The Case Where $\Delta\beta \neq 0$

Here we consider the case when a phase mismatch is present. To solve the coupled equations, we multiply the first equation of (12.4-28) with $\exp(-i\,\Delta\beta\, z)$ on both sides. This leads to

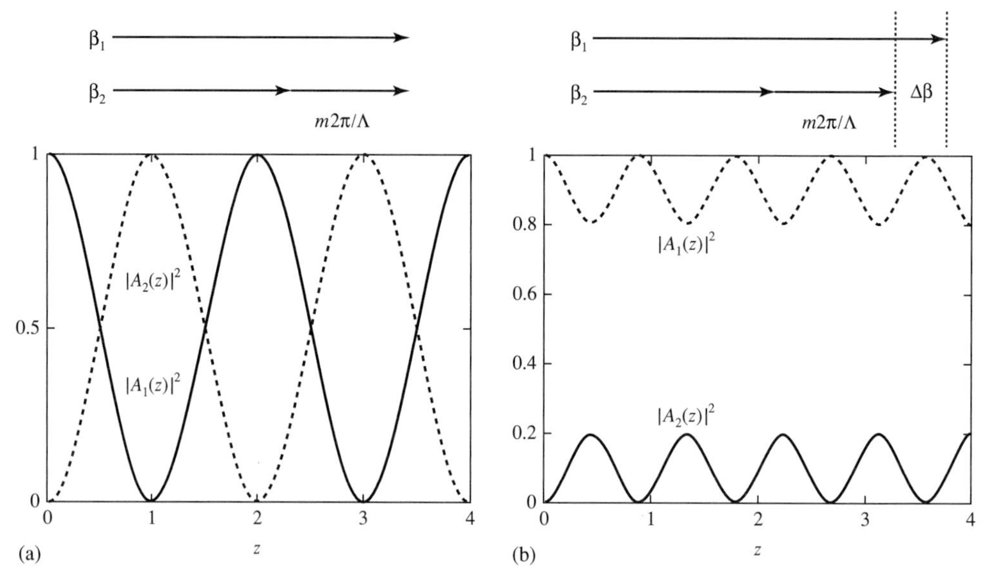

Figure 12.16 Mode powers $|A_1|^2$ and $|A_2|^2$ versus z, with $\kappa = \pi/2$, where, without loss of generality, we assume a real coupling constant κ. The vectors above the figures show the momentum diagrams of the codirectional coupling. (a) $\Delta\beta = 0$, (b) $\Delta\beta = 4\kappa$.

$$e^{-i\Delta\beta z}\frac{d}{dz}A_1 = -i\kappa A_2(z) \tag{12.4-37}$$

Carrying out a differentiation with respect to z on both sides of Equation (12.4-37), and using (12.4-28) to eliminate A_2, we arrive at the following differential equation:

$$\frac{d^2}{dz^2}A_1 - i\,\Delta\beta\frac{d}{dz}A_1 + |\kappa|^2 A_1 = 0 \tag{12.4-38}$$

Equation (12.4-38) is a second-order ordinary differential equation. The general solution for A_1 can easily be obtained. Once A_1 is obtained, the general solution for A_2 can then be obtained directly from Equation (12.4-37). After a few steps of algebra, the general solutions can be written

$$\begin{aligned} A_1(z) &= e^{i(\Delta\beta/2)z}\left[\left(\cos sz - i\frac{\Delta\beta}{2}\frac{\sin sz}{s}\right)A_1(0) - i\kappa\frac{\sin sz}{s}A_2(0)\right] \\ A_2(z) &= e^{-i(\Delta\beta/2)z}\left[-i\kappa^*\frac{\sin sz}{s}A_1(0) + \left(\cos sz + i\frac{\Delta\beta}{2}\frac{\sin sz}{s}\right)A_2(0)\right] \end{aligned} \tag{12.4-39}$$

where

$$s^2 = \kappa^*\kappa + \left(\frac{\Delta\beta}{2}\right)^2 \tag{12.4-40}$$

and $A_1(0)$ and $A_2(0)$ are the mode amplitudes at $z = 0$.

In the case of single-beam incidence at $z = 0$ (i.e., $A_2(0) = 0$), the solutions become

$$A_1(z) = e^{i(\Delta\beta/2)z}\left(\cos sz - i\frac{\Delta\beta}{2}\frac{\sin sz}{s}\right)A_1(0)$$
$$A_2(z) = e^{-i(\Delta\beta/2)z}\left(-i\kappa^*\frac{\sin sz}{s}\right)A_1(0) \tag{12.4-41}$$

We note that

$$|A_1(z)|^2 + |A_2(z)|^2 = |A_1(0)|^2 \tag{12.4-42}$$

which means that the total power flow along the z direction is conserved. The coupling efficiency (or diffraction efficiency) is defined as

$$\eta = \left|\frac{A_2(L)}{A_1(0)}\right|^2 \tag{12.4-43}$$

where L is the interaction length. Using Equations (12.4-41) and (12.4-40), we obtain

$$\eta = \frac{|\kappa|^2}{|\kappa|^2 + (\Delta\beta/2)^2}\sin^2|\kappa|L\sqrt{1+\left(\frac{\Delta\beta}{2|\kappa|}\right)^2} \tag{12.4-44}$$

Inspection of Equation (12.4-44), or (12.4-41), reveals that the power transfer or coupling efficiency can never be complete, except in the phase-matched case, $\Delta\beta = 0$. Figure 12.16b plots the mode powers $|A_1|^2$ and $|A_2|^2$ as a function of interaction distance z. The incomplete power transfer between the modes is a result of the phase mismatch. The maximum coupling efficiency can be written, according to Equation (12.4-44),

$$\eta_{\max} = \frac{|\kappa|^2}{|\kappa|^2 + (\Delta\beta/2)^2} \tag{12.4-45}$$

The maximum coupling efficiency drops to $\frac{1}{2}$ when $\Delta\beta = 2|\kappa|$. Figure 12.17 plots the coupling efficiency as a function of the phase mismatch $\Delta\beta\, L$ for the case when $|\kappa|L = \pi/2$. We note that the coupling efficiency is 100% when $\Delta\beta = 0$, and the efficiency drops as the phase mismatch $\Delta\beta$ increases.

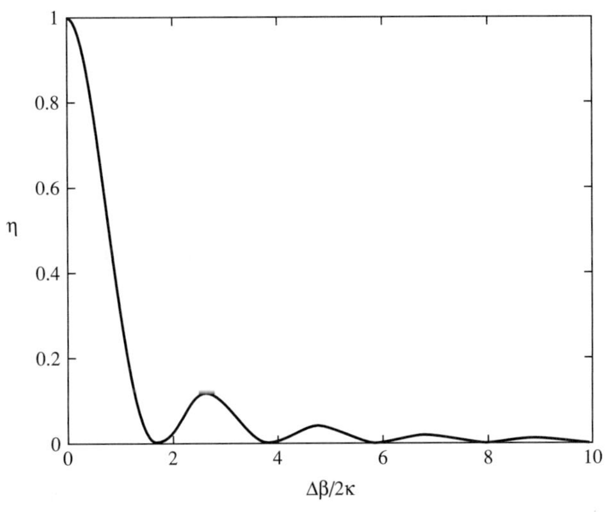

Figure 12.17 Coupling efficiency η as a function of the phase mismatch, with $\kappa L = \pi/2$, where, without loss of generality, we assume a real coupling constant κ.

Contradirectional Coupling ($\beta_1\beta_2 < 0$)

When the coupled modes are propagating in opposite directions, say, $\beta_1 > 0$ and $\beta_2 < 0$, the sign factor $\beta_1/|\beta_1|$ and $\beta_2/|\beta_2|$ become 1 and −1, respectively. The coupled equations (12.4-24) in this case become

$$\frac{d}{dz}A_1 = -i\kappa A_2(z)e^{i\Delta\beta z}$$
$$\frac{d}{dz}A_2 = i\kappa^* A_1(z)e^{-i\Delta\beta z} \tag{12.4-46}$$

where the coupling constant κ is again given by Equation (12.4-29). The net power flow in the $+z$ direction for this case is $|A_1|^2 - |A_2|^2$. The coupled equations are again consistent with the conservation of energy (see Problem 12.11), which requires that

$$\frac{d}{dz}(|A_1|^2 - |A_2|^2) = 0 \tag{12.4-47}$$

We now solve (12.4-46) for the mode amplitudes.

The Case Where $\Delta\beta = 0$

In the case of perfect phase matching, the coupled equations (12.4-46) become

$$\frac{d}{dz}A_1 = -i\kappa A_2$$
$$\frac{d}{dz}A_2 = i\kappa^* A_1 \tag{12.4-48}$$

The above equation can easily be integrated by first eliminating A_2. The solutions with the boundary condition $A_2(L) = 0$ can be written

$$A_1(z) = \frac{\cosh|\kappa|(L-z)}{\cosh|\kappa|L}A_1(0)$$
$$A_2(z) = -i\frac{\kappa^*}{|\kappa|}\frac{\sinh|\kappa|(L-z)}{\cosh|\kappa|L}A_1(0) \tag{12.4-49}$$

We note that

$$|A_1(z)|^2 - |A_2(z)|^2 = |A_1(0)|^2 - |A_2(0)|^2 = |A_1(L)|^2 = \frac{|A_2(0)|^2}{\cosh^2|\kappa|L} \tag{12.4-50}$$

which indicates that the net power flow along the z axis is conserved. Figure 12.18 shows the mode power $|A_1|^2$ and $|A_2|^2$ as functions of z. We note that the power of mode 1 is a decreasing function of interaction distance z, as the mode penetrates the periodic medium. The loss of mode power is converted into the power of mode 2, which increases as it propagates along the $-z$ direction.

The coupling efficiency defined as

$$\eta = \frac{|A_2(0)|^2}{|A_1(0)|^2} \tag{12.4-51}$$

is given, according to (12.4-49), by

$$\eta = \tanh^2|\kappa|L \tag{12.4-52}$$

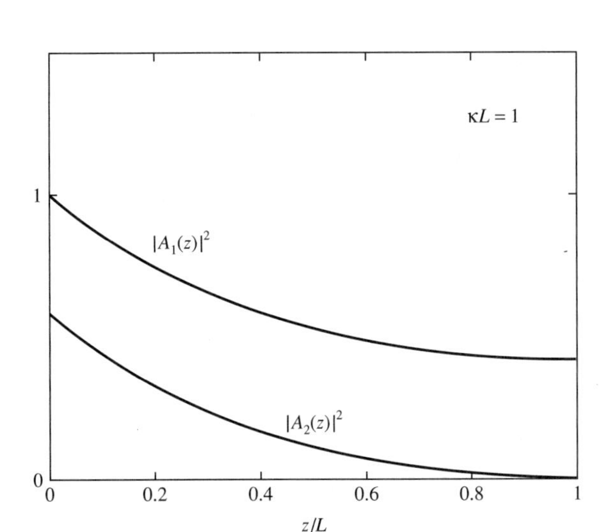

Figure 12.18 Mode power $|A_1|^2$ and $|A_2|^2$ versus z in phase-matched Bragg scattering in a contradirectional interaction. The mode power of the incident wave decays exponentially, donating power to the scattered mode, which grows exponentially along the $-z$ direction.

We find that the coupling efficiency is an increasing function of $|\kappa|L$ and approaches unity as $|\kappa|L$ becomes infinity.

The Case Where $\Delta\beta \neq 0$

We now consider the case when a phase mismatch is present. The solutions of the coupled equations (12.4-46) subject to the boundary condition $A_2(L) = 0$ can be written

$$A_1(z) = e^{i(\Delta\beta/2)z}\frac{s\cosh s(L-z) + i(\Delta\beta/2)\sinh s(L-z)}{s\cosh sL + i(\Delta\beta/2)\sinh sL}A_1(0)$$

$$A_2(z) = e^{-i(\Delta\beta/2)z}\frac{-i\kappa^*\sinh s(L-z)}{s\cosh sL + i(\Delta\beta/2)\sinh sL}A_1(0) \tag{12.4-53}$$

where

$$s^2 = \kappa^*\kappa - \left(\frac{\Delta\beta}{2}\right)^2 \tag{12.4-54}$$

From the above solution $A_2(z)$, it can be seen that the power transfer between the two modes in the region between $z = 0$ and $z = L$ is given by

$$\eta = \frac{|A_2(0)|^2}{|A_1(0)|^2} = \frac{|\kappa|^2\sinh^2 sL}{s^2\cosh^2 sL + (\Delta\beta/2)^2\sinh^2 sL} \tag{12.4-55}$$

Again, we notice that the coupling efficiency decreases as $\Delta\beta$ increases. A complete power transfer for contradirectional coupling, however, only occurs when phase matching is satisfied ($\Delta\beta = 0$) and the interaction distance L is infinite. This situation is somewhat different from that of the codirectional coupling, where power is exchanged back and forth between the two modes and complete power transfer happens periodically in space provided $\Delta\beta = 0$. Figure 12.19 plots the coupling efficiency as a function of the phase mismatch $\Delta\beta$. Aside

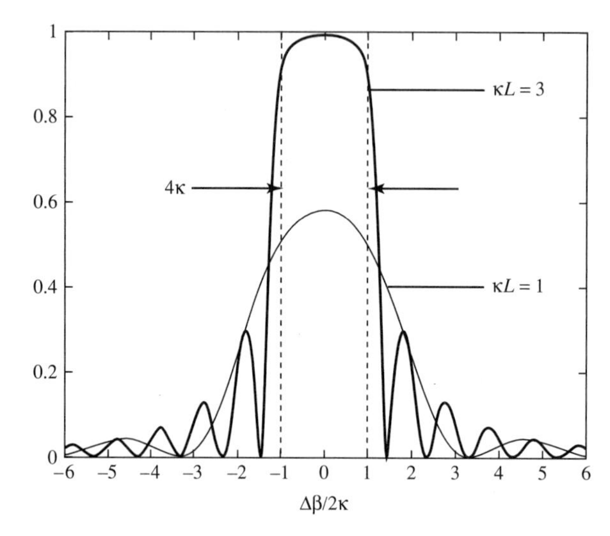

Figure 12.19 Coupling efficiency as a function of the phase mismatch in contradirectional coupling.

from the main peak at $\Delta\beta = 0$, the coupling spectrum also consists of a series of sidelobes on both sides of the main peak. The width of the main peak is in the spectral range where

$$-2|\kappa| < \Delta\beta < 2|\kappa| \tag{12.4-56}$$

This is also the width of the photonic bandgap. These sidelobes peak approximately at $sL = i(p + 1/2)\pi$ with $p = 1, 2, 3, \ldots$, which corresponds to $\Delta\beta = \pm 2\sqrt{\kappa^*\kappa + (p + 1/2)^2(\pi/L)^2}$. The peak coupling of these sidelobes is given, according to Equation (12.4-55), by

$$\eta = \frac{|\kappa L|^2}{\pi^2(p + 1/2)^2 + |\kappa L|^2} \tag{12.4-57}$$

These sidelobes become appreciable when $|\kappa L| > \pi/2$. In fact, the peak coupling efficiency at the first sidelobe reaches 10% when $|\kappa L| = \pi/2$, at which point the coupling efficiency at the main peak is only 84%. Zero coupling efficiency occurs at $sL = iq\pi$ (with $q = 1, 2, 3, \ldots$), which corresponds to $\Delta\beta = \pm 2\sqrt{\kappa^*\kappa + (q\pi/L)^2}$.

In the general case when both modes are present initially (i.e., $A_1(0) \neq 0$ and $A_2(L) \neq 0$), the solutions are

$$\begin{aligned}
A_1(z) &= e^{i(\Delta\beta/2)z}\left(\frac{s\cosh s(L-z) + i(\Delta\beta/2)\sinh s(L-z)}{s\cosh sL + i(\Delta\beta/2)\sinh sL}A_1(0)\right.\\
&\qquad\left. + \frac{-i\kappa e^{i(\Delta\beta/2)L}\sinh sz}{s\cosh sL + i(\Delta\beta/2)\sinh sL}A_2(L)\right)\\
A_2(z) &= e^{-i(\Delta\beta/2)z}\left(\frac{-i\kappa^*\sinh s(L-z)}{s\cosh sL + i(\Delta\beta/2)\sinh sL}A_1(0)\right.\\
&\qquad\left. + e^{i(\Delta\beta/2)L}\frac{s\cosh sz + i(\Delta\beta/2)\sinh sz}{s\cosh sL + i(\Delta\beta/2)\sinh sL}A_2(L)\right)
\end{aligned} \tag{12.4-58}$$

These equations will be useful in the investigation of dispersion properties of fiber Bragg gratings, particularly in the case when there are several segments of gratings in series.

Inside the photonic bandgap where $-2|\kappa| < \Delta\beta < 2|\kappa|$, s is a real number. The solutions of the wave equation are exponential functions of z. These are the so-called evanescent waves.

The incidence of a beam of light in this case will be totally reflected, provided the periodic medium is semi-infinite. In the spectral regime where $2|\kappa| < |\Delta\beta|$, s is a pure imaginary number. The solutions of the wave equation are sinusoidal functions of z. The waves are propagating in the periodic medium.

EXAMPLE: CONTRADIRECTIONAL COUPLING IN VOLUME INDEX GRATINGS

Referring to Figure 12.20, we consider an example of a volume grating whose index of refraction is given by

$$n(z) = n_0 + n_1 \cos(Kz) \tag{12.4-59}$$

where n_0, n_1, and K are constants. The constant n_0 is the averaged index of refraction of the medium, n_1 may be regarded as the depth of the sinusoidal index modulation, and K is related to the period of the index grating Λ by

$$K = \frac{2\pi}{\Lambda} \tag{12.4-60}$$

The dielectric perturbation for such a volume index grating is given by

$$\Delta\varepsilon(z) = \varepsilon_0 \, \Delta n^2(z) = 2\varepsilon_0 n_0 n_1 \cos(Kz) \tag{12.4-61}$$

Since both the dielectric perturbation and the unperturbed dielectric constant are scalars, mode coupling between TE waves (s-waves) and TM waves (p-waves) does not exist. Consequently, only coupling between waves of the same polarization state can take place. This is possible only in the contradirectional coupling, because the phase-matching condition for codirectional coupling can never be satisfied.

The mode coupling for TE waves and for TM waves are similar. The only difference is the coupling constant (12.4-27). Let θ be the angle between the wavevector $\mathbf{k}$ of the incident beam and the z axis, and $\mathbf{k}'$ be the wavevector of the reflected beam as shown in Figure 12.20. The coupling constants, according to Equation (12.4-27), are given by

$$\kappa = C^{(1)} = \frac{\omega^2\mu}{2\beta}\mathbf{p}'\cdot\varepsilon_1\mathbf{p} = \frac{\omega^2\mu\varepsilon_0 n_0}{2\beta}n_1\mathbf{p}'\cdot\mathbf{p} = \begin{cases} \dfrac{\omega^2\mu\varepsilon_0 n_0}{2\beta}n_1, & \text{TE} \\ \dfrac{\omega^2\mu\varepsilon_0 n_0}{2\beta}n_1\cos 2\theta, & \text{TM} \end{cases} \tag{12.4-62}$$

where $\mathbf{p}$ and $\mathbf{p}'$ are the polarization states of the beams, $\varepsilon_1 = \varepsilon_0 n_0 n_1$ is the first Fourier expansion coefficient, and $\beta = k\cos\theta$ is the z component of the wavevector $\mathbf{k}$. Note that the

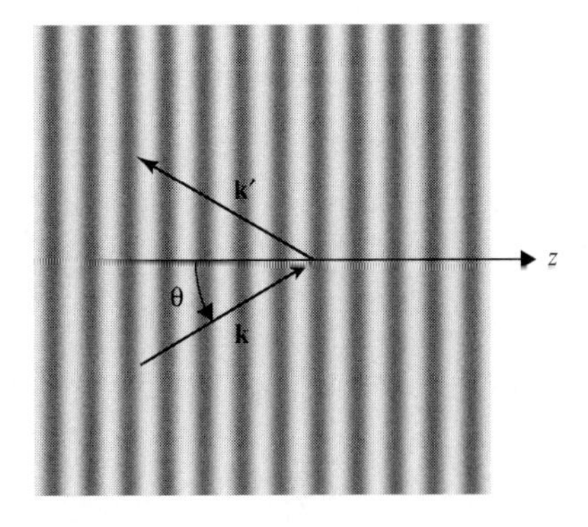

Figure 12.20 A volume index grating as a Bragg reflector, with an off-axis incident beam.

difference between the two coupling constants is only a directional factor of cos 2θ, which is the cosine of the angle between the electric field vector of the coupled TM waves. According to Equation (12.4-62), the coupling constant vanishes at $\theta = 45°$ for TM waves. This corresponds to zero reflection of TM waves at the Brewster angle. (Note: For $n_1 \ll n_0$, the Brewster angle is 45°.) For normal incidence with $\theta = 0$, the coupling constant becomes

$$\kappa = \frac{\omega}{2c} n_1 = \frac{\pi}{\lambda} n_1 \tag{12.4-63}$$

The phase mismatch factor $\Delta\beta$ is given by

$$\Delta\beta = 2k \cos\theta - \frac{2\pi}{\Lambda} = 2n_0 \frac{2\pi}{\lambda} \cos\theta - \frac{2\pi}{\Lambda} \tag{12.4-64}$$

where k is the wavenumber of the optical beam and Λ is the period of the index grating. We note that phase matching occurs when the period equals a half-wave length (i.e., $\Lambda = \lambda/2n_0$).

The reflectance spectrum is given, according to Equation (12.4-55), by

$$\eta = \frac{|A_2(0)|^2}{|A_1(0)|^2} = \frac{|\kappa|^2 \sinh^2 sL}{s^2 \cosh^2 sL + (\Delta\beta/2)^2 \sinh^2 sL} \tag{12.4-65}$$

Figure 12.19 shows the reflectance spectrum as a function of $\Delta\beta$. Let ω_0 be the frequency at which the phase-matching condition (Bragg condition) is satisfied. Then $\Delta\beta$ can be written

$$\Delta\beta = \frac{2n_0}{c}(\omega - \omega_0) \cos\theta \tag{12.4-66}$$

The reflectance spectrum consists of a main peak with a sharp cutoff and a series of sidelobes. The bandwidth of the main peak is given approximately by

$$\Delta\beta = 4|\kappa| \tag{12.4-67}$$

because at $\Delta\beta = \pm 2|\kappa|$, the parameter s becomes zero and the wavefunction changes from exponential to sinusoidal. Thus the spectral regime $\Delta\beta = 4|\kappa|$ is often called the "stopband" or, equivalently, the photonic bandgap. The size of the photonic bandgap in terms of frequency is given by

$$\frac{\Delta\omega_{gap}}{\omega_0} = \frac{\Delta\lambda_{gap}}{\lambda_0} = \frac{\Delta\nu_{gap}}{\nu_0} = 2\left|\frac{n_1}{n_0}\right| \tag{12.4-68}$$

where λ_0 is the wavelength at the center of the main peak, and ν_0 is the corresponding frequency in units of hertz.

To illustrate the use of a volume index grating as a Bragg reflector, we consider a volume index grating with the following parameters: $n_0 = 1.5$, $n_1 = 0.001$, and $\lambda_0 = 1550$ nm. To ensure a peak reflection at $\lambda_0 = 1550$ nm, the period of the index grating must be $\Lambda = \lambda_0/2n_0 = 517$ nm.

If we desire a reflectance of 99% at $\lambda_0 = 1550$ nm, then the interaction length must be such that $\kappa L = 3.0$, according to (12.4-52). Thus the interaction length is

$$L = \frac{3.0}{\kappa} = 3.0 \frac{\lambda_0}{\pi n_1} = 1.5 \text{ mm}$$

The bandwidth of the reflectance peak is, according to Equation (12.4-68),

$$\Delta\lambda_{\text{gap}} = 2\lambda_0 \left|\frac{n_1}{n_2}\right| = 2.07 \text{ nm}, \quad \text{or equialently}, \quad \Delta\nu_0 = 258 \text{ GHz}$$

12.5 PERIODIC WAVEGUIDES

Referring to Figure 12.21a, we consider a periodic dielectric waveguide in which the periodic dielectric perturbation is due to a small sinusoidal variation of the index of refraction of the guiding layer of a dielectric slab waveguide. The index variation can be obtained, for example, by the diffusion of metal ions into the guiding layer (e.g., Ti in $LiNbO_3$). A similar type of index variation in the core of a single-mode silica fiber can be obtained by UV exposures. A periodic dielectric perturbation can also be created by having a surface corrugation on the guiding layer. The surface corrugation (or surface relief) can be obtained, for example, by using photolithographic techniques. These periodic waveguides are used for spectral optical filters [1], mode converters, as well as in distributed feedback lasers [2–4]. These applications will be described further later on.

Using the perturbation approach, the dielectric constant of the periodic waveguide structure shown in Figure 12.21 can be written

$$\varepsilon(\mathbf{r}) = \varepsilon_a(\mathbf{r}) + \Delta\varepsilon(\mathbf{r}) \tag{12.5-1}$$

with $\varepsilon_a(\mathbf{r})$ being the dielectric constant for the unperturbed waveguide structure:

$$\varepsilon_a(\mathbf{r}) = \begin{cases} \varepsilon_0 n_1^2, & 0 < x \\ \varepsilon_0 n_2^2, & -t < x < 0 \\ \varepsilon_0 n_3^2, & x < -t \end{cases} \tag{12.5-2}$$

where n_1, n_2, n_3 are the indices of refraction of the layers of the dielectric slab waveguide as discussed in Chapter 3, t is the thickness of the guiding layer, and $\Delta\varepsilon(\mathbf{r})$ is the periodic dielectric perturbation. For the structure shown in Figure 12.21a, the periodic dielectric perturbation can be written

$$\Delta\varepsilon(\mathbf{r}) = \begin{cases} 0, & 0 < x \\ \varepsilon_0 \, \Delta n_{20}^2 \cos(Kz), & -t < x < 0 \\ 0, & x < -t \end{cases} \tag{12.5-3}$$

where Δn_{20}^2 is a constant, and K is a constant related to the period Λ by $K = 2\pi/\Lambda$. For the structure shown in Figure 12.21b, the periodic dielectric perturbation due to the periodic surface corrugation can be written

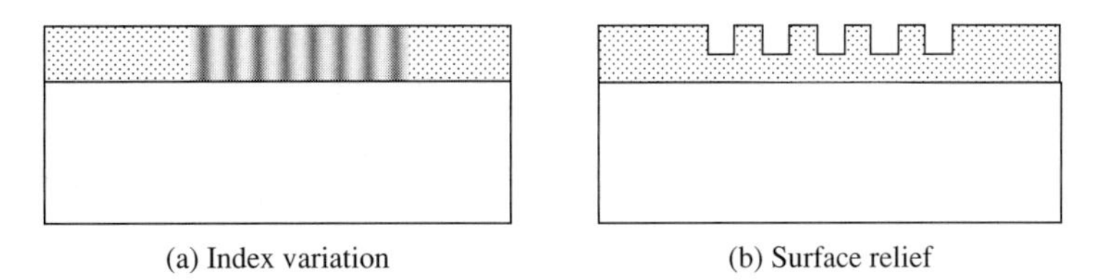

Figure 12.21 Schematic drawing of periodic dielectric waveguides. The periodic dielectric perturbation can be obtained by either changing the index of refraction of the guiding layer (or cladding layer) as shown in (a), or by etching the surface of the guiding layer (or cladding layer) as shown in (b).

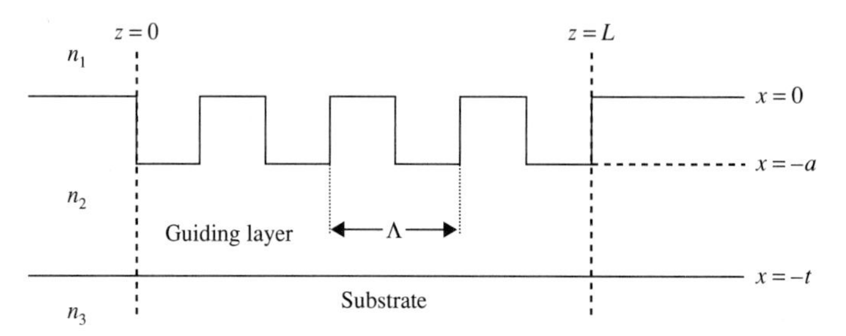

Figure 12.22 A corrugated periodic waveguide.

$$\Delta\varepsilon(\mathbf{r}) = \begin{cases} 0, & 0 < x \\ \varepsilon_0(n_1^2 - n_2^2)S(z), & -a < x < 0 \\ 0, & x < -a \end{cases} \tag{12.5-4}$$

where a is the depth of the surface corrugation (see Figure 12.22), and $S(z)$ is a square-wave periodic function given by

$$S(z) = \begin{cases} 1, & 0 < z < \Lambda/2 \\ 0, & \Lambda/2 < z < \Lambda \end{cases} \quad \text{with} \quad S(z + \Lambda) = S(z) \tag{12.5-5}$$

Since $\Delta n^2(\mathbf{r})$ is a scalar for the structures shown in Figure 12.21, it follows, from Equation (12.4-18), that the index variation or the corrugation couples only TE to TE modes and TM to TM modes, but not TE to TM. For multimode waveguides, the coupling between the fundamental TE mode and high-order TE modes is possible. Similarly, the coupling between the fundamental TM mode and high-order TM modes is also possible. The coupling constants can be evaluated by using Equation (12.4-18). In the following, we illustrate the calculation of the coupling constants for TE waves. For the index grating shown in Figure 12.21a, the dielectric perturbation (12.5-3) can be rewritten

$$\Delta\varepsilon(\mathbf{r}) = \begin{cases} 0, & 0 < x \\ \varepsilon_0 \; \Delta n_{20}^2 \; \dfrac{1}{2}(e^{-iKz} + e^{+iKz}), & -t < x < 0 \\ 0, & x < -t \end{cases} \tag{12.5-6}$$

We note that, in periodic dielectric perturbation due to a sinusoidal index grating, the Fourier expansion consists only of the fundamental orders ($-K$ and $+K$). All higher order Fourier components ($m \neq 1$) are zero. The coupling constant is thus given by

$$C_{kn}^{(1)} = \frac{\omega}{4}\langle k|\varepsilon_1(x, y)|n\rangle = \frac{\omega}{4}\int \mathbf{E}_k^* \cdot \varepsilon_1(x, y)\mathbf{E}_n \, dx \, dy = \frac{\omega}{8}\varepsilon_0 \; \Delta n_{20}^2 \int_{-t}^{0} E_{ky}^*(x)E_{ny}(x) \, dx \tag{12.5-7}$$

where $E_{ky}(x)$ and $E_{ny}(x)$ are wavefunctions of the kth and nth TE modes.
We note that the coupling constant is directly proportional to the depth of the index modulation.

For the periodic dielectric perturbation due to surface corrugation, we can expand the square-wave function (12.5-4) as a Fourier series and rewrite the dielectric perturbation as

$$\Delta\varepsilon(\mathbf{r}) = \begin{cases} 0, & 0 < x \\ \varepsilon_0(n_1^2 - n_2^2) \displaystyle\sum_{m=-\infty}^{\infty} b_m e^{-im(2\pi/\Lambda)z}, & -a < x < 0 \\ 0, & x < -a \end{cases} \tag{12.5-8}$$

where b_m are the Fourier expansion coefficients. For the case of a square-wave surface corrugation, the Fourier expansion of the periodic function $S(z)$ can be written

$$S(z) = \sum_{m=-\infty}^{\infty} b_m e^{-im(2\pi/\Lambda)z} = \sum_{m=-\infty}^{\infty} e^{im\pi/2} \frac{\sin(m\pi/2)}{m\pi} e^{-im(2\pi/\Lambda)z} \tag{12.5-9}$$

so that

$$b_m = e^{im\pi/2} \frac{\sin(m\pi/2)}{m\pi} \tag{12.5-10}$$

or equivalently,

$$b_m = \begin{cases} \frac{1}{2}, & m = 0 \\ 0, & m = \text{even} \\ \dfrac{i}{m\pi}, & m = \text{odd} \end{cases} \tag{12.5-11}$$

The coupling constant due to the mth component of the periodic dielectric perturbation is thus written

$$\kappa = C_{kn}^{(m)} = \frac{\omega}{4} \int \mathbf{E}_k^*(x) \cdot \varepsilon_m(x) \mathbf{E}_n(x)\, dx = \frac{\omega}{4} \varepsilon_0 b_m (n_1^2 - n_2^2) \int_{-a}^{0} E_{ky}^*(x) \cdot E_{ny}(x)\, dx \tag{12.5-12}$$

We note that the magnitude of the coupling constant is an increasing function of the depth of corrugation a (for $a \ll t$).

Figure 12.23 shows several schematic drawings of periodic dielectric waveguides for applications in Bragg reflectors, filters, and couplers. In what follows, we discuss some of these device applications.

Bragg Reflectors (Broadband)

In broadband Bragg reflector applications, we need a strong dielectric perturbation, which provides a large coupling constant for the contradirectional coupling. The dielectric perturbation provides the coupling between an incoming mode and the same mode that propagates in the opposite direction. In multimode waveguides, a strong periodic dielectric perturbation may cause coupling between modes of different spatial wavefunctions. In other words, it may couple a mode $E_k(x) \exp[i(\omega t - \beta_k z)$ to a different mode $E_n(x) \exp[i(\omega t + \beta_n z)$ propagating in the opposite direction. The situation in a single-mode waveguide is relatively simple.

We first consider the simple case of a TE mode propagation in a single-mode slab waveguide structure. Furthermore, we are interested in the spectral regime where the forward propagating TE mode $A\mathbf{E}_0(x) \exp[i(\omega t - \beta z)]$ is only coupled with the backward wave $B\mathbf{E}_0(x) \exp[i(\omega t + \beta z)]$, where A and B are constants, and $\mathbf{E}_0(x)$ is the TE mode wavefunction. Based on the discussion in the last section, an efficient contradirectional coupling occurs only when $\beta - (-\beta) = 2\beta = m2\pi/\Lambda$ for some integer m. The coupling between the backward $B\mathbf{E}_0(x) \exp[i(\omega t + \beta z)]$ and the forward $A\mathbf{E}_0(x) \exp[i(\omega t - \beta z)]$ by the mth Fourier component of the periodic dielectric perturbation can thus be described by, according to Equation (12.4-46),

$$\begin{aligned} \frac{d}{dz} A &= -i\kappa B(z) e^{i\Delta\beta z} \\ \frac{d}{dz} B &= i\kappa^* A(z) e^{-i\Delta\beta z} \end{aligned} \tag{12.5-13}$$

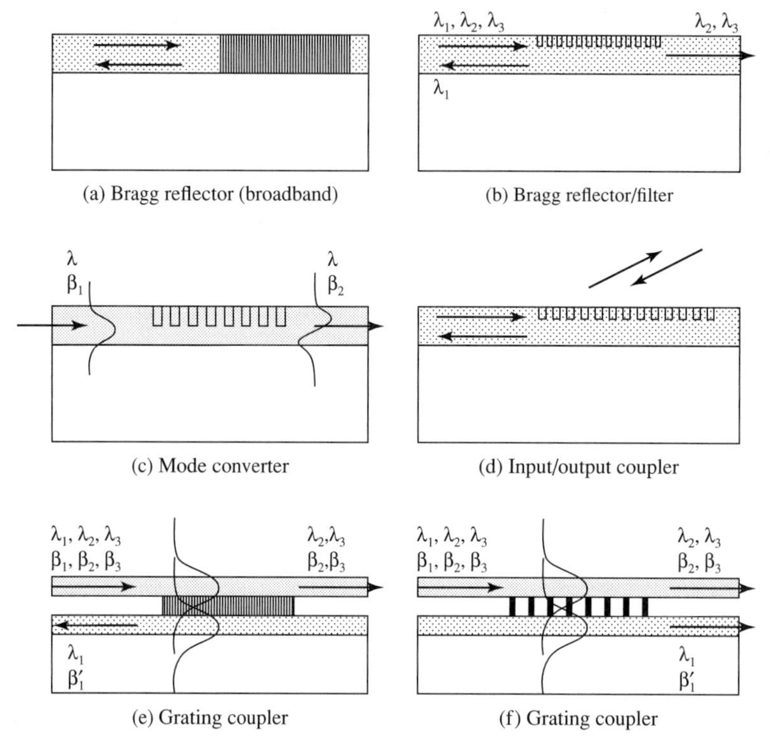

Figure 12.23 Schematic drawing of various photonic devices based on periodic dielectric perturbation.

where the coupling constant κ, for the case of periodic surface corrugation, is given by, according to Equation (12.5-12),

$$\kappa = C_{00}^{(m)} = \frac{\omega}{4}\int \mathbf{E}_0^*(x)\cdot \varepsilon_m(x)\mathbf{E}_0(x)\,dx\ \frac{\omega}{4}\varepsilon_0 b_m \int_{-a}^{0} (n_1^2 - n_2^2)|E_{0y}(x)|^2\,dx \qquad (12.5\text{-}14)$$

where $E_{0y}(x)$ is the wavefunction of the fundamental TE mode. We note that the magnitude of the coupling constant is an increasing function of the depth of the corrugation.

The momentum mismatch is given by

$$\Delta\beta = \beta - (-\beta) - m\frac{2\pi}{\Lambda} = 2\beta - m\frac{2\pi}{\Lambda} \equiv 2(\beta - \beta_0) \qquad (12.5\text{-}15)$$

where $\beta_0 = m\pi/\Lambda$. For the case of the fundamental TE mode in a single-mode waveguide, we may define n_{eff} as the effective index of the mode such that

$$\beta = n_{\text{eff}}\frac{\omega}{c} = n_{\text{eff}}\frac{2\pi}{\lambda} \qquad (12.5\text{-}16)$$

The momentum mismatch in the contradirectional coupling can thus be written

$$\Delta\beta = 2\beta - m\frac{2\pi}{\Lambda} \equiv \frac{2n_{\text{eff}}}{c}(\omega - \omega_0) \qquad (12.5\text{-}17)$$

where ω_0 is the center frequency where the Bragg condition is satisfied.

Following the mathematical steps described in the last section, the coupled equations can be solved analytically. The general solutions are written

$$A(z) = C_1 e^{i(\Delta\beta/2)z - sz} + C_2 e^{i(\Delta\beta/2)z + sz}$$
$$B(z) = \frac{i}{\kappa}\frac{d}{dz}A(z) \tag{12.5-18}$$

where C_1 and C_2 are constants, and the parameter s is given by

$$s = \sqrt{|\kappa|^2 - \left(\frac{\Delta\beta}{2}\right)^2} = \sqrt{|\kappa|^2 - [\beta(\omega) - \beta_0]^2} \tag{12.5-19}$$

The solutions subject to the boundary condition $B(L) = 0$ can be written, according to Equations (12.4-53),

$$A(z) = e^{i(\Delta\beta/2)z}\frac{s\cosh s(L-z) + i(\Delta\beta/2)\sinh s(L-z)}{s\cosh sL + i(\Delta\beta/2)\sinh sL}A(0)$$
$$B(z) = e^{-i(\Delta\beta/2)z}\frac{-i\kappa^*\sinh s(L-z)}{s\cosh sL + i(\Delta\beta/2)\sinh sL}A(0) \tag{12.5-20}$$

The reflectance (or diffraction efficiency) is given by

$$R = \left|\frac{B(0)}{A(0)}\right|^2 = \frac{|\kappa|^2\sinh^2 sL}{s^2\cosh^2 sL + (\Delta\beta/2)^2\sinh^2 sL} \tag{12.5-21}$$

Maximum reflectance occurs at $\Delta\beta = 0$, where the reflectance is given by

$$R_{\max} = \tanh^2|\kappa|L \tag{12.5-22}$$

We note that the reflectance is an increasing function of $|\kappa|L$. The width of the main peak can be estimated as follows. Significant contradirectional coupling occurs in the spectral regime, where $-2|\kappa| < \Delta\beta < 2|\kappa|$. In this regime, the parameter s is real and the general solutions (12.5-18) are exponential, indicating a strong exchange of energy between the forward mode and the backward mode. This spectral regime is equivalent to a frequency range of, according to Equations (12.5-16) and (12.5-17),

$$-\frac{|\kappa|c}{n_{\text{eff}}} < \omega - m\frac{\pi c}{n_{\text{eff}}\Lambda} < \frac{|\kappa|c}{n_{\text{eff}}} \quad (m = 1, 2, 3, \ldots) \tag{12.5-23}$$

The spectral width of the photonic bandgap where peak reflectance occurs is thus given by

$$\Delta\omega_{\text{gap}} = 2\frac{|\kappa|c}{n_{\text{eff}}} \tag{12.5-24}$$

where we recall that n_{eff} is the effective index of the mode. It is important to note that the coupling constant depends on the Fourier expansion order m. An effective broadband reflector must have a high reflectance. Since the peak reflectance is given by $\tanh^2\kappa L$, according to Equation (12.4-52), an efficient broadband reflector requires $\kappa L \gg 1$. Furthermore, the useful bandwidth for reflection is limited by the photonic bandgap $\Delta\omega_{\text{gap}}$. Thus a large coupling constant κ is desirable for a broadband reflector.

We now consider the general solutions and their resemblance to the Bloch waves. According to Equations (12.5-18), we can rewrite the general solution of the contradirectional coupling as

$$E(z) = [C_1 e^{i(\Delta\beta/2)z - sz} + C_2 e^{i(\Delta\beta/2)z + sz}]e^{-i\beta z} + [C_3 e^{-i(\Delta\beta/2)z - sz} + C_4 e^{-i(\Delta\beta/2)z + sz}]e^{+i\beta z} \tag{12.5-25}$$

where C_1, C_2, C_3, and C_4 are constants. Only C_1 and C_2 are independent. The above solution can further be written

$$E(z) = B_1 e^{-i\beta' z} + B_2 e^{+i\beta' z} \tag{12.5-26}$$

where B_1 and B_2 are constants, and

$$\beta' = \beta - \frac{\Delta\beta}{2} \pm is = m\frac{\pi}{\Lambda} \pm i\sqrt{|\kappa|^2 - \left(\frac{\Delta\beta}{2}\right)^2} = m\frac{\pi}{\Lambda} \pm i\sqrt{|\kappa|^2 - [\beta(\omega) - \beta_0]^2} \tag{12.5-27}$$

where we used $\Delta\beta \equiv 2(\beta - \beta_0)$ with $\beta_0 \equiv m\pi/\Lambda$.

We note that for a range of frequencies such that $|\Delta\beta(\omega)| < 2|\kappa|$, β' has an imaginary part. This is the so-called photonic bandgap (or "forbidden" region) in which the evanescent behavior shown in Figure 12.18 occurs and which is formally analogous to the energy gap in semiconductors where the periodic crystal potential causes the electron propagation constants to become complex. Note that for each value of m, $m = 1, 2, 3 \ldots$, there exists a gap whose center frequency ω_{0m} satisfies $\beta(\omega_{0m}) = m\pi/\Lambda$. The exceptions are values of m for which κ is zero. We can approximate $\beta(\omega)$ near its Bragg value $(m\pi/\Lambda)$ by $\beta(\omega) \sim (\omega/c)n_{\text{eff}}$ (n_{eff} is an effective index of refraction). The result is, according to Equation (12.5-17),

$$\beta' \cong m\frac{\pi}{\Lambda} \pm i\left[|\kappa|^2 - \left(\frac{n_{\text{eff}}}{c}\right)^2 (\omega - \omega_0)^2\right]^{1/2} \tag{12.5-28}$$

where ω_0 is the center frequency where the Bragg condition is satisfied. In other words, ω_0 is the value of ω for which the unperturbed β is equal to $\beta_0 \equiv m\pi/\Lambda$.

A plot of Re β' and Im β' (for $m = 1$) versus ω, based on (12.5-28), is shown in Figure 12.24. We note that the width of the "forbidden" frequency zone is

$$(\Delta\omega)_{\text{gap}} = \frac{2|\kappa|c}{n_{\text{eff}}} \tag{12.5-29}$$

where κ is according to Equation (12.5-17) a function of the integer m. It follows from Equation (12.5-28) that

$$(\text{Im}\beta')_{\max} = |\kappa| \tag{12.5-30}$$

The complex wavenumber β' obtained here by using coupled-wave analysis is indeed the Bloch wavenumber.

In solid-state physics, it is well known that the behavior of electrons is described by means of electron wavefunctions. The wavefunctions in a periodic electron potential are of the form (Bloch form)

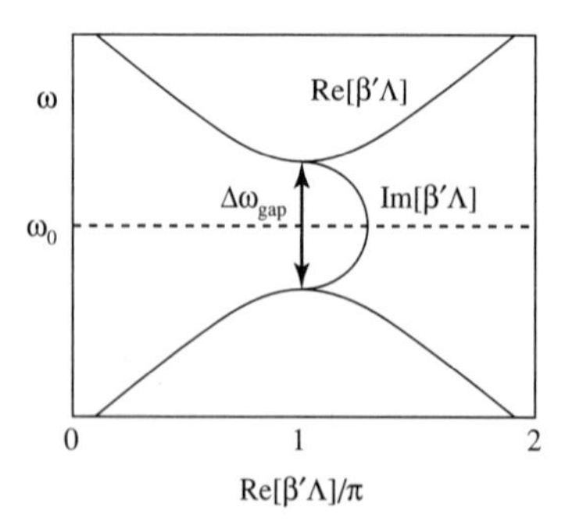

Figure 12.24 Dependence of the real and imaginary parts of the propagation constant, β', of the modes in a periodic waveguide. At frequencies $|\omega - \omega_0| < \Delta\omega_{\text{gap}}/2$, $\text{Im}(\beta') \neq 0$ and the modes are evanescent. At these frequencies, Re $\beta' = m\pi/\Lambda$ (with $m = 1$ in the figure).

$$\Psi_i = u_i(\mathbf{r}) \exp\left(-i\frac{E_i t}{\hbar} + i\mathbf{k}_i \cdot \mathbf{r}\right) \tag{12.5-31}$$

There exist regions of electron energy E_i where the propagation constant $\mathbf{k}_i$ is complex independently of the direction of $\mathbf{k}_i$ in complete formal analogy with Equation (12.5-28). These are the *forbidden energy gaps* of the crystal. Recent proposals and experiments [1, 5] suggest that it should be possible to engineer "optical crystals" that have a two-dimensional (2-D) or three-dimensional (3-D) periodicity that will possess a forbidden frequency gap for which optical propagation will be evanescent (i.e., with a complex propagation constant that is a two-dimensional or three-dimensional generalization of Equation (12.5-28)). A simple example of 2-D photonic crystals will be discussed in Section 12.8.

Bragg Reflectors/Filters

For spectral filter applications, it is often desirable to have a spectral filter that provides a narrow bandwidth. In the area of optical communications, spectral filters with narrow bandwidth of a few gigahertz are often needed. These filters are employed to block undesirable radiation and to transmit only the radiation in the desired spectral regime. A weak periodic perturbation with a small coupling constant is desirable for this application. Based on the results obtained for the volume index gratings described in the last section, the bandwidth of high reflectance is given by

$$\frac{\Delta\omega_{\text{gap}}}{\omega_0} = \frac{\Delta\lambda_{\text{gap}}}{\lambda_0} = \frac{\Delta\nu_{\text{gap}}}{\nu_0} = 2\left|\frac{n_1}{n_0}\right| \tag{12.5-32}$$

For example, a bandwidth of 25 GHz at $\lambda = 1550$ nm would require an index modulation of $n_1/n_0 \approx 6 \times 10^{-5}$. As noted earlier, a narrowband filter requires a small coupling constant κ. For a high-efficiency filter, a long interaction is often needed to achieve $\kappa L \gg 1$. The reflectance spectrum of a Bragg reflector is given, according to Equation (12.4-55), by

$$R = \frac{|\kappa|^2 \sinh^2 sL}{s^2 \cosh^2 sL + (\Delta\beta/2)^2 \sinh^2 sL} \tag{12.5-33}$$

It is important to note that the reflectance spectrum as shown in Figure 12.19 exhibits several sidelobes on both sides of the photonic bandgap. These sidelobes result from the impedance mismatch at the edge of the periodic structure. For spectral filter applications, these sidelobes are very undesirable. Spatial apodization, which will smooth over the transition from the homogeneous region to the periodic region, can be employed to eliminate the sidelobes. This subject will be discussed later.

Mode Converters

To illustrate the principle of operation of the mode converter shown in Figure 12.23c, we consider a two-mode waveguide. The waveguide supports two confined modes of propagation. A periodic dielectric perturbation can be introduced to provide the coupling between these two modes. Both codirectional and contradirectional couplings are possible, depending on the period of the periodic perturbation. Let the propagation constants of these two modes be β_1 and β_2. The Bragg conditions for the coupling are given as follows:

$$\Delta\beta = \begin{cases} \beta_1 - \beta_2 - m(2\pi/\Lambda) = 0, & \text{codirectional} \\ \beta_1 + \beta_2 - m(2\pi/\Lambda) = 0, & \text{contradirectional} \end{cases} \tag{12.5-34}$$

where m is an integer. We note that contradirectional coupling requires high-frequency gratings (short-period gratings), whereas codirectional coupling requires long-period gratings. For the case of dielectric perturbation via periodic surface corrugation, the coupling constant for TE modes is given, according to Equation (12.5-12), by

$$\kappa = C_{12}^{(m)} = \frac{\omega}{4}\int \mathbf{E}_1^*(x)\cdot\varepsilon_m(x)\mathbf{E}_2(x)\,dx = \frac{\omega}{4}\varepsilon_0 b_m(n_1^2 - n_2^2)\int_{-a}^{0} E_{1y}^*(x)\cdot E_{2y}(x)\,dx \qquad (12.5\text{-}35)$$

where we recall that b_m is the Fourier expansion coefficient of the surface corrugation, and a is the depth of corrugation. We note that the coupling constant is the same for both codirectional coupling and contradirectional coupling.

Input/Output Couplers

Mode coupling between a guided mode and a radiation mode is possible provided the Bragg condition is satisfied. Depending on the period of the grating, there are several possibilities. To illustrate this, we assume a single-mode dielectric slab waveguide with a substrate index of n_3 and a guiding layer of index n_2. Let β_0 be the propagation constant of the guided mode, and let β be the propagation constant of the scattered wave (radiation mode). An efficient Bragg scattering requires that

$$\beta_0 - \beta - m(2\pi/\Lambda) = 0 \qquad (12.5\text{-}36)$$

or equivalently,

$$\beta = \beta_0 - m(2\pi/\Lambda) \qquad (12.5\text{-}37)$$

for some integer m. Depending on the grating period Λ, the propagation constant of the scattered wave may vary over a large range. The magnitude of the propagation constant β determines whether the scattered wave will be radiating toward the air (index n_1) or the substrate (index n_3). Table 12.1 lists the possibilities, where n_{eff} is the effective index of the guided mode. Figure 12.25 shows the momentum diagrams for the two forward output couplings. Generally, the guided mode can couple with many radiation modes, according to Equation (12.5-37). Let the electric field of the waves be written

$$E(x, z, t) = \left\{A_0(z)E_0(x)e^{-i\beta_0 z} + \sum_{\rho=1,2}\int [a_{\rho\beta}(z)E_{\rho\beta}(x)e^{-i\beta z}]d\beta\right\}e^{i\omega t} \qquad (12.5\text{-}38)$$

where A_0 and $a_{\rho\beta}$ are constants, $\rho = 1$ stands for air radiation mode, $\rho = 2$ stands for substrate radiation mode, $E_0(x)$ is the wavefunction of the guided mode, and $E_{\rho\beta}(x)$ represents

TABLE 12.1 Dependence of β on Grating Period

	β	Radiation Modes	Spectral Regime/Grating
I	$n_1 k < \beta < n_3 k$	Forward substrate mode	$(n_{\text{eff}} - n_3)\omega < cK < (n_{\text{eff}} - n_1)\omega$
II	$0 < \beta < n_1 k$	Forward air mode (and substrate mode)	$(n_{\text{eff}} - n_1)\omega < cK < n_{\text{eff}}\omega$
III	$\beta = 0$	Vertical air mode (and substrate mode)	$n_{\text{eff}}\omega = cK$
IV	$-n_1 K < \beta < 0$	Backward air mode (and substrate mode)	$n_{\text{eff}}\omega < cK < (n_{\text{eff}} + n_1)\omega$
V	$-n_3 K < \beta < -n_1 K$	Backward substrate mode	$(n_{\text{eff}} + n_1)\omega < cK < (n_{\text{eff}} + n_3)\omega$

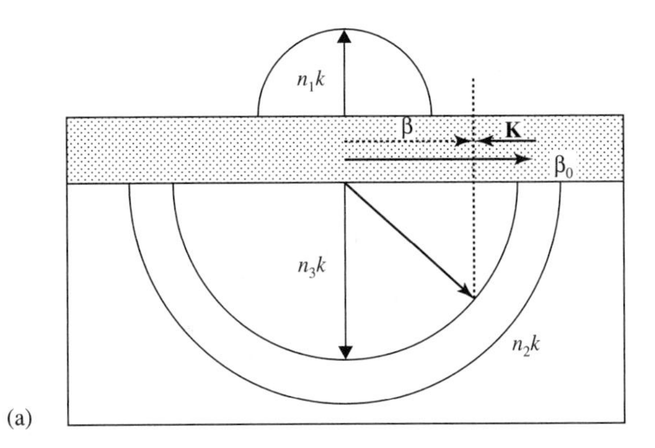

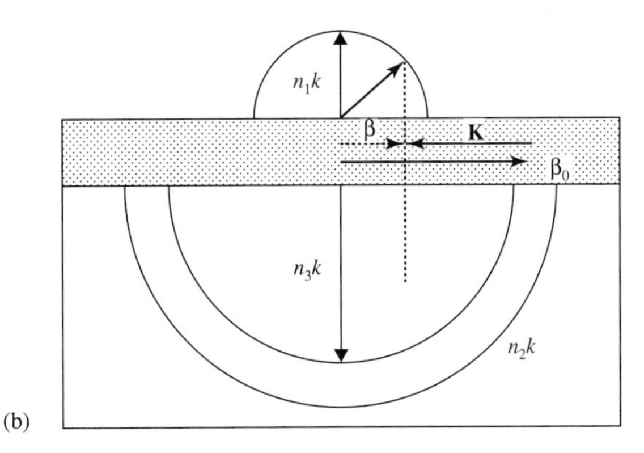

Figure 12.25 Schematic drawing of the momentum diagrams for the two forward output couplings. (a) Forward radiation into the substrate mode. (b) Forward radiation into the air mode.

the wavefunction of the radiation modes. Let $\beta_0 > 0$; the coupled equations for this case are written, according to Equation (12.4-17),

$$\frac{d}{dz}A_0 = -i\sum_{\rho=1,2}\int C_{0\rho\beta}^{(m)}a_{\rho\beta}(z)e^{i(\beta_0-\beta-m2\pi/\Lambda)z}d\beta \tag{12.5-39}$$

$$\frac{d}{dz}a_{\rho\beta} = -i\frac{\beta}{|\beta|}C_{0\rho\beta}^{(m)*}A_0(z)e^{-i(\beta_0-\beta-m2\pi/\Lambda)z} \tag{12.5-40}$$

where the coupling coefficient $C_{0\rho\beta}^{(m)}$ is written

$$C_{0\rho\beta}^{(m)} = \frac{\omega}{4}\langle 0|\varepsilon_m(x,y)|\rho\beta\rangle = \frac{\omega}{4}\int \mathbf{E}_0^*\cdot\varepsilon_m(x,y)\mathbf{E}_{\rho\beta}\,dx\,dy \tag{12.5-41}$$

The coupled equations can be solved as follows. For weak dielectric perturbation, we may assume that the amplitude of the guided mode $A_0(z)$ is a constant when we integrate Equation (12.5-40). This leads to

$$a_{\rho\beta}(z) = \begin{cases} -iA_0(z)C_{0\rho\beta}^{(m)*}\dfrac{\sin(\Delta\beta/2)z}{\Delta\beta/2}e^{-i(\Delta\beta/2)z}, & 0<\beta \\ +iA_0(z)C_{0\rho\beta}^{(m)*}\dfrac{\sin(\Delta\beta/2)(z-L)}{\Delta\beta/2}e^{-i(\Delta\beta/2)(z+L)}, & \beta<0 \end{cases} \tag{12.5-42}$$

where L is the interaction length, and $\Delta\beta = \beta_0 - \beta - m2\pi/\Lambda$. Substituting Equation (12.5-42) into Equation (12.5-39), we obtain

$$\frac{d}{dz}A_0 = -A_0(z)|C_{0\rho\beta}^{(m)}|^2 \int_{\beta>0} \frac{\sin(\Delta\beta/2)z}{\Delta\beta/2} e^{i(\Delta\beta/2)z} d\beta + A_0(z)|C_{0\rho\beta}^{(m)}|^2 \int_{\beta<0} \frac{\sin(\Delta\beta/2)(z-L)}{\Delta\beta/2} e^{i(\Delta\beta/2)(z-L)} d\beta$$

In arriving at the above equation, we assume that the coupling constant is a slowly varying function of β, and the couplings are dominated by the modes that satisfy the Bragg condition ($\Delta\beta = 0$). In the case of forward radiation mode with $\beta > 0$, the first integral yields π, while the second integral yields 0 (see Problem 12.13). In the case of backward radiation mode with $\beta < 0$, the first integral yields 0, while the second integral yields $-\pi$. Thus we obtain

$$\frac{d}{dz}A_0 = -A_0(z)\pi|C_{0\rho\beta}^{(m)}|^2 \quad \text{with } \beta = \beta_0 - m(2\pi/\Lambda) \tag{12.5-43}$$

The above equation is valid for either forward or backward radiation mode and is for a given integer m. If we consider the contribution of all orders of the Fourier components of the dielectric perturbation, then the equation for $A_0(z)$ can be written

$$\frac{d}{dz}A_0 = -A_0(z)\sum_{\rho=1,2}\sum_{m} \pi|C_{0\rho\beta}^{(m)}|^2 = -\alpha A_0(z) \quad \text{with } \beta = \beta_0 - m(2\pi/\Lambda) \tag{12.5-44}$$

where

$$\alpha = \sum_{\rho=1,2}\sum_{m} \pi|C_{0\rho\beta}^{(m)}|^2 = \pi \sum_{\rho=1,2}\sum_{m} \left| \frac{\omega}{4} \int \mathbf{E}_0^* \cdot \varepsilon_m(x, y)\mathbf{E}_{\rho\beta}\, dx\, dy \right|^2 \tag{12.5-45}$$

is the radiation decay constant for the guided mode, with $\beta = \beta_0 - m(2\pi/\Lambda)$. The above equation can be employed to evaluate the radiation decay constant. For the case of square-wave surface corrugation as shown in Figure 12.22, the radiation constant for the fundamental TE mode can be written

$$\alpha = \sum_{\rho=1,2}\sum_{m} \left| \frac{\omega}{4} \varepsilon_0 b_m (n_1^2 - n_2^2) \int_{-a}^{0} E_{0y}^*(x) \cdot E_{\rho\beta y}(x)\, dx \right|^2 \tag{12.5-46}$$

The radiation decay constant can further be calculated by substituting the wavefunctions of the modes.

12.6 SPECTRAL FILTERS AND FIBER BRAGG GRATINGS

One of the most interesting and important applications of spatially periodic optical waveguides is their use as optical reflectors [1], filters, and dispersion compensators. These applications owe much of their impetus to advances in fabricating high-efficiency index gratings in spatially doped (and treated) silica fibers by means of exposure to standing-wave patterns of ultraviolet light [2–4, 6, 7]. Figure 12.26 shows a schematic drawing of the process of recording the index grating in an optical fiber.

The index perturbation in the silica fiber after the UV exposure is in the form

$$\Delta n(x, y, z) = \Delta n_0 \sin \frac{2\pi}{\Lambda} z \tag{12.6-1}$$

with

$$\Lambda = \frac{\lambda}{2\sin\theta} \tag{12.6-2}$$

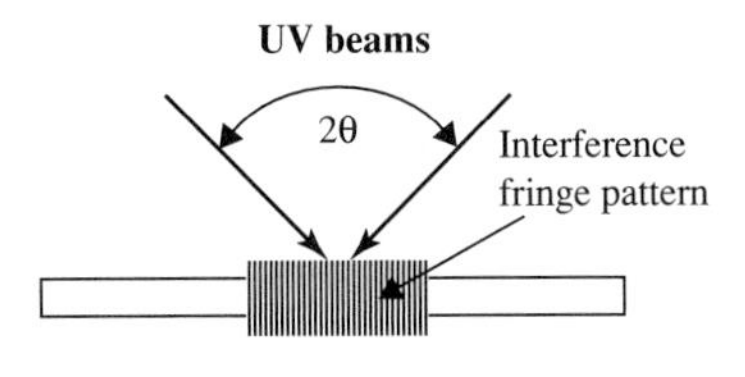

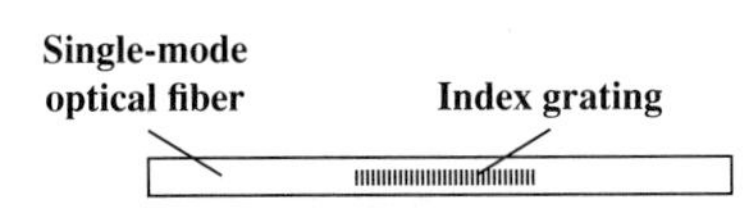

Figure 12.26 The irradiation of a silica fiber by the interferometric standing-wave pattern of UV beams causes a chemically induced "permanent," periodic index perturbation in the fiber. The period is $\Lambda = \lambda/(2 \sin \theta)$, where θ is the half-angle between the interfering beams, and λ is the wavelength of the beams. Doping with phosphorus and molecular hydrogen-loading of the fiber increase the sensitivity of the fiber to the ultraviolet radiation.

where Δn_0 is a constant, θ is the incidence angle of the two interfering beams, and λ is the wavelength of the beams. Practical systems employ excimer lasers ($\lambda = 0.248$ μm) or doubled-argon lasers ($\lambda = 0.244$ μm) as the radiation source. The basic feature of the periodic waveguide is that at frequencies near the Bragg frequency ω_0, an incident mode is strongly reflected as indicated by the reflectance spectrum shown in Figure 12.19. For frequencies outside the photonic bandgap at ω_0, the optical beam is transmitted with minimum loss due to reflection sidelobes and possibly coupling to radiation modes. The amplitude reflection coefficient of a periodic waveguide of length L is obtained from Equations (12.5-20).

$$r(\omega) = \frac{B(0)}{A(0)} = \frac{-i\kappa^* \sinh sL}{s \cosh sL + i(\Delta\beta/2) \sinh sL} \tag{12.6-3}$$

with

$$\Delta\beta = \beta - (-\beta) - m\frac{2\pi}{\Lambda} = 2\beta - m\frac{2\pi}{\Lambda} \equiv \frac{2n_{\text{eff}}}{c}(\omega - \omega_0) \tag{12.6-4}$$

$$s = \sqrt{|\kappa|^2 - \left(\frac{\Delta\beta}{2}\right)^2} = \sqrt{|\kappa|^2 - \left[\frac{n_{\text{eff}}}{c}(\omega - \omega_0)\right]^2} \tag{12.6-5}$$

where n_{eff} is the effective index of refraction of the mode of propagation in the fiber.

We now discuss the optical properties of a fiber Bragg grating (FBG). Without loss of generality, we may assume that the coupling constant κ is real and positive. This is legitimate provided we take a proper choice of the origin $z = 0$. The amplitude reflection coefficient r is, in general, a complex number. We define the phase shift ϕ of the reflection as

$$r = |r|e^{-i\phi} = \sqrt{R}\, e^{-i\phi} \tag{12.6-6}$$

where R is the intensity reflectance. Using Equations (12.6-3) and (12.6-6), we obtain

$$R(\omega) = \left|\frac{B(0)}{A(0)}\right|^2 = \frac{\kappa^2 \sinh^2 sL}{s^2 \cosh^2 sL + (\Delta\beta/2)^2 \sinh^2 sL} \tag{12.6-7}$$

and

$$\phi = \frac{\pi}{2} + \tan^{-1}\left(\frac{\Delta\beta}{2s} \tanh sL\right) \tag{12.6-8}$$

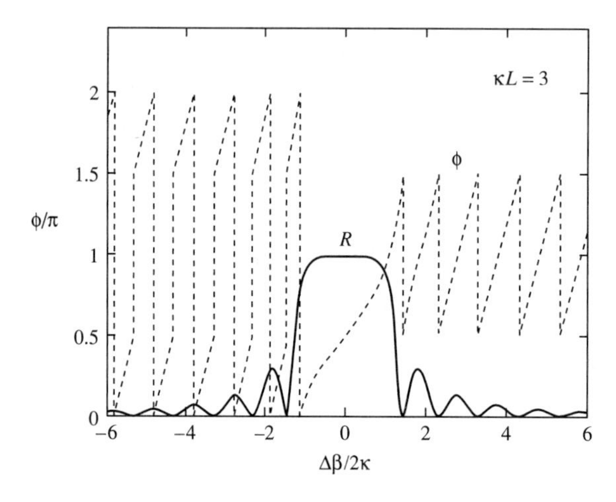

Figure 12.27 The reflectance spectrum and phase shift of a periodic waveguide with $\kappa L = 3$. The dark solid curve is the reflectance R as a function of $\Delta\beta$, and the light solid curve is for the phase shift ϕ. The region between the vertical lines at $\Delta\beta = \pm 2\kappa$ indicates the photonic bandgap.

The analytical expression of the phase shift (12.6-8) yields a result within 0 and π. This is a result of the principal value of the arctangent function. Generally, the phase shift of the complex reflection coefficient can be anywhere between 0 and 2π. The multivalued nature of trigonometric functions does not create problems as most physically measurable parameters are dependent on the derivative of the phase shift (e.g., group delay equals the frequency derivative of the phase shift).

We note that the reflectance is maximum at the center of the photonic bandgap where $\Delta\beta = 0$. The phase shift is $\pi/2$ at $\Delta\beta = 0$, assuming a real κ. The coupling constant is, in general, a complex number, depending on the choice of the origin. Figure 12.27 shows a calculated plot of the reflectance $|r(\omega)|^2$ versus $\Delta\beta$ (which is proportional to the frequency detuning $(\omega - \omega_0)/2\pi$). We note the extremely high reflectance that can be obtained. The vertical lines indicate the edge of the photonic bandgap. The figure also shows a plot of the phase shift of the Bragg reflection. For practical applications with interaction lengths on the order of 1 mm, the reflectance is near unity within the photonic bandgap. The phase shift, however, exhibits a strong dependence on the frequency detuning. We note that the phase shift varies from $-\pi/2$ to $3\pi/2$ between the two reflectance minima (zero) around the photonic bandgap. Outside the photonic bandgap, the phase shift varies over a range of π between any neighboring pair of reflectance minima (zero). We also note that the phase shift is discontinuous at frequencies where the reflectance vanishes. This discontinuity of π is actually not a problem as the phase shift of a zero field amplitude is actually undefined at these frequencies. The phase information of a Bragg reflector is of importance in applications such as vertical cavity lasers, which will be discussed in Chapter 16, and in calculating the effect of the filter on incident optical pulses.

By differentiating the phase shift with respect to the angular frequency, we obtain an effective group delay, according to Equation (12.6-8),

$$\tau = \frac{d\phi}{d\omega} = \frac{\kappa^2 \dfrac{\sinh sL}{sL} \cosh sL - \left(\dfrac{\Delta\beta}{2}\right)^2}{\kappa^2 \cosh^2 sL - \left(\dfrac{\Delta\beta}{2}\right)^2} \tau_0 \tag{12.6-9}$$

where $\tau_0 = d(\beta L)/d\omega$ is the group delay for the beam to traverse through the medium (of length L) in the absence of the periodic index variation. Physically, τ is the time delay of beam propagation due to a penetration into the periodic medium and then a reflection from the same medium. At the center of the photonic bandgap ($\Delta\beta = 0$), the group delay reaches its minimum of

$$\tau_{\min} = \frac{\tanh \kappa L}{\kappa L} \tau_0 \tag{12.6-10}$$

The minimum group delay at the center of the photonic bandgap is consistent with the minimum penetration of the optical beam, which decays exponentially in the periodic medium. As the optical frequency is detuned from the center of the photonic bandgap, the exponential decay constant is decreasing, and thus the optical beam penetrates further into the periodic medium. This is consistent with the increase of the group delay. The minimum group delay at the center of the photonic bandgap is a decreasing function of κL, according to Equation (12.6-10). This is also consistent with the decrease of the penetration depth as κL increases. The effective group delay reaches τ_0 at the bandedges, where $s = 0$.

Figure 12.28 shows the normalized effective group delay as well as the intensity reflectance as functions of the normalized frequency $\Delta\beta/(2\kappa)$. We note that in the forbidden gap ($-2\kappa < \Delta\beta < 2\kappa$), the intensity reflectance is indeed very high, reflecting the nature of the photonic bandgap (also known as the stopband). The effective group delay inside the photonic bandgap is less than τ_0. Outside the photonic bandgap, however, the normalized effective group delay oscillates around unity. In the immediate vicinity of the photonic bandedge, where $sL = i\pi$, the intensity reflectance is zero, and the effective group delay is

$$\tau = \frac{\kappa^2 L^2 + \pi^2}{\pi^2} \tau_0 = \left(1 + \frac{\kappa^2 L^2}{\pi^2}\right) \tau_0 \tag{12.6-11}$$

which can be very large as κL becomes very large. The oscillating behavior of the effective group delay is a result of the interference due to a reflected wave at both ends of the periodic medium. It is important to note that the effective group delay exhibits a strong dispersion. In other words, the effective group delay is a fast varying function of the frequency detuning outside the photonic bandgap, especially for periodic media with large κL. The strong oscillation in the group delay is a result of the impedance mismatch at the edge of the periodic medium, which creates reflections.

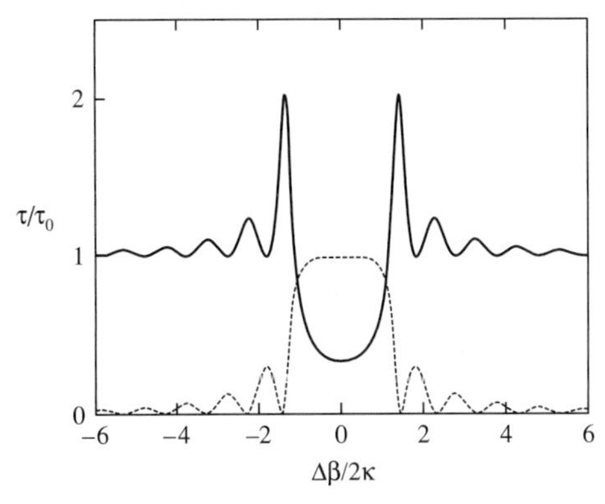

Figure 12.28 Normalized effective group delay τ/τ_0 and the reflectance R of a finite segment of a periodic medium as functions of $\Delta\beta/2\kappa$. We assume $\kappa L = 3.0$ in this plot. The photonic bandgap is between $-2\kappa < \Delta\beta < 2\kappa$. The dashed curve is the intensity reflectance.

EXAMPLE: COUPLING COEFFICIENT

To appreciate the significance of $\Delta n \leq 10^{-3}$, which is achievable in a periodic index optical fiber in which the index perturbation is caused by ultraviolet exposure, we will calculate the coupling coefficient κ that results from an index perturbation

$$n(x, y, z) = n_0 + n_1 \sin \frac{2\pi}{\Lambda} z$$

where n_0 is the average index of refraction of the fiber, and n_1 is the depth of the index modulation. The coupling constant according to Equation (12.4-63) is given by

$$\kappa = \frac{\omega}{2c} n_1 = \frac{\pi}{\lambda} n_1$$

If the filter is to be used at the communication wavelength of $\lambda = 1.55\ \mu\text{m}$ in a fiber with $n_1 = 10^{-3}$, the result is $\kappa = (10^{-3})/(1.55 \times 10^{-4}) = 6.5\ \text{cm}^{-1}$. A length of a periodic index fiber 6 mm long can thus result in a reflection of

$$|r(\omega)|^2 = |\tanh(\kappa L)|^2 = 0.998$$

The spectral bandwidth of the photonic bandgap is, according to Equation (12.4-68),

$$\Delta\lambda_{\text{gap}} = 2\left|\frac{n_1}{n_0}\right|\lambda_0 = 2.07\ \text{nm}$$

Distributed Feedback Lasers

If a periodic medium is provided with sufficient gain at frequencies near the Bragg frequency ω_0 (where $m\pi/\Lambda \approx \beta$), oscillation can result without the benefit of end reflectors. The mirror feedback of the conventional Fabry–Perot type laser is provided here by the continuous coherent backscattering from the periodic dielectric perturbation. In the following discussion we will consider two generic cases: (1) the bulk properties of a gain medium are perturbed periodically (e.g., with index variation) [2] and (2) the boundary of a waveguide laser is perturbed periodically (e.g., with surface corrugation) [3]. Both cases will be found to lead to the same set of coupled equations. Let g be the intensity gain of the optical beam in the medium; then the coupled modes can be written $A(z) \exp(-i\beta z + gz/2)$ and $B(z) \exp(+i\beta z - gz/2)$. If we define a complex propagation constant

$$\beta' = \beta - ig/2 \tag{12.6-12}$$

and an effective Bragg mismatch of

$$\Delta\beta' = 2\beta - m\frac{2\pi}{\Lambda} = \Delta\beta + ig \tag{12.6-13}$$

then the coupled equations are identical to the passive case we derived earlier. Thus the reflection coefficient obtained earlier is still valid. In other words, we can write the reflection coefficient as

$$r(\omega) = \frac{B(0)}{A(0)} = \frac{-i\kappa^* \sinh s'L}{s' \cosh s'L + i(\Delta\beta'/2) \sinh s'L} \tag{12.6-14}$$

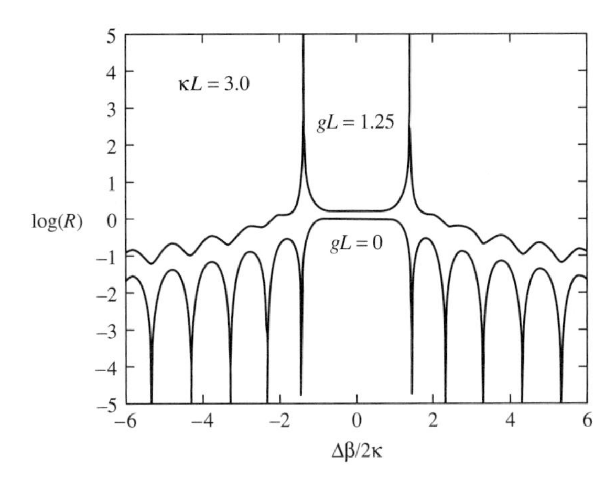

Figure 12.29 Reflectance spectrum of a Bragg reflector with $gL = 0$, and with $gL = 1.25$. The vertical lines indicate the location of the edges of the photonic bandgap.

with

$$s' = \sqrt{|\kappa|^2 - \left(\frac{\Delta\beta + ig}{2}\right)^2} \tag{12.6-15}$$

By examining Equation (12.6-14), we note that the reflectance can be greater than unity. Furthermore, the reflectance can reach infinity when the denominator vanishes. This is the situation of oscillation when a finite output is obtained without any input (zero input). Figure 12.29 plots the reflectance spectrum of a Bragg reflector with gain. In the example shown in the figure, a gain coefficient of $gL = 1.25$ and a coupling constant of $\kappa L = 3$ can lead to oscillations outside the photonic bandgap (near the edges of photonic bandgap).

At the center of the photonic bandgap, the reflection coefficient can be written

$$r(\omega) = \frac{B(0)}{A(0)} = \frac{-i\kappa^* \sinh s_0 L}{s_0 \cosh s_0 L - (g/2) \sinh s_0 L} \tag{12.6-16}$$

with

$$s_0 = \sqrt{|\kappa|^2 + (g/2)^2} \tag{12.6-17}$$

We note that the denominator in Equation (12.6-16) is always positive. Thus oscillation will never occur at the center of the photonic bandgap regardless of the magnitude of the gain. Semiconductor lasers with built-in monolithic gratings, the *distributed feedback lasers* (DFB lasers), are the topic of Chapter 16.

12.7 CHIRPED AND TAPERED INDEX GRATINGS

In Section 12.6 we discussed optical properties of an index grating with a dielectric constant that is a periodic function of position. In this section, we consider a more general case of index gratings in which either the amplitude of the index modulation and/or the periodicity varies with position. For example, let us consider an index grating given by

$$n(z) = n_0 + n_1(z) \cos\left(\frac{2\pi}{\Lambda(z)} z\right) \tag{12.7-1}$$

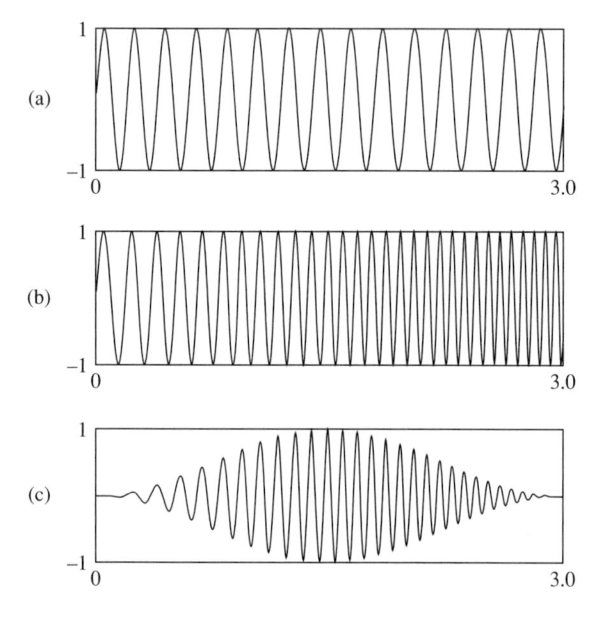

Figure 12.30 Schematic drawing of various index gratings: (a) a simple index grating with a constant period and a constant amplitude of index modulation, (b) a chirped index grating with a constant amplitude of index modulation, and (c) a chirped index grating with a tapered amplitude of index modulation.

where the amplitude of the index modulation $n_1(z)$ is a function of position z, and the period $\Lambda(z)$ is also a function of position z. In the case of a chirped index grating, the period of the grating $\Lambda(z)$ can be an increasing (or decreasing) function of position. In the case of a tapered index grating, the amplitude of the index modulation $n_1(z)$ is a smooth function of position. Figure 12.30 shows schematic drawings of some examples of these index gratings.

The coupled equations derived earlier in this chapter can be employed to find the reflection coefficient. We rewrite the coupled equations as

$$\begin{aligned} \frac{d}{dz} A &= -i\kappa(z) B e^{i\Delta\beta z} \\ \frac{d}{dz} B &= i\kappa^*(z) A e^{-i\Delta\beta z} \end{aligned} \tag{12.7-2}$$

where the coupling constant $\kappa(z)$, which is related to n_1, is a function of z, and the wavenumber mismatch (or momentum mismatch) $\Delta\beta$ is also a function of z and is given by

$$\Delta\beta = 2\beta - \frac{2\pi}{\Lambda(z)} \tag{12.7-3}$$

where we consider only the fundamental Fourier component ($m = 1$), and the period Λ is a function of z. The field amplitudes A and B can be obtained numerically by integrating the coupled equations from $z = L$ to $z = 0$ with the boundary condition of $B(L) = 0$. Once the field amplitudes are obtained, the reflection coefficient is given by $r = B(0)/A(0)$.

In addition to the numerical integration of the coupled differential equations, the reflection coefficient r can also be obtained by using a matrix method. In a chirped or tapered grating, the coupling constant κ and the grating period Λ are functions of position z. If we divide the chirped or tapered grating into a large number of segments, then we may treat each segment as a simple index grating. In this case, the results obtained in the previous sections can be used for each of the segments.

General solutions of the coupled equations for a simple index grating (with constant n_1 and Λ) can be written

$$A(z) = \left[\left(\cosh sz - i\frac{\Delta\beta}{2s}\sinh sz\right)A(0) - i\frac{\kappa}{s}\sinh szB(0)\right]e^{i(\Delta\beta/2)z}$$
$$B(z) = \left[i\frac{\kappa^*}{s}\sinh szA(0) + \left(\cosh sz + i\frac{\Delta\beta}{2s}\sinh sz\right)B(0)\right]e^{-i(\Delta\beta/2)z} \tag{12.7-4}$$

where $A(0)$ and $B(0)$ are arbitrary field amplitudes at $z = 0$. We define the actual electric field of the waves as

$$a(z) = A(z)e^{-i\beta z}$$
$$b(z) = B(z)e^{+i\beta z} \tag{12.7-5}$$

Then the general solutions can be written

$$a(z) = \left[\left(\cosh sz - i\frac{\Delta\beta}{2s}\sinh sz\right)a(0) - \left(i\frac{\kappa}{s}\sinh sz\right)b(0)\right]e^{-iKz/2}$$
$$b(z) = \left[\left(i\frac{\kappa^*}{s}\sinh sz\right)a(0) + \left(\cosh sz + i\frac{\Delta\beta}{2s}\sinh sz\right)b(0)\right]e^{+iKz/2} \tag{12.7-6}$$

or, equivalently, in a matrix form as

$$\begin{bmatrix} a(z)\exp(+iKz/2) \\ b(z)\exp(-iKz/2) \end{bmatrix} = \begin{bmatrix} \cosh sz - i\frac{\Delta\beta}{2s}\sinh sz & -i\frac{\kappa}{s}\sinh sz \\ i\frac{\kappa^*}{s}\sinh sz & \cosh sz + i\frac{\Delta\beta}{2s}\sinh sz \end{bmatrix} \begin{bmatrix} a(0) \\ b(0) \end{bmatrix} \tag{12.7-7}$$

where $K = 2\pi/\Lambda$, and $a(0)$ and $b(0)$ are the field amplitudes at $z = 0$. We note that the matrix is unimodular. By inverting the above matrix equation, we obtain

$$\begin{bmatrix} a_0 \\ b_0 \end{bmatrix} = \begin{bmatrix} \cosh sz + i\frac{\Delta\beta}{2s}\sinh sz & i\frac{\kappa}{s}\sinh sz \\ -i\frac{\kappa^*}{s}\sinh sz & \cosh sz - i\frac{\Delta\beta}{2s}\sinh sz \end{bmatrix} \begin{bmatrix} a(z)\exp(+iKz/2) \\ b(z)\exp(-iKz/2) \end{bmatrix} \tag{12.7-8}$$

To treat a general index grating with chirping and/or tapering, we divide the grating into a large number of segments. Figure 12.31 shows the division of a chirped grating into N segments

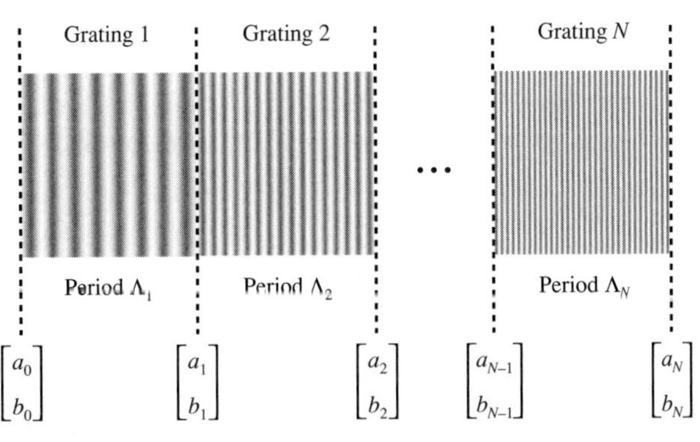

Figure 12.31 Schematic drawing of the division of a general chirped and tapered index grating into N segments. Each of the N segments can be treated as a simple index grating.

of gratings. For sufficiently large N, we may treat each segment as a simple homogeneous index grating. Let us represent the field amplitude at the end of the mth segment as a column vector

$$\begin{bmatrix} a_m \\ b_m \end{bmatrix} \tag{12.7-9}$$

Without loss of generality, we may select the segments so that each segment consists of an integral number of grating periods. In this case, ($z = M\Lambda$, M = integer) we can ignore the exponential factors $\exp(iKz)$ and $\exp(-iKz)$, which are either 1 or -1. The matrix relationship for the first segment can thus be written

$$\begin{bmatrix} a_0 \\ b_0 \end{bmatrix} = \begin{bmatrix} \cosh s_1\ell_1 + i\dfrac{\Delta\beta_1}{2s_1}\sinh s_1\ell_1 & i\dfrac{\kappa_1}{s_1}\sinh s_1\ell_1 \\ -i\dfrac{\kappa_1^*}{s_1}\sinh s_1\ell_1 & \cosh s_1\ell_1 - i\dfrac{\Delta\beta_1}{2s_1}\sinh s_1\ell_1 \end{bmatrix} \begin{bmatrix} a_1 \\ b_1 \end{bmatrix} \tag{12.7-10}$$

where ℓ_1 is the length of the first segment, and

$$s_1 = \sqrt{|\kappa_1|^2 - (\Delta\beta_1/2)^2} \tag{12.7-11}$$

with

$$\Delta\beta_1 = 2\beta - \frac{2\pi}{\Lambda_1} \tag{12.7-12}$$

We define a matrix

$$M_1 = M(s_1, \Delta\beta_1, \ell_1) = \begin{bmatrix} \cosh s_1\ell_1 + i\dfrac{\Delta\beta_1}{2s_1}\sinh s_1\ell_1 & i\dfrac{\kappa_1}{s_1}\sinh s_1\ell_1 \\ -i\dfrac{\kappa_1^*}{s_1}\sinh s_1\ell_1 & \cosh s_1\ell_1 - i\dfrac{\Delta\beta_1}{2s_1}\sinh s_1\ell_1 \end{bmatrix} \tag{12.7-13}$$

for the first segment. Then the matrix relationship can be written

$$\begin{bmatrix} a_0 \\ b_0 \end{bmatrix} = M_1 \begin{bmatrix} a_1 \\ b_1 \end{bmatrix} \tag{12.7-14}$$

This matrix M_1 is unimodular, and it can be shown that

$$M(s, \Delta\beta, \ell_1 + \ell_2) = M(s, \Delta\beta, \ell_1)M(s, \Delta\beta, \ell_2) \tag{12.7-15}$$

provided the two segments have the same period Λ and coupling constant κ.

We now combine the matrix relations for each of the segments and obtain

$$\begin{bmatrix} a_0 \\ b_0 \end{bmatrix} = M_1M_2M_3 \cdots M_N \begin{bmatrix} a_N \\ b_N \end{bmatrix} = M \begin{bmatrix} a_N \\ b_N \end{bmatrix} = \begin{bmatrix} M_{11} & M_{12} \\ M_{21} & M_{22} \end{bmatrix} \begin{bmatrix} a_N \\ b_N \end{bmatrix} \tag{12.7-16}$$

with M_m being the matrix for the mth segment,

$$M_m = M(s_m, \Delta\beta_m, \ell_m) = \begin{bmatrix} \cosh s_m\ell_m + i\dfrac{\Delta\beta_m}{2s_m}\sinh s_m\ell_m & i\dfrac{\kappa_m}{s_m}\sinh s_m\ell_m \\ -i\dfrac{\kappa_m^*}{s_m}\sinh s_m\ell_m & \cosh s_m\ell_m - i\dfrac{\Delta\beta_m}{2s_m}\sinh s_m\ell_m \end{bmatrix} \tag{12.7-17}$$

where ℓ_m is the length of the mth segment, $\Delta\beta_m = 2\beta - 2\pi/\Lambda_m$, κ_m is the corresponding coupling constant for the mth segment, $s_m = \sqrt{|\kappa_m|^2 - (\Delta\beta_m/2)^2}$, and M_{11}, M_{12}, M_{21}, and M_{22} are the matrix elements of the overall matrix M.

The Bragg reflection coefficient can thus be written in terms of the matrix elements of the overall matrix:

$$r = \left(\frac{b_0}{a_0}\right)_{b_N=0} = \frac{M_{21}}{M_{11}} \tag{12.7-18}$$

We will now investigate the optical properties of tapered and chirped index gratings. Specifically, we discuss the possibility of suppressing sidelobes in the reflection spectrum of a simple Bragg reflection filter, and suppressing ripples and spikes in the group delay spectrum.

Flat-Top Spectral Filters

As discussed in this chapter, the dispersion relationship of an index grating exhibits a photonic bandgap, where the wavenumber of propagation is a complex number. The unique property of the photonic bandgap can be employed for high-reflectance mirrors for laser cavities. Although a single periodic medium like a simple index grating or a quarter-wave stack can be employed as a narrowband filter, there are too many sidelobes that render these filters useless for many applications, including dense wavelength division multiplexing (DWDM) optical networks. It is known that multiple quarter-wave stacks separated by half-wave layers can be employed for narrowband filters, which exhibit flat-top passbands. However, strong ripples exist in these filters.

It is known that the sidelobes of a Bragg reflector can be eliminated by using chirping and tapering. In the following, we show an example of tapering that can be employed to eliminate the sidelobes and obtain a flat-top passband required in many applications, including DWDM optical networks. From the fabrication point of view, tapered index gratings are relatively easy to achieve via holographic interference. Specifically, we consider a tapered index grating given by

$$n(x) = n_0 + n_1 \sin^2\frac{\pi x}{L} \cos(Kx) \tag{12.7-19}$$

where n_0 is the background index of refraction, n_1 is the amplitude of the index modulation, K is the grating wavenumber, and L is the interaction length of the index grating. Figures 12.32 and 12.33 show the reflection spectra of two tapered index gratings with different coupling strengths. We note that the sidelobes can be reduced to below 50 dB, or even 75 dB.

Chirped Fiber Bragg Gratings for Dispersion Compensation

As discussed earlier in this chapter, a simple index grating is a dispersive optical element. The group delay upon reflection from a simple index grating exhibits a minimum when the frequency of the input beam is tuned to the center of the photonic bandgap. This is consistent with minimum penetration of optical energy into the index grating. For practical applications, particularly in long-distance transmission of signals through optical fibers, a dispersion compensation module must exhibit a group delay that is a linear function of the frequency in the spectral regime of interest. This can be achieved by using a chirped index grating. The basic concept is illustrated as follows. Referring to Figure 12.34, we consider a chirped index grating that can be viewed as a number of simple index gratings in sequence. In this case, the grating period is a decreasing function of position. As a result of the Bragg matching condition,

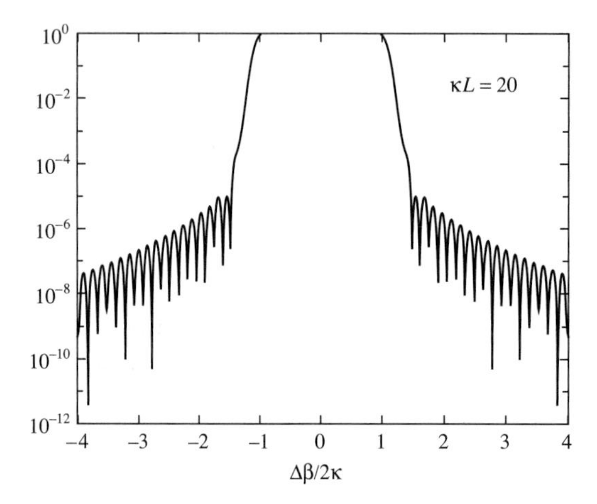

Figure 12.32 Reflection spectrum of a tapered index grating with $\kappa L = \pi n_1 L/\lambda = 20$.

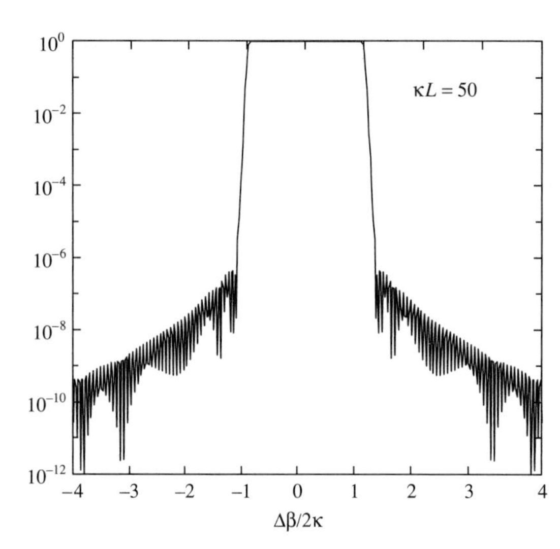

Figure 12.33 Reflection spectrum of a tapered index grating with $\kappa L = \pi n_1 L/\lambda = 50$.

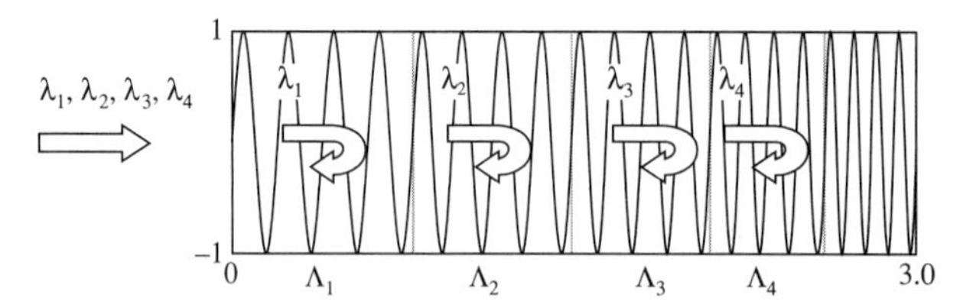

Figure 12.34 Schematic drawing of a chirped index grating for dispersion compensation. Optical waves at wavelength λ_m will be reflected at the position where the grating period is $\Lambda_m = \lambda_m/2n$.

$$\Lambda_m = \frac{\lambda_m}{2n} \qquad (12.7\text{-}20)$$

where n is the index of refraction of the medium, optical waves at wavelength λ_m will be reflected at the position where the grating period satisfies the above condition. As a result of the spatial chirping of the index grating, a group delay dispersion is created. In other words,

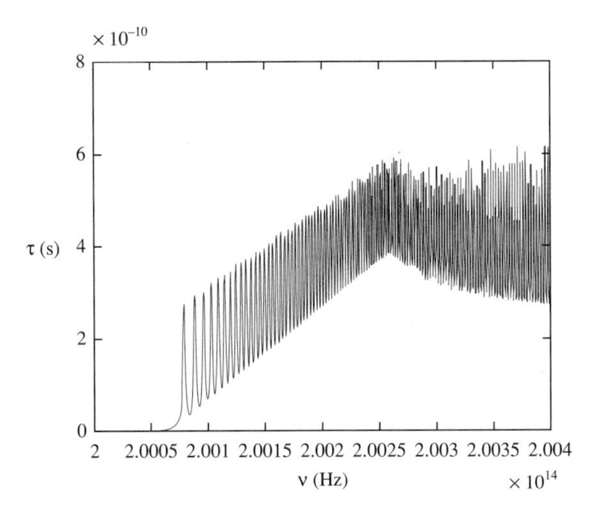

Figure 12.35 Calculated group delay as a function of frequency for a linearly chirped index grating.

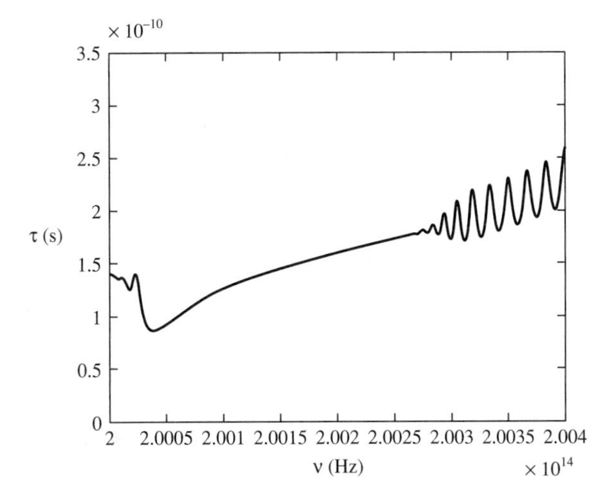

Figure 12.36 Calculated group delay as a function of frequency for a linearly chirped index grating with a Gaussian tapering.

optical waves at λ_4 will have a larger group delay as compared with optical waves at λ_1. In principle, a linear chirping will lead to a group delay that is a linear function of wavelength (or frequency) in a limited spectral regime of interest.

The group delay of a chirped index grating can be obtained by solving the coupled equations (12.7-2) from $z = L$ to $z = 0$ with the boundary condition of $B(L) = 0$ at each frequency. Once the field amplitude $B(0)$ is obtained, we can calculate the group delay by taking the derivative of the phase shift with respect to the angular frequency. Similarly, the group delay can be obtained by using the matrix method described earlier in this section. Figure 12.35 shows a calculated group delay as a function of frequency for a linearly chirped index grating. We notice that strong ripples (spikes) exist in the group delay. The strong ripples are a result of the sharp ending of the index grating at $z = 0$ and $z = L$. The ripples can be reduced or eliminated by using tapering or apodization. Figure 12.36 shows the result of a Gaussian tapering. The tapering provides a smooth ending to the index grating at $z = 0$ and $z = L$. We note that the tapering of the chirped index grating leads to the elimination of the group delay spikes.

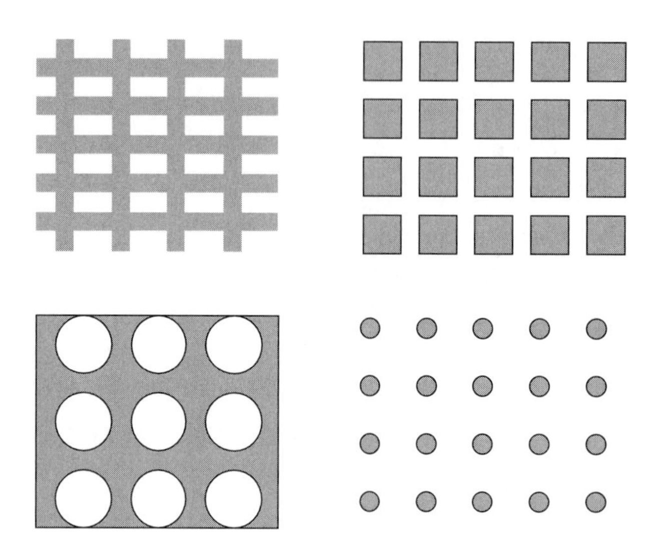

Figure 12.37 Schematic drawings of some 2-D periodic structures. In addition to rectangular lattices, hexagonal lattices can also be constructed.

12.8 2-D AND 3-D PERIODIC MEDIA (PHOTONIC CRYSTALS)

We have so far discussed one-dimensional (1-D) periodic media whose index of refraction varies periodically along one direction (z). For guided waves in single-mode fibers, the 1-D periodic medium is adequate to stop the propagation at some particular frequency band, as the direction of propagation is limited to only one direction (along the fiber axis). In free space, optical waves can propagate in many directions. Although a photonic bandgap may exist for any given direction of propagation in 1-D periodic media, these bandgaps occur at different frequencies. As a result, there's no unique frequency region where the propagation of optical waves can be completely stopped. This is illustrated in Figure 12.5 for the case of TE waves in a periodic layered medium (1-D). To really stop the propagation at a given frequency for all directions of propagation, a 2-D or 3-D periodic variation of the index of refraction may be needed. Figure 12.37 shows schematic drawings of some 2-D periodic media, involving a 2-D periodic array of cylinders in air (or voids in a host material).

Generally, solution of the Bloch waves in 2-D or 3-D periodic media requires numerical techniques and a computer program. To illustrate the concept of photonic bandgaps in 2-D or 3-D periodic media, we consider a simple case of a 2-D periodic medium whose dielectric constant is separable. In other words, the dielectric constant of the 2-D periodic structure can be written

$$\varepsilon(x, y) = \varepsilon_H(x) + \varepsilon_V(y) = \varepsilon_0 n_H^2(x) + \varepsilon_0 n_V^2(y) \tag{12.8-1}$$

where $\varepsilon_H(x) = \varepsilon_0 n_H^2(x)$ and $\varepsilon_V(y) = \varepsilon_0 n_V^2(y)$ are one-dimensional periodic functions. Figure 12.38 shows a 2-D periodic structure that satisfies the separable condition (12.8-1). The structure can be separated into the sum of two periodic layered media with index profile given by

$$n_H^2(x) = \begin{cases} n_1^2, & 0 < x < a \\ n_2^2, & a < x < a + b \equiv \Lambda \end{cases}$$
$$n_V^2(y) = \begin{cases} n_1^2, & 0 < y < a \\ n_2^2, & a < y < a + b \equiv \Lambda \end{cases} \tag{12.8-2}$$

where a and b are thicknesses of the layers, $\Lambda = a + b$ is the period, and n_1 and n_2 are the refractive indices of the layers, respectively.

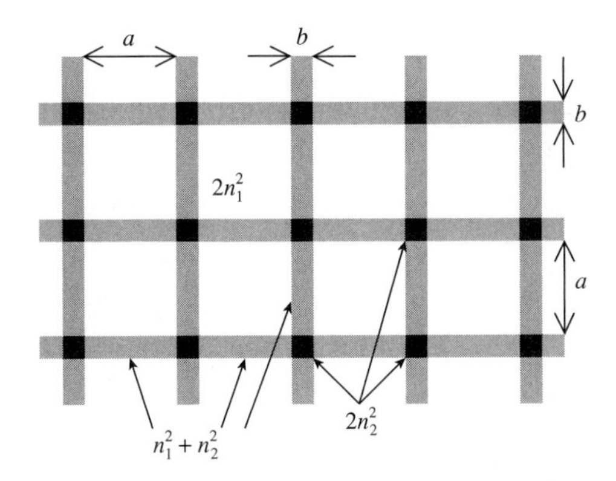

Figure 12.38 Schematic drawing of a separable 2-D periodic medium. The dielectric constants involved are $\varepsilon_0(2n_1^2)$, $\varepsilon_0(n_1^2+n_2^2)$, and $\varepsilon_0(2n_2^2)$. The big white square regions (usually air) have a dielectric constant of $\varepsilon_0(2n_1^2)$. The small dark square regions have a dielectric constant of $\varepsilon_0(2n_2^2)$. The rest of the structure has a dielectric constant of $\varepsilon_0(n_1^2 + n_2^2)$.

For simplicity in illustrating the concept, we consider propagation in the xy plane. The wave equation in this case can be written

$$\left(\frac{\partial^2}{\partial x^2} + \frac{\partial^2}{\partial y^2} + \omega^2\mu\varepsilon_0 n_H^2(x) + \omega^2\mu\varepsilon_0 n_V^2(y)\right)\mathbf{E}(x, y) = 0 \tag{12.8-3}$$

We further consider propagation of the electromagnetic wave with the $\mathbf{E}$ vector perpendicular to the xy plane and assume a solution in the form

$$\mathbf{E}(x, y) = \hat{\mathbf{z}}F(x)G(y) \tag{12.8-4}$$

The wave equation (12.8-3) becomes

$$\left(\frac{1}{F}\frac{\partial^2}{\partial x^2}F + \omega^2\mu\varepsilon_0 n_H^2(x)\right) + \left(\frac{1}{G}\frac{\partial^2}{\partial y^2}G + \omega^2\mu\varepsilon_0 n_V^2(y)\right) = 0 \tag{12.8-5}$$

Since the above equation is the sum of an x-dependent part and a y-dependent part, it follows that

$$\begin{aligned}\left(\frac{1}{F}\frac{\partial^2}{\partial x^2}F + \omega^2\mu\varepsilon_0 n_H^2(x)\right) &= \beta^2 \\ \left(\frac{1}{G}\frac{\partial^2}{\partial y^2}G + \omega^2\mu\varepsilon_0 n_V^2(y)\right) &= -\beta^2\end{aligned} \tag{12.8-6}$$

where β^2 is an arbitrary constant. Thus the 2-D problem is reduced to two one-dimensional problems. The results obtained in Section 12.2 can now be employed to obtain the Bloch waves and the dispersion relationship. The solutions in Bloch form can be written

$$\begin{aligned}F(x) &= e^{-iK_x x} f_{K_x}(x) \\ G(x) &= e^{-iK_y y} g_{K_y}(y)\end{aligned} \tag{12.8-7}$$

where K_x and K_y are components of the Bloch wavevector, and $f_{Kx}(x)$ and $g_{Kx}(y)$ are periodic functions. The dispersion relationship for the 2-D periodic medium can be written, according to Equations (12.2-27),

$$\cos K_x\Lambda = \cos k_{1x}a \cos k_{2x}b - \frac{1}{2}\left(\frac{k_{2x}}{k_{1x}} + \frac{k_{1x}}{k_{2x}}\right)\sin k_{1x}a \sin k_{2x}b, \tag{12.8-8}$$

$$\cos K_y\Lambda = \cos k_{1y}a \cos k_{2y}b - \frac{1}{2}\left(\frac{k_{2y}}{k_{1y}} + \frac{k_{1y}}{k_{2y}}\right)\sin k_{1y}a \sin k_{2y}b \tag{12.8-9}$$

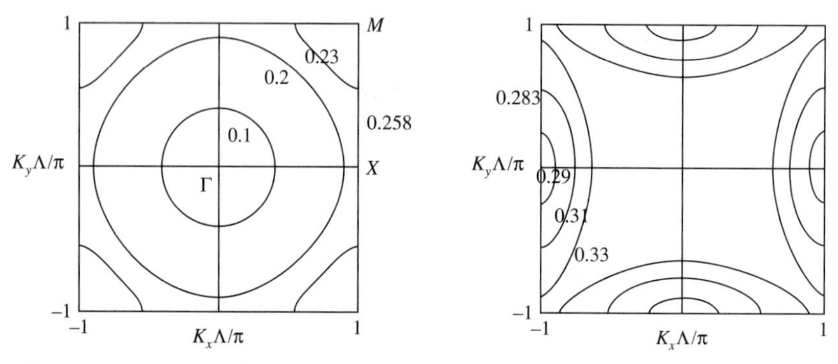

Figure 12.39 Normal surface in the K_xK_y plane of a 2-D periodic medium at various normalized frequencies: $\omega\Lambda/(2\pi c) = \Lambda/\lambda = 0.1, 0.2, 0.23, 0.258, 0.283, 0.29, 0.31, 0.33$. The normal surfaces plotted in the reduced Brillouin zone exhibit a fourfold symmetry. The parameters are $n_1^2 = 0.5$, $n_2^2 = 8.2$, $a = 0.8\Lambda$, and $b = 0.2\Lambda$. The corresponding values of the dielectric constants of the structure are 1.0, 8.7, and 16.4.

where k_{1x}, k_{1y}, k_{2x}, and k_{2y} are given, according to Equations (12.8-2) and (12.8-6), by

$$
\begin{aligned}
k_{1x} &= \sqrt{(n_1\omega/c)^2 - \beta^2} \\
k_{2x} &= \sqrt{(n_2\omega/c)^2 - \beta^2} \\
k_{1y} &= \sqrt{(n_1\omega/c)^2 + \beta^2} \\
k_{2y} &= \sqrt{(n_2\omega/c)^2 + \beta^2}
\end{aligned}
\qquad (12.8\text{-}10)
$$

The above equations can now be employed to obtain the relationship between ω and (K_x, K_y), and the photonic band structures. To obtain the dispersion relationship $\omega = \omega(K_x, K_y)$, we must eliminate β^2 from Equations (12.8-8) and (12.8-9). This is usually done as follows. For each given frequency ω, Equations (12.8-8) and (12.8-9) can be employed to obtain a Bloch wavevector (K_x, K_y) for each value of β^2. By connecting all the points in the K_xK_y plane, the normal surface for the given frequency in the K_xK_y plane is obtained. Figure 12.39 shows the normal surface for various normalized frequencies $\omega\Lambda/(2\pi c) = \Lambda/\lambda$ in the reduced Brillouin zone ($-1 < K_x\Lambda/\pi < 1$, $-1 < K_y\Lambda/\pi < 1$). At a low normalized frequency (0.1), the normal surface is almost a circle. At this frequency, the wavelength is 10 times the period, and the periodic structure behaves like a homogeneous medium. At a higher normalized frequency (0.2), the normal surface is no longer circular, reflecting the effect of the periodic structure. At an even higher normalized frequency (0.23), a partial photonic bandgap occurs where $K_x\Lambda = \pi$ or $K_y\Lambda = \pi$ and parts of the normal surface coincide with the edge of the Brillouin zone. At an even higher frequency (0.258), the normal surface is an exact square, which coincides completely with the edge of the Brillouin zone. At this frequency, the Bloch wave is evanescent for all directions of propagation, and we have a photonic bandgap. The photonic bandgap occupies the normalized frequency region between (0.258, 0.283). The normal surfaces for all frequencies within the photonic bandgap are all square. At normalized frequencies higher than the upper edge of the photonic bandgap (0.29, 0.31, 0.33), parts of the normal surfaces coincide with the edge of the Brillouin zone. The normal surfaces of these frequencies are folded back into the reduced Brillouin zone for convenience.

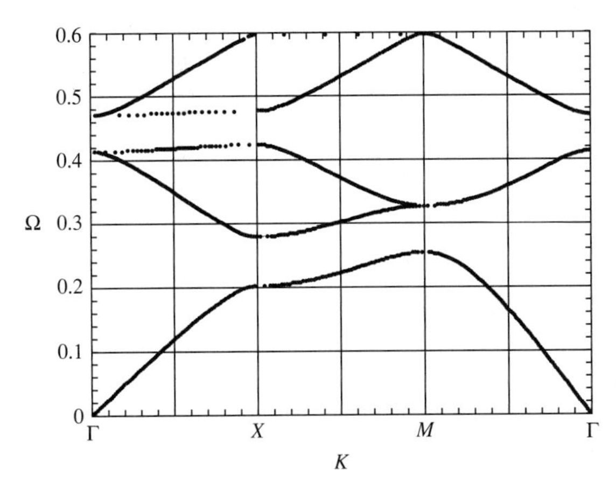

Figure 12.40 Band structure of a 2-D periodic medium. The vertical axis is the normalized frequency $\Omega \equiv \omega\Lambda/(2\pi c) = \Lambda/\lambda$. The parameters are $n_1^2 = 0.5$, $n_2^2 = 8.2$, $a = 0.8\Lambda$, and $b = 0.2\Lambda$. The corresponding values of the dielectric constants of the structure are 1.0, 8.7, and 16.4.

Equations (12.8-8) and (12.8-9) can also be employed to plot the band structure (ω versus K). For a square lattice, the band structure is often plotted in three major directions in the Brillouin zone. For convenience, the center of the Brillouin zone is designated as Γ ($K_x = 0$, $K_y = 0$), point X is located at ($K_x\Lambda = \pi$, $K_y = 0$), and point M is located at ($K_x = 0$, $K_y\Lambda = \pi$) (see Figure 12.39). To plot the band structure along ΓX, we set $K_y = 0$ in Equation (12.8-9). For a given frequency ω, setting $K_y = 0$ leads to a value for β^2. Substituting this value of β^2 into Equation (12.8-8), we obtain K_x. Similarly, for the band structure along XM, we set $K_x\Lambda = \pi$. The band structure along ΓM can easily be obtained by setting $\beta^2 = 0$. This is a result of symmetry. Figure 12.40 shows the band structure of a periodic medium. For this example, we note that a photonic bandgap exists at a normalized frequency $\omega\Lambda/(2\pi c) = \Lambda/\lambda$ between (0.258, 0.283). Another photonic bandgap also exists in the normalized frequency region between (0.430, 0.476).

The discussion above is for TE waves with **E** vector perpendicular to the plane of propagation (xy plane). The band structure for TM waves can be quite different from that of the TE wave.

The separable case discussed involves a periodic dielectric structure that has three different regions, each with a unique refractive index. Such a structure is difficult to manufacture. In practice, most 2-D and 3-D periodic structures are made of a high-index material (e.g., GaAs, Si, Ge) by creating a periodic array of voids. The index of refraction of the voids is 1. So the whole structure involves two different refractive indices. A large difference in the refractive indices between the voids and the host material is important to support a true photonic bandgap. Furthermore, a proper choice of the fill factor is also important to ensure the existence of a true photonic bandgap. Further investigation of this subject can be found in References [8–10].

Bragg Reflection Waveguides

As discussed in Chapter 3, total internal reflection can be employed for supporting confined propagation in dielectric media. The electromagnetic radiation is confined by the high refractive index of the guiding layer (core), as the electromagnetic wave undergoes total internal

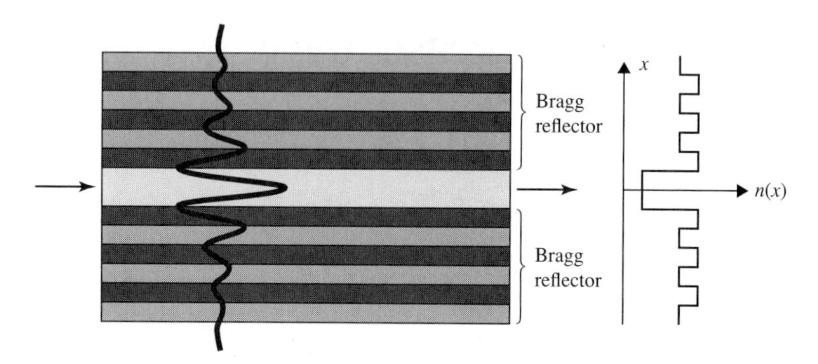

Figure 12.41 Schematic drawing of a Bragg reflection waveguide, consisting of a dielectric slab sandwiched between periodic layered structures.

reflection at the core–clad interfaces. Confined propagation can also be achieved by employing Bragg reflections at the core–clad interfaces. With the availability of Bragg reflection, a high refractive index is no longer required. This opens up the possibility of confined propagation in low-index media (including air and vacuum). This type of waveguide is known as a Bragg reflection waveguide (or photonic crystal waveguide) [11, 12].

To illustrate the concept, we consider a Bragg reflection waveguide as shown in Figure 12.41. A low-index slab is sandwiched between two Bragg reflectors. The Bragg reflectors can be made of periodic layered structures. The electromagnetic wave with a frequency in the photonic bandgap will undergo Bragg reflection from the bounding periodic layered structures. With a sufficient number of periods in the Bragg reflectors, the reflectivity can approach 100%. Thus the energy is trapped in the region bounded by the Bragg reflectors.

The mode condition for Bragg reflection waveguides can be obtained by solving the wave equation. For the simple case as shown in Figure 12.41, we may assume a plane wave solution in the dielectric slab and evanescent Bloch waves in the bounding layered media. To illustrate this, we consider a Bragg reflection waveguide with an index profile given by

$$n(x) = \begin{cases} n_1, & -b - d/2 < x < -b - a - d/2 \\ n_2, & -d/2 < x < -b - d/2 \\ n_c, & |x| < d/2 \\ n_2, & d/2 < x < b + d/2 \\ n_1, & b + d/2 < x < b + a + d/2 \end{cases} \tag{12.8-11}$$

where n_c is the core index, n_1 and n_2 are the refractive indices of the layers in the Bragg reflectors, d is the thickness of the core layer (guiding layer), and a and b are the thicknesses of the layers in the Bragg reflectors. For simplicity, we assume the Bragg reflectors have an infinite number of periods. The electric field of the symmetric waveguide modes can be written $E(x)\exp(-i\beta z)$, where β is the propagation constant along the z axis, and $E(x)$ can be written

$$E(x) = \begin{cases} E_K(x)e^{iKx}, & x < -d/2 \\ c_1 \cos(k_c x), & |x| < d/2 \\ E_K(x)e^{-iKx}, & x > d/2 \end{cases} \tag{12.8-12}$$

where $E_K(x)e^{-iKx}$ is the Bloch wavefunction given by Equations (12.2-24) and (12.2–26) derived earlier in this chapter, c_1 is a constant, and k_c is the x component of the wavevector in the guiding layer,

$$k_c = \sqrt{(\omega n_c/c)^2 - \beta^2} \tag{12.8-13}$$

For TE modes, we obtain the following mode condition, after matching the boundary conditions (Problem 12.12)

$$k_c \tan\left(k_c \frac{d}{2}\right) = -ik_{1x} \frac{e^{iK\Lambda} - A - B}{e^{iK\Lambda} - A + B} \quad \text{(even modes)} \tag{12.8-14}$$

where A and B are given in Equations (12.2-12), Λ is the period of the Bragg reflectors, and k_{1x} is given by

$$k_{1x} = \sqrt{(\omega n_1/c)^2 - \beta^2} \tag{12.8-15}$$

A similar derivation for the antisymmetric modes will yield the following mode condition:

$$k_c \cot\left(k_c \frac{d}{2}\right) = ik_{1x} \frac{e^{iK\Lambda} - A - B}{e^{iK\Lambda} - A + B} \quad \text{(odd modes)} \tag{12.8-16}$$

In the spectral regime where K is a complex number, the guided modes exhibit an oscillatory decay in the Bragg reflectors (Figure 12.41). For further details, interested readers are referred to Reference [11].

Waveguiding in a Bragg reflection waveguide can also be understood from the "defect" point of view. The waveguide shown in Figure 12.41 can be viewed as a periodic medium with a defect at the location of the guiding layer. The electron wavefunctions in a perfect crystal are Bloch waves whose probability distribution is uniform in the crystal. A defect (or disorder) in solids has a tendency to trap electrons. As a result, the electron wavefunctions are localized around the defect, provided the defect is strong enough. This was first pointed out by P. W. Anderson in 1958 and is known as Anderson localization [13]. Similarly, a defect in photonic periodic media has a tendency to trap electromagnetic waves. A waveguide is formed if the defects form a line (say, along the z axis) along the direction of propagation. The surface of a periodic layered medium can be considered as an extreme case of defect. Thus it is reasonable to believe that the surface of a Bragg reflector can also support a surface wave [14].

Two-dimensional and three-dimensional periodic media can also be employed as Bragg reflection waveguides. This is done by creating a defect in the periodic structure. Figure 12.42 shows some examples. In these examples, propagation in the xy plane consists of evanescent Bloch waves for frequencies inside the photonic bandgap. The structure is homogeneous in the z axis. Under the appropriate conditions, such structures can support confined propagation along the z axis.

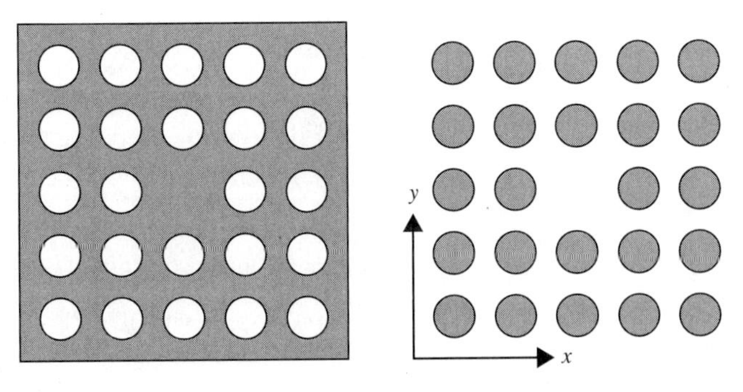

Figure 12.42 Schematic drawings of Bragg reflection waveguides using a defect in 2-D periodic structure to confine the propagation of electromagnetic waves along the z axis.

PROBLEMS

12.1 Show that the Bragg law of diffraction can be written

$$\mathbf{k}' - \mathbf{k} = m\mathbf{K}, \quad m = 1, 2, 3, \ldots$$

where $\mathbf{k}$ is the wavevector of the incident wave, $\mathbf{k}'$ is the wavevector of the scattered wave, and $\mathbf{K}$ is a reciprocal lattice vector.

12.2

(a) Consider a periodic array of slits with a period of Λ. Let θ be the angle of incidence. Show that diffraction occurs at angles θ' that satisfy the following equation:

$$\Lambda \sin\theta + \Lambda \sin\theta' = m(\lambda/n), \quad m = 1, 2, 3, \ldots$$

(b) Let z be the direction of periodicity. The grating equation can be written

$$k_z' - k_z = m(2\pi/\Lambda), \quad m = 1, 2, 3, \ldots$$

12.3 Find the matrix elements of the unit-cell translation matrix by deriving Equations (12.2-12).

12.4 Show that the unimodular nature of the translation matrix is consistent with the conservation of energy.

12.5 Find the matrix elements of the unit-cell translation matrix for the TM waves by deriving Equations (12.2-14).

12.6 Solve the eigenvalue problem of Equation (12.2-21) and derive the eigenvalues and eigenvectors. Derive the Bloch wave of Equation (12.2-26).

12.7

(a) Derive the following equation,

$$\cos K\Lambda = \cos k_1 a \cos k_2 b - \frac{1}{2}\left(\frac{n_2}{n_1} + \frac{n_1}{n_2}\right)\sin k_1 a \sin k_2 b$$

and plot $K\Lambda$ versus frequency ω for a quarter-wave stack.

(b) Show that the fundamental bandgap is given by

$$\Delta\omega_{\text{gap}} = \omega_0 \frac{4}{\pi}\sin^{-1}\left|\frac{n_2 - n_1}{n_2 + n_1}\right|$$

(c) Show that the even bandgap vanishes in a quarter-wave stack (see Figure 12.7).

(d) Plot the field pattern for frequency at the center of the fundamental bandgap.

12.8 The component of the Poynting vector in the direction of periodicity is zero inside the bandgap. Derive Equation (12.2-46). [Hint: $A + A^* = 2\cos K\Lambda = (-1)^m(e^{-K_i\Lambda} + e^{K_i\Lambda})$, $|e^{iK\Lambda} - A|^2 = e^{-2K_i\Lambda} + AA^* - (A + A^*)(-1)^m e^{-K_i\Lambda}$.]

12.9 Derive the following expression for the reflectivity of a Bragg reflector:

$$|r_N|^2 = \frac{|C|^2}{|C|^2 + (\sin K\Lambda/\sin NK\Lambda)^2}$$

12.10

(a) Show that $d(|A_1|^2 + |A_2|^2)/dz = 0$, where A_1 and A_2 satisfy the following coupled equations:

$$\frac{d}{dz}A_1 = -i\kappa A_2(z)e^{i\Delta\beta z}$$

$$\frac{d}{dz}A_2 = -i\kappa^* A_1(z)e^{-i\Delta\beta z}$$

(b) Find the general solution by deriving Equations (12.4-39).

12.11

(a) The net power flow in the $+z$ direction for contradirectional coupling is $|A_1|^2 - |A_2|^2$. Show that $d(|A_1|^2 + |A_2|^2)/dz = 0$ where A_1 and A_2 satisfy the following coupled equations:

$$\frac{d}{dz}A_1 = -i\kappa A_2(z)e^{i\Delta\beta z}$$

$$\frac{d}{dz}A_2 = i\kappa^* A_1(z)e^{-i\Delta\beta z}$$

(b) Find the general solution by deriving Equations (12.4-58).

12.12 Consider the symmetric Bragg reflection waveguide shown in Section 12.8.

(a) Show that the Bloch wave at the boundary $x = d/2$ can be written

$$E(x) = a_0 e^{-ik_{1x}(x-d/2)} + b_0 e^{+ik_{1x}(x-d/2)}$$

where a_0 and b_0 are given by Equation (12.2-24).

(b) Derive the mode conditions (12.8-14) and (12.8-16).

12.13 Let $u = \Delta\beta/2$, with $\Delta\beta = \beta_0 - \beta - m2\pi/\Lambda$. Show that the following integral,

$$I = \int_{\beta>0} \frac{\sin(\Delta\beta/2)z}{\Delta\beta/2} e^{i(\Delta\beta/2)z}\, d\beta$$

can be written

$$I = 2\int_{u_1}^{u_0} \frac{\sin uz}{u} e^{iuz} du, \quad \text{with } u_0 = (\beta_0 - m2\pi/\Lambda)/2,$$

$$u_1 = (\beta_0 - \beta_{max} - m2\pi/\Lambda)/2$$

where β_{max} is the maximum propagation constant of the radiation mode. The integral can be approximated by

$$I \approx 2\int_{-\infty}^{+\infty} \frac{\sin uz}{u} e^{iuz} du = 2\int_{0}^{+\infty} \frac{\sin 2uz}{u} du = \pi$$

provided z is positive.

REFERENCES

1. Yablonovitch, E. Inhibited spontaneous emission in solid state physics and electronics. *Phys. Rev. Lett.* **58**:2059 (1987).
2. Hill, K. O., Y. Fujii, D. C. Johnson, and B. Kawasaki, Photosensitivity in optical fiber waveguides: application to reflection filter fabrications. *Appl. Phys. Lett.* **32**:646 (1978).
3. Meltz, G., W. W. Morey, and W. H. Glenn, Formation of Bragg gratings in optical fibers by a transverse holographic method. *Opt. Lett.* **14**:823 (1989).
4. Hill, K. O., B. Malo, F. Bilodeau, and D. C. Johnson, Photosensitivity in optical fibers. *Ann. Rev. Mater. Sci.* **23**:125 (1993).
5. Joanopoulos, J. D., R. D. Meade, and J. N. Winn, *Photonic Crystals: Molding in the Flow of Light.* Princeton University Press, Princeton, NJ, 1995.
6. Archambault, J.-L., L. Reekie, and P. St. J. Russell, 100% reflectivity Bragg reflectors produced in optical fibres by single excimer laser pulses. *Electron Lett.* **29**:453 (1993).
7. Morey, W. W., G. A. Ball, and G. Meltz, Photoinduced Bragg gratings in optical fibers. *Optics Photonics News*, **Feb.**:8 (1994).
8. Yablonovitch, E., and T. J. Gmitter, Photonic band structures: the face-center cubic case. *Phys. Rev. Lett.* **63**:1950 (1989).
9. Yablonovitch, E., T. J. Gmitter, and K. M. Leung, Photonic band structures: the face-center cubic case employing non-spherical atoms. *Phys. Rev. Lett.* **67**:2295 (1991).
10. Joannopoulos, J. D., R. D. Meade, and J. N. Winn, *Photonic Crystals.* Princeton University Press, Princeton, NJ, 1995.
11. Yeh, P., and A. Yariv, Bragg reflection waveguides. *Opt. Commun.* **19**:427 (1976).
12. Yeh, P., A. Cho, and A. Yariv, Observation of confined propagation in Bragg waveguides. *Appl. Phys. Lett.* **30**:471 (1977).
13. Anderson, P. W., *Phys. Rev.* **109**:1492 (1958).
14. Yeh, P., A. Cho, and A. Yariv, Optical surface waves in periodic layered media. *Appl. Phys. Lett.* **32**:104 (1978).

CHAPTER 13

WAVEGUIDE COUPLING

13.0 INTRODUCTION

In this chapter we consider the coupling between two or more dielectric waveguides. We start from a discussion of the fundamental properties of guided modes in general dielectric waveguides, including orthogonality among the modes and power flows. Dielectric perturbation theory is then introduced to describe the coupling between two parallel waveguides, and the coupling of N parallel identical waveguides. The formalism is applied in analyzing a number of important applications, including electro-optic coupling and directional couplers.

13.1 GENERAL PROPERTIES OF MODES

The general requirement for a waveguide of electromagnetic radiation is that there be a flow of energy only along the guiding structure and not perpendicular to it. This means that the electromagnetic fields of guided modes will be appreciable only in the immediate neighborhood of the guiding structure. A dielectric cylinder of arbitrary cross section, such as the one shown in Figure 13.1, can serve as a waveguide provided its dielectric constant is large enough. For optical waves, this means that the index of refraction of the core of the waveguide must be larger than its surroundings. Generally, a beam of light propagating in a transversely inhomogeneous medium tends to bend toward the high-index region, according to the ray equation. Thus the higher index of refraction in the core of the guiding structure has an effect similar to that of a converging lens. Under the appropriate conditions, this converging effect due to the higher core index may cancel out exactly the spreading due to diffraction. When this happens, a guided mode is supported by the dielectric structure.

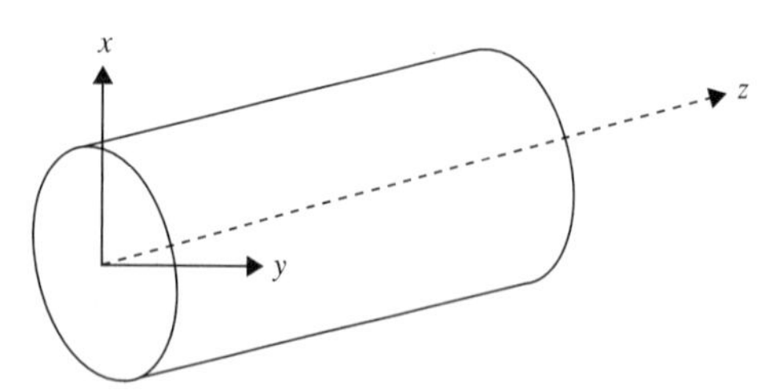

Figure 13.1 A section of a general cylindrical dielectric waveguide.

We begin by considering a waveguide with an arbitrary cross section as illustrated in Figure 13.1. The axis of the guide will be taken as the z axis, and the time variation of the modes is of the form $\exp(i\omega t)$. Maxwell's equations can be written

$$\nabla \times \mathbf{H} = i\omega\varepsilon_0 n^2 \mathbf{E} \tag{13.1-1}$$

$$\nabla \times \mathbf{E} = -i\omega\mu_0 \mathbf{H} \tag{13.1-2}$$

where n is the refractive index distribution of the dielectric structure and is a function of x and y only. Since the whole dielectric waveguide is homogeneous along the z axis, solutions to the wave equations (13.1-1) and (13.1-2) can be taken as

$$\mathbf{E} = \mathscr{E}(x, y) \exp[(i\omega t - \beta z)] \tag{13.1-3}$$

$$\mathbf{H} = \mathscr{H}(x, y) \exp[(i\omega t - \beta z)] \tag{13.1-4}$$

where β is the propagation constant to be determined from Maxwell's equations. We limit ourselves to dielectric structures that consist of piecewise homogeneous and isotropic materials, or materials with a small gradient in the distribution of the refractive index so that the wave equations (13.1-1) and (13.1-2) reduce to the scalar wave equation:

$$\nabla^2 \mathbf{E} + \left(\frac{\omega}{c}\right)^2 n^2(x, y)\mathbf{E} = 0 \tag{13.1-5}$$

Substitution of Equation (13.1-3) for $\mathbf{E}$ in the above equation yields

$$\nabla_t^2 \mathscr{E} + \left[\left(\frac{\omega}{c}\right)^2 n^2(x, y) - \beta^2\right]\mathscr{E} = 0 \tag{13.1-6}$$

where $\nabla_t^2 = \nabla^2 - \partial^2/\partial z^2$ is the transverse Laplacian operator. Equation (13.1-6) governs the transverse behavior of the field. In the case of piecewise homogeneous dielectric structures (e.g., single-mode fibers), Equation (13.1-6) holds separately in each homogeneous region. Therefore the field must be solved for separately in each region, and then the tangential components of the field must be matched at each interface. Another important boundary condition for confined modes is that the field amplitudes are zero at infinity. The axial propagation constant β must be the same throughout the waveguide structure in order to satisfy the boundary conditions at all points on the interfaces of those homogeneous media.

The wave equation (13.1-6) can be viewed as an eigenvalue problem, with $\mathscr{E}$ being the eigenfunctions and β^2 the eigenvalues. The eigenfunctions must satisfy the continuity condition at the interfaces, and the boundary conditions at infinity. Given a refractive index profile $n^2(x, y)$, there are in general an infinite number of eigenvalues, corresponding to an infinite number of modes. However, normally only a finite number of these modes are confined near the core of the dielectric structure and will propagate freely along the guide. One of the necessary conditions for a guided mode is that the fields fall off exponentially outside the core of the waveguide structure. Consequently, the quantity $(\omega/c)^2 n^2(x, y) - \beta^2$ must be negative in the region far away from the guiding region (core) and at infinity. In other words, the propagation constant β of a confined mode must be such that

$$\frac{\omega^2}{c^2} n^2(\infty) < \beta^2 \tag{13.1-7}$$

where $n^2(\infty)$ is the index of refraction of the waveguide structure at infinity ($\sqrt{x^2 + y^2} \rightarrow \infty$).

On the other hand, if the fields vanish at infinity (i.e., $|E(\infty)|^2 = 0$), the continuity of the fields requires that the magnitude of the field $|E(x, y)|$ attain a maximum value at some point in the xy plane. Normally, $E(x, y)$ is a smooth function of space. The existence of a maximum requires that the Laplacian of the field be negative. In other words, the propagation constant β of a confined mode must be such that

$$\beta^2 < \frac{\omega^2}{c^2} n^2(x, y) \tag{13.1-8}$$

in some region of the xy plane (usually in the core of the waveguide structure). In particular, if n_c is the maximum value of the refractive index profile $n^2(x, y)$, then the propagation constant of a confined mode must satisfy the following condition:

$$\beta^2 < \frac{\omega^2}{c^2} n_c^2 \tag{13.1-9}$$

In the region where condition (13.1-8) is satisfied, solutions to the wave equation are oscillatory. These oscillatory solutions must be matched to the exponential solutions at the boundary of the dielectric interfaces. Therefore not all the β's satisfying conditions (13.1-7) and (13.1-9) are legitimate eigenvalues of the confined modes. Often, only a discrete set of eigenvalues are legitimate for confined modes. This is in agreement with the results obtained in Chapter 3.

The dielectric structure also supports unconfined modes that are solutions of the wave equation. The propagation constants β of these unconfined modes do not have to satisfy the above conditions. These modes are often referred to as the radiation modes.

Orthogonality of Modes

The waveguide modes supported by an arbitrary dielectric structure have an important and useful orthogonality property. Let $(\mathbf{E}_1, \mathbf{H}_1)$ and $(\mathbf{E}_2, \mathbf{H}_2)$ be the fields of two linearly independent solutions to Maxwell's equations (13.1-1) and (13.1-2). Since there is no source (charge density and electrical current) in the dielectric waveguide structure, the following relation holds:

$$\nabla \cdot (\mathbf{E}_1 \times \mathbf{H}_2 - \mathbf{E}_2 \times \mathbf{H}_1) = 0 \tag{13.1-10}$$

The above equation is known as the Lorentz reciprocity theorem and is valid provided both ε and μ are symmetric tensors. If we replace the operator ∇ by $\nabla_t + \mathbf{a}_z\, \partial/\partial z$ and assume that the fields $(\mathbf{E}_1, \mathbf{H}_1)$ and $(\mathbf{E}_2, \mathbf{H}_2)$ are of the forms (13.1-3) and (13.1-4) with propagation constants β_1 and β_2, respectively, then the reciprocity relation reduces to

$$\nabla_t \cdot (\mathbf{E}_1 \times \mathbf{H}_2 - \mathbf{E}_2 \times \mathbf{H}_1) - i(\beta_1 + \beta_2)\mathbf{a}_z \cdot (\mathbf{E}_1 \times \mathbf{H}_2 - \mathbf{E}_2 \times \mathbf{H}_1) = 0 \tag{13.1-11}$$

where we recall that ∇_t is the transverse gradient operator and $\mathbf{a}_z$ is a unit vector along the z axis. Using the two-dimensional form of the divergence theorem, we obtain

$$\iint_S \nabla_t \cdot (\mathbf{E}_1 \times \mathbf{H}_2 - \mathbf{E}_2 \times \mathbf{H}_1) da = \oint_C (\mathbf{E}_1 \times \mathbf{H}_2 - \mathbf{E}_2 \times \mathbf{H}_1) \cdot \mathbf{n}\, dl$$

$$= i(\beta_1 + \beta_2) \iint_S (\mathbf{E}_1 \times \mathbf{H}_2 - \mathbf{E}_2 \times \mathbf{H}_1) \cdot \mathbf{a}_z\, da \tag{13.1-12}$$

where S denotes an arbitrary surface in the xy plane, C denotes the boundary of the surface, and $\mathbf{n}$ is a unit vector normal to curve C and $\mathbf{a}_z$. If we take S as the entire xy plane, then the contour integral in the above equation vanishes, since the fields vanish at infinity. Thus we obtain

$$(\beta_1 + \beta_2) \iint (\mathbf{E}_1 \times \mathbf{H}_2 - \mathbf{E}_2 \times \mathbf{H}_1) \cdot \mathbf{a}_z\, da = 0 \tag{13.1-13}$$

where the integral is carried over the entire xy plane.

Substitution of Equations (13.1-3) and (13.1-4) for **E** and **H** in Equation (13.1-13) yields

$$(\beta_1 + \beta_2) \iint (\mathcal{E}_1 \times \mathcal{H}_2 - \mathcal{E}_2 \times \mathcal{H}_1) \cdot \mathbf{a}_z \, da = 0 \tag{13.1-14}$$

since the common exponential terms can be canceled. To show that each term in Equation (13.1-14) vanishes separately, we consider the two solutions $(\mathbf{E}_1, \mathbf{H}_1)$ and $(\mathbf{E}'_2, \mathbf{H}'_2)$, where $(\mathbf{E}'_2, \mathbf{H}'_2)$ is the mirror transform of the mode $(\mathbf{E}_2, \mathbf{H}_2)$ with respect to the plane $z = 0$. Because of the symmetry of the dielectric structure, the transformed mode is also a solution to Maxwell's equations (13.1-1) and (13.1-2). This mirror transform corresponds to a reversal of the direction of propagation, and the directions of the longitudinal component of the electric field and the transverse component of the magnetic field are also reversed:

$$\begin{aligned}
\mathbf{E}'_{2t} &= \mathcal{E}_{2t}(x, y) \exp[(i\omega t + \beta_2 z)] \\
E'_{2z} &= -\mathcal{E}_{2z}(x, y) \exp[(i\omega t + \beta_2 z)] \\
\mathbf{H}'_{2t} &= -\mathcal{H}_{2t}(x, y) \exp[(i\omega t + \beta_2 z)] \\
H'_{2z} &= \mathcal{H}_{2z}(x, y) \exp[(i\omega t + \beta_2 z)]
\end{aligned} \tag{13.1-15}$$

where the subscript t stands for the transverse component. Since only the transverse components of the field contribute to the integrals (13.1-13) and (13.1-14), the equation corresponding to Equation (13.1-14) for this case $((\mathbf{E}_1, \mathbf{H}_1)$ and $(\mathbf{E}'_2, \mathbf{H}'_2))$ is

$$(\beta_1 - \beta_2) \iint (-\mathcal{E}_1 \times \mathcal{H}_2 - \mathcal{E}_2 \times \mathcal{H}_1) \cdot \mathbf{a}_z \, da = 0 \tag{13.1-16}$$

Addition and subtraction of Equations (13.1-14) and (13.1-16) yield

$$\iint (\mathcal{E}_1 \times \mathcal{H}_2) \cdot \mathbf{a}_z \, da = \iint (\mathcal{E}_2 \times \mathcal{H}_1) \cdot \mathbf{a}_z \, da = 0 \tag{13.1-17}$$

If we further assume that the dielectric structure is lossless (i.e., ε and μ are real tensors), a similar derivation leads to

$$\iint (\mathcal{E}_1 \times \mathcal{H}_2^*) \cdot \mathbf{a}_z \, da = \iint (\mathcal{E}_2^* \times \mathcal{H}_1) \cdot \mathbf{a}_z \, da = 0 \tag{13.1-18}$$

By including the time and z dependence, the orthogonal relationship in lossless dielectric waveguide structures can be written

$$\iint (\mathbf{E}_1 \times \mathbf{H}_2^*) \cdot \mathbf{a}_z \, da = 0 \tag{13.1-19}$$

where $\mathbf{H}_2^*$ is the complex conjugate of $\mathbf{H}_2$. This latter relationship shows that power flow in a lossless waveguide is the sum of the power carried by each mode individually. If the power is normalized to 1 W, the orthonormalization of the modes can be written

$$\frac{1}{2} \iint (\mathbf{E}_m \times \mathbf{H}_n^*) \cdot \mathbf{a}_z \, da = \delta_{mn} \tag{13.1-20}$$

where m, n are mode subscripts, and δ_{mn} is the Kronecker delta. For radiation modes, δ is the delta function.

For transverse electric (TE) or transverse magnetic (TM) modes, the orthonormality (13.1-20) reduces to (see Problem 13.7)

$$\frac{\beta_m}{2\omega\mu}\iint (\mathbf{E}_m \cdot \mathbf{E}_n^*)\, da = \delta_{mn} \qquad \text{(TE)} \tag{13.1-21}$$

$$\frac{\beta_m}{2\omega}\iint \left(\mathbf{H}_m \cdot \frac{1}{\varepsilon}\mathbf{H}_n^*\right) da = \delta_{mn} \qquad \text{(TM)} \tag{13.1-22}$$

Energy Transport

In a lossless dielectric waveguide, each mode carries power and propagates along the guide independently of the presence of other modes. This is apparent from the orthogonality of the modes expressed by Equation (13.1-20). The transport of power is given by the real part of the complex Poynting vector over the entire xy plane. Because of the orthogonality, we are able to treat the power transport of one mode at a time. For a given mode of propagation, the time-averaged power flow is given by

$$P = \tfrac{1}{2}\mathrm{Re}\iint (\mathbf{E} \times \mathbf{H}^*) \cdot \mathbf{a}_z\, da \tag{13.1-23}$$

The field energy per unit length of the mode is given by

$$U = \tfrac{1}{4}\mathrm{Re}\iint (\mathbf{E} \cdot \varepsilon\mathbf{E}^* + \mathbf{H} \cdot \mu\mathbf{H}^*)\, da \tag{13.1-24}$$

where we assume that both ε and μ are real so that the integral is also real. Since Maxwell's equations are linear, the power flow P and the energy density U are proportional. The constant of proportionality has the dimension of velocity and is called the velocity of energy transport:

$$v_e = \frac{P}{U} \tag{13.1-25}$$

This energy velocity can be shown to equal the group velocity, which is defined as

$$v_g = \frac{\partial\omega}{\partial\beta} \tag{13.1-26}$$

In a dielectric waveguide the propagation constant β for each mode is a function of ω. When a light pulse with a finite spread in frequency of $\delta\omega$ is propagating in a waveguide, it is possible that the power of each spectral component is carried by only one mode. If $\delta\beta$ is the corresponding spread in the propagation constant of the mode, the velocity of the pulse is given by Equation (13.1-26).

To prove that the energy velocity v_e and the group velocity v_g are equal, we start from Maxwell's equations by substituting $\nabla_t + \mathbf{a}_z\, \partial/\partial z$ for the operator ∇ in Equations (13.1-1) and (13.1-2).

$$\nabla_t \times \mathbf{H} + \mathbf{a}_z \times \frac{\partial}{\partial z}\mathbf{H} = i\omega\varepsilon\mathbf{E} \tag{13.1-27}$$

$$\nabla_t \times \mathbf{E} + \mathbf{a}_z \times \frac{\partial}{\partial z}\mathbf{E} = -i\omega\mu\mathbf{H} \tag{13.1-28}$$

where $\varepsilon = \varepsilon_0 n^2$, and we assume a more general permeability tensor μ. We recall that $\mathbf{a}_z$ is a unit vector along the waveguide. Since the z dependence of the fields is in the form of Equations (13.1-3) and (13.1-4), the above equations can be written

$$\nabla_t \times \mathbf{H} - i\beta \mathbf{a}_z \times \mathbf{H} = i\omega\varepsilon \mathbf{E} \tag{13.1-29}$$

$$\nabla_t \times \mathbf{E} - i\beta \mathbf{a}_z \times \mathbf{E} = -i\omega\mu \mathbf{H} \tag{13.1-30}$$

Suppose now that β is changed by an infinitesimal amount $\delta\beta$. If $\delta\omega$, $\delta\mathbf{E}$, and $\delta\mathbf{H}$ are the corresponding changes in ω, $\mathbf{E}$, and $\mathbf{H}$, respectively, we may follow the derivation in Chapter 1 and obtain the following equation:

$$\nabla_t \cdot \mathbf{F} - i4\delta\beta \,\mathrm{Re}[(\mathbf{E} \times \mathbf{H}^*) \cdot \mathbf{a}_z] = -i2\delta\omega[\mathbf{E} \cdot \varepsilon\mathbf{E}^* + \mathbf{H} \cdot \mu\mathbf{H}^*] \tag{13.1-31}$$

where $\mathbf{F}$ is given by

$$\mathbf{F} = \delta\mathbf{E} \times \mathbf{H}^* + \delta\mathbf{H}^* \times \mathbf{E} + \mathbf{H} \times \delta\mathbf{E}^* + \mathbf{E}^* \times \delta\mathbf{H} \tag{13.1-32}$$

If we perform an integration over the entire xy plane on Equation (13.1-31), and use the two-dimensional divergence theorem

$$\iint_S \nabla_t \cdot \mathbf{F}\, da = \oint_C \mathbf{F} \cdot \mathbf{n}\, dl \tag{13.1-33}$$

we obtain

$$\oint_C \mathbf{F} \cdot \mathbf{n}\, dl - i8\delta\beta P = -i8\delta\omega U \tag{13.1-34}$$

where C is a contour at infinity, and P and U are given by Equations (13.1-23) and (13.1-24), respectively. The contour integral vanishes because the field amplitudes of the confined modes are zero at infinity. This leads to

$$\delta\beta\, P = \delta\omega\, U \tag{13.1-35a}$$

By using the definition of the group velocity and energy velocity, Equation (13.1-35a) can be written

$$v_e\, \delta\beta = \delta\omega = \frac{\partial\omega}{\partial\beta}\delta\beta = v_g\, \delta\beta \tag{13.1-35b}$$

Since $\delta\beta$ is an arbitrary infinitesimal number, we conclude that

$$v_e = v_g \tag{13.1-36}$$

for a confined mode in a dielectric waveguide. In the above derivation, we assume that both ε and μ are independent of the frequency ω. So, strictly speaking, the equality of the energy velocity and the group velocity is valid only in the spectral regime where the chromatic dispersion of the material can be neglected.

13.2 DIELECTRIC PERTURBATION THEORY AND MODE COUPLING

In the preceding section we derived some general properties of dielectric waveguide modes, including the orthogonality and energy transport. Confined modes can actually be excited and propagate along the axis of the waveguide structure independently provided the dielectric constant $\varepsilon(x, y) = \varepsilon_0 n^2(x, y)$ remains independent of z. In the case when there is a dielectric perturbation $\Delta\varepsilon(x, y, z)$ due to waveguide imperfections, bending, surface corrugations, or the like, the modes of propagation are coupled to each other. In other words, if a pure mode is excited at the beginning of the waveguide, some of its power may be transferred to other modes. The detail of the coupling and the energy exchange among the modes due to a periodic dielectric perturbation $\Delta\varepsilon(x, y, z)$ were described in Section 12.4.

Here we consider a special case in which the dielectric perturbation is independent of z. In other words, the dielectric perturbation $\Delta\varepsilon(x, y, z) = \Delta\varepsilon(x, y)$ is a function of x and y only (i.e., $\partial\Delta\varepsilon/\partial z = 0$). There are many practical situations where it is desirable or necessary to obtain the propagation constants and the wavefunctions of the modes of the whole waveguide structure including the dielectric perturbation $\Delta\varepsilon(x, y)$. In the event that the wave equation is difficult to solve (or cannot be directly solved), perturbation theory provides a method of treating such problems. Often the perturbation $\Delta\varepsilon(x, y)$ is chosen so that the unperturbed waveguide structure can easily be solved or has known solutions. Let the dielectric constant be written

$$\varepsilon(x, y) = \varepsilon_a(x, y) + \Delta\varepsilon(x, y) \tag{13.2-1}$$

where $\varepsilon_a(x, y)$ is the dielectric constant of the unperturbed waveguide. Let the unperturbed modes of propagation be

$$\mathbf{E}_m = \mathscr{E}_m(x, y) \exp[i(\omega t - \beta_m z)] \tag{13.2-2}$$

whose transverse wavefunctions $\mathscr{E}_m$ satisfy the unperturbed wave equation

$$[\nabla_t^2 + \omega^2\mu\varepsilon_a(x, y)][\mathscr{E}_m(x, y) = \beta_m^2\mathscr{E}_m(x, y) \tag{13.2-3}$$

The modes form a complete orthogonal set, as discussed earlier, and obey the orthonormalization relation

$$\frac{\beta_m}{2\omega\mu}\iint \mathscr{E}_m \cdot \mathscr{E}_n^* \, dx\, dy = \delta_{mn} \tag{13.2-4}$$

We now consider the effect of a dielectric perturbation $\Delta\varepsilon(x, y)$ that is small compared with $\varepsilon_a(x, y)$. We assume that the application of such a small perturbation will only cause small changes in the mode wavefunctions and propagation constants. Let $\delta\mathscr{E}_m$ and $\delta\beta_m^2$ be the changes in the mode wavefunctions and propagation constants, respectively. The true wave equation now takes the form

$$[\nabla_t^2 + \omega^2\mu\varepsilon_a(x, y) + \omega^2\mu\, \Delta\varepsilon(x, y)](\mathscr{E}_m + \delta\mathscr{E}_m) = (\beta_m^2 + \delta\beta_m^2)(\mathscr{E}_m + \delta\mathscr{E}_m) \tag{13.2-5}$$

If we neglect the second-order terms $\Delta\varepsilon\, \delta\mathscr{E}_m$ and $\delta\beta_m^2\, \delta\mathscr{E}_m$ and use Equation (13.2-3), then Equation (13.2-5) simplifies to

$$[\nabla_t^2 + \omega^2\mu\varepsilon_a(x, y)]\delta\mathscr{E}_m + \omega^2\mu\, \Delta\varepsilon\, \mathscr{E}_m = \beta_m^2\, \delta\mathscr{E}_m + \delta\beta_m^2\, \mathscr{E}_m \tag{13.2-6}$$

To solve this equation we expand $\delta\mathscr{E}_m$ in terms of the unperturbed wavefunctions of the modes:

$$\delta\mathscr{E}_m(x, y) = \sum_n a_{mn}\, \mathscr{E}_n(x, y) \tag{13.2-7}$$

where a_{mn} are constants. Substituting Equation (13.2-7) for $\delta\mathscr{E}_m$ in Equation (13.2-6), and using Equation (13.2-3), we obtain

$$\sum_n a_{mn}(\beta_n^2 - \beta_m^2)\mathscr{E}_n(x, y) = (\delta\beta_m^2 - \omega^2\mu\, \Delta\varepsilon)\mathscr{E}_m(x, y) \tag{13.2-8}$$

If we now scalar-multiply the above equation by $\mathscr{E}_m^*$ and integrate over the entire xy plane, we observe that the expression on the left vanishes because of the orthogonal property (13.2-4). Thus we obtain the following equation:

$$\iint \mathscr{E}_m^* \cdot (\delta\beta_m^2 - \omega^2\mu\,\Delta\varepsilon)\mathscr{E}_m\,dx\,dy = 0 \tag{13.2-9}$$

Since $\delta\beta_m^2$ is a constant, this can be written

$$\delta\beta_m^2 = \frac{\iint \mathscr{E}_m^* \cdot \omega^2\mu\,\Delta\varepsilon\,\mathscr{E}_m\,dx\,dy}{\iint \mathscr{E}_m^* \cdot \mathscr{E}_m\,dx\,dy} \tag{13.2-10}$$

This equation gives the first-order correction to the propagation constant β_m^2. Using Equation (13.2-4) and $\delta\beta_m^2 = 2\beta_m\,\delta\beta_m$, Equation (13.2-10) can also be written

$$\delta\beta_m = \frac{\omega}{4}\iint \mathscr{E}_m^* \cdot \Delta\varepsilon\,\mathscr{E}_m\,dx\,dy \tag{13.2-11}$$

To obtain the correction $\delta\mathscr{E}_m$ for the mode wavefunction, we scalar-multiply each side of Equation (13.2-8) by $\mathscr{E}_n^*$ (with $n \neq m$) and integrate over the entire xy plane. This leads to

$$a_{mn} = \frac{\omega\beta_n}{2(\beta_m^2 - \beta_n^2)}\iint \mathscr{E}_n^* \cdot \Delta\varepsilon(x, y)\,\mathscr{E}_m\,dx\,dy, \quad n \neq m \tag{13.2-12}$$

The value a_{mm} is not given by such a process; it is to be chosen so as to normalize the resultant mode wavefunction according to Equation (13.2-4) and is given by

$$a_{mm} = -\frac{1}{4}\frac{\delta\beta_m^2}{\beta_m^2} = -\frac{1}{2}\frac{\delta\beta_m}{\beta_m} \tag{13.2-13}$$

Using the expression for $\delta\beta_m$, the coefficient a_{mm} can thus be written

$$a_{mm} = -\frac{\omega}{8\beta_m}\iint \mathscr{E}_m^* \cdot \Delta\varepsilon\,\mathscr{E}_m\,dx\,dy \tag{13.2-14}$$

It is convenient to define the "coupling coefficients" as

$$\kappa_{nm} = \frac{\omega}{4}\iint \mathscr{E}_n^* \cdot \Delta\varepsilon\,\mathscr{E}_m\,dx\,dy \tag{13.2-15}$$

so that the expression for the first-order correction to the mode wavefunction can be written, according to Equation (13.2-12) and (13.2-7),

$$\delta\mathscr{E}_m(x, y) = \sum_{n\neq m}\frac{2\beta_n}{\beta_m^2 - \beta_n^2}\kappa_{nm}\,\mathscr{E}_n(x, y) - \frac{\kappa_{mm}}{2\beta_m}\mathscr{E}_m(x, y) \tag{13.2-16}$$

The correction to the propagation constant $\delta\beta_m$ can thus be written

$$\delta\beta_m = \kappa_{mm} \tag{13.2-17}$$

The results obtained here can be used to evaluate the attenuation coefficient of the modes when linear absorption is present in the waveguide structure. The key result (13.2-10) is often written

$$\delta\beta_m^2 = \frac{\omega^2}{c^2}\frac{\iint \mathscr{E}_m^* \cdot \Delta n^2(x, y)\,\mathscr{E}_m\,dx\,dy}{\iint \mathscr{E}_m^* \cdot \mathscr{E}_m\,dx\,dy} = \frac{\omega^2}{c^2}\langle \Delta n^2(x, y)\rangle \tag{13.2-18}$$

where $\Delta n^2(x, y) = \Delta\varepsilon(x, y)/\varepsilon_0$, and $\langle \Delta n^2(x, y)\rangle$ denotes the statistically averaged value of the index perturbation. Equation (13.2-18) is also known as the Hellman–Feynman theorem in quantum mechanics.

It is important to remember that Equation (13.2-18) is only a first-order approximation. There are situations where the dielectric perturbation is particularly strong, or where the off-diagonal elements are such that $2\beta_n\kappa_{nm}$ is no longer much smaller than $|\beta_m^2 - \beta_n^2|$. This can lead to a significant correction of the wavefunctions, according to Equation (13.2-16). For situations like this, we need to express the wavefunctions in terms of a linear combination of the unperturbed wavefunctions,

$$\psi(x, y) = \sum_m c_m \mathscr{E}_m(x, y) \tag{13.2-19}$$

where c_m are constants to be determined so that $\psi(x, y)$ satisfies the following wave equation:

$$[\nabla_t^2 + \omega^2\mu\varepsilon_a(x, y) + \omega^2\mu\, \Delta\varepsilon(x, y)]\psi = \beta^2\psi \tag{13.2-20}$$

where $\Delta\varepsilon(x, y)$ is the dielectric perturbation. This approach is particularly useful for situations where the total number of modes involved is small (e.g., 2, 3 or 4). Substituting Equation (13.2-19) into Equation (13.2-20), we obtain

$$[\nabla_t^2 + \omega^2\mu\varepsilon_a(x, y) + \omega^2\mu\, \Delta\varepsilon(x, y)] \sum_m c_m \mathscr{E}_m(x, y) = \beta^2 \sum_m c_m \mathscr{E}_m(x, y) \tag{13.2-21}$$

Using Equation (13.2-3), the above equation becomes

$$\sum_m (\beta_m^2 + \omega^2\mu\, \Delta\varepsilon(x, y)) c_m \mathscr{E}_m(x, y) = \beta^2 \sum_m c_m \mathscr{E}_m(x, y) \tag{13.2-22}$$

We now scalar-multiply the above equation by $\mathscr{E}_n^*$ $(n = 1, 2, 3, 4, \ldots)$ and integrate over the entire xy plane. We obtain, after using the orthonormal relationship (13.2-4),

$$\begin{aligned}
(\beta_1^2 + 2\beta_1\kappa_{11})c_1 + 2\beta_1\kappa_{12}c_2 + 2\beta_1\kappa_{13}c_3 + 2\beta_1\kappa_{14}c_4 + \cdots &= \beta^2 c_1 \\
2\beta_2\kappa_{21}c_1 + (\beta_2^2 + 2\beta_2\kappa_{22})c_2 + 2\beta_2\kappa_{23}c_3 + 2\beta_2\kappa_{24}c_4 + \cdots &= \beta^2 c_2 \\
2\beta_3\kappa_{31}c_1 + 2\beta_3\kappa_{32}c_2 + (\beta_3^2 + 2\beta_3\kappa_{33})c_3 + 2\beta_3\kappa_{34}c_4 + \cdots &= \beta^2 c_3 \\
2\beta_4\kappa_{41}c_1 + 2\beta_4\kappa_{42}c_2 + 2\beta_4\kappa_{43}c_3 + (\beta_4^2 + 2\beta_4\kappa_{44})c_4 + \cdots &= \beta^2 c_4 \\
&\vdots
\end{aligned} \tag{13.2-23}$$

or equivalently, in matrix form,

$$\begin{bmatrix}
\beta_1^2 + 2\beta_1\kappa_{11} & 2\beta_1\kappa_{12} & 2\beta_1\kappa_{13} & 2\beta_1\kappa_{14} & \cdots \\
2\beta_2\kappa_{21} & \beta_2^2 + 2\beta_2\kappa_{22} & 2\beta_2\kappa_{23} & 2\beta_2\kappa_{24} & \cdots \\
2\beta_3\kappa_{31} & 2\beta_3\kappa_{32} & \beta_3^2 + 2\beta_3\kappa_{33} & 2\beta_3\kappa_{34} & \cdots \\
2\beta_4\kappa_{41} & 2\beta_4\kappa_{42} & 2\beta_4\kappa_{43} & \beta_4^2 + 2\beta_4\kappa_{44} & \cdots \\
\vdots & \vdots & \vdots & \vdots & \vdots
\end{bmatrix}
\begin{bmatrix} c_1 \\ c_2 \\ c_3 \\ c_4 \\ \vdots \end{bmatrix}
= \beta^2 \begin{bmatrix} c_1 \\ c_2 \\ c_3 \\ c_4 \\ \vdots \end{bmatrix} \tag{13.2-24}$$

where κ_{nm} is given by Equation (13.2-15). The wavefunctions and the corresponding eigenvalues β^2 can thus be obtained by solving the eigenvalue problem (13.2-23). To illustrate the use of the above method, we consider the following example.

EXAMPLE: STRESS-INDUCED BIREFRINGENCE IN SINGLE-MODE FIBER

Under the influence of a shear strain S_6, the two modes, LP_{01x} and LP_{01y}, are coupled. The dielectric perturbation due to the shear strain is given, according to Equations (9.6-17) and (9.6-34), by

$$\Delta\varepsilon = \varepsilon_0 n^4 \begin{bmatrix} 0 & p_{44}S_6 \\ p_{44}S_6 & 0 \end{bmatrix} \quad (13.2\text{-}25)$$

where p_{44} is a photoelastic coefficient of the material (silica). For this case, the Hellman–Feynman theorem leads to a zero correction to the propagation constant for the LP_{01} mode.

The propagation constants can be obtained by solving the following eigenvalue problem according to Equation (13.2-24),

$$\begin{bmatrix} \beta_0^2 + 2\beta_0\kappa_{11} & 2\beta_0\kappa_{12} \\ 2\beta_0\kappa_{21} & \beta_0^2 + 2\beta_0\kappa_{22} \end{bmatrix} \begin{bmatrix} c_1 \\ c_2 \end{bmatrix} = \beta^2 \begin{bmatrix} c_1 \\ c_2 \end{bmatrix} \quad (13.2\text{-}26)$$

where β_0 is the propagation of the mode before the perturbation, and κ_{11}, κ_{22}, κ_{12}, and κ_{21} are given by

$$\begin{aligned}
\kappa_{11} &= \frac{\omega}{4}\iint \mathscr{E}_0^* \cdot \Delta\varepsilon_{11}\mathscr{E}_0\, dx\, dy = 0 \\
\kappa_{22} &= \frac{\omega}{4}\iint \mathscr{E}_0^* \cdot \Delta\varepsilon_{22}\mathscr{E}_0\, dx\, dy = 0 \\
\kappa_{12} &= \frac{\omega}{4}\iint \mathscr{E}_0^* \cdot \Delta\varepsilon_{12}\mathscr{E}_0\, dx\, dy = \frac{2\omega\mu}{\beta_0}\frac{\omega}{4}\varepsilon_0 n^4 p_{44}S_6 \\
\kappa_{21} &= \frac{\omega}{4}\iint \mathscr{E}_0^* \cdot \Delta\varepsilon_{21}\mathscr{E}_0\, dx\, dy = \frac{2\omega\mu}{\beta_0}\frac{\omega}{4}\varepsilon_0 n^4 p_{44}S_6
\end{aligned} \quad (13.2\text{-}27)$$

Using Equation (13.2-26), Equation (13.2-26) becomes

$$\begin{bmatrix} \beta_0^2 - \beta^2 & 2\beta_0\kappa_{12} \\ 2\beta_0\kappa_{21} & \beta_0^2 - \beta^2 \end{bmatrix} \begin{bmatrix} c_1 \\ c_2 \end{bmatrix} = 0 \quad (13.2\text{-}28)$$

The eigenvalues are given by

$$\beta^2 = \beta_0^2 \pm 2\beta_0\kappa_{12} = \beta_0^2 \pm \omega^2\mu\varepsilon_0 n^4 p_{44}S_6 \quad (13.2\text{-}29)$$

The strain-induced birefringence is thus given by

$$\Delta n = \tfrac{1}{2}n^3 p_{44}S_6 \quad (13.2\text{-}30)$$

Using $p_{44} = -0.07$ for silica, $n = 1.45$, and a shear strain of $S_6 = 10^{-7}$, we obtain a stress-induced birefringence of $\Delta n = 2 \times 10^{-8}$.

13.3 COUPLING OF TWO PARALLEL WAVEGUIDES—DIRECTIONAL COUPLER

We now consider the coupling between the modes of two parallel waveguides separated by a finite distance. If a physical overlap of the mode wavefunctions is present, the modes may be coupled. Exchange of power between guided modes of adjacent waveguides is known as directional coupling. This phenomenon is similar to the electron motion in a two-atom

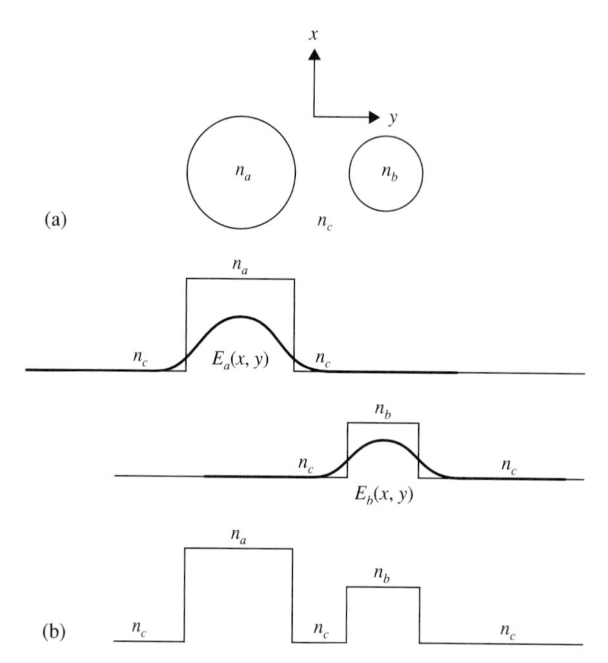

Figure 13.2 Examples of waveguide structures consisting of two parallel cylindrical dielectric waveguides: (a) two parallel circular dielectric waveguides separated by a finite distance and (b) two parallel dielectric slab waveguides separated by a finite distance. n_a and n_b are the core indices, and n_c is the clad index.

molecule. Waveguide directional couplers perform a number of useful functions in optical communications, including power division, power coupling, and switching.

In this section, we treat the coupling between the modes of two parallel waveguides by using the coupled-mode theory. Referring to Figure 13.2, we consider the case of two cylindrical dielectric waveguides separated by a finite distance. Let $\mathscr{E}_a(x, y) \exp[i(\omega t - \beta_a z)]$ and $\mathscr{E}_b(x, y) \exp[i(\omega t - \beta_b z)]$ be the modes of propagation of the individual waveguides when they are far apart. When these two waveguides are separated by a finite distance, the electric field of a general wave propagation in the coupled-waveguide structure can be approximated by

$$\mathbf{E}(x, y, z, t) = A(z)\mathscr{E}_a(x, y) \exp[i(\omega t - \beta_a z)] + B(z)\mathscr{E}_b(x, y) \exp[i(\omega t - \beta_b z)] \qquad (13.3\text{-}1)$$

provided the two waveguides are not too close to each other and $\mathscr{E}_a(x, y) \exp[i(\omega t - \beta_a z)]$ and $\mathscr{E}_b(x, y) \exp[i(\omega t - \beta_b z)]$ are the only confined modes of the waveguides. In the absence of coupling—hat is, if the distance between waveguides a and b is infinite—$A(z)$ and $B(z)$ do not depend on z and will be independent of each other, since each of the two terms on the right side of Equation (13.3-1) satisfies the wave equation separately.

Let $n^2(x, y)$ be the refractive index distribution of the composite waveguide structure shown in Figure 13.2. It can be written

$$n^2(x, y) = \begin{cases} n_a^2, & \text{core } a \\ n_b^2, & \text{core } b \\ n_c^2, & \text{elsewhere} \end{cases} \qquad (13.3\text{-}2)$$

For the purpose of mathematical convenience in the mode coupling, we define

$$\Delta n_a^2(x, y) = \begin{cases} n_a^2 - n_c^2, & \text{core } a \\ 0, & \text{elsewhere} \end{cases}$$

$$\Delta n_b^2(x, y) = \begin{cases} n_b^2 - n_c^2, & \text{core } a \\ 0, & \text{elsewhere} \end{cases} \tag{13.3-3}$$

$$n_s^2(x, y) = n_c^2$$

The index profile of the composite waveguide structure can thus be written

$$n^2(x, y) = n_s^2(x, y) + \Delta n_a^2(x, y) + \Delta n_b^2(x, y) \tag{13.3-4}$$

where $n_s^2(x, y)$ represents the refractive index distribution of the region outside the cores of the two waveguides, $\Delta n_a^2(x, y)$ represents the presence of waveguide a, and $\Delta n_b^2(x, y)$ represents the presence of waveguide b. With this definition, the index profile of waveguide a is $n_a^2(x, y) = n_s^2(x, y) + \Delta n_a^2(x, y)$, and the index profile of waveguide b is $n_b^2(x, y) = n_s^2(x, y) + \Delta n_b^2(x, y)$. It is thus obvious that the individual waveguide modes $\mathscr{E}_\alpha(x, y)$ (with $\alpha = a, b$) satisfy the equation

$$\left(\frac{\partial^2}{\partial x^2} + \frac{\partial^2}{\partial y^2} + \frac{\omega^2}{c^2}[n_s^2(x, y) + \Delta n_\alpha^2(x, y)]\right)\mathscr{E}_\alpha(x, y) = \beta_\alpha^2\, \mathscr{E}_\alpha(x, y) \tag{13.3-5}$$

The presence of waveguide b imposes a dielectric perturbation $\Delta n_b^2(x, y)$ on the propagation of the modes $\mathscr{E}_\alpha(x, y) \exp[i(\omega t - \beta_a z)]$ and vice versa. The electric field of the true wave propagation as given by Equation (13.3-1) must obey the wave equation

$$\left(\frac{\partial^2}{\partial x^2} + \frac{\partial^2}{\partial y^2} + \frac{\partial^2}{\partial z^2} + \frac{\omega^2}{c^2}[n_s^2(x, y) + \Delta n_a^2(x, y) + \Delta n_b^2(x, y)]\right)\mathbf{E} = 0 \tag{13.3-6}$$

The problem at hand is to solve for the mode amplitudes $A(z)$ and $B(z)$. We substitute Equation (13.3-1) for $\mathbf{E}$ in Equation (13.3-6) and use the assumption of slow variation of the mode amplitudes over z. After using Equation (13.3-5), we obtain

$$\begin{aligned} &-2i\beta_a \frac{dA}{dz}\mathscr{E}_a e^{i(\omega t - \beta_a z)} - 2i\frac{dB}{dz}\mathscr{E}_b e^{i(\omega t - \beta_b z)} \\ &= -\frac{\omega^2}{c^2}\Delta n_b^2(x, y) A \mathscr{E}_a e^{i(\omega t - \beta_a z)} - \frac{\omega^2}{c^2}\Delta n_a^2(x, y) B \mathscr{E}_b e^{i(\omega t - \beta_b z)} \end{aligned} \tag{13.3-7}$$

We now take the scalar product of Equation (13.3-7) with $\mathscr{E}_a^*(x, y)$ and $\mathscr{E}_b^*(x, y)$, respectively, and integrate over the entire xy plane. The result, using the normalization condition (13.2-4), is

$$\begin{aligned} \frac{dA}{dz} &= -i\kappa_{ab} B e^{i(\beta_a - \beta_b)z} - i\kappa_{aa} A \\ \frac{dB}{dz} &= -i\kappa_{ba} A e^{-i(\beta_a - \beta_b)z} - i\kappa_{bb} B \end{aligned} \tag{13.3-8}$$

where

$$\begin{aligned} \kappa_{ab} &= \frac{\omega}{4}\varepsilon_0 \iint \mathscr{E}_a^* \cdot \Delta n_a^2(x, y)\, \mathscr{E}_b \, dx\, dy \\ \kappa_{ba} &= \frac{\omega}{4}\varepsilon_0 \iint \mathscr{E}_b^* \cdot \Delta n_b^2(x, y)\, \mathscr{E}_a \, dx\, dy \end{aligned} \tag{13.3-9}$$

$$\kappa_{aa} = \frac{\omega}{4}\varepsilon_0 \iint \mathscr{E}_a^* \cdot \Delta n_b^2(x, y)\mathscr{E}_a \, dx \, dy$$
$$\kappa_{bb} = \frac{\omega}{4}\varepsilon_0 \iint \mathscr{E}_b^* \cdot \Delta n_a^2(x, y)\mathscr{E}_b \, dx \, dy \tag{13.3-10}$$

In arriving at Equation (13.3-8), we use the assumption that the waveguides are not too close, so that the overlap integral of the mode functions is small, that is,

$$\iint \mathscr{E}_a^* \cdot \mathscr{E}_b \, dx \, dy \ll \iint \mathscr{E}_a^* \cdot \mathscr{E}_a \, dx \, dy \tag{13.3-11}$$

The terms with κ_{aa} and κ_{bb} result from the dielectric perturbations to one of the waveguides due to the presence of the other waveguide, and represent only a small correction to the propagation constants β_a and β_b, respectively (see Equations (13.2-11) and (13.2-17)). The terms with κ_{ab} and κ_{ba} represent the exchange coupling between the two waveguides. Despite their different appearance in Equation (13.3-9), these two coupling constants (κ_{ab}, κ_{ba}) actually form a complex conjugate pair. In other words, $\kappa_{ba} = \kappa_{ab}^*$. Such a relationship is important to ensure the conservation of energy (see Problem 13.3). So if we take the total field as

$$\mathbf{E}(x, y, z, t) = A(z)\mathscr{E}_a e^{i[(\omega t-(\beta_a+\kappa_{aa})z]} + B(z)\mathscr{E}_b e^{i[(\omega t-(\beta_b+\kappa_{bb})z]} \tag{13.3-12}$$

instead of Equation (13.3-1), the coupled equations (13.3-8) become

$$\frac{dA}{dz} = -i\kappa_{ab} B e^{i2\delta z}$$
$$\frac{dB}{dz} = -i\kappa_{ba} A e^{-i2\delta z} \tag{13.3-13}$$

where 2δ is the phase mismatch given by

$$2\delta = (\beta_a + \kappa_{aa}) - (\beta_b + \kappa_{bb}) \tag{13.3-14}$$

Solutions of Equations (13.3-13) subject to a single input at waveguide a (i.e., $A(0) = A_0$, $B(0) = 0$) are given by

$$A(z) = A_0 e^{i\delta z}\left(\cos sz - i\frac{\delta}{s}\sin sz\right)$$
$$B(z) = -iA_0 e^{-i\delta z}\frac{\kappa}{s}\sin sz \tag{13.3-15}$$

where we assume $\kappa_{ab} = \kappa_{ba} = \kappa$ and

$$s = \sqrt{\kappa^2 + \delta^2} \tag{13.3-16}$$

In terms of powers $P_a(z) = |A(z)|^2$ and $P_b(z) = |B(z)|^2$ in the two waveguides, the solutions become

$$P_a(z) = P_0 - P_b(z)$$
$$P_b(z) = P_0 \frac{\kappa^2}{\kappa^2 + \delta^2}\sin^2\sqrt{\kappa^2 + \delta^2}\, z \tag{13.3-17}$$

where $P_0 = |A(0)|^2 = A_0^2$ is the input power to waveguide a at $z = 0$. Complete power transfer occurs in a distance $L = \pi/2\kappa$ provided $\delta = 0$; that is, the phase velocities of the two modes are equal. For $\delta \neq 0$, the maximum fraction of power that can be transferred is, from Equations (13.3-17),

$$\frac{\kappa^2}{\kappa^2 + \delta^2} \tag{13.3-18}$$

where the coupling constant $\kappa = \kappa_{ab} = \kappa_{ba}$ is given by Equations (13.3-9). For the case of parallel slab waveguides, the coupling constant can be evaluated by using the explicit expressions of the field obtained in Chapter 3. In the special case of two identical symmetric slab waveguides, the coupling constant for the same TE modes of propagation is

$$\kappa = \frac{2h^2 q e^{-qs}}{\beta(t + 2/q)(h^2 + q^2)} \left(\frac{2\pi}{\lambda}\right)^2 (n_2^2 - n_1^2) \tag{13.3-19}$$

where t is the waveguide width, s is the separation, and q and h are given by Equations (3.1-9) in Chapter 3.

The extension to two parallel channel waveguides that are confined in the vertical direction, as well as in the horizontal direction, is straightforward. In the well-confined case $t \gg 2/q$, Equation (13.3-19) simplifies to

$$\kappa = \frac{2h^2 q e^{-qs}}{\beta t(h^2 + q^2)} \left(\frac{2\pi}{\lambda}\right)^2 (n_2^2 - n_1^2) \tag{13.3-20}$$

A typical value of κ obtained at $\lambda \sim 1$ μm with t, $s \sim 3$ μm, and $\Delta n \sim 5 \times 10^{-3}$ is $\kappa \sim 5$ cm^{-1} so that coupling distances are of the order of magnitude of $\kappa^{-1} \sim 2$ mm.

A form of an electro-optic switch based on directional coupling [1] is as follows. The length L of the coupler is chosen so that for $\delta = 0$ (i.e., synchronous case) $\kappa L = \pi/2$. From Equations (13.3-17) it follows that all the input power to guide a exits from guide b at $z = L$. The switching is achieved by applying an electric field to guide a (or b) in such a way as to change its propagation constant until

$$\delta L = \tfrac{1}{2}(\beta_a - \beta_b)L = \frac{\sqrt{3}}{2}\pi \tag{13.3-21}$$

that is, $\delta = \sqrt{3}\kappa$. It follows from Equations (13.3-17) that at this value of δ

$$P_a = P_0 \quad \text{and} \quad P_b = 0 \tag{13.3-22}$$

that is, the power reappears at the output of guide a. A control of δ can thus be used to achieve any division of the powers between the outputs of guides a and b.

In practice, a convenient way to control δ is to fabricate the directional coupler in an electro-optic crystal (e.g., $LiNbO_3$). In this case, according to Equation (9.1-17), the application of an electric field E across one of the two waveguides will cause the index of refraction to change by

$$\Delta n \propto n^3 rE \tag{13.3-23}$$

where r is the appropriate electro-optic tensor element. The change Δn will give rise to a change in propagation constant

$$\delta \sim \frac{\omega}{c}\Delta n \sim \frac{\omega}{c} n^3 rE \tag{13.3-24}$$

The control of the power output from both arms of a directional coupler by means of an applied voltage is illustrated in Figure 13.3. The electrode geometry for applying a field to the waveguide is illustrated in Figure 13.4.

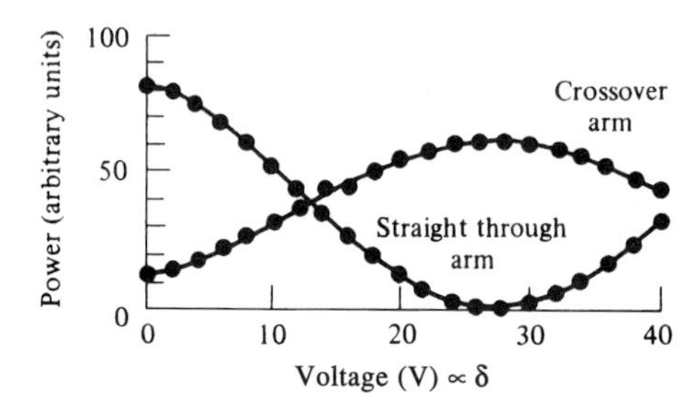

Figure 13.3 The dependence of the power output from the arms of a directional coupler on the (voltage-controlled) phase constant mismatch δ. (After Reference [2].)

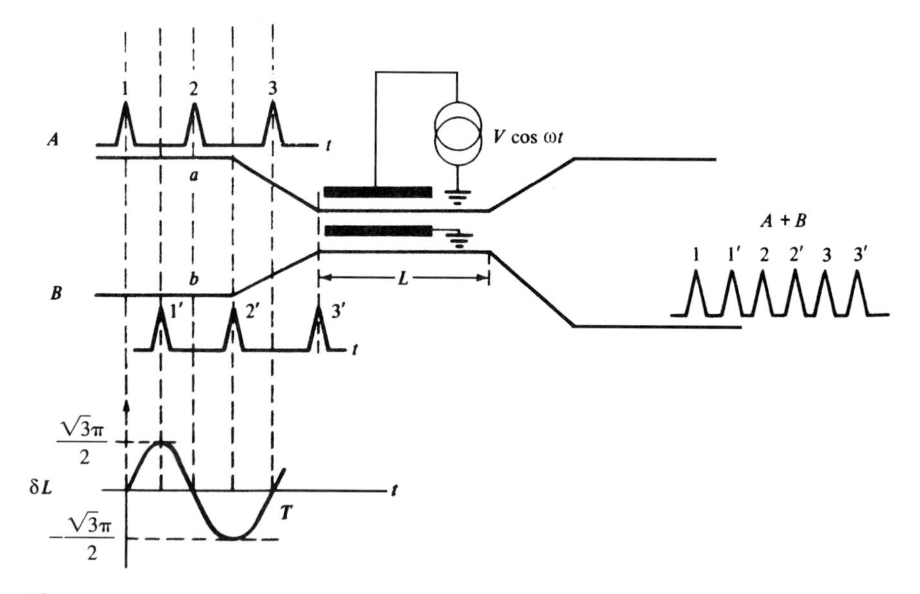

Figure 13.4 Multiplexing and demultiplexing in a directional coupling configuration. The two waveguides are indicated by the solid lines a and b and are brought into close proximity in the region between $z = 0$ and $z = L$. A pair of electrodes are deposited around waveguide a, so that a phase mismatch can be produced by an applied electric field.

One of the interesting applications for electro-optically switched directional couplers is in the area of very high-frequency ($>5 \times 10^9$ Hz) sampling and of multiplexing and demultiplexing of optical binary pulse trains. An example of the latter is demonstrated by Figure 13.4. Two independent, but synchronized, data pulse trains A and B are fed into legs a and b, respectively, of a directional coupler. The length of the coupling section satisfies the power transfer condition $\kappa L = \pi/2$. The phase mismatch δ between the two waveguides is controlled, as discussed above, by an electric field applied across one of the waveguides. This electric field is due to a microwave signal at a frequency ω_m. The resulting peak phase constant mismatch, which occurs at the maxima and minima of the applied voltage, satisfies the condition (13.3-21),

$$|\delta_{\text{max}}| = \frac{\sqrt{3}\pi}{2L} \tag{13.3-25}$$

so that the B pulses, which are synchronized to arrive during the extrema of the microwave signal, exit from arm b. Pulses A, on the other hand, arrive when the applied field, hence δ, is zero and, since $\kappa L = \pi/2$, cross over and exit from guide b. The result is that both pulse trains A

and B are interleaved or, in electrical engineering parlance, multiplexed in the output of guide b. The (combined) output from b can be fed into the input of a second directional coupler fed with a signal at $2\omega_m$ and multiplexed thereby with a second data train, and so on.

The device can, of course, be operated in reverse, right to left in the figure, and act as a demultiplexer for separating the dense bit train $A + B$ entering b into the individual trains A and B.

Approximate Wavefunctions and Propagation Constants

In addition to coupled-mode analysis, perturbation theory can also be employed to find approximate solutions to the wave equation that governs the propagation of electromagnetic radiation in the composite structure. In other words, the approximation of the actual normal modes of propagation can be obtained by carrying out a perturbation analysis. This technique is particularly useful when the wavefunctions as well as the eigenvalues (propagation constants) of the individual waveguides are known, and the closed-form solutions of the wavefunctions and eigenvalues of the composite waveguide structure are not available.

Again, let $\mathscr{E}_a(x, y) \exp[i(\omega t - \beta_a z)]$ and $\mathscr{E}_b(x, y) \exp[i(\omega t - \beta_a z)]$ be the fundamental modes of propagation of the individual waveguides when they are apart. The waveguide modes of the composite structure can be approximated by

$$\mathbf{E}(x, y)e^{-i\beta z} = [C_a\mathscr{E}_a(x, y) + C_b\mathscr{E}_b(x, y)] \exp[-i\beta z)] \tag{13.3-26}$$

where C_a and C_b are constants. The approximation is legitimate provided that the two waveguides are not too close to each other. To calculate the propagation constant β and to evaluate the constants C_a and C_b, we decompose the refractive index profile into three parts, according to Equation (13.3-4). Substituting Equation (13.3-26) into the wave Equation (13.3-6) and using Equation (13.3-5), we obtain

$$C_a\left(\beta_a^2 + \frac{\omega^2}{c^2}\Delta n_b^2(x, y) - \beta^2\right)\mathscr{E}_a + C_b\left(\beta_b^2 + \frac{\omega^2}{c^2}\Delta n_a^2(x, y) - \beta^2\right)\mathscr{E}_b = 0 \tag{13.3-27}$$

We now take the scalar product of Equation (13.3-27) with $\mathscr{E}_a^*(x, y)$ and $\mathbf{E}_b^*(x, y)$, respectively, and integrate over the entire xy plane. These lead to the following linear equation for C_a and C_b:

$$\begin{bmatrix} \beta_a^2 - \beta^2 + K_a & J_a + I^*(\beta_b^2 - \beta^2) \\ J_b + I(\beta_a^2 - \beta^2) & \beta_b^2 - \beta^2 + K_b \end{bmatrix}\begin{bmatrix} C_a \\ C_b \end{bmatrix} = 0 \tag{13.3-28}$$

where I, J_a, J_b, K_a, and K_b are constants given by

$$\begin{aligned} I &= \iint \mathscr{E}_b^* \cdot \mathscr{E}_a \, dx\, dy \\ J_a &= \left(\frac{\omega}{c}\right)^2 \iint \mathscr{E}_a^* \cdot \Delta n_a^2(x, y)\, \mathscr{E}_b \, dx\, dy \\ J_b &= \left(\frac{\omega}{c}\right)^2 \iint \mathscr{E}_b^* \cdot \Delta n_b^2(x, y)\, \mathscr{E}_a \, dx\, dy \\ K_a &= \left(\frac{\omega}{c}\right)^2 \iint \mathscr{E}_a^* \cdot \Delta n_b^2(x, y)\, \mathscr{E}_a \, dx\, dy \\ K_b &= \left(\frac{\omega}{c}\right)^2 \iint \mathscr{E}_b^* \cdot \Delta n_a^2(x, y)\, \mathscr{E}_b \, dx\, dy \end{aligned} \tag{13.3-29}$$

and we have used the following normalization,

$$\iint \mathscr{E}_a^* \cdot \mathscr{E}_a \, dx\, dy = \iint \mathscr{E}_b^* \cdot \mathscr{E}_b \, dx\, dy = 1 \qquad (13.3\text{-}30)$$

These integrals are carried out over the entire xy plane. Here $\Delta n_a^2(x, y)$ and $\Delta n_b^2(x, y)$ vanish except at the cores a and b, respectively. Physically, I is the overlap integral of the two individual wavefunctions that are not orthogonal to each other, K_a and K_b are, respectively, the dielectric perturbation to one of the waveguides due to the presence of the other waveguide, and J_a and J_b represent the exchange coupling between the two waveguides.

The eigenvalues β_1 and β_2 are the solutions of the secular equation

$$\begin{vmatrix} \beta_a^2 - \beta^2 + K_a & J_a + I^*(\beta_b^2 - \beta^2) \\ J_b + I(\beta_a^2 - \beta^2) & \beta_b^2 - \beta^2 + K_b \end{vmatrix} = 0 \qquad (13.3\text{-}31)$$

This is a quadratic equation in β^2 and, in general, will yield two desired propagation constants β_1 and β_2. The corresponding wavefunctions can be obtained by solving for the two constants C_a and C_b.

In the special case when the two waveguides are identical—that is, $\beta_a^2 = \beta_b^2 = \beta_0^2$, $K_a = K_b = K$, and $J_a = J_b = J$—the propagation constants and the corresponding wavefunctions are given by

$$\begin{aligned} \beta_1^2 &= \beta_0^2 + \frac{K + J}{1 + I} \\ \beta_2^2 &= \beta_0^2 + \frac{K - J}{1 - I} \end{aligned} \qquad (13.3\text{-}32)$$

$$\begin{aligned} \mathbf{E}_1 &= \frac{1}{\sqrt{2(1 + I)}} (\mathscr{E}_a + \mathscr{E}_b) \\ \mathbf{E}_2 &= \frac{1}{\sqrt{2(1 - I)}} (\mathscr{E}_a - \mathscr{E}_b) \end{aligned} \qquad (13.3\text{-}33)$$

The mode with wavefunction $\mathbf{E}_1$ and propagation constant β_1 is the fundamental mode with a symmetric wavefunction. The mode with wavefunction $\mathbf{E}_2$ and propagation constant β_2 is the high-order mode with an antisymmetric wavefunction. We note that these wavefunctions are mutually orthogonal.

13.4 COUPLING OF *N* PARALLEL IDENTICAL WAVEGUIDES—SUPERMODES

The modes of propagation of a periodic array of waveguides, which consists of N identical waveguides, are often called supermodes in laser diode arrays. A schematic drawing of a laser diode array is shown in Figure 13.5. The array consists of three parallel waveguides on a common substrate. Although closed-form solutions for the case of N identical slab dielectric wavegudies can be obtained, numerical analysis is often needed to find the modes of propagation of a general array of waveguides.

In the previous section, the perturbation approach was employed to obtain an approximate solution for the propagation constants and the wavefunctions of two parallel waveguides. We now extend this approach to the case of N identical waveguides. Figure 13.6 shows a simple example of N parallel identical slab waveguides.

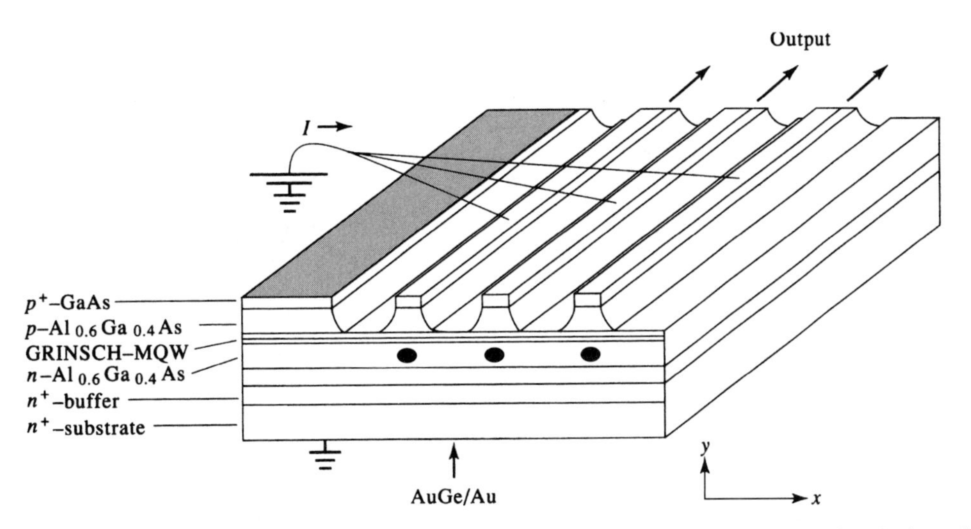

Figure 13.5 A three-channel semiconductor laser array. The elliptical spots correspond to the intensity pattern of the individual wavefunction of the confined mode of propagation. (After Reference [3]).

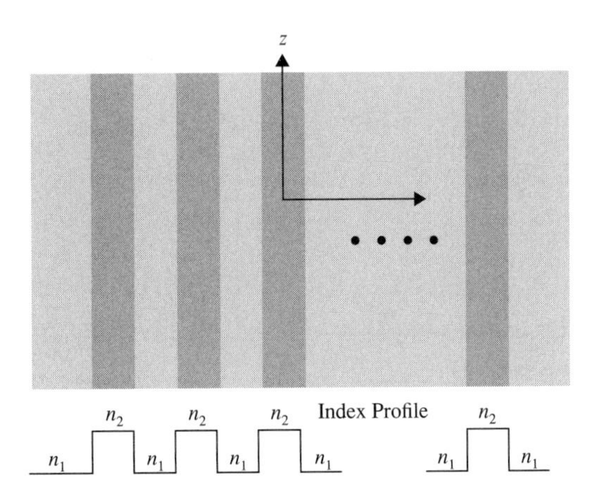

Figure 13.6 A simple example of N parallel identical slab waveguides. The structure consists of N parallel dielectric slabs separated by an equal distance. The lower figure shows the refractive index profile (with $n_1 < n_2$).

For the sake of simplicity in illustrating the concept, we assume that each of the individual waveguides supports only one confined mode. In this approach, the wavefunctions of the supermodes are written as linear combinations of the unperturbed wavefunctions of the individual waveguides,

$$\mathbf{E}(x, y)e^{-i\beta z} = \left(\sum_{n=1}^{N} C_n \mathscr{E}_n(x, y) \right) e^{-i\beta z} \tag{13.4-1}$$

where C_n is a constant, β is the propagation constant to be determined, and $\mathscr{E}_n(x, y)$ is the unperturbed wavefunction of the mode supported by the nth waveguide when the separation between the waveguides is infinite. In this expansion, we neglect the coupling with other modes of different propagation constants (e.g., radiation modes). We now follow the approach that leads to Equation (13.3-28) and assume that only nearest-neighbor coupling is present. This leads to

$$\begin{bmatrix} K-\Delta & J-I\Delta & 0 & 0 & \cdots & 0 & 0 & 0 \\ J-I\Delta & K-\Delta & J-I\Delta & 0 & \cdots & 0 & 0 & 0 \\ 0 & J-I\Delta & K-\Delta & J-I\Delta & \cdots & 0 & 0 & 0 \\ 0 & 0 & J-I\Delta & K-\Delta & \cdots & 0 & 0 & 0 \\ \vdots & \vdots & \vdots & \vdots & \cdots & \vdots & \vdots & \vdots \\ 0 & 0 & 0 & 0 & \cdots & K-\Delta & J-I\Delta & 0 \\ 0 & 0 & 0 & 0 & \cdots & J-I\Delta & K-\Delta & J-I\Delta \\ 0 & 0 & 0 & 0 & \cdots & 0 & J-I\Delta & K-\Delta \end{bmatrix} \begin{bmatrix} C_1 \\ C_2 \\ C_3 \\ C_4 \\ \vdots \\ C_{N-2} \\ C_{N-1} \\ C_N \end{bmatrix} = 0 \qquad (13.4\text{-}2)$$

where I, J, and K are the integrals defined in Equations (13.3-29) for a pair of neighboring waveguides, and Δ is given by

$$\Delta = \beta^2 - \beta_0^2 \qquad (13.4\text{-}3)$$

where β_0 is the propagation constant of the mode of a single individual unperturbed waveguide. In arriving at Equation (13.4-2), we used $I^* = I$, $J_a = J_b = J$, and $K_a = K_b = K$.

We notice that all the elements of the matrix in Equation (13.4-2) are zero except the diagonal elements and those immediately next to the diagonal elements. This is a result of the nearest-neighbor coupling. In addition, all the diagonal elements are identical ($K - \Delta$), and all the nonzero off-diagonal elements are also identical ($J - I\Delta$). This is a result of the fact that all the individual waveguides are identical and equally spaced. Closed-form solutions exist for such a simple and symmetric matrix equation.

The matrix equation (13.4-2) can be rewritten in recurrence form as

$$(J - I\Delta)C_n + (K - \Delta)C_{n+1} + (J - I\Delta)C_{n+2} = 0 \qquad (13.4\text{-}4)$$

with the boundary condition

$$C_{N+1} = C_0 = 0 \qquad (13.4\text{-}5)$$

The solution to Equation (13.4-4) subject to the preceding boundary condition can be written

$$C_n = \sin\frac{ns\pi}{N+1}, \quad n = 1, 2, 3, \ldots, N \qquad (13.4\text{-}6)$$

where s is an arbitrary integer given by

$$s = 1, 2, 3, \ldots, N \qquad (13.4\text{-}7)$$

We note that there are N independent solutions, each corresponding to an integer s. Substituting Equation (13.4-6) for C_n in Equation (13.4-4), we obtain the following closed-form solution for the propagation constants of the sth supermode:

$$\beta_s^2 - \beta_0^2 = \Delta = \frac{K + 2J\cos[s\pi/(N+1)]}{1 + 2I\cos[s\pi/(N+1)]}, \quad s = 1, 2, 3, \ldots, N \qquad (13.4\text{-}8)$$

where we recall that β_0 is the propagation constant of the confined mode of a single waveguide when the separation between the waveguides is infinite, and I, J, and K are the integrals defined in Equations (13.3-29) for a pair of neighboring waveguides. If the overlap integral I is much less than 1 ($I \ll 1$), Equation (13.4-8) can be written approximately

$$\beta_s^2 - \beta_0^2 = \Delta = K + 2J\cos\frac{s\pi}{N+1}, \quad s = 1, 2, 3, \ldots, N \qquad (13.4\text{-}9)$$

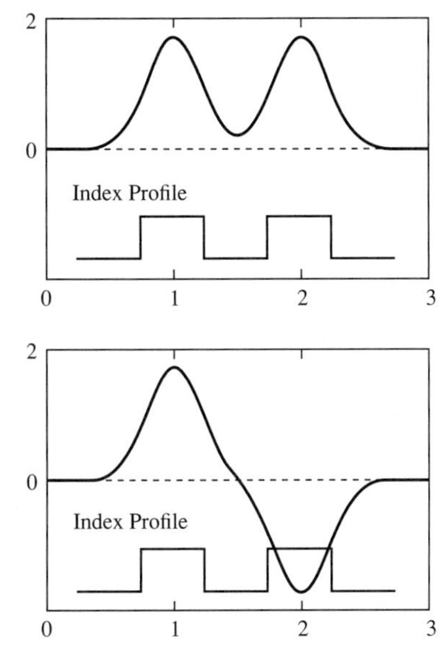

Figure 13.7 Wavefunctions of the supermodes of two parallel identical waveguides.

We notice that there are N independent supermodes. The propagation constants span a region of $K \pm 2J$. We consider the situation when N approaches infinity. In this case, the propagation constants span a continuous band with a width of $4J$.

To compare with earlier results for the case of two coupled waveguides, let us put $N = 2$ in Equation (13.4-8). The propagation constants according to Equation (13.4-8) and the wavefunctions according to Equations (13.4-1) and (13.4-6) are identical to the results obtained earlier. Figure 13.7 shows the wavefunctions for the case when $N = 2$.

To illustrate further the use of the perturbation approach, we consider the following example.

EXAMPLE: A FIVE-CHANNEL WAVEGUIDE

Consider the coupling of five identical parallel waveguides that are equally spaced. Let us assume that the individual waveguide supports only one confined mode of propagation. Find the propagation constants and the wavefunctions of the supermodes. Taking $N = 5$, and putting $s = 1, 2, 3, 4$, and 5 in Equation (13.4-9), we obtain the following propagation constants:

$$\beta_1^2 = \beta_0^2 + K + \sqrt{3}J$$

$$\beta_2^2 = \beta_0^2 + K + J$$

$$\beta_3^2 = \beta_0^2 + K$$

$$\beta_4^2 = \beta_0^2 + K - J$$

$$\beta_5^2 = \beta_0^2 + K - \sqrt{3}J$$

The wavefunctions can be obtained by using Equations (13.4-1) and (13.4-6) with the substitution of $N = 5$ and $s = 1, 2, 3, 4, 5$. The results are

$$\mathbf{E}_1(x, y)e^{-i\beta_1 z} = \left(\frac{1}{2}\mathscr{E}_1(x, y) + \frac{\sqrt{3}}{2}\mathscr{E}_2(x, y) + \mathscr{E}_3(x, y) + \frac{\sqrt{3}}{2}\mathscr{E}_4(x, y) + \frac{1}{2}\mathscr{E}_5(x, y)\right)e^{-i\beta_1 z}$$

$$\mathbf{E}_2(x, y)e^{-i\beta_2 z} = (\mathscr{E}_1(x, y) + \mathscr{E}_2(x, y) - \mathscr{E}_4(x, y) - \mathscr{E}_5(x, y))e^{-i\beta_2 z}$$

$$\mathbf{E}_3(x, y)e^{-i\beta_3 z} = (\mathscr{E}_1(x, y) - \mathscr{E}_3(x, y) + \mathscr{E}_5(x, y))e^{-i\beta_3 z}$$

$$\mathbf{E}_4(x, y)e^{-i\beta_4 z} = (\mathscr{E}_1(x, y) - \mathscr{E}_2(x, y) + \mathscr{E}_4(x, y) - \mathscr{E}_5(x, y))e^{-i\beta_4 z}$$

$$\mathbf{E}_5(x, y)e^{-i\beta_5 z} = \left(\frac{1}{2}\mathscr{E}_1(x, y) - \frac{\sqrt{3}}{2}\mathscr{E}_2(x, y) + \mathscr{E}_3(x, y) - \frac{\sqrt{3}}{2}\mathscr{E}_4(x, y) + \frac{1}{2}\mathscr{E}_5(x, y)\right)e^{-i\beta_5 z}$$

Note that these wavefunctions are not normalized. They are, however, mutually orthogonal. Figure 13.8 shows the wavefunctions of these five modes.

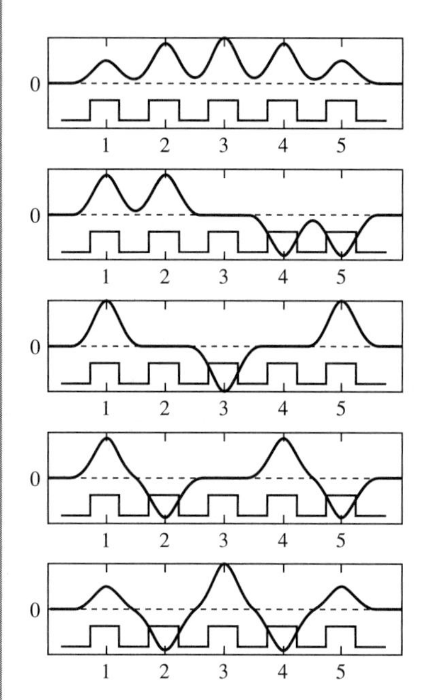

Figure 13.8 Wavefunctions of the supermodes of five parallel identical waveguides.

13.5 PHASE MATCHING AND FREQUENCY SELECTIVE COUPLING—MULTIPLEXING

According to Equation (13.3-18), the maximum efficiency of energy coupling between two different parallel waveguides is

$$\frac{\kappa^2}{\kappa^2 + \delta^2} \tag{13.5-1}$$

where κ is the coupling constant, and δ is proportional to the difference of the propagation constants. The coupling efficiency can be very low when $\kappa \ll \delta$. The coupling between two different waveguides can be made highly efficient and even frequency selective provided a periodic dielectric perturbation is introduced between the waveguides.

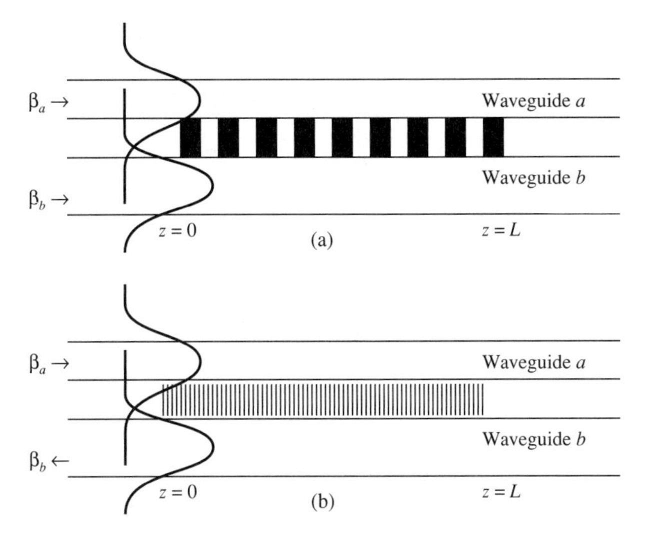

Figure 13.9 (a) Codirectional coupling between two different waveguides. (b) Contradirectional coupling between two different waveguides.

Referring to Figure 13.9, we consider the coupling between the modes of two different waveguides. The general analysis introduced in Chapter 12 can now be applied to treat the coupling between two different waveguides in the presence of a periodic dielectric perturbation.

We assume that the propagation constants of the modes of the two waveguides are sufficiently different (i.e., $\kappa_{ab} \ll |\beta_b - \beta_a|$) so that coupling is negligible in the absence of the periodic perturbation. Figure 13.9a shows the case of codirectional coupling, which often involves long-period gratings. Figure 13.9b is for contradirectional coupling, which involves short-period (or high-frequency) gratings.

Codirectional Grating Couplers

The electric field of the optical propagation in this case can be written

$$\mathbf{E}(x, y, z) = A(z)\mathscr{E}_a(x, y)e^{-i\beta_a z} + B(z)\mathscr{E}_b(x, y)e^{-i\beta_b z} \tag{13.5-2}$$

where $A(z)$ and $B(z)$ are mode amplitudes as functions of z. In the presence of the periodic dielectric perturbation, the general solutions can be written, according to Equations (12.4-39),

$$\begin{aligned} A(z) &= e^{i(\Delta\beta/2)z}\left[\left(\cos sz - i\frac{\Delta\beta}{2}\frac{\sin sz}{s}\right)A(0) - i\kappa\frac{\sin sz}{s}B(0)\right] \\ B(z) &= e^{-i(\Delta\beta/2)z}\left[-i\kappa^*\frac{\sin sz}{s}A(0) + \left(\cos sz + i\frac{\Delta\beta}{2}\frac{\sin sz}{s}\right)B(0)\right] \end{aligned} \tag{13.5-3}$$

with

$$s^2 = \kappa^*\kappa + \left(\frac{\Delta\beta}{2}\right)^2 \tag{13.5-4}$$

$$\kappa = \frac{\omega}{4}\int \mathscr{E}_a^*(x, y)\cdot\varepsilon_m(x, y)\,\mathscr{E}_b\,dx\,dy \tag{13.5-5}$$

$$\Delta\beta = \beta_a - \beta_b - m\frac{2\pi}{\Lambda} = 0 \tag{13.5-6}$$

where Λ is the period of the dielectric perturbation, $\varepsilon_m(x, y)$ is the mth Fourier component of the periodic dielectric perturbation, and $A(0)$ and $B(0)$ are the mode amplitudes at $z = 0$.

In the case of a single beam incidence at $z = 0$ (i.e., $B(0) = 0$), the solutions become

$$A(z) = e^{i(\Delta\beta/2)z}\left(\cos sz - i\frac{\Delta\beta}{2}\frac{\sin sz}{s}\right)A(0)$$
$$B(z) = e^{-i(\Delta\beta/2)z}\left(-i\kappa^*\frac{\sin sz}{s}\right)A(0) \tag{13.5-7}$$

The coupling efficiency (or diffraction efficiency) is defined as

$$\eta = \left|\frac{B(L)}{A(0)}\right|^2 \tag{13.5-8}$$

where L is the interaction length. Using Equations (13.5-7), we obtain

$$\eta = \frac{|\kappa|^2}{|\kappa|^2 + (\Delta\beta/2)^2}\sin^2|\kappa|L\sqrt{1+\left(\frac{\Delta\beta}{2|\kappa|}\right)^2} \tag{13.5-9}$$

We now examine the coupling efficiency as a function of wavelength. Let λ_0 be the center wavelength where the phase-matching condition is satisfied. In other words,

$$\Delta\beta = \beta_a - \beta_b - m\frac{2\pi}{\Lambda} = n_a\frac{2\pi}{\lambda_0} - n_b\frac{2\pi}{\lambda_0} - m\frac{2\pi}{\Lambda} = 0, \quad \text{at } \lambda = \lambda_0 \tag{13.5-10}$$

where n_a and n_b are the effective indices of the modes of propagation of the individual waveguides. According to Equation (13.5-10), the phase matching for codirectional coupling requires a grating period of

$$\Lambda = m\frac{\lambda_0}{|n_a - n_b|}, \quad m = 1, 2, 3, \ldots \tag{13.5-11}$$

If we assume that the effective indices are insensitive to the variation in wavelength as compared with $1/\lambda$, then the phase mismatch $\Delta\beta$ for a general wavelength can be written

$$\Delta\beta = (n_a - n_b)\left(\frac{2\pi}{\lambda} - \frac{2\pi}{\lambda_0}\right) \tag{13.5-12}$$

We further assume that the interaction length is such that $|\kappa|L = \pi/2$ so that the coupling efficiency is 100% at $\lambda = \lambda_0$. According to Equation (13.5-9), the coupling efficiency drops to 50% when $\Delta\beta = 1.60|\kappa|$ (assuming $|\kappa|L = \pi/2$). Using Equations (13.5-9) and (13.5-12), we obtain a FWHM bandwidth of

$$\Delta\lambda = 1.60\frac{\lambda_0^2}{\pi|n_a - n_b|}|\kappa| \tag{13.5-13}$$

Consider an example of two parallel waveguides with $n_a = 1.50$, $n_b = 1.51$, $|\kappa| = 5\ \text{cm}^{-1}$, and $\lambda_0 = 1\ \mu\text{m}$. The above formula yields $\Delta\lambda = 25$ nm. The grating period needed for phase matching is an integral multiple of 100 μm, according to Equation (13.5-11). Generally, the bandwidth of codirectional coupling is relatively broad. The bandwidth is inversely proportional to the difference of the effective indices, $|n_a - n_b|$, and is directly proportional to the coupling constant. So if a narrow bandwidth of coupling is desired (e.g., in the case of frequency-selective couplers), the waveguides should be designed to be sufficiently different, and with a small coupling constant. It is important to note that a small coupling constant requires a long interaction length ($|\kappa|L = \pi/2$) to achieve 100% coupling.

Using $|\kappa|L = \pi/2$, and the grating period of Equation (13.5-11), the coupling bandwidth can also be written (assuming $m = 1$)

$$\Delta\lambda = 0.8\frac{\lambda_0}{N} \tag{13.5-14}$$

where N is the number of periods in the grating ($N = \Lambda/L$). We note that the bandwidth of grating coupling is inversely proportional to the number of periods N.

Contradirectional Grating Couplers

The electric field of the optical propagation in this case can be written

$$\mathbf{E}(x, y, z) = A(z)\mathscr{E}_a(x, y)e^{-i\beta_a z} + B(z)\mathscr{E}_b(x, y)e^{+i\beta_b z} \tag{13.5-15}$$

where $A(z)$ and $B(z)$ are mode amplitudes as functions of z. Note that the mode in waveguide b, $\mathscr{E}_b(x, y)e^{+i\beta_b z}$, is propagating in the $-z$ direction. The general solutions for the case of contradirectional coupling subject to the boundary condition $B(L) = 0$ according to Section 12.4, can be written

$$\begin{aligned} A(z) &= e^{i(\Delta\beta/2)z}\frac{s\cosh s(L-z) + i(\Delta\beta/2)\sinh s(L-z)}{s\cosh sL + i(\Delta\beta/2)\sinh sL}A(0) \\ B(z) &= e^{-i(\Delta\beta/2)z}\frac{-i\kappa^*\sinh s(L-z)}{s\cosh sL + i(\Delta\beta/2)\sinh sL}A(0) \end{aligned} \tag{13.5-16}$$

with

$$s^2 = \kappa^*\kappa - \left(\frac{\Delta\beta}{2}\right)^2 \tag{13.5-17}$$

$$\Delta\beta = \beta_a + \beta_b - m\frac{2\pi}{\Lambda} = 0 \tag{13.5-18}$$

where κ is the coupling constant given by Equation (13.5-5), and $A(0)$ is the input mode amplitude at $z = 0$. From the above solution $B(z)$, it can be seen that the energy coupling efficiency between the two modes in the region between $z = 0$ and $z = L$ can be written

$$\eta = \frac{|B(0)|^2}{|A(0)|^2} = \frac{|\kappa|^2\sinh^2 sL}{s^2\cosh^2 sL + (\Delta\beta/2)^2\sinh^2 sL} \tag{13.5-19}$$

We now examine the frequency-selective nature of the contradirectional grating coupler. Again, let λ_0 be the wavelength where $\Delta\beta = 0$. The phase-matching condition in contradirectional coupling requires a grating with a period of

$$\Lambda = m\frac{\lambda_0}{|n_a + n_b|}, \quad m = 1, 2, 3, \ldots \tag{13.5-20}$$

The coupling efficiency at this wavelength is given by

$$\eta(\lambda_0) = \frac{|B(0)|^2}{|A(0)|^2} = \tanh^2|\kappa|L \tag{13.5-21}$$

Following the discussion earlier, the phase mismatch as the wavelength is varied from λ_0 can be written

$$\Delta\beta = (n_a + n_b)\left(\frac{2\pi}{\lambda} - \frac{2\pi}{\lambda_0}\right) \tag{13.5-22}$$

where n_a and n_b are the effective indices of the modes of the waveguides.

According to Figure 12.19 and the discussion in Section 12.4, the bandwidth of contradirectional coupling is given by

$$\Delta\beta = 4|\kappa| \tag{13.5-23}$$

Using Equations (13.5-22) and (13.5-23), we obtain the bandwidth of contradirectional coupling as

$$\Delta\lambda = \frac{2\lambda_0^2}{\pi(n_a + n_b)}|\kappa| \tag{13.5-24}$$

As a result of the contradirectional coupling, we note that the denominator is proportional to the sum of the two effective indices. This leads to a much narrower bandwidth—and thus the advantages of strong frequency selectivity.

Using the same example of two parallel waveguides with $n_a = 1.50$, $n_b = 1.51$, $|\kappa| = 5\ \text{cm}^{-1}$, and $\lambda_0 = 1\ \mu\text{m}$, the above formula yields $\Delta\lambda = 0.11$ nm for contradirectional coupling. The grating period needed for phase matching is an integral multiple of 0.33 μm, according to Equation (13.5-20). Generally, the bandwidth of contradirectional coupling is relatively narrow. The bandwidth is inversely proportional to the sum of the effective indices, $(n_a + n_b)$, and is directly proportional to the coupling constant $|\kappa|$.

Using $|\kappa|L = \pi/2$, which yields a peak coupling efficiency of 84% according to Equation (13.5-21), and the grating period of Equation (13.5-20), the coupling bandwidth can also be written (assuming $m = 1$)

$$\Delta\lambda = \frac{\lambda_0}{N} \tag{13.5-25}$$

where N is the number of periods in the grating ($N = \Lambda/L$). We note that the bandwidth of grating coupling in this case is also inversely proportional to the number of periods N. It is important to note that the expression is valid for the case when $|\kappa|L$ if of the order of unity. For cases when $|\kappa|L \gg 1$, we must resort to Equation (13.5-24) for the bandwidth. In these cases, any further increases in the interaction length L (or the number of periods) will not decrease the bandwidth, according to Equation (13.5-24). This is a direct result of the exponentially decaying nature of the mode amplitudes $A(z)$ and $B(z)$ along the z axis.

The frequency-selective nature of contradirectional grating coupling is desirable in optical communications using dense wavelength division multiplexing (DWDM) techniques. In DWDM optical communications systems, many different frequency channels are employed to increase the transmission capacity of the networks. Figure 13.10 shows a schematic drawing of contradirectional grating couplers for use in DWDM optical communications systems.

13.6 MODE CONVERTERS

Here we consider the coupling between modes in a multimode waveguide, and the coupling between the two polarization modes of a single-mode waveguide (e.g., single-mode fiber). Let $\mathbf{E}_1(x, y)\exp(-i\beta_1 z)$ and $\mathbf{E}_2(x, y)\exp(-i\beta_2 z)$ be the electric fields of the modes of propagation in the waveguide. The general electric field in the waveguide can be written

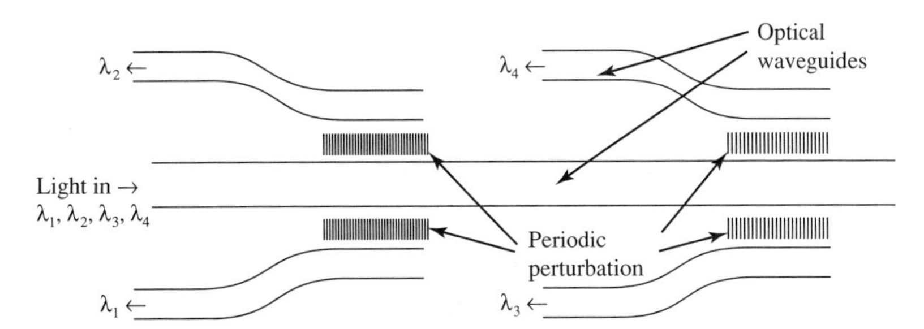

Figure 13.10 Frequency-selective contradirectional couplers for use in DWDM optical communications systems [4].

$$\mathbf{E}(x, y, z) = A_1\mathbf{E}_1(x, y)\exp(-i\beta_1 z) + A_2\mathbf{E}_2(x, y)\exp(-i\beta_2 z) \qquad (13.6\text{-}1)$$

where A_1 and A_2 are mode amplitudes, and β_1 and β_2 are the propagation constants. In a perfect waveguide without any perturbation, the mode amplitudes are constants.

In a mode converter, a perturbation is introduced that can cause the coupling and the exchange of energy between the modes. As a result, the mode amplitudes A_1 and A_2 will be dependent on the distance of propagation z. In what follows, we first consider the coupling between modes in a multimode waveguide, and then the coupling between the two polarization modes of a single-mode waveguide (e.g., single-mode fiber).

Mode Converters in Multimode Waveguides

Referring to Figure 13.11, let us focus our attention on the coupling between two modes with different propagation constants ($\beta_1 \neq \beta_2$). Based on the earlier discussion, an efficient coupling and exchange of energy will occur provided the following conditions are met:

(a) A periodic perturbation is present with a period Λ such that $\beta_1 - \beta_2 = m(2\pi/\Lambda)$, where m is a nonzero integer.

(b) The two wavefunctions $\mathbf{E}_1(x, y)$ and $\mathbf{E}_2(x, y)$ and the mth Fourier component of the dielectric perturbation $\varepsilon_m(x, y)$ have sufficient overlap so that the coupling constant

$$\kappa = C_{12}^{(m)} = \frac{\omega}{4}\int \mathbf{E}_1^*(x, y)\cdot\varepsilon_m(x, y)\mathbf{E}_2(x, y)\,dx \qquad (13.6\text{-}2)$$

defined in Equation (12.4-18) is nonzero.

As the two modes couple in the waveguide, the energy is transferred from one mode to the other and vice versa. To ensure that the energy is completely transferred from one mode to the other, a proper choice of the coupling–length product (e.g., $|\kappa|L = \pi/2$) is important.

One of the most important applications of mode converters is in a special scheme of chromatic dispersion compensation in optical communications. As discussed in Chapter 3, single-mode fibers made of silica exhibit a chromatic dispersion of about $D = 17$ ps/nm-km in the spectral regime around $\lambda = 1550$ nm. In multimode silica fibers (e.g., step-index silica fibers with $V > 2.405$), some of the high-order modes (e.g., LP_{02} mode) may exhibit negative chromatic dispersion as large as 500 ps/nm km. Such a negative dispersion can be employed for chromatic dispersion compensation in optical communications. Figure 13.12 shows a schematic drawing of the basic principle of dispersion compensation using a multimode fiber that supports a high-order mode (e.g., LP_{02} mode) with a huge negative chromatic dispersion.

For practical applications in optical communications, the multimode fiber shown in Figure 13.12 must be designed to support very few confined modes. This will eliminate the

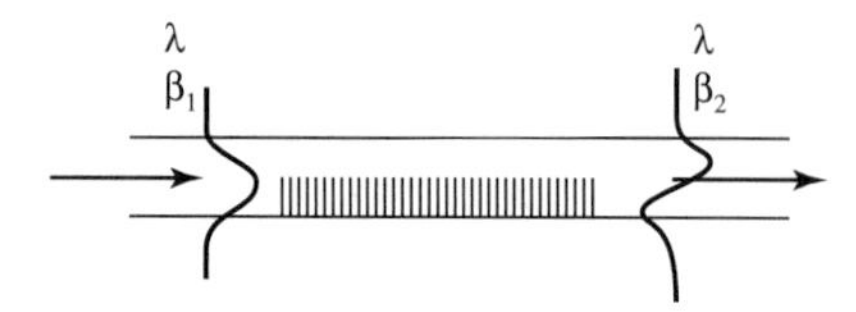

Figure 13.11 Schematic drawing of a mode converter that converts the energy of mode 1 to that of mode 2 in a dielectric slab waveguide.

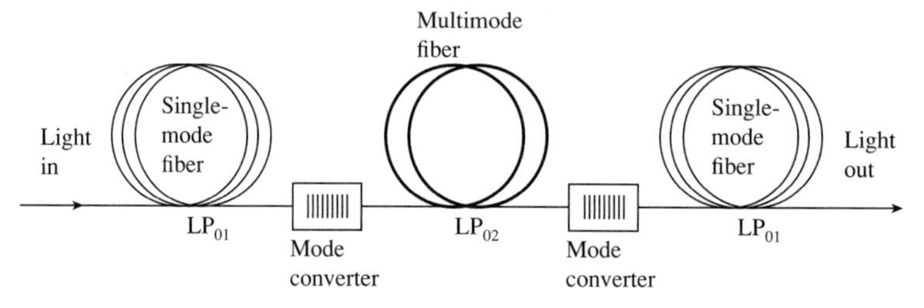

Figure 13.12 Schematic drawing of a method of dispersion compensation using the negative dispersion of a high-order mode of propagation in multimode fibers. The mode converters are employed to convert the optical energy from LP_{01} mode to LP_{02} mode, and then back to LP_{01} mode after the propagation in the multimode fiber.

problem of modal scrambling when too many modes are present with small differences in their propagation constants. Furthermore, the multimode fiber must have a core diameter similar or identical to that of the single-mode fibers. This will ensure an efficient coupling with minimum insertion loss at the junction. If the core diameters are well matched, the energy in the single-mode fiber will excite the fundamental mode (LP_{01}) of the multimode fiber. A mode converter is now needed to convert the optical energy in the fundamental mode to the desired high-order mode (e.g., LP_{02}). A second mode converter is needed to convert the energy back to the fundamental mode (LP_{01}) so that an efficient coupling back to the single-mode fiber can be achieved.

Polarization Mode Converters

Generally, all waveguide modes are polarized. In planar waveguides (e.g., slab waveguides), there are TE modes and TM modes, which are mutually orthogonal. The propagation constants of these two types of modes are usually different, even in a waveguide that supports only one confined TE mode and one confined TM mode. In circular dielectric waveguides (e.g., silica fibers), there are linearly polarized modes (LP_{lmx} and LP_{lmy} modes), which are also mutually orthogonal. As a result of the circular symmetry, these two modes (with the same lm) have the same propagation constant. In a single-mode optical fiber, there are actually two polarization modes. They are the LP_{01x} and LP_{01y} modes propagating along the axis of the fiber (the z axis). In a polarization-maintaining fiber (or a birefringent fiber), the circular symmetry is no longer present. As a result, these two modes (LP_{01x} and LP_{01y}) have different propagation constants. Based on the earlier discussion, an efficient coupling and exchange of energy between the polarization modes will occur provided the following conditions are met:

(a) A periodic perturbation with a period Λ is still required if the two polarization modes are sufficiently different in their propagation constants. The period must be such that $\beta_1 - \beta_2 = m(2\pi/\Lambda)$, where m is a nonzero integer.

(b) The mth Fourier component of the dielectric perturbation $\varepsilon_m(x, y)$ must have off-diagonal tensor elements so that the coupling constant defined in Equation (13.6-2) is nonzero.

In what follows, we examine the coupling between the polarization modes. We start from the coupled equations (12.4-28),

$$\frac{d}{dz}A_1 = -i\kappa A_2(z)e^{i\Delta\beta z}$$
$$\frac{d}{dz}A_2 = -i\kappa^* A_1(z)e^{-i\Delta\beta z} \tag{13.6-3}$$

where A_1 is the amplitude of one polarization mode, A_2 is the amplitude of the other polarization mode, κ is the coupling constant due to the dielectric perturbation $\Delta\varepsilon(x, y, z)$, and

$$\Delta\beta = \beta_1 - \beta_2 \tag{13.6-4}$$

Here we assume a general dielectric perturbation with the coupling constant given by

$$\kappa = C_{12}^{(m)} = \frac{\omega}{4}\int \mathbf{E}_1^*(x, y)\cdot \varepsilon(x, y)\mathbf{E}_2(x, y)\, dx \tag{13.6-5}$$

To illustrate the polarization mode coupling, we consider the case when a particular polarization mode is excited in a circular single-mode fiber with $\Delta\beta = 0$. In this case, the boundary condition is $A_2(0) = 0$. Solutions of the coupled equations can be written

$$A_1(z) = A_1(0)\cos|\kappa z|$$
$$A_2(z) = -i\frac{\kappa^*}{|\kappa|}A_1(0)\sin|\kappa z| \tag{13.6-6}$$

where we assume that κ is a constant, and $A_1(0)$ is the mode amplitude of the polarized input at $z = 0$. Since both modes have the same propagation constant, the energy is coupled back and forth between the two polarization modes. In optical fibers with a sufficient length, the polarization modes are coupled due to a dielectric perturbation resulting from external perturbations such as bending and stress. As a result, the coupling constant κ is a slowly but randomly varying function of z. In this case, the solutions can be written approximately as

$$A_1(z) = A_1(0)\cos|X|$$
$$A_2(z) = -i\frac{X^*}{|X|}A_1(0)\sin|X| \tag{13.6-7}$$

where

$$X = \int_0^z \kappa\, dz \tag{13.6-8}$$

Depending on the nature of the coupling, X is in general a complex number. For a circular fiber with a fixed length and a randomly varying coupling constant, X is also a random complex variable. In this case, the output polarization is also randomly varying.

To avoid random coupling between the two polarization modes, some special fibers can be made with an elliptical core (see Figure 13.13). Such fibers support two polarization modes with different propagation constants. For the elliptical core shown in Figure 13.13, the long axis of the core ellipse is parallel to the y axis. It can be shown that the propagation constant of the LP_{01y} mode is greater than that of the LP_{01x} mode. In other words, the fundamental mode (mode with the largest β) is polarized along the long axis of the elliptical core. In practice, an external perturbation (e.g., stress) can be applied to a circular fiber so that the fiber is effectively an elliptical fiber.

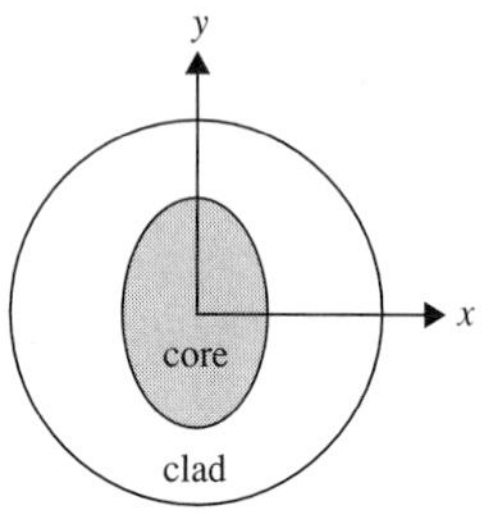

Figure 13.13 An optical fiber with an elliptical core. The propagation constants of the polarization modes are such that $\beta_{LP_{01x}} < \beta_{LP_{01y}}$.

If the core of the fiber is made to have a large ellipticity, then the difference ($\Delta\beta = \beta_{LP_{01x}} - \beta_{LP_{01y}}$) in the propagation constants can be sufficiently large ($|\kappa| \ll \Delta\beta$) to suppress the energy coupling between the polarization modes. These fibers are known as polarization-maintaining fibers (PMFs). The fraction of energy that can be coupled between the two polarization modes is given by

$$\eta = \frac{\kappa^2}{\kappa^2 + (\Delta\beta/2)^2} \tag{13.6-9}$$

PROBLEMS

13.1 Derive Equation (13.2-18):

$$\delta\beta_m^2 = \frac{\omega^2}{c^2}\frac{\iint \mathscr{E}_m^* \cdot \Delta n^2(x,y)\mathscr{E}_m\, dx\, dy}{\iint \mathscr{E}_m^* \cdot \mathscr{E}_m\, dx\, dy} = \frac{\omega^2}{c^2}\langle \Delta n^2(x,y)\rangle$$

The correction to the propagation constant β^2 is the expectation value of the perturbation.

13.2 Show that in a lossless waveguide the total power flow is the sum of the individual modal power flows.

13.3 Rewrite Equation (13.3-5) using Dirac notation and operators as

$$(L + \Delta n_\alpha^2)|\alpha\rangle = \beta_\alpha^2|\alpha\rangle, \quad \alpha = a, b$$

where L is a linear Hermitian operator representing $\nabla_t^2 + \omega^2[n_s^2(x,y)]/c^2$, with $\nabla_t^2 = \partial^2/\partial x^2 + \partial^2/\partial y^2$.

(a) Show that with this notation, the coupling constants (13.3-9) become

$$\kappa_{ab} = \langle a|\Delta n_a^2|b\rangle\omega\varepsilon_0/4$$

$$\kappa_{ba} = \langle b|\Delta n_b^2|a\rangle\omega\varepsilon_0/4$$

Show that $\kappa_{ba} = \kappa_{ab}^*$. Show that the forms of the coupled equations (13.3-13) are consistent with the conservation of the modes' power.

(b) Derive the solutions (13.3-15).

(c) Show that the solutions are consistent with the conservation of energy.

13.4

(a) Show that the wavefunctions in Equation (13.3-33) are mutually orthogonal.

(b) Sketch the wavefunctions.

13.5 Referring to Figure P13.5, we consider a dual-clad circular fiber with an additional cladding of refractive index $n_c < n_2$.

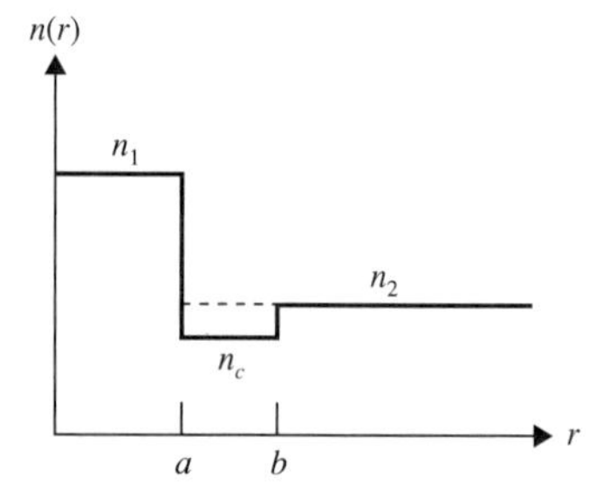

Figure P13.5

(a) Treat the additional cladding as a dielectric perturbation. Write down the perturbation $\Delta n^2(r)$.

(b) Using the mode of the step-index fiber as the basis, find the correction to the propagation constant due to the presence of the dielectric perturbation.

13.6 Referring to Figure P13.6, we consider a fiber with an elliptical core with $\delta = a - b \ll a$.

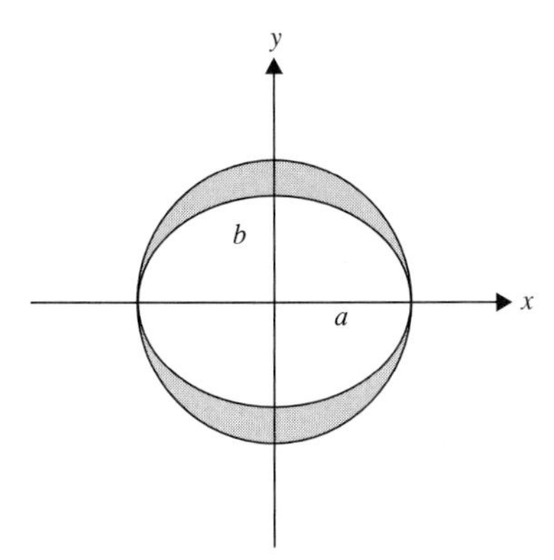

Figure P13.6

(a) Treat the shaded region as a dielectric perturbation. Write down the perturbation $\Delta n^2(x, y)$.

(b) Using the fundamental mode (LP_{01}) of the step-index circular fiber as the basis, show that the field intensity inside the core of the unperturbed fiber can be written

$$|E|^2 = \begin{cases} |A|^2 J_0^2(hr) + \left|\dfrac{hA}{2\beta}\right|^2 4J_1^2(hr)\cos^2\phi, & LP_{01x} \\ |A|^2 J_0^2(hr) + \left|\dfrac{hA}{2\beta}\right|^2 4J_1^2(hr)\sin^2\phi, & LP_{01y} \end{cases}$$

If we neglect the small term $(hA/2\beta)^2$, show that the integral of the field intensity is

$$\langle E|E\rangle = \pi a^2 |A|^2 \left(1 + \frac{h^2}{q^2}\right) J_1^2(ha)$$

(c) Using the Hellman–Feynman theorem (Problem 13.1), show that

$$\Delta\bar{n} = \bar{n}_{LP_{01x}} - \bar{n}_{LP_{01y}} = \frac{q^2h^2}{2\beta^2(q^2+h^2)}\frac{a-b}{a}(n_1 - n_2)$$

assuming $\delta = a - b \ll a$. You may assume the field value remains the same in the shaded region as the field value at $r = a$.

(d) Calculate the birefringence for a single-mode fiber SMF-28 with $n_2 = 1.4600$, $n_1 = 1.4628$, $a = 4.7$ μm, and $\lambda = 1.55$ μm. You need to solve for β, q, and h.

13.7 Let (E_m, H_m) and (E_n, H_n) be two arbitrary modes of a lossless dielectric waveguide, and let I_{mn} be

$$I_{mn} = \frac{1}{2}\iint (\mathbf{E}_m \times \mathbf{H}_n^*)\cdot\hat{\mathbf{z}}\, da = \delta_{mn}$$

(a) Show that

$$I_{mn} = \frac{\beta_n}{2\omega\mu}\iint (\mathbf{E}_m\cdot\mathbf{E}_n^*)\,da + \frac{i}{2\omega\mu}\iint (\mathbf{E}_m\cdot\nabla)\mathbf{E}_{nz}^*\,da$$

This shows that the orthonormality for TE modes reduces to

$$\frac{\beta_n}{2\omega\mu}\iint (\mathbf{E}_m\cdot\mathbf{E}_n^*)\,da = \delta_{mn} \quad (TE)$$

[Hint: Use the vector identity $[\mathbf{A}\times(\nabla\times\mathbf{B})]\cdot\mathbf{C} = \mathbf{A}\cdot(\mathbf{C}\cdot\nabla)\mathbf{B} - \mathbf{C}\cdot(\mathbf{A}\cdot\nabla)\mathbf{B}$.]

(b) Show that from (a) and integration by parts

$$2\delta_{mn} = I_{mn}^* + I_{nm} = \frac{\beta_n+\beta_m}{2\omega\mu}\iint (\mathbf{E}_m\cdot\mathbf{E}_n^*)\,da + \frac{i}{2\omega\mu}\iint (\mathbf{E}_{nz}^*\nabla\cdot\mathbf{E}_m - \mathbf{E}_{mz}^*\nabla\cdot\mathbf{E}_n)\,da$$

Thus either $\nabla\cdot\mathbf{E} = 0$ or $\hat{\mathbf{z}}\cdot\mathbf{E} = 0$ is sufficient for the orthonormality relationship in (a).

(c) Follow the procedures in (a) and (b) and use the same definition for I_{mn}. Show that

$$I_{mn} = \frac{\beta_n}{2\omega}\iint \left(\mathbf{H}_m\cdot\frac{1}{\varepsilon}\mathbf{H}_n^*\right)da + \frac{i}{2\omega}\iint \frac{1}{\varepsilon}(\mathbf{H}_m\cdot\nabla)\mathbf{H}_{nz}^*\,da$$

Thus, for TM modes, $H_z = 0$, the orthonormality relationship reduces to

$$\frac{\beta_m}{2\omega}\iint \left(\mathbf{H}_m\cdot\frac{1}{\varepsilon}\mathbf{H}_n^*\right)da = \delta_{mn} \quad (TM)$$

(d) Show that either $\mathbf{H}\cdot\nabla\varepsilon = 0$ or $\hat{\mathbf{z}}\cdot\mathbf{H} = 0$ is sufficient for the orthonormality relationship in (c).

13.8 Show that in the case of electro-optic mode coupling in which $\beta_m^{TM} \neq \beta_m^{TE}$, one can use a periodic electro-optic constant or electric field along the direction of propagation to obtain phase-matched operation. How would you accomplish this in practice? (Be bold and invent freely.)

13.9 In the case of coupling between modes that carry power in opposite directions, the coupled equations are written

$$\frac{dA}{dz} = \kappa_{ab}Be^{-i(\beta_b-\beta_a)z}, \quad \frac{dB}{dz} = \kappa_{ab}^*Ae^{-i(\beta_b-\beta_a)z}$$

(a) Show that the conservation of energy becomes

$$\frac{d}{dz}(|A|^2 - |B|^2) = 0$$

(b) Show that the solutions in Equation (13.5-16) satisfied the above condition.

13.10 Assume that the two channel waveguides are different. Solve the eigenvalue problem of Equation (13.3-28):

$$\begin{bmatrix} \beta_a^2 - \beta^2 + K_a & J_a + I^*(\beta_b^2 - \beta^2) \\ J_b + I(\beta_a^2 - \beta^2) & \beta_b^2 - \beta^2 + K_b \end{bmatrix} \begin{bmatrix} C_a \\ C_b \end{bmatrix} = 0$$

(a) Find the eigenvalues and the eigenmodes assuming real propagation constants β_a and β_b.

(b) Assume the presence of individual exponential gain constants γ_a and γ_b in each guide. Find the eigenvalues and the eigenmodes.
[Hint: Use complex propagation constants $\beta_a' = \beta_a + i\gamma_a/2$ and $\beta_b' = \beta_b + i\gamma_b/2$.]

13.11 Referring to Figure 13.5, assume that the regions between the optical waveguides are highly absorbing. Explain qualitatively why the desirable $(+ + \cdots +)$ supermode has a smaller modal gain at a given injection current than the $(+ - + - \cdots +)$ supermode.
[Hint: Use Equation (13.4-6).]

13.12 Consider the linear array as shown in Figure 13.4-2. Let the waveguides be separated by a distance (center to center) of Λ in the x-direction. The field at the output end ($z = 0$) can be written

$$E(x, y) = \sum_{n=1}^{N} C_n \mathcal{E}_n(x, y) = \sum_{n=1}^{N} C_n \mathcal{E}_0(x - n\Lambda, y)$$

where $\mathcal{E}_0(x, y)$ is the wavefunction of the modes of the waveguides.

Using Equation (2.12-32), the far-field intensity distribution can be written

$$F(x_1, y_1) \propto \iint E(x, y) \exp[ik(xx_1 + yy_1)/L]\, dx\, dy$$

where L is the distance of observation measured from the output end, (x_1, y_1) is the coordinate of the observation plane. Define $\theta = x_1/L$ as the angle of observation.

(a) Show that

$$F(x_1, y_1) \propto \sum_{n=1}^{N} C_n \exp(ikn\Lambda\theta) \iint \mathcal{E}_0(x, y) \exp[ik(xx_1 + yy_1)/L]\, dx\, dy$$

If $\mathcal{E}_0(x, y)$ is approximately a delta function $\delta(x, y)$, show that

$$F(\theta) \propto \sum_{n=1}^{N} C_n \exp(ikn\Lambda\theta)$$

(b) Evaluate the summation for the supermode with $C_n = 1$ for all n.

(c) Evaluate the summation for the supermode with $C_n = (-1)^n$.

13.13 The integrated magnetic energy is the same as the integrated electric energy for a confined mode in dielectric waveguides.

(a) Starting from Equations (13.1-1) and (13.1-2), show that

$$i\omega(\mathbf{H}^* \cdot \mu\mathbf{H} - \mathbf{E}^* \cdot \varepsilon\mathbf{E}) = \nabla \cdot (\mathbf{E} \times \mathbf{H}^*)$$

(b) Show that for guided waves of the form (13.1-3, 4)

$$\int \mathbf{H}^* \cdot \mu\mathbf{H}\, dx\, dy = \int \mathbf{E}^* \cdot \varepsilon\mathbf{E}\, dx\, dy$$

where the integrals are carried out over the entire xy-plane. In other words, the integrated magnetic energy is the same as the integrated electric energy.

REFERENCES

1. Somekh, S., E. Garmire, A. Yariv, H. L. Garvin, and R. G. Hunsperger, Channel optical waveguide directional couplers. *Appl. Phys. Lett.* **27**:327 (1975).
2. Campbell, J. C., F. A. Blum, D. W. Shaw, and K. L. Lawley, GaAs electro-optic directional-coupler switch. *Appl. Phys. Lett.* **27**:202 (1975).
3. Kapon, E., C. Lindsey, J. Katz, S. Margalit, and A. Yariv, Coupling mechanism of gain guided laser arrays. *Appl. Phys. Lett.* **44**:389 (1984).
4. Yeh, P., and H. F. Taylor, Contradirectional frequency-selective couplers for guided-wave optics. *Appl. Opt.* **19**:2848 (1980).

CHAPTER 14

NONLINEAR OPTICAL EFFECTS IN FIBERS

14.0 INTRODUCTION

Most optical fibers that are used in optical communications are made of fused silica (SiO_2), which exhibits a very low attenuation in the spectral regime around $\lambda = 1550$ nm. The nonlinear optical effect in silica fibers is relatively small. The Kerr coefficient n_2 is ~ 3×10^{-20} m^2/W. However, the length of interaction can be relatively long (e.g., kilometers). As a result, nonlinear optical effects at high optical power levels cannot be ignored in a long segment of fiber. The nonlinear optical effect has become important recently in wavelength division multiplexed (WDM) optical networks, where erbium-doped fiber amplifiers (EDFAs) are employed to amplify the power levels of transmission. In dense wavelength division multiplexed (DWDM) optical networks where the channel wavelengths, more precisely the frequencies, are located at values corresponding to the International Telecom Union (ITU) grids, the channel spacings are identical in frequency (e.g., 100 GHz or 50 GHz). The presence of many channels at equal and close frequency spacing, 25 GHz and 50 GHz are typical, can lead to crosstalk among the neighboring channels via nonlinear optical four-wave mixing (FWM). Furthermore, some of the nonlinear optical effects increase with the total power of all the channels. Among most useful optical materials, silica has the lowest nonlinear optical effects. Despite this, nonlinear optical effects in DWDM optical networks are routinely observed even at milliwatt power levels due to the high intensities achieved in the small diameter fibers and the long interaction lengths. In this chapter we describe various nonlinear optical effects in single-mode optical fibers. These include four-wave mixing (FWM), cross-phase modulation (XPM), self-phase modulation (SPM), phase conjugation, wavelength conversion, as well the propagation of optical solitons.

14.1 KERR EFFECT AND SELF-PHASE MODULATION

We consider the propagation of a single beam of light inside a single-mode fiber. Let z be the axis of the fiber. As a result of the optical Kerr effect described in Chapter 8, the refractive index of bulk silica is often written

$$n = n_0 + n_2 I \tag{14.1-1}$$

where I is the intensity of the beam, n_0 is the index of refraction in the limit of infinitesimal low intensities, and n_2 is the optical Kerr coefficient (in units of m^2/W). In a single-mode silica fiber, both n_0 and the intensity I are functions of space (x, y, z). For linearly polarized modes (LP_{01}), the intensity is written, according to Equations (3.3-33) or (1.4-21)

$$I = \frac{\beta}{2\omega\mu}|\mathbf{E}(x, y)|^2 \tag{14.1-2}$$

where β is the propagation constant, and $\mathbf{E}(x, y)$ is the wavefunction of the mode of propagation as described in Chapter 3. As a result of the position dependence, it is appropriate to rewrite Equation (14.1-1) as

$$n = n_0(x, y) + n_2 I(x, y) \tag{14.1-3}$$

where $n_0(x, y)$ is the index profile of the fiber. The nonlinear contribution to the index of refraction can be treated as a small perturbation. Using the perturbation theory described in Chapter 13, we can calculate the effect of the nonlinear term $n_2 I(x, y)$ on the propagation constant. According to Equation (13.2-18), the correction to the propagation constant due to the Kerr effect can be written

$$\delta\beta^2 = \frac{\omega^2}{c^2}\frac{\iint \mathbf{E}^* \cdot \Delta n^2(x, y)\mathbf{E}\, dx\, dy}{\iint |\mathbf{E}(x, y)|^2\, dx\, dy} = \frac{\omega^2}{c^2}\langle \Delta n^2(x, y)\rangle \tag{14.1-4}$$

For typical beam intensities, $n_2 I(x, y) \ll n_0$ and thus $\delta\beta \ll \beta$. Using Equation (14.1-3), the above equation can be written

$$2\beta\,\delta\beta = \frac{\omega^2}{c^2}\frac{\iint \mathbf{E} \cdot 2n_0 n_2 I(x, y)\mathbf{E}\, dx\, dy}{\iint |\mathbf{E}(x, y)|^2\, dx\, dy} \tag{14.1-5}$$

For most step-index fibers used in optical communications, the index step ($n_{core} - n_{clad}$) is very small compared with the core index n_{core}. Under this circumstance, Equation (14.1-5) can be written, according to Equation (14.1-2),

$$2\beta\,\delta\beta = 2n_{eff}n_2\frac{\beta}{2\omega\mu}\frac{\omega^2}{c^2}\frac{\iint |\mathbf{E}(x, y)|^4\, dx\, dy}{\iint |\mathbf{E}(x, y)|^2\, dx\, dy} \tag{14.1-6}$$

where n_{eff} is the effective index as defined by

$$\beta = n_{eff}\frac{\omega}{c} \tag{14.1-7}$$

In arriving at Equation (14.1-6), we also assumed that n_2 is a constant throughout the fiber structure. This is legitimate as both the core and the clad of most fibers are made of silica. Using Equations (14.1-7) and (14.1-6), we obtain the following expression for the correction to the effective index of the mode of propagation due to the Kerr effect:

$$\delta n_{\text{eff}} = n_2 \frac{\beta}{2\omega\mu} \frac{\iint |\mathbf{E}(x, y)|^4 \, dx \, dy}{\iint |\mathbf{E}(x, y)|^2 \, dx \, dy} \tag{14.1-8}$$

We now define an effective area of the mode of propagation in the fiber as

$$A_{\text{eff}} = \frac{\left(\iint |\mathbf{E}(x, y)|^2 \, dx \, dy\right)^2}{\iint |\mathbf{E}(x, y)|^4 \, dx \, dy} \tag{14.1-9}$$

where $\mathbf{E}(x, y)$ is the wavefunction of the mode. The correction to the effective index (14.1-8) can then be written

$$\delta n_{\text{eff}} = n_2 \frac{\frac{\beta}{2\omega\mu} \iint |\mathbf{E}(x, y)|^2 \, dx \, dy}{A_{\text{eff}}} \tag{14.1-10}$$

We note that the numerator in the above equation is the power of the optical beam P. Thus the correction to the effective index can be written

$$\delta n_{\text{eff}} = n_2 \frac{P}{A_{\text{eff}}} \tag{14.1-11}$$

where P is the power of the mode and A_{eff} is the effective area of the mode of propagation in the fiber. The effective area A_{eff} is a useful parameter for the mode of propagation in fiber optics. The parameter is related to the wavefunction of the mode of propagation, which depends on the waveguide structure. For the special case of a wavefunction with a Gaussian distribution,

$$\mathbf{E}(x, y) = \exp\left(-\frac{x^2 + y^2}{a^2}\right) \tag{14.1-12}$$

The effective area is

$$A_{\text{eff}} = \pi a^2 \tag{14.1-13}$$

The effective area A_{eff} is a measure of the physical area occupied by the mode. For a well-confined mode, the effective area is approximately the core of the waveguide. In the extreme case when the wavefunction is uniform inside the core and zero elsewhere, the effective area is exactly the area of the core.

EXAMPLE: KERR-INDUCED CHANGE IN EFFECTIVE INDEX

Consider the transmission of 10 mW of optical power inside a single mode fiber with an effective area of 100 μm^2 for the mode of propagation. The Kerr-induced change in the effective index of the mode of propagation is

$$P = 10 \text{ mW}$$

$$A_{\text{eff}} = 100 \ \mu\text{m}^2$$

$$n_2 = 3 \times 10^{-20}\ \text{m}^2/\text{W}$$

$$\delta n_{\text{eff}} = n_2 I_{\text{eff}} = n_2 \frac{P}{A_{\text{eff}}} = 3 \times 10^{-12}$$

The change in the effective index is thus very small. However, in a fiber of length L, the change of phase shift due to the change of the effective index is given by

$$\delta\phi = \frac{2\pi}{\lambda}\delta n_{\text{eff}} L \tag{14.1-14}$$

where L is the length of the fiber and λ is the wavelength of light. For $\lambda = 1.5\ \mu\text{m}$ and L in the range of 100 km, the phase shift is of the order of 1, which is a very significant phase shift.

When the beam of light inside the fiber consists of a series of pulses, representing the signals, the Kerr-induced phase shift can be time dependent. This follows directly from Equation (14.1-11). In fact, the intrinsic optical Kerr effect is relatively fast. As a result, the change in the effective index follows almost instantaneously the variation of the local intensity. Thus, for high-frequency (or broadband) signals, the Kerr-induced phase shift is also a fast varying function of time. The time-varying Kerr-induced phase shift leads to phenomena known as self-phase modulation (SPM) and cross-phase modulation (XPM).

Self-Phase Modulation

We consider the transmission of a beam of light through a single-mode fiber of length L. At the output end ($z = L$) of the fiber, the electric field of the beam can be written

$$E(x, y, L, t) = A_0\psi(x, y)\exp[i(\omega t - kL)] = A_0\psi(x, y)\exp[i(\omega t - \phi)] \tag{14.1-15}$$

where A_0 is a constant, ω is the frequency, and $\psi(x, y)$ is the wavefunction of the mode of propagation. The phase shift ϕ is written

$$\phi = \frac{2\pi}{\lambda} n_{\text{eff}} L \tag{14.1-16}$$

Using Equations (14.1-16) and (14.1-11), we can write the Kerr-induced phase shift as

$$\delta\phi(t) = \frac{2\pi}{\lambda}\delta n_{\text{eff}} L = \frac{2\pi}{\lambda} n_2 L \frac{P(t)}{A_{\text{eff}}} \equiv \frac{2\pi}{\lambda} n_2 L I_{\text{eff}}(t) \tag{14.1-17}$$

where $I_{\text{eff}}(t)$ is the effective intensity ($P(t)/A_{\text{eff}}$). This is the phase shift of a beam of light inside a single-mode fiber due to the instantaneous intensity of the same beam. If the phase of a beam varies with time, we may define the "local frequency" as

$$\omega'(t) = \frac{\partial}{\partial t}[\omega - \phi(t)] = \omega - \frac{\partial}{\partial t}\delta\phi(t) = \omega - \frac{2\pi}{\lambda} n_2 L \frac{\partial}{\partial t} I(t) \tag{14.1-18}$$

where ω is the original frequency of the beam and, for simplicity, we drop the subscript eff. The Kerr-induced (or self-induced) change in frequency is thus

$$\Delta\omega = \omega' - \omega = -\frac{\partial}{\partial t}\delta\phi(t) = -\frac{2\pi}{\lambda} n_2 L \frac{\partial}{\partial t} I(t) \tag{14.1-19}$$

(a)

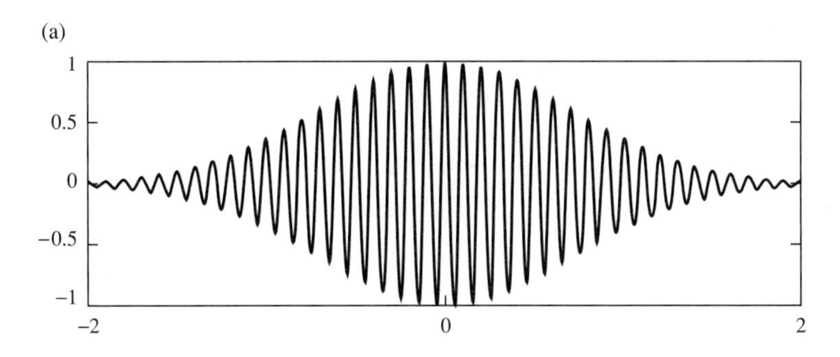

(b)

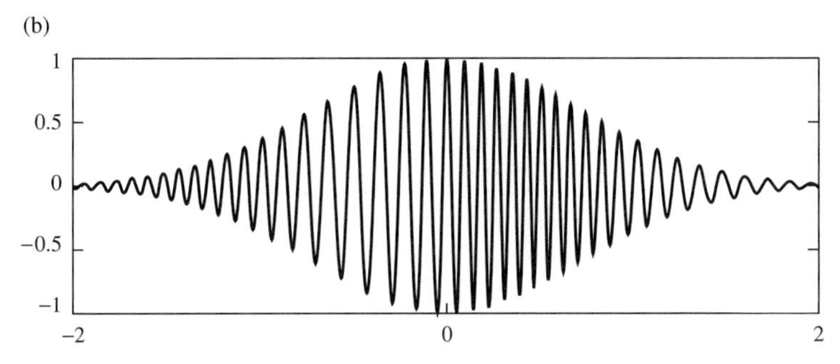

Figure 14.1 Electric field as a function of time: (a) at the input end of the fiber and (b) at the output end of the fiber. The parameters used are $A_{\text{eff}} = 100\ \mu\text{m}^2$, $P = 5$ mW, $L = 2500$ km, $\tau = 0.05$ ps, and $n_2 = 3 \times 10^{-20}\ \text{m}^2/\text{W}$.

To illustrate the self-phase modulation and its effect on the frequency, we consider a simple example of a pulse with a Gaussian temporal intensity distribution,

$$I(t) = I_0 \exp(-t^2/\tau^2) \tag{14.1-20}$$

where I_0 is a constant and τ is a measure of the pulse duration. Substituting Equation (14.1-20) for the intensity $I(t)$ in Equation (14.1-19), we obtain

$$\Delta\omega = \omega'(t) - \omega = \frac{2\pi}{\lambda} n_2 L \frac{2t}{\tau^2} I_0 \exp\left(\frac{-t^2}{\tau^2}\right) \tag{14.1-21}$$

By examining the above equation, we note that the leading edge ($t < 0$) of the pulse exhibits a decrease in the local frequency, whereas the trailing edge exhibits an increase in the local frequency. Thus, as a result of the transmission through a single-mode fiber, a Gaussian pulse acquires a time-varying frequency chirp due to the Kerr effect. Figure 14.1 shows the electric field as a function of time at the output end of the fiber. We note the frequency chirping in the output pulse (b). $\Delta\omega$ in Equation (14.1-21) is also a measure of the frequency broadening as a result of the nonlinear effect. The frequency broadening becomes significant for intense short pulses in a long segment of fiber.

14.2 CROSS-PHASE MODULATION—POLARIZATION

In wavelength division multiplexing (WDM), there are many channels of communication within a single fiber in optical networks. Each of the channels has a unique carrier frequency (or wavelength). As a result of the Kerr-induced change in the refractive index, the temporal variation of the intensity of one channel can cause the phase modulation of other channels. This is known as the cross-phase modulation (XPM). To illustrate the concept, we consider the propagation of two waves having different frequencies ω_1 and ω_2 inside a single-mode fiber. For simplicity, we assume that both waves are polarized. For example, they can correspond to the LP_{01x} mode of the fiber (see Chapter 3). Let the electric field of the optical beams be written

$$\mathbf{E}(x, y, z, t) = \tfrac{1}{2}\hat{\mathbf{x}}(E_{01}(x, y)e^{i(\omega_1 t-\beta_1 z)} + E_{02}(x, y)e^{i(\omega_2 t-\beta_2 z)} + \text{c.c.}) \tag{14.2-1}$$

where $E_{01}(x, y)$ and $E_{02}(x, y)$ are the wave functions of LP_{01} modes, and β_1 and β_2 are the propagation constants of the two waves, respectively.

According to Equation (8.6-4), the third-order nonlinear polarization due to the presence of the two fields is also polarized along the x axis and can be written

$$P_x^{(3)}(x, y, z, t) = 4\chi_{1111}E_xE_xE_x \tag{14.2-2}$$

where E_x is the x component of the total field, χ_{1111} is a third-order nonlinear susceptibility, and the superscript (3) indicates third-order nonlinearity.

We now substitute Equation (14.2-1) for E_x into the above equation. This leads to

$$P_x^{(3)}(x, y, z, t) = \tfrac{4}{8}\chi_{1111}(E_{01}(x, y)e^{i(\omega_1 t-\beta_1 z)} + E_{02}(x, y)e^{i(\omega_2 t-\beta_2 z)} + \text{c.c.})^3 \tag{14.2-3}$$

There are in general 64 terms in the above equation. Here we are interested in terms with frequencies ω_1 and ω_2 only since these are the only terms that can affect our input field. Neglecting terms with frequencies $3\omega_1$, $3\omega_2$, $(2\omega_1 - \omega_2)$, $(2\omega_2 - \omega_1)$, we obtain

$$P_x^{(3)}(x, y, z, t) = \tfrac{1}{2}(P_{NLx}^{(\omega_1)}e^{i(\omega_1 t-\beta_1 z)} + P_{NLx}^{(\omega_2)}e^{i(\omega_2 t-\beta_2 z)} + \text{c.c.}) \tag{14.2-4}$$

where the amplitudes of the nonlinear polarizations are given by

$$\begin{aligned} P_{NLx}^{(\omega_1)} &= \chi_{\text{eff}}^{(3)}(|E_{10}|^2 + 2|E_{20}|^2)E_{10} \\ P_{NLx}^{(\omega_2)} &= \chi_{\text{eff}}^{(3)}(|E_{20}|^2 + 2|E_{10}|^2)E_{20} \end{aligned} \tag{14.2-5}$$

with

$$\chi_{\text{eff}}^{(3)} = 3\chi_{1111} \tag{14.2-6}$$

We note that the first term in Equation (14.2-5) is consistent with Equation (8.6-6). This term is responsible for the self-phase modulation described in Section 14.1. The second term in Equation (14.2-5) is responsible for the cross-phase modulation.

If we include the nonlinear polarization, then the total polarization at frequencies ω_1 and ω_2 can be written

$$\begin{aligned} P_x(x, y, z, t) = \tfrac{1}{2}\{&[\varepsilon_0 n^2(\omega_1) + \chi_{\text{eff}}^{(3)}(|E_{10}|^2 + 2|E_{20}|^2)]E_{10}e^{i(\omega_1 t-\beta_1 z)} \\ &+ [\varepsilon_0 n^2(\omega_2) + \chi_{\text{eff}}^{(3)}(|E_{20}|^2 + 2|E_{10}|^2)]E_{20}e^{i(\omega_2 t-\beta_2 z)} + \text{c.c.}\} \end{aligned} \tag{14.2-7}$$

We note that the nonlinear polarizations are equivalent to a slight change in the index of refraction for each of the frequencies. These changes can be written

$$\begin{aligned} \Delta n(\omega_1) &= \frac{3\chi_{xxxx}}{2n\varepsilon_0}(|E_{10}|^2 + 2|E_{20}|^2) = n_2(I_1 + 2I_2) \\ \Delta n(\omega_2) &= \frac{3\chi_{xxxx}}{2n\varepsilon_0}(|E_{20}|^2 + 2|E_{10}|^2) = n_2(I_2 + 2I_1) \end{aligned} \tag{14.2-8}$$

where we ignore a slight dependence of n_2 on the frequency. For the modes of propagation, the intensities I_1 and I_2 are both functions of space (x, y). Following the discussions of last section, and using the definition of effective areas of the modes, we can write the change in the effective index of the modes as

$$\delta n_{\text{eff}}(\omega_1) = n_2(I_{1\text{eff}} + 2I_{2\text{eff}}) = n_2\left(\frac{P_1}{A_{1\text{eff}}} + 2\frac{P_2}{A_{2\text{eff}}}\right)$$
$$\delta n_{\text{eff}}(\omega_2) = n_2(I_{2\text{eff}} + 2I_{1\text{eff}}) = n_2\left(\frac{P_2}{A_{2\text{eff}}} + 2\frac{P_1}{A_{1\text{eff}}}\right) \tag{14.2-9}$$

where P_1 and P_2 are powers, and $A_{1\text{eff}}$ and $A_{2\text{eff}}$ are effective areas of the modes, respectively. The Kerr-induced phase shifts can now be written

$$\delta\phi_1 = \frac{2\pi}{\lambda_1}L\delta n_{\text{eff}}(\omega_1) = \frac{2\pi}{\lambda_1}Ln_2(I_{1\text{eff}} + 2I_{2\text{eff}}) = \frac{2\pi}{\lambda_1}Ln_2\left(\frac{P_1}{A_{1\text{eff}}} + 2\frac{P_2}{A_{2\text{eff}}}\right)$$
$$\delta\phi_2 = \frac{2\pi}{\lambda_2}L\delta n_{\text{eff}}(\omega_2) = \frac{2\pi}{\lambda_2}Ln_2(I_{2\text{eff}} + 2I_{1\text{eff}}) = \frac{2\pi}{\lambda_2}Ln_2\left(\frac{P_2}{A_{2\text{eff}}} + 2\frac{P_1}{A_{1\text{eff}}}\right) \tag{14.2-10}$$

Here we note that the first term is responsible for the self-phase modulation (SPM), whereas the second term is responsible for the cross-phase modulation (XPM). In the case when there are more than two modes of propagation, the cross-phase modulation must be summed over all other modes of propagation. The phase shift of mode m due to cross-phase modulation is thus written

$$\delta\phi_{m\text{XPM}} = \frac{4\pi}{\lambda_m}Ln_2\sum_{i\neq m} I_{i\text{eff}} = \frac{4\pi}{\lambda_m}Ln_2\sum_{i\neq m}\frac{P_i}{A_{i\text{eff}}} \tag{14.2-11}$$

where the summation is over all modes $(i \neq m)$ in the same fiber, λ_m is the wavelength of the mth mode, P_i is the power of the ith mode, and $A_{i\text{eff}}$ is the effective area of the ith mode.

The Kerr-induced change in frequency of the mth mode due to all other modes $(i \neq m)$ is thus

$$\Delta\omega = \omega' - \omega = -\frac{\partial}{\partial t}\delta\phi_{m\text{XPM}}(t) = -\frac{4\pi}{\lambda_m}n_2L\sum_{i\neq m}\frac{\partial}{\partial t}I_{i\text{eff}}(t) \tag{14.2-12}$$

In a WDM optical network where many channels, each with its own wavelength, share the same fiber, the cross-phase modulation will in general increase with the number of channels. In the following, we consider the nonlinear optical coupling between the two polarization components of the same wave.

Coupling Between Polarization Components of Same Wave

Let the electric field of the optical wave be written

$$\mathbf{E}(x, y, z, t) = \tfrac{1}{2}(\hat{\mathbf{x}}E_{0x}(x, y)e^{i(\omega t-\beta_x z)} + \hat{\mathbf{y}}E_{0y}(x, y)e^{i(\omega t-\beta_y z)} + \text{c.c.}) \tag{14.2-13}$$

where $E_{0x}(x, y)$ and $E_{0y}(x, y)$ are the wavefunctions of LP_{01} modes, and β_x and β_y are the propagation constants. In the case of circular fibers, $\beta_x = \beta_y$. For simplicity, we rewrite Equation (14.2-13) as

$$\mathbf{E}(x, y, z, t) = (\hat{\mathbf{x}}E_x + \hat{\mathbf{y}}E_y) \tag{14.2-14}$$

We note that E_x and E_y are simply the x and y components of the electric field, respectively.
According to Equation (8.6-4), the third-order nonlinear polarization can be written

$$P_i^{(3)}(x, y, z, t) = 4\chi_{ijkl}E_jE_kE_l \tag{14.2-15}$$

where E_j is the jth component (x or y) of the electric field given by Equation (14.2-13), and a summation over repeated indices is assumed. We now examine the x and y components of the nonlinear polarization of Equation (14.2-15). Using the approximation of the linearly polarized modes (LP modes), the electric field has only x and y components. The x and y components of the nonlinear polarization can thus be written

$$\begin{aligned} P_x^{(3)}(x, y, z, t) &= 4\chi_{1111}E_xE_xE_x + 4\chi_{1122}E_xE_yE_y + 4\chi_{1212}E_yE_xE_y + 4\chi_{1221}E_yE_yE_x \\ &= 4\chi_{1111}E_xE_xE_x + 4(\chi_{1122} + \chi_{1212} + \chi_{1221})E_xE_yE_y \end{aligned} \tag{14.2-16}$$

$$\begin{aligned} P_y^{(3)}(x, y, z, t) &= 4\chi_{2222}E_yE_yE_y + 4\chi_{2211}E_yE_xE_x + 4\chi_{2121}E_xE_yE_x + 4\chi_{2112}E_xE_xE_y \\ &= 4\chi_{2222}E_yE_yE_y + 4(\chi_{2211} + \chi_{2121} + \chi_{2112})E_yE_xE_x \end{aligned} \tag{14.2-17}$$

where we keep only the terms with nonzero nonlinear coefficients. For silica, which is an isotropic medium, there are 21 nonzero elements of the third-order nonlinear optical susceptibility tensor χ_{ijkl} of which only three are independent. A symmetry argument leads to (see Problem 14.8)

$$\chi_{1111} = \chi_{1122} + \chi_{1212} + \chi_{1221} \quad \text{and} \quad \chi_{1111} = \chi_{2222} \tag{14.2-18}$$

Thus the nonlinear polarization can be written

$$P_x^{(3)}(x, y, z, t) = 4\chi_{1111}E_xE_xE_x + 4\chi_{1111}E_xE_yE_y = 4\chi_{1111}E_x(E_xE_x + E_yE_y) \tag{14.2-19}$$

$$\begin{aligned} P_y^{(3)}(x, y, z, t) &= 4\chi_{2222}E_yE_yE_y + 4\chi_{2222}E_yE_xE_x \\ &= 4\chi_{1111}E_yE_yE_y + 4\chi_{1111}E_yE_xE_x = 4\chi_{1111}E_y(E_yE_y + E_xE_x) \end{aligned} \tag{14.2-20}$$

where we have also used the symmetry $\chi_{1122} = \chi_{2211}$, $\chi_{1212} = \chi_{2121}$, $\chi_{1221} = \chi_{2112}$.
Substituting Equation (14.2-13) for E_x, E_y in Equations (14.2-19) and (14.2-20), and limiting ourselves to the amplitudes of the nonlinear terms that oscillate at the same frequency as the input optical beam ω, we obtain

$$\begin{aligned} P_x^{(3)}(x, y, z, t) &= \chi_{xxxx}\{(\tfrac{3}{2}|E_{0x}|^2 + \tfrac{2}{2}|E_{0y}|^2)E_{0x}e^{i(\omega t-\beta_x z)} + \tfrac{1}{2}E_{0y}^2E_{0x}^*e^{i(\omega t-\beta' z)}\} + \text{c.c.} \\ P_y^{(3)}(x, y, z, t) &= \chi_{xxxx}\{(\tfrac{3}{2}|E_{0y}|^2 + \tfrac{2}{2}|E_{0x}|^2)E_{0y}e^{i(\omega t-\beta_y z)} + \tfrac{1}{2}E_{0x}^2E_{0y}^*e^{i(\omega t-\beta'' z)}\} + \text{c.c.} \end{aligned} \tag{14.2-21}$$

where β' and β'' are given by

$$\begin{aligned} \beta' &= 2\beta_y - \beta_x \\ \beta'' &= 2\beta_x - \beta_y \end{aligned} \tag{14.2-22}$$

We note that the last terms have different propagation constants (β', β''). Their contributions to the phase modulation do not add up cumulatively in a long segment of birefringent fiber (when $|\beta_x - \beta_y|L \gg 1$). In this case, the Kerr-induced change in the index of refraction can be written

$$\begin{aligned} \Delta n_x &= \frac{3\chi_{xxxx}}{2n\varepsilon_0}(|E_{0x}|^2 + \tfrac{2}{3}|E_{0y}|^2) = n_2(I_{0x} + \tfrac{2}{3}I_{0y}) \\ \Delta n_y &= \frac{3\chi_{xxxx}}{2n\varepsilon_0}(|E_{0y}|^2 + \tfrac{2}{3}|E_{0x}|^2) = n_2(I_{0y} + \tfrac{2}{3}I_{0x}) \end{aligned} \tag{14.2-23}$$

We note that both self-phase modulation and cross-phase modulation between the polarization components are present. The phase changes due to the nonlinear Kerr effect can be obtained in a similar way. They are given by

$$\delta\phi_x = \frac{2\pi}{\lambda} L n_2 \left(I_{0x\text{eff}} + \frac{2}{3} I_{0y\text{eff}}\right) = \frac{2\pi}{\lambda} L n_2 \left(\frac{P_{0x}}{A_{0x\text{eff}}} + \frac{2}{3}\frac{P_{0y}}{A_{0y\text{eff}}}\right)$$
$$\delta\phi_y = \frac{2\pi}{\lambda} L n_2 \left(I_{0y\text{eff}} + \frac{2}{3} I_{0x\text{eff}}\right) = \frac{2\pi}{\lambda} L n_2 \left(\frac{P_{0y}}{A_{0y\text{eff}}} + \frac{2}{3}\frac{P_{0x}}{A_{0x\text{eff}}}\right) \tag{14.2-24}$$

where $I_{0x\text{eff}}$ and $I_{0y\text{eff}}$ are the effective intensities, P_{0x} and P_{0y} are the powers, and $A_{0x\text{eff}}$ and $A_{0y\text{eff}}$ are the effective areas of the polarization components, respectively.

The Kerr-induced phase retardation can thus be written

$$\delta\phi_y - \delta\phi_x = \frac{2\pi}{3\lambda} L n_2 (I_{0y\text{eff}} - I_{0x\text{eff}}) = \frac{2\pi}{3\lambda} L n_2 \left(\frac{P_{0y}}{A_{0y\text{eff}}} - \frac{P_{0x}}{A_{0x\text{eff}}}\right) \tag{14.2-25}$$

The induced phase retardation can contribute to the change of the polarization state as a result of the nonlinear Kerr coupling between the polarization components (see also Problem 8.14).

For circular fibers ($\beta_x = \beta_y = \beta$), the nonlinear polarization (14.2-21) becomes

$$P_x^{(3)}(x, y, z, t) = \chi_{xxxx}\{(\tfrac{3}{2}|E_{0x}|^2 + \tfrac{2}{2}|E_{0y}|^2)E_{0x} + \tfrac{1}{2}E_{0y}^2 E_{0x}^*\}e^{i(\omega t - \beta z)} + \text{c.c.}$$
$$P_y^{(3)}(x, y, z, t) = \chi_{xxxx}\{(\tfrac{3}{2}|E_{0y}|^2 + \tfrac{2}{2}|E_{0x}|^2)E_{0y} + \tfrac{1}{2}E_{0x}^2 E_{0y}^*\}e^{i(\omega t - \beta z)} + \text{c.c.} \tag{14.2-26}$$

The amplitude of the nonlinear polarization can thus be written

$$P_{\text{NL}x}^{(\omega)} = \chi_{xxxx}\{(3|E_{0x}|^2 + 2|E_{0y}|^2)E_{0x} + E_{0y}^2 E_{0x}^*\}$$
$$P_{\text{NL}y}^{(\omega)} = \chi_{xxxx}\{(3|E_{0y}|^2 + 2|E_{0x}|^2)E_{0y} + E_{0x}^2 E_{0y}^*\} \tag{14.2-27}$$

which is consistent with Equation (8.6-11). For a beam of polarized light in the fiber, the nonlinear polarization leads to a slight change of the effective index of refraction of the mode of propagation (the optical Kerr effect) as discussed in Section 8.6 (see also the discussion in Problem 8.14).

14.3 NONDEGENERATE FOUR-WAVE MIXING

We have so far discussed self-phase modulation (SPM) and cross-phase modulation (XPM) due to the optical Kerr effect in single-mode fibers. In WDM optical networks, there are many channels of optical waves, each with its own unique frequency of propagation. Under the appropriate conditions, energy exchange among different channels may occur. The main such exchange is due to third-order optical nonlinearity and involves four waves. This process is termed optical four-wave mixing (FWM). We have discussed optical four-wave mixing and phase conjugation in bulk nonlinear media in Section 8.8. Here in this section we consider the same process in single-mode fibers that are widely used in optical networks. Optical four-wave mixing is particularly important when the frequency spacing between communication channels is identical (e.g., 100 GHz) [1, 2]. The optical four-wave mixing can lead to undesirable cross talk among channels of communication. Under the appropriate conditions, optical four-wave mixing can be employed for temporal pulse restoration (phase conjugation) as well as wavelength conversion. Generally, the pulse undergoes a phase conjugation that is manifested as a reversal in the sign of the chirp. This leads to the restoration of the original pulse shape as the pulse propagates through another segment of fiber of the same length and undoes all the chromatic dispersion experienced by the original pulse.

Generally, a nonlinear optical four-wave mixing process involves the participation of four optical waves. Such a general third-order process can be viewed as the generation of a fourth optical wave with an input of three optical waves. Let the frequencies and wavevectors of the three input waves be (ω_1, β_1), (ω_2, β_2), and (ω_3, β_3); then the frequencies and wavevectors of the fourth optical wave (ω_4, β_4) can be written

$$\omega_4 = \omega_1 \pm \omega_2 \pm \omega_3 \tag{14.3-1}$$

$$\beta_4 = \beta_1 \pm \beta_2 \pm \beta_3 \tag{14.3-2}$$

where we note that all wavevectors are along the axis of the fiber (z axis). In general, there are many possible combinations of the nonlinear optical four-wave mixing. Each of these processes leads to a new wave at a new frequency and a new wavevector. For optical communications in silica fibers, we are interested in the case when all four frequencies are in the same or similar band of communication channels. In other words, we are interested in the case when the frequency differences among the waves are small compared with the frequencies of the waves (i.e., $|\omega_m - \omega_n| \ll \omega_n$, $m, n = 1, 2, 3, 4$). Thus we limit our discussion to the case when

$$\omega_4 = \omega_1 + \omega_2 - \omega_3 \tag{14.3-3}$$

$$\beta_4 = \beta_1 + \beta_2 - \beta_3 \tag{14.3-4}$$

Figure 14.2 shows the relative position of the frequencies in optical four-wave mixing. In general, all four frequencies can be different. This is nondegenerate four-wave mixing (NDFWM). The special case of $\omega_1 = \omega_2$ is known as partially degenerate four-wave mixing (PDFWM).

These two equations can be viewed as the conservation of energy and momentum. Since the wavevector (β_4) is related to the frequency (ω_4) by the dispersion relationship, these two equations may not be satisfied simultaneously. This is the issue of phase matching discussed earlier in Section 8.8. Thus nonlinear optical four-wave mixing in single-mode fibers occurs only in some particular input configurations when the above two equations are satisfied simultaneously. In this section, we will discuss a very interesting and useful scheme of nonlinear optical four-wave mixing in single-mode fibers in which the phase-matching condition is satisfied. In particular, we will be discussing the nondegenerate and the partially degenerate four-wave mixing processes, which have important applications, including wavelength conversion, phase conjugation, and spectral reversal.

We first consider the case of nondegenerate four-wave mixing in which three input optical waves with frequencies ω_1, ω_2, and ω_3 mix in a single-mode fiber and produce a fourth optical wave at $\omega_1 + \omega_2 - \omega_3$ in the same fiber (see Figures 14.2 and 14.3). For simplicity, we

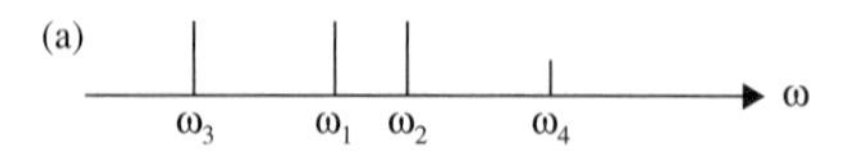

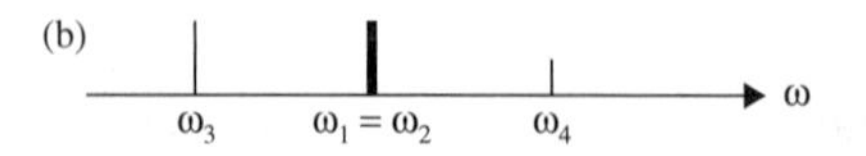

Figure 14.2 Relative position of the frequencies in (a) nondegenerate four-wave mixing (NDFWM) and (b) partially degenerate four-wave mixing (PDFWM).

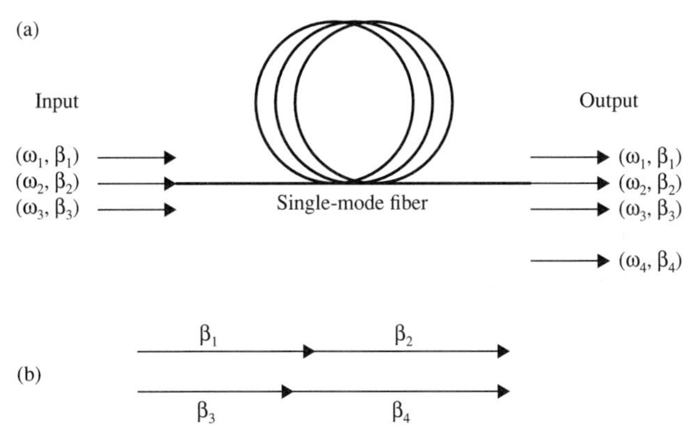

Figure 14.3 (a) Schematic drawing of nondegenerate four-wave mixing in a single-mode fiber. At the input end, three waves at (ω_1, β_1), (ω_2, β_2), and (ω_3, β_3) are coupled into a single-mode fiber. A new wave at (ω_4, β_4) is generated. (b) Vector diagram of the conservation of momentum.

assume that all three waves are polarized (e.g., LP_{01x} mode). Let the three input optical waves inside the fiber be written

$$\begin{aligned}\mathbf{E}_1(\mathbf{r}, t) &= \tfrac{1}{2}\hat{\mathbf{x}}A_1E_{10}(x, y)e^{i(\omega_1 t-\beta_1 z)} + \text{c.c.}\\ \mathbf{E}_2(\mathbf{r}, t) &= \tfrac{1}{2}\hat{\mathbf{x}}A_2E_{20}(x, y)e^{i(\omega_2 t-\beta_2 z)} + \text{c.c.}\\ \mathbf{E}_3(\mathbf{r}, t) &= \tfrac{1}{2}\hat{\mathbf{x}}A_3E_{30}(x, y)e^{i(\omega_3 t-\beta_3 z)} + \text{c.c.}\end{aligned} \tag{14.3-5}$$

where $E_{10}(x, y)$, $E_{20}(x, y)$, and $E_{30}(x, y)$ are the wavefunctions of the LP_{01} modes, A_1, A_2, and A_3 are amplitudes of the waves, and β_1, β_2, and β_3 are the corresponding propagation constants. According to the third-order term in Equation (8.1-1), a nonlinear (NL) polarization at $\omega_1 + \omega_2 - \omega_3$ is generated,

$$\begin{aligned}P_x^{(\text{NL})}(\mathbf{r}, t) &= 3\chi_{1111}^{(3)}A_1A_2A_3^*E_{10}E_{20}E_{30}^*e^{i[(\omega_1+\omega_2-\omega_3)t-(\beta_1+\beta_2-\beta_3)z]} + \text{c.c.}\\ &= \text{Re}\,[6\chi_{1111}^{(3)}A_1A_2A_3^*E_{10}E_{20}E_{30}^*e^{i[(\omega_1+\omega_2-\omega_3)t-(\beta_1+\beta_2-\beta_3)z]}]\end{aligned} \tag{14.3-6}$$

The nonlinear polarization is actually a volume distribution of oscillating dipoles that can radiate a new wave at the frequency $\omega_4 = \omega_1 + \omega_2 - \omega_3$, provided the wavevector of the new wave satisfies the momentum conservation condition $\beta_4 = \beta_1 + \beta_2 - \beta_3$. In a long segment of fiber, the newly generated wave at frequency ω_4 and wavevector β_4 will also mix with any two of the other three waves. This will lead to the generation of waves at frequencies ω_1, ω_2, and ω_3, and wavevectors β_1, β_2, and β_3. Thus all four waves are coupled with energy exchange among the waves as they propagate along the fiber. In the following, we derive the coupled equations for the amplitude of the waves.

Coupled-Mode Formulation of Nondegenerate Four-Wave Mixing

In the following we will consider the four-wave mixing process shown in Figure 14.3a. Three waves of different frequencies $(\omega_1, \omega_2, \omega_3)$ are launched at the input end of a single-mode fiber. The new wave generated by the nonlinear polarization is labeled as wave 4. We will derive the coupled equations that govern the coupling of these four waves in the fiber. Let the electric fields of the waves be written

$$\begin{aligned}\mathbf{E}_1(\mathbf{r},t) &= \tfrac{1}{2}\hat{\mathbf{x}}A_1(z)E_{10}(x,y)e^{i(\omega_1 t-\beta_1 z)} + \text{c.c.}\\ \mathbf{E}_2(\mathbf{r},t) &= \tfrac{1}{2}\hat{\mathbf{x}}A_2(z)E_{20}(x,y)e^{i(\omega_2 t-\beta_2 z)} + \text{c.c.}\\ \mathbf{E}_3(\mathbf{r},t) &= \tfrac{1}{2}\hat{\mathbf{x}}A_3(z)E_{30}(x,y)e^{i(\omega_3 t-\beta_3 z)} + \text{c.c.}\\ \mathbf{E}_4(\mathbf{r},t) &= \tfrac{1}{2}\hat{\mathbf{x}}A_4(z)E_{40}(x,y)e^{i(\omega_4 t-\beta_4 z)} + \text{c.c.}\end{aligned} \tag{14.3-7}$$

where the mode amplitudes $A_1(z)$, $A_2(z)$, $A_3(z)$, and $A_4(z)$ are functions of z, reflecting the coupling as they propagate along the z axis. For simplicity, we assume that all four waves are in the LP_{01x} mode.

We now examine the nonlinear polarizations that are relevant to the interaction among the four waves. Particularly, we are interested in the nonlinear polarizations that are oscillating at the frequencies of the waves (ω_1, ω_2, ω_3, ω_4). A general expression of the third-order nonlinear polarization can be written, according to Equation (8.6-1),

$$\begin{aligned}P_x^{(\text{NL})}(\mathbf{r},t) = 4\chi_{1111}^{(3)}\tfrac{1}{8}[&A_1(z)E_{10}(x,y)e^{i(\omega_1 t-\beta_1 z)} + A_2(z)E_{20}(x,y)e^{i(\omega_2 t-\beta_2 z)}\\ &+ A_3(z)E_{30}(x,y)e^{i(\omega_3 t-\beta_3 z)} + A_4(z)E_{40}(x,y)e^{i(\omega_4 t-\beta_4 z)} + \text{c.c.}]^3\end{aligned} \tag{14.3-8}$$

where we note that even in the simple case of identical polarization (LP_{01x}), there are 512 terms in the third-order nonlinear polarization. The choice of the relevant terms in $P_x^{(\text{NL})}$ requires some judicious reasoning. First, we ignore the third-harmonic terms with frequencies of $3\omega_n$ ($n = 1, 2, 3, 4$), and terms with frequencies $2\omega_n + \omega_m$ ($m, n = 1, 2, 3, 4$). These terms, if phase matched, will radiate into waves in the spectral regimes, which are outside the band of communication channels. Second, we also ignore nonlinear terms with frequencies $2\omega_n - \omega_m$ ($m, n = 1, 2, 3, 4$). Thus keeping only the terms that oscillate at the frequencies of the waves (ω_1, ω_2, ω_3, ω_4), we obtain

$$\begin{aligned}P_x^{(\text{NL}\omega_1)}(\mathbf{r},t) = \tfrac{1}{2}\chi_{1111}^{(3)}\times\{&6A_3A_4A_2^*E_{30}E_{40}E_{20}^*e^{i[(\omega_3+\omega_4-\omega_2)t-(\beta_3+\beta_4-\beta_2)z]} + [6|A_2E_{20}|^2A_1E_{10}\\ &+ 6|A_3E_{30}|^2A_1E_{10} + 6|A_4E_{40}|^2A_1E_{10} + 3|A_1E_{10}|^2A_1E_{10}]e^{i(\omega_1 t-\beta_1 z)} + \text{c.c.}\}\\ P_x^{(\text{NL}\omega_2)}(\mathbf{r},t) = \tfrac{1}{2}\chi_{1111}^{(3)}\times\{&6A_3A_4A_1^*E_{30}E_{40}E_{10}^*e^{i[(\omega_3+\omega_4-\omega_1)t-(\beta_3+\beta_4-\beta_1)z]} + [6|A_1E_{10}|^2A_2E_{20}\\ &+ 6|A_3E_{30}|^2A_2E_{20} + 6|A_4E_{40}|^2A_2E_{20} + 3|A_2E_{20}|^2A_2E_{20}]e^{i(\omega_2 t-\beta_2 z)} + \text{c.c.}\}\\ P_x^{(\text{NL}\omega_3)}(\mathbf{r},t) = \tfrac{1}{2}\chi_{1111}^{(3)}\times\{&6A_1A_2A_4^*E_{10}E_{20}E_{40}^*e^{i[(\omega_1+\omega_2-\omega_4)t-(\beta_1+\beta_2-\beta_4)z]} + [6|A_1E_{10}|^2A_3E_{30}\\ &+ 6|A_2E_{20}|^2A_3E_{30} + 6|A_4E_{40}|^2A_3E_{30} + 3|A_3E_{30}|^2A_3E_{30}]e^{i(\omega_3 t-\beta_3 z)} + \text{c.c.}\}\\ P_x^{(\text{NL}\omega_4)}(\mathbf{r},t) = \tfrac{1}{2}\chi_{1111}^{(3)}\times\{&6A_1A_2A_3^*E_{10}E_{20}E_{30}^*e^{i[(\omega_1+\omega_2-\omega_3)t-(\beta_1+\beta_2-\beta_3)z]} + [6|A_1E_{10}|^2A_4E_{40}\\ &+ 6|A_2E_{20}|^2A_4E_{40} + 6|A_3E_{30}|^2A_4E_{40} + 3|A_4E_{40}|^2A_4E_{40}]e^{i(\omega_4 t-\beta_4 z)} + \text{c.c.}\}\end{aligned} \tag{14.3-9}$$

We note that there are many nonlinear terms with the dependence $\exp[i(\omega_n t - \beta_n z)]$ ($n = 1, 2, 3, 4$). These terms are responsible for self-phase modulations and cross-phase modulations, as discussed earlier. They are responsible for a slight modification of the propagation constants β_n. Here we ignore these terms and concentrate our attention on the terms that are responsible for the four-wave mixing. Thus we can write

$$\begin{aligned}P_x^{(\text{NL})}(\mathbf{r},t) = \tfrac{1}{2}\chi_{1111}^{(3)}\times\{&6A_3A_4A_2^*E_{30}E_{40}E_{20}^*e^{i[(\omega_3+\omega_4-\omega_2)t-(\beta_3+\beta_4-\beta_2)z]}\\ &+ 6A_3A_4A_1^*E_{30}E_{40}E_{10}^*e^{i[(\omega_3+\omega_4-\omega_1)t-(\beta_3+\beta_4-\beta_1)z]}\\ &+ 6A_1A_2A_4^*E_{10}E_{20}E_{40}^*e^{i[(\omega_1+\omega_2-\omega_4)t-(\beta_1+\beta_2-\beta_4)z]}\\ &+ 6A_1A_2A_3^*E_{10}E_{20}E_{30}^*e^{i[(\omega_1+\omega_2-\omega_3)t-(\beta_1+\beta_2-\beta_3)z]} + \text{c.c.}\}\end{aligned} \tag{14.3-10}$$

We now substitute **E** (the sum of all four waves) from Equation (14.3-7), and the nonlinear polarization from Equation (14.3-10) into the following wave equation (from Equation (8.3-2)):

$$\nabla^2 \mathbf{E} = \mu_0 \frac{\partial^2}{\partial t^2}(\varepsilon_0 \mathbf{E} + \mathbf{P}) = \mu_0 \varepsilon \frac{\partial^2 \mathbf{E}}{\partial t^2} + \mu_0 \frac{\partial^2}{\partial t^2} \mathbf{P}_{\mathrm{NL}}$$

We note that each of the waves with a constant amplitude is a solution of the wave equation in the absence of nonlinear coupling. The last equation must be satisfied separately at each of the four frequencies. By assuming slowly varying amplitudes (i.e., $\partial^2 A_n(z)/\partial^2 z \ll \beta_n \partial A_n(z)/\partial z$), we obtain, after a few steps of algebra,

$$\begin{aligned}
E_{10} \frac{dA_1}{dz} &= -i3\chi_{1111}^{(3)} \frac{\omega_1^2}{\beta_1} \mu_0 E_{30} E_{40} E_{20}^* A_3 A_4 A_2^* \exp(i\,\Delta\beta\, z) \\
E_{20} \frac{dA_2}{dz} &= -i3\chi_{1111}^{(3)} \frac{\omega_2^2}{\beta_2} \mu_0 E_{30} E_{40} E_{10}^* A_3 A_4 A_1^* \exp(i\,\Delta\beta\, z) \\
E_{30} \frac{dA_3}{dz} &= -i3\chi_{1111}^{(3)} \frac{\omega_3^2}{\beta_3} \mu_0 E_{10} E_{20} E_{40}^* A_1 A_2 A_4^* \exp(-i\,\Delta\beta\, z) \\
E_{40} \frac{dA_4}{dz} &= -i3\chi_{1111}^{(3)} \frac{\omega_4^2}{\beta_4} \mu_0 E_{10} E_{20} E_{30}^* A_1 A_2 A_3^* \exp(-i\,\Delta\beta\, z)
\end{aligned} \tag{14.3-11}$$

where $\Delta\beta$, describing the phase mismatch in NDFWM, is given by

$$\Delta\beta = \beta_1 + \beta_2 - \beta_3 - \beta_4 \tag{14.3-12}$$

In arriving at Equations (14.3-11), we use the frequency condition (14.3-3) and assume that $\chi_{1111}^{(3)}$, which is proportional to the Kerr coefficient, is a constant in the spectral regime of interest. We now multiply both sides of the first equation in (14.3-11) by $E_{10}(x, y)$ and then integrate over the entire xy plane. This leads to

$$\frac{dA_1}{dz} = -i3\chi_{1111}^{(3)} \frac{\omega_1^2}{\beta_1} \mu_0 A_2^* A_3 A_4 \frac{\int E_{10}^* E_{20}^* E_{30} E_{40}\, dx\, dy}{\int |E_{10}|^2 dx\, dy} \exp(i\,\Delta\beta\, z) \tag{14.3-13}$$

We now define

$$f_1 = \frac{\int E_{10}^* E_{20}^* E_{30} E_{40}\, dx\, dy}{\int |E_{10}|^2\, dx\, dy \int |E_{10}|^2\, dx\, dy} \tag{14.3-14}$$

Equation (14.3-13) can be written

$$\frac{dA_1}{dz} = -i n_2 \frac{2\omega_1}{c} A_2^* A_3 A_4 f_1 \exp(i\,\Delta\beta\, z) \tag{14.3-15}$$

where n_2 is the Kerr coefficient. In arriving at Equation (14.3-15), we use the relationship from Equation (8.6-7),

$$n_2 = \frac{3c\mu_0 \chi_{1111}^{(3)}}{n_{\mathrm{eff}}^2 \varepsilon_0} \tag{14.3-16}$$

where n_{eff} is the effective index of the mode of propagation, as well as the mode normalization from Equation (13.2-4)

$$\int |E_{m0}|^2 \, dx \, dy = \frac{2\omega_m \mu_0}{\beta_m}, \quad m = 1, 2, 3, 4 \tag{14.3-17}$$

In Equation (14.3-16), we ignore the slight variation of the effective index among the four waves.

Similarly, we can obtain the coupled equations for the other three waves. The coupled equations can thus be written

$$\begin{aligned}
\frac{dA_1}{dz} &= -in_2 \frac{2\omega_1}{c} A_2^* A_3 A_4 f_1 \exp(i \,\Delta\beta\, z) - \frac{\alpha}{2} A_1 \\
\frac{dA_2}{dz} &= -in_2 \frac{2\omega_2}{c} A_1^* A_3 A_4 f_2 \exp(i \,\Delta\beta\, z) - \frac{\alpha}{2} A_2 \\
\frac{dA_3}{dz} &= -in_2 \frac{2\omega_3}{c} A_1 A_2 A_4^* f_3 \exp(-i \,\Delta\beta\, z) - \frac{\alpha}{2} A_3 \\
\frac{dA_4}{dz} &= -in_2 \frac{2\omega_4}{c} A_1 A_2 A_3^* f_4 \exp(-i \,\Delta\beta\, z) - \frac{\alpha}{2} A_4
\end{aligned} \tag{14.3-18}$$

where the mode overlap constants are given by

$$\begin{aligned}
f_2 &= \frac{\int E_{10}^* E_{20}^* E_{30} E_{40} \, dx \, dy}{\int |E_{20}|^2 \, dx \, dy \int |E_{20}|^2 \, dx \, dy} \\
f_3 &= \frac{\int E_{10} E_{20} E_{30}^* E_{40}^* \, dx \, dy}{\int |E_{30}|^2 \, dx \, dy \int |E_{30}|^2 \, dx \, dy} \\
f_4 &= \frac{\int E_{10} E_{20} E_{30}^* E_{40}^* \, dx \, dy}{\int |E_{40}|^2 \, dx \, dy \int |E_{40}|^2 \, dx \, dy}
\end{aligned} \tag{14.3-19}$$

In Equations (14.3-18), we have added the linear attenuation coefficient α for the mode propagation in the fiber. We have also assumed a constant n_2 in the spectral regimes of the four waves.

In the case when $\omega_1 \approx \omega_2 \approx \omega_3 \approx \omega_4$, all four mode overlap constants f_n ($n = 1, 2, 3, 4$) are equal to $1/A_{\text{eff}}$, where A_{eff} is the effective area of the modes of propagation (LP_{01x}) (see Equation (14.1-9)). This is the case where analytical solutions of the coupled equations are available. We first consider the case where phase matching is satisfied.

The Case Where $\Delta\beta = 0$

Let ω_0 be the zero-dispersion point in the single-mode fiber. For silica fibers (e.g., SMF-28), the zero dispersion occurs at around $\lambda_0 = 1.30$ μm ($\omega_0 = 2\pi c/\lambda_0$). The propagation constant β of the modes can be written approximately as

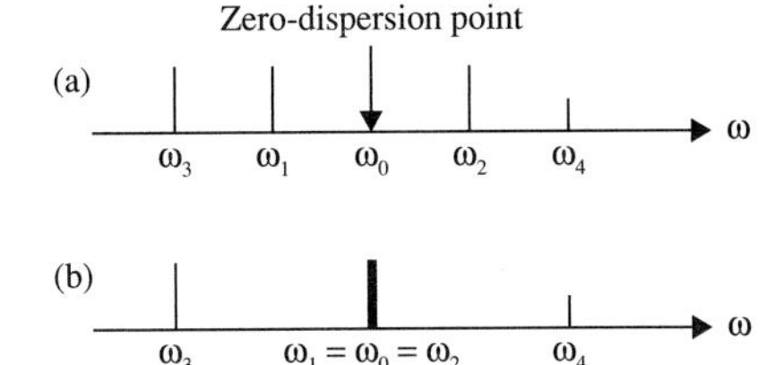

Figure 14.4 Relative position of the frequencies in the case of phase-matched four-wave mixing. The frequencies are positioned symmetrically with respect to the zero-dispersion point ω_0. (a) Nondegenerate four-wave mixing and (b) partially degenerate four-wave mixing.

$$\beta_n = \beta(\omega_n) = \beta(\omega_0) + \left(\frac{\partial\beta}{\partial\omega}\right)_{\omega_0}(\omega_n - \omega_0) + \frac{1}{6}\left(\frac{\partial^3\beta}{\partial\omega^3}\right)_{\omega_0}(\omega_n - \omega_0)^3 + \cdots \tag{14.3-20}$$

where the quadratic term $(\partial^2\beta/\partial\omega^2)_{\omega_0}(\omega_n - \omega_0)^2/2$ vanishes, as ω_0 is the zero-dispersion point. According to Equation (14.3-20), phase matching ($\beta_1 + \beta_2 = \beta_3 + \beta_4$) can be achieved when the frequencies of the four waves are positioned symmetrically with respect to the zero-dispersion point. In other words,

$$\omega_0 - \omega_1 = \omega_2 - \omega_0 \quad \text{and} \quad \omega_0 - \omega_3 = \omega_4 - \omega_0 \tag{14.3-21}$$

This is shown in Figure 14.4.

It is important to note that we have ignored the modifications to the propagation constants due to self-phase modulation and cross-phase modulation. These nonlinear contributions $(\Delta\beta)_{\text{NL}}$ can be written, according to Equations (14.2-9),

$$(\Delta\beta)_{\text{NL}} = \frac{n_2}{A_{\text{eff}}}\left(P_4\frac{\omega_4}{c} + P_3\frac{\omega_3}{c} - P_2\frac{\omega_2}{c} - P_1\frac{\omega_1}{c}\right) \tag{14.3-22}$$

where A_{eff} is the effective area of the modes. We note that the nonlinear contributions depend on the mode powers and may not be ignored, particularly when the mode powers are high. In some cases, the nonlinear contributions $(\Delta\beta)_{\text{NL}}$ may cancel the linear $\Delta\beta$ under the appropriate conditions.

When the Kerr coefficient is expressed in units of m^2/W, the mode amplitudes can be conveniently written

$$A_n = \sqrt{P_n}\exp(i\phi_n) = \sqrt{Q_n}\exp(-\alpha z/2)\exp(i\phi_n), \quad n = 1, 2, 3, 4 \tag{14.3-23}$$

where the P_n are the corresponding powers of the modes in units of W, the ϕ_n are the corresponding phases, and the Q_n are the corresponding powers of the modes in the absence of linear attenuation. Using these new notations, the four complex coupled equations can thus be written in terms of the following eight real equations:

$$\begin{aligned}
\frac{dQ_1}{dz} &= -4\gamma\sqrt{Q_1Q_2Q_3Q_4}\exp(-\alpha z)\sin\Delta\phi \\
\frac{dQ_2}{dz} &= -4\gamma\sqrt{Q_1Q_2Q_3Q_4}\exp(-\alpha z)\sin\Delta\phi \\
\frac{dQ_3}{dz} &= +4\gamma\sqrt{Q_1Q_2Q_3Q_4}\exp(-\alpha z)\sin\Delta\phi \\
\frac{dQ_4}{dz} &= +4\gamma\sqrt{Q_1Q_2Q_3Q_4}\exp(-\alpha z)\sin\Delta\phi
\end{aligned} \tag{14.3-24}$$

and

$$\frac{d}{dz}\phi_1 = -2\frac{\gamma}{Q_1}\sqrt{Q_1Q_2Q_3Q_4}\,\exp(-\alpha z)\cos\Delta\phi$$
$$\frac{d}{dz}\phi_2 = -2\frac{\gamma}{Q_2}\sqrt{Q_1Q_2Q_3Q_4}\,\exp(-\alpha z)\cos\Delta\phi$$
$$\frac{d}{dz}\phi_3 = -2\frac{\gamma}{Q_3}\sqrt{Q_1Q_2Q_3Q_4}\,\exp(-\alpha z)\cos\Delta\phi \quad (14.3\text{-}25)$$
$$\frac{d}{dz}\phi_4 = -2\frac{\gamma}{Q_4}\sqrt{Q_1Q_2Q_3Q_4}\,\exp(-\alpha z)\cos\Delta\phi$$

where γ, which is a constant describing the four-wave mixing gain, is given by

$$\gamma = \frac{n_2\omega_0}{cA_{\text{eff}}} = \frac{2\pi}{\lambda_0 A_{\text{eff}}}n_2 \quad (14.3\text{-}26)$$

where λ_0 is the wavelength corresponding to the frequency ω_0. In arriving at Equations (14.3-24) and (14.3-25), we assumed $|\omega_n - \omega_0| \ll \omega_0$. In typical optical communications networks, the frequency difference between channels is in the range of 100 GHz, whereas the frequencies (ω_1, ω_2, ω_3, ω_4) of the waves are in the range of 200 THz.

From the four equations for the phases, we obtain

$$\frac{d}{dz}\Delta\phi = 2\gamma\left(\frac{1}{Q_3}+\frac{1}{Q_4}-\frac{1}{Q_1}-\frac{1}{Q_2}\right)\sqrt{Q_1Q_2Q_3Q_4}\,\exp(-\alpha z)\cos\Delta\phi \quad (14.3\text{-}27)$$

where $\Delta\phi$ is given by

$$\Delta\phi = \phi_1 + \phi_2 - \phi_3 - \phi_4 \quad (14.3\text{-}28)$$

By examining the coupled equations for the mode powers (14.3-24), and Equation (14.3-27), we note that there exist four independent invariants (constants of integration). They are

$$\frac{d}{dz}(Q_1 + Q_2 + Q_3 + Q_4) = 0$$
$$\frac{d}{dz}(Q_1 - Q_2) = 0$$
$$\frac{d}{dz}(Q_3 - Q_4) = 0 \quad (14.3\text{-}29)$$
$$\frac{d}{dz}(\sqrt{Q_1Q_2Q_3Q_4}\cos\Delta\phi) = 0$$

Equations (14.3-24) and (14.3-27) are similar to those of phase locking in coupled oscillators. As a result of the coupling, the relative phases of the waves are locked. By examining Equation (14.3-27), we note that the phases are locked at

$$\Delta\phi = \pm\pi/2 \quad (14.3\text{-}30)$$

These are solutions of Equation (14.3-27). They can also easily be obtained for the special case when $P_4(0) = 0$ (i.e., the input power of the fourth wave is zero). Using $P_4(0) = 0$ and the fourth invariant in (14.3-29), we obtain $Q_1Q_2Q_3Q_4 \cos\Delta\phi = 0$ for all z. Since all mode powers are finite, we conclude that $\cos\Delta\phi = 0$ for all z. This leads to the solutions of Equation (14.3-30). It is important to note that the solutions of Equation (14.3-30) are valid in the general case when $P_4(0) \neq 0$.

We now consider a practical case in lightwave systems designed for either phase conjugation or wavelength conversion in which the fourth wave is absent at the input ($z = 0$). According to the fourth equation in (14.3-24) and the conditions that $Q_4(0) = 0$ and $Q_4(z) \geq 0$ for all $z > 0$, we find that $\Delta\phi = +\pi/2$ is a proper solution. In this case, the coupled equations become

$$\begin{aligned}
\frac{dQ_1}{dz} &= -4\gamma\sqrt{Q_1Q_2Q_3Q_4}\,\exp(-\alpha z) \\
\frac{dQ_2}{dz} &= -4\gamma\sqrt{Q_1Q_2Q_3Q_4}\,\exp(-\alpha z) \\
\frac{dQ_3}{dz} &= +4\gamma\sqrt{Q_1Q_2Q_3Q_4}\,\exp(-\alpha z) \\
\frac{dQ_4}{dz} &= +4\gamma\sqrt{Q_1Q_2Q_3Q_4}\,\exp(-\alpha z)
\end{aligned} \tag{14.3-31}$$

The above coupled equations can be integrated in the case of equal pump powers, that is, $P_1(0) = P_2(0) = P_p(0)$. Here we designate waves 1 and 2 as the pump waves. With the help of the invariants (14.3-29) and the boundary conditions, the solutions of the coupled equations can be written

$$\begin{aligned}
Q_1 &= P_p(0)U(z) \\
Q_2 &= P_p(0)U(z) \\
Q_3 &= P_p(0)[1 - U(z)] + P_3(0) \\
Q_4 &= P_p(0)[1 - U(z)]
\end{aligned} \tag{14.3-32}$$

where $U(z)$ is a positive real function in the range of $[0, 1]$ that satisfies the following differential equation:

$$\frac{dU}{U\sqrt{[(1-U) + P_3(0)/P_p(0)](1-U)}} = -4\gamma P_p(0)\exp(-\alpha z)\,dz \tag{14.3-33}$$

A direct integration of the above equation yields

$$U = \frac{1+\sigma}{1+\sigma\cosh^2[g(z)\sqrt{1+\sigma}]} \tag{14.3-34}$$

where σ is a constant, representing the ratio of the mode powers at the input

$$\sigma = \frac{P_3(0)}{P_p(0)} \tag{14.3-35}$$

and $g(z)$ is given by

$$g(z) = 2\gamma P_p(0)\frac{1-e^{-\alpha z}}{\alpha} = \frac{4\pi}{\lambda_0}\frac{P_p(0)}{A_{\text{eff}}}\frac{1-e^{-\alpha z}}{\alpha}n_2 \tag{14.3-36}$$

By examining the function $U(z)$, we note that it is a monotonically decreasing function of z. As z approaches infinity, U reaches a minimum. In the hypothetical case of lossless transmission ($\alpha = 0$), U reaches a minimum of zero. For a finite segment of fiber, $(1 - U)$ is a measure of the fractional power transfer among the waves. Figure 14.5 shows the mode powers as a function of interaction length in a single-mode silica fiber. At a power level of 20 mW for waves 1 and 2, and a power of 15 mW for wave 3, the gain due to FWM is not

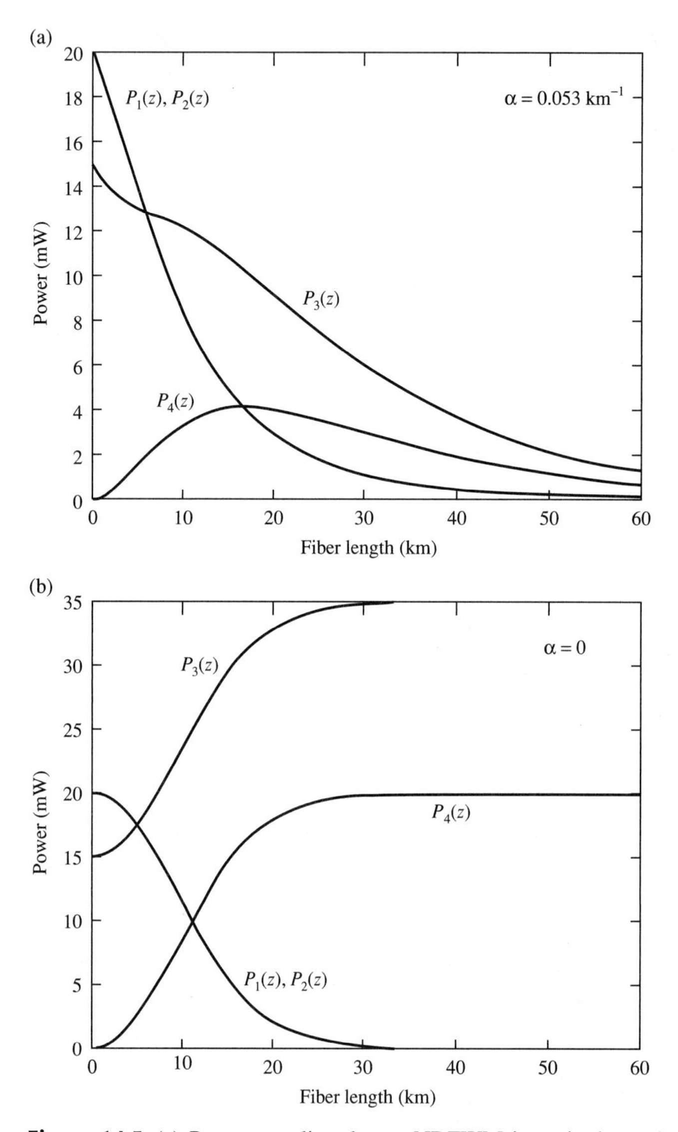

Figure 14.5 (a) Power coupling due to NDFWM in a single-mode silica fiber for a special case of $P_1(0) = P_2(0) = P_p(0) = 20$ mW, and $P_3(0) = 15$ mW. The parameters are wavelength $\lambda_0 = 1300$ nm, a Kerr coefficient of $3 \times 10^{-20}\ \text{m}^2/\text{W}$, a linear attenuation coefficient of $\alpha = 0.053\ \text{km}^{-1}$ $A_{\text{eff}} = 70\ \mu\text{m}^2$. (b) Power coupling due to nondegenerate FWM in a single mode silica fiber for the hypothetical case of lossless transmission ($\alpha = 0$).

sufficient to overcome the linear attenuation of $\alpha = 0.053\ \text{km}^{-1}$. So $P_3(z)$ is a decreasing function of z. The power of wave 4, $P_4(z)$, increases initially due to gain from FWM. After reaching a maximum, the power starts to decay due to the linear attenuation. Figure 14.5b shows the hypothetical case of a lossless transmission ($\alpha = 0$). In this case, we note that the powers of waves 3 and 4 are increasing functions of z, while the powers of waves 1 and 2 are decreasing functions of z. In the NDFWM process, waves 1 and 2 donate energy to waves 3 and 4. Figure 14.6 shows the power coupling for the case when the power levels of waves 1 and 2 are increased to 40 mW. In this case, the gain due to NDFWM can be greater than the

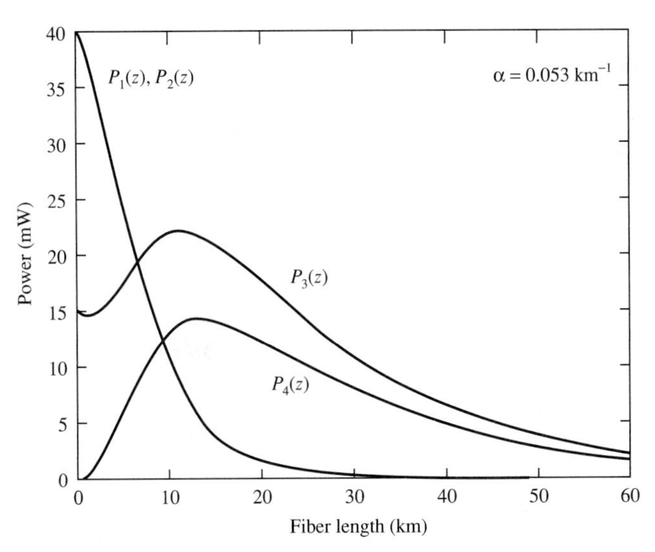

Figure 14.6 Power coupling due to NDFWM in a single-mode silica fiber for a special case of $P_1(0) = P_2(0) = P_p(0) = 40$ mW, and $P_3(0) = 15$ mW. The parameters are wavelength $\lambda_0 = 1300$ nm, a Kerr coefficient of 3×10^{-20} m^2/W, a linear attenuation coefficient of $\alpha = 0.053$ km^{-1}, and $A_{\text{eff}} = 70\ \mu\text{m}^2$.

linear attenuation. The power of wave 3, $P_3(z)$, can be amplified after an initial drop due to linear attenuation. The amplification is a result of the gain due to NDFWM, which occurs as soon as the power level of wave 4 is built up from the mixing process. After reaching their maxima, the power levels of both waves 3 and 4 decrease due to linear attenuation. In this regime, the powers of all waves decrease to a level where the nonlinear gain becomes smaller than the linear attenuation.

The Case Where $\Delta\beta \neq 0$

We now consider the case of phase mismatch. When $\Delta\beta \neq 0$, the coupled equations become

$$
\begin{aligned}
\frac{dQ_1}{dz} &= -4\gamma\sqrt{Q_1Q_2Q_3Q_4}\,\exp(-\alpha z)\sin(\Delta\phi - \Delta\beta\, z) \\
\frac{dQ_2}{dz} &= -4\gamma\sqrt{Q_1Q_2Q_3Q_4}\,\exp(-\alpha z)\sin(\Delta\phi - \Delta\beta\, z) \\
\frac{dQ_3}{dz} &= +4\gamma\sqrt{Q_1Q_2Q_3Q_4}\,\exp(-\alpha z)\sin(\Delta\phi - \Delta\beta\, z) \\
\frac{dQ_4}{dz} &= +4\gamma\sqrt{Q_1Q_2Q_3Q_4}\,\exp(-\alpha z)\sin(\Delta\phi - \Delta\beta\, z)
\end{aligned}
\qquad (14.3\text{-}37)
$$

and

$$
\frac{d}{dz}\Delta\phi = 2\gamma\left(\frac{1}{Q_3} + \frac{1}{Q_4} - \frac{1}{Q_1} - \frac{1}{Q_2}\right)\sqrt{Q_1Q_2Q_3Q_4}\,\exp(-\alpha z)\cos(\Delta\phi - \Delta\beta\, z) \qquad (14.3\text{-}38)
$$

In the case of phase mismatch, the locking of phases among the four waves is no longer present. In other words, $\Delta\phi =$ constant is not a solution of Equation (14.3-38). The solution of $\Delta\phi = \pi/2$ can be considered valid provided $\Delta\beta$ is small enough such that $\Delta\beta\, L \ll 1$, where L is the length of interaction. This occurs when the frequencies of the waves are near the zero dispersion point, or when the frequency difference between the waves are very small. In these cases, the solutions are still given by Equations (14.3-32) and (14.3-34), provided $g(z)$ is given by

$$g(z) = 2\gamma P_p(0)\frac{\Delta\beta\, e^{-\alpha z}\sin(\Delta\beta\, z) + \alpha[1 - e^{-\alpha z}\cos(\Delta\beta\, z)]}{(\Delta\beta)^2 + \alpha^2} \tag{14.3-39}$$

In the case of $\alpha \ll |\Delta\beta|$, $g(z)$ becomes, according to Equation (14.3-39),

$$g(z) \approx 2\gamma P_p(0)\frac{\sin(\Delta\beta\, z)}{\Delta\beta} \tag{14.3-40}$$

As a result of the sinusoidal function in $g(z)$, the powers of all four waves exhibit an oscillatory decaying behavior. In the case of phase mismatch, the optimum coupling length is

$$L_{\text{optimum}} = \frac{\pi}{2(\Delta\beta)} = \frac{L_c}{4} \tag{14.3-41}$$

where L_c is the coherent length defined as $L_c = 2\pi/(\Delta\beta)$. After this distance, the powers of waves 3 and 4 will flow back to waves 1 and 2. To get an idea about the magnitude of the phase mismatch, we consider the following examples.

EXAMPLE: FWM WITH LARGE PHASE MISMATCH

We consider a single-mode silica fiber in the spectral regime around ν_0 = 193.5 THz (λ_0 = 1550 nm), where the dispersion is about 17 ps/nm-km, which is equivalent to $(\partial^2\beta/\partial\omega^2)$ = 2.2×10^{-23} s^2/km. Let the frequencies of the waves be

$$\nu_0 - \nu_1 = \nu_2 - \nu_0 = 100 \text{ GHz}$$

$$\nu_0 - \nu_3 = \nu_4 - \nu_0 = 200 \text{ GHz}$$

The phase mismatch is given by

$$\Delta\beta = (\beta_1 + \beta_2 - \beta_3 - \beta_4) = \frac{1}{2}\left(\frac{\partial^2\beta}{\partial\omega^2}\right)_{\omega_0}[(\omega_1 - \omega_0)^2 + (\omega_2 - \omega_0)^2 - (\omega_3 - \omega_0)^2 - (\omega_4 - \omega_0)^2] \tag{14.3-42}$$

Using the parameters above, we obtain $\Delta\beta$ = 17.4 km^{-1}. This is a large phase mismatch. Four-wave mixing with such a large phase mismatch is almost negligible in fibers with a length over a kilometer.

EXAMPLE: FWM WITH SMALL PHASE MISMATCH

Consider the same fiber in the above example, with the following channel spacings:

$$\nu_0 - \nu_1 = \nu_2 - \nu_0 = 100 \text{ MHz}$$

$$\nu_0 - \nu_3 = \nu_4 - \nu_0 = 200 \text{ MHz}$$

In this case, the frequency spacing is 1000 times smaller. The phase mismatch is a million times smaller. In other words, $\Delta\beta = 1.74 \times 10^{-5}$ km^{-1}. For a fiber with L = 100 km, $\Delta\beta\, L = 1.74 \times 10^{-3}$, which is much less than 1. Thus four-wave mixing under these conditions can be significant provided the power levels are high enough.

14.4 PARTIALLY DEGENERATE FOUR-WAVE MIXING

For wavelength conversion in optical fibers, it is convenient to use partially degenerate four-wave mixing (PDFWM), where $\omega_1 = \omega_2$, and only three distinct waves are present. Let the electric fields be written

$$\begin{aligned} \mathbf{E}_1(\mathbf{r}, t) &= \tfrac{1}{2}\hat{\mathbf{x}}A_1E_{10}(x, y)e^{i(\omega_1 t-\beta_1 z)} + \text{c.c.} \\ \mathbf{E}_3(\mathbf{r}, t) &= \tfrac{1}{2}\hat{\mathbf{x}}A_3E_{30}(x, y)e^{i(\omega_3 t-\beta_3 z)} + \text{c.c.} \\ \mathbf{E}_4(\mathbf{r}, t) &= \tfrac{1}{2}\hat{\mathbf{x}}A_4E_{40}(x, y)e^{i(\omega_4 t-\beta_4 z)} + \text{c.c.} \end{aligned} \tag{14.4-1}$$

where $E_{10}(x, y)$, $E_{30}(x, y)$, and $E_{40}(x, y)$ are the wavefunctions of the LP_{01} modes, A_1, A_3, and A_4 are amplitudes of the waves, ω_1, ω_3 and ω_4 are the corresponding frequencies, and β_1, β_3, and β_4 are the corresponding propagation constants. Figure 14.7 is a schematic drawing of partially degenerate four-wave mixing in a single-mode fiber.

According to the third-order term in Equation (8.1-1), a nonlinear (NL) polarization at $2\omega_1 - \omega_3$ is generated,

$$\begin{aligned} P_x^{(NL)}(\mathbf{r}, t) &= \tfrac{3}{2}\chi_{1111}^{(3)}A_1^2A_3^*E_{10}^2E_{30}^*e^{i[(2\omega_1-\omega_3)t-(2\beta_1-\beta_3)z]} + \text{c.c.} \\ &= \text{Re}[3\chi_{1111}^{(3)}A_1^2A_3^*E_{10}^2E_{30}^*e^{i[(2\omega_1-\omega_3)t-(2\beta_1-\beta_3)z]}] \end{aligned} \tag{14.4-2}$$

As in the case of nondegenerate four-wave mixing, the nonlinear polarization is a volume distribution of oscillating dipoles, which can radiate a new wave at the frequency $\omega_4 = 2\omega_1 - \omega_3$, provided the wavevector of the new wave satisfies the momentum conservation condition $\beta_4 = 2\beta_1 - \beta_3$. In a long segment of fiber, the newly generated wave at frequency ω_4 and wavevector β_4 will also mix with the other two waves. This will lead to the generation of waves at frequencies ω_1 and ω_3 and wavevectors β_1 and β_3. Thus all three waves are coupled

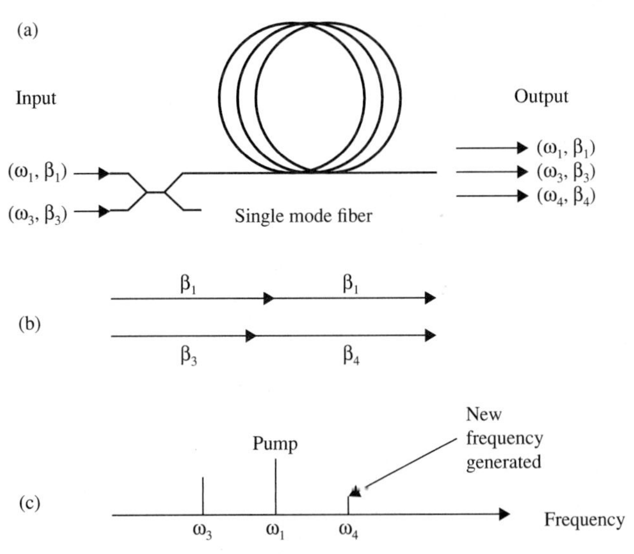

Figure 14.7 (a) Schematic drawing of partially degenerate four-wave mixing in a single-mode fiber. (b) Vector diagram of the conservation of momentum. (c) Frequency diagram of the waves. Note that frequencies ω_3 and ω_4 are equidistant from the pump frequency ω_1, which is located at (or near) the zero-dispersion point.

with energy exchange among the waves as they propagate along the fiber. The conservation of photon energy and momentum, in this case, becomes

$$\omega_4 = 2\omega_1 - \omega_3$$
$$\beta_4 = 2\beta_1 - \beta_3 \tag{14.4-3}$$

We note that frequencies ω_3 and ω_4 are equidistant from the pump frequency ω_1. Since the wavevector (β_4) is related to the frequency (ω_4) by the dispersion relationship, these two equations may not be satisfied simultaneously. Thus partially degenerate nonlinear optical four-wave mixing in single-mode fibers occurs significantly only when the pump frequency ω_1 is located at the zero-dispersion point.

Coupled-Mode Formulation of Partially Degenerate Four-Wave Mixing

As a result of the coupling via the third-order nonlinear polarization, $A_1(z)$, $A_3(z)$, and $A_4(z)$ are functions of z, reflecting the energy exchange as they propagate along the z axis. For simplicity, we again assume that all four waves are in the LP_{01x} mode. The coupled equations in this case are given by

$$\frac{dA_1}{dz} = -in_2 \frac{2\omega_1}{c} A_1^* A_3 A_4 f_1 \exp(i\,\Delta\beta\, z) - \frac{\alpha}{2} A_1$$
$$\frac{dA_3}{dz} = -in_2 \frac{\omega_3}{c} A_1^2 A_4^* f_3 \exp(-i\,\Delta\beta\, z) - \frac{\alpha}{2} A_3 \tag{14.4-4}$$
$$\frac{dA_4}{dz} = -in_2 \frac{\omega_4}{c} A_1^2 A_3^* f_4 \exp(-i\,\Delta\beta\, z) - \frac{\alpha}{2} A_4$$

where n_2 is the Kerr coefficient, f_1, f_3, and f_4 are the mode wavefunction overlap constants defined in Section 14.3, α is the linear attenuation coefficient, and $\Delta\beta$, describing the phase mismatch in PDFWM, is given by

$$\Delta\beta = 2\beta_1 - \beta_3 - \beta_4 \tag{14.4-5}$$

In arriving at Equations (14.4-4), we use the frequency condition (14.4-3) and again assume that $\chi^{(3)}_{1111}$, which is proportional to the Kerr coefficient, is a constant in the spectral regime of interest. In the case when $\omega_1 \approx \omega_3 \approx \omega_4$, all three mode overlapping constants are equal to $1/A_{\text{eff}}$, where A_{eff} is the effective area of the modes of propagation (LP_{01x}). This is the case where analytical solutions of the coupled equations are available.

For the case of phase matching, $\Delta\beta = 0$, the solution of the mode powers can be obtained using the method employed in nondegenerate four-wave mixing (NDFWM) in the last section. With the following boundary conditions,

$$P_1(z=0) = P_p(0)$$
$$P_3(z=0) = P_3(0) \tag{14.4-6}$$
$$P_4(z=0) = 0$$

the solutions for the mode powers can be written

$$P_1(z) = P_p(0)V(z)\exp(-\alpha z)$$
$$P_3(z) = \{\tfrac{1}{2}P_p(0)[1 - V(z)] + P_3(0)\}\exp(-\alpha z) \tag{14.4-7}$$
$$P_4(z) = \tfrac{1}{2}P_p(0)[1 - V(z)]\exp(-\alpha z)$$

where $V(z)$ is given by

$$V(z) = \frac{1 + 2\sigma}{1 + 2\sigma \cosh^2[\frac{1}{2} g(z)\sqrt{1 + 2\sigma}]} \tag{14.4-8}$$

where σ is a constant, again representing the ratio of the mode powers at the input

$$\sigma = \frac{P_3(0)}{P_p(0)} \tag{14.4-9}$$

and $g(z)$ is again given by

$$g(z) = 2\gamma P_p(0)\frac{1 - e^{-\alpha z}}{\alpha} = \frac{4\pi}{\lambda_0}\frac{P_p(0)}{A_{\text{eff}}}\frac{1 - e^{-\alpha z}}{\alpha}n_2 \tag{14.4-10}$$

with

$$\gamma = \frac{n_2\omega_0}{cA_{\text{eff}}} = \frac{2\pi}{\lambda_0 A_{\text{eff}}}n_2 \tag{14.4-11}$$

In Equations (14.4-10) and (14.4-11), λ_0 is the wavelength corresponding to the frequency of the pump wave $\omega_0 = \omega_1$ (Note: $\omega_1 \approx \omega_3 \approx \omega_4$.)

Here again, the function $V(z)$ is a monotonically decreasing function of z. At $z = 0$, $V(0) = 1$. As z approaches infinity, V reaches a minimum. In the hypothetical case of lossless transmission ($\alpha = 0$), $g(z) = 2\gamma P_p(0)z$. The function $V(z)$ reaches a minimum of zero, representing a complete transfer of power from the pump wave (wave 1) to waves 3 and 4. For a finite segment of fiber, $(1 - V)$ is a measure of the fractional power transfer among the waves. In practice, linear attenuation in fiberoptic transmission cannot be neglected. In this case, the function $V(z)$ has the functional form $(1 - e^{-\alpha z})/\alpha$. We can conclude that the power coupling among the three waves will become insignificant in the region where $z \gg 1/\alpha$. Figure 14.8 shows the mode powers as a function of interaction length in a single-mode silica fiber. At a power level of 40 mW for the pump wave (wave 1), and a power of 15 mW for wave 3, the gain due to FWM is not sufficient to overcome the linear attenuation of $\alpha = 0.053$ km^{-1}. So $P_3(z)$ is a decreasing function of z. The power of wave 4, $P_4(z)$, increases initially due to gain from PDFWM. After reaching a maximum, the power starts to decay due to the linear attenuation. Figure 14.8b shows the hypothetical case of lossless transmission ($\alpha = 0$). In this case, we note that the powers of waves 3 and 4 are increasing functions of z, while the power of the pump wave (wave 1) is a decreasing function of z. The power of the pump wave is distributed equally to waves 3 and 4. In the PDFWM process, the pump wave (wave 1) donates energy to waves 3 and 4.

Figure 14.9 shows the power coupling for the case when the pump power level is increased to 60 mW. In this case, the gain due to PDFWM can be greater than the linear attenuation. The power of wave 3, $P_3(z)$, can be amplified after an initial drop due to linear attenuation. The amplification is a result of the gain due to PDFWM, which occurs as soon as the power level of wave 4 is built up from the mixing process. After reaching their maxima, the power levels of both waves 3 and 4 decrease due to linear attenuation. In this regime, the powers of all waves decrease to a level where the nonlinear gain becomes smaller than the linear attenuation. For the purpose of wavelength conversion, an optimum choice of the length of fiber will be at the location where the power of wave 4 reaches its maximum.

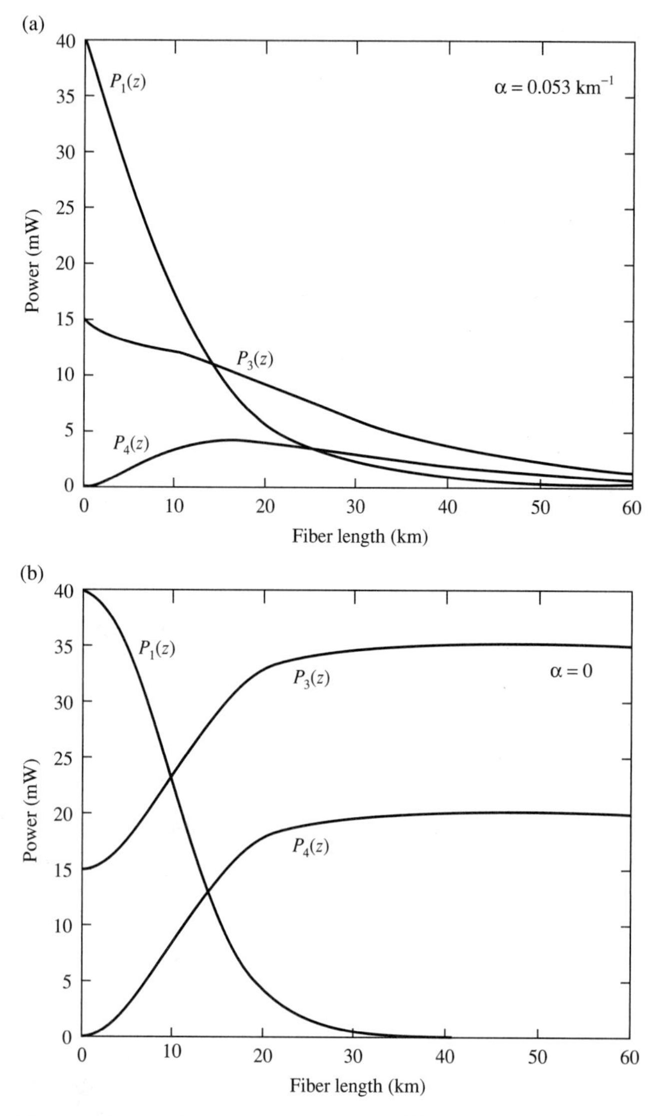

Figure 14.8 (a) Power coupling due to PDFWM in a single-mode silica fiber for a special case of $P_1(0) = P_p(0) = 40$ mW, and $P_3(0) = 15$ mW. The parameters are wavelength $\lambda_0 = 1300$ nm, a Kerr coefficient of 3×10^{-20} m^2/W, a linear attenuation coefficient of $\alpha = 0.053$ km^{-1}, and $A_{\text{eff}} = 70$ μm^2. (b) Power coupling due to the same PDFWM in a single-mode silica fiber for the hypothetical case of lossless transmission ($\alpha = 0$).

The optimum length of interaction for wavelength conversion can be derived by requiring $dP_4(z)/dz = 0$. Using Equations (14.4-7), we obtain the condition of optimum length of interaction

$$\frac{dV}{dz} = -\alpha(1 - V) \tag{14.4-12}$$

where $V(z)$ is given by Equation (14.4-8). Substituting Equation (14.4-8) for V in Equation (14.4-12), we obtain

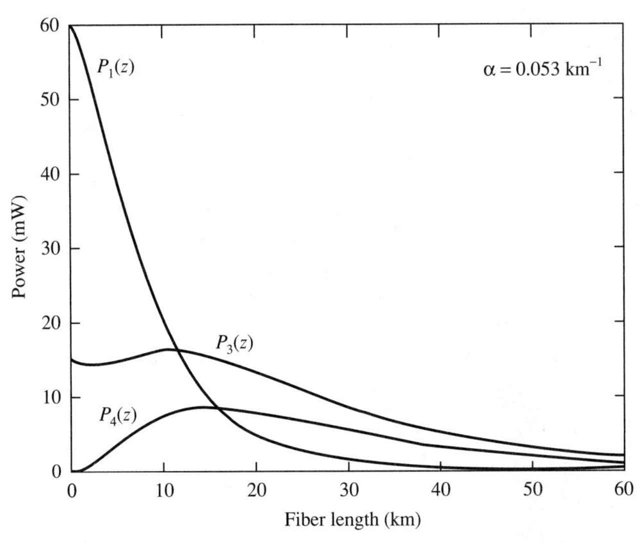

Figure 14.9 Power coupling due to PDFWM in a single-mode silica fiber for a special case of $P_1(0) = P_p(0) = 60\ \text{mW}$, and $P_3(0) = 15\ \text{mW}$. The parameters are wavelength $\lambda_0 = 1300\ \text{nm}$, a Kerr coefficient of $3 \times 10^{-20}\ \text{m}^2/\text{W}$, a linear attenuation coefficient of $\alpha = 0.053\ \text{km}^{-1}$, and $A_{\text{eff}} = 70\ \mu\text{m}^2$.

$$V\frac{dg(z)}{dz} = \frac{\alpha}{\sqrt{1+2\sigma}} \tanh\left[\tfrac{1}{2}\, g(z)\sqrt{1+2\sigma}\right] \tag{14.4-13}$$

where $g(z)$ is given by Equation (14.4-10). If $\gamma P_p(0)$ is not very large such that $\gamma P_p(0)/\alpha \ll 1$, then $g(z) \ll 1$ and $V(z) \approx 1$, as well as $\tanh\,[g(z)\sqrt{1+2\sigma}/2] \approx g(z)\sqrt{1+2\sigma}/2$. The optimum fiber length for wavelength conversion is thus given, according to Equation (14.4-13), by

$$L_{\text{optimum}} = \frac{\ln 3}{\alpha} \tag{14.4-14}$$

For the example shown in Figure 14.8, the optimum length is about 20 km.

In wavelength conversion via PDFWM, the pump wave (wave 1) is donating energy to both waves 3 and 4. The conversion efficiency can thus be defined as

$$\eta = \frac{\Delta P_3(z) + \Delta P_4(z)}{P_1(0)} = \frac{P_3(z) - P_3(0)e^{-\alpha z} + P_4(z)}{P_1(0)} \tag{14.4-15}$$

Using Equations (14.4-7), the conversion efficiency can be written

$$\eta = (1 - V)\exp(-\alpha z) = \frac{2\sigma \sinh^2[\tfrac{1}{2}\, g(z)\sqrt{1+2\sigma}]}{1 + 2\sigma \cosh^2[\tfrac{1}{2}\, g(z)\sqrt{1+2\sigma}]}\exp(-\alpha z) \tag{14.4-16}$$

The optimum conversion efficiency for the case when $\gamma P_p(0)/\alpha \ll 1$, $g(z) \ll 1$, and $V(z) \approx 1$ can be written, according to Equations (14.4-14) and (14.4-16),

$$\eta_{\max} = \frac{8\gamma^2 P_3(0) P_p(0)}{27\alpha^2} \tag{14.4-17}$$

Generally, the conversion efficiency is an increasing function of the pump power $P_p(0)$.

EXAMPLE: OPTIMUM CONVERSION EFFICIENCY

Consider an example of a phase-matched PDFWM with wavelength λ_0 = 1300 nm, a Kerr coefficient of 3×10^{-20} m^2/W, a linear attenuation coefficient of α = 0.053 km^{-1}, and A_{eff} = 70 μm^2. The coupling constant γ is given by Equation (14.4-11),

$$\gamma = \frac{n_2\omega_0}{cA_{eff}} = \frac{2\pi}{\lambda_0 A_{eff}} n_2 = 0.0021 \text{ m/W}$$

With power levels of $P_3(0) = P_p(0)$ = 10 mW, the optimum conversion efficiency is given by Equation (14.4-17),

$$\eta_{max} = \frac{8\gamma^2 P_3(0) P_p(0)}{27\alpha^2} = 0.045$$

With power levels of $P_3(0)$ = 30 mW and $P_p(0)$ = 10 mW, the optimum conversion efficiency becomes

$$\eta_{max} = \frac{8\gamma^2 P_3(0) P_p(0)}{27\alpha^2} = 0.136$$

The optimum interaction length is given by Equation (14.4-14)

$$L_{optimum} = \frac{\ln 3}{\alpha} = 20.7 \text{ km}$$

Figures 14.10 and 14.11 show the conversion efficiency as a function of the interaction length for various input power levels. We note that both Equations (14.4-14) and (14.4-17) are good approximations for the optimum interaction length and the optimum conversion efficiency in the low-efficiency regime ($V \approx 1$, or $\eta \ll 1$).

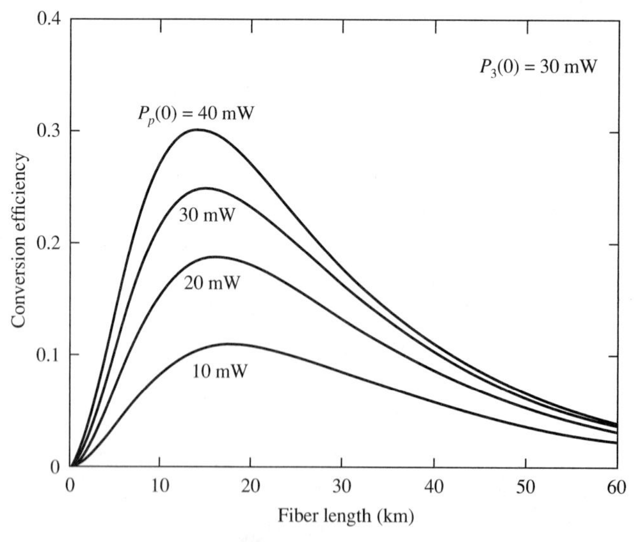

Figure 14.10 Wavelength conversion efficiency via PDFWM in a single-mode silica fiber for a special case of $P_3(0)$ = 30 mW and various pump power levels. The parameters are wavelength λ_0 = 1300 nm, a Kerr coefficient of 3×10^{-20} m^2/W, a linear attenuation coefficient of α = 0.053 km^{-1}, and A_{eff} = 70 μm^2.

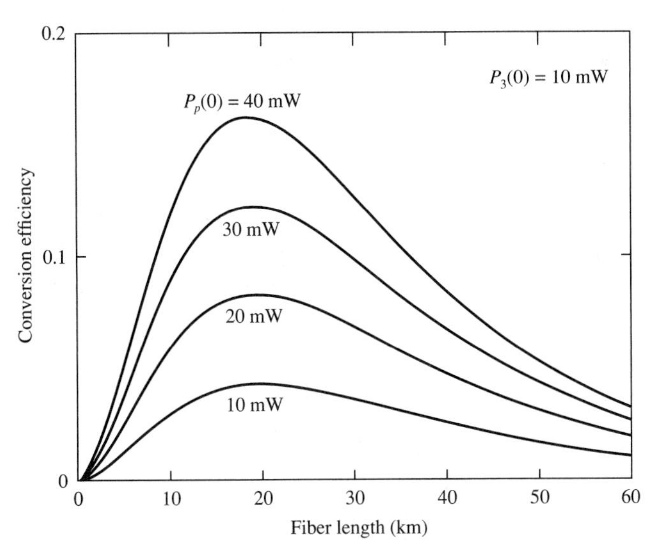

Figure 14.11 Wavelength conversion efficiency via PDFWM in a single-mode silica fiber for a special case of $P_3(0) = 10$ mW and various pump power levels. The parameters are wavelength $\lambda_0 = 1300$ nm, a Kerr coefficient of 3×10^{-20} m^2/W, a linear attenuation coefficient of $\alpha = 0.053$ km^{-1}, and $A_{\text{eff}} = 70$ μm^2.

Phase Conjugation and Frequency Chirp Reversal

We now discuss an important property of FWM in single-mode fibers. Particularly, we are interested in the relationship between the temporal profiles of waves 3 and 4. By including the temporal profiles, the electric field amplitudes can be written

$$\mathbf{E}_1(\mathbf{r}, t) = \tfrac{1}{2}\hat{\mathbf{x}}A_1(z)E_{10}(x, y)e^{i(\omega_1 t - \beta_1 z)} + \text{c.c.}$$
$$\mathbf{E}_3(\mathbf{r}, t) = \tfrac{1}{2}\hat{\mathbf{x}}A_3(z)p_3(t)E_{30}(x, y)e^{i(\omega_3 t - \beta_3 z)} + \text{c.c.} \qquad (14.4\text{-}18a)$$
$$\mathbf{E}_4(\mathbf{r}, t) = \tfrac{1}{2}\hat{\mathbf{x}}A_4(z)p_4(t)E_{40}(x, y)e^{i(\omega_4 t - \beta_4 z)} + \text{c.c.}$$

where $p_3(t)$ and $p_4(t)$ are the temporal profiles of waves 3 and 4, respectively. In Equations (14.4-18a), we assume that the pump wave is a continuous wave (cw).

According to the nonlinear polarization (14.4-2), or the third coupled equation in (14.4-4), in PDFWM, the temporal profile of the fourth wave is proportional to the phase conjugate of the temporal profile of the third wave. Without loss of generality, we can write, according to Equation (14.4-2)

$$p_4(t) = p_3^*(t) \qquad (14.4\text{-}18b)$$

provided the pump wave (wave 1) is a continuous wave. In other words, the temporal profile of wave 4 is a phase conjugate of the temporal profile of wave 3. As a result of the phase conjugation, the spectral distribution of wave 4 is an inversion of that of wave 3 about the pump frequency ω_1. This interesting property of spectral inversion is illustrated in Figure 14.12.

The spectral inversion about the pump frequency ω_1 can also be seen from the frequency relationship (14.4-3) among the waves,

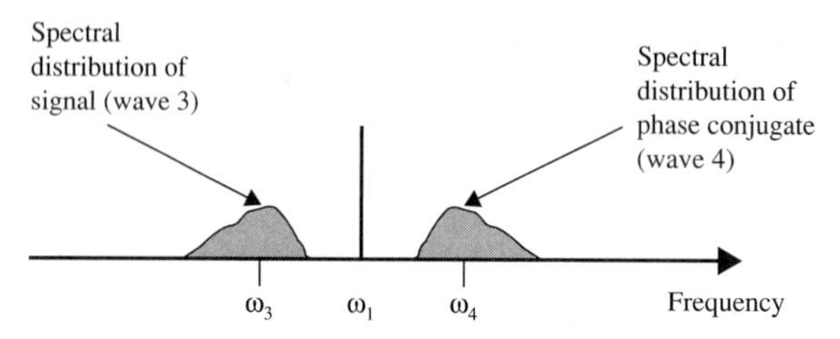

Figure 14.12 Schematic drawing of spectral inversion about the pump frequency (ω_1) in partially degenerate four-wave mixing (PDFWM) in a single mode fiber. The shaded regions indicate the normalized spectral distribution of waves 3 and 4, respectively.

$$\omega_4 = 2\omega_1 - \omega_3 \tag{14.4-19}$$

Let the signal wave (wave 3) be written in terms of a linear summation of all its spectral components,

$$p_3(t)\exp(i\omega_3 t) = \exp(i\omega_3 t)\int q_3(\Omega)e^{i\Omega t}\, d\Omega \tag{14.4-20}$$

where $q_3(\Omega)$ is the amplitude of the frequency component at $(\omega_3 + \Omega)$. Each frequency component $(\omega_3 + \Omega)$ of the signal wave (wave 3) will undergo PDFWM with the pump wave (wave 1) at frequency ω_1, producing a new wave at frequency $[2\omega_1 - (\omega_3 + \Omega)] = (\omega_4 - \Omega)$ with an amplitude of $q_3^*(\Omega)$, according to Equation (14.4-19). The new wave at $(\omega_4 - \Omega)$ is a frequency component of the phase conjugate wave (wave 4). We now sum up all frequency components of the newly generated waves. This is the phase conjugate wave (wave 4). Thus we can write

$$\begin{aligned} p_4(t)\exp(i\omega_4 t) &\equiv \exp(i\omega_4 t)\int q_4(\Omega)e^{i\Omega t}\, d\Omega \\ &= \exp(i\omega_4 t)\int q_3^*(\Omega)e^{-i\Omega t}\, d\Omega \end{aligned} \tag{14.4-21}$$

and consequently,

$$q_4(\Omega) = q_3^*(-\Omega) \tag{14.4-22}$$

where $q_4(\Omega)$ is the normalized amplitude of the frequency component of wave 4 at $(\omega_4 + \Omega)$. This result is consistent with Equation (14.4-18b), and the spectral inversion shown in Figure 14.12.

The unique property of spectral inversion in PDFWM can be employed for the restoration of pulses (or dispersion compensation) in optical communications links. As discussed in Chapter 7, an optical pulse will undergo pulse spreading and frequency chirping in single-mode fibers, which exhibit group velocity dispersion (GVD). As a result of the group velocity dispersion, different frequency components of an optical pulse travel at different speeds inside the fibers. In positive dispersion fibers ($D > 0$), the high-frequency components travel faster than the low-frequency components. This leads to the spreading and a negative chirping of the pulse. Here we define positive chirping for the case when the local frequency of an optical pulse increases with time.

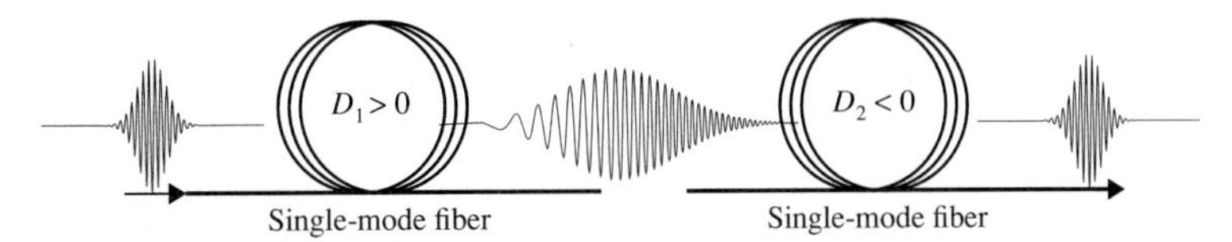

Figure 14.13 Schematic drawing of a conventional scheme of dispersion compensation in single-mode fibers. An input optical pulse without chirping is transmitted through a positive dispersion fiber with $D_1 > 0$. At the exit end of this fiber, the output pulse is broadened and chirped, with high-frequency components in the leading edge. The original pulse can be restored if the second fiber has a negative group velocity dispersion such that $D_1L_1 + D_2L_2 = 0$, where L_1 and L_2 are the lengths of the fibers.

In conventional dispersion compensation, an output pulse from a positive dispersion fiber ($D_1 > 0$) is made to propagate through a segment of negative dispersion fiber ($D_2 < 0$). In the first fiber, the high-frequency components of the pulse travel faster than those near the center frequency and thus appear in the leading edge, whereas the low-frequency components travel slower and appear in the trailing edge. When such a broadened and chirped pulse is transmitted through a negative dispersion fiber, the low-frequency components in the trailing edge can catch up with the high-frequency components, provided the dispersion–length product of the second fiber matches with that of the first fiber. In other words,

$$D_1L_1 + D_2L_2 = 0 \tag{14.4-23}$$

When the above condition is satisfied, the pulse spreading and chirping are eliminated. The concept is illustrated in Figure 14.13. The drawback of such a scheme of dispersion compensation is the requirement of a negative dispersion fiber, which often exhibits high loss of transmission.

The spectral inversion of PDFWM in single-mode fibers also produces a chirp reversal of an optical pulse. This is illustrated in Figure 14.14. A chirped optical pulse is fed into a single-mode fiber, which is pumped by a cw pump wave. If the single-mode fiber has zero dispersion, the pulse will remain undistorted as it propagates in the fiber. However, as a result of the spectral inversion, the high-frequency leading edge is shifted in frequency and becomes a low-frequency leading edge. Similarly, the low-frequency trailing edge becomes a high-frequency trailing edge. The net result is a chirp reversal.

Chirp reversal can be employed for elimination of the spreading and chirping of the optical pulse by using only positive dispersion fibers. The basic concept is illustrated in

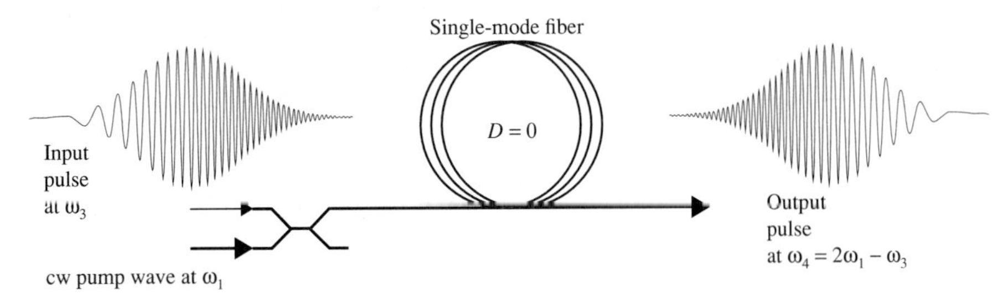

Figure 14.14 Schematic drawing of chirp reversal in partially degenerate four-wave mixing in a sigle-mode fiber with a cw pump wave at frequency ω_1.

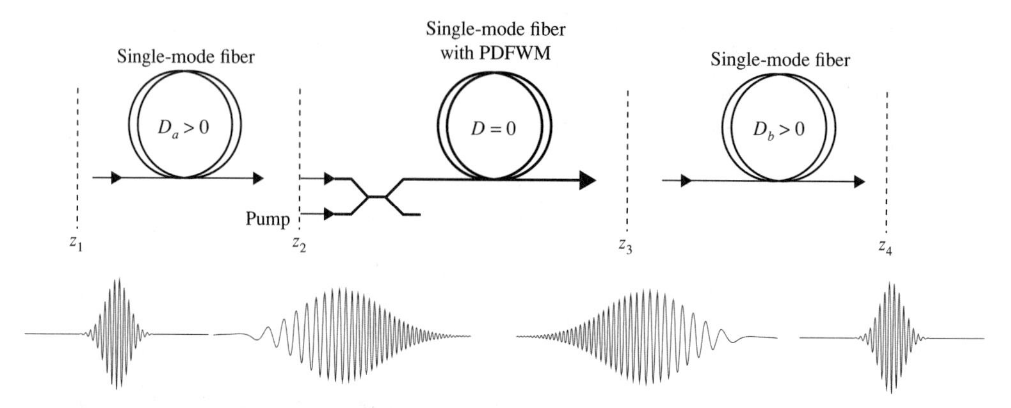

Figure 14.15 Schematic drawing of a scheme of dispersion compensation via PDFWM in single-mode fibers. The pulse profiles are examined at four locations: z_1, z_2, z_3, and z_4.

Figure 14.15 and involves two positive dispersion fibers with an interposing phase conjugator. At the input end z_1 of the first fiber (with $D_a > 0$), an unchirped pulse is launched into the fiber. The output pulse at location z_2 is broadened and chirped. The output pulse is now coupled into a phase conjugator, which consists of PDFWM in a single-mode fiber (with $D = 0$). As a result of the spectral inversion, the output pulse at location z_3 has a reversed chirp. This pulse is now transmitted through another positive dispersion fiber ($D_b > 0$) with a properly chosen length. The original pulse shape is restored at the final output location z_4, provided $D_aL_a = D_bL_b$, where L_a and L_b are the lengths of the fibers.

The dispersion compensation scheme as illustrated by Figure 14.15 can also be explained in terms of the Fourier spectra of the waves. Let us follow the evolution of an optical pulse envelope through the system and examine the Fourier transform of the optical pulse envelope at each of the four locations (z_1, z_2, z_3, z_4):

1. Wave 3 at $z = z_1$: $q_{31}(\Omega) = q_3(\Omega)$
2. Wave 3 at $z = z_2$: $q_{32}(\Omega) = q_{31}(\Omega)\exp(-i\beta_3 L_a)$
$$= q_3(\Omega)\exp(-i\beta_{30}L_a - i\Omega\tau_a - i\delta_a L_a\Omega^2) \tag{14.4-24}$$
3. Wave 4 at $z = z_3$: $q_{41}(\Omega) = q_{32}^*(-\Omega)$
4. Wave 4 at $z = z_4$: $q_{42}(\Omega) = q_{41}(\Omega)\exp(-i\beta_4 L_b) = q_{41}(\Omega)\exp(-i\beta_{40}L_b - i\Omega\tau_b - i\delta_b L_b\Omega^2)$

where L_a and L_b are the lengths of the fibers, β_3 and β_4 are the propagation constants of waves 3 and 4, respectively, $\beta_{30} = \beta_3(\Omega = 0)$, $\beta_{40} = \beta_4(\Omega = 0)$, and τ_a, τ_b, δ_a, and δ_b are given by

$$\begin{aligned}
\tau_a &= \left(\frac{\partial\beta_3}{\partial\omega}\right)_{\Omega=0} L_a = \frac{L_a}{v_g(\omega_3)} \\
\tau_b &= \left(\frac{\partial\beta_4}{\partial\omega}\right)_{\Omega=0} L_b = \frac{L_b}{v_g(\omega_4)} \\
\delta_a &= \frac{1}{2}\left(\frac{\partial^2\beta_3}{\partial\omega^2}\right)_{\Omega=0} \\
\delta_b &= \frac{1}{2}\left(\frac{\partial^2\beta_4}{\partial\omega^2}\right)_{\Omega=0}
\end{aligned} \tag{14.4-25}$$

Physically, τ_a and τ_b are the propagation delays in fibers a and b, respectively, and δ_a and δ_b are the group velocity dispersion (GVD) in fibers a and b, respectively. It follows that the Fourier transform of the output pulse can be written, according to Equations (14.4-24),

$$q_{42}(\Omega) = q_{41}(\Omega)\exp(-i\beta_{40}L_b - i\Omega\tau_b - i\delta_b L_b\Omega^2)$$
$$= q_3^*(-\Omega)\exp(+i\beta_{30}L_a - i\Omega\tau_a + i\delta_a L_a\Omega^2)\exp(-i\beta_{40}L_b - i\Omega\tau_b - i\delta_b L_b\Omega^2) \quad (14.4\text{-}26)$$

If we choose the two fibers such that

$$\delta_a L_a = \delta_b L_b \quad (14.4\text{-}27)$$

then

$$q_{42}(\Omega) = q_3^*(-\Omega)\exp(-i\phi - i\Omega\tau) \quad (14.4\text{-}28)$$

where ϕ is a constant phase, and $\tau = \tau_a + \tau_b$ is the total group delay. We now examine the temporal profile at the output of the system,

$$p_4(t) = \int q_{42}(\Omega)e^{i\Omega t}d\Omega = \int q_3^*(-\Omega)e^{-i\phi - i\Omega\tau}e^{i\Omega t}d\Omega = e^{-i\phi}p_3^*(t-\tau) \quad (14.4\text{-}29)$$

The (squared magnitude) output envelope is thus identical to that of the input, except with a group delay of τ. The dispersion compensation scheme using PDFWM in single-mode fibers described earlier was first proposed in 1978 [3]. An experimental demonstration of such a dispersion compensation scheme using PDFWM in a dispersion-shifted fiber (with a zero dispersion at $\lambda_0 = 1549$ nm) was carried out in 1993 [4].

14.5 OPTICAL SOLITONS

We have demonstrated that self-phase modulation (SPM) due to the optical Kerr effect in single-mode fibers can lead to frequency chirping. The chromatic dispersion in single-mode fiber can also lead to frequency chirping (and pulse broadening). In 1973 Hasegawa [5] proposed that the balance between the pulse broadening due to self-phase modulation and the compression due to group velocity dispersion (GVD) in a low-loss fiber can lead to the evolution of self-sustaining propagating pulses, the so-called optical solitons. As a result of the Kerr effect described by the relation $n = n_0 + n_2 I$, an optical pulse experiences a self-phase modulation as it propagates in the nonlinear medium. The self-phase modulation in conjunction with the intensity variation of the pulse leads to a frequency chirping with the low-frequency components in the leading edge of the pulse. If the medium also exhibits group velocity dispersion characterized by the parameter $\beta'' = (\partial^2\beta/\partial^2\omega)$, and as described in Chapter 7, then it is possible that the frequency-chirped pulse may be compressed under the appropriate conditions. As we know, the group velocity dispersion alone will cause transform-limited pulses to broaden. The combination of self-phase modulation due to the Kerr effect and the group velocity dispersion can lead to self-sustaining pulses (the solitons) provided the two effects balance out. It is interesting to note that solitons are not limited to optics. As a matter of historical interest, a solitary water wave was observed and first described by John Scott Russel in a barge canal in Great Britain in 1834 [6].

The Mathematical Description of Solitons

To illustrate qualitatively the basic physics of this phenomenon, we consider first in Figure 14.16 what happens to a transform-limited optical pulse as it propagates in a fiber whose index of refraction depends on the field intensity I,

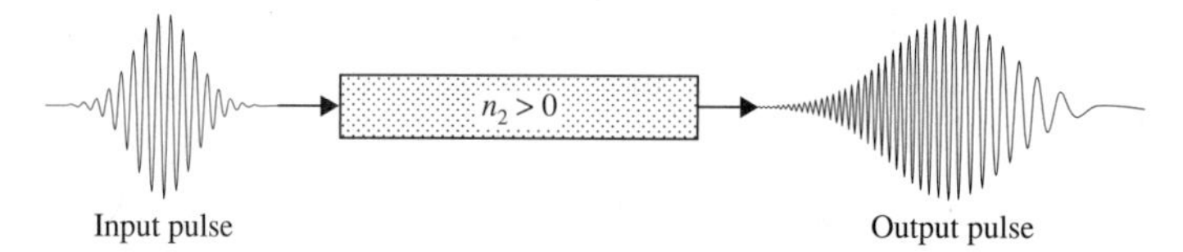

Figure 14.16 Spatial profiles of input and output pulses. The input pulse is unchirped. The output pulse is chirped due to self-phase modulation in a nonlinear medium with $n_2 > 0$.

$$n = n_0 + n_2 I \tag{14.5-1}$$

where n_0 is the refractive index at zero intensity, and n_2 is the Kerr coefficient. To be specific, we will assume a Gaussian pulse

$$E_{\text{in}} = A_0 \exp\left(i\omega_0 t - \frac{\alpha t^2}{2}\right) \tag{14.5-2}$$

where A_0 is a constant that does not play a role in this discussion and α is also a constant. After a short propagation distance, ΔL, the pulse will emerge as

$$E_{\text{out}} = E_{\text{in}} \exp(-ik\,\Delta L) = A_0 \exp\left[i\left(\omega_0 t - \frac{\omega_0}{c} n(I)\,\Delta L\right)\right] \exp\left(-\frac{\alpha t^2}{2}\right) \tag{14.5-3}$$

where c is the speed of light in vacuum.

Its output phase is thus

$$\phi(\Delta L, t) = \omega_0 t - \frac{\omega_0}{c}[n_0 + n_2 I_0 \exp(-\alpha t^2)]\,\Delta L \tag{14.5-4}$$

while the instantaneous frequency (or local frequency) is

$$\omega(\Delta L, t) = \frac{d\phi}{dt} = \omega_0 + 2\frac{\omega_0}{c} n_2 I_0\,\Delta L\,\alpha t \exp(-\alpha t^2) \tag{14.5-5}$$

where I_0 is the instantaneous intensity at the center of the pulse ($t = 0$).

The local frequency is plotted in Figure 14.16 for the case $n_2 > 0$. Over the central, and most important portion of the pulse, it is chirped positively, that is, $d\omega/dt > 0$. If we examine the chirped pulse in the space domain, we note that the leading edge ($t < 0$) consists of low-frequency components of the wave, whereas the trailing edge ($t > 0$) consists of high-frequency components of the wave. It follows directly that if a pulse with, say, a positive chirp, $d\omega/dt > 0$, enters a linear dispersive fiber with

$$\beta'' \equiv \frac{d^2\beta}{d\omega^2} = \left(-\frac{1}{v_g^2}\frac{dv_g}{d\omega}\right) < 0 \quad (\text{or } D > 0)$$

it will narrow with propagation distance since the late arriving high-frequency components will be sped up during their transit relative to the lower frequencies. The position along the fiber where all frequency components "catch up" with each other is where the minimum pulse width occurs. Beyond this point the pulse will be broadened.

Imagine a hypothetical fiber made up of alternating short segments, half of which have $\beta'' < 0$ and $n_2 = 0$, while in the remainder $\beta'' = 0$ and $n_2 > 0$. A section of such a fiber with three segments is shown in Figure 14.17. We start with a transform-limited pulse in plane 1, which by plane 2 becomes broadened and chirped due to the dispersion $\beta'' \neq 0$, as discussed in Chapter 7. Propagation in the nonlinear fiber ($n_2 > 0$) results in a reversal of the sign of the

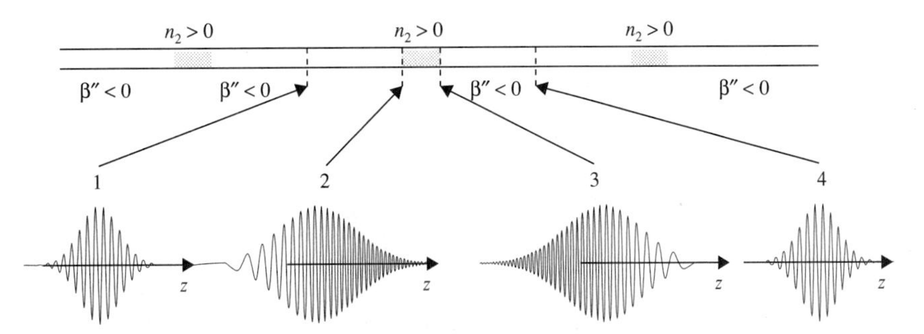

Figure 14.17 A representation of the soliton physics in which the simultaneous effects of linear dispersion due to β'' and nonlinear chirping due to n_2 are shown, for the sake of illustration as operating on the propagating pulse in succession. Spatial profiles of the optical pulse at four points of one unit cell of the periodic fiber are shown.

chirp at plane 3 without a change in the pulse length. The relative slowing down of the low frequencies at the leading edge occurring in the last segment of the dispersive fiber in the unit cell ($\beta'' < 0$) gives rise to a pulse with the original shape and zero chirp at 4. In an actual fiber, both effects (i.e., n_2 and β'') exist simultaneously. This results in steady-state nonspreading pulses—solitons.

To find the properties of the soliton, we will derive first the equation governing propagation in a dispersive nonlinear medium.

The Nonlinear Schrödinger Equation

We are particularly interested in the evolution of the pulse shape as it propagates in the nonlinear and dispersive medium. Consider the propagation of an optical pulse along the z axis. The electric field of the pulse can be written

$$E(z, t) = \int_{-\infty}^{\infty} S(\omega) e^{i(\omega t - \beta z)} d\omega \tag{14.5-6}$$

where $S(\omega)$ is the Fourier spectrum of the pulse and β is the propagation constant. In a dispersive medium, the propagation constant can be written

$$\beta(\omega) = \beta_0 + \beta'(\omega - \omega_0) + \tfrac{1}{2}\beta''(\omega - \omega_0)^2 + \cdots \tag{14.5-7}$$

where ω_0 is a reference frequency, which is often taken as the carrier frequency of the pulse, and

$$\beta_0 = \beta(\omega_0), \quad \beta' = \left(\frac{\partial \beta}{\partial \omega}\right)_{\omega=\omega_0}, \quad \beta'' = \left(\frac{\partial^2 \beta}{\partial \omega^2}\right)_{\omega=\omega_0} \tag{14.5-8}$$

Using Equation (14.5-7), the electric field (14.5-6) can be written

$$E(z, t) = e^{i(\omega_0 t - \beta_0 z)} \int_{-\infty}^{\infty} S(\omega_0 + \Omega) e^{i(\Omega t - \beta' \Omega z - \beta'' \Omega^2 z/2)} d\Omega \tag{14.5-9}$$

where Ω is the frequency measured from the reference frequency ω_0,

$$\Omega = \omega - \omega_0 \tag{14.5-10}$$

If we write

$$E(z, t) = A(z, t)e^{i(\omega_0 t - \beta_0 z)} \tag{14.5-11}$$

then the amplitude (or the envelope function) of the pulse can be written

$$A(z, t) = \int_{-\infty}^{\infty} S(\omega_0 + \Omega)e^{i(\Omega t - \beta'\Omega z - \beta''\Omega^2 z/2)}d\Omega \tag{14.5-12}$$

From this integral form, we can derive a differential equation for the evolution (or propagation) of the amplitude as follows. Taking partial derivatives in Equation (14.5-12), we obtain

$$\frac{\partial}{\partial t}A(z, t) = i\int_{-\infty}^{\infty} \Omega S(\omega_0 + \Omega)e^{i(\Omega t - \beta'\Omega z - \beta''\Omega^2 z/2)}d\Omega$$

$$\frac{\partial^2}{\partial t^2}A(z, t) = -\int_{-\infty}^{\infty} \Omega^2 S(\omega_0 + \Omega)e^{i(\Omega t - \beta'\Omega z - \beta''\Omega^2 z/2)}d\Omega \tag{14.5-13}$$

$$\frac{\partial}{\partial z}A(z, t) = -i\int_{-\infty}^{\infty} (\beta'\Omega + \beta''\Omega^2/2)S(\omega_0 + \Omega)e^{i(\Omega t - \beta'\Omega z - \beta''\Omega^2 z/2)}d\Omega$$

According to Equations (14.5-13), the envelope wave equation for $A(z, t)$ in linear media can be written

$$\frac{\partial A}{\partial z} + \frac{1}{v_g}\frac{\partial A}{\partial t} - i\frac{\beta''}{2}\frac{\partial^2 A}{\partial t^2} = 0 \tag{14.5-14}$$

where v_g is the group velocity given by

$$v_g = \frac{1}{\beta'} = \left(\frac{\partial\omega}{\partial\beta}\right)_{\omega=\omega_0} \tag{14.5-15}$$

If the power level is high enough, then we must consider the correction to the propagation constant β due to the Kerr effect. This can be done by adding a correction term to the propagation constant in Equations (14.5-9) or (14.5-12). We obtain

$$A(z, t) = \int_{-\infty}^{\infty} S(\omega_0 + \Omega)e^{i(\Omega t - \beta'\Omega z - \beta''\Omega^2 z/2)}e^{-i\beta_2 I z}d\Omega \tag{14.5-16}$$

where I is the local intensity of the wave given by

$$I = \tfrac{1}{2}c\varepsilon_0 n_0 |A(z, t)|^2 \tag{14.5-17}$$

and β_2 is given by

$$\beta_2 = \frac{\omega_0}{c}n_2 \tag{14.5-18}$$

As a result of this correction term, the spatial derivative of the amplitude becomes, approximately,

$$\frac{\partial}{\partial z}A(z, t) = -i\int_{-\infty}^{\infty} (\beta_2 I + \beta'\Omega + \beta''\Omega^2/2)S(\omega_0 + \Omega)e^{i(\Omega t - \beta'\Omega z - \beta''\Omega^2 z/2)}d\Omega \tag{14.5-19}$$

The equation for the field amplitude (envelope function), including the Kerr effect, thus becomes, according to Equations (14.5-17) and (14.5-19),

$$\frac{\partial A}{\partial z} + \frac{1}{v_g}\frac{\partial A}{\partial t} - i\frac{\beta''}{2}\frac{\partial^2 A}{\partial t^2} = -i\tfrac{1}{2}\omega_0\varepsilon_0 n_0 n_2 |A|^2 A \tag{14.5-20}$$

We find it useful to transform to a coordinate system moving at the group velocity v_g. Thus we define

$$\tau = t - \frac{z}{v_g} \tag{14.5-21}$$

and rewrite Equation (14.5-20) as

$$-i\frac{\partial A}{\partial z} - \frac{\beta''}{2}\frac{\partial^2 A}{\partial \tau^2} + \frac{1}{2}\omega_0\varepsilon_0 n_0 n_2 |A|^2 A = 0 \tag{14.5-22}$$

This equation governs the evolution of the optical pulse in a nonlinear dispersive medium. Equation (14.5-22), known as the nonlinear Schrödinger equation, is the main result. A solution (the fundamental solution) of Equation (14.5-22) in the case of $\beta'' < 0$ is

$$A(z, \tau) = A_0 \frac{e^{-i\gamma z}}{\cosh(\tau/\tau_0)} = A_0 \operatorname{sech}(\tau/\tau_0)\exp(-i\gamma z) \tag{14.5-23}$$

where A_0 is a real constant, and τ_0 and γ, are constants given by

$$\begin{aligned} \frac{1}{\tau_0} &= \left(\frac{\omega_0\varepsilon_0 n_0 n_2}{-2\beta''}\right)^{1/2} A_0 \\ \gamma &= -\frac{\beta''}{2\tau_0^2} = \frac{\omega_0\varepsilon_0 n_0 n_2}{4} A_0^2 \end{aligned} \tag{14.5-24}$$

We note that both constants τ_0 and γ are positive in the case of $\beta'' < 0$ and $n_2 > 0$. In terms of the original (z, t) variables, the fundamental soliton solution is

$$A(z, t) = A_0 \operatorname{sech}\left(\frac{t - z/v_g}{\tau_0}\right)\exp(-i\gamma z) \tag{14.5-25}$$

The intensity pattern of the fundamental soliton is illustrated in Figure 14.18. The full width at half maximum of the intensity pattern is

$$\tau_{FWHM} = 1.76\tau_0 \tag{14.5-26}$$

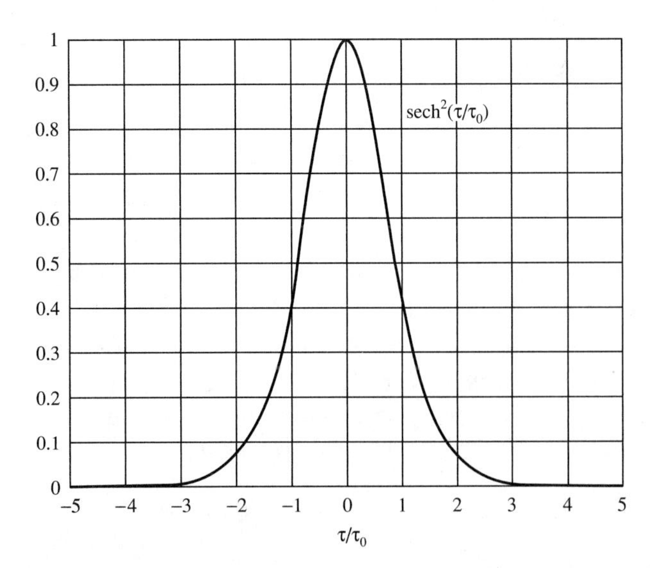

Figure 14.18 The intensity pattern of the fundamental soliton.

We may view the soliton width τ_0 as an independent parameter that uniquely characterizes the soliton in a fiber with given n_2 and β''. It follows from (14.5-24) that for a given τ_0 the two remaining constants of $A(z, t)$, A_0 and γ, are determined. It is important to note that the amplitude A_0 is not an arbitrary constant. Recall that the pulse energy is given by

$$U = \tfrac{1}{2} c n_0 \varepsilon_0 A_{\text{eff}} \int |E(z, t)|^2 \, d\tau \tag{14.5-27}$$

where A_{eff} is the effective area of the mode of propagation in the fiber.

Using $\int_{-\infty}^{\infty} \text{sech}^2(x)\, dx = 2$, we can write the pulse energy as

$$U = \tfrac{1}{2} c n_0 \varepsilon_0 A_{\text{eff}} \tau_0 A_0^2 \tag{14.5-28}$$

or equivalently, according to Equations (14.5-24),

$$U = \frac{-\beta'' c}{\tau_0 \omega_0 n_2} A_{\text{eff}} \tag{14.5-29}$$

which is positive for $\beta'' < 0$. Employing the commonly used dispersion parameter D (in units of ps/nm-km), according to Equation (7.1-17),

$$D = -\frac{2\pi c}{\lambda_0^2} \beta'' \tag{14.5-30}$$

the pulse energy can be written

$$U = \frac{\lambda_0^3 D}{4\pi^2 c \tau_0 n_2} A_{\text{eff}} \tag{14.5-31}$$

We note that the energy of the fundamental soliton is determined as soon as the pulse width is fixed. The pulse energy is inversely proportional to the pulse width so that lower-energy soliton pulses are broader. This causes pulse broadening due to attenuation in long fibers, a problem that can be remedied with the use of optical amplifiers. We note that, in a given fiber (fixed β'' and n_2), the peak electric field A_0 of the fundamental soliton is fixed by the relation (14.5-24). Since the field is limited essentially to the time duration within the pulse, the peak power in a soliton is

$$P_{\text{soliton}} = \frac{U}{\tau_{\text{FWHM}}} = \frac{\lambda_0^3 D}{4\pi^2 c \tau_0 \tau_{\text{FWHM}} n_2} A_{\text{eff}} \tag{14.5-32}$$

The numerical example that follows will help us get some appreciation of the soliton properties and the range of soliton powers.

EXAMPLE: OPTICAL SOLITONS IN SILICA FIBERS

We will use the following data: $\lambda_0 = 1.55 \times 10^{-6}$ m, $n_0 = 1.45$, $n_2 = 3 \times 10^{-20}$ m^2/W, $D =$ 18 ps/nm-km, the fiber effective area of $A_{\text{eff}} = 100\ \mu\text{m}^2$. We will assume a data transmission rate of 10^{10} bits/s and a corresponding $\tau_{\text{FWHM}} = 3 \times 10^{-11}$ s, $\tau_0 = \tau_{\text{FWHM}}/1.76 = 1.76 \times 10^{-11}$ s. The pulse energy, according to Equation (14.5-31) is given by

$$U = \frac{\lambda_0^3 D}{4\pi^2 c \tau_0 n_2} A_{\text{eff}} = 1.1 \times 10^{-12} \text{ joule}$$

The peak pulse power in the fiber soliton mode is thus

$$P_{\text{soliton}} = \frac{U}{\tau_{\text{FWHM}}} = \frac{\lambda_0^3 D}{4\pi^2 c \tau_0 \tau_{\text{FWHM}} n_2} A_{\text{eff}} = 0.036 \text{ watt}$$

We note that if we wish to communicate at a higher bit rate, that is, $f_{\text{bit}} > 10^{10}$ bit/s, we need to use smaller values of τ_0. This will increase the peak power, which is proportional to τ_0^{-2}. At power levels exceeding, say, 100 mW in our fiber, the corresponding optical intensity, in a typical fiber, will be near 10^5 W/cm^2. This intensity is sufficiently elevated so that nonlinear effects such as Raman and four-wave mixing become important and lead to degradation [2].

Much of the recent interest in solitons for optical fiber communications is a result of a series of experimental demonstrations by L. Mollenauer [7–14], which have established its viability for long-distance transmission (thousands of kilometers).

High-Order Solitons

The nonlinear Schrödinger (NLS) equation also supports high-order modes of propagation. For an input pulse of the form

$$A(z = 0, \tau) = NA_0 \operatorname{sech}(\tau/\tau_0) \tag{14.5-33}$$

where N is an integer known as the order of the soliton, the analytical solutions of the NLS equation for $N = 1, 2$ are given by [15–17]

$$A_1(z, \tau) = A_0 \frac{1}{\cosh(\tau/\tau_0)} \exp(-i\gamma z) \tag{14.5-34}$$

$$A_2(z, \tau) = A_0 \frac{4[\cosh(3\tau/\tau_0) + 3e^{-i8\gamma z} \cosh(\tau/\tau_0)]}{[\cosh(4\tau/\tau_0) + 4\cosh(2\tau/\tau_0) + 3\cos(8\gamma z)]} \exp(-i\gamma z) \tag{14.5-35}$$

The first-order solution (the fundamental soliton) given by Equation (14.5-34) has a temporal intensity profile of $\operatorname{sech}^2(\tau/\tau_0)$, which is independent of position z in the fiber. This means the fundamental soliton keeps its initial pulse shape during propagation throughout the fiber. However, the pulse shape of the second-order soliton ($N = 2$) is not independent of position z in the fiber. In fact, the pulse shape is a periodic function of position z in the fiber. The period of oscillation, which is the same for all solitons, is given by $\gamma z = \pi/4$ or, equivalently,

$$z_{\text{period}} = \frac{\pi}{4\gamma} = -\frac{\pi\tau_0^2}{2\beta''} \tag{14.5-36}$$

where we note that $\beta'' < 0$. The period can also be written in terms of the dispersion parameter D (ps/nm-km):

$$z_{\text{period}} = \frac{\pi^2 c \tau_0^2}{D\lambda_0^2} \tag{14.5-37}$$

For the example of a soliton in silica fiber given above, the soliton period is $z_{\text{period}} = 21$ km (with a FWHM pulse width of 30 ps). The soliton period is proportional to the square of the pulse width. As a result, the period becomes very small for subpicosecond solitons. The

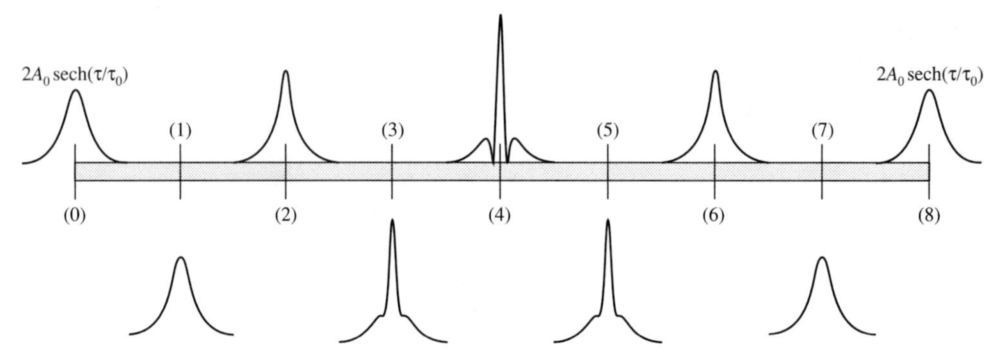

Figure 14.19 Temporal pulse profiles $|A_2|$ of $N = 2$ soliton at various locations in a fiber: (0) $\gamma z = 0$, (1) $\gamma z = \pi/32$, (2) $\gamma z = 2\pi/32$, (3) $\gamma z = 3\pi/32$, (4) $\gamma z = 4\pi/32$, (5) $\gamma z = 5\pi/32$, (6) $\gamma z = 6\pi/32$, (7) $\gamma z = 7\pi/32$, (8) $\gamma z = 8\pi/32$.

evolution of higher-order solitons is symmetric around $\gamma z = \pi/8$. For high-order modes with $N > 2$, analytic solutions are not available. Figure 14.19 is a plot of the pulse shape of $N = 2$ soliton at various positions in the fiber. We note that the pulse shape undergoes periodic compression and broadening as the pulse propagates inside the fiber.

PROBLEMS

14.1 Using Equation (14.1-21),

$$\Delta\omega = \omega'(t) - \omega = \frac{2\pi}{\lambda} n_2 L \frac{2t}{\tau^2} I_0 \exp\left(-\frac{t^2}{\tau^2}\right)$$

(a) Sketch $\Delta\omega$ as a function of time, and find the maximum of $\Delta\omega$.

(b) Estimate the spectral broadening due to self-phase modulation and show that

$$\Delta\omega_{\text{broadening}} = \frac{8\pi}{\sqrt{2e}} n_2 \frac{I_0 L}{\tau\lambda}$$

where L is the length of propagation and $e = 2.71828$.

14.2 Show that the coupled equations (14.3-18) for nondegenerate four-wave mixing are consistent with the conservation of energy,

$$\frac{d}{dz}(|A_1|^2 + |A_2|^2 + |A_3|^2 + |A_4|^2) = 0$$

provided $\alpha = 0, f_1 = f_2 = f_3 = f_4 = f$, and the frequencies satisfy Equation (14.3-3).

14.3 In a medium where the dispersive (β'') effects can be neglected, the envelope equation (14.5-20) becomes

$$\frac{\partial A}{\partial z} + \frac{1}{v_g}\frac{\partial A}{\partial t} = -i\tfrac{1}{2}\omega_0\varepsilon_0 n_0 n_2 |A|^2 A$$

Show that for short distances, a pulse with an input envelope

$$A(t) = A_0 \exp(-\alpha t^2)$$

propagating in a nondispersive ($\beta'' = 0$) fiber becomes

$$A(z,t) = A\left(t - \frac{z}{v_g}\right)\exp\left(-i\frac{\omega_0\varepsilon_0 n_0 n_2 z}{2}|A(z,t)|^2\right) \qquad (14.\text{P-1})$$

14.4 Show that for a Gaussian optical pulse with an envelope $A(t) = A_0 \exp(-\alpha t^2)$, the intensity (in W/m^2) can be written

$$I(t) = \frac{|A(t)|^2}{2\eta} = \tfrac{1}{2}c\varepsilon_0 n_0 |A(t)|^2 = I_0 \exp(-2\alpha t^2),$$
$$I_0 = \frac{A_0^2}{2\eta} = \tfrac{1}{2}c\varepsilon_0 n_0 A_0^2 \qquad (14.\text{P-2})$$

Show that the propagation in a dispersive ($\beta'' \neq 0$) and nonlinear ($n_2 \neq 0$) fiber results in a chirped output pulse whose envelope near the peak can be approximated by

$$A(z,t) \cong A\left(t - \frac{z}{v_g}\right)\exp\left[i\frac{2\omega_0 n_2 \alpha I_0 z}{c}\left(t - \frac{z}{v_g}\right)^2\right]$$

where an uninteresting time-independent phase shift was left out.

14.5 Consider the propagation of a pulse through a periodic fiber such as shown in Figure 14.17. Assume that in plane 1 the pulse envelope is a chirpless Gaussian pulse $A(t) = A_0 \exp(-\alpha t^2)$. Show, using the results of Section 7.2 [Equation (7.2-13) is especially relevant] and Problem 14.2, that the chirping of the pulse due to propagation between planes 1 and 2 can be reversed between planes 2 and 3 so that the pulse returns to its original envelope at 4 as shown in the figure. Derive the conditions necessary for the pulse envelope to repeat itself between 1 and 4. [Hint: You should find that the condition for phase reversal and pulse restoration between planes 1 and 4 is

$$I_0 = \frac{\lambda_0 |\beta''| \alpha}{\pi |n_2|} \left(\frac{L}{L_{NL}} \right) \sqrt{1 + (2\beta'' L\alpha)^2} \tag{3}$$

where $2L$ is the length of the dispersive fiber segments (Figure 14.17 shows a unit cell that contains two dispersive sections of length L straddling a nonlinear fiber of length L_{NL})].

14.6 Show that in the case of $L = (L_{NL})$ and in the limit of $4|\beta''|^2\alpha^2L^2 \ll 1$, expression (14.P-3) agrees with the results (14.5-24), if we take $2\alpha = 1/\tau_0^2$.

14.7 Consider a sech pulse with a temporal profile $A(t) = A_0 \operatorname{sech}(t/\tau_0)$. Using the following integral identity,

$$\int_{-\infty}^{\infty} \operatorname{sech}(\alpha t) e^{-i\omega t} dt = \frac{2\pi}{\alpha} \operatorname{sech}\left(\frac{\pi}{2\alpha} \omega \right)$$

show that

$$\Delta t_{1/2}\, \Delta\omega_{1/2} = 1.97$$

where $\Delta t_{1/2}$ and $\Delta\omega_{1/2}$ are full width at half-maximum of the temporal profile and the spectrum, respectively. The Fourier transform of a sech pulse has a sech profile.

14.8 Consider the nonlinear polarizations given by Equations (14.2-16) and (14.2-17). For isotropic media, the direction of the nonlinear polarization should be parallel to the electric field vector.

(a) Show that

$$4\chi_{1111}E_xE_x + 4(\chi_{1122} + \chi_{1212} + \chi_{1221})E_yE_y$$
$$= 4\chi_{2222}E_yE_y + 4(\chi_{2211} + \chi_{2121} + \chi_{2112})E_xE_x$$

(b) Using the symmetry between x and y, show that

$$\chi_{1122} = \chi_{2211}, \quad \chi_{1212} = \chi_{2121}, \quad \chi_{1221} = \chi_{2112}$$

(c) Show that

$$\chi_{1111} = \chi_{1122} + \chi_{1212} + \chi_{1221} \quad \text{and} \quad \chi_{1111} = \chi_{2222}$$

[Hint: The equation in (a) must be valid for all E_x and E_y.]

14.9

(a) Derive Equations (14.3-11).
(b) Derive Equations (14.3-18).
(c) Derive Equations (14.3-24), (14.3-25), and (14.3-29).
(d) Derive the solutions (14.3-32).
(e) Derive the optimum coupling length when $\alpha \ll |\Delta\beta|$, Equation (14.3-41).

14.10

(a) Derive Equations (14.4-4).
(b) Derive the solutions (14.4-7).
(c) Derive Equations (14.4-12) and (14.4-14).
(d) Derive Equation (14.4-17).

14.11 Show that Equation (14.5-35) reduces to (14.5-33) at $z = 0$.

REFERENCES

1 Wu, W., P. Yeh, and S. Chi, Phase conjugation by four-wave mixing in single mode fibers. *IEEE Photonics Tech. Lett.* **6**(12):1448 (1994).

2 Agrawal, G. P., *Nonlinear Fiber Optics*. Academic Press, London, 1989.

3 Yariv, A., D. Fekete, and D. M. Pepper, Compensation for channel dispersion by nonlinear optical phase conjugation. *Opt. Lett.* **4**:52 (1979).

4 Watanabe, S., T. Naito, and T. Chikama, Compensation of chromatic dispersion in a single mode fiber by optical phase conjugation. *IEEE Photonics Technol. Lett.* **5**:92 (1993).

5 Hasegawa, A., and F. D. Tappert, Transmission of stationary nonlinear optical pulses in dispersive dielectric fibers. I. Anomalous dispersion. *Appl. Phys. Lett.* **23**:142 (1984).

6 Dodd, R. K., J. C. Eilbeck, J. D. Gibbon, and H. C. Morris, *Solitons and Nonlinear Wave Equations*. Academic Press, London, 1982.

7 Mollenauer, L. F., R. H. Stolen, and J. P. Gordon, *Phys. Rev. Lett.* **45**:1095 (1980).

8 Mollenauer, L. F., and R. H. Stolen, *Opt. Lett.* **9**:13 (1984).

9 Mollenauer, L. F., and J. P. Gordon, Long-distance, high-bit-rate transmission using solitons in optical fibers. In: *The Froehlic/Kent Encyclopedia of Telecommunication.* Marcel Dekker, New York, 1995, p. 329. Excellent semipopular review article.

10 Gordon, J. P., and H. A. Haus, Random walk of coherently amplified solitons in optical fiber. *Opt. Lett.* **11**:665 (1986).

11 Mecozzi, A., J. D. Morres, H. A. Haus, and Y. Lai, Solition transmission control. *Opt. Lett.* **16**:1841 (1991).

12 Kodama, Y., and A. Hasegawa, Generation of asymptotically stable optical solitons and suppression of the Gordon–Haus effect. *Opt. Lett.* **17**:33 (1992).

13 Mollenauer, L. F., J. P. Gordon, and S. G. Evangelides, The sliding frequency guiding filter: an improved form of soliton jitter control. *Opt. Lett.* **17**:1675 (1992).

14 Mollenauer, L. F., E. Lichtman, M. J. Neubelt, and G. T. Harvey, Demonstration using sliding-frequency guiding filters of error-free soliton transmission over more than 20 Mm at 10 Gbit/s single channel and over more than 13 Mm at 20 Gbit/s in a two channel WDM. *Electtron. Lett.* **29**:910 (1993).

15 Zakharov, V. E., and A. B. Schabat, *Sov. Phys. JETP* **34**:62 (1972).

16 Satsuma, J., and N. Yajima, *Prog. Theor. Phys. Suppl.* **N55**:284 (1974).

17 Ozyazici, M. S., and M. Sayin, Effect of loss and pulse width variation on soliton propagation. *J. Optoelectron. Adv. Mater.* **5**:(2): 447 (2003).

CHAPTER 15

SEMICONDUCTOR LASERS—THEORY AND APPLICATIONS

15.0 INTRODUCTION

We have discussed the basic principles of operation of lasers, using discrete quantum states of atomic systems for the sake of simplicity in illustrating the concept. Examples of these lasers include argon ion lasers and ruby lasers. The laser emission originates from the stimulated transition from a higher quantum state to a lower quantum state in an atomic system with a population inversion. In this chapter we will discuss an important class of lasers, the semiconductor lasers. In this class of lasers, the coherent optical emission originates from the stimulated transition from a higher energy band (conduction band) to a lower energy band (valence band) in a semiconductor.

The semiconductor laser invented in 1961 [1–3] is the first laser to make the transition from a research topic and specialized applications to the mass consumer market. The semiconductor laser is now widely used in many household products (e.g., CD players) and optical communications networks. It is, by economic standards and the degree of its applications, the most important of all lasers. The main features that distinguish the semiconductor laser are:

1. Small physical size (300 μm × 10 μm × 50 μm) that enables it to be incorporated easily into other instruments.
2. Its direct pumping by low-power electric current (15 mA at 2 V is typical), which makes it possible to drive it with conventional transistor circuitry.
3. Its efficiency in converting electric power to light. Actual operating efficiencies exceed 50%.
4. The ability to modulate its output by direct modulation of the pumping current at rates exceeding 20 GHz. This is of major importance in high-data-rate optical communications systems.
5. The possibility of integrating it *monolithically* with electronic field effect transistors, microwave oscillators, bipolar transistors, and optical components in III–V semiconductors to form integrated optoelectronic circuits.
6. The semiconductor-based manufacturing technology, which lends itself to mass production.

7. The compatibility of its output beam dimensions with those of typical silica-based optical fibers and the possibility of tailoring its output wavelength to the low-loss ($\lambda = 1.5$ μm), low-dispersion ($\lambda = 1.3$ μm) region of such fibers.
8. The possibility of tailing its output wavelength to the blue spectral regime for display and the readout of high-density optical storage.

From the pedagogic point of view, understanding how a modern semiconductor laser works requires, in addition to the basic theory of the interaction of radiation with electrons that was developed in Chapter 5, an understanding of dielectric waveguiding [4, 5] (Chapter 3) and elements of solid-state theory of semiconductors [6, 7]. The latter theory will be taken up in the next few sections.

15.1 SOME SEMICONDUCTOR PHYSICS BACKGROUND

In this section we will briefly develop some of the basic background material needed to understand semiconductor lasers. The student is urged to study the subject in more detail, using any of the numerous texts dealing with the wave mechanics of solids (e.g., Reference [6]). The main difference between electrons in semiconductors and electrons in other laser media is that in semiconductors all the electrons occupy, thus share, the whole crystal volume, while in a conventional laser medium, ruby, for example, the Cr^{3+} electrons are localized to within 1 or 2 Å of their parent Cr^{3+} ion. For a typical Cr doping level of 2×10^{19} cm^{-3}, the average separation between neighboring Cr^{3+} ions is about 20 Å. The electrons on a given ion do not communicate with those on other ions. In other words, the electrons are very well localized and the electronic energy states are discrete.

In a semiconductor, on the other hand, because of the proximity of the neighboring atoms, the electrons are no longer localized. Consider the case of silicon crystals. When the Si atoms are infinitely apart, the electron states within each individual Si atom are discrete, and the wavefunctions of the electrons are well localized around the nucleus of the Si atom. As the Si atoms are brought closer as in the case of silicon crystals, the electron states are broadened into electron energy bands (see Figure 15.1). Depending on the width of the energy bands, gaps may exist between the energy bands. These gaps are usually smaller than the original separation between the neighboring energy levels of the individual atoms.

In addition to the energy bands, the wavefunctions of the electrons are no longer localized as the atoms are brought closer to form a solid crystal. As a matter of fact, the spread of the wavefunctions encompasses the whole crystal. As a result of the physical overlap of the wavefunctions, no two electrons in a crystal can be placed in the same quantum state, that is, possess the same eigenfunction. This is the *Pauli exclusion principle*, which is one of the more important axiomatic foundations of quantum mechanics. Each electron thus must possess a unique spatial wavefunction and an associated eigenenergy (the total energy associated with the state). If we plot a horizontal line, as in Figure 15.1, *for* each allowed electron energy (eigenenergy), we will discover that the energy levels cluster within bands that are separated by "energy gaps" ("forbidden" gaps). These energy bands collapse into discrete energy levels when the atoms are separated infinitely apart. A schematic description of the energy level spectrum of electrons in a crystal is shown in Figure 15.1. Table 15.1 lists some semiconductor materials and their properties.

The manner in which the available energy states are occupied determines the conduction properties of the crystal. If the electrons exactly fill one or more energy bands while the next

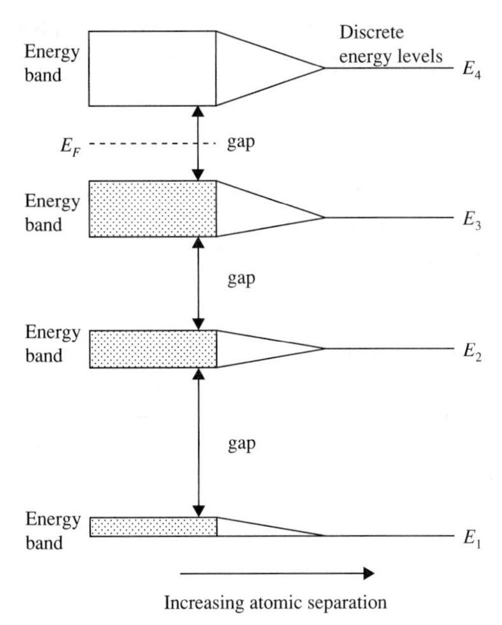

Figure 15.1 The energy bands of electrons in a crystal. The discrete energy states are broadened into energy bands as atoms are brought closer to form a solid crystal. In a given material these bands are usually occupied, in the ground state, up to some uppermost level. The energy E_F that marks, in the limit of $T \to 0$, the transition from fully occupied electron states ($E < E_F$) to empty states ($E > E_F$) is called the *Fermi energy*. It does not, except accidentally, correspond to an eigenenergy of an electron in the crystal.

Table 15.1 Properties of Selected Semiconductors

Material	Bandgap (eV)	Band	Mobility @300 K (cm^2/V-s)		Effective Mass		Dielectric Constant, $\varepsilon/\varepsilon_0$	Refractive Index @ $h\nu \approx E_g$
			Electrons	Holes	Electrons (long/trans)	Holes (heavy/light)		
C	5.47	*Indirect*	2,000	2100	1.4/0.36	1.08/0.36	5.7	—
Si	1.124	*Indirect*	1,450	505	0.92/0.19	0.54/0.15	11.9	3.5
AlN	6.2	Direct	—	14	—	—	9.14	2.7
AlP	2.41	*Indirect*	60	450	3.61/0.21	0.51/0.21	9.8	—
AlAs	2.15	*Indirect*	294	—	1.1/0.19	0.41/0.15	10	3.2
AlSb	1.61	*Indirect*	200	400	1.8/0.26	0.33/0.12	12	3.6
GaN	3.44	Direct	440	130	0.22	0.96	10.4	—
GaP	2.27	*Indirect*	160	135	4.8/0.25	0.67/0.17	11.1	3.45
GaAs	1.424	Direct	9,200	320	0.063	0.5/0.076	12.4	3.6
GaSb	0.75	Direct	3,750	680	0.0412	0.28/0.05	15.7	3.8
InN	1.89	Direct	250	—	0.12	0.5/0.17	9.3	—
InP	1.34	Direct	5,900	150	0.079	0.56/0.12	12.6	3.4
InAs	0.353	Direct	33,000	450	0.021	0.35/0.026	15.1	3.5
InSb	0.17	Direct	77,000	850	0.0136	0.34/0.0158	16.8	4.2

higher band is completely empty, then the crystal will be an insulator, provided the energy gap is large enough, say, ~3 eV. We recall that the thermal energy k_BT at room temperature is only about 0.026 eV. With such a wide bandgap, the thermal excitation across the energy gap is negligible. An external electric field will not cause the flow of current. Under the action of a constant force, the momentum of an electron, and thus the energy, may increase with time. However, in a completely filled energy band, there is simply no energy states to accommodate the "accelerated" electron. Since every accessible state is filled, nothing changes when the electric field is applied. This is a consequence of the Pauli exclusion principle.

If the gap between the uppermost filled band (the valence band) and the next higher band (the conduction band) is small, say, <2 eV, then thermal excitation can cause the elevation of some electrons at room temperature from the valence band to the conduction band where they are able to be accelerated into empty states and the crystal can conduct electricity. Such crystals are called *semiconductors.* Their degree of conductivity can be controlled not only by the temperature but also by "doping" them with impurity atoms. The wavefunction of an electron in a given band, say, the valence band, is characterized by a vector **k** and a corresponding (Bloch) wavefunction

$$\psi(\mathbf{r}) = u_{\mathbf{k}}(\mathbf{r})e^{i\mathbf{k}\cdot\mathbf{r}} \tag{15.1-1}$$

where **k** is the wavevector (also known as the Bloch wavevector).

The function $u_{\mathbf{k}}(\mathbf{r})$ is periodic and possesses the same periodicity as the crystal lattice. The factor $\exp(i\mathbf{k}\cdot\mathbf{r})$ is responsible for the wave nature of the electronic motion in the crystal. In the free-electron approximation, we simply set $u_{\mathbf{k}}(\mathbf{r}) = 1$ for all **k**. In this case, the wavefunctions are simply plane waves. The magnitude of the wavevector is related to the de Broglie wavelength λ_e of the electron by

$$\lambda_e = \frac{2\pi}{|\mathbf{k}|} = \frac{2\pi}{k} \tag{15.1-2}$$

The vector **k** can only possess a prescribed set of values (i.e., it is quantized), which are obtained by requiring that the total phase shift $\mathbf{k}\cdot\mathbf{r}$ across a crystal with dimensions L_x, L_y, L_z be some integral multiple of 2π,

$$k_i = \frac{2\pi}{L_i}s, \quad s = 1, 2, 3, \ldots \tag{15.1-3}$$

where $i = x, y, z$. We can thus divide the total volume in **k** space into cells, each with a volume

$$\Delta V_k \equiv \Delta k_x\, \Delta k_y\, \Delta k_z = \frac{(2\pi)^3}{L_xL_yL_z} = \frac{(2\pi)^3}{V} \tag{15.1-4}$$

where V is the volume of the crystal, so that we may associate with each such differential volume in **k** space a quantum state (two states when we allow for the two intrinsic spin states of each electron). The number of such states within a spherical shell (in **k** space) of radial thickness dk and radius k is then given by the volume of the shell divided by the volume (15.1-4) ΔV_k per state

$$\rho(k)\, dk = \frac{k^2V}{\pi^2}\, dk \tag{15.1-5}$$

so that $\rho(k)$ is the number of states per unit volume of **k** space. (A factor of 2 for spin was included to account for the fact that each given (spatial) state can accommodate two electrons, each in a unique spin "up" or "down" state.) The energy, measured from the bottom of the band, of an electron with wavevector **k** in, say, the conduction band (indicated henceforth by a subscript c) is

$$E_c(\mathbf{k}) = \frac{\hbar^2k^2}{2m_c} \tag{15.1-6}$$

where m_c is the effective mass of an electron in the conduction band. In the simplest and idealized case, which is the one we are considering here, the energy depends only on the

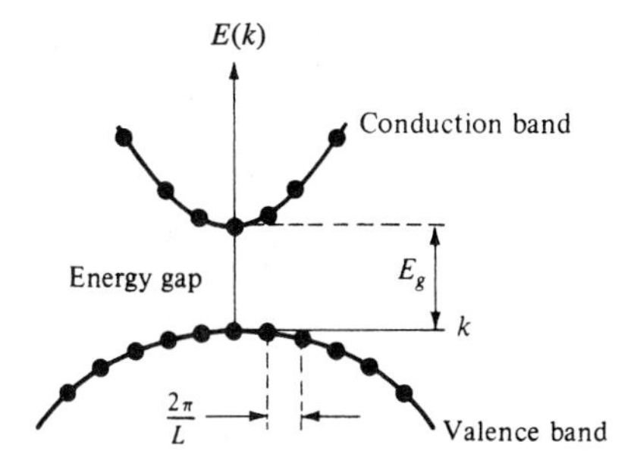

Figure 15.2 A typical energy band structure for a direct gap semiconductor with $m_c < m_v$. The uniformly spaced dots correspond to electron states.

magnitude k of the electron propagation vector and not its direction. We recall $\hbar k$ is the momentum of the electron in the band. Thus the energy in Equation (15.1-6) may be viewed as the kinetic energy of the electron in the band.

We often need to perform electron number counting, not in **k** space but as a function of the energy E. The density of states function $\rho(E)$ (the number of electronic states per unit energy interval per unit crystal volume) is determined from the conservation of states relation

$$\rho(E)\, dE = \frac{1}{V}\rho(k)\, dk$$

which with the use of Equations (15.1-5) and (15.1-6) leads to

$$\rho_c(E) = \frac{1}{2\pi^2}\left(\frac{2m_c}{\hbar^2}\right)^{3/2} E^{1/2}$$

or

$$\rho_c(\omega) = \hbar\rho_c(E) = \frac{1}{2\pi^2}\left(\frac{2m_c}{\hbar}\right)^{3/2} \omega^{1/2} \tag{15.1-7}$$

where $\hbar\omega = E$. A similar expression but with m_c replaced by m_v, the effective mass in the valence band, applies to the valence band. Table 15.1 lists some important parameters of selected semiconductors.

Figure 15.2 depicts the energy–k relationship of a direct gap semiconductor, that is, one where the conduction band minimum and the valence band maximum occur at the same value of **k**. The dots represent allowed (not necessarily occupied) electron energies. Note that, following Equation (15.1-3), these states are spaced uniformly along the k axis.

The Fermi–Dirac Distribution Law

The probability that an electron state at energy E is occupied by an electron is given by the Fermi–Dirac distribution function [6, 7]

$$f(E) = \frac{1}{e^{(E-E_F)/k_BT} + 1} \tag{15.1-8}$$

where E_F is the Fermi level, k_B is the Boltzmann constant ($k_B = 1.38066 \times 10^{-23}$ J/K), and T is the temperature. It is important to note that the Fermi level is a function of the temperature. The Fermi level at $T = 0$ K is often called the *Fermi energy*. For electron energies well below the Fermi level such that $E_F - E \gg kT, f(E) \to 1$ and the electronic states are fully occupied, while well above the Fermi level $E - E_F \gg kT$, $f(E) \propto \exp(-E/kT)$ and approaches the

Boltzmann distribution. At $T = 0$ K, $f(E) = 1$, for $E < E_F$, and $f(E) = 0$, for $E > E_F$ so that all levels below the Fermi energy are occupied while those above it are empty. In thermal equilibrium, a single Fermi level applies to both the valence and conduction bands. Under conditions in which the thermal equilibrium is disturbed, such as in a *p-n* junction with a current flow or a bulk semiconductor in which a large population of conduction electrons and holes is created by photoexcitation, separate Fermi levels called *quasi-Fermi levels* are used for each of the bands. The concept of quasi-Fermi levels in excited systems is valid whenever the carrier scattering time ($\sim 10^{-12}$ s) within a band is much shorter than the equilibration time ($\sim 10^{-9}$–10^{-3} s) between bands. This is usually true at the large carrier densities used in *p-n* junction lasers.

The Fermi level in semiconductors also depends on the density of impurity doping. In very highly doped semiconductors, the Fermi level is forced either (1) into the conduction band for donor impurity doping or (2) into the valence band for acceptor impurity doping. This situation is demonstrated by Figure 15.3. According to Equation (15.1-8), at $T = 0$ K, all the states below E_F are filled while those above it are unoccupied as shown in the figure. In this respect the highly doped semiconductor (also known as degenerate semiconductor) behaves like a metal, in which case the conductivity does not disappear at very low temperatures. In a degenerate semiconductor, the carrier concentration is so high that the Boltzmann approximation for the occupancy of the states is no longer valid. The unoccupied states in the valence band (unshaded area in Figure 15.3b) are called *holes*, and they are treated exactly like electrons except that their charge, corresponding to an electron deficiency, is positive and their energy increases downward in the diagram. The number of holes in the semiconductor depicted by Figure 15.3b is the number of electron states falling within the unshaded area at the top of the valence band. The process of exciting an electron from state *a* to state *b* (Figure 15.3b) in the valence band can also be viewed as one whereby a hole is excited from *b* to *a*. The advantage of this point of view is the symmetry in the language and mathematical description that it brings to the discussions of current flow due to electrons in the conduction band and those in the valence band.

To better appreciate the role of the quasi-Fermi level, consider a nonthermal equilibrium situation in which electrons are excited into the conduction band of a degenerate *p*-type semiconductor at a very high rate. This can be done by injecting electrons into the *p* region

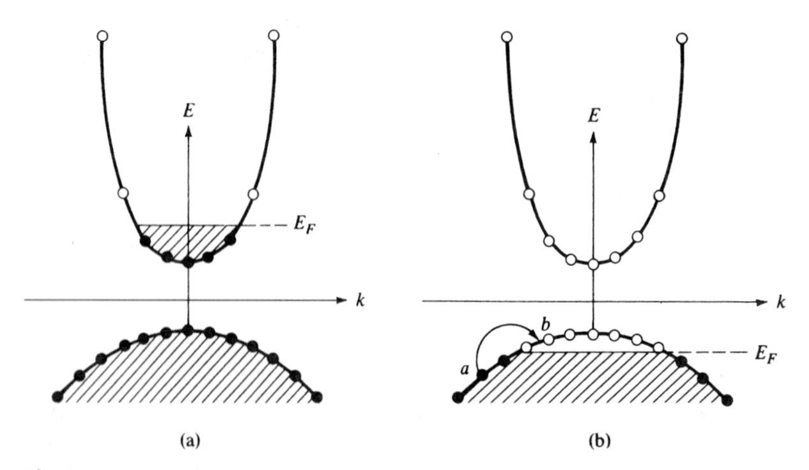

Figure 15.3 (a) Energy band of a degenerate *n*-type semiconductor at 0 K. (b) A degenerate *p*-type semiconductor at 0 K. The cross-hatching represents regions in which all the electron states are filled. Empty circles indicate unoccupied states (holes).

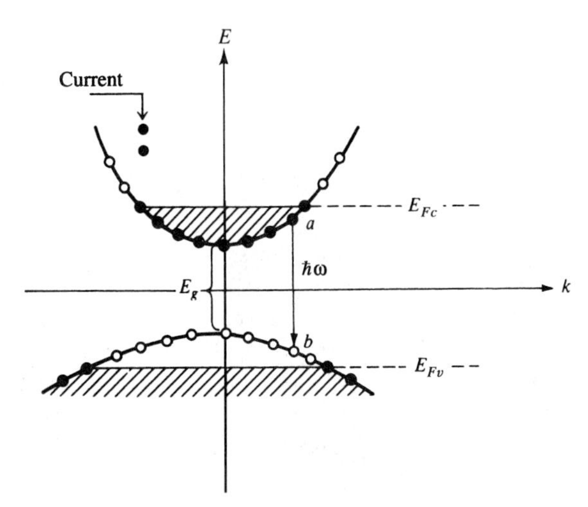

Figure 15.4 Electrons are injected at a rate of I/Ve per unit volume (I = total current) into the conduction band of a semiconductor.

across a *p-n* junction or by subjecting the semiconductor to an intense light beam with $h\nu > E_g + E_{Fc} + E_{Fv}$, (see Figure 15.4) so that for each absorbed photon an electron is excited into the conduction band from the valence band. This situation is depicted in Figure 15.4. Following this excitation, electrons relax, by emitting optical and acoustic phonons, to the bottom of the conduction band in times of $\sim 10^{-12}$ s while their relaxation across the gap back to the valence band—a process referred to as electron–hole recombination—is characterized by a time constant of

$$\tau \sim (3\text{–}4) \times 10^{-9} \text{ s}$$

It is important in analyzing the process of light amplification in semiconductors to determine the quasi-Fermi level E_{Fc} for a given rate of excitation. Assuming that the relaxation to the bottom of the band into which the carriers are excited is instantaneous, we have

$$\frac{N_c}{\tau} = \frac{I}{Ve} \tag{15.1-9}$$

where N_c is the density (m^{-3}) of electrons in the conduction band, I is the injection current (in amperes), τ is the electron relaxation time back to the valence band (electron–hole recombination time), and V is the volume into which the electrons are confined following the injection.

The density of electrons with energies between E and $E + dE$ is the product of $\rho_c(E)$—the density of allowed electron states—and the occupation probability $f_c(E)$ of these states. Using Equations (15.1-7) and (15.1-8),

$$N_c = \frac{I\tau}{eV} = \int_0^\infty \rho_c(E) f_c(E)\, dE$$

$$= \frac{1}{2\pi^2}\left(\frac{2m_c}{\hbar^2}\right)^{3/2} \int_0^\infty \frac{E^{1/2}}{e^{(E-E_{Fc})/k_BT} + 1}\, dE \tag{15.1-10}$$

For a given injection current I the only unknown quantity in (15.1-10) is the conduction quasi-Fermi level E_{Fc}. We can thus invert, in practice by numerical methods, (15.1-10) and solve it for $E_{Fc}(T)$ as a function of I, or equivalently of N_c (see Appendix E). We shall make use later of this fact. At $T = 0$ the integral is replaced by

$$\int_0^{E_{Fc}} E^{1/2} dE = \tfrac{2}{3} E_{Fc}^{3/2}$$

yielding

$$E_{Fc}(T = 0) = (3\pi^2)^{2/3} \frac{\hbar^2}{2m_c} N_c^{2/3} \tag{15.1-11}$$

Another fact that we need before proceeding to the subject of optical gain in semiconductors is that when an electron makes a transition (induced or spontaneous) between a conduction band state and the valence band, the two states involved must have the same **k** vector. This is due to the fact that, according to quantum mechanics, the rate of such a transition is always proportional to an integral over the crystal volume that involves the product of the initial state wavefunction and the complex conjugate of that of the final state. Such an integral would, according to (15.1-1), be vanishingly small except when the condition

$$\mathbf{k}_f = \mathbf{k}_i \tag{15.1-12}$$

is satisfied. In band diagrams such as that of Figure 15.4, the transitions are consequently described by vertical arrows. Equation (15.1-12) is consistent with the fact that the momentum of the photon is negligible relative to that of the electron.

15.2 GAIN AND ABSORPTION IN SEMICONDUCTOR (LASER) MEDIA

Consider the semiconductor material depicted in Figure 15.5 in which by virtue of electron pumping (e.g., via current injection) a nonthermal equilibrium steady state is obtained in which *simultaneously* large densities of electrons and holes coexist in the *same* space. These are characterized by quasi-Fermi levels E_{Fc} and E_{Fv}, respectively, as shown. These quasi-Fermi levels depend on the charge carrier densities, as well as the temperature.

Let an optical beam at a (radian) frequency ω_0 travel through the crystal. Let a represent an electronic state in the conduction band, and b represent an electronic state in the valence band. This beam will induce downward $a \to b$ transitions that lead to amplification as well as $b \to a$ absorbing transitions. Net amplification of the beam results if the rate of $a \to b$ transitions exceeds that of $b \to a$.

As discussed in the previous section, only transitions in which the upper and lower electron states have the same **k** vector are allowed. The pair of levels a and b in Figure 15.5

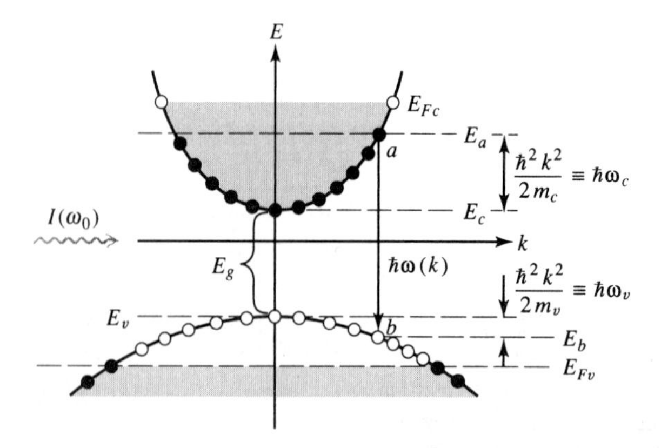

Figure 15.5 An optical beam at ω_0 with intensity $I(\omega_0)$ is incident on a pumped semiconductor medium characterized by quasi-Fermi levels E_{Fc} and E_{Fv}. A single level pair $a \to b$ with the same **k** value is shown. The induced transition $a \to b$ contributes one photon to the beam.

are thus characterized by some **k** value. Let us consider a group of such levels with nearly the same **k** value and hence with nearly the same transition energy,

$$\hbar\omega(\mathbf{k}) = E_g + \frac{\hbar^2 k^2}{2m_c} + \frac{\hbar^2 k^2}{2m_v} \tag{15.2-1}$$

(In the following, the **k** dependence of ω will be omitted but understood.) The density of such level pairs whose **k** values fall within a spherical shell of thickness dk is, according to Equation (15.1-5), $\rho(k)\, dk/V$.

Before proceeding, let us remind ourselves of some results developed in connection with conventional laser media. The gain constant $\gamma(\omega_0)$ is given by Equation (5.6-25) as

$$\gamma(\omega_0) = -\frac{k}{n^2}\chi''(\omega_0), \quad k = \frac{2\pi n}{\lambda} \tag{15.2-2}$$

where $\chi''(\omega_0)$, the imaginary part of the electric susceptibility, is given, according to Equation (5.6-24), by

$$\chi''(\omega_0) = \frac{(N_1 - N_2)\lambda_0^3}{8\pi^3 t_{\text{spont}}\, \Delta\nu\, n} \frac{1}{1 + 4(\nu - \nu_0)^2/(\Delta\nu)^2} \tag{15.2-3}$$

Combining the last two equations and defining the "relaxation time" T_2 by $T_2 = (\pi\, \Delta\nu)^{-1}$ leads to

$$\gamma(\omega_0) = \frac{(N_1 - N_2)\lambda_0^2}{4\pi^2 t_{\text{spont}}} \frac{T_2}{\pi[1 + (\omega - \omega_0)^2 T_2^2]} \tag{15.2-4}$$

In semiconductors T_2 is the mean lifetime for coherent interaction of electrons with a monochromatic field and is of the order of the phonon–electron collision time. Numerically, $T_2 \sim 10^{-12}$ s for typical semiconductors. Given an electron in an upper state a, the lower state b with the same **k** value may be occupied by another electron. The downward rate of transitions is thus proportional to

$$R_{a\to b} \propto f_c(E_a)[1 - f_v(E_b)]$$

that is, to the product of the probabilities $f_c(E_a)$ that the upper (conduction) state is occupied and the probability $(1 - f_v)$ that the lower (valence) state is empty. The functions $f_{v,c}(E)$ are given, according to Equation (15.1-8), by

$$f_c(E_a) = \frac{1}{e^{(E_a - E_{Fc})/k_B T} + 1} \tag{15.2-5}$$

$$f_v(E_b) = \frac{1}{e^{(E_b - E_{Fv})/k_B T} + 1} \tag{15.2-6}$$

allowing for the fact that under pumping conditions $E_{Fc} \neq E_{Fv}$. In typical semiconductor lasers, the quasi-Fermi level for conduction electrons is near the bottom of the conduction band, and the quasi-Fermi level for valence electrons is near the top of the valence band.

In translating to the case of semiconductors the results that were developed for conventional lasers, the population inversion density $(N_2 - N_1)$ is thus replaced by the effective inversion due to electrons and holes within dk.

$$N_2 - N_1 \to \frac{\rho(k)\, dk}{V}\{f_c(E_a)[1 - f_v(E_b)] - f_v(E_b)[1 - f_c(E_a)]\}$$

$$= \frac{\rho(k)\, dk}{V}[f_c(E_a) - f_v(E_b)] \tag{15.2-7}$$

where we recall that the energy of the photon involved is given by

$$E_a - E_b \equiv \hbar\omega = E_g + \frac{\hbar^2 k^2}{2m_c} + \frac{\hbar^2 k^2}{2m_v} \tag{15.2-8}$$

Equation (15.2-7) is of central importance and is a capsule statement of the difference between the population inversion in a conventional laser medium where the level occupation probability obeys Boltzmann statistics and that of a semiconductor medium governed by Fermi–Dirac statistics.

Returning to the gain expression (15.2-4), we use Equation (15.2-7) to rewrite it as

$$d\gamma(\omega_0) = \frac{\rho(k)\,dk}{V}(f_c - f_v)\frac{\lambda_0^2}{4n^2\tau}\left(\frac{T_2}{\pi[1 + (\omega - \omega_0)^2 T_2^2]}\right)$$

where ω_0 is the frequency of the photon, and $\hbar\omega = \hbar\omega(k)$ is the energy difference of the electron states involved in the transition, as given in Equation (15.2-1). The differential designation $d\gamma(\omega_0)$ is to remind us that only electrons with **k** vectors within dk are included here. We have also replaced, to agree with popular usage, the term spontaneous lifetime (t_{spont}) by the recombination lifetime τ for an electron in the conduction band with a hole in the valence band. To obtain the gain constant, we must add up the contributions from all the electrons. This leads to

$$\gamma(\omega_0) = \int_0^\infty \frac{dk\,\rho(k)}{V}[f_c(\omega) - f_v(\omega)]\frac{\lambda_0^2}{4n^2\tau}\left(\frac{T_2}{\pi[1 + (\omega - \omega_0)^2 T_2^2]}\right) \tag{15.2-9}$$

We will find it easier to carry out the indicated integration in Equation (15.2-9) in the ω domain ($\hbar\omega$ being the separation $E_a(\mathbf{k}) - E_b(\mathbf{k})$). From Equation (15.2-1), we rewrite the energy as

$$\hbar\omega = E_g + \frac{\hbar^2 k^2}{2m_r} \tag{15.2-10}$$

where m_r is the reduced effective mass given by

$$\frac{1}{m_r} = \frac{1}{m_v} + \frac{1}{m_c} \tag{15.2-11}$$

Using Equation (15.2-10), we obtain

$$d\omega = \frac{\hbar}{m_r}k\,dk$$

$$k = (\hbar\omega - E_g)^{1/2}\left(\frac{2m_r}{\hbar^2}\right)^{1/2}$$

Thus the expression (15.2-9) for the gain coefficient $\gamma(\omega_0)$ becomes

$$\gamma(\omega_0) = \int_0^\infty (\hbar\omega - E_g)^{1/2}\left(\frac{2m_r}{\hbar^2}\right)^{1/2}\frac{m_r\lambda_0^2[f_c(\omega) - f_v(\omega)]}{4n^2\pi^2\hbar\tau}\,\frac{T_2}{\pi[1 + (\omega - \omega_0)^2 T_2^2]}\,d\omega \tag{15.2-12}$$

In most situations we can replace the normalized function

$$\frac{T_2}{\pi[1 + (\omega - \omega_0)^2 T_2^2]} \to \delta(\omega - \omega_0)$$

which is merely a statement of the fact that its width $\Delta\omega \sim T_2^{-1}$ is narrower than other spectral features of interest. In this case the integration (15.2-12) becomes

$$\gamma(\omega_0) = \frac{\lambda_0^2}{8\pi^2 n^2 \tau}\left(\frac{2m_c m_v}{\hbar(m_v + m_c)}\right)^{3/2}\left(\omega_0 - \frac{E_g}{\hbar}\right)^{1/2}[f_c(\omega_0) - f_v(\omega_0)] \tag{15.2-13}$$

where we recall that ω_0 is the frequency of the photon and $E_a - E_b = \hbar\omega_0$. The condition for net gain $\gamma(\omega_0) > 0$ is thus

$$f_c(\omega_0) > f_v(\omega_0) \tag{15.2-14}$$

which is the equivalent, in a semiconductor, of the conventional inversion condition $N_2 > N_1$. Using Equations (15.2-5) and (15.2-6), the gain condition (15.2-14) becomes

$$\frac{1}{e^{(E_a - E_{Fc})/k_B T} + 1} > \frac{1}{e^{(E_b - E_{Fv})/k_B T} + 1} \tag{15.2-15}$$

Recalling that $E_a - E_b = \hbar\omega_0$, condition (15.2-15) is satisfied provided

$$\hbar\omega_0 < E_{Fc} - E_{Fv} \tag{15.2-16}$$

so that only frequencies whose photon energies $\hbar\omega_0$ are smaller than the quasi-Fermi level separations are amplified. Condition (15.2-16) was first derived by Basov et al. [1] and Bernard and Duraffourg [8]. The general features of the gain dependence $\gamma(\omega_0)$ on the frequency ω_0 are illustrated by Figure 15.6. The gain is zero at $\hbar\omega_0 < E_g$, since no electronic

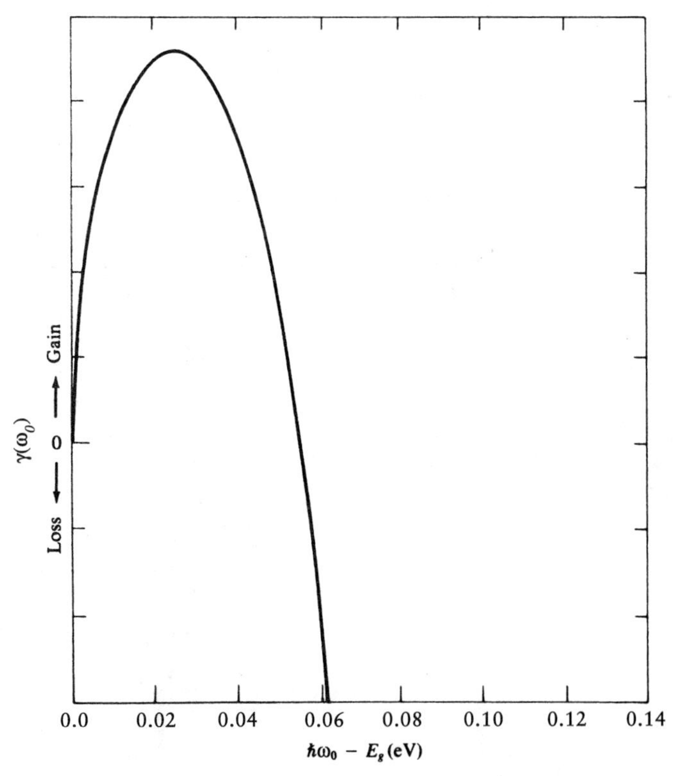

Figure 15.6 A typical plot of gain $\gamma(\omega_0)$ as a function of frequency for a fixed pumping level of the density of the injected electrons N. (After Reference [9].)

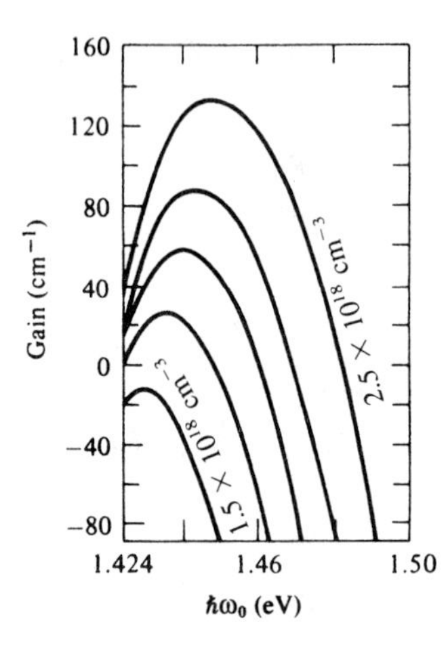

Figure 15.7 A plot based on Equation (15.2-12) of the photon energy dependence of the optical gain (or loss = negative gain) of GaAs with the injected carrier density as a parameter. (After Reference [9].)

transitions exist at these energies. As the photon energy becomes greater than the bandgap, the gain starts to increase. The gain becomes zero again at the frequency where $\hbar\omega_0 = E_{Fc} - E_{Fv}$. The spectral profile of the gain coefficient is dominated by the Fermi–Dirac distribution of the electrons. At higher frequencies the semiconductor absorbs.

Figure 15.7 shows calculated plots based on Equation (15.2-12) with the density of the (injected) electrons as a parameter. The curves are based on the following physical constants of GaAs: $m_c = 0.067m_e$, $m_v = 0.48m_e$, $T_2 \sim 0.5$ ps, $\tau \sim 3 \times 10^{-9}$ s, $E_g = 1.43$ eV (with $m_e = 9.910953 \times 10^{-31}$ kg). We note that the minimum density to achieve transparency ($\gamma = 0$) is $N_{tr} \sim 1.55 \times 10^{18}$ cm^{-3}. The peak gain corresponding to a given inversion density N_c is plotted in Figure 15.8.

It follows from Figure 15.8 that semiconductor media are capable of achieving very large gains ranging up to a few hundred cm^{-1}. In a laser the amount of gain that actually prevails is clamped by the phenomenon of saturation (see Section 5.7) to a value equal to the loss. In a typical semiconductor laser, this works out to $20 < \gamma < 80$ cm^{-1}. In this region we can approximate the plot of Figure 15.8 by a linear relationship

$$\gamma_{max} = B(N - N_{tr}) \tag{15.2-17}$$

The constant B fitting the data of Figure 15.8 is $B \sim 1.5 \times 10^{-16}$ cm^2 and is typical of GaAs/GaAlAs lasers at 300 K. The gain constant B decreases with the increase of the temperature T. This is due to the broadening of the transition regions of the Fermi functions $f_c(\omega)$ and $f_v(\omega)$ in Equation (15.2-12). At 77 K, $B \sim 5 \times 10^{-16}$ cm^2. Figure 15.8 shows that the semiconductor diode is capable of producing extremely large incremental gains, with only moderate increases of the inversion density, hence the current, above the transparency value ($N_{tr} \sim 1.55 \times 10^{18}$ cm^{-3} in the figure). It is thus possible to obtain oscillation in a semiconductor laser with active regions that are only a few tens of microns long. Commercial diode lasers have typical lengths of ~250 μm. Figure 15.9 shows the gain coefficient versus photon energy over the bandgap at $T = 0$, 77, 300, 400 K. We note that both the gain bandwidth and the peak gain decrease with the increase of operation temperatures.

For additional background material on semiconductor lasers, the student is advised to consult References [10–12].

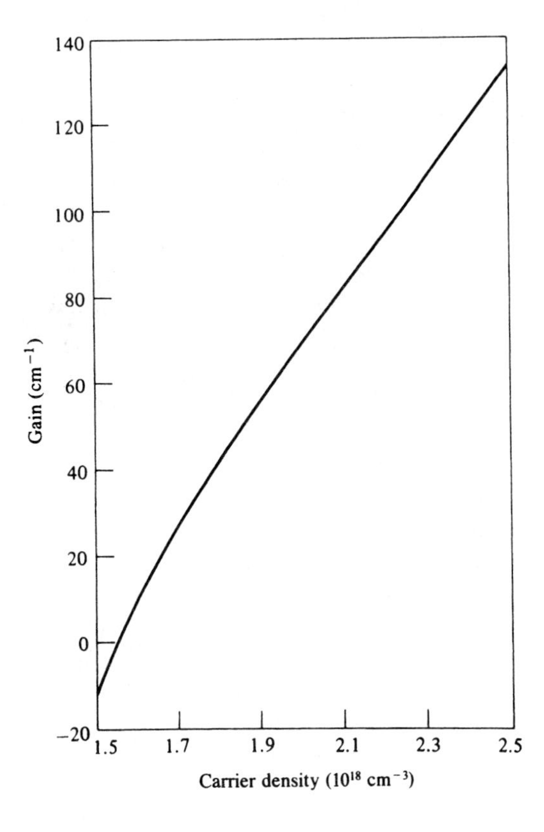

Figure 15.8 A plot of the peak gain γ_{max} of Figure 15.7 as a function of the inversion density at $T = 300$ K.

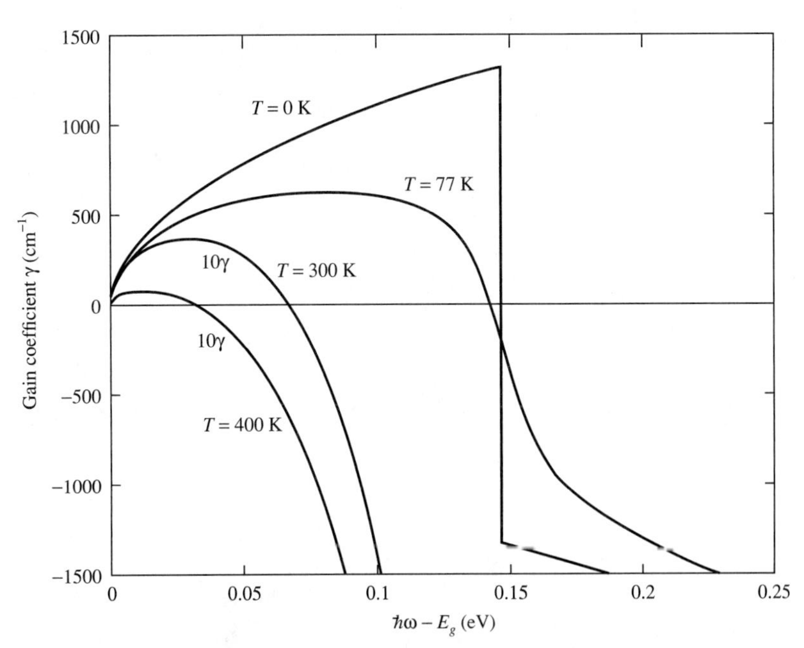

Figure 15.9 An example of the gain coefficient of a GaAs laser as a function of the photon energy at various temperatures. The curves for $T = 300$ K and 400 K are 10 γ. In this case, the carrier density leads to quasi-Fermi levels of $E_{Fc} - E_{Fv} \approx E_g + 0.15$ eV.

15.3 GaAs/$Ga_{1-x}Al_xAs$ LASERS

According to Table 15.1, we note that the energy gap of binary III–V semiconductors covers a broad range (from a fraction of 1 eV to a few eV) for applications. Since the bandgap translates to the emission photon energy of lasers, a wide emission spectrum is covered. In practice, however, operational devices are limited by the availability of substrate materials and the possibility of lattice matching in the crystal growth of the laser structures. The most common substrates that are available with relatively low defect density are GaAs and InP crystals. In the early stage of the crystal growth development, it was realized that a solid solution of two binary compound semiconductors can form ternary or quaternary alloy semiconductors. For example, the solid solution of GaAs and AlAs can form $Ga_{1-x}Al_xAs$ semiconductors. The two most important classes of semiconductor lasers are those that are based on III–V semiconductors that can be grown on GaAs or InP substrates.

The first system is based on GaAs and $Ga_{1-x}Al_xAs$ semiconductor crystals. In this case, the active region is either GaAs or $Ga_{1-x}Al_xAs$. Since an AlAs semiconductor crystal has a larger bandgap than that of GaAs, the ternary compound crystal $Ga_{1-x}Al_xAs$ has a bandgap between those of GaAs ($x = 0$, $E_{gap} = 1.43$ eV) and AlAs ($x = 1$, $E_{gap} = 2.1$ eV). The subscript x indicates the fraction of Ga atoms in GaAs that are replaced by Al. The resulting lasers emit (depending on the active region molar fraction x and its doping) at $0.75\ \mu m < \lambda < 0.88\ \mu m$. This spectral region is convenient for short-haul (2 km) optical communications in silica fibers.

The second system has $Ga_{1-x}In_xAs_{1-y}P_y$ as its active region. The lasers emit in the range $1.1\ \mu m < \lambda < 1.6\ \mu m$ depending on x and y. The region near 1.55 μm is especially favorable, since, as shown in Figure 3.21, optical fibers are available with losses as small as 0.15 dB/km at this wavelength, making it extremely desirable for long-distance optical communications. In this section we will consider GaAs/$Ga_{1-x}Al_xAs$ lasers. A generic laser of this type, depicted in Figure 15.10, has a thin (0.1–0.2 μm) region of GaAs sandwiched between two regions of GaAlAs. It is consequently called a *double heterostructure* laser. The basic layered structure

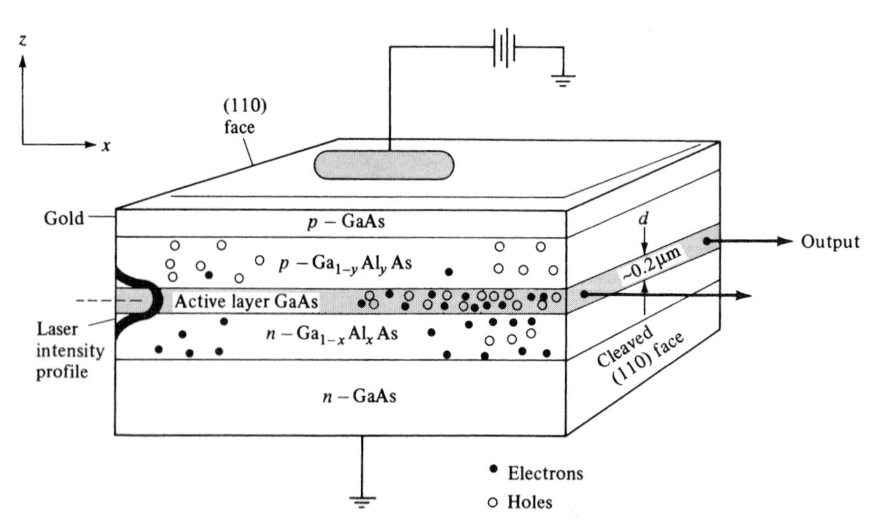

Figure 15.10 A typical double heterostructure GaAs/GaAlAs laser. Electrons and holes are injected into the active GaAs layer from the n and p GaAlAs. Photons with frequencies near $\nu = E_g/h$ are amplified by stimulating electron–hole recombination.

is grown epitaxially on a crystalline GaAs substrate so that it is uninterrupted crystallographically. The favored crystal growth techniques are liquid-phase epitaxy and chemical vapor deposition using metallo-organic reagents (MOCVD) [11, 13, 14]. Another important technique—molecular beam epitaxy (MBE) [11, 13, 15, 16]—uses atomic beams of the crystal constituents in ultrahigh vacuum to achieve extremely fine thickness and doping control.

The thin active region is usually undoped while one of the bounding $Ga_{1-x}Al_xAs$ layers is doped heavily n-type and the other p-type. The difference

$$n_{\mathrm{GaAs}} - n_{\mathrm{Ga}_{1-x}\mathrm{Al}_x\mathrm{As}} \cong 0.62x$$

between the indices of refraction of GaAs and the ternary crystal with a molar fraction x gives rise to a three-layered dielectric waveguide of the type illustrated in Figure 3.1. At this point the student should review the basic modal concepts discussed in Chapter 3. The lowest-order (fundamental) confined mode has its energy concentrated mostly in the GaAs (high-index) layer. The index distribution and a typical modal intensity plot for the lowest-order mode are shown in Figure 15.11. When a positive bias is applied to the device, electrons are injected from the n-type $Ga_{1-x}Al_xAs$ into the active GaAs region while a density of holes equal to that of the electrons in the active region is caused by injection from the p side. The density of holes must equal that of the electrons to achieve charge neutrality.

The electrons that are injected into the active region are prevented from diffusing out into the p region by means of the potential barrier ΔE_c due to the difference ΔE_g between the energy gaps of GaAs and $Ga_{1-x}Al_xAs$. The x dependence of the energy gap of $Ga_{1-x}Al_xAs$ is approximated by [13]

$$E_g(x < 0.37) = (1.424 + 1.247x)\ \mathrm{eV}$$

and is plotted in Figure 15.12.

The total discontinuity ΔE_g of the energy gap at a GaAs/GaAlAs interface is taken up mostly (60%) by the conduction bandedge, that is, $\Delta E_c = 0.6\,\Delta E_g$, while 40% is left to the valence band, $\Delta E_v = 0.4\,\Delta E_g$, so that both holes and electrons are effectively confined to the active region. This double confinement of injected carriers as well as of the optical mode energy to the same region is probably the single most important factor responsible for the successful realization of low-threshold continuous semiconductor lasers [17–19]. Under these conditions, we expect the gain experienced by the mode to vary as $1/d$, where d is the thickness of the active (GaAs) layer, since at a given total current, the carrier density, hence the gain, will be proportional to $1/d$. To further discuss the dependence on the thickness of the active region d, we start with the basic definition of the modal gain,

$$g = \frac{\text{Power generated per unit length (in } x)}{\text{Power carried by beam}}$$

$$= \frac{-\int_{-\infty}^{-d/2} \alpha_n |E|^2 dz + \int_{-d/2}^{d/2} \gamma |E|^2 dz - \int_{d/2}^{\infty} \alpha_p |E|^2 dz}{\int_{-\infty}^{\infty} |E|^2 dz} \qquad (15.3\text{-}1)$$

where γ is the gain coefficient experienced by a plane wave in a medium whose inversion density is equal to that of the active medium. The gain coefficient γ is given by Equations (15.2-12) and (15.2-17). The parameter α_n is the loss constant of the unpumped n-$Ga_{1-x}Al_xAs$ and is due mostly to free electron absorption. α_p is the loss (by free holes) in the bounding p-$Ga_{1-y}Al_yAs$ region. We note that as $d \to \infty$, $g \to \gamma$.

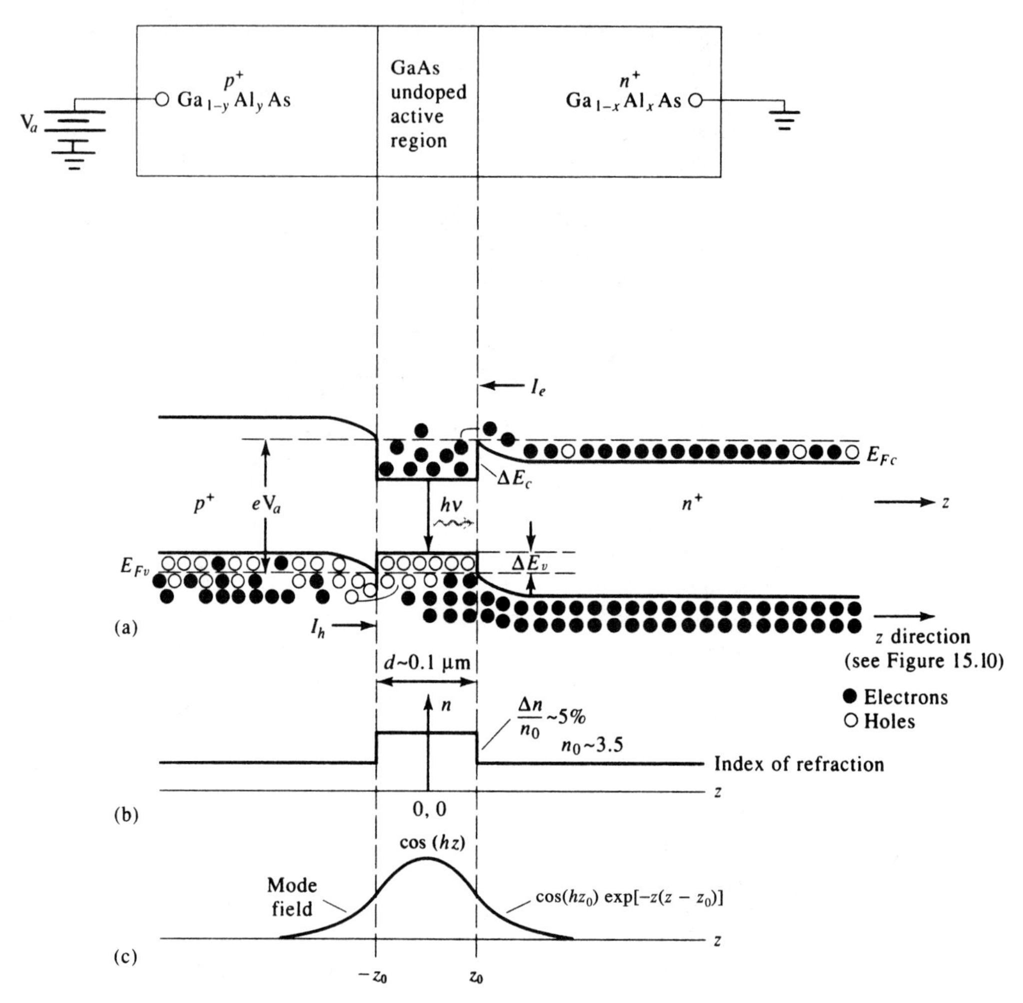

Figure 15.11 (a) The energy bandedges of a strongly forward-biased (near-flattened) double heterostructure GaAs/GaAlAs laser diode. Note the trapping of electrons (holes) in the potential well formed by the conduction (valence) bandedge energy discontinuity $\Delta E_c(\Delta E_v)$. (b) The spatial (z) profile of the index of refraction which is responsible for dielectric waveguiding in the high-index (GaAs) layer. (c) The mode profile of the fundamental optical mode in a slab waveguide.

In the case when γ is a constant over $-d/2 < z < d/2$. It is convenient to rewrite Equation (15.3-1) as

$$g = \gamma\Gamma_a - \alpha_n\Gamma_n - \alpha_p\Gamma_p \tag{15.3-2}$$

with

$$\Gamma_a = \frac{\int_{-d/2}^{d/2} |E|^2 dz}{\int_{-\infty}^{\infty} |E|^2 dz}, \quad \Gamma_n = \frac{\int_{-\infty}^{-d/2} |E|^2 dz}{\int_{-\infty}^{\infty} |E|^2 dz}, \quad \Gamma_p = \frac{\int_{d/2}^{\infty} |E|^2 dz}{\int_{-\infty}^{\infty} |E|^2 dz} \tag{15.3-3}$$

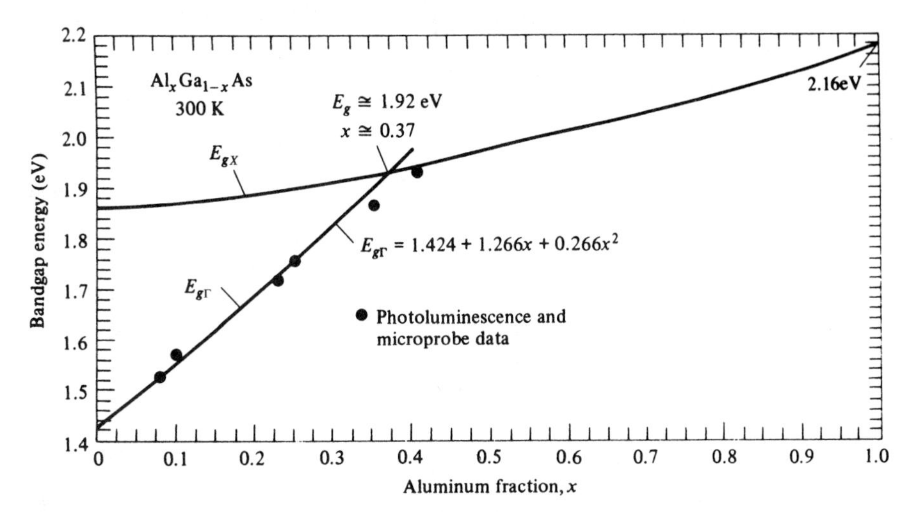

Figure 15.12 The magnitude of the energy gap in Ga$_{1-x}$Al$_x$As as a function of the molar fraction x. For $x > 0.37$ the bandgap is indirect. (After Reference [11].)

where Γ_a is very nearly the fraction of the mode power carried within the active GaAs layer, while Γ_n and Γ_p are, respectively, the fraction of the power in the n and p regions. As long as $\Gamma_a \sim 1$, that is, most of the mode energy is in the active region, the gain g is approximately equal to γ, which is inversely proportional to the active region thickness d. As d decreases further, an increasing fraction of the mode intensity is carried outside the active region as can be seen from the modal waveguide solution plotted in Figure 15.13 [11]. The portion of optical mode intensity outside the active region does not get amplified, and even suffers attenuation.

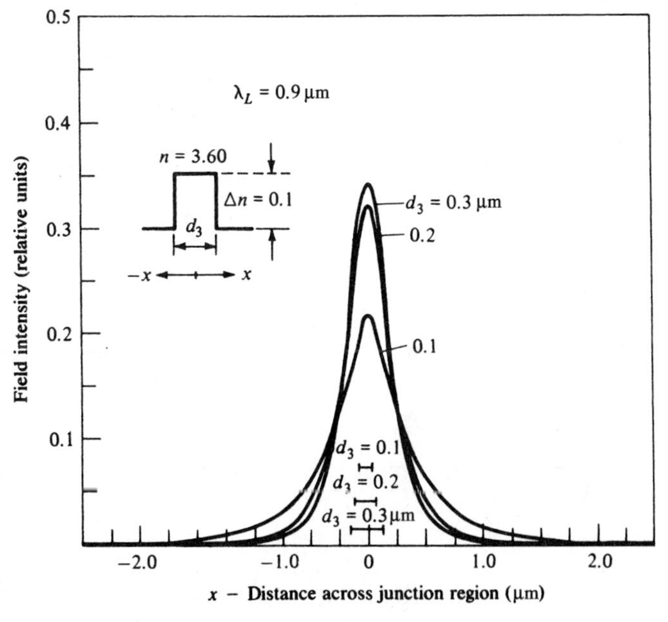

Figure 15.13 Calculated near-field intensity distribution of the step discontinuity waveguide for various values of the guiding layer thickness. (After Reference [11].)

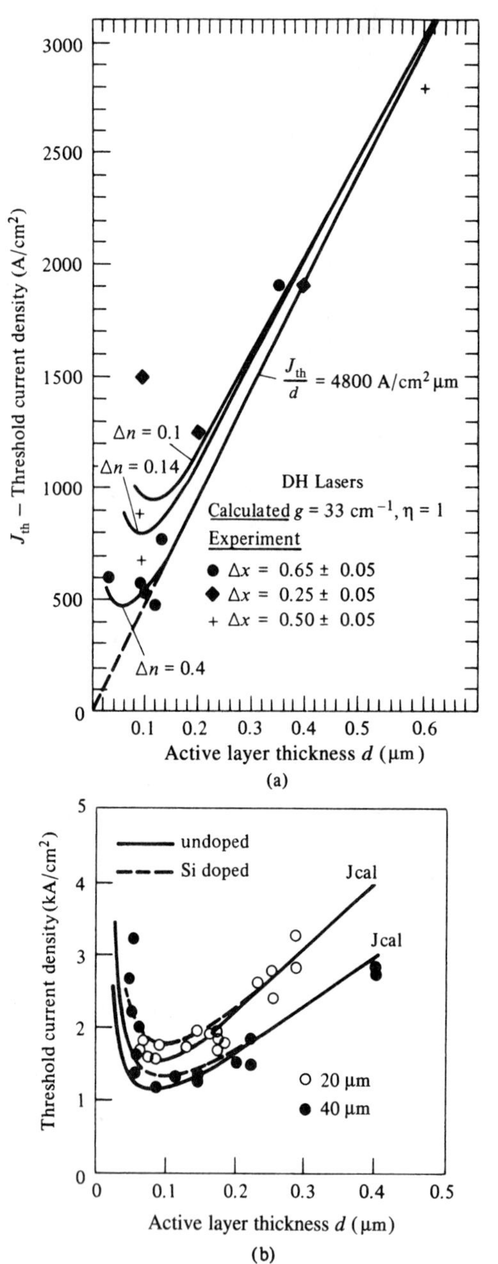

Figure 15.14 (a) Calculated and experimental values of the threshold current density as a function of the active layer thickness d for broad-area 500 μm long AlGaAs DH diode lasers of "undoped" active layers. Notable exceptions are the experimental data for $\Delta x = 0.25$, which were obtained from diodes with heavy-Ge-doped active layers. (After Reference [11].) (b) Calculated and experimental values of the threshold current density as a function of active layer thickness d for stripe-geometry (20 and 40 μm wide stripe contacts) 300 μm long AlGaAs DH diode lasers ($\Delta x = 0.25$) of "undoped" and low-Si-doped active layers ($x = 0.05$). (After Reference [21].)

The resulting decrease of the confinement factor Γ_a eventually dominates over the d^{-1} dependence and the gain begins to decrease with further decrease of d [20]. A plot of the threshold current dependence on d is depicted in Figure 15.14. The bottoming out and eventual increase of J_{th} for $d \lesssim 0.1$ μm is due to the decrease of the confinement factor Γ_a and the increase of the relative role of the losses in the p- and n-GaAlAs bounding layer as Γ_n and Γ_p increase, that is, as an increasing fraction of the mode intensity is carried within these lossy unpumped regions as shown in Figure 15.13.

EXAMPLE: THRESHOLD CURRENT DENSITY IN DOUBLE HETEROSTRUCTURE LASERS

Consider the case of a GaAs/GaAlAs laser of the type illustrated in Figure 15.11. We will use the following parameters: $\tau \sim 4 \times 10^{-9}$ s, $L = 500$ μm. The threshold gain condition (15.3-2) is

$$\gamma\Gamma_a = \alpha_n\Gamma_n + \alpha_p\Gamma_p - \frac{1}{L}\ln R + \alpha_s \tag{15.3-4}$$

where R is the reflectivity of the cavity mirrors and the term α_s accounts for scattering losses (mostly at heterojunction interfacial imperfections). The largest loss term in lasers with uncoated faces is usually L^{-1} ln R. In our case, taking $R = 0.31$ as due to Fresnel reflectivity at a GaAs($n = 3.5$)/air interface, we obtain

$$-\frac{1}{L}\ln R = 23.4 \text{ cm}^{-1}$$

The rest of the loss terms are assumed to add up to ~10 cm^{-1}, so that taking $\Gamma_a \sim 1$, the total gain needed is 33.4 cm^{-1}. This requires, according to Figure 15.8, an injected carrier density of $N \sim 1.7 \times 10^{18}$ cm^{-3}. Under steady-state conditions, the rate at which carriers are injected into the active region must equal the electron–hole recombination rate,

$$\frac{J}{e} = \frac{Nd}{\tau}$$

Using the above data we obtain

$$\frac{J}{d} = \frac{eN}{\tau} \sim 6.8 \times 10^3 \text{ A/cm}^2\text{-}\mu\text{m}$$

This value of J/d is in reasonable agreement with the measured value of $\sim 5 \times 10^3$ in Figure 15.14. If we use this value to estimate the lowest threshold current density, which from Figure 15.14 occurs when $d \sim 0.08$ μm, we obtain

$$J_{\min} = 0.68 \times 10^4 \times 0.08 = 544 \text{ A/cm}^2$$

again, close to the range of observed values.

The successful epitaxial growth of thin layers of $Ga_{1-x}Al_xAs$ on top of GaAs (and vice versa), which is the main reason for the success of double heterostructure lasers, is due to the fact that their lattice constants are the same, to within a fraction of a percent, over the range $0 \le x \le 1$. This can be seen from the plot of Figure 15.15, which shows the lattice constant corresponding to various compositions of III–V semiconductors as a function of the bandgap energy. We note that the line connecting the AlAs ($x = 1$) and the GaAs ($x = 0$) is nearly horizontal, which corresponds to a (very nearly) constant lattice constant over this compositional range.

15.4 SOME REAL LASER STRUCTURES

The double heterostructure lasers discussed in Section 15.3 lack the means for confining the current and the radiation in the lateral (y) direction. As a result, typical broad-area lasers can support more than one transverse (y) mode, resulting in unacceptable mode hopping as well

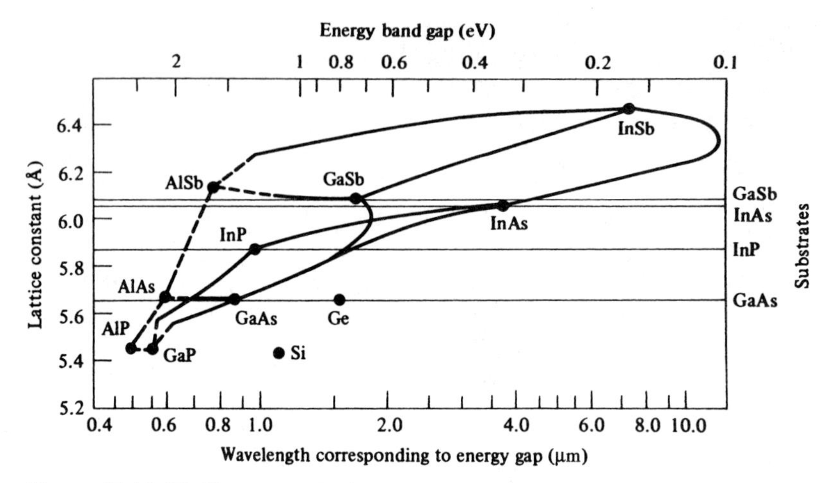

Figure 15.15 III–V compounds: lattice constants versus energy bandgaps and corresponding wavelengths. The solid lines correspond to direct-gap materials and the dashed lines to in-direct-gap materials. The binary-compound substrates that can be used for lattice-matched growth are indicated on the right. (After Reference [11].)

as spatial and temporal instabilities. To overcome these problems, modern semiconductor lasers employ some form of transverse optical and carrier confinement. A typical and successful example of this approach is the buried heterostructure laser [22] shown in Figure 15.16. To fabricate these lasers, the first three layers—n-$Ga_{1-x}Al_xAs$, GaAs, and p-$Ga_{1-y}Al_yAs$—are grown on an n-GaAs crystalline substrate by one of the epitaxial crystal growth techniques described earlier. The structure is then etched through a mask down to the substrate level, leaving stand a thin (~3 μm) rectangular mesa composed of the original layers. A "burying" $Ga_{1-z}Al_zAs$ layer is then regrown on both sides of the mesa, resulting in the structure shown in Figure 15.16.

The most important feature of the buried heterostructure laser is that the active GaAs region is surrounded on *all* sides by the lower index GaAlAs, so that the structure is that of a rectangular dielectric waveguide for optical propagation. The transverse dimensions of the

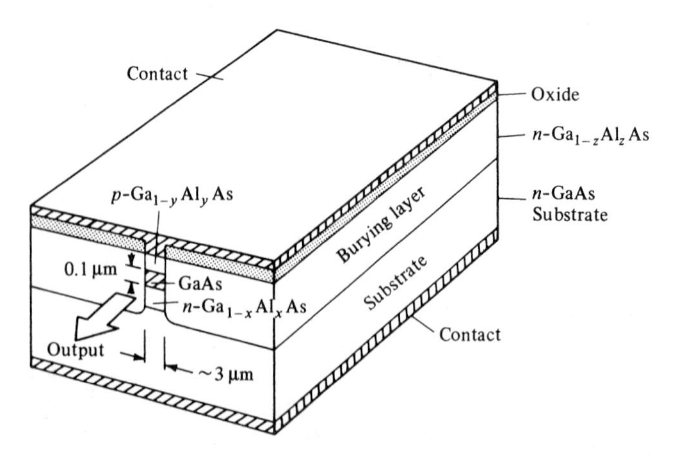

Figure 15.16 A buried heterostructure laser [22].

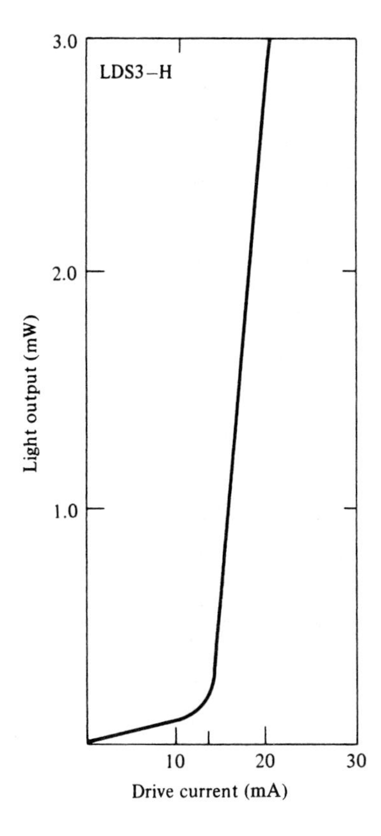

Figure 15.17 Power versus current plot of a low-threshold (~14 mA) commercial DH GaAs/GaAlAs laser. (After Reference [24].)

active region and the index discontinuities (i.e., the molar fractions x, y, and z) are so chosen that only the lowest-order transverse mode can propagate in the laser waveguide. Another important feature of this laser is the confinement of the injected carriers by the boundaries of the active region due to the energy band discontinuity at the GaAs/GaAlAs interfaces as discussed in the last section. These act as potential barriers inhibiting carrier escape out of the active region. GaAs semiconductor lasers utilizing this structure have been fabricated, see Chapter 16, with threshold currents of less than 1 milliampere [23]; more typical lasers have thresholds of ~20 milliamperes. A typical power versus current plot of a commercial laser is shown in Figure 15.17, while the far-field angular intensity distribution is shown in Figure 15.18.

Quaternary GaInAsP Semiconductor Lasers

Optical fiber communications over long distances (say, >10 km) uses, almost exclusively, lasers emitting in spectral regions near 1.3 μm and 1.55 μm. The 1.3 μm lasers are important because the group velocity dispersion of silica-based fibers at this wavelength is very small. The first-order group velocity dispersion parameter D, defined by Equation (3.5-20), is plotted in Figure 3.19 and is zero at $\lambda = 1.3$ μm, so that optical pulses at this wavelength undergo, according to Equation (3.5-22), minimal spread with distance. The wavelength region around 1.55 μm is where the optical absorption coefficient of silica fibers reaches a minimum, which makes it a favorite for long-haul links. Lasers in these wavelength regions [25] are fabricated

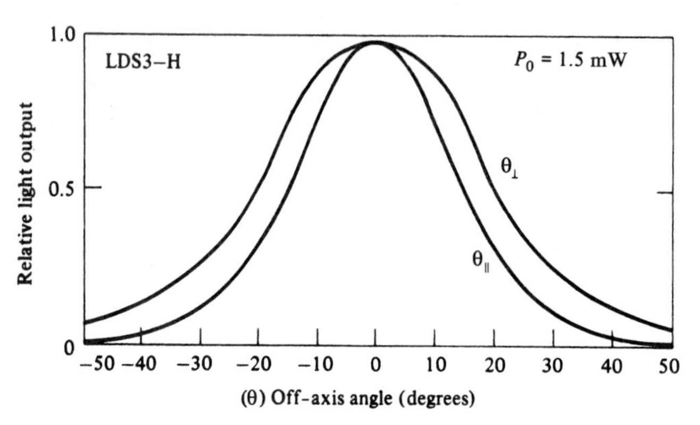

Figure 15.18 Far-field angular intensity distribution of a low-threshold commercial DH GaAs/GaAlAs laser. (After Reference [24].)

using active layers of $GA_{1-x}In_xAs_{1-y}P_y$. From Figure 15.15, we find that such lasers spanning the 0.9 μm < λ < 1.7 μm region can be lattice-matched to InP, which possesses a lower index of refraction to produce dielectric waveguides in which the InP epitaxial layers act as cladding layers to the quaternary $Ga_{1-x}In_xAs_{1-y}P_y$ active layer. The quaternary layer plays in this system the role played by GaAs in the GaAs/GaAlAs laser depicted in Figure 15.10. A typical quaternary laser structure is shown in Figure 15.19. Modern versions of these laser systems employ active regions with thicknesses in the 50–100 Å range. These are the *quantum well lasers*. These lasers possess lower threshold currents and have a larger modulated bandwidth compared to earlier generations employing "thick" (~1000 Å) active regions. They are discussed in detail in Chapter 16. Propagation in optical fibers at 1.55 μm without repeaters has been demonstrated experimentally at distances of ~150 km [26–29].

Power Output of Injection Lasers

The considerations of saturation and power output in an injection laser are basically the same as that of conventional lasers, which were described in Sections 5.7 and 6.4. As the injection current is increased above the threshold value, the laser oscillation intensity builds up. The resulting stimulated emission shortens the lifetime of the inverted carriers to the point where the magnitude of the inversion is clamped at its threshold value. Let η_i be the probability that an injected carrier recombines radiatively within the active region. We can write the following expression for the power emitted by stimulated emission:

$$P_e = \frac{(I - I_t)\eta_i}{e} h\nu \tag{15.4-1}$$

where I is the electric current and I_t is the threshold electric current. The reason for an internal quantum efficiency η_i that is less than unity is, mostly, the existence of a leakage current component that bypasses the active *p-n* junction region. Part of the emitted power P_e is dissipated inside the laser resonator, and the rest is coupled out through the end reflectors. These two powers are, according to Equation (15.3-4), proportional to the effective internal loss $\alpha \equiv \alpha_n\Gamma_n + \alpha_p\Gamma_p + \alpha_s$ and to the mirror loss $-L^{-1}\ln R$, respectively, where α_s is the attenuation coefficient due to other loss mechanisms such as scattering loss and absorption due to impurities, and α_n and α_p are absorption loss (due to carriers) in the n and p layers. We can thus write the output power as

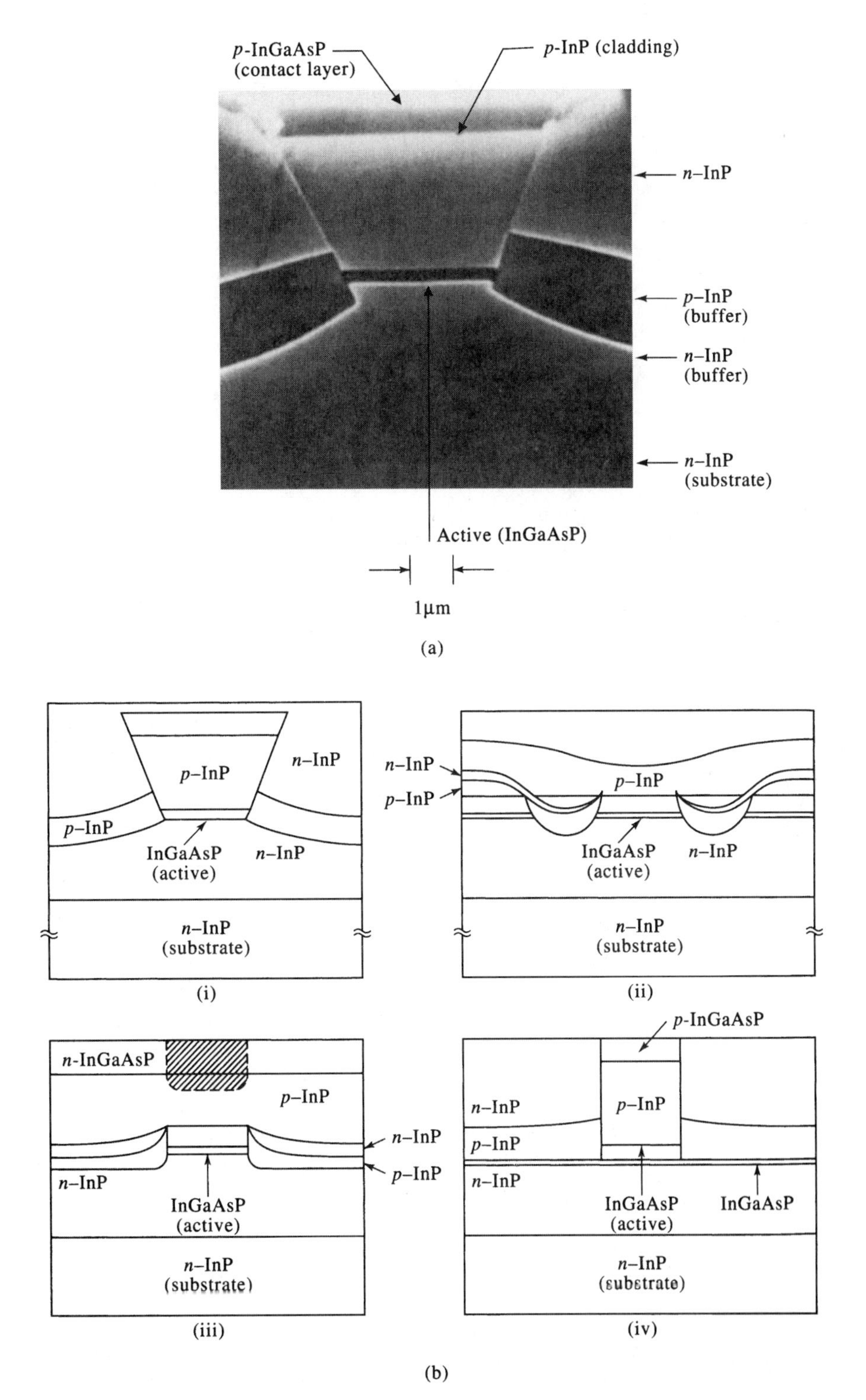

Figure 15.19 Typical structures of buried active region InP/GaInAsP diode lasers: (a) SEM [28, 30]; (b) drawing of different structures [27].

$$P_0 = \frac{(I - I_t)\eta_i h\nu}{e} \frac{(1/L)\ln(1/R)}{\alpha + (1/L)\ln(1/R)} \tag{15.4-2}$$

The external differential quantum efficiency η_{ex} is defined as the ratio of the photon output rate that results from an increase in the injection rate (carriers per second) to the increase in the injection rate:

$$\eta_{ex} = \frac{d(P_0/h\nu)}{d[(I - I_t)/e]} \tag{15.4-3}$$

Using Equation (15.4-2), we obtain

$$\eta_{ex}^{-1} = \eta_i^{-1}\left(\frac{\alpha L}{\ln(1/R)} + 1\right) \tag{15.4-4}$$

By plotting the dependence of η_{ex} on L we can determine η_i, which in GaAs is around 0.9–1.0.

Since the incremental efficiency of converting electrons into useful output photons is η_{ex}, the main remaining loss mechanisms degrading the conversion of electrical to optical power is the small discrepancy between the energy eV_{appl} supplied to each injected carrier and the photon energy $h\nu$. This discrepancy is due mostly to the series resistance of the laser diode. The efficiency of the laser in converting electrical power input to optical power is thus

$$\eta = \frac{P_0}{IV_{appl}} = \eta_i \frac{I - I_t}{I} \frac{h\nu}{eV_{appl}} \frac{\ln(1/R)}{\alpha L + \ln(1/R)} \tag{15.4-5}$$

where I is the electric current and V_{appl} is the applied voltage. In practice, $eV_{appl} \sim 1.4E_g$ and $h\nu \sim E_g$. Values of $\eta \sim 30\%$ at 300 K have been achieved.

15.5 DIRECT-CURRENT MODULATION OF SEMICONDUCTOR LASERS

Since the main application of semiconductor lasers is as sources for optical communications systems, the problem of high-speed modulation of their output by the high-data-rate information is one of great technological importance. A unique feature of semiconductor lasers is that, unlike other lasers that are modulated externally (see Chapter 9), the semiconductor laser can be modulated directly by modulating the injection current. This is especially important in view of the possibility of monolithic integration of the laser and the modulation electronic circuit, as will be discussed in Section 15.7. The following treatment follows closely that of Reference [31].

If we denote the photon density inside the active region of a semiconductor laser by P and the injected electron (and hole) density by N, then we can write

$$\begin{aligned} \frac{dN}{dt} &= \frac{I}{Ve} - \frac{N}{\tau} - A(N - N_{tr})P \\ \frac{dP}{dt} &= A(N - N_{tr})P\Gamma_a - \frac{P}{\tau_p} \end{aligned} \tag{15.5-1}$$

where I is the total injection current, V is the volume of the active region, τ is the spontaneous recombination lifetime, and τ_p is the photon lifetime as limited by absorption in the bounding media, scattering and coupling through the output mirrors. The term $A(N - N_{tr})P$ is the net rate per unit volume of induced transitions. N_{tr} is the inversion density needed to achieve transparency as defined by Equation (15.2-17). The constant A is a temporal growth coefficient that by definition is related to the constant B defined by Equation (15.2-17) by the relation $A = Bc/n$, with n being the index of refraction and c being the speed of light in vacuum. Γ_a is the mode confinement factor defined by Equation (15.3-3), and its presence here is to ensure that the total number, rather than the density variables used in (15.5-1), of electrons undergoing stimulated transitions is equal to the number of photons emitted. The contribution of spontaneous emission to the photon density is neglected since only a very small fraction ($\sim 10^{-4}$) of the spontaneously emitted power enters the lasing mode.

By setting the left sides of Equations (15.5-1) equal to zero, we obtain the steady-state solutions N_0 and P_0

$$\begin{aligned} 0 &= \frac{I_0}{Ve} - \frac{N_0}{\tau} - A(N_0 - N_{tr})P_0 \\ 0 &= A(N_0 - N_{tr})P_0\Gamma_a - \frac{P_0}{\tau_p} \end{aligned} \tag{15.5-2}$$

The above equations can be solved for the steady-state electron density N_0 and photon density P_0.

For the purpose of studying the current injection, we consider the case where the current is made up of dc and ac components

$$I = I_0 + i_1 e^{i\omega_m t} \tag{15.5-3}$$

where ω_m is the modulation frequency, and I_0 and i_1 are constants. Under the circumstance, we define the small-signal modulation response n_1 and p_1 by

$$N = N_0 + n_1 e^{i\omega_m t} \quad \text{and} \quad P = P_0 + p_1 e^{i\omega_m t} \tag{15.5-4}$$

where N_0 and P_0 are the steady-state solutions of (15.5-2), and n_1 and p_1 are constants.

Using Equations (15.5-3) and (15.5-4), and the result $A(N_0 - N_{tr}) = (\tau_p\Gamma_a)^{-1}$ from (15.5-2) in (15.5-1), we obtain the following algebraic equations for n_1 and p_1:

$$\begin{aligned} -i\omega_m n_1 &= -\frac{i_1}{Ve} + \left(\frac{1}{\tau} + AP_0\right)n_1 + \frac{1}{\tau_p\Gamma_a}p_1 \\ i\omega_m p_1 &= -AP_0\Gamma_a n_1 \end{aligned} \tag{15.5-5}$$

Our main interest is in the modulation response of the photon density $p_1(\omega_m)/i_1(\omega_m)$ so that from (15.5-5) we obtain

$$p_1(\omega_m) = \frac{-(i_1/Ve)AP_0\Gamma_a}{\omega_m^2 - i\omega_m/\tau - i\omega_m AP_0 - AP_0/\tau_p} \tag{15.5-6}$$

A typical measurement of $p_1(\omega_m)$ is shown in Figure 15.20b. The response curve (normalized) is flat at low frequencies, peaks at the "relaxation resonance frequency" ω_R, and then drops steeply. The expression for the peak frequency is obtained by minimizing the magnitude of the denominator of Equation (15.5-6):

$$\omega_R = \sqrt{\frac{AP_0}{\tau_p} - \frac{1}{2}\left(\frac{1}{\tau} + AP_0\right)^2} \tag{15.5-7}$$

In a typical semiconductor laser with $L = 300\ \mu\text{m}$, we have from Equation (4.7-3) $\tau_p = (n/c)[\alpha - (1/L)\ln R]^{-1} \sim 10^{-12}$ s, $\tau \sim 4 \times 10^{-9}$ s, and $AP_0 \sim 10^9\ \text{s}^{-1}$ so that to a very good accuracy

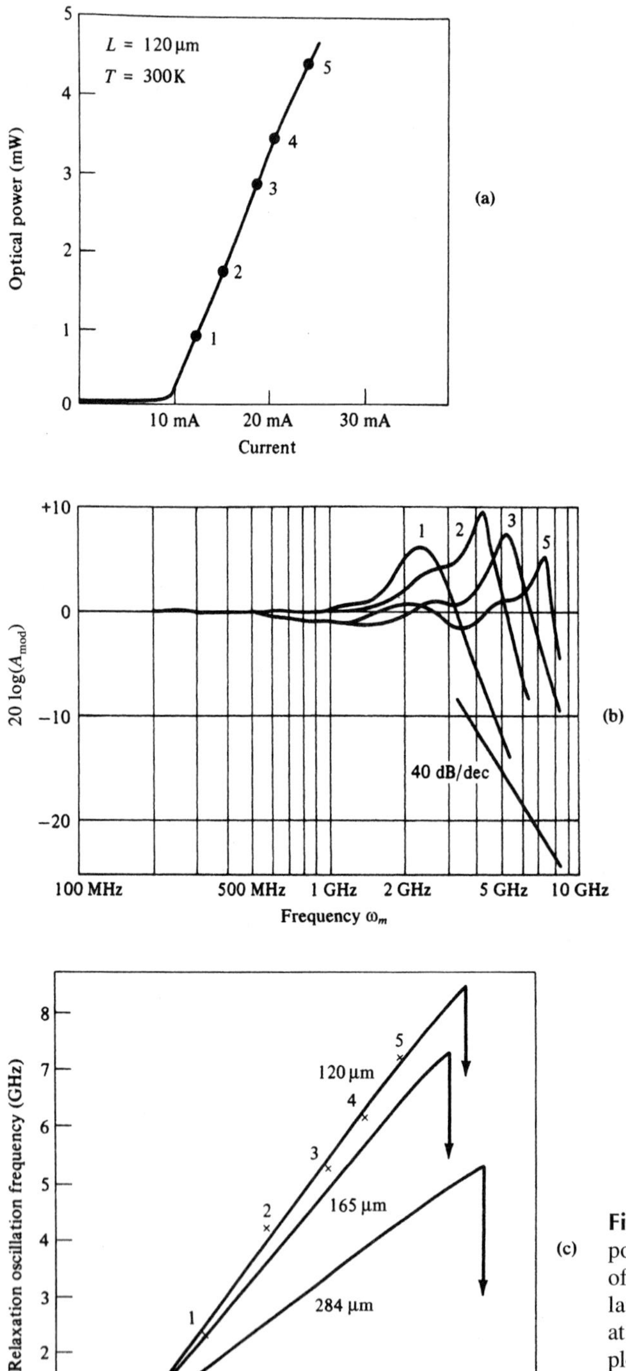

Figure 15.20 (a) The cw light output power versus current characteristic of a laser of length 120 μm. (b) Modulation characteristics of this laser at various bias points indicated in the plot. (c) Measured relaxation oscillation resonance frequency of lasers of various cavity lengths as a function of $\sqrt{P}$, where P is the cw output optical power. The points of catastrophic damage are indicated by downward pointing arrows. (After Reference [31].)

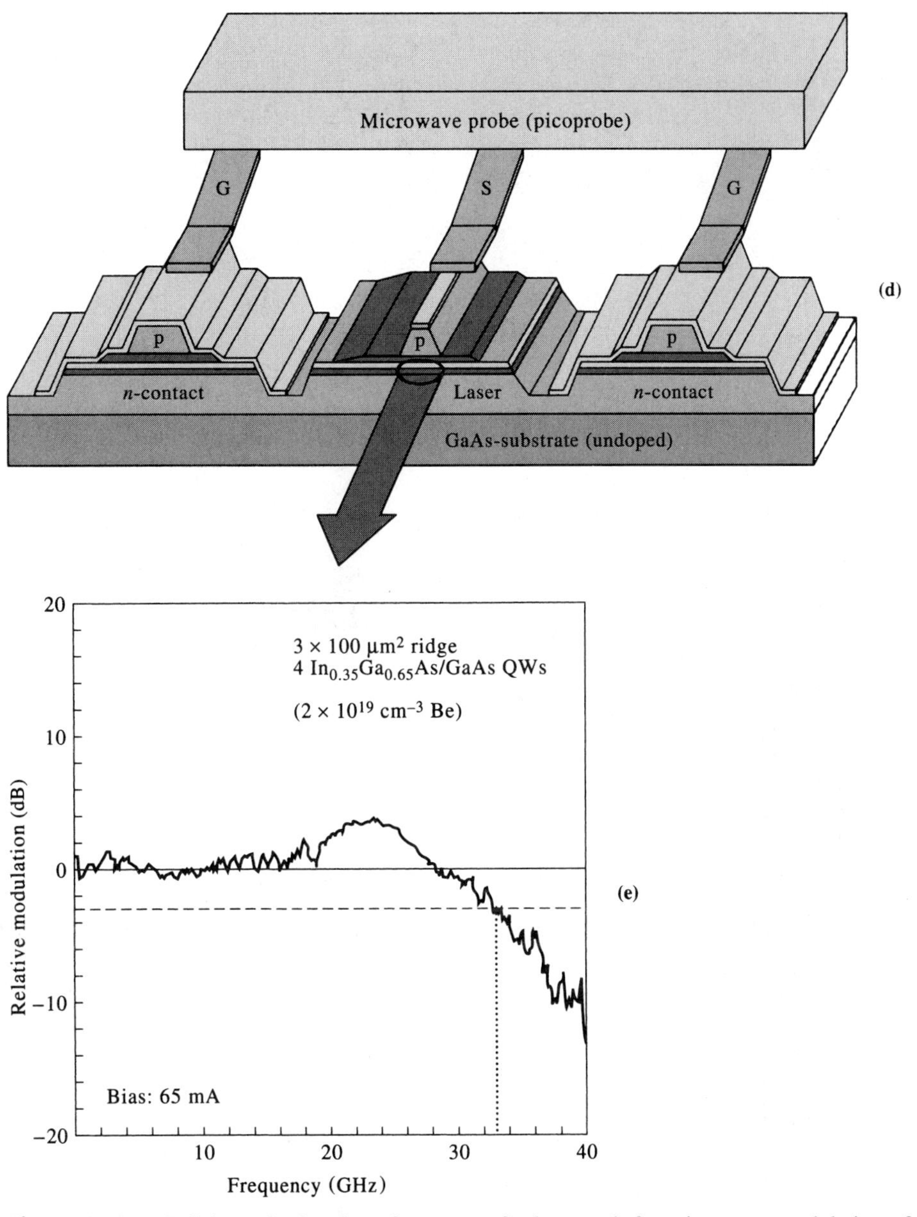

Figure 15.20 (d) Schematic drawing of a current feed network for microwave modulation of high-speed lasers. (e) The corresponding frequency response. (After Reference [28].)

$$\omega_R = \sqrt{\frac{AP_0}{\tau_p}} \qquad (15.5\text{-}8)$$

The last result is extremely useful, since it suggests that to increase ω_R and thus increase the useful linear region of the modulation response $p_1(\omega_m)/i_1(\omega_m)$, we need to increase the optical gain coefficient A, decrease the photon lifetime τ_p, and operate the laser at as high internal photon density P_0 as possible. The observed linear dependence of the modulation resonance frequency ω_R on the square root of the power output $\sqrt{P}$ is demonstrated in Figure 15.20c for lasers of various lengths. A detailed discussion of the optimum strategy for maximizing ω_R is given in Reference [31]. Figure 15.20d shows the schematic drawing of a microwave current feeding electrodes for high-frequency modulation, and Figure 15.20e shows the corresponding frequency response.

It is somewhat tedious but straightforward to show that Equation (15.5-8) can also be written

$$\omega_R = \sqrt{\frac{1 + A\tau_p\Gamma_a N_{tr}}{\tau\tau_p}\left(\frac{I_0}{I_{th}} - 1\right)} \tag{15.5-9}$$

where I_{th} is the threshold current and I_0 is the operation current.

EXAMPLE: MODULATION BANDWIDTH IN GaAs/GaAlAs LASERS

Here, using Equation (15.5-8), we will estimate the uppermost useful modulation frequency ω_R of a typical GaAs/GaAlAs laser. We shall assume a typical laser emitting 5×10^{-3} watt from a single face with an active area cross section of 3 μm × 0.1 μm, a facet reflectivity of $R = 0.31$, and an index of refraction $n_0 = 3.5$. Solving for P_0 from the following relationship,

$$\frac{(1-R)P_0 ch\nu}{n_0} = \frac{\text{Power}}{\text{Area}}$$

we obtain $P_0 = 1.21 \times 10^{15}$ photons/cm^3 for the photon density in the laser cavity. The constant A has a typical value of 2×10^{-6} cm^3/s. (This can be checked against the relationship $A = Bc/n_0$, where B is the spatial gain parameter of Equation (15.2-17).) The photon lifetime τ_p is obtained from Equation (4.7-3),

$$\tau_p = \frac{n_0}{c}\left(\alpha - \frac{1}{L}\ln R\right)^{-1}$$

which for $L = 120$ μm, $\alpha = 10$ cm^{-1}, and $R = 0.31$ yields $\tau_p \sim 1.08 \times 10^{-12}$ s. Combining these results gives

$$\nu_R \equiv \frac{\omega_R}{2\pi} = \frac{1}{2\pi}\sqrt{\frac{AP_0}{\tau_p}} = \frac{1}{2\pi}\sqrt{\frac{2 \times 10^{-6} \times 1.2 \times 10^{15}}{1.08 \times 10^{-12}}}$$
$$= 7.53 \times 10^{9} \text{ Hz}$$

This value is in the range of the experimental data shown in Figure 15.20, which was obtained on a laser with characteristics similar to that used in our example. The square root law dependence of ω_R on the photon density (or power output) predicted by Equation (15.5-8) is verified by the data of Figure 15.20c.

15.6 GAIN SUPPRESSION AND FREQUENCY CHIRP IN CURRENT-MODULATED SEMICONDUCTOR LASERS

In Section 15.5, we solved for the modulation of the optical power output (or, equivalently, the photon density inside the laser resonator), which is due to a modulation of the current flowing through the laser. The discussion leads to an understanding of the useful range of the modulation speed. Here, in this section, we discuss the effect of current modulation on the carrier density in the laser. The current is taken as

$$I(t) = I_0 + i_1(\omega_m) \exp(i\omega_m t) \tag{15.6-1}$$

while the photon density inside the laser, which is proportional to the optical power output, is

$$P(t) = P_0 + p_1(\omega_m) \exp(i\omega_m t) \tag{15.6-2}$$

where I_0, P_0, i_1, and p_1 are constants, and ω_m is the modulation frequency. Physically, $i_1(\omega_m)$ is the amplitude of the current modulation, while $p_1(\omega_m)$ is the modulation amplitude of the photon density. We also take the carrier density in the active regions (the inverted population) as

$$N(t) = N_0 + n_1(\omega_m) \exp(i\omega_m t) \tag{15.6-3}$$

where N_0 and n_1 are constants.

Ideally we would like $p_1(\omega_m)/i_1(\omega_m)$, the frequency modulation response, to be a constant independent of ω_m and, above threshold, we will expect that $n_1(\omega_m) = 0$, indicating perfect gain clamping. As we shall find out, neither expectation is realized fully. As a matter of fact, if we solve Equations (15.5-5) for $n_1(\omega_m)$, the result is

$$n_1(\omega_m) = -i\left(\frac{i_1}{Ve}\right)\frac{\omega_m}{\omega_m^2 - \dfrac{AP_0}{\tau_p} - i\omega_m\left(\dfrac{1}{\tau} + AP_0\right)} \tag{15.6-4}$$

where we recall that τ is the spontaneous recombination lifetime, τ_p is the photon cavity lifetime, and A is a constant proportional to the constant B in Equation (15.2-17). We thus find that, under dynamic conditions, the carrier density, hence the gain, is not clamped at the threshold value N_0 but has an oscillating component whose amplitude $n_1(\omega_m)$ is given by Equation (15.6-4). Its general features are depicted in Figure 15.21. We note a peak at ω_R, which is also the resonance frequency for the amplitude modulation response $p_1(\omega_m)/i_1(\omega_m)$, as given by Equation (15.5-7). It is known that injected electrons in semiconductors have a contribution to the magnitude of the index of refraction. Thus the modulation of the carrier density is accompanied by a modulation of the index of refraction, leading to a frequency

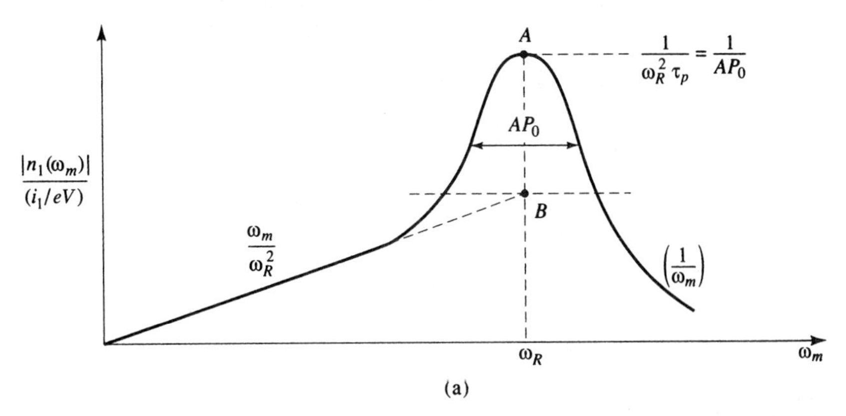

Figure 15.21 (a) A theoretical plot of the carrier density modulation n_1 as a function of the current modulation frequency ω_m. (b) (1) A scanning Fabry–Perot spectrum of a GaInAsP ($\lambda = 1.31$ μm) DFB laser with no current modulation. (2) The spectrum of the same laser when the current is modulated at $f_m = 550$ MHz, horizontal scale = 1 GHz/div. (Courtesy of H. Blauvelt, P. C. Chen, and N. Kwong of ORTEL Corporation, Alhambra, California)

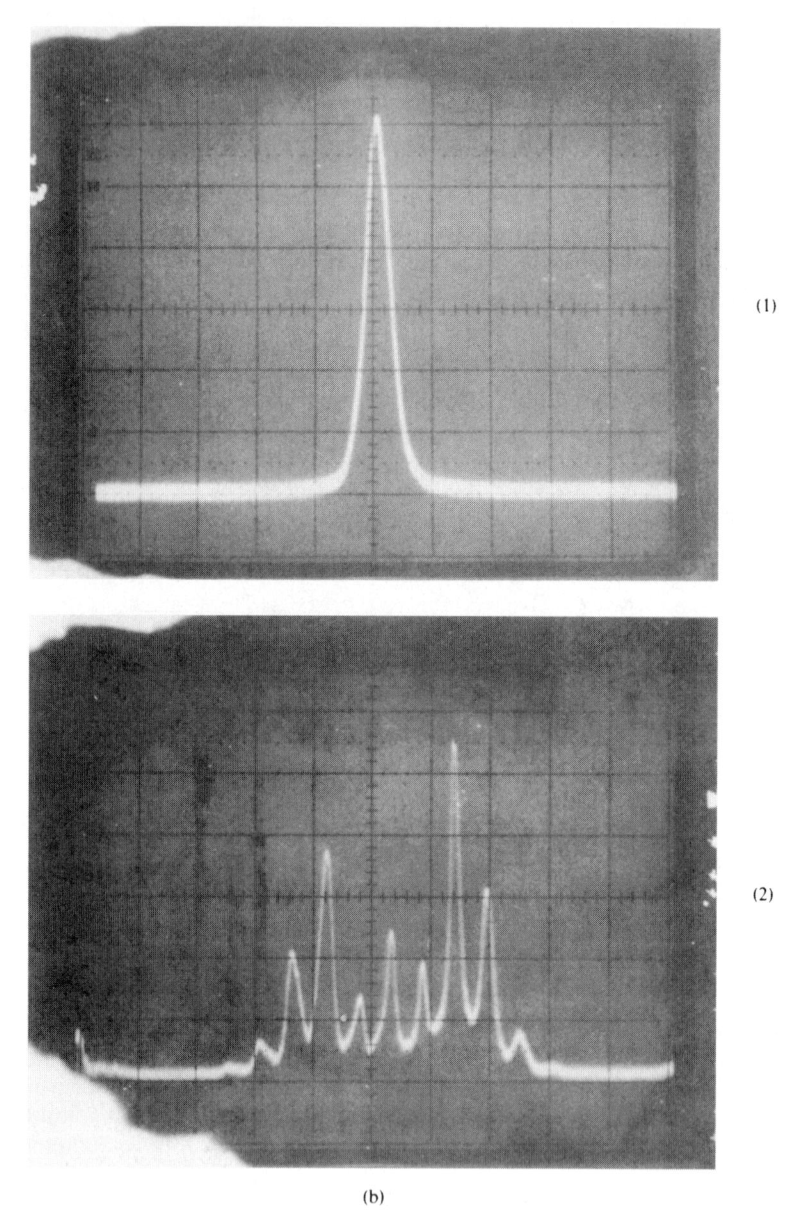

Figure 15.21 *(continued)*

modulation (FM) of the output optical field. This parasitic FM, most of the time undesired, has a number of important consequences. The most important of these is the resulting spectral broadening of the laser field that, in dispersive fibers ($D \neq 0$), leads to an increase in the spreading of optical pulses with distance, as discussed in Section 3.5.

Before embarking on the analysis of the parasitic frequency modulation, we need to introduce two new physical concepts: (1) the gain suppression effect and (2) the amplitude–phase coupling effect.

Gain Suppression

The gain experienced in an inverted semiconductor laser medium by an optical wave is invariably lower the higher the optical intensity. This is due partly to optical gain saturation and partly to gain suppression. The first effect reflects the drop of the total electron population density N in (15.5-1) with the increase in photon density P. This effect is accounted for properly by the coupled rate equations (15.5-1). The second mechanism reducing the gain takes place even when the total carrier density N is constant and reflects the reduction of the density of *resonant carriers* (electrons and holes) in the *immediate vicinity* of points a and b in Figure 15.5, which contribute to the gain. This is due both to spectral "hole burning," as discussed in Section 6.10 and illustrated in Figure 6.37, and to an increase in the effective electron temperature by the optical field. By electron "temperature" we mean the temperature used in the Fermi function (15.1-8). Under dynamic nonthermal equilibrium conditions, this temperature may differ from the lattice temperature. Such an increase causes, according to the discussion in Sections 15.1 and 15.2, the electron (and hole) distribution to broaden. Thus more carriers are spread to higher energies, reducing in the process the density of resonant carriers that contribute to the gain. This effect asserts itself with a time constant of $<10^{-12}$ s, characteristic of electron–electron and electron–phonon collisions, and for system applications involving modulation at $\omega_m/2\pi < 3 \times 10^{10}$ Hz, it can be considered as responding instantaneously to the optical field.

Our main departure from the analyses of Section 15.5 is to take the optical gain constant as

$$G(N, P) = G(N)(1 - \varepsilon P) \approx G(N_{\text{th}}) + A(N - N_{\text{th}}) - \varepsilon G(N_{\text{th}})P \tag{15.6-5}$$

where N is the carrier density, P is the photon density, and the gain suppression (due to photon density) is represented by the factor $(1 - \varepsilon P)$. The constant ε is called the *gain suppression factor*. We can view Equation (15.6-5) as a Taylor series expansion about the threshold point $N = N_{\text{th}}$, $P = 0$. The numerical value of ε can be estimated theoretically in a given system but more often is evaluated experimentally [29]. We now rewrite the rate equations (15.5-1) as

$$\begin{aligned} \frac{dN}{dt} &= \frac{I}{Ve} - \frac{N}{\tau} - G(N, P)P \\ \frac{dP}{dt} &= \Gamma_a G(N, P)P - \frac{P}{\tau_p} \end{aligned} \tag{15.6-6}$$

At steady state, $d/dt = 0$, we obtain

$$0 = \frac{I_0}{Ve} - \frac{N_0}{\tau} - [G(N_{\text{th}}) + A(N_0 - N_{\text{th}}) - \varepsilon G(N_{\text{th}})P_0]P_0$$

$$0 = \Gamma_a[G(N_{\text{th}}) + A(N_0 - N_{\text{th}}) - \varepsilon G(N_{\text{th}})P_0]P_0 - \frac{P_0}{\tau_p}$$

The value of $G(N_{\text{th}})$ is obtained from the second equation evaluated at threshold ($N_0 = N_{\text{th}}$, $P_0 = 0$)

$$G(N_{\text{th}}) = \frac{1}{\Gamma_a \tau_p}$$

We use the last result to simplify the last two equations

$$0 = \frac{I_0}{Ve} - \frac{N_0}{\tau} - \frac{P_0}{\Gamma_a \tau_p}$$
$$0 = A(N_0 - N_{\text{th}}) - \frac{\varepsilon P_0}{\Gamma_a \tau_p} \tag{15.6-7}$$

Performing a "small signal" expansion as in Equations (15.5-3) and (15.5-4) leads to

$$i\omega_m n_1 = \frac{i_1}{Ve} - \left(\frac{1}{\tau} + AP_0\right) n_1 - \frac{(1 - \varepsilon P_0)}{\tau_p \Gamma_a} p_1$$
$$i\omega_m p_1 = \Gamma_a A P_0 n_1 - \frac{\varepsilon P_0}{\tau_p} p_1 \tag{15.6-8}$$

We note that for the case $\varepsilon = 0$, these equations reduce to Equations (15.5-5). Solving (15.6-8) for $p_1(\omega_m)$ yields

$$p_1(\omega_m) = \frac{-\Gamma_a A P_0 \left(\dfrac{i_1}{Ve}\right)}{\omega_m^2 - i\omega_m \left(\dfrac{1}{\tau} + AP_0 + \dfrac{\varepsilon P_0}{\tau_p}\right) - \left(\dfrac{AP_0}{\tau_p} + \dfrac{\varepsilon P_0}{\tau \tau_p}\right)} \tag{15.6-9}$$

Under typical conditions such as those of the earlier example, we have

$P_0 = 1.2 \times 10^{21}$ photons/m^3

$A = 2 \times 10^{-12}$ m^3/s $\qquad \tau = 4 \times 10^{-9}$ s

$\tau_p = 10^{-12}$ s $\qquad \varepsilon = 10^{-23}$ m^3

so that $\varepsilon P_0/\tau_p = 1.2 \times 10^{10}\ \text{s}^{-1} \gg AP_0$.

Using the inequalities $\varepsilon P_0/\tau_p \gg AP_0$ and $\varepsilon/\tau_p \gg A$, we find that the peak modulation response occurs at

$$\omega_R \approx \sqrt{\frac{AP_0}{\tau_p} - \frac{\varepsilon^2 P_0^2}{2\tau_p^2}}$$

A comparison of this last result to Equation (15.5-8) shows that when $\varepsilon \neq 0$, that is, gain suppression is considered, the modulation resonant frequency, ω_R, does not increase indefinitely with P_0 but reaches a maximum value at a photon density of

$$(P_0)_{\max} = \frac{A\tau_p}{\varepsilon^2} \tag{15.6-10}$$

The maximum value of ω_R, which obtains when $P_0 = (P_0)_{\max}$, is

$$(\omega_R)_{\max} = \frac{A}{\sqrt{2}\varepsilon} \tag{15.6-11}$$

Using the typical numerical constants given following Equation (15.6-9) and in the example of Section 15.5, we estimate

$$(P_0)_{\max} = \frac{A\tau_p}{\varepsilon^2} = \frac{2 \times 10^{-12} \times 10^{-12}}{10^{-46}} = 2 \times 10^{22}\ \text{photons/m}^2$$

which, using the expressions given in the example of Section 15.5, corresponds to a power output of ~80 mW per facet. The corresponding maximal resonant modulation frequency is

$$\left(\frac{\omega_R}{2\pi}\right)_{\max} = \left(\frac{1}{2\pi}\right)\frac{A}{\sqrt{2\varepsilon}} = \frac{2\times 10^{-12}}{2\pi\sqrt{2}\times 10^{-23}} = 2.24\times 10^{10}\ \text{Hz}$$

The last result points out the role of the gain suppression in placing a practical upper limit on the modulation bandwidth that can be achieved by current modulation of semiconductor lasers. To modulate at significantly higher frequencies, $\omega_m > \omega_R$, one needs to employ external modulators, such as those that are discussed in Chapter 9.

Amplitude–Phase coupling

Amplitude–phase coupling, which causes a second effect that is attendant upon the carrier density modulation ($n_1(\omega_m) \neq 0$), is the frequency modulation ("chirp") of the laser output field. The amplitude $n_1(\omega_m)$ of the carrier density fluctuation is obtained by solving (15.6-8):

$$n_1(\omega_m) = \frac{-\left(i\omega_m + \dfrac{\varepsilon P_0}{\tau_p}\right)\left(\dfrac{i_1}{Ve}\right)}{\omega_m^2 - \left(\dfrac{AP_0}{\tau} + \dfrac{\varepsilon P_0}{\tau\tau_p}\right) - i\omega_m\left(\dfrac{1}{\tau} + AP_0 + \dfrac{\varepsilon P_0}{\tau_p}\right)} \tag{15.6-12}$$

A comparison of Equations (15.6-12) and (15.6-9) shows that

$$n_1(\omega_m) = \frac{\left(i\omega_m + \dfrac{\varepsilon P_0}{\tau_p}\right)}{\Gamma_a A P_0} p_1(\omega_m) \tag{15.6-13}$$

Since for our assumed $\exp(i\omega_m t)$ time dependence, $i\omega_m = d/dt$, we use Equation (15.6-13) to write

$$\Delta N(t) = \frac{1}{\Gamma_a A}\left(\frac{1}{P_0}\frac{dP}{dt} + \frac{\varepsilon}{\tau_p}\Delta P(t)\right) \tag{15.6-14}$$

where $N(t) = N_0 + \Delta N(t)$ and $P(t) = P_0 + \Delta P(t)$. Equation (15.6-14) applies to any arbitrary photon density modulation, not necessarily harmonic.

Our next task is to find the effect of the carrier density modulation $\Delta N(t)$ on the laser frequency. As discussed in Chapter 5, the index of refraction of a gain medium is a complex number. We write

$$n_0(t) = n_0'(t) - in_0''(t) \tag{15.6-15}$$

where the imaginary part n_0'' accounts for the gain or absorption, and the time dependence reflects the dependence of the real part n_0' and the imaginary part n_0'', on the time-modulated carrier density $\Delta N(t)$.

The imaginary part of the index n_0 is related to the spatial exponential gain constant of the laser medium since the spatial dependence of the field is

$$E(z) \propto E_0 \exp\left(-\frac{i\omega}{c}(n_0' - in_0'')z\right) = E_0\exp\left(-\frac{i\omega n_0'}{c}z\right)\exp\left(-\frac{\omega n_0''}{c}z\right)$$

Since our rate equations (15.6-6) and (15.6-8) are in the time domain, we need to convert the spatial growth parameter $-\omega n_0''/c$ to a temporal one. We use

$$\frac{d|E|}{dt} = \frac{\partial |E|}{\partial z}\frac{dz}{dt} \cong -\left(\frac{\omega n_0''}{c}\right)\frac{c}{n_0'}|E| = -\frac{\omega n_0''}{n_0'}|E|$$

where we recall n_0'' is negative in a gain medium. It thus follows that the exponential gain constant $G(N, P)$ of (15.6-6) is related to n_0'' by

$$G = -\frac{2\omega n_0''}{n_0'} = -\frac{4\pi\nu n_0''}{n_0'} \tag{15.6-16}$$

where the factor of 2 accounts for the fact that G is the temporal growth constant of the photon density ($\propto$ optical intensity), that is, of $|E|^2$.

From the last equation, it follows that

$$\frac{\partial n_0''}{\partial N} = -\frac{n_0'}{4\pi\nu}\frac{\partial G}{\partial N} = -\frac{n_0'}{4\pi\nu}A \tag{15.6-17}$$

where we used Equation (15.6-5) to substitute $\partial G/\partial N = A$. A modulation $\Delta N(t)$ of the carrier density causes, according to Equation (15.6-17), a corresponding perturbation

$$\Delta n_0'' = -\frac{n_0'}{4\pi\nu} A\, \Delta N(t) \tag{15.6-18}$$

Our next task is to obtain the dependence of the real index perturbation $\Delta n_0'$ on $\Delta N(t)$. Now $\Delta n_0'$ and $\Delta n_0''$ are related through Kramers–Kroning relations, which are discussed in Appendix C. It has, however, proved useful to relate $\Delta n_0'$ and $\Delta n_0''$ by means of a parameter, the so-called phase–amplitude coupling constant, the α parameter [32, 33].

$$\alpha = \frac{\Delta n_0'}{\Delta n_0''} \tag{15.6-19}$$

The α parameter (also known as the Henry α-factor [33]) is a function of the carrier density N_0 of the semiconductor laser medium, and typical values for it range between 3 and 5 [32, 34]. It is important not to confuse it with the attenuation coefficient. The same α-factor is also responsible for the enhancement of linewidth in semiconductor lasers (see Equation (10.7-31)).

Combining Equations (15.6-19) and (15.6-18) results in

$$\Delta n_0' = -\frac{\alpha n_0' A}{4\pi\nu}\Delta N(t) \tag{15.6-20}$$

A perturbation $\Delta n_0'$ due to carrier density modulation causes a perturbation $\Delta\nu$ of the laser frequency

$$\frac{\Delta\nu}{\nu} = -\frac{\Delta n_0'}{n_0'}\Gamma_a = \frac{\alpha\Gamma_a A}{4\pi\nu}\Delta N(t) \tag{15.6-21}$$

The first equality of (15.6-21) follows directly from the basic Fabry–Perot resonance frequency relation, Equation (6.2-2). We assume that the index change of $\Delta n_0'$ due to injected carriers occurs only in the active region. The confinement factor Γ_a accounts for (possibly) partial filling of the resonator by the active medium. Substitution in the last equation for $\Delta N(t)$ from Equation (15.6-14) [32, 35] gives

$$\Delta\nu(t) = \frac{\alpha}{4\pi}\left(\frac{1}{P_0}\frac{dP}{dt} + \frac{\varepsilon}{\tau_p}\Delta P(t)\right) \tag{15.6-22}$$

for the laser frequency chirp. We note that the frequency chirp is proportional to the phase–amplitude coupling parameter α. The contribution involving dP/dt is called the *transient chirp*, while that which is proportional to $\Delta P(t)$ is termed the *adiabatic chirp*. The latter involves the gain suppression parameter ε and is usually dominant at low ($\leq 10^8$ Hz) frequencies, while the transient term dominates at typical microwave ($>10^9$ Hz) frequencies.

The Field Spectrum of a Chirping Laser

In the development leading to Equation (15.6-22), we showed that any modulation of the power of a semiconductor laser by means of current modulation causes a frequency chirp. In the following, we will derive the spectrum of the output optical field of a laser with sinusoidal power modulation in order to obtain an appreciation of the expected order of magnitude of the effect, especially the amount of spectral spread [36, 37]. We take the optical field of the laser as

$$E(t) = \left(E_0 + \frac{s}{2} E_0 \sin \omega_m t \right) \exp\{i[2\pi\nu_0 t + \phi(t)]\} \tag{15.6-23}$$

where E_0 and s are constants, ω_m is the modulation frequency, $\phi(t)$ is a time-varying phase, and ν_0 is the average optical frequency. The dimensionless constant s is a measure of the depth of modulation. For a field with a time-varying phase, the local carrier frequency $\nu(t)$ of the above field is $\nu(t) = \nu_0 + (1/2\pi)(d\phi/dt)$ and corresponds to the number of optical cycles per second at time t. The phase $\phi(t)$ in Equation (15.6-23) is related to the chirp $\Delta\nu(t)$ of Equation (15.6-22) by the general result

$$\phi(t) = 2\pi \int_0^t \Delta\nu(t')\, dt' = \frac{\alpha}{2}\left(\frac{1}{P_0} P(t) + \frac{\varepsilon}{\tau_p} \int_0^t \Delta P(t')\, dt' \right) \tag{15.6-24}$$

If $\phi(t)$ were a constant, which would be the case when $\alpha = 0$, the spectrum of the topical field $E(t)$ would consist of a carrier of amplitude E_0 at ν_0 and two sidebands with amplitudes $s/4$. When $\phi(t)$ is not constant, additional sidebands appear and the optical spectrum broadens. This chirp-induced broadening is of special concern in applications that involve high-data-rate communications in fibers, since, as shown in Sections 1.5 and 3.5, the temporal broadening of pulses in dispersive fibers is directly proportional to the product of the spectral width of the light and the propagation length. Any increase in the laser's spectral width would thus limit the rate at which the data can be transmitted in such a fiber. In the case considered here, we can write the photon density $P(t)$ as

$$P(t) \propto \langle |E(t)|^2 \rangle = E_0^2 \left(1 + \frac{s^2}{4} \right) + sE_0^2 \sin \omega_m t = P_0 + P_1 \sin \omega_m t \tag{15.6-25}$$

where the angular brackets < > indicate averaging over a few optical periods. Substitution of the last expression into Equation (15.6-24) leads to

$$\phi(t) = \frac{\alpha}{2} \left[\underset{\text{transient}}{\frac{P_1}{P_0} \sin \omega_m t} \quad \underset{\text{adiabatic}}{- \frac{\varepsilon P_1}{\omega_m \tau_p} \cos \omega_m t} \right]$$

where we left out time-independent terms that correspond to unimportant fixed phase shifts. At high modulation frequencies such that $\omega_m \gg \varepsilon P_0/\tau_p$ (this corresponds, using the numerical data used earlier in this section, to $\omega_m/2\pi \gg 2 \times 10^9$ Hz), the first term ("transient") in the square brackets dominates, so that the optical phase can be taken as

$$\phi(t) = \frac{m_1 \alpha}{2} \sin \omega_m t$$
$$m_1 \equiv \frac{P_1}{P_0} = \frac{s}{1 + s^2/4} \approx s \tag{15.6-26}$$

when s (and m_1) $\ll 1$. The total optical field, Equation (15.6-23), assumes the form

$$E(t) = \left(E_0 + \frac{m_1}{2} E_0 \sin(\omega_m t)\right) \exp\left[i\left(\omega_0 t + \frac{m_1 \alpha}{2} \sin(\omega_m t)\right)\right] \tag{15.6-27}$$

where we used $s \sim m_1$ ($s \ll 1$).

We can use the Bessel function identity

$$e^{i\delta \sin x} = \sum_{n=-\infty}^{\infty} J_n(\delta) e^{inx} \tag{15.6-28}$$

to rewrite Equation (15.6-27):

$$\frac{E(t)}{E_0} = \sum_{n=-\infty}^{\infty} J_n(\delta) \exp[i(\omega_0 + n\omega_m)t] - i\frac{m_1}{4} \sum_{n=-\infty}^{\infty} J_n(\delta) \exp[i(\omega_0 + (n+1)\omega_m)t]$$

$$+ i\frac{m_1}{4} \sum_{n=-\infty}^{\infty} J_n(\delta) \exp[i(\omega_0 + (n-1)\omega_m)t]$$

Some of the sidebands are:

$$\begin{aligned}
&\text{at } \omega_0, && E_0\left(J_0(\delta) + i\frac{m_1}{2} J_1(\delta)\right) \exp(i\omega_0 t) \\
&\text{at } \omega_0 + \omega_m, && E_0\left(J_1(\delta) - i\frac{m_1}{4} J_0(\delta) + i\frac{m_1}{4} J_2(\delta)\right) \exp[i(\omega_0 + \omega_m)t] \\
&\text{at } \omega_0 - \omega_m, && E_0\left(-J_1(\delta) + i\frac{m_1}{4} J_0(\delta) - i\frac{m_1}{4} J_2(\delta)\right) \exp[i(\omega_0 - \omega_m)t] \\
&\text{at } \omega_0 + 2\omega_m, && E_0\left(J_2(\delta) - i\frac{m_1}{4} J_1(\delta) + i\frac{m_1}{4} J_3(\delta)\right) \exp[i(\omega_0 + 2\omega_m)t] \\
&\text{at } \omega_0 - 2\omega_m, && E_0\left(J_2(\delta) - i\frac{m_1}{4} J_1(\delta) + i\frac{m_1}{4} J_3(\delta)\right) \exp[i(\omega_0 - 2\omega_m)t]
\end{aligned} \tag{15.6-29}$$

where $\delta = m_1\alpha/2$ is the phase modulation index of the optical field. The amplitudes of the sidebands at $\omega_0 \pm n\omega_m$, in this case, have the same magnitude. This is a consequence of the form of (15.6-27) but is not generally true, so that the optical sidebands, in general, for m_1 and δ not zero, are not symmetric about ω_0. This is considered in Problem 15.11. An experimental graph showing the spectrum of the output field of a laser whose current is modulated is given in Figure 15.21b. The spectrum can be fit well with an adiabatic chirp spectrum (i.e., phase modulation 90° out of phase with current), corresponding to a field

$$E(t) = \left(E_0 + \frac{m_1}{2} E_0 \sin(\omega_m t)\right) \exp[i(2\pi\nu_0 t + \delta \cos \omega_m t)] \tag{15.6-30}$$

with $m_1 = 0.2$ and $\delta = 3.3$.

In the transient case, we found (see Equation (15.6-27)) that the phase modulation index δ, the amplitude of the phase excursion, is equal to $m_1\alpha/2$ and that it can be determined from a fit to the experimental sideband distribution. Since the intensity modulation index m_1 can be determined straightforwardly from a spectral analysis of the laser intensity, the combination of the field spectrum, obtainable with a scanning Fabry–Perot etalon, and the intensity spectrum, obtained from a spectral analysis of the detected photocurrent, can be used to determine the amplitude–phase coupling constant α [32].

15.7 INTEGRATED OPTOELECTRONICS

In one of its rare moments of cooperative spirit, nature has endowed the III–V semiconductors based on GaAs/GaAlAs and InP/GaInAsP with a double gift. These are, as discussed earlier, the materials of choice for semiconductor lasers; but in addition it is possible to use them, especially InGaAs/GaAs and GaAs/GaAlAs, as base materials for electronic circuits in a manner similar to that in silicon. A completely new electronic technology based on GaAs/GaAlAs is now emerging [36]. It takes advantage of the large mobility of electrons in GaAs for very high switching speeds.

It was pointed out in 1971 [37] that it should be possible to bring together monolithically in a III–V semiconductor the two principal actors of the modem communication era—the transistor and the laser—in new integrated optoelectronic circuits. This new technology is now taking its first tentative steps from the laboratory to applications.

The basic philosophy, as well as an example of an integrated optoelectronic device, is shown in Figure 15.22, which shows a buried heterostructure GaAs/GaAlAs laser, similar to that illustrated in Figure 15.16, fabricated monolithically on the same crystal as a field-effect transistor (FET). The output current of the FET (see arrows) supplies the electron injection to the active region of the laser. This current and thus the laser power output can be controlled by a bias voltage applied to the gate electrode.

An example of a feasibility model of an integrated optoelectronic optical repeater, which incorporates a detector, a FET current preamplifier, a FET laser driver, and a laser, is shown in Figure 15.23. The main reason for the accelerating drive toward an integrated optoelectronic circuit technology [40] derives from the reduction of parasitic reactances that are always associated with conventional wire interconnections, plus the compatibility with the integrated electronic circuits technology that makes it possible to apply the advanced techniques of the latter to this new class of devices. More examples of optoelectronic integrated circuits are demonstrated in Figures 15.24 and 15.25.

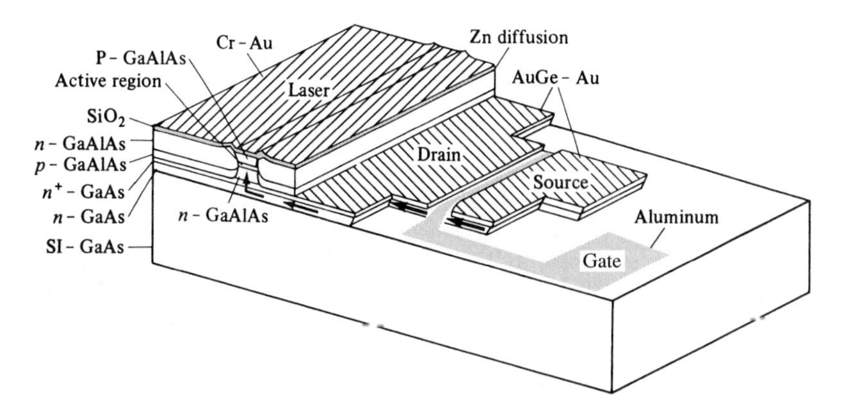

Figure 15.22 A GaAs n-channel field-effect transistor integrated monolithically with a buried heterostructure GaAs/GaAlAs laser. The application of a gate voltage is used to control the bias current of the laser. This voltage can oscillate and modulate the light at frequencies >10 GHz. (After Reference [38].)

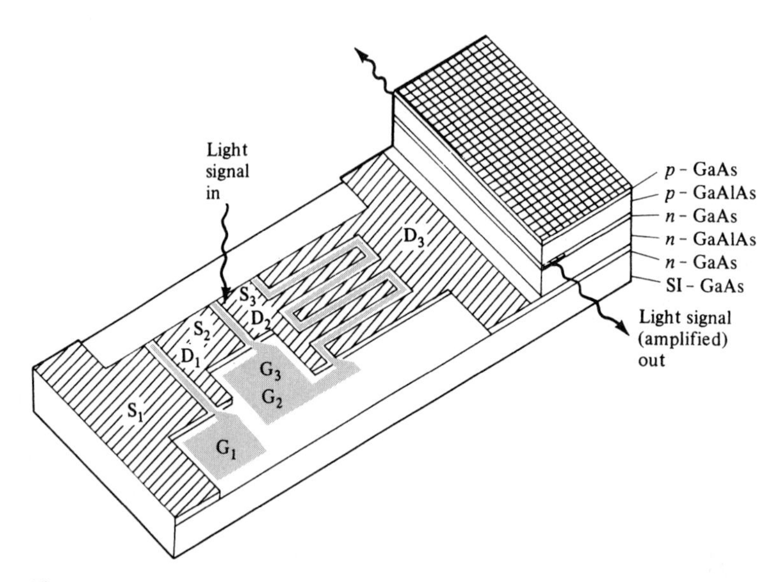

Figure 15.23 A monolithically integrated optoelectronic repeater containing a detector, transistor current source, a FET amplifier, and a laser on a single-crystal GaAs substrate. (After Reference [39].)

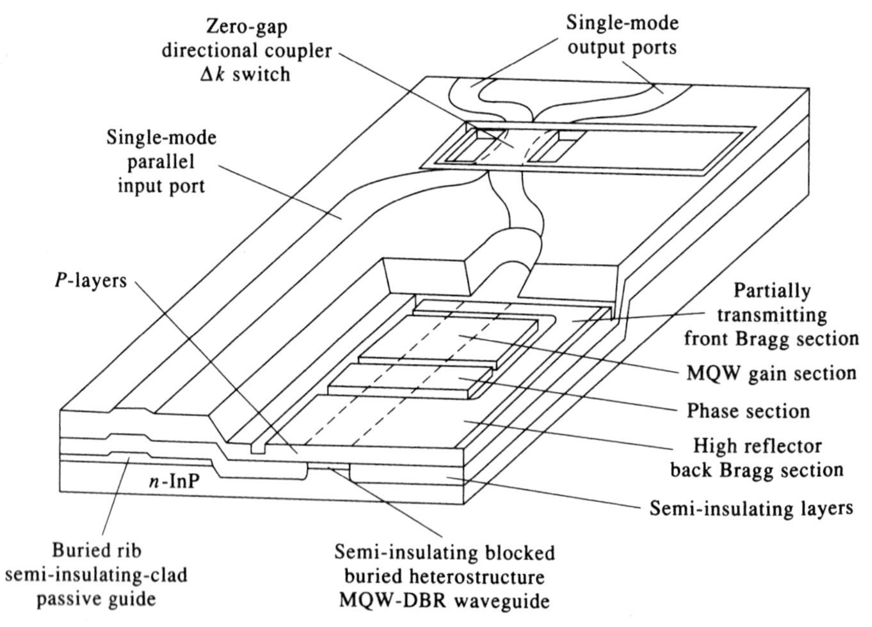

Figure 15.24 A monolithic circuit containing a tunable multisection InGaAsP/InP 1.55 μm laser employing multiquantum well (MQW) gain section, a passive waveguide for an external input optical wave, and a directional coupler switch for combining the laser output field and that of the external input at the output ports. (After Reference [41].)

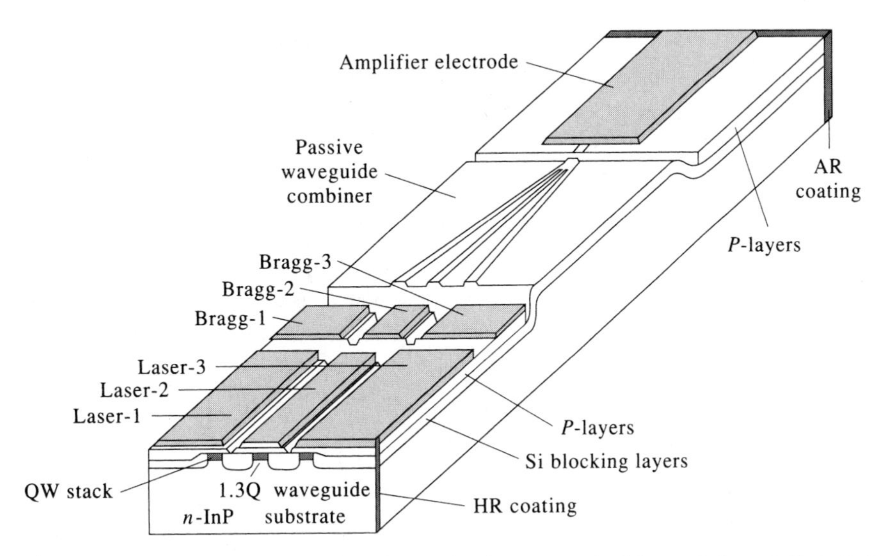

Figure 15.25 An optoelectronic integrated circuit composed of three ~1.5 μm InGaAs/InP distributed feedback lasers each tuned to a slightly different wavelength. The three wavelengths are fed into a single waveguide and amplified in a single amplifying section. (After Reference [42].)

PROBLEMS

15.1 Derive Equation (15.6-12).

15.2 Derive Equations (15.5-7) and (15.5-8).

15.3 Assume a fiber with $L = 10$ km and a group velocity dispersion parameter of 10 ps/nm-km (see Section 3.4). Calculate the maximum data rate through the fiber in bits/s if we use a semiconductor laser with characteristics similar to those used in the example of Section 15.6. For the purpose of this calculation, assume that a data rate of N bits/s is equivalent to a current modulation frequency of $\omega_m/2\pi = N$.

15.4 Find the frequency ω_R that maximizes $p_i(\omega_m)$ as given by Equation (15.6-9) and, using the approximations given, derive Equation (15.6-11).

15.5

(a) Evaluate and plot the gain $\gamma(\omega)$ of an inverted GaAs crystal under the following conditions:

$$N_{\text{elec}} = N_{\text{hole}} = 3 \times 10^{18}\ \text{cm}^{-3}$$

$$m_c - 0.07\ m_{\text{electron}}$$

$$m_h = 0.4\ m_{\text{electron}}$$

$$T = 0\ \text{K}$$

$$E_g = 1.45\ \text{eV}$$

$$T_2 = \infty$$

(b) Comment qualitatively on the changes in $\gamma(\omega)$ as the temperature is raised.

(c) What is the effect of a finite T_2 on $\gamma(\omega)$?

15.6 Consider the effect on the modulation response $p_1(\omega_m)/i_1(\omega_m)$ of the inclusion of a nonlinear gain term bP in the rate equations (15.5-1)

$$\frac{dN}{dt} = \frac{I}{eV} - \frac{N}{\tau} - A(1 - bP)(N - N_{\text{tr}})P$$

$$\frac{dP}{dt} = A(1 - bP)(N - N_{\text{tr}})P\Gamma_a - \frac{P}{\tau_p}$$

where $bP \ll 1$. Show that the main effect is a damping of the resonance peak at ω_R.

15.7 Solve for the carrier density modulation $N = N_0 + N_1 e^{i\omega_m t}$ in a semiconductor laser whose current is modulated at

$$I = I_0 + I_1 e^{i\omega_m t} \tag{15.P-1}$$

$$\omega_m = \text{Modulation frequency} \ll \omega_{\text{opt}} \tag{15.P-2}$$

(See Section 15.5.)

15.8 Assume $\varepsilon = \varepsilon_0 - aN$, where a is a constant, and that the instantaneous frequency of the semiconductor laser obeys

$$\frac{\Delta\nu}{\nu} = -\frac{\Delta\varepsilon}{\varepsilon} \qquad (15.P\text{-}3)$$

Find the form of the laser optical field due to the current modulation. What is the (phase) modulating index of the field?

15.9 Using the data of Figure 15.7, what is the total current needed to render the active medium of a semiconductor laser transparent? Assume an active volume of $300 \times 2 \times 0.2\ \mu\text{m}^3$ and a recombination lifetime of $\tau = 3 \times 10^{-9}$ s.

15.10 If the thickness of the active region in Problem 15.9 were reduced to 100 Å, can we obtain enough gain from a semiconductor laser to overcome a distributed loss constant of $\alpha = 20\ \text{cm}^{-1}$ and $R = 0.9$? What will be the transparency current? What will be the threshold current? Assume a mode height normal to the interfaces of $t = 4000$ Å and

$$\Gamma_a \sim \frac{d(\text{active region})}{t(\text{mode height})}$$

15.11 Plot the optical spectrum of a wave with simultaneous amplitude modulation (AM) and frequency modulation (FM):

$$E(t) = \left(E_0 + \frac{m_1}{2} E_0 \sin(\omega_m t)\right) \exp\{i[\omega_0 t + \delta \sin(\omega_m t + \alpha)]\}$$

for $\alpha = 0, \pi/6, \pi/4, \pi/2$.

REFERENCES

1. Basov, N. G., O. N. Krokhin, and Y. M. Popov, Production of negative temperature states in *p-n* junctions of degenerate semiconductors, *Sov. Phys. JETP* **40**:1320 (1961).
2. Hall, R. N., G. E. Fenner, J. D. Kingsley, T. J. Soltys, and R. O. Carlson, Coherent light emission from GaAs junctions. *Phys. Rev. Lett.* **9**:366 (1962).
3. Nathan, M. I., W. P. Dumke, G. Burns, F. H. Dills, and G. Lasher, Stimulated emission of radiation from GaAs *p-n* junctions. *Appl. Phys. Lett.* **1**:62 (1962).
4. Yariv, A., and R. C. C. Leite, Dielectric waveguide mode of light propagation in *p-n* junctions. *Appl. Phys. Lett.* **2**:55 (1963).
5. Anderson, W. W., Mode confinement injunction lasers. *IEEE J. Quantum Electron.* **1**:228 (1965).
6. Kittel, C., *Introduction to Solid State Physics*, 5th ed. Wiley, New York, 1982.
7. Yariv, A., *Introduction to the Theory and Applications of Quantum Mechanics*. Wiley, New York, 1982.
8. Bernard, M. G., and G. Duraffourg, Laser conditions in semiconductors. *Phys. Status Solidi* **1**:699 (1961).
9. Vahala, K., L. C. Chiu, S. Margalit, and A. Yariv, On the linewidth enhancement factor *a* in semiconductor injection lasers. *Appl. Phys. Lett.* **42**:631 (1983).
10. Stern, F., Semiconductor lasers: theory. In: *Laser Handbook*, F. T. Arecchi and E. O. Schultz Du Bois (eds.). North Holland, Amsterdam, 1972.
11. Kressel, H., and J. K. Butler, *Semiconductor Lasers and Heterojunction LEDS*. Academic Press, New York, 1977.
12. Yariv, A., *Quantum Electronics*, 2nd ed. Wiley, New York, 1975, p. 219.
13. Casey, H. C., and M. B. Panish, *Heterostructure Lasers*. Academic Press, New York, 1978.
14. Dupuis, R. D., and P. D. Dapkus, *Appl. Phys. Lett.* **31**:466 (1977).
15. Cho, A. Y., and J. R. Arthur, in *Progress in Solid State Physics*, Vol. 10, J. O. McCaldin and G. Somoraj (eds.). Pergamon Press, Elmsford, NY, p. 157.
16. Tsang, W. T. and A. Y. Cho, Growth of GaAs/GaAlAs by molecular beam epitaxy. *Appl. Phys. Lett.* **30**:293 (1977).
17. Hayashi, J., M. B. Panish, and P. W. Foy, A low-threshold room-temperature injection laser. *IEEE J. Quantum Electron.* **5**:211 (1969).

18. Kressel, H., and H. Nelson, Close confinement gallium arsenide *p-n* junction laser with reduced optical loss at room temperature. *RCA Rev.* **30**:106 (1969).
19. Alferov, Zh. I., et al., *Sov. Phys. Semicond.* **4**:1573 (1971).
20. Botez, D., and G. J. Herskowitz, Components for optical communication systems: a review. *Proc. IEEE* **68** (1980).
21. Chinone, N., H. Nakashima, I. Ikushima, and R. Ito, Semiconductor lasers with a thin active layer (≪0.1 IJ-m) for optical communication. *Appl. Opt.* **17**:311 (1978).
22. Tsukada, T., GaAs-$Ga_{1-x}Al_xAs$ buried heterostructure injection lasers. *J. Appl. Phys.* **45**:4899 (1974).
23. Derry, P., et al., Ultra low threshold graded-index separate confinement single quantum well buried heterostructure (Al, Ga) as lasers with high reflectivity coatings. *Appl. Phys. Lett.* **50**:1773 (1987).
24. Ortel Corp., Alhambra, CA. Product Data Sheets.
25. Hsieh, J. J., J. A. Rossi, and J. P. Donnelly, Room temperature CW operation of GalnAsP/InP double heterostructure diode lasers emitting at 1.1 μm. *Appl. Phys. Lett.* **28**:709 (1976).
26. Suematsu, Y., Long wavelength optical fiber communication. *Proc. IEEE* **71**:692 (1983).
27. A comprehensive book dealing with long-wavelength lasers is G. P. Agrawal and N. K. Dutta, *Long Wavelength Semiconductor Lasers*, Van Nostrand, New York, 1986.
28. Ralston, J. D., S. Weisser, K. Eisele, R. E. Sah, E. C. Larkins, J. Rosenzweig, J. Fleissner, and K. Bender, Low-bias-current direct modulation up to 33 GHz in InGaAs/AlGaAs pseudomorphic MQW ridge-waveguide lasers. *IEEE Photonics Technol.* **6**:1076 (1994).
29. Coldren, L. A., and S. W. Corzine, *Diode Lasers and Photonic Integrated Circuits.* Wiley, New York, 1995, p. 195.
30. Hirao, M., S. Tsuji, K. Mizushi, A. Doi, and M. Nakamura, *J. Opt. Commun.* **1**:10 (1980).
31. Lau, K. T., N. Bar-Chaim, I. Ury, and A. Yariv, Direct amplitude modulation of semiconductor GaAs lasers up to X-band frequencies. *Appl. Phys. Lett.* **43**:11 (1983).
32. Harder, C., K. Vahala, and A. Yariv, Measurement of the linewidth enhancement factor *a* of semiconductor lasers, *Appl. Phys. Lett.* **42**:428 (1983).
33. Henry, C. H., Line broadening of semiconductor lasers in coherence amplification and quantum effects. In: Semiconductor Lasers, Y. Yamamoto (ed.). Wiley-Interscience, New York, 1991, Chapter 2.
34. Vahala, K., L. C. Chiu, S. Margalit, and A. Yariv, On the linewidth factor *a* in semiconductor injection lasers. *Appl. Phys. Lett.* **42**:631 (April 1983).
35. Koch, T., and J. Bowers, Nature of wavelength chirping in directly modulated semiconductor lasers. *Electron Lett.* **20**:1038 (1984).
36. See, for example, J. P., Bailbe, A. Marty, P. H. Hiep, and G. E. Rey, Design and fabrication of high speed GaAlAs/GaAs heterojunction transistors. *IEEE Trans. Electron. Dev.* **ED-27**:1160 (1980).
37. Yariv, A., Active integrated optics. In: *Fundamental and Applied Laser Physics*, Proc. ESFAHAN Symposium, Aug. 29, 1971, M. S. Feld, A. Javan, and N. Kurnit (eds.). Wiley, New York, 1972.
38. Ury, I., K. Y. Lau, N. Bar-Chaim, and A. Yariv, Very high frequency GaAlAs laser-field effect transistor monolithic integrated circuit. *Appl. Phys. Lett.* **41**:126 (1982).
39. Yust, M., et al., A monolithically integrated optical repeater. *Appl. Phys. Lett.* **10**:795 (1979).
40. Bar-Chaim, N., I. Ury, and A. Yariv, Integrated optoelectronics. *IEEE Spectrum* **May**:38 (1982).
41. Hernandez-Gil, F., T. L. Koch, U. Koren, R. P. Gnall, and C. A. Burrus, Tunable MQW-DBR laser with a monolithically integrated InGaAsP/InP directional coupler switch. Paper PD 17, *Conference on Laser Engineering and Optics* (*CLEO*), (1989).
42. Koren, U., et al., Wavelength division multiplexing light source with integrated quantum well tunable lasers and optical amplifiers. *Appl. Phys. Lett.* **54**:2056 (May 1989).

CHAPTER 16

ADVANCED SEMICONDUCTOR LASERS

16.0 INTRODUCTION

In this chapter we discuss several different types of semiconductor lasers, including quantum well (QW) lasers, distributed feedback (DFB) lasers, and vertical cavity surface emitting lasers (VCSELs). The student is assumed to have a basic knowledge of elementary quantum mechanics. The quantum well laser [1] is similar in most respects to the conventional double heterostructure laser of the type shown in Figure 15.11 except for the thickness of the active layer. In the quantum well it is ~50–100 Å, while in conventional lasers it is ~1000 Å. This feature leads to profound differences in performance. The main advantage to derive from the thinning of the active region is almost too obvious to state—a decrease in the threshold current that is nearly proportional to the thinning. This reduction can be appreciated directly from Figure 15.7. The carrier density in the active region needed to render the active region transparent is $\sim 10^{18}\ \text{cm}^{-3}$. It follows that just to reach transparency we must maintain a total population of $N_{\text{transp}} \sim V_a \times 10^{18}$ electrons (holes) in the conduction (valence) band of the active region where V_a (in units of cm^3) is the volume of the active region. The injection current to sustain this population is approximately

$$I_{\text{transp}} \sim \frac{e \times 10^{18}}{\tau} V_a \qquad (16.0\text{-}1)$$

and is proportional to the volume of the active region. A thinning of the active region thus reduces V_a and I_{transp} proportionately. In a properly designed laser, the sum of the free carrier, scattering, and mirror (output) coupling can be made small enough so that the increment of current, above the transparency value, needed to reach threshold is small in comparison to I_{transp}. The reduction of the transparency current that results from a small V_a thus leads to a small threshold current [2–6]. Quantum well active regions are also used as the amplifying medium in distributed feedback lasers and in vertical cavity surface emitting lasers. Both of these classes of important semiconductor lasers are discussed in separate sections of this chapter.

16.1 CARRIERS IN QUANTUM WELLS (ADVANCED TOPIC)

The essential difference between the gain of a pumped quantum well semiconductor medium and that of a bulk semiconductor laser has to do with the densities of states in both of these media. The density of states of a bulk semiconductor was derived in Section 15.1 and is given by Equation (15.1-7). It was used in Equation (15.2-9) to derive an expression for the gain $\gamma(\omega_0)$ of that medium. In this section we will repeat this procedure using the density of states function of a quantum well. This is one of the few places in this book where we need to turn to quantum mechanics. The student without a quantum mechanical background but with a good electromagnetic preparation can simply think of the electron using the de Broglie picture as a wave obeying, not Maxwell's equations, but the Schrödinger equation. The traveling-wave solutions of this equation are of the form of modes $\psi_m(x, z) = u_m(\mathbf{r}) \exp(-iE_m t/\hbar)$, where $\hbar$ is Planck's constant divided by 2π, E_m is the quantized energy of an electron in state m, and $u_m(\mathbf{r})$ is the eigenfunction.

Referring to Figure 16.1, we consider the motion of an electron in the conduction band of a quantum well. The quantum well consists of a thin layer (about 100 Å) of GaAs sandwiched between two $Ga_{1-x}Al_xAs$ semiconductors. The whole structure shares a single crystal lattice. Since $Ga_{1-x}Al_xAs$ has a larger bandgap than GaAs, potential energy barriers exist at the interface between GaAs and $Ga_{1-x}Al_xAs$. The electron in the potential well is free (with an effective mass m_c) to move in the x and y directions, but is confined in the z direction (normal to the junction planes) due to the presence of the potential barriers at the junction between GaAs and GaAlAs. The potential barrier ΔE_c was given in Section 15.3 for the GaAs/$Ga_{1-x}Al_xAs$ system as $\Delta E_c \sim 0.75$ eV.

For the sake of simplification, we shall take ΔE_c as infinite. (This is a close approximation for barriers $\gtrsim 100$ Å and aluminum molar fraction $x \gtrsim 0.3$.) The wavefunction $u(\mathbf{r})$ of the electrons in this potential well obeys the time-independent Schrödinger equation [7]:

$$V(z)u(\mathbf{r}) - \frac{\hbar^2}{2m_c}\left(\frac{\partial^2}{\partial z^2} + \frac{\partial^2}{\partial x^2} + \frac{\partial^2}{\partial y^2}\right)u(\mathbf{r}) = Eu(\mathbf{r}) \tag{16.1-1}$$

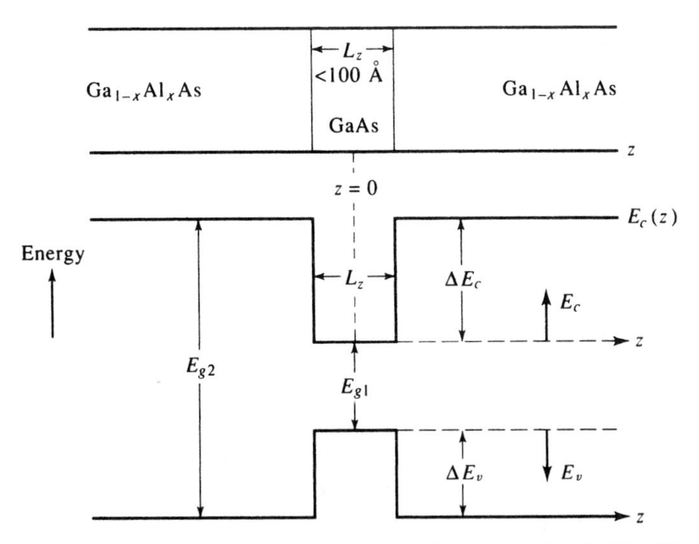

Figure 16.1 The layered structure and the bandedges of a GaAlAs/GaAs/GaAlAs quantum well.

where E is the energy of the electron while $V(z) = E_c(z)$ is the potential energy function confining the electrons in the z direction. For convenience, we will measure the energy relative to that of an electron at the bottom of the conduction band in the GaAs active region as shown in Figure 16.1. Since the quantum potential well $V(z)$ is a function of z only, the solution of the time-independent Schrödinger equation can be written

$$u(\mathbf{r}) = \psi_k(\mathbf{r}_\perp)u(z) = u_{\mathbf{k}_\perp}(\mathbf{r}_\perp)e^{i\mathbf{k}_\perp \cdot \mathbf{r}_\perp}u(z) \tag{16.1-2}$$

where $u_{\mathbf{k}_\perp}(\mathbf{r}_\perp)e^{i\mathbf{k}_\perp \cdot \mathbf{r}_\perp}$ is the Bloch wavefunction of electron motion in the xy plane with $u_{\mathbf{k}_\perp}(\mathbf{r}_\perp)$ possesses the crystal periodicity, $\mathbf{r}_\perp$ represents a coordinate in the xy plane, $\mathbf{k}_\perp$ is an arbitrary constant representing the Bloch wavevector for motion in the xy plane, and $u(z)$ is the wavefunction describing motion in the z direction. We note that the electron wavefunction in the xy plane is a Bloch wave, as the medium is "homogeneous" in the xy plane (i.e., no potential barriers). Substituting Equation (16.1-2) in (16.1-1), we obtain

$$\begin{aligned} -\frac{\hbar^2}{2m_c}\left(\frac{\partial^2}{\partial x^2} + \frac{\partial^2}{\partial y^2}\right)u_{\mathbf{k}_\perp}(\mathbf{r}_\perp)e^{i\mathbf{k}_\perp \cdot \mathbf{r}_\perp} &= \frac{\hbar^2 k_\perp^2}{2m_c}u_{\mathbf{k}_\perp}(\mathbf{r}_\perp)e^{i\mathbf{k}_\perp \cdot \mathbf{r}_\perp} \\ \left(V(z) - \frac{\hbar^2}{2m_c}\frac{\partial^2}{\partial z^2}\right)u(z) = \left(E - \frac{\hbar^2 k_\perp^2}{2m_c}\right)u(z) &= E_z u(z) \end{aligned} \tag{16.1-3}$$

where E_z is a separation constant to be determined. By separating the wavefunction into the form given by Equation (16.1-2), the electron energy is written

$$E = \frac{\hbar^2 k_\perp^2}{2m_c} + E_z \tag{16.1-4}$$

where the first part can be viewed as the kinetic energy of the electron motion in the conduction band along the xy direction, while E_z is the energy for the electron motion in the z direction. For simplicity, the energy of the electron at the bottom of the conduction band is taken to be zero. For a potential well with infinite barrier, $u(z)$ must vanish at $z = \pm L_z/2$. Taking $V(z) = 0$ inside the potential well, Equations (16.1-3) yield simple sinusoidal solutions for the the wavefuntion $u(z)$,

$$u_l(z) = \begin{cases} \cos l\dfrac{\pi}{L_z}z, & l = 1, 3, 5, \ldots \\ \sin l\dfrac{\pi}{L_z}z, & l = 2, 4, 6, \ldots \end{cases} \tag{16.1-5}$$

where L_z is the width of the potential well. Substitution of Equation (16.1-5) into (16.1-3), yields the energy of the electron,

$$E_c(\mathbf{k}_\perp, l) = \frac{\hbar^2 k_\perp^2}{2m_c} + E_z = \frac{\hbar^2 k_\perp^2}{2m_c} + l^2\frac{\hbar^2\pi^2}{2m_c L_z^2} \equiv \frac{\hbar^2 k_\perp^2}{2m_c} + E_{lc}, \tag{16.1-6}$$

where E_z is given by

$$E_z \equiv E_{lc} = l^2\frac{\hbar^2\pi^2}{2m_c L_z^2} = l^2 E_{1c}, \quad l = 1, 2, 3, \ldots \tag{16.1-7}$$

We put a subscript c to indicate the conduction band. For a potential well with infinite barriers, all wavefunctions (with an arbitrary l) are confined inside the well. In other words, electrons are confined by the potential well regardless of their energy. A similar set of wavefunction and energy equations can be obtained for electrons in the valence band. In

Equation (16.1-6), E_{1c} is the energy of the lowest state ($l = 1$). The lowest state (ground state) has a symmetric wavefunction, reflecting the symmetry of the potential well. We recall that the zero energy is taken as the bottom of the conduction band. It is important to note that the wavefunction (16.1-5) is only an approximation. In the approximation, we assume that the electron motion in the z direction is dominated by the potential well. Strictly speaking, the wavefunction $u(z)$ should also be a Bloch wave, reflecting the crystal periodicity in the z direction.

Similar results with $m_c \rightarrow m_v$ apply to the holes in the valence band. We recall that the hole energy E_v is measured downward in our electronic energy diagrams so that

$$E_v(\mathbf{k}_\perp, l) = \frac{\hbar^2 k_\perp^2}{2m_v} + l^2 \frac{\hbar^2 \pi^2}{2m_v L_z^2} = \frac{\hbar^2 k_\perp^2}{2m_v} + E_{lv}, \quad l = 1, 2, 3, \ldots \tag{16.1-8}$$

measured (downward) from the top of the valence band. The complete wavefunctions are then

$$\psi_c(\mathbf{r}) = \sqrt{\frac{2}{L_z}}\, \psi_{\mathbf{k}_\perp c}(\mathbf{r}_\perp) CS\left(l \frac{\pi}{L_z} z\right) \tag{16.1-9}$$

for electrons in the conduction band and

$$\psi_v(\mathbf{r}) = \sqrt{\frac{2}{L_z}}\, \psi_{\mathbf{k}_\perp v}(\mathbf{r}_\perp) CS\left(l \frac{\pi}{L_z} z\right) \tag{16.1-10}$$

for holes in the valence band. We defined $CS(x) \equiv \cos(x)$ or $\sin(x)$ in accordance with Equation (16.1-4), and $\psi_{\mathbf{k}_\perp c}(\mathbf{r}_\perp)$ and $\psi_{\mathbf{k}_\perp v}(\mathbf{r}_\perp)$ are Bloch wavefunctions for electron motion in the xy plane. The lowest-lying electron and hole wavefunctions are

$$\psi_c(\mathbf{r})_{\text{ground state}} = \sqrt{\frac{2}{L_z}} \cos\left(\frac{\pi}{L_z} z\right) \psi_{\mathbf{k}_\perp c}(\mathbf{r}_\perp) \tag{16.1-11}$$

$$\psi_v(\mathbf{r})_{\text{ground state}} = \sqrt{\frac{2}{L_z}} \cos\left(\frac{\pi}{L_z} z\right) \psi_{\mathbf{k}_\perp v}(\mathbf{r}_\perp) \tag{16.1-12}$$

and are shown along with the next higher level in Figure 16.2. In a real quantum well made of semiconductors, the height ΔE_c of the confining well (see Figure 15.11) is finite. The finite potential well can still support confined states, provided the barrier height ΔE_c is sufficient (usually at least as big as E_{1c}). The wavefunction of confined states in a finite potential well consists of a sinusoidal function inside the well and an exponential decaying function in the z direction outside the well. The number of confined states depends on the height of the energy barrier ΔE_c. It is interesting to note that the electron wavefunctions $u(z)$ are identical to those of TE confined modes of a symmetric slab waveguide. The mathematical procedure for solving (16.1-3) for a finite potential well is similar to that used in Chapter 3 to obtain the TE modes of a dielectric waveguide that obeys a Schrödinger-like equation (3.1-4). As a matter of fact, to determine the number of confined eigenmodes as well as their eigenvalues we use a procedure identical to that of Figure 3.2.

The Density of States

The considerations applying here are similar to those of Section 15.1. Since the electron is "free" in the x and y directions, we apply two-dimensional quantization by assuming the

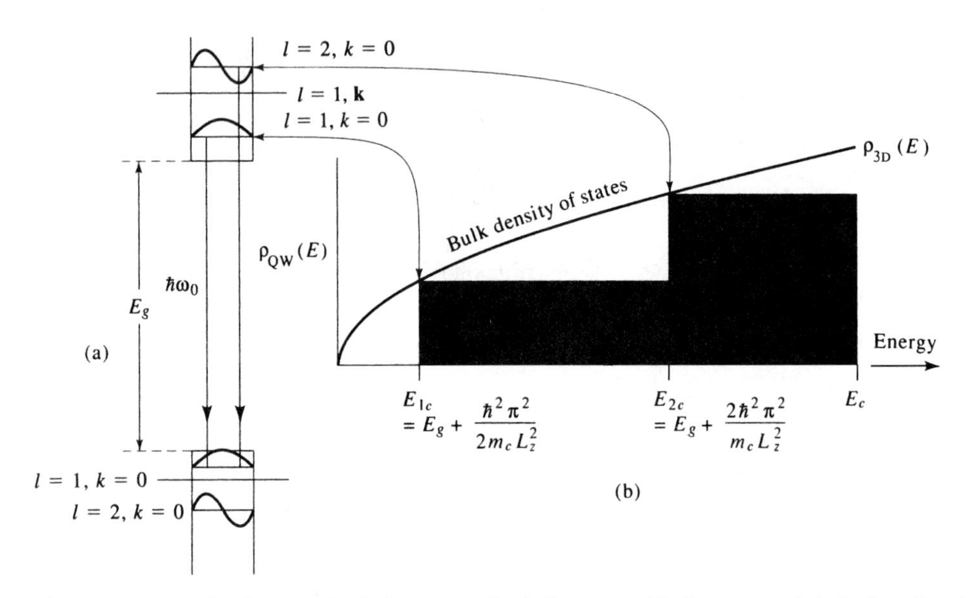

Figure 16.2 (a) The first two $l = 1$, $l = 2$ quantized electron and hole states and their eigenfunctions in an infinite potential well. (b) A plot of the volumetric density of states $(1/AL_z)/[dN(E)/dE]$ (i.e., the number of states per unit area (A) per unit energy divided by the thickness z of the active region) of electrons in a quantum well and of a bulk semiconductor. (Courtesy of M. Mittelstein, California Institute of Technology, Pasadena, California.)

electrons are confined to a rectangle L_xL_y (with L_x and L_y much larger than L_z). This leads, as in Equation (15.1-3), to a quantization of the components of the **k** vectors.

$$k_x = n\frac{\pi}{L_x}, \quad n = 1, 2, \ldots; \qquad k_y = m\frac{\pi}{L_y}, \quad m = 1, 2, \ldots \tag{16.1-13}$$

The area in $\mathbf{k}_\perp$ space per one eigenstate is thus $A_{\text{state}} = \pi^2/L_xL_y \equiv \pi^2/A_\perp$. We will drop the subscript $\perp$ from now on so that $k \equiv k_\perp$. The number of states with transverse values of k' less than some given k is obtained by counting the number of all positive lattice points (m, n), according to Equations (16.1-13). Dividing the area $\pi k^2/4$ by A_{state} (the factor $\frac{1}{4}$ is due to the fact that we only count the positive lattice points according to Equations (16.1-13)), we obtain

$$N(k) = \frac{k^2 A_\perp}{2\pi} \tag{16.1-14}$$

where $A_\perp = L_xL_y$ is the area of the quantum well in the xy direction, and a factor of 2 for the two spin orientations of each electron was included. This is the total number of states available for electrons with transverse wavenumber less than k.

The number of states between k and $k + dk$ is

$$\rho(k)dk = \frac{dN(k)}{dk}dk = A_\perp \frac{k}{\pi} dk \tag{16.1-15}$$

and is the same for the conduction or valence band. The total number of states with total energies between E and $E + dE$ is

$$\frac{dN(E)}{dE}dE = \frac{dN(k)}{dk}\frac{dk}{dE}dE = A_\perp \frac{k}{\pi} dk \tag{16.1-16}$$

The number of states per unit energy per unit area is thus

$$\frac{1}{A_\perp}\frac{dN(E)}{dE} = \frac{k}{\pi}\frac{dk}{dE} \tag{16.1-17}$$

Using Equation (16.1-6) with $\mathbf{k}_\perp \to \mathbf{k}$, $l = 1$, and limiting the discussion to the conduction band, we obtain the relation between the electron energy at the lowest state $l = 1$ and its **k** value as

$$k = \sqrt{\frac{2m_c}{\hbar^2}}(E - E_{1c})^{1/2} \tag{16.1-18}$$

Substituting Equation (16.1-18) for k in Equation (16.1-17), we obtain the two-dimensional density of states (per unit energy and unit area),

$$\rho_{\text{QW}}(E) \equiv \frac{1}{A_\perp}\frac{dN(E)}{dE} = \frac{m_c}{\pi\hbar^2} \tag{16.1-19}$$

Recall that this is the density of electron states. The actual density of electrons depends on the details of occupancy of these states according to the Fermi–Dirac distribution function as is discussed in the next section. We note that the density of states is a constant for electrons that are free to move in the xy plane, but are confined in a potential well in the z direction. An expression similar to (16.1-19) but with $m_c \to m_v$ applies to the valence band. In the reasoning leading to Equation (16.1-19), we considered only one transverse $u(z)$ quantum state with a fixed l quantum number (see Equation (16.1-19)). But once $E > E_{2c}$, as an example, an electron of a given total energy E can be found in either the $l = 1$ or $l = 2$ state so that the density of states at $E = E_{2c}$ doubles. At $E = E_{3c}$ it triples, and so on. This leads to a staircase density of states function. The total density of states thus increases by $m_c/\pi\hbar^2$ at each of the energies E_{lc} of Equation (16.1-6), which is expressed mathematically as

$$\rho_{\text{QW}}(E) = \sum_{n=1}^{\text{all states}} \frac{m_c}{\pi\hbar^2} H(E - E_{nc}) \tag{16.1-20}$$

where $H(x)$ is the Heaviside function that is equal to unity when $x > 0$ and is zero when $x < 0$.

The first two steps of the staircase density of states are shown in Figure 16.2. In the figure we plotted the volumetric density of states of the quantum well medium ρ_{QW}/L_z so that we can compare it to the bulk density of states in a conventional semiconductor medium. It is a straightforward exercise to show that, in this case, the QW volumetric density of states equals the bulk value $\rho_{3\text{D}}(E)$ at each of the steps, as shown in the figure.

The Selection Rules

Consider an amplifying electron transition from an occupied state in the conduction band to an unoccupied state in the valence band. The states $l = 1$ in the conduction band have the highest electron population. (The Fermi law (15.1-8) shows how the electron occupation drops with energy.) The same argument shows that the highest population of holes is to be found in the $l = 1$ valence band state. It follows that, as far as populations are concerned, the highest optical gain will result from the transition from an $l = 1$ state in the valence band to an $l = 1$ state in the conduction band. According to the elementary theory of quantum mechanics, the transition rate between two quantum states (initial state u_i and final state u_f) is given by

$$R_{f\leftarrow i} \propto \int u_f^*(\mathbf{r}) e\mathbf{r} \cdot \mathbf{E} u_i(\mathbf{r})\, d^3\mathbf{r} \tag{16.1-21}$$

where $\mathbf{E}$ is the electric field vector of the optical wave and $\mathbf{r}$ is the position of the electron.

The gain constant is also proportional to the rate of traisntion $R_{f\leftarrow i}$. Using Equations (16.1-11) and (16.1-12) for the lowest lying electron and hole wavefunctions, the transition rate between these two states becomes

$$R_{f\leftarrow i} \propto \int dx\, dy \int_{-L_z/2}^{L_z/2} e\mathbf{r} \cdot \mathbf{E} \cos^2 \frac{\pi z}{L_z}\, dz \tag{16.1-22}$$

Since

$$\int_{-L_z/2}^{L_z/2} z \cos^2 \frac{\pi z}{L_z}\, dz = 0 \tag{16.1-23}$$

it follows that an allowed transition occurs only when the optical field $\mathbf{E}$ is polarized in the xy plane. The optical $\mathbf{E}$ vector thus must lie in the plane of the quantum well (xy plane). A field polarized along the z direction does not stimulate any transitions between the two lowest lying levels and thus does not exercise gain (or loss) for these two states.

16.2 GAIN IN QUANTUM WELL LASERS

To obtain an expression for the gain of an optical wave confined (completely) within a quantum well medium [8], we follow a procedure identical to that employed in the case of a bulk semiconducting medium that was developed in Section 15.2. An amplifying transition at some frequency $\hbar\omega_0$ is shown in Figure 16.2. The upper electron state and the lower hole state (the unoccupied electron state in the valence band) have the same l and $\mathbf{k}$ values (see discussion of selection rules in Section 16.1) so that the transition energy is

$$\begin{aligned} \hbar\omega &= E_c - E_v = E_g + E_c(\mathbf{k}, l) + E_v(\mathbf{k}, l) \\ &= E_g + \left(\frac{1}{m_c} + \frac{1}{m_v}\right)\frac{\hbar^2}{2}\left(k^2 + l^2\frac{\pi^2}{L_z^2}\right) \\ &= E_g + \frac{\hbar^2}{2m_r}\left(k^2 + l^2\frac{\pi^2}{L_z^2}\right) \end{aligned} \tag{16.2-1}$$

where m_r is the reduced mass

$$m_r = \frac{m_c m_v}{m_c + m_v} \tag{16.2-2}$$

and $l = 1, 2, \ldots$ is the quantum number of the z-dependent eigenfunction $u_l(z)$ as in Equation (16.1-5). We start again with Equation (15.2-4) but this time in the correspondence of Equation (15.2-7) replace $\rho(k)/V$ by the equivalent quantum well $\rho_{QW}(\mathbf{k})/L_z$ volumetric carrier density,

$$N_1(m^{-3}) \rightarrow \frac{\rho_{QW}(\mathbf{k})}{L_z} dk\, f_v(E_v)[1 - f_c(E_c)] = \frac{k}{\pi L_z}(f_v - f_v f_c)\, dk$$

$$N_2(m^{-3}) \rightarrow \frac{\rho_{QW}(\mathbf{k})}{L_z} dk\, f_c(E_c)[1 - f_v(E_v)] = \frac{k}{\pi L_z}(f_c - f_c f_v)\, dk$$

where $\rho_{QW}(\mathbf{k})$ is given by Equation (16.1-19) and is independent of $\mathbf{k}$. The effective inversion population density due to carriers between k and $k + dk$ is thus

$$N_2 - N_1 \rightarrow \frac{k\,dk}{\pi L_z}[f_c(E_c) - f_v(E_v)] \tag{16.2-3}$$

The division of ρ_{QW} by L_z is due to the need, in deriving the gain constant, to use the *volumetric* density of inverted population consistent with the definition of N_1 and N_2 in Equation (15.2-4). E_c and E_v are, respectively, the upper and lower energies of the carriers involved in a transition. We use Equations (15.2-4) and (16.2-3) to write the contribution to the gain due to electrons within dk and in a single, say, $l = 1$, subband as

$$d\gamma(\omega_0) = \frac{k\,dk}{\pi L_z}[f_c(E_c) - f_v(E_v)]\frac{\lambda_0^2}{4n^2\tau}\left(\frac{T_2}{\pi[1 + (\omega - \omega_0)^2 T_2^2]}\right) \tag{16.2-4}$$

where T_2 is the coherence collision time of the electrons and τ is the electron–hole recombination lifetime assumed to be a constant. We find it more convenient to transform from the k variable to the transition frequency ω (see Equation 16.2-1)). From (16.2-1) it follows that

$$dk = \frac{m_r}{\hbar k}\,d\omega$$

so that (16.2-4) becomes

$$\gamma(\omega_0)_{l=1} = \frac{m_r \lambda_0^2}{4\pi\hbar L_z n^2 \tau}\int_0^\infty [f_c(\hbar\omega) - f_v(\hbar\omega)]\frac{T_2\,d\omega}{\pi[1 + (\omega - \omega_0)^2 T_2^2]} \tag{16.2-5}$$

where ω_0 is the frequency of the optical wave, and we used the convention that $f_c(\hbar\omega)$ is the Fermi function for the conduction electron at the upper transition (electron) state E_c, while $f_v(\hbar\omega)$ is the Fermi function for the valence electron at the lower transition state. To include, as we should, the contributions from all other subbands ($l = 2, 3, \ldots$) we replace, using Equation (16.1-20),

$$\frac{m_r}{\pi\hbar^2} \rightarrow \frac{m_r}{\pi\hbar^2}\sum_{l=1}^{\infty} H(\omega - \omega_l) \tag{16.2-6}$$

where $\hbar\omega_l$ is the energy difference between the bottom of the l subband in the conduction band and the l subband in the valence band,

$$\hbar\omega_l = E_g + l^2\frac{\hbar^2\pi^2}{2m_r L_z^2} \tag{16.2-7}$$

To get an analytic form for Equation (16.2-5) we will assume that the phase coherence "collision" time T_2 is long enough so that

$$\frac{T_2}{\pi[1 + (\omega - \omega_0)^2 T_2^2]} \rightarrow \delta(\omega - \omega_0) \tag{16.2-8}$$

which simplifies Equation (16.2-5) to

$$\gamma(\omega_0) = \frac{m_r \lambda_0^2}{4\pi\hbar L_z n^2 \tau}[f_c(\hbar\omega_0) - f_v(\hbar\omega_0)]\sum_{l=1}^{\infty} H(\hbar\omega_0 - \hbar\omega_l) \tag{16.2-9}$$

Equations (16.2-5) and (16.2-9) constitute our key result. They contain most of the basic physics of gain in quantum well media. Consider, first, the dependence of the gain on the Fermi functions f_c and f_v. An increase in the pumping current leads to an increase in the density of

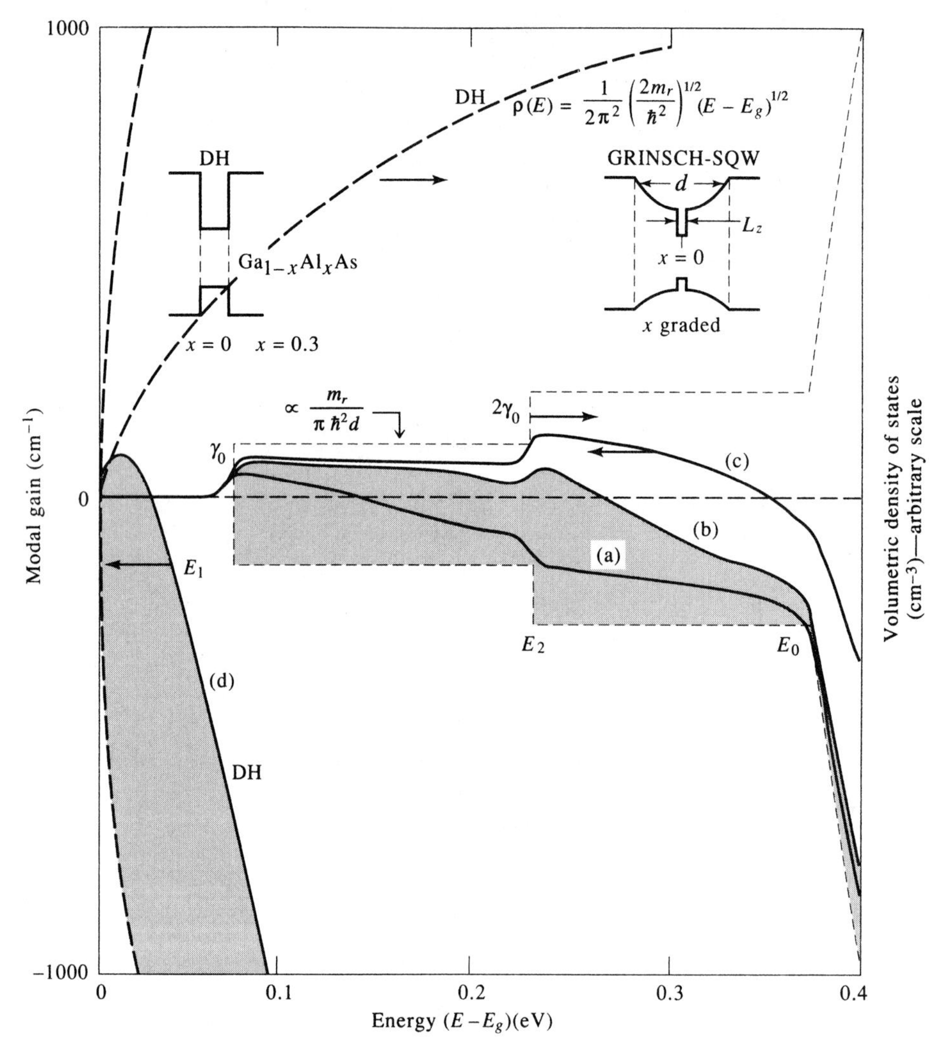

Figure 16.3 Gain (solid curves) and the joint density of states function (dashed lines) in a graded index, separate confinement heterostructure single quantum well (GRINSCH-SQW) laser, and a conventional double heterostructure (DH) laser. The gain curves (a), (b), and (c) are for successively larger injection current densities, and curve (d) applies to the DH with the same current density as the QW laser curve (a). To meaningfully compare the density of states of a quantum well laser and the bulk (DH) laser we divided the former by the width $W = 4L_z$ of the optical confinement distance. This, in addition to rendering the dimensions identical, makes both curves proportional to the maximum (available) modal gain. (Courtesy of D. Mehuys, California Institute of Technology, Pasadena, California.)

injected carriers in the active region and, with it, to an increase in the quasi-Fermi energies E_{F_c} and E_{F_v}. This leads to a larger spectral region of ω_0 where the gain condition (Equation (15.2-14))

$$f_c(\hbar\omega_0) - f_v(\hbar\omega_0) > 0 \tag{16.2-10}$$

is satisfied. This situation is depicted in Figure 16.3. The solid curves (a), (b), and (c) show the modal gain of a typical GaAs quantum well laser at three successively increasing current densities. The modal gain is equal to the medium gain $\gamma(\omega_0)$ of Equation (16.2-5) multiplied by the optical confinement factor $\Gamma_a \sim L_z/d$, where d is the effective width of the optical mode. The dashed curve corresponds to the gain available at infinite current density ($f_v(\hbar\omega_0) = 0$, $f_c(\hbar\omega_0) = 1$) and, thus, the gain in this case according to Equation (16.2-9), is proportional to the density of states function

$$\rho_{QW}(\hbar\omega_0) = \sum_{l=1}^{\infty} \frac{m_r}{\pi\hbar^2} H(\hbar\omega_0 - \hbar\omega_l) \tag{16.2-11}$$

The first frequency to experience transparency, then gain, as the current is increased, according to the idealized staircase density of states model, is ω_0, where

$$\hbar\omega_0 = E_g + E_{1c} + E_{1v} = E_g + \frac{\hbar^2\pi^2}{2m_r L_z^2} \tag{16.2-12}$$

$\hbar\omega_0$ is thus the energy difference between the $l = 1$ ($k = 0$) conduction band state and the $l = 1$ ($k = 0$) valence band state. The inversion factor $f_c(\hbar\omega_0) - f_v(\hbar\omega_0)$ is always larger at this frequency than at higher frequencies. As the current is increased, and with it the density of electrons (holes) in the conduction (valence) band, the quasi-Fermi levels (E_{F_c}, E_{F_v}) move deeper into their respective bands. There now exists a range of frequencies between the value given by Equation (16.2-12) and $\omega_0 = (E_g + E_{F_c} + E_{F_v})$, where the gain condition (16.2-10) is satisfied. At even higher pumping the contribution from the $l = 2$ subband (see Figure 16.3, curve (b)) adds to that from $l = 1$ and the maximum available gain doubles to $2\gamma(\omega_0)$. Curve (d) in Figure 16.3 shows the gain of a conventional double heterostructure laser. We note that equal increments of current will yield larger increments of gain in the SQW case, that, at low currents, the SQW gain tends to saturate at a constant value γ_0, and that the width of the spectral region that experiences gain is much larger in the SQW case compared to DH lasers.

The gain expression $\gamma(\omega_0)$ derived as Equation (16.2-9) is that of the quantum well medium. It is the gain experienced by a wave that is completely confined to the quantum well. Since the quantum well thickness if typically 50 Å $< L_z <$ 100 Å, while the mode width (height) is typically $d \sim$ 1000 Å, the actual gain experienced by the mode—modal gain—is given as in Equations (15.3-2) and (15.3-3) by

$$g_{\text{modal}} = \gamma\Gamma_a \approx \gamma\frac{L_z}{d} \tag{16.2-13}$$

When we use the expression (16.2-9) for γ, we find that the modal gain at a given areal density of carriers is independent of the thickness L_z of the quantum well and is inversely dependent on the optical mode width (height) d.

Multiquantum Well Laser

The small thickness of the quantum well relative to that of the optical mode width (height) ($L_z/d = 2 \times 10^{-2}$ typically) makes it practical to employ more than one quantum well as the active region. To first approximation, the total electronic inversion is divided equally among the quantum wells, and the total modal gain is the sum of the individual modal gains of each well. The advantage of multiquantum well lasers is that, as shown in Figure 16.4, the gain from a single well tends to saturate with carrier density, hence with current, because of the flat-top nature of the density of states. The use of multiple quantum wells enables each well to operate much within its linear gain–current region, thus extracting the maximum modal gain at a given total injected carrier density. This effect also results in a large differential gain $A \equiv \partial g/\partial N$, which, as shown in Section 15.5, leads to a higher laser modulation bandwidth.

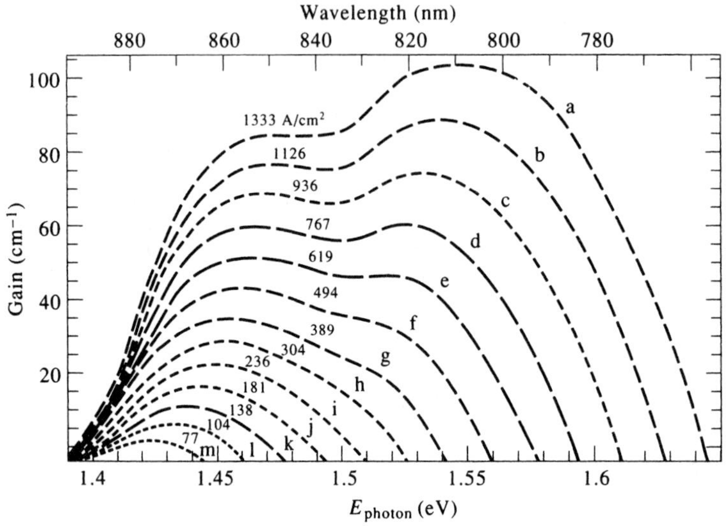

Figure 16.4 Theoretical plot of the exponential (modal) gain constant versus wavelength of a quantum well laser. (Courtesy of Michael Mittelstein, California Institute of Technology, Pasadena, California.)

This behavior is illustrated by Figure 16.5. We see that the optimal number of wells in a given laser depends on the requisite modal gain which, at oscillation, is equal to the laser losses. A laser with an effective loss constant of $d_{\text{eff}} = \alpha - 1/L \ln R$ (α = loss constant, R is the mirrors' reflectivity) of, say, 10 cm^{-1}, will have, according to the figure, the lowest threshold current with one ($N = 1$) quantum well.

A theoretical plot of the exponential (modal) gain constant as a function of photon energy, or wavelength, is shown in Figure 16.4. The parameter is the injection current density. The interesting feature is the leveling off of the gain at the lower photon energies with increasing current and the appearance of a second peak at the higher current due to the population of the $l = 2$ state of the quantum well.

Figure 16.6 shows the layered structure of a single quantum well GaAs/GaAlAs laser. The 80 Å wide quantum well is bounded on each side with a graded index region. This graded index (and graded energy gap) region is grown by tapering the Al concentration from 0% to 60% in a gradual fashion as shown. The graded region functions as both a dielectric waveguide and as a funnel for the injected electrons and, not shown, the holes, herding them into the quantum well.

16.3 DISTRIBUTED FEEDBACK LASERS

All laser oscillators employ optical feedback. By the word *feedback* we mean a means for ensuring that part of the optical field passing through a given point returns to the point repeatedly. If the delay is equal to an integral number of optical periods, this leads, in the presence of gain, to a sustained self-consistent oscillating mode where the field stimulated by atoms at any moment adds up coherently and *in phase* to those emitted earlier. In the laser resonators studied so far in this book, the feedback was provided by two oppositely facing reflectors (Fabry–Perot lasers). Feedback can also be achieved in a traveling wave folded-path geometry (e.g., a ring resonator). In distributed feedback (DFB) lasers [9–11], the reflection

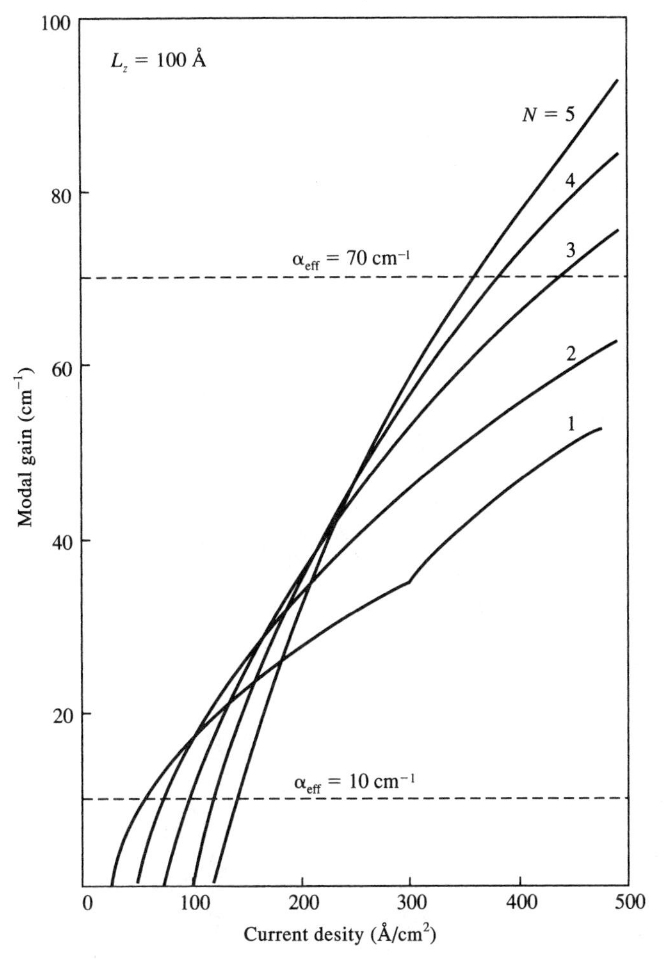

Figure 16.5 The modal gain g_{modal} ($= \gamma\Gamma$) as a function of the injected current with the number N of quantum wells as a parameter. (After Reference [8].)

feedback of forward into backward waves, and vice versa, takes place not at the end reflectors but continuously throughout the length of the resonator. This coupling is due to a spatially periodic modulation of the index of refraction of the medium or of its optical gain. These lasers enjoy a wavelength stability that is far superior to those of ordinary Fabry–Perot lasers. This stability is due to the fact that the laser mode prefers to oscillate at a frequency such that the spatial period A of the index perturbation is equal to some (usually small) integer (ℓ) number of half-wavelengths in the medium (or waveguide):

$$\Lambda = \ell \frac{\lambda_g}{2}, \quad \ell = 1, 2, 3, \ldots \tag{16.3-1}$$

with

$$\lambda_g = \frac{2\pi}{\beta} = \frac{\lambda}{n_{\text{eff}}} \tag{16.3-2}$$

where β is the propagation constant of the optical field in the waveguide, λ is the wavelength in vacuum, and n_{eff} is the effective index of the mode of propagation in the waveguide. This

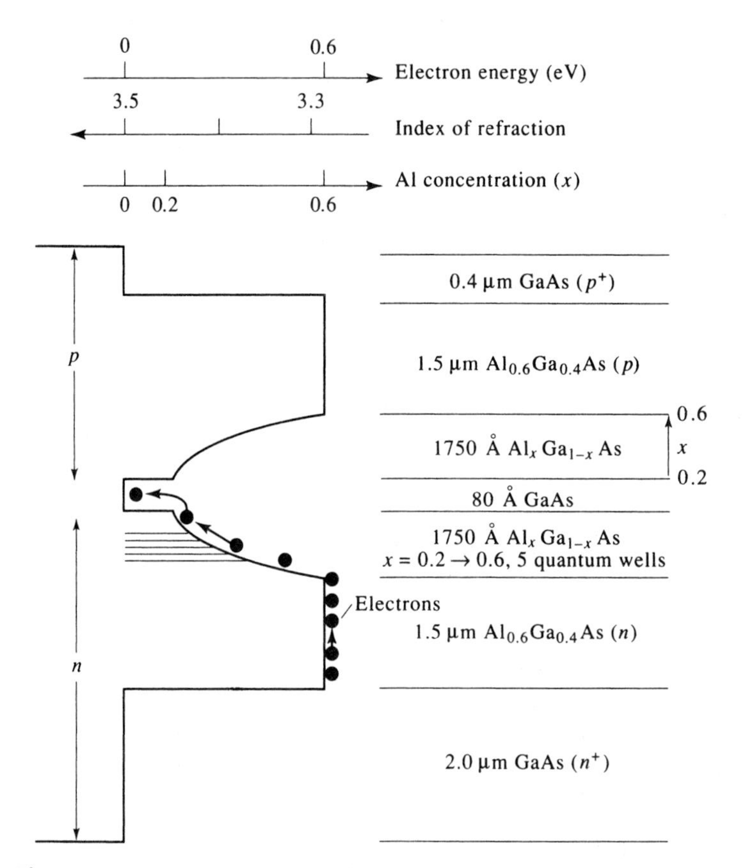

Figure 16.6 Schematic drawing of the conduction bandedge and doping profile of a single quantum well, graded index separate confinement heterojunction (GRINSCH) laser. (Courtesy of H. Chen, California Institute of Technology, Pasadena, California.)

condition (16.3-1), which ensures that reflections from different unit cells of the periodic perturbation add up in phase, is referred to as the *Bragg condition*. This is in analogy with the, formally similar, phenomenon of X-ray diffraction from the periodic lattice of crystals. Optical properties of periodically perturbed dielectric waveguides were discussed earlier in Chapter 12 (Section 12.5). The Bragg condition (16.3-1) is equivalent to $\Delta\beta = 0$ in Equation (12.5-15). Equation (16.3-1) enables the laser designer, through a choice of the period Λ, to "force" the laser to oscillate at any predetermined wavelength, provided that the amplifying medium is capable of providing sufficient gain at that wavelength. This property is especially important in semiconductor lasers used in optical fiber communications. Such lasers are often required to operate within narrow, prescribed wavelength regions to minimize pulse spreading by chromatic (group velocity) dispersion or to avoid crosstalk from other laser beams at different wavelengths sharing the same fiber. In dense wavelength division multiplexed (DWDM) optical networks, different laser carriers are separated at a frequency spacing of about 100 GHz. In these systems, the laser frequencies need to be stabilized with a drift of less than a few gigahertz.

Distributed feedback (DFB) lasers were discussed briefly in Section 12.6. Here we will start our extensive treatment of the distributed feedback laser with a derivation of the relevant coupled-mode equations. The essence of the DFB laser is a dielectric waveguide with a spatially periodic perturbation, and *with gain*. Let z be the axis along the direction of the waveguide; the field inside the waveguide consists of a forward propagating wave (along $+z$ direction) and a backward propagating wave (along $-z$ direction). The field can be written

$$\mathbf{E}(z) = A(z)\mathbf{E}_0(x, y) \exp[i(\omega t - \beta z)] + B(z)\mathbf{E}_0(x, y) \exp[i(\omega t + \beta z)] \tag{16.3-3}$$

where ω is the frequency, β is the propagation constant of the mode, and $A(z)$ and $B(z)$ are the mode amplitudes. Here we assume that both the forward and backward propagating waves are of the same spatial mode $\mathbf{E}_0(x, y)$ (usually the fundamental mode) of the waveguide. The z dependence of the mode amplitudes reflects the coupling between the waves. Waveguides with periodic dielectric perturbation were discussed in Chapter 12. The results obtained in Chapter 12 can now be employed for the discussion of distributed feedback lasers. By adding gain terms in the coupled equations, we obtain, according to Equations (12.5-13)

$$\begin{aligned} \frac{d}{dz} A &= -i\kappa B(z) e^{i2\Delta\beta z} + \gamma A \\ \frac{d}{dz} B &= i\kappa^* A(z) e^{-i2\Delta\beta z} - \gamma B \end{aligned} \tag{16.3-4}$$

where γ is the amplitude gain coefficient for the mode, $2\Delta\beta$ is the phase mismatch given by

$$2\Delta\beta = 2\beta - \ell\frac{2\pi}{\Lambda} = 2n_{\text{eff}}\frac{2\pi}{\lambda} - \ell\frac{2\pi}{\Lambda} \equiv 2(\beta - \beta_0) \tag{16.3-5}$$

and the coupling constant κ is given by

$$\kappa = \frac{\omega}{4}\int \mathbf{E}_0^*(x, y) \cdot \varepsilon_\ell(x, y)\mathbf{E}_0(x, y)\, dx \tag{16.3-6}$$

where $\varepsilon_\ell(x, y)$ is the ℓth Fourier component of the periodic dielectric perturbation. The coupling constant κ is, in general, a complex number. The phase of the coupling constant depends on the choice of the location of the origin ($z = 0$). For a corrugated periodic waveguide as shown in Figure 12.22, the choice of origin as shown in the figure leads to a pure imaginary coupling constant as given by Equations (12.5-11) and (12.5-12). In the event when the coupling constant κ vanishes, the mode amplitudes become

$$\begin{aligned} A(z) &= A(0)e^{\gamma z} \\ B(z) &= B(0)e^{-\gamma z} \end{aligned} \tag{16.3-7}$$

which agrees with the linear amplification of the modes in the gain medium. When κ is finite, the coupled equations can easily be solved by defining $A'(z)$ and $B'(z)$ as

$$\begin{aligned} A(z) &= A'(z)e^{\gamma z} \\ B(z) &= B'(z)e^{-\gamma z} \end{aligned} \tag{16.3-8}$$

The result is

$$\begin{aligned} \frac{d}{dz} A' &= -i\kappa B'(z) e^{i2(\Delta\beta + i\gamma)z} \\ \frac{d}{dz} B' &= i\kappa^* A'(z) e^{-i2(\Delta\beta + i\gamma)z} \end{aligned} \tag{16.3-9}$$

in a form identical to (12.5-13), provided we replace $\Delta\beta$ with $2(\Delta\beta + i\gamma)$. With this substitution, we can thus use Equations (12.5-20) to write down directly the solutions for the complex mode amplitudes of the forward propagating wave $A(z) = A'(z)e^{\gamma z}$ (the incident wave) and the backward propagating wave $B(z) = B'(z)e^{-\gamma z}$ (the reflected wave), respectively, inside the amplifying periodic dielectric waveguide. Assuming a single forward propagating wave of mode amplitude $A(0)$ incident from the left ($z = 0$), the boundary condition for the backward propagating wave is $B(L) = 0$. The solutions for the mode amplitudes subject to the boundary condition $B(L) = 0$ can be written

$$A(z) = e^{i(\Delta\beta)z} \frac{S \cosh S(L - z) + i(\Delta\beta + i\gamma) \sinh S(L - z)}{S \cosh SL + i(\Delta\beta + i\gamma) \sinh SL} A(0)$$

$$B(z) = e^{-i(\Delta\beta)z} \frac{-i\kappa^* \sinh S(L - z)}{S \cosh SL + i(\Delta\beta + i\gamma) \sinh SL} A(0) \tag{16.3-10}$$

where S and $\Delta\beta$ are given by

$$S^2 = \kappa^*\kappa - (\Delta\beta + i\gamma)^2 = |\kappa|^2 + (\gamma - i\,\Delta\beta)^2 \tag{16.3-11}$$

$$\Delta\beta = \beta(\omega) - \ell\frac{\pi}{\Lambda} \equiv [\beta(\omega) - \beta_0] \tag{16.3-12}$$

The fact that S now is complex makes for a major qualitative difference between the behavior of the passive periodic guide (12.5-20) and the periodic guide with gain (16.3-10). To demonstrate this difference, consider the case when the condition

$$S \cosh SL + i(\Delta\beta + i\gamma) \sinh SL = 0 \tag{16.3-13}$$

is satisfied. It follows from Equation (16.3-10) that both the reflectance, $B(0)/A(0)$, and the transmittance, $A(L)/A(0)$, become infinite. *The device acts as an oscillator*, since it yields finite output fields $B(0)$ and $A(L)$ with no input $[A(0) = 0]$. Condition (16.3-13) is thus the oscillation condition for a distributed feedback laser. For the case of $\gamma = 0$, it follows, from Equation (12.5-20), that $|B(0)/A(0)| < 1$, and $|A(L)/A(0)| < 1$ as appropriate to a passive device with no internal gain.

Let ω_0 be the frequency where $\beta(\omega_0) = \beta_0$ so that $\Delta\beta = [\beta(\omega_0) - \beta_0] = 0$. This is known as the Bragg frequency. According to Equation (16.3-5), the Bragg frequency can be written $\omega_0 = \pi c/(\Lambda n_{\text{eff}})$. For frequencies very near the Bragg frequency ω_0 ($\Delta\beta \cong 0$) and for sufficiently high-gain constant γ so that condition (16.3-13) is nearly satisfied, the guide acts as a high-gain amplifier. The amplified output is available either in reflection with a mode amplitude gain

$$\frac{B(0)}{A(0)} = \frac{-i\kappa^* \sinh SL}{S \cosh SL + i(\Delta\beta + i\gamma) \sinh SL} \tag{16.3-14}$$

or in transmission with a mode amplitude gain

$$\frac{A(L)}{A(0)} = e^{i\Delta\beta L} \frac{S}{S \cosh SL + i(\Delta\beta + i\gamma) \sinh SL} \tag{16.3-15}$$

The behavior of the incident and reflected fields for a high-gain case is sketched in Figure 16.7. Note the qualitative difference between this case and the passive (no gain) one depicted in Figure 12.18.

The reflection power gain, $|B(0)/A(0)|^2$, and the transmission power gain, $|A(L)/A(0)|^2$, are plotted in Figure 16.8 as a function of $\Delta\beta$ and γ. Each plot contains four infinite gain singularities at which the oscillation condition (16.3-13) is satisfied. These are four longitudinal laser modes. Higher orders exist but are not shown.

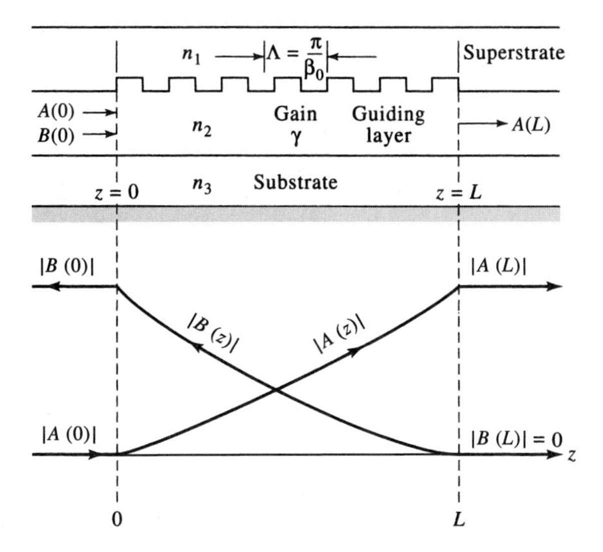

Figure 16.7 Incident and reflected fields inside an amplifying periodic waveguide.

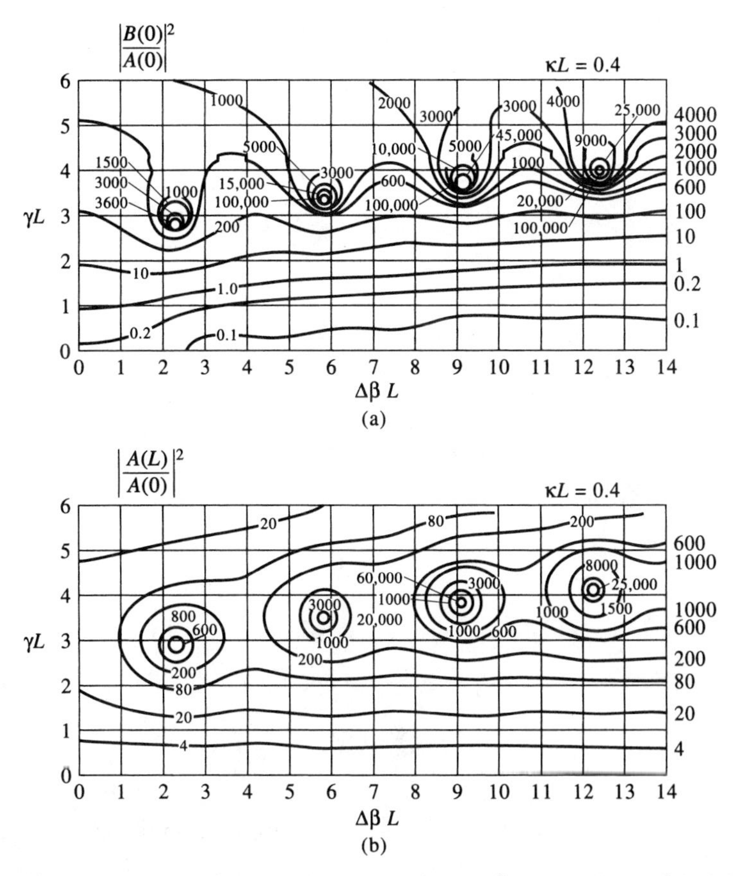

Figure 16.8 (a) Reflection gain contours in the $\Delta\beta\, L - \gamma L$ plane. $\Delta\beta$ is defined by Equation (16.3-12) and is proportional to the deviation of the frequency $\omega_0 \equiv \pi c/\Lambda n$. The plots are symmetric about $\Delta\beta$, so that only one-half ($\Delta\beta > 0$) of the plots are shown. (b) Transmission gain. (Courtesy of H. W. Yen.)

Oscillation Condition

The oscillation condition (16.3-13) can be written

$$S + i(\Delta\beta + i\gamma) \tanh SL = 0 \tag{16.3-16}$$

This is a transcendental equation of the complex variable $(\Delta\beta + i\gamma)$. In general, one has to resort to a numerical solution to obtain the threshold values of $\Delta\beta$ and γ for oscillation [12]. In some limiting cases, however, we can obtain approximate solutions. In the high-gain $\gamma \gg |\kappa|$ case, we have from the definition of $S^2 = |\kappa|^2 + (\gamma - i\,\Delta\beta)^2$

$$S \approx (\gamma - i\,\Delta\beta)\left(1 + \frac{|\kappa|^2}{2(\gamma - i\,\Delta\beta)^2}\right), \quad \gamma \gg |\kappa| \tag{16.3-17}$$

By using this approximation, the oscillation condition (16.3-16) becomes

$$1 - \tanh SL = -\frac{|\kappa|^2}{2(\gamma - i\,\Delta\beta)^2} \tag{16.3-18}$$

Since $\gamma \gg |\kappa|$, tanh SL is near unity and therefore can be given approximately as

$$\tanh SL \approx 1 - 2e^{-2SL} \tag{16.3-19}$$

Thus the oscillation condition (16.3-16) becomes

$$\frac{4(\gamma - i\,\Delta\beta)^2}{|\kappa|^2} e^{-2SL} = -1 \tag{16.3-20}$$

Equating the phases on both sides of (16.3-20) results in

$$-2\tan^{-1}\frac{\Delta\beta}{\gamma} + 2\Delta\beta\,L - \frac{\Delta\beta\,L|\kappa|^2}{\gamma^2 + (\Delta\beta)^2} = (2m + 1)\pi \tag{16.3-21}$$

where m is an integer.

For oscillation near the Bragg frequency ω_0 (i.e., the frequency at which $\Delta\beta = 0$) and in the limit $\gamma \gg |\Delta\beta|, |\kappa|$, the oscillating mode frequencies are given by

$$(\Delta\beta_m)L \cong \left(m + \frac{1}{2}\right)\pi, \quad m = 0, \pm 1, \pm 2, \pm 3, \ldots \tag{16.3-22}$$

and since $\Delta\beta \equiv \beta - \beta_0 \approx (\omega - \omega_0)n_{\text{eff}}/c$,

$$\omega_m = \omega_0 + \left(m + \tfrac{1}{2}\right)\frac{\pi c}{n_{\text{eff}}L} \tag{16.3-23}$$

We note that no oscillation can take place exactly at the Bragg frequency ω_0. The mode frequency spacing is

$$\omega_{m+1} - \omega_m \cong \frac{\pi c}{n_{\text{eff}}L} \tag{16.3-24}$$

which is approximately the same as in a two-reflector resonator of length L.

The threshold gain value γ_m is obtained by equating the amplitudes in Equation (16.3-20). This leads to the following equation:

$$\frac{e^{2\gamma_m L}}{\gamma_m^2 + (\Delta\beta)_m^2} = \frac{4}{|\kappa|^2} \tag{16.3-25}$$

This equation indicates an increase in threshold gain value γ_m with increasing mode number m (i.e., increasing $\Delta\beta_m$, according to Equation (16.3-22)). This is also evident from the numerical gain plots (Figure 16.8). An important feature that follows from Equation (16.3-25) is that the threshold gain for modes with the same $|\omega - \omega_0|$, or equivalently the same $|\Delta\beta|$, is the same. Thus two modes will exist with the lowest threshold, one on each side of ω_0. This property of DFB lasers is usually undesirable, and methods for obtaining single-mode operation are discussed in the last part of this section. High-speed (data rate) optical communications in fibers require that the optical source put out a single frequency in order to minimize the temporal spread of the optical pulses with distance, which is caused by group velocity dispersion. The periodic perturbation in semiconductor DFB lasers is achieved by incorporating a grating, usually in the form of a rippled interface, in the laser structure. This is achieved by interrupting the crystal growth at the appropriate stage and wet-chemical etching a corrugation into the topmost layer by using an interferometrically produced photoresist mask [10]. Growth of a layer with a different index of refraction, or optical absorption on top of the rippled surface, results in the desired spatial modulation.

A diagram of a distributed feedback laser using a GaAs–GaAlAs structure is shown in Figure 16.9. The waveguiding layer, as well as that providing the gain (active layer), is that of p-GaAs. The feedback is provided by corrugating the interface between the p-$Ga_{0.93}Al_{0.07}As$ and p-$Ga_{0.7}Al_{0.3}As$, where the main index discontinuity responsible for the guiding occurs. Figure 16.12 shows an example of a periodic gain grating. The laser in this example is based on the quaternary $Ga_{1-x}In_xAs_{1-y}P_y$ as the active region and InP as the high-energy gap, low-index cladding layer. The feedback is achieved by growing an extra-absorbing (i.e., low-energy gap) layer and then etching through a mask to leave behind a periodic array of absorbing islands.

The increase in threshold gain with the longitudinal mode index m predicted by Equation (16.3-25) and by the plots of Figure 16.8 manifests itself in a high degree of mode discrimination in the distributed feedback laser.

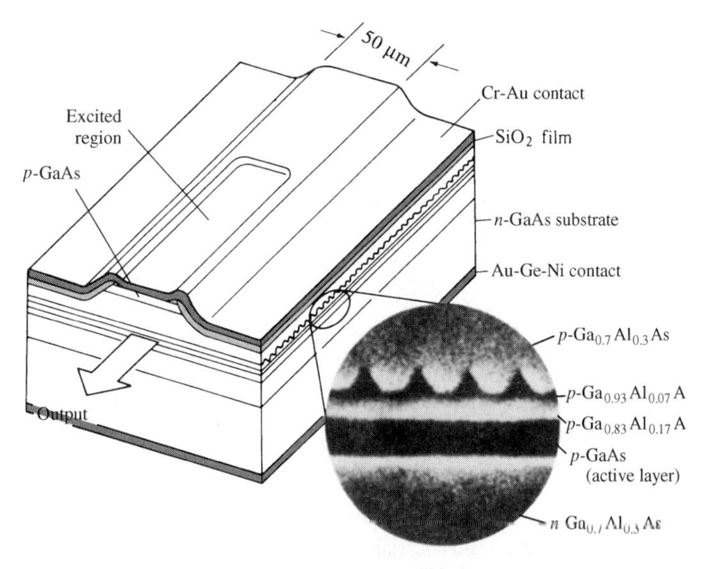

Figure 16.9 A GaAs–GaAlAs cw injection laser with a corrugated interface. The insert shows a scanning electron microscope photograph of the layered structure. The feedback is in third order ($\ell = 3$) and is provided by a corrugation with a period $\Lambda = 3\lambda_g/2 = 0.345$ μm. The thin (0.2 μm) p-$Ga_{0.83}A_{0.17}As$ layer provides a potential barrier that confines the injected electrons to the active (p-GaAs) layer, thus increasing the gain. (After Reference [11].)

It follows from Equations (16.3-23) and (16.3-25) that the two lowest threshold modes are those with $m = 0$ and $m = -1$ and that they are situated symmetrically on either side of the Bragg frequency ω_0 just outside the bandgap (see also Figure 12.29).

To understand why the basic DFB laser of Figure 16.7, in which the index of refraction is spatially periodic, does not oscillate at the Bragg frequency, consider Figure 16.10a. For convenience, we pick the origin $z = 0$ as shown in the figure. Let the reflection coefficient of a wave (at ω) incident from the left ($z < 0$), on the plane $z = 0$, be r_2, and for a wave incident from the right, r_1. The mode amplitude coefficient r_2 is given according to Equation (16.3-14) by

$$r_2 = \frac{-i\kappa^* \sinh SL_1}{S \cosh SL_1 + i(\Delta\beta + i\gamma) \sinh SL_1} \tag{16.3-26}$$

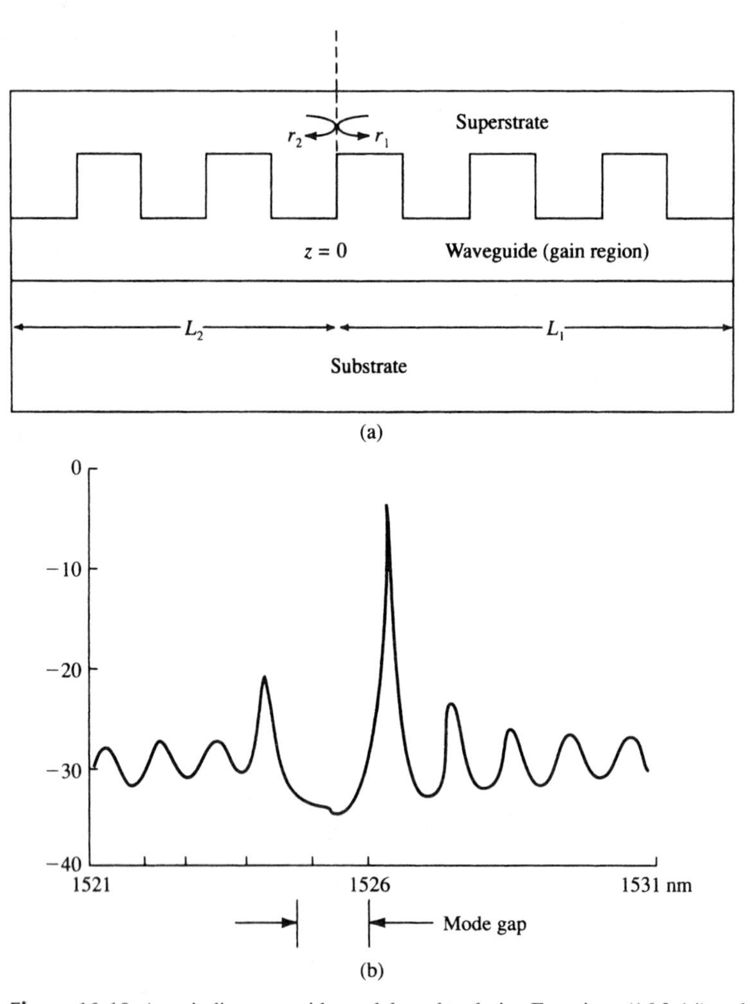

Figure 16.10 A periodic waveguide model used to derive Equations (16.3-14) and (16.3-15). (a) A periodic (DFB) GaInAsP waveguide laser. The origin $z = 0$ is located at the dashed line. (b) The spontaneous emission spectrum below, but near, threshold of a DFB laser showing the mode gap. (c) A DFB laser with a phase shift section. (d) A quarter-wavelength ($\lambda/4$) shifted DFB laser. (e) The spontaneous emission spectrum below threshold of a $\lambda/4$-shifted DFB laser. (Courtesy of P. C. Chen, ORTEL Corporation.)

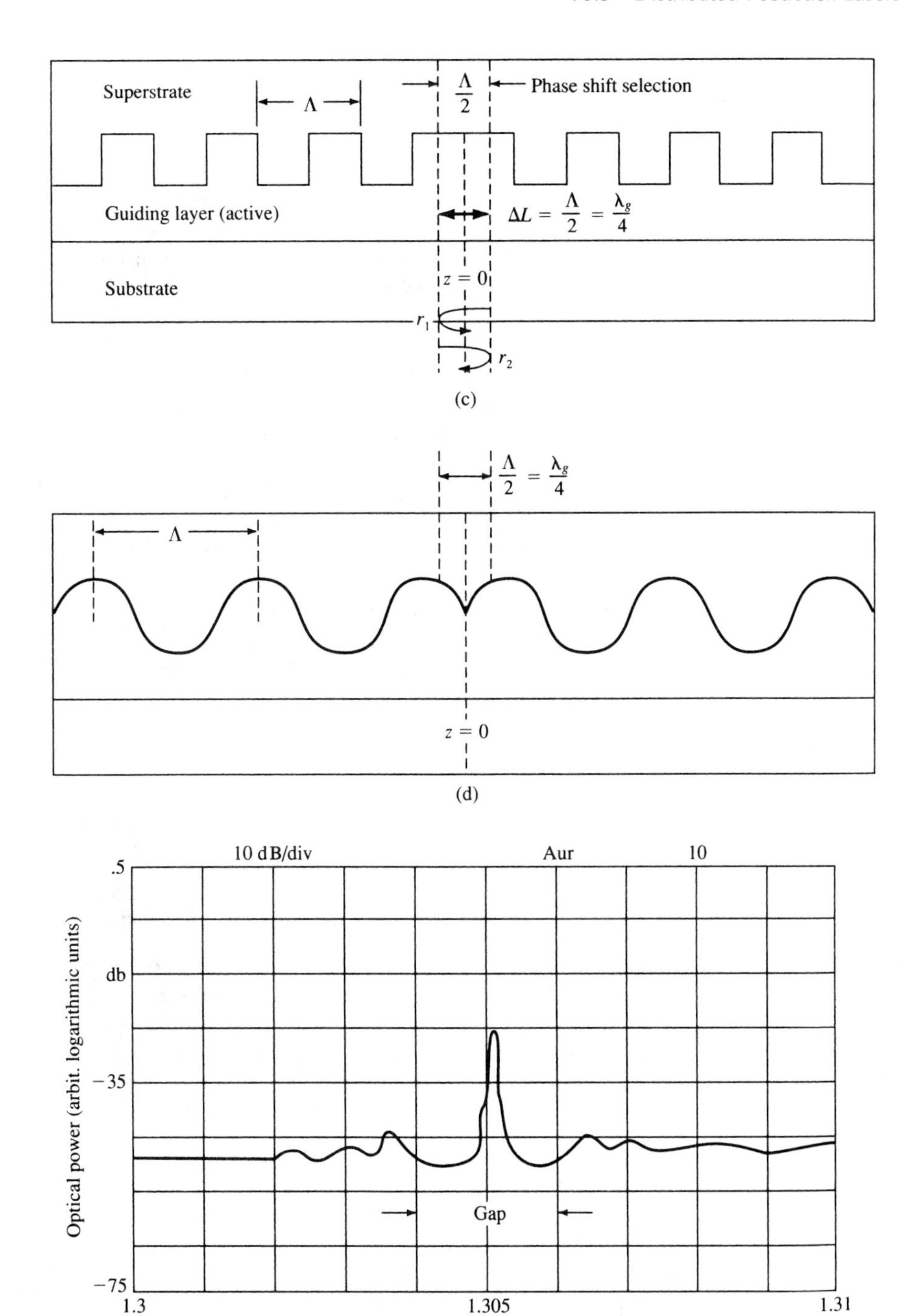

Figure 16.10 *(continued)*

with

$$S = \sqrt{\kappa^*\kappa + (\gamma - i\,\Delta\beta)^2}$$
$$\Delta\beta = \frac{\omega}{c} n_{\text{eff}} - \ell \frac{\pi}{\Lambda} = (\omega - \omega_0)\frac{n_{\text{eff}}}{c} \tag{16.3-27}$$

where κ is the coupling constant, γ is the mode amplitude gain coefficient, and L_1 is the segment of the periodic waveguide on the right side of the origin ($0 < z$). As we indicated earlier, the phase of the complex coupling constant κ depends on the choice of the origin. The choice of origin as shown in Figure 16.10 leads to a real coupling constant (i.e., $\kappa^* = \kappa$). The mode amplitude reflection coefficient r_1, "looking" to the left, can be obtained by solving the coupled equation with the boundary condition of $A(-L_2) = 0$, and a finite input of $B(0)$ at $z = 0$. The reflection coefficient in this case in defined as $A(0)/B(0)$. Using Equations (12.7-4) with the substitution of $\Delta\beta$ with $2(\Delta\beta + i\gamma)$, as well as using $A(-L_2) = 0$, we obtain

$$r_1 = \frac{-i\kappa \sinh SL_2}{S \cosh SL_2 + i(\Delta\beta + i\gamma) \sinh SL_2} \tag{16.3-28}$$

At the Bragg frequency (center of the photonic bandgap) ω_0, where $\Delta\beta = 0$, the two mode amplitude reflection coefficients become

$$r_1 = \frac{-i\kappa \sinh SL_2}{S \cosh SL_2 - \gamma \sinh SL_2}$$
$$r_2 = \frac{-i\kappa^* \sinh SL_1}{S \cosh SL_1 - \gamma \sinh SL_1} \tag{16.3-29}$$

and their product is written

$$r_1 r_2 = \frac{-\kappa^*\kappa \sinh SL_1 \sinh SL_2}{(S \cosh SL_2 - \gamma \sinh SL_2)(S \cosh SL_1 - \gamma \sinh SL_1)} \tag{16.3-30}$$

This result is actually independent of the choice of the location of the origin ($z = 0$). By examining the above equation, we note that the denominator is a positive real number, as both S and γ are real. It follows that at the Bragg frequency ($\Delta\beta = 0$), $r_1 r_2$ = negative number. The oscillation condition for a laser, on the other hand, is

$$r_1(\omega) r_2(\omega) = 1 \tag{16.3-31}$$

It follows that the periodic index DFB laser cannot oscillate at the Bragg frequency ω_0, where $\Delta\beta = 0$. The oscillation condition (16.3-31) was discussed in Chapter 6. This is just a sophisticated way of saying that, at steady state, the oscillation condition is equivalent to demanding that a wave launched, say, to the right, returns after one round trip with the same amplitude and the same phase (modulo $m2\pi$). In fact, condition (16.3-31) is consistent with (16.3-16), if we use the general expressions for r_1 and r_2 as given by Equations (16.3-26) and (16.3-28), respectively (see Problem 16.7).

Oscillation thus takes place at the symmetrically situated frequencies shown in Figure 16.8. The two oscillation frequencies nearest the Bragg frequency require the lowest gain and are given, according to Equation (16.3-23), by

$$\omega = \omega_0 \pm \frac{\pi c}{2 n_{\text{eff}} L} \tag{16.3-32}$$

The threshold gains for oscillation at these two frequencies are, according to Equation (16.3-25) and Figure 16.8, equal, so that they are in practice equally likely to oscillate. This situation is highly undesirable in practice, since it results in wavelength instabilities and spectral broadening. This is unacceptable, for example, in long-haul, high-data-rate fiber links, where the increased spectral width due to multiwavelength oscillation was shown in Chapter 3 to limit the data rate due to pulse broadening by group velocity dispersion.

The existence of two such oscillating wavelengths is shown in the spectrum of Figure 16.10b as the two peaks on either side of the "gap."

A widely employed method [13] for forcing the DFB laser to oscillate preferentially at a single midgap frequency is shown in Figures 16.10c and 16.10d. An extra section of length $\lambda_g/4$ is inserted at the center of the laser (λ_g is the "guide" wavelength, $\lambda_g = \lambda/n_{\text{eff}}$). The mode amplitude reflection coefficients r_1 and r_2 "looking" to the left and right, respectively, from the midplane are now given by their previous values (i.e., those corresponding to Figure 16.10a), each multiplied by $\exp[-i(\pi/2)]$ to account for the added propagation delay in the $\lambda_g/4$ section. At $\omega = \omega_0$ ($\Delta\beta = 0$), r_1 becomes $r_1 \exp[-i(\pi/2)]$, and r_2 becomes $r_2 \exp[-i(\pi/2)]$. The product of the two reflection coefficients $r_1' r_2'$ is now $-r_1 r_2$, which is a *positive* number, so that oscillation at ω_0 is possible. This is illustrated in Figure 16.10e.

Gain-Coupled Distributed Feedback Lasers

Another type of distributed feedback laser is one where the periodic modulation is not of the index of refraction but of the gain or losses of the medium [14]. To analyze this situation, we remind ourselves that gain or loss can be represented by taking the dielectric constant ε of a medium as complex. It is a straightforward matter to show (see Problem 16.6) that ε can be expressed as

$$\varepsilon = \varepsilon_0 n^2 \left(1 + i\frac{2\gamma}{k_0 n}\right) \tag{16.3-33}$$

where $k_0 = 2\pi/\lambda = \omega\sqrt{\mu\varepsilon_0}$, n is the index of refraction, and γ ($\ll k_0$) is the exponential gain constant of the field amplitude. In a lossy medium, $\gamma < 0$.

In the case where n and γ are periodic, we can write

$$\begin{aligned} n(z) &= n_0 + n_1 \cos\frac{2\pi}{\Lambda}z \\ \gamma(z) &= \gamma_0 + \gamma_1 \cos\frac{2\pi}{\Lambda}z \end{aligned} \tag{16.3-34}$$

where n_0, n_1, γ_0, and γ_1 are constants, and Λ is the period of the periodic modulation. Limiting ourselves to the case $n_1 \ll n_0$, $\gamma_1 \ll \gamma_0$, we can write, according to Equation (16.3-33),

$$\omega^2\mu\varepsilon(z) = \left[k_0^2 n_0^2 + i2k_0 n_0\gamma_0 + 4k_0 n_0\left(\frac{\pi n_1}{\lambda} + i\frac{\gamma_1}{2}\right)\cos\frac{2\pi}{\Lambda}z\right] \tag{16.3-35}$$

or equivalently,

$$n(z) = n_0 + i\frac{\gamma_0}{k_0} + \left(n_1 + i\frac{\gamma_1}{k_0}\right)\cos\frac{2\pi}{\Lambda}z \tag{16.3-36}$$

The coupled equations (16.3-4) were derived based on a real periodic dielectric perturbation. With the presence of periodic gain modulation as given by Equation (16.3-36), the coupled equations can be derived in a similar manner (see Section 12.4). They are given by

$$\begin{aligned} \frac{d}{dz}A &= -i\kappa B(z)e^{i2\Delta\beta z} + \gamma_0 A \\ \frac{d}{dz}B &= i\kappa A(z)e^{-i2\Delta\beta z} - \gamma_0 B \end{aligned} \tag{16.3-37}$$

where the coupling constant is given by

$$\kappa = \frac{\pi}{\lambda}\left(n_1 + i\frac{\gamma_1}{k_0}\right) = \frac{\pi n_1}{\lambda} + i\frac{\gamma_1}{2} \tag{16.3-38}$$

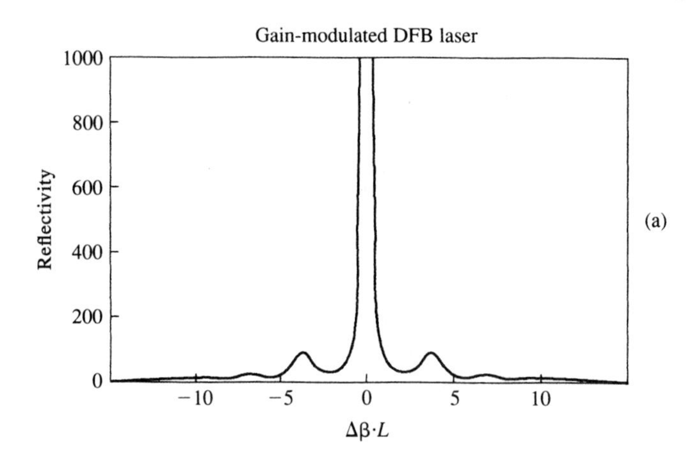

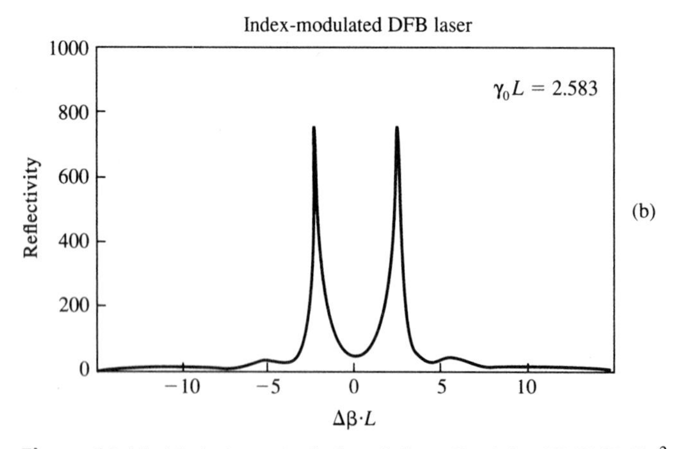

Figure 16.11 (a) A theoretical plot of the reflectivity $|E_r(0)/E_i(0)|^2$ of a waveguide with a gain $\gamma = \gamma_0 + \gamma_1 \sin(2\pi/\Lambda)z$. (b) A similar plot for an index modulated waveguide, $\gamma = \gamma_0$, $n = n_0 + n_1 \sin(2\pi/\Lambda)z$. (c) The measured oscillation spectrum of a GaInAsP distributed feedback laser with gain coupling: $\lambda = 1.5427$ μm. A single oscillating mode is present at the Bragg wavelength. Higher-order modes have output powers that are down by a factor of >45 dB (i.e., >32,000), compared to that of the fundamental mode. (Parts (a) and (b) courtesy of M. McAdam, California Institute of Technology, Pasadena, California. Part (c) courtesy of Dr. P. C. Chen, ORTEL Corporation.)

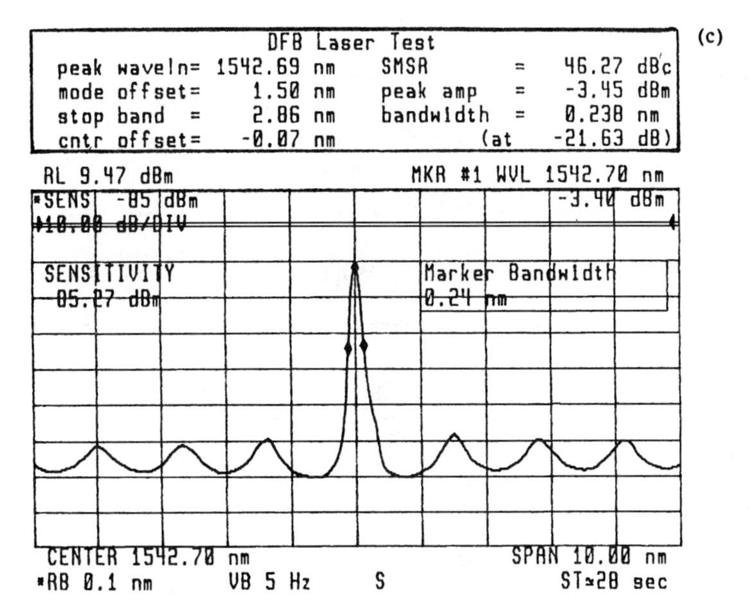

Figure 16.11 (*Continued*)

When $\gamma_0 = \gamma_1 = 0$, the coupled equations (16.3-37) reduce to (12.4-46). We can view Equation (16.3-38) as a generalization of Equation (12.4-63). The solutions in this case are given by

$$r_1 = \frac{-i\kappa \sinh SL_2}{S \cosh SL_2 + i(\Delta\beta + i\gamma_0) \sinh SL_2}$$
$$r_2 = \frac{-i\kappa \sinh SL_1}{S \cosh SL_1 + i(\Delta\beta + i\gamma_0) \sinh SL_1} \qquad (16.3\text{-}39)$$

with

$$S = \sqrt{\kappa^2 + (\gamma_0 - i\,\Delta\beta)^2} = \sqrt{\left(\frac{\pi n_1}{\lambda} + i\frac{\gamma_1}{2}\right)^2 + (\gamma_0 - i\,\Delta\beta)^2} \qquad (16.3\text{-}40)$$

We now consider the special case of pure gain modulation (i.e., $\gamma_1 \neq 0$ and $n_1 = 0$). At the Bragg frequency ω_0, where $\Delta\beta = 0$, the two reflection coefficients become

$$r_1 = \frac{-i\kappa \sinh SL_2}{S \cosh SL_2 - \gamma_0 \sinh SL_2}$$
$$r_2 = \frac{-i\kappa \sinh SL_1}{S \cosh SL_1 - \gamma_0 \sinh SL_1} \qquad (\omega = \omega_0) \qquad (16.3\text{-}41)$$

where κ becomes pure imaginary, $\kappa = i\gamma_1/2$, and S becomes positive and real, $S = \sqrt{\gamma_0^2 + (\gamma_1/2)^2}$. Examining the reflection coefficients (16.3-41), we note that the denominators are all positive. As a result, the product $r_1 r_2$ at ω_0 becomes a *positive* number, so that laser oscillation can now take place at the exact Bragg frequency ω_0. If we plot $|r^2|$ versus $\Delta\beta$ for this case (pure gain modulation), we obtain the result shown in Figure 16.11a. For comparison, we show in Figure 16.11b a plot of the reflectivity in the case of index modulation, which shows two

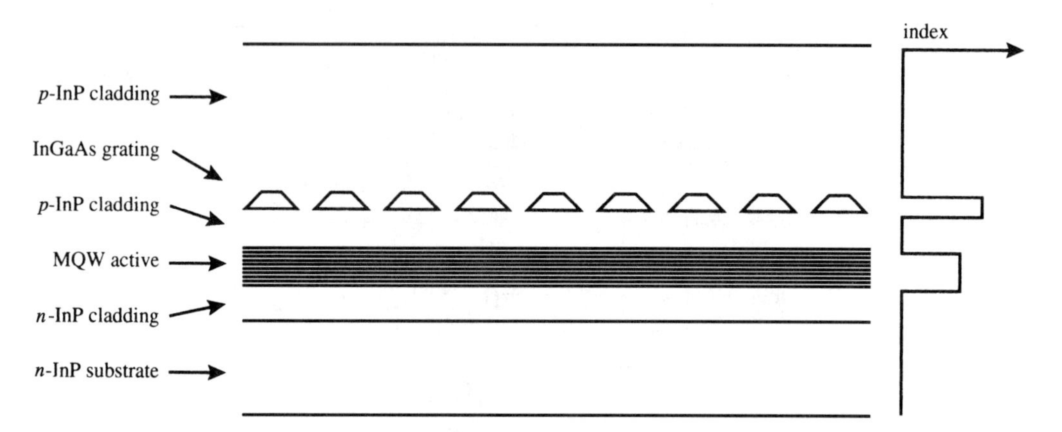

Figure 16.12 A periodic lossy layer, that is, a layer with an energy gap smaller than $\hbar\omega_{oscill}$, provides the periodic gain coupling in a semiconductor DFB laser. (Courtesy of Dr. P. C. Chen, ORTEL Corporation.)

modes situated symmetrically about ω_0. The experimental oscillation spectrum of a gain-coupled laser is shown in Figure 16.11c. It demonstrates the strong suppression of higher-order modes.

A cross section of a commercial gain-coupled DFB laser is shown in Figure 16.12. The periodic modulation of the gain is achieved by photolithographic corrugation [14] of an absorbing layer near the active region. The layer is incorporated for this purpose in the epitaxially grown laser structure. Additional layers grown epitaxially result in "burying" the periodic loss layer.

16.4 VERTICAL CAVITY SURFACE EMITTING SEMICONDUCTOR LASERS

Vertical cavity surface emitting semiconductor lasers (VCSELs) differ from their more conventional relatives in that the optical beam travels at right angles to the active region instead of in the plane of the active regions [15–17]. We recall that the active region is usually a thin slab of a few hundred nanometers in thickness. A typical VCSEL structure is illustrated in Figure 16.13. The VCSEL consists of an active layer sandwiched between two Bragg reflectors. A laser cavity in the vertical direction is formed by the two Bragg reflectors. The top and bottom Bragg reflectors consist, each, of alternating layers of semiconductor $GaIn_xAl_yAs_{1-x-y}$ with different x and y compositions. The difference in the index of refraction between adjacent layers gives rise to a high reflectivity (>99%) at the vicinity of the Bragg frequency ω_0 from each such "stack." For optimum mirror reflectivity, these stacks are usually made of alternating quarter-wave layers. The thickness of each layer is $\lambda/4n$, where n is the index of refraction of the layer. These mirror layers are grown epitaxially along with the rest of the laser layers. The laser biasing current flows through the mirrors so that they are highly doped to reduce the series resistance. The gain is provided by a small number, typically 1 to 4, of quantum wells that are placed near a maximum of the standing-wave pattern in the cavity to maximize the stimulated emission rate into the oscillation field. The total length of the spacer region, ② and ③, that straddles the active region is typically $L = \lambda'$, where λ' is the wavelength in the medium ($\lambda' = \lambda/n$, where n is the index of refraction). This

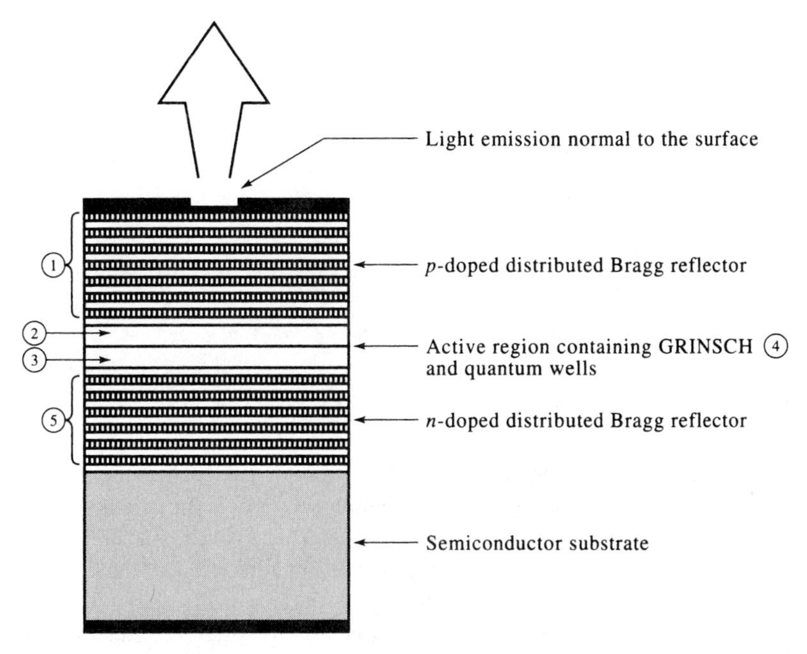

Figure 16.13 A schematic cross section of a vertical cavity surface emitting semiconductor laser based on the GaInAlAs alloy system.

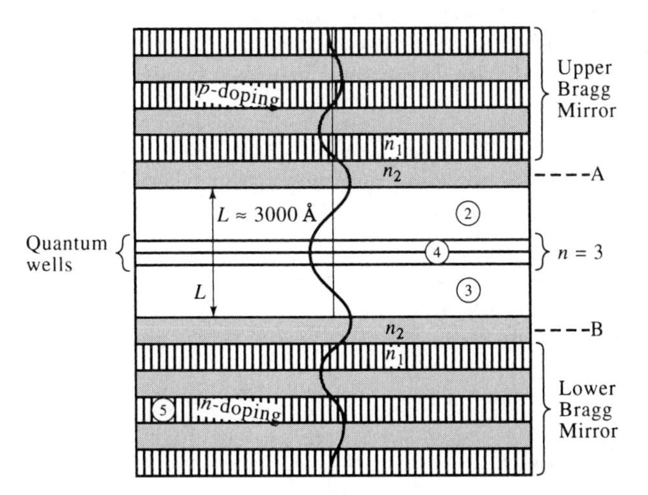

Figure 16.14 The field distribution of the laser mode inside a vertical cavity laser with $L = \lambda/n$ with three quantum wells. Note the evanescent decay of the field envelope inside the Bragg mirrors and the constant-amplitude standing wave between the mirrors.

translates, near $\lambda = 1\ \mu m$, to $L = 0.3\ \mu m$. Typical mode diameters are in the range of 3–10 μm. A typical Bragg stack consisting of, say, 15 quarter-wave layers is 2 μm thick.

The field distribution inside a vertical cavity laser is shown in Figure 16.14. We note that inside the Bragg mirror the optical wave amplitude undergoes exponential evanescence. This is in agreement with Equations (12.4-49) and Figure 12.18, which describe the evanescent decay of an optical wave inside a periodic medium for optical frequencies within the "forbidden" frequency gap [12].

Since the distance L_z traveled in the amplifying medium is small (approximately 100 Å per quantum well), the gain per pass is very small, and laser oscillation is made possible by the extremely high reflectance (>99%) of the Bragg mirror and the very low losses in regions ② and ③. Figure 16.14 conveys the relative scale of the key layer thicknesses.

The Oscillation Condition of a Vertical Cavity Laser

The oscillation condition of the VCSEL can be written

$$r_1(\omega)r_2(\omega)\exp\left(2\sum_{m=1}^{N}\gamma_m(\omega)L_z - i2\frac{\omega}{c}nL\right) = 1 \tag{16.4-1}$$

where $r_1(\omega)$ and $r_2(\omega)$ are amplitude reflection coefficients of the Bragg mirrors, γ_m is the amplitude gain coefficient of the mth quantum well, N is the number of quantum wells, L_z is the thickness of each of the quantum wells, L is the cavity spacing (i.e., space between the Bragg reflectors), and n is the averaged index of refraction of the cavity medium along the light path. Equation (16.4-1) is consistent with the oscillation condition of the Fabry–Perot laser described in Equation (6.1-12). Physically, it is merely a statement of the requirement that after one round trip a wave returns to its, arbitrary, starting plane with the same amplitude and, to within an integral multiple of 2π, the same phase. The factor of 2 in the exponent accounts for the round-trip propagation in the cavity.

In what follows, we will assume that each quantum well N contributes equally to the gain so that $\sum_{m=1}^{N}\gamma_m(\omega)L_z \equiv N\gamma(\omega)L_z$. We also assume identical Bragg reflectors so that we can write $r_1(\omega) = r_2(\omega) = r(\omega)$. Taking the absolute value of Equation (16.4-1), we obtain

$$|r(\omega)|^2 = \exp(-2N\gamma(\omega)L_z) \tag{16.4-2}$$

This is the mirror reflectivity needed to achieve oscillation. Since the optical wave travels at right angles to the plane of the quantum wells, the gain γ is not the modal gain g_{modal} of Equation (16.2-13) but the bulk gain γ of a quantum well medium. We note that according to Equation (16.2-9), the product $\gamma(\omega)L_z$ is independent of L_z. (This is strictly true when L_z is sufficiently small so that contributions to the gain from excited states ($l > 1$, see Equation (16.1-6)) in the quantum well are negligible. In practice, this is satisfied at room temperature for $L_z < 70$Å.) From experimental data of edge emitting quantum well lasers, we determine that for $L_z = 70$ Å the maximum gain due to the $l = 1$ energy level of the quantum well with a fully inverted population is $\gamma(\omega_0) = 5 \times 10^3$ cm^{-1}. Using this value in Equation (16.4-2) and taking $L_z = 70$ Å, we obtain

$$|r(\omega)|^2 = \exp[-2N \times 5 \times 10^3 \times 7 \times 10^{-7}]$$

for the reflectivity needed for oscillation.

$$N = 1, \quad |r(\omega)|^2 = 0.993$$

$$N = 2, \quad |r(\omega)|^2 = 0.986$$

$$N = 3, \quad |r(\omega)|^2 = 0.979$$

$$N = 4, \quad |r(\omega)|^2 = 0.972$$

where N is the number of quantum wells so that reflectivities around $R(\equiv |r(\omega)|^2) = 98\%$ are required of the Bragg reflectors. We will next take a small detour to review the optical properties of these reflectors that are relevant to VCSELs.

The Bragg Mirror

The analysis of the Bragg mirror is an excellent example of the power of the coupled-mode formalism developed in Chapter 12. In the case when the mirrors are made of alternating layers of two different materials of uniform refractive index, an exact analysis using the 2×2 matrix method can be carried out to obtain the reflectivity of the mirror (see Section 12.2). The periodic perturbation of the index of refraction couples, exactly as in the case of the DFB laser, two waves propagating in opposite directions. The coupling is described by the coupled equations (12.4-46). To obtain the coupling constant κ, we write the dielectric constant of the Bragg reflector as

$$\varepsilon(x, y, z) = \varepsilon_0 n^2(x, y, z) = \tfrac{1}{2}(n_1^2 + n_2^2)\varepsilon_0 + \tfrac{1}{2}(n_2^2 - n_1^2)\varepsilon_0 f(z) \tag{16.4-3}$$

where n_1 and n_2 are the indices of refraction of the two alternating layers, and $f(z)$ is a periodic square wavefunction of unity amplitude defined as

$$f(z) = \begin{cases} 1, & 0 < z < \Lambda/2 \\ -1, & \Lambda/2 < z < \Lambda \end{cases} \tag{16.4-4}$$

The periodic square wavefunction can be written as a Fourier series,

$$f(z) = \sum_{\ell} a_\ell e^{-i\ell \frac{2\pi}{\Lambda} z}, \quad a_\ell = i\frac{1 - e^{i\pi\ell}}{\ell\pi} = i\frac{1 - \cos(\ell\pi)}{\ell\pi} \tag{16.4-5}$$

We now treat the second term in Equation (16.4-3) as a dielectric perturbation $\Delta\varepsilon$,

$$\Delta\varepsilon(x, y, z) = \sum_{\ell \neq 0} \varepsilon_\ell(x, y) \exp\left(-i\ell \frac{2\pi}{\Lambda} z\right) \tag{16.4-6}$$

Then the ℓth Fourier component of the dielectric perturbation can be written

$$\varepsilon_\ell = i\varepsilon_0 \frac{1 - \cos(\ell\pi)}{\ell\pi} \frac{1}{2}(n_1^2 - n_2^2) \tag{16.4-7}$$

Based on the discussion in Section 12.4, the coupling constant can be written, according to Equations (12.4-27) and (12.4-29),

$$\kappa = \frac{\omega^2\mu}{2\bar{n}k_0}\varepsilon_\ell = i\frac{\omega^2\mu\varepsilon_0}{4\bar{n}k_0}\frac{1 - \cos(\ell\pi)}{\ell\pi}(n_2^2 - n_1^2) \tag{16.4-8}$$

where $k_0 = \omega/c$, and $\bar{n} = \sqrt{n_1^2 + n_2^2}/\sqrt{2}$ is the average index of refraction in the Bragg mirror. When $\ell = 1$ we have

$$\kappa = i\frac{\sqrt{2}}{\lambda}\frac{(n_2^2 - n_1^2)}{\sqrt{n_1^2 + n_2^2}} \tag{16.4-9}$$

For the case when $n_1 \approx n_2$, the coupling constant can be written

$$\kappa = i\frac{2\Delta n}{\lambda} \tag{16.4-10}$$

where $\Delta n = n_2 - n_1$ is the index difference of the adjacent layers. The mirror reflectivity at the Bragg frequency is given, according to Equation (12.4-52), by

$$R(\omega_0) = |r(\omega_0)|^2 = \tanh^2|\kappa L_m| \tag{16.4-11}$$

where L_m is the total thickness of the stack of alternating layers in the Bragg mirror.

To obtain an appreciation for the magnitude of reflectivities that we may expect in a typical Bragg mirror, we will design a Bragg mirror to operate at a center wavelength of $\lambda_0 = 0.875$ μm. The unit cell consists of a pair of epitaxially grown $Ga_{0.8}Al_{0.2}As$ and AlAs layers. The index of refraction difference is $\Delta n = n_{Ga_{0.8}Al_{0.2}As} - n_{AlAs} = 0.55$. The average index is $n = 3.3$. Since the thickness of a unit cell is $\lambda_0/2n$, the length of the Bragg mirror with N_m periods is $L = N_m\lambda_0/2n$. The result in the case of $N_m = 15$ is

$$R(\omega_0) = |r(\omega_0)|^2 = \tanh^2\left(N_m \frac{\Delta n}{n}\right) = \tanh^2\left(\frac{15 \times 0.55}{3.3}\right) = 0.973$$

This value is sufficient, according to the discussion following Equation (16.4-2), to satisfy the oscillation conditions in vertical cavity lasers with more than four inverted ($N \geq 4$) quantum wells.

For frequency in the vicinity of the Bragg frequency, the amplitude reflection coefficient is obtained, according to Equation (12.4-53),

$$r(\omega) = \frac{-i\kappa^* \sinh sL}{s \cosh sL + i\,\Delta\beta \sinh sL} \tag{16.4-12}$$

where

$$2\Delta\beta = 2\beta - \ell\frac{2\pi}{\Lambda} = 2\bar{n}\frac{2\pi}{\lambda} - \ell\frac{2\pi}{\Lambda} \equiv 2(\beta - \beta_0) \tag{16.4-13}$$

$$s^2 = \kappa^*\kappa - (\Delta\beta)^2 \tag{16.4-14}$$

A plot of the reflectivity $|r(\omega)|^2$ based on Equation (16.4-12) and the experimental parameters of the above example is shown in Figure 16.15a. An experimental plot of a Bragg mirror with the same parameters is shown in Figure 16.15b. The phase shift $\phi(\omega)$ of the complex reflectance $r(\omega) = |r(\omega)|\exp[-i\phi(\omega)]$ is shown in Figure 16.15c. Assuming $\Delta n < 0$, the amplitude reflection coefficient at the Bragg frequency $r(\omega_0)$ is real and positive (with $\phi(\omega_0) = 0$). For a more detailed treatment of Bragg mirrors and light propagation in stratified media, the reader is referred to Section 12.2 and Reference [12].

The Oscillation Frequencies

The phase part of Equation (16.4-1) can be employed to obtain an expression for the oscillation frequencies of a surface emitting Bragg mirror laser. For simplicity, we assume two identical Bragg mirrors with $r_1(\omega) = r_2(\omega) = |r(\omega)|e^{-i\phi(\omega)}$. The phase condition is

$$2\phi(\omega) + 2\frac{\omega}{c}nL = 2m\pi, \quad m = 1, 2, \ldots \tag{16.4-15}$$

Let us denote the two neighboring oscillation frequencies corresponding to m and $m + 1$ as ω_m and ω_{m+1}, respectively:

$$\begin{aligned} \phi(\omega_m) + \frac{\omega_m}{c}nL &= m\pi \\ \phi(\omega_{m+1}) + \frac{\omega_{m+1}}{c}nL &= (m+1)\pi \end{aligned} \tag{16.4-16}$$

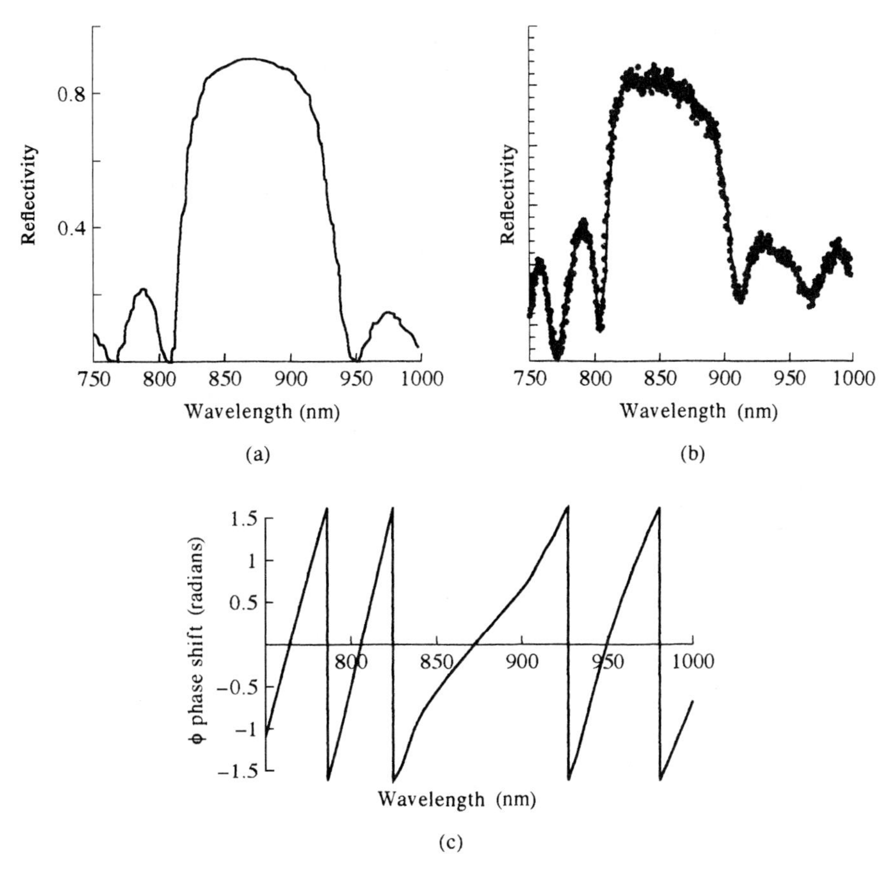

Figure 16.15 Calculated (a) and measured (b) reflectivity of a 15-period $Al_{0.2}GA_{0.8}As/AlAs$ distributed Bragg reflector. The calculated phase shift $\phi(\omega)$ is plotted in (c) (see also Figure 12.27 and the discussion in Section 12.6). (Courtesy of J. Obrien, California Institute of Technology, Pasadena, California.)

so that

$$\left(\phi(\omega_{m+1}) - \phi(\omega_m) + \frac{\omega_{m+1} - \omega_m}{c} nL\right) = \pi \tag{16.4-17}$$

The phase $\phi(\omega)$ in the immediate vicinity of the Bragg frequency is approximately a linear function of the frequency and can be written

$$\phi(\omega) \cong \tau(\omega - \omega_0) + \phi_0 \tag{16.4-18}$$

where ϕ_0 is a constant and τ is the group delay of reflection at the Bragg frequency (see Section 12.6). According to Equation (12.6-10), the group delay τ is given by

$$\tau = \frac{\tanh|\kappa|L_m}{|\kappa|L_m}\tau_0 \approx \frac{1}{|\kappa|L_m}\tau_0 - \frac{n}{|\kappa|c} \tag{16.4-19}$$

where we assumed $|\kappa|L_m \gg 1$ so that $\tanh|\kappa|L_m \approx 1$; $\tau_0 = d(\beta L_m)/d\omega = nL_m/c$ is the group delay for the beam to traverse through the medium of Bragg reflector (of length L_m) in the absence of the periodic index modulation.

Using Equations (16.4-18) and (16.4-19), Equation (16.4-17) becomes

$$2\pi\Delta\nu = (\omega_{m+1} - \omega_m) = \frac{\pi c}{n(L + 1/|\kappa|)} \tag{16.4-20a}$$

for the intermode frequency interval, or equivalently,

$$\Delta\nu = (\nu_{m+1} - \nu_m) = \frac{c}{2nL(1 + 1/|\kappa|L)} \tag{16.4-20b}$$

The effective length of the Bragg mirror resonator at the Bragg frequency is thus not the mirror spacing L but

$$L_{\text{eff}} = L + \frac{1}{|\kappa|} \tag{16.4-21}$$

The contribution $1/|\kappa|$ is due to the evanescent penetration of the oscillating laser field into the Bragg mirrors, as illustrated in Figure 16.14. Since two Bragg mirrors are assumed in the analysis, the penetration distance into a single Bragg mirror is $1/(2|\kappa|)$.

We recall that the field behavior inside the periodic Bragg mirror (at the Bragg frequency ω_0) is given, according to Equation (12.5-28), by

$$\exp(-i\beta' z) = \exp\left(-i\frac{\pi}{\Lambda}z\right)\exp(-|\kappa|z) \tag{16.4-22}$$

and an intensity of

$$I(z) = I(0)\exp(-2|\kappa|z) \tag{16.4-23}$$

which corresponds to an effective energy penetration distance of $\sim 1/(2|\kappa|)$ in each of the Bragg mirrors in agreement with the value of $1/(|\kappa|)$ of Equation (16.4-21).

EXAMPLE: INTERMODE FREQUENCY SEPARATION

To obtain an appreciation for the intermode frequency spacing of (16.4-20a), we will consider the laser depicted in Figure 16.14. The data for the Bragg mirror is the same as used in the example following Equation (16.4-11). The basic parameters are

$$\lambda = 1\ \mu\text{m}, \quad L = \lambda = 1\ \mu\text{m}$$

$$|\kappa| = \frac{2\Delta n}{\lambda} = \frac{2 \times 0.55}{1} = 1.1\ \mu\text{m}^{-1}$$

$$L_{\text{eff}} = L + \frac{1}{|\kappa|} = (1 + 0.91)\ \mu\text{m} = 1.91\ \mu\text{m}$$

$$2\pi\Delta\nu = (\omega_{m+1} - \omega_m) = \frac{\pi}{n(L + 1/|\kappa|)}$$

(Note that the penetration depth, 0.91 μm, is about the same as the intermirror spacing, L.)

$$\Delta\nu \equiv \frac{\omega_{m+1} - \omega_m}{2\pi} = \frac{c}{2nL_{\text{eff}}} = \frac{3 \times 10^{10}}{2 \times 3.3 \times 1.91 \times 10^{-4}} = 2.38 \times 10^{13}\ \text{Hz} = 793\ \text{cm}^{-1}$$

The high-reflectivity region of the Bragg mirror (the photonic bandgap) is given, according to Equation (12.5-24), by

$$(\Delta\nu)_{\text{Bragg}} = \frac{(\Delta\omega)_{\text{gap}}}{2\pi} \approx \frac{|\kappa| c}{\pi n} = \frac{1.1 \times 10^4 \ (\text{cm}^{-1}) \times 3 \times 10^{10}}{\pi \times 3.3} = 3.47 \times 10^{13} \text{ Hz}$$

This number is comparable to the intermode spacing $\Delta\nu = 2.38 \times 10^{13}$ Hz, so that only one mode at a time will experience high reflectivity and will satisfy the oscillation condition. This leads in most cases to a single-mode oscillation. This is to be contrasted with more conventional cleaved-facet-reflectors, edge-emitting semiconductor lasers, where $L = 300$ μm and the mode spacing is correspondingly shorter.

We conclude by showing in Figure 16.16 a photograph of a two-dimensional array of surface emitting lasers.

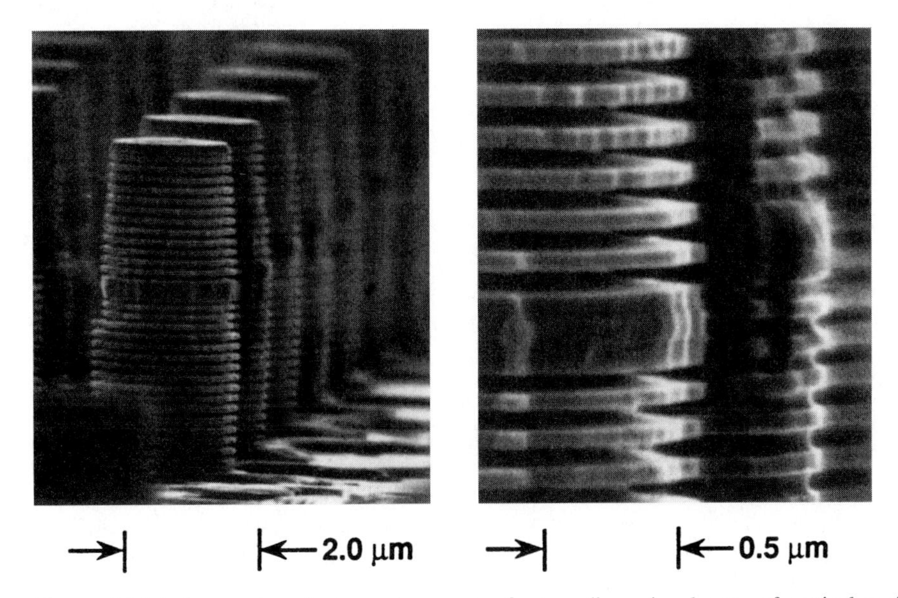

Figure 16.16 A scanning electron micrograph of a two-dimensional array of vertical cavity surface emitting lasers. The multilayered structure of alternating GaAs and AlAs layers is demonstrated by partial preferential etch of the AlAs layers. (Courtesy of *Scientific American* magazine and A. Scherer, California Institute of Technology, Pasadena, California.)

PROBLEMS

16.1 Solve the one-dimensional Schrödinger equation (16.1-3) in the case of a simple square potential well, where

$$V(z) = \begin{cases} -V_0, & -L/2 < z < L/2 \\ 0, & \text{elsewhere} \end{cases}$$

16.2 Assume that as we scale the length L of a quantum well laser we maintain the differential quantum efficiency η_{ex} constant by increasing R.

(a) Derive the expression relating R (mirror reflectivity) to L.

(b) Show that $I_{\text{threshold}}$ is proportional to L.

16.3 Show qualitatively that for a given m_c and injection current the maximum gain obtains when $m_v = m_c$.

16.4 Estimate the coupling constant κ of the DFB laser whose spontaneous emission spectrum is given in Figure 16.10e.

16.5

(a) Using a computer program, plot the magnitude of the reflection coefficient (see Equation (16.3-14)) of a periodic amplifying waveguide as a function of $\Delta\beta L$, assuming $\kappa L = 0.4$. Let γL be the parameter, and generate plots with $\gamma L = 2$, 2.9, 3.5, and 3.8.

(b) Plot the equi-gain contours in the γL–$\Delta\beta L$ plane as in Figure 16.12.

16.6

(a) Derive the coupled-mode equations of a DFB laser with a periodic modulation of its losses. The spatial periodic loss can be accounted for according to Maxwell's equations by taking the dielectric constant of the waveguide as

$$\varepsilon'(\mathbf{r}) = \varepsilon(\mathbf{r})\left(1 - i\frac{\sigma(\mathbf{r})}{\omega\varepsilon(\mathbf{r})}\right), \quad \text{with} \quad \sigma(\mathbf{r}) = \sigma_0 + \sigma_1(x)\cos\frac{2\pi}{\Lambda}z$$

where $\sigma(\mathbf{r})$ is the medium conductivity.

(b) Compare the coupling coefficient κ in this case to that of index modulation (see Equation (12.5-14)).

(c) Estimate the magnitude of κ in the case of a loss-modulated waveguide where the effective index is $n_{\text{eff}} = 3.5$, $\Lambda = 0.22$ μm, and $\lambda \sim 1.55$ μm. The lossy layer has an absorption coefficient of $\alpha = 300\ \text{cm}^{-1}$ and a thickness of 1000 Å. It is situated at the center of the guiding layer. Assume that the waveguide mode is highly confined to the inner layer ($n = 3.51$).

16.7 Starting from Equations (16.3-26) and (16.3-28),

$$r_2 = \frac{-i\kappa^* \sinh SL_1}{S\cosh SL_1 + i(\Delta\beta + i\gamma)\sinh SL_1}, \quad \text{and}$$

$$r_1 = \frac{-i\kappa \sinh SL_2}{S\cosh SL_2 + i(\Delta\beta + i\gamma)\sinh SL_2}$$

show that

$$r_1 r_2 = \frac{-\kappa^*\kappa \sinh SL_1 \sinh SL_2}{S[S\cosh SL + i(\Delta\beta + i\gamma)\sinh SL] - \kappa^*\kappa \sinh SL_1 \sinh SL_2}$$

where $L = L_1 + L_2$ is the total length of the periodic waveguide. Note that $r_1 r_2 = 1$ when $S\cosh SL + i(\Delta\beta + i\gamma)\sinh SL = 0$.

16.8 Examine the phase ϕ of $r_1 r_2$ in Problem 16.7. This phase is the "roundtrip phase shift" of light in DFB lasers. At oscillation this phase shift must be zero or an integral multiple of 2π. Show that in the limit $\gamma \to 0$, the phase shift ϕ is within $(0, 2\pi)$ for frequencies in the photonic bandgap. In other words, $0 < \phi < 2\pi$ within the bandgap. Thus, oscillation in DFB lasers cannot occur within the photonic bandgap (see Figures 12.29 and 16.8).

REFERENCES

1. van der Ziel, J. P., R. Dingle, R. C. Miller, W. Wiegmann, and W. A. Nordland, Jr., Laser oscillation from quantum states in very thin GaAs-$Al_{0.2}Ga_{0.8}$sAs multi-layer structures. *Appl. Phys. Lett.* **26**:463 (1975). See also R. D. Dupuis and P. D. Dapkus, *IEEE J. Quantum Electron.* **QE-16**:170 (1980).
2. Mittelstein, M., *Theory and Experiments on Unstable Resonators and Quantum Well GaAssGaAlAs Lasers*. Ph.D. thesis in Applied Physics, California Institute of Technology, Pasadena, CA, 1989, p. 54.
3. Tsang, W. T., Extremely low threshold (AlGa)As modified multi-quantum well heterostructure laser grown by MBE. *Appl. Phys. Lett.* **39**:786 (1981).

4. Mehuys, D., *Linear, Nonlinear and Tunable Guided Wave Modes for High Power (GaAl)As Semiconductor Lasers*. Ph.D. thesis, California Institute of Technology, Pasadena, CA, June 1989.
5. Derry, P., et al., Ultra low threshold graded-index separate confinement single quantum well buried heterostructure (Al, Ga) as lasers with high reflectivity coatings. *Appl. Phys. Lett.* **50**:1773 (1987).
6. Eng, L. E., et al., Submilliampere threshold current pseudomorphic InGaAs/AlGaAs buried heterostructure quantum well lasers grown by molecular beam epitaxy. *Appl. Phys. Lett.* **55**:1378 (Oct. 1989).
7. Dingle, R., W. Wiegmann, and C. H. Henry, Quantum states of confined carriers in very thin $Al_xGat_{1-x}As$-GaAs-$Al_xGA_{1-x}As$ heterostructures. *Phys. Rev. Lett.* **33**:827 (1974). Also see G. Bastard and J. A. Brum, Electronic states in semiconductor heterostructures. *IEEE J. Quantum Electron.* **QE-22**:1625 (1986).
8. Arakawa, Y., and A. Yariv, Theory of gain, modulation response and spectral linewidth in A1GaAs quantum-well lasers. *IEEE J. Quantum Electron.* **QE-21**:1666 (1985).
9. Kogelnik, H., and C. V. Shank, Coupled wave theory of distributed feedback lasers. *J. Appl. Phys.* **43**:2328 (1972).
10. Nakamura, M., A. Yariv, H. W. Yen, S. Somekh, and H. L. Garvin, Optically pumped GaAs surface laser with corrugation feedback. *Appl. Phys. Lett.* **22**:515 (1973).
11. Aiki, K., M. Nakmura, J. Umeda, A. Yariv, A. Katzir, and H. W. Yen, GaAs–GaAlAs distributed feedback laser with separate optical and carrier confinement. *Appl. Phys. Lett.* **27**:145 (1975).
12. Yariv, A., and P. Yeh, *Optical Waves in Crystals*. Wiley-Interscience, New York, 1984.
13. Haus, H. A., and C. V. Shank, Antisymmetric taper of distributed feedback lasers. *IEEE J. Quantum Electron.* **QE-12**:532 (1976).
14. Nakano, Y., Y. Luo, and K. Tada, Facet reflection independent, single longitudinal mode oscillation in a GaAlAs/GaAs distributed feedback laser equipped with a gain-coupling mechanism. *Appl. Phys. Lett.* **55**:16 (1989).
15. Iga, K., S. Ishikawa, S. Ohkouchi, and T. Nishimura, Room-temperature pulsed oscillation of GaAlAs/GaAs surface emitting injection laser. *Appl. Phys. Lett.* **45**:348 (1984).
16. Jewell, J. L., J. P. Harbison, A. Scherer, Y. H. Lee, and L. T. Florez, Vertical cavity surface emitting lasers. *IEEE J. Quantum Electron.* **27**:1332 (1991).
17. Jewell, J. L., J. P. Harbison, and A. Scherer, Microlasers. *Sci. Am.* p. **86** (November 1991).

CHAPTER 17

OPTICAL AMPLIFIERS

17.0 INTRODUCTION

Linear amplifiers are essential elements in any point-to-point system. In optical networks, there are many loss mechanisms, including insertion loss, branching loss, and propagation attenuation in silica fibers. Linear optical amplifiers are useful in restoring the power levels of optical signals in the optical domain. Optical amplification in fiber links [1–3] has been recognized as having major system implications for very long distance transmission of information (> 1000 km) using optical fibers and for local distribution systems involving a large number of subscribers. The conventional way of compensating for optical loss in light-wave communication systems has been the rather costly and cumbersome procedure of electronic regeneration at the repeater stations. The regeneration process includes photon–electron conversion, electrical amplification, retiming, pulse reshaping, and finally electron–photon conversion. In dense wavelength division multiplexed (DWDM) optical networks, there are many frequency channels in a single optical fiber. The conventional way requires the separation of the signals of all the channels for the regeneration, and then recombining all the channels. This is a very expensive approach, particularly for DWDM optical networks. Optical amplifiers are capable of amplifying the power levels of all the channels simultaneously in the optical domain in a manner that is transparent to the modulation format, provided the gain bandwidth is wide enough. This eliminates the need of costly optical-to-electrical and electrical-to-optical conversion at the repeater stations, and provides a simple and economical means of bandwidth upgrade in optical networks. The raison d'être for the optical amplifiers is that they make it possible to maintain the optical power at sufficiently high levels along the path so that the signal-to-noise ratio (SNR) degradation due to signal shot noise and receiver noise is reduced to practical inconsequence.

Depending on the role of optical amplifiers in the optical networks, amplifiers are generally classified into the following three categories:

1. *Booster amplifier*. Many lasers, particularly tunable ones, are designed to provide low optical output power levels for stability purposes. These lasers must be immediately followed by an optical amplifier to boost the power level.
2. *In-line amplifier*. As a signal's power level decreases due to propagation, branching, or tap losses, the in-line amplifier allows the signal power level to be amplified within the signal path (fiber). Good gain flatness is often required as many amplifiers may be cascaded.

3. *Preamplifier*. For the purpose of achieving high signal-to-noise ratio in the detection, preamplifiers are employed to amplify weak signals before the receivers. High gain and low noise performance are essential.

In this chapter, we discuss various optical amplifiers, including semiconductor optical amplifiers (SOAs), erbium-doped fiber amplifiers (EDFAs), and Raman optical amplifiers. We also take up the important subject of amplified spontaneous emission noise introduced by the optical amplifier [4, 5] and the signal-to-noise ratio (SNR) in fiber links, where amplifiers are cascaded.

17.1 SEMICONDUCTOR OPTICAL AMPLIFIERS

In the last two chapters, we discussed various semiconductor lasers. A semiconductor optical amplifier (SOA) is based on the same technology as a Fabry–Perot diode laser. As discussed in Chapters 15 and 16, such a laser consists of an amplifying medium located inside a resonant (Fabry–Perot type) cavity. The amplification (or gain) is achieved by electric carrier injection into a semiconductor to provide a population inversion. In order to get only the amplification function, it is necessary to prevent self-oscillation (lasing) of the device. This is accomplished by eliminating cavity reflections through the use of an antireflection (AR) coating and angle cleaving the chip facets. Figure 17.1 shows a schematic drawing of a semiconductor optical amplifier.

The main advantage of a SOA is the possibility of very large gains (>20 dB) in a short (less than ~400 μm) semiconductor chip. The main disadvantage of the SOA compared to the erbium-doped fiber amplifier (EDFA) is the short lifetime of the carrier recombination. The short recombination lifetime in the SOA can cause crosstalk between communication channels. In addition, the presence of residual reflections from the facets of the SOA requires the need for optical isolators. The presence of even minute reflection ($R < 10^{-5}$) can give rise to instabilities and excess noise in the source laser oscillator. Impressive results, however, have been demonstrated. Unlike EDFAs, which are optically pumped, SOAs are electrically pumped by injected current. The discussion in this section follows closely that in References [6–8].

Depending on the level of residual reflectivity of the facet, SOAs can be divided into two categories: resonant or Fabry–Perot amplifiers (FPAs) and traveling-wave amplifiers (TWAs). In resonant amplifiers, the facet reflectivities are lower than that of a laser so that self-oscillation is not possible. The reflectivities are, however, significant enough to provide multiple reflections in the active region. This gives rise to a resonant cavity and the gain spectrum consists of a series of peaks corresponding to the longitudinal Fabry–Perot resonances

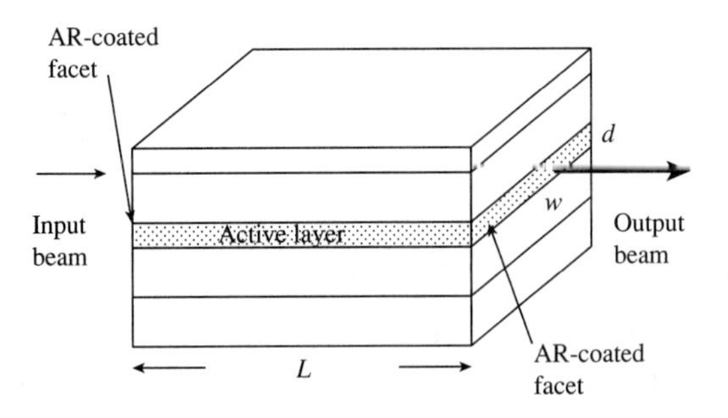

Figure 17.1 Schematic drawing of a semiconductor optical amplifier (SOA) with an active layer of thickness d, width w, and a length L. The facets are antireflection coated to prevent self-oscillation.

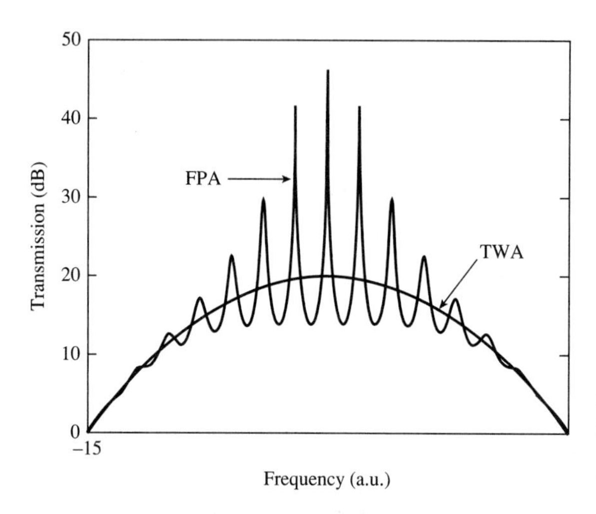

Figure 17.2 Transmission spectrum of a semiconductor optical amplifier with a facet reflectivity of $R = 0.01$, and a single-pass peak intensity gain of 20 dB ($G_{peak} = 100$). The smooth curve is the extreme case when $R = 0$.

of the amplifier chip. In the spectral regimes around the peaks, the gains are much higher. Resonant SOAs are manufactured using an AR coating with a reflectivity around 10^{-2}. They typically feature a gain ripple of 10–20 dB and a useful bandwidth of 2–10 GHz around each peak. TWA devices incorporate a coating with a reflectivity of less than 10^{-4} (see Figure 17.2). They show a gain ripple of a few decibels and a bandwidth better than 5 THz (e.g., $\Delta\lambda = 40$ nm in the $\lambda = 1550$ nm window).

The basic structure of a SOA consists of an active region whose two facets have been antireflection coated. Assuming an identical facet reflectivity R, the transmission of such an active Fabry–Perot cavity is given by, according to the discussion in Chapter 4,

$$T = \frac{(1 - R)^2 G}{(1 - GR)^2 + 4GR \sin^2(2\pi n_{eff} L/\lambda)} \tag{17.1-1}$$

where R is the facet reflectivity, G is the internal single-pass intensity gain of the guided mode, n_{eff} is the effective index of the guided mode, L is the cavity length, and λ is the wavelength. In the event when the facet reflectivity reduces to zero, the transmission becomes the single-pass intensity gain G. With the presence of a residual facet reflectivity ($R \neq 0$), the transmission spectrum deviates from the single-pass gain spectrum due to multiple reflections inside the cavity. Figure 17.2 shows the transmission of a SOA. The transmission spectrum consists of resonance peaks whose transmittances are much higher than the single-pass gain. These resonance peaks occur when the cavity length is an integral number of half-wavelengths, (i.e., $L = m(\lambda/2n_{eff})$, with $m = 1, 2, 3, \ldots$). The maximum and minimum transmissions are given by

$$T_{max} = \frac{(1 - R)^2 G}{(1 - GR)^2} \tag{17.1-2}$$

$$T_{min} = \frac{(1 - R)^2 G}{(1 + GR)^2} \tag{17.1-3}$$

Operation of the amplifier in the FPA or TWA modes will depend on the value of the round-trip amplitude GR. For TWAs, a gain ripple of less than 1 dB would require the product GR to be less than 0.058 (see Problem 17.1). For a single-pass gain of 30 dB ($G = 1000$), this means a facet reflectivity of less than 5.8×10^{-5}, not an easy task to achieve. Furthermore,

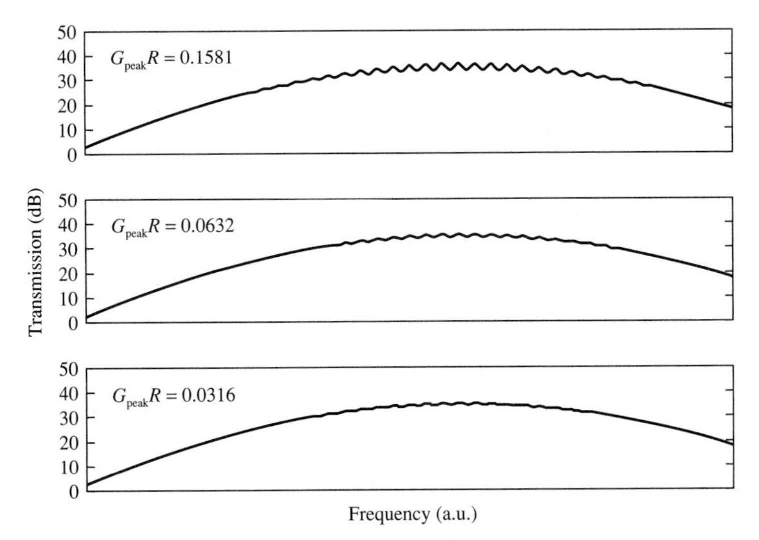

Figure 17.3 Gain spectrum of a semiconductor optical amplifier with various facet reflectivities R. The gain ripple will be less than 1 dB provided GR is less than 0.058, where G is a single-pass intensity gain.

most antireflection coatings are optimized for a particular wavelength and often have low reflectivity only over a limited spectral regime. A semiconductor optical amplifier will, in general, operate both as a resonant amplifier (FPA) and traveling-wave amplifier (TWA) in different parts of the gain spectrum. Figure 17.3 shows the transmission ripples of a SOA at various values of the facet reflectivities.

For SOAs, the challenges in the early days of development included the realization of low-facet reflectivites and low insertion loss. In addition to the facet residual reflectivity problem, the SOA is also plagued with the problem of strong polarization dependence. The polarization dependence has two origins. First, virtually all conventional semiconductor lasers have unequal mode confinement factor for the TE and TM modes. Second, the TE and TM modes have different effective indices (n_{eff}). This leads to different facet reflectivities for the TE and TM modes. In addition, the maxima and minima of the transmission of these two modes (TE and TM) do not overlap. A number of schemes have been proposed to eliminate the polarization-dependence problem. For example, two mutually perpendicular SOAs in series can be employed to neutralize the problem. A double-pass configuration with the help of a reflector and a 90° polarization rotator can also be employed to solve the problem. In practice, a single polarization-independent SOA is more desirable. This can be achieved by designing the waveguide with nearly equal mode confinement factor for the TE and TM modes, and nearly equal effective index. For example, a waveguide with a square cross section would provide the symmetry needed.

The gain spectrum of the SOA is basically the same as that of the semiconductor laser. As discussed in Chapters 15 and 16, the gain bandwidth depends strongly on the pumping (current injection) and is limited by the difference between the carrier's quasi-Fermi levels. Like all gain media, the gain of a SOA will also saturate. The saturation output power of an optical amplifier is a very important parameter for in-line applications when numerous amplifiers have to be cascaded (e.g., in long-haul optical networks). The saturation power is usually defined as the output power for which the gain is reduced by 3 dB from the small-signal gain. As discussed in Chapter 5, the gain saturation is a result of the decrease of the population inversion caused by the increased stimulated emission recombination at high optical powers. The saturation also depends on other carrier recombination rates (radiative and non-radiative) at a given pumping rate, as well as on waveguide geometry and confinement factor.

The gain saturation can cause crosstalk between two channels of signals being amplified by the same optical amplifier when the total amplifier output power approaches the saturation output power. In this case, the gain of one channel depends on the presence or absence of optical power in the other channel (or channels). At a carrier lifetime of only a few hundred picoseconds in the SOA, the gain can easily be modulated by a signal with a bit rate up to a few gigahertz. The modulation of the gain can also be caused by the beating frequency of two amplified signals. As a result of the beating, the modulated gain then creates a new frequency when a third signal is being amplified, resulting in gain-induced four-wave mixing. Consider the presence of two signals at frequencies ω_1 and ω_2. The beating of these two signals causes the SOA gain to be modulated, through saturation, at frequency $\Omega = \omega_2 - \omega_1$. A signal at frequency ω_3 will be amplified with the modulated gain, thus acquiring sidebands at $\omega_3 \pm \Omega$. This is the gain-induced four-wave mixing. The gain modulation is strongly dependent on the beat frequency (or the frequency separation) of the two amplified signals and thus places stringent limitations on the channel spacing in multichannel systems (WDM) with optical amplifiers.

In addition to the gain-induced four-wave mixing, self-phase modulation (SPM) and cross-phase modulation (XPM) also occur in SOAs. As discussed earlier in Chapter 15, the refractive index of the active layer depends on the carrier density. A signal with high bit rate (a few gigahertz) can cause a temporal variation of the carrier density during propagation of the pulses. This leads to a temporal variation of the refractive index of the active layer, and thus the phase modulation. As a result, the amplification of optical pulses can be accompanied by a considerable spectral distortion even when the pulse shape remains undistorted immediately after the SOA. The spectral distortion (usually chirping and broadening) will cause pulse distortion and signal fading in optical fibers.

17.2 ERBIUM-DOPED FIBER AMPLIFIERS

In Section 6.11, we briefly discussed Er^{3+}:silica fiber lasers. The Er^{3+}:silica laser oscillators consist of a section of Er^{3+}-doped silica fiber bounded by fiber Bragg gratings (FBGs) functioning as mirrors. Again, if the reflectors (mirrors) are removed, the erbium-doped silica fiber becomes an optical amplifier. The energy levels involved in the pumping and amplification transitions in a typical Er^{3+}:silica fiber laser are shown in Figure 6.53. Er^{3+} is a rare earth ion from the lanthanide family. The erbium ion Er^{3+} has 11 electrons in the 4f orbital. It is an element of choice for amplification in the 1.55 μm spectral regime, due to its extremely wide gain bandwidth resulting from the transition between the states ${}^4I_{15/2}$ and ${}^4I_{13/2}$. In the host medium (silica), all $(2J + 1)$ components of both multiplets of the erbium ion are split and homogeneously broadened by the environmental perturbations due to the neighboring oxygen ions in the silica glass (SiO_2). In other words, the ${}^4I_{15/2}$ level is split into 16 sublevels ($2J + 1$, with $J = 15/2$), and the ${}^4I_{13/2}$ level is split into 14 sublevels ($2J + 1$, with $J = 13/2$). As a result of the random environments (in fused silica) for each of the erbium ions, all these sublevels are further broadened due to inhomogeneous broadening. This leads to the extremely wide gain bandwidth, which is desirable for broadband optical communications applications. There are two strong peaks in absorption cross section of the erbium-doped fiber, at 1480 nm and 980 nm, corresponding to the transitions (${}^4I_{15/2} \rightarrow {}^4I_{11/2}$) and (${}^4I_{15/2} \rightarrow {}^4I_{13/2}$), respectively.

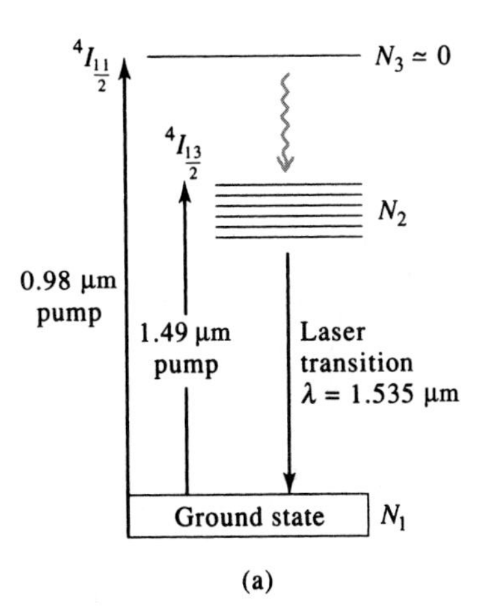

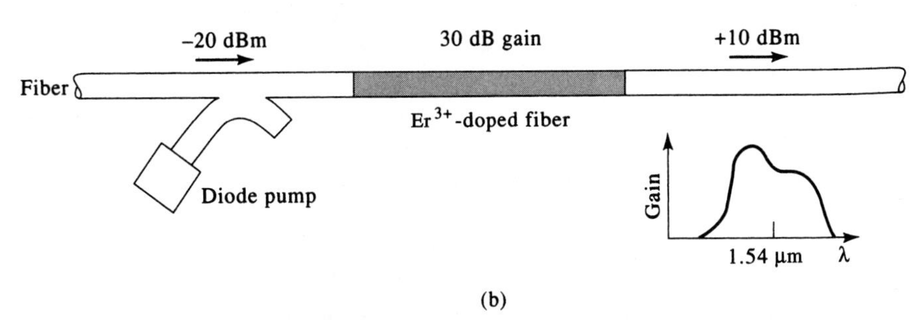

Figure 17.4 (a) The pertinent energy diagram of Er^{3+} in silica for pumping at $\lambda = 0.98$ μm (preferred). (b) Schematic drawing showing the amplifying fiber spliced into the transmission fiber and the method for coupling the pump radiation into the fiber.

Figure 17.4 shows a schematic drawing of an erbium-doped fiber amplifier and the relevant energy levels involved. Population inversion can be obtained via optical pumping by radiation at $\lambda = 980$ nm or $\lambda = 1480$–1490 nm. The radiation at 1480–1490 nm directly pumps the electrons to the state ($^4I_{13/2}$) with population N_2. Indirect optical pumping is achieved by radiation at 980 nm, which lifts electrons to the upper state ($^4I_{11/2}$). The electron at this state has a lifetime of about 1 μs and then decays to the intermediate state ($^4I_{13/2}$). The intermediate state ($^4I_{13/2}$) is metastable with a spontaneous lifetime of about $t_2 = 10.2 \times 10^{-3}$ s. Each decay results in the (spontaneous) emission of a photon, so $t_2 = t_{spont}$. Such a long spontaneous lifetime is responsible for the lack of crosstalk between communication channels. The pumping field is usually obtained from semiconductor lasers and is coupled into the amplifying fiber whose length is typically between a few meters and a few tens of meters. The fiber amplifying section can be spliced smoothly into the fibers in optical networks, with virtually zero reflection. This eliminates the possibility of self-oscillation. Furthermore, the circular symmetry of the ideal fiber eliminates the polarization dependence. Thus an erbium-doped fiber amplifier (EDFA) works as a polarization-insensitive traveling-wave amplifier (TWA), with a typical gain in the range of 20–40 dB from a few meters of fiber. A typical gain spectrum is also shown in Figure 17.4. We note that the gain of the EDFA varies significantly with wavelength. This is a problem in WDM optical networks when the number of channels covers a wide spectral regime, and when the amplifiers are cascaded. Passive optical filters can be employed to flatten the gain spectrum.

As a result of the combination of both homogeneous and inhomogeneous broadening, the gain spectrum can cover the spectral range from $\lambda = 1500$ nm to $\lambda > 1600$ nm. There are two spectral windows that are of particular importance. They are the C-band (1530–1560 nm) and the L-band (1560–1610 nm).

It is important to note that both the pump beam intensity and the population inversion decrease along the fiber due to linear attenuation. At the same time, the signal-carrying beam is being amplified. This causes the signal beam intensity to continue to increase until the point where the pump beam is depleted by the combination of linear attenuation and stimulated emission. Beyond this point, the signal intensity starts to decrease with distance. Consequently, there is an optimal fiber length where the gain is maximized. The EDFAs feature lower noise levels and lower crosstalk noise in multichannel amplification. These desirable features are a result of the much longer recombination times for the excited states.

EDFAs provide advantages over regenerative repeaters as well as other amplification systems. Once installed, EDFAs leave room for expansion into high-speed systems as well as allowing manufacturers to use the same erbium amplifier repeater for different types of systems with differing bit rates. With the wide gain bandwidth, EDFAs provide simultaneous amplification of all the wavelengths. EDFAs also demonstrate advantages over traveling-wave semiconductor laser amplifiers and Raman amplifiers (discussed in the last section). SOAs provide high gain, large bandwidth, and lower current consumption with a high reliability and a small size. SOAs are bidirectional amplifiers and can be used only as lumped amplifiers. Disadvantages of SOAs as compared to EDFAs involve the interference between adjacent pulses in the saturation regime due to the short gain recovery time. The short gain recovery time allows the amplification condition of a given pulse to affect only a relatively small number of the following time slots, which results in a pulse-pattern dependent intersymbol interference (ISI). This limits the maximum bit rate and reduces the acceptable input signal power. SOAs produce a significant pulse-pattern distortion in the output signal at 25 Gbps where no significant degradation was observed with EDFAs over 100 Gbps [9]. An additional disadvantage of SOAs is a large coupling loss between the SOA and the optical fiber. The losses degrade both the fiber-to-fiber gain and the effective noise figure. Losses can reach up to 10 dB [10]. Both Raman amplifiers and EDFAs amplify optical signals in the fiber

TABLE 17.1 Selected Properties of SOAs and EDFAs

Properties of Amplifiers	Erbium-Doped Fiber Amplifiers (EDFAs)	Semiconductor Optical Amplifers (SOAs)
Active medium	Er^{3+} ion in silica	Electron–hole in semiconductors
Typical length	Few meters	500 μm
Pumping	Optical	Electrical current
Gain spectrum	$\lambda = 1500$–1600 nm	$\lambda = 1300$–1500 nm
Gain bandwidth	25–35 nm	100 nm
Relaxation time	0.1–1 ms	<10–100 ps
Maximum gain	30–50 dB	25–30 dB
Saturation power	>10 dBm	0–10 dBm
Crosstalk	—	For bit rate <10 GHz
Polarization	Insensitive	Sensitive
Noise figure	3–4 dB	6–8 dB
Insertion loss	<1 dB	4–6 dB
Optics	Pump laser diode couplers, fiber splice	Antireflection coatings, fiber-waveguide coupling
Optoelectronic integration	No	Yes

by transferring energy from a pump beam to the signal beam. In Raman amplifiers, energy is transferred through an effect in the fiber known as stimulated Raman scattering (SRS). Raman amplifiers have a low noise figure, low connection loss, high gain, high output power, and broad bandwidth. The major drawback of Raman amplifiers as compared to EDFAs is their low efficiency (see Section 17.5) [11]. Table 17.1 summarizes some of the important properties of both EDFAs and SOAs.

The gain of EDFAs can be studied using the basic rate equations described in Section 6.4. Several other models have been developed for different applications. Interested readers are referred to References [12–18].

17.3 AMPLIFIED SPONTANEOUS EMISSION

As with any amplifier, both SOAs and EDFAs add noise to the signal that is being amplified. The amplification of the input signal is a result of the stimulated emission. In addition to the process of stimulated emission, the spontaneous emission of photons also occurs. These spontaneous photons are also amplified by the active medium of the amplifier. The amplified spontaneous emission (ASE) from the active medium is a fundamental source of noise. The noise added by the optical amplifiers can lead to poor signal reception in analog communications or high bit error rate (BER) in digital transmission systems. For electronic amplifiers, it is common to quantify the noise added by the amplifier by a parameter known as the noise figure. In this chapter we will investigate the noise in optical amplifiers and define an appropriate noise figure to quantify the added noise. The noise figure of an optical amplifier is defined as

$$F = \frac{(SNR)_{\text{input}}}{(SNR)_{\text{output}}} = \frac{S_i / N_i}{S_o / N_o} = \frac{N_o}{GN_i} = \frac{GN_i + N_{\text{amp}}}{GN_i} = 1 + \frac{N_{\text{amp}}}{GN_i} \tag{17.3-1}$$

where SNR stands for signal-to-noise ratio, S_i is the input signal power, S_o is the output signal power, N_i is the noise power of the signal at input, N_o is the noise power at output, G is the gain of the amplifier ($G = S_o/S_i$), and N_{amp} is the noise power added by the amplifier at the output. To find the noise figure of an optical amplifier, we need to calculate the noise power added to the signal beam by the amplifier. In Chapter 11, we discussed noise in photodiodes where the added noise was that of Johnson noise (thermal noise). In optical amplifiers, the noise originates mainly from the spontaneous emission in the gain medium (which is a medium not in thermal equilibrium). In Section 10.6 we discussed the effect of spontaneous emission power on the spectral linewidth of the laser output. In this section we will derive the effect of spontaneous emission noise in a traveling-wave optical amplifier in which the gain medium, with no mirrors, is used to amplify a weak input field. The basic engineering problem is to find the degradation of the signal-to-noise power that is caused by the (inevitable) addition of some spontaneous emission (noise) power to the amplified signal. We will return to the issue of signal-to-noise ratio and noise figure after evaluating the power of the amplified spontaneous emission at the end of this section. The spontaneous emission noise is also referred to as quantum noise. It is thus of fundamental nature and cannot be avoided.

To evaluate the added noise power from spontaneous emission, we consider an inverted atomic medium with population densities N_2 and N_1 in the upper and lower transition levels, respectively. The inverted medium occupies the space between $z = 0$ and $z = L$ (see Figure 17.5), with a transverse dimension of a and b. An optical beam with power P is propagating through

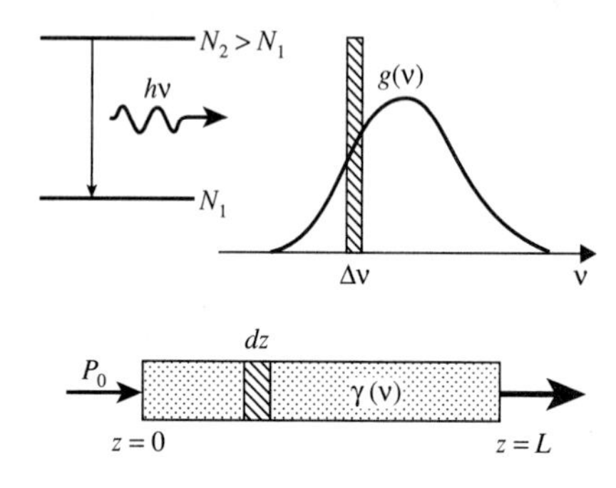

Figure 17.5 Schematic drawing of an optical amplifier of length L and cross section $A = ab$, with an intensity gain coefficient $\gamma(\nu)$ and a lineshape function $g(\nu)$.

the gain medium. The coherent amplification of the input beam power P due to stimulated emission is given by

$$dP = \gamma P \, dz \tag{17.3-2}$$

where γ, the intensity gain coefficient, is given by Equation (5.6-19) as

$$\gamma(\nu) = (N_2 - N_1)\frac{c^2 g(\nu)}{8\pi n^2 \nu^2 t_{\text{spont}}} \tag{17.3-3}$$

where we recall that $g(\nu)$ is the normalized lineshape function, t_{spont} is the spontaneous lifetime of the upper state, and n is the index of refraction of the medium. As photons are emitted via the spontaneous emission process in the medium, they are also being amplified as they propagate. We will next calculate the power of the amplified spontaneous emission at the output of the optical amplifier. The amplified spontaneous emission will mix with the amplified signal beam at the detector, creating electrical intensity noise that leads to a degradation of the signal-to-noise ratio.

Referring to Figure 17.5, we consider a differential volume element $A\,dz$ at z, where A is the cross-sectional area of the amplifier medium. The number of excited atoms (in the upper state) inside the volume is $N_2 A\,dz$. Thus the total optical power due to spontaneous emission from this volume element can be written

$$P_N = \frac{N_2 h\nu A \, dz}{t_{\text{spont}}} \tag{17.3-4}$$

where $h\nu$ is the photon energy. This power covers the whole gain spectrum as described by the normalized lineshape function of $g(\nu)$. Only the fraction of the spontaneous emission power that falls in the spectral regime of $(\nu, \nu + \Delta\nu)$ is of interest to us, where ν is the optical signal carrier frequency and $\Delta\nu$ is the optical bandwidth of interest. Physically, $\Delta\nu$ can be the bandwidth of a bandpass filter in front of the detector, or the bandwidth of the input signal. Thus the noise power due to spontaneous $2 \to 1$ transitions that fall within the optical bandwidth of $\Delta\nu$ is

$$\Delta P_N = \frac{N_2 h\nu A \, dz}{t_{\text{spont}}} g(\nu)\,\Delta\nu \tag{17.3-5}$$

We note this power is emitted into many transverse spatial modes. Only a small fraction is emitted into the transverse spatial mode of the input wave. For all practical cases, we assume that the input wave is a single transverse spatial mode (e.g., a polarized Gaussian beam in free space, or a single guided mode in a waveguide), propagating in the $+z$ direction. The issue now is to find the fraction of power that is emitted into the transverse spatial mode of the signal wave along the $+z$ direction. Assuming a bandwidth of $\Delta\nu$, the total number of spatial modes (including transverse and longitudinal modes) in free space is given, according to Equation (4.0-12), by

$$N_{\text{modes}} \cong \frac{8\pi\nu^2 n^3 V}{c^3}\Delta\nu$$

where n is the index of refraction of the medium, and $V = AL$ is the volume of the amplifying medium. If each of the modes receives an equal share of the energy of the spontaneous emission, then the fraction of spontaneous radiation emitted into one of the modes is given by

$$\alpha = \frac{1}{N_{\text{modes}}} \cong \frac{c^3}{8\pi\nu^2 n^3 V\,\Delta\nu} \tag{17.3-6}$$

This parameter is also known as the spontaneous emission factor of a laser cavity of volume $V = AL$. For the spontaneous emission of an optical amplifier, we are interested in the fraction of power going into a single transverse spatial mode defined by the input signal beam. For each spatial transverse mode, there are many longitudinal modes of traveling waves that are separated by a frequency spacing of c/nL. Within a spectral bandwidth of $\Delta\nu$, the number of longitudinal modes of traveling waves is given by $\Delta\nu/(c/nL)$ (see Problem 17.4). Thus the fraction of spontaneous emission power going into this group of longitudinal modes all corresponding to a single spatial transverse mode (one direction of propagation, one polarization) is given by

$$\beta = \alpha\frac{\Delta\nu}{(c/nL)} = \frac{\lambda^2}{8\pi n^2 A} = \frac{c^2}{8\pi n^2\nu^2 A} \tag{17.3-7}$$

where A is the area of the amplifying medium. This parameter is also called the β-factor of spontaneous emission. Thus the spontaneous emission power that is emitted into the transverse spatial mode of the input wave is written

$$dP_N = \beta\frac{N_2 h\nu A\,dz}{t_{\text{spont}}}g(\nu)\,\Delta\nu \tag{17.3-8}$$

or equivalently, using Equation (17.3-7),

$$dP_N = \frac{c^2}{8\pi\nu^2 n^2}\frac{N_2 h\nu\,dz}{t_{\text{spont}}}g(\nu)\,\Delta\nu \tag{17.3-9}$$

This is the amount of spontaneous emission power received by a transverse spatial mode (one polarization, one direction of propagation) due to a volume element of the gain medium. In waveguide structures, the modes (including confined modes and radiation modes) may have different spatial overlap with the active atoms. In this case, the fraction of spontaneous emission power can be dependent on the wavefunction of the specific transverse spatial mode involved. Here we consider an optical amplifier in free space of volume $V = AL$, where all transverse spatial modes receive an equal share of the spontaneously emitted power.

Using Equation (17.3-3) for the intensity gain coefficient, Equation (17.3-9) can be written

$$dP_N = \frac{N_2}{N_2 - N_1}\gamma(\nu)h\nu\,\Delta\nu\,dz \tag{17.3-10}$$

where we recall that $\Delta\nu$ is the bandwidth of detection. For the total increment of optical power due to stimulated emission and spontaneous emission into the transverse spatial mode of the input wave, we combine Equations (17.3-2) and (17.3-10). (If the two contributions were coherent, we would add their fields.) This leads to

$$dP = \gamma P \, dz + \frac{N_2}{N_2 - N_1} \gamma(\nu) h\nu \, \Delta\nu \, dz \tag{17.3-11}$$

where the second term is due to spontaneous emission from the volume element $A\,dz$. The differential equation of the power of the transverse spatial mode is thus given by

$$\frac{dP}{dz} = \gamma P + \frac{N_2}{N_2 - N_1} \gamma h\nu \, \Delta\nu \tag{17.3-12}$$

The solution of Equation (17.3-12) subject to the boundary condition $P(0) = P_0$ is

$$P(z) = P_0 e^{\gamma z} + \mu h\nu \, \Delta\nu (e^{\gamma z} - 1) \tag{17.3-13}$$

where P_0 is the input signal power, and

$$\mu \equiv \frac{N_2}{N_2 - N_1} \tag{17.3-14}$$

is the population inversion factor. We note that the total power in Equation (17.3-13) consists of an amplified signal power ($P_0 e^{\gamma z}$) and a noise power due to amplified spontaneous emission. The power of amplified spontaneous emission (ASE) is thus given by

$$P_{\text{ASE}} = \mu h\nu \, \Delta\nu (e^{\gamma L} - 1) = \mu h\nu \, \Delta\nu (G - 1) \tag{17.3-15}$$

where $G = \exp(\gamma L)$ is the intensity gain of the amplifier.

The signal-to-noise power ratio at the output of the optical amplifier (in the optical domain) is

$$\left(\frac{S}{N}\right)_{\text{output}} = \frac{GS_i}{GN_i + N_{\text{amp}}} = \frac{GP_0}{GN_i + \mu h\nu \, \Delta\nu (G - 1)} \tag{17.3-16}$$

where N_i is the noise of the signal at input, N_{amp} is the noise power added by the amplifier, and $G \equiv \exp(\gamma L)$ is the one-pass intensity gain. From the point of view of power bookkeeping, the effect of spontaneous emission is seen to be equivalent to a noise input power

$$N_{i\ \text{amp}} = \mu h\nu \, \Delta\nu \left(1 - \frac{1}{G}\right) \tag{17.3-17a}$$

provided the first term in the denominator can be neglected (i.e., $GN_i \ll \mu h\nu \, \Delta\nu(G - 1)$). For an ideal four-level gain medium ($\mu = 1$) with a high gain ($G \gg 1$), the equivalent input noise power becomes

$$N_{i\ \text{amp}} = h\nu \, \Delta\nu \tag{17.3-17b}$$

If the optical amplifier were to be employed as a preamplifier in an optical receiver, then the minimum detectable power in the sense defined in Section 11.4 is given by

$$(P_2)_{\min} \sim h\nu \, \Delta\nu \tag{17.3-18}$$

which is the same as that obtained in Equation (11.4-10) in the case of a heterodyne detection scheme with unity quantum efficiency ($\eta = 1$). The optical preamplifier is thus an "ideal" quantum limited receiver. To achieve the minimum detectable signal power, we need, according to Equation (17.3-18), to reduce the bandwidth $\Delta\nu$ as much as possible. This is more easily done at the radio frequencies of the heterodyne signal than at optical frequencies. The practical advantage thus lies with heterodyne reception. In practice, $\Delta\nu$ should be chosen to match the bandwidth of the signal beam. As discussed later in this section and the next section, we will use a definition of SNR that is based on the signal and noise powers following detection at the receiver.

The approach leading to Equation (17.3-12) is quite general and should apply also to an atomic medium that is in thermal equilibrium (at temperature T) and hence is absorbing. We can use Equation (17.3-12) in this case, provided we put $\gamma(\nu) \rightarrow -\alpha(\nu)$ (α being the medium absorption coefficient) and $(N_2/N_1) = \exp(-h\nu/kT)$, where $k = k_B$ is the Boltzmann constant, as appropriate to a medium in thermal equilibrium.

The result is

$$\frac{dP}{dz} = -\alpha P + \frac{\alpha h\nu \, \Delta\nu}{e^{h\nu/kT} - 1} \tag{17.3-19}$$

whose solution is

$$P(z) = P_0 e^{-\alpha z} + \frac{h\nu \, \Delta\nu}{e^{h\nu/kT} - 1}(1 - e^{-\alpha z}) \tag{17.3-20}$$

If the medium is "black", $e^{-\alpha L} \ll 1$ (i.e., all incident radiation is absorbed), the output power is

$$P(L) = \frac{h\nu}{e^{h\nu/kT} - 1}\Delta\nu \tag{17.3-21}$$

independent of L and the input. This result is the same as the Johnson noise formula (10.5-6), which was obtained using quite a different point of view. For electromagnetic waves with frequencies in the microwave regime, $h\nu \ll kT$ at room temperature, the noise power (17.3-21) becomes $kT\,\Delta\nu$, which is the expression for the thermal (Johnson) noise.

If the laser medium contains a transition, other than that responsible for the gain, that causes an absorption coefficient α and is characterized by a temperature T (this would be the temperature appearing in the Boltzmann ratio of the populations involved), then we must add the spontaneous emission from the upper to lower level of this transition to that from the upper, amplifying, laser level. Using Equations (17.3-12) and (17.3-19), we obtain

$$\frac{dP}{dz} = (\gamma - \alpha)P + \frac{N_2}{N_2 - N_1}\gamma h\nu \, \Delta\nu + \frac{\alpha h\nu \, \Delta\nu}{e^{h\nu/kT} - 1} \tag{17.3-22}$$

whose solution is

$$P(z) = P(0)e^{[\gamma(\nu)-\alpha(\nu)]z} + h\nu \, \Delta\nu \left(\frac{\alpha}{e^{h\nu/kT} - 1} + \frac{\gamma}{1 - N_1/N_2} \right) \frac{e^{(\gamma-\alpha)z} - 1}{\gamma - \alpha} \tag{17.3-23}$$

At optical frequencies where $h\nu \gg kT$, the contribution of the first term in the large parentheses, which represents spontaneous emission due to atomic levels involved in the absorbing transitions, is, in most cases, negligible compared to the second term, which is due to spontaneous emission in the lasing transition. At lower, say, microwave, frequencies the loss contribution may become appreciable.

Signal-to-Noise Ratio and Noise Figure

We now return to the issue of signal-to-noise ratio (SNR) and noise figure (NF) for optical amplifiers. As defined in Equation (17.3-1), the noise figure is the ratio of the input SNR to output SNR. Thus the noise figure depends on the definition of SNR of optical amplifiers, particularly the definition of noise power. For electrical signals, both the signal and noise are measured in terms of electrical power (or energy).

For electronic amplifiers, the input noise power is the thermal noise (Johnson noise) $k_B T_0 \Delta f$, where k_B is Boltzmann's constant, T_0 is the standard "room" temperature of 290 K, and Δf is the electronic signal bandwidth. The noise figure for the electronic amplifier is often written (see Section 11.7, footnote 15)

$$F = \frac{(SNR)_{\text{input}}}{(SNR)_{\text{output}}} = \frac{S_i/N_i}{S_o/N_o} = 1 + \frac{\text{Noise power added by amplifer}}{G k_B T_0 \ \Delta f} = 1 + \frac{T_A}{290} \qquad (17.3\text{-}24)$$

where S_i is the input signal power, S_o is the output signal power, N_i is the noise power at input, N_o is the noise power at output, and T_A is the effective "noise temperature," so that $k_B T_A \Delta f$ is the noise power added by the amplifier. With this definition, a hypothetical noiseless amplifer would be characterized by $T_A = 0$. The noise figure definition in Equation (17.3-1) or (17.3-24) is independent of the signal. Furthermore, in the event when cascading with another amplifier is necessary, the noise figure of the cascade can easily be obtained by the following cascading formula:

$$F - 1 = (F_1 - 1) + \frac{(F_2 - 1)}{G_1} \qquad (17.3\text{-}25)$$

where F_2 is the noise figure of the second amplifier and G_1 is the gain of the first amplifier.

For optical amplifiers, we need to generalize the definition of the noise figure. Optical amplifiers exhibit some features not shared with electronic amplifiers. These include the fundamental source of noise due to amplified spontaneous emission. For optical signals, there are two possibilities in the definition (and measurement) of the signal power and the noise power. In the optical domain, the power (or energy) is measured in terms of the mean square of the field amplitudes (e.g., $|\mathbf{E}|^2$, where $\mathbf{E}$ is the electric field of the optical waves). The square of the field amplitude is proportional to the photon number. However, in the electronic domain, we often define the power (or energy) as the mean square of the electric current at the detector ($\overline{i^2}$), which is proportional to the square of the photon number. These two definitions of the signal power and noise power lead to different forms of noise figure [19]. In this book, we will adopt the convention of using mean square of photocurrent at the detector ($\overline{i^2}$) for both the signal and noise power levels since it is the one connected most directly to communications system considerations. Although the noise figure resulting from this definition is not signal independent and the cascading formula does not apply, the approach is consistent with test instruments in use by major optical amplifier equipment manufacturers [20]. We will address the issue of amplifier cascading in the next section.

Thus, for optical amplifiers, the input signal-to-noise ratio (SNR) is defined as

$$(SNR)_{\text{input}} = \frac{S_i}{N_i} = \frac{(\overline{i^2_{\text{signal}}})}{\overline{(\Delta i)^2}} = \frac{(\overline{i^2_{\text{signal}}})}{(\overline{i^2_{\text{noise}}})} \qquad (17.3\text{-}26)$$

where $\overline{i^2_{\text{signal}}}$ is the mean square of the photocurrent i_{signal} generated by the optical signal of power level S_0, and $\overline{i^2_{\text{noise}}}$ is the mean square (variance) of the noise current. For a shot-noise limited signal and an ideal receiver with unity quantum efficiency and negligible thermal noise, the signal current is $(\overline{i^2_{\text{signal}}}) = (\bar{I})^2 = (eS_0/h\nu)^2$. Thus, the SNR input becomes, according to Equation (10.4-9),

$$(SNR)_{\text{input}} = \frac{S_i}{N_i} = \frac{\left(\frac{eS_0}{h\nu}\right)^2}{2e\left(\frac{eS_0}{h\nu}\right)\Delta f} \tag{17.3-27}$$

or equivalently,

$$(SNR)_{\text{input}} = \frac{S_i}{N_i} = \frac{S_0}{2h\nu\,\Delta f} \tag{17.3-28}$$

where S_0 is the input signal power S_i of the optical beam and Δf is the electronic bandwidth of photodetection.

The calculation of SNR_{output} requires evaluation of the noise current created by the presence of the amplified spontaneous beam at the receiver. This is discussed in the next section.

17.4 OPTICAL AMPLIFICATION IN FIBER LINKS

As discussed earlier, optical amplifiers can be employed in optical networks to maintain the optical signal power at sufficiently high levels along the path so that the signal-to-noise ratio (SNR) degradation due to signal shot noise and receiver noise is reduced to practical inconsequence. An undesirable side effect of the optical amplifier on the SNR of the detected signal is to add, upon detection, a noise power component, at frequencies near that of the signal power. We will derive the noise power by considering the mean square current caused by the beating between the optical field of the amplified (optical) spontaneous emission power of the amplifier and the signal optical field. Both the amplified spontaneous emission (ASE) noise power and the gain $\exp(\gamma l)$ play important roles in the noise figure of the optical amplifier. Figure 17.6 shows a typical plot of the gain spectrum $\gamma(\lambda)$ and the noise figure as functions of the wavelength.

Figure 17.7 shows the two spectral windows of the amplified output spontaneous emission power that beat (at the detector) with the optical signal field S_0 at frequency ω_0 to generate an output noise current at some arbitrary frequency ω_m. This mechanism thus gives rise to a spectral continuum of RF noise current extending from dc to approximately $\Delta\omega_{\text{gain}}$, the width

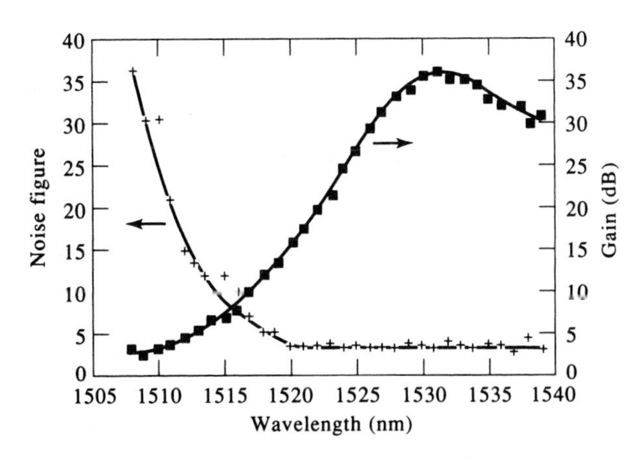

Figure 17.6 Noise factor and gain spectrum of the silica Er^{3+} fiber amplifier for a constant pump power of 34.2 mW at 0.98 μm. (After Reference [5].)

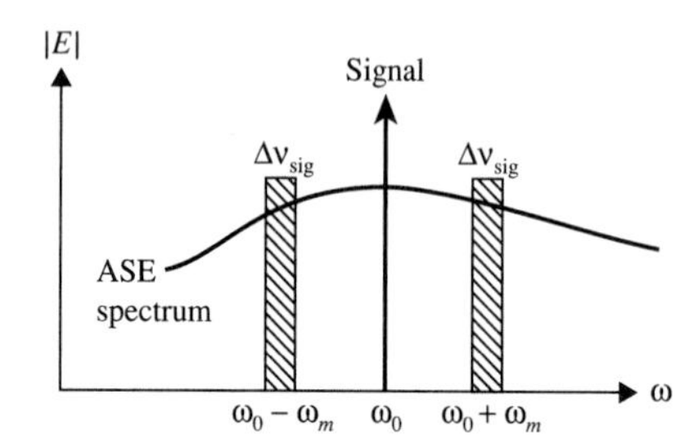

Figure 17.7 At the output of the amplifier, the amplified spontaneous emission fields near $\omega_0 + \omega_m$ and near $\omega_0 - \omega_m$ will each beat (mix) at the detector with the amplified optical signal field at ω_0 to yield radio frequency (RF) currents with frequencies near ω_m. These currents, which occupy a spectral width of $\Delta\nu_{sig}$, cannot be separated from those due to intentional (signal) modulation of the intensity at ω_m and thus constitute RF noise.

of the (amplified) spontaneous emission spectrum. To estimate this current, we first need an expression for the optical spontaneous emission power at the output of an optical amplifier. According to Equation (17.3-15), the amplified spontaneous emission (ASE) power in a *single* transverse spatial mode (one polarization, one direction $+z$) within a spectral bandwidth $\Delta\nu_{opt}$ at the output of an optical amplifier is

$$N_a = P_{ASE} = \mu h\nu\, \Delta\nu_{opt}(G - 1) \tag{17.4-1}$$

where $\Delta\nu_{opt}$ is the optical bandwidth, $G = \exp(\gamma l)$ is the single-pass power gain of the optical amplifier with a distributed intensity gain coefficient γ (see Equation (17.3-3)) and length l, and

$$\mu = \frac{N_2}{N_2 - N_1} \tag{17.4-2}$$

is the population inversion factor of the transition. It accounts for the larger value of N_2, and hence larger spontaneous emission power, in atomic (amplifier) systems in which $N_1 \neq 0$. In an optical amplifier, the power gain per pass is given by $G = \exp(\gamma l)$, where the intensity gain coefficient γ is proportional to $N_2 - N_1$. A large N_1 thus causes a larger N_2 for a given gain γ. The spontaneous emission power is proportional to N_2.

Referring to Figure 17.7 again, we denote the output optical power at ω_0 as S and that of the amplified spontaneous emission at each of the spectral windows ($\omega_0 - \omega_m$ and $\omega_0 + \omega_m$) as N_a. Following the discussion in Section 11.11, we can write the optical field (sum of signal field and noise field) arriving at the detector as

$$E = \sqrt{S}e^{i(\omega_0 t + \Phi_S)} + \sqrt{N_a}\,e^{i[(\omega_0 - \omega_m)t + \Phi_1]} + \sqrt{N_a}\,e^{i[(\omega_0 + \omega_m)t + \Phi_2]} \tag{17.4-3}$$

where S is the amplified optical signal power level, N_a is the ASE power at each of the spectral windows shown in Figure 17.7 (i.e., $N_a = P_{ASE} = \mu h\nu\, \Delta\nu_{sig}(G - 1)$), and Φ_S, Φ_1, and Φ_2 are the phases. Due to the random nature of the amplified spontaneous emission, the phases Φ_1 and Φ_2 are uniformly distributed over (0 and 2π). The detector current is proportional to the absolute square of the sum of the signal field plus the noise field. Using Equation (17.4-3), the photoelectric current can be written

$$i = \frac{Se\eta}{h\nu}\left(1 + 2\frac{N_a}{S} + 2\sqrt{\frac{N_a}{S}}\cos(\omega_m t + \Phi_2 - \Phi_S) + 2\sqrt{\frac{N_a}{S}}\cos(-\omega_m t + \Phi_1 - \Phi_S)\right) \tag{17.4-4}$$

where η is the quantum efficiency of the optoelectronic conversion. The mean square beat current at ω_m is then

$$(\overline{i^2})_{\text{ASE-signal}} = 4\left(\frac{e\eta}{h\nu}\right)^2 SN_a = \frac{4e^2\eta^2 S(G - 1)\mu\, \Delta\nu_{sig}}{h\nu} \tag{17.4-5}$$

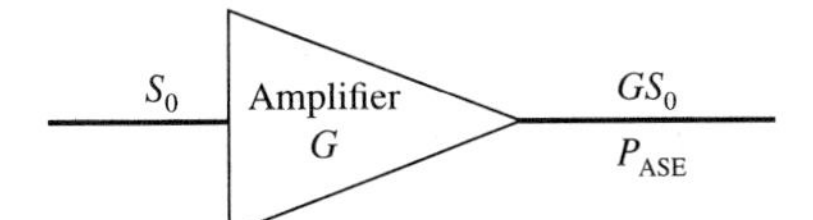

Figure 17.8 An optical amplifier with a power gain G and an input signal power S_0. P_{ASE} is the total power of the amplified spontaneous emission (ASE) at the output of the amplifier in the appropriate bandwidth $\Delta\nu$.

where we have taken $N_a = P_{ASE} = \mu h\nu\ \Delta\nu_{sig}(G-1)$ as the noise power from each of the two ASE frequency bands, each with a width $\Delta\nu_{sig}$ (signal bandwidth), one above and one below the signal frequency, which contribute incoherently to the beat current. This optical signal bandwidth $\Delta\nu_{sig}$ is also the electronic bandwidth Δf_{sig} in the photodetector. In the remainder, we will drop the subscript "sig" and use Δf or $\Delta\nu$ only.

Consider an optical in-line amplifier as shown in Figure 17.8. The input signal power is S_0, and it enters the amplifier in a *single* transverse (usually the fundamental) fiber mode. The amplified output signal power is $S = GS_0$, while N_a, as given by Equation (17.4-1), represents the (optical) amplified spontaneous emission power P_{ASE} at the output, which is generated within the amplifier within an optical bandwidth of $\Delta\nu$. If we were to detect the signal at the input to the amplifier, the main noise contribution would, in an ideal case (i.e., a noiseless receiver), be that of the signal shot noise so that the signal-to-noise power ratio (SNR) at the input to the amplifier is, according to Equation (10.4-9), or Equations (17.3-27) and (17.3-28),

$$SNR_{in} = \frac{(\overline{i^2})_{signal}}{(\overline{i^2})_{noise}} = \frac{\left(\dfrac{S_0 e}{h\nu}\right)^2}{2e^2\dfrac{S_0}{h\nu}\Delta f} = \frac{S_0}{2h\nu\ \Delta f} \tag{17.4-6}$$

where we assume 100% detection efficiency $\eta = 1$ for simplicity. The detected signal "power" (mean square of photocurrent) at the output of the optical amplifier is

$$(\overline{i^2})_{out} = \left(\frac{GS_0 e}{h\nu}\right)^2 \tag{17.4-7}$$

while the noise power is that of the ASE-signal noise (17.4-5) and the shot noise (10.4-9)

$$\left(\overline{i^2_{shot}}\right)_{out} = 2e\bar{I}\ \Delta f = 2e\frac{eGS_0}{h\nu}\Delta f \tag{17.4-8}$$

where $\bar{I} = eGS_0/h\nu$ is the average photocurrent. The noise current component that is due to beating of ASE frequencies with themselves is proportional to P_{ASE}^2 and can be made negligible compared to the ASE-signal current if the signal power $S(z)$ is not allowed to drop too low due to attenuation and/or optical filtering. We have neglected for similar reasons the shot noise due to the ASE. The SNR at the output of the amplifier is thus

$$SNR_{out} = \frac{\left(\dfrac{GS_0 e}{h\nu}\right)^2}{\dfrac{2e^2GS_0}{h\nu}\Delta f + \dfrac{4e^2G(G-1)S_0\mu\ \Delta f}{h\nu}} \tag{17.4-9}$$

where we assumed a 100% detector quantum efficiency. The denominator consists of mean square noise current from the shot noise and the mean square noise current from the beating between the signal and the amplified spontaneous emission.

Using Equations (17.4-9) and (17.4-6), we can write the noise figure as

$$F = \frac{SNR_{\text{in}}}{SNR_{\text{out}}} = \frac{G + 2G(G-1)\mu}{G^2} = \frac{1}{G} + 2\mu\left(1 - \frac{1}{G}\right) \tag{17.4-10}$$

For large gain $G \gg 1$, the second term in the denominator of Equation (17.4-9) dominates, and the noise figure becomes

$$F = \frac{SNR_{\text{in}}}{SNR_{\text{out}}} \approx 2\mu$$

which in an ideal four-level ($N_1 = 0$, $\mu = 1$) amplifier is equal to 2 (or 3 dB). The single high-gain optical amplifier will thus degrade the SNR of the detected output by a factor of 2 (or 3 dB). We recall that this degradation is tolerated only in order to save the signal from the far worse fate of succumbing, in its attenuated state, to the noise of the receiver. An experimental verification of the 3 dB limit is shown in Figure 17.6.

In a very long (100 km or longer) fiber link, we will need to amplify the signal a number of times. We will consequently develop in what follows a formalism for treating systematically cascades of amplifiers. A generalization of the expression (17.4-9) for the SNR of the detected signal at an arbitrary point z along the link is to write

$$SNR(z) = \frac{\left[\dfrac{eS(z)}{h\nu}\right]^2}{\dfrac{2e^2S(z)\,\Delta f}{h\nu} + \dfrac{4e^2N_a(z)S(z)}{(h\nu)^2} + \dfrac{4kT_e\,\Delta f}{R}} \tag{17.4-11}$$

where $S(z)$ is the amplified optical signal power, $N_a(z)$ is the noise power of the amplified spontaneous emission, and the last term in the denominator represents the mean square thermal noise current of the receiver (at point z) whose effective noise temperature is T_e. R is the output impedance of the detector including the receiver's input impedance. Equation (17.4-11) neglects, again, the shot noise due to the ASE, the ASE–ASE beat noise, and the intensity fluctuation noise of the source laser. If the signal power $S(z)$ can be maintained above a certain level by repeated amplification, we can neglect the receiver noise term. Under these realistic circumstances, the SNR expression (17.4-11) becomes

$$SNR(z) = \frac{S^2(z)}{2S(z)h\nu\,\Delta f + 4N_a(z)S(z)} \tag{17.4-12}$$

$S(z)$ is the signal power at z, while $N_a(z)$ is the total noise power of ASE at z originating in *all* the preceding amplifiers ($z' < z$).

Let us next consider the realistic scenario of a long fiber with amplifiers employed serially at fixed and equal intervals (z_0), as illustrated in Figure 17.9.

The signal power level $S(z)$ at the fiber input and at the output of each amplifier is S_0. The signal is attenuated by a factor of $L \equiv \exp(-\alpha z_0)$ in the distance z_0 between amplifiers and is boosted back up by the gain $G = L^{-1} = e^{\alpha z_0}$ at each amplifier to the initial level S_0. The spontaneous emission power $N_a(z)$ is attenuated by a factor L between two neighboring amplifiers boosted to its original value by the amplifier and increases by an increment of P_{ASE} at the output of each amplifier. We employ Equation (17.4-1) for P_{ASE} to calculate the SNR of the detected current at the output of the nth amplifier. Assuming $G \gg 1$, the result is

$$SNR_n = \frac{S_0}{2h\nu\,\Delta f[1 + 2n\mu(e^{\alpha z_0} - 1)]} \tag{17.4-13}$$

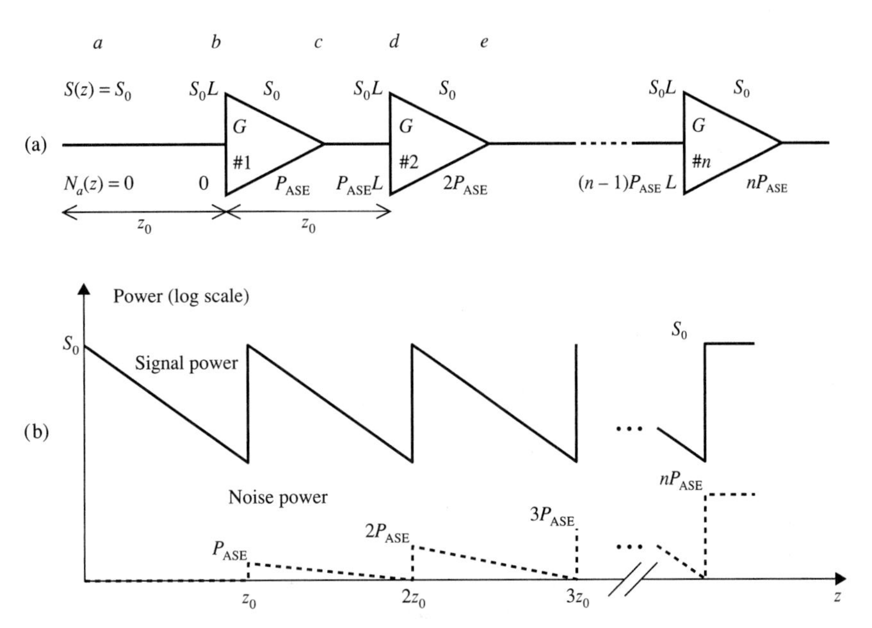

Figure 17.9 (a) A fiber link with periodic amplification. The spontaneous emission power $N_a(z)$ and the signal power $S(z)$ at the amplifiers' input and output planes are indicated. The amplifiers are separated by a distance z_0 so that the gain from each amplifier balances the propagation loss L. In other words, $GL = 1$. P_{ASE} is the ASE power added by each amplifier. The total ASE power at the end of the nth amplifier is nP_{ASE}. (b) The figure shows the signal power and the noise power as functions of the distance of propagation in the fiber. At the locations of the amplifiers ($z_0, 2z_0, 3z_0, \ldots$), the signal power is restored to the original power S_0, and the ASE power is increased by P_{ASE}.

We note that the power of cumulative ASE at the output of the nth amplifier is nP_{ASE}. When $\exp(\alpha z_0) = G \gg 1$, we find a z^{-1} (more exactly an n^{-1}) dependence of the SNR rather than the $\exp(-\alpha z)$ dependence of a fiber without amplification in which the main noise mechanism is shot noise. The physical reason for this difference is that the repeated amplification keeps the signal level high as well as the level of the signal–ASE beat noise. The latter is kept well above the signal shot noise. A fixed amount of beat noise power is thus added at each stage, leading to the inverse distance dependence of the SNR.

Equation (17.4-13) suggests that the SNR at z can be improved by reducing z_0, that is, by using smaller intervals between the amplifiers, which, of course, entails reducing the gain $G = \exp(\alpha z_0)$ of each. Let us take the limit of Equation (17.4-13) as $z_0 \to 0$, that is, the separation between amplifiers tends to zero. In this limit the whole length of the fiber acts as a distributed amplifier with a gain constant $\gamma = \alpha$, just enough to maintain the signal at a constant value. Since $S(z)$ is a constant, we need only evaluate the ASE optical power $N_a(z)$ in order to obtain, using Equation (17.4-12), an expression for the SNR at z. To find how much noise power is added by the amplifying fiber, we consider a differential length dz. It may be viewed as a discrete amplifier with a gain of $\exp(\gamma\, dz)$ so that its contribution to ASE power $N_a(z)$ is given by Equation (17.4-1) as

$$dN_a = (e^{\gamma(dz)} - 1)\mu h\nu\, \Delta f \tag{17.4-14}$$

or

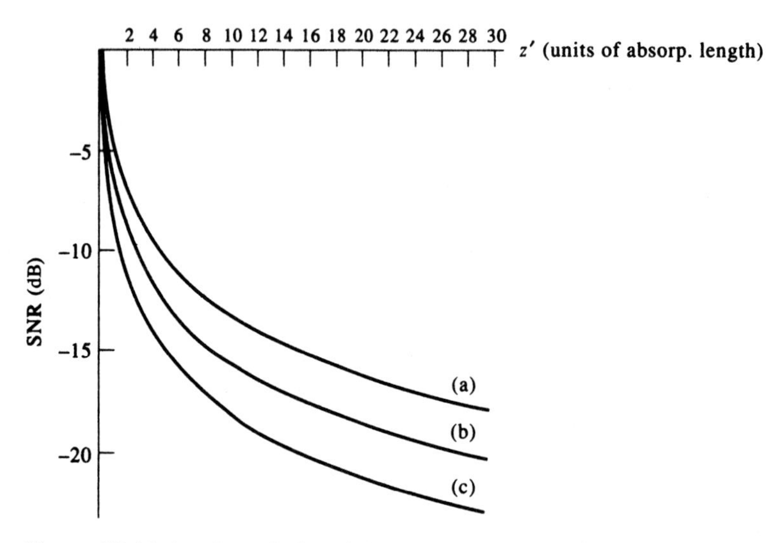

Figure 17.10 A universal plot of the degradation of the SNR compared to the initial ($z = 0$) value in the cases of (a) continuous amplification ($\gamma = \alpha$), ($\mu = 1$); (b) periodic amplification every $z_0 = \alpha^{-1}$ ($z' = 1, 2, 3, \ldots$; $\mu = 1$; curve is to be read only at $z' = 1, 2, 3, \ldots$); and (c) periodic amplification every $z_0 = 2\alpha^{-1}$ ($z' = 2, 4, 6, \ldots$; $\mu = 1$; curve is to be read only at $z' = 2, 4, 6, \ldots$).

$$\frac{dN_a}{dz} = \gamma\mu h\nu\,\Delta f, \quad N_a(z) = \gamma\mu h\nu\,\Delta f z \tag{17.4-15}$$

where, since no spontaneous emission is present at the input, we used $N_a(0) = 0$. We note that the ASE power is a linear function of z. Using Equation (17.4-15) in (17.4-11) and taking $S(z) = S_0$, $\gamma = \alpha$ results in

$$SNR(z) = \frac{S_0}{2(1 + 2\mu\alpha z)h\nu\,\Delta f} \tag{17.4-16}$$

We can also obtain Equation (17.4-16) as the limit of (17.4-13) when $z_0 \to 0$. It is interesting to compare the (ideal) distributed amplifier to the discrete amplifier case of Equation (17.4-13)

$$(SNR)(z) = \frac{S_0}{2[1 + 2(z/z_0)\mu(e^{\alpha z_0} - 1)]h\nu\,\Delta f} \tag{17.4-17}$$

where we used $G = \exp(\alpha z_0)$ and $n = z/z_0$.

Figure 17.10 shows plots of the ideal continuous amplification case described by Equation (17.4-16) as well as two cases of discrete amplifier cascades (Equation (17.4-13)). The advantage of continuous amplification compared to, say, amplification every α^{-1} is seen to be less than 2 dB so that the latter may be taken as a practical optimum configuration. In a low-loss optical fiber, say, with $\alpha = 0.2$ dB/km, the distance between amplifiers that are placed every α^{-1} km would be 21.7 km. Figure 17.11 shows the SNR of the detected signal along a realistic link for the case of (a) continuous amplification, (b) discrete amplifiers spaced by $z_0 = \alpha_0^{-1}$, and (c) no amplification at all. The launched power is $P_0 = 5$ mW, $\lambda = 1.55$ μm, $\Delta f = 10^9$ Hz, and $\alpha = 0.2$ dB/km. Curve (b) is to be read only at multiples of $z = \alpha^{-1} = 21.7$ km, which are the output planes of the optical amplifiers. Curve (c) assumes detection with a receiver with a noise figure of 4 dB ($F = 4$ dB) and an input impedance of 1000 Ω.

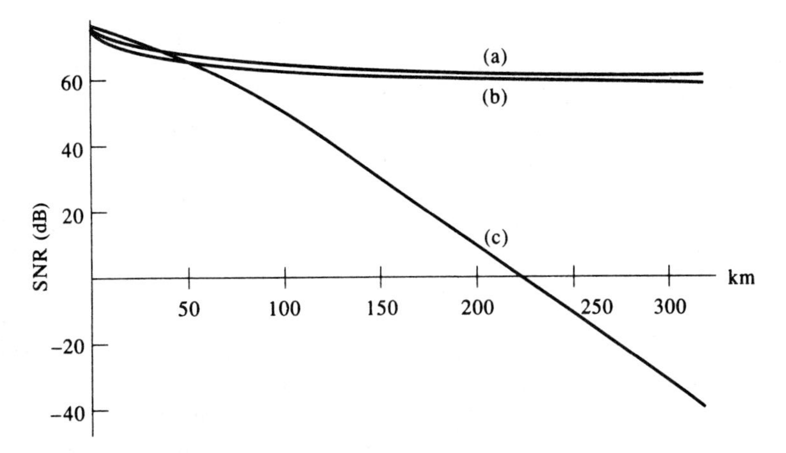

Figure 17.11 SNR of detected signal in a fiber link with (a) a continuous amplifier $\gamma = \alpha$ ($\mu = 1$); (b) discrete amplifiers employed every absorption length $\alpha^{-1} = 21.7$ km (0.2 dB/km fiber loss; $\mu = 1$; curve is to be read only at multiples of 21.7 km); and (c) no optical amplification and detection with a receiver with a noise figure of 4 dB. The power launched into the fiber is 5 mW, the fiber loss is 0.2 dB/km, $\lambda = 1.55$ μm, the detection bandwidth is $\Delta f = 10^9$ Hz, and the detector load impedance is 1000 ohms.

We note that if, for example, we need to maintain a SNR exceeding 50 dB, we must use a fiber link shorter than 100 km if no amplifier is used, but if laser amplifiers are used every, say, $z_0 = 2\alpha^{-1}$ (= 21.7 km), fiber length in excess of 1000 km can be employed.

The above discussion centers on the use of optical amplification in long-distance transmission of data. A second class of applications, no less important, is that of distribution systems with a very large number of subscribers. The use of optical amplifiers makes it possible to maintain the power arriving at a subscriber's premises at sufficiently high levels so as not to be degraded by the receiver noise. The number of subscribers that can thus be served by a single laser can be increased by anywhere from 1 to 3 orders of magnitude.

17.5 RAMAN OPTICAL AMPLIFIERS

We have discussed semiconductor optical amplifiers (SOAs) and erbium-doped fiber amplifiers (EDFAs). These amplifiers are the so-called lumped amplifiers in which the gain is lumped at a point in the transmission line. As discussed in the last section, distributed amplification can provide better system performance especially in terms of the signal-to-noise ratio (SNR). In this section, we discuss fiber Raman amplifiers, which are by their nature a distributed amplifier. The following discussion follows closely that in Reference [21].

As discussed in Section 8.6, the Raman effect, discovered in 1928 by Sir Chandrasekhera Venkata Raman, is an inelastic scattering of light that is accompanied by elementary excitations in the medium. The principle of Raman amplification is based on the phenomenon of stimulated Raman scattering (SRS), which was discussed in Section 8.6. Raman amplification in optical fibers was first observed and measured by Stolen and Ippen [22]. Their measurement showed a frequency shift of approximately 13.2 THz in a silica fiber. Although Raman amplifiers were demonstrated in experiments using solid-state lasers, they have not yet been deployed in real field systems. With the availability of high-power diode pump lasers, the feasibility of Raman amplifiers has increased accordingly. It is important to point out some unique features of Raman amplifiers. Raman amplifiers provide low noise amplification and offer arbitrary gain band (i.e., the gain band is determined by the pump frequency).

In a Raman amplifier, the basic stimulated scattering process can be described in the following. A pump photon at frequency ω_L interacts with some atomic entity in the medium (e.g., SiO_2 molecule in silica); the atomic entity absorbs the photon with energy $\hbar\omega_L$ and is thus elevated to a virtual state (i.e., not an eigenstate) and then relaxes to some excited eigenstate. In the silica fiber this is an excited vibrational state of the silica molecule. In the process, a photon at a lower frequency ω_S is emitted. The energy deficiency $\hbar(\omega_L - \omega_S)$ equals the energy of the first vibrational state to which the molecule is excited. The above discussion is centered around a single atomic entity (e.g., a molecule) in a medium. When the pump beam is propagating in the medium, many of the molecules along the beam path will be excited to their vibrational states. The collective vibrational motion of the molecules in the medium at a frequency $\omega_A = (\omega_L - \omega_S)$ is physically equivalent to the propagation of an elastic wave (or acoustic wave) at frequency ω_A in the medium. Using the particle picture of an acoustic wave discussed in Section 9.7, the acoustic wave consists of phonons with an energy given by $\hbar\omega_A$. As a result of the stimulated Raman scattering, a laser beam with photon energy $\hbar\omega_L$ can donate energy to another laser beam at a lower photon energy $\hbar\omega_S$, while at the same time creating an acoustic wave with phonon energy $\hbar\omega_A$ in the medium. The rate at which this process takes place is proportional to the intensity of the field at ω_S. This leads to an exponential amplification as a function of distance, that is, exponential gain at ω_S.

In an optical fiber with a long interaction length, the laser beam at frequency ω_S can be amplified significantly. Figure 17.12 illustrates the energy levels involved in stimulated Raman scattering. As indicated, the SRS process is insensitive to the frequency of the pump beam. In other words, the gain band is determined by the frequency of the pump laser. This property is desirable in fiberoptic communications.

Generally, both Stokes and anti-Stokes processes are possible in SRS (see Figure 8.8). In anti-Stokes Raman transitions, the molecules are initially in the vibrational excited state. The molecules relax to the ground state after the Raman transition. For silica fibers, the Raman energy shift $\hbar(\omega_L - \omega_S) = \hbar\omega_A$ is about 55 meV, which is also the energy of the first excited state in the molecule. Since the excited states are populated only slightly at room temperature

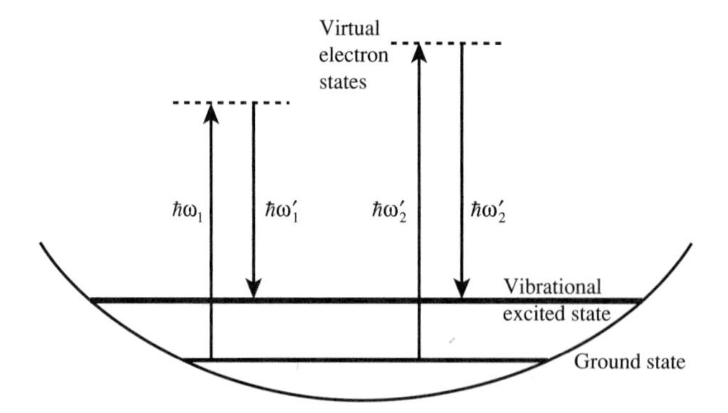

Figure 17.12 Schematic drawing of the vibrational energy levels and the Stokes Raman transitions ($\omega_1' < \omega_1$, $\omega_2' < \omega_2$). In the Stokes process, the molecular system is excited from the ground state to an excited state (vibrational excited state). The process is insensitive to the frequency of the pump beam. The frequency shift of the photons is related to the vibration frequency of the molecules. In other words, $\omega_1 - \omega_1' = \omega_2 - \omega_2' = \omega_A$, where ω_A is the molecular vibration frequency. The waves at ω_1' (or ω_2') are amplified by those at ω_1 (or ω_2).

($k_B T \approx 26$ meV), the anti-Stokes intensities are relatively small compared with the Stokes intensities.

At a given pump laser frequency ω_L, the gain for the Stokes line is centered at the frequency $\omega_S = \omega_L - \omega_A$, where ω_A is the vibrational frequency of the molecules. The gain bandwidth is determined by the linewidth of the first excited state, which is related to the decay lifetime of the state. In Raman spectroscopy, the lifetime of the molecular vibration is often deduced from the width of the Raman line. Like the Kerr effect, Raman scattering is also relatively weak in most media. In optical fibers, however, the nonlinear effect due to Raman scattering can be significant due to the high optical intensity of a beam of laser in the core and the long interaction lengths (in the range of 100 km) afforded by these waveguides.

For a Raman amplifier at $\lambda = 1500$ nm, the Stokes wave is shifted from the pump wave by about $\Delta\lambda = 100$ nm and continues to propagate forward along the fiber together with the pump wave. If the pump is actually one channel of a multiwavelength division multiplexed communications system, then its Stokes wave may overlap with other channels at longer wavelengths—leading to crosstalk and Raman amplification. Raman amplification causes shorter wavelength channels to experience power depletion and act as a pump for the amplification of longer wavelength channels. This can skew the power distribution among the WDM channels—reducing the signal-to-noise ratio of the lowest frequency channels and introducing crosstalk on the high-frequency channels. Both of these effects can lower the information-carrying capacity of the optical system. Thus a proper bandwidth and channel allocation is important when Raman amplifiers are employed.

We now consider the energy coupling between the signal and the pump in a fiber Raman amplifier. According to Equations (8.6-24) and (8.6-25), the Raman coupling between two waves is described by

$$\begin{aligned} \frac{d}{dz}\phi_L &= -\alpha_L\phi_L - G_R'\,\phi_L\phi_S \\ \frac{d}{dz}\phi_S &= -\alpha_S\phi_S + G_R'\,\phi_L\phi_S \end{aligned} \tag{17.5-1}$$

with ϕ_L and ϕ_S being the photon fluxes given by

$$\begin{aligned} \phi_L(x, y, z) &= \frac{I_L}{\hbar\omega_L} = \frac{1}{\hbar\omega_L}\frac{k_L}{2\mu_0\omega_L}|E_L(x, y, z)|^2 \\ \phi_S(x, y, z) &= \frac{I_S}{\hbar\omega_S} = \frac{1}{\hbar\omega_S}\frac{k_S}{2\mu_0\omega_S}|E_S(x, y, z)|^2 \end{aligned} \tag{17.5-2}$$

where z is an axis along the direction of propagation in the fiber, α_L and α_S are the linear attenuation coefficients, I_L and I_S are the beam intensities of the waves in the fiber, k_L and k_S are the beam wavenumbers (propagation constants) in the fiber, and E_L and E_S are the electric field amplitudes of the waves. Photon flux is a measure of the number of photons per unit area per unit time. The constant G_R' is the Raman gain coefficient for photon fluxes given by

$$G_R' = \frac{12\hbar\mu_0^2\omega_L^2\omega_S^2}{k_L k_S}\,\mathrm{Im}(\chi_{R1}^{(3)}) = -\frac{12\hbar\mu_0^2\omega_L^2\omega_S^2}{k_L k_S}\,\mathrm{Im}(\chi_{R2}^{(3)}) \tag{17.5-3}$$

where $\chi_{R1}^{(3)} = \chi_{R2}^{*(3)}$ are the third-order nonlinear susceptibilities responsible for Raman scattering, as discussed in Section 8.6. We note that the coupled equations in (17.5-1) are consistent with the conservation of photon flux in the event when the linear attenuation coefficients vanish ($\alpha_L = \alpha_S = 0$). In other words, one laser photon (at ω_L) is converted into one Stokes photon (at ω_S).

For fiber Raman amplifiers, both the pump and the signal are guided waves of the fiber (e.g., LP_{01}). The power coupling between the signal wave and the pump wave can be obtained by integrating Equations (17.5-1) over the *xy* plane. According to the definitions in Equations (17.5-2), the pump power and the signal power can be written

$$
\begin{aligned}
P_L(z) &= \hbar\omega_L \int \phi_L(x, y, z)\, dx\, dy = \frac{k_L}{2\mu_0\omega_L} \int |E_L(x, y, z)|^2\, dx\, dy \\
P_S(z) &= \hbar\omega_S \int \phi_S(x, y, z)\, dx\, dy = \frac{k_S}{2\mu_0\omega_S} \int |E_S(x, y, z)|^2\, dx\, dy
\end{aligned}
\tag{17.5-4}
$$

where the integrations are over the entire *xy* plane. We now perform similar integrations over the *xy* plane for the coupled equations (17.5-1). These lead to the following coupled equations for the modal power in the fiber:

$$
\begin{aligned}
\frac{d}{dz} P_L &= -\alpha_L P_L - \frac{G_R'}{\hbar\omega_S A_{\text{eff}}^R} P_L P_S \\
\frac{d}{dz} P_S &= -\alpha_S P_S + \frac{G_R'}{\hbar\omega_L A_{\text{eff}}^R} P_L P_S
\end{aligned}
\tag{17.5-5}
$$

where A_{eff}^R is the effective area for the Raman coupling given by

$$
A_{\text{eff}}^R = \frac{\int |E_L(x, y)|^2 dx\, dy \cdot \int |E_S(x, y)|^2 dx\, dy}{\int |E_L(x, y)|^2 |E_S(x, y)|^2 dx\, dy}
\tag{17.5-6}
$$

Note that since $\omega_L > \omega_S$, the power gain by the signal beam at ω_S is less than that lost by the pump beam. The difference represents the power going to the molecular excitation at $\omega_A = (\omega_L - \omega_S)$. The effective area for Raman coupling is different from the effective area of each of the modes. In fact, these effective areas satisfy the following relationship, according to the Schwartz inequality,

$$
\sqrt{A_{\text{eff}}^L A_{\text{eff}}^S} \le A_{\text{eff}}^R
\tag{17.5-7}
$$

where A_{eff}^L and A_{eff}^S are the effective areas of the pump wave and the signal wave, respectively. The equality holds only when the mode wavefunctions of the two waves are identical. Depending on how far the frequencies of the waves are from the cutoff, the wavefunctions of these two waves can be quite different.

For fiber Raman amplifiers, the coupled equations (17.5-5) are often written

$$
\begin{aligned}
\frac{d}{dz} P_p &= -\alpha_p P_p - g_R \frac{\omega_p}{\omega_s} P_p P_s \\
\frac{d}{dz} P_s &= -\alpha_s P_s + g_R P_p P_s
\end{aligned}
\tag{17.5-8}
$$

where $P_p = P_L$ is the pump power, $P_s = P_S$ is the signal power that is being amplified, α_p and α_s are the linear attenuation coefficients, ω_p and ω_s are the frequencies of the waves, and the Raman gain coefficient g_R is given by

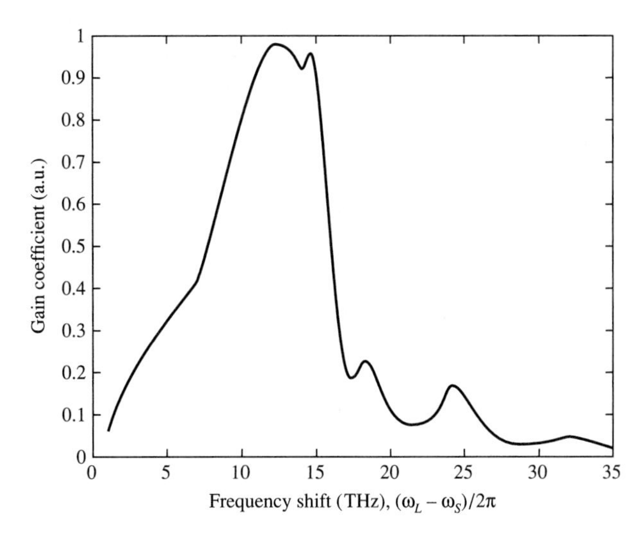

Figure 17.13 Typical gain spectrum of a Raman amplifier in silica fiber as a function of the frequency shift between the pump and signal frequencies.

$$g_R = \frac{G'_R}{\hbar\omega_L A^R_{\text{eff}}} = \frac{12\mu_0^2\omega_L\omega_S^2}{A^R_{\text{eff}}k_L k_S}\,\text{Im}(\chi^{(3)}_{R1}) = -\frac{12\mu_0^2\omega_L\omega_S^2}{A^R_{\text{eff}}k_L k_S}\,\text{Im}(\chi^{(3)}_{R2}) \tag{17.5-9}$$

or equivalently,

$$g_R = \frac{24\pi}{c\varepsilon_0^2 A^R_{\text{eff}} n^2 \lambda_S}\,\text{Im}(\chi^{(3)}_{R1}) = -\frac{24\pi}{c\varepsilon_0^2 A^R_{\text{eff}} n^2 \lambda_S}\,\text{Im}(\chi^{(3)}_{R2}) \tag{17.5-10}$$

where $n^2 = n_L n_S \approx n_S^2$, with n_L and n_S being the indices of refraction. We note that the Raman gain coefficient (in units of $\text{m}^{-1}\text{W}^{-1}$) has a wavelength dependence of $1/\lambda$, in addition to the dependence of $\chi^{(3)}_{R1}$ and $\chi^{(3)}_{R2}$ on the wavelength. Figure 17.13 shows a typical gain spectrum of a fiber Raman amplifier.

As shown in Figure 17.13, the Raman gain spectrum of the optical fiber exhibits a broad continuum shape mainly due to the amorphous nature of the material. The actual profile of the gain spectrum depends on the presence of various dopants, including GeO_2. As shown in the figure, the full width at half maximum (FWHM) gain bandwidth is approximately $\Delta\lambda = 50$ nm in the spectral regime around $\lambda = 1500$ nm. In DWDM optical network applications, the gain flatness among all the channels is an important issue. This can be achieved by using multiple pump beams at proper frequency spacings [23–25].

In the event when the difference between the two linear attenuation coefficients can be neglected (i.e., $\alpha_p = \alpha_s = \alpha$), exact solutions of the coupled equations (17.5-8) are available. We first rewrite the coupled equations in terms of photon rates (i.e., number of photons per second). They are written, according to Equations (17.5-8),

$$\begin{aligned}\frac{d}{dz}\left(\frac{P_p}{\hbar\omega_p}\right) &= -\alpha\left(\frac{P_p}{\hbar\omega_p}\right) - g_R\hbar\omega_p\left(\frac{P_p}{\hbar\omega_p}\right)\left(\frac{P_s}{\hbar\omega_s}\right)\\ \frac{d}{dz}\left(\frac{P_s}{\hbar\omega_s}\right) &= -\alpha\left(\frac{P_s}{\hbar\omega_s}\right) + g_R\hbar\omega_p\left(\frac{P_p}{\hbar\omega_p}\right)\left(\frac{P_s}{\hbar\omega_s}\right)\end{aligned} \tag{17.5-11}$$

where we assume $\alpha = \alpha_p = \alpha_s$. The solutions of (17.5-11) are written, according to Equations (8.6-27) and (8.6-28),

$$P_p(z) = P_p(0)\frac{1+m^{-1}}{1+m^{-1}\exp\left(\frac{\gamma}{\alpha}(1-e^{-\alpha z})\right)}\exp(-\alpha z)$$

$$P_s(z) = P_s(0)\frac{1+m}{1+m\exp\left(-\frac{\gamma}{\alpha}(1-e^{-\alpha z})\right)}\exp(-\alpha z) \tag{17.5-12}$$

where the Raman exponential gain coefficient γ is given by

$$\gamma = g_R\left(P_s(0)\frac{\omega_p}{\omega_s} + P_p(0)\right) = \frac{G'_R}{\hbar\omega_p A^R_{\text{eff}}}\left(P_s(0)\frac{\omega_p}{\omega_s} + P_p(0)\right) \tag{17.5-13}$$

and m is the photon rate ratio at input ($z = 0$)

$$m = \frac{P_p(0)}{P_s(0)}\frac{\omega_s}{\omega_p} \tag{17.5-14}$$

In the above, $P_p(0)$ and $P_s(0)$ are the powers of the pump and signal beams at input ($z = 0$), respectively. The parameter m is the ratio of photon rates (in photons/s) at the input (the photon rate is defined as the number of photons per second crossing the input plane $z = 0$). For most Raman amplifiers in the spectral regime around 1500 nm, the frequency ratio (ω_s/ω_p) is about 0.93. So the parameter m is approximately the ratio of pump power to signal power at input.

For Raman amplifiers with a gain of at least 10 dB, the parameter m is much greater than 1. In other words, the pump power is significantly higher than the signal power in Raman amplifiers. In this case, the signal power can be written, according to Equations (17.5-12),

$$P_s(z) = P_s(0)\exp\left(\frac{\gamma}{\alpha}(1-e^{-\alpha z})\right)\exp(-\alpha z) \tag{17.5-15}$$

where $P_s(0)$ is the signal power at input ($z = 0$). Figure 17.14 shows the pump power and signal power as a function of interaction distance z in the fiber. We note that the pump power is a decreasing function of z as it donates energy to the signal beam and suffers the propagation attenuation loss. The signal power increases with z initially as it gains energy from the pump beam. The signal power reaches a maximum at a point where the gain due to Raman amplification equals the loss due to linear attenuation. This location is given, according to Equation (17.5-15), by

$$z = L_{\text{eff}} = \frac{1}{\alpha}\ln\left(\frac{\gamma}{\alpha}\right) \tag{17.5-16}$$

where L_{eff} is often called the effective length of the Raman amplifier. For distance $z > L_{\text{eff}}$, the signal power starts to decline. For the example in Figure 17.14, the effective fiber length is approximately 40 km. In this case a gain of over 15 dB can be obtained by using a pump power of 500 mW in a single-mode fiber. It is important to note that the effective fiber length depends on the power of the pump beam, as γ is dependent on the power of the pump beam, according to Equation (17.5-13).

Using Equations (17.5-15) and (17.5-16), the maximum signal power is given by

$$P_s(z)_{\max} = P_s(0)\frac{\alpha}{\gamma}\exp\left(\frac{\gamma}{\alpha}-1\right) \cong P_s(0)\frac{\alpha}{g_R P_p(0)}\exp\left(\frac{g_R P_p(0)}{\alpha}-1\right) \tag{17.5-17}$$

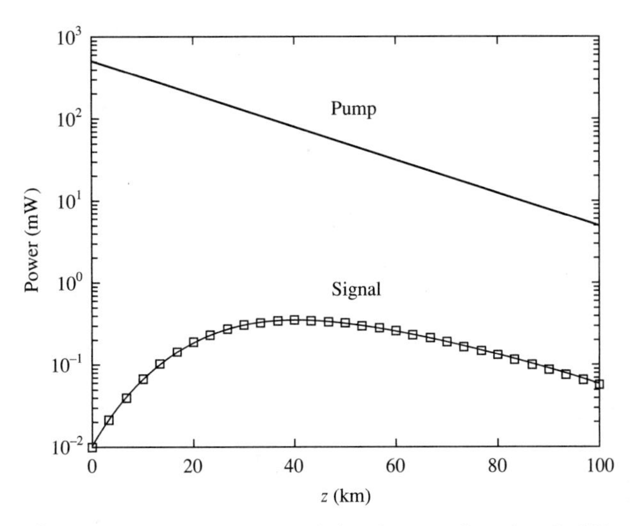

Figure 17.14 Pump power and signal power (in units of mW) versus z in a fiber Raman amplifier. The parameters used are $P_p(0) = 500$ mW, $P_s(0) = 0.01$ mW, linear attenuation $\alpha = 0.046$/km (corresponds to 0.2 dB/km), $\lambda_p = 1500$ nm, and Raman gain coefficient $g_R = 5.9 \times 10^{-4}$ m^{-1}W^{-1}. The solid curves are from Equations (17.5-12). The dotted curve (with □) is obtained by using the approximation (17.5-15).

The maximum Raman power gain is thus given by

$$G_{\max} = \frac{P_s(z)_{\max}}{P_s(0)} = \frac{\alpha}{\gamma}\exp\left(\frac{\gamma}{\alpha} - 1\right) \cong \frac{\alpha}{g_R P_p(0)}\exp\left(\frac{g_R P_p(0)}{\alpha} - 1\right) \tag{17.5-18}$$

Figure 17.15 shows an example of the maximum Raman gain as a function of the pump power $P_p(0)$. We note that a gain of 30 dB (10^3) would require a pump power of around 800 mW.

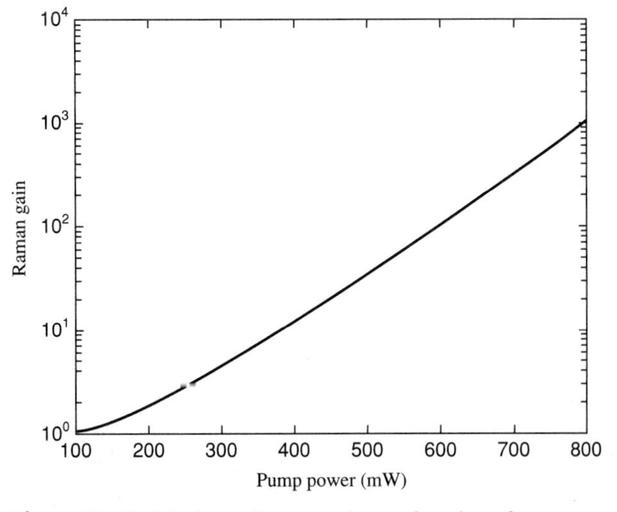

Figure 17.15 Maximum Raman gain as a function of pump power in a silica fiber. The parameters used are $\alpha = 0.046$/km, $\lambda_p = 1500$ nm, and Raman gain coefficient $g_R = 5.9 \times 10^{-4}$ m^{-1}W^{-1}.

EXAMPLE: MAXIMUM GAIN IN SILICA FIBER

Consider a single-mode silica fiber with a linear attenuation of 0.2 dB/km. The attenuation coefficient is $\alpha = 0.046$/km. We assume a Raman gain coefficient $g_R = 5.9 \times 10^{-4}\ \mathrm{m^{-1}W^{-1}}$ and a pump power of 500 mW. The exponential gain coefficient γ is, according to Equation (17.5-13),

$$\gamma = g_R\left(P_s(0)\frac{\omega_p}{\omega_s} + P_p(0)\right) \approx g_R P_p(0) = 0.295/\mathrm{km}$$

The effective length of the Raman amplifier is, according to Equation (17.5-16),

$$L_{\mathrm{eff}} = \frac{1}{\alpha}\ln\left(\frac{\gamma}{\alpha}\right) = 40.4\ \mathrm{km}$$

The maximum gain is, according to Equation (17.5-17),

$$\frac{P_s(z)_{\max}}{P_s(0)} = \frac{\alpha}{\gamma}\exp\left(\frac{\gamma}{\alpha} - 1\right) = 35\ (\text{or } 15\ \mathrm{dB})$$

PROBLEMS

17.1 The gain ripple in units of dB is given by

$$\rho = 10\log_{10}(T_{\max}/T_{\min})$$

Show that $\rho = 1$ dB requires $GR = 0.0575$.

17.2

(a) Show that the FWHM gain bandwidth of resonant transmission in a Fabry–Perot amplifier is given approximately by

$$\Delta\nu_{\mathrm{FWHM}} = \frac{c}{2n_{\mathrm{eff}}L}\frac{1-GR}{\pi\sqrt{GR}}$$

The increase in the resonant transmission comes with a decrease of the gain bandwidth.

(b) For a SOA with $L = 500\ \mu$m, $n_{\mathrm{eff}} = 3.6$, $G = 100$, and $R = 0.008$, find the maximum gain at resonance and the gain bandwidth at resonance.

17.3 Consider a transverse spatial mode inside a box of cross-sectional area A and length L. The power of the zero-point fluctuation can be obtained by multiplying the energy density with the speed of light. Show that the power is

$$P_{\mathrm{zero}} = \tfrac{1}{2}h\nu\,\Delta\nu$$

[Note: The number of longitudinal modes within $\Delta\nu$ is $\Delta\nu/(c/nL)$.]

17.4

(a) The spontaneous emission factor α is defined as the fraction of power emitted into a single cavity mode (usually the lasing mode). Show that if all modes are treated equally, the spontaneous emission factor α is given by

$$\alpha = \frac{c^3}{8\pi\nu^2 n^3 V\,\Delta\nu}$$

where $\Delta\nu$ is the optical bandwidth.

(b) In an optical amplifier, the cavity is eliminated via antireflection at the facets (or ends). The spontaneous emission factor β is defined as the fraction of power emitted into a single spatial transverse mode (usually the spatial mode defined by the input beam). Show that if all modes are treated equally, the spontaneous emission factor β is given by

$$\beta = \alpha\frac{\Delta\nu}{(c/nL)} = \frac{c^2 L}{8\pi\nu^2 n^2 V} = \frac{c^2}{8\pi\nu^2 n^2 A}$$

where $\Delta\nu$ is the optical bandwidth. This approach is legitimate only for the lower order spatial modes whose frequency spacing is c/nL. For high-order spatial modes, the frequency spacing is larger than c/nL by a directional factor $1/\cos\theta$, where θ is the effective ray angle of the spatial mode. For the fundamental spatial mode (and lower order spatial modes) the ray angle θ is near zero.

(c) Using the argument of Section 4.0, the number of spatial modes can be obtained by the following formula:

$$N_{\text{trans}} = \frac{1}{\beta} = 2 \times 2 \frac{\frac{1}{4}\pi k^2}{\left(\frac{\pi}{a}\frac{\pi}{b}\right)} = \frac{k^2}{\pi} A = \frac{4\pi n^2 \nu^2}{c^2} A$$

The β factor obtained in this way is different from that of (b) by a factor of 2. Try to explain the difference.

17.5 Consider a dielectric waveguide that supports a discrete set of confined modes of propagation as well as a continuum of radiation modes. All these modes are mutually orthogonal. For the purpose of simplicity, the modes are normalized to an energy of one photon in a segment of length L along the direction of propagation. Consider the presence of an atom in the excited state. The rate of spontaneous emission into a confined mode (mode m) is given by the Fermi golden rule

$$R_m = \frac{2\pi}{\hbar} |\langle f | \mu \cdot E_m(x_d, y_d) | i \rangle|^2 \rho = \frac{2\pi}{\hbar} |\mu_{fi}|^2 |E_m(x_d, y_d)|^2 \rho$$

where (i, f) represents the initial and final states of the transition, (x_d, y_d) is the location of the dipole in the waveguide, $\mu_{fi} = \langle f | \mu | i \rangle$ is the matrix element of the dipole moment μ of the atom, ρ is the density of photon states, and E_m is the normalized electric field vector. Using the notation, the spontaneous emission factor can be written

$$\beta_m = \frac{R_m}{\sum_m R_m} = R_m t_{\text{spont}}$$

where the spontaneous lifetime is defined as

$$\frac{1}{t_{\text{spont}}} = \sum_m R_m$$

We note that R_m depends on the location of the dipole (excited atom). In the case when the waveguide is filled with a spatial distribution of radiating dipoles (excited atoms), we can perform a spatial average to obtain the spontaneous emission factor.

(a) Show that the density of states for traveling waves with a given transverse spatial mode (one direction and one polarization) in a waveguide segment of length L is given by

$$\rho = \frac{dN}{h\, d\nu} = \frac{n_g L}{hc}$$

where n_g is the group index.

(b) Let $P(x, y)$ be the spatial distribution of the excited atoms. The average rate of spontaneous emission into the confined mode is given by

$$\bar{R}_m = \frac{2\pi}{\hbar} |\mu_{fi}|^2 \rho \frac{\iint_\infty |E_m(x_d, y_d)|^2 P(x_d, y_d)\, dx_d\, dy_d}{\iint_\infty P(x_d, y_d)\, dx_d\, dy_d}$$

Define an effective area and an effective index as

$$n_{\text{eff}}^2 = \frac{\iint_\infty n^2(x, y) |E_m(x, y)|^2 dx\, dy}{\iint_\infty |E_m(x, y)|^2 dx\, dy} \quad \text{and}$$

$$A_{\text{eff}} = \frac{\iint_\infty |E_m(x, y)|^2 dx\, dy \iint_\infty P(x, y)\, dx\, dy}{\iint_\infty |E_m(x, y)|^2 P(x, y)\, dx\, dy}$$

Show that $\bar{R}_m$ can be written

$$\bar{R}_m = \frac{2\pi}{\hbar} |\mu_{fi}|^2 \frac{2\nu}{\varepsilon_0 n_{\text{eff}}^2 A_{\text{eff}}} \frac{n_g}{c}$$

(c) Assuming that the total spontaneous emission rate remains the same as that of a homogeneous medium with an index of refraction of n_{eff},

$$\frac{1}{t_{\text{spont}}} = \frac{2\pi}{\hbar} |E|^2 |\mu_{fi}|^2 \frac{8\pi}{c^3} n_{\text{eff}}^2 n_g \nu^2 L^3 \frac{1}{h}$$

where h is Planck's constant, and E is the amplitude of a plane wave with the normalization

$$\tfrac{1}{2} \varepsilon_0 n_{\text{eff}}^2 |E|^2 L^3 = h\nu$$

show that

$$\frac{1}{t_{\text{spont}}} = \frac{2\pi}{\hbar} |\mu_{fi}|^2 \frac{8\pi}{c^3} n_{\text{eff}}^2 n_g \nu^2 L^3 \frac{1}{h} \frac{2h\nu}{\varepsilon_0 n_{\text{eff}}^2 L^3}$$
$$= \frac{2\pi}{\hbar} |\mu_{fi}|^2 \frac{16\pi}{\varepsilon_0 c^3} n_g \nu^3$$

Strictly speaking, this assumption is valid only when the transverse dimension of the waveguide is much larger than the wavelength. For details on the total spontaneous emission rate for dipoles embedded in dielectric waveguides, interested students are referred to References [26–28].

(d) Using (b) and (c), show that the spontaneous emission factor β can be written

$$\beta = \bar{R}_m t_{\text{spont}} = \frac{c^2}{8\pi n_{\text{eff}}^2 \nu^2 A_{\text{eff}}}$$

This is the fraction of spontaneous emission energy into a single spatial mode (one direction and one polarization).

(e) Show that for the special case of a uniform distribution of excited atoms in a rectangular volume with cross-sectional area A and length L, the effective area can be written

$$A_{\text{eff}} = \frac{\int_{-\infty}^{\infty} n^2(x, y)|E_m(x, y)|^2 dx\, dy}{n_{\text{eff}}^2 \int_A |E_m(x, y)|^2 dx\, dy} A = \frac{A}{\Gamma}$$

where Γ is the confinement factor (fraction of mode energy inside the active region A),

$$\Gamma = \frac{n_{\text{eff}}^2 \int_A |E_m(x, y)|^2 dx\, dy}{\int_{-\infty}^{\infty} n^2(x, y)|E_m(x, y)|^2 dx\, dy} = \frac{\int_A |E_m(x, y)|^2 dx\, dy}{\int_{-\infty}^{\infty} |E_m(x, y)|^2 dx\, dy}$$

In this case, show that the spontaneous emission factor can be written

$$\beta = \bar{R}_m t_{\text{spont}} = \frac{c^2}{8\pi n_{\text{eff}}^2 \nu^2 A} \Gamma$$

17.6 Assume a fiber distribution network fed by a single semiconductor laser at $\lambda = 1.55$ μm with a power output $P_0 = 10$ mW. The power is divided into N branches, amplified by an optical fiber amplifier (in each branch) and then divided again into M branches. Determine the maximum number of "subscribers" NM that can be serviced by the system assuming: $\Delta f = 10^9$ Hz; R (receiver input impedance) is 10^3 ohms, $T_e = 1000$ K; and a minimum SNR at the subscriber of 42 dB. The maximum power level at the output of the amplifiers is 10 mW.

17.7 Derive Equation (17.3-28) by adding the so-called zero-point fluctuation energy of $h\nu/2$ per mode to a classical signal $\sqrt{S_0}$ normalized such that S_0 is the optical power level. The power of the zero-point fluctuation of a transverse spatial mode within a bandwidth of $\Delta\nu$ is $P_{\text{zero}} = h\nu\, \Delta\nu/2$ according to Problem 17.3. There are two spectral windows of the zero-point fluctuation that beat with the optical signal field at frequency ω_0 to generate a noise current at the modulation frequency ω_m.

(a) Show that the relevant optical field can be written

$$E = \sqrt{S_0}\, e^{i(\omega_0 t + \Phi_S)} + \sqrt{N_0}\, e^{i[(\omega_0 - \omega_m)t + \Phi_1]} + \sqrt{N_0}\, e^{i[(\omega_0 + \omega_m)t + \Phi_2]}$$

where S_0 is the optical signal power level, N_0 is the power of the zero-point fluctuation (i.e., $N_0 = P_{\text{zero}} = h\nu\, \Delta\nu_{\text{sig}}/2$), and Φ_S, Φ_1, and Φ_2 are the phases.

(b) Due to the random nature of the zero-point fluctuation, the phases Φ_1 and Φ_2 are uniformly distributed over (0 and 2π). Show that the photoelectric current can be written

$$i = \frac{S_0 e\eta}{h\nu}\left(1 + 2\frac{N_0}{S_0} + 2\sqrt{\frac{N_0}{S_0}} \cos(\omega_m t + \Phi_2 - \Phi_S) + 2\sqrt{\frac{N_0}{S_0}} \cos(-\omega_m t + \Phi_1 - \Phi_S)\right)$$

where η is the quantum efficiency of the optoelectronic conversion, and the mean square beat current at ω_m is

$$\langle \overline{i_N^2} \rangle = 4\left(\frac{e\eta}{h\nu}\right)^2 S_0 N_0 = 2\left(\frac{e\eta}{h\nu}\right)^2 S_0 (h\nu\, \Delta\nu_{\text{sig}})$$

where $N_0 = h\nu\, \Delta\nu_{\text{sig}}/2$ with $\Delta\nu_{\text{sig}}$ being the optical signal bandwidth.

(c) Show that

$$(SNR)_{\text{input}} = \frac{S_i}{N_i} = \frac{S_0}{2h\nu\, \Delta\nu_{\text{sig}}}$$

which is identical to Equation (17.3-28). [Note: The optical signal bandwidth $\Delta\nu_{\text{sig}}$ is the electronic bandwidth of photodetection: $\Delta\nu_{\text{sig}} = \Delta f$].

REFERENCES

1 Simon, J. C., Semiconductor laser amplifier for single mode optical fiber communications. *J. Opt. Commun.* **4**:51 (1983).

2 Mears, R. J., L. Reekie, I. M. Jauncey, and D. N. Payne, Low noise erbium-doped fiber amplifier operating at 1.54 mm. *Electron. Lett.* **23**:1026 (1987).

3 Hagimoto, K., et al., A 212 km non-repeatered transmission experiment at 1.8 Gb/s using LD pumped Er^{3+}-doped fiber amplifiers in an IM/direct-detection repeater system. In: *Proceedings of the Optical Fiber Conference*, Houston, TX, postdeadline Paper PDI5, 1989.

4 Olshansky, R., Noise figure for Er-doped optical fibre amplifiers. *Electron. Lett.* **24**:1363 (1988).

5 Payne, D. N., Tutorial session abstracts. In: *Optical Fiber Communication (OFC 1990) Conference*, San Francisco. Optical Society of America, Washington, DC, 1990.

6 Olsson, N. A., Semiconductor optical amplifiers. *Proc. IEEE* **80**:375 (1992).

7 Simon, J. C., GaInAsP semiconductor laser amplifiers for single mode fiber communications. *J. Lightwave Technol.* **5**:1286 (1987).

8 Saitoh, T., and T. Mukai, 1.5 mm GaInAsP traveling wave semiconductor laser amplifier. *IEEE J. Quantum Electron.* **QE-23**:1010 (1987).

9 Mukai, T., K. Inoue, and T. Saito, Homogenous gain saturation in 1.5 μm InGaAsP traveling-wave semiconductor laser amplifiers. *Appl. Phys. Lett.* **51**:381 (1987).

10 Digonnet, M. J. F. (ed.), *Rare Earth Doped Fiber Lasers and Amplifiers*. Stanford University/Marcel Dekker, New York, 1993.

11 Edagawa, N., K. Mochizuki, S. Ryu, and H. Wakabayashi, Amplification characteristics of fiber Raman amplifiers. ICICE Technical Report, 1988, translated from Japanese.

12 Armitage, J. R., Three-level fiber laser amplifier: a theoretical model. *Appl. Opt.* **27**(23):4831 (1988).

13 Morkel, P. R., and R. I. Laming, Theoretical modeling of erbium-doped fiber amplifiers with excited-state absorption. *Opt. Lett.* **14**(19):1062 (1989).

14 Desurvire, E., Analysis of erbium-doped fiber optical amplifiers pumped in the $^4I_{15/2}$–$^4I_{13/2}$ band. *IEEE Photon. Technol. Lett.* **1**:293 (Oct. 1989).

15 Giles, C. R., and D. DiGiovanni, Spectral dependence of gain and noise in erbium-doped fiber optical amplifiers. *IEEE Photon. Technol. Lett.* **2**:797 (Nov. 1990).

16 Giles, C. R., and E. Desurvire, Propagation of signal and noise in concatenated erbium-doped fiber optical amplifiers. *J. Lightwave Technol.* **9**:147 (Feb. 1991).

17 Giles, C. R., and E. Desurvire, Modeling erbium-doped fiber amplifiers. *J. Lightwave Technol.* **9**:271 (Feb. 1991).

18 Desurvire, E., *Erbium-Doped Fiber Amplifiers*. Wiley, New York, 1994.

19 Haus, H. A., The noise figure of optical amplifiers. *IEEE Photonics Technol. Lett.* **10**(11):1602 (1998).

20 Baney, D. M., P. Gallion, and R. Tucker, Theory and measurement techniques for the noise figure of optical amplifiers. In: *Optical Fiber Technology*, Vol. 6. Academic Press, New York, 2000, pp. 122–154.

21 Namiki, S., and Y. Emori, Ultrabroad-band Raman amplifiers pumped and gain-equalized by wavelength-division-multiplexed high-power laser diodes. *IEEE J. Selected Topics Quantum Electron.* **7**(1):3 (2001).

22 Stolen, R. H., and E. P. Ippen, Raman gain in glass optical waveguides. *Appl. Phys. Lett.* **22**:276 (1973).

23 Islam, M. N., Raman amplifiers for telecommunications. *IEEE J. Selected Topics Quantum Electron.* **8**(3):548 (2002).

24 Hu, J., B. S. Marks, and C. R. Menyuk, Flat-gain fiber Raman amplifiers using equally spaced pumps. *J. Lightwave Technol.* **22**(6):1519 (2004).

25 Perlin, V. E., and H. G. Winful, Optimal design of flat-gain wide-band fiber Raman amplifiers. *J. Lightwave Technol.* **20**(2):250 (2002).

26 Khosravi, H., and R. Loudon, Vacuum field fluctuation and spontaneous emission in a dielectric slab. *Proc. R. Soc. London A* **436**:373 (1992).

27 Bjork, G., S. Machida, Y. Yamamoto, and K. Igeta, Modification of spontaneous emission rate in planar dielectric microcavity structures. *Phys. Rev. A* **44**:669 (1991).

28 Chance, R. R., A. Prock, and R. Silbey, Molecular fluorescence and energy transfer near interfaces. In: *Advances in Chemical Physics*, Vol. 37, I. Prigogine and S. A. Rice (eds.). Wiley, New York, 1978, pp. 1–65.

CHAPTER 18

CLASSICAL TREATMENT OF QUANTUM NOISE AND SQUEEZED STATES

18.0 INTRODUCTION

In this book we have, in a number of locations, discussed the subject of noise. This was done mostly in the context of optical detection, lasers, and optical parametric amplification. In the last chapter, we discussed the amplified spontaneous emission (ASE) and its impact on the uncertainty in the detection of light. Some of the most important and elegant phenomena of optics, including the zero-point fluctuation energy ($h\nu/2$), related to optical waves and their detection cannot be predicted and thus cannot be treated by the classical theory. They can be explained only by quantizing the electromagnetic field (i.e., applying the formalism of quantum mechanics to optics). The resulting discipline is known as *quantum optics*. Important topics that fall within this discipline include amplitude and phase noise (fluctuations), the statistics of photogenerated electrons, and the field of nonlinear optical squeezing. These areas are just too important to forego. In this chapter, by assuming just *one* result from quantum mechanics (the uncertainty principle), we attempt to treat all the above-mentioned phenomena classically and obtain results that agree with those of quantum optics.

18.1 THE UNCERTAINTY PRINCIPLE AND QUANTUM NOISE

One of the better-known uncertainties of quantum mechanics relates to the simultaneous measurement of the position (x) and momentum (p) of a particle and decrees that the product of the uncertainties Δp and Δx must obey [1]

$$\Delta p \, \Delta x \geq \hbar/2 \tag{18.1-1}$$

where $\hbar \equiv h/2\pi$ and $h = 6.62377 \times 10^{-34}$ J-s is Planck's constant. These fundamental uncertainties extend to optical measurements such as measurements of the amplitudes and phases of optical fields. Their proper study involves the elegant formalism of quantum optics [1–3]. Since we have foresworn quantum mechanics in this book, we cannot approach this subject from first principles. We can, however, appreciate many of the consequences and even obtain numerically correct results for the important scenarios by accepting just *one* basic consequence of quantum mechanics: that of the *uncertainty principle*.

The Uncertainty Principle

Let us represent the classical monochromatic electric field of some mode, say, one in a resonator, as

$$e(t) = |E| \cos(\omega t + \beta) = \mathrm{Re}[E \exp(i\omega t)] \tag{18.1-2}$$

where β is a phase angle and

$$E = |E| \exp(i\beta) = E_1 + iE_2 \tag{18.1-3}$$

is the complex phasor representing the field. It is shown in Figure 18.1a as a vector in the complex E plane of length $|E|$ and projections E_1 and E_2 along the real and imaginary axes, respectively.

According to quantum mechanics [1, 2], the complex amplitude E in Equation (18.1-3) *cannot be* specified exactly. This uncertainty is represented in Figure 18.1b by means of the "uncertainty circle." The most probable position of the tip of the phasor E, on measurement, will be found near the center of the circle and is extremely unlikely to be found outside it. We recall here that phasor does not represent the spatial direction of the field vector but rather the relation between the real (in-phase component) and the imaginary (quadrature component) of the complex field amplitude. The field phasor corresponding to the center of this circle is denoted as $\langle E \rangle$, the "expectation value" of E. The expectation value corresponds to the quantum mechanical ensemble average, that is, to the average of a large number of independent field determinations (measurements) under *identical* conditions. This expectation value obeys in all respects the same (Maxwell's) equations as its classical counterpart. There is even a theorem in quantum mechanics, Ehrenfest's theorem [2], to prove it.

Repeated measurements of the projections of E, $E_1(t)$, and $E_2(t)$ will, in general, yield different results, and the results for $E(t) = E_1(t) + iE_2(t)$ will tend to cluster about the center of the circle, which is a graphical way to describe the most probable region in which the tip of $E(t)$ will fall. The uncertainty that results from this inherent quantum-imposed spread in the values of $E(t)$ can be thought of as *quantum noise*. We shall devote the rest of this chapter to a consideration of the consequences of this noise in optical measurements. A classical *approximation* of quantum physics is to write the basic monochromatic field of an electromagnetic mode as

$$e(t) = \mathrm{Re}[E \exp(i\omega t)] = E_1(t) \cos \omega t - E_2(t) \sin \omega t \tag{18.1-4}$$

where the complex amplitude $E(t)$ is

$$E(t) = E_1(t) + iE_2(t) \tag{18.1-5}$$

Our discussion in this chapter is restricted to electromagnetic fields that are classical analogies of quantum coherent states (e.g., output of ideal lasers). Since the complex amplitude cannot be specified exactly, according to quantum mechanics, we need to treat $E(t)$ as a random complex variable. Thus we can write $E_1(t)$ and $E_2(t)$ as

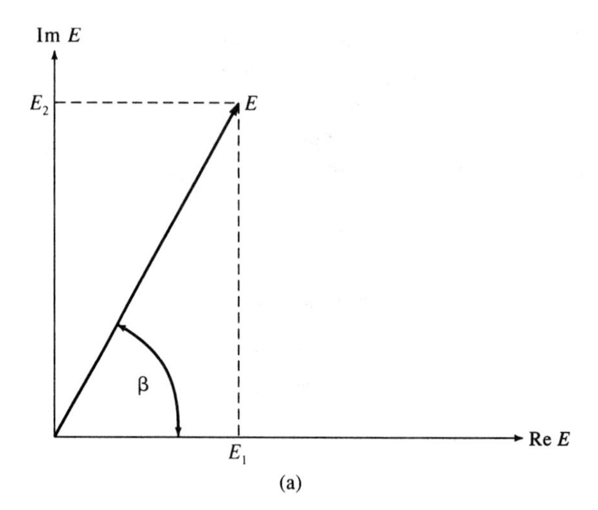

(a)

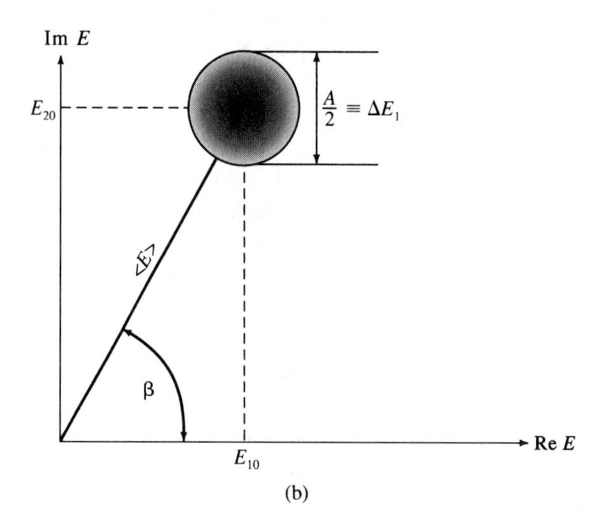

(b)

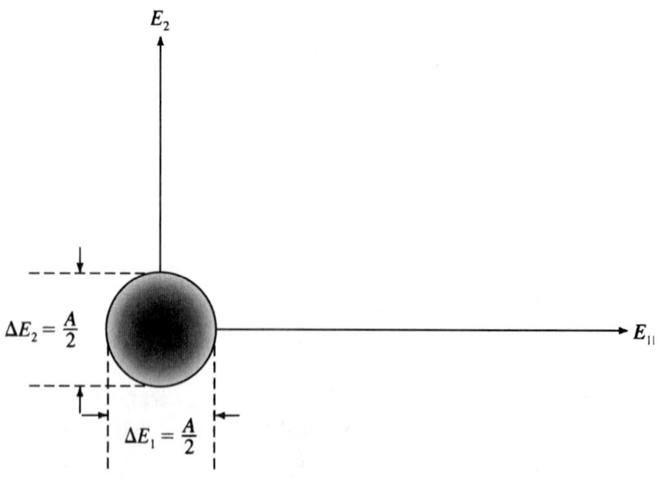

(c)

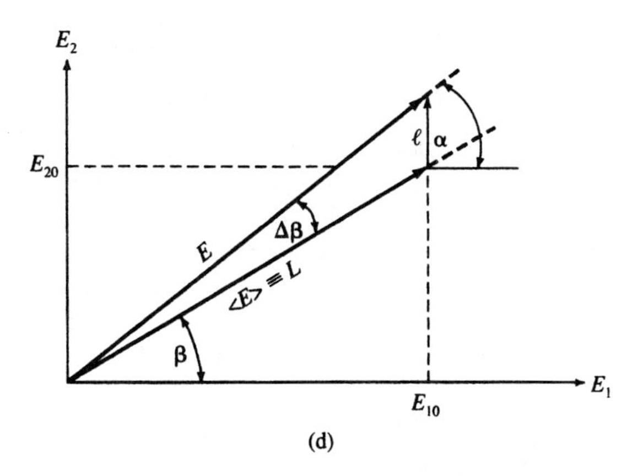

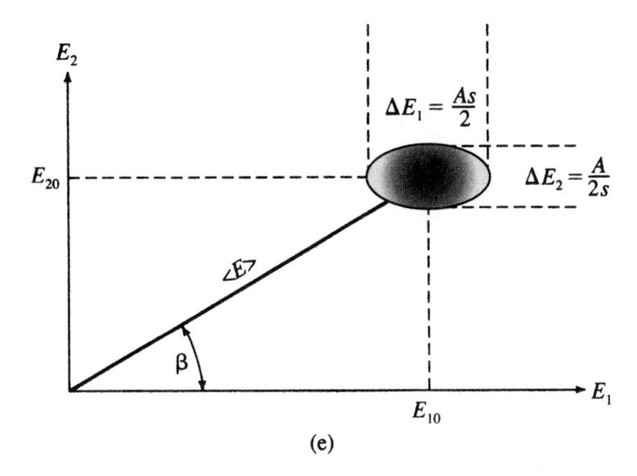

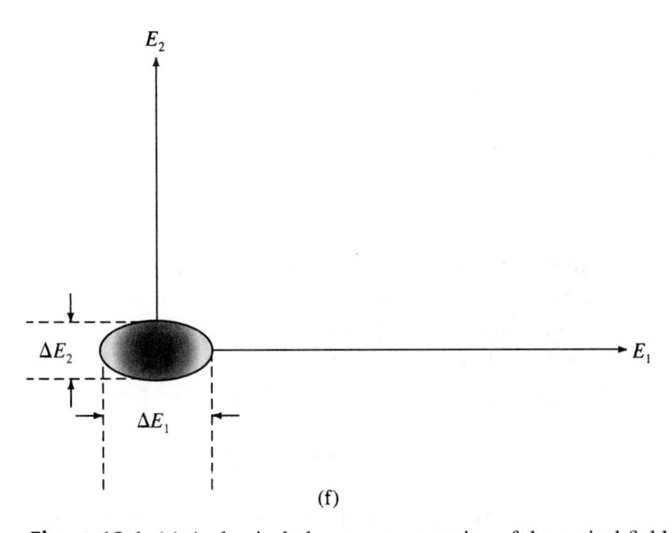

Figure 18.1 (a) A classical phasor representation of the optical field. (b) A representation of a coherent field that is consistent with quantum optics. In this special case, $\Delta E_1 = \Delta E_2$. (c) The electromagnetic field representation of an unexcited vacuum ($\langle E \rangle = 0$) optical mode. (d) An equivalent representation of the field with a random phasor added, vectorially, to the tip of the classical phasor. (e) A "squeezed" field. (f) A squeezed vacuum ($\langle E \rangle = 0$) mode. The squeezing factor (defined in Equation (18.2-8)) is $s > 1$.

$$\begin{aligned}E_1(t) &= \langle E_1(t)\rangle + \delta E_1(t) = E_{10} + \delta E_1(t)\\ E_2(t) &= \langle E_2(t)\rangle + \delta E_2(t) = E_{20} + \delta E_2(t)\end{aligned} \tag{18.1-6}$$

where $\langle E_{1,2}(t)\rangle \equiv E_{10,20}$ is the expectation value of $E_{1,2}(t)$, and $\delta E_1(t)$ and $\delta E_2(t)$ are random variables representing the fundamental quantum mechanical uncertainties (or statistical fluctuations). They have zero mean, $\langle \delta E_{1,2}(t)\rangle = 0$, and are uncorrelated, $\langle \delta E_1(t)\delta E_2(t)\rangle = 0$. Here the angular brackets $\langle\ \rangle$ indicate ensemble averaging. For time-invariant (or stationary) cases, the ensemble average can also be represented by the temporal average.

Before proceeding with a description of these statistical fluctuations, we will, as is the practice in quantum optics, find it useful to relate the *mean* electric field of a mode to the mean number n of optical photons (quanta) in the mode. This is legitimate provided $|\delta E_{1,2}| \ll |E_{10,20}|$ (see Problem 18.1). Taking the mode volume as V, the dielectric constant of the medium as ε, and using Equation (1.4-22), we obtain

$$\frac{\varepsilon}{2}|\langle E\rangle|^2 V = \frac{\varepsilon}{2}(E_{10}^2 + E_{20}^2)V = n\hbar\omega = \text{Field energy within volume } V \tag{18.1-7}$$

so that the mean field amplitude is expressible in terms of the mean photon number n as

$$|\langle E\rangle| = \sqrt{E_{10}^2 + E_{20}^2} = \left(\frac{2\hbar\omega}{\varepsilon V}\right)^{1/2}\sqrt{n} \equiv A\sqrt{n} \tag{18.1-8}$$

where A is given by

$$A \equiv \left(\frac{2\hbar\omega}{\varepsilon V}\right)^{1/2} \tag{18.1-9}$$

With this definition, the constant A is physically the absolute magnitude of the mean electric field amplitude for a single photon with energy $\hbar\omega$ inside a space of volume V. The mean electric field amplitude is proportional to the square root of the mean photon number n.

If we define the measure of the fundamental quantum mechanical uncertainties as

$$\begin{aligned}\Delta E_1 &= \langle (E_1 - E_{10})^2\rangle^{1/2} = \langle (\delta E_1(t))^2\rangle^{1/2}\\ \Delta E_2 &= \langle (E_2 - E_{20})^2\rangle^{1/2} = \langle (\delta E_2(t))^2\rangle^{1/2}\end{aligned} \tag{18.1-10}$$

then according to quantum mechanics [2]

$$\Delta E_1\,\Delta E_2 \geq \frac{A^2}{4} \tag{18.1-11}$$

In other words, a simultaneous measurement of the in-phase (E_1) and quadrature (E_2) components of the electric field must obey the uncertainty principle. This is the only quantum mechanical result that we will use in this chapter.

The output field of most laser oscillators is in the *coherent state* [1, 2] in which the uncertainty is divided equally between the two quadrature components E_1 and E_2:

$$\Delta E_1 = \Delta E_2 = \frac{A}{2} \tag{18.1-12}$$

For mathematical simplicity, we define the normalized dimensionless field as

$$\begin{aligned}x_1 &\equiv E_1/A\\ x_2 &\equiv E_2/A\\ x &\equiv E/A = x_1 + ix_2\end{aligned} \tag{18.1-13}$$

Physically, the complex random number x is a measure of the photon number amplitude. In fact, it can be shown that $|\langle x\rangle|^2$ is the mean photon number, according to Equation (18.1-8). Using these dimensionless field amplitudes, we can rewrite the uncertainty relationship (18.1-11) as

$$\Delta x_1\,\Delta x_2 \geq \tfrac{1}{4} \tag{18.1-14a}$$

and for the *coherent state* field

$$\Delta x_1 = \Delta x_2 = \tfrac{1}{2} \tag{18.1-14b}$$

A profound consequence of the uncertainty relation, Equation (18.1-11), is that it applies even to a mode that, classically, is *not excited*, that is, $n = 0$. This *vacuum state*, which corresponds to (18.1-6) with $E_{10} = E_{20} = 0$, is illustrated by Figure 18.1c. Electromagnetic fields in which the uncertainties of the two quadrature components are unequal, that is, $\Delta x_1 \neq \Delta x_2$, are called *squeezed states*. Such states have been produced experimentally using nonlinear optical techniques [4, 5]. A squeezed electromagnetic field is illustrated in Figure 18.1e. For squeezed states, the uncertainty circle becomes an ellipse. A squeezed vacuum field is shown in Figure 18.1f.

Instead of representing the field in terms of the quadrature amplitudes E_1 and E_2, we can use the description of Figure 18.1d in which the field $E(t)$ is written as a vector sum of a mean phasor $\langle E\rangle$ and a random phasor $\delta E = E(t) - \langle E\rangle$. Let this random phasor be written

$$\delta E = \ell(t) = |\ell|e^{i\alpha} = \delta E_1(t) + i\delta E_2(t) \tag{18.1-15}$$

where the phase angle α of this fluctuation phasor is measured from $\langle E\rangle$. For a coherent field, the phase angle α is uniformly distributed (is equally likely to occur) between 0 and 2π. Thus we have

$$\langle|\ell|^2\rangle = \langle(\delta E_1(t))^2\rangle + \langle(\delta E_2(t))^2\rangle = 2\langle(\delta E_1(t))^2\rangle = \frac{A^2}{2} \tag{18.1-16}$$

Some of the consequences of the uncertainty relation (18.1-11) are explored in what follows.

The Energy of an Electromagnetic Mode

The energy of an electromagnetic field $E(t)$ is given classically by

$$U = \tfrac{1}{2}\varepsilon V(EE^*) \tag{18.1-17}$$

We choose, without loss of generality, the direction of the mean value $\langle E\rangle$ as the real axis so that the complex field amplitude can be written

$$E = L + |\ell|\cos\alpha + i|\ell|\sin\alpha \tag{18.1-18}$$

where we recall ℓ is the complex random phasor responsible for the statistical fluctuation, and L is the mean value of the complex field amplitude, $L \equiv \langle E\rangle$. Since we are treating E as a random variable, the energy of the electromagnetic field is also a random variable. Using Equation (18.1-18), the mean value of the field energy is written

$$\begin{aligned}\langle U\rangle &= \frac{\varepsilon V}{2}\langle L^2 + 2L|\ell|\cos\alpha + |\ell|^2\cos^2\alpha + |\ell|^2\sin^2\alpha\rangle \\ &= \frac{\varepsilon V}{2}\left(L^2 + \langle|\ell|^2\rangle\right) = \frac{\varepsilon V}{2}\left(A^2 n + \tfrac{1}{2}A^2\right) = \hbar\omega\left(n + \tfrac{1}{2}\right)\end{aligned} \tag{18.1-19}$$

where we used the definition of A in Equation (18.1-8) and $\langle \sin^2\alpha \rangle = \langle \cos^2\alpha \rangle = \frac{1}{2}$, as well as the fact that $\langle \cos\,\alpha \rangle = 0$, since α is distributed uniformly between 0 and 2π. Although Equation (18.1-19) is derived for coherent states, it is valid for a general quantum field.

It follows that in the case when the classical field is zero, that is, $\langle E \rangle$ and n, according to Equation (18.1-8), are zero, the mode energy is $\hbar\omega/2$. This is the *zero-point vibration energy of the mode*. It is one of the main consequences of quantum mechanics and it does *not* have a classical counterpart. The only assumption used in the derivation of the zero-point fluctuation is the uncertainty principle.

Uncertainty in Energy

As we indicated, the field energy is a random variable. The uncertainty (or variance) of the mode energy can be defined by

$$\langle (\Delta U)^2 \rangle = \langle U - \langle U \rangle^2 \rangle = \langle U^2 \rangle - \langle U \rangle^2 \tag{18.1-20}$$

Using Equation (18.1-17) for U, taking $\langle \cos^m\alpha \rangle = 0$ for m odd, and neglecting terms $O(\ell^4/L^4)$, we obtain

$$\Delta U \equiv \langle (\Delta U)^2 \rangle^{1/2} = \hbar\omega\sqrt{n} \tag{18.1-21}$$

or in terms of the number of photons, N, in the resonator

$$\Delta N = \frac{\Delta U}{\hbar\omega} = \sqrt{n} \equiv \sqrt{\langle N \rangle} \tag{18.1-22}$$

and

$$(\Delta N)^2 = \langle N \rangle = n \tag{18.1-23}$$

where $\Delta N \equiv \langle (N - n)^2 \rangle^{1/2}$.

The relation (18.1-23) between the mean square uncertainty in the number of photons N and the average number $\langle N \rangle$ is consistent with a Poissonian *probability distribution of the photon number* N [2, 3]

$$p(N, n) = \frac{n^N e^{-n}}{N!} \tag{18.1-24}$$

for the number of photons N in the resonator. For an electromagnetic field with an average number of photons n, the probability of finding N photons is given by $p(N, n)$.

Phase Uncertainty

Since E is a complex random variable, the phase β of the electric field is also a random variable. The uncertainty $\Delta\beta$ in the value of the field phase can be obtained as follows. We define a new set of random variables δF_1 and δF_2,

$$\begin{aligned} \delta F_1 &= \delta E_1 \cos\beta + \delta E_2 \sin\beta \\ \delta F_2 &= -\delta E_1 \sin\beta + \delta E_2 \cos\beta \end{aligned} \tag{18.1-25}$$

where β is the phase angle of the complex field E. With this definition, the random variable δF_1 is parallel to the mean field $\langle E \rangle$, whereas the random variable δF_2 is perpendicular to the mean field $\langle E \rangle$. Equations (18.1-25) can be viewed as a coordinate rotation. It can easily be shown that

$$\Delta F_1 \, \Delta F_2 = \Delta E_1 \, \Delta E_2 \geq \frac{A^2}{4} \tag{18.1-26}$$

The uncertainty ΔF_1 is related to the uncertainty of the mean field amplitude, and thus the photon number, whereas the uncertainty ΔF_2 is related to the phase angle uncertainty $\Delta\beta$. Thus we have

$$\begin{aligned} \Delta\beta &= \frac{\Delta F_2}{|\langle E\rangle|} \\ \Delta N &= \left(\frac{|\langle E\rangle + \Delta F_1|^2}{A^2}\right) - \left(\frac{|\langle E\rangle|^2}{A^2}\right) \cong 2\frac{|\langle E\rangle|\Delta F_1}{A^2} \end{aligned} \tag{18.1-27}$$

provided $\Delta F_1 \ll |\langle E\rangle|$ and $\Delta F_2 \ll |\langle E\rangle|$. In other words, (18.1-27) is valid when the mean photon number is much greater than 1 (or equivalently, $1 \ll n$). Combining Equations (18.1-26) and (18.2-27), we obtain

$$\Delta N \, \Delta\beta \geq \tfrac{1}{2} \tag{18.1-28}$$

where the equality holds for a coherent field. This is an important result and states the fundamental quantum mechanical limit on the simultaneous measurement of the phase (β) and excitation level N (number of photons) of a quantized electromagnetic field.

Fluctuation of Photoelectron Number

If an optical wave is incident on a perfect photodetector whose area is A_{rea}, then for each incident and absorbed photon, ideally, one photoelectron is emitted. The resulting current is thus

$$i = \frac{ec\varepsilon|E|^2 A_{\text{rea}}}{2\hbar\omega} \tag{18.1-29}$$

where e is the absolute value of the electronic charge, and the optical power incident on the detector is $c\varepsilon|E|^2 A_{\text{rea}}/2$. Since E is a complex random variable, the electric current is also a random variable. If we use the dimensionless field $x \equiv E/A$ (A is defined in Equations (18.1-13)), the expression for the current becomes

$$i = \frac{ecA_{\text{rea}}}{V}(x_1^2 + x_2^2) \cong \frac{ecA_{\text{rea}}}{V}(x_{10}^2 + x_{20}^2 + 2x_{10}\,\Delta x_1 + 2x_{20}\,\Delta x_2) = i_0 + \Delta i \tag{18.1-30}$$

where the random variables are written $x_1 = x_{10} + \Delta x_1$ and $x_2 = x_{20} + \Delta x_2$, and the mean current and the current fluctuation are written

$$i_0 = \frac{ecA_{\text{rea}}}{V}(x_{10}^2 + x_{20}^2) = \frac{ec}{L_R} n \tag{18.1-31}$$

$$\Delta i = 2\frac{ec}{L_R}(x_{10}\,\Delta x_1 + x_{20}\,\Delta x_2) \tag{18.1-32}$$

where we used $x_{10}^2 + x_{20}^2 = |\langle E\rangle|^2/A^2 = n$, and $L_R = V/A_{\text{rea}}$ is the length of quantization volume used in Equation (18.1-9).

The total number of photoelectrons generated during a time interval $\tau = L_R/c$ (corresponding to a detection bandwidth $B = 1/\tau$) is

$$N_e = \frac{i\tau}{e} = \frac{[i_0 + \Delta i(t)]\tau}{e} \tag{18.1-33}$$

The average number of photoelectrons is then given by

$$\langle N_e \rangle = \frac{i_0 \tau}{e} = n \tag{18.1-34}$$

which is exactly the mean photon number. The variance of the photoelectrons is given by

$$\begin{aligned}\langle \Delta N_e^2 \rangle &= \frac{\langle (\Delta i)^2 \rangle \tau^2}{e^2} = 4\langle (x_{10}\,\Delta x_1 + x_{20}\,\Delta x_2)^2 \rangle = 4\,[x_{10}^2 \langle (\Delta x_1)^2 \rangle + x_{20}^2 \langle (\Delta x_2)^2 \rangle] \\ &= 4(x_{10}^2 + x_{20}^2)\langle (\Delta x_1)^2 \rangle = n \end{aligned} \tag{18.1-35}$$

where we used $\langle \Delta x_1\, \Delta x_2 \rangle = 0$, Equation (18.1-18), and $\langle \Delta i \rangle = 0$ as well as

$$\langle (\Delta x_1)^2 \rangle = \langle (\Delta x_2)^2 \rangle \equiv \langle (\Delta x)^2 \rangle = \tfrac{1}{4} \tag{18.1-36}$$

We note that the above equation holds for a coherent state. It follows that

$$\langle \Delta N_e^2 \rangle = \langle N_e \rangle \tag{18.1-37}$$

for the photoelectrons generated by incoming optical radiation in a coherent state. This is in agreement with a Poisson distribution for the photoelectron number N_e. Poissonian statistics apply to the case where each event (electron emission in this case) is completely random and independent so that there exist no correlations between individual emission events. This is exactly the scenario shown in Sections 10.3 and 10.4, which leads to shot noise in the current spectrum

$$S_i(\nu) = \frac{\overline{i_N^2(\nu)}}{\Delta \nu} = 2ei_0 \tag{18.1-38}$$

where $S_i(\nu)$ is the spectral density of the photocurrent at the radio frequency ν. We have thus demonstrated that *the electronic shot noise in the photocurrent can be attributed to the quantum field fluctuations.*

Minimum Detectable Optical Power Increment

Most of the methods used to measure the power of an electromagnetic wave employ detectors that convert absorbed optical power to a proportional output current. In a perfect detector, which releases one electron into the external circuit for each absorbed photon, we have

$$i = \frac{Pe}{\hbar\omega} \tag{18.1-39}$$

where P is the optical power to be measured. Let our measurement of the current consist of accumulating it for T seconds, which results in a total number of collected electrons (or holes)

$$N_e(T) = \frac{iT}{e} = \frac{PT}{\hbar\omega} \quad \Rightarrow \quad P = \frac{\hbar\omega}{T} N_e(T) \tag{18.1-40}$$

The mean squared uncertainty in the power measurement is thus given by

$$\langle (\Delta P)^2 \rangle = \left(\frac{\hbar\omega}{T}\right)^2 \langle (\Delta N_e)^2 \rangle = \left(\frac{\hbar\omega}{T}\right)^2 \langle N_e(T) \rangle \tag{18.1-41}$$

where, in the last equality, we used Equation (18.1-37) by assuming a coherent state for the incoming radiation. The average number of collected electrons during the observation interval T is given by

$$\langle N_e(T)\rangle = \frac{\langle P\rangle}{\hbar\omega}T \tag{18.1-42}$$

so that the mean square uncertainty in the measurement of power is

$$\langle(\Delta P)^2\rangle = \frac{\hbar\omega}{T}\langle P\rangle \tag{18.1-43}$$

Defining, arbitrarily, the minimum detectable power as that power at which the root mean squared fluctuation is equal to the average, that is,

$$\langle P\rangle_{\text{min}} = \langle(\Delta P)^2\rangle^{1/2} \tag{18.1-44}$$

we obtain

$$P_{\text{min}} = \hbar\omega B = \hbar\omega\,\Delta\nu \tag{18.1-45}$$

$B = \Delta\nu = 1/T$ is the bandwidth of the current integrating system. One often refers to the quantity $\hbar\omega B$ as the minimum detectable power. It is treated in some detail in Section 11.4. The presence of $\hbar$ indicates that the power uncertainty $\langle(\Delta P)^2\rangle^{1/2}$ can be considered as quantum noise.

18.2 SQUEEZING OF OPTICAL FIELDS

It is possible to take a coherent optical field, such as the output of a laser, and reduce the fluctuation of one of its quadrature components, say, ΔE_1, at the expense of ΔE_2, or vice versa. The resulting uncertainty diagram becomes elliptical, while the product $\Delta E_1\,\Delta E_2$ retains its initial value of $A^2/4$. This is referred to as *squeezing* [4, 5]. Squeezing is accomplished, usually, by a nonlinear optical operation on the field. One of the most common methods of achieving squeezing employs degenerate optical parametric amplification, which is discussed in Section 8.4. In this case (the degenerate case), the "pump" frequency is twice that of the "signal," $\omega_{\text{pump}} = 2\omega_{\text{signal}}$. To demonstrate how squeezing is accomplished in this situation, we will need, first, to revisit the topic of parametric amplification.

Optical parametric amplification (OPA) is described by Equations (8.4-10). In the case of a degenerate phase-matched parametric amplifier, we have

$$\omega_1 = \omega_2 \equiv \omega \quad \text{and} \quad \omega_3 = 2\omega$$

If we designate the complex amplitude of the field at frequency ω as E, the parametric amplifier equations become, according to Equations (8.4-10) and (8.4-11),

$$\frac{dE}{dz} = -i\frac{g}{2}E^* \tag{18.2-1}$$

with a parametric amplification coefficient given by

$$g = \frac{\omega d}{n_0}\sqrt{\mu/\varepsilon_0}\,E_3 \tag{18.2-2}$$

where d is the effective nonlinear susceptibility, E_3 is the electric field amplitude of the pump wave (at frequency 2ω), and n_0 is the index of refraction. The coupling constant g is complex, since it is proportional to the complex pump amplitude E_3. Without loss of generality, we can take the pump field at $z = 0$ as

$$E_3(t) = -|E_3| \sin 2\omega t = \frac{i|E_3|}{2}(e^{i2\omega t} - e^{-i2\omega t}) = \mathrm{Re}(E_3 e^{i2\omega t}) \tag{18.2-3}$$

This choice determines the time reference. In this case, $E_3 = +i|E_3|$, the coupling constant g is imaginary, and Equation (18.2-1) assumes the form

$$\frac{dE}{dz} = \tfrac{1}{2}|g|E^* \tag{18.2-4}$$

where

$$|g| = \frac{\omega d}{n_0}\sqrt{\mu/\varepsilon_0}\,|E_3|$$

It is convenient to express the "signal" field at ω as in Equation (18.1-2) in terms of its quadrature amplitudes, E_1 and E_2,

$$E = (E_1 + iE_2), \quad E_1 = \tfrac{1}{2}(E + E^*), \quad E_2 = \frac{-i}{2}(E - E^*) \tag{18.2-5}$$

so that the time-dependent ("signal") field is given by

$$e(t) = \mathrm{Re}[E \exp(i\omega t)] = \mathrm{Re}[(E_1 + iE_2)\exp(i\omega t)] = E_1 \cos \omega t - E_2 \sin \omega t \tag{18.2-6}$$

If we substitute the first of Equations (18.2-5) in (18.2-4), we obtain

$$\begin{aligned} \frac{dE_1}{dz} &= \frac{|g|}{2}E_1 \\ \frac{dE_2}{dz} &= -\frac{|g|}{2}E_2 \end{aligned} \tag{18.2-7}$$

so that at output of the parametric amplifier $z = L$,

$$\begin{aligned} E_1(L) &= E_1(0)\exp\left(\frac{|g|}{2}L\right) = E_1(0)s \\ E_2(L) &= E_2(0)\exp\left(-\frac{|g|}{2}L\right) = \frac{E_1(0)}{s} \end{aligned} \tag{18.2-8}$$

where the squeezing factor s is defined, in accordance with Figure 18.1e, by

$$s = \exp(|g|L/2)$$

Here in this section L is the length of interaction in the parametric amplifier. Degenerate parametric amplification is thus seen to lead to amplification of one quadrature component (E_1) and to the attenuation of the other component (E_2). The choice of which component is amplified is determined by the phase of the pump E_3. This is illustrated by Figure 18.2. If we now express the field amplitudes as in (18.1-3), including their quasi-quantum mechanical fluctuations, the last two equations become

$$\begin{aligned} E_1(L, t) &= [E_{10}(0) + \delta E_1(0, t)]\exp\left(\frac{|g|L}{2}\right) \\ E_2(L, t) &= [E_{20}(0) + \delta E_2(0, t)]\exp\left(-\frac{|g|L}{2}\right) \end{aligned} \tag{18.2-9}$$

where δE_1 and δE_2 represent the quantum fluctuations of the input field. The mean fields E_{10} and E_{20} *as well as the fluctuations* $\delta E_1(t)$ and $\delta E_2(t)$ are thus found to be amplified (attenuated)

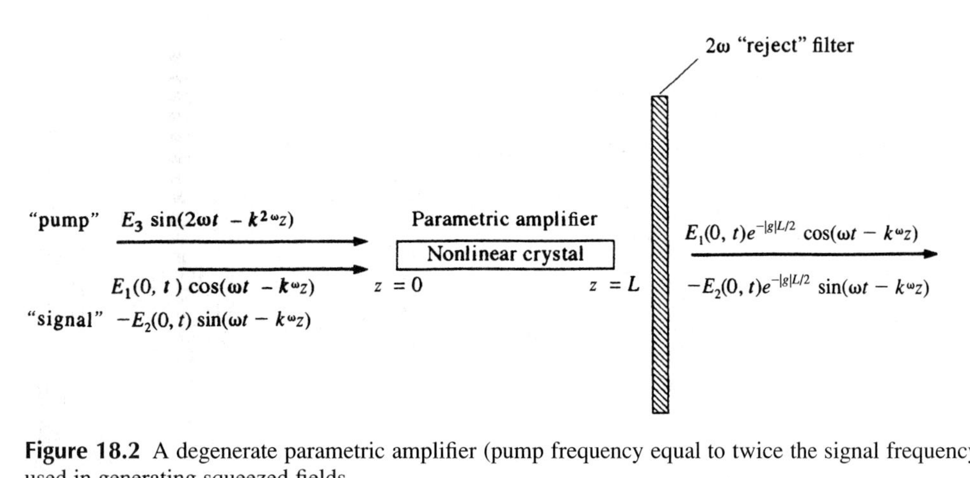

Figure 18.2 A degenerate parametric amplifier (pump frequency equal to twice the signal frequency) used in generating squeezed fields.

by the nonlinear parametric interaction. The standard deviations of the output ($z = L$) fluctuations are

$$\Delta E_1(L) \equiv \langle(\delta E_1(L, t))^2\rangle^{1/2} = \Delta E_1(0)\exp\left(\frac{|g|L}{2}\right) = \frac{A}{2}\exp\left(\frac{|g|L}{2}\right)$$
$$\Delta E_2(L) \equiv \langle(\delta E_2(L, t))^2\rangle^{1/2} = \Delta E_2(0)\exp\left(-\frac{|g|L}{2}\right) = \frac{A}{2}\exp\left(-\frac{|g|L}{2}\right) \tag{18.2-10}$$

where A is given by Equation (18.1-9).

The uncertainty product

$$\Delta E_1(L)\,\Delta E_2(L) = \Delta E_1(0)\,\Delta E_2(0) = \frac{A^2}{4} \tag{18.2-11}$$

remains unchanged, although the uncertainty area is now elliptical rather than circular. This is illustrated in Figure 18.3a for the case of $\exp(|g|L/2) = 2$.

The case of a parametric amplifier with no input is particularly interesting. Classically, we expect no output. Quantum mechanically, however, there exists an input field, the so-called zero-point fluctuation or *vacuum field*, represented by the origin-centered circle of Figure 18.1c. This field can be represented classically by Figure 18.1a with $\langle E_1\rangle = \langle E_2\rangle = 0$. We write the input field (at $z = 0$) as

$$e_{\text{in}}(t) = \text{Re}\{[\delta E_1(0, t) + i\delta E_2(0, t)]\exp(i\omega t)\} = \delta E_1(0, t)\cos\omega t - \delta E_2(0, t)\sin\omega t$$
$$\langle\delta E_1(0, t)\rangle = \langle\delta E_2(0, t)\rangle = 0 \tag{18.2-12}$$
$$\langle(\delta E_1(0, t))^2\rangle = \langle(\delta E_2(0, t))^2\rangle = \frac{A^2}{4}$$

The resulting output is given by (18.2-9) with $E_{10}(0) = E_{20}(0) = 0$ and is

$$E_1(L, t) \equiv \delta E_1(0, t)\exp\left(\frac{|g|L}{2}\right)$$
$$E_2(L, t) \equiv \delta E_2(0, t)\exp\left(-\frac{|g|L}{2}\right) \tag{18.2-13}$$

so that the output field

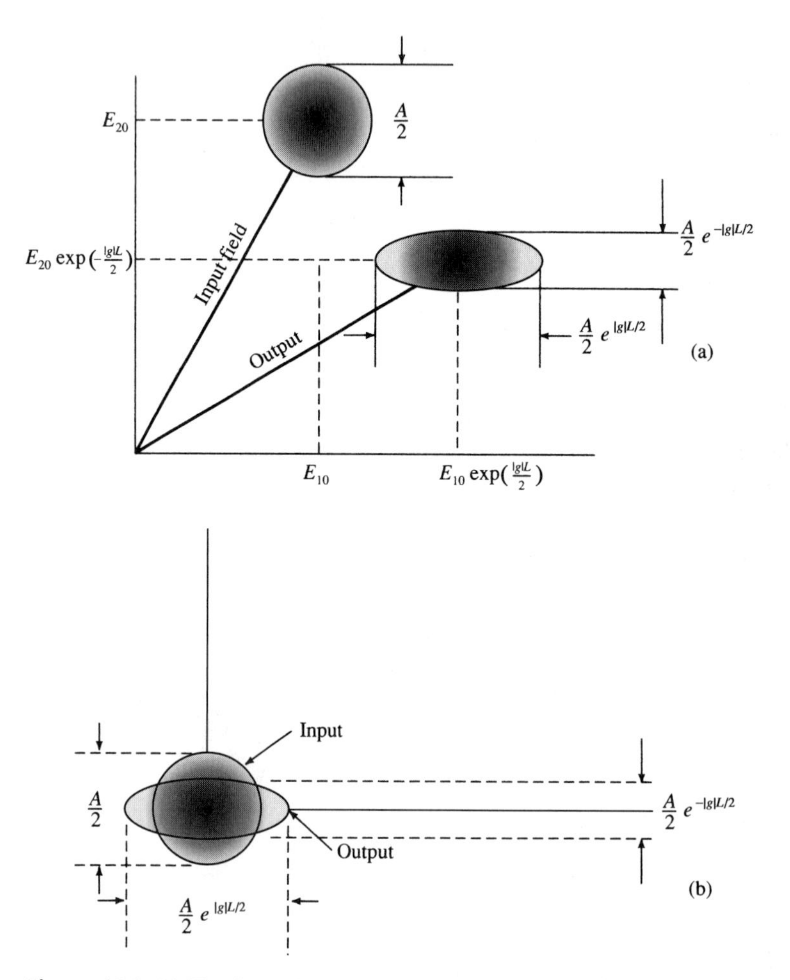

Figure 18.3 (a) The input field to a degenerate parametric amplifier shown in Figure 18.2 and the squeezed output fields for the case $\exp(|g|L/2) = 2$ (6 dB squeezing). (b) Same as (a) except that the input field is zero.

$$e_{\text{out}}(t) = \delta E_1(0, t) \exp\left(\frac{|g|L}{2}\right) \cos \omega t - \delta E_2(0, t) \exp\left(-\frac{|g|L}{2}\right) \sin \omega t$$

corresponding to a field with a zero mean but with squeezed vacuum fluctuations—the so-called *squeezed vacuum.* The input (circle) and output (ellipse) fluctuations in this case are depicted by Figure 18.3b for a parametric gain $\exp(|g|L) = 4$.

Experimental Demonstration of Squeezing

Referring to Figure 18.4, we consider an experimental setup that is used often to demonstrate squeezing [5]. A nonlinear crystal is employed as a parametric amplifier, which amplifies an incoming electromagnetic wave $E \exp(i\omega t)$. An optical filter located at the end of the parametric amplifier is employed to block the pump beam. The output of the optical parametric amplifier at ω is then combined in a balanced homodyne receiver with the strong local oscillator, also of frequency ω, which is *coherent* with that of the pump field at 2ω. (Usually these two fields are derived from the same master laser oscillator at ω.) The two combined fields, whose

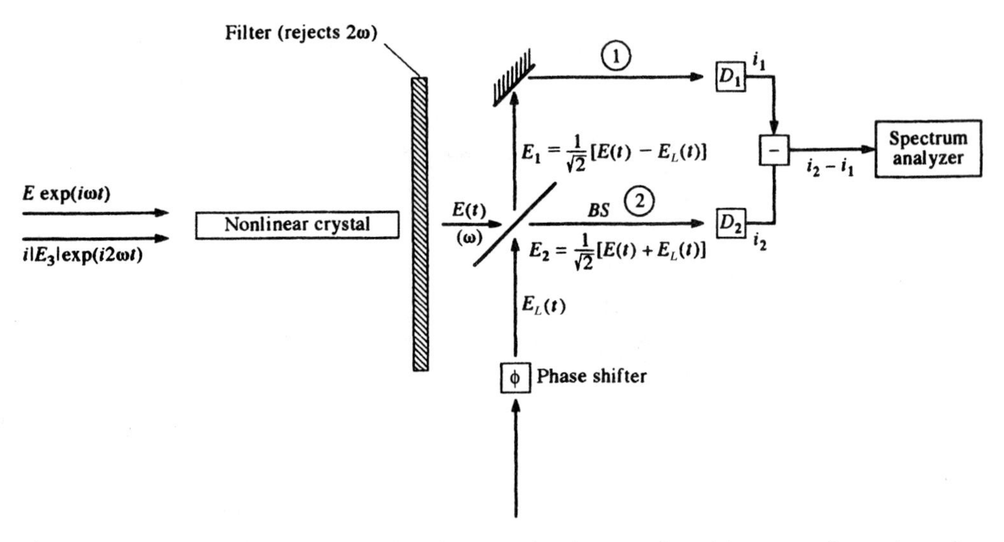

Figure 18.4 A balanced homodyne receiver for measuring the squeezing of the vacuum fluctuations of an electromagnetic field. The squeezing is caused by degenerate parametric amplification. Only the complex amplitudes are marked in the homodyne receiver section since the factor $\exp(i\omega t)$ has been dropped.

complex amplitudes are $E_1 = 1/\sqrt{2}[E(t) - E_L(t)]$ and $E_2 = 1/\sqrt{2}[E(t) + E_L(t)]$, are detected, respectively, by photodetectors D_1 and D_2. Here $E_L(t)$ is the electric field of the local oscillator. The resulting currents i_1 and i_2 are then subtracted from each other in the electronic circuit. The net current $i_2 - i_1$ is fed to a spectrum analyzer that displays the spectral density of $(i_2 - i_1)$, a quantity that according to Equations (10.2-5) and (10.2-7) is proportional to $\langle (i_2 - i_1)^2 \rangle$. The detected photocurrents are given, according to Equation (11.1-2), by

$$
\begin{aligned}
i_1 &= KE_1(t)E_1^*(t) = \frac{K}{2}[E(t) - E_L(t)][E^*(t) - E_L^*(t)] \\
i_2 &= \frac{K}{2}[E(t) + E_L(t)][E^*(t) + E_L^*(t)]
\end{aligned}
\tag{18.2-14}
$$

where K is a detector constant. The result of the electronic subtraction is thus

$$
i_2 - i_1 = K[E_L^*(t)E(t) + \text{c.c.}] \tag{18.2-15}
$$

This result illustrates the raison d'être for the balanced homodyne receiver. Since the output $(i_2 - i_1)$ contains only mixed ("signal" × local oscillator) product terms, fluctuations $\Delta E_L(t)$ of the local oscillator field, which lead to "large" terms $E_L\,\Delta E_L^*(t)$ in the photocurrents i_1 and i_2, cancel out in the current subtraction. These terms would, in the case of a single detector receiver, mask the signal term $E_L^*E(t)$.

In most of the experiments [4, 5] demonstrating squeezing, there exists no input to the parametric amplifier. In this case, the output of the parametric amplifier is given by Equations (18.2-9) with $E_{10}(0) = E_{20}(0) = 0$:

$$
E(t) = \delta E_1(0, t) \exp\left(\frac{|g|L}{2}\right) + i\delta E_2(0, t) \exp\left(-\frac{|g|L}{2}\right) \tag{18.2-16}
$$

that is, the squeezed vacuum field. The complex amplitude of the local oscillator field at the beam splitter is taken as the sum of the average field plus a fluctuation term. The fluctuation may be due to basic quantum causes or any other cause. A phase factor $\exp(i\phi)$ accounts for the phase shifter

$$E_L(t) = [E_{L0} + \delta E_L(t)] \exp(i\phi) \tag{18.2-17}$$

E_{L0} can, without any loss of generality, be taken as a real number. Substituting the last two equations for $E_L(t)$ and $E(t)$ in Equation (18.2-15) and neglecting the terms involving $\delta E_L(t)$, since $\delta E_L(t) \ll E_{L0}$, results in

$$(i_2 - i_1) = 2KE_{L0}\left[\delta E_1(0, t) \exp\left(\frac{|g|L}{2}\right) \cos\phi - i\delta E_2(0, t) \exp\left(-\frac{|g|L}{2}\right) \sin\phi\right] \tag{18.2-18}$$

Since both $\langle \delta E_1(t) \rangle$ and $\langle \delta E_2(t) \rangle$ are zero, the time-averaged $(i_2 - i_1)$, the quantity that normally will be registered by a sensitive ammeter, is zero. This problem is avoided by squaring $(i_2 - i_1)$. This is accomplished usually [5] by the spectrum analyzer, which displays the spectral density $S_f(\Omega)$ of the input $f(t)$, where

$$\langle f^2(t) \rangle = \int_0^\infty S_f(\Omega)\, d\Omega$$

as discussed in Section 10.2. In our case, the input is $f(t) = i_1(t) - i_2(t)$, so that the output of the spectrum analyzer is proportional, in the case of a constant $S_{(i_2 - i_1)}(\Omega)$, to

$$\begin{aligned} S_{(i_2-i_1)}(\Omega) \propto \langle (i_2 - i_1)^2 \rangle &= 4K^2E_{L0}^2[\langle(\delta E_1(0, t))^2\rangle \exp(|g|L) \cos^2\phi + \langle(\delta E_1(0, t))^2\rangle \exp(-|g|L) \sin^2\phi] \\ &= K^2E_{L0}^2A^2[\exp(|g|L) \cos^2\phi + \exp(-|g|L) \sin^2\phi] \end{aligned} \tag{18.2-19}$$

where we used (18.1-6) and $\langle \delta E_1(t)\delta E_2(t) \rangle = 0$.

A typical result of such an experiment is shown in Figure 18.5. The observed dependence of the photocurrent fluctuations on the phase ϕ of the local oscillator field is in agreement with Equation (18.2-19) and constitutes a dramatic verification of squeezing of the vacuum fluctuations. The dashed horizontal line is the result when the optical parametric amplifier is blocked.

It is instructive to view the squeezed states by plotting, in Figure 18.6, the actual sinusoidal optical fields corresponding to six representative points inside the uncertainty ellipse of

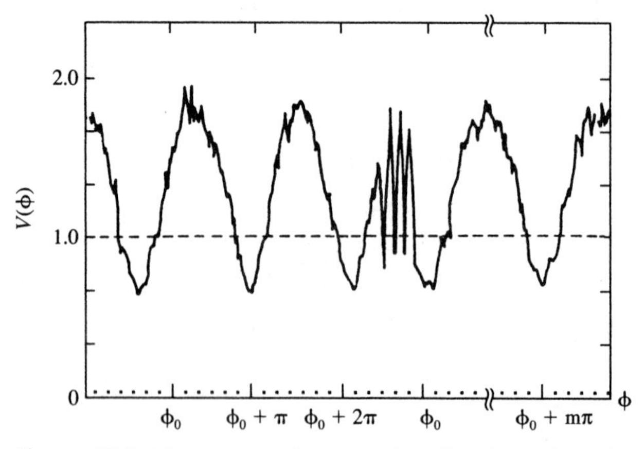

Figure 18.5 Measurement demonstrating the phase dependence of the quantum fluctuations in a squeezed state. The squeezing is achieved by degenerate parametric amplification. The phase dependence of the rms voltage from a balanced homodyne receiver is displayed versus the local oscillator phase ϕ. The noise voltage is centered on $\nu = 1.8$ MHz. With the parametric amplifier blocked, $|g| = 0$, the output is given by the dashed horizontal line with no ϕ dependence. The dips represent 50% of the electronic noise power relative to that of unsqueezed vacuum (i.e., $|g| = 0$) input [5].

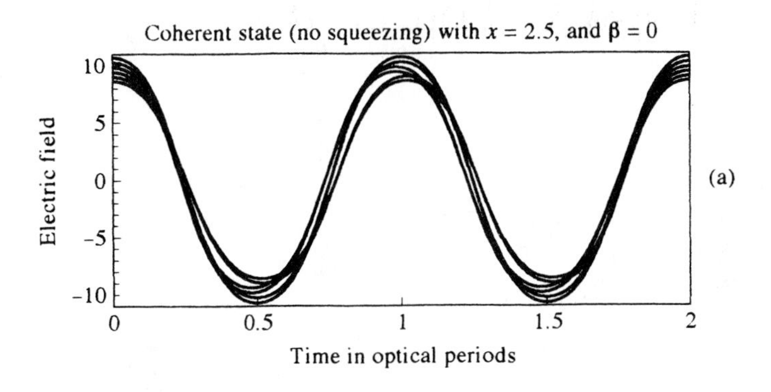

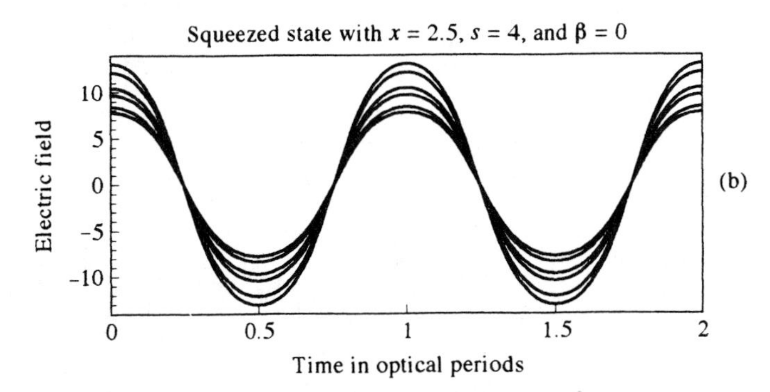

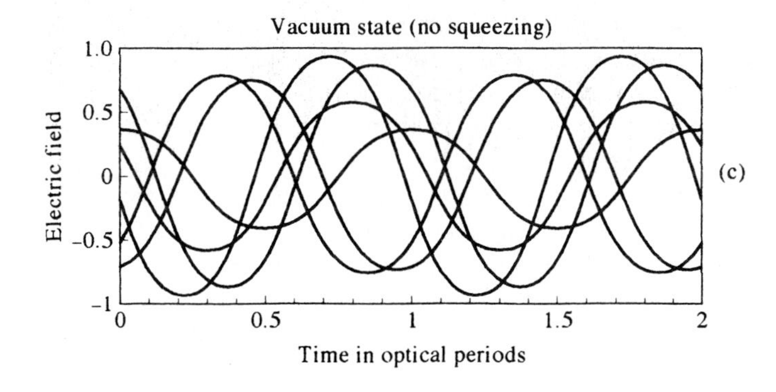

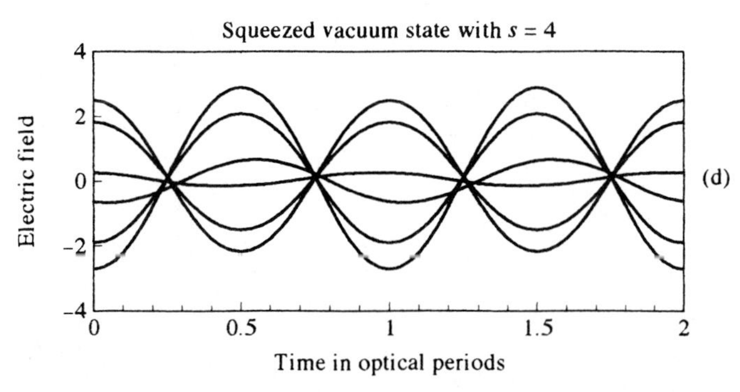

Figure 18.6 Representation of (a) unsqueezed electric field, (b) squeezed ($s = 4$) state, (c) unsqueezed "vacuum" field, and (d) squeezed ($s = 4$) vacuum field.

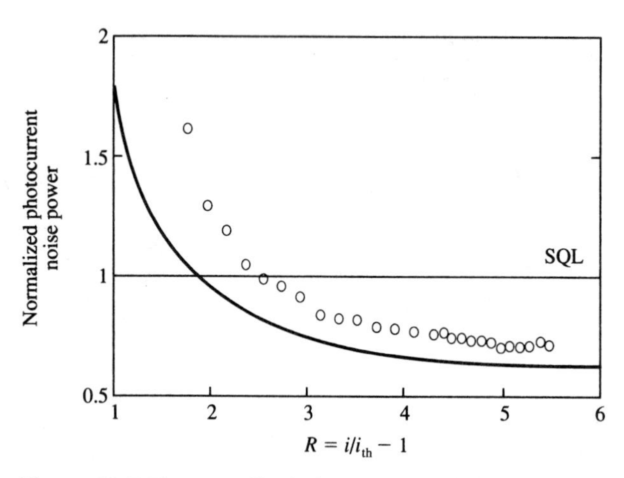

Figure 18.7 The normalized photocurrent noise spectral density at $\nu = 29$ MHz as a function of the injection current to a semiconductor laser. i_{th} is the oscillation threshold current. SQL—the standard quantum limit—is the level corresponding to shot noise. (After Reference [7].)

Figure 18.1e. We recall that each such point represents a possible realization of the field (complex) amplitude. The case of no squeezing ($s = 0$) is shown in Figure 18.6a, while a squeezed state with $x = 2.5$, $s = 4$, $\beta = 0$ is shown in Figure 18.6b. We note that in the squeezed state we trade an increase in the accuracy of measuring the frequency (or phase) for an increased amplitude fluctuation in qualitative agreement with Equation (18.1-15). The vacuum state ($\langle E \rangle = 0$) is shown in Figure 18.6c, which contains plots of representative points from Figure 18.1c, while the squeezed vacuum (Figure 18.1f) is shown in Figure 18.6d.

Another type of squeezing, "number squeezing," which results in a photocurrent noise level below that of shot noise, can be exhibited in semiconductor diode lasers [6]. According to Equation (18.1-28), the photon number uncertainty and the phase uncertainty are related by $\Delta N\, \Delta\beta = \frac{1}{2}$. A squeezing in the photon number uncertainty will lead to a bigger uncertainty in the phase angle. For photodetectors that are phase insensitive, the number squeezing leads to a lower noise. This squeezing results when the injection current to the lasers is highly constant and/or when proper frequency-dependent feedback is employed [7]. Such lasers may find practical uses in communications [8] and atomic measurements [9], for example. Experimental data showing such squeezing is shown in Figure 18.7.

In conclusion, it is worthwhile to remind ourselves that the classical treatment of this chapter seems to do a good job in representing the results of the rigorous quantum treatment—but only up to second-order electric field products. If we were to ask some more difficult questions, say, those involving expectation values of the electric field raised to the third power or higher, the classical approach fails.

PROBLEMS

18.1

(a) Compare $|\langle E\rangle|^2$ and $\langle |E|^2\rangle$, where E is given by Equations (18.1-6) and show that

$$\langle |E|^2\rangle = E_{10}^2 + E_{20}^2 + \langle(\delta E_1)^2\rangle + \langle(\delta E_2)^2\rangle$$
$$= |\langle E\rangle|^2 + \langle(\delta E_1)^2\rangle + \langle(\delta E_2)^2\rangle$$

For random variables with positive variances, $|\langle E\rangle|^2 < \langle |E|^2\rangle$.

(b) Using the phase angle β, we can write

$$E_1 = |E|\cos\beta \quad \text{and} \quad E_2 = |E|\sin\beta$$

Show that for an electromagnetic field with a well-defined photon number (i.e., $\Delta N = 0$)

$$E_{10} = E_{20} = 0$$

Quantum electromagnetic fields with a well-defined photon number are known as number states. The phases of these states are completely undetermined. For these states, Equation (18.1-7) is not appropriate.

18.2 In quantum optics, the Hamiltonian of an electromagnetic field is written

$$H = \hbar\omega(a^\dagger a + \tfrac{1}{2})$$

where $a^\dagger$ and a are creation operator and annihilation operator, respectively. These operators satisfy the following relations:

$$a^\dagger|n\rangle = \sqrt{n+1}\,|n+1\rangle \quad \text{and} \quad a|n\rangle = \sqrt{n}\,|n-1\rangle$$

where the $|n\rangle$ are eigenstates of the Hamiltonian. These eigenstates of the Hamiltonian are known as the number states, which have well-defined photon numbers. The energy of an electromagnetic field is exactly like that of a quantum harmonic oscillator. In other words,

$$H|n\rangle = E_n|n\rangle = \hbar\omega(n + \tfrac{1}{2})|n\rangle$$

A coherent state $|\alpha\rangle$ is defined as an eigenstate of the annihilation operator. In other words,

$$a|\alpha\rangle = \alpha|\alpha\rangle$$

where α is an eigenvalue. Such a state can be expressed in terms of a linear combination of the number states,

$$|\alpha\rangle = \sum_n c_n|n\rangle$$

(a) Show that

$$c_n = c_0\frac{\alpha^n}{\sqrt{n!}}, \quad \text{where} \quad |c_0|^2 = \exp(-|\alpha|^2)$$

In other words, a normalized coherent state can be written

$$|\alpha\rangle = \sum_n c_n|n\rangle = \sum_n c_0\frac{\alpha^n}{\sqrt{n!}}|n\rangle = e^{-|\alpha|^2/2}\sum_n\frac{\alpha^n}{\sqrt{n!}}|n\rangle$$

It is interesting to note that the coherent state with $\alpha = 0$ is the vacuum state, that is,

$$|\alpha = 0\rangle = |0\rangle$$

(b) The number operator can be written $\mathbf{N} = a^\dagger a$. Show that

$$\mathbf{N}|n\rangle = a^\dagger a|n\rangle = n|n\rangle$$
$$\langle N\rangle \equiv \langle\alpha|\mathbf{N}|\alpha\rangle = |\alpha|^2$$
$$\langle\alpha|a|\alpha\rangle = \alpha,\ \langle\alpha|a^\dagger|\alpha\rangle = \alpha^*$$
$$\langle\alpha|a^2|\alpha\rangle = \alpha^2,\ \langle\alpha|a^{\dagger 2}|\alpha\rangle = \alpha^{*2}$$
$$\langle\alpha|a^\dagger a|\alpha\rangle = \langle\alpha|aa^\dagger|\alpha\rangle - 1 = |\alpha|^2$$

(c) The probability of finding N photons in a coherent state is

$$p(N) = |c_N|^2 = |\langle N|\alpha\rangle| = e^{-|\alpha|^2}\frac{|\alpha|^{2n}}{N!}$$

Show that

$$\langle N\rangle \equiv \sum_n np_n = |\alpha|^2$$

$$(\Delta N)^2 \equiv \langle(N - \langle N\rangle)^2\rangle \equiv \sum_n (n - \langle N\rangle)^2 p_n = |\alpha|^2 = \langle N\rangle$$

18.3 We quantize the field and write the electric field as

$$e(t) = \mathrm{Re}(Ee^{i\omega t}) = E_1\cos(\omega t) - E_2\sin(\omega t) = \frac{A}{2}(a^\dagger e^{i\omega t} + ae^{-i\omega t})$$

where $a^\dagger$ and a are creation operator and annihilation operator, respectively.

(a) Show that E_1 and E_2 are now written

$$E_1 = \frac{A}{2}(a^\dagger + a) \quad \text{and} \quad E_2 = \frac{A}{2i}(a^\dagger - a)$$

and

$$\langle\alpha|E_1|\alpha\rangle = \frac{A}{2}(\alpha + \alpha^*) \quad \text{and} \quad \langle\alpha|E_2|\alpha\rangle = \frac{A}{2i}(\alpha^* - \alpha)$$

The eigenvalue α is related to the mean field $\langle E\rangle$. Find the relationship.

(b) Show that

$$\langle\alpha|E_1^2|\alpha\rangle = \frac{A^2}{4}\langle\alpha|(a^{\dagger 2} + a^2 + a^\dagger a + aa^\dagger)|\alpha\rangle$$

$$= \frac{A^2}{4}(\alpha^2 + \alpha^{*2} + 2\alpha\alpha^* + 1)$$

$$\langle\alpha|E_2^2|\alpha\rangle = \frac{A^2}{4}\langle\alpha|(-a^{\dagger 2} - a^2 + a^\dagger a + aa^\dagger)|\alpha\rangle$$

$$= \frac{A^2}{4}(-\alpha^2 - \alpha^{*2} + 2\alpha\alpha^* + 1)$$

$$(\Delta E_1)^2 = \langle\alpha|E_1^2|\alpha\rangle - \langle\alpha|E_1|\alpha\rangle^2 = \frac{A^2}{4}$$

$$(\Delta E_2)^2 = \langle\alpha|E_2^2|\alpha\rangle - \langle\alpha|E_2|\alpha\rangle^2 = \frac{A^2}{4}$$

This proves Equation (18.1-12) for the coherent states.

18.4 Two coherent states with different eigenvalues α and α' are only approximately orthogonal. Show that

$$|\langle\alpha|\alpha'\rangle|^2 = \exp(-|\alpha - \alpha'|^2)$$

18.5 For a monochromatic field with a well-defined frequency ω, an uncertainty in the phase angle $\Delta\beta$ can be translated into an uncertainty in time. In other words,

$$\Delta\beta = \omega\,\Delta t$$

where Δt is an uncertainty in time. Show that $\Delta N\,\Delta\beta \geq \frac{1}{2}$ is a direct consequence of the following uncertainty:

$$\Delta U\,\Delta t \geq \hbar/2$$

where U is the energy of the field.

REFERENCES

1. See, for example, A. Yariv, *Quantum Electronics*, 4th ed. Wiley, New York, 1988, p. 13.
2. Glauber, R. J., Coherent and incoherent states of radiation fields. *Phys. Rev.* **131**:2776 (1963).
3. Louisell, W. H., *Radiation and Noise in Quantum Electronics*. McGraw-Hill, New York, 1964.
4. Yuen, H. P., and J. H. Shapiro, Generation and detection of two-photon coherent states in degenerate four wave mixing. *Opt. Lett.* **4**:334 (1979).
5. Wu, L., J. H. Kimble, J. L. Hall, and H. Wu, Generation of squeezed states by parametric down conversion. *Phys. Rev. Lett.* **57**:2520 (1986).
6. Yamamoto, Y., S. Machida, and O. Nilsson, Amplitude squeezing in a pump-noise-suppressed laser oscillator. *Phys. Rev. A* **34**:4025 (1986).
7. Kitching, J., A. Yariv, and Y. Shevy, Room temperature generation of squeezed light from a semiconductor laser with weak optical feedback. *Phys. Rev. Lett.* **74**:3372 (1995).
8. Saleh, B. E. A., and M. C. Teich, Information transmission with number-squeezed light. *Proc. IEEE* **80**:451 (1992). Also by the same authors, *Fundamentals of Photonics*, Wiley, New York, 1991, pp. 414–416.
9. Wieman, C. E., and L. Hollberg, Using diode lasers for atomic physics. *Rev. Sci. Instrum.* **62**:1 (1991).

APPENDIX A

WAVE EQUATION IN CYLINDRICAL COORDINATES AND BESSEL FUNCTIONS

Here we present the wave equation in cylindrical coordinates and solutions of the Bessel functions. In cylindrical coordinates (r, ϕ, z), the vector differential operators and the Laplacian operator take the following forms:

$$\Delta\psi = \mathbf{u}_r \frac{\partial\psi}{\partial r} + \mathbf{u}_\phi \frac{1}{r}\frac{\partial\psi}{\partial\phi} + \mathbf{u}_z \frac{\partial\psi}{\partial z}$$

$$\nabla \cdot \mathbf{A} = \frac{1}{r}\frac{\partial}{\partial r}(rA_r) + \frac{1}{r}\frac{\partial}{\partial\phi}A_\phi + \frac{\partial}{\partial z}A_z$$

$$\nabla \times \mathbf{A} = \mathbf{u}_r\left(\frac{1}{r}\frac{\partial}{\partial\phi}A_z - \frac{\partial}{\partial z}A_\phi\right) + \mathbf{u}_\phi\left(\frac{\partial}{\partial z}A_r - \frac{\partial}{\partial r}A_z\right) + \mathbf{u}_z\frac{1}{r}\left(\frac{\partial}{\partial r}(rA_\phi) - \frac{\partial}{\partial\phi}A_r\right)$$

$$\nabla^2\psi = \frac{1}{r}\frac{\partial}{\partial r}\left(r\frac{\partial\psi}{\partial r}\right) + \frac{1}{r^2}\frac{\partial^2\psi}{\partial\phi^2} + \frac{\partial^2\psi}{\partial z^2}$$

where ψ is a scalar function and $\mathbf{A}$ is a vector function.

Since the refractive index profiles $n(r)$ of most optical fibers are cylindrically symmetric, it is convenient to use the cylindrical coordinate system. The field components are E_r, E_ϕ, E_z, H_r, H_ϕ, and H_z. Since the unit vectors $\mathbf{u}_r$ and $\mathbf{u}_\phi$ are not constant vectors, the wave equations involving the transverse components (E_r, E_ϕ, H_r, H_ϕ) are very complicated. The wave equation for the z component of the field vectors, however, remains simple,

$$(\nabla^2 + k^2)\begin{bmatrix} E_z \\ H_z \end{bmatrix} = 0 \tag{A-1}$$

where $k^2 = \omega^2 n^2/c^2$ and ∇^2 is the Laplacian operator given by

$$\nabla^2 = \frac{\partial^2}{\partial r^2} + \frac{1}{r}\frac{\partial}{\partial r} + \frac{1}{r^2}\frac{\partial^2}{\partial\phi^2} + \frac{\partial^2}{\partial z^2}$$

The problems of wave propagation in a cylindrical structure are usually approached by solving for E_z and H_z first and then expressing E_r, E_ϕ, H_r, and H_ϕ in terms of E_z and H_z. Since we are concerned with propagation along the axis of the waveguide (z axis), we assume

$$\begin{bmatrix} \mathbf{E}(\mathbf{r}, t) \\ \mathbf{H}(\mathbf{r}, t) \end{bmatrix} = \begin{bmatrix} \mathbf{E}(r, \phi) \\ \mathbf{H}(r, \phi) \end{bmatrix} \exp[i(\omega t - \beta z)] \tag{A-2}$$

that is, every component of the field vector assumes the same z and t dependence of $\exp[i(\omega t - \beta z)]$. Maxwell's curl equations are now written in terms of the cylindrical components and are given by

$$i\omega\varepsilon E_r = i\beta H_\phi + \frac{1}{r}\frac{\partial}{\partial\phi} H_z \tag{A-3a}$$

$$i\omega\varepsilon E_\phi = -i\beta H_r - \frac{\partial}{\partial r} H_z \tag{A-3b}$$

$$i\omega\varepsilon E_z = -\frac{1}{r}\frac{\partial}{\partial\phi} H_r + \frac{1}{r}\frac{\partial}{\partial r}(rH_\phi) \tag{A-3c}$$

and

$$-i\omega\mu H_r = i\beta E_\phi + \frac{1}{r}\frac{\partial}{\partial\phi} E_z \tag{A-4a}$$

$$-i\omega\mu H_\phi = -i\beta E_r - \frac{\partial}{\partial r} E_z \tag{A-4b}$$

$$-i\omega\mu H_z = -\frac{1}{r}\frac{\partial}{\partial\phi} E_r + \frac{1}{r}\frac{\partial}{\partial r}(rE_\phi) \tag{A-4c}$$

Using Equations (A-3a), (A-3b), (A-4a), and (A-4b), we can solve for E_r, E_ϕ, H_r, and H_ϕ in terms of E_z and H_z. The results are

$$\begin{aligned} E_r &= \frac{-i\beta}{\omega^2\mu\varepsilon - \beta^2}\left(\frac{\partial}{\partial r}E_z + \frac{\omega\mu}{\beta}\frac{\partial}{r\partial\phi}H_z\right) \\ E_\phi &= \frac{-i\beta}{\omega^2\mu\varepsilon - \beta^2}\left(\frac{\partial}{r\partial\phi}E_z - \frac{\omega\mu}{\beta}\frac{\partial}{\partial r}H_z\right) \end{aligned} \tag{A-5}$$

$$\begin{aligned} H_r &= \frac{-i\beta}{\omega^2\mu\varepsilon - \beta^2}\left(\frac{\partial}{\partial r}H_z - \frac{\omega\varepsilon}{\beta}\frac{\partial}{r\partial\phi}E_z\right) \\ H_\phi &= \frac{-i\beta}{\omega^2\mu\varepsilon - \beta^2}\left(\frac{\partial}{r\partial\phi}H_z + \frac{\omega\varepsilon}{\beta}\frac{\partial}{\partial r}E_z\right) \end{aligned} \tag{A-6}$$

These relations show that it is sufficient to determine E_z and H_z in order to specify uniquely the wave solution. The remaining components can be calculated from Equations (A-5) and (A-6). With the assumed z dependence of Equation (A-2), the wave equation (A-1) becomes

$$\left(\frac{\partial^2}{\partial r^2} + \frac{1}{r}\frac{\partial}{\partial r} + \frac{1}{r^2}\frac{\partial^2}{\partial\phi^2} + (k^2 - \beta^2)\right)\begin{bmatrix} E_z \\ H_z \end{bmatrix} = 0 \tag{A-7}$$

This partial differential equation is separable, and the solution takes the form

$$\begin{bmatrix} E_z \\ H_z \end{bmatrix} = \psi(r)\exp(\pm il\phi) \tag{A-8}$$

where $l = 0, 1, 2, 3, \ldots$, so that E_z and H_z are single-valued functions of ϕ. Then Equation (A-7) becomes

$$\frac{\partial^2 \psi}{\partial r^2} + \frac{1}{r}\frac{\partial \psi}{\partial r} + \left(k^2 - \beta^2 - \frac{l^2}{r^2}\right)\psi = 0 \tag{A-9}$$

where $\psi = E_z, H_z$.

Equation (A-9) is the Bessel differential equation, and the solutions are called Bessel functions of order l. If $k^2 - \beta^2 > 0$, the general solution of Equation (A-9) is

$$\psi(r) = c_1 J_l(hr) + c_2 Y_l(hr) \tag{A-10}$$

where $h^2 = k^2 - \beta^2$, c_1 and c_2 are constants, and J_l and Y_l are Bessel functions of the first and second kind, respectively, of order l. If $k^2 - \beta^2 < 0$, the general solution of Equation (A-9) is

$$\psi(r) = c_1 I_l(qr) + c_2 K_l(qr) \tag{A-11}$$

where $q^2 = \beta^2 - k^2$, c_1 and c_2 are constants, and I_l and K_l are the modified Bessel functions of the first and second kind, respectively, of order l.

To proceed with physical solutions, we need the asymptotic forms of these functions for small and large arguments. Only leading terms will be given for simplicity.

For $x \ll 1$:

$$\begin{aligned}
&J_l(x) \to \frac{1}{l!}\left(\frac{x}{2}\right)^l \\
&Y_0(x) \to \frac{2}{\pi}\left(\ln\frac{x}{2} + 0.5772\ldots\right) \\
&Y_l(x) \to -\frac{(l-1)!}{\pi}\left(\frac{2}{x}\right)^l \qquad l = 1, 2, 3, \ldots \\
&I_l(x) \to \frac{1}{l!}\left(\frac{x}{2}\right)^l \\
&K_0(x) \to -\left(\ln\frac{x}{2} + 0.5772\ldots\right) \\
&K_l(x) \to \frac{(l-1)!}{2}\left(\frac{2}{x}\right)^l
\end{aligned} \tag{A-12}$$

For $x \gg 1, l$:

$$\begin{aligned}
&J_l(x) \to \left(\frac{2}{\pi x}\right)^{1/2}\cos\left(x - \frac{l\pi}{2} - \frac{\pi}{4}\right) \\
&Y_l(x) \to \left(\frac{2}{\pi x}\right)^{1/2}\sin\left(x - \frac{l\pi}{2} - \frac{\pi}{4}\right) \\
&I_l(x) \to \left(\frac{1}{2\pi x}\right)^{1/2} e^x \\
&K_l(x) \to \left(\frac{\pi}{2x}\right)^{1/2} e^{-x}
\end{aligned} \tag{A-13}$$

In these formulas l is assumed to be a nonnegative integer. The transition from the small x behavior to the large x asymptotic form occurs in the region of $x \sim l$.

For traveling waves in the radial direction, it is often convenient to introduce Hankel functions defined as

$$
\begin{aligned}
H_m^{(1)}(x) &= J_m(x) + iY_m(x) \\
H_m^{(2)}(x) &= J_m(x) - iY_m(x)
\end{aligned}
\tag{A-14}
$$

where $H_m^{(1)}(x)$ is the Hankel function of the first kind, and $H_m^{(2)}(x)$ is the Hankel function of the second kind. For $x \gg 1, l$,

$$
\begin{aligned}
H_m^{(1)}(x) &\to \left(\frac{2}{\pi x}\right)^{1/2} \exp\left[j\left(x - \frac{m\pi}{2} - \frac{\pi}{4}\right)\right] \\
H_m^{(2)}(x) &\to \left(\frac{2}{\pi x}\right)^{1/2} \exp\left[-j\left(x - \frac{m\pi}{2} - \frac{\pi}{4}\right)\right]
\end{aligned}
\tag{A-15}
$$

With the time dependence of Equation (A-2), $H_m^{(2)}(x)$ is a wave traveling in the direction of positive x. In other words, $H_m^{(2)}(r)$ is an outward traveling wave in the radial direction (r).

Some low-order Bessel functions are plotted in Figures A.1 to A.4 for reference.

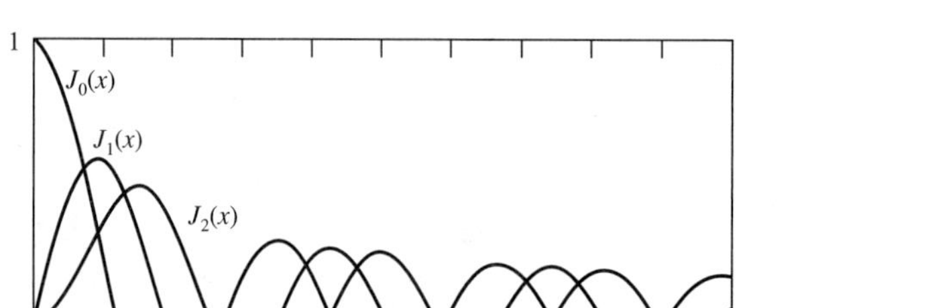

Figure A.1 Bessel functions of the first kind.

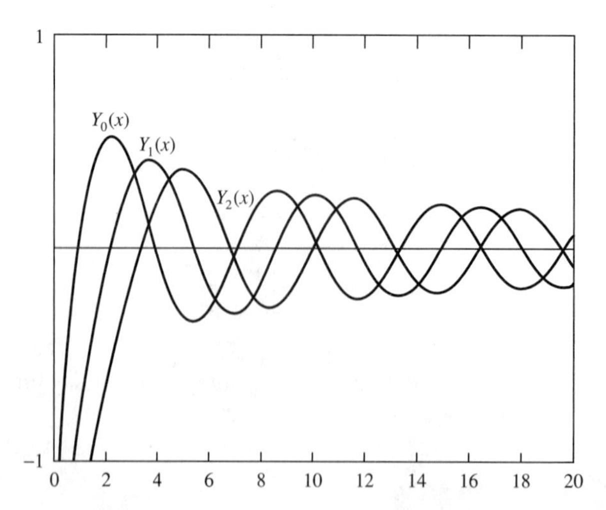

Figure A.2 Bessel functions of the second kind.

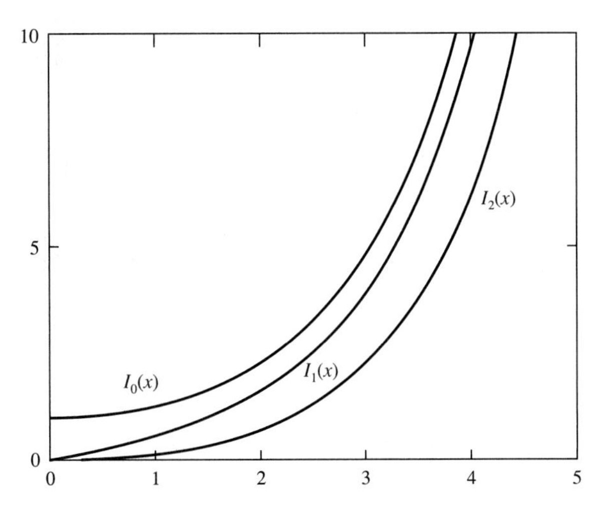

Figure A.3 Modified Bessel functions of the first kind.

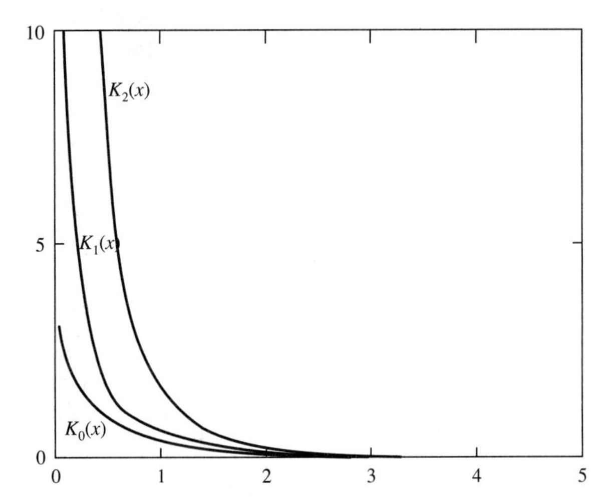

Figure A.4 Modified Bessel functions of the second kind.

APPENDIX B

EXACT SOLUTIONS OF THE STEP-INDEX CIRCULAR WAVEGUIDE

The geometry of the step-index circular waveguide is shown in Figure B.1. It consists of a core of refractive index n_1 and radius a, and a cladding of refractive index n_2 and radius b. The radius b of the cladding is usually chosen to be large enough so that the field of confined modes is virtually zero at $r = b$. In the calculation below we will put $b = \infty$; this is a legitimate assumption in most waveguides, as far as confined modes are concerned.

The radial dependence of the fields E_z and H_z is given by Equation (A-10) or (A-11) in Appendix A, depending on the sign of $k^2 - \beta^2$. For confined propagation, β must be larger than $n_2\omega/c$ (i.e., $\beta > n_2k_0 = n_2\omega/c$). This ensures that the wave is evanescent in the cladding region, $r > a$. The solution is thus given by Equation (A-11) with $c_1 = 0$. This is evident from the asymptotic behavior for large r given by (A-13). The evanescent decay of the field also ensures that the power flow is confined around the z axis, that is, no radial power flow exists. Thus the fields of a confined mode in the cladding ($r > a$) are given by

$$\begin{aligned} E_z(r, t) &= CK_l(qr) \exp[i(\omega t + l\phi - \beta z)] \\ H_z(r, t) &= DK_l(qr) \exp[i(\omega t + l\phi - \beta z)] \end{aligned} \qquad r > a \qquad \text{(B-1)}$$

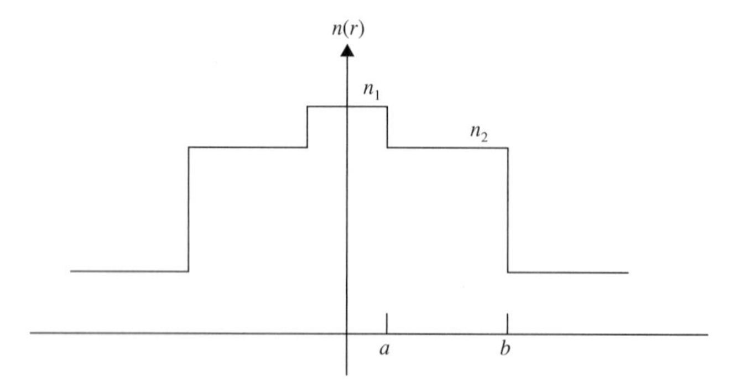

Figure B.1 Structure and index profile of a step-index circular waveguide.

where C and D are two arbitrary constants, and q is given by

$$q^2 = \beta^2 - n_2^2 k_0^2, \quad k_0 = \frac{\omega}{c} \tag{B-2}$$

For the fields in the core, $r < a$, we must consider the behavior of the fields as $r \to 0$. According to (A-12), Y_l and K_l are divergent as $r \to 0$. Since the fields must remain finite at $r = 0$, the proper choice for the fields in the core ($r < a$) is Equation (A-10) with $c_2 = 0$. This becomes evident only when matching, at the interface $r = a$, the tangential components of the field vectors **E** and **H** in the core with the cladding field components derived from (B-1); we are unable to accomplish this if the radial dependence of the core fields is given by I_l. Thus the propagation constant β must be less than $n_1 k_0$ and the core fields are given by

$$\left.\begin{aligned} E_z(r, t) &= AJ_l(hr) \exp[i(\omega t + l\phi - \beta z)] \\ H_z(r, t) &= BJ_l(hr) \exp[i(\omega t + l\phi - \beta z)] \end{aligned}\right\} \quad r < a \tag{B-3}$$

where A and B are two arbitrary constants, and h is given by

$$h^2 = n_1^2 k_0^2 - \beta^2 \tag{B-4}$$

In the field expressions (B-1) and (B-3), we have taken a "+" sign in front of $l\phi$ in the exponents. A negative sign would yield a set of independent solutions, but with the same radial dependence. Physically, l plays a role similar to the quantum number describing the z component of the orbital angular momentum of an electron in a cylindrically symmetric potential field. Thus, if the positive sign in front of $l\phi$ corresponds to a clockwise "circulation" of the Poynting vector about the z axis, the negative sign would correspond to a counterclockwise "circulation" of the Poynting vector around the axis. This can also be seen by examining the ϕ component of the Poynting vector (S_ϕ). Since the fiber itself does not possess any preferred sense of rotation, these two states are degenerate. In other words, they have exactly the same propagation constant.

Equations (B-1) and (B-3) together require that $h^2 > 0$ and $q^2 > 0$, which translates to

$$n_1 k_0 > \beta > n_2 k_0 \tag{B-5}$$

which can be regarded as a necessary condition for confined modes to exist. This is identical to the condition discussed in Chapter 3 for the dielectric slab waveguide and can be expected on intuitive grounds from our discussions of total internal reflection at a dielectric interface.

Using Equations (B-1) and (B-3) in conjunction with Equations (A-5) and (A-6), we can calculate all the field components in both the cladding and the core regions. The results follow.

Core ($r < a$):

$$\begin{aligned} E_r &= \frac{-i\beta}{h^2}\left(AhJ_l'(hr) + \frac{i\omega\mu l}{\beta r} BJ_l(hr)\right) \exp[i(\omega t + l\phi - \beta z)] \\ E_\phi &= \frac{-i\beta}{h^2}\left(\frac{il}{r} AJ_l(hr) - \frac{\omega\mu}{\beta} BhJ_l'(hr)\right) \exp[i(\omega t + l\phi - \beta z)] \\ E_z &= AJ_l(hr) \exp[i(\omega t + l\phi - \beta z)] \end{aligned} \tag{B-6}$$

$$H_r = \frac{-i\beta}{h^2}\left(BhJ_l'(hr) - \frac{i\omega\varepsilon_1 l}{\beta r}AJ_l(hr)\right)\exp[i(\omega t + l\phi - \beta z)]$$

$$H_\phi = \frac{-i\beta}{h^2}\left(\frac{il}{r}BJ_l(hr) + \frac{\omega\varepsilon_1}{\beta}AhJ_l'(hr)\right)\exp[i(\omega t + l\phi - \beta z)] \tag{B-7}$$

$$H_z = BJ_l(hr)\exp[i(\omega t + l\phi - \beta z)]$$

where $J_l'(hr) = dJ_l(hr)/d(hr)$, and $\varepsilon_1 = \varepsilon_0 n_1^2$.

Cladding ($r > a$):

$$E_r = \frac{i\beta}{q^2}\left(CqK_l'(qr) + \frac{i\omega\mu l}{\beta r}DK_l(qr)\right)\exp[i(\omega t + l\phi - \beta z)]$$

$$E_\phi = \frac{i\beta}{q^2}\left(\frac{il}{r}CK_l(qr) - \frac{\omega\mu}{\beta}DqK_l'(qr)\right)\exp[i(\omega t + l\phi - \beta z)] \tag{B-8}$$

$$E_z = CK_l(qr)\exp[i(\omega t + l\phi - \beta z)]$$

$$H_r = \frac{i\beta}{q^2}\left(DqK_l'(qr) - \frac{i\omega\varepsilon_2 l}{\beta r}CK_l(qr)\right)\exp[i(\omega t + l\phi - \beta z)]$$

$$H_\phi = \frac{i\beta}{q^2}\left(\frac{il}{r}DK_l(hr) + \frac{\omega\varepsilon_2}{\beta}CqK_l'(hr)\right)\exp[i(\omega t + l\phi - \beta z)] \tag{B-9}$$

$$H_z = DK_l(qr)\exp[i(\omega t + l\phi - \beta z)]$$

where $K_l'(qr) = dK_l(qr)/d(qr)$, and $\varepsilon_2 = \varepsilon_0 n_2^2$.

These fields must satisfy the boundary conditions that E_ϕ, E_z, H_ϕ, and H_z be continuous at $r = a$. This leads to

$$\begin{aligned}
&AJ_l(ha) - CK_l(qa) = 0\\
&A\left(\frac{il}{h^2 a}J_l(ha)\right) + B\left(-\frac{\omega\mu}{h\beta}J_l'(ha)\right) + C\left(\frac{il}{q^2 a}K_l(qa)\right) + D\left(-\frac{\omega\mu}{q\beta}K_l'(qa)\right) = 0\\
&BJ_l(ha) - DK_l(qa) = 0\\
&A\left(\frac{\omega\varepsilon_1}{h\beta}J_l'(ha)\right) + B\left(\frac{il}{h^2 a}J_l(ha)\right) + C\left(\frac{\omega\varepsilon_2}{q\beta}K_l'(qa)\right) + D\left(\frac{il}{q^2 a}K_l(qa)\right) = 0
\end{aligned} \tag{B-10}$$

where the primes on J_l and K_l again refer to differentiation with respect to their arguments ha and qa, respectively. Equations (B-10) yield nontrivial solutions for A, B, C, and D, provided the determinant of their coefficients vanishes. This requirement yields the following mode condition that determines the propagation constant β:

$$\left(\frac{J_l'(ha)}{haJ_l(ha)} + \frac{K_l'(ha)}{qaK_l(qa)}\right)\left(\frac{n_1^2 J_l'(ha)}{haJ_l(ha)} + \frac{n_2^2 K_l'(ha)}{qaK_l(qa)}\right) = l^2\left[\left(\frac{1}{qa}\right)^2 + \left(\frac{1}{ha}\right)^2\right]^2\left(\frac{\beta}{k_0}\right)^2 \tag{B-11}$$

Equation (B-11), together with (B-4) and (B-2), is a transcendental function of β for each l. The function $J_l'(x)/xJ_l(x)$ in Equation (B-11) is a rapidly varying oscillatory function of $x = ha$. Therefore Equation (B-11) may be considered roughly as a quadratic equation in

$J_l'(ha)/haJ_l(ha)$. For a given l and a given frequency ω, only a finite number of eigenvalues β can be found that satisfy Equations (B-11) and (B-5). Once the eigenvalues have been found, we employ (B-10) to solve for the ratios B/A, C/A, and D/A that determine the six field components of the mode corresponding to each propagation constant β. These ratios are, from (B-10),

$$\frac{C}{A} = \frac{J_l(ha)}{K_l(qa)}$$

$$\frac{B}{A} = \frac{i\beta l}{\omega\mu}\left(\frac{1}{q^2a^2} + \frac{1}{h^2a^2}\right)\left(\frac{J_l'(ha)}{haJ_l(ha)} + \frac{K_l'(qa)}{qaK_l(qa)}\right)^{-1} \tag{B-12}$$

$$\frac{D}{A} = \frac{J_l(ha)}{K_l(qa)}\frac{B}{A}$$

The quantity B/A is of particular interest because it is a measure of the relative amount of E_z and H_z in a mode (i.e., $B/A = H_z/E_z$). Note that E_z and H_z are out of phase by $\pi/2$.

MODE CHARACTERISTICS AND CUTOFF CONDITIONS

In the treatment of slab waveguide modes in Chapter 3, we show that the solutions are easily separated into two classes—the TE and TM modes. In the circular waveguide, the solutions also separate into two classes. However, these are not in general TE or TM, each having in general nonvanishing E_z, H_z, E_ϕ, H_ϕ, E_r, and H_r components. The two classes in solutions can be obtained by noting that Equation (B-11) is quadratic in $J_l'(ha)/haJ_l(ha)$, and when we solve for this quantity, we obtain two different equations corresponding to the two roots of the quadratic equation. The eigenvalues resulting from these two equations yield the two classes of solutions that are designated conventionally as the *EH* and *HE* modes.

By solving Equation (B-11) for $J_l'(ha)/haJ_l(ha)$, we obtain

$$\frac{J_l'(ha)}{haJ_l(ha)} = -\left(\frac{n_1^2 + n_2^2}{2n_1^2}\right)\frac{K_l'}{qaK_l} \pm\left[\left(\frac{n_1^2 - n_2^2}{2n_1^2}\right)^2\left(\frac{K_l'}{qaK_l}\right)^2 + \frac{l^2}{n_1^2}\left(\frac{\beta}{k_0}\right)^2\left(\frac{1}{q^2a^2} + \frac{1}{h^2a^2}\right)^2\right]^{1/2} \tag{B-13}$$

where the arguments of K_l' and K_l are qa. We now use the Bessel function relations

$$J_l'(x) = -J_{l+1}(x) + \frac{l}{x}J_l(x)$$
$$J_l'(x) = J_{l-1}(x) - \frac{l}{x}J_l(x) \tag{B-14}$$

and Equation (B-13) becomes

EH modes:

$$\frac{J_{l+1}(ha)}{haJ_l(ha)} = \left(\frac{n_1^2 + n_2^2}{2n_1^2}\right)\frac{K_l'(qa)}{qaK_l(qa)} + \left(\frac{1}{(ha)^2} - R\right) \tag{B-15a}$$

HE modes:

$$\frac{J_{l-1}(ha)}{haJ_l(ha)} = -\left(\frac{n_1^2 + n_2^2}{2n_1^2}\right)\frac{K_l'(qa)}{qaK_l(qa)} + \left(\frac{1}{(ha)^2} - R\right) \tag{B-15b}$$

where

$$R = \left[\left(\frac{n_1^2 - n_2^2}{2n_1^2}\right)^2\left(\frac{K_l'(qa)}{qaK_l(qa)}\right)^2 + \left(\frac{l\beta}{n_1k_0}\right)^2\left(\frac{1}{q^2a^2} + \frac{1}{h^2a^2}\right)^2\right]^{1/2} \tag{B-16}$$

Equations (B-15a) and (B-15b) can be solved graphically by plotting both sides as functions of ha, letting $(qa)^2 = (n_1^2 - n_2^2)k_0^2 - (ha)^2$ on the right-hand side.

We consider first the special case when $l = 0$. At $l = 0$ we have $\partial/\partial\phi = 0$, and all the field components of the modes are radially symmetric. There are two families of solutions that correspond to (B-15b) and (B-15a). In the first case, the *HE* mode condition (B-15b) becomes

$$\frac{J_1(ha)}{haJ_0(ha)} = -\frac{K_1(qa)}{qaK_0(qa)} \quad \text{(TE)} \tag{B-17a}$$

where we used $K_0'(x) = -K_1(x)$ and $J_{-1}(x) = -J_1(x)$. Under condition (B-17a), the constants A and C vanish according to (B-10) or (B-12). By substituting $A = C = 0$ and $l = 0$ in Equations (B-6) through (B-9), we find that the only nonvanishing field components are H_r, H_z, and E_ϕ. These solutions are thus referred to as TE modes. If the eignevalues are β_m, $m = 1, 2, 3, \ldots$, the TE modes are designated as TE_{0m}, $m = 1, 2, 3, \ldots$, where the first subscript is $l = 0$.

In the second case, the *EH* mode condition, Equation (B-15a) at $l = 0$ becomes

$$\frac{J_1(ha)}{haJ_0(ha)} = -\frac{n_2^2K_1(qa)}{qan_1^2K_0(qa)} \quad \text{(TM)} \tag{B-17b}$$

where we used $K_0'(x) = -K_1(x)$. In this case the constants B and D vanish according to (B-10) or (B-12). By substituting $B = D = 0$ and $l = 0$ in Equations (B-6) through (B-9), we find that the only nonvanishing field components are E_r, E_z, and H_ϕ. These solutions are thus referred to as TM modes and are designated as TM_{0m}.

Now consider the graphical solution of Equations (B-17a) and (B-17b). Confined modes require that q be real to achieve the exponential decay of the field in the cladding. Thus we need only consider ha in the range $0 \le ha \le V \equiv k_0a(n_1^2 - n_2^2)^{1/2}$. The right-hand sides of Equations (B-17a) and (B-17b) are always negative. Starting from $-K_1(V)/VK_0(V)$ for TE modes at $ha = 0$, the right side of Equation (B-17a) is a monotonically decreasing function of ha and becomes asymptotical, according to (A-12)

$$-\frac{K_1(qa)}{qaK_0(qa)} \xrightarrow[ha\to V]{} \frac{2}{(V^2 - h^2a^2)\ln(V^2 - h^2a^2)} \tag{B-18}$$

which diverges to $-\infty$ at $ha = V$. The right side of Equation (B-17b) for TM modes behaves identically except for a factor of n_2^2/n_1^2. On the left sides of Equations (B-17a) and (B-17b), $J_1(ha)/haJ_0(ha)$ starts from $\frac{1}{2}$ at $ha = 0$ and increases monotonically until it diverges to ∞ at $ha = 2.405$, which is the first zero of $J_0(ha)$. Beyond $ha = 2.405$, $J_1(ha)/haJ_0(ha)$ varies from $-\infty$ to $+\infty$ between the zeros of $J_0(ha)$. For large values of ha, $J_1(ha)/haJ_0(ha)$ is a function resembling $-(ha)^{-1}\tan(ha - \pi/4)$, according to Figure B.2, which shows the two curves describing the right and left sides of Equation (B-17a). The normalized frequency $V = k_0a(n_1^2 - n_2^2)^{1/2}$ is assumed to be high enough so that two modes, marked by the circles at the intersection of the two curves, exist. The vertical asymptotes are given by the roots of $J_0(ha) = 0$. If the

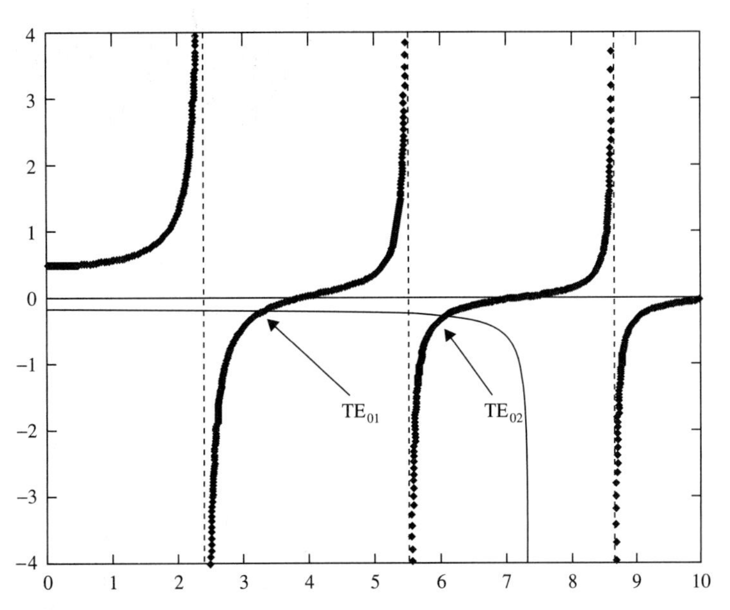

Figure B.2 Graphical determination of the propagation constants of TE modes ($l = 0$) for a step-index waveguide, with $n_1 = 1.4628$, $n_2 = 1.4600$, $a = 20$ μm, and $V = 7.334$.

maximum value of ha, $(ha)_{max} = V$, is smaller than the first root of $J_0(x)$, 2.405, there can be no intersection of the two curves for rea1 β. If V is between the first and the second zero of $J_0(x)$, there will be exactly one intersection of the two curves. Thus the cutoff value (a/λ) for TE_{0m} (or TM_{0m}) waves is given by

$$\left(\frac{a}{\lambda}\right)_{0m} = \frac{x_{0m}}{2\pi(n_1^2 - n_2^2)^{1/2}} \tag{B-19}$$

where x_{0m} is the mth zero of $J_0(x)$. The first three zeros, as shown in Figure B.3, are

$$x_{01} = 2.405, \quad x_{02} = 5.520, \quad x_{03} = 8.654$$

For higher zeros, the asymptotic formula

$$x_{0m} \approx (m - \tfrac{1}{4})\pi$$

gives adequate accuracy (to at least three figures).

When $l \neq 0$ in Equations (B-1), the modes are no longer TE or TM but become the EH or HE modes of the waveguide. These can still be solved graphically in a manner similar to that outlined for the $l = 0$ case. For $l = 1$, the two curves representing the two sides of the EH mode condition (B-15a) are shown in Figure B.4. The normalized frequency $V = k_0 a(n_1^2 - n_2^2)^{1/2}$ is assumed to be 7.334, so that there are two intersections. These are the EH_{11} and EH_{12} modes. The vertical asymptotes are given by the roots of $J_1(x) = 0$. Figure B.5 shows those of the HE modes. At the same value of $V = 7.334$ there are three intersections that correspond to HE_{11}, HE_{12}, and HE_{13} modes, respectively. The vertical asymptotes are also given by the roots of $J_1(x) = 0$. Note that, as shown in Figure B.5, the intersection for HE_{11} mode always exists regardless of the value of V. This mseans the HE_{11} mode does not have a cutoff. All other HE_{1m}, EH_{1m} modes have cutoff values of a/λ given by

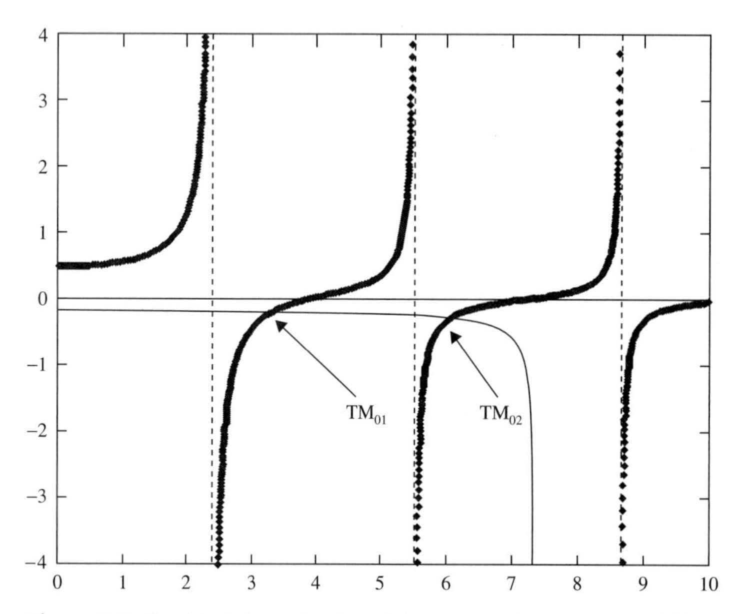

Figure B.3 Graphical determination of the propagation constants of TM modes ($l = 0$) for a step-index waveguide, with $n_1 = 1.4628$, $n_2 = 1.4600$, $a = 20$ μm, and $V = 7.334$.

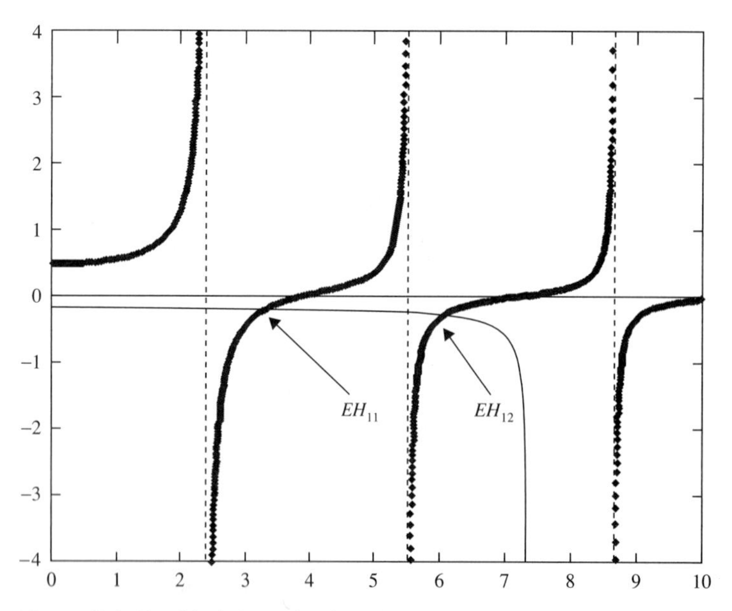

Figure B.4 Graphical determination of the propagation constants of *EH* modes ($l = 1$) for a step-index waveguide, with $n_1 = 1.4628$, $n_2 = 1.4600$, $a = 20$ μm, and $V = 7.334$.

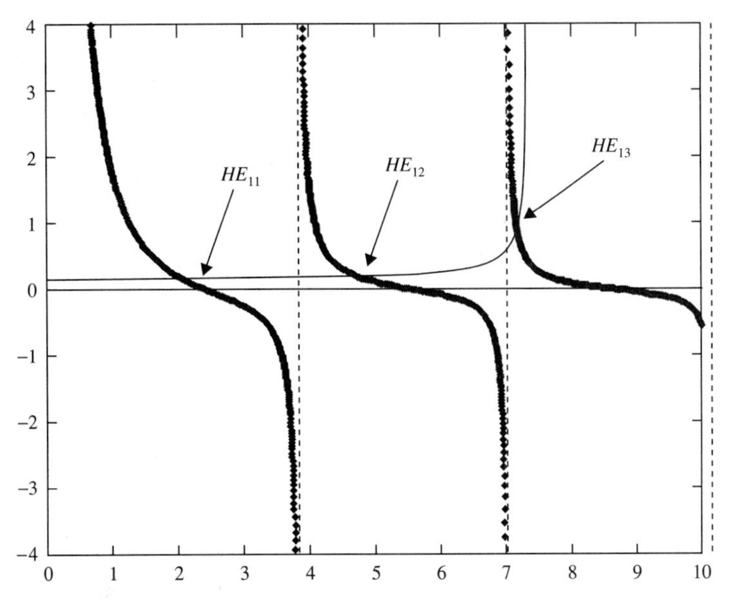

Figure B.5 Graphical determination of the propagation constants of HE modes ($l = 1$) for a step-index waveguide, with $n_1 = 1.4628$, $n_2 = 1.4600$, $a = 20$ μm, and $V = 7.334$.

$$\left(\frac{a}{\lambda}\right)_{1m} = 0, \quad \frac{3.832}{2\pi(n_1^2 - n_2^2)^{1/2}}, \quad \frac{7.016}{2\pi(n_1^2 - n_2^2)^{1/2}}, \quad \ldots \quad (HE \text{ modes})$$
$$\left(\frac{a}{\lambda}\right)_{1m} = \frac{3.832}{2\pi(n_1^2 - n_2^2)^{1/2}}, \quad \frac{7.016}{2\pi(n_1^2 - n_2^2)^{1/2}}, \quad \frac{10.173}{2\pi(n_1^2 - n_2^2)^{1/2}}, \quad \ldots \quad (EH \text{ modes}) \tag{B-20}$$

Note that the zeros of $J_1(x)$ are 3.832, 7.016, 10.173, 13.323. For higher zeros, the asymptotic formula

$$x_{0m} \approx (m + \tfrac{1}{4})\pi$$

gives adequate accuracy (to at least three figures).

For $l > 1$, the cutoff values for a/λ are given by [1]

$$\left(\frac{a}{\lambda}\right)_{lm}^{EH} = \frac{x_{lm}}{2\pi(n_1^2 - n_2^2)^{1/2}} \tag{B-21}$$

$$\left(\frac{a}{\lambda}\right)_{lm}^{HE} = \frac{z_{lm}}{2\pi(n_1^2 - n_2^2)^{1/2}} \tag{B-22}$$

where x_{lm} is the mth zero of $J_l(x) = 0$, and z_{lm} is the mth root of

$$zJ_l(z) = (l - 1)\left(1 + \frac{n_1^2}{n_2^2}\right)J_{l-1}(z), \quad l > 1 \tag{B-23}$$

If we substitute the propagation constant β for $l > 1$ into Equations (B-12), we find that B/A is neither zero nor infinite. This means that both E_z and H_z are present in these modes. The designation of these hybrid modes is based on the relative contribution of E_z and H_z to a transverse component (e.g., E_r or E_ϕ) of the field at some reference point. If E_z makes the larger contribution, the mode is considered TM-dominant and designated HE_{lm}. The fundamental

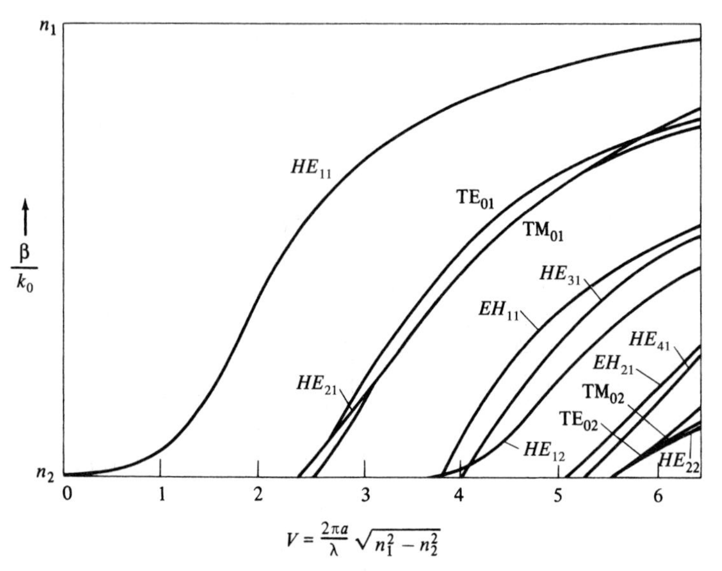

Figure B.6 Normalized propagation constant as a function of V parameter for a few of the lowest-order modes of a step-index waveguide [2].

mode HE_{11}, for example, is TM-dominant. On the other hand, if H_z makes the larger contribution, the mode is considered TE-dominant and designated EH_{lm}. The mode HE_{11} can propagate at any wavelength, as noted earlier, since $(a/\lambda)_{11}^{HE} = 0$. The next modes that can propagate, according to Equation (B-19) are the TE_{01} and TM_{01} modes. Since x_{lm} or z_{lm} forms an increasing sequence for fixed l and increasing m, or for fixed m and increasing l, the number of allowed modes increases as the square of a/λ.

For many applications, the important characteristic of a mode is the propagation constant β as a function of the frequency ω (or normalized frequency V). This information is often presented as the mode index of the confined mode

$$n = \frac{\beta}{k_0} \tag{B-24}$$

as a function of $V = k_0 a(n_1^2 - n_2^2)^{1/2}$; here $k_0 = \omega/c$. Since the phase velocity of a mode is ω/β, n is the ratio of the speed of light in vacuum to the mode phase velocity (n is also called the effective mode index). Figure B.6 shows n for a number of the low-order modes of the step-index circular waveguide [2]. We note that at cutoff, each mode has a value of $(\beta/k_0) = n_2$. We can easily understand this by recalling that as the mode approaches cutoff, the fields extend well into the cladding layer. Thus near cutoff the modes are poorly confined and most of the energy propagates in medium 2 and thus $n = n_2$. By similar reasoning, for frequencies far above cutoff, the mode is tightly confined to the core, and n approaches n_1. As discussed earlier, for $V < 2.405$, only the fundamental HE_{11} mode can propagate. This is an important result, since for many applications single-mode propagation is required. These applications include interferometry that calls for well-defined stationary phase fronts and optical communications by transmission of very short optical pulses. In the latter case, the excitation of many modes would lead to pulse broadening, since the different modes possess different group velocities. This limits the number of pulses (i.e., bits) that can be packed into a given time slot and still be separable on the receiving end.

EXAMPLE: HYBRID MODES IN STEP-INDEX FIBERS

Consider a single-mode silica fiber with core index of $n_1 = 1.4628$, and a clad index of $n_2 = 1.4600$, with a core radius of $a = 4.7$ μm and an optical wave of wavelength = 1.55 μm. These parameters lead to $V = 1.723$. By solving numerically the mode condition equation, we obtain the following parameter for the HE_{11} mode:

$$ha = 1.4235, \quad \beta = 1.4609(\omega/c), \quad \mu|B|^2/\varepsilon_1|A|^2 = 0.9956$$

We note that the HE_{11} mode is almost an equal mixture of TE and TM modes, but TM dominates the HE_{11} mode.

For the same index profile with a core radius of $a = 15$ μm and wavelength = 1.55 μm, the V parameter becomes $V = 5.501$. There are three modes for $l = 1$. They are HE_{11}, EH_{11}, and HE_{12}. The solution of the mode condition yields

$$HE_{11}, \quad ha = 2.0276, \quad n_{\text{eff}} = 1.4624, \quad \mu|B|^2/\varepsilon_1|A|^2 = 0.9972$$

$$EH_{11}, \quad ha = 4.2697, \quad n_{\text{eff}} = 1.4611, \quad \mu|B|^2/\varepsilon_1|A|^2 = 1.0100$$

$$HE_{12}, \quad ha = 4.5463, \quad n_{\text{eff}} = 1.4609, \quad \mu|B|^2/\varepsilon_1|A|^2 = 0.9908$$

where $n_{\text{eff}} = \beta/(\omega/c)$.

REFERENCES

1. Snitzer, E., Cylindrical dielectric waveguide modes. *J. Opt. Soc. Am.* **51**:491 (1961).
2. Keck, D. B., *Fundamentals in Optical Fiber Communications*, M. K. Barnoski (ed.). Academic Press, New York, 1976, Chapter 1.

KRAMERS–KRONIG RELATIONS

The dielectric function $\varepsilon(\omega)$ or refractive index $n(\omega)$ depends on the response of a material to electromagnetic radiation. According to the discussion in Chapter 5, the dielectric function $\varepsilon(\omega)$ depends sensitively on the electronic structure of the medium and is very useful in the determination of the electronic structure of a material. The real and imaginary parts of the complex dielectric function $\varepsilon(\omega)$ are both functions of frequency ω. The Kramers–Kronig relations enable us to find the real part of $\varepsilon(\omega)$ if we know the imaginary part of $\varepsilon(\omega)$ at all frequencies, and vice versa. Such relations were first introduced by Kramers and Kronig in 1926 to study the dielectric constant of a substance.

The dielectric function derived in Chapter 5 relates the monochromatic components of the displacement and electric vectors in the following form:

$$D(\omega) = \varepsilon(\omega)E(\omega) \tag{C-1}$$

As a consequence of the frequency dependence of $\varepsilon(\omega)$, there is a temporally nonlocal connection between the displacement $D(t)$ and the electric field $E(t)$. Such a connection can easily be constructed by Fourier transform. Let us put

$$D(t) = \int_{-\infty}^{\infty} D(\omega)e^{i\omega t}d\omega \tag{C-2}$$

$$E(t) = \int_{-\infty}^{\infty} E(\omega)e^{i\omega t}d\omega \tag{C-3}$$

Substitution of Equation (C-1) for $D(\omega)$ in Equation (C-2) gives

$$D(t) = \int_{-\infty}^{\infty} \varepsilon(\omega)E(\omega)e^{i\omega t}d\omega \tag{C-4}$$

If we invert the Fourier integral in Equation (C-3), we obtain

$$E(\omega) = \frac{1}{2\pi}\int_{-\infty}^{\infty} E(t)e^{-i\omega t}dt \tag{C-5}$$

We now substitute Equation (C-5) for $E(\omega)$ in Equation (C-4) and obtain

$$D(t) = \frac{1}{2\pi}\int_{-\infty}^{\infty} \varepsilon(\omega)e^{i\omega t}d\omega \int_{-\infty}^{\infty} E(t')e^{-i\omega t'}dt' \tag{C-6}$$

By rearranging the order of integration and letting $\tau = t - t'$, the last equation can be written

$$D(t) = \frac{1}{2\pi}\int_{-\infty}^{\infty} d\tau \int_{-\infty}^{\infty} \varepsilon(\omega)e^{i\omega\tau}d\omega\, E(t-\tau) \tag{C-7}$$

We now define a function $F(\tau)$ as

$$F(\tau) = \frac{1}{2\pi}\int_{-\infty}^{\infty} [\varepsilon(\omega) - \varepsilon_0]e^{i\omega\tau}d\omega \tag{C-8}$$

so that, Equation (C-7) can be written

$$D(t) = \varepsilon_0 E(t) + \int_{-\infty}^{\infty} F(\tau)E(t-\tau)\, d\tau \tag{C-9}$$

Equation (C-9) gives a nonlocal relation between $D(t)$ and $E(t)$ in which the displacement $D(t)$ at time t depends on the electric field E at times other than t. If the dielectric function $\varepsilon(\omega)$ is independent of ω (i.e., $\varepsilon(\omega)$ = constant), Equation (C-8) yields $F(\tau) = \varepsilon_0\chi\delta(\tau)$, where χ is the electric susceptibility, and the temporal nonlocal response disappears. But if $\varepsilon(\omega)$ varies with ω, $F(\tau)$ is no longer a delta function, and the nonlocal response results.

Using $[\varepsilon(\omega) - \varepsilon_0] = \varepsilon_0[n^2(\omega) - 1]$ and Equation (5.4-10a) for $n^2(\omega)$, a contour integration of Equation (C-8) yields

$$F(\tau) = 0 \quad \text{for } \tau < 0 \tag{C-10}$$

This result agrees with the principle of causality, which states that the displacement $D(t)$ at time t depends on the value of the electric field E at all earlier times. Using Equation (C-10), Equation (C-9) becomes

$$D(t) = \varepsilon_0 E(t) + \int_{0}^{\infty} F(\tau)E(t-\tau)\, d\tau \tag{C-11}$$

If we allow $F(\tau)$ to be an arbitrary real function of τ such that the integral in Equation (C-11) always converges, this equation can be viewed as the most general relation between D and E in a uniform isotropic medium since only the causality condition is assumed. We now start from Equation (C-11) and substitute Equations (C-3) and (C-4) for $E(t)$ and $D(t)$, respectively, and we get

$$\varepsilon(\omega) = \varepsilon_0 + \int_{0}^{\infty} F(\tau)e^{-i\omega\tau}\, d\tau \tag{C-12}$$

This formula can be used to determine the frequency dependence of the dielectric constant, provided $F(\tau)$ is known. In general, $\varepsilon(\omega)$ is complex. However, since $F(\tau)$ is a real function, $\varepsilon(\omega)$ satisfies the following symmetry relation:

$$\varepsilon^*(\omega) = \varepsilon(-\omega) \tag{C-13}$$

If we separate the real and imaginary parts through the relation

$$\varepsilon(\omega) = \varepsilon'(\omega) - i\varepsilon''(\omega) \tag{C-14}$$

we obtain

$$\varepsilon'(-\omega) = \varepsilon'(\omega) \tag{C-15}$$

$$\varepsilon''(-\omega) = -\varepsilon''(\omega) \tag{C-16}$$

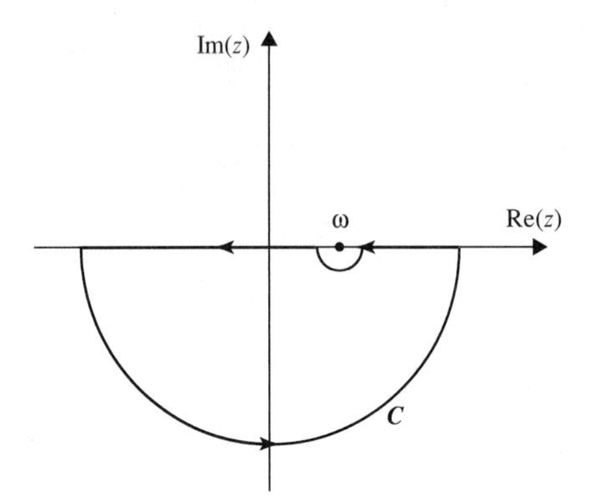

Figure C.1 Contour of the integration in Equation (C-18).

which shows that the real part of the dielectric constant is an even function and the imaginary part is an odd function of the frequency ω.

The relation (C-12) determines the dielectric constant as a function of frequency ω, which is a real variable. Let us now view ω as a function of a complex variable and rewrite Equation (C-12) as

$$\varepsilon(z) = \varepsilon_0 + \int_0^{\infty} F(\tau)e^{-iz\tau}\, d\tau \tag{C-17}$$

where z is a complex variable. Such an $\varepsilon(z)$ is an analytic function of z in the lower half of the complex plane. In the upper half-plane where $z = \omega + i\gamma$ with $\gamma > 0$, the integral in Equation (C-17) diverges.

We now consider the contour integral

$$J = \oint \frac{\varepsilon(z) - \varepsilon_0}{z - \omega}\, dz \tag{C-18}$$

in which the contour of integration is a counterclockwise path that consists of the real axis going around the point $z = \omega$ from below, along an infinitesimal semicircle, and being closed by an infinite semicircle lying in the lower half-plane (see Figure C.1). Note that the integrand of Equation (C-18) has no singularity in the region surrounded by the contour C.

According to Cauchy's theorem, the integral is zero (i.e., $J = 0$). Furthermore, the value of $\varepsilon(z) - \varepsilon_0$ tends to zero as $z \to \infty$. Therefore the integration over the infinite semicircle gives no contribution, and Equation (C-18) can be written

$$0 = J = \lim_{a \to 0}\left(\int_{\infty}^{\omega + a} \frac{\varepsilon(z) - \varepsilon_0}{z - \omega}\, dz + \int_{\omega - a}^{-\infty} \frac{\varepsilon(z) - \varepsilon_0}{z - \omega}\, dz \right) - i\pi[\varepsilon(\omega) - \varepsilon_0] \tag{C-19}$$

where a is the radius of the infinitesimal semicircle, and the term $i\pi[\varepsilon(\omega) - \varepsilon_0]$ is a result of the integration over the infinitesimal semicircle.

The above equation can be rewritten

$$\varepsilon(\omega) - \varepsilon_0 = \frac{i}{\pi} P \int_{-\infty}^{\infty} \frac{\varepsilon(z) - \varepsilon_0}{z - \omega}\, dz \tag{C-20}$$

where P indicates that the integral is a principal-value integral. Since the integration is over the real axis, we may replace z by ω' and Equation (C-20) becomes

$$\varepsilon(\omega) - \varepsilon_0 = \frac{i}{\pi} P \int_{-\infty}^{\infty} \frac{\varepsilon(\omega') - \varepsilon_0}{\omega' - \omega}\, d\omega' \tag{C-21}$$

Using Equation (C-14) and equating the real and imaginary parts of both sides of Equation (C-21), we obtain the Kramers–Kronig relations:

$$\varepsilon'(\omega) - \varepsilon_0 = \frac{1}{\pi} P \int_{-\infty}^{\infty} \frac{\varepsilon''(\omega')}{\omega' - \omega} d\omega' \tag{C-22}$$

$$\varepsilon''(\omega) = \frac{1}{\pi} P \int_{-\infty}^{\infty} \frac{\varepsilon'(\omega') - \varepsilon_0}{\omega' - \omega} d\omega' \tag{C-23}$$

Recalling that

$$0 = \frac{1}{\pi} P \int_{-\infty}^{\infty} \frac{\varepsilon_0}{\omega' - \omega} d\omega' \tag{C-24}$$

Equation (C-21) can also be written

$$\varepsilon''(\omega) = \frac{1}{\pi} P \int_{-\infty}^{\infty} \frac{\varepsilon'(\omega')}{\omega' - \omega} d\omega' \tag{C-25}$$

Using Equation (C-24), we can remove the singularity in the integrand of Equations (C-22) and (C-23) by rewriting the Kramer–Kronig relationship as

$$\varepsilon'(\omega) - \varepsilon_0 = \frac{1}{\pi} \int_{-\infty}^{\infty} \frac{\varepsilon''(\omega') - \varepsilon''(\omega)}{\omega' - \omega} d\omega' \tag{C-26}$$

$$\varepsilon''(\omega) = \frac{1}{\pi} \int_{-\infty}^{\infty} \frac{\varepsilon'(\omega') - \varepsilon'(\omega)}{\omega' - \omega} d\omega' \tag{C-27}$$

The symmetry properties (C-15) and (C-16) can be used to rewrite Equations (C-22), (C-23), (C-26), and (C-27) in the following forms, which contain only positive frequencies:

$$\varepsilon'(\omega) - \varepsilon_0 = \frac{2}{\pi} P \int_{0}^{\infty} \frac{\omega'\varepsilon''(\omega')}{\omega'^2 - \omega^2} d\omega' \tag{C-28}$$

$$\varepsilon''(\omega) = \frac{2\omega}{\pi} P \int_{0}^{\infty} \frac{\varepsilon'(\omega')}{\omega'^2 - \omega^2} d\omega' \tag{C-29}$$

or

$$\varepsilon'(\omega) - \varepsilon_0 = \frac{2}{\pi} \int_{0}^{\infty} \frac{\omega'\varepsilon''(\omega') - \omega\varepsilon''(\omega)}{\omega'^2 - \omega^2} d\omega' \tag{C-30}$$

$$\varepsilon''(\omega) = \frac{2\omega}{\pi} \int_{0}^{\infty} \frac{\varepsilon'(\omega') - \varepsilon'(\omega)}{\omega'^2 - \omega^2} d\omega' \tag{C-31}$$

The Kramers–Kronig relations enable us to calculate the function $\varepsilon'(\omega)$ if we know the function $\varepsilon''(\omega)$, and vice versa. They are of very general validity and are derived directly from Equation (C-11), which contains only the assumption of causality between the displacement D and electric field E. As an example of using the Kramers–Kronig relations, let us consider the attenuation of electromagnetic radiation determined by the imaginary part of the dielectric constant (i.e., $\varepsilon''(\omega)$).

According to Equation (C-24), the integral (C-25) vanishes identically provided $\varepsilon'(\omega)$ = constant. It follows directly from Equation (C-25) that in a dispersionless medium, that is, when $\varepsilon'(\omega)$ = constant, the imaginary part of the dielectric constant vanishes. In other words, any dispersive medium is at the same time also an absorbing medium. Or equivalently, any dispersionless medium is lossless. According to Equation (C-26), we obtain $\varepsilon'(\omega) = \varepsilon_0$ provided $\varepsilon''(\omega) = 0$. In other words, the only truly dispersionless medium is a vacuum.

Consider now, as a second example, the real part of the dielectric constant of an atomic system for which the damping constant γ_j is small and negligible. From Equation (5.5-10a), we have

$$\varepsilon'(\omega) - \varepsilon_0 = \frac{Ne^2}{m} \sum_j \frac{f_j}{(\omega_j^2 - \omega^2)} \tag{C-32}$$

This equation can be rewritten

$$\varepsilon'(\omega) - \varepsilon_0 = \frac{2}{\pi} P \int_0^\infty \sum_j \frac{f_j \delta(\omega' - \omega_j)}{\omega'^2 - \omega^2} d\omega' \tag{C-33}$$

where δ is the Dirac delta function. Comparing Equation (C-33) with Equation (C-28), we obtain an expression for the imaginary part of the dielectric constant in terms of the oscillator strengths,

$$\varepsilon''(\omega) = \frac{\pi N e^2}{2\omega m} \sum_j f_j \delta(\omega - \omega_j) \tag{C-34}$$

Using the sum rule (5.4-12) and integrating Equation (C-32), we obtain the sum rule

$$\int_0^\infty \varepsilon''(\omega)\omega \, d\omega = \frac{\pi e^2}{2m} NZ \tag{C-35}$$

where Z is the number of electrons per atom and N is the number of atoms per unit volume.

TRANSFORMATION OF A COHERENT ELECTROMAGNETIC FIELD BY A THIN LENS

In this appendix we will derive one of the most important results of the theory of coherent optics, which deals with the transformation of a coherent monochromatic field by a lens.

Consider the propagation of an optical beam

$$E = \mathrm{Re}[u(x, y, z)e^{i\omega t}]$$

from an "input" plane $z = 0$ through a lens at z and then to the back focal "output" plane at $z + f$ as shown in Figure D.1. $u(\mathbf{x})$ is thus the complex amplitude of the field, and f is the focal length of the lens.

We use Equations (2.12-15) and (2.12-16) to transform the input beam at $z = 0$ to the plane 1.

$$u_1(x_1, y_1) = \frac{i}{\lambda z}\iint_{\Sigma_0} u(x, y)e^{-ikr}\,dx\,dy \tag{D-1}$$

where $u(x, y) \equiv u(x, y, z = 0)$ and

$$r = \sqrt{(x_1 - x)^2 + (y_1 - y)^2 + z^2} \approx z + \frac{(x_1 - x)^2}{2z} + \frac{(y_1 - y)^2}{2z} \tag{D-2}$$

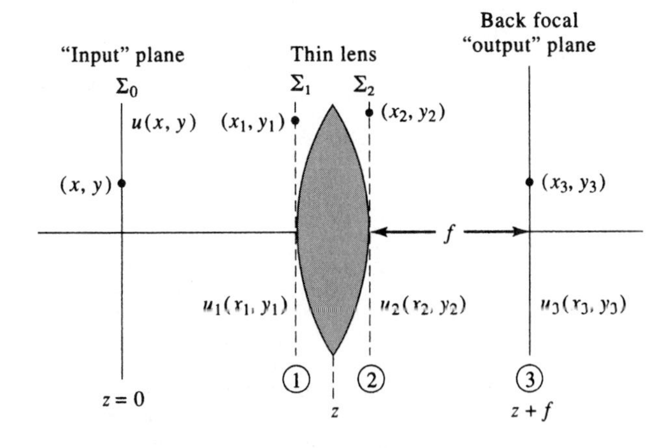

Figure D.1 A lens at z transforms an "input" beam $u(x, y)$ at $z = 0$ to an "output" at $z + f$. The lens plane is designated as Σ_1 and assumed infinite in its transverse dimensions so that truncation effects are neglected.

We rewrite Equation (D-1) as

$$u_1(x_1, y_1) = \frac{ie^{-ikz}}{\lambda z} \iint_{\Sigma_0} u(x, y) e^{-(ik/2z)[(x_1-x)^2 + \text{s.y.}]}\, dx\, dy \tag{D-3}$$

In Equation (D-2) and the rest of this appendix, s.y. stands for "similar terms with $x \to y$." As an example $[(x_1 - x)^2 + \text{s.y.}] \equiv [(x_1 - x)^2 + (y_1 - y)^2]$. Relation (D-3), which results from using the approximate form of (D-2), is called the Kirchhoff diffraction integral. The approximation is valid when the neglected terms in Equation (D-2) multiplied by $k(= 2\pi/\lambda)$ are small compared to 2π. The field at plane Σ_2 is obtained by multiplying the field at u_1 by the lens transfer function (2.6-24)

$$\begin{aligned} u_2(x_2, y_2) &= u_1(x_2, y_2) e^{(ik/2f)(x_2^2 + y_2^2)} \\ &= \frac{ie^{-ikz}}{\lambda z} \iint_{\Sigma_0} dx\, dy\, u(x, y) e^{-i[(k/2z)(x_2 - x)^2 - (k/2f)x_2^2 + \text{s.y.}]} \end{aligned} \tag{D-4}$$

Next, we apply Equation (D-3) again to "propagate" from plane 2 to plane 3

$$\begin{aligned} u_3(x_3, y_3, z + f) = &-\frac{e^{-ik(z+f)}}{\lambda^2 zf} \iint_{\Sigma_2} dx_2\, dy_2\, e^{-i(k/2f)[(x_3 - x_2)^2 + \text{s.y.}]} \\ &\times \left[\iint_{\Sigma_0} dx\, dy\, u(x, y) e^{-i[(k/2z)(x_2 - x)^2 - (k/2f)x_2^2 + \text{s.y.}]} \right] \end{aligned} \tag{D-5}$$

We rearrange the sum of *all* of the terms in the exponents of Equation (D-5). The result is

$$\begin{aligned} k(z + f) &+ \frac{k}{2f}[(x_3 - x_2)^2 + \text{s.y.}] + \left[\frac{k}{2z}(x_2 - x)^2 - \frac{k}{2f}x_2^2 + \text{s.y.} \right] \\ &= k(z + f) + k\frac{(x_3^2 + y_3^2)}{2f}\left(1 - \frac{z}{f}\right) - \frac{k}{f}(xx_3 + yy_3) \\ &\quad + \frac{k}{2z}\left\{ \left[x_2 - \left(x + \frac{z}{f}x_3 \right) \right]^2 + \left[y_2 - \left(y + \frac{z}{f}y_3 \right) \right]^2 \right\} \end{aligned}$$

Changing the order of integration, we rewrite Equation (D-5) as

$$\begin{aligned} &u_3(x_3, y_3, z + f) \\ &= -\frac{\exp\left\{ -ik\left[(z + f) + \dfrac{x_3^2 + y_3^2}{2f}\left(1 - \dfrac{z}{f}\right) \right] \right\}}{\lambda^2 zf} \cdot \iint_{\Sigma_0} dx\, dy\, u(x, y) \exp\left[i\left(\frac{kxx_3}{f} + \frac{kyy_3}{f} \right) \right] \\ &\quad \iint_{\Sigma_2} dx_2\, dy_2 \exp\left(-\frac{ik}{2z}\left\{ \left[x_2 - \left(x + \frac{z}{f}x_3 \right) \right]^2 + \left[y_2 - \left(y + \frac{z}{f}y_3 \right) \right]^2 \right\} \right) \end{aligned} \tag{D-6}$$

Considering the apertures Σ_0, Σ_2 as infinite and using the definite integral

$$\int_{-\infty}^{\infty} e^{-ia^2 x^2}\, dx = \frac{\sqrt{\pi}}{|a|} e^{-i\pi/4}$$

the integral over Σ_2 is equal to $-i2\pi z/k$ so that recalling that $k\lambda = 2\pi$, Equation (D-6) becomes

$$u_3(x_3, y_3, z + f) = \frac{ie^{-ik(z+f)}}{\lambda f} \exp\left[i\frac{k}{2f}\left(\frac{z}{f} - 1\right)(x_3^2 + y_3^2)\right] \iint_{\Sigma_0} u(x, y) \exp\left[ik\left(\frac{xx_3}{f} + \frac{yy_3}{f}\right)\right] dx\, dy \tag{D-7}$$

Recalling the definition of the Fourier transform, we can rewrite Equation (D-7) as

$$u_3(x_3, y_3, z + f) = i\frac{\exp\left[-ik(z + f) + \frac{ik}{2f}\left(\frac{z}{f} - 1\right)(x_3^2 + y_3^2)\right]}{\lambda f}(2\pi)^2 F\{u(x, y)\}_{p=-kx_3/f, q=-ky_3/f} \tag{D-8}$$

$F\{u(x, y)\}$ is the double (x, y) Fourier transform of $u(x, y)$ and is a function of the variables p and q. An especially simple form results if the plane Σ_0 is the front focal plane, that is, $z = f$. In this case,

$$u_3(x_3, y_3, 2f) = i\frac{4\pi^2 e^{-i2kf}}{\lambda f} F\{u(x, y)\}_{p=-kx_3/f, q=-ky_3/f} \tag{D-9}$$

The output field $u_3(x_3, y_3, 2f)$ is thus the (scaled) Fourier transform of the input field $u(x, y)$.

FERMI LEVEL AND ITS TEMPERATURE DEPENDENCE

As discussed in Chapter 15, the probability that a quantum state at energy E is occupied by an electron is governed by the Fermi–Dirac distribution function

$$f(E) = \frac{1}{e^{(E-E_F)/k_BT} + 1} \tag{E-1}$$

where E_F is the Fermi level. We define the Fermi energy as

$$E_{F0} = E_F(T = 0\text{ K}) \tag{E-2}$$

In the regime where $\exp[(E - E_F)/k_BT] \gg 1$, the occupancy is much less than unity, the Fermi–Dirac distribution reduces to the Boltzmann distribution. This is the classical regime (or the nondegenerate regime).

At 0 K, the occupancy probability is 1 for all states with energy below the Fermi energy, and 0 for all states with energy above the Fermi energy. So the Fermi energy is related to the electron density by the following equation:

$$n = \frac{N}{V} = \frac{1}{2\pi^2}\left(\frac{2m}{\hbar^2}\right)^{3/2}\int_0^{E_{F0}} \sqrt{E}\,dE = \frac{1}{3\pi^2}\left(\frac{2mE_{F0}}{\hbar^2}\right)^{3/2} \tag{E-3}$$

where m is the electron mass, N is the total number of electrons, and V is the volume of space occupied by the electrons. According to Equation (E-3), the Fermi energy is an increasing function of the electron density. Because of the Pauli exclusion principle, each state can only accommodate two electrons. More states are needed for more electrons. Since the highest occupied state has energy E_{F0}, the Fermi energy must increase with the electron density in order to accommodate all electrons.

At temperature above absolute zero, evaluation of the occupancy requires knowledge of the Fermi level. The temperature dependence of the Fermi level can be obtained from the following integral:

$$n = \frac{N}{V} = \frac{1}{2\pi^2}\left(\frac{2m}{\hbar^2}\right)^{3/2}\int_0^{\infty} \frac{E^{1/2}}{e^{(E-E_F)/k_BT} + 1}\,dE \tag{E-4}$$

where the integration over all states must yield the electron density. This equation must be inverted in order to obtain the Fermi level as a function of the temperature and the electron density. Unfortunately, the integral on the right side of Equation (E-4) is not invertible analytically.

In the following, we will describe two approximations. Using the following definitions

$$x = E/k_BT \quad \text{and} \quad \eta = E_F/k_BT \tag{E-5}$$

we can write Equation (E-4) as

$$\frac{2}{3}\left(\frac{E_{F0}}{k_BT}\right)^{3/2} = \int_0^\infty \frac{\sqrt{x}}{e^{x-\eta}+1}dx \tag{E-6}$$

or equivalently,

$$n = \frac{1}{3\pi^2}\left(\frac{2mE_{F0}}{\hbar^2}\right)^{3/2} = \frac{1}{2\pi^2}\left(\frac{2mk_BT}{\hbar^2}\right)^{3/2}\int_0^\infty \frac{\sqrt{x}}{e^{x-\eta}+1}dx \equiv n_Q\frac{2}{\sqrt{\pi}}\int_0^\infty \frac{\sqrt{x}}{e^{x-\eta}+1}dx \tag{E-7}$$

where n_Q is the so-called quantum concentration

$$n_Q = 2\left(\frac{mk_BT}{2\pi\hbar^2}\right)^{3/2} \tag{E-8}$$

Numerical integration of the right side of Equation (E-6) can be carried out for each given value of η. We note η is related to the Fermi level E_F through Equation (E-3). The temperature is then obtained by equating the result of the integration to the left side of Equation (E-6). If we do this for many different values of η, we will then obtain the temperature dependence of the Fermi level.

Figures E.1 and E.2 show the normalized Fermi level as a function of normalized temperature. The numerical result is compared with two approximations. They are the Sommerfeld approximation and the Joyce–Dixon approximation [1]:

$$\frac{E_F}{E_{F0}} = \left[1 - \frac{\pi^2}{12}\left(\frac{k_BT}{E_{F0}}\right)^2\right] \quad \text{(Sommerfeld)} \tag{E-9}$$

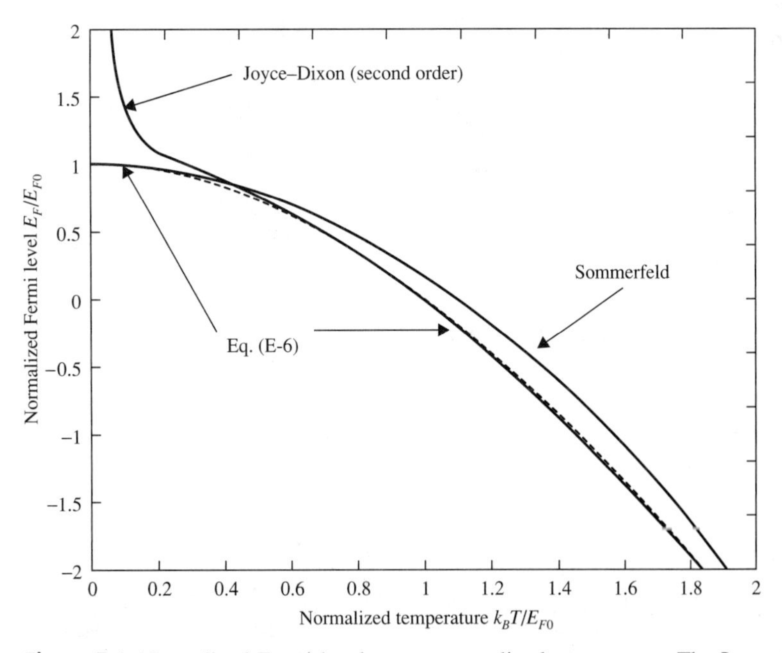

Figure E.1 Normalized Fermi level versus normalized temperature. The Joyce–Dixon approximation in this figure is calculated up to the second order.

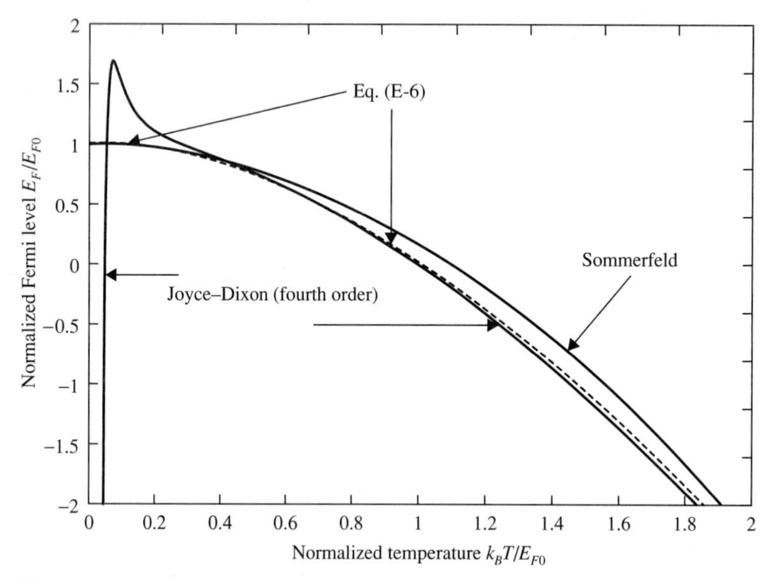

Figure E.2 Normalized Fermi level versus normalized temperature. The Joyce–Dixon approximation in this figure is calculated up to fourth order.

$$\frac{E_F}{k_BT} = \ln r + \frac{1}{\sqrt{8}}r - \left(\frac{3}{16} - \frac{\sqrt{3}}{9}\right)r^2 + a_3r^3 + a_4r^4 + \cdots \quad \text{(Joyce–Dixon)} \tag{E-10}$$

with r given by

$$r = \frac{n}{n_Q} = \frac{4}{3\sqrt{\pi}}\left(\frac{E_{F0}}{k_BT}\right)^{3/2} \tag{E-11}$$

and a_3 and a_4 given by [2]

$$a_3 = 0.000148386 \quad \text{and} \quad a_4 = -0.00000442563 \tag{E-12}$$

We notice that the Sommerfeld approximation gives better results for normalized temperature below 0.42 ($k_BT < 0.42E_{F0}$), whereas the Joyce–Dixon approximation provides better results for normalized temperature above 0.42 ($k_BT > 0.42E_{F0}$). At low temperatures, the parameter r becomes a large number. The Joyce–Dixon approximation does not converge well. This is the regime where the Sommerfeld approximation provides better results. For semiconductor applications, the above results can be applied for the electrons as well as the holes individually, provided they can be treated in separate equilibrium states. In this situation, the Fermi levels are the so-called quasi-Fermi levels, and the effective masses for the electrons and holes must be used.

REFERENCES

1. Joyce, W. B., and R. W. Dixon, *Appl. Phys. Lett.* **31**:354 (1977).
2. See, for example, C. Kittel and H. Kroemer, *Thermal Physics*, 2nd ed. Freeman, New York, 1980.

APPENDIX F

ELECTRO-OPTIC EFFECT IN CUBIC $\bar{4}3M$ CRYSTALS

As an example of transverse modulation[1] and of the application of the electro-optic effect, we consider the case of crystals of the $\bar{4}3m$ symmetry group. Examples of this group are InAs, CuCl, GaAs, and CdTe. The last two are used for modulation in the infrared since they remain transparent beyond 10 μm. These crystals are cubic and have axes of fourfold symmetry along the cube edges ($\langle 100 \rangle$ directions) and threefold axes of symmetry along the cube diagonals $\langle 111 \rangle$.

To be specific, we apply the field in the $\langle 111 \rangle$ direction—that is, along a three fold-symmetry axis. Taking the field magnitude as E, we have

$$\mathbf{E} = \frac{E}{\sqrt{3}} (\mathbf{e}_1 + \mathbf{e}_2 + \mathbf{e}_3) \tag{F-1}$$

where $\mathbf{e}_1$, $\mathbf{e}_2$, and $\mathbf{e}_3$ are unit vectors directed along the cube edges x, y, and z, respectively. The three nonvanishing electro-optic tensor elements are, according to Table 9.1 (see $\bar{4}3m$ tensor), r_{41}, $r_{52} = r_{41}$, and $r_{63} = r_{41}$. Thus, using Equations (9.1-9) and (9.1-10), with

$$\left(\frac{1}{n^2}\right)_1 = \left(\frac{1}{n^2}\right)_2 = \left(\frac{1}{n^2}\right)_3 \equiv \frac{1}{n_o^2}$$

we obtain

$$\frac{x^2 + y^2 + z^2}{n_o^2} + \frac{2r_{41}E}{\sqrt{3}}(xy + yz + xz) = 1 \tag{F-2}$$

[1] *Transverse modulation* is the term applied to the case when the field is applied normal to the direction of propagation.

as the equation of the index ellipsoid. One can proceed formally at this point to derive the new directions x', y', and z' of the principal axes of the ellipsoid. A little thought, however, will show that the $\langle 111 \rangle$ direction along which the field is applied will continue to remain a threefold-symmetry axis, whereas the remaining two orthogonal axes can be chosen *anywhere* in the plane normal to $\langle 111 \rangle$. Thus (F-2) is an equation of an ellipsoid of revolution about $\langle 111 \rangle$. To prove this we choose $\langle 111 \rangle$ as the z' axis, so

$$z' = \frac{1}{\sqrt{3}}x + \frac{1}{\sqrt{3}}y + \frac{1}{\sqrt{3}}z \tag{F-3}$$

and take

$$\begin{aligned} x' &= \frac{1}{\sqrt{2}}y - \frac{1}{\sqrt{2}}z \\ y' &= \frac{-2}{\sqrt{6}}x + \frac{1}{\sqrt{6}}y + \frac{1}{\sqrt{6}}z \end{aligned} \tag{F-4}$$

Therefore

$$\begin{aligned} x &= -\frac{2}{\sqrt{6}}y' + \frac{1}{\sqrt{3}}z' \\ y &= \frac{1}{\sqrt{2}}x' + \frac{1}{\sqrt{6}}y' + \frac{1}{\sqrt{3}}z' \\ z &= -\frac{1}{\sqrt{2}}x' + \frac{1}{\sqrt{6}}y' + \frac{1}{\sqrt{3}}z' \end{aligned} \tag{F-5}$$

Substituting (F-5) in (F-2), we obtain the equation of the index ellipsoid in the x', y', z' coordinate system as

$$(x'^2 + y'^2)\left(\frac{1}{n_o^2} - \frac{r_{41}E}{\sqrt{3}}\right) + z'^2\left(\frac{1}{n_o^2} + \frac{2r_{41}}{\sqrt{3}}E\right) = 1 \tag{F-6}$$

so the principal indices of refraction become

$$\begin{aligned} n_{y'} &= n_{x'} = n_o + \frac{n_o^3 r_{41} E}{2\sqrt{3}} \\ n_{z'} &= n_o - \frac{n_o^3 r_{41} E}{\sqrt{3}} \end{aligned} \tag{F-7}$$

It is clear from Equation (F-6) that other choices of x' and y', as long as they are normal to z' and to each other, will work as well since x' and y' enter Equation (F-6) as the combination $x'^2 + y'^2$, which is invariant to rotations about the z' axis. The principal axes of the index ellipsoid (F-6) are shown in Figure F.1.

An amplitude modulator based on the foregoing situation is shown in Figure F.2. The fractional intensity transmission is given by Equation (9.2-10) as

$$\frac{I_o}{I_i} = \sin^2\frac{\Gamma}{2}$$

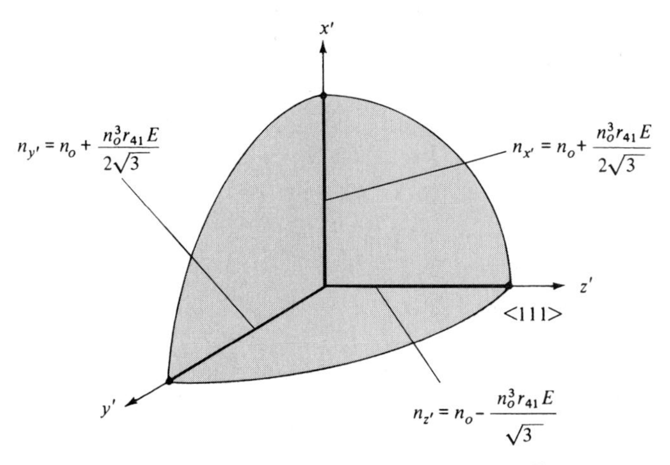

Figure F.1 The intersection of the index ellipsoid of $\bar{4}3m$ crystals (with **E** parallel to $\langle 111 \rangle$) with the planes $x' = 0$, $y' = 0$, $z' = 0$. The principal indices of refraction for this case are $n_{x'}$, $n_{y'}$, and $n_{z'}$.

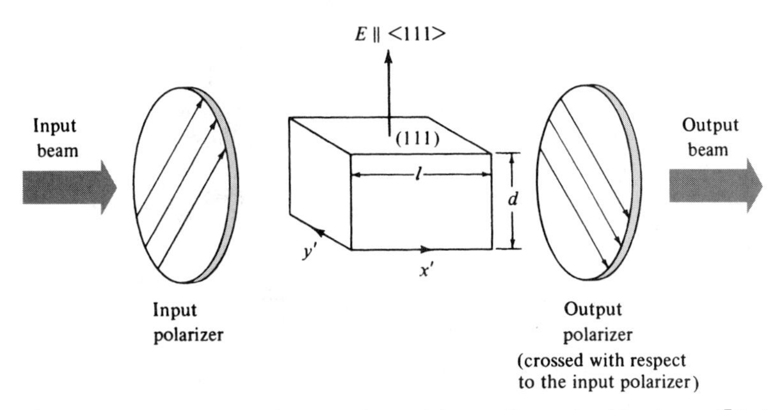

Figure F.2 A transverse electro-optic modulator using a zinc-blend-type ($\bar{4}3m$) crystal with **E** parallel to the cube diagonal $\langle 111 \rangle$ direction.

where the retardation, using (F-7), is

$$\Gamma = \phi_{z'} - \phi_{y'}$$
$$= \frac{(\sqrt{3}\pi) n_o^3 r_{41}}{\lambda_0} \left(\frac{Vl}{d} \right) \quad \text{(F-8)}$$

An important difference between this case where the electric field is applied normal to the direction of propagation and the longitudinal case (9.2-3) is that here Γ is proportional to the crystal length l.

A complete discussion of the electro-optic effect in $\bar{4}3m$ crystals is given in C. S. Namba, *J. Opt. Soc. Am.* **51**:76 (1961). A summary of his analysis is included in Table F.1.

TABLE F.1 Electro-optical Properties and Retardation in $\bar{4}3m$ (Zinc-Blende Structure) Crystals for Three Directions of Applied Field

	$E_{\perp}(001)$ Plane	$E_{\perp}(110)$ Plane	$E_{\perp}(111)$ Plane
	$E_x = E_y = 0, E_z = E$	$E_x = E_y = \frac{E}{\sqrt{2}}, E_z = 0$	$E_x = E_y = E_z = \frac{E}{\sqrt{3}}$
Index ellipsoid	$\frac{x^2+y^2+z^2}{n_o^2} + 2r_{41}Exy = 1$	$\frac{x^2+y^2+z^2}{n_o^2} + \sqrt{2}r_{41}E(yz+zx) = 1$	$\frac{x^2+y^2+z^2}{n_o^2} + \frac{2}{\sqrt{3}}r_{41}E(yz+zx+xy) = 1$
n'_x	$n_o + \frac{1}{2}n_o^3 r_{41}E$	$n_o + \frac{1}{2}n_o^3 r_{41}E$	$n_o + \frac{1}{2\sqrt{3}}n_o^3 r_{41}E$
n'_y	$n_o - \frac{1}{2}n_o^3 r_{41}E$	$n_o - \frac{1}{2}n_o^3 r_{41}E$	$n_o + \frac{1}{2\sqrt{3}}n_o^3 r_{41}E$
n'_z	n_o	n_o	$n_o - \frac{1}{\sqrt{3}}n_o^3 r_{41}E$
$x'\ y'\ z'$ Coordinates	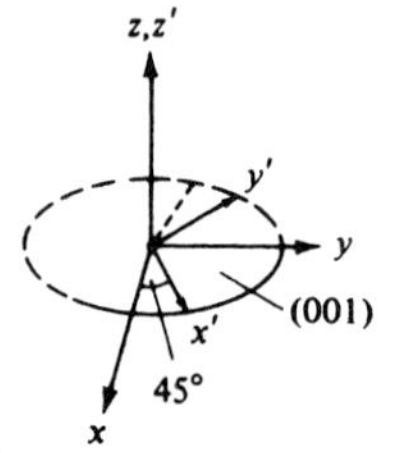	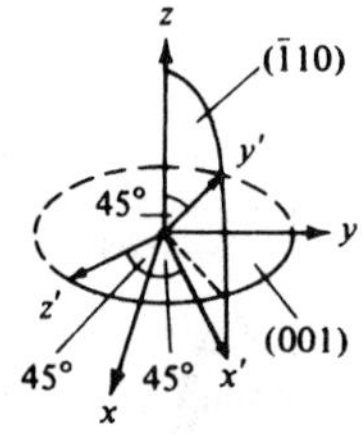	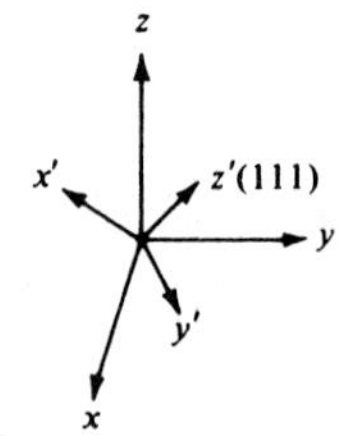
Directions of optical path and axes of crossed polarizer	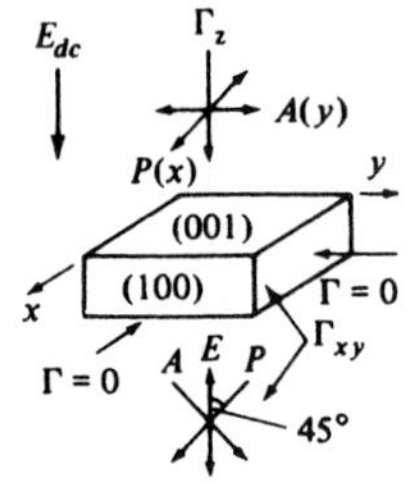	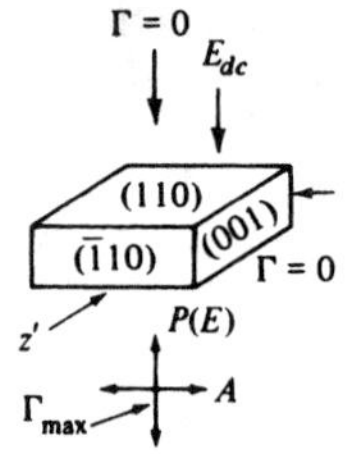	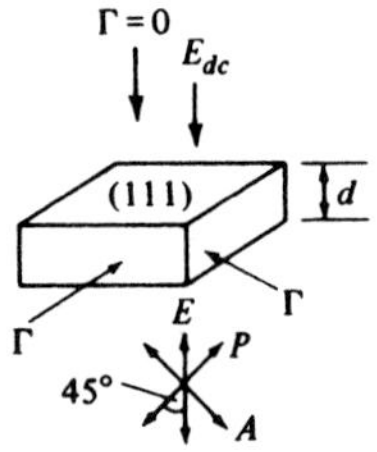
Retardation phase difference $\Gamma(V = Ed)$	$\Gamma = \frac{2\pi}{\lambda}n_o^3 r_{41}V$; $\Gamma_{xy} = \frac{\pi}{\lambda}\frac{l}{d}n_o^3 r_{41}V$	$\Gamma_{max} = \frac{2\pi}{\lambda}\frac{l}{d}n_o^3 r_{41}V$	$\Gamma = \frac{\sqrt{3}\pi}{\lambda}\frac{l}{d}n_o^3 r_{41}V$

After C. S. Namba, *J. Opt. Soc. Am.* **51**:76 (1961).

CONVERSION FOR POWER UNITS AND ATTENUATION UNITS

In optical networks, the power units and propagation attenuation units are often described in terms of dBm and dB/km, respectively. Here we present the conversion tables for ready reference.

The optical power in units of dBm is given by

$$P(\text{dBm}) = 10 \log_{10} P(\text{mW})$$

Table G.1 lists some common power levels.

The attenuation in units of dB/km is given by

$$\alpha(\text{dB/km}) = (10 \log_{10} e)\alpha(\text{km}^{-1}) = 4.343\ \alpha(\text{km}^{-1})$$

Table G.2 lists some common attenuation levels.

TABLE G.1 Power Equivalents

Power in Units of dBm	Power in Units of mW
−20 dBm	0.01 mW
−10 dBm	0.1 mW
0 dBm	1 mW
1 dBm	1.26 mW
2 dBm	1.58 mW
3 dBm	2 mW
7 dBm	5 mW
10 dBm	10 mW
11.8 dBm	15 mW
15 dBm	31.6 mW
20 dBm	100 mW

TABLE G.2 Attenuation Equivalents

Linear Attenuation α in Units of dB/km	Linear Attenuation α in Units of km^{-1}
0.043 dB/km	0.01 km^{-1}
0.22 dB/km	0.05 km^{-1}
0.43 dB/km	0.1 km^{-1}
1 dB/km	0.23 km^{-1}
4.3 dB/km	1.0 km^{-1}
10 dB/km	2.3 km^{-1}
43 dB/km	10 km^{-1}

Author Index

Subject Index